Lecture Notes in Control and Information Sciences

Edited by A. V. Balakrishnan and M. Thoma

Vol. 1: Distributed Parameter Systems: Modelling and Identification
Proceedings of the IFIP Working Conference, Rome, Italy, June 21–26, 1976
Edited by A. Ruberti
V, 458 pages. 1978

Vol. 2: New Trends in Systems Analysis
International Symposium, Versailles, December 13–17, 1976
Edited by A. Bensoussan and J. L. Lions
VII, 759 pages. 1977

Vol. 3: Differential Games and Applications
Proceedings of a Workshop, Enschede, Netherlands, March 16–25, 1977
Edited by P. Hagedorn, H. W. Knobloch, and G. J. Olsder
XII, 236 pages. 1977

Vol. 4: M. A. Crane, A. J. Lemoine
An Introduction to the Regenerative Method for Simulation Analysis
VII, 111 pages. 1977

Vol. 5: David J. Clements, Brian D. O. Anderson
Singular Optimal Control: The Linear Quadratic Problem
V, 93 pages. 1978

Vol. 6: Optimization Techniques
Proceedings of the 8th IFIP Conference on Optimization Techniques, Würzburg, September 5–9, 1977
Part 1
Edited by J. Stoer
XIII, 528 pages. 1978

Vol. 7: Optimization Techniques
Proceedings of the 8th IFIP Conference on Optimization Techniques, Würzburg, September 5–9, 1977
Part 2
Edited by J. Stoer
XIII, 512 pages. 1978

Vol. 8: R. F. Curtain, A. J. Pritchard
Infinite Dimensional Linear Systems Theory
VII, 298 pages. 1978

Vol. 9: Y. M. El-Fattah, C. Foulard
Learning Systems:
Decision, Simulation, and Control
VII, 119 pages. 1978

Vol. 10: J. M. Maciejowski
The Modelling of Systems with Small Observation Sets
VII, 241 pages. 1978

Vol. 11: Y. Sewaragi, T. Soeda, S. Omatu
Modelling, Estimation, and Their Applications for Distributed Parameter Systems
VI, 269 pages. 1978

Vol. 12: I. Postlethwaite, A. G. J. McFarlane
A Complex Variable Approach to the Analysis of Linear Multivariable Feedback Systems
IV, 177 pages. 1979

Vol. 13: E. D. Sontag
Polynomial Response Maps
VIII, 168 pages. 1979

Vol. 14: International Symposium on Systems Optimization and Analysis
Rocquentcourt, December 11–13, 1978;
IRIA LABORIA
Edited by A. Bensoussan and J. Lions
VIII, 332 pages. 1979

Vol. 15: Semi-Infinite Programming
Proceedings of a Workshop, Bad Honnef, August 30 – September 1, 1978
V, 180 pages. 1979

Vol. 16: Stochastic Control Theory and Stochastic Differential Systems
Proceedings of a Workshop of the „Sonderforschungsbereich 72 der Deutschen Forschungsgemeinschaft an der Universität Bonn"
which took place in January 1979 at Bad Honnef
VIII, 615 pages. 1979

Vol. 17: O. I. Franksen, P. Falster, F. J. Evans
Qualitative Aspects of Large Scale Systems
Developing Design Rules Using APL
XII, 119 pages. 1979

Vol. 18: Modelling and Optimization of Complex Systems
Proceedings of the IFIP-TC 7 Working Conference
Novosibirsk, USSR, 3–9 July, 1978
Edited by G. I. Marchuk
VI, 293 pages. 1979

Vol. 19: Global and Large Scale System Models
Proceedings of the Center for Advanced Studies (CAS)
International Summer Seminar
Dubrovnik, Yugoslavia, August 21–26, 1978
Edited by B. Lazarević
VIII, 232 pages. 1979

Vol. 20: B. Egardt
Stability of Adaptive Controllers
V, 158 pages. 1979

Vol. 21: Martin B. Zarrop
Optimal Experiment Design for Dynamic System Identification
X, 197 pages. 1979

For further listing of published volumes please turn over to inside of back cover.

Lecture Notes in Control and Information Sciences

Edited by A.V. Balakrishnan and M. Thoma

38

System Modeling and Optimization

Proceedings of the 10th IFIP Conference
New York City, USA, August 31 – September 4, 1981

Edited by R.F. Drenick and F. Kozin

Springer-Verlag
Berlin Heidelberg GmbH 1982

Editors

R.F. Drenick
Department of EE & CS
Polytechnic Institute of New York
333 Jay Street
Brooklyn, N.Y. 11201
USA

F. Kozin
Department of EE & CS
Polytechnic Institute of New York
333 Jay Street
Brooklyn, N.Y. 11201
USA

ISBN 978-3-540-11691-2 ISBN 978-3-540-39459-4 (eBook)
DOI 10.1007/978-3-540-39459-4

Originally published by Springer-Verlag Berlin Heidelberg New York in 1982

Lecture Notes in Control and Information Sciences

Edited by A.V. Balakrishnan and M. Thoma

38

System Modeling and Optimization

Proceedings of the 10th IFIP Conference
New York City, USA, August 31 – September 4, 1981

Edited by R.F. Drenick and F. Kozin

Springer-Verlag
Berlin Heidelberg GmbH 1982

Editors

R.F. Drenick
Department of EE & CS
Polytechnic Institute of New York
333 Jay Street
Brooklyn, N.Y. 11201
USA

F. Kozin
Department of EE & CS
Polytechnic Institute of New York
333 Jay Street
Brooklyn, N.Y. 11201
USA

ISBN 978-3-540-11691-2 ISBN 978-3-540-39459-4 (eBook)
DOI 10.1007/978-3-540-39459-4

Originally published by Springer-Verlag Berlin Heidelberg New York in 1982

FOREWORD

These Proceedings are made up of papers which were selected from those presented at the 10th IFIP Conference on System Modeling and Optimization in New York City, Aug. 31 to Sept. 4, 1981.

The Conference was organized on behalf of the Technical Committee of the International Federation for Information Processing by the Polytechnic Institute of New York. Its aim was to bring together scientists engaged in studies concerned with the modeling and the optimization of socio-economic, managerial, biological, and technological systems. It was attended by two hundred seventy-one persons of whom more than half came from outside the United States. In all, thirty different countries were represented.

The organization of a major international conference in a city of the size of New York is a complex and expensive undertaking. In fact it would have been impossible without the generous financial support of:

The USAF Office of Scientific Research
The National Science Foundation
The International Business Machines Corporation.

This support is gratefully acknowledged by the conference co-organizers.

The papers which appear in this volume were selected by a three-step reviewing process. A preliminary screening of the proposed contributions was made by the members of the Conference Organizing Committee, namely:

J.J. Bongiorno, Polytechnic Institute of New York
E.A. Cherniavsky, the Brookhaven National Laboratory
E.G. Coffman, Jr., Bell Telephone Laboratories
J. Cullum, IBM Watson Research Center
J.J. Golembeski, Bell Telephone Laboratories
P. Green, IBM Watson Research Center
R.A. Haddad, Polytechnic Institute of New York
P.J. Kolesar, Columbia University
M. Overton, New York University
P.E. Sarachik, Polytechnic Institute of New York
J. Traub, Columbia University

with the assistance of N. Hauser, J. Kao, C.W. Marshall, L. Shaw, and P. Sen, all associated with the Polytechnic Institute, and R.R. Mohler, Oregon State University.

A second round of reviewing was performed by members of the International Program Committee which consisted of

J. Stoer FRG, Chairman

A.V. Balakrishnan, USA
G.B. Dantzig, USA
H. Freeman, USA
M. Iri, Japan
K. Malanowski, Poland
E. Rofman, France
P.D. Lax, USA
W. Leontief, USA
J.L. Lions, France
G.I. Marchuk, USSR
M.J.D. Powell, UK
A. Ruberti, Italy

A third series of reviews took place during the conference itself which relied, in addition to those already mentioned, on

L. Arnold, USA
B. Asselmeyer, FRG
L.N. Belykh, USSR
F. Boesch, USA
V. Chichinadze, USSR
F. Clarke, Canada
Y. Ermolyev, USSR
D. Feigenbaum-Cleiman, Brazil
W. Gersch, USA
J.O. Gray, UK
V. Haas, USA
U. Heilemann, FRG
J.B. Hiriart-Urruty, France
P. Kall, Switzerland
H. Kano, Japan
A. Kershenbaum, USA
W. Krabs, FRG
L. Kurz, USA
I. Lasiecka, USA
W.S. Levine, USA
M. Lucertini, Italy
D.H. Martin, South Africa
J.B. Mockus, USSR
C. Paige, Canada
O. Pironneau, France
J.Ch. Pomerol, France
Z. Rekasius, USA
M. Robin, France
E. Roxin, USA
J. Rumbaugh, USA
E. Sachs, FRG
C. Saguez, France
M. Shooman, USA
L. Slominski, Poland
S. Sugimoto, Japan
Y. Sunahara, Japan
P. Thoft-Christensen, Denmark
A.L. Tits, USA
C. VanNuffelen, Netherlands
N.E. Walker, USA
S. Walukiewicz, Poland
R.L. Williams, USA
P. Wolfe, USA
K. Yajima, Japan
A. Zemanian, USA

The assistance of these reviewers is gratefully acknowledged. Equally grateful acknowledgments are due to Ms. E. Pignataro who organized the non-technical part of the conference program, and to the conference secretaries, Ms. B. Johnson, Ms. M. Hitchcock, and Ms. E. Samsen who oversaw the administrative aspects of the conference and the preparation of these proceedings.

R.F. Drenick F. Kozin

Conference Co-Chairmen

TABLE OF CONTENTS

INVITED PLENARY LECTURES

INVITED TOPICAL LECTURES

CONTROL THEORY

IDENTIFICATION AND ESTIMATION

CONTROL APPLICATIONS

DISTRIBUTED PARAMETER SYSTEMS

STOCHASTIC CONTROL

THEORY AND APPLICATIONS OF PROGRAMMING

PROGRAMMING ALGORITHMS

MANAGEMENT SCIENCE

BIOLOGICAL MODELS

COMPUTER MODELS

COMPUTATIONAL PROBLEMS IN MAGNETIC FUSION RESEARCH

John Killeen
National Magnetic Fusion Energy Computer Center
Lawrence Livermore National Laboratory
P.O. Box 5509
Livermore, California 94550

ABSTRACT

Numerical calculations have had an important role in fusion research since its beginning, but the application of computers to plasma physics has advanced rapidly in the last few years. One reason for this is the increasing sophistication of the mathematical models of plasma behavior, and another is the increased speed and memory of the computers which made it reasonable to consider numerical simulation of fusion devices. The behavior of a plasma is simulated by a variety of numerical models. Some models used for short times give detailed knowledge of the plasma on a microscopic scale, while other models used for much longer times compute macroscopic properties of the plasma dynamics.

The computer models used in fusion research are surveyed. One of the most active areas of research is in time-dependent,three-dimensional, resistive magnetohydrodynamic models. These codes are reviewed briefly.

1. INTRODUCTION

For the purpose of discussing the physical models, and also the numerical methods, it is convenient to consider the following categories of computer codes used to model the physics of fusion devices.

1. Time-dependent magnetohydrodynamics
2. Plasma transport in a magnetic field
3. MHD and guiding-center equilibria
4. MHD stability of confinement systems
5. Vlasov and particle models
6. Multi-species Fokker-Planck codes
7. Hybrid codes

In a short paper it is impossible to review all of the above topics, so a brief description of the models is given, followed by a review of time-dependent, three-dimensional, resistive magnetohydrodynamic codes.

1.1 Time-dependent magnetohydrodynamics

Detailed comparison of experimental data from pulsed high-beta devices with theory, taking due account of experimental complications depends on the application of 2-D and 3-D (two- and three-dimensional) versions of codes analogous to the 1-D Hain-Roberts code [1]. Most of the physical phenomena important here lie in the fast MHD time scale (nonoseconds to microseconds). For Tokamak configurations, the corresponding effects occur on longer time scales-milliseconds; however, the questions of stability of Tokamak discharges toward MHD modes are very important. One example of great interest is the study of the early stages of a Tokamak discharge and the formation and destruction of magnetic surfaces.

Fast time scale MHD codes are typically used to investigate the time dependent behavior of instabilities. The main question to be answered is whether or not a particular MHD mode will be unstable, and if so, how fast will it grow and what is its structure. Although linear MHD stability calculations have made a significant contribution to our understanding of plasma phenomena, nonlinear MHD problems, including the effects of resistivity, rely totally on computers. In order to analyze nonlinear resistive instabilities, the time dependent MHD equations of motion must be solved. The most advanced resistive MHD stability codes are nonlinear and three dimensional. This degree of generality is necessary in order to study the coupling of modes.

There are a great variety of MHD codes being developed. Within the fluid theory various degrees of complexity are considered. The so-called ideal MHD is an infinite conductivity approximation. The more realistic models include the transport coefficients, e.g., thermal conductivity and electrical resistivity, and these can be scalars or tensors [1,2]. Two-dimensional codes are now fairly standard and there are several three-dimensional codes. In some cases perturbation theory is used and the equations are then linearized and Fourier analyzed in one or two coordinates.

The choice of coordinate system varies among these codes. A fixed Eulerian grid is the usual choice, but Lagrangian descriptions, particularly using magnetic flux surfaces as coordinate surfaces [3], are proving useful in certain problems as are particle-in-cell methods [4]. In the work of Brackbill [5] a moving grid is used which is not a Lagrangian grid. A variety of difference schemes are being used, ranging from fully explicit using a Lax-Wendroff or a leap-frog scheme, to implicit methods employing the ADI scheme or "splitting" (the method of fractional time steps). In a later section of this paper we shall describe the variety of numerical methods used in three-dimensional codes.

1.2 Plasma transport in a magnetic field

In order to simulate the transport of a plasma in a magnetic confinement device over most of its lifetime--from tens to hundreds of milliseconds--a set of partial differential equations of the diffusion type must be solved. Typical dependent variables are the number densities and temperatures of each particle species, current densities, and magnetic fields. The transport coefficients such as thermal conductivity, electrical resistivity, and diffusion coefficients are obtained from the best available theories, but the codes also have the capability of easily changing the form of the coefficients in order to develop phenomenological models. In the past years, a considerable effort has been devoted to the numerical solution of these equations for toroidal plasmas which provides an excellent means of comparing theory with experiment. Reviews of these models and their application to Tokamaks are given in Refs. [6,7]. In all of these codes implicit difference methods are used for the solution of the coupled diffusion equations.

1.3 MHD and guiding-center equilibria

It is necessary to develop time-independent codes to support the design and operation of each major fusion experiment. These include codes used to compute and study prospective equilibrium plasma configurations. Experimental devices incorporating the idea of axial symmetry in a torus appear to be capable of plasma confinement for times which are of great interest. One reason for this result is the assurance of equilibria in such devices, as predicted by MHD and guiding-center theories. The computation of these two-dimensional equilibria involves the solution of an elliptic partial differential equation. Iterative methods such as SOR, ADI, and ICCG, and direct methods including cyclic reduction and FFT have all been used in these codes. Equilibrium computational methods are reviewed in Refs. [8, 9].

1.4 MHD stability of confinement systems

Another important class of time-independent code is used to determine the ideal MHD stability of equilibrium configurations by calculating eigenvalues of the linearized perturbed equations. Variational methods are used, and finite elements and Fourier series are used to represent eigenmodes. Reviews of these computations have been given by Grimm et al [10] and Gruber et al [11].

1.5 Vlasov and particle models

Particle codes are fundamental in that they compute in detail the motion of particles under the influence of their self-consistent electric and magnetic fields, as well as any externally imposed fields. These codes give phase-space distribution functions, fluctuation and wave spectra, and orbits of individual particles. They are ideal for providing detailed information on the growth and saturation of strong instabilities and the effects of turbulence.

Particle codes are usually classified as either "electrostatic" or "electromagnetic". In the first type only the self-consistent electric field is computed via Poisson's equation and the magnetic field is either absent or constant in time. Recent methods and results are reviewed in Ref. [12]. In the last five years there has been a considerable development in electromagnetic codes. They are either relativistic and fully electromagnetic, i.e., the particle equations of motion are relativistic and the electric and magnetic fields are obtained from the full Maxwell equations (wave equations) as in Ref. [13] or they are in the nonradiative limit where the equations are nonrelativistic and displacement currents are neglected as in Ref. [14]. Recent research [15-19] in particle codes has concentrated on developing implicit, orbit-averaging and moment algorithms for moving the particles and computing their charge and current densities. These results are of great importance to the fusion program, as they will allow particle codes to be used to simulate plasma phenomena on time scales much longer than previously considered.

1.6 Multi-species Fokker-Planck codes

In the simulation of magnetically confined plasmas where the ions are not Maxwellian and where a knowledge of the distribution functions is important, kinetic equations must be solved. At number densities and energies typical of mirror machines, the end losses are due primarily to the scattering of charged particles into the loss cones in velocity space by classical Coulomb collisions. The kinetic equation describing this process is the Boltzmann equation with Fokker-Planck collision terms [20]. The use of this equation is not restricted to mirror systems. The heating of plasmas by energetic neutral beams, the thermalization of α-particles in DT plasmas, the study of runaway electrons and ions in Tokamaks, and the performance of two-energy component fusion reactors are other examples where the solution of the Fokker-Planck equation is required [21].

The first injection of neutral beams into Tokamak plasmas took place at the Culham, Princeton, and Oak Ridge laboratories in 1972-73. The injected ions were studied with linearized Fokker-Planck models [22,23,24] and the expected plasma heating was observed experimentally.

With the advent of much more powerful neutral beams, it is now possible to consider neutral-beam-driven Tokamak fusion reactors. For such devices, three operating regimes [25] can be considered: (1) the beam-driven thermonuclear reactor, (2) the two-energy component torus (TCT), and (3) the energetic-ion-reactor e.g., the counterstreaming ion torus (CIT). In order to study reactors in regimes (2) or (3), a non-linear Fokker-Planck model must be used because most of the fusion energy is produced by beam-beam or beam-plasma reactions. Furthermore, when co and counter injection are used, or major radius compression is employed, a two velocity-space dimensional Fokker-Planck operator is required. A non-linear, two-dimensional, multi-species Fokker-Planck model [21] developed for the mirror program was applied successfully to several scenarios of TCT operation [26,27,28]. The successful application of the two-dimensional Fokker-Planck model to the energy multiplication studies of TCT led to the formulation of a more complete model of beam-driven Tokamak behavior [29,30].

Neutral beam heated Tokamaks [25] are characterized by a warm Maxwellian background plasma, whose evolution can be described by a set of macroscopic transport equations, and one or more energetic species which are quite non-Maxwellian, whose evolution should be represented by Fokker-Planck equations. The coupling of these systems is by means of particle and energy sources in the multispecies transport equations and a Maxwellian target plasma in the multi-species Fokker-Planck equations.

The Fokker-Planck/Transport (FPT) Code [29,30] models the time-dependent behavior of such a system. The model assumes the existence of an arbitrary number of Maxwellian warm ion species which are described by their individual densities $n_a(\rho,t)$ and by a common temperature profile $T_i(\rho,t)$ where ρ is the average radius of a flux surface. The electrons are described by a separate temperature profile $T_e(\rho,t)$ and their density is determined by quasineutrality. The energetic species are represented by velocity space distribution functions $f_b(v,\theta,\rho,t)$, where v is speed and θ is pitch angle. In addition to one-dimensional radial transport equations for the bulk plasma densities and temperatures, and nonlinear Fokker-Planck equations in two-dimensional velocity space for the energetic ion distribution functions; neutral beam deposition and neutral transport are modeled using Monte Carlo codes [31,32].

1.7 Hybrid codes

There is a need for codes which can best be described as Hybrid Codes; these are codes which combine the good features of fluid codes with the good features of particle codes. The advantage of a particle code is that it contains the most complete treatment of the physics. Its disadvantage also stems from this feature because it is forced to follow the development of the plasma on the fastest time scale and shortest space scale at which significant plasma phenomena occur. These scales are typically much shorter than the time and size scale in fusion devices. The feature of fluid codes which is attractive is that they treat the plasma on a coarser scale and hence need many fewer time steps and spatial points; however, the motions of certain classes of particles are often crucial. It is clear that proper treatment of such phenomena requires an accurate description of the important class of particles. On the other hand, it is possible to treat the rest of the plasma by means of fluid equations.

Another class of hybrid code is the coupling of a Fokker-Planck code to a plasma transport (diffusion) code which was described in the last subsection.

2. TIME-DEPENDENT, THREE-DIMENSIONAL, RESISTIVE MAGNETOHYDRODYNAMICS

In order to achieve the high densities and temperatures required for a successful fusion reactor, a plasma must be confined by a magnetic field for a sufficiently long time. In the attempts to achieve this confinement, the problem of stability has emerged as one of the most important. The most dangerous types of instabilities are the magnetohydrodynamic (MHD) instabilities in which the plasma is assumed to behave as a conducting fluid and the instabilities involve displacement of macroscopic portions of the plasma. It is a particular MHD instability, the resistive instability which is considered in this section.

Resistivity can destroy the stabilization achieved by the shearing of the lines of force. In the case of a magnetic field which has shear or which changes direction, the magnetic energy can be reduced by allowing the fields to mix and annihilate. This is prevented in a perfectly conducting plasma, but with finite conductivity an instability can develop in which the magnetic lines of force are torn into "islands". This type of resistive instability is known as a resistive tearing mode [33].

There are three types of resistive modes: (1) the rippling mode, which is driven by a gradient in the resistivity and is usually not important when large temperature gradients are unlikely; (2) the gravitational mode (g-mode) which is the resistive equivalent of the interchange instability and is important in sheared systems; and (3) the tearing mode, which is the resistive equivalent of the kink mode and involves displacement of the whole plasma.

The modes grow on a time scale intermediate between the resistive diffusion time $\tau_R = 4\pi a^2/\eta c^2$ and the hydromagnetic transit time $\tau_H = a(4\pi\rho)^{1/2}B^{-1}$ where 'a' is a characteristic dimension of the plasma layer, η is the resistivity, ρ is the mass density of the plasma, B is the magnetic field, and c is the speed of light.

Due to the many possible equilibrium configurations and the many approximations necessary to make the problem analytically tractable, it is *usually not possible* to analytically describe the general parameter dependence of the growth rates. In order to obtain results for specific and wide choices of equilibrium magnetic fields and boundary conditions, numerical models have been developed to study these resistive instabilities. Some codes employ a set of "reduced" MHD equations which are appropriate for large aspect ratio, low β, Tokamak plasmas.

The development of three-dimensional, initial-value codes for the solution of MHD equations has evolved from a few exploratory studies in the 1970's to several production type codes, which at the present time are making important contributions to the understanding of resistive instabilities in toroidal plasmas. This evolution is examined from several points of view. It is useful to consider the following options for initial-value MHD code development:

- primitive MHD equations, reduced MHD equations
- compressible, incompressible
- ideal MHD, dissipative MHD
- one, two, three dimensions
- linear, nonlinear
- Eulerian, Lagrangian, ALE, dynamical grid
- finite differences, finite elements, expansions
- explicit difference equations, implicit difference equations

The codes considered in this review will be categorized according to the above options. In addition the applications of the codes will be discussed, as these usually explain the choice of options.

2.1 Primitive, incompressible, linear, resistive MHD codes

The numerical study of linear MHD stability has followed two paths. In the first, the linearized MHD equations (either resistive or ideal) are treated as an initial value problem [34-43]. The equilibrium state of the system is specified and is given a perturbation. The MHD equations are advanced forward in time to trace the evolution of the plasma. The fastest growing instability eventually dominates over all other motions. The second approach which has been used very successfully is to utilize the energy principle [10,11] to gain information on the full spectrum of ideal MHD instabilities.

The resistive instability of an incompressible plasma was first analytically investigated by Furth, Killeen, and Rosenbluth [33]. They used the plane slab model, in which the equilibrium depends only on y, the magnetic field is $\hat{x}\, B_{xo} + \hat{z}\, B_{zo}$, and $\vec{v}_0 = 0$. In that paper perturbations of the form

$f_1(y) \exp [i(k_x x + k_z z) + \omega t]$ are assumed, and the problem is to solve an eigenvalue problem for ω, the growth rate of the instability. In order to solve the problem the plasma is divided into two regions, a narrow inner region about the plane for which the wave vector is perpendicular to the zero-order magnetic field ($\vec{k} \cdot \vec{B}_0 = 0$) and an outer region where the infinite conductivity equations hold. By matching the solutions within the resistive layer to the outer ideal MHD solutions, FKR found resistive tearing modes with growth rates, $p = \omega\tau_R$, proportional to $S^{2/5}$, where $S = \tau_R/\tau_H$.

We can also assume an arbitrary time-dependence and the problem becomes an initial-value problem. Two regions are not used, i.e., the same equations hold throughout the plasma. The initial-value problem is then solved numerically. This method of solution was developed [34] for the linear model simultaneously with the analytic technique and is described in Ref. [35]. The initial-value codes, RIPPLE, use the same basic equations and assumptions as FKR and are capable of finding tearing, rippling, and gravitational modes as well as mixed modes.

We assume that the hydromagnetic approximation is valid, and the ion pressure and inertia terms are neglected in Ohm's law. An isotropic resistivity is assumed, the fluid is assumed to be incompressible, and perturbations in resistivity result only from convection. The basic equations are:

$$\frac{\partial \vec{B}}{\partial t} = \text{curl } (\vec{v} \times \vec{B}) - \text{curl } (\frac{\eta}{4\pi} \text{curl } \vec{B}) \quad , \tag{2.1}$$

$$\text{div } \vec{B} = 0 \quad , \qquad \text{div } \vec{v} = 0 \quad , \tag{2.2}$$

$$\text{curl } (\rho \frac{d\vec{v}}{dt}) = \text{curl } (\frac{1}{4\pi} \text{curl } \vec{B} \times \vec{B}), \tag{2.3}$$

$$\frac{\partial \eta}{\partial t} + \vec{v} \cdot \nabla \eta = 0 \quad . \tag{2.4}$$

In Eqs. (2.1) - (2.3) we consider $\vec{B} = \vec{B}_0 + \vec{B}_1$ and $\vec{v} = \vec{v}_0 + \vec{v}_1$, where $\vec{B}_0$ and $\vec{v}_0$ are given and the subscript 1 denotes perturbed quantities. We obtain, to first order, the following set of linearized equations:

$$\frac{\partial \vec{B}_1}{\partial t} = \text{curl } (\vec{v}_0 \times \vec{B}_1 + \vec{v}_1 \times \vec{B}_0) - \frac{1}{4\pi} \text{curl } (\eta_0 \text{ curl } \vec{B}_1 + \eta_1 \text{ curl } \vec{B}_0) \tag{2.5}$$

$$\text{div } \vec{B}_1 = 0 \ , \ \text{div } \vec{v}_1 = 0 \quad , \tag{2.6}$$

$$\rho_0 \text{ curl} \left[\frac{\partial \vec{v}_1}{\partial t} + (\vec{v}_0 \cdot \nabla) \vec{v}_1 \right] = \frac{1}{4\pi} \text{ curl } \left[(\vec{B}_0 \cdot \nabla) \vec{B}_1 + (\vec{B}_1 \cdot \nabla) \vec{B}_0 \right] \quad , \tag{2.7}$$

with η_1 determined from the first-order version of Eq. (2.4).

In order to consider more realistic equilibrium magnetic fields, a cyclindrical model, RIPPLE IV, was developed [35], which also used the above equations of incompressible magnetohydrodynamics. This model has been extensively applied to the study of tearing modes in reversed field pinch [36,37], and tokamak [37] equilibria. Dibiase developed a new cylindrical model [37,39] which includes the effects of compressibility, viscosity, and thermal conductivity along with finite resistivity. This model has also been applied to the RFP [37-39].

The linear model given by Eqs. (2.5) - (2.7) is applied in RIPPLE IV in order to study specific diffuse pinch configurations. The equilibrium is given by $\vec{B}_0(r) = \hat{\theta}\, B_{\theta 0}(r) + \hat{z}\, B_{z0}\,(r)$, $\vec{v}_0 = 0$, and $\eta_0 = \eta_0\,(r)$. These functions are chosen to describe a particular experiment, and the stabilizing effect of the location of the conducting walls (R_w) with reference to the singular surface can be determined. We assume perturbations of the form $f_1(r,t)\,\exp\,[i(m\theta + k_z z)]$. We can find a consistent system of four equations involving the components B_{r_1}, B_{θ_1}, v_{r_1}, v_{θ_1} and an equation for η_1. These equations are solved by implicit difference methods in the r direction for each mode (m, k_z).

Recently there has been interest in the effect of equilibrium flow on the tearing mode [44,45]. We have developed a new linear initial-value code, RIPPLE V, [40] to study this problem. For this work we have gone back to the plane slab model using the incompressible MHD equations. We have also applied the linear slab model to the double tearing mode [46]. In this case there are two neighboring singular surfaces, i.e., surfaces for which $\vec{k} \cdot \vec{B}_0 = 0$. If these surfaces lie close to one another, the modes at each singular surface may interact leading to an enhanced growth rate.

In all of the above linear models the initial-value problems solved are one-dimensional, i.e., the zero-order fields are given by $\vec{B}_0(y)$ or $\vec{B}_0(r)$ and the perturbed variables take the form

$$f_1(y,t)\,\exp\,[i(k_x x + k_z z)] \text{ or } f_1(r,t)\,\exp\,[i(m\theta + k_z z)]$$

In many toroidal confinement devices it is not possible to specify the equilibrium fields as functions of one variable. In tokamaks and compact torii the zero-order field can be specified by $\vec{B}_0(r,z)$. To study tearing modes in such configurations we have developed a new two-dimensional, linear code, RIPPLE VI, [41] in which the perturbations are of the form $f_1(r,z,t)\exp\,[in\phi]$, $n \geq 1$. We use the incompressible MHD equations to derive a set of eight coupled linear partial differential equations. For the case, n = 0, we have developed a 2D axisymmetric linear code (ALIMO) [47],which makes use of field and velocity stream functions, resulting in a system of four equations.

These linear codes are used for extensive parameter studies of prospective equilibria. Stable and unstable regions of wave number space, growth rates of exponentially growing modes and their mode structure are calculated. In order to study the longtime, large amplitude behavior of these modes, and to simulate experimental devices in controlled fusion research, the non-linear fluid equations must be solved. In general, such a calculation requires the simultaneous advancement in time of eight non-linear partial differential equations in several spatial dimensions.

2.2 Primitive, compressible, nonlinear, resistive MHD codes

After the linear modes have been determined, the resultant eigenmodes may be used to start a nonlinear calculation which can then follow their growth to some final state. These nonlinear calculations have all been with initial-value codes in two and three space dimensions. The nonlinear codes are not limited to stability studies. They are also used to simulate total experiments. By incorporating time dependent boundary conditions, such as circuit equations, a simulation can start with a neutral gas, and model its ionization, pinching, and subsequent relaxation.

The primitive resistive magnetohydrodynamic (MHD) equations relate the electromagnetic fields $\underline{E}$ and $\underline{B}$ to the fluid velocity $\underline{v}$ and the thermodynamic variables (the pressure, P, and mass density, ρ). In terms of the nondimensional variables

$$\underline{x}/a \rightarrow \underline{x} \qquad t/t_H \rightarrow t \qquad \underline{B}/B_0 \rightarrow \underline{B}$$
$$\underline{v}/v_A \rightarrow \underline{v} \qquad \rho/\rho_0 \rightarrow \rho \qquad P/P_0 \rightarrow P$$
$$\eta/\eta_0 \rightarrow \eta$$

these equations take the form

$$\frac{\partial \underline{B}}{\partial t} - \nabla \times \left(\underline{v} \times \underline{B} - \frac{\eta}{S} \nabla \times \underline{B}\right) = 0 \tag{2.8a}$$

$$\frac{\partial \rho \underline{v}}{\partial t} + \nabla \cdot (\rho \underline{v}\, \underline{v}) + \frac{1}{2} \nabla P + \underline{B} \times (\nabla \times \underline{B}) = 0 \tag{2.8b}$$

$$\frac{\partial \rho}{\partial t} + \nabla \cdot (\rho \underline{v}) = 0 \tag{2.8c}$$

$$\frac{\partial P}{\partial t} + \nabla \cdot (\gamma P \underline{v}) - (\gamma - 1) \left\{ \underline{v} \cdot \nabla P + \frac{2\eta}{S} (\nabla \times \underline{B})^2 \right\} = 0 \tag{2.8d}$$

where η is the electrical resistivity of the fluid, and we have assumed Ohm's law in the form

$$\underline{E} + \underline{v} \times \underline{B} = \frac{\eta}{S} \underline{j} \tag{2.9}$$

We assume the perfect gas law is valid so that $\rho e = P/(\gamma - 1)$, where e is the specific internal energy and $\gamma = C_p/C_v$ is the ratio of specific heats for the fluid.

In the normalization described above subscripts $(\ )_0$ refer to characteristic values of various quantities, a is a characteristic length for changes in the magnetic field, $v_A = B_0/(4\pi\rho_0)^{1/2}$ is the Alfvén velocity, and $t_H = a/v_A$ is the Alfvén transit time. The normalization of the thermodynamic quantities is chosen such that $P_0 = B_0^2 / 8\pi = \rho_0 v_A^2 / 2$. The quantity $S = t_R/t_H = 4\pi a v_A/c^2\eta_0$ is the magnetic Reynolds number, and is the ratio of the two characteristic time scales of Eq. (2.8): the resistive diffusion time $t_R = 4\pi a^2/c^2\eta_0$, and the Alfvén transit time t_H.

Equations (2.8) are a set of eight nonlinear equations in the eight unknowns $\underline{B}$, $\rho\underline{v}$, ρ, and P. The above set of equations are not all in conservation form, Equation (2.8c) is in conservation form, and Eq. (2.8a) is in the form of a pseudovector conservation law. Equation (2.8b) could be converted to conservation form by adding $\underline{B}\ (\nabla \cdot B) = 0$ to the equation. This addition has not been done because on a finite difference mesh $\nabla \cdot \underline{B}$ is not necessarily zero. The pressure equation (2.8d) is also not in conservation form. If we define in our normalized variables a total energy density w as

$$w \equiv \frac{1}{2} \rho v^2 + B^2 + P/(\gamma - 1) \quad , \tag{2.10}$$

an equation in conservation form for w can be obtained [48]. When conservation form is used the pressure becomes a subsidiary variable. When the plasma β ($=P/B^2$) is small, the error in the pressure calculation becomes equal to or greater than the calculated pressure. In this form, then, low β plasmas cannot be simulated.

At this point we should consider the question of why use the primitive MHD model, which leads to eight coupled equations, when the reduced MHD model is available, leading to only two equations. The reduced equations employ Tokamak ordering, i.e. β is of order ε^2, where $\varepsilon = \frac{a}{R} \ll 1$, is the aspect ratio of the torus, and the toroidal field is much larger than the poloidal field. The toroidal field is assumed constant in the reduced model. If we wish to consider high β Tokamaks, or the reversed field pinch (RFP), where the toroidal and poloidal field are comparable, or a compact torus (small aspect ratio) then the reduced equations are not applicable and we must use the primitive MHD model.

One of the earliest 3D resistive MHD codes was TRINITY developed by Roberts and Boris [49]. It was an Eulerian code using an explicit leap-frog difference scheme with the resistive diffusion terms employing the DuFort and Frankel algorithm. Similar 3D codes subsequently developed were those of Wooten et al (ideal) [50], Cochran et al [51], and Pritchett et al (ideal) [52]. These codes have been used to study internal kink and interchange instabilities in simple geometries.

Sykes and Wesson have also developed a 3D Eulerian code which includes a tensor resistivity and thermal conductivity [53-55]. Their code employs a Lax-Wendroff scheme and the diffusion terms are also treated by an explicit method. This code has been used to study internal sawtooth oscillations in Tokamaks [53], spontaneous toroidal field reversal in the RFP [54], and high beta internal sawtooth oscillations in toroidal geometry [55].

All of the above codes must satisfy a Courant time-step condition. Generally this is much smaller than one imposed by the time scale of the physical instability being studied. To avoid this limitation Finan [56,57] developed a 3D Eulerian code which uses the Douglas-Gunn algorithm for Alternating-Direction Implicit temporal advancement. The eight equations are solved simultaneously to avoid syncronization errors. The resulting finite difference equations are a coupled system of nonlinear algebraic equations which are solved by the Newton-Raphson iteration technique. Time steps 10-50 times the Courant condition are used in long time simulations. The Finan code, IMP, generalizes to three dimensions techniques that have proved successful in two dimensional resistive MHD calculations [2,46,48,58]. The code has been used to model the tearing mode and internal kink instability in cartesian geometry and the internal kink in toroidal geometry [56,57]. The code has also successfully modeled spontaneous field reversal in the RFP [56].

All of the above 3D nonlinear codes use finite differences in all three directions. In order to obtain the required spatial resolution for high S resistive instabilities, and simultaneously to decrease the computer time for the simulation of these instabilities, recent 3D MHD code developments have used expansions in one or two directions with finite differences in the remaining direction(s).

Hender [59] has developed a 3D Eulerian code to study the coupling of resistive "g" modes in reversed field pinches. It assumes cylindrical geometry and starts with an equilibrium which is a function of r only. The linear codes RIPPLE 4a [35,36] or RESTAB [39] are used to calculate unstable modes which are parametrized by m and k_z. The 3D nonlinear code uses finite differences in the r direction and a Fourier expansion in θ and z. The finite difference method is a mixture of implicit and explicit algorithms.

Mirin [60] has developed a 3D Eulerian code to study the nonlinear growth of MHD instabilities in axisymmetric toroidal equilibria which are functions of r and z. The linear code RIPPLE VI [41] calculates unstable modes, periodic in ϕ, which can be used as initial values for the 3D nonlinear code. The nonlinear code uses an explicit Lax-Wendroff method in r and z with a Fourier expansion in ϕ.

Turner and Wesson are developing a similar code [61] for 3D MHD simulations of JET.

2.3 Reduced, nonlinear, resistive MHD codes

In order to study kink modes in Tokamaks, a set of reduced equations was derived [62] and solved in two dimensions (cylindrical coordinates with helical symmetry) by several groups. Strauss [63,64] has extended the two dimensional model [62] to three dimensions, and solves the equations on an Eulerian mesh with a leap-frog time advancement algorithm. Strauss [63] uses this 3D code to study internal kinks in a square pipe.

The resistive MHD equations in three dimensions can be similarly simplified [65] by the application of Tokamak ordering. The first assumption is $B_\theta/B_\zeta \sim \varepsilon$, where B_θ is the poloidal field, B_ζ is the toroidal field, and $\varepsilon = a/R_0 \ll 1$ is the aspect ratio. Here a is the minor radius and R_0 is the major radius of the torus. A cylindrical coordinate system (r, θ, ζ) is employed, where $r (0 \le r \le a)$ is the minor radius, $\theta (0 \le \theta \le 2\pi)$ is the poloidal angle, and $\zeta (0 \le \zeta \le 2\pi)$ is the toroidal angle. This ordering has the important effect that the time variation of the toroidal magnetic field, B_ζ, can be ignored and, consequently, the fastest time scale of the equations, the time for the propagation of Alfvén waves across the magnetic field, is removed from the dynamics. The fastest time-scale remaining in the equations is the time for Alfvén waves to propagate along the magnetic field. The slow timescale is the resistive diffusion time. The second assumption is to consider only a low β plasma, where β is the ratio of plasma pressure to magnetic field pressure. Specifically, $\beta \sim \varepsilon^2$ is assumed.

The above assumptions yield two scalar partial differential equations which in dimensionless form are [65]

$$\frac{D\psi}{Dt} = \eta J_\zeta - \frac{\partial \phi}{\partial \zeta} - E_\zeta^W \qquad (2.11a)$$

and

$$\frac{DU}{Dt} = -S^2 \left[\hat{\zeta} \cdot (\nabla\psi \times \nabla J_\zeta) + \frac{\partial J_\zeta}{\partial \zeta}\right] \tag{2.11b}$$

where

ψ is the poloidal flux function (normalized to $a^2 B_{\zeta o}$) defined by
$\underline{B} = (-\varepsilon \nabla\psi \times \hat{\zeta} + \hat{\zeta}) B_{\zeta o}$,

$\frac{D}{Dt}$ is the convective derivative $\frac{\partial}{\partial t} + \underline{v}_\perp \cdot \nabla$,

$\underline{v}$ is the fluid velocity in units of a/τ_r,

$\perp$ denotes perpendicular to $\hat{\zeta}$,

η is the resistivity normalized to unity at the magnetic axis,

$J_\zeta = \nabla_\perp^2 \psi$ is the toroidal component of the plasma current density normalized to $\mu_o R_o / B_{\zeta o}$,

ϕ is the velocity stream function, $\underline{v}_\perp = \nabla\phi \times \hat{\zeta}$,

E_ζ^W is the equilibrium toroidal electric field at the wall, and
$U \equiv \nabla_\perp^2 \phi$ is the toroidal vorticity.

The Oak Ridge code RS3 [66] solves Eq. (2.11) using differences in all three directions with a mixed explicit-implicit scheme. This code has been superceded by RSF [67] which uses finite differences in r and expansions in the two periodic directions, and runs considerably faster for the same accuracy. In order for a code such as RSF to model three dimensional behavior, care must be taken to insure that a sufficient number of modes are chosen for possible coupling and excitation. RSF typically uses 35 modes, but can handle 80.

The code has been applied to the feedback stabilization of the 2/1 tearing mode [68], the study of Mirnov oscillations [69], and an extensive study of major disruptions under varying conditions [70-73]. The explanation of disruptions found in these investigations involves a coupling of tearing modes, in particular 2/1 and 3/2 modes. Such studies are only possible with three dimensional codes.

Another 3D reduced MHD code which makes use of an expansion in the two periodic directions is that of Edery et al [74,75]. Similar codes are also in operation at the Princeton Plasma Physics Laboratory [76] and in Japan at JAERI [77].

2.4 Summary

A number of codes which solve either the primitive or reduced equations of resistive magnetohydrodynamics in three dimensions, as initial-value problems, have been examined. Because of the nature of resistive instabilities e.g., tearing and reconnection, the Eulerian representation is generally used. Recent developments favor the use of expansions in one or two directions because of increased speed and accuracy, as long as the number of modes considered is not too large. The codes which employ expansions are also more compatible with the linear codes [34-43] which solve each mode separately. These modes can be combined to start the nonlinear problems.

Work performed under the auspices of the U.S. Department of Energy by the Lawrence Livermore National Laboratory under Contract W-7405-ENG-48.

REFERENCES

1. K. Hain, G. Hain, K. V. Roberts, S. J. Roberts, and W. Köppendörfer, *Z. Naturforsch*, 15a, 1039 (1960).

2. I. Lindemuth and J. Killeen, *J. Comput. Phys.* 13, (1973) 181.

3. David Potter, in *Methods in Computational Physics* (Academic Press, New York, 1976), Vol. 16, 43-84.

4. R. Morse, in *Methods in Computational Physics* (Academic Press, New York, 1970), Vol. 9, pp. 213-240.

5. J. U. Brackbill, in *Methods in Computational Physics* (Academic Press, New York, 1976), Vol. 16, pp. 1-41.

6. J. T. Hogan, ibid., pp. 131-165.

7. M. L. Watkins, M. H. Hughes, P. M. Keeping, K. V. Roberts, and J. Killeen, ibid., pp. 166-210.

8. Brendan McNamara, ibid., pp. 211-252.

9. K. Lackner, *Computer Physics Comm.* 12 (1976) 33.

10. R. C. Grimm, J. M. Greene, and J. L. Johnson, in *Methods in Computational Physics* (Academic Press, New York, 1976), Vol. 16, pp. 253-281.

11. R. Gruber, F. Troyon, D. Berger, L. C. Bernard, S. Rousset, R. Schreiber, W. Kerner, W. Schneider and K. V. Roberts, *Computer Physics Comm.* 21 (1981) 323-371.

12. J. M. Dawson, H. Okuda, and B. Rosen, in *Methods in Computational Physics* (Academic Press, New York, 1976), Vol. 16, pp. 282-326.

13. A. B. Langdon and B. Lasinski, ibid., pp. 327-366.

14. C. W. Nielson and H. R. Lewis, ibid., pp. 367-388.

15. R. J. Mason, *J. Comput. Phys.* 41 (1981) 233.

16. J. Denavit, *J. Comput. Phys.* 42 (1981) 337.

17. B. I. Cohen, A. B. Langdon, and A. Friedman, accepted by *J. Comput. Phys.*

18. B. I. Cohen, R. P. Freis, and V. Thomas, accepted by *J. Comput. Phys.*

19. J. U. Brackbill and D. W. Forslund, submitted to *J. Comput. Phys.*

20. J. Killeen and K. D. Marx, in *Methods in Computational Physics* (Academic Press, New York, 1970), Vol. 9, 421-489.

21. J. Killeen, A. A. Mirin, and M. E. Rensink, in *Methods in Computational Physics* (Adademic Press, New York, 1976), Vol. 16, pp. 389-432.

22. J. G. Cordey and W.G.F. Core, *Phys. Fluids*, 17, (1974) 1626.

23. J. D. Gaffey, *J. Plasma Phys.* 16, (1976) 149.

24. J. D. Callen, et al., *Plasma Physics and Controlled Nuclear Fusion Research* (IAEA, Vienna) I, (1975) 645.

25. D. L. Jassby, *Nuclear Fusion,* 17, (1977) 309.

26. J. Killeen, et al., Lawrence Livermore Laboratory Report UCID-16530 (1974).

27. H. L. Berk, et al., *Plasma Physics and Controlled Nuclear Fusion Research,* (IAEA, Vienna) III, (1974) 569.

28. J. Killeen, et al., *Seventh European Conference on Controlled Fusion and Plasma Physics* 1, (1975) 22.

29. A. A. Mirin, J. Killeen, K. D. Marx, and M. E. Rensink, *J. Comput. Phys.* 23, (1977) 23.

30. J. Killeen, A. A. Mirin, and M. G. McCoy in *Modern Plasma Physics* (IAEA Vienna 1981) p. 395.

31. G. G. Lister, et al., *Plasma Heating in Toroidal Devices*, (Varenna, Italy) (1976) 303.

32. M. H. Hughes and D. E . Post, *J. Comput. Phys.* 28, (1978) 43.

33. H. P. Furth, J. Killeen, and M. N. Rosenbluth, *Phys. Fluids* 6, (1963) 459.

34. J. Killeen and H. P. Furth, *Bull. Am. Phys. Soc.* 6, (1961) 193.

35. J. Killeen in *Physics of Hot Plasmas* edited by B. J. Rye and J. C. Taylor, Plenum, New York, 1970, p. 202.

36. J. E. Crow, J. Killeen and D. C. Robinson, *Sixth European Conference on Controlled Fusion and Plasma Physics*, Moscow (1973), 269.

37. J. A. Dibiase, PhD Thesis, Univ. of Calif., Davis, UCRL-51591 (1974).

38. J. A. Dibiase, J. Killeen, D. C. Robinson, D. Schnack, in *Third Topical Conference on Pulsed High Beta Plasmas,* (Culham 1975) Pergamon Press, Oxford (1976) 283-289.

39. J. Dibiase and J. Killeen, *J. Comp. Phys.* 24, (1977) 158.

40. J. Killeen and A. I. Shestakov, *Phys. Fluids* 21, (1978) 1746.

41. A. I. Shestakov, J. Killeen and D. D. Schnack, *J. Comp. Phys.* (Accepted 1981).

42. G. Batemen, W. Schneider, and W. Grossmann, *Nuclear Fusion,* 14 (1974) 669.

43. A. Sykes and J. Wesson, ibid., 645.

44. D. Dobrott, S. C. Prager, and J. B. Taylor, *Phys. Fluids* 20, (1977) 1850.

45. R. K. Pollard and J. B. Taylor, *Phys. Fluids* 22, (1979) 126.

46. D. Schnack and J. Killeen, in *Theoretical and Computational Plasma Physics,* (Trieste, 1977), IAEA Vienna(1978) 337-360.

47. A. I. Shestakov, D. D. Schnack, and J. Killeen, *Ninth Conference on Numerical Simulation of Plasmas,* Evanston (1980) OA-6.

48. D. Schnack and J. Killeen, *J. Comput. Phys.* 35, (1980) 110.

49. K. V. Roberts and J. P. Boris, in *Proceedings of the Culham Conference on Computational Physics* 2, (1969) 44.

50. J. Wooten, H. R. Hicks, G. Bateman and R. A. Dory, ORNL/TM-4784 (1974).

51. F. L. Cochran, P. C. Liewer and G. Bateman in *Proceedings of the Eighth Conference on Numerical Simulation of Plasmas*, Paper PB-3, LLNL Report CONF-780614 (1978).

52. P. L. Pritchett, C. C. Wu, and J. M. Dawson, *Phys. Fluids* 21 (1978), 1543.

53. A. Sykes and J. A. Wesson, *Phys. Rev. Letters* 37 (1976) 140.

54. A. Sykes and J. A. Wesson, *Eighth European Conference on Controlled Fusion and Plasma Physics*, Prague, (1977).

55. A. Sykes in "Proceedings of the 1981 Wildhaus Workshop on Computational MHD Models of the Behavior of Magnetically Confined Plasmas" to be published in *Computer Physics Comm.*

56. C. H. Finan III, "The Alternating-direction Implicit Numerical Solution of the Time-dependent, Three-dimensional, Single Fluid, Resistive Magnetohydrodynamic Equations," PhD Thesis, UCRL-53086, (December 1980).

57. C. H. Finan III and J. Killeen in "Proceedings of the 1981 Wildhaus Workshop on Computational MHD Models of the Behavior of Magnetically Confined Plasmas" to be published in *Computer Physics Comm.*

58. D. Schnack and J. Killeen, *Nuclear Fusion*, 19 (1979) 877.

59. T. C. Hender and D. C. Robinson in "Proceedings of the 1981 Wildhaus Workshop on Computational MHD Models of the Behavior of Magnetically Confined Plasmas" to be published in *Computer Physics Comm.*

60. A. A. Mirin and A. I. Shestakov, *Bull. Am. Phys. Soc.* 26 (1981) 956.

61. M. F. Turner and J. A. Wesson, in "Proceedings of the 1981 Wildhaus Workshop on Computational MHD Models of the Behavior of Magnetically Confined Plasmas" to be published in *Computer Physics Comm.*

62. R. White, D. Monticello, M. N. Rosenbluth, H. Strauss and B. B. Kadomtsev, *Plasma Physics and Controlled Nucl. Fusion*, Tokyo, IAEA, Vienna (1975) Vol I, 495.

63. H. R. Strauss, *Phys. Fluids* 19 (1976) 134.

64. H. R. Strauss, *Phys. Fluids* 20 (1977) 1354.

65. B. V. Waddell, B. Carreras, H. R. Hicks and J. A. Holmes, *Phys. Fluids* 22 (1979) 896.

66. H. R. Hicks, B. Carreras, J. A. Holmes, D. K. Lee, S. J. Lynch, and B. V. Waddell, in *Proceedings of the Eighth Conference on Numerical Simulation of Plasmas*, Paper PA-7, Monterey, CA., LLNL Report CONF-780614 (1978).

67. H. R. Hicks, J. A. Holmes, D. K. Lee, B. Carreras and B. V. Waddell, (1981). Accepted in *J. Comput. Phys.*

68. J. A. Holmes, B. Carreras, H. R. Hicks, S. J. Lynch, B. V. Waddell, *Nuclear Fusion*, 19 (1979) 1333.

69. B. Carreras, B. V. Waddell, H. R. Hicks, ibid., 1423.

70. H. R. Hicks, J. A. Holmes, B. A. Carreras, D. J. Tetreault, G. Berge, J. P. Freidberg, P. A. Politzer, D. Sherwell, *Eighth International Conference on Plasma Physics and Controlled Nuclear Fusion Research,* IAEA Vienna (1981) Vol I, 259.

71. B. Carreras, H. R. Hicks, J. A. Holmes, B. V. Waddell, *Phys. Fluids* 23 (1980) 1811.

72. B. Carreras, J. A. Holmes, H. R. Hicks, V. E. Lynch, *Nuclear Fusion,* 21 (1981) 511.

73. B. Carreras, H. R. Hicks, D. K. Lee, *Phys. Fluids* 24 (1981) 66.

74. D. Edery, et al., in *Plasma Physics and Controlled Nuclear Fusion Research,* IAEA Vienna (1981) Vol I, 269.

75. D. Edery, R. Pellat, J. L. Soulé "Proceedings of the 1981 Wildhaus Workshop on Computational MHD Models of the Behavior of Magnetically Confined Plasmas" to be published in *Computer Phys. Comm.*

76. H. R. Strauss, W. Park, D. A. Monticello, R. B. White, S. C. Jardin, M. S. Chance, A.M.M. Todd, A. H. Glasser, *Nuclear Fusion,* 20 (1980) 638.

77. G. Kurita, et al., JAERI-M 9788 (1981).

A SIMPLE MODEL OF A CAPITAL STOCK IN EQUILIBRIUM WITH THE TECHNOLOGY AND THE PREFERENCES

Tjalling C. Koopmans, Yale University

A great variety of fields of inquiry and of techniques are discussed at this meeting of IFIP. Among the thirteen invited speakers alone I find that eight treat applications of system analysis while five are more theory-oriented. The situation in economics is similar in some respects and different in others. It is my impression that the professions that you are working for, or giving advice to, apparently do not feel that you are trying to reform their fields or perhaps even inject irrelevancies into it. In economics we have two groups of people. There is one which is similar to this audience. These are the mathematical economists and econometricians who are highly competent theoreticians. The other group consists of people who know the economy very thoroughly, its history and its facts, and who have developed a certain wisdom that derives from their experience. There is an overlap between two groups, of course, but it is quite small, and otherwise there is more standoffishness between them than is desirable. There is a need for more communication between them.

I have exerted a modest effort in that direction during the last twenty-five years since I became an economic theorist. I have always followed a certain schedule in my work. I would first do a study of some problem in mathematical economics and seek to make sure that my conjectures could be proved. I would then follow up with an article addressed to, or intended to engage the interest of, a broader group of economists. This type of article then sought to bring out the aims of mathematical economists in that particular subject. I find this is necessary. If mathematical economists want to work on worthwhile problems they should at least make clear what the problems are when they publish their work, so they obtain a feedback that tells them whether their work is regarded as pertinent.

My address here is an example of this process. The original study is discussed in a technical paper, jointly with Terje Hansen from the School of Business in Bergen, Norway [1]. Later on I developed a much simpler example. Instead of n capital goods, and m consumption goods and ℓ resources, as in the original work, the example assumes just one capital good, two consumption goods, and one resource (labor). Nevertheless it retains the structure and ideas of the more general problem. It was originally prèsented at a conference of the International Economic Association in S'Agaro, Spain, in 1975 and it appeared in the conference proceedings [2]. I will present here a somewhat expanded version of that example [3].

The example is based on a linear model, such as those used in linear programming or in activity analysis. That model is defined by Table 1, which is to be read in the following way.

TABLE 1: Illustrative Three-Sector Model

	Brief Notations	Activity Levels x_i and Technical Coefficients x_1	x_2	x_3	Availabilities and Total "Outputs" $\geq$	
(1)	f_1 f_2 f_3	$-a_1$	$-a_2$	$-a_3$	$-z^1$	(capital input)
(2)		b_1	b_2	b_3	z^2	(capital "output")
(3)	-1 -1 -1	-1	-1	-1	-1	(labor)
(4)	d_1 d_2 0	1	0	0	y_1	(cons. good "1")
(5)		0	1	0	y_2	(cons. good "2")

There are three processes, represented by the columns headed by x_1, x_2, x_3. Rows (1) and (2) stand for the inequality constraints

$$-a_1x_1 - a_2x_2 - a_3x_3 \geq -z^1 \tag{1}$$

$$b_1x_1 + b_2x_2 + b_3x_3 \geq z^2 \tag{2}$$

The first expresses the fact that a capital input of at least z^1 is required if the system is to run at the activity levels x_1, x_2, x_3. Row (2) is similarly interpreted. Row (3) is used for normalization: Because all entries are -1, the unit of labor is the total available, and is also the unit in which the activity levels are measured. The technical coefficients a_i and b_i represent how much capital input is required at the beginning of the period to sustain activity i over one period at unit level, and how much capital is returned intact at the end of that period.

Note that one must have $a_1 > b_1$ in order to express the fact that there will be some loss of capital effectiveness through the use of the first process. Similarly $a_2 > b_2$, but $a_3 < b_3$ which expresses that at the end of one period some new capital must have been produced during that period to make up for the loss of effectiveness through use.

Rows (4) and (5) represent the inequalities

$$x_1 \overset{(>)}{=} y_1 \ , \tag{3}$$

$$x_2 \overset{(>)}{=} y_2 \ , \tag{4}$$

They specify that the first process operated at unit level produces one unit of the first consumption good, the second process one unit of the other, and that no less will be consumed than is produced. Evidently, one must also have

$$x_i \geq 0 \ , \qquad y_i \geq 0 \ , \qquad z^t \geq 0 \ . \tag{6}$$

Two problems can be formulated for this system. The first is this.

P_1: for given capital stock z^1, available at the beginning of the i-th period,

$$\max \sum_{t=1}^{\infty} (\alpha)^{t-1} u(y_1^t, y_2^t) , \qquad 0 < \alpha < 1 , \tag{7}$$

subject to the above inequality constraints.

In this expression, (u_1^t, y_2^t) represents the utility attributed to the two consumption levels y_1, y_2 in the year t, and α is a discount factor. That is, $\alpha = 1/(1+\rho)$, where ρ is the discount rate. The maximand is thus the total discounted utility for all times in the future.

The origin of this problem formulation is with Frank Ramsey, a co-worker of John Maynard Keynes in the 30's [4]. Ramsey died young, but by the time of his death he had published four articles, each fundamental yet each in a different field. The one I am citing here is on optimal economic growth, and differs from P_1 mainly by the use of a continuous time variable. Another article was on optimal taxation. Both of these were in mathematical economics but there was a third article by him on the foundations of mathematics. Probably some of you here are familiar with it. A fourth article set an important precedent for statisticians. It dealt with personal or subjective probability, as it is now sometimes called.

The problem P_1 now is to determine how, beginning with a first-period starting level z^1, the capital stock will grow, or diminish, or perhaps fluctuate, if the discounted sum of future utilities is maximized. The utility function is assumed to be differentiable as needed, strictly concave, and to be increasing in each variable y_i. In other words, the greater the consumption, the greater the utility. This last assumption has an immediate consequence. The inequalities (4) and (5) must be satisfied as strict equalities, for if ever $y_i < x_i$ the utility is unnecessarily diminished, and so is the maximum in (7). Therefore, instead of (4) and (5), one will always have

$$x_1 = y_1 , \qquad x_2 = y_2 . \tag{8}$$

The second problem that can be considered presupposes that Problem P_1 can be solved. It is the following.

P_2: Find z^1 such that $\hat{z}^t = z^1$, $\hat{y}_i^t = \hat{y}_i$, for all $t > 1$, where $\hat{z}^t$, y_i^t solve P_1.

This is the problem I will discuss here. It requires you to find an initial capital stock such that the solution to Problem P_1 will lead to an indefinite repetition of that same capital stock. Each period's capital input will then be the same period's capital output. In other words, given a discount rate α (which presupposes a given interest rate $\rho = (1-\alpha)/\alpha$, what is the capital stock that will preserve itself under this optimization?

In Fig. 1a, the output levels y_1 and y_2 are plotted along two coordinate axes. According to (8), consumption and output levels are the same. The x_i and y_i can thus share the axes as shown. Since x_1 and x_2 are measured in units of labor, neither can exceed that unit. They range from 0 to 1 along each axis. The feasible region is thus the triangle which is denoted with $\mathcal{X}$ in the figure. The utility function may be defined more widely than on the limited feasible region. In the figure it is represented by the "level (or indifference) curves" $u(y_1,y_2)$ = const. Any two points on the same level curve produce the same utility. The figure also shows the symbols d_1 and d_2 introduced in Table 1.

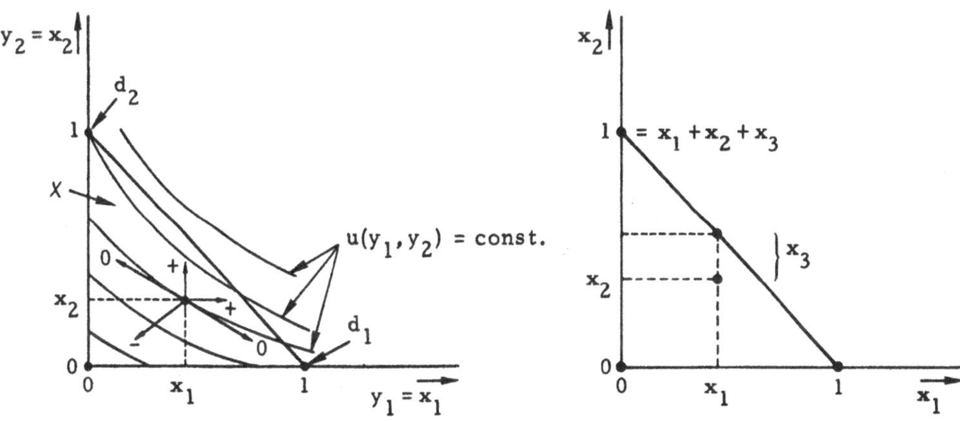

Fig. 1a Level curves for single-period utility function

Fig. 1b Activity levels under full-labor constraint

Note that the quantity x_3 is not shown in Fig. 1a. It is however brought out in Fig. 1b which displays a pair of values x_1, x_2, i.e., of the labor that goes into producing goods 1 and 2, along with the values of x_3, i.e., along with the labor that remains available for capital formation.

Figure 1b assumes that there is no slack in the labor constraint, i.e., that

$$x_1 + x_2 + x_3 = 1 \;, \tag{9}$$

rather than ≤ 1, which means that all labor is fully used. I will make this assumption, also. It will in fact be convenient for me to assume even more, namely, that there is no slack in the constraints (1) and (2) as well, i.e., that

$$-a_1x_1 - a_2x_2 - a_3x_3 = -z^1 \tag{10}$$

$$b_1x_1 + b_2x_2 + b_3x_3 = z^2 \;, \tag{11}$$

in place of (1) and (2). These assumptions are discussed in detail in the paper [3]. I will briefly discuss the condition under which they hold in a moment.

Observe now that each triplet (x_1,x_2,x_3) corresponds in a 1-to-1 way with a pair (z^1,z^2) of capital input and capital output values. The mapping between them is linear since it is defined by the linear eqs. (9), (10), and (11), and it carries the triangular feasible region $\mathcal{X}$ of Fig. 1a into a corresponding region in the $(-z^1,z^2)$-quadrant. This region is again a triangle and denoted in Fig. 2 with $\mathcal{Z}$. The two rectangular x-axes are mapped into the skew axes shown in the figure. The third x-axis is implied by (9). Conversely, to every point $(-z^1,z^2)$ in the set $\mathcal{Z}$ of feasible capital input and capital output values, there corresponds exactly one pair (x_1,x_2) in $\mathcal{X}$.

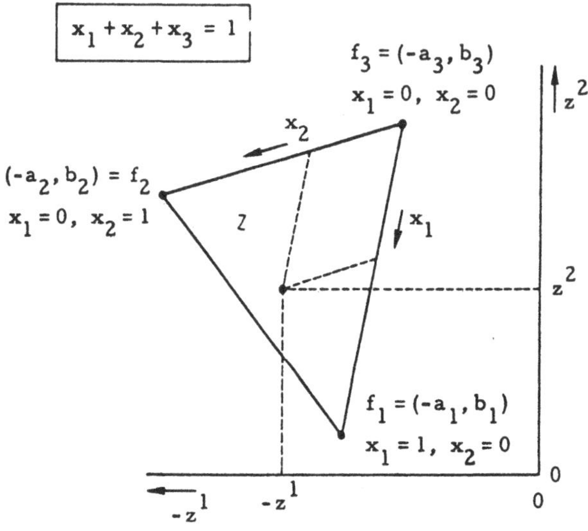

Fig. 2 Attainable set $\mathcal{Z}$ in the capital input and output space

The mapping also carries the utility function $u(y_1,y_2)$ into a function $\phi(z^1,z^2)$. The level lines of u are transformed into those of ϕ, as suggested in Fig. 3. Given two values z^1 and z^2, for instance the two shown in the figure, the pair (x_1,x_2) which corresponds to it in the region $\mathcal{X}$ may or may not be optimal. It certainly is optimal if there are no slacks in the constraints (1), (2), and (3), as I have assumed here so far, but it may not be if there are. The question therefore is this. Under what conditions does it remain optimal even when slacks are present?

It will not be possible to derive these conditions here. (They are derived in [3].) I can however state them here. They are the following. The two capital constraints (1) and (2), first of all, will have no slacks if the tangent to the level curve passing through the point $(-z^1,z^2)$ has a finite negative slope. The point in Fig. 3 clearly has this property.

The condition for the labor constraint is slightly different. The magnitude of the slope of the tangent must be less than that of the line $\overline{Of_3}$.

These conditions can be combined into the following. If $s(-z^1,z^2)$ is the slope of the tangent to the level curve through $(-z^1,z^2)$, then

$$-\frac{b_3}{a_3} \leq s(-z^1,z^2) < 0 \tag{12}$$

is sufficient for the non-optimality of any combination of slacks in the three constraints. If the "<" sign on the right is replaced with "≤", the condition is necessary.

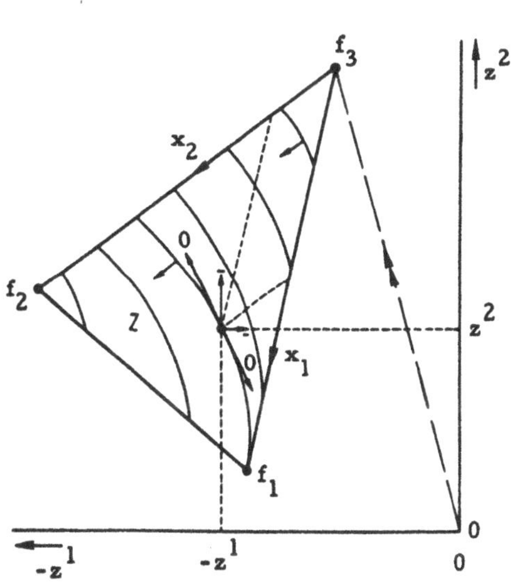

Fig. 3 Test of no-slack constraints for capital input and output

We now have the solution to the optimization problem for all points $(-z^1,z^2)$ for which this no-slack condition holds. It will become clear presently that this solution is all we will need to have in order to solve the problem we have called P_2 above, namely the conditions under which an invariant capital stock, i.e., one that preserves itself, is optimal. First, for it to be invariant, the capital input for each period must be equal to the capital output: $z^1 = z^2 = z$. This is accordingly a necessary condition for invariance. The points which satisfy this condilie along a straight line through the origin with slope (-1) and denoted with $\mathcal{L}$ in Fig. 4. Feasibility is a second necessary condition. It restricts the stocks z to the points lying inside the triangle $\mathcal{Z}$, i.e., to the segment $\mathcal{S}$ in which $\mathcal{L}$ intersects $\mathcal{Z}$. (This intersection is non-empty because of the inequalities between a_i and b_i I assumed earlier.)

The third condition is that a point (-z,z) on $\mathcal{S}$ will qualify as representing the initial value for an invariant capital stock if the level curve of the utility function ϕ through that point has the slope $(-1/\alpha)$ where α is the discount factor introduced earlier in Eq. (7). This condition is, surprisingly, again a slope criterion. It is derived in the following way.

Consider just two successive terms in the sum in (7) which is

$$U = \sum_{t=1}^{\infty} (\alpha)^{t-1}u(x_1^t,x_2^t) = \ldots(\alpha)^{t-1}\phi(-z^t,z^{t+1}) + (\alpha)^t\phi(z^{t+1},z^{t+2}) + \ldots \quad (13)$$

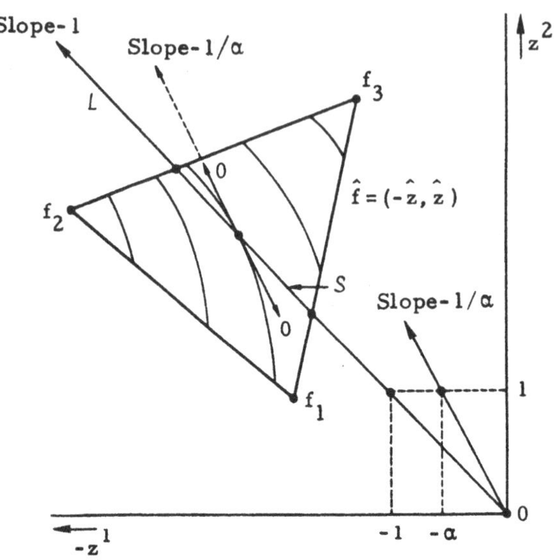

Fig. 4 Condition for an invariant optimal capital stock in an interior point of $\mathcal{Z}$

A necessary condition for U to be maximized by setting $z^t = \hat{z}$ for all t is then that, for each t, the sum of the terms

$$(\alpha)^{t-1}\phi(-\hat{z},z^{t+1}) + (\alpha)^t\phi(-z^{t+1},\hat{z})$$

be maximized with respect to z^{t+1} by setting

$$z^{t+1} = \hat{z} .$$

This condition evidently implies that the "slope condition"

$$\frac{\partial\phi(-\hat{z},z)}{\partial z} + \alpha\,\frac{\partial\phi(-z,\hat{z})}{\partial z} = 0$$

hold for $z = \hat{z}$, or in terms of the slope $s(-z^1,z^2)$ introduced above

$$s(-\hat{z},\hat{z}) = -1/\alpha ,$$

which is the third condition I have just mentioned.

It is of interest to know how such a stock depends on the value of the discount factor α. In Fig. 5, I have diagrammed part of the segment $\mathcal{S}$. It shows two points, both labeled $\hat{f}(\alpha)$, in which the level curves have the same slope $(-1/\alpha)$, and between them a third point $f(\alpha')$ with a different slope $(-1/\alpha')$ which corresponds to a larger discount factor $\alpha' > \alpha$. Such patterns are quite consistent with strict concavity of the function ϕ.

This observation is of some importance because it bears on a discussion among economists which has quite a long history. The issue is this. If the interest rate changes, does the invariant capital stock change? I myself think so because a smaller interest rate, hence a larger value of α, would mean giving more weight to the receipts expected in the future. Therefore the discount rate should, via the discount factor, have an influence on the value of the self-preserving capital stock. Intuitively, most economists would have said, the lower the discount rate (hence the larger α) the larger the self-preserving capital. We have a counter-intuitive case here, however. In going from point $\hat{f}(\alpha')$ in Fig. 5 to $\hat{f}(\alpha)$ in the upper left, the capital stock $\hat{z}$ increases, but the discount factor decreases from α' to α. This result has come as somewhat of a shock to economists reasoning from intuition.

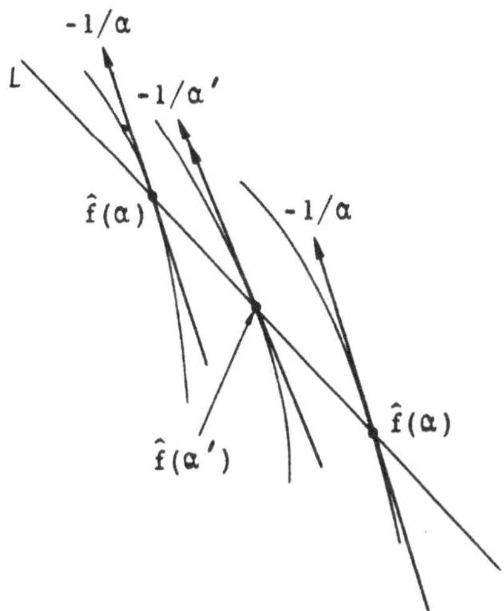

Fig. 5 Case of more than one invariant capital stock for a given α

There is however an economic situation in which this disturbing phenomenon cannot take place, namely, when the two consumption goods are "normal goods." This situation prevails when both of the two optimal consumptions y_1 and y_2, which are attainable within a given budget, increase as the budget is increased. It is the "normal" case because it describes a situation in which, if you get richer you consume more goods. But not all situations are normal. If the two goods are meat and potatoes, and you grow richer, you are likely to consume more meat, but fewer potatoes.

In the normal case the magnitude of the slope $s(y_1,y_2)$ of the level curve through (y_1,y_2), namely,

$$|s(y_1,y_2)| = \frac{\partial u(y_1,y_2)}{\partial y_1} \quad \frac{\partial u(y_1,y_2)}{\partial y_2} \tag{14}$$

is monotonically increasing in y_2, for fixed y_1, and monotonically decreasing in y_1, for fixed y_2. Consider now a straight line in the y-space (or x-space), such as $d_1'd_2'$ in Fig. 6a. As you proceed from the point $(d_1',0)$ towards $(0,d_2')$, the magnitude of the slope in (14) decreases. This should be quite clear from the figure.

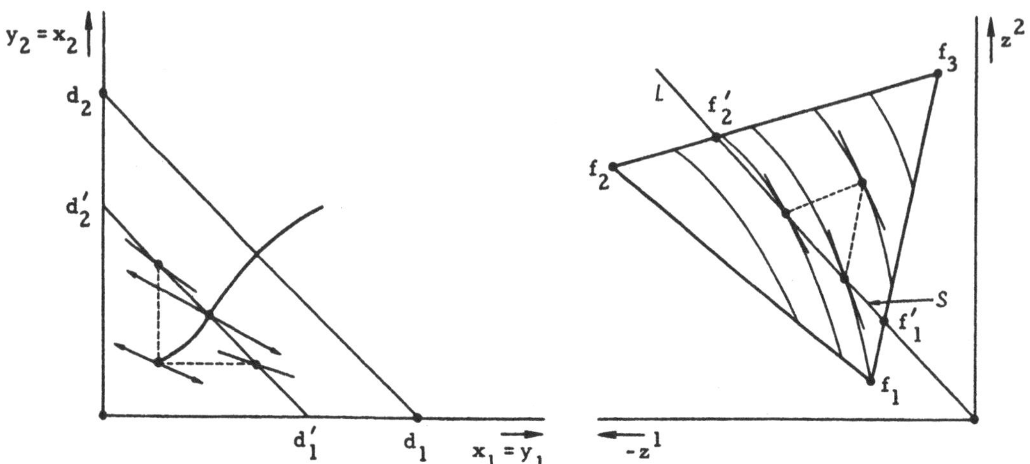

Fig. 6a The case of normal consumption goods

Fig. 6b Invariant capital stock increases with α

Now let the segment $\overline{d_1'd_2'}$ be so chosen that it maps into the segment in the z-space. The point $(d_1',0)$ will then be carried into one such as f_1' in Fig. 6b, and $(d_2',0)$ into f_2'. It is evident that the slope of the level curves decreases in magnitude, as you proceed on $\mathcal{S}$ from the lower right to the upper left, and hence, as α increases. The capital stock $\hat{z}$ however increases also. Consequently, in this case, the self-preserving capital stock behaves in the intuitively expected way.

A question of interest in the normal case is the range of invariant capital stocks that are achievable as α varies over the interval (0,1). Let $\bar{z}$ and $\underline{z}$ be the largest and smallest value which $\hat{z}$ can attain in this way. It is clearly possible that, as you proceed along in Fig. 6b towards f_2' you approach a point short of f_2' at which the slope s attains the value (-1), hence α approaches its upper bound $\bar{\alpha} = 1$. When this happens invariant capital stocks $z > \hat{z}(1)$ cannot be realized. On the other hand, if the magnitude $|s|$ of the slope at f_2' is still greater than 1, $|s| = 1/\alpha > 1$, then f_2' is an upper limit on the invariant capital stock attainable by increasing the discount factor. This is illustrated in Fig. 7a.

Going in the other direction along $\mathcal{S}$, namely towards f_1', α decreases towards the smallest value $\underline{\alpha} = a_3/b_3$ permitted by (12), i.e., without slack in the labor constraint. It is then possible that the invariant capital stock $\hat{z}(\underline{\alpha})$ is represented by a point in the interior of $\mathcal{S}$, as also indicated in Fig. 7a. If α is decreased further, this invariance cannot be enforced any longer (unless slack is allowed in the labor constraint. The input and output capital stocks then diverge, as indicated in Fig. 7a by the curve passing through $\hat{f}$. Its image in the x-space is sketched in Fig. 7b.

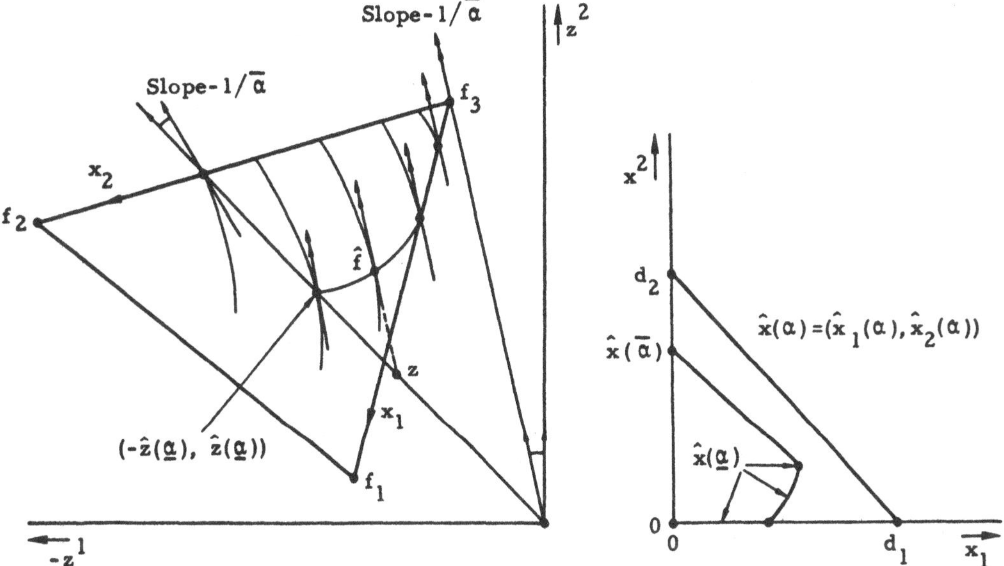

Fig. 7a Locus of the invariant capital stock for $1 > \alpha > 0$

Fig. 7b Locus of associated consumption vector for $1 > \alpha > $)

I will now turn to the final topic of this presentation. This is the question of the stability of the self-preserving capital stock, that is, if $\hat{z}$ is invariant and if the initial stock z^1 differs slightly from $\hat{z}$, will the sequence $z^1, z^2, \ldots$ obtained by the optimization of (13) converge to $\hat{z}$? The same question can also be asked regarding small changes in the discount factor α. Both questions were answered by an analysis carried out rigorously by my colleague K. Iwai. It establishes a link with "catastrophe theory." I will however give only a brief account of it here.

Iwai's analysis shows that instability prevails in the case I have called counter-intuitive earlier and which is illustrated in Fig. 8a. The slopes $(-1/\alpha)$ of the level curves (not shown in the figure) decrease in absolute value as you travel along the segment $\mathcal{S}$ from f_2' to f_1' (and α increases from $\bar{\alpha}$ to α'). In such a case, according to Iwai's results, if for some α in the interval $\bar{\alpha} > \alpha > \alpha'$ the initial capital stock z is even only slightly larger than the self-preserving $\hat{z}(\alpha)$ for that point, that stock will continue to grow larger as time goes on; and if it is even only slightly smaller it will continue to grow smaller. This is indicated in Fig. 8b. Our result probably is the based for the economists' intuition. I am here concerned not so much with the relationship between discount rate and the self-preserving capital stock, as with the stability of that relationship.

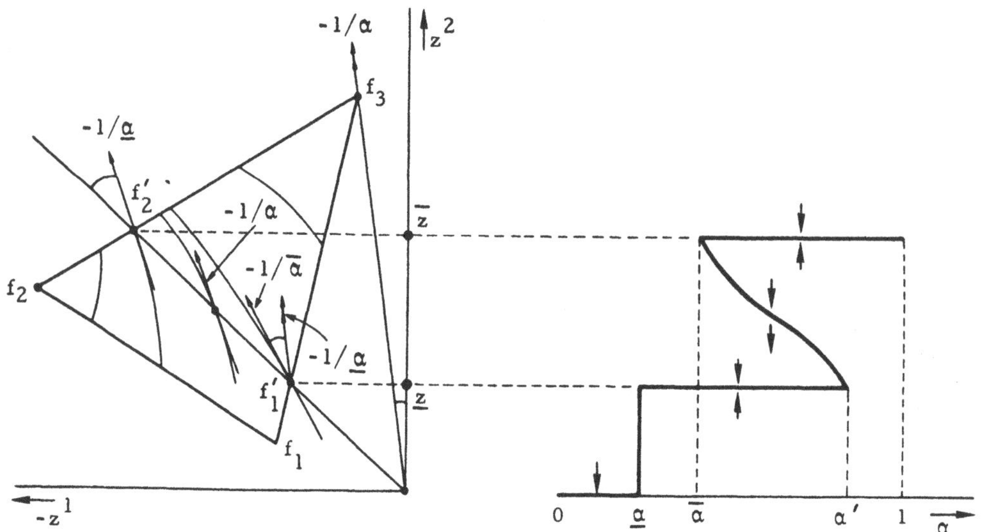

Fig. 8a An unstable invariant stock $\hat{z}(\alpha)$ Fig. 8b "Catastrophes" at $\bar{\alpha}$ and at α'

The situation is similar for small changes in α. Suppose (see Fig. 8b) that the self-preserving level $\bar{z}$ of capital stock has been established for some α slightly in excess of $\bar{\alpha}$, and that thereupon the population becomes less patient, i.e., wants an α smaller than $\bar{\alpha}$. In this case, the "optimal" z^t will soon begin its decline as t increases and a kind of catastrophic drop of z^t down to $\underline{z}$ will take place. Conversely, if the public has an $\alpha > \alpha'$, in equilibrium with a stock $z = \underline{z}$, and then grows more concerned with the future and becomes more parsimonious, increasing α slightly above α', then, z^t will work itself up to the level $\bar{z}$ in the figure, at a high cost of transition.

Here, therefore, is the point of contact with catastrophe theory. I confess that I have in the past been unappreciative of that theory. I have heard several of Professor Zeeman's lectures [6], and whenever his examples were drawn from a field with which I am not familiar my reaction was one of great interest. But when his example belonged to a field I think I know something about I had some reservations. Now I must apologize to Professor Zeeman. The kind of catastrophe he writes about can take place under assumptions frequently made in my own field, and in a very definite well formulated problem area. The way a slightly raised impatience of the public can lead to a catastrophic fall in the self-preserving capital stock is a case in point.

In conclusion I would like to indicate the nature of the mathematical problem that results if there are many consumption goods, capital goods, and resources. (In the problem discussed here labor was the only resource.) In the more general problem, the question of self-preservation concerns a vector of capital stocks. This question is equivalent to that of the existence of a fixed point for a continuous mapping of a convex set into itself. In addition to the existence, there is also the problem of the computation of such a fixed point. It is a problem that has occupied economists for the last ten to twenty years. My colleague Herbert Scarf at Yale is one of the leading contributors to the development of such algorithms. His algorithm in fact provides the solution to the more general problem [5]. Its application to the problems discussed in this lecture was developed by Terje Hansen [1].

REFERENCES

1. T. Hansen and T.C. Koopmans, Definition and Computation of a Capital Stock Invariant Under Optimization, Jour. of Economic Theory, Vol. 5(1972) pp. 487-523.

2. T.C. Koopmans, Examples of Production Relations Based on Microdata, in G.C. Harcourt (ed.), the Microeconomic Foundations of Macroeconomics, Macmillan Press (1977).

3. T.C. Koopmans, Examples of Production Relations Based on Microdata, Cowles Foundation Paper No. 455, Yale Univ. (1978).

4. F. Ramsey, A Mathematical Theory of Saving, Economic Jour. Vol. 38 (1928) pp.534-459.

5. H.E. Scarf, The Computation of Economic Equilibria, Yale Univ. Press (1973).

6. E. Zeeman, Catastrophe Theory - Selected Papers, Addison Wesley (1977).

SOME NUMERICAL PROBLEMS

ARISING FROM LINEAR SYSTEMS

Chris Paige
School of Computer Science
McGill University
Montreal, Quebec, Canada H3A 2K6

Abstract

Mathematically equivalent but computationally different ways of computing a result can lead to totally different answers. Some simple problems in linear systems will be used to illustrate this difficulty. The reasons for the failure of some well known algorithms will be given, along with insights into the design of good numerical algorithms in this and related areas.

1. Introduction

Many mathematically correct algorithms do not work well on computers, giving answers which can have unnecessary inaccuracies or even be meaningless. Here we comment on some well known algorithms for solving some problems in linear systems theory. We show that some of these do not work satisfactorily on a computer (they are "numerically unstable") and where possible we indicate why this unnecessary loss of accuracy occurs. It is known that certain general techniques are numerically risky, and so the knowledge gained from the individual algorithms here will be useful in avoiding such dangers when designing other algorithms. On the positive side, we know that certain practices contribute to numerically reliable results, and where possible we will indicate some of these as ways of avoiding the aforementioned dangers. This will provide some useful tools for good algorithm design, especially for control problems.

Numerically stable and efficient algorithms are not yet available for many interesting problems in systems theory, and we point out some such problems here that are quite basic. This lack of numerically reliable algorithms combined with the ever increasing use of digital computers in the design and application of control systems makes some understanding of the computational behaviour of algorithms both important and interesting for workers in this area. We hope this brief exposition will serve as an introduction to the subject for those who have not studied more substantial material, such as the numerical work of Wilkinson (see [13]), or the application of good numerical algorithms to linear systems (see for example [5]).

2. Computing Exponentials

The linear constant coefficient ordinary differential equation

$$\dot{x}(t) = Ax(t) \ , \quad x(0) = x_o \ ,$$

occurs regularly in modelling linear systems. Here A is a given constant $n \times n$ matrix, and the solution is $x(t) = \exp(tA)\, x_o$ where

$$\exp(tA) = I + tA + \frac{t^2A^2}{2!} + \ldots$$

For this and other reasons computing the matrix exponential is a basic problem in linear systems, and yet the very thorough and informative paper by Moler and Van Loan [11] concludes that none of the many methods they studied is entirely satisfactory. We are still searching for a general and efficient numerically stable method for computing the matrix exponential. This does not mean that in any given case we cannot compute a satisfactory answer, and in fact several of the available methods will give good answers most of the time, and some of the methods will tell us when the answer is suspect. However it does mean that we do not yet know of an algorithm for computing the exponential of a matrix that will always return as much information as the data and the precision of the computer will allow.

One general numerical danger can be exhibited by considering one of the most obvious, if inefficient, ways of computing $\exp(A)$. Let

$$T_k(A) = I + A + \frac{A^2}{2!} + \ldots + \frac{A^k}{k!}$$

be a Taylor series approximation to $\exp(A)$. The following computer algorithm could be used to approximate $\exp(A)$.

```
EXPA  := I;  (to contain T_k(A))
AKOKF := I;  (to contain A^k/k!)
FOR k := 1,2,.. UNTIL NO CHANGE IN EXPA DO
    AKOKF := AKOKF * A/k ;
    EXPA  := EXPA + AKOKF ;
```

This increases the order of the Taylor series approximation until, because of rounding, the next term does not alter the computed result of the last line. It can be shown that this is a reasonable way to stop if the computer uses rounding by adding, and that the resulting *truncation* error can safely be ignored. The algorithm is very simple, can it go wrong? First we give a brief introduction to rounding errors.

Let ε be the relative precision of our computer. Effectively this means that if we have a real number α and store it in our computer, we will know that the resulting floating point computer number $\tilde{\alpha}$ obtained by rounding α to fit the computer word length will satisfy

$$\tilde{\alpha} = \alpha(1 + \varepsilon_1) \ , \quad |\varepsilon_1| \le \varepsilon \ ,$$

and ε is the smallest such number for which we can always say this. For this exposition we will just consider rounding errors and ignore the dangers of overflow or underflow, something we could not afford to do in practice.

If a given algorithm for computing $\exp(\alpha)$ for any given real scalar α could be relied on to give an answer EXPA such that

$$\text{EXPA} = \exp(\alpha + \delta\alpha)$$

with

$$|\delta\alpha| \le \phi(k).\varepsilon.|\alpha| \; ,$$

where $\phi(k)$ is some low order polynomial in k, the number of steps our algorithm required, then we would say the algorithm is *numerically stable*. We see this means that we would be sure of obtaining *the exact answer* for *slightly perturbed initial data*. A rounding error analysis following the work of Wilkinson (see for example [13]) would be required to obtain such a bound on $\delta\alpha$, and for a good numerically stable algorithm we could hope for such a bound with something like $\phi(k) = 3k$. Note that the resulting bound on this equivalent error in α would then be only $3k$ times the bound on the error obtained just by storing α. As a result we would be content that the algorithm would return just about as much information as the data and precision of computation would allow.

With this background let us now try our Taylor series algorithm taking the matrix A to be a scalar. For our computed result EXPA we compute the *equivalent data perturbation*

$$\delta A = \ln(\text{EXPA}) - A$$

such that $\text{EXPA} = \exp(A + \delta A)$ is the *exact* answer for perturbed data $A + \delta A$. If $|\delta A/A|$ is small the algorithm could be numerically stable, otherwise it is not. On an IBM computer with relative precision $\varepsilon = 16^{-5} \cong 10^{-6}$ we tried $A = 7.7$, and obtained convergence in 27 steps giving $\text{EXPA} = 2208.340$ $(\exp(7.7) = 2208.348)$, and $\delta A = -.3\times10^{-5}$. This looked good and suggested the algorithm could be stable. Next we tried $A = -7.7$, obtaining convergence in 36 steps with $\text{EXPA} = .0005929$ $(\exp(-7.7) = .0004528)$, and $\delta A = .27$, which is a hopeless result, showing the algorithm is not even stable for *scalars* A, and so it will not be stable for matrices A.

The difficulty here is that unless $|A|$ is small we get a large growth in magnitude of the numbers we compute. In the present case this is not so important if the final result is large, as when we took $A = 7.7$, but it is catastrophic when the final result is small, and in general can lead to unnecessary loss of accuracy even when the final result is not small. Here our computations gave

$$T_6(-7.7) = 157.237 \quad , \quad T_7(-7.7) = -161.185 \; , \quad T_8(-7.7) = 145.296$$

which, because of their size and our precision of computation, almost certainly have *absolute errors* of about the same size as our final computed *result* $T_{36}(-7.7) = .0005929$. Thus the contributions of these earlier values to the final result are effectively obscured by the rounding errors, and the final result has no figures accurate. Here there was no large cancellation in any one step, the largest relative decrease in size being from $T_{22}(-7.7) = -.007752$ to $T_{23}(-7.7) = -.001727$, a factor of less than 5 ; instead the size is brought down over many steps.

For a negative scalar A we would get a much improved result by applying the Taylor series algorithm to -A and inverting the result. If a matrix A had all negative eigenvalues we could do the same thing, but then we could have difficulties from errors introduced by the inversion. For general matrices it is wise to avoid solution of systems of equations or inversion unless absolutely necessary. Again this danger is often related to large increases in size, for example the bound on the relative error in the computed inverse (using a numerically stable algorithm, see [13]) of a matrix A is proportional to $\chi(A).\varepsilon$ where $\chi(A) = ||A||.||A^{-1}||$ is called the condition number of A for solution of equations. We see $\chi(A) \geq 1$, and if $\chi(A)$ is large then we can have large relative errors introduced by forming A^{-1}. However for some matrices we know $\chi(A)$ is small, and then we need have no fears about solving equations with them. An important class of such matrices is that of orthogonal matrices, then if $Q^TQ = QQ^T = I$ we have $Q^{-1} = Q^T$ and $\chi(Q) = 1$ for the correct norm. Transformations with orthogonal matrices are favoured by numerical analysts because they do not lead to the changes in size that were seen to cause difficulties in this Section.

We return to the problem of computing exp(A) . In parallel with the discussion for scalar α we would like an efficient algorithm which for any given real $n\times n$ matrix A could be relied on to give a computed matrix EXPA such that

$$\text{EXPA} = \exp(A + \delta A)$$

with
$$||\delta A|| \leq \phi(k,n).\varepsilon.||A||$$

for some matrix norm $||.||$. Here $\phi(k,n)$ is some low order polynomial in n and the number of steps k the algorithm requires to compute EXPA . The algorithm would then be numerical stable, but we know of no such algorithm. For more details on this problem and related insights on numerical computations see [11].

3. Linear Equations

In solving the system of linear algebraic equations

$$Ax = b\ , \quad \text{nonsingular } \underset{n\times n}{A} \text{ and } \underset{n\times 1}{b} \text{ given,}$$

we have numerically stable algorithms [13] which can be relied on to give a computed result $\tilde{x}$ satisfying

$$(A + \delta A)\tilde{x} = b\ , \quad ||\delta A|| \leq \phi(n).\varepsilon.||A||\ .$$

Thus $\tilde{x}$ will be the exact answer for nearby data. An interesting point is that the residual r for such a result satisfies

$$||r|| = ||b-A\tilde{x}|| = ||\delta A\tilde{x}|| \le \phi(n).\varepsilon.||A||.||\tilde{x}|| ,$$

and $||r||/(||A||.||\tilde{x}||)$ will necessarily be small. Thus computing this ratio is a nice way to check on the stability of an algorithm for solving linear systems. If the ratio is not small, then the algorithm is not numerically stable.

Solving the matrix Lyapunov equation

$$AX + XA^T + H = 0$$

appears to be a basic numerical problem in linear-quadratic theory. Here A and $H=H^T$ are given real $n\times n$ matrices, and this is a linear system of equations for the unknown matrix X . The method of Bartels and Stewart [1] is popular for solving this problem. This method first reduces A to Schur form

$$Q^TAQ = R , \; Q^TQ=QQ^T=I ,$$

where R is upper triangular, except for 2×2 blocks on the diagonal corresponding to the complex conjugate pairs of eigenvalues of A . The method then solves

$$RY + YR^T + Q^THQ = 0$$

for Y and then computes $X = QYQ^T$.

Partly because it uses nice orthogonal transformations Q , the method used for reducing A to Schur form is known to be numerically stable. Also the solution for Y is a straightforward solve if R is upper triangular, and the formation of X is numerically stable. Even so the resulting method is apparently *not* numerically stable. Belanger and McGillivray [2] carried out computations with this method on a problem with $n=24$ arising from the study of a magnetic levitation vehicle suspension. They carried out their computations on an IBM 360 Model 75 in double precision ($\varepsilon = 16^{-13} \cong 2\times 10^{-16}$) with $H=I$, and obtained a computed solution X_1 such that the largest off-diagonal element of $AX_1+X_1A^T$ was 1.08×10^{-4} . This means that their residual

$$||AX_1 + X_1A^T + H|| > 10^{-4} ,$$

and from our discussion on residuals of linear equations we would expect a numerically stable algorithm working with this precision to give a far smaller residual than this.

By carrying out one step of iterative refinement, i.e. by solving

$$A\delta X_1 + \delta X_1A^T + (H + AX_1 + X_1A^T) = 0$$

for δX_1 and forming $X_2 = X_1 + \delta X_1$, they found that the largest off-diagonal element of $AX_2 + X_2A^T$ was 2.22×10^{-16} , (a stunning improvement for one step of refinement) thus showing the size of residual that could be obtained for this particular problem. Even with this numerical instability and iterative refinement, the speed and final accuracy of the result were both favorable compared with the other methods they considered.

We thus have a popular method which although numerically unstable, appears to work well if iterative refinement is used when the residual is too large. This is quite a good method, but it has dangers. For example on a computer with less precision this problem could have led to a meaningless result which no amount of iterative refinement would have improved.

The algorithm in [1] was designed to solve

$$AX + XB = C \;; \quad \underset{m\times m}{A}\,, \quad \underset{n\times n}{B}\,, \text{ and } \underset{m\times n}{C} \text{ given .}$$

For this problem an algorithm that could be relied on to give a computed solution $\tilde{X}$ which satisfies

$$(A+\delta A)\tilde{X} + \tilde{X}(B+\delta B) = C+\delta C$$

$$||\delta F|| \le \phi_F(m,n).\varepsilon.||F||, \quad \text{for} \quad F = A,\ B, \text{ and } C\ ,$$

would be numerically stable. The algorithm in [1] does not do this, though it is not clear where the difficulty lies. It may be the sensitivity of the eigenvalues obtained via the Schur reduction that causes the trouble. The Schur reduction effectively computes the eigenvalues, and then these play a key role in the solution of equations. Whatever the reason, we still appear to need an efficient numerically stable algorithm for solving this problem.

For the matrix Lyapunov equation $AX + XA^T + H = 0$ with $H = H^T$ we often have positive definite H and stable A, and then X is also symmetric positive definite. In this case it makes sense to work from the factor L of $H = LL^T$ and compute the factor U of $X = U^TU$, thus avoiding unnecessary squaring of matrices with its concomitant loss of accuracy. The recent work of Hammarling [7] gives a nice way of doing this, but like the method in [1] it is based on the Schur reduction of A, and could suffer similar numerical instability. Computational results would be most useful here.

4. Sensitivity of Eigenvalues

In Section 3 we suggested that the sensitivity of the eigenvalues of a matrix could have led to the poor performance of a numerical algorithm. Here we give a useful example of such sensitivity. Consider the $n\times n$ upper triangular matrix

$$A = \begin{bmatrix} 1 & -1 & \cdot & -1 & -1 \\ & 1 & \cdot & -1 & -1 \\ & & \cdot & \cdot & \cdot \\ & & & 1 & -1 \\ & & & & 1 \end{bmatrix},$$

which has n eigenvalues of unity and determinant $\det(A) = 1$. It is a favourite among numerical analysts and appears to be quite far from being singular. However noting that $1 + 2^1 + 2^2 + .. + 2^r = 2^{r+1} - 1$ for any integer $r \ge 0$, we see

$$(1\ ,\ 2^1\ ,\ 2^2\ ,\ldots,\ 2^{n-1})\begin{bmatrix} 1 & -1 & \cdot & -1 & -1 \\ & 1 & \cdot & -1 & -1 \\ & & \cdot & \cdot & \cdot \\ & & & 1 & -1 \\ \alpha & \alpha & \cdot & \alpha & 1+\alpha \end{bmatrix} = 0$$

when $\alpha = -2^{1-n}$. So for large n we see that only a very small perturbation is needed to make A singular. Thus a very small perturbation has moved an eigenvalue from 1 to 0 . The matrix is not as far from being singular as first sight suggests. We introduce this example for a specific purpose, but in general matrices do not have to be large to have sensitive eigenvalues.

The idea of sensitivity as used here has nothing to do with how the eigenvalues are computed. One of the essential points in understanding numerical algorithms is to separate the ideas of sensitivity of problems and stability of algorithms. We see that sensitivity of the eigenvalues is just a *mathematical* property of the given matrix alone. And no matter what the sensitivities of the eigenvalues λ_i are for a given matrix, we know that a numerically stable algorithm will give us computed eigenvalues $\tilde{\lambda}_i$ which are *exact* for a *nearby* matrix $A + \delta A$, see [13]. We see this is a property of the *algorithm* alone. However if the eigenvalues *are* sensitive, then we may have *large* $|\tilde{\lambda}_i - \lambda_i|$ even with a numerically stable algorithm: this is no reflection on the algorithm, it has done all that could be desired of it.

As a result of this discussion it is reasonable to conclude that although we have excellent numerically stable algorithms for computing eigensolutions (see [13]), it is wise to avoid using these as substeps early in a larger computation if we can do so. This is because sensitive eigenvalues may cause unnecessary loss of accuracy in later computations, and this may well be the difficulty with the algorithm in [1] discussed in Section 3.

5. Computing Controllability

Several methods have been proposed for determining if a given linear system is controllable or not, and the numerical properties of some of these have been discussed in [12]. For exposition we consider the single input time invariant linear system

$$\dot{x} = Ax + bu\ ;\quad \underset{n\times n}{A} \ \text{ and } \ \underset{n\times 1}{b} \ \text{ given .}$$

Wonham [14] showed that such a system is controllable if and only if a vector f can be chosen so that $A + bf^T$ has any given set of eigenvalues. Based on this the following easily implementable method has been suggested in [4]. Take a random n-vector f . If A and $A + bf^T$ have no two eigenvalues equal then the pair (b,A) is controllable.

A possible danger with such an approach should be fairly obvious from the last section. Suppose the pair (b,A) is uncontrollable, but the eigenvalues of A are very sensitive. Even with a numerically stable algorithm the computed eigenvalues $\tilde{\lambda}_i$ for A are liable to be quite different from the true eigenvalues of A, and different from the theoretically fixed eigenvalues of $A+bf^T$. Thus it would be possible to conclude that an uncontrollable system is controllable - an error we would prefer not to make. An example of this is given in [12].

A more promising approach to testing controllability is to transform A and b with an orthogonal matrix Q so that

$$\tilde{b} = Q^T b = \begin{bmatrix} \alpha_{10} \\ 0 \\ 0 \\ \cdot \\ 0 \end{bmatrix}, \quad \tilde{A} = Q^T A Q = \begin{bmatrix} \alpha_{11} & \alpha_{12} & \cdot & \alpha_{1,n-1} & \alpha_{1,n} \\ \alpha_{21} & \alpha_{22} & \cdot & \alpha_{2,n-1} & \alpha_{2,n} \\ 0 & \alpha_{32} & \cdot & \alpha_{3,n-1} & \alpha_{3,n} \\ 0 & 0 & \cdot & \cdot & \cdot \\ 0 & 0 & 0 & \alpha_{n,n-1} & \alpha_{n,n} \end{bmatrix}. \tag{5.1}$$

This avoids finding eigenvalues, and maintains the size of the matrix and vector. If we define $\tilde{x} = Q^T x$ then the system

$$\dot{\tilde{x}} = \tilde{A}\tilde{x} + \tilde{b}u$$

is equivalent to the earlier system. This is a simple direct computation, see for example [10], and was originally proposed in [6]. It is known [8] that the system is controllable if and only if

$$\text{rank } (b, A-\lambda I) = n \quad \text{for all } \lambda .$$

It is then obvious from the transformed system that the system is controllable if and only if

$$\alpha_{10}\alpha_{21}\cdots\alpha_{n,n-1} \neq 0 .$$

Since it can be shown that the computed $\tilde{A}$ and $\tilde{b}$ are exact for nearby $A+\delta A$ and $b+\delta b$ this looks like a good algorithm, and it was used to give good results in [12].

Unfortunately the numerical problem of determining if a given system is controllable or not does not appear to have been satisfactorily solved. One problem lies in the yes-no nature of the question. What happens if the system is controllable, but very close to an uncontrollable system? This may not be obvious from the previous computation. For example the pair

$$(b,A) = \left[\begin{array}{c|cccc} 1 & -1 & \cdot & -1 & -1 \\ & 1 & \cdot & -1 & -1 \\ & & \cdot & \cdot & \cdot \\ & & & 1 & 1 \end{array}\right]$$

has $\alpha_{10}\alpha_{21}\cdots\alpha_{n,n-1} = 1$, and so is controllable, and looks clearly so. Yet we have seen in Section 4 that

$$(b',A') = \left[\begin{array}{c|cccc} 1 & -1 & \cdot & -1 & -1 \\ & 1 & \cdot & -1 & -1 \\ & & \cdot & \cdot & \cdot \\ \alpha & \alpha & \cdot & 1+\alpha & 1+\alpha \end{array}\right]$$

has rank less than n when $\alpha = -2^{1-n}$. This means rank $(b',A'-0.I) < n$ and the pair (b',A') is uncontrollable. But for large n this pair is extremely close to (b,A) , and so it can be difficult to tell when a system is close to an uncontrollable system. I believe such problems are considered in [3].

It was pointed out in [12] that it is desirable to know just how close a given system is to the nearest uncontrollable system, and a suggested approach to this was to compute

$$\mu(A,b) = \text{minimum} \; ||(\delta b,\delta A)||$$

such that $(b+\delta b, A+\delta A)$ is an uncontrollable pair. This has been considered in [9] but has not been satisfactorily solved, and appears to be another interesting problem in this area.

The reduction in (5.1) promises to be a very useful tool in this area. Several important computations involving transformations similar to (5.1) are given in the very comprehensive paper by Van Dooren [5], while use is made of (5.1) for the pole assignment problem in [9] and [10].

References

[1] Bartels R.H. and Stewart G.W., Solution of the matrix equation AX+XB=C . Commun. ACM 15, 820-826 (1972)

[2] Belanger P.R. and McGillivray T.P., Computational experience with the solution of the matrix Lyapunov equation. IEEE Trans. automatic Control AC-21, 799-800 (1976)

[3] Boley D., Computing the controllability-observability of linear time-invariant systems. A numerical approach. Ph.D. dissertation, Computer Science, Stanford University, Stanford, California (1981)

[4] Davison E.J., Gesing W. and Wang S.H., An algorithm for obtaining the minimal realization of a linear time-invariant system and determining if a system is stabilizable-detectable. IEEE Trans. automatic Control AC-23, 1048-1054 (1978)

[5] Van Dooren P.M., The generalized eigenstructure problem in linear system theory. IEEE Trans. automatic Control AC-26, 111-129 (1981)

[6] Van Dooren P.M., Emami-Naeini A. and Silverman L., Stable extraction of the Kronecker structure of pencils. Proc. 17th IEEE Conf. Decision Contr., 521-524 (1979)

[7] Hammarling S.J., The numerical solution of the Lyapunov equation. NPL, Teddington, England, Rep. DNACS 49/81 (Aug. 1981)

[8] Hautus M.L.J., Controllability and observability conditions of linear autonomous systems. Proc. Kon. Ned. Akad. Wetensh. Ser. A. 72, 443-448 (1969)

[9] Miminis G.S., Numerical algorithms for controllability and eigenvalue allocation. M.Sc. dissertation, Computer Science, McGill University, Montreal, Quebec (1981)

[10] Miminis G.S. and Paige C.C., An algorithm for pole assignment of time invariant linear systems. Internat. J. Control, to appear (1982)

[11] Moler C.B. and Van Loan C., Nineteen dubious ways to compute the exponential of a matrix. SIAM Review 20, 801-836 (1978)

[12] Paige C.C., Properties of numerical algorithms related to computing controllability. IEEE Trans. automatic Control AC-26, 130-138 (1981)

[13] Wilkinson J.H., The Algebraic Eigenvalue Problem. London: Oxford Univ. Press (1965)

[14] Wonham W.M., On pole assignment in multi-input controllable linear systems. IEEE Trans. automatic Control AC-12, 660-665 (1967)
[15] Golub G.H., Nash S. and Van Loan C., A Hessenberg-Schur method for the problem AX+XB=C. IEEE Trans. automatic Control AC-24, 909-913 (1979)

Note added in printing: It has been brought to my attention by S. Hammarling that the residual using the algorithm in [1] *will* be well-behaved. This is shown in [15], which also gives an improved algorithm along the lines of that in [1]. These algorithms are thus effectively stable for the general AX + XB = C problem, and the algorithm in [7] will be excellent for the matrix Lyapunov equation. The discussion in Section 3 on possible large residuals for these equations is incorrect. I have since found that the code used in [2] was faulty, leading to an incorrect result for the initial residual. The algorithms in [1], [7], and [15] are then even better than suggested in [2], and give numerically satisfactory results without requiring iterative refinement.

Chris Paige

OPTIMAL SHAPE DESIGN FOR ELLIPTIC SYSTEMS

O. PIRONNEAU
University of Paris 13 and INRIA

Plan

1. Definition and Applications
2. Solutions by optimization techniques
3. Optimality conditions
4. Discretization
5. Examples
6. Open problems.

1. Definition and Applications

Let A be an unbounded operator which defines a function Φ from $\Omega \subset R^n$ into R^m. Let $E(\Phi,\Omega)$ be a real valued functional of Φ and Ω. Let $\mathcal{O}$ be a set of admissible Ω. Problems of the type

$$\min_{\Omega \in \mathcal{O}} \{E(\Phi,\Omega) : A(\Phi,\Omega) = 0\} \tag{1}$$

are called optimal shape design problems.

More generally, we may extend this definition to any optimization problem with a criteria that involves a function defined by Partial Differential Equations and with optimization variables which are geometrical elements of the problem (domains, subdomains, curves, points with Dirac masses ...).

There are many concrete problems of this type relevant to industry. In fact, almost all distributed systems are simulated numerically in industry for optimization and very often the optimization variable is geometrical. Let us list any examples.

Structural design :

Minimization of the veight with respect to geometry under constraints on the stresses. Applications are in architecture (shells) ship building , large airplanes ...

Engine design :

Optimization of performance with respect to the shape of the combustion chamber for jet engine or motors of cars.

Airplane design :

Optimization of aerodynamic performance (drag, pressure distribution) with respect to the shape (usually of the wing or some part like the cockpit).

Ship design :

Similar problems for the front and near of a ship.

Magnetostatic design :

Optimization of rotors of electric motors or of electromagnets.

2. Solution by optimization techniques

The list of examples in the previous paragraph demonstrates that optimal shape design is not only a mathematical game and that in many instances engineers are in fact simulating their systems only to be able to optimize it.

It is commonly admitted that a good engineer with some experience of the system can arrive within 5 % of an optimum by purely intuitive methods provided that the system has less than, say, 5 parameters.
Beyong these numbers one must use more rationnal methods.

Now, if $z = \{z_1,\ldots,z_N\}$ denotes the variables with respect to which $E(z)$ is to be minimized one method that can always be used is to compute

$$\frac{\partial E}{\partial z_i} \simeq \frac{E(z_i + k) - E(z_i)}{k} \tag{2}$$

and change z_i into $z_i + k$ if $E(z_i + k) < E(z_i)$ and so on for $i = 1\ldots N$. This so called method of local variationscan be further refined by computing $E(z)$ for $N^2/2 + 2N+1$ different z and taking the minimum in z of the quadratic approximation of E from these values of $E(z)$. These methods are certainly very simple but they are very expensive computationnally so if there is a problem of computing time (used there usually is) one must call on some more elaborate method .

One other temptation is to use a fixed point technique. This again is best understood on an example.
Suppose we want to solve

$$\min_{\Omega} \int_{\Gamma} |\nabla\Phi - u_d|^2 \, d\Gamma \quad , \quad \Gamma = \text{boundary of } \Omega \tag{3}$$

where Φ is the solution of $A(\Phi,\Omega) = 0$. The fixed point algorithm would be to move Γ of a quantity proportionnal to $|\nabla\Phi - u_d|^2$ on the basis that the bigger this is the more Γ has to be moved.
There are many instances where this technique works [6] [7] but there are an equal number of instances where it does not work. So in the end (or on the average) the engineer might find that technique too risky.

The right way to compute the minimum in $z \in R^N$ of a function $E(z)$ is to compute its gradient Grad $E(z)$ analytically if possible and to approach the minimum as the limit of a convergent sequence (or subsequence) $\{z^m\}$. The simplest is the method of steepest descent :

$$z^{m+1} = z^m - \rho \text{ Grad } E(z^m) \qquad (\rho \text{ small positive number}) \tag{4}$$

where Grad $E(z^m)$ is defined by

$$E(z^m + \delta z) = E(z^m) + \langle \text{Grad } E(z^m), \delta z\rangle + 0(\|\delta z\|) \tag{5}$$

In (5) $\langle , \rangle$ is the scalar product, $\| \quad \|$ is the norm of the space of z.

For optimal control, it is usually much faster with the conjugate Gradient Method (see [15] for instance) :

$$z^{m+1} = z^m + \rho^m h^m \tag{6}$$

$$\rho^m = \arg \min_{\rho} E(z^m + \rho h^m) \tag{7}$$

$$h^m = - \text{ Grad } E(z^m) + \gamma^m h^{m-1} \tag{8}$$

$$\gamma^m = \|\text{Grad } E(z^m)\|^2 / \|\text{ Grad } E(z^{m-1})\| \tag{9}$$

In the next paragraph we shall show how one can compute the gradient of a functional with respect to a geometrical element but let us notice right now that the above gradient methods require $z \in Z$, Z vector space while the sum of two geometrical elements is not defined off-hand. So even though optimal shape problems may be well defined from the physical point of view they require some transcription to become an usual optimization problem. This can be summarized into the diagram of Figure 1.

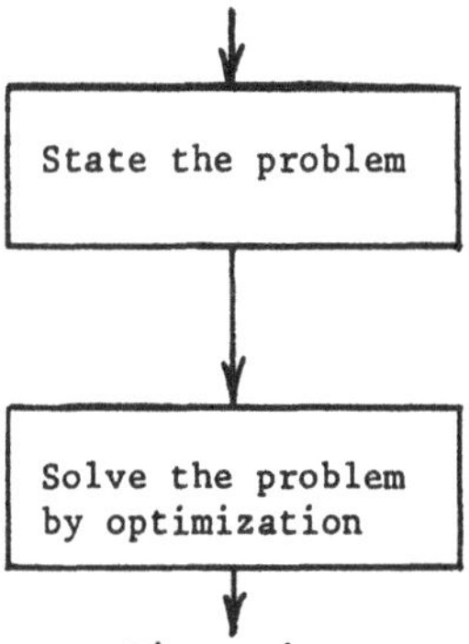

State the problem

↓

Solve the problem by optimization

Figure 1
A too ideal solution

Well, unfortunately, one usually ends up by solving many optimization problems according to the diagram of Figure 2.

What happens is that an engineer often does not know "a priori" what is an admissible shape for his problem : he solves his problem then find that the solution has an angle or is too thin ... so he must add this new constraint into the problem

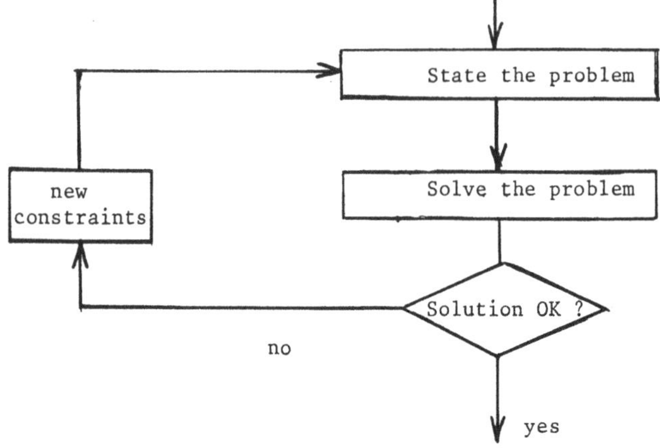

Figure 2
A more realistic view of things

For this reason optimum shape design is an expensive business ; many industrial laboratories may find it too costly, consequently perhaps too technocratic : intuitive optimization is enough

For this case also however the calculus of variations can be of great help to the intuition of the engineer.

The analytical expression of the gradient of E with respect to the shape tells the engineers in what direction and where it is best to modify his present design.

To summarize we shall make the 2 statements

. 3-d shape optimizations can rarely be done by intuition

. there are two ways to use the gradient of the criteria with respect to the shape : As a qualitative information for the modification of the actual design(formula (4) ρ intuited) ; or quantitatively inside a conjugate gradient algorithm (see (6), (9)).

3. Optimality conditions

3.1. The method of mapping

Consider the typical problem

$$\min_{\Omega\in\mathcal{O}} E(\Omega) = \int_D |\nabla\Phi(\Omega) - u_d|^2 \, dx \tag{10}$$

with $\Phi(\Omega)$ the solution of

$$-\Delta\Phi = f \quad \text{in } \Omega \qquad \Phi|_\Gamma = 0 \qquad \Gamma = \text{boundary of } \Omega \tag{11}$$

and

$$\mathcal{O} = \{\Omega \text{ , open set of } R^n : \Omega \supset D\} \qquad (12\text{-}$$

Let $C_{0,1}$ be the unit cercle centered at the origin and let

$$\mathcal{T} = \{T \text{ one to one mapping : } T(C_{0,1}) \in \mathcal{O}\}$$

For a given $T \in \mathcal{T}$ the integral in (10) can be expressed in term of the new variable $X = T^{-1}(x)$:

$$E(T(C_{0,1})) = \int_{T^{-1}(D)} |T_x^{-1} \nabla \hat{\Phi} - \hat{u}_d|^2 \; |T_x^{-1}| \, dX \tag{13}$$

where T_x^{-1} is the inverse of the Jacobian matrix T_x and $|T_x^{-1}|$ is its determinant. The $\hat{}$ recalls that the functions are evaluated at $T(X)$ for example

$$\nabla\hat{\Phi}(X) = \left\{ \frac{\partial\Phi}{\partial y_1}(y), \ldots, \frac{\partial\Phi}{\partial y_m}(y) \right\}_{y=T(X)} \quad ; \; \hat{u}_d(X) = u_d(T(X)) \tag{14}$$

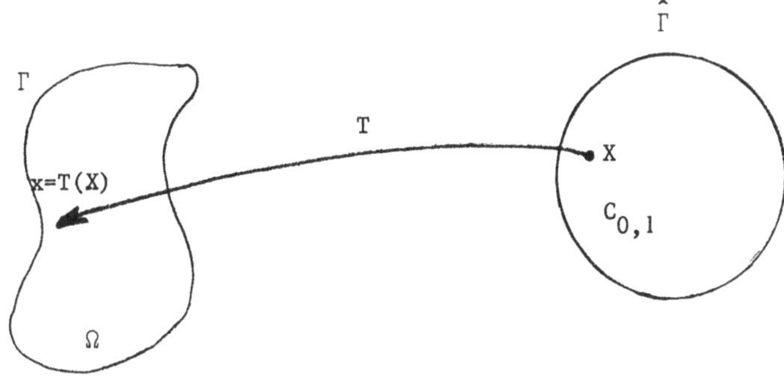

Figure 3

Similarly (11) can be evaluated with respect to X :

$$\nabla.(|T_x^{-1}|(T_x^{-1})^t T_x^{-1}\nabla\,\hat{\Phi}) = \hat{f} \quad \text{in} \quad C_{0,1} \quad ; \quad \hat{\Phi}|_{\hat{\Gamma}} = 0 \tag{15}$$

Thus problem (10) is a minimization of (13) with respect to $T \in \mathcal{T}$, and the techniques of optimal control, with control in the coefficients, can be applied (see [9] , [4]). This method is good on problems for which the explicit construction of T is easy. For example the 3 cases of figure 4 can be mapped each respectively from a rectangle, a cercle, a shell with constant thickness, all by affine transformations.

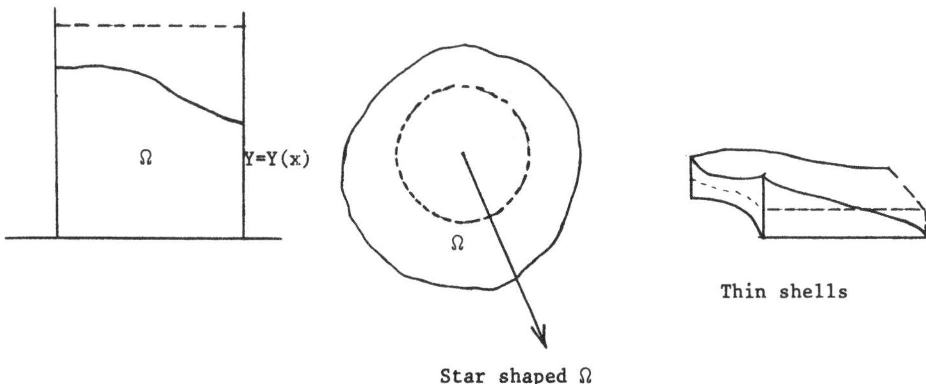

Figure 4

For numerical tests in these lines see [2][11] .
From the theoretical point of view this method seems also to be quite good to obtain optimality conditions with a set of minimal hypothesis on the regularity of the data (see [12]).

3.2. The method of local variations

From the practical point of view it is not really necessary to put a vector space structure on $\mathcal{O}$. One only needs to know how to construct a set of domains of $\mathcal{O}$ "close" to a given $\Omega \in \mathcal{O}$ (the largest hyperplane to $\mathcal{O}$ in some sense).

So let $\Omega \in \mathcal{O}$ with boundary Γ and let Ω_α be the domain with boundary Γ_α constructed as follows

$$\Gamma_\alpha = \{x + \alpha(x)n(x) : x \in \Gamma\} \tag{16}$$

where n(x) is the normal vector of Γ at x and α is a given real valued function on Γ.

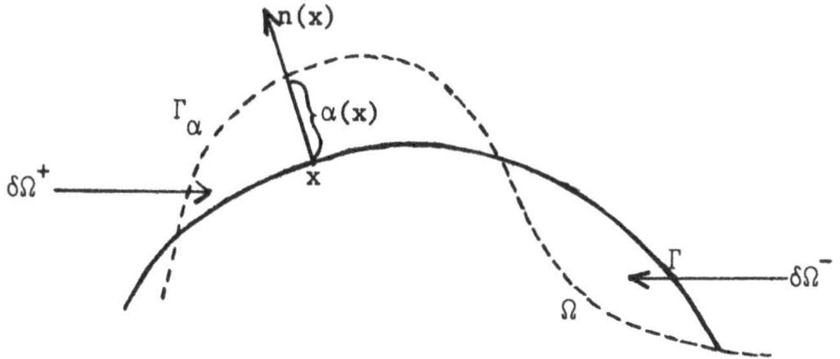

Figure 5

Proposition 1 :

If the data are smooth and Γ is C^1, then refering to (10)-(11) :

$$E(\Omega_\alpha) - E(\Omega) = \int_\Gamma \alpha \frac{\partial\Phi}{\partial n}(\Omega) \frac{\partial p}{\partial n} d\Gamma + O(\|\alpha\|)_{C^1} \tag{17}$$

where p is the solution of

$$-\Delta p = -2\nabla.[I_D(\nabla\Phi(\Omega) - U_d)] \quad \text{in } \Omega \quad p|_\Gamma = 0 \tag{18}$$

Proof :

From (11) we have

$$\Delta\delta\Phi = 0 \quad \text{in} \quad \Omega \cap \Omega_\alpha \quad ; \delta\Phi = \Phi(\Omega_\alpha) - \Phi(\Omega) \tag{19}$$

By a Taylor expansion

$$\Phi(\Omega_\alpha)|_x = \Phi(\Omega_\alpha)|_{x+\alpha n} - \alpha \frac{\partial\Phi}{\partial n}|_{x+\alpha n} + O(\|\alpha\|) = -\alpha \frac{\partial\Phi}{\partial n}|_{x+\alpha n} + O(\|\alpha\|)$$

It is possible to show that $\Phi(\Omega_\alpha) \to \Phi(\Omega)$ when $\|\alpha\|_{C^1} \to 0$ so

$$\delta\Phi|_\Gamma = -\alpha \frac{\partial\Phi}{\partial n}(\Omega) + O(\|\alpha\|) \tag{20}$$

Now from (10)

$$\delta E = E(\Omega_\alpha) - E(\Omega) = 2\int_D (\nabla\Phi(\Omega) - u_d)\nabla\delta\Phi \, dx + O(\|\delta\Phi\|) \tag{21}$$

So (18) implies that (Green's formula)

$$\delta E = \int_\Omega \nabla p \nabla\delta\Phi \, dx - \int_\Gamma \frac{\partial p}{\partial n}\delta\Phi \, d\Gamma \tag{22}$$

Now (19)-(20) yield (17).

Now let us investigate the case of Neuman conditions

$$\Phi - \Delta\Phi = f \quad \text{in} \quad \Omega \qquad \frac{\partial\Phi}{\partial n}\Big|_{\Gamma} = 0 \tag{23}$$

Proposition 2

Refering to (10)-(22), for smooth data one has

$$E(\Omega_\alpha) - E(\Omega) = \int_\Gamma (p\Phi + \nabla p\nabla\Phi - f\Phi)\alpha \, d\Gamma + 0(\|\alpha\|_{C^1}) \tag{24}$$

where p is the solution in $H^1(\Omega)$ of

$$\int_\Omega (pw + \nabla p \; \nabla w)dx = 2\int_D (\nabla\Phi(\Omega)-u_d) \; \nabla w \; dx \quad \forall w \in H^1(\Omega) \tag{25}$$

Proof :

In variational form(23) is

$$\int_\Omega (\Phi w + \nabla\Phi \; \nabla w - f \; w)dx = 0 \qquad \forall w \in H^1(\Omega) \tag{26}$$

Let $\tilde{w}$ be any smooth extention of w outside Ω and let us use the following symbolism (see figure 5).

$$\int_{\delta\Omega} = \int_{\delta\Omega^+} - \int_{\delta\Omega^-} \text{ where } \delta\Omega^+ = \Omega_\alpha - \Omega \cap \Omega_\alpha \text{ ,} \delta\Omega^- = \Omega - \Omega \cap \Omega_\alpha \tag{27}$$

then (25) implies

$$\int_{\delta\Omega} (\Phi w + \nabla\Phi\nabla w - fw)dx + \int_\Omega (\delta\Phi w + \nabla\delta\Phi\nabla w)dx = 0 \quad ; \tag{28}$$

therefore (29) used with $\delta\Phi$ in place of w gives

$$2\int_D (\nabla\Phi - u_d)\nabla\delta\Phi \; dx = -\int_{\delta\Omega} (\Phi p + \nabla\Phi\nabla p - fp)dx \tag{29}$$

Now the element of volume in $\delta\Omega$ is approximatly dx = $\alpha d\Gamma$ so (28) and (29) lead to (24).

4. Discretization

4.1. Discretization by FEM

Let us consider again the model problem (10)-(22). To discretize it by the finite element method one may divide Ω into triangles $\{T_k\}_1^M$ of average size h and set $\Omega = \cup T_k$ and

$$H_h = \{w_h \in C^o(\Omega) : w_h|_{T_k} \in P^1 \quad k = 1,\dots,M\} \tag{30}$$

Here $C^o(\Omega)$ is the space of continuous functions, P^1 the set of Polynomials of degree 1. It is easy to show that the functions of H_h are completly determined from their values $\{\Phi_i\}$ at the vertices $\{q_i\}^N$ of the triangles ($\Phi_h(x)$ is the linear interpolation of Φ_i, Φ_j, Φ_k where q^i, q^j, q^k are the 3 vertices of the triangle which contains x) Another way of saying the same thing is

$$\Phi_h(x) = \sum_{i=1}^{N} \Phi_i \, w^i(x) \tag{31}$$

$$\text{where } w^i(.) \in H_h, \quad w^i(q^j) = \delta_{ij} \tag{32}$$

With these notations, problem (10)-(22) can be approximated by

$$\min_{\Omega} E(\Omega) = \int_D |\nabla \Phi_h - u_d|^2 \, dx \tag{33}$$

where Φ_h is the solution of

$$\int_\Omega (\nabla\Phi_h \nabla w_h + \Phi_h w_h - f w_h) dx = 0 \quad \forall w_h \in H_h \; ; \quad \Phi_h \in H_h \tag{34}$$

Notice that now the optimization variables are $\{q^i\}_1^N$.

<u>Proposition 3</u> :

If each vertex q^k is moved by δq^k then Φ_h is modified by $\delta\Phi_h$ solution, up to higher order terms δq of

$$\int_\Omega (\nabla\delta\Phi_h . \nabla w_h + \delta\Phi_h \, w_h) dx = \int_\Omega [\nabla\Phi_h.(\nabla\delta q.\nabla w_h) - \nabla.\delta q(\nabla\Phi_h.\nabla w_h + w_h\Phi_h) - \delta q \, w_h \Phi_h] \, dx \tag{35}$$

$$\text{where } \delta q(x) = \sum_1^N w^k(x) \, \delta q^k . \tag{36}$$

Corollary

If Ω' is obtained from Ω by moving its vertices then

$$E'(\Omega)-E(\Omega) = \int_{\Omega} [\nabla\Phi_h.(\nabla\delta q.\nabla p_h)-\nabla.\delta q(\nabla\Phi_h \nabla p_h+\Phi_h p_h) -$$

$$- \delta q\, p_h\Phi_h]\, dx + O(\delta q) \tag{37}$$

where p_h is the solution of

$$\int_{\Omega} \nabla p_h \nabla w_h\, dx = 2\int_{D} (\nabla\Phi_h-u_d) \nabla w_h dx \quad \forall w_h \in H_h \; ; \; p_h \in H_h \tag{38}$$

Proofs

The proofs are a little technical so the reader is sent to [10]. These results enable us to use a gradient type algorithm to solve (33)-(34). For example :

Algorithm :

0. Choose an initial Ω^o, choose K, choose ρ

1. For k = 0.,K do
   - solve (34) with $\Omega = \Omega^k$
   - solve (38) with $\Omega = \Omega^k$
   - compute

$$\chi_i^j = \int_{\Omega^k} [\nabla\Phi_h \nabla w^j \frac{\partial p_h}{\partial x_i} - \frac{\partial w^k}{\partial x_i}(\nabla\Phi_h \nabla p_h + \Phi_h p_h) - w^k \Phi_h p_h]\, dx \tag{39}$$

   - change $q_i^j \to q_i^j + \rho\, \chi_i^j$

4.2. Discretization by other methods

Naturally in conjunction with the method of mapping one can use the method of finite differences : there are no difficulties but the formulas are rather long.

If a good precision is not required one may use a penalty method to avoid working with moving grids.
For example following [8] . One can approximate (11) by

$$-\Delta\Phi^\varepsilon + \frac{1}{\varepsilon} I_\Omega \; \Phi^\varepsilon = f \quad \text{in} \quad C =]0,1[^{\,2} \supset \Omega \; ; \quad \Phi^\varepsilon|_\Gamma = 0 \tag{40}$$

Then (40) is approximated by finite differences (see figure 6), for example

$$- \Delta ij \quad \Phi + \frac{1}{2\varepsilon} [1 + \min \{1, | \frac{Y_i}{h} - j|\} \operatorname{sign}(\frac{Y_i}{h} - j)] \quad \Phi_{ij} = F_{ij} \quad (41)$$

where Δij is the usual 5 points formula.

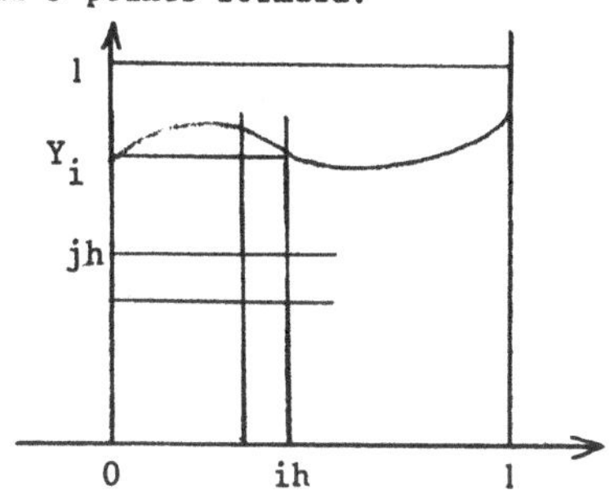

Figure 6

So again the problem has been transformed into an optimal control problem where the controls appear in the coefficient of the PDE (here Y_i).

5. Applications

5.1. Optimization of a Nozzle (Angrand [1])

The problem is

$$\min_{\Omega \supset D} \{ \int_D |\nabla\Phi - u_d|^2 \, dx \;:\; \Delta\Phi = 0 \text{ in } \Omega \;,\; \Phi|_\Gamma = \Phi_\Gamma \} \tag{42}$$

it corresponds to the design of a nozzle at low speed with a required velocity u_d of air in some prescribed region D ; Φ is the potential of the flow Φ_Γ is non zero only at the entrance and exit of the nozzle.
Figure 7 shows the initial Ω^o, figure 8, 10, 12 several computed nozzles and figure 9, 11 the intermediate Ω^k, $k < K$.

5.2. Design of an electro magnet (Marrocco-Pironneau [10])

The problem is

$$\min_{\Omega} \{ \int_D |\nabla\Phi - u_d|^2 \, dx \;:\; \nabla.[\mu \; |\nabla\Phi|^2)\nabla\Phi] = j \;\text{ in }\; C,\; \Phi|_{\partial C} = 0 \;\} \tag{43}$$

where $\mu(.)$ is a non linear function of its argument in $\Omega \subset C$ and a given other function in $A \subset C - \Omega$.
This corresponds to the design of an electro magnet with constant interpolar magnetic field u_d ; Φ is the magnetic potential, μ the reluctance, j the currents.
The non linear μ in the iron region Ω is shown in figure 13, the general shape of Ω is shown in figure 14.
Figures 15, 16, 17 show the triangulation and how it is moved when the iron region is deformed. Figures 18, 19, 20 show the equipotential lines along the iterations and Figure 21 displays the corresponding decrease of the cost function.
Notice that oscillations occur and that a smoothing causes the cost function to increase.

5.3. Airfoil design (Angrand [1])

The stream function Ψ in R^2 around an airfoil S at subsonic speed is solution of

$$\nabla.(K-|\nabla\Psi|^2)^{2.5} = 0 \;\text{ in }\; \Omega - \overset{\circ}{S} \;;\; \Psi|_{\Gamma_\infty} = \Psi_\infty \;,\; \Psi|_S = C_z \tag{44}$$

where K is a numerical constant. The constant C_z is proportional to the lift of the airfoil and determined from the Joukowski condition at the trailing edge TE

$$J(s,\alpha) = \nabla\Psi(TE)^{+} = \nabla\Psi(TE)^{-} = 0 \tag{45}$$

If U_∞ is the velocity at infinity and α is the incidence angle of the airfoil

$$\Psi_\infty = |U_\infty|(x_2 \cos \alpha - x_1 \sin \alpha) \tag{46}$$

To avoid separation of the boundary layer one may solve

$$\min_S E(\Psi(S,\alpha)) = \int_S \left| |\nabla\Psi|^2 - p_d \right|^2 d\Gamma \tag{47}$$

subject to (44)-(45)-(46).

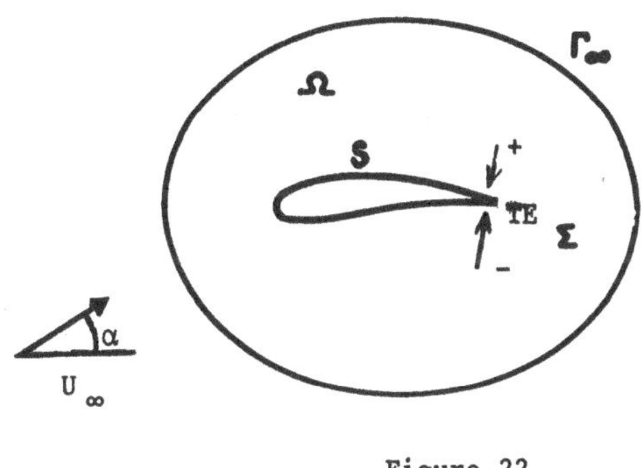

Figure 22

Figure 23 shows the result of an optimization together with the intermediate shapes.

As in the 2 other examples the FEM has been used.

Computing time is reasonable for the first two examples (about 1 min of IBM 3033) while the last example in an expensive calculation because many points are needed to represent properly such flows and the answer is needed with a good precision so about 40 iterations (intermediate shapes) are necessary.

6. Open problems

6.1. Mathematical questions

The problem of existence of solution was studied by Chesnay [5]; the result shows that if the boundaries are uniformly lipchitz the solution exists. Thus one may wonder what happens if there is no such regularity. Murat-Cioranescu [13] show that the question is linked to the theory of Homogeneization. They constructed a sequence Ω^m of domains converging to Ω (but with oscillating boundaries) with the following property.

If ϕ^m is the solution of

$$- \Delta \phi^m = f \text{ in } \Omega^m , \quad \phi^m|_{\Gamma^m} = 0 \tag{48}$$

then $\phi^m \to \Psi$ solution of

$$- \Delta\Psi + \alpha\Psi = f \text{ in } \Omega , \quad \Psi|_\Gamma = 0 , \quad \alpha > 0 \tag{49}$$

This therefore shows that the problem

$$\min_\Omega \{ \int_\Omega |\nabla(\phi-\Psi)|^2 dx : -\Delta\phi=f \text{ in } \Omega ; \quad \phi|_\Gamma = 0 \}$$

has no solution.

6.2. Algorithm problems

From the computational side there are a number of problems. First a problem of topology of the solution. Suppose that the solution Ω has two holes but Ω^o has one hole only then $\{\Omega^m\}$ will never converge to Ω but only to a suboptimum of the problem. Some solution to this difficulty is given in [3].
A major problem comes from the moving meshes care must be taken that the meshes always stay regular ; this difficulty is time consuming from the programming point of view only.
Thirdly one may wish to use higher order isoparametric elements ; for these the theoretical formula are still to be worked out.
Finally many industrial design problems have a non differentiable criteria ; thus all the difficulties of non differentiable optimization are attached to these.

6.3. Other optimal shape design

Naturally many important design problems have non elliptic PDEs.

Such a problem with a parabolic equation was studied in [14] it is of the type : what is the best way of swimming or flying for a deformable body ?

In the class of hyperbolic systems there are many problems connected with accoustics (minimization of jet noise, optimization of concert halls) and optics. To the knowledge of the author these problems have not been studied.

7. Conclusion

The techniques of the calculus of variation and of optimization proved to be successful for several optimal shape design problems however these remain expensive both in the qualification of the engineers required to understand the method and in computing time. However it seems difficult to do without such techniques for 3-dimensional optimizaation problems. The field is well studied from the mathematical point of view but still in its beginnings from the industrial implementation side.

References

[1] F. ANGRAND : Méthodes numériques pour la conception optimale en aérodynamique, 3 cycle thesis, University of Paris 6, June 1980.

[2] D. BEGIS, R. GLOWINSKI : Application de la méthode des éléments finis à l'approximation d'un domaine optimal. Appl. Math. and Opt. 2 n° 2 (1975).

[3] J. CEA, A. GIAN, J. MICHEL : Quelques résultats sur l'identification de domaine. Calcolo III-IV (1973).

[4] G. CHAVENT : Analyse fonctionnelle et identification de coefficients répartis. Thèse d'Etat, University of Paris 6 (1971).

[5] D. CHESNAY : On the existence of a solution in a domain identification problem, J. of Math. Anal. and Appl. 52, 189-289 (1975).

[6] C. CRYER : On the approximate solution of free boundary problems, J. Assoc. Comp. Mach. (1970), pp. 397-411.

[7] A. DERVIEUX : Résolution de Problèmes à frontières libres, University of Paris 6 Thèse d'Etat, June 1981.

[8] H. KAWARADA : Numerical methods for free surface problems by means of penalty. Lecture notes in Math. Springer 704 (1979).

[9] J.L. LIONS : Optimal Control of Systems governed by partial differential equations North Holland (1970).

[10] A. MARROCCO, O. PIRONNEAU : Optimum design with Lagrangian finite elements. Comp. Meth. in Appl. Mech. and Eng. 15 (1978) pp. 277-308.

[11] Ph. MORICE : Une méthode d'optimisation de forme de domaine. Proc. IFIP-IRIA, Versailles (1974) Springer, pp. 454-467.

[12] F. MURAT, J. SIMON. Sur le contrôle par un domaine géométrique. University of Paris 6, Doct. Thesis (1977).

[13] F. MURAT, G. CIORANESCU : to appear.

[14] O. PIRONNEAU, D. KATZ : Optimal swimming of flagellated micro-organism. J. Fluid. Mech. 66 (1974), pp. 391-415.

[15] E. POLAK : Computational method in optimization. Academic Press (1971).

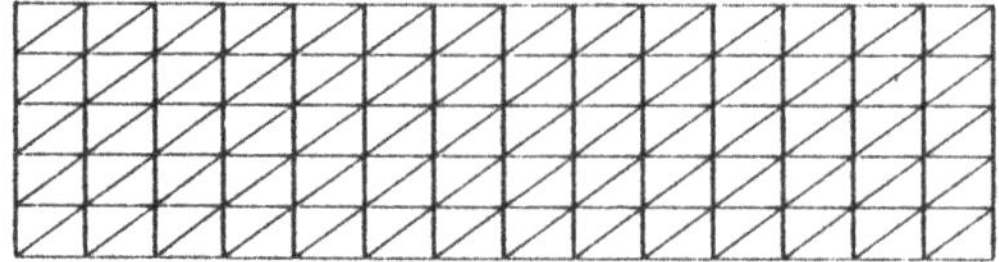

Figure 7

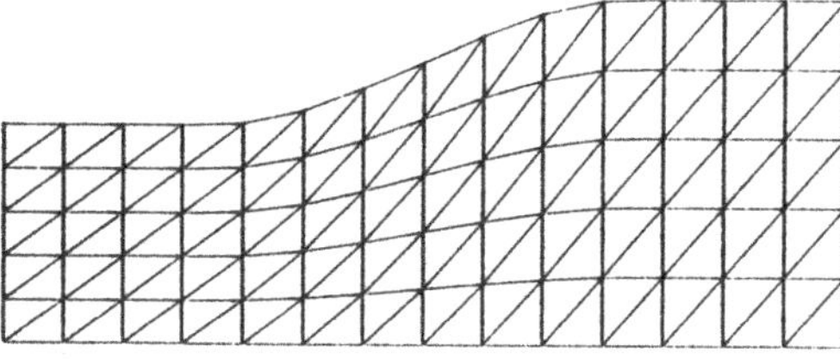

Figure 8

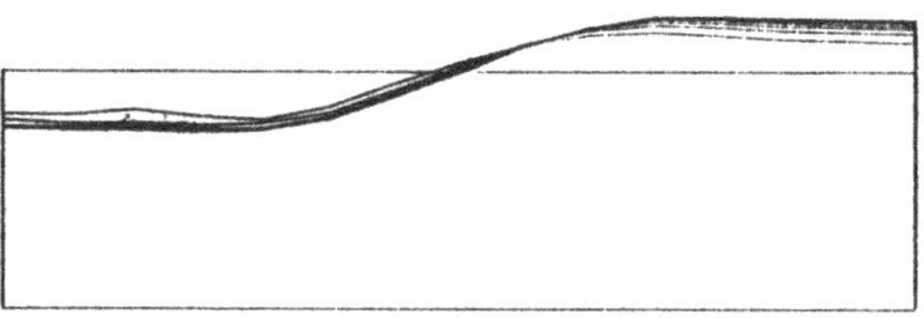

Figure 9

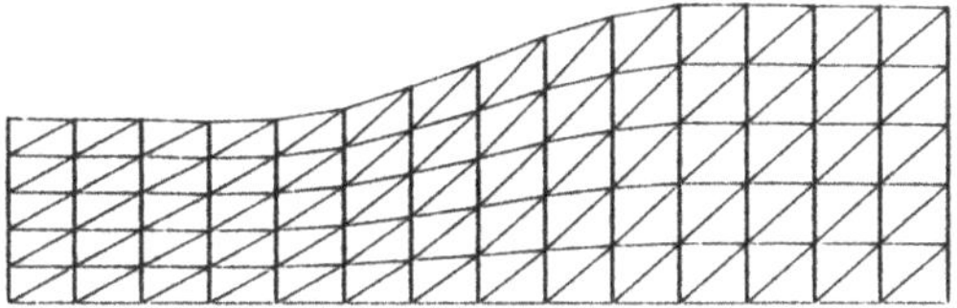

Figure 10

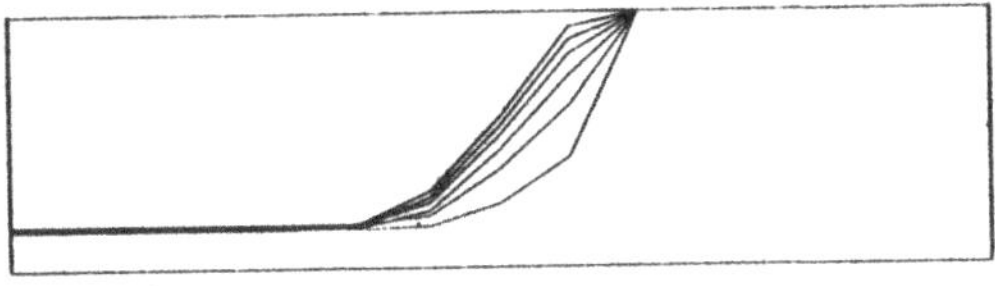

Figure 11

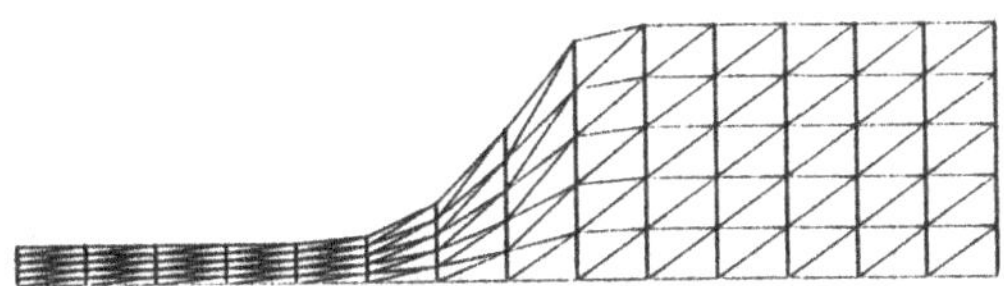

Figure 12

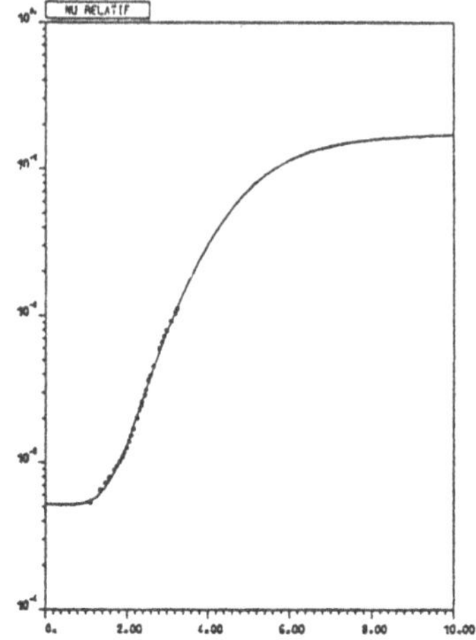

Figure 13

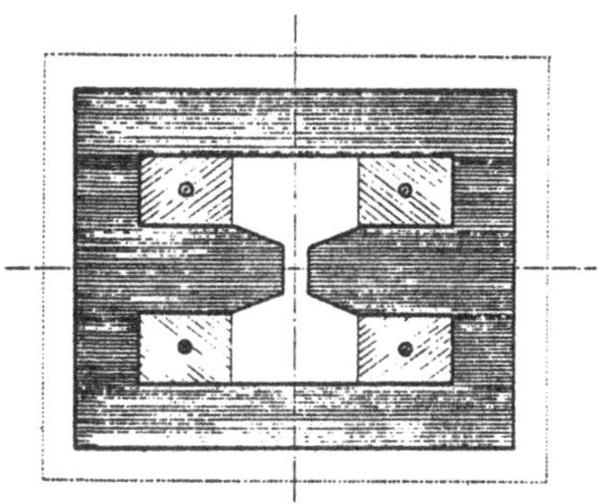

Figure 14

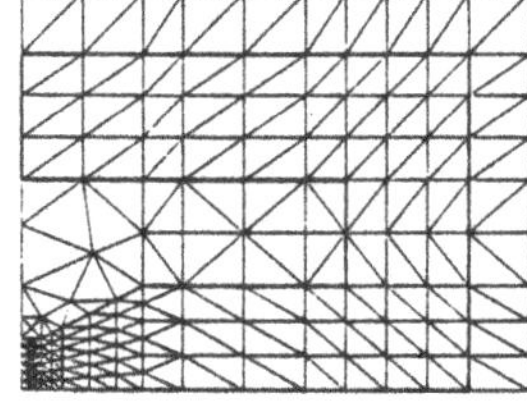

Figure 15

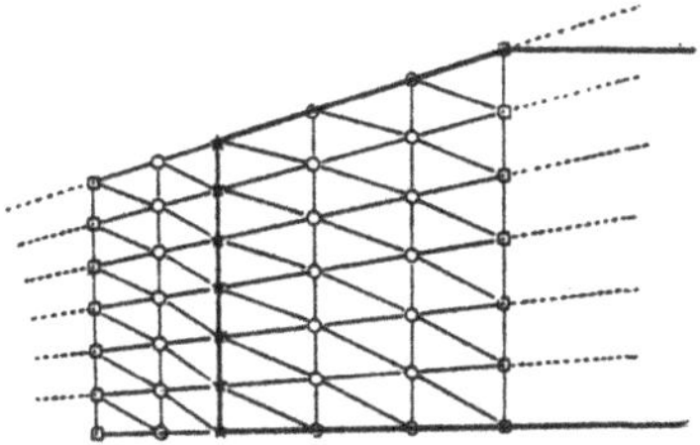

Figure 16

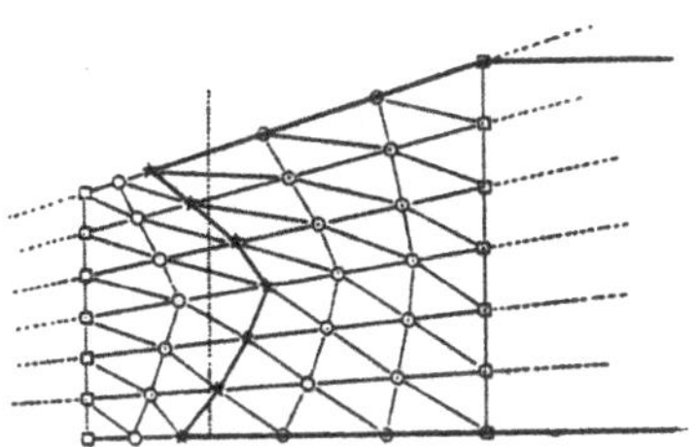

Figure 17

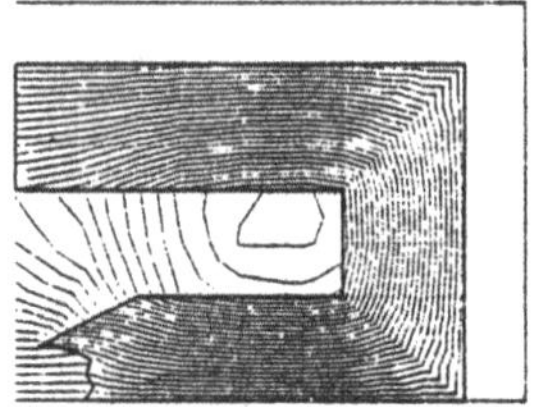

Figure 18

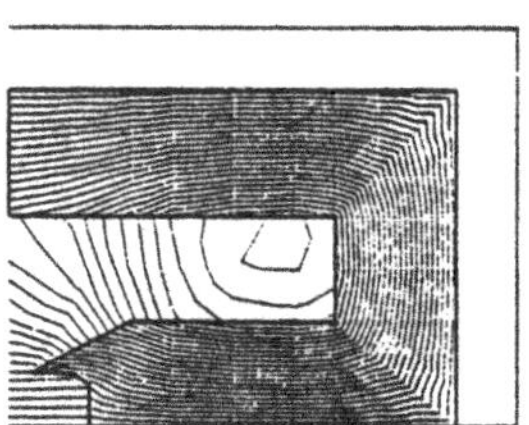

Figure 19

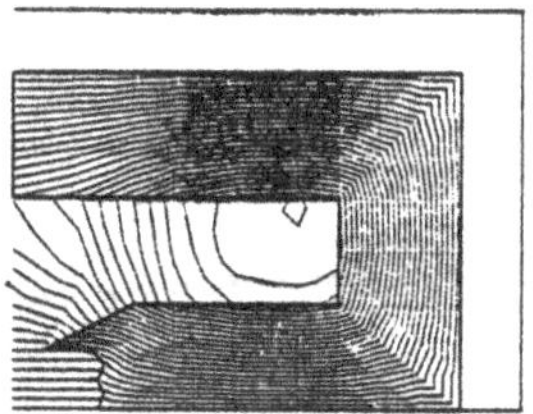

Figure 20

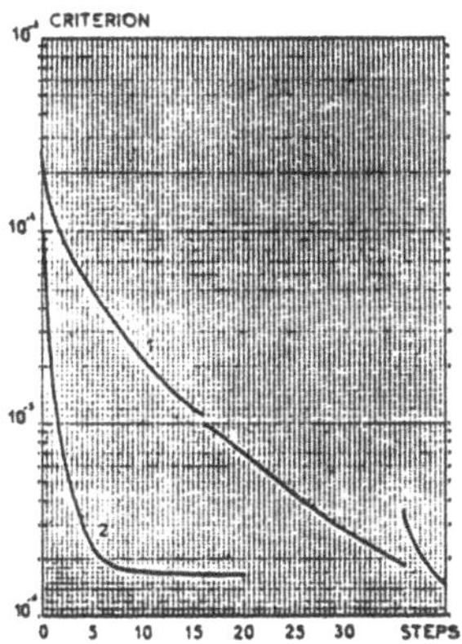

Figure 21

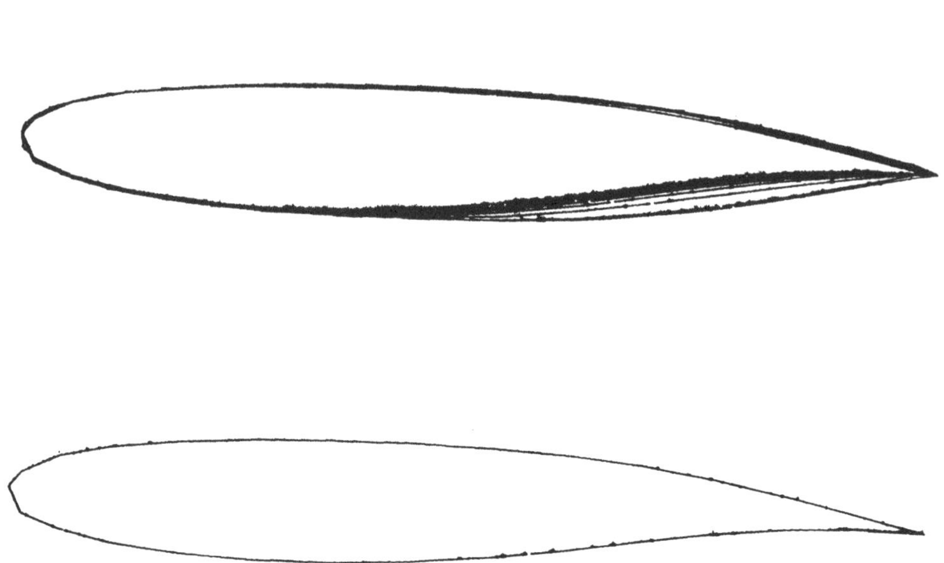

Figure 23

URBAN SYSTEMS ANALYSIS AND URBAN POLICY

E. S. Savas
Assistant Secretary

Office of Policy Development and Research

U.S. Department of Housing and Urban Development

Washington, D.C. 20410

September 4, 1981

In the spirit of improving the usefulness and extending the scope of urban systems analysis, I'd like to make three points today:

1. Urban systems analysis is encumbered with certain dysfunctional myths, which inhibit its effective use in policy making;
2. It is applicable at different levels, which present different environments and call for different methods of analysis;
3. There are other competing and complementary disciplines that can profitably be embraced by systems analysis, and the education of systems analysts need reform in order to embrace them.

Let me start with the myths and go through them.

THE MYTHS

The Myth of Decision-Making

Perhaps the most mischievous myth that afflicts the field of urban systems analysis is the very notion of decision-making. As conventionally taught, alternative choices are analyzed, and the consequences are arrayed before the decision-maker, whereupon decision-making occurs. In this idealistic scenario, decision-making is a neat, crisp act which follows quite automatically from the analysis. In fact, however, in the urban policy arena polices are rarely made; at best they emerge from a vague, prolonged, diffuse, pluralistic and evolutionary process. They are often the accumulation of sequential ad hoc choices in environments characterized by a multiplicity of conflicting sources of partial information and disparate judgements. They require persuasion and consensus-building, not merely decisions. The situation is described eloquently by Kash (1968):

> "Decision-making . . . is a process." Phrases like the "decision-making process" and the "process of policy formulation" are not mere incantation: they refer to the continuous flow of decisions, large and small, that make up the seamless web of policy formation and administrative action . . . government. The dynamic flux of the policy process makes the

job of the advisor particularly difficult. It means that there is no orderly procedure whereby the advisor can state his views or explain his research, and then retire from the scene, confident that his advice will receive systematic consideration. There are numerous distractions and competing demands on the decision-maker's time and span of attention. Decisions once made can become unmade a week later. The advisor may face a difficult task to secure a full hearing for his views in the first place, and then must struggle to keep attention focused on his recommendations for a long enough period to assure action of some kind. Continuity is thus an essential attribute of effective communication of policy-oriented research.

A corollary of this is that the advice cannot simply be given to the top levels, if favourable decision and effective implementation of advice is desired. Consider the case of a high level decision-maker accepting the recommendation of an advisory group and making a "policy" decision designed to implement the advice. Unless the subordinates carry out the decision effectively, the whole intent can be defeated. Comprehension of the basis for the decision reached at the higher level can be a vital factor in winning the consent and enthusiasm of those who must execute the decision and, in doing so, make a myriad of other decisions, which can determine the success or failure of the original decision.

The Myth of Problem-Solving

A second major myth inherent in the conventional view of systems analysis is the idea of problem solving. The quantitative roots of systems analysis nurture the belief that problems and solutions constitute a dyad; after all, every problem in mathematics textbooks comes with a matching solution, usually in the back of the book. Use of the right technique will produce the right answer. Given this view of the world, urban issues are treated as problems with answers, and the entire armamentarium of problem-solving weaponry is wheeled into position and brought to bear to "solve the problem." Again, this is a false and dysfunctional view of urban policy issues. More often than not, it is far more useful to approach these issues as conditions to cope with, or situations to be ameliorated, rather than as problems to be solved. (For example, poverty is not a "problem" that can be solved; some people will always be labelled poor, with all the stigma that implies, regardless of their absolute level of income.) Incremental improvement, and even "muddling through," are not to be sneered at when "solutions" are unattainable. As Wildavsky (1974) points out, problems involving government aren't solved, they're merely superseded.

The Myth of Optimality

A third myth that obstructs effective urban systems analysis is the myth of optimality. Of course, optimization is a useful construct, but it plays a surprisingly small role in the exasperating urban environment where, in linear programming parlance, there is usually no *feasible solution*. In this complex setting, optimization consists of trying to determine which of the multitudinous constraints can be forced back, at minimal political cost, far enough to create one feasible--hence, optimal--point.

The Myth of the Moon

The fourth myth holds that if we could successfully accomplish something as difficult as getting to the moon, we should be able to . . . r evitalize cities, solve social ills, etc.

The journey to the moon was indeed a triumph of systems analysis, but it offers little encouragement to urban analysts. Getting to the moon was relatively easy, for two important reasons: (1) there are no constituents on the moon; (2) there was no identifiable constituency on earth that could be organized to oppose the effort of going to the moon.

The total absence of a lunar constituency meant that the public did not care where the astronaunts landed, whether in the Sea of Tranquilty or on the Heights of Assininity, or what they did once that got there. Therefore, the decisions about lunar landing sites could be based entirely on scientific and technical considerations. However, with respect to more terrestrial decisions, the public cares a great deal indeed about where a waste dispoal site is to be located, for example, or where an airport is to be built, and therefore, technical analysis plays a relatively minor role in such decisions. An algorithm which seems to prove that the "best" place for a drug-treatment centre is just down the block from a councilman's home, proves instead that the analyst is a poor one who should be ignored, or worse.

No group felt that its vital interests were threatened by a large-scale national effort to send a man to the moon and bring him back alive. On the other hand, many interest groups saw positive benefits to themselves from such a programme. Under such circumstances, a national consensus was easy to achieve, and a massive programme could be mounted without opposition.

But this fortunate conjunction of circumstances rarely arises on the urban scene. In particular, there is no consensus about constructing transportation networks, building subsidized housing, or improving the productivity of municipal workers.

The Myth of Attributable Achievement

When a course of action--recommended after analysis of a complex issue--is adopted and proves successful, many forces will have conspired and converged to bring about the observed outcome. Not all of them can be identified, nor can their impacts readily be assessed. The causal chain cannot be traced back, for it is more like a net than a chain. Therefore, each of the many participants can legitimately claim credit, for each can orient the net uniquely, and sight along one particular thread. In short, success has many fathers (while failure is an orphan). The analyst who was trained to revere the creativity of the solitary scientist, and who seeks his professional rewards through attributable achievement, is likely to be frustrated in this setting.

The Myth of Model-Building

This myth, the sixth in our list, also merits comment. The builder of a mathematical model sees his work as culminating in a product, a theoretically sound and comprehensive tool of quite general applicability. But, if the activity that the model presumes to represent is a very complex one, such as an urban system, a product, no matter how complex--and particularly if complex--is inadequate for improving that activity ("solving the problem"): a process is needed.

Ideally, model-building is called for--nay, demanded--as part of the improvement process. Too often, unfortunately, a model is constructed under the builder's implicit assumption that its mere presence will initiate the necessary process. While the existence of such a tool is necessary--but not sufficient--for its use, it is <u>neither</u> necessary <u>nor</u> sufficient for initiating the improvement process.

Certainly a model can be useful in the process, under favourable circumstances. Sometimes a model can even be used to initiate this processs.[1] However, there are also other possible initiators of the change process. Therefore, the question can rightly be asked for any given issue: are there more cost-effective ways to initiate or accelerate the change process than by constructiong a costly mathematical model? Rarely is this question asked.

The Myth of Planning

How could any rational person, let alone a systems analyst, oppose

---

[1]For example, the model of world dynamics (Meadows, 1972), while of limited scientific merit, was a very effective tool which successfully captured public attention and focused it on the issue of the limits to growth in a world of finite resources. In this respect, it has probably hastened the process of change--vis., societal acceptance of this newly-perceived condition and gradual public adjustment to it.

planning? While I would be among the last to deny the virtues of planning--I do attempt to practice it myself--it does not merit quite such an exalted position. In the first place, emphasizing it in this way painfully accentuates the distinction between planning and accomplishing. Secondly, the emphasis on planning exaggerates our ability to foresee and forfend.

With respect to the well-known abyss between planning and doing, it is well to remember that problems not perceived by the public are problems not acted upon. A problem has to exist before it can be addressed effectively (i.e., coped with). Detection of a problem by an intellectual elite, such as analysts, is not sufficient to induce action, for the public does not reward long-range political decisions which prevent problems from arising, nor does it punish its politicians for short-sightedness.

As an example, consider the issue of birth control. For years demographers had been vainly calling attention to the problem of overpopulation that they foresaw. Yet, in 1960, the President of the United States enunciated the policy that family planning was a private matter, and government would have no role in it; public funds were not to be used, even for research on fertility and birth control. Just ten years later, however, abortion was available virtually on demand in New York hospitals, paid for by public funds. What had happened in the interim was a raising of the public consciousness to the "world population bomb," the spectre of a "population explosion," and other such popularizations of what had hitherto been exclusively a concern of an obscure scientific community. The change in public awareness made possible this dramatic reversal of public policy.

As to our ability to anticipate and correct, it is important to note with humility that our predictive models for urban phenomena are poor, and are not likely to become dramatically better. It is difficult to forecast the effect of a disturbance on a system, and it is difficult to calculate the kind and quantity of anticipatory corrective action that should be taken to counteract that effect. The combination of this factor, and the preceding one, conspires to limit the applicability of feed-forward control to urban problems. In other words, planning is of limited potential in this arena and we should not expect too much of it.

As a modest example, consider the following, which is also drawn from the New York experience. It was predicted that legislation to provide free medical care to low-income families would have a certain budgetary impact. The prediction was wrong, and the cost to the taxpayer turned out to be much greater than had been expected. The reason: planners assumed a relatively low enrollment rate, due to ignorance and apathy, but welfare rights groups, neighbourhood groups, and legal clinics, in poverty areas were effective in making contact with, and educating, eligible patients, and helping them enroll in the programme.

In addition, actions in urban government often have quite unintended consequences. For example, consider the convoluted workings of New York City's civil service system, which was designed to hire the most meritorious job applicants into the public service. In fact, the process results in an _inverse_ merit system: the _lower_ an applicant scores on an entrance examination, the _more_ likely he or she is to be hired, and the _higher_ the score, the _less_ the probability of being hired! This perverse and unintended consequence comes about becuase of the long, drawn out, bureaucratic procedures that are built into the hiring and selection process to assure that the merit principle is followed scrupulously. The end result was a median delay of seven months between the time someone applied for a job and someone was hired to fill it. What happened was that the better people, who scored well, found jobs elsewhere, while those lower down were still available when the inexorable, but ponderous, mechanism of the "merit system" (finally produced a job offer (Savas and Ginsburg, 1973).

The Myth of Technical Elegance

This eighth myth results in excessive value being accorded to technical elegance by systems analysts. This is a well-known perversion that afflicts many service professions, wherein the norms and values of the insiders come to dominate those of the public at large. The irrelevance of technical elegance is abundantly obvious to political leaders and government executives, if not yet to analysts.

The Myth of Decomposition

The ninth myth on this list derives from the fact that the analytic disciplines are very good at abstracting a problem from its natural state, dissecting it, idealizing it, applying simplifying assumptions, and then forcing it into a well-defined category. Unfortunately, the distinguishing characteristic of complex social systems, such as urban systems, is that they are aggregates with strong bonds between components, and with values playing major roles. Such systems incorporate--totally and indistinguishably--elements which historically have been aggregated artificially into the man-made intellectual compartments that we call the fields of knowledge. Urban policy issues do not lend themselves readily to such decomposition, for they are robust and resistant to abstraction and fractionation.

The Myth of Irrational Politics

A common lament by urban systems analysts is that decision-making is too political, and not sufficiently rational. All this reveals is the analyst's failure to understand the objectives of the decision influencers. The implicit objective functions of the latter include terms, often unarticulated, that the analyst is either unaware of or unable to handle, and therefore, prefers to ignore.

LEVELS OF APPLICATION

It is useful to differentiate between two different levels of application, the operational level and the institutional or strategic level. (Other, finer distinctions and sub-levels could be identified, but it is not necessary for the purpose here to do so.)

The operational level can be illustrated by an analysis to determine the appropriate replacement policy for fire trucks. Given certain well-known data on purchase costs, repair costs, repair rates, etc., one can readily utilize the standard tools of systems analysis to arrive at a reasonable technical conclusion, even an optimal one. Nevertheless, even in this case, there is no simple, universally acceptable, technical "solution." Vendors have a major stake in the outcome, and will attempt to influence it. The fireman's association can also be expected to press for early replacement. On the other hand, political leaders campaigning on a platform of fiscal austerity are likely to favour a longer replacement cycle. Other agencies may see themselves as competing with the fire department for scare resources, and will, no doubt, suggest better uses of the municipal funds. Thus, even at this simplest and purest technical level, the factors cited above come into play and complicate the analysis, or, more precisely, affect the policy outcome which the technical analysis was intended to inform.

At the broadest level of institutional or strategic analysis, profound societal issues are involved. For example, it has been argued that the relatively high crime rate in American cities is an inevitable consequence of the high status we accord individual, as opposed to group, rights, and of the weakness of our informal institutions (family, church, neighbours) in controlling deviant behavior (Bayley, 1980). Reducing urban crime through a gradual societal change at this level is very different from a programme to reduce crime by better allocation of police resources; it requires a very different sort of analysis; one fraught with the difficulties, and subject to all the dysfunctional myths, cited above.

The difference between the different levels of systems analysis can be further illustrated by reference to residential refuse collection. At the operational level, one can conduct an analysis to choose the most economical vehicle for the job; one whose capacity is properly matched to the route and the trip to the disposal site. (If the vehicle is too large, there will not be enough working time available to fill the truck completely, and hence, a smaller, less expensive truck would have been adequate. Conversely, if the truck is too small, much of its time will be spent travelling to and from the disposal site, and therefore, the working time of its crew will be squandered; the larger the crew, the more costly this option.)

At the institutional level, one can conduct an analysis to determine whether eliminating the city agency and turning over the entire function to a private, profit-making contractor is best; there is impressive evidence from three different countries which indicates quite conclusively that substantial economies can generally be obtained by this action (Savas, 1979a).

Both of these approaches have the potential of improving this service. Clearly, the operational analysis requires more data, lends itself more readily to modelling, is more complicated technically, and yields a relatively unambiguous answer. On the other hand, the question of contracting for service is enveloped in ideological issues, abounds in intangibles and incommensurables, is perceived as a major threat by established bureaucracies, poses transitional problems, and appears to be risky in that it requires a new, untried relationship. Technical analysis plays a relatively minor role in this decision, and aside from some simple economic calculations--but not <u>too</u> simple (Savas, 1979b)--there is no need for formal models, elegant constructs, or extensive data collection.

The foregoing examples have been drawn from the subject area of public services planning, but other examples could just as easily have been drawn from physical network planning. For instance, deciding whether or not to build a bridge is a strategic question; designing it is an operational one.

EDUCATING FOR URBAN SYSTEMS ANALYSIS

In urban systems analysis applied at the strategic level, one may confront such fundamental issues as intergovernmental relations, the appropriate degree of centralization and decentralization in municipal government, the role of neighbourhoods vis-a-vis regional government, and the choice of delivery systems for providing services: government, private contractors, the marketplace, franchises, grants, vouchers or voluntary citizen associations. It is evident that, at this level, the tools and training required by the systems analyst will differ greatly from those he needs to work effectively at the more technical, operational level.

At the strategic level, there are other competing and complementary disciplines and perspectives that can be and are employed. To name but a few, they include law, political science, political economy, economics, sociology, organizational behaviour and psychology. These provide alternative decision-making paradigms, and it behooves the urban analyst to appreciate and be conversant with, if not fluent in, these approaches. Systems analysis should embrace the aspects of these fields which are particularly appropriate for urban policy and planning purposes. Analysts should be made sensitive to conflicting value judgements. The conventional education of systems analysts is sadly lacking in this respect.

As part of his training, the urban analyst should be made aware of his exposed role; he is a participant and not a detached observer. Thus, the traditional analyst, while comfortable with the abstract concepts of resource allocation, is frequently disconcerted when he finds himself caught up in the unfamiliar political turmoil of which group gets what, when and where. After all, this is what the allocation of public resources is all about.

Another role in which the urban analyst must become more comfortable through better training is that of a change agent. The analyst rarely recommends maintaining the status quo; almost invariably he recommends change. But at the strategic level in the urban policy arena, change means altering institutional relationships, work patterns, and the relative power of people. Such changes are bitterly opposed and difficult to effect. Nor surprisingly, the role of change agent in this setting is a hazardous one. Socrates, an early analyst who questioned the established ways of thinking, was prosecuted by a public official, and given poison by a civil servant.

SUMMARY

A number of dysfunctional myths have the effect of obscuring the proper role of systems analysis in urban policy-making, and inhibiting its effective use for that purpose. These are the myths of:

1. decision-making

2. problem-solving
3. optimality
4. the moon
5. attributable achievement
6. model-building
7. planning
8. technical elegance
9. decomposition
10. irrational politics

One must recognize that systems analysis can be applied at two very different levels: the operational level and the institutional or strategic level. The environments at these two levels differ greatly and require different tools, approaches and emphasis. The education of urban systems analysts should stress these distinctions, and should familiarize the analyst with other complementary disciplines and viewpoints, such as those of law, political science, economics, sociology and psychology.

REFERENCES

Bayley, David H. (1980) "Ironies of American Law Enforcement." The Public Interest 59 (Spring): 45-56.

Kash, D. E. (1968) "Research and Development of the University," Science 160 (June 21): 1313-1318.

Savas, E. S. (1979a) "Public vs. Private Refuse Collection: A Critical Review of the Evidence," Urban Analysis 6: 1-13.

Savas, E. S. (1979b) "How Much Do Government Services Really Cost?" Urban Affairs Quarterly 15 (September): 23-38.

Savas, E. S., and Sigmund Ginsburg (1973) "The Civil Service: A Meritless System?" The Public Interest 32 (Summer): 70-85.

Wildavsky, Aaron (1974) The Politics of the Budgetary Process, Little, Brown and Company, Boston, Mass.

ON THE TREATMENT OF CHRONIC FORMS OF A DISEASE ACCORDING TO A MATHEMATICAL MODEL.

G.I. Marchuk, L.N. Belykh

Institute of Numerical Mathematics of the USSR Academy of Sciences

Moscow / USSR

1. Model of Infectious Disease.

The mathematical model of an infectious disease based on immunological assumptions was advanced by G.I. Marchuk in 1975. The model is the system of the ordinary differential equations of the following kind:

$$\begin{aligned}
\frac{dV}{dt} &= (\beta - \gamma F)V \\
\frac{dF}{dt} &= \rho C - \eta \gamma F V - \mu_f F \\
\frac{dC}{dt} &= \xi(m)\alpha V(t-\tau)F(t-\tau)\theta(t-\tau) - \mu_c (C - C^*) \\
\frac{dm}{dt} &= \sigma V - \mu_m m
\end{aligned} \qquad (1)$$

with the initial conditions at t=0

$$V(0) = V^0 > 0,\ F(0) = F^0 \geq 0,\ C(0) = C^0 \geq 0,\ m(0) = m^0 \geq 0. \qquad (2)$$

Here we distinguish the following most essential characteristics of a disease:

V - concentration of viruses (pathogenic multiplying antigens)

F - concentration of antibodies (immune substrates neutralizing the viruses)

C - concentration of plasma cells (antibody-producents)

m - characteristics (mass or area) of an organ damaged by viruses

It should be noted that $\xi(m)$ is a continuous non - increasing function describing the immune system failure due to the damage of the organ (for a healthy organ $\xi(m) = 1$, for an entirely damaged one $\xi(m) = 0$) and $\theta(t)$ is a step function $\theta(t)=1 \quad t \geq 0, \quad \theta(t)=0 \quad t<0$.

According to the model (1)-(2) the disease process is described as the following. At t=0 the initial population of viruses penetrates into the body, where it starts to multiply and to injure the organ cells. In the blood viruses collide with the receptors of immunocompetent cells (antibodies).

That brings about the immune system stimulation. After a certain period of time τ, necessary for a division and differentiation of ICC, a numerous population (clone) of plasma cells appears in the body. Their main function is the antibody production. Antibodies bind the viruses and the outcome of a disease depends on the outcome of the struggle between them. If the viruses damage the organ seriously, the general state of a body grows worse, the consequence being that the immune system failure takes place.

The analytical and numerical research has shown the validity of the following statements:

Theorem 1. For all $t \geq 0$ there exists the single solution of model (1)-(2).

Theorem 2. Provided the initial conditions (2) are non-negative the solution of (1)-(2) is non-negative either for all $t \geq 0$.

Theorem 3. A sufficient condition for the asymptotic stability of stationary solution

$$V_1 = 0, \quad F_1 = \rho C^*/\mu_f = F^*, \quad C_1 = C^*, \quad m_1 = 0 \tag{3}$$

is inequality $\beta < \gamma F^*$, it being known that if inequality $0 < V^0 < IB$ is valid, then $V(t) < V^0 e^{-at}$, where

$$IB = \mu_f(\gamma F^* - \beta)/(\beta\eta\gamma), \quad a = \gamma\rho C^*/(\mu_f + \eta\gamma V^0) - \beta > 0. \tag{4}$$

Theorem 4. At $\xi(m) \equiv 1$ and $\mu_c \tau \leq 1$ the sufficient condition for the asymptotic stability of stationary solution

$$V_2 = \frac{\mu_c(\mu_f\beta - \gamma\rho C^*)}{\beta(\alpha\rho - \mu_c\eta\gamma)} > 0, \quad F_2 = \beta/\gamma,$$
$$C_2 = \frac{\alpha\mu_f\beta - \eta\mu_c\gamma^2 C^*}{\gamma(\alpha\rho - \mu_c\eta\gamma)}, \quad m_2 = \sigma V_2/\mu_m \tag{5}$$

is the inequality

$$0 < \frac{f-d}{a-g\tau} < b-g-f\tau \tag{6}$$

where

$$a = \mu_f + \mu_c + \eta\gamma V_2,$$
$$b = \mu_c(\eta\gamma V_2 + \mu_f) - \eta\gamma\beta V_2$$
$$d = \gamma\mu_c\eta\beta V_2,$$
$$g = \alpha\rho V_2,$$
$$f = \beta\alpha\rho V_2$$

it being known that in case $\alpha \to \infty$ condition (6) might come to the form

$$0 < \beta - \gamma F^* < \frac{1}{\tau + \frac{1}{\mu_c + \mu_f}} \tag{7}$$

Theorem5. In case $\xi(m) \equiv 1$, $\beta > \gamma F^*$, $F^\circ = F^*$, $C^\circ = C^*$, $m^\circ = 0$ and small V^0 the maximum value of V (t) is independent of V° and may be estimated by the value

$$V_{max} = \frac{(\beta - \gamma F^*)(\mu_f + a)}{\gamma(\rho g - \eta\gamma f)}$$

where

$$f = (\beta + \gamma F^*)/2,\ a = \beta - \gamma f,\ g = \alpha f e^{-a\tau}/(\mu_c + a)$$

The stationary solution (3) is interpreted as a healthy body state and a value IB as an immunological barrier of the body.

We simulate the infection of a healthy body by a small dose of viruses $V^0 > 0$ ($F^0 = F^*, C^0 = C^*$, m = 0), that corresponds to natural conditions. Simulation has shown that there exists four types of solution qualitatively different from each other which might be interpreted as forms of a disease (a subclinical form, an acute form with the recovery, a chronic form and a lethal form). A subclinical form arises when the conditions of theorem 3 are valid,and is characterized by the stable removal of the viruses from the body,independent of the immune response effectiveness. An acute form is characterized by the pronounced dynamics of viruses (rapid growth and an abrupt fall to the small value practically to zero). The prolonged persistention of viruses in the body is characteristic of chronic forms, and stable chronic forms arise in particular under the conditions of theorem 4. The lethal outcome is caused by the unlimited growth of viruses. It is possible for a disease to transit from an acute form to the chronic one and then to lethal due to the serious damage of the organ.

The main biological conclusion is the following. Arising of any disease form is independent of an infection dose but depends on the immunological status of an organism specific for a given type of viruses (the set of model parameters).

The following biological corollaries - hypothesis have been formulated.

1. The increase in the immunological barrier IB due to the level of immunocompetent cells in the healthy body C^*, for example with the help of vaccination, is a good way of disease prophylaxis (theorem 3).

2. In the body with the arbitrarily high sensitivity of the immune system ($\alpha \to \infty$) the small population of viruses might be stablely presented (theorem 4).

3. It is favourable to treat the acute form of a disease using the drugs with only antipathogenic properties.

2. The Nature of the Chronic Forms Origin and the Ways of Their Treatment.

We advanced the following hypothesis of the immunological nature of the chronic forms origin of a disease.

Chronic forms of a disease are determined by the week stimulation of the immune system.

We formulated the hypothesis using the following speculations. According to model (1)-(2) the outcome of a disease (recovery or chronization) depends on the width of the interval (t_1,t_2) where the inequality $F(t) > \beta/\gamma$ is valid and consequantly the concentration of viruses V(t) decreases. If this interval is sufficiently wide, then V(t) gets very small (practically approaches zero), that being regarded as the recovery.

Otherwise V(t) "fails" to get small, reaches its minimum at $t=t_2$, starts growing again and chronization arises. If we assume that the width of the interval (t_1,t_2) is determined by the number of antibodies produced (the more antibodies the wider the interval) and if we take into consideration that the antibody production is determined by the immune system stimulation then we come to the given hypothesis.

One of the simulation results has shown that if the infectious dose is largely increased (in thousands times), then the chronic form might be transited to an acute one. Moreover, it gave us an idea of the chronic form treatment - the widening of the interval (t_1, t_2) due to the strengthening of the immune response. The idea is the main in the ways of treatment which we have offered and, in particular, in the biostimulation theory.

The essence of the biostimulation or aggravation theory is the following. In a body subject to the stable chronic form a new non-multiplying non-pathogenic antigen (a biostimulator) is injected with increasing doses in a certain interval Δt. The immune system begins to form a response against the new antigen. The injection of high doses of the biostimulator might lead to the situation when due to the macrophage competition between two antigens (viruses and a biostimulator) the immune response specific for the viruses will be blocked, i.e. the immune system mill react only against biostimulators and will forget the disease agents, the result being that the viruses concentration in the body increases. After a certain period of time the biostimulator injections are over and the biostimulators are speedily removed from the organism.The immune system is again face to face with viruses. But the situation has entirely changed .For the time when the biostimulators were in the body the concentration of viruses has increased greatly and acquired the ability of the effective stimulation of the immune system, that brings about the entire removal of the viruses from the body and an increase of the immunological status specific for given viruses due to memory cells,i.e. the next contacts with viruses will lead to the more favourable forms of a disease (see biological corollary 1). Apparently , the great increase in the viruses concentration might lead to a serious damage of an organ. Then it is necessary to combine biostimulation and antipathogenic therapy. Thus, the artificial aggravation of a chronic disease due to biostimulators leads to an entire self - treatment of the body.

According to the theory and on the basis of models (1)-(2) the biostimulation model has been constructed. With the help of inequality (7) we have estimated that several weeks are required for the viruses to achieve concentration effectively stimulating the immune system. The simulation has shown that the estimate is good at Δt = 1 day. The clinical application of the aggravation theory for the pneumonia treatment testifies in its favour.

It should be noted that in our opinion antibioticotherapy (even in case of the entire removal of viruses) cannot lead to the entire recovery because next contacts with the same antigen lead to chronization. Moreover, the antibiotics are immunodepressants and their application brings about a decrease in the immunological status. The next contacts,therefore, might intensify the disease gravity,whereas in case of the biostimulation we have quite the opposite situation.

Except biostimulation and antibiotics therapy we simulate the treatment of chronic forms with the help of the nonspecific factor SAP founded recently(an antibody producents stimulator). The injection of it into the body calls forth the situation when the quantity of antibody increases threefold on the peak of the immune response.

The mechanism of a factor action and its influence on the immunological status are unknown yet but a certain part of immunologists hopes for its succesful application for the chronic forms treatment. The threefold increase of maximum number of antibodies, obviously , calls forth an increase in the width of the interval (t_1 ,t_2) and indeed we may hope that there will be a recovery. The simulation has shown that such situation is possible in some cases but the exact answer may be given by a clinical verification only.

3. Hypotheses, Facts and Suppositions.

Thus, according to our theory, the chronic forms of a disease are determined by the weak stimulation of the immune system. On the other hand, some experimental facts show that the chronic forms are determined by the T - immunodeficiency. There is no contradiction here because T - immunodeficiency calls forth the failure of T-B cooperation, which results in the decrease of B-lymphocytes stimulation and of the immune response effectivity both quantitavely and qualitatively (only Ig M production).

Moreover, general T - immunodeficiency might also mean the deficiency of T - suppressors and, thus, that of tolerancy as well. According to our results this can lead to chronization due to the fact that the body is permitted to react on small doses of viruses (see the simulation results in section 2) These facts, therefore, are in agreement with our theory and the curious supposition arises: the low zone tolerancy is expedient from the chronic forms preventing point of view. Moreover, the same can be said about the antibody production chain IgM → IgG → IgA . It is known that Ig M - antibodies arise at the beginning of the immune response, then IgG achieve their peak and the response is finished by IgA production. Such a timedifferent antibodies action appears to call forth the widening of the interval(t_1, t_2) where $dV/dt < 0$.

In this case the absence of any element in the antibodies production chain will make it possible for a disease to be chronic and the chronic forms might be determined by, for example, the absence of IgA.

Using the antibody equation structures it is possible to show that from the point of view of the chronic forms preventing the optimal strategy for the body to product antibodies is the chain - short lifetime IgM → long lifetime IgG.

Thus, we have explained the practical meaning of our term "weak stimulation" and the possible chronization due to immunodeficiencies . The chronic forms treatment appears to be the subproblem of the immunodeficiencies treatment problem.

4. Computational Aspects.

The hypotheses,corollaries and conclusions advanced above are mainly based on a great number of simulation results of models of (1)-(2) type the general form of which is

$$\frac{dx}{dt} = G(x, x(t-\tau), t) \tag{8}$$

with the initial condition

$$x(t) = \varphi(t) \qquad t \in [t^0 - \tau, t^0] \tag{9}$$

where functions G, φ satisfy the conditions of the existence and the uniqness of the solution. The feature of the systems of such kind is a well-known fact that the solution $x(t)$ is a piece-wise smooth function. In other words, at the points $t_K = t^0 + K\tau$ (k = 0,1,2, ... 3,...) the solution derivatives $x^{(i)}(t) = d^i x / dt^i$ (i=1,2,3,...) undergo,generally speaking, discontinuances $\Delta x_K^i = x^{(i)}(t_K+0) - x_K^{(i)}(t_K-0)$ But L-order approximation of the solution,in case classical computational methods are used,is based on Taylor expansion of solution and consequently on the existence of the continuous derivatives $x^{(i)}(t)$ $(i = \overline{1, L+1})$ for all t $\geqslant$ t° including the points t_K $(K = 0, 1, 2, \dots)$. Thus, an application of classical schemes to system (8)-(9),generally speaking, leads to the loss of approximation order with all arising from that mathematical consequencies which in its turn might call forth the false biological interpretations. To avoid such a situation we have described piece-wise smooth functions from the point of view of the generalized functions theory /3/ and have proved the theorem which enables us to apply a "smooth apparatus" to a piece-wise smooth function and, in particular , to use classical schemes for systems of (8)-(9) type.

Theorem 6. For any given in (a,b) piece-wise smooth function $f(t)$ which has the ν derivatives ($\nu > 1$) on the differentiation in - tervals and at the points $\{t_K\}$ from (a,b)($K = \overline{1, N}$) undergoes its finite discontinuances or those of their derivatives $\Delta f_K^{(i)}$ $(i = \overline{0, \nu};\ K = \overline{1, N})$ there exists the unique function (g(t) computed according to the formula:

$$g(t)=\sum_{K=1}^{N}\sum_{i=0}^{\nu-1}\Delta f_K^{(i)}\frac{(t-t_K)^i}{i!}\,\theta(t-t_K)$$

such that the difference $u(t)=f(t)-g(t)$ is $(\nu-1)$ times continuously differentiable function in (a,b).

Using the theorem or more exactly applying the smooth apparatus to the smooth function $u(t)$ and making, therefore , conclusions with respect to the piece-wise smooth function $f(t)$ due to the relationship $f(t)=u(t)+g(t)$ we have constructed for the function $f(t)$ the generalized Taylor series, the generalized interpolation polynom and Runge-Kutta schemes with given approximation order for systems (8)-(9). The solution of test examples confirm the advanced approach.

It is obvious that on the basis of theorem 6 one may construct the difference schemes for the systems of neutral type or for those with discontinuance of the right-hand part.

References:

I. Г.И. Марчук. Простейшая математическая модель вирусного заболевания.-Новосибирск,1975 - 22с. Препринт ВЦ СО АН СССР.
2. Г.И. Марчук. Математические модели в иммунологии.-М., Наука.1980 - 264с.
3. В. С. Владимиров. Обобщенные функции в математической физике. - М.,Наука, 1976 - 280с.
4.G. I. Marchuk,L.N. Belykh. Mathematical Model of Infec - tious Disease. - Calcolo, 1979,vol XYI,p. 399-414.

THE APPLICABILITY OF THE HAMILTON-JACOBI VERIFICATION TECHNIQUE

Frank H. Clarke

Department of Mathematics
University of British Columbia
Vancouver, B.C. V6T 1Y4
Canada

ABSTRACT

We discuss an extended (nonsmooth) version of the classical Hamilton-Jacobi verification technique. It turns out that for "normal" optimal control problems, the conditions of this method are necessary and sufficient.

1. THE PROBLEM AND THE METHOD

Let us consider the following standard dynamics of control theory:

$$\dot{x}(t) = f(x(t),u(t)) , \qquad 0 \le t \le 1 \tag{1}$$

and the constraints

$$x(0) = x_0 , \quad x(1) \in C , \quad u(t) \in U \text{ a.e.} , \tag{2}$$

where $x(\cdot)$ and $u(\cdot)$ assume values in R^n and R^m respectively. The problem is that of finding the (measurable) control u and corresponding state x which satisfy (1) (2) and which maximize $g(x(1))$. For ease of exposition, we adopt throughout this article the assumptions that U is compact and C closed, that f and g are locally Lipschitz, and that for each x the set $f(x,U)$ is convex. In the parlance of the trade, we are dealing with the relaxed optimal control problem of Mayer, in the absence of unilateral (or explicit) state constraints.

The issue that we address is that of confirming the (local) optimality of an admissible control-state pair (u,x) that we have somehow managed to identify. There exists an old, useful, and periodically rediscovered technique for doing this. Various names and labels have been attached to it, among them "the royal road of Caratheodory", "dynamic programming", "the Bellman equation", "Krotov functions", and our own choice: "the Hamilton-Jacobi equation". (See R.B. Vinter and R.M. Lewis [7] for more discussion and for references.) The technique may be seen in action, for example, in [1] (as the basis for a simple proof of an inequality in mathematical physics) and in [2] (where it confirms the synthesis of an optimal control problem arising in renewable resources).

Although the basic idea remains the same in various treatments, not all its implementations are strictly equivalent. The one we are about to discuss retains a high degree of generality while being applicable "in all reasonable problems", a phrase that we shall explain later on.

We require some notation. An <u>arc</u> is an absolutely continuous function from $[0,1]$ to R^n. An <u>ε-tube</u> about an arc $x(\cdot)$ is the subset

$$\{(t,y) : 0 \le t \le 1 , |x(t) - y| < \varepsilon\} ,$$

where $\varepsilon > 0$. The (true) <u>Hamiltonian</u> of the problem is the function H defined by

$$H(x,p) = \max\{<p,f(x,u)> : u \in U\} . \tag{3}$$

The (classical) <u>Hamilton-Jacobi equation</u> is the relationship

$$W_t(t,x) + H(x,W_x(t,x)) = 0 \tag{4}$$

where W is a function of (t,x), and W_t, W_x denote partial derivatives. We shall say that W satisfies the <u>boundary condition</u> provided

$$W(1,\cdot) \text{ and } g(\cdot) \text{ agree on } C . \tag{5}$$

Here, in the raiment we have chosen for it, is the old and useful idea:

<u>PROPOSITION 1</u> *Suppose that* (u,x) *is a given control-state pair, and that on some* ε*-tube about* x *there is a continuously differentiable solution* W *of the Hamilton-Jacobi equation which satisfies the boundary condition as well as the condition*

$W(0,x_0) = g(x(1))$. *Then* (u,x) *is a local solution to the optimal control problem.*

Proof: We shall show that if (v,y) is any admissible control-state pair for which $(t,y(t))$ lies in the ε-tube about x, one has $g(y(1)) \le g(x(1))$. To see this, we write

$$g(y(1)) = W(1,y(1)) = W(0,y(0)) + \int_0^1 \frac{d}{dt} W(t,y(t))dt$$

$$= W(0,x_0) + \int_0^1 \{W_t(t,y) + \langle W_x(t,y),\dot{y}\rangle\}dt$$

$$\le g(x(1)) + \int_0^1 \{W_t + H(y,W_x)\}dt \quad (\text{since } \dot{y} = f(y,v))$$

$$= g(x(1)) . \qquad \text{QED}$$

We see then that one way to verify the optimality of a given pair (u,x) is to exhibit a function W which satisfies the hypotheses of Proposition 1 ; hence the term "verification technique". But if (u,x) is a solution, is it necessarily the case that there exists a function W to confirm it in this office? There may not be such a W, as evidenced by the following.

EXAMPLE 1 We set $n = 2$, and we denote by (x,y) points in R^2. Let

$$U = [-1,1], \quad f(x,y,u) = (0,xu), \quad x_0 = (0,0), \quad C = R^2, \quad g(a,b) = b .$$

Note that the state $(0,0)$, with any control u, is optimal for this version of the problem. The function W of Proposition 1, if it exists, satisfies

$$W(0,0,0) = 0, \quad W(1,a,b) = b \tag{6}$$

$$W_t(t,x,y) + |xW_y(t,x,y)| = 0 , \tag{7}$$

where (7) is valid in an ε-tube about the arc $(0,0)$. Since $W_t \le 0$, we derive from (6) that

$$W(t,0,0) = 0 . \tag{8}$$

Again from (6), one has $W_y(1,0,0) = 1$, so that for some δ in $(0,\varepsilon)$ one has

$$W_y(t,x,0) \ge 1/2 \quad \text{for} \quad 1-\delta \le t \le 1, \quad |x| < \delta .$$

Combining this last conclusion with (7) yields

$$W_t(t,x,0) \le |x|/2 \quad \text{for} \quad 1-\delta \le t \le 1, \quad |x| < \delta .$$

Setting $\tau = 1-\delta$, this leads to

$$W(t,x,0) \le W(\tau,x,0) - (t-\tau)|x|/2 \quad \text{for} \quad \tau \le t \le 1, \quad |x| < \delta . \tag{9}$$

Let t lie in $(\tau,1)$, and consider now the quantity $q = W_x(t,0,0)$. One has, using (9) and (8):

$$q = \lim_{x \downarrow 0} [W(t,x,0) - W(t,0,0)]/x$$

$$\le \lim_{x \downarrow 0} [W(\tau,x,0) - (t-\tau)|x|/2]x = W_x(\tau,0,0) - (t-\tau)/2 .$$

Similarly,

$$q = \lim_{x \downarrow 0} [W(t,-x,0) - W(t,0,0)]/(-x) \geq W_x(\tau,0,0) + (t-\tau)/2 .$$

This contradiction shows that no such W exists.

The technique of Proposition 1 is thus seen to be inapplicable in some cases. An evident way to react to this setback is to extend the technique to greater levels of generality, in the hope of capturing more cases within its scope. One such effort is due to R.B. Vinter and R.M. Lewis [7], who consider sequences of continuously differentiable functions satisfying, among other things, a certain inequality related to the Hamilton-Jacobi equation. We shall take a different tack in which no sequences intervene, but in which we depart from considering only smooth functions. Our guide will be the calculus of generalized gradients [3] [5].

2. THE MODIFIED HAMILTON-JACOBI EQUATION

Let $\phi(\cdot)$ be a locally Lipschitz function on R^n. The generalized gradient of ϕ at x, denoted $\partial\phi(x)$, is the set

$$\text{co}\ \{\lim_{i\to\infty} \nabla\phi(x_i) : x_i \to x\} ,$$

where all sequences $\{x_i\}$ which converge to x and along which ϕ is differentiable, and such that $\lim_{i\to\infty} \nabla\phi(x_i)$ exist, are considered, and where "co" denotes "convex hull". The generalized gradient is defined for other classes of functions beside the locally Lipschitz ones, and an extensive calculus has been developed; we refer the reader to [5] for details and references.

A locally Lipschitz function W of (t,x) is said to be a solution of the (modified) Hamilton-Jacobi equation provided that (in some suitable region) one has

$$\max \{\alpha + H(x,\beta) : (\alpha,\beta) \in \partial W(t,x)\} = 0 . \qquad (10)$$

(Note that if W is continuously differentiable, then $\partial W(t,x)$ reduces to the singleton $\{[W_t(t,x), W_x(t,x)]\}$, so that (4) and (10) coincide.) With this terminology established, one can now assert

PROPOSITION 2 *Proposition 1 remains true if the words "continuously differentiable" are replaced by "locally Lipschitz".*

The proof, which we omit, is based upon elementary properties of generalized gradients.

Since the class of verification functions is larger than it was, we may now hope to find a verification function W in Example 1 where before none existed. The reader may verify that $W(t,x,y) = y + (1-t)|x|$ is such a one.

Is the verification technique, as extended by Proposition 2, applicable to all cases? We adduce the following to show that it is not.

EXAMPLE 2 We have $n = 2$ as in example 1, and we define U = unit ball in R^2,

$f(x,y,u) = u$, $x_0 = (0,0)$, $C = \{1\} \times R$, $g(a,b) = b$. Note that the only admissible control-state pair (and so the optimal one) is $u \equiv (1,0)$, $(x,y)(t) = (t,0)$. A verification function W would satisfy

$$W(0,0,0) = 0, \quad W(1,1,b) = b, \quad \text{and}$$

$$\alpha + |(\beta,\gamma)| \le 0 \quad \text{for all} \quad (\alpha,\beta,\gamma) \in \partial W(t,x,y), \tag{11}$$

where the latter holds for all (t,x,y) in an ε-tube about the arc $(t,0)$. Now let (x',y') be any arc originating at $(0,0)$, corresponding to a control u' in U, and lying in the ε-tube about $(t,0)$. Then, by the Mean Value Theorem for generalized gradients,

$$W(1,x'(1),y'(1)) = W(1,x'(1),y'(1)) - W(0,0,0)$$
$$\in [1,x'(1),y'(1)] \cdot \partial W(t,tx'(1),ty'(1)) \le 0 \qquad \text{(by (11))}.$$

If K is a Lipschitz constant for W, one also deduces

$$y'(1) = W(1,1,y'(1)) \le W(1,x'(1),y'(1)) + K|1 - x'(1)|$$
$$\le K|1 - x'(1)| ,$$

in view of the preceding inequality. Applied to the arc $x'(t) = st$, $y'(t) = (1-s^2)^{1/2}t$ (for any $s < 1$), this last conclusion yields

$$s \le (K^2 - 1)/(K^2 + 1) ,$$

a contradiction which establishes the non-existence of a Lipschitz verification function in this example.

If, as the above shows, the extended verification technique of Proposition 2 still fails to apply to all cases, what grounds could we have to be satisfied with it? The answer lies in the fact that, as we shall see, the method applies to all "nondegenerate" problems. (Thus example 2 will turn out to be a pathological, or degenerate, problem.) To make this notion precise, we must introduce some terminology from the theory of necessary conditions.

3. EXTREMALS AND NORMALITY

An arc x is said to be an <u>extremal</u> provided there exists an arc p such that

$$(-\dot{p},\dot{x}) \in \partial H(x,p) \qquad \text{for almost all} \quad t \quad \text{in} \quad [0,1] . \tag{12}$$

The arc p is referred to as the <u>coextremal</u>. The <u>transversality condition</u> is

$$p(1) \in \lambda\partial g(x(1)) - N_C(x(1)) , \tag{13}$$

where λ is a scalar and N_C denotes the normal cone to C (at $x(1)$); see [3]. These constructs appear in the theory of necessary conditions, which includes the assertion [4] that if x solves the problem (locally), then there exists a coextremal p for x and a nonnegative scalar λ for which the transversality condition is satisfied, and such that $|p(t)| + \lambda$ is never 0.

If f happens to admit a derivative f_x continuous in (x,u), then it can be shown that (12) holds iff there is a point u in U such that

$$\dot{x} = f(x,u) , \quad -\dot{p} = f_x^*(x,u)p$$

$$\max \{\langle p,f(x,w)\rangle : w \in U\} = \langle p,f(x,u)\rangle .$$

Of course, we recognize in this case the familiar conditions of the Pontryagin maximum principle.

The degenerate case of these necessary conditions is that in which they can hold for $\lambda = 0$, for then the objective function g plays no part in them. Borrowing the analogous term from the calculus of variations, we accordingly refer to the problem as normal at x provided that there is no nonvanishing coextremal p for x which satisfies $-p(1) \in N_C(x(1))$. We call the problem normal provided that, for all local solutions x, the problem is normal at x. The following points out one special class of normal problems.

PROPOSITION 3 *A free endpoint problem (i.e. one for which $C = R^n$) is normal.*

This follows immediately from the fact that, in this case, one has $N_C(x(1)) = \{0\}$. Note that the problem of example 1 is normal, being a free endpoint one. That of example 2 is not, since (as the reader may verify) the arc $(t,0)$ admits the (abnormal) coextremal $(1,0)$.

The following asserts that the verification technique of Proposition 2 applies (at least) whenever the necessary conditions are nondegenerate (i.e. when the problem is normal):

THEOREM [6] *In order that the control-state pair (u,x) solve the problem (locally) it is sufficient, and if the problem is normal, also necessary, that there exist on a tube about x a locally Lipschitz solution W of the Hamilton-Jacobi equation satisfying the boundary condition and also the condition $W(0,x_0) = g(x(1))$.*

The above provides grounds for being satisfied with Proposition 2 as a "sufficiently applicable" verification technique. We shall not give the proof, which applies to problems more general than the one considered here; somewhat involved, it is based upon a formula for the generalized gradient of a certain "value function", the function $V(u)$ which is the value of the problem whose terminal constraint is $x(1) \in C + u$. Besides providing a means of constructing the verification function W, the full development establishes the following interesting sensitivity result: the function V is Lipschitz (near the origin) when the problem is normal.

REFERENCES

1. R.A. Adams, F.H. Clarke, Gross' logarithmic Sobolev inequality: a simple proof, Amer. Journal of Math. 101 (1979) 1265-1269.

2. C.W. Clark, F.H. Clarke, G. Munro, The optimal exploitation of renewable resource stocks, Econometrica 47 (1979) 25-47.

3. F.H. Clarke, Generalized gradients and applications, Transactions Amer. Math. Soc. 205 (1975) 247-262.

4. __________, Extremal arcs and extended Hamiltonian systems, Transactions Amer.

Math. Soc. 231 (1977) 349-367.

5. __________, Generalized gradients of Lipschitz functionals, Advances in Mathematics 40 (1981) 52-67.

6. __________, "Nonsmooth Analysis and Optimization", monograph, forthcoming.

7. R.B. Vinter, R.M. Lewis, "A verification theorem which provides a necessary and sufficient condition for optimality", IEEE Trans. on Automatic Control AC-25 (1980) 84-89.

STATIONARY DETERMINISTIC FLOWS IN DISCRETE SYSTEMS : I

E. Gelenbe
ERA 452 du CNRS
LRI Bât. 490, Université Paris-Sud
91405 ORSAY (FRANCE)

ABSTRACT : We consider a deterministic system whose state space is the (n-dimensional) first orthant : one can view it as a Petri net, a vector addition system, or a deterministic network of queues evolving with time. Our purpose is to develop a quantitative measurement oriented theory of such systems. The objective is therefore similar to that of operational analysis. The system is examined over an infinite time interval and assumptions are made concerning the flows of significant events. Under these assumptions several basic theorems are proved in the scalar case, where flows in and out of a subsystem are being considered. In particular the following *asymptotic* results are established : the difference between state frequency measurements at instants of arrival and of departure are identical as $t \to \infty$; furthermore as $t \to \infty$ the system satisfies equations similar to the familiar "birth and death" equations. Little's formula as such does *not* hold however, though a related theorem can be established. The generality of our assumptions is comparable to that of the point process approach of Franken, König and others, although our approach is strictly deterministic (i.e. single sample path in stochastic terms).

1 - INTRODUCTION

1.1 - General considerations

Complex concurrent systems are conveniently described by vector addition systems (VAS) which originated with the work of Karp and Miller on parallel program schemata [1]. These models are essentially equivalent to the nets introduced by C.A. Petri which were originally suggested as models for management systems, but which are now very often effectively used to describe computer operating system behaviour. Petri nets and VAS are frequently used to prove qualitative properties of complex concurrent systems, for instance in the area of computer communication protocols, such as the non-existence of deadlock between parallel processes. They are also used as formal specification tools, as models in the design of large hardware/software systems, and for building standards of communication protocols.
It would be most convenient if essentially the same models could incorporate properties which would make them suitable for performance evaluation. Such an extension would have obvious practical advantages for the system designer. Indeed such extensions have been suggested by Sifakis [2], and others, but have not led to the development of a substantial quantitative theory useable for performance evaluation and prediction.
In the area of computer performance evaluation, probabilistic queueing network models (see for instance [3]) have gained wide popularity. More recently, more direct relationships between models of this type and measurement data was made via "operational models" introduced by Buzen [4] and developed by Buzen and Denning [5], without probabilistic assumptions. However, a direct link between the formalisms used in queueing networks and those used in Petri nets has not been available, except for the most obvious analogy which would associate customers and servers on the one hand with tokens and transitions on the other hand.
This paper is an attempt to develop a "measurement" oriented theory within a formal context very close to that of VAS. The existence of such a theory would allow the analysis of both qualitative and quantitative aspects within one formal framework. Indeed conventional VAS are imbedded in the model we use. We may also view this work as an attempt to deal with very general sample paths of a stochastic queueing system ; in this aspect we deal with a generality comparable with that of [7].

1.2 - The formal model

Let $\underline{k}(t)$ be an ℓ-dimensional vector of natural numbers defined for each real valued $t \geq 0$. $\underline{k}(t)$ will be a left continuous function of t. $\underline{k}(0) = \underline{k}_0$ is given.
Let $T = \{0 = t_0 < t_1 < t_2 < \cdots < t_i < t_{i+1} < \cdots\}$ be a countable infinite subset of the positive real line. We shall say that T is the set of transition times of $K = \{\underline{k}(t), t \geq 0\}$; namely, we suppose that

$$k(t) = \underline{k}(t_i^+) \text{ for } t \in]t_i, t_{i+1}], \text{ for each } i = 0,1,2,\ldots$$

This means that changes, or "jumps" in K can only occur at the instants included in

T. This, of course, does not imply that there will be a jump at each instant which is an element of T.

A subset S of $\{1,2,\dots,\ell\}$ will be called a cut, and for a given S we shall define for each $t \geq 0$

$$N(t) = \sum_{i \in S} k_i(t)$$

where $k_i(t)$ is the i-th component of the vector k(t). The cut S will also give rise to a restriction T_S of T :

$$T_S = \{t_i : t_i \in T \text{ and } N(t_i^+) \neq N(t_i)\} .$$

We can view T_S as the set of jump instants of $\{N(t),\ t \geq 0\}$, or as the jump instants in T which can be detected by observing N(t).

1.3 - Relationship with vector addition systems

The relationships between the model described in the preceding section and VAS is obvious.

A VAS is defined by an ℓ-dimensional vector of natural numbers $\underline{k}_0$, and by a finite set $R = \{\underline{r}_1,\dots,\underline{r}_k\}$ of integer valued ℓ-dimensional vectors. From $\underline{k}_0$ we can construct some ℓ-dimensional vector as follows :

$$\underline{k}^0 = \underline{k}_0$$

$$\underline{k}^m = \underline{k}^{m-1} + \underline{r}^m, \quad \underline{r}^m \in R$$

as long as each $\underline{k}_i$, $1 \leq i \leq m$, is composed of natural numbers. That is, we use the elements of R to transform $\underline{k}_0$ by successive additions with the only restriction that each of the resulting vectors remain in N^ℓ (the ℓ-dimensional space of natural numbers)

With respect to the model presented in Section 1.2, we can simply consider that

$$\underline{k}(t_m^+) = \underline{k}^m$$

Thus our model differs from a VAS by the introduction of time.

1.4 - Relationship with Petri nets

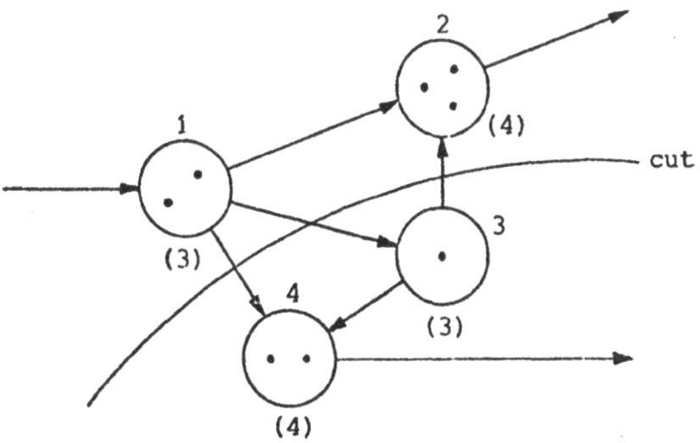

Figure 1 : Example of a Petri net.

Consider the Petri net shown in Fig. 1. Arrows represent the flow of tokens among the four places or transitions. The incoming arrow to transition n° 1 indicates that it may receive tokens from the "outside world", while tokens may go the "outside world" from transitions n° 2 and 4. The system may be represented by the vector

$$\underline{k} = (k_1, k_2, k_3, k_4)$$

where k_i is the number of tokens present at position i. In Figure 1, we show the initial vector which is

$$\underline{k}_0 = (2,3,1,2)$$

and the numbers in parentheses next to each transition represent the number of tokens necessary to fire that transition. Thus $\underline{k}_0$ is a stable state of the system.

Suppose now that a time $t_1 > 0$ one token arrives to transition 1 which will then fire and realease its three tokens to each of 2,3,4 which will in turn fire, etc. leading to a new stable state

$$\underline{k}(t_1^+) = (0,1,1,0)$$

Let the cut S be the set {3,4} as shown in Figure 1. Then

$$N(0) = 3$$

and

$$N(t_1^+) = 1$$

An external arrival of a token at time t_2 will provoke a state transition of the system to

$$\underline{k}(t_2^+) = (1,1,1,0)$$

and so on. Thus the formal model described in Section 1.2, differs from a conventional Petri net only in the introduction of the time parameter t, as was the case with respect to VAS.

1.5 - Relationship with probabilistic queueing models

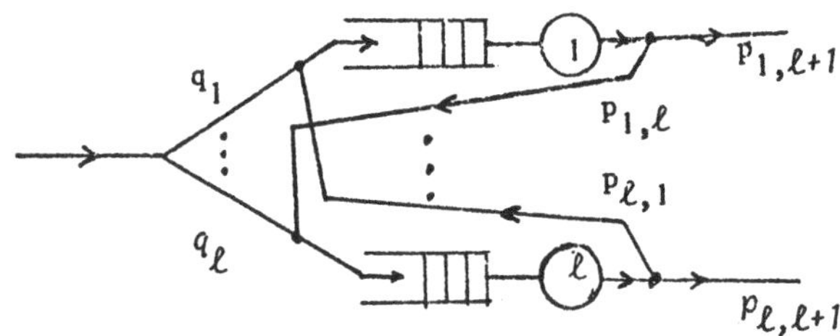

Figure 2 : Probabilistic network of queues

Consider the probabilistic network of queues shown in Figure 2. Customers arriving to the system are routed initially to service station i, $1 \le i \le \ell$, with probability q_i, and after some queueing and service time at that station will either leave the system with probability $p_{i,\ell+1}$ or enter queue j with probability $p_{i,j}$ and so on.

We shall denote by ω a complete history of the system and Ω will be the set of all such histories. Any ω can be constructed as follows, supposing that these histories begin at time $t=0$.

Let successive customers arrive at instants

$$A = \{0 \le a_1 \le a_2 \le a_3 \le \cdots \le a_i \le a_{i+1} \le \cdots\}$$

and for the i-th customer in the sequence let

$$y_i = \{y_{i1}, Y_{i2}, \cdots, Y_{ij}, Y_{i,j+1} \cdots\}$$

be the sequence of service stations visited, at which it sill receive services of duration

$$Z_i = \{Z_{i1}, Z_{i2}, \cdots, Z_{ij}, Z_{i,j+1}, \cdots\}$$

Then each ω will be completely defined if we specify the three sets A, $\{Y_i\}$, $\{Z_i\}$. For a given ω; however, we may only be interested in the movements of the customers in the system, in which case we can construct the function

$$\underline{k}(\omega,t) = (\underline{k}_1(\omega,t), \cdots, k_\ell(\omega,t)), \qquad t \ge 0$$

when $k_j(\omega,t)$ is the number of customers in queue j at time t for history ω. Similarly from history ω we can construct the jump instants

$$0 \le t_1(\omega) \le t_2(\omega) \le \cdots \le t_i(\omega) \le \cdots$$

Thus the $\underline{k}(t)$ constructed in Section 1.2, corresponds to the $\underline{k}(\omega,t)$ of this section. In the probabilistic framework, one most often works directly with probability distributions, rather than sith sample paths, even through the probability distributions are initially assigned to each history ω in Ω.

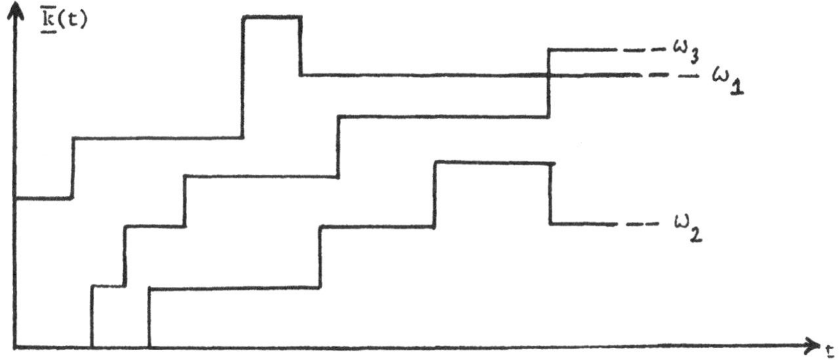

Figure 3 : Trajectories or histories associated with the queueing network.

Thus for some time $\tilde{t}$, one tends to compute quantities related to $p(\underline{k}(\tilde{t}) \in A)$; which is the probability that the (random process) $\bar{\underline{k}}(t)$[1] is in the subset A of its set of possible states at time $\tilde{t}$; but this is in fact

$$p(\bar{\underline{k}}(\tilde{t}) \in A) = \sum_{\omega \in \Omega(\tilde{t},A)} p(\omega) \tag{1}$$

where

$$\Omega(\tilde{t},A) = \{\omega \mid \underline{k}(\omega,\tilde{t}) \in A\} \ . \tag{2}$$

---

1) $\bar{\underline{k}}(t) = \{\{\underline{k}(\omega,t) \ : \ \omega \in \Omega\}, \ t \ge 0\}$.

Thus, results concerning $\underline{k}(\omega,t)$ for one ω or for a family of ω's may be carried over to the probability distributions of a probabilistic model via (1) and (2).

2 - PROPERTIES OF FLOWS

Let us now return to the properties of the formal model defined in Section 1.2 as follows :

- $\underline{k}(t) \in \mathbb{N}^{\ell}$ (the ℓ-domensional space of natural numbers) defined for each t ; $\underline{k}(t) = (k_1(t),\cdots,k_{\ell}(t))$.
- $T = \{0<t_1<t_2<\cdots<t_i<t_{i+1}<\cdots\}$ the set of jump instants of $\underline{k}(t)$;
- S a subset of $\{1,\ldots,\ell\}$ which we call a cut, and which gives rise to

$$N(t) = \sum_{i\in S} k_i(t), \qquad t \geq 0$$

and to $T_S = \{t_i \ : \ t_i \in T \ \text{ and } \ N(t_i^+) \neq N(t_i)\}$.

All these quantities are to be viewed as being deterministic. One may also view $\{\underline{k}(t),\ t\geq 0\}$ as being in fact identical to some sample path $\{\underline{k}(\omega,t),\ t\geq 0\}$ associated with a given history ω of the random process $\{k(t),t\geq 0\} = \{\ \underline{k}(\omega,t) \ : \ \omega \in \Omega \ ,\ t \geq 0\}$.

2.1 - Flows associated with the cut S

$T_S \subseteq T$ is obtained by considering only those time instants at which N(t) changes values. We shall define two subsets Λ and Γ of T_S such that :

$$\Lambda \cap \Gamma = \emptyset,\ \Lambda \cup \Gamma = T_S$$

where

$$\Lambda = \{t_i \ : \ t_i \in T_S \text{ and } N(t_i^+) > N(t_i)\}$$

$$\Gamma = \{t_i \ : \ t_i \in T_S \text{ and } N(t_i^+) < N(t_i)\}$$

and write

$$\Lambda = \{0<a_1<a_2<\cdots<a_i<a_{i+1}<\cdots\}$$

$$\Gamma \quad \{0<d_1<d_2<\cdots<d_i<d_{i+1}<\cdots\}$$

which we shall view as the arrival and departure instants, respectively, to and from the cut S.

2.2 - Flow assumptions

Some restrictive assumptions will now be made concerning the flows which we consider.

2.1.1- Initial condition on N(t)

$$N(0) = 0$$

2.2.2- One step transitions of N(t)

For any $t_i \in T_S$, $N(t_i^+) = N(t_i) \pm 1$

2.2.3- Causal flows

$$\sum_{i=1}^{\infty} 1(a_i\ t) \geq \sum_{i=1}^{\infty} 1(d_i<t) \qquad \text{for all } t \geq 0.$$

2.2.4- Stationary flows

There exist real numbers $0<\lambda<\infty$, $0<\gamma<\infty$ such that

$$\lim_{i\uparrow\infty} \frac{1}{i} \left|a_i - \frac{i}{\lambda}\right| = 0 \qquad \lim_{i\uparrow\infty} \frac{1}{i} \left|d_i - \frac{i}{\gamma}\right| = 0$$

Under these assumptions, we can now present some useful general results.

3 - PROPERTIES OF ARRIVAL AND DEPARTURE INSTANT MEASURES

From a theoretical standpoint, Λ and Γ define sequences of significant instants of the formal system. Thus properties of observations of the system at those instants are of interest. If one takes a practical view-point arising from the necessity to carry out measurements, the Λ or Γ instants can be used to trigger data collection. Our first result states that the sequences Λ and Γ are equivalent with respect to the measures they offer from observations of the state of the cut S, namely of $\{N(t),t\geq 0\}$.
Let us define for each $t\geq 0$ and natural number $n\geq 0$,

$$U(n,t) = \sum_{i=1}^{\infty} 1(N(a_i) = n, a_i<t) \Big/ \sum_{i=1}^{\infty} 1(a_i<t)$$

$$V(n,t) = \sum_{i=1}^{\infty} 1(N(d_i^+) = n, d_i<t) \Big/ \sum_{i=1}^{\infty} 1(d_i<t)$$

We can say that $U(n,t)$ is the relative frequency of observing that $N(t) = n$ at arrival instants, and that $V(n,t)$ is the corresponding quantity just after departure instants.

Theorem 3.1 : Under assumptions 2.2.1 to 2.2.4,

$$\lim_{t\uparrow\infty} |U(n,t) - V(n,t)| = 0 \qquad \text{if } \lambda = \gamma .$$

Analogous theorems are often established for various queueing theoretic models under probabilistic assumptions.
The proof of this theorem is simplified by the use of the following lemma.

Lemma 3.2 : Let $0<x_1<x_2<\cdots<x_i<x_{i+1}<\cdots$

be an infinite sequence such that $x_i<\infty$ for $i<\infty$, and

$$\lim_{i\uparrow\infty} \frac{1}{i} \left|x_i - \frac{i}{\nu}\right| = 0$$

for some real number $0<\nu<\infty$. Denote by $X(t)$ the integer

$$X(t) = \sum_{i=1}^{\infty} 1(x_i<t), \quad \text{for } t>0 .$$

Then

(a) $$x_{X(t)} = \frac{X(t)}{\nu} + \varepsilon_x(t)$$

where

$$\lim_{t\uparrow\infty} \frac{|\varepsilon_x(t)|}{X(t)} = 0$$

(b) $$X(t) = \theta t + \nu_x(t)$$

where

$$\lim_{t\uparrow\infty} \frac{|\theta_x(t)|}{t} = 0$$

(c) Hence $$x_{X(t)} = t + \varepsilon_x(t) + \frac{1}{\nu}\theta_x(t)$$

where

$$\lim_{t\uparrow\infty} \frac{|x_{X(t)} - t|}{X(t)} = 0 \quad .$$

Proof of Lemma 3.2 : Clearly $i\to\infty$ implies $x_i\to\infty$; since $i<\infty$ implies $x_i<\infty$, it follows that

$$i\to\infty \text{ if and only if } x_i\to\infty \ .$$

By definition,

$$x_{X(t)} < t \le x_{X(t)+1}$$

since

$$x_{X(t)}\to\infty \text{ if and only if } X(t)\to\infty$$

and because $t\to\infty$ implies that $x_{X(t)}\to\infty$, it follows that $t\to\infty$ implies $X(t)\to\infty$.

By assumption

$$\lim_{X(t)\to\infty} \frac{1}{X(t)} \left|x_{X(t)} - \frac{X(t)}{\nu}\right| = 0 \ .$$

Therefore, if we define

$$\varepsilon_x(t) = x_{X(t)} - \frac{X(t)}{\nu}$$

then

$$\lim_{t\to\infty} \frac{|\varepsilon_x(t)|}{X(t)} = 0$$

which establishes (a).

Thus we can write :

$$\frac{X(t)}{\nu} + \varepsilon_x(t) < t \le \frac{X(t)+1}{\nu} + \varepsilon'_x(t)$$

where

$$\varepsilon'_x(t) = x_{X(t)+1} - \left(\frac{X(t)+1}{\nu}\right) \ .$$

Thus

$$-(1+\nu\varepsilon'_x(t)) \le X(t) - \nu t < -\nu\varepsilon_x(t)$$

where we know that by (a)

$$\lim_{t\to\infty} \frac{|\varepsilon_x(t)|}{X(t)} = 0 \qquad \lim_{t\to\infty} \frac{|\varepsilon'_x(t)|}{X(t)+1} = 0 \quad .$$

Clearly we have :

$$\lim_{t\to\infty} \frac{|\varepsilon'_x(t)|}{X(t)} = 0$$

and

$$|X(t) - \nu t| \le \max(\nu|\varepsilon_x(t) \ , \ |1+\nu\varepsilon'_x(t)|) \ .$$

Therefore if we write

$$X(t) = \nu t + \theta_x(t)$$

we will have

$$\lim_{t\uparrow\infty} \frac{|\theta_x(t)|}{X(t)} = 0, \quad \text{or} \quad \lim_{t\uparrow\infty} \frac{|\theta_x(t)|}{t+ \ _x(t)} = 0$$

which necessarily implies $\lim_{t\uparrow\infty} \frac{|\theta_x(t)|}{t} = 0$, completing the proof of (b).

Assertion (c) is an immediate consequence of (a) and (b). Q.E.D.

Corollary 3.3 : As an immediate consequence, we have :

$$N(t) \to +\infty \text{ as } t \to +\infty \quad \text{if } \lambda = \gamma$$

and

$$\lim_{t\uparrow\infty} \frac{N(t)}{t} = 0 \qquad \text{if } \lambda = \gamma \ .$$

Proof of Theorem 3.1. : For each $d_i < t$ construct $a_{k(i)}$ such that

$$a_{k(i)} = \sup \{a_j : a_j < d_i \quad \text{and} \quad N(a_j) = N(d_i^+)\}$$

by assumption 2.2.1 and 2.2.2. We can then write for each $n \geq 0$

$$\sum_{i=1}^{\infty} 1(N(d_i^+) = n,\ d_i < t) = \sum_{i=1}^{\infty} 1(N(a_{k(i)}) = n,\ d_i < t) = \sum_{j=1}^{\infty} 1(N(a_j) = n,\ a_j < t)$$
$$- A(n,t)$$

where

$$A(n,t) = \sum_{j=1}^{\infty} 1(N(a_j) = n,\ a_j < t,\ a_j \neq a_{k(i)} \quad \text{for some } d_i < t \text{ with } N(d_i^+) = n).$$

Using the definition of $U(n,t)$ we can write

$$U(n,t) = \sum_{j=1}^{\infty} 1(a_j < t) - A(n,t) = V(n,t) \sum_{i=1}^{\infty} 1(d_i < t)$$

since $\sum_{i=1}^{\infty} 1(d_i < t) = \sum_{j=1}^{\infty} 1(a_j < t) - N(t)$

we have

$$[U(n,t) - V(n,t)] \sum_{j=1}^{\infty} 1(a_j < t) = A(n,t) - V(n,t)\, N(t)$$

or

$$U(n,t) - V(n,t) = \frac{N(t)}{\sum_{j=1}^{\infty} 1(a_j < t)} \left[\frac{A(n,t)}{N(t)} - 1\right] .$$

Notice that

$$N(t) = \sum_{n=0}^{\infty} A(n,t), \text{ hence } A(n,t) \leq N(t)$$

Furthermore by Lemma 3.1 (b)

$$N(t) = \lambda t - \gamma t + \theta_a(t) - \theta_d(t)$$

and

$$\sum_{j=1}^{\infty} 1(a_j < t) = \lambda t + \theta_a(t)$$

Therefore,

$$|U(n,t) - V(n,t)| \leq \frac{\lambda t - \gamma t + \theta_a(t) - \theta_d(t)}{\lambda t + \theta_a(t)} \quad .$$

Recalling Lemma 3.2 (b), we see that when $\lambda = \gamma$

$$\lim_{t\uparrow\infty} |U(n,t) - V(n,t)| = 0 \qquad \text{Q.E.D.}$$

4 - THE RATE EQUATIONS

We shall now prove a relationship akim to the birth and death equations for certain continuous time markovian systems. Of course, our context remains strictly deterministic and governed only by the Flow Assumptions of Section 2.2.

Define the following quantities

$$Q(n,t) = \frac{1}{t}\int_0^t 1(N(\tau) = n)\, d \ , \ n \geq 0$$

$$\lambda(n,t) = \sum_{i=1}^{\infty} 1(a_i<t, N(a_i) = n) \Big/ \int_0^t 1(N(\tau) = n)\, d\tau, \qquad n \geq 0$$

$$u(n+\ ,t) \equiv \sum_{i=1}^{\infty} 1(d_i<t, N(d_i^+) = n) \Big/ \int_0^t 1(N(\tau) = n+1)\, d\tau, \qquad n \geq 0$$

Theorem 4.1 : If $\lambda = \gamma$, then

$$\lim_{t\uparrow\infty} |\lambda(n,t)\ Q(n,t) - \mu(n+1,t)\ Q(n+1,t)| = 0$$

Proof : We can write as a consequence of Theorem 3.1 that

$$U(n,t) = V(n,t) + h(t)$$

where

$$\lim_{t\uparrow\infty} h(t) = 0 \ .$$

However, by definition

$$U(n,t) = \lambda(n,t)\cdot t\ Q(n,t) \Big/ \sum_{i=1}^{\infty} 1(a_i<t)$$

$$V(n,t) = \mu(n+1,t)\cdot t\ Q(n+1,t) \Big/ \sum_{i=1}^{\infty} 1(d_i<t) \quad .$$

Therefore

$$\lambda(n,t)\ Q(n,t) = \mu(n+1,t)\ Q(n+1,t) \quad \sum_{i=1}^{\infty} 1(a_i<t) \Big/ \sum_{i=1}^{\infty} 1(d_i<t) + \frac{h(t)}{t} \sum_{i=1}^{\infty} 1(a_i<t)$$

or

$$\lambda(n,t)\ Q(n,t)-\mu(n+1,t)\ Q(n+1,t) = \mu(n+1,t)\ Q(n+1,t) \frac{N(t)}{\sum_{i=1}^{\infty} (d_i<t)} + h(t) \frac{\sum_{i=1}^{\infty} 1(a_i<t)}{t}$$

$$= V(n,t) \frac{N(t)}{t} + h(t) \frac{\sum_{i=1}^{\infty} 1(a_i<t)}{t} \ .$$

Now using the fact that $\lambda = \gamma$ and Lemma 3.2 (b) we have :

$$\lim_{t\uparrow\infty} \frac{N(t)}{t} = \lim_{t\uparrow\infty} \frac{\theta_a(t)-\theta_b(t)}{t} = 0 \qquad \text{(already mentioned in Corollary 3.3)}$$

$$\lim_{t\uparrow\infty} \frac{1}{t} \sum_{i=1}^{\infty} 1(a_i<t) = \lambda + \lim_{t\uparrow\infty} \frac{\theta_a(t)}{t} = \lambda \quad .$$

Also $0 \le V(n,t) \le 1$ by definition. Therefore since $\lim_{t\uparrow\infty} h(t) = 0$ we have

$$\lim_{t\uparrow\infty} |\lambda(n,t)\, Q(n,t) - \mu(n+1,t)\, Q(n+1,t)| = 0 \qquad \text{Q.E.D.}$$

5 - A MODIFIED "LITTLE'S FORMULA"

A well-know formula in queueing theory, initially proved by J.D.C. Little [3,4,7] states that "$\bar{N} = \lambda T$" or that the average number in the system is given by the product of the arrival rate and of the response time. Under our assumptions (2.2.1 to 2.2.4) this formula does not hold in its standard form as we shall see below.

Define

$$T(t) = \sum_{i=1}^{\infty} (d_i - a_i)\, 1(d_i<t) \Big/ \sum_{i=1}^{\infty} 1(d_i<t) \;.$$

This quantity can be viewed as the average of the time intervals separating arrival and departure intervals. If the arrival instant a_i in some sense provokes the departure instant d_i, then it can be interpreted as an average response time of the system in the time interval $[0,t[$.

Theorem 5.1 : If $\lambda = \gamma$ then

$$\lim_{t\uparrow\infty} \frac{1}{t} \left| \frac{1}{t} \int_0^t N(\tau)\, d\tau - \lambda T(t) \right| = 0 \qquad .$$

Remark 5.2 : In many queueing systems one proves that "the average number in the system is equal to the arrival rate multiplied by the average response time", although cases in which this result (Little's formula) does not hold can be constructed [9]. Theorem 5.1 states that under assumptions 2.2.1 to 2.2.4, these quantities differ at most by a function of t which increases strictly more slowly than t.

Proof of Theorem 5.1 : Let us define for each d_i :

$$a_{j(i)} = \sup\, \{a_j : a_j < d_i,\ N(a_j) = N(d_i^+)\}.$$

Then clearly

$$\int_0^t N(\tau)\, d\tau = \sum_{i=1}^{\infty} (d_i - a_{j(i)})\, 1(d_i<t) + \int_{d_{D(t)}}^t N(\tau)\, d\tau$$

where we recall that

$$d_{D(t)} = \sup\, \{d_i : d_i<t\} \;.$$

We must have

$$\sum_{i=1}^{\infty} (d_i - a_{j(i)})\, 1(d_i<t) = \sum_{i=1}^{\infty} (d_i - a_i)\, 1(d_i<t)$$

because by assumption

$$\sum_{i=1}^{\infty} 1(a_i<t) \ge \sum_{i=1}^{\infty} 1(d_i<t)$$

so that for each $d_i<t$ there must at least one $a_j<t$. Therefore :

$$\int_0^t N(\tau)\, d\tau = T(t) \sum_{i=1}^{\infty} 1(d_i<t) + \int_{d_{D(t)}}^t N(\tau)\, d\tau \qquad .$$

In the interval $]d_{D(t)},t[$, $N(\tau)$ is a non-decreasing function of τ, therefore

$$\int_{d_{D(t)}}^{t} N(\tau)\, d\tau \le [t-d_{D(t)}]\, N(t) \quad .$$

Hence using Lemma 3.2 (a) and (c), we have

$$d_{D(t)} = \frac{D(t)}{\gamma} + \varepsilon_d(t) = t + \varepsilon_d(t) + \theta_d(t)/\gamma$$

so that

$$0 \le \frac{1}{t}\int_{d_{D(t)}}^{t} N(\tau)\, d\tau \le \frac{\gamma\,\varepsilon_d(t) + \theta_d(t)}{\gamma t}\cdot[\theta_a(t) - \theta_d(t)]$$

where we have used (for $\lambda = \gamma$) Lemma 3.2 (c) :

$$N(t) = \theta_a(t) - \theta_d(t) \ .$$

Thus since

$$\sum_{i=1}^{\infty} 1(d_i<t) = \gamma t + \theta_d(t) = \lambda t + \theta_d(t) \ ,$$

We have :

$$\frac{1}{t}\left|\frac{1}{t}\int_0^t N(\tau)\, d\tau - \lambda T(t)\right| \le T(t)\,\frac{\theta_d(t)}{t} + \frac{(\lambda\varepsilon_d(t)+\theta_d(t))(\theta_a(t)-\theta_d(t))}{\lambda\, t^2}$$

Since as $t\to\infty$, $\theta_d(t)/t \to 0$, $\theta_a(t)/t\to 0$, $\varepsilon_d(t)/t\to 0$, and because

$$T(t) \le d_{D(t)} = t+\varepsilon_d(t)+\theta_d(t)/\gamma \ .$$

We have

$$\lim_{t\to\infty} \frac{1}{t}\left|\frac{1}{t}\int_0^t N(\tau)\, d\tau - \lambda T(t)\right| = 0 \qquad \text{Q.E.D.}$$

6 - CONCLUSION

In this paper we have set the framework for the study of systems governed by flows of events more general than those examined in the usual queueing theoretical structure.

As such our flow assumptions can be best compared to those made in recent work such as [7] in which stochastic queueing models are examined under general (non-independent) arrival and departure streams.

Our approach is strictly deterministic and thus can be viewed as a single sample path analysis in a stochastic framework.

It is easy to see that our assumptions hold in very general cases. Nevertheless some general theorems can be proved.

We show in particular that the empirical state frequencies at arrival and departure instants are assymptotically identical. Furthermore we exhibit equations similar to those of birth and death processes. We show that a variant of Little's formula is satisfied. Several examples illustrating the class of systems we consider are exhibited.

Subsequent papers shall deal with the vector discrete space case, as well as with cases where the state space is continuous.

REFERENCES

[1] R. KARP and R. MILLER : Properties of a model of parallel computation, SIAM J. of App. Math., vol. 14, Nov. 1966.

[2] J. SIFAKIS : Thèse de Docteur Ingénieur, Université de Grenoble, 1974.

[3] E. GELENBE and I. MITRANI : Analysis and synthesis of computer system models, Academic Press, 1980.

[4] J.P. BUZEN : Fundamental operational laws of computer system performance, Acta Informatica, vol. 7, n° 2, p. 167-182, 1976.

[5] P.J. DENNING, J.P. BUZEN : The operational analysis of queueing network models, A.C.M. Computing Surveys, p. 225-262, 1978.

[6] P. FRANKEN : A New Approach to Investigation of Stationary Queueing Systems, Sektion Mathematik der Humboldt-Universität zu Berlin, Preprint Nr. 6/76, Nov. 1976.

[7] P. FRANKEN, D. KÖNIG, U. ARNDT, V. SCHMIDT : Queues and Point Processes, Akademie-Verlag, Berlin, 1979.

[8] S. STIDHAM : Queueing systems where $L \neq \lambda W$. NCSU Techn. Report n° 78-6, August, 1978.

RECENT TRENDS OF THE OPTIMAL CONTROL FOR STOCHASTIC DISTRIBUTED PARAMETER SYSTEMS

by

Yoshifumi SUNAHARA
Professor

Faculty of Polytechnic Sciences
Kyoto Institute of Technology
Matsugasaki, Kyoto 606, Japan

Abstract

A survey is presented of some recent developments of the optimal control theory for stochastic distributed parameter systems within the framework of stochastic process theory in the function space. First, various kinds of mathematical models of practical systems are exhibited, where the significance of uncertainties in the system structure and/or the environment is taken into account. Then, a promising approach of application is discussed which treats in a unified manner the optimal control for a class of distributed parameter systems involving a stochastic parameter. A representative result of digital simulation experiments is also included. The paper ends with conclusions and recommendations for future works in the field.

1. Introduction

Recognizing the fact that dynamic behaviors of all real physical systems are, in fact, distributed, recent advances in control sciences have stirred a great deal of enthusiasm in the development of state estimate and optimal control of stochastic distributed parameter systems (SDPSs). In the 1960's, very little work of the optimal control has been done in distributed parameter systems. This was due to the fact that the theoretical aspect of distributed parameter systems (DPSs) is much more difficult than for lumped parameter systems. In spite of a wide range of recent applications of the DPSs, it has been only in the past decade that a high rate of publishing papers in this field has appeared and that readily applicable approaches have, in particular, been developed to estimate the system state and to determine unknown parameters using noisy observation data. A familiar fashion of the optimal control for DPSs can be found in the form of the modal optimal control [1], [2]. Since the major body of the optimal control theory of DPSs is beautifully summarized in Ref. [3], the reappearance of historical aspects is abandoned in this paper. However, during recent years, many theoretical works in the context of DPSs have centered on the function space concept, where abstract versions of the state estimate and optimal control have been shown in a suitable Hilbert, Banach or Sobolev space. For a recent survey of the aspect of function space methods, Ref. [4] brings us a useful information with many references included. A fruitful line of theoretical development was initiated with the terminology, "variational inequality" [5]. Many excellent results are collected in Ref. [6] and this is one of the most important publications in the field of DPSs. Similar problems were studied in Ref. [7] where the notion of a feedback control was introduced.

This survey is organized as follows: Various kinds of mathematical models of SDPSs are exhibited for the purpose of showing the existing status of uncertainties in Section II. With this exhibition as the first step, in order to guarantee the mathematical security of system models in Section III, an example is briefly described of giving conditions for existence of a unique continuous solution to a stochastic partial differential equation. Section IV is devoted to describing a brief survey of studies on the optimal control for SDPSs and to presenting particular results developed by the author and his colleagues, including digital simulation experiments. Discussions on problems which are opened to future studies close this survey with the list of references.

II. Mathematical Models for Stochastic Distributed Parameter Systems

We know that the answer for the question whether a system should be modeled mathematically as a DPS or not is not simple, and that a general and unique model which is capable of expressing various kinds of DPSs can not be made. However, several mathematical models described below are undoubtedly important in the stochastic distributed parameter class of systems of interest.

Let $u(t,x)$ be the scalar system state.

(M.1) SDPS with an additive noise:

A mathematical model of a general class of nonlinear SDPSs is given by

$$\frac{\partial u(t,x)}{\partial t} = F(t,x,u,u_x,u_{xx},\cdots) + f(t,x) + G(t,x)\gamma(t,x),\ t \in]0,t_f[,\ x \in D \tag{2.1}$$

where the subscripts express the partial derivatives. In Eq. (2.1), F is a conventional nonlinear function but this is assumed to be sufficiently smooth in order to guarantee the existence of a unique continuous solution, f is considered to be the distributed control signal, G is a known function and $\gamma(t,x)$ is the distributed noise process which is, in many cases, assumed to be the zero-mean white noise process with respect to t. With a well-posed linear spatial differential operator L_x, a linear version of Eq. (2.1) takes a very familiar form as

$$\frac{\partial u(t,x)}{\partial t} = L_x u(t,x) + f(t,x) + G(t,x)\gamma(t,x),\ t \in]0,t_f[,\ x \in D \tag{2.2}$$

Examples of the initial and boundary conditions are respectively as follows:

I.C. $u(0,x) = u_0(x)$, $x \in D$ (2.3a)

B.C. $B_x u(t,x) = \gamma_b(t,x)$, $t \in]0, t_f[$,

$x \in \partial D$ (the boundary at the system state) (2.3b)

In (2.3a), u_0 is the initial state function which is usually assumed to be Gaussian and this is independent of $\gamma(t,x)$. In Eq. (2.3b), B_x is also a well-posed linear spatial differential operator and $\gamma_b(t,x)$ expresses an uncertainty as the boundary input of the system.

(M.2) DPS with stochastic parameters

In recent years, a strong interest has been in the field of nuclear or chemical plants that fall into the category of DPSs with stochastic parameters. This implies that one or more coefficients in differential operators of the mathematical model are considered to be stochastic variables. Since measurements of physical properties of the system considered inherently exhibit remarkable uncertainties, the system modeling by a class of partial differential equations with stochastic parameters is more realistic. An example of the general form expressed by using the parabolic type is

$$\frac{\partial u(t,x)}{\partial t} = A(t,x,\omega;D_x)u(t,x) + f(t,x), \quad t \in]0, t_f[, \ x \in D \tag{2.4a}$$

together with the same initial condition as given by (2.3a) and with the boundary condition

$$B_j(t,x;D_x)u(t,x) = 0, \quad t \in]0, t_f[, \ x \in \partial D, \ j = 1,2,\cdots,m \tag{2.4b}$$

where A is a partial differential operator involving stochastic coefficients, B_j is a boundary operator with deterministic nonvanishing coefficient. In Eqs. (2.4), the symbol ω expresses the generic point of the sample space and this will be omitted to write in the sequel because no confusions will result. Since differential operators are stochastic, the problem of finding a solution to Eq. (2.4a) is included in the mathematical literature of stochastic eigenvalue problems.

(M.3) DPS with space dependent parameters

Among many practical systems, there has recently been much practical interest shown in problems of identifying unknown system parameters which depend on the system state. Recently, motivated by the oil exploration survey, the following mathematical model described by a class of partial differential equations of hyperbolic type is frequently adopted:

$$a_r(x)\frac{\partial^2 u(t,x)}{\partial t^2} - \frac{\partial}{\partial x}\{a_e(x)\frac{\partial u(t,x)}{\partial x}\} = 0, \quad t > 0, \ x > 0 \tag{2.5a}$$

with the initial and boundary conditions

$$u(0,x) = \frac{\partial u(0,x)}{\partial t} = 0, \qquad x > 0, \tag{2.5b}$$

$$- a_e(0)\frac{\partial u(t,0)}{\partial x} = g(t), \qquad t > 0, \tag{2.5c}$$

where both $a_r(x)$ and $a_e(x)$ are respectively unknown coefficients to be determined and the function g(t) is a stochastic input at the boundary. According to the motivation, the model (2.5) is used for estimating unknown parameter $\{a_r(x), a_e(x)\}$ based on the boundary measurement $y(t) = u(t,0)$. Consequently, use is in the future area of the model (2.5) for the aspect of controls.

(M.4) DPS with the moving boundary

A large number of systems of practical interest involve boundaries moving by phase change such as melting or solidification, chemical reaction and so on. An example of the mathematical model is

$$\frac{\partial u(t,x)}{\partial t} = a\frac{\partial^2 u(t,x)}{\partial x^2}, \quad 0 < x < y(t), \quad 0 < t < t_f \tag{2.6a}$$

with the initial condition

$$u(0,x) = u_0(x) \geq 0, \quad 0 < x < y(0) \tag{2.6b}$$

In Eq. (2.6a), the function y(t) is the moving boundary and a is a positive constant. Taking into account physical properties of the system considered, the boundary condition can be formulated. An illustrative example can be seen in Ref. [8]. Very little work has been done in the optimal control scheme of this class of DPSs, particularly compared with what has been accomplished for DPSs with fixed boundaries. In many cases, uncertainties appear at the boundary.

(M.5) DPS with a semi-permeable wall

Mathematical models are omitted to describe here because of the space limit and because the mathematical model including the control aspect is not available at the present time (for more detail, please see Ref. [9]).

III. Uncertainties and Mathematical Security of System Models

Of all the aspects of the subject considered in this paper, the most popular has been the development of the feedback configuration of the optimal control for DPSs. Stochastic versions of the optimal control problem for SDPSs were reported in Refs. [10] to [13] where all consider the optimal stochastic control problem along the extended line of attack on lumped parameter concepts and all of those works were concerned only with the mathematical model (M-1), as might be expected.

In modeling SDPSs, different authors assume different classes of parameter models, each one representing a particular case adapted to a specific practical system. Although such types as hyperbolic or parabolic partial differential equations are representative of system models, on the other hand, they can be grouped into four different classes associated with the existing status of uncertainties.

Class S_1: Involves uncertainties in the interior region of the system.

Class S_2: Involves uncertainties at the system boundary.

Class-S_3: Involves uncertainties simultaneously in the interior region and at the boundary of the system.

Class-S_4: Operates in the stochastic environment.

There are a number of ways in which a piece of work in the field of the optimal control for SDPSs could be specified. One would be by the control system configuration. Can the system be configurated in a form of the optimal feedback control or the optimal open-loop control? Does the system have distributed or boundary controls? What kinds of mechanisms are adopted to measure the system state? How are the time and space evolutions of the system measured; throughout the whole region D of the system state u(t,x), or only at specific points on the system boundary ∂D? Although no classification method is satisfactory, the principal one selected here deals with the class of controls as follows:

Class-C_1: Involves the distributed control where controllers are set in the spatial domain.

Class-C_2: Involves the pointwise control where controllers are set at a finite number of spatial points.

Class-C_3: Involves the boundary control where controllers are set at the system boundary.

Figure 1 shows an illustration of the situation of control classes C_1, C_2 and C_3.

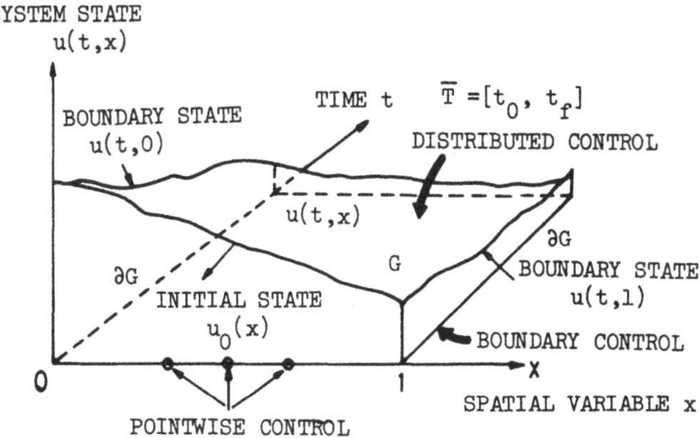

Fig. 1 Illustration of the situation of control

Although there is a sizable body of theory and of computational method for the stochastic version of optimal control for DPSs, the principal activities in this study are concerned with (1) evolution operator ideas associated with the semi-group as, for example in Ref. [13] to [15] (2) method of variational forms as, for example in Refs. [6] and [17]. However, stochastic problems involving partial differential equations with stochastic parameter have not had a great deal of attention in the research area of controls. Very recently, in remarkable papers [18], [19], a general theory in this context was developed.

It has become customary to begin a survey paper in the field of DPSs involving stochastic parameters by remarking on the following two fundamental items:

(Item-1) Problems involved with the field mentioned above are much more difficult than those with SDPSs with constant parameters.

(Item-2) Computational difficulty arises when we are dealing with numerical methods for mathematical models.

In fact, these remarks still remain valid. However they are now being resolved and now have much less significance than they had in 1970's. For the purpose of explaining this circumstance, in this survey paper, the optimal control problem of SDPSs with stochastic coefficients is separated into a number of subproblems which allow the problem to be solved in a successive manner. The subproblems involved in the stochastic optimal control are as follows:

Step-1: Write the mathematical description containing uncertainties of the physical system under consideration and choose the control class.

Step-2: Decide a measuring method and an allocation of sensors in the spatial domain.

Step-3: Examine the mathematical security of the mathematical model adopted in Step-1.

Step-4: Choose a criterion of control performance.

Step-5: Define a class of admissible control.

Step-6: Find the optimal control satisfying the control criterion adopted in Step-4.

Step-7: Consider numerical aspects of computing and generating the optimal control signal obtained in Step-6.

In the sequel, the steps outlined above are explored along an illustrative example.

Step-1. Mathematical model: We shall consider a SDPS described by

$$\frac{\partial u(t,x)}{\partial t} - a\frac{\partial^2 u(t,x)}{\partial x^2} + b\frac{\partial u(t,x)}{\partial x}\dot{w}(t) = f(t,x),\ t\in T =]0,t_f[,\ x\in D =]0,1[\tag{3.1a}$$

with the initial and boundary conditions

$$\text{I.C.}\quad u(0,x) = u_0(x), \qquad x\in D, \tag{3.1b}$$

$$\text{B.C.}\quad u(t,0) = u(t,1) = 0,\ t\in]0,t_f[, \tag{3.1c}$$

where a, b are positive constants, respectively, $\dot{w}(t)$ is a formal derivative of the Brownian motion process, i.e., the Gaussian white noise and f(t,x) denotes the spatially distributed control signal.

Step-2. Observation mechanism: For simplicity, the author is excluding a number of interesting problems relating to the optimal choice of observation mechanisms and to the optimal sensor allocation.

Step-3. Mathematical security of the system model: First of all, the (closed) time interval $\bar{T} = [0,t_f]$ is partitioned into $0 = t_0^{(m)} < t_1^{(m)} <\cdots< t_m^{(m)} = t_f$, where $\{t_i^{(m)}\}\subset\{t_i^{(m+1)}\}$. The Brownian motion process w(t) is also approximated by

$$w^{(m)}(t) = w(t_i^{(m)}) + \frac{\Delta_i w^{(m)}}{\tau_i^{(m)}}(t - t_i^{(m)}), \tag{3.2}$$

where $\tau_i^{(m)} \triangleq t_{i+1}^{(m)} - t_i^{(m)}$ and

$$\Delta_i w^{(m)} = w(t_{i+1}^{(m)}) - w(t_i^{(m)}). \tag{3.3}$$

It is easy to show that $w^{(m)}(t)$ converges uniformly to sample values of $w(t)$ w.p.1 as $\tau^{(m)} = \max_i\{\tau_i^{(m)}\} \to 0$ [20], [21]. Thus, the $w^{(m)}(t)$-process defined by (3.2) almost surely has its piecewise continuous derivative, Eq. (3.1) can be approximated by

$$\frac{\partial u^{(m)}(t,x)}{\partial t} - a\frac{\partial^2 u^{(m)}(t,x)}{\partial x^2} + b\frac{dw^{(m)}}{dt}\frac{\partial u^{(m)}(t,x)}{\partial x} = f(t,x), \quad (t,x) \in T \times D \tag{3.4a}$$

together with the initial and boundary conditions

$$\text{I.C. } u^{(m)}(0,x) = u_0(x), \quad x \in D \; : \; \text{B.C. } u^{(m)}(t,0) = u^{(m)}(t,1) = 0, \quad t \in T. \tag{3.4b}$$

Let H be a class of functions which are square integrable, i.e., $H = L^2(D)$. We need the following basic assumptions:

(H-1) For the initial condition, we assume that

$$u_0 \in L^2(\Omega;H), \tag{3.5}$$

where we write the well-known probability space by (Ω,F,μ).

(H-2) For the control signal, we assume that

$$f \in L^2(\Omega;L^2(T,H)). \tag{3.6}$$

Furthermore, we introduce the linear operator A_D from $V(= H_0^1(D))$ into $V'(= H^{-1}(D))$ which is defined by

$$\langle A_D\psi,\phi\rangle \triangleq \int_0^1 a\frac{\partial\psi(x)}{\partial x}\frac{\partial\phi(x)}{\partial x}dx, \text{ for } {}^\forall\psi,\phi \in V. \tag{3.7}$$

Along a similar notion to (3.7), we shall introduce another linear operator A_s from V into $L(R^1,H)$ as

$$A_s\psi\gamma \triangleq b\frac{\partial\psi(x)}{\partial x}\gamma, \text{ for } {}^\forall\psi \in V, \ {}^\forall\gamma \in R^1 \tag{3.8}$$

where L is the space of linear continuous mappings of R^1 into H. With (3.7) and (3.8), from Eq. (3.4a), it is a simple exercise to show that

$$(u^{(m)}(t),\psi) + \int_0^t \langle A_D u^{(m)}(s),\psi\rangle ds + \int_0^t (A_s u^{(m)}(s),\psi)dw^{(m)}(s)$$

$$= (u_0,\psi) + \int_0^t (f(s),\psi)ds, \text{ for } {}^\forall\psi \in V. \tag{3.9}$$

Basically, the main difficulty for a generalized class of DPSs described by a partial differential equation is due to the infinite dimensionality of the state space. From this point of view, it is extremely difficult to show that $u^{(m)}(t)$ determined by Eq. (3.9) converges to $u(t)$ in some sense, as $m \to \infty$. One method is to introduce the Galerkin's method [22] by noting that V is separable. That is, we may choose $\{e_i\}_{i=1}^\infty$ as an orthonormal basis of H. Define

$$V_k = H_k = \text{span}[e_1,e_2,\cdots,e_k]. \tag{3.10a}$$

$$\Pi_k(\cdot) = \sum_{i=1}^{k} (\cdot, e_i)e_i; \; \tilde{\Pi}_k(\phi) = \sum_{i=1}^{k} \langle\phi, e_i\rangle e_i, \quad \text{for } \phi \in V'. \tag{3.10b}$$

Then, by using the same procedure as in Ref. [22], Eq. (3.9) can be approximated by

$$u_k^{(m)}(t) + \int_0^t A_{Dk} u_k^{(m)}(s)ds + \int_0^t A_{sk} u_k^{(m)}(s)dw^{(m)}(s) = u_{0k} + \int_0^t f_k(s)ds, \tag{3.11}$$

where $u_k^{(m)}(t) = \Pi_k u^{(m)}(t)$, $f_k(t) = \Pi_k f(t)$, $A_{Dk}\phi = \tilde{\Pi}_k A_D \phi$ and $A_{sk}\phi = \Pi_k A_s \phi$. The following lemma states the existence of $u_k(t)$ as $m \to \infty$.

[Lemma-1] [23] With (H-1) and (H-2), we can extract a subsequence $u_k^{(m)}(t)$ such that, for a finite fixed k, $\text{l.i.m.}_{m\to\infty} u_k^{(m)}(t) = u_k(t)$, where $u_k(t)$ is a solution to

$$u_k(t) + \int_0^t A_{Dk} u_k(s)ds + \int_0^t A_{sk} u_k(s)dw(s) + \frac{1}{2}\int_0^t A_{sk} u_k(s) A_{sk}^*(s)ds = u_{0k} + \int_0^t f_k(s)ds, \tag{3.12}$$

where

$$\frac{1}{2} A_{sk}\phi A_{sk}^* = \frac{1}{2}\sum_{i=1}^{k} (A_{sk}\phi, A_{sk}e_i)e_i, \text{ for } {}^{\forall}\phi \in V_k. \tag{3.13}$$

We are now ready to describe an important theorem stating the existence and uniqueness of a solution of the stochastic partial differential equation considered here.

[Theorem-1] [23] With (H-1) and (H-2), Eq. (3.12) has a unique continuous solution

$$u_k \in L^2(\Omega; L^2(T; V_k) \cap C(\bar{T}; H_k)), \tag{3.14}$$

where C is the space consisting of a continuous function ϕ defined in $\bar{T}$.

Furthermore, by a similar statement to Lemma-1, $\lim_{k\to\infty} u_k = u$ weakly in $L^2(\Omega; L^2(T,V))$, where, for $\phi \in V$, u(t) satisfies

$$(u(t),\phi) + \int_0^t \{\langle A_D u(s),\phi\rangle + \frac{1}{2}(A_s u(s), A_s\phi)\}ds$$
$$+ \int_0^t (A_s u(s),\phi)dw(s) = (u_0,\phi) + \int_0^t (f(s),\phi)ds. \tag{3.15}$$

There exists a unique continuous solution to Eq. (3.15) such that $u \in L^2(\Omega; L^2(T,V)) \cap C(\bar{T}; H))$.

Since the proofs of Lemma-1 and Theorem-1 are rather lengthy, the descriptions are omitted (see Ref. [23]). However, it should not be overlooked that we do not need the condition $a - (b^2/2) > 0$ but require only a simple condition $a > 0$.

IV. Optimal Control

We shall proceed to find the optimal control signal in this section. Before doing this, Steps 4 and 5 should be performed.

Step-4. Control criterion: Define the functional J by

$$J(f) \triangleq \frac{1}{2}E\{\int_0^{t_f} [(u(t), Mu(t)) + N(f(t), f(t))]dt\}, \tag{4.1}$$

where $M \in L^2(T;H\otimes_2 H)$ and N is a positive constant. Thus, the problem is to find the feedback optimal control f^o in Eq. (3.15) in such a way that the functional $J(f)$ becomes minimal with respect to $f \in W_{ad}$, where W_{ad} denotes a class of admissible control and this definition is given in the next step.

Step-5. A class of admissible control: The class W_{ad} in Step-4 is defined by

$$W_{ad} = \{f | f \in L^2(\Omega;L^2(T,H)) \text{ and } f. \text{ is } F.\text{-measurable}\}, \tag{4.2}$$

where F_t is a σ-algebra generated by the solution process $\{u(s);\ 0 \le s \le t\}$.

Step-6. Optimal control strategy: By using the calculus of variations, it can be proved that the control $f^o \in W_{ad}$ is optimal, if and only if

$$(\delta J(f^o),\ f - f^o) \ge 0, \text{ for } {}^\forall f \in W_{ad} \tag{4.3}$$

where δJ denotes the Gateaux differential with respect to f, [22]. Noting that the functional $J(f)$ defined by (4.1) is a quadratic form, the inequality (4.3) is expressed in the form [22]

$$E\{\int_0^{t_f} [(u^o(t), Mv(t)) + N(f^o(t), f(t) - f^o(t))]dt\} \ge 0 \tag{4.4}$$

where $u^o(t)$ is the solution to Eq. (3.15) corresponding to the optimal control f^o and $v(t) = u(t) - u^o(t)$ which is the solution to

$$(v(t),\phi) + \int_0^t \langle A_D v(s), \phi\rangle ds + \frac{1}{2}\int_0^t (A_s v(s), A_s\phi)ds$$
$$+ \int_0^t (A_s v(s), \phi)dw(s) = \int_0^t (f(s) - f^o(s), \phi)ds, \text{ for } {}^\forall\phi \in V. \tag{4.5}$$

Now, we introduce the following adjoint system,

$$(P(t)u^o(t), \phi) - \int_0^t \{\langle A_D^* P(s)u^o(s), \phi\rangle + \frac{1}{2}(A_s P(s)u^o(s), A_s\phi)$$
$$- (P(s)A_s u^o(s), A_s\phi)\}ds + \int_0^t (P(s)A_s u^o(s), \phi)dw(s)$$
$$= (P(0)u^o(0), \phi) - \int_0^t (Mu^o(s), \phi)ds, \text{ for } {}^\forall\phi \in V \tag{4.6}$$

where P is a deterministic self-adjoint Hilbert Schmidt operator such that

$$P \in C(\bar{T}, H\otimes_2 H) \cap L^2(T;V\otimes_2 H) \cap L^2(T;H\otimes_2 V) \tag{4.7}$$

with $P(t_f) = 0$ and $V\otimes_2 H$ denotes the class of Hilbert Schmidt operator of $\mathcal{L}(H;V)$. It can easily be proved that there exists a unique solution to Eq. (4.6) and that [23]

$$Pu^o \in L^2(\Omega;C(\bar{T};H) \cap L^2(T;V)). \tag{4.8}$$

With this result, the following optimality theorem is stated:

[Theorem-2] With (4.8), a necessary and sufficient condition for existing the optimal control f^o is found to be

$$(P(t)E\{u^o(t)|F_t\} + Nf^o(t),\ f(t) - f^o(t)) \geq 0, \text{ for } t \in T,\ \forall f \in W_{ad}, \text{ a.e.} \tag{4.9}$$

Furthermore, the optimal control $f^o(t)$ is given by

$$f^o(t) = -N^{-1}P(t)\bar{u}_t \tag{4.10}$$

where $\bar{u}_t = E\{u^o(t)|F_t\}$ and $P(t)$ satisfies the following operator Riccati equation:

$$\begin{aligned}(\dot{P}(t)\phi_1, \phi_2) - \langle A_D\phi_1, P(t)\phi_2\rangle - \langle A_D{}^*P(t)\phi_1, \phi_2\rangle \\ - \frac{1}{2}(A_s\phi_1, A_sP(t)\phi_2) - \frac{1}{2}(A_sP(t)\phi_1, A_s\phi_2) + (P(t)A_s\phi_1, A_s\phi_2) \\ - (P(t)N^{-1}P(t)\phi_1, \phi_2) + (M\phi_1, \phi_2) = 0\end{aligned} \tag{4.11a}$$

with the terminal condition

$$(P(t_f)\phi_1, \phi_2) = 0. \tag{4.11b}$$

The proof has been completed in Ref. [23].

Step-7. Numerical aspect: Both time and spatial variables were partitioned as $t_i = i\Delta t$ $(i = 0,1,\cdots,n-1)$ and $x_j = j\Delta x$ $(j = 0,1,\cdots,m-1)$ for computer implementation. Parameter values were set as $a = 0.1$, $b = 1.5$, $M(\cdot) = 143\int_0^1 \exp[-11\{(x-0.5)^2 + (y-0.5)^2\}](\cdot)dy$ and $N = 0.01$. With $\Delta t = 0.0001$ and $\Delta x = 0.1$, Eq. (3.15) was simulated on a digital computer where $f(t,x) = 0$. This purpose is to emphasize an advantage of applying the control $f(t,x)$ in the subsequent experiment. Figure 2 shows a sample run of the $u(t,x)$-process without the control. Equation (3.15) was simulated once again on a digital computer with the optimal control f^o determined by (4.10) and (4.11). A representative of sample runs of the $u^o(t,x)$-process is depicted in Fig. 3.

V. Conclusions

The elegant body of mathematical theory pertaining to finite-dimensional stochastic systems and its successful application to many fundamentally linear problems can be extended to infinite-dimensional systems with stochastic inputs. A serious difficulty arising from the infinite dimensional system with stochastic parameters concerns the theory of operator Riccati equation in function spaces. Such theoretical methods as necessary conditions, existence of solutions, uniqueness, etc. are of undoubted importance. These are not only of academic interests but of contribution to better understanding of the problems being considered. However, a number of problems remain open in the field of the control of SDPSs. Numerical algorithms which are capable of saving computer time are also required.

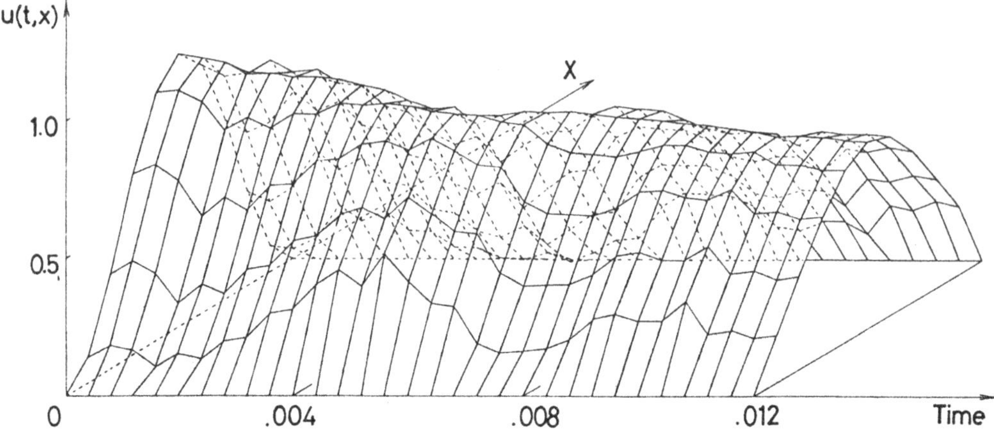

Fig. 2 A control-free sample run of the u(t,x)-process

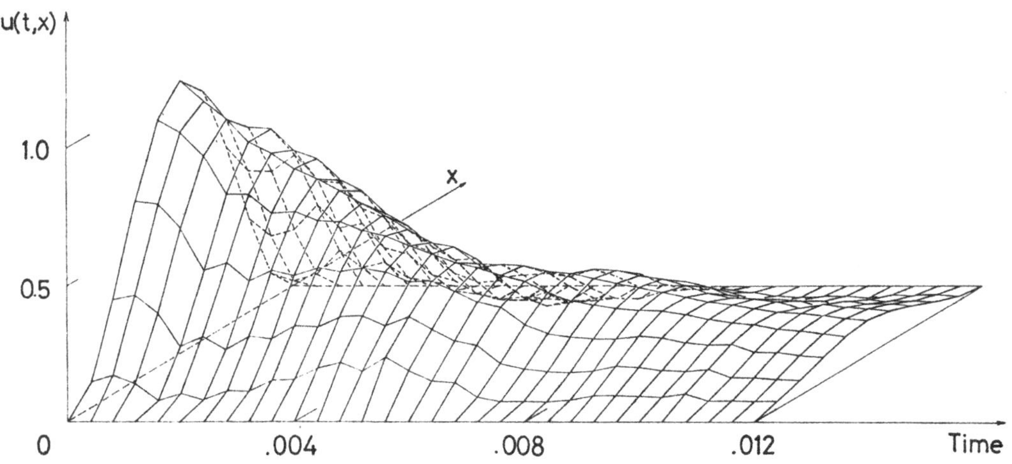

Fig. 3 A controlled sample run of the u(t,x)-process

Acknowledgment

The author would like to thank Dr. ShinIchi Aihara for his assistance. This work was supported in part by a grant of the Educational Ministry of Japan, No. 56550289.

References

1) Balas, M.J.: Modal Control of Certain Flexible Dynamic Systems, SIAM J. Control and Optimization, Vol. 16, No. 3, pp. 450-463 (1978)

2) Gibson, J.S.: An Analysis of Optimal Modal Regulation, Convergence and Stability, SIAM J. Control and Optimization, Vol. 19, No. 5, pp. 686-707 (1981)

3) Robinson, A.C.: A Survey of Optimal Control of Distributed-Parameter Systems, Automatica, Vol. 7, pp. 371-388 (1971)

4) Mitter, S.K.: Optimal Control of Distributed Parameter Systems, Preprints of Joint Automatic Control Conference, Boulder, Colorado (1969)

5) Lions, J.L. and Stampacchia, G.: Variational Inequalities, Commun. Pure Appl. Math., Vol. 20, pp. 493-519 (1967)

6) Lions, J.L.: Optimal Control of Systems Governed by Partial Differential Equations, (Trans. by Mitter, S.K.), Springer-Verlag, N.Y. (1971)

7) Balakrishnan, A.V.: Optimal Control Problems in Banach Spaces, SIAM J. Control Series A, Vol. 3, pp. 152-179 (1965)

8) Sunahara, Y.: Identification of Distributed Parameter Systems, Chapter 2, Distributed Parameter Control Systems, ed. by Tzafestas, S.G., Pergamon Press, London (to appear)

9) Sunahara, Y., Aihara, Sh. and Ishikawa, M.: On the State Estimates for a Class of Stochastic Distributed Parameter Systems with a Semi-Permeable Wall,: Preprints for the 20th IEEE Conference on Decision and Control, San Diego, Calif. (1981)

10) Kushner, H.J.: On the Optimal Control of a System Governed by a Linear Parabolic Equation with White Noise Inputs, SIAM J. Control, Vol. 6, pp. 596-614 (1968)

11) Bensoussan, A.: Control of Stochastic Partial Differential Equations, Chapter 4, Distributed Parameter Systems, ed. by Ray, W.H. and Lainiotis, D.G., Marcel Dekker, N.Y. (1978)

12) Balakrishnan, A.V.: Identification and Stochastic Control of a Class of Distributed Systems with Boundary Noise, Lecture Notes in Economics and Mathematical Systems, Vol. 107, Springer-Verlag, N.Y. (1974)

13) Curtain, R.F. and Pritchard, A.J.: Infinite Dimensional Linear Systems Theory, Lecture Notes in Control and Information Sciences, Vol. 8, Springer-Verlag, N.Y. (1978)

14) Balakrishnan, A.V.: An Operator Theoretic Formulation of a Class of Control Problems and the Steepest Descent Method of Solution, SIAM J. Control, Vol. 1, pp. 109-127 (1963)

15) Falb, P.L.: Infinite Dimensional Control Problems-I; On the Closure of the Set of Attainable States for Linear Systems, J. Math. Anal. Applic, Vol. 9, pp. 12-22 (1964)

16) Bensoussan, A. and Viot, M.: Optimal Control of Stochastic Linear Distributed Parameter Systems, SIAM, J. Cont., Vol. 13, pp. 904-926 (1975)

17) Bensoussan, A. and Lions, J.L.: Applications des Inéquations Voriationnelles en Contrôle Stochastique, Dunod (1978)

18) Ahmed, N.U.: Stochastic Control on Hilbert Space for Linear Evolution Equations with Random Operator-Valued Coefficients, SIAM J. Control and Optimization, Vol. 19, pp. 401-430 (1981)

19) Pardoux, E.: Equationes aux Dérivées Partielles Stochastiques non Linéaires Monotones Thèse a L'université de Paris (1975)

20) Wong, E. and Zakai, M.: On the Relation between Ordinary and Stochastic Differential Equations, Int. J. Eng. Scie., Vol. 3, pp. 213-229 (1965)

21) McShane, E.J.: Stochastic Calculus and Stochastic Models, Academic Press, N.Y. (1974)

22) Lions, J.L.: Quelques Méthodes de Résolution des Problèmes aux Limites Nonlinéaires, Dunod, Paris (1969)

23) Sunahara, Y., Aihara, Sh. and Kojima, F.: On the Optimal Control for Distributed Parameter Systems with Stochastic Coefficients, Preprints of 8th SICE Symposium on Control Theory, pp. 233-238 (1979)

MODELS IN THE POLICY PROCESS: PAST, PRESENT, AND FUTURE

Warren E. Walker
The Rand Corporation
Santa Monica, CA 90406

In 1976, Greenberger, Crenson, and Crissey, in their book Models in the Policy Process [9, p. 26] wrote:

> Both model designers and sponsors share a general impression that the actual uses of modeling in government have fallen short of expectations. The gap between expectation and achievement is widest in the policy applications of modeling.

I am afraid that, if they were writing their book today—5 years later—they would come to the same conclusion. However, even though the conclusion might be the same, the intervening years have brought some important changes in the types of models being developed and how they are being used. These trends, coupled with major technological developments, portend significant changes in the use of models in the policy process. As a result, I believe that the gap between expectation and achievement will become considerably narrowed.

In this talk, I will describe the evolution of policy modeling. For expository purposes, I will divide the evolution into three periods:

- o the early years—the period from the late 1960's through the early 1970's
- o the recent past—the period since the early 1970's
- o the future

Forgive me if I include some simplifications and broad generalizations in order to make my points stark and clear. I have no doubt that each of you will have counterexamples for some of my statements. However, I hope you will agree that, overall, I have captured the essence of the field's past and future.

When I say policy models, I mean models that can be used to evaluate the consequences of alternative decisions that might be made by a policymaker, who is typically a public official. These models are usually designed, built, and used by policy analysts as part of a project to find solutions to problems confronting the policymaker.

EARLY YEARS

Policy modeling is not a new endeavor. It can be traced directly back to the operational analyses performed for the military during World War II. By the end of the war, a total of about 700 scientists had participated in studies that had come to be called operations research [20, p. 140]. After the war, operations research techniques were applied to a wide variety of systems. At first, most of the settings for these applications were in the private sector. Typical applications were in inventory control, production planning, and facilities location. The immediate clients for most of these early studies were lower-level managers who had operational responsibilities.

Gradually, as the methodological tools were improved and computers grew more powerful, models were developed to support higher managerial levels. By the early 1970's, analysts in the private sector had begun to build strategic planning models designed to support the decisionmaking responsiblities of top corporate management.

The use of quantitative analysis in the non-military sector of government lagged behind its use in industry by about ten years. Analysts began to apply mathematical models to the problems of state and local governments in the late 1960's. As in industry, most of the early efforts in the public sector focused on attempts to increase efficiency and effectiveness in situations where it was fairly clear what these terms meant and how they could be measured--such as in dispatching fire companies, designing police patrol areas, and scheduling hospital admissions (although there were some well publicized attempts to build comprehensive urban planning models, none of which was particularly successful*).

Even with the limited scope and objectives of these early policy studies, very few of them led to the implementation of new policies in the client agencies--the key measure of the success of a policy study. (I maintain that, even if the models used are elegant and the analysis impeccable, a policy study cannot be considered successful if it has no influence on policy decisions.) It is generally acknowledged that the process by which most of the analysis was carried out during this period

*See [16].

was responsible in large part for this notable lack of success. The relationship most commonly found among the analyst, the model, and the policymaker is illustrated in a simplified and exaggerated way in Fig. 1.

The model was central to the process. Often large and complex, the model required considerable amounts of data and was expensive to run. In many cases it was an optimization model or a comprehensive simulation model. The inner workings of the model were seldom understood by anyone but its builders. The analyst might spend a small amount of time with the policymaker at the beginning of the project defining the problem and identifying data sources, but after this modest interaction they would have little or no contact until the end of the project. After the analysis was completed, the analyst would present the results to the decisionmaker in the form of a briefing and/or a final report. More often than not the report remained on the shelf, and the results of the study were never used.

This process, viewed from the perspective of the roles and interactions of the analyst and policymaker in the various stages of the project, is depicted in Fig. 2. Here we see that the analyst and policymaker interact only at the very beginning and very end of the project. The analyst, with little or no input from the policymaker, builds the model, runs the model, and analyzes the results. He then presents his findings to the policymaker. By this time, the original problem may have changed considerably, the alternative solutions examined might no longer be viable, or the policymaker may have entirely forgotten about the project. Heiss [10] summed up this situation when he wrote (in 1974):

> The urban researcher has limited impact since, as related to the urban decisionmaking process, he comes late, leaves early, and does not get involved in implementation.

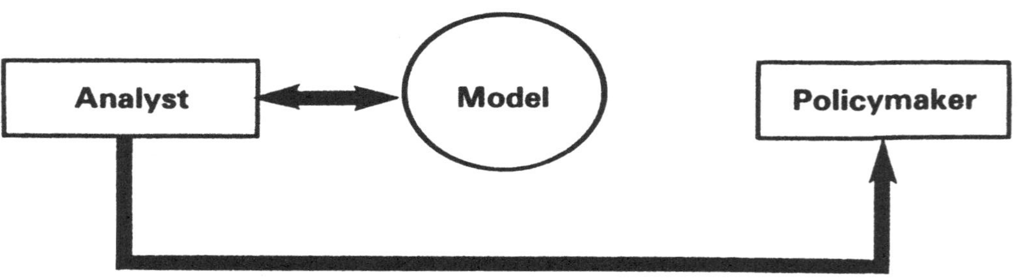

Fig. 1--Relationships among policymaker, analyst, and model: early years

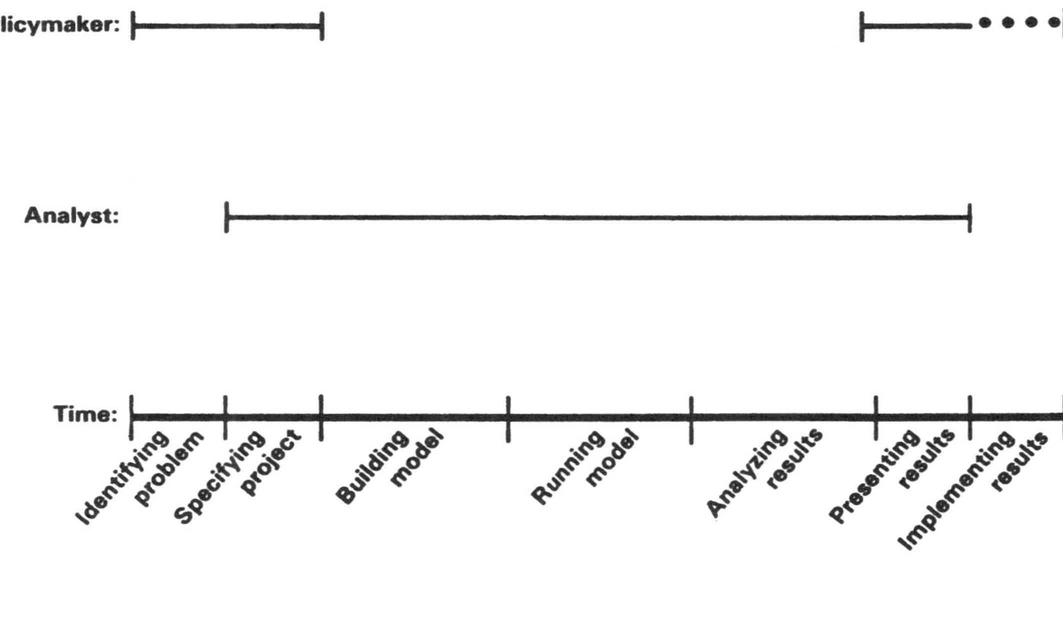

Fig. 2—Roles of policymaker and analyst: early years

RECENT PAST

The last few years have been a period of reassessment and revision. Analysts began to realize that models are only one element in a policy analysis study, and that most of the other elements play a more important role in determining the study's success. In particular, there has been a growing appreciation that, in order to make the analysis relevant, useful, and implementable, the analyst must interact much more closely with the policymaker and must obtain a much better understanding of the way he makes his decisions. As a result, much closer working relationships have developed between analysts and policymakers. As Roberts wrote [25]:

> The client is the boss ... [He must] be persuaded that you have properly taken into account his issues, his questions, his level of concerns. Otherwise he will not believe the model you have built, he will not accept it, and he will not use it.

In part because the focus has shifted from the model toward implementation of results, we have not recently witnessed any major methodological developments to match some of the breakthroughs in policy modeling that occurred in the earlier period. Instead, there has been a broadening of scope along several dimensions:

- in application areas
- in the level of policymaker who is the client for the analysis
- in the range of performance measures considered
- in the academic disciplines represented

In addition, the approach being used to model large systems is to build many interrelated and interacting smaller models instead of a single large model.

We discuss each of these developments in turn.

Application Areas

Policy models had their earliest successes when used to analyze the operating problems of government agencies. In recent years, practically no area of endeavor has escaped the eye of the modeler. For example, policy modeling has been successfully applied in such diverse areas as blood-banking [3], managing the spruce budworm in North American forests [1], locating rural social service centers in India [23], and protecting a Dutch estuary from floods [8].

Level of Policymaker

Policy studies using sophisticated modeling tools are being performed for policymakers at increasingly higher levels of government. This trend involves both a shift in the type of client for policy research (from lower-level managers and agency heads to top-level government officials) and in the types of problems being addressed (from the operational and tactical problems of an agency to strategic planning for an entire jurisdiction).

The primary models currently being used by top-level government officials are planning and budgeting models (see, for example, [21] and [5]). These models are used in a number of ways. For example:

- to help analyze the fiscal impacts of local government development policies
- to evaluate alternative economic, educational, social, and environmental policies before they are implemented
- to perform revenue and expenditure forecasting
- for goal setting and problem definition

The development and use of these planning and budgeting models parallels the development and use (about ten years earlier) of corporate planning models in the private sector.

Comprehensive policy analysis models for high-level government policymaking have also recently been constructed to analyze other areas, such as energy (see [6]), water management (see [8]), and the environment (see [11, p. 236]).

Performance Measures

Typically, early policy studies focused on one or two quantifiable criteria (such as cost, travel time to fires, tons of refuse collected, etc.) that were related to the stated objective of the study. However, there has been increasing acknowledgment by analysts that (1) there are usually multiple, and often conflicting, objectives in public-sector planning problems, (2) many policy impacts are not quantifiable, and (3) some of the policymaker's objectives are unstated and only dimly understood even by him.

These factors pose serious problems for the analyst who is trying to help a policymaker choose a course of action. Because of this, much attention in recent years has focused on how to handle them. One approach that has been developed for assessing the many projected impacts from a policy is called "decision analysis" (see [14]). In decision analysis, all of the impacts are quantified, and a weighted combination taken to produce a single measure of value, which can be used to rank the alternative policies. The weights are based on the value system or preference structure of the policymaker.

An alternative approach, which maintains the disaggregated information on individual impacts and handles qualitative as well as quantitative impacts, is to present the impacts in the form of a matrix called a scorecard. Figure 3 is a sample scorecard that presents selected results from a study that compared three alternative ways to protect a Dutch estuary from flooding [7]. The entries in each column represent the consequences associated with a particular alternative--in this case (1) permanent closure with a dam, (2) temporary closure with a storm surge barrier (SSB), and (3) leaving the estuary open but increasing the height of the surrounding dikes. The entries in a row show how a particular consequence varies from alternative to alternative. Each impact is expressed in terms of the natural units commonly used to characterize it (e.g., hectares, kilometers, number of beach visits per year). Qualitative impacts can also be shown (e.g., none, minor, or major for the impact on the attractiveness of the area), which enables the consideration of issues such as the equity of a policy or its impact on the quality of life. A color or shading scheme such as that shown here is normally used to indicate the relative rankings

	Alternatives		
	Closed case	SSB case	Open case
SECURITY			
Land flooded (ha) in 1/4000 storm (90% prob.)	0	0	400
Technical uncertainty	None	Scour	Dikes
Expected land flooded during transition pd. (ha)	430	200	530
RECREATION			
Added shoreline (km)	17	11	6
Added sea beach visits (1000/yr)	338	0	0
Decrease in attractiveness of area	None	Minor	Major
Major tourist site created?	No	Yes	No
Decrease in salt-water fish quantity (%)	75	0	25
NATIONAL ECONOMY (PEAK YEAR)			
Jobs	5800	9000	5700
Imports (DFL million)	110	200	130
Production (DFL million)	580	940	560

Rankings: Best Intermediate Worst

Fig. 3—Sample scorecard from policy analysis of the Oosterschelde (Polano) project

of the policies for each impact, but the process of making comparisons, tradeoffs, and selections among the policies is left to the decisionmaker's judgment and intuition. Removing the reliance on weights allows the analyst to present results that are relatively value-free. In addition, it makes it possible for different interest groups to agree on a single alternative (perhaps for different reasons), while they might be unable to agree on weights to assign to the various performance measures.

In addition to these efforts to determine how to combine, contrast, and/or present information on stated objectives to the policymaker, analysts have begun to address the more difficult issue of how to include a policymaker's hidden or unstated objectives (e.g. political and organization considerations) in the analysis. One potentially fruitful line of inquiry is to use models to generate a number of very different policies each of which performs about as well with respect to the stated objectives.* The policymaker can then choose among them based on factors other than those calculated by the models. Of course, the scorecard approach also lets the policymaker factor in the unstated objectives for each of the alternatives examined in the analysis.

Academic Disciplines

Broadening the scope of policy modeling in the areas of application, level of policymaking, and impacts considered has brought a concomitant increase in the number of disciplines represented on a policy study team. In the early period, although lip-service was given to interdisciplinary teams, most policy studies that used mathematical models were staffed almost exclusively with technical specialists--operations researchers, statisticians, and computer scientists.

In recent years, the staffing of many policy studies has been broadened to include a wide range of disciplines. Most staff members are experts in the various aspects of the system being studied. For example, an interdisciplinary project team at the International Institute for Applied Systems Analysis (IIASA) built a set of models to be used in

---

*See, for example, [2] and [26].

planning integrated regional development [17]. The agriculture module alone drew upon agronomists for information about crops, geographers for data about soil and climate, engineers for agricultural technologies, and economists for cost and resource allocation questions. Also included on the project team were hydrologists, demographers, and urban planners.

The increased focus on implementation of the results of a policy study have also led to the inclusion of psychologists, sociologists, and political scientists on the team. Their job is to understand the policy-maker, the organizational environment, and the political environment within which the policy must be accepted, implemented, and operated.

Many Small Models

The IIASA models mentioned above are a good example of the trend in modeling toward analyzing a large complex system by building an interlinked system of small models (or modules) rather than building a single, comprehensive, complicated, and expensive large model. In the modular approach (which some call the "Tinker-Toy" approach [11, p. 216]) each module simulates in sufficient detail the behavior of one aspect of the system. The module can be used separately to study the impacts of a proposed policy on a specific portion of the system, or interactively with other modules to study the behavior of the entire system. The output from one module can be used directly as input to another module, or tabulated, analyzed, and combined with outputs from other modules to form an input data set for a subsequent model.

In the IIASA model, each of five important aspects of a region's development was represented as a separate module: industry, agriculture, water, population, and migration (see Fig. 4). Certain data and values are shared or flow among them: prices, wages, water demand and cost, and labor availability. A central integration model allocates capital and labor among the sectors; and the linked models work out the consequences of alternative allocations.

A similar modeling approach was recently used in a policy analysis of the water-management system in the Netherlands [8] -- the study was dubbed PAWN. In PAWN, the Dutch water-management system was divided into 12 sectors, each of which represented a major suppier or user of water

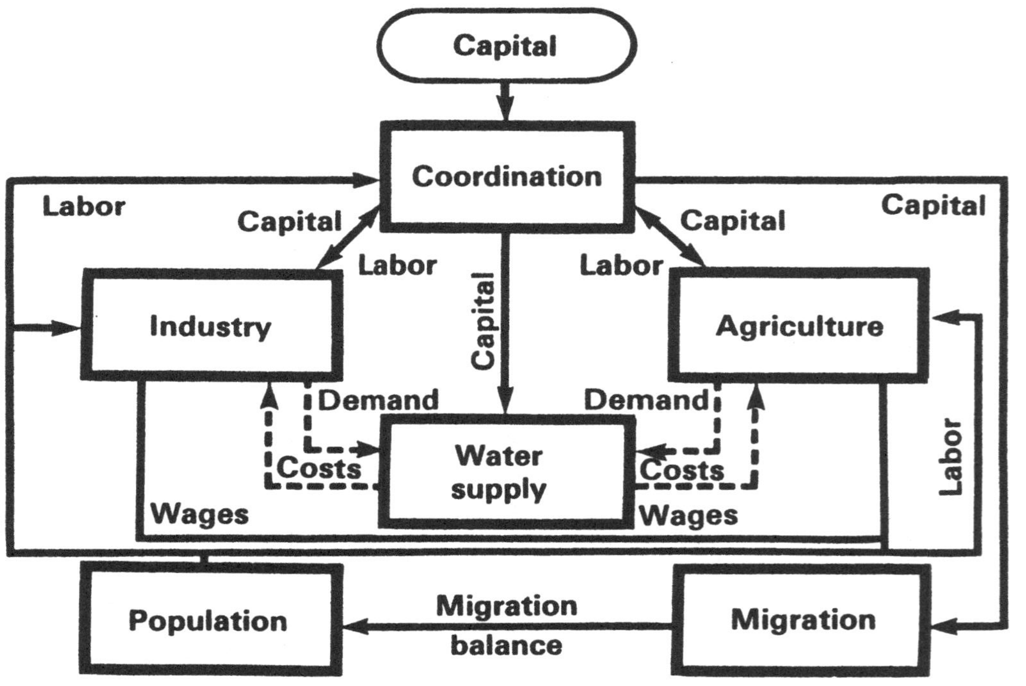

Fig. 4--IIASA system of models for regional development

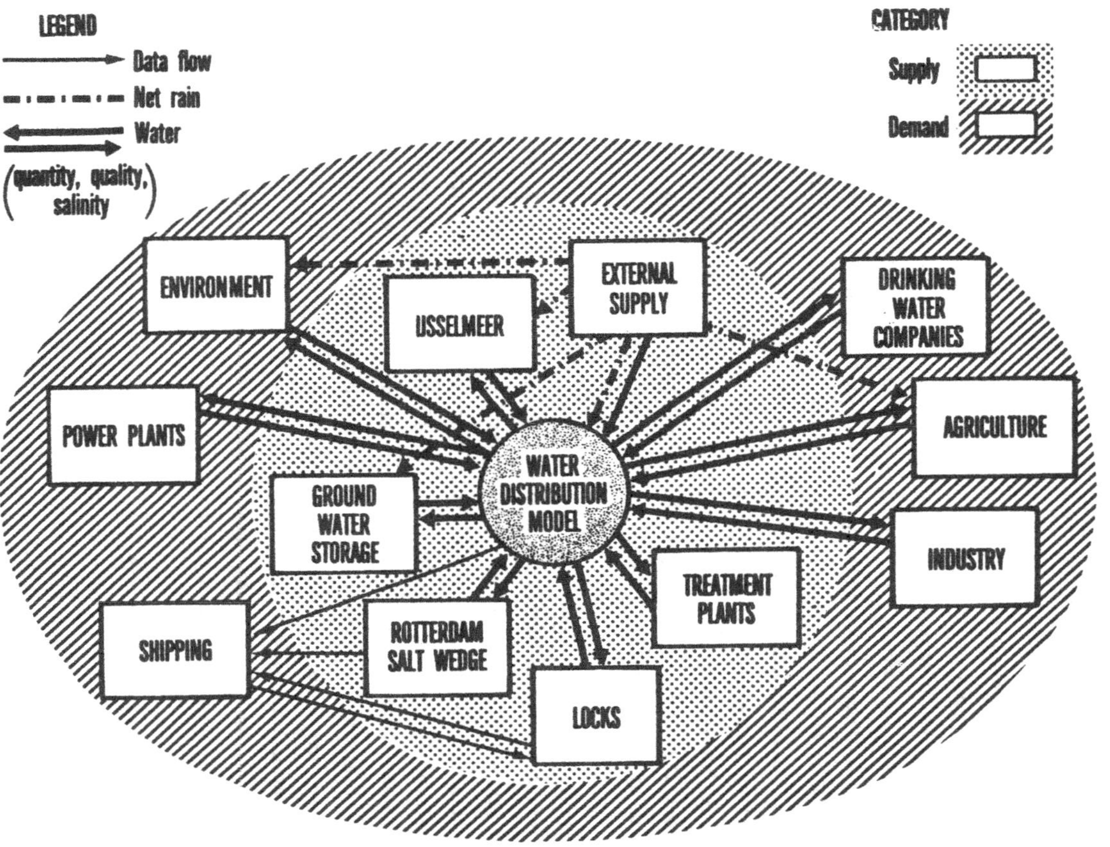

Fig. 5--PAWN system diagram

(see Fig. 5). Each of the sectors was modeled separately. Most of the models were then run separately, with outputs from one sometimes used as inputs to another. In some, cases, however, a subset of the models was run interactively, with one module calling another as a subroutine, and information being passed back and forth among the modules. Figure 6 illustrates the relationship among the Distribution Model (the central integration model, which simulates the flow of water throughout the country), the external supply sector (which supplies input to all the models), and a number of models representing the agriculture sector.

The modular approach to modeling is attractive for a variety of reasons. In addition to mitigating the problems inherent in building a single large model, it provides flexibility and convenience for the analysts, and facilitates communication with the policymaker. As summarized by Kunreuther et al. [15, pp. 21-22] the modular approach also makes it "relatively easy to adapt to a wide variety of circumstances, availability of data, and types of analyses without having to incur large amounts of time, skill, and confusion in reprogramming".

Highlights of Recent Past

Figures 7 and 8 highlight three of the most important recent trends in policy modeling. Perhaps the most important development, in terms of improving the chances for successful implementation, is the trend toward making a policy analysis study a joint effort of the analyst and the policymaker. Many recent studies have included the policymaker and members of his staff as full partners on the project team. The policymaker, therefore, has become involved in all phases of the project's work except for the actual running of the model.

Second, as Fig. 7 makes clear, studies of large complex systems have increasingly used several small models instead of a single, monolithic model.

Finally, as suggested by Fig. 8, analysts have increasingly recognized that the study does not end with the preparation of the final report. They have become increasingly concerned with the implementation phase. Their experiences in the early years made it clear that good analytical results do not necessarily lead to successful implementation (see, for

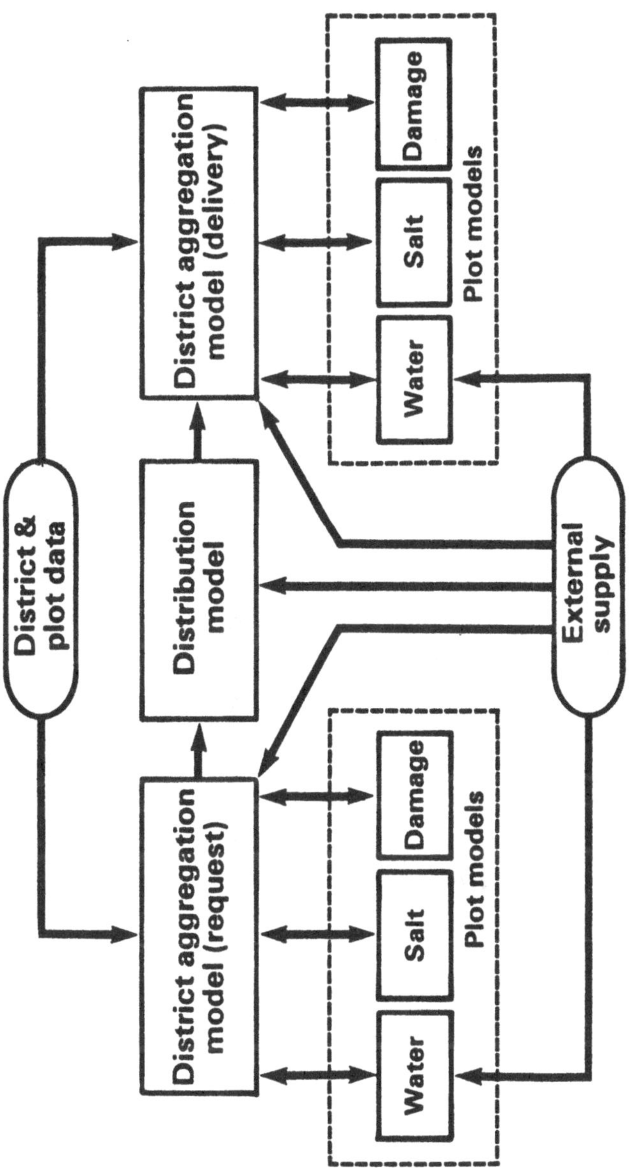

Fig. 6—Interactions among PAWN agriculture models

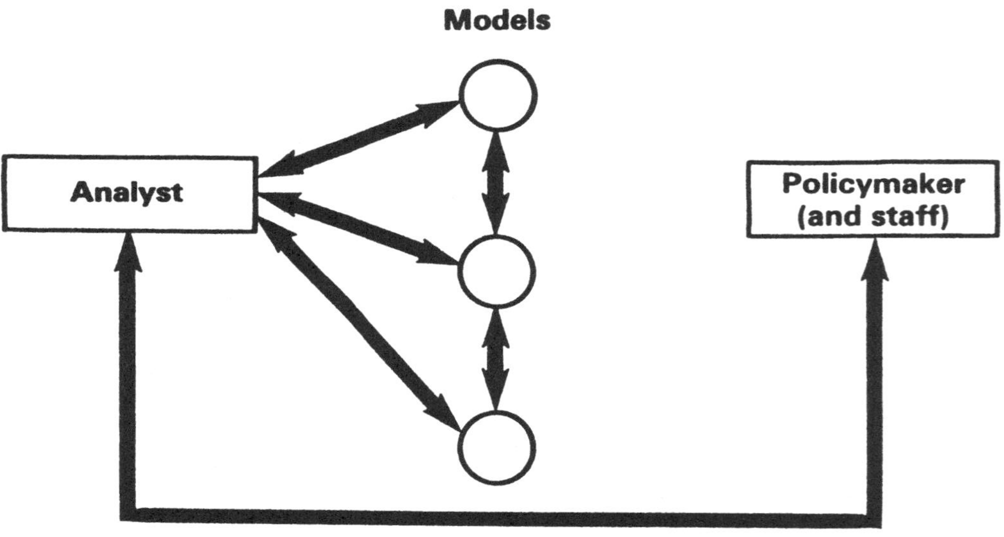

Fig. 7--Roles of policymaker and analyst: recent past

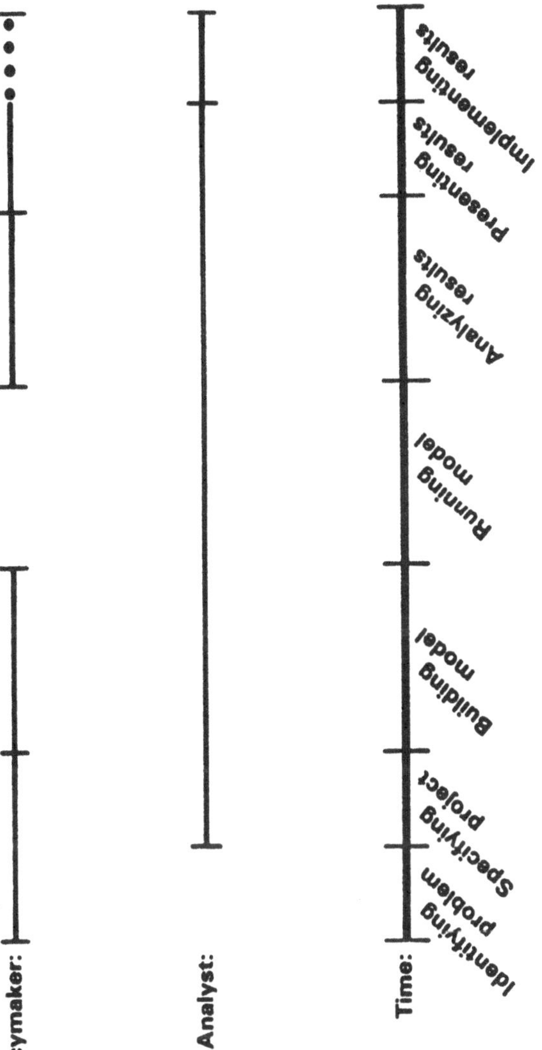

Fig. 8--Relationships among policymaker, analyst, and models: recent past

example, [28]). Thus, attention is now being paid to implementation during all stages of the study. Implementation costs and political and organizational problems are factored into the analysis; the study often includes the development of an implementation plan; and people knowledgeable in organizational behavior and the process of planned change are sometimes included on the project team.

THE FUTURE

The development of commercial time-sharing services in the late 1960's began a movement that is still accelerating--toward personalized computer systems, direct access to and interaction with models and data, and decentralization of computer resources. The availability of micro-computers, interactive terminals, and data communications networks has been growing exponentially. I believe that these developments have profound implications for the use of models in the policy process.

In 1977, there were about 200,000 microcomputers in use in the United States. A jump to three million operating units by 1984 is forecast [22]. In the not too distant future policymakers and members of their staffs will have computer terminals in their offices, much as they now have calculators. These terminals will give them direct and immediate access to policy models, and provide the potential for significant changes in the nature of policy analysis and the roles and interactions of the various participants in a policy analysis study.

The broad outlines of what may be in store can be seen in recent developments in the private sector, where increasing attention is being paid to on-line interactive systems that assist managers at all levels of a corporation in making their decisions. Such systems are broadly called "decision support systems" or DSSs [12].

Basically, a DSS embeds decision models in a management information system (MIS), and provides the decisionmaker with on-line access to both the information in the MIS and the outputs from the various models. The major elements of a DSS in a public sector setting are illustrated in Fig. 9.

At the heart of a DSS is the policymaker (not the policy model). The DSS's primary purpose is to support a policymaker in making decisions--to act as an extension of his own decisionmaking process, or, as others have phrased it, as "an executive mind-support system" [13].

Through use of a simple, forgiving, English-like command language, the policymaker interacts with both an integrated data base and an inter-linked system of policy models. The command language acts as a buffer between the policymaker and the computer, and allows a "conversation" based on the policymaker's concepts, vocabulary, and definition of the decision problem.

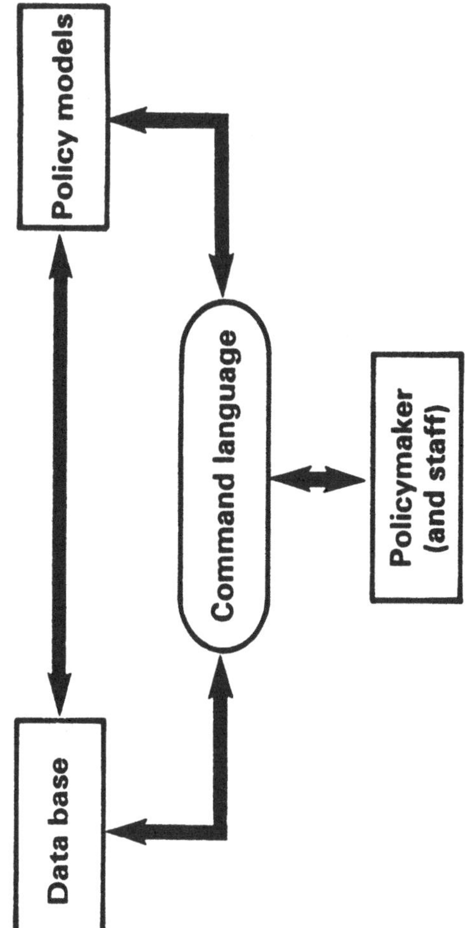

Fig. 9--Major elements of a decision support system

The data base retains all the relevant information about the policy area in an organized, systematic manner, and is continually updated. Policy models often fall into disuse because the input data gradually become out of date, and it is costly and inconvenient to collect the required new data on an ad-hoc basis. In the case of a DSS, the updating of input data is automatic and institutionalized.

The set of interlinked policy models (or modules) in a DSS is not unlike the sets of models we said were currently being developed for policy analysis studies. The individual modules are likely to be even smaller than the modules in current studies, and more easily combined to produce models that can analyze new situations or answer new questions in a dynamic environment. The models will draw the majority of their inputs from the data base and place much of their output back onto the data base. This output can then be used by other models as a source of input data.

In order to make the models attractive to policymakers and guard against their falling into disuse, they will tend to be self-documenting, easily updated,* and so easy to use that they will become a natural part of the policymaker's decision process. They will also be problem-oriented, and relevant to the real problems confronting the policymaker. The idea is <u>not</u> to automate the decision process or capture the essence of the decision process in one or more models. Instead, each module gives the policymaker information about those parts of the system that are structured and can be modeled. The policymaker then combines these outputs with personal knowledge, understanding, and judgments about those aspects of the situation that are not taken into account by the models, to reach a decision about the best course of action. While the process may require running some or all modules a number of times under various sets of assumptions, they are run under the control of the policymaker or his staff, to supply information he has requested and not information an outside analyst (or the management information department) thinks he should see.

---

*Updating procedures will be incorporated in the routine maintenance of the system so that changes are made to the models to match changes in the environment. Some changes would be made automatically--e.g. changes in the input data and new parameter values that can be calculated from information in the (constantly updated) data base.

The new technology that makes decision support systems possible also makes it possible to display information for the policymaker in ways that capture his attention and facilitate his understanding. Even now, the output of models can be displayed at terminals in the form of colored maps, pie-charts, histograms, etc. Graphic displays can even show the simulated behavior of the system over time.

In the future, "situation rooms" for policymakers might be developed, which would utilize large displays driven by the computer. House and McLeod [11, p. 96], in discussing this idea, said that the room should be designed tc make the presentation of simulation results more intelligible and dramatic to policymakers, community groups, and concerned citizens:

> To this end, there would be calibrated dials labeled with the names of several of the more important exogenous variables and parameters that would be under the control of the experimenter. Furthermore the computer would be programmed to run in "rep-op," a mode in which a complete simulation is run and automatically repeated and displayed at a rate fast enough to change in apparently real time with the movement of the dials. Thus, serious researchers and the curious public alike could immediately see the probable long-term impact of proposed policy changes.

There are many documented examples of decision support systems that have already been developed and successfully used. Among the private sector examples are a corporate planning system for Xerox of Canada Limited [27], a planning and reporting system for Liberty National Bank [4], and a financial planning system for a large scientific research organization [18].

There are also a surprising number of public-sector decision support systems already in operation. For example, computer-aided dispatching systems for fire departments, police departments, and emergency medical services are decision support systems. The New York City Fire Department's Management Information and Control System (MICS) includes sophisticated algorithms for helping the dispatcher make rapid decisions concerning (1) how many fire companies to send to an incoming alarm, and which specific companies to send (see [24, Chap. 11]), and (2) how best to provide coverage to an area of the city when all of its fire companies are busy fighting fires (see [24, Chap. 12]).

The future role of models in the policy process that I have just sketched out is an extrapolation of earlier trends. As shown in Fig. 10, decision support systems continue the trend toward expanding the policymaker's role in a policy analysis study and making the study more responsive to his needs. Because the policymaker will be interacting directly with the models, he will have a better understanding of the meaning of the numbers produced by the models, a better feeling for the differences among the policies, and more confidence in the results. All of these factors should increase the chances for successful implementation.

Figure 11 indicates that the future policymaker will become a full partner in a policy study. He will be involved in every step of the process: from problem identification through implementation of results. Maintenance of a close working relationship between the analyst and the policymaker throughout the study will also do much to increase the chances that the study will be a success.

The fact that the policymaker and his staff will be playing a more active role in building policy models, running them, and analyzing the results, does not mean that there will be a declining need for policy analysts in the future. It only means that there will be a greater need for analysts to work within public agencies. In fact, since I foresee more widespread use of policy modeling, I expect that there will actually be a greater total number of policy analysts in the future. Many of them will continue to be employed by private research corporations and management consulting firms. Someone has to build the models, and I believe that most of them will continue to be built by outside parties, albeit with the active participation of representatives of the government agency.

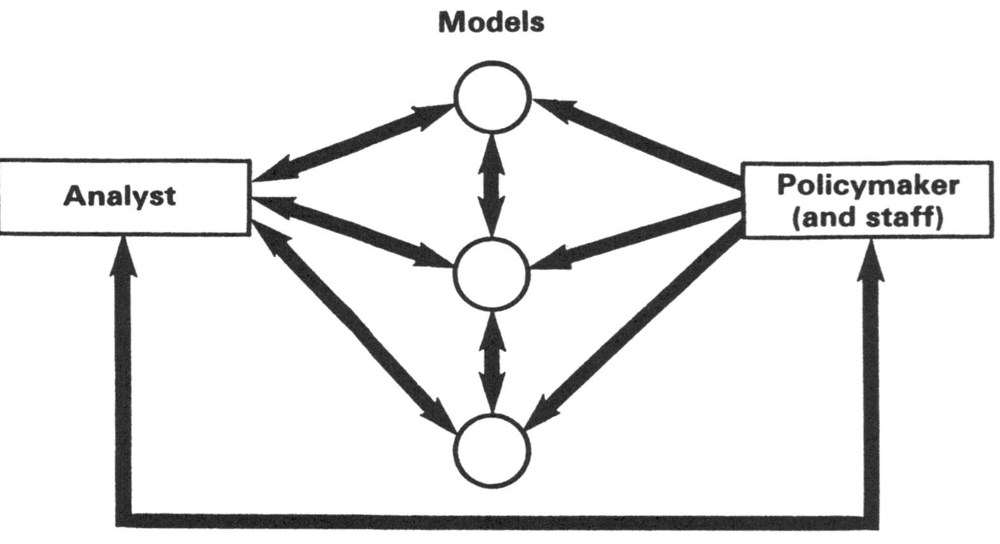

Fig. 10—Relationships among policymaker, analyst, and models: future

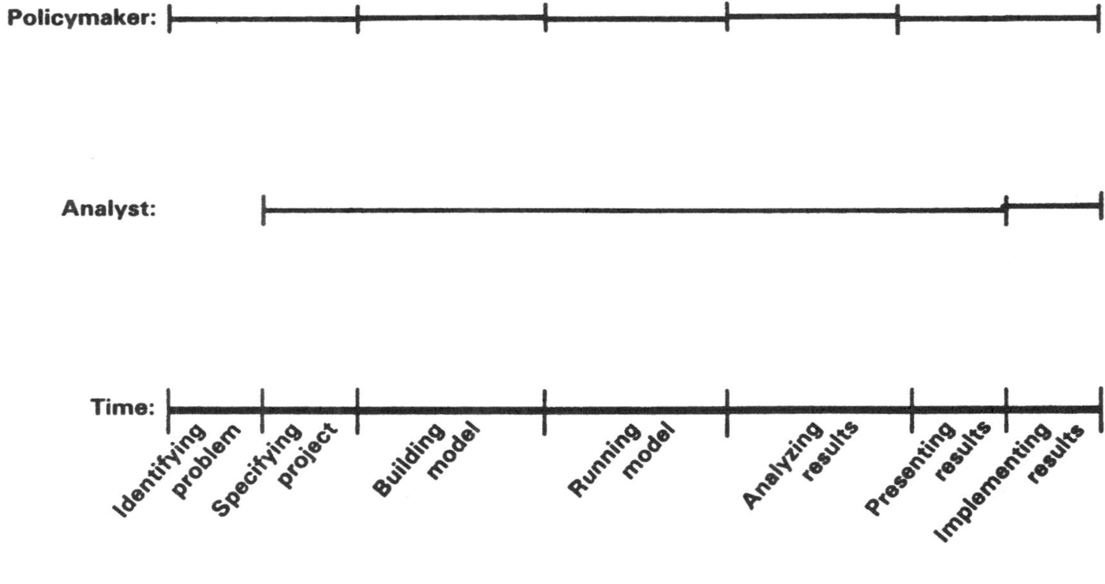

Fig. 11--Roles of policymaker and analyst: future

CONCLUSIONS

What does this excursion into the past, present, and future of policy modeling imply for the policy analyst and model builder? Based on evidence from the early years and even the recent past one might conclude that the use of models in the process of public sector decisionmaking has been tried for twenty years and has been found wanting. In other words, one might conclude that the public sector has had a brief fling with using computer models, and that their use is unlikely to grow much in the future.

However, I believe that the confluence of several independent developments has set the stage for a dramatic increase in the use of models in the policy process.

- o First, there are the technological developments. In the past few years science fiction has become science fact. Microcomputers are being mass produced that have the capabilities of computers that used to fill a room, but at a fraction of the cost. Through telecommunications, these computers can be connected to others located almost anywhere. Interactive terminals give the user direct and immediate access to the computer. Data base management systems, query languages, and graphical displays provide information in a form that can be used directly by a decisionmaker.
- o Second, decision support systems have gained acceptance as management tools in the _private_ sector. While public agencies generally lag the private sector in the use of such tools, they eventually proceed along similar paths.
- o Last, but perhaps most important, public agencies are beginning to experience unprecedented fiscal pressures. These pressures are likely to motivate the search for new, creative solutions to public sector management and service delivery problems.

Even though each of these developments has been evolutionary--the result of gradual changes in technology, managerial procedures, and fiscal conditions--their confluence portends a revolutionary change in the use of models in the public sector. In contrast to the negative tone of my opening remarks, my scenario for the future suggests that we may be on the threshold of a new era in the use of models for governmental planning and policy analysis.

REFERENCES

1. Baskerville, G., W. Clark, C. Holling, D. Jones, and C. Miller, Ecological Policy Design: A Case Study of Forests, Insects, and Managers, John Wiley & Sons, London (1980).

2. Brill, E.D., Jr., S. Chang, and L.D. Hopkins, "Modeling to Generate Alternatives: The HSJ Approach and an Illustration Using a Problem in Land Use Planning," Management Science, forthcoming.

3. Brodheim, E., and G.P. Prastacos, "The Long Island Blood Distribution System as a Prototype for Regional Blood Management," Interfaces, Vol. 9, No. 5, pp. 3-20 (November 1979).

4. Bruce, J.W., "Management Reporting System: A New Marriage Between Management and Financial Data Through Management Science," Interfaces, Vol. 6, No. 1, Part 2, pp. 54-63 (November 1975).

5. Dutton, W.H. and M.S. Hollis, Fiscal Impact Budgeting Systems, Public Policy Research Organization, University of California, Irvine (November 1979).

6. Federal Energy Administration, Project Independence Report, Washington, D.C. (1974).

7. Goeller, B.F., A.F. Abrahamse, J.H. Bigelow, J.G. Bolten, D.M. de Ferranti, J.C. De Haven, T.F. Kirkwood, and R.L. Petruschell, Protecting an Estuary From Floods: A Policy Analysis of the Oosterschelde, Report No. R-2121/1-NETH, The Rand Corporation, Santa Monica, California (December 1977).

8. Goeller, B.F., et al., Policy Analysis of Water Management For The Netherlands: Vol. I, Summary Report, Report No. R-2500/1-NETH, The Rand Corporation, Santa Monica, California (forthcoming).

9. Greenberger, M., M.A. Crenson, and B.L. Crissey, Models in the Policy Process, Russell Sage Foundation, New York (1976).

10. Heiss, F.W., Urban Research and Urban Policymaking: An Observatory Perspective, Bureau of Governmental Research and Service, University of Colorado, Boulder (1974).

11. House, P.W., and J. McLeod, Large-Scale Models for Policy Evaluation, John Wiley & Sons, New York (1977).

12. Keen, P.G.W. and M.S.S. Morton, Decision Support Systems: An Organizational Perspective, Addison-Wesley Publishing Co., Reading, Massachusetts (1978).

13. Keen, P.G.W. and G.R. Wagner, "DSS: An Executive Mind-Support System," Datamation, Vol. 25, No. 12, pp. 117-122 (November 1979).

14. Keeney, R.L. and H. Raiffa, Decisions With Multiple Objectives: Preference and Value Tradeoffs, John Wiley & Sons, New Nork (1976).

15. Kunreuther, H., J. Lepore, L. Miller, J. Vinso, J. Wilson, B. Borkan, B. Duffy, and N. Katz, An Interactive Modeling System For Disaster Policy Analysis, Program on Technology, Environment and Man, Monograph #26, Institute of Behavioral Science, University of Colorado, Boulder (1978).

16. Lee, D.B., Jr., "Requiem For Large-Scale Models," Journal of the American Institute of Planners, Vol. 39, pp. 163-178 (1973).

17. Levien, R.E., "Applied Systems Analysis: From Problem Through Research to Use," IIASA Reports, Vol. 3, No. 1, pp. 19-37 (January-March 1981).

18. Meador, C.L., and D.N. Ness, "Decision Support Systems: An Application to Corporate Planning," Sloan Management Review, Vol. 15, No. 2, pp. 51-68 (Winter 1974).

19. Mohan, J.J., "Starfire, F.D.N.Y.: Communications Computer Network," With New York Firemen, First Issue 1980, pp. 12-13 and 23.

20. Miser, H.J., "Operations Research and Systems Analysis," Science, Vol. 209, pp. 139-146 (July 4, 1980).

21. Naylor, T.H., H.E. Glass, J. Wall, and D.N. Milstein, SIMPLAN: A Computer Based Planning System for Government, Duke University Press, Durham, North Carolina (1977).

22. Palmer, S.D., "The Microcomputer Explosion," ICP Interface: Administrative and Accounting, Vol. 6, No. 1, p. 12 (Spring 1981).

23. Patel, N.R., "Locating Rural Social Service Centers in India," Management Science, Vol. 25, No. 1, pp. 22-30 (January 1979).

24. Rand Fire Project (W.E. Walker, J.M. Chaiken, and E.J. Ignall, eds.), Fire Department Deployment Analysis, Elsevier North Holland, Inc., New York (1979).

25. Roberts, E.B., in Simulation in the Service of Society, pp. 6-7 (January 1975).

26. Schilling, D.A., A. McGarity, and C. ReVelle, "Hidden Attributes and the Display of Information in Multiobjective Analysis," Management Science, forthcoming,

27. Seaberg, R.A. and C. Seaberg, "Computer Based Decision Systems in Xerox Corporate Planning," Management Science, Vol. 20, No. 4, Part II, pp. 575-584 (December 1973).

28. Walker, W.E., Changing Fire Company Locations: Five Implementation Case Studies, Stock No. 023-000-00456-9, U.S. Government Printing Office, Washington, D.C. (January 1978).

A RITZ-TYPE APPROACH TO THE CALCULATION OF OPTIMAL CONTROL FOR NONLINEAR, DYNAMIC SYSTEMS

B. Asselmeyer
Coordinated Science Laboratory
University of Illinois
1101 W. Springfield Ave.
Urbana, Illinois 61801

A Ritz-type parameterization of the control functions is used to simplify the calculation of optimal control functions. From the well-known gradient technique for problems constrained by differential equations, an iteration formula for the parameters is derived. This approach allows the efficient solution of fairly general problems even if only small computers (like desk-top calculators, personal computers or micro/mini computers) are used.

I. Introduction

The problem considered in this paper is the numerical calculation of optimal control functions for dynamic systems. For many technical applications, the problem can be formulated as follows: We consider a dynamic system, which is described by a set of nonlinear, explicit, differential equations of first order (state equations)

$$\dot{\underline{x}} = \underline{g}(\underline{x},\underline{u},t) \qquad \begin{aligned} \underline{x} &= (n\times 1) \text{ state vector} \\ \underline{u} &= (\ell\times 1) \text{ control vector} \\ t &= \text{time} \\ \underline{g} &= (n\times 1) \text{ function vector.} \end{aligned} \tag{1}$$

The aim is to bring the states of this system from a given initial state at time t_0

$$\underline{x}(t_0) = \underline{a} \qquad \underline{a} = (n\times 1) \text{ constant vector} \tag{2}$$

to a goal, described by a set of terminal conditions at time t_f

$$\underline{q}(\underline{x}(t_f),t_f) = 0 \qquad \underline{q} = (r\times 1) \text{ function vector}, \quad r \le n \tag{3}$$

(where the elements q_i generally are nonlinear functions). For that the control vector $\underline{u}$ is determined to minimize the value of a performance criterion

$$P(\underline{x}(t_f),t_f) \Rightarrow \min \tag{4}$$

while the elements u_i of $\underline{u}$ are separately bounded

$$u_{i\,min} \le u_i \le u_{i\,max}. \tag{5}$$

The solution of such an optimization problem can generally be found only numerically. Several different approaches are known (summarized for instance in Ref. 1), but all require quite considerable computational effort. For the application of such optimization schemes with small computers therefore a simplification of these general procedures is necessary, in order to reduce execution time as well as program length and amount of data to be stored.

In this paper a simplification of the well-known gradient procedure for dynamic optimization problems is presented. Following an idea of Ritz (Ref. 2), a parameterization of the control functions is introduced,

$$u_i = \sum_{i=1}^{m_i} p_{ij} f_{ij}(t) \tag{6}$$

where $f_{ij}(t)$ are suitable chosen, known functions of time and p_{ij} are the parameters to be determined. Following a suggestion of Bryson (Ref. 3) an iteration formula for these parameters is derived. It will be shown that this formula corresponds to the one known from ordinary optimization problems, where each function evaluation there corresponds to a complete integration of the systems differential equations here and the final conditions are seen as equality constraints. Application of a similar procedure had been proposed in Ref. 4 and discussed for certain examples in Ref. 5-Ref. 7. Lately the convergence properties of such methods have been explored in Ref. 8.

Although only suboptimal solutions can be computed that way, the application of such a simplified optimization scheme is advantageous especially for technical problems. Since the computation uses a mathematical model of the real system (which seldom has an accuracy better than a few percent), it is not very reasonable to compute the optimal solution for the model with a very high accuracy, if the control functions found in such a way are to be applied to the real system. In addition, a suboptimal solution, which improves the behavior of the system, is useful for application, even if it is only a crude approximation of the optimal one. Since with such simplified procedures solutions can be found relatively fast, they can be used for on-line applications as discussed in Ref. 9.

II. Derivation of the Iteration Formula for the Optimal Control Problem

The adjoint system to (1) is defined by

$$\underline{\lambda} = \frac{\partial g}{\partial \underline{x}} \underline{\lambda} \qquad \lambda = (n\times 1) \text{ adjoint state vector.} \tag{7}$$

These adjoint variables can be used to compute the change of either the cost criterion or one of the terminal conditions induced by a change in the control functions $\delta\underline{u}$

$$dF|_{t_f} = \int_{t_o}^{t_f} \left(\frac{\partial g}{\partial \underline{u}} \underline{\lambda}\right)^T \delta\underline{u}\, dt \tag{8}$$

$$F = P_1, q_1, \ldots, q_\ell$$

where $\frac{\partial g}{\partial \underline{u}}$ is the $(\ell\times n)$ matrix of partial derivatives of the systems functions $\underline{g}$ with respect to the control functions. The terms in brackets are called influence functions (see Ref. 1)

$$\underline{\lambda}_u^F = \frac{\partial g}{\partial \underline{u}} \underline{\lambda} \tag{9}$$

where F defines the appropriate boundary conditions for $\underline{\lambda}$.

Using the definition of the control function in (6) the change δu_i in one of the control functions is given by

$$\delta u_i = \sum_{j=1}^{m_i} \delta p_{ij} f_{ij}(t). \tag{10}$$

If only a differential change of one parameter is considered, it follows from (8) that

$$\frac{\partial F}{\partial p_{ij}} = \int_{t_o}^{t_f} \lambda_{u_i}^F f_{ij}(t)dt. \tag{11}$$

If $f_{ij}(t)$ is a member of a complete set of orthonormal functions, then (11) can be seen to determine the coefficients of a series expansion of $\lambda_{u_i}^F$ using just these functions. An approximation for $\lambda_{u_i}^F$ can therefore be defined as

$$\lambda_{u_i}^F \approx \hat{\lambda}_{u_i}^F = \sum_{j=1}^{m_i} \frac{\partial F}{\partial p_{ij}} f_{ij}(t). \tag{12}$$

Using a steepest descent technique for the determination of the change in the control function, this change according to Ref. 3 can be assumed to be

$$\delta u_i = \beta\lambda_{u_i}^P + \sum_{j=1}^{r} \gamma_j \lambda_{u_i}^{q_j} \tag{13}$$

where β is a stepsize factor and γ_j are yet to be determined weighting coefficients, which take into account the need to fulfill the terminal conditions while minimizing the cost criterion. These parameters can be determined as follows.

For the change of a terminal condition, (8) can be written using (13) and replacing all the influence functions there by (12)

$$dq_i = \beta \int_{t_o}^{t_f} (\underline{\lambda}_u^{q_i})^T \underline{\lambda}_u^P dt + \sum_{n=1}^{r} \gamma_n \int_{t_o}^{t_f} (\underline{\lambda}_u^{q_i})^T \underline{\lambda}_u^{q_n} dt$$

$$\approx \beta \sum_{k=1}^{\ell} \sum_{j=1}^{m_\ell} \frac{\partial P}{\partial p_{kj}} \int_{t_o}^{t_f} (\lambda_{u_k}^{q_i} f_{kj}(t))dt + \sum_{n=1}^{r} \gamma_n \sum_{k=1}^{\ell} \sum_{j=1}^{m_\ell} \frac{\partial q_n}{\partial p_{kj}} \int_{t_o}^{t_f} (\lambda_{u_k}^{q_i} f_{kj})dt. \tag{14}$$

We combine all parameters into one vector

$$\underline{P} = (p_{11},\ldots,p_{1m_1},\ldots,p_{\ell m_\ell})^T \tag{15}$$

and define the partial derivatives accordingly

$$\frac{\partial F}{\partial \underline{p}} = \text{grad } F = (\frac{\partial F}{\partial p_{11}},\ldots,\frac{\partial F}{\partial p_{\ell m_\ell}})^T. \tag{16}$$

Using (11) for the integral expressions in (14), we can write this after rearranging and using the abbreviations (16)

$$dq_i = \beta \text{ grad } P^T \text{ grad } q_i + \sum_{n=1}^{r} \gamma_n \text{ grad } q_n^T \text{ grad } q_i. \tag{17}$$

Combining grad q_j into a matrix G_q

$$G_q = (\text{grad } q_1,\ldots,\text{grad } q_r) \tag{18}$$

and prescribing changes Δq_j for all the terminal conditions, (17) represents a set of r equations for the determination of the parameters γ_j. Solving these equations yields

$$\underline{\gamma} = (G_q^T G_q)^{-1} \Delta \underline{q} - \beta (G_q^T G_q)^{-1} G_q^T \text{ grad } P. \tag{19}$$

Matrix $(G_q^T G_q)$ is invertible, if the total number of parameters is greater or equal the number of terminal conditions (this can be obtained by suitable choice of $\underline{u}$ according to (6)) and if the grad q_j are linearly independent (this can be obtained by suitable choice of the functions $f_{ij}(t)$ in (6), if the terminal conditions of the original problem are also linearly independent).

For the change $\delta \underline{p}$ of the parameters, which determines the change of the control functions according to (10), it follows from (13) using (12) and (19)

$$\delta \underline{p} = \underbrace{\beta \text{ grad } P}_{1} + \underbrace{G_q (G_q^T G_q)^{-1} \Delta \underline{q}}_{2} - \underbrace{\beta G_q (G_q^T G_q)^{-1} G_q^T \text{ grad } P}_{3}. \tag{20}$$

With (20) an iterative computation of $\underline{u}_{opt}$ can be defined:

Step 0: Initialize parameter vector $\underline{p}$ and define $\underline{u}^{(1)}$. Let n=1.

Step 1: Compute $P^{(1)}$ and $\underline{q}^{(1)}$ by integration of system (1) using $\underline{u}^{(1)}$.

Step 2: Change each parameter one at a time and integrate system (1). Compute grad $P^{(n)}$ and $G_q^{(n)}$ by finite difference approximation.

Step 3: Determine $\delta \underline{p}$ from (20) and therewith $\delta \underline{u}$ from (10).

Step 4: Compute $P^{(n+1)}$ and $\underline{q}^{(n+1)}$ by integration of system (1) using $\underline{u}^{(n+1)} = u^{(n)} + \delta \underline{u}$.

Step 5: Check if iteration has converged (for instance if $|P^{(n+1)} - P^{(n)}| < \varepsilon_p$). If not, let n: = n+1 and continue at step 2.

Note that during this iteration no use of the adjoint system has been made. Only the original system (1) is integrated. Adjoint variables and influence functions were only used for the derivation of (20).

III. Interpretation and Extension

The iteration formula (20) is the same one found in Ref. 10 for the ordinary optimization problem with equality constraints. These equality constraints are here the final conditions of the original dynamic problem, which can be determined by integrating system (1) using the control functions defined by (6) with the presently valid set of parameters. The value of the cost criterion has to be determined the same way and corresponds to the value of the function to be optimized there.

The iteration formula contains three terms. The first one induces a change in the performance criterion; if a minimum is sought, then the stepsize parameter β is negative (and can be determined for instance by an one-dimensional search). The second term induces desired changes Δq in the terminal conditions. The third term linearly compensates changes in the terminal conditions induced by term one.

For the actual computation it may be advantageous first to find a solution which fulfills the terminal conditions by just using term two in (20). This means that before the actual optimization starts, first an admissible solution is found through a Newton-type iteration. During the second stage then the fulfillment of the terminal conditions can be retained by adding a correction term in (20). This correction term

$$\Delta p^{corr} = \sum_j G_q (G_q^T G_q)^{-1} \Delta q^{(j)} \tag{21}$$

is the result of an subsidiary iteration, which according a Newton-scheme reduces the actual deviations in the terminal conditions; it uses as many terms as are necessary to reach

$$\| \underline{q} \| < \varepsilon_q . \tag{22}$$

It is known that the simple gradient procedure yields slow convergence. It can be improved by using for instance conjugate gradient or variable metric methods as summarized in Ref. 11. All these methods can be used in a similar way by introducing a matrix K and modifying (20) to

$$\Delta \underline{p} = \beta K \text{ grad } P + K G_q (G_q^T K G_q)^{-1} \Delta \underline{q} - K G_q (G_q^T K G_q)^{-1} G_q^T K \text{ grad } P \tag{23}$$

while choosing K according to the desired method. Using second derivatives, K can be seen as the inverse Hessian matrix of P with respect to $\underline{p}$. (23) is just stated here, a derivation can be found in Ref. 12.

For the derivation of (20) the functions $f_{ij}(t)$ for the series expansion of each control function were assumed to be out of a complete, orthonormal set. For the application of this formula in an iterative optimization problem, however, it is sufficient for these functions to be linearly independent only.

It must be pointed out that of course only suboptimal solutions are found that way. How closely they approximate the true optimal solution depends on the choice of the function f_{ij} as well as on their number. Convergence for this type of control function parameterization is treated in Ref. 8.

IV. Example

To demonstrate that the arbitrary choice of the type of functions $f_{ij}(t)$ and of the number of parameters does not pose a big difficulty in many cases, the application of the method described here to an example from a technical background is given.

We consider a stirred-tank reactor in a chemical plant, as it is described in Fig. 1. A very simple model for this reactor has been given in Ref. 13. It has 2 state equations given here in a dimensionless, normalized form, one for the concentration of the reaction product in the reactor

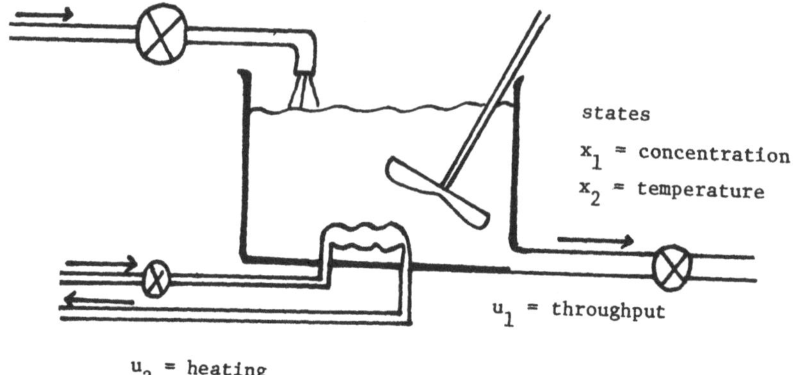

Fig. 1. Systems configuration.

$$\dot{x}_1 = u_1\cdot(5-x_1)/12.5 - x_1\cdot 18{,}828\cdot 10^{33}\cdot \exp-(75{,}23/x_2) \tag{24}$$

the other for the temperature there

$$\dot{x}_2 = \frac{1}{400}\,(u_1(24-32x_2) + 28{,}8u_2(3{,}73-4x_2)/(5+0{,}92u_2) + 242{,}88\cdot 10^{33}\cdot \exp-(75{,}23/x_2)). \tag{25}$$

The control function u_1 is the throughput, u_2 is the external heating. These equations are highly nonlinear due to the modeling of the reaction constants as exponential functions of the temperature. In addition, each of the control functions must be nonnegative and has an upper bound too.

Normally a system like that is just controlled around its steady state. The problem considered here is to determine an optimal control function u_1, if for u_2 a time profile is prescribed, for instance an interruption of the heating due to a maintenance cycle. In this example a fixed time interval of 10 units is considered at the end of which the states should be at their steady state values again, while u_2 is prescribed to be zero for 5 time units in the middle of the interval and at its steady state value all the other times. The cost criterion is to maximize the gain during this period, given by

$$J = \int_{t_o}^{t_f} (r_1u_1x_1 - r_2u_1 - r_3u_2)dt \rightarrow \max \tag{26}$$

where r_1 represents the gain from the product, r_2 and r_3 are the costs of the used material and the heating, respectively.

Figure 2 shows the results of several computations. In the left part, the state

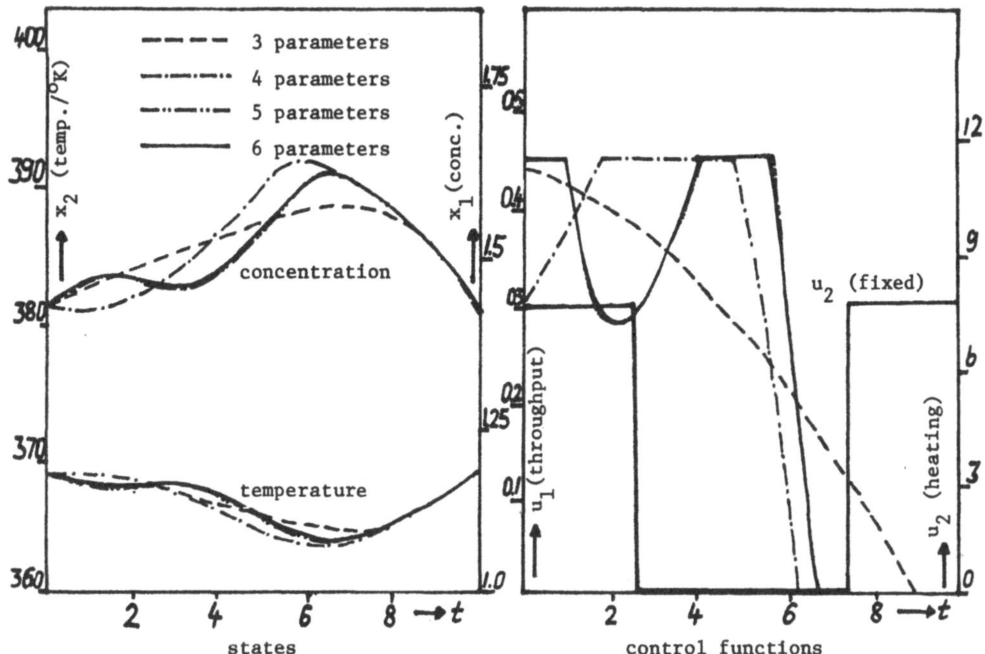

Fig. 2. Results of optimization with different numbers of parameters.

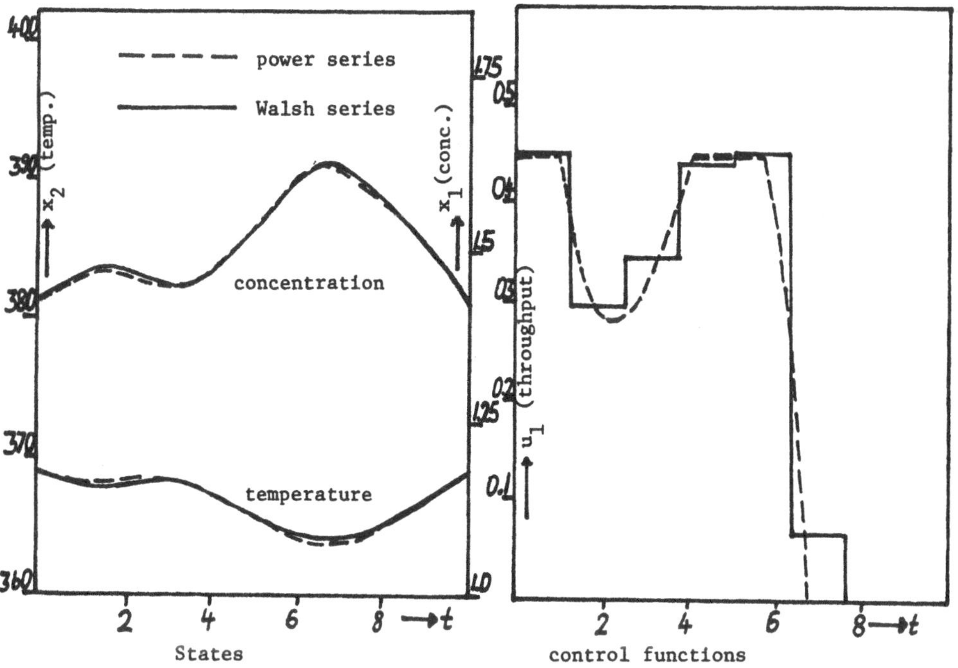

Fig. 3. Results of optimization with different sets of basis functions.

trajectories for state x_1 (concentration of product in the vessel) and state x_2 (temperature in the vessel) are plotted over the considered time interval of 10 units. In the right part the corresponding optimal control functions, u_1 and u_2 are plotted. The control function u_1 (which is determined by the optimization procedure) is assumed to be

$$u_1 = \sum_{j=1}^{m} p_j \left(\frac{t}{10}\right)^{(m-1)} . \tag{27}$$

In Fig. 2 the influence of the number of parameters on the solution is shown; although for 3 and 4 parameters a solution is found, increasing the number to 5 still yields a significant change in the control function and the state trajectories. For 6 and more parameters, approximately the same trajectory for the control function is found as with 5 parameters. To ensure the existence of all the matrices for the iteration, only two parameters would have been necessary, since two final conditions exist; for an sufficient approximation of the optimal control function, however, at least 5 parameters were necessary in this example.

Figure 3 indicates that the choice of the basis functions $f_{ij}(t)$, which is arbitrary also, should not be very difficult, provided a sufficient number of free parameters is considered. Here the solutions found for a power series approximation as in Fig. 2 and an approximation by the first 8 elements of a Walsh series are shown. Both yield very similar state trajectories and the control functions are as similar as they can be regarding the different function bases.

Although these results are no proof, they indicate that choice of number of parameters and type of functions for the approximation of the control function are normally not very difficult. Use of this method for different examples (as done in Ref. 12) yields about the same experiences. Also the initial guess for the parameters as needed in step 0 of the iteration scheme is not very difficult, expecially for technical systems, where physical insight often gives a clear indication of what the control function could look like. The used optimization method shows in addition a wide region of convergence, which makes the choice of these initial values not very difficult either.

V. Conclusions

An optimization method for dynamic systems based on the steepest descent technique has been simplified using a Ritz-type approximation of the control functions to be determined. By reducing the dynamic optimization problem to a parameter optimization typically about 90% of the computation time as well as of the program length can be saved, thus allowing the treatment of technically relevant system on small computers.

Due to the approximation of the control function, this procedure gives only suboptimal solutions. The degree of optimality obtainable depends on the number of parameters and on the type of functions used for approximation. Although both of these are arbitrary (with some restrictions), a number of examples which this method was used on, showed it not to be very difficult to find a suitable approximation which yielded satisfactory results, at least for technical systems where the model to be optimized corresponds to the real systemoonly with an accuracy of a few percent anyway.

References

1. H. Tolle: Optimization methods, Springer-Verlag, 1971.

2. W. Ritz: Über eine neue Methode zur Lösung gewisser Variationsprobleme der mathematischen Physik. J. reine angew. Math., Vol. 135, 1908, pl

3. A. Bryson, W. Denham, E. Caroll, K. Mikami: Determination of the lift or drag program that minimizes reentry heating with acceleration or range constraints using a steepest descent procedure. JAS-Paper Nr. 61-6.

4. H. Rosenbrock, C. Storey: Computational techniques for chemical engineers. Pergamon Press, 1966.

5. G. Hicks, W. Ray: Approximation methods for optimal control synthesis. Can. J. Chem. Eng., Vol. 49, 1971, p. 522.

6. W. Williamson: Use of polynomial approximations to calculate suboptimal controls. J. Am. Inst. Aero. Astron., Vol. 9, 1971, Nr. 11, p. 2271.

7. R. Brusch: A nonlinear programming approach to space-shuttle trajectory optimization. J. Opt. Theor. Appl., Vol. 13, 1974, Nr. 1, p. 94.

8. H. Sirisena, A. Chou: Convergence of the control parameterization Ritz method for nonlinear optimal control problems. J. Opt. Theor. Appl., Vol. 29, 1979, Nr. 3, p. 369.

9. B. Asselmeyer: A two level optimal final-value control system for nonlinear plants realized with mini/micro computers. J. Opt. Contr. Appl. Meth., Vol. 2, to appear 1982.

10. J. Rosen: The gradient projection method for nonlinear programming, Part I: Linear constraints. SIAM J., Vol. 8, 1960, Nr. 1, p. 181.

11. S. Jacoby, J. Kowalik, J. Pozzo: Iterative methods for nonlinear optimization problems. Prentica Hall, Inc., 1972.

12. B. Asselmeyer: Zur optimalen Endwertregelung nichtlinearer Systeme mit Hilfe Kleiner Rechner, Darmstädter Dissertation, 1980.

13. J. Douglas: Process dynamics and control, Vol. 1, Analysis of dynamic systems. Prentice Hall, Inc., 1972.

ON LOWER CLOSURE AND LOWER SEMICONTINUITY IN THE EXISTENCE THEORY FOR OPTIMAL CONTROL

E.J. Balder
Mathematical Institute, University of Utrecht
Utrecht, the Netherlands

1. Introduction.

Lower closure and closure problems form very useful stepping stones to obtain existence results for optimal control problems. An excellent introduction to the subject of lower closure is given in the first part of [12a]. A distinction should be made between the lower closure problems which have additional weak convergence conditions for the "derivative functions", and those which have not.

In the absence of additional weak convergence, the orientor field usually must have a rather strong upper semicontinuity property, property (Q), jointly in time and state variable [3a-b,4a]. We remark that property (Q) is a natural generalization of the classical notion of semi-normality in the calculus of variations.

The situation is different when the lower closure problem has weakly converging derivative functions. Then, namely, the orientor field merely needs to have an upper semicontinuity property in the sense of Kuratowski, property (K), and this only with respect to the state variable [3d,4a]. It should be mentioned that in earlier work such problems were still treated by working with property (Q) type semicontinuity [2,3c]. Later, it was recognized that property (K) type semicontinuity is enough for such problems. A notable exception to this rule is formed by the class of lower closure problems with infinite - dimensional, weakly converging derivative functions. To deal with these problems, it seems unavoidable to impose a property (Q) type semicontinuity condition in the state variable [10,4b].

The role of lower closure in the existence theory for optimal control is parallelled in the calculus of variations by that of lower semicontinuity of integral functionals [1b,3d,8,12b-c,14]. At present there is a strong tendency in important branches of optimal control theory to reduce optimal control problems to variational problems by means of a so-called deparametrization or reduction procedure [5,6,9,13c,15]. It therefore seems relevant to ask to what extent lower closure problems can be reduced to lower semicontinuity problems, as was done by the present author [1c-f]. Here at least one difference with the original lower semicontinuity problems in the calculus of variations can be expected: by the very nature of deparametrization, the integrands must be allowed to take infinite values.

Globally speaking, one proceeds as follows: (i) one formulates a suitable Lagrangian for deparametrizing the lower closure problem; (ii) since the Lagrangian has to figure as the integrand in the associated integral functional, one must make sure that it has the semicontinuity, convexity and measurability properties required to apply the lower semicontinuity result one has in mind; (iii) one applies said result; (iv) one "reparametrizes" by applying a standard argument involving a measurable implicit function result.

The effect of this approach is to simplify the subject of lower closure to a considerable extent, and to provide excellent starting points for reaching more general lower closure and existence results [1c-f].

The objective of this note is merely to illustrate the above steps. A new, useful Lagrangian and Hamiltonian will be presented in section 2. There, we also discuss the relations between semicontinuity of the orientor field and semicontinuity of the Lagrangian and Hamiltonian. In section 3 we present an apparently new lower semicontinuity result for integral functionals and a lower closure result which it implies.

2. Semicontinuity of orientor fields, Lagrangians and Hamiltonians.

Let G be a subset of $\mathbb{R}$, A a subset of $G\times\mathbb{R}^n$ and $\approx{Q}$ a multifunction from A into $\mathbb{R}^m$ with nonempty values; here G represents the time domain, $\mathbb{R}^n$ the state space and $\tilde{\tilde{Q}}$ the orientor field. For $t\in G$ we denote the section of A at t by $A(t)$. We shall suppose that $A(t)$ is closed for each $t \in G$.

The orientor field $\tilde{\tilde{Q}}$ is said to have <u>property</u> (K) with respect to A at $(t,x)\in A$ if

$$\tilde{\tilde{Q}}(t,x) = \bigcap_{\gamma>0} \operatorname{cl} \tilde{\tilde{Q}}(t,x;\gamma),$$

where

$$\tilde{\tilde{Q}}(t,x;\gamma) \equiv \cup \{\tilde{\tilde{Q}}(t,x'):x' \in A(t),\ |x'-x|<\gamma\}.$$

We say that $\tilde{\tilde{Q}}$ has <u>property</u> (Q) with respect to A at $(t,x)\in A$ [3a] if

$$\tilde{\tilde{Q}}(t,x) = \bigcap_{\gamma>0} \operatorname{cl}\operatorname{co} \tilde{\tilde{Q}}(t,x;\gamma).$$

The <u>modified Lagrangian</u> $L:G\times\mathbb{R}^{n+m+1}\to(-\infty,+\infty]$ corresponding to $\tilde{\tilde{Q}}$ is defined as follows [1c]:

$$L(t,x,\xi,\lambda) \equiv \begin{cases} \inf\{\eta: \eta \geq \lambda, (\xi,\lambda) \in \tilde{\tilde{Q}}(t,x)\} \text{ if } (t,x) \in A, \\ +\infty \quad \text{if } (t,x) \notin A. \end{cases}$$

The modified Hamiltonian $H: G \times \mathbb{R}^{n+m+1} \to [-\infty,+\infty]$ corresponding to $\tilde{\tilde{Q}}$ is defined as follows [1e]:

$$H(t,x,p,q) \equiv \begin{cases} \sup\{\langle p,\xi\rangle+(q-1)\eta: (\xi,\eta)\in\tilde{\tilde{Q}}(t,x)\} \text{ if } (t,x)\in A,\ q \geq 0, \\ +\infty \text{ if } (t,x) \in A,\ q < 0, \\ -\infty \text{ if } (t,x) \notin A. \end{cases}$$

Note here that we follow the usual convention "$\inf \phi \equiv +\infty$ ". It is easy to verify that $H(t,x,.,.)$ is the Fenchel-conjugate $L^*(t,x,.,.)$ of $L(t,x,.,.)$.

The semicontinuity properties (K) and (Q) of the orientor field turn out to be equivalent to certain properties of L and H, which involve semicontinuity and conjugation.

Theorem 2.1. The orientor field $\tilde{\tilde{Q}}$ has property (K) with respect to A at $(t,x)\in A$ if and only if for each $(\xi,\lambda)\in\mathbb{R}^{m+1}$ $L(t,.,.,.)$ is lower semicontinuous at (x,ξ,λ).

Theorem 2.2. The orientor field $\tilde{\tilde{Q}}$ has property (Q) with respect to A at $(t,x)\in A$ if and only if

$$L(t,x,.,.) = \bar{H}^*(t,x,.,.),$$

where $\bar{H}$ is defined by

$$\bar{H}(t,x,p,q) \equiv \limsup_{x'\to x} H(t,x',p,q).$$

Partial results in the direction of this characterization of property (Q) were obtained in [7,12a]. Theorem 2.2 itself has been proven in [1e].

3. Lower closure and lower semicontinuity: an example.

All results in [1c-d,1f] were derived from a single, classical lower semicontinuity result [3d,8,12b-c,14]. In [1e] results on conjugate integral functionals [13a], together with theorem 2.2, play a central role in proving lower semicontinuity. Now the classical result mentioned above also follows as a special case from [1b]. Here we will present another lower semicontinuity result that follows from [1b]. It will be discussed more extensively in forthcoming work by the author.

Let us begin with some preparations. Let G be the interval $[0,T]$ for some $T > 0$. We shall denote the Lebesgue σ-algebra on G by $\mathcal{L}$ and the Borel σ-algebra on $\mathbb{R}^r$ by $\mathcal{B}^r$, etc. Let $\{x_k\}_1^\infty$ be a sequence of absolutely continuous functions from G into $\mathbb{R}^n$ and $x_0: G\to\mathbb{R}^n$ an

L-measurable function such that for each $t \in G$

$$\{x_k(t)\}_1^\infty \text{ converges to } x_0(t). \qquad (3.1)$$

Let $1 \leq \alpha \leq n$. For any $x \in \mathbb{R}^n$ we describe the partitioning of x into its projections on $\mathbb{R}^\alpha$ and $\mathbb{R}^{n-\alpha}$ by $x \equiv (x^{(1)}, x^{(2)})$, etc. We shall suppose that $x_0^{(1)}$ (composed of the first α component functions of x_0) is absolutely continuous. The absolutely continuous part (by Lebesgue decomposition) of $x_0^{(2)}$ will be denoted as $\dot{X}_0^{(2)}$. Also, we suppose for the derivatives that

$$\{\dot{x}_k^{(1)}\}_1^\infty \text{ converges weakly to } \dot{x}_0^{(1)}, \qquad (3.2)$$

in the sense of the $\sigma(L_1^\alpha(G), L_\infty^\alpha(G))$-topology, and that for each $k \geq 1$

$$\dot{x}_k^{(2)}(t) \geq 0 \text{ a.e.} \qquad (3.3)$$

<u>Theorem</u> 3.1. Suppose that (3.1)-(3.3) hold. Let $l: G \times \mathbb{R}^{2n} \to (-\infty, +\infty]$ be such that for each $t \in G$

$l(t,.,.)$ is lower semicontinuous on $\mathbb{R}^{2n}$,

$l(t, x_0(t), .)$ is convex on $\mathbb{R}^n$

$l(t, x_0(t), \xi^{(1)}, .)$ is nonincreasing on $\mathbb{R}^{n-\alpha}$ (coordinatewise) for each $\xi^{(1)} \in \mathbb{R}^\alpha$, and that

$\{l^-(\cdot, x_k(\cdot), \dot{x}_k(\cdot))\}_1^\infty$ is uniformly integrable,

where $l^- \equiv \max(-l, 0)$. Then

$$\liminf_{k \to \infty} \int_G l(t, x_k(t), \dot{x}_k(t))\, dt \geq \int_G l(t, x_0(t), \dot{x}_0^{(1)}(t), \dot{X}_0^{(2)}(t))\, dt.$$

The result is obtained by specializing the generalized convexity requirement - in the sense of [1a,11] - in [1b] to that of ordinary convexity for the $\xi^{(1)}$ -variable and monotone convexity for the $\xi^{(2)}$ -variable.

We shall now describe the lower closure problem. Let A be as in section 2 and let U be a multifunction from A into $\mathbb{R}^p$ with $L \times B^{n+p}$-measurable graph M. By M(t) we denote the section of M at t, $t \in G$. Let $f: M \to \mathbb{R}^n$, $f_0: M \to (-\infty, +\infty]$ be $L \times B^{n+p}$ -measurable functions. Let $\{u_k\}_1^\infty$ be a sequence of L-measurable control functions from G into $\mathbb{R}^p$ such that for each $k \geq 1$ and for a.e.t

$$x_k(t) \in A(t),\ u_k(t) \in U(t, x_k(t)),\ \dot{x}_k(t) = f(t, x_k(t), u_k(t)) \qquad (3.4)$$

We shall set

$$\tilde{\tilde{Q}}(t,x) \equiv \{(\xi,\eta) \in \mathbb{R}^{n+1} : \xi^{(1)} = f^{(1)}(t,x,u),\ \xi^{(2)} \geqslant f^{(2)}(t,x,u),$$
$$\eta \geqslant f_0(t,x,u),\ u \in U(t,x)\},\ (t,x) \in A. \tag{3.5}$$

Theorem 3.2. Suppose that (3.1)-(3.4) hold and that for each $t \in G$
$A(t)$ is closed,
$M(t)$ is closed,
$f^{(1)}(t,.,.)$ is continuous on $M(t)$,
$f^{(2)}(t,.,.)$ is (componentwise) lower semicontinuous on $M(t)$,
$f_0(t,.,.)$ is lower semicontinuous on $M(t)$,
$\tilde{\tilde{Q}}$ has property (K) at $(t,x_0(t))$ with respect to $A(t)$,
$\tilde{\tilde{Q}}(t,x_0(t))$ is convex,
$\{f_0^-(\cdot,x_k(\cdot),u_k(\cdot))\}^\infty$ is uniformly integrable.

Then, if $i \equiv \liminf_{k\to\infty} \int f_0(t,x_k(t),u_k(t))\,dt < +\infty$, there exists an L-measurable function $u_* : G \to \mathbb{R}^p$ such that for a.e.t

$$x_0(t) \in A(t), u_*(t) \in U(t,x_0(t)),\ \dot{x}_0^{(1)}(t) = f^{(1)}(t,x_0(t),u_*(t)),$$
$$\dot{x}_0^{(2)}(t) \geqslant f^{(2)}(t,x_0(t),u_*(t)).$$

Note that the above lower closure result can easily be used to derive existence results; cf. [3b]. The proof of Theorem 3.2 by way of Theorem 3.1 uses the modified Lagrangian L, corresponding to (3.5), in that one takes l in Theorem 3.1 to be the lower semicontinuous hull $\bar{L}$ of L with respect to the (x,ξ,λ) -variable. Measurability of $\bar{L}$ follows from [13b], (3.5) and our basic measurability assumptions; cf.[1f].

References

[1a] E.J. Balder, An extension of duality-stability relations to non-convex optimization problems, SIAM J. Control Optimization 15 (1977), 329-343.

[1b] ——, Lower semicontinuity of integral functionals with nonconvex integrands by relaxation-compactification, ibid. 19 (1981), 533-542.

[1c] ——, Lower closure problems with weak convergence conditions in a new perspective, ibid. 20 (1982), No. 2, to appear.

[1d] ——, On existence problems for the optimal control of certain nonlinear integral equations of Urysohn type, J. Optimization Theory Appl., to appear.

[1e] ——, The complete Hamiltonian characterization of property (Q)

and its application to infinite dimensional lower closure problems, submitted.

[1f] ——, An existence result for optimal control problems with unbounded time domain and its application to optimal economic growth problems, submitted.

[2] L.D. Berkovitz, Existence and lower closure theorems for abstract control problems, SIAM J. Control 12 (1974), 27-42.

[3a] L. Cesari, Existence theorems for weak and usual optimal solutions in Lagrange problems with unilateral constraints I, Trans. Amer. Math. Soc. 124 (1966), 369-412.

[3b] ——, Existence theorems for optimal controls of the Mayer type, SIAM J. Control 6 (1968), 517-552.

[3c] ——, Closure problems for orientor fields and weak convergence, Arch. Rational Mech. Anal. 55 (1974), 332-356.

[3d] ——, Lower semicontinuity and lower closure without seminormality conditions, Ann. Mat. Pura Appl. 98 (1974), 381-397.

[4a] L. Cesari and M.B. Suryanarayana, Convexity and property (Q) optimal control theory, SIAM J. Control 12 (1974), 705-720.

[4b] ——, Existence theorems for Pareto optimization; multivalued and Banach space valued functionals, Trans. Amer. Math. Soc. 244 (1978), 37-65.

[5] F. Clarke, The generalized problem of Bolza, SIAM J. Control Optimization 14 (1976), 682-699.

[6] I. Ekeland and R. Temam, Convex Analysis and Variational Problems, Dunod, Paris, 1972, English transl., North-Holland, Amsterdam, 1976.

[7] G.S. Goodman, The duality of convex functions and Cesari's property (Q), J. Optimization Theory Appl. 19 (1976), 17-23.

[8] A.D. Ioffe, On lower semicontinuity of integral functions I, SIAM J. Control Optimization 15 (1977), 521-538.

[9] A.D. Ioffe and V.M. Tichomirov, Theorie der Extremalaufgaben, Nauka, Moscow, 1974, German transl., Deutscher Verlag der Wissenschaften, Berlin, 1979.

[10] P.J. Kaiser and M.B. Suryanarayana, Orientor field equations in Banach spaces, J. Optimization Theory Appl. 19 (1976), 141-164.

[11] J.J. Moreau, Fonctionelles Convexes, Séminaire sur les Equatiens aux Dérivées Partielles, Collége de France, Paris, 1966.

[12a] C. Olech, Existence theory in optimal control problems- the underlying ideas, International Conference on Differential Equations, H.A. Antosiewicz, ed., Academic Press, New York, 1975, 612-629.

[12b] C. Olech, Weak lower semicontinuity of integral functions, J. Optimization Theory Appl. 19 (1976), 3-16.

[12c] ——, A characterization of L_1-weak lower semicontinuity of integral functionals, Bull. Acad. Pol. Sc. 25 (1977), 135-142.

[13a] R.T. Rockafellar, Integrals which are convex functionals II, Pacific J. Math. 39 (1971), 439-469.

[13b] ——, Integral functionals, normal integrands and measurable selections, Nonlinear Operators and the Calculus Of Variations, J.P. Gossez ed., Lecture Notes in Mathematics No.543, Springer, Berlin, 1976, 157-207.

[13c] ——, Duality in optimal control, Mathematical Control Theory, W.A. Coppel, ed., Lecture Notes in Mathematics No. 680, Springer, Berlin, 1978, 219-257.

[14] L. Tonelli, Opere Scelte, Cremonese, Rome, 1962.

[15] L.E. Zachrisson, Deparametrization of the Pontryagin maximum principle, Mathematical Theory of Control, A.V. Balakrishnan and L.W. Neustadt, ed., Academic Press, New York, 1967, 234-245.

A DISCRETE MAXIMUM PRINCIPLE CONCERNING THE OPTIMAL COST OF DETERMINISTIC CONTROL PROBLEMS

R. GONZALEZ*, E. ROFMAN**

*Electronic Department - University of Rosario
Pellegrini 250 - 2000 Rosario - Argentina

**I.N.R.I.A.
Rocquencourt BP 105 - 78153 Le Chesnay - France

INTRODUCTION :

Previously [1], we have dealt with the numerical solutions of some optimal deterministic control problems, using as a basic tool of analysis the characterization (introduced in [2], [3]) of the optimal cost function as the maximum element of a suitable set of subsolutions of the associated Hamilton-Jacobi equation. In this paper, to compute that maximum element, we present a new algorithm who makes possible to solve non-trivial problems in computers of small central memories.

In part I we study the stationary case. In I.1 is introduced a control problem to be optimized that considers the use in each strategy of stopping times, continuous and impulsive controls. So $V(x) = \inf_{\tau, u(.), z(.)} J(x, \tau, u(.), z(.))$. We present a fitted set W of subsolutions and it is shown that V(x) is the unique solution of the equivalent problem (P): Find the maximum element of the set W.

In I.2 we consider the discretized problem (P_h), its solution, the algorithm to compute it and its properties. Using a particular scheme to discretize the partial derivatives of the functions considered, we are enabled to define an algorithm that, iterative and succesively, increases the values of these functions in the vertices of the triangulation employed, until the approximate solution $\bar{w}^h$ is found.

In I.3 it is proved that the solutions $\bar{w}^h$ of the problems (P_h) converge to the solution V(x) of (P).

In Part II we consider the non stationary case. In particular, we give a solution to the problem of optimal control of an electricity production system, applying the methodology described in this paper. Systems with a significant number of thermal generators may be optimized in this form.

This work has been partly supported by D.O.E., Office of Electric Energy Systems, under contract 01-80RA-50154.

I. THE STATIONARY PROBLEM.

I.1. The optimal control problem and an equivalent formulation (the problem (P)).

To control a system with trajectories in a bounded set $\Omega \subset \mathbb{R}^n$, we use stopping time control, impulsive control and continuous control.

In the intervals of time free of the action of impulsive controls, the trajectories of the system satisfies the differential equation:

$$\begin{cases} \frac{dy}{dt} = f(y,u) \\ y(0) = x \qquad x \in \Omega \subset \mathbb{R}^n \end{cases} \tag{1.1}$$

where u(.) is a measurable function of time, with values in a compact set $U \subset \mathbb{R}^m$.

At times θ_ν $(0 \leq \theta_1 < \theta_2 < \dots)$, impulses $z(\theta_\nu)$ are applied; they produce jumps of amplitude g_ν, in the trajectory of the system:

$$y(\theta_\nu^+) = y(\theta_\nu^-) + g(y(\theta_\nu^-), z(\theta_\nu)) \tag{1.2}$$

$y(\theta_\nu^+)$, $(y(\theta_\nu^-))$ is the right limit (left) of the trajectory y(.). The set Z of impulsive controls is a compact set of $\mathbb{R}^p$.

The control strategy is determined by the stopping time $\tau \geq 0$, the function u(.), the times $\{\theta_\nu\}$ and the impulses $z(\theta_\nu)$.

We assign to each strategy of control $(\tau, u(.), z(.))$, the cost value J:

$$J(x,\tau,u(.),z(.)) = \int_0^\tau e^{-\alpha s} l(y(s),u(s))\,ds + \varphi(y(\tau))e^{-\alpha\tau} + \\ + \sum_\nu e^{-\alpha\theta_\nu} \cdot q(y(\theta_\nu^-), z(\theta_\nu)) \tag{1.3}$$

being l the instantaneous cost, φ the final cost, $\alpha > 0$ the instantaneous actualization factor and $q > 0$ the cost of application of an impulse.

We are trying to find, $\forall\, x \in \Omega$ the value of the optimal cost function V(x) defined by:

$$V(x) = \inf_{\tau, u(.), z(.)} J(x,\tau,u(.),z(.)) \tag{1.4}$$

In the following, we shall suppose that $\forall\, t$, $y(t) \in \Omega$, and that f, l, φ, g, q are continuous and bounded functions (and lipschitzean functions of y). When suitable conditions are verified by the constants of lipschitzeannity, it is proved that (see [1], [4]); $\exists\, L > 0 \,/\, \forall x, x' \in \Omega$, $|V(x) - V(x')| \leq L\,|x - x'|$ (1.5)

and in this case we can prove the following

Theorem 1.1: (Characterization of V(x))

Let be $W = \{ w : \Omega \longrightarrow \mathbb{R} / (1.7), (1.8), (1.9), (1.10) \}$ (1.6)

where

w is a lipschitzean function (1.7)

$$w(x) \leq \min_{z \in Z} \left(w(x+g(x,z)) + q(x,z) \right) \quad \forall\, x \in \Omega \tag{1.8}$$

$$w(x) \leq \varphi(x) \quad \forall\, x \in \Omega \tag{1.9}$$

$$\min_{u \in U} \left(\frac{\partial w(x)}{\partial x} f(x,u) + l(x,u) - \alpha w(x) \right) \geq 0 \quad \text{a.e.} x \in \Omega \tag{1.10}$$

then

$$V(x) \in W \tag{1.11}$$

$$V(x) \geq w(x) \quad \forall\, x \in \Omega,\ \forall\, w \in W \tag{1.12}$$

For the proof, see [4].

Theorem 1.1 makes possible to find the optimal cost function defined in (1.4), solving the equivalent problem:

(P): Find the maximum element of the set W defined by (1.6).

I.2. The discretized problem (P_h).

2.0. Preliminary comments.

In this chapter we shall introduce sets W^h, finite dimensional approximation of W, looking for a numerical device to compute V(x). Following this idea, after a discretization Ω^h of the set Ω, we shall define W^h by functions w^h verifying properties related to (1.7)-(1.10). The main difficulty of this approach is the choice of W^h having maximum element $\bar{w}^h$. In fact, after introducing in W^h the natural partial order: $w_1 \leq w_2$ iff $w_1(x_i^h) \leq w_2(x_i^h)$ $\forall\, x_i^h$ vertex of Ω^h, it is not possible, in general, to insure the existence of $\bar{w}^h$. We show, in what follows, that thanks to a criterium used in the discretization of the derivatives that appear in (1.10), (see (2.3)) we obtain:

a) There exist an unique maximal element in W^h, furthermore, this maximal element is also the maximum element $\bar{w}^h$.

b) A characterization of $\bar{w}^h$ that enables us to compute it with an iterative algorithm of relaxation type.

2.1. Description of the discretization procedure.

a) The set Ω is approximated with a triangulation Ω^h (union of simplices).

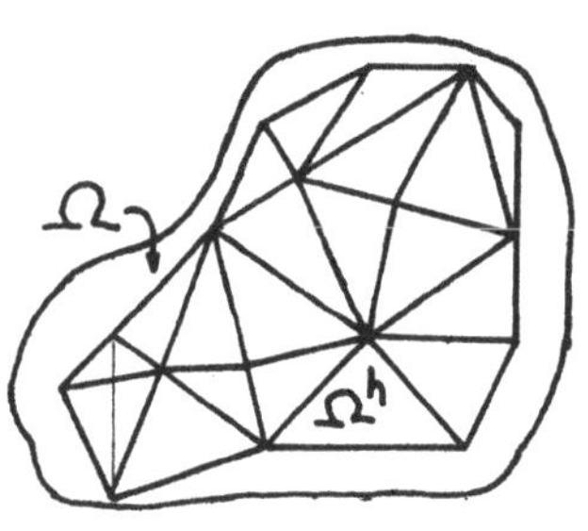

b) We consider in place of W, the set W^h of functions w^h: $\overline{\Omega}^h \dashrightarrow \mathbb{R}$ (w^h continuous in $\overline{\Omega}^h$, $\frac{\partial w^h}{\partial x}$ is constant in the interior of each simplex of Ω^h; i.e. w^h are linear finite elements) that satisfy $\forall\ x_i^h$ vertex of Ω^h (2.1), (2.2), (2.3), discretizations of the restrictions (1.8), (1.9), (1.10).

$$w^h(x_i^h) \leq w^h(x_i^h + g(x_i^h, z)) + q(x_i^h, z) \qquad (2.1)$$

$\forall\ z \in Z^h \subset Z$, finite set that approximates the set of impulsive controls Z.

$$w^h(x_i^h) \leq \varphi(x_i^h) \qquad (2.2)$$

$$\frac{\partial w^h}{\partial x}(x_i^h; u) \, . \, \|f(x_i^h, u)\| + l(x_i^h, u) - \alpha\, w^h(x_i^h) \geq 0 \qquad (2.3)$$

$\forall\ u \in U^h \subset U$, finite set that approximates the set of admissible continuous controls U. With $\frac{\partial w^h}{\partial x}(x_i^h; u)$ we denote the derivative of w^h in the direction of f, i.e.

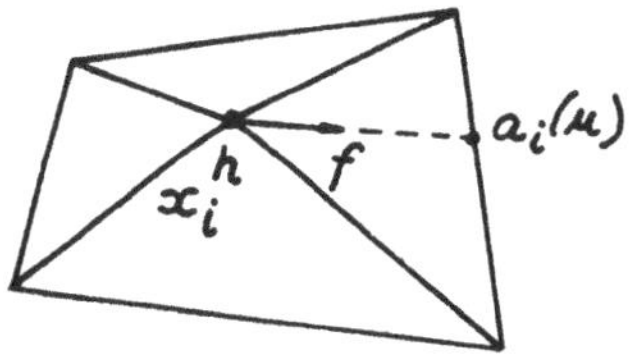

$$\left.\begin{aligned} &\frac{\partial w^h}{\partial x_f}(x_i^h; u) \, . \, \|f(x_i^h, u)\| = \frac{w^h(a_i(u)) - w^h(x_i^h)}{\|a_i(u) - x_i^h\|} \, . \, \|f(x_i^h, u)\| \\ &\text{if} \quad f(x_i^h, u) \neq 0 \\ &\frac{\partial w^h}{\partial x_f}(x_i^h; u) \, . \, \|f(x_i^h, u)\| = 0 \quad \text{if} \ f(x_i^h, u) = 0 \end{aligned}\right\} \qquad (2.4)$$

c) We introduce the problem (P_h): discretization of problem (P). (P_h): Find the maximum element $\bar{w}^h$ of the set W^h, considering in W^h the partial order $w^h \leq \tilde{w}^h$ iff $w^h(x_i^h) \leq \tilde{w}^h(x_i^h)$ $\forall\ x_i^h$ vertex of Ω^h.

2.2. Analysis of Problem (P_h)

In the set W^h is defined only a partial order and in consequence, it is not obvious, a priori, the existence of a maximum element. However, by virtue of the properties of W^h, we shall prove that there exists a unique maximal element in W^h that is also the maximum element of W^h.

To do this, we transform the restrictions (2.1), (2.3) in more useful equivalent relations ((2.6), (2.7)). Considering that $a_i(u)$ and

$(x_i^h+g(x_i^h,z))$ are convex combinations of the vertices of the simplices to whom they belong, we have:

$$\begin{cases} a_i(u) = \sum_{j=1}^{n_h} \lambda_j(x_i^h,u).x_j^h \ , \ \lambda_j \geqslant 0 \ , \ \sum_{j=1}^{n_h} \lambda_j = 1 \\ x_i^h+g(x_i^h,z) = \sum_{j=1}^{n_h} \lambda'_j(x_i^h,z).x_j^h \ , \ \lambda'_j \geqslant 0 \ , \ \sum_{j=1}^{n_h} \lambda'_j = 1 \end{cases} \tag{2.5}$$

Being w^h an affine function, (2.1), (2.3) are equivalent to:

$$w^h(x_i^h) \leqslant \min_{u \in U^h} \left[\beta(x_i^h,u).\left(\sum_{j=1}^{n_h} \lambda_j(x_i^h,u).w^h(x_j^h)\right) + \Delta t(x_i^h,u).1(x_i^h,u)\right] \tag{2.6}$$

$$w^h(x_i^h) \leqslant \min_{z \in Z^h} \left[\sum_{j=1}^{n_h} \lambda'_j(x_i^h,u).w^h(x_j^h)+q(x_i^h,z)\right] \tag{2.7}$$

where:

$$\begin{cases} \beta(x_i^h,u) = 0 \quad \text{if} \quad f(x_i^h,u) = 0 \\ \beta(x_i^h,u) = \|f(x_i^h,u)\| \,/\, (\|f(x_i^h,u)\| + \alpha\|a_i(u)-x_i^h\|) \quad \text{if} \quad f(x_i^h,u) \neq 0 \end{cases} \tag{2.8}$$

$$\begin{cases} \Delta t(x_i^h,u) = 1/\alpha \quad \text{if} \quad f(x_i^h,u) = 0 \\ \Delta t(x_i^h,u) = \|a_i(u)-x_i^h\| \,/\, (\|f(x_i^h,u)\| + \alpha\|a_i(u)-x_i^h\|) \\ \qquad\qquad\qquad\qquad\qquad\qquad \text{if} \quad f(x_i^h,u) \neq 0 \end{cases} \tag{2.9}$$

W^h may now be defined as the set of linear finite elements on Ω^h that satisfy (2.2), (2.6), (2.7) and we prove the following:

Theorem 2.1:
There exists $\bar{w}^h$, maximum element of W^h.
Proof: Let be

$$\hat{w}^h(x_i^h) = \sup\left\{w^h(x_i^h) / w^h \in W^h\right\} \tag{2.10}$$

$\hat{w}^h$ is well defined by virtue of (2.2). From (2.2), (2.10) it follows that $\hat{w}^h$ verifies (2.2). We shall prove that $\hat{w}^h$ satisfies (2.6), (2.7). In (2.6), (2.7) the factors that multiply $w^h(x_j^h)$ are non-negative, then by virtue of (2.6), (2.10) we have:

$$w^h(x_i^h) \leqslant \min_{u \in U^h} \left[\beta(x_i^h,u).\left(\sum_{j=1}^{n_h} \lambda_j(x_i^h,u).\hat{w}^h(x_j^h)\right) + \Delta t(x_i^h,u).1(x_i^h,u)\right]$$

and, as w^h is an arbitrary element of W^h, we obtain:

$$\hat{w}^h(x_i^h) \leqslant \min_{u \in U^h} \left[\beta(x_i^h,u).\left(\sum_{j=1}^{n_h} \lambda_j(x_i^h,u).\hat{w}^h(x_j^h)\right) + \Delta t(x_i^h,u).1(x_i^h,u)\right]$$

i.e., $\hat{w}^h$ satisfies (2.6). In a similar way, it is proved that $\hat{w}^h$ verifies (2.7) and in consequence, $\hat{w}^h \in W^h$; by virtue of (2.10), $\hat{w}^h$ is the maximum element of W^h, i.e. $\hat{w}^h = \bar{w}^h$.

2.3. Characterization of the maximum element $\bar{w}^h$.

We define the operator $M: \mathbb{R}^{nh} \longrightarrow \mathbb{R}^{nh}$ in the following form:

$$(Mw^h)(x_i^h) = \min \Big\{ \varphi(x_i^h), \min_{z\in Z^h}\Big[\Big(\sum_{j=1}^{nh} \lambda'_j(x_i^h,z).w^h(x_j^h)\Big)+q(x_i^h,z)\Big], \quad (2.11)$$
$$\min_{u\in U^h}\Big[\beta(x_i^h,u).\Big(\sum_{j=1}^{nh} \lambda_j(x_i^h,u).w^h(x_j^h)\Big)+\Delta t(x_i^h,u).1(x_i^h,u)\Big]\Big\}$$

and we obtain the following characterization of the maximum element of W^h.

Theorem 2.2:
$\bar{w}^h$ is the maximum element of W^h if and only if $\bar{w}^h \equiv M\,\bar{w}^h$

Proof of the necessary condition:
Let be $\bar{w}^h$ the maximum element of W^h and suppose that $\exists\, i_o,\ \varepsilon > 0$ /

$$\bar{w}^h(x_{i_o}^h) + \varepsilon \leq (M\,\bar{w}^h)(x_{i_o}^h) \quad (2.12)$$

We define $\check{w}^h(x_i^h) = \bar{w}^h(x_i^h)\ \ \forall\, i\neq i_o$, $\check{w}^h(x_{i_o}) = \bar{w}^h(x_{i_o}^h)+\varepsilon$, then we have:

$$\begin{cases} \check{w}^h(x_{i_o}^h) \leq (M\,\bar{w}^h)(x_{i_o}^h) \leq (M\,\check{w}^h)(x_{i_o}^h) \\ \check{w}^h(x_i^h) = \bar{w}^h(x_i^h) \leq (M\bar{w}^h)(x_i^h) \leq (M\,\check{w}^h)(x_i^h) \quad \forall\, i\neq i_o \end{cases}$$

by virtue of (2.12) and the monotony of M. In consequence, $\check{w}^h \in W^h$ and $\check{w}^h > \bar{w}^h$; this contradiction has the origin in (2.12), then $\bar{w}^h(x_i^h) = (M\,\bar{w}^h)(x_i^h)\ \ \forall\, i$, i.e. $\bar{w}^h \equiv M\,\bar{w}^h$.

The proof of the sufficient conditions is given in [4]. It uses essentially the positivity of q(x,z) and the following Dicrete Maximum Principle:

DMP: Let be C^h a subset of vertices of Ω^h and S^h the set of vertices not included in C^h. If $\forall\, x_i^h \in C^h$

$$\min_{u\in U^h}\Big(\frac{\partial w^h}{\partial x_f}(x_i^h;u).\,\|f(x_i^h,u)\| - \alpha\, w^h(x_i^h)\Big) \geq 0 \quad (2.13)$$

then, $\exists\ 0 \leq \gamma < 1$ such that

$$\max_{x_i^h\in C^h} w^h(x_i^h) \leq \gamma\Big[\Big(\max_{x_j^h\in S^h} w^h(x_j^h)\Big) \vee 0\Big] \quad (2.14)$$

2.4. Algorithm to compute $\bar{w}^h$.

Making use of the characterization of the maximum element $\bar{w}^h$, we define an algorithm that generates an increasing sequence of functions w^h_ν. These functions converge to a function satisfying $M\, w^h = w^h$, i.e. to $\bar{w}^h$.

Algorithm:

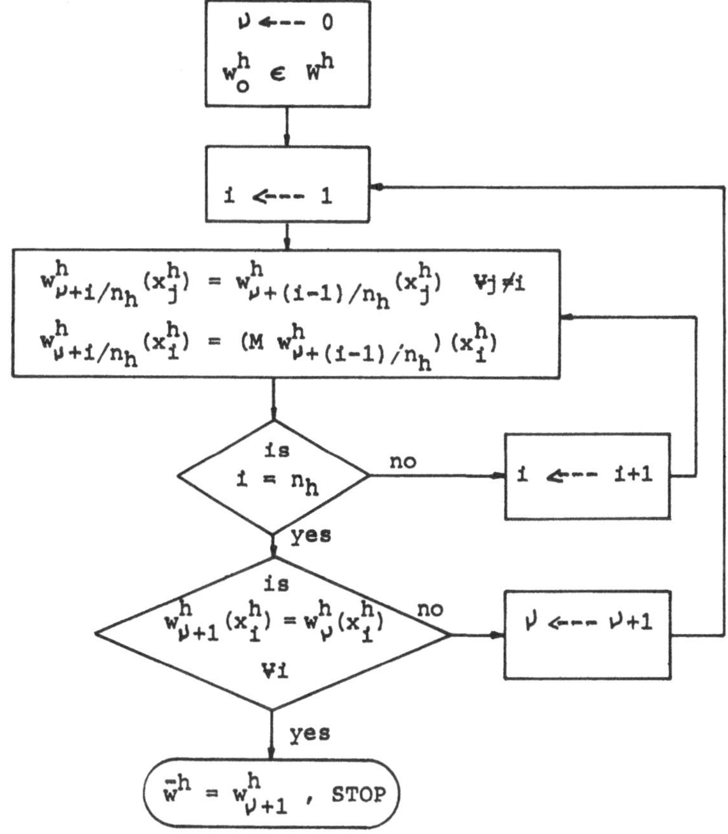

The algorithm gives the solution of problem P_h in the following sense.

Theorem 2.3:

The algorithm finishes at $\bar{w}^h$ in a finite number of steps or generates a sequence $\{w^h_\nu\}$ that converges to $\bar{w}^h$.

Proof:

If the algorithm finishes in a finite number of steps ($\bar{\nu}$), that means that $w^h_{\bar{\nu}}$ is not modified in the last loop, i.e. $w^h_{\bar{\nu}}(x^h_i) = M w^h_{\bar{\nu}}(x^h_i)$ $\forall\, i$, and then by virtue of Theorem 2.2, $w^h_{\bar{\nu}} = \bar{w}^h$.

By definition of the operator M, we have:

$$w^h \in W^h \Longleftrightarrow w^h \leq M\, w^h \quad \text{and} \quad w^h \leq \tilde{w}^h \Longrightarrow M\, w^h \leq M\, \tilde{w}^h$$

As $w^h_o \in W^h$, it results by induction: $w^h_{\nu+i/n_h} \in W^h$ and

$$w^h_\nu \leq \ldots\ldots\ldots \leq w^h_{\nu+(i-1)/n_h} \leq w^h_{\nu+i/n_h} \leq \ldots\ldots \leq w^h_{\nu+1} \leq \ldots\ldots \leq \varphi \quad \text{then}$$

$$\exists\, \check{w}^h / \lim_{\nu \to \infty} w^h_{\nu+i/n_h} = \check{w}^h \quad \forall\, i=1,\ldots,n_h \tag{2.16}$$

From (2.15) we have: $w^h_{\nu+i/n_h}(x^h_i) = (M\ w^h_{\nu+(i-1)/n_h})(x^h_i)$, then, by (2.16) we obtain: $\check{w}^h(x^h_i) = (M\ \check{w}^h)(x^h_i)$.

In consequence, by Theorem 2.2 it follows that: $\lim_{\nu \to \infty} w^h_\nu = \bar{w}^h$.

Remarks.

1) The algorithm needs only the values of the function $w^h_{\nu+(i-1)/n_h}$ at the points $a_i(u)$ and $(x^h_i+g(x^h_i,z))$, to compute $w^h_{\nu+i/n_h}(x^h_i)$. Then, in general it is only necessary to have in the central memory of the computer, the values of $w^h_{\nu+(i-1)/n_h}(x^h_j)$ for a little number of vertices x^h_j. This property allows the application of this algorithm in computers with small central memories (minicomputers).

2) As theorem 2.3 shows, the convergence of the algorithm does not depend on the order of the vertices x^h_j of the triangulation Ω^h; however, a careful choice of that order may allow:

a) An easy retrieval of the information needed for the computation, from the masive memory to the central memory of the computer.

b) An acceleration of the convergence of the algorithm.

3) The algorithm implies the saturation of at least one of the restrictions (2.2), (2.6), (2.7) in each iteration. In practice, the convergence will not be lost if that saturation is omitted in some steps.

I.3. Convergence of $\bar{w}^h$ to V(x).

We can proove that $\lim_{h \to 0} \bar{w}^h(x) = V(x)\quad \forall\ x \in \Omega$ (being $\|h\|$ the maximum size of the simplices of Ω^h). The proof of the inequality $V(x) \leqslant \lim_{\|h\| \to 0} \bar{w}^h(x)$ is obtained following the techniques used in [1]. To obtain the inequality $V(x) \geqslant \overline{\lim}_{\|h\| \to 0} \bar{w}^h(x)$, we make use of the discrete maximum principle introduced in Theorem 2.2 (this DMP gives also implicitly the stability of the method). The complete proof of the convergence is included in [4].

II.1. The non-stationary case.

The method developed in I, can be easily extended to the non-stationary case. Here, we shall limit ourselves to show the restrictions that replace (2.2), (2.6), (2.7):

$$w^h(x^h_i,\mu\Delta) \leqslant \varphi(x^h_i,\mu\Delta) \qquad (2.2)'$$

$$w^h(x_i^h,\mu\Delta) \leqslant \frac{1}{(1+\alpha\Delta)} \min_{u \in U^h} \left\{ \sum_{j=1}^{n_h} \lambda_j (x_i^h,\mu\Delta,u) . w^h(x_j^h,(\mu+1)\Delta) + 1(x_i^h,\mu\Delta,u)\Delta \right\} \quad (2.6)'$$

$$w^h(x_i^h,\mu\Delta) \leqslant \min_{z \in Z^h} \left\{ \sum_{j=1}^{n_h} \lambda_j^{\prime} (x_i^h,\mu\Delta,z) . w^h(x_j^h,\mu\Delta) + q(x_i^h,\mu\Delta,z) \right\} \quad (2.7)'$$

There is also a new "final condition":

$w^h(x_i^h,\bar{\mu}\Delta) = \varphi(x_i^h,\bar{\mu}\Delta)$ where $\bar{\mu}\Delta = T$ (final time)

In the non-stationary case it is not necessary to be $\alpha > 0$. The complete study of the problem is developed in [4].

II.2. Application to a problem of power generation.

We have applied the method developed, to minimize the operative cost of an electric power production system, that comprises an hydraulic generator and two termic generators. In this case, we have to deal with a system of four quasi-variational inequalities (IQV), concerning the optimal cost $V_i(x,t)$; $(i=1,4)$:

$$\begin{cases} \dfrac{\partial V_i(x,t)}{\partial t} + \min\limits_{(P_1,P_2,P_h)} \left(\dfrac{\partial V_i(x,t)}{\partial x} . (A-P_h) + C_1P_1 + C_2P_2 + C_3(D-P_1-P_2-P_h)^+ + C_h.P_h \right) \geqslant 0 \\ V_i(x,t) \leqslant V_j(x,t) + k_j^i \quad \forall\ j \neq i \end{cases} \quad (3.1)$$

We have obtained the numerical solution of (3.1) using a suitable modification of the method introduced in II.1. We have studied the optimization of the system during a period of 192 hours and as a result, we have obtained feedback policies giving, with a discretization step length of 1 hour, the power of operation of each generator. For the numerical computations, we have used the minicomputer PDP 11/23.

REFERENCES

The complete list of references is included in [4].

[1] R. GONZALEZ: Sur la résolution de l'équation de Hamilton-Jacobi du contrôle déterministe. Cahier de Mathematique de la Decision 8029-8029 bis. Ceremade. Université de Paris-Dauphine, France, 1980.

[2] R. GONZALEZ: Sur la l'existence d'une solution maximale de l'équation de Hamilton-Jacobi, C.R.A.S. Paris (1976), Serie A, 1287-1290.

[3] R. GONZALEZ, E. ROFMAN: An algorithm to obtain the maximum solution of the Hamilton-Jacobi equation. Lectures Notes in Control and Information Sciences, Vol. 6, Springer-Verlag, 1978.

[4] R. GONZALEZ, E. ROFMAN: Rapport de Recherche, INRIA - Rocquencourt-France, to appear in 1982.

On the Computational Complexity of Clustering and Related Problems

Teofilo F. Gonzalez
Programs in Mathematical Sciences
The University of Texas at Dallas
Richardson, Texas 75080/USA

Abstract

The problem of clustering a set of n points into k groups under various objective functions is studied. It is shown that under some objective functions clustering problems are NP-hard even when the points to be grouped are restricted to lie in the two dimensional euclidean space. Our results can be extended to show that their corresponding approximation problems are also NP-hard. It is shown that some restricted graph partition problems are also NP-hard. Keywords: NP-complete problems, approximation algorithms, clustering problems.

I. INTRODUCTION

The problem of clustering a set of objects arises in many disciplines. Because of the wide range of applications, there are many variations of this problem. The main difference between these clustering problems is in the objective function. Research in different fields of study during the past thirty years has produced a long list of clustering algorithms. However, very little is known about the merits of these algorithms. Even simple questions regarding to the computational complexity of most clustering problems have not yet been answered. In this paper, we study the computational complexity of typical clustering problems.

In what follows, we define some of the typical clustering problems we are interested in studying. Let $G=(V,E,W)$ be a weighted undirected graph with vertex set V, edge set E and a disimilarity or weight function $W:E \to R_0^+$ (the set of non-negative reals). A k-split of the set of vertices V is a set of nonempty vertex subsets $B_1,B_2,\ldots,B_k$ such that $\cup B_i=V$. The sets B_i in a k-split are called clusters. The clusters are said to be nonoverlapping when $\sum|B_i| = |V|$. In what follows, we shall concentrate only on nonoverlapping clustering problems. An objective function, f: $B_1,B_2,\ldots,B_k \to R_0^+$, is defined for each k-split. For k-split $B_1,B_2,\ldots,B_k$, we define S_ℓ as the sum of the weights assigned to the edges adjacent to any pair of nodes in set B_ℓ, i.e., $S_\ell = \sum_{\substack{i,j \in B_\ell \\ \{i,j\} \in E}} W(\{i,j\})$. M_ℓ denotes the maximum weight assigned to an edge whose endpoints are vertices in cluster B_ℓ, i.e., $M_\ell = \max_{\substack{i,j \in B_\ell \\ \{i,j\} \in E}} \{W(\{i,j\})\}$. Some typical objective functions are shown in Table 1.

$f(B_1,B_2,\ldots,B_k$	
$(\sum\sum)$	$\sum_{\ell=1}^{k} S_\ell$
$(\sum 1/\|\cdot\|\sum)$	$\sum_{\ell=1}^{k} S_\ell/\|B_\ell\|$
$(M\sum)$	$\max_{1\le\ell\le k}\{S_\ell\}$
$(\sum M)$	$\sum_{\ell=1}^{k} M_\ell$
$(M\,M)$	$\max_{1\le\ell\le k}\{M_\ell\}$
(D)	$\sum_{\ell=1}^{k}\sum_{i\in B_\ell}\|\|x_i - m_\ell\|\|^2$, where the set of points (vertices) are in m-dimensional space, $\|\|x_k\|\| = \sqrt{{}_\ell\Sigma((x_k)_\ell)^2}$ and m_ℓ is the centroid, i.e., $m_\ell = (1/\|B_\ell\|)\sum_{i\in B_\ell} x_i$.

Table 1. Objective Functions

A clustering problem has one of the following forms:

(P1) Given a graph G, an objective function f and an integer k, find a k-split with least objective function value, i.e., find a k-split $(B_1^*,B_2^*,\ldots,B_k^*)$ such that $f(B_1^*,B_2^*,\ldots,B_k^*) = \min\{f(B_1,B_2,\ldots,B_k)\,|\,(B_1,B_2,\ldots,B_k)$ is a k-split for G$\}$.

(P2) Given a graph G, an objective function f and a real w, find for the least value of k a k-split with objective function value less than or equal to w, i.e., find a k-split $(B_1{}^*,B_2^*,\ldots,B_k^*)$ such that $f(B_1^*,B_2^*,\ldots,B_k^*) \le w$ and $f(B_1,B_2,\ldots,B_{k'}) > w$ for all k'-splits with $k' < k$.

(P3) Given a graph G, an objective function f, an integer k' and a real w. Is there a k-split $(B_1,B_2,\ldots,B_k)$ with objective function value $\le w$ for some $k \le k'$?

It can be easily shown that the decision problem P3 is computationally not harder than P1 and P2, i.e., any algorithm which solves P1 or P2 can be used to solve P3. This relation implies that if problem P3 is NP-complete then both P1 and P2 are NP-hard. In what follows when we refer to optimization clustering problems, it is implied that we refer to problems of the form P1. Whenever we wish to consider problems in the form P2, we shall state it explicitly.

An m-dimensional clustering problem is one in which the vertices of G are points in the m-dimensional euclidean space, the set of edges is complete and the weight of each edge is given by the euclidean distance between the two points it joins, i.e., $W(x_i,x_j) = ||x_i - x_j||$ where $||x_k|| = \sqrt{{}_\ell\Sigma\,((x_k)_\ell)^2}$.

We shall refer to a clustering problem as an α - $\beta\gamma$ problem, where $\alpha \in \{2,3,\ldots,k\}$ is the number of clusters; β means that it is either a β-dimensional clustering problem ($\beta \in \{1,2,\ldots,m\}$) or that the problem has been defined over an arbitrarily weighted graph (β=g); and $\gamma \in \{\Sigma\Sigma,\Sigma 1/|\cdot|\Sigma,M\Sigma,\Sigma M,MM,D\}$ is the objective function (see Table 1). For example, k-2ΣΣ indicates that the number of clusters k is an input to the problems; it is a 2-dimensional euclidean problems; and the objective function is ΣΣ (see Table 1). Note that any algorithm which solves the k-2ΣΣ problem will also solve the 2-2ΣΣ problem, but the converse is not true. In the 2-2ΣΣ problem, the set of vertices in G is always partitioned into two clusters whereas in the k-2ΣΣ problem the set of vertices in G will be partitioned into k clusters, where k could be any integer greater than 1.

Let us now define the k-maxcut problem. This problem is similar to the k-gΣΣ problem, but instead of finding a nonoverlapping k-split minimizing the sum of the weights of the edges inside a cluster, the objective is to find a nonoverlapping k-split maximizing the sum of the weights of the edges between clusters [SG,K and GJS].

A reader not familiar with NP-complete problems and approximate solutions is referred to [HS,GJ2 and K]. Our notation is that of [HS].

It is simple to prove that for any k, the k-gΣΣ problem is computationally identical to the k-maxcut problem, i.e., any algorithm solving one of these problems will also solve the other problem. The k-maxcut problem for k = 2 was shown to be NP-hard in [K]; in [SG] it was shown to be NP-hard for k > 2; and in [GJS] it was shown to be NP-hard for k = 2 even when the weight of every edge is zero or one. Hence, k-gΣΣ is NP-hard. Sahni and Gonzalez [SG] showed that there is an efficient (1/k)-approximation algorithm for the k-maxcut problem, whereas the k-gΣΣ ε-approximation problem is NP-hard. Using the same approach as the one in [SG], one can show that k-gΣM, k-gMΣ, k-gMM, k-gΣ1/|•|Σ and their corresponding ε-approximation problems are also NP-hard.

Fisher [F] showed that the k-1D problem can be solved in polynomial time. This was shown by first proving that in every problem instance there exists an optimal solution with the property that the convex hulls of every pair of distinct clusters are disjoint. This reduces the problem to one that can be solved by dynamic programming procedures. Bodin [Bd] extended this approach to solve other clustering problems. A similar approach was used by Brucker [Br] to show that the k-1ΣM, k-1MM and k-1Σ1/|•|Σ can be solved in polynomial time. The k-1ΣM problem can also be solved by reducing it to the problem of finding the largest k gaps [Br], which can be solved in O(n log n) time. When k is some fixed constant, finding the largest k gaps can be solved in linear time [G1]. Bock [Bk] showed that the 2-mΣ1/|•|Σ problem can be solved in polynomial time. This was shown by first proving that every instance of the k-mΣ1/|•|Σ problem has an optimal solution with the property that the convex hulls of every pair of distinct clusters are disjoint. The 2-gMM problem can be solved efficiently by reducing the problem to that of testing whether a graph is bipartite or not [Br]. Gonzalez [G2] showed that the k-2(*) problem can be solved efficiently when k is some fixed constant and (*) represents objective functions with some given properties.

For general graphs, most clustering problems are NP-hard. On the other hand, 1-dimensional clustering problem can be solved efficiently. The complexity of most 2-dimensional clustering problems is not known. In this paper, we study the computational complexity of exact and approximate solutions to these problems.

For optimization problems of the form P2, one can show that the k-gΣΣ ε-approximation problem is computationally identical to the k-maxcut ε-approximation problem. For k-gΣΣ, k-maxcut, k-gΣ1/|•|Σ, k-gMΣ, k-gΣM and k-gMM the 1-approximation problem is NP-hard. The proof follows the same approach as the one in [SG] but uses the result in [GJ1], which states that the 1-approximation problem for graph coloration is NP-hard.

Algorithms for other clustering problems appear in [AM], [JL], [S1], [S2], [FV], [DH], [M], [Sh] and [R].

In section II, we show that the k-2MM problem is NP-hard. The same reduction is then used in section III to show that the following problems are also NP-hard: k-2MΣ, k-2MM 1.36-approximation and k-2MΣ 1.16-approximation.

II. The Complexity of the k-2MM Decision Problem

In this section it is shown that the k-2MM decision problem is NP-complete. This result is obtained by reducing a restricted version of the exact cover by three sets problem to it.

The exact cover by three sets (XC3) problem was shown to be NP-complete in [GJ3] and is defined as follows:

> Exact Cover by Three Sets(XC3): Given a finite set of elements $X=\{x_1,x_2,\ldots x_{3q}\}$ and a collection of 3-element subsets of X, $C=\{(x_{i_\ell},x_{j_\ell},x_{k_\ell})\mid 1 \le \ell \le m\}$, in which no element in X appears in more than three subsets. The problem consists of determining whether C has an exact cover for X, i.e., a subcollection $C' \subseteq C$ such that every element in X occurs in exactly one member of C'.

The restricted version of this problem, to be used in our reduction, is denoted RXC3. This problem is exactly like the XC3 problem, except that each element in X appears in exactly three subsets of C. RXC3 is shown to be NP-complete in [G2].

In order to simplify the presentation of our result, we begin by showing that the k-gMM decision problem is NP-complete (lemma 1). The construction used in this lemma is then modified to show thet the k-gMM decision problem is NP-complete even when the input graph, after deleting all edges with weight different than one, is planar and no node is of degree greater than six (lemma 2). We then show how this result can be used to prove that the k-2MM decision problem is NP-complete (theorem 1). The reduction RXC3 α k-gMM is identical to the one in [GJ2], which was used to show that partition of a graph into triangles is NP-complete.

Lemma 1: The decision problem k-gMM is NP-complete.

Proof: It is simple to show that the decision problem k-gMM can be solved in nondeterministic polynomial time. We now show that RXC3 α k-gMM.

Given an instance, (X,C), of the restricted exact cover by 3-sets problem, we construct an instance of the k-gMM decision problem which we denote KG. KG=(G=(V,E, W),k,d) is defined as follows:

Vertex set: There is a vertex (v_i) for each element of set X and nine vertices, $(a_{\ell,1},b_{\ell,1},c_{\ell,1},\ldots,a_{\ell,3},b_{\ell,3},c_{\ell,3})$, are introduced for each 3-element subset of X in C.

Edge set: The set of edges is complete,i.e., for every pair of vetices $i\neq j$ edge $\{i,j\}$ is in E.

Weights: For each 3-element subset of X in C, eighteen edges will get a weight of one. The edges introduced for $(x_{i_{\ell,1}},x_{i_{\ell,2}},x_{i_{\ell,3}})$ C are shown in figure 1. All other edges are given the weight of two.

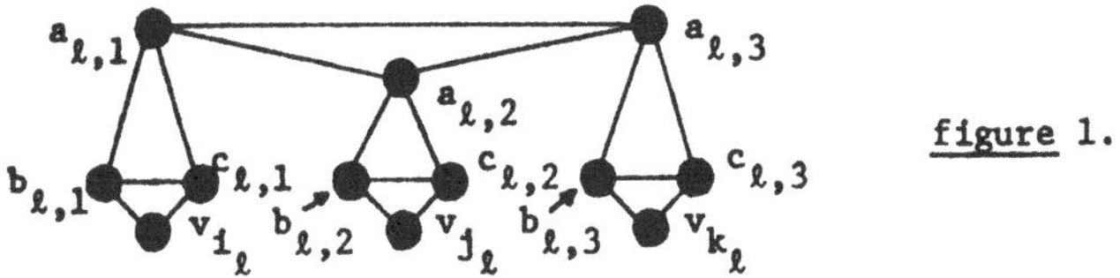

figure 1.

The maximum number of clusters, k, is 3m+q. The maximum weight for an edge inside a cluster, d, is one.

In order to complete the proof of the lemma it is only required to show that KG has a k-split with objective function value $\leq$ d iff (X,C) has an exact cover, since the construction of KG can be carried out in polynomial time.

Claim: KG has a k-split with objective function value $\leq$ d iff (X,C) has an exact cover.

Proof: First of all it is shown that if (X,C) has an exact cover, then KG has a k-split $(B_1,B_2,\ldots,B_k)$ with objective function value $\leq d=1$. Let C' be any exact cover for (X,C). Assume without loss of generality that $(x_{i_{1,1}},x_{i_{1,2}},x_{i_{1,3}}),\ldots,$ $(x_{i_{q,1}},x_{i_{q,2}},x_{i_{q,3}})$ are the elements in C which form an exact cover C'. Let $B_{\ell,j}=\{b_{\ell,j},c_{\ell,j},v_{i_{\ell,j}}\}$ for $1\leq j\leq 3$ and $1\leq \ell\leq q$; let $B_{\ell,4}=\{a_{\ell,1},a_{\ell,2},a_{\ell,3}\}$ for $1\leq \ell\leq q$; and let $B_{\ell,j}=\{a_{\ell,j},b_{\ell,j},c_{\ell,j}\}$ for $1\leq j\leq 3$ and $q+1\leq \ell\leq m$. It is simple to show that $(B_{1,1},\ldots,B_{q,4},B_{q+1,1},\ldots,B_{m,3})$ is a k-split with objective function value equal to d for KG.

In order to complete the proof of the claim it is only required to show that if KG has a k-split with objective function $\leq d=1$, then (X,C) has an exact cover. Let $B_1,B_2,\ldots,B_k$ be a k-split with objective function value $\leq$d for KG. Since no four nodes are completly connected by edges with a weight of one and since the number of nodes in KG is 3*k, we have that each B_i must have exactly three nodes. Let

$\gamma_z=(\ell,j)$, for $z=1,\ldots,3q$, if vertex v_{i_z} is in the same cluster with $b_{\ell,j}$ and $c_{\ell,j}$. It can be easily shown that: i) If $\gamma_z=(\ell,j)$ for some z, then $a_{\ell,j}$ is not in the same cluster with $b_{\ell,j}$ and $c_{\ell,j}$.

and ii) If for all z $\gamma_z\neq(\ell,j)$ then $a_{\ell,j}, b_{\ell,j}$ and $c_{\ell,j}$ are in the same cluster.

We now show that if for some z, $\gamma_z=(\ell,j)$ then there exists z_i, $1\leq i\leq 3$, such that $\gamma_{z_i}=(\ell,i)$. Let j_1 and j_2 be such that $\{j_1,j_2,j\}=\{1,2,3\}$. The construction rules together with the fact that each cluster has exactly three nodes implies that $a_{\ell,j}$ can be in a cluster only with either $b_{\ell,j}$ and $c_{\ell,j}$ or a_{ℓ,j_1} and a_{ℓ,j_2}. Since i) holds true for $\gamma_z=(\ell,j)$, it must then be that $a_{\ell,1}, a_{\ell,2}$ and $a_{\ell,3}$ are in the same cluster. This fact together with i) and ii) imply that there exists z_i, $1\leq i\leq 3$, such that $\gamma_{z_i}=(\ell,i)$.

Now, let $A=\{\ell\,|\,\gamma_z=(\ell,j)\}$. Clearly $|A|=q$. Also, it is simple to see that $C'=\{(x_{i_{\ell,1}},x_{i_{\ell,2}},x_{i_{\ell,3}})\,|\,\ell\in A\}$ is an exact cover for C. This completes the proof of the claim and the lemma. □

Before proving our next result, we outline the construction to be used in it. First of all, the construction in lemma 1 (figure 1) is replaced by the one given in figure 2.

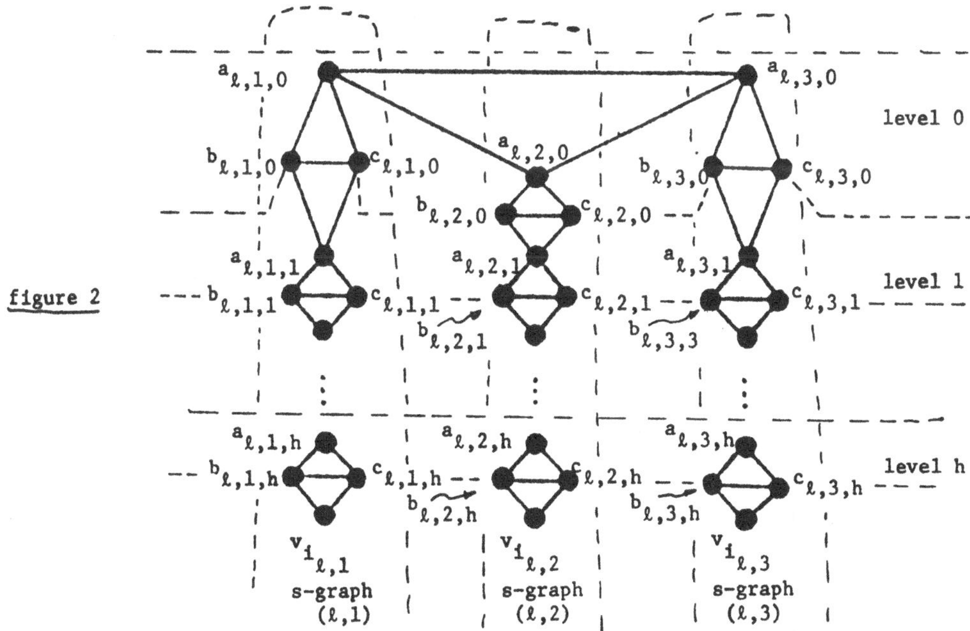

figure 2

The subgraph induced by the set of nodes $a_{\ell,j,z}, b_{\ell,j,z}, c_{\ell,j,z}, v_{i_{\ell,j}}$ $0\leq z\leq h$ is called s-graph(ℓ,j). For $z=0,1,\ldots,h$, nodes $a_{\ell,j,z}, b_{\ell,j,z}$ and $c_{\ell,j,z}$ are said to be in level z. The weight assigned to all edges introduced by the rule implied in by figure 2 is one.

It is simple to show that not all the graphs constructed by using the above rule, starting with an instance of RXC3, are planar. In order to guarantee planarity, we shall modify our construction rule. h is selected in such a way that at each level z ($z \geq 1$) only two adjacent s-graphs cross and **after** level h all the s-graphs that include node v_j are adjacent to each other. The crossing of the two s-graphs at level z is handled by applying the transformation shown in figure 3.

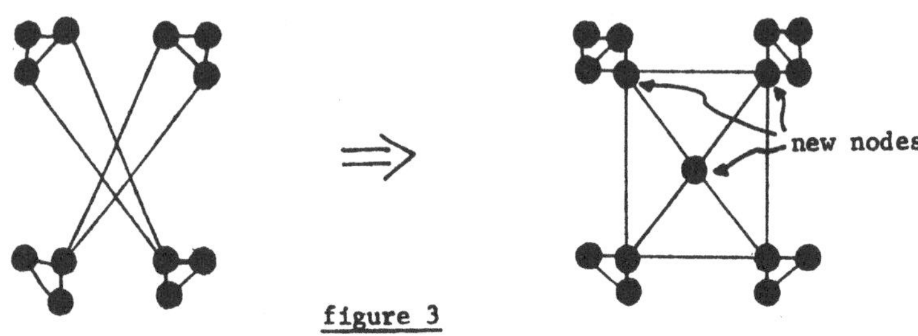

figure 3

Lemma 2: The decision problem k-gMM is NP-complete even when the input graph, after deleting all the edges with a weight different than one, is planar and no node is of degree greater than six.

Proof: The construction is as outlined above and the proof is similar to the proof of lemma 1. □

The subgraphs, in figure 2, consisting of two triangles placed side by side are called diamonds. The ends are the two nodes of degree two in it. It should be clear that two diamonds connected in series can replace any diamond and the resulting construction can also be used in lemma 2. This transformation can be carried out any number of times, as long as the total transformation takes polynomial time.

In the final transformation we replace the constructions implied in figures 1,2 and 3 by the one in figure 4.

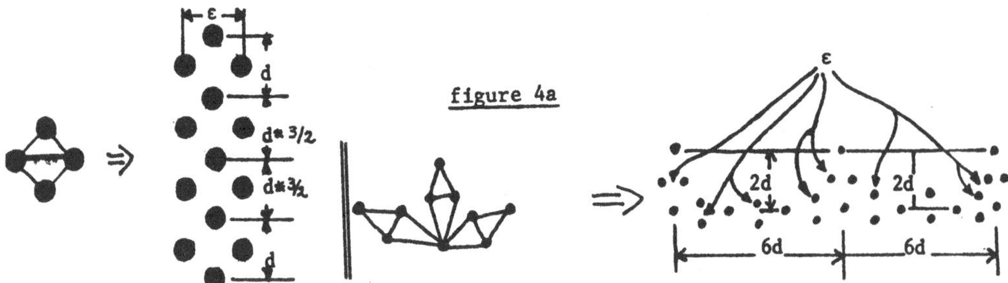

figure 4a

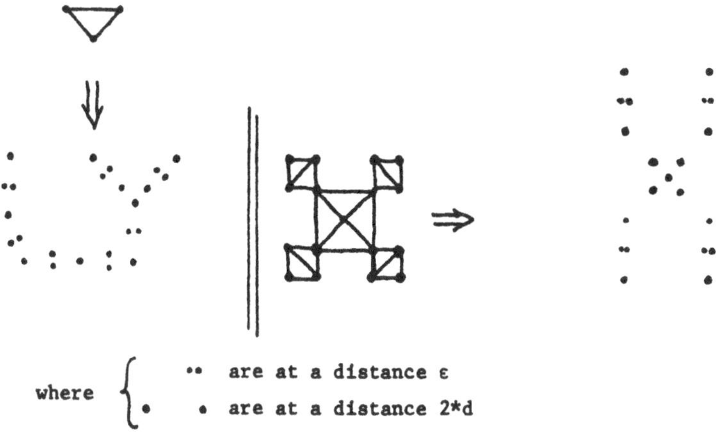

figure 4b.

After taking care of some simple details, one can show that two points are at a distance $\leq$ d+ε iff these two points had an edge between them with a weight of one in the construction used in lemma 2 (after adding several diamonds as shown in figure 4).

Theorem 1: The k-2MM problem is NP-complete.

Proof: The construction used in this proof follows the rules shown in figure 4 and the proof follows the same lines as the proof of lemma 2. □

III. The Complexity of Related Problems.

After a careful examination of the construction rules shown in figure 4, one can show that the closest three points not at a distance $\leq$ d + ε of each other, are at least $1/\sqrt{2}$ units apart. Using this fact together with the techniques used in [SG], one can prove the following theorem.

Theorem 2: The k-2MM $(1/\sqrt{2})$-approximation problem is NP-hard. □

The proofs and constructions of the next two theorems are similar to the ones in theorems 1 and 2. For brevity they will not be included.

Theorem 3: The k-2MΣ decision problem is NP-complete. □

Theorem 4: The k-2MΣ (1.16)-approximation problem is NP-hard. □

The construction used in section II can be easily adapted to show that partition of a graph into triangles is NP-complete even when the graphs to be partitioned are planar and no node is of degree ≥ 6.

(The formal proofs of our theorems will appear in a subsequent paper.)

References

[AM] Augustson, J. G. and J. Minker, "An Analysis of Some Graph Theoretical Cluster Techniques," J.ACM, 17,571-588,(October 1970).

[Bk] Bock, H. H., "Automatische Klassifikation," Vandenhoek und Ruprecht, Gottingen, 1974.

[Bd] Bodin, L. D., "A Graph Theoretic Approach to the Grouping of Ordering Data," Networks, 2, 307-310, (1972).

[Br] Brucker, P. "On the Complexity of Clustering Problems," in R. Henn, B. Korte and W. Oletti (eds), Optimiening and Operations Research, Lecture Notes in Economics and Mathematical Systems, Springer, Berlin (1977).

[DH] Duda, R. and P. Hart, "Pattern Classification and Scene Analysis," John Wiley and Sons, New York, 1973.

[FV] Fisher, L. and J. Van Ness, "Admissible Clustering Procedures," Biometrica, 58:91-104, 1971.

[F] Fisher, W. D., "On Grouping for Maximum Homogeneity, "JASA, 53:789-798,1958.

[G1] Gonzalez, T., "Algorithms on Sets and Related Problems," Technical Report 75-15, The University of Oklahoma, 1975.

[G2] Gonzalez, T.,Manuscript in preparation.

[GJ1] Garey, M. R. and D. S. Johnson, "The Complexity of Near-Optimal Graph Coloring," JACM, 23, 1, 43-69, (Jan 1976).

[GJ2] Garey, M. R. and D. S. Johnson, "Computers and Intractability: A Guide to the Theory of NP-Completeness," W. H. Freeman and Company, San Francisco, 1980.

[GJ3] Garey, M. R. and D. S. Johnson, Unpublished results referenced in [GJ2].

[HS] Horowitz, E. and S. Sahni, "Fundamentals of Computer Algorithms," Computer Science Press, Inc., 1978.

[JL] Johnson, D. B. and J. M. Lafuente, "Controlled Single Pass Classification Algorithm with applications to Multilevel Clustering," Scientific Report #ISR-18, Information Science and Retreival, Cornell University, Oct 1970.

[R] Rohlf, F. J. "Single Link Clustering Algorithms," RC 8569 (#37332) Research Report, IBM, T. J. Watson Research Center, Nov. 1980.

[K] Karp, R. M., "Reducibility Among Combinatorial Problems," In Complexity of Computer Computations, R. E. Miller and J. W. Thatcher, Eds, Plenum Press, N. Y. 1972, p.p. 85-104.

[M] Meisel, W. S., "Computer-Oriented Approaches to Pattern Recognition," Academic Press, New York, 1972.

[SG] Sahni, S. and T. Gonzalez, "P-Complete Approximation Problems," JACM, 23, 555-565, 1976.

[S1] Salton, G. "The Smart Retreival System, Experiments in Automatic Document Processing," Prentice-Hall, New Jersey (1971).

[S2] Salton, G., "Dynamic Information and Library Processing," Prentice-Hall, New Jersey (1975).

[Sh] Shamos, M.I., "Geometry and Statistics: Problems at the Interface," in J. F. Traub (ed), Algorithms and Complexity: New Directions and Recent Results, Academic Press, New York, 251-280, 1976.

THE APPLICATION OF VECTOR MINIMISATION TECHNIQUES IN THE ANALYSIS OF MULTILOOP NONLINEAR FEEDBACK SYSTEMS.

J.O. Gray and N.B. Nakhla
Dept of Electrical Engineering
University of Salford, Salford M5 4WT,
ENGLAND.

1. INTRODUCTION

Harmonic linearisation procedures of first order have been extensively used for the analysis and design of nonlinear multivariable systems [1,2,3,4,5], based on the assumption that the input to each nonlinearity is a pure sinusoid. This requirement limits the field of applicability of such elementary describing function methods with their useful graphical interpretations to systems that strictly satisfy the low pass frequency characteristic.

Recently [6,8] , higher order solutions have been derived for a limited class of multivariable systems containing a set of bounded separable nonlinear elements,where the effect of a single superharmonic signal component is included. A successive approximation technique is used to determine restricted regions in the parameter space, for the prediction of progressively higher order harmonic balance points. The method used is first to obtain approximate values for the fundamental frequencies and amplitudes of any possible oscillations, by employing single harmonic approximants to the nonlinear elements when seeking solutions to the linearised system characteristic equation in the vicinity of the $j\omega$ axis. These data are subsequently used to derive estimates of the corresponding superharmonic signal content, in the determination of overall harmonic balance.

The object of this paper is to discuss the numerical procedures incorporated for the estimation of a single superharmonic signal as well as outlining possible strategies for the extension to higher dimensionality problems, either in terms of a larger number of loops or when more superharmonics are included. A graphical procedure is presented, which yields information concerning loop interaction effects at or near possible harmonic balance points. These results are particularly significant because the exact contribution from both the diagonal and off diagonal system elements now involve the effects of the superharmonic signals and thus represent an improvement on earlier results [4]. The procedure is made robust by allowing for possible parameter variations as well as errors related to the data intervals chosen, which errors in this case inherently account for the presence of harmonics.

2. A NUMERICAL APPROACH FOR THE ANALYSIS OF HARMONIC BALANCE CONDITIONS

Consider the multivariable nonlinear system configuration of Fig.(1). A successive approximation technique [8] has proved successful in the prediction of oscillatory modes when the elementary single harmonic describing function analysis fails. A main feature of the procedure adopted is the use of an optimisation strategy for the determination of possible solutions to the harmonic balance equations. The solution is assumed to be of the form of a truncated Fourier series, some of the coefficients of which are adjusted using an iterative algorithm to minimise certain error functions. The number of terms assumed in the series relates to the order of the solution required. In the implementation of an optimisation procedure, the two essential aspects are:

a) The definition of appropriate error functions to be minimised;

b) The choice of a suitable optimisation algorithm;

A detailed discussion of each aspect is now given.

a) In the estimation of the superharmonic signal content when seeking possible harmonic balance conditions, the error functions used for minimisation are derived [6] with reference to a graphical interpretation of the equation set

$$\min. \{(1+t_{ii}^{(r)}) + \sum_{\substack{k=1\\k\neq i}}^{2} \left| t_{ik}^{(r)} \frac{\tilde{a}_k^{(r)}}{\tilde{a}_i^{(r)}} \right| \varepsilon^{j\Psi_k^{(r)}} \} = 0 \qquad \ldots\ldots\ldots (1)$$

$$\forall \Psi_k^{(r)} \qquad 0 \leqslant \Psi_k^{(r)} < 2\pi$$

for $i = 1,2$ and $r = 1,p$;
which is taken initially for ease of computation, as an approximation to the exact harmonic balance equation set

$$(1+t_{ii}^{(r)})\ \tilde{a}_i^{(r)} + \sum_{\substack{k=1\\k\neq i}}^{2} t_{ik}^{(r)}\ \tilde{a}_k^{(r)} = 0 \qquad \ldots\ldots\ldots (2)$$

for $i = 1,2$ and $r = 1,p$, where moduli is taken to simplify the computation and $t_{ik}^{(r)}$ are elements of the system metrix $T^{(r)}$ and $\tilde{a}_k^{(r)}$ are signal values. These parameters are completely defined in the list of symbols given below.

From Fig. (2) the error functions $f_i^{(r)}$ and $h_i^{(r)}$ are defined by

$$f_i^{(r)} = u_i^{(r)} \cos \emptyset_i^{(r)}$$
$$h_i^{(r)} = u_i^{(r)} \sin \emptyset_i^{(r)} \qquad \ldots\ldots\ldots (3)$$

Solutions are sought, which minimise the summation

$$S_1 = \sum_{\substack{r=1\\r=p}} \sum_{i=1}^{2} \{ (f_i^{(r)})^2 + (h_i^{(r)})^2 \} \qquad \ldots\ldots\ldots (4)$$

The use of both $f_i^{(r)}$ and $h_i^{(r)}$ as opposed to the possible minimisation of $u_i^{(r)}$ tends to ensure higher efficiency in the operation of the iterative procedure since it imposes a restriction on the area within which the vectors $U_i^{(r)}$ exist between two successive iterations, thus resulting in a significant limitation on the allowable parameter variations. This is illustrated in Fig. (3) in which ${}_jU_i^{(r)}$ is taken to be the error vector at the jth iteration, ${}_ju_i^{(r)}$ is its magnitude and ${}_jf_i^{(r)}$, ${}_jh_i^{(r)}$ are its components. Now, if the criterion is to minimise $u_i^{(r)}$, then the error vector in the (j+1)th iteration, given by ${}_{j+1}U_i^{(r)}$ can be anywhere within the circle of radius ${}_jU_i^{(r)}$ as shown in Fig. (3), whereas seeking the minimisation of both $f_i^{(r)}$ and $h_i^{(r)}$ restricts the area that can contain ${}_{j+1}U_i^{(r)}$, to the rectangle with dimensions $(2{}_jf_i^{(r)} \times 2{}_jh_i^{(r)})$ contained within that circle.
The inconsistency which may arise from there being solutions to equation (1) which are not possible solutions to equation (2) can be overcome by seeking roots of the characteristic equation of the second order linearised system. Those solutions of equation (1) for which there are roots in the vicinity of the $j\omega$ axis, are then retained for future consideration.

In the determination of solutions to the exact harmonic balance equation set (2), the error functions are defined as:

$$v_i^{(r)} = \mathrm{Re}\,\{(1+t_{ii}^{(r)}) + \sum_{\substack{k=1\\k\neq i}}^{2} t_{ik}^{(r)} \frac{\tilde{a}_k^{(r)}}{\tilde{a}_i^{(r)}}\}$$

$$w_i^{(r)} = \mathrm{Im}\,\{(1+t_{ii}^{(r)}) + \sum_{\substack{k=1\\k\neq i}}^{2} t_{ik}^{(r)} \frac{\tilde{a}_k^{(r)}}{\tilde{a}_i^{(r)}}\} \qquad \ldots\ldots\ldots (5)$$

and the solution is determined, to minimise the sum of squares

$$S_2 = \sum_{\substack{r=1\\r=p}} \sum_{i=1}^{2} \{ (v_i^{(r)})^2 + (w_i^{(r)})^2 \} \qquad \ldots\ldots\ldots (6)$$

The definition of the error functions in equation (5) as well as equation (3) signifies the fact that they are related to certain error vectors in the complex frequency domain and therefore both parameters which define those vectors must be taken into account, for efficient optimisation purposes.

b) The minimisation procedure used is based on a method by Peckham [7], which minimises the sum of squares of nonlinear functions as defined above without the explicit evaluation of any derivatives of those functions; the latter is an advantage whenever the functions are not analytically defined. In general terms, a solution requires the determination of a vector of variables $\underline{R} = (R_1, R_2, \ldots R_n)$ which minimises the sum of squares S defined by

$$S(\underline{R}) = \sum_{k=1}^{M} \{F_k(\underline{R})\}^2 \qquad \ldots\ldots\ldots\ldots (7)$$

In the estimation of the superharmonic signal content i.e. when S_1 is minimised in equation (4), the vector $\underline{R}$ corresponds to the set of variables $\{a_\ell^{(r)}, \psi_\ell^{(r)};\ \ell=1,2\}$ i.e. n=4 and the set of error functions $\{\ f_\ell^{(r)},\ h_\ell^{(r)};\ell = 1,2$ and $r=1,p\}$ constitutes the set of functions $\{F_k(\underline{R});\ k=1,2,\ldots M\ \}$ where M = 8.

Alternatively, in the minimisation of equation (6), the 8 functions are defined by the set of error functions $\{v_\ell^{(r)},\ w_\ell^{(r)}, \ell=1,2$ and $r = 1,p\}$ which are to be minimised using 8 variables.

If each $F_k(\underline{R})$ is written in a linear approximation as

$$F_k \simeq H_k + \sum_{i=1}^{n} J_{ki}\, R_i \qquad \ldots\ldots\ldots (8)$$

where H_k is a constant and J_{ki} is a Jacobian element given by $J_{ki} = \partial F_k/\partial R_i$, then the M-element vector of functions $\underline{F}$ is given by:

$$\underline{F} = \underline{H} + J\underline{R} \qquad \ldots\ldots\ldots (9)$$

in which $\underline{H}$ is a vector of constants and J is the Jacobian matrix.

In the determination of the minimum $\underline{Y}$ given by the equation

$$J^T J \underline{Y} = -J^T \underline{H} \qquad \ldots\ldots\ldots (10)$$

a direct evaluation of J and $\underline{H}$ is avoided by using the function values at a set of $P \geqslant (n+1)$ points, as well as the corresponding vectors of variables at those points. If $\underline{R}^{(L)}$ is defined as the Lth such vector and $F_k^{(L)}$ is the value of kth function at this point, then formulae can be obtained for the coefficients of the linear approximation, J and $\underline{H}$, in terms of the $\underline{R}^{(L)}$ and $F_k^{(L)}$ (L=1,2...P and k=1,2...M), through a minimisation of the weighted sum of squares over this point set, of the difference between the linear approximation and the actual function values, given by

$$\sum_{L=1}^{P} W_L^2 \left(H_k + \sum_{i=1}^{n} J_{ki}\, R_i^{(L)} - F_k^{(L)}\right)^2 \qquad \ldots\ldots\ldots (11)$$

for k=1,2,...M; W_L is a weighting function chosen to give more emphasis to function values near the minimum and is defined by:

$$W_L^2 = 1 \Big/ \sum_{k=1}^{M} (F_k^{(L)})^2 \qquad \ldots\ldots\ldots (12)$$

This estimate of the position of the minimum then replaces the point of the set with the highest function value $S(\underline{R})$ and this constitutes an iteration. Initially, the set of P points consists of (n+1) points which correspond to that point defined by the initial estimates for the n variables as well as another n points generated around the latter, along each of the n dimensions.

3. STRATEGY FOR HIGHER DIMENSIONALITY PROBLEMS

In the following, a brief outline is given on the extension of the procedure to more general problems, namely that of a 2-loop system where 2 superharmonics need be considered as well as a 3-loop system with a single pth harmonic. In a third order analysis which involves two superharmonics, ranges of amplitudes of the fundamental ($a_\ell^{(1)}$, $\ell = 1,2$) are defined around the solution points at the second level of approximation. These data can be used for the estimation of the corresponding harmonic signals, using an indirect approach where the Fourier coefficients of the harmonics alone are adjusted and now 8 variables are used to minimise 12 error functions. This analysis is performed in 2 stages; in the first only the dominant superharmonic coefficients are iteratively determined to minimise all $f_i^{(r)}$ and $h_i^{(r)}$ for the fundamental and that harmonic alone i.e. 4 variables are used to minimise 8 error functions. In the second stage, the parameters related to this harmonic are adjusted iteratively around their values determined above, while seeking the parameter values for the next significant harmonic, to minimise all $f_i^{(r)}$ and $h_i^{(r)}$ for the fundamental and both harmonics simultaneously. This two-stage least squares estimation ensures the efficiency of the iterative procedure.

In the analysis of a 3-loop system, an approximate equation set similar to equation set (1), does not have the simple graphical interpretation of Fig.(2) and therefore cannot be successfully used for the superharmonic estimation. However, with reference to the harmonic balance system matrix equations, given by

$$\left[I + T^{(r)}\right] \hat{a}^{(r)} = 0 \qquad \ldots\ldots (13)$$

for r=1,p and if $D_{ki}^{(r)}$ is taken to be the cofactor of the element (k,i) in $\left[I+T^{(r)}\right]$, then the following relationships [9] hold at the point of harmonic balance.

$$\frac{\tilde{a}_1^{(r)}}{D_{i1}^{(r)}} = \frac{\tilde{a}_2^{(r)}}{D_{i2}^{(r)}} \qquad \frac{\tilde{a}_3^{(r)}}{D_{i3}^{(r)}} \qquad \ldots\ldots (14)$$

for i = 1,2,3.

The combination of 2 terms at a time in each row i so that $D_{ii}^{(r)}$ is a common term in the 2 resulting equations, yields a set of 6 equations. However, since these equations are intended for use within a hill climbing procedure where best results can be effected by keeping the error functions as independent of each other as possible, then the equations in that set can be reduced to the following:

$$\begin{bmatrix} -D_{13}^{(r)} & 0 & D_{11}^{(r)} \\ D_{22}^{(r)} & -D_{21}^{(r)} & 0 \\ 0 & D_{33}^{(r)} & -D_{32}^{(r)} \end{bmatrix} \begin{bmatrix} \tilde{a}_1^{(r)} \\ \tilde{a}_2^{(r)} \\ \tilde{a}_3^{(r)} \end{bmatrix} = 0 \qquad \text{.....(15)}$$

So, in the superharmonic estimation, values of $a_\ell^{(p)}$ and $\psi_\ell^{(p)}$ (ℓ=1,2,3) are sought, which give solutions to the equation set:

$$\min \{ D_{ii}^{(r)} + \left| D_{ik}^{(r)} \frac{\tilde{a}_i^{(r)}}{\tilde{a}_k^{(r)}} \right| \varepsilon^{j\Psi_i^{(r)}} \} = 0 \qquad \text{....(16)}$$

$$\forall \Psi_i^{(r)} \quad 0 \leq \Psi_i^{(r)} < 2\pi$$

for i=1,2,3 (respectively k=3,1,2) and r=1,p. Each equation in this set has the graphical interpretation shown in Fig.(4). In this case, there are 12 error functions to be minimised using 6 variables.

Initial approximations are determined for the parameters of all possible oscillations, which can be used in a direct optimisation procedure to seek solutions to equation (13), through a minimisation of the sum of squares.

$$S_1 = \sum_{\substack{r=1 \\ r=p}} \sum_{i=1}^{3} \{ (v_i^{(r)})^2 + (w_i^{(r)})^2 \} \qquad \text{.....(17)}$$

where $v_i^{(r)}$ and $w_i^{(r)}$ are defined by

$$v_i^{(r)} = \operatorname{Re} \{ (1+t_{ii}^{(r)}) + \sum_{\substack{j=1 \\ j\neq i}}^{3} t_{ij}^{(r)} \frac{\tilde{a}_j^{(r)}}{\tilde{a}_i^{(r)}} \}$$

$$w_i^{(r)} = \operatorname{Im} \{ (1+t_{ii}^{(r)}) + \sum_{\substack{j=1 \\ j\neq i}}^{3} t_{ij}^{(r)} \frac{\tilde{a}_j^{(r)}}{\tilde{a}_i^{(r)}} \} \qquad \text{.....(18)}$$

In this case 12 functions are to be minimised using 12 varibles. However, because the initial estimates are generally quite close to the final solution point, a problem of this order can be solved by the existing routines.

4. A COMPUTATIONAL PROCEDURE WITH A GRAPHICAL INTERPRETATION

It can be shown that the harmonic effects computed above can be incorporated within a sequential loop balance computational procedure similar to that of [4,5], to render higher accuracy limit cycle predictions and yield graphical plots of a similar nature in the frequency domain.

In the modified form of the sequential loop balance procedure, for a 2-loop system with a single pth significant harmonic, solutions are now sought to the harmonic balance equation set which is a subset of equation (2), that for which r=1. The equations are tested individually in a sequential manner, over a grid of parameters $(a_1^{(1)}, a_2^{(1)})$ which is the same as that used when seeking solutions to equation(1). Use is also made of the sets of matrices $T^{(1)}$ produced in that earlier computation, using corresponding values of $a_k^{(p)}$ and $\psi_k^{(p)}$ which minimise equation (4). Allowances are made within the algorithm for the possible parameter variations and the use of discrete data.

A useful graphical interpretation in the frequency domain can be given, using the $h_i^{(1)}$ - loci, where $h_i^{(1)}$ is the open loop transfer function for the fundamental signal in loop i, when all the other loops are closed and there are specified amounts of the pth superharmonic in circulation. Each point on the $h_i^{(1)}$ - loci is plotted as

$$h_i^{(1)} = t_{ii}^{(1)} + \sum_{\substack{j=1\\ j\neq i}}^{2} t_{ij}^{(1)} \frac{\tilde{a}_j^{(1)}}{\tilde{a}_i^{(1)}} \qquad \ldots.\ (19)$$

where the $t_{ij}^{(1)} \frac{\tilde{a}_j^{(1)}}{\tilde{a}^{(1)}}$ interaction vector represent the computed values for all loops $j\neq i$, of the interaction effects at or near balance condition. The procedure adopted in the determination of the $T^{(1)}$ matrices used to produce the displays, ensures that all the vectors in the critical regions of the frequency domain are displayed and hence the resultant plots should prove useful for design in a manner analogous to those given elsewhere [4].

5. EXAMPLE OF USE

Consider the feedback system shown in Fig.(5). In the determination of possible oscillatory modes, the sequential loop balance method based on the sinusoidal input describing function approximation does not yield any predictions, which result is supported by the displays of Fig.(6).

The successive approximation analysis however yields the results shown in Table(1), where the parameter values determined via a digital simulation are also given for comparison.

The computational procedure of section (4) yields the numerical results of Table(2) and the associated $h_i^{(1)}$ loci are shown in Fig.(8).

In Table(1) it is noted that the inclusion of the third harmonic effects results in accurate limit cycle predictions. The appropriate adjustments on the elements of $T^{(1)}$ due to these harmonics are reflected in the accurate predictions of Table(2) with a high nearest point index, which indicates high confidence in the result. This prediction is a major improvement on the sequential loop balance results that do not make the correct prediction as far as the existence of the limit cycle are concerned. This is to be expected because of the high amount of third harmonic which is in circulation in loop 2.

The interaction vector displays of Fig.(8) include the harmonic effects in the circulating signals and emphasise the large degree of signal interaction in loop 2. The major limit cycle occurs in loop 1 where there is an insignificant superharmonic component while the major superharmonic content is in loop 2. The plots suggest the use of a single phase - lead compensator in loop 1, to avoid possible region of limit cycle operation and such a design strategy would be a useful first step in any approach to system stability and the satisfaction of further closed loop system specifications.

6. CONCLUSION

The complex problem of predicting oscillatory modes in nonlinear multivariable systems has been successfully approached by a procedure of successive approximations where major use is made of combining the philosophies of harmonic balance and an optimisation technique. Several aspects involved in the implementation of the procedure have been considered in detail, as well as the strategies adopted in the extension of the method to the solution of more complex problems. The use of these accurate procedures in the enhancement of a computer graphical method has been discussed as a step towards their incorporation within the computer aided design suite for nonlinear systems which is currently under development on a Prime 550 computer at Salford University.

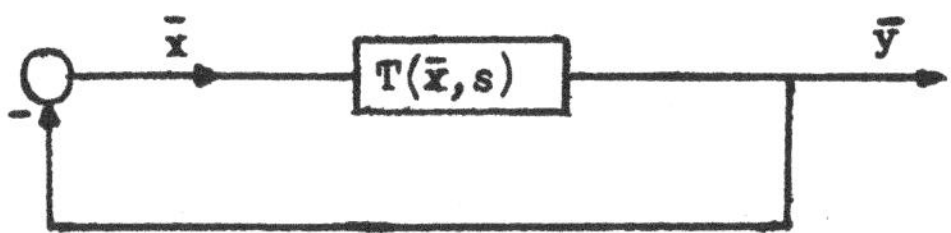

Fig.(1) Nonlinear system structure

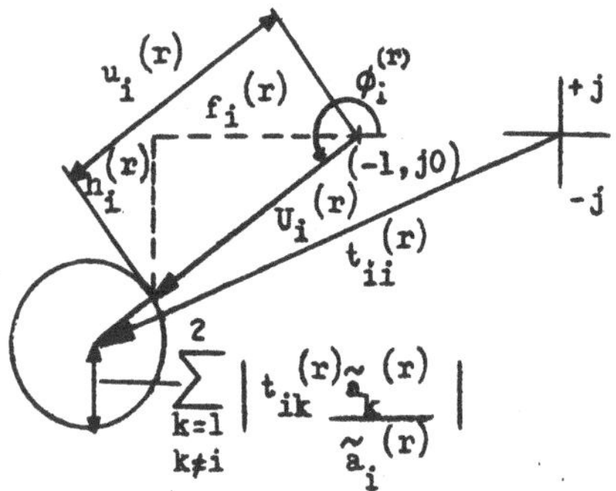

Fig.(2) Graphical Interpretation of equation (1)

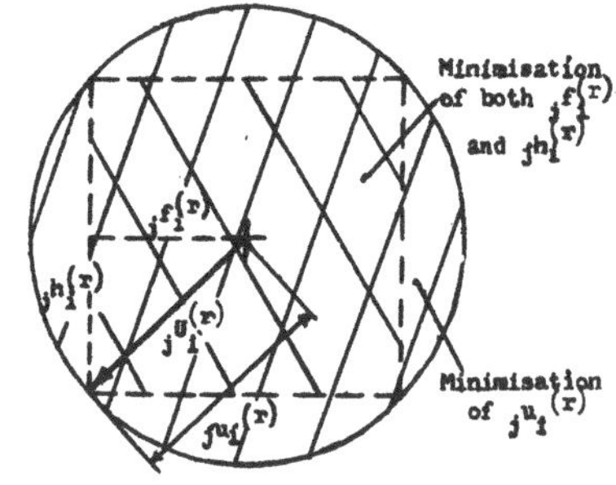

Fig.(3) Area for possible trajectories of $U_i^{(r)}$, based on two different criteria

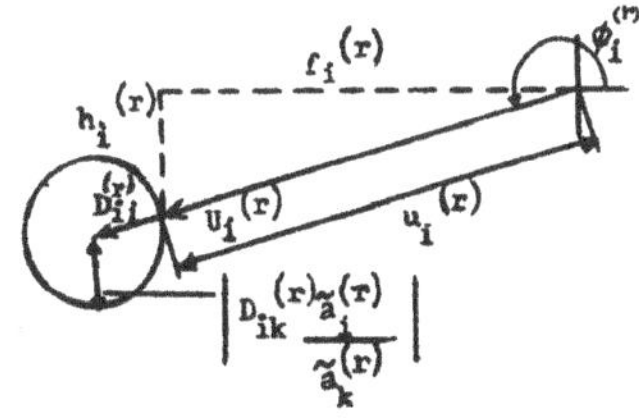

Fig.(4) Graphical Interpretation of equation (16)

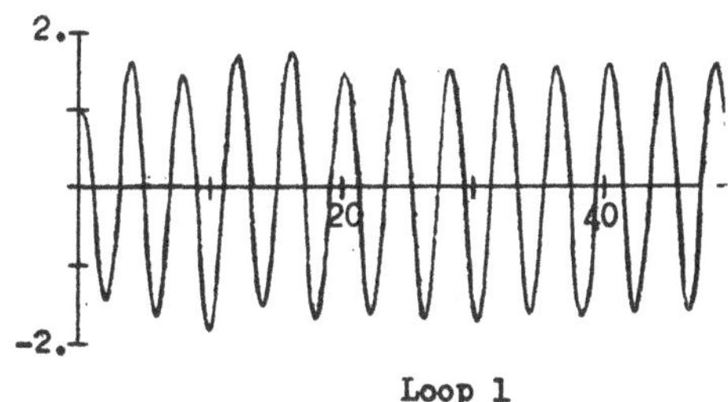

Loop 1

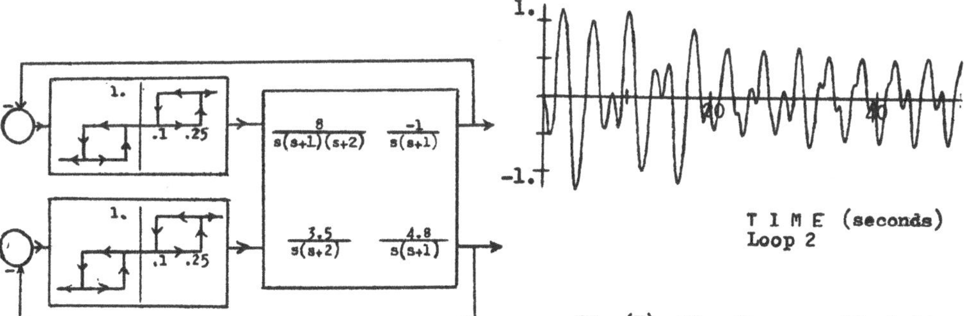

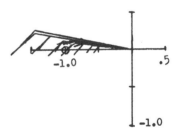

T I M E (seconds)
Loop 2

Fig.(7) Time Response Simulation Result

Fig.(5) System Configuration for the Example

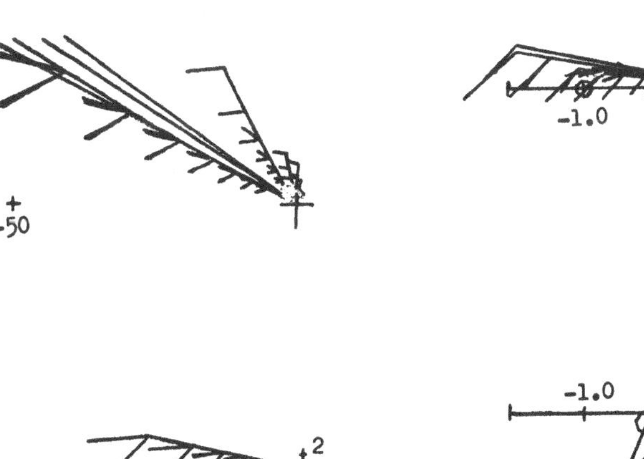

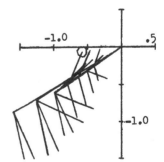

Fig.(6) Nyquist Interaction vector display using the sequential computational procedure; Single harmonic solution

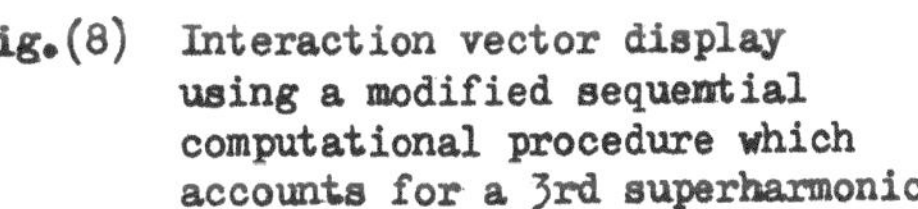

Fig.(8) Interaction vector display using a modified sequential computational procedure which accounts for a 3rd superharmonic

Parameter	Computed Result Based on the 3rd Harmonic Presence	Simulated Result
Frequency(rad/s)	1.55	1.547
Loop 1		
Amplitude of fundamental($a_1^{(1)}$)	1.608	1.606
% 3rd. harmonic	2.24	2.47
Relative phase of 3rd harmonic w.r.t. fundamental (rad.)	-1.53	-0.88
Loop 2		
Amplitude of fundamental($a_2^{(1)}$)	0.38	0.32
% 3rd. harmonic	32.4	25.7
Relative phase of 3rd harmonic w.r.t. fundamental (rad.)	2.17	2.6
Relative phase of $a_2^{(1)}$ w.r.t. $a_1^{(1)2}$ (rad.)	-2.65	-2.78

TABLE (1)

COMPARISON OF COMPUTED RESULTS AND SIMULATED RESULT FOR THE EXAMPLE

ω = 1.5 rad/s.	Minimum	Maximum	Nearest
Amplitude 1	1.44	2	1.728
Amplitude 2	0.26	1.94	0.325
Nearest point index = 83.7 %			

TABLE (2)

PREDICTED PARAMETERS FOR LIMIT CYCLE OPERATION, USING A SEQUENTIAL COMPUTATIONAL PROCEDURE WHICH ACCOUNTS FOR THE PRESENCE OF A 3RD SUPER-HARMONIC SIGNAL, IN THE EXAMPLE

List of Symbols [6] :

x_k periodic input signal to nonlinear elements n_{ik}, where

$$x_k = \sum_{r=1}^{q} a_k^{(r)} \varepsilon^{j(r\omega t+\theta_k^{(r)})}$$

$\tilde{a}_k^{(r)}$ the rth harmonic component of x_k, where

$$\tilde{a}_k^{(r)} = a_k^{(r)} \sin(r\omega t+\theta_k^{(r)})$$

$\theta_k^{(r)}$ the relative phase of $\tilde{a}_k^{(r)}$ w.r.t. an arbitrarily chosen reference which is usually taken to be $\tilde{a}_1^{(1)}$.

$t_{ik}^{(r)}$ the transmittance between the kth input and the ith output, determined using the matrices $G(jr\omega)$ and $N^{(r)}$, where the former corresponds to $G(s)$ at $s=jr\omega$ and the ikth element of the latter is given by

$$n_{ik}^{(r)} = \frac{C_{ik}^{(r)}}{a_k^{(r)}\varepsilon^{j\psi_k^{(r)}}}\{(a_k^{(1)},0),(a_k^{(p)},\psi_k^{(p)})\}$$

in the presence of a single pth superharmonic; $C_{ik}^{(r)}$ is the complex Fourier coefficient of the rth harmonic component at the output of nonlinearity n_{ik}.

$\hat{a}^{(r)}$ vector of the rth harmonic sinusoids, given by

$$\hat{a}^{(r)} = (\tilde{a}_1^{(r)}, \tilde{a}_2^{(r)}, \ldots \tilde{a}_m^{(r)})^\tau$$

τ indicates transposition.

REFERENCES

1. Mees, A.I. (1973). 'Describing Functions, Circle Criteria and Multiloop Feedback Systems'. Proc. IEEE, 120(1), 126-130.
2. Ramani, N. and Atherton, D.P. (1973). 'Frequency Response Methods for Nonlinear Systems'. Proc. Canadian Conf. Auto. Control, 9.2-1-9.2-15.
3. Atherton, D.P. (1981). 'Analysis and Design of Nonlinear Feedback Systems'. IEE Proc., 128, Pt. D, No. 5, Sept, 173-180.
4. Gray, J.O. and Taylor, P.M. (1979). 'Computer Aided Design of Multivariable Nonlinear Systems'. Automatica, 15, 281-297.
5. Gray, J.O. and Taylor, P.M. (1977). 'Frequency Response Methods in the Design of Multivariable Nonlinear Feedback Systems'. Proc. IFAC Symposium on Multivariable Technological Systems, Fredericton.
6. Gray, J.O. and Nakhla, N.B. (1981). 'Prediction of Limit Cycles in Multivariable Nonlinear Systems'. IEE Proc.,128, Pt.D,5,Sept,233-241.
7. Peckham, G. (1970). 'A New Method for Minimising a sum of Squares Without Calculating Gradients'. The Computer Journal, Vol. 13(4), 418-420.
8. Gray, J.O. and Nakhla, N.B. (1981). 'A Numerical Method for the Analysis of Harmonic Balance Conditions in Multiloop Nonlinear Systems'. Proc. IEE, Int. Conf. Control and Appl., Warwick, U.K., 301-306.
9. Wylie, C.R. 'Advanced Engineering Mathematics', McGraw-Hill, 1966, 3rd.Ed.,400-455.

SINGULAR STEADY STATE LQG PROBLEMS: ESTIMATION AND OPTIMIZATION

Violet B. Haas*
School of Electrical Engineering
Purdue University
West Lafayette, IN 47907, U.S.A.

Abstract. The duality principle that links linear quadratic control with state estimation is extended to include steady state singular problems.

1. Introduction. A signal process $z(\cdot)$ is modeled as the output of a linear, time invariant, dynamical system driven by stationary white noise having sample functions $w(\cdot)$:

$$\dot{x}(t) = A'x(t) + D'w(t),$$
$$z(t) = B'x(t), \quad t \geq 0. \tag{1.1}$$

Primes denote matrix transposition, $x(t)\varepsilon\mathcal{R}^n$, $w(t)\varepsilon\mathcal{R}^q$ and $z(t)\varepsilon\mathcal{R}^m$. The initial state is a random variable and for $s \geq 0$, $t \geq 0$, we suppose that

$$E[w(t)] = 0,\ E[x(0)] = 0,\ E\left[w(t)w'(s)\right] = I\delta(t-s),\ E\left[x(0)x'(0)\right] = P_0.$$

The measurement process is corrupted by stationary white noise with sample functions $v(\cdot)$, and is represented by

$$y(t) = z(t) + v(t), \tag{1.2}$$

where, $E[v(t)] = 0$, $E\left[v(t)v'(s)\right] = R\delta(t-s)$ and the noise processes $w(\cdot)$ and $v(\cdot)$ are uncorrelated with each other and each is uncorrelated with the initial state. The functions $v(\cdot)$ and $w(\cdot)$ are random elements of $L^2_m[0,T]$ and $L^2_q[0,T]$ respectively, for any $T > 0$. $L^2_m[0,T]$ is the set of functions square integrable on $[0,T]$ with values in $\mathcal{R}^m$. Matrices A, B, D, R, are constant and R is nonnegative definite and symmetric. Given the measurements $y(s)$ for $0 \leq s < t$, we seek that linear estimate, $\hat{x}(t)$ of $x(t)$ which provides the minimum, stationary, mean squared error of the state vector.

The assumption of additive white noise is a convenience. Frequently one desires to perform state estimation on a system containing random measurement errors which

*This research was supported by NSF grant ECS7918885.

are not white. In this event the noise can often be simulated by a "shaping filter," which is itself a dynamical system containing additive white noise, and the augmented system contains some output measurements which are noise free. Estimation and filtering problems in such cases have been considered by several authors, including Bryson and Johansen [1], Tse and Athans [2], Fairman [3] and Fogel and Huang [4]. These authors employ a variety of coordinate transformations to obtain solutions or redefine the estimator so it is not strictly a Kalman-Bucy filter. Because of this these authors could not see the duality principle relating this problem with an optimal control problem. The duality was found for time-varying systems by Ju and Haas [5]. Here we shall uncover this duality for the steady state.

The estimate $\hat{x}(t)$ is to be the state of a dynamical system,

$$\dot{\hat{x}}(t) = A'\hat{x}(t) + K\left[y(t) - B'\hat{x}(t)\right], \tag{1.3}$$

and the error of this estimate, $\tilde{x}(t) = x(t) - \hat{x}(t)$, satisfies,

$$\dot{\tilde{x}}(t) = A'\tilde{x}(t) + D'w(t) - K\left[y(t) - B'\hat{x}(t)\right]. \tag{1.4}$$

We wish to choose K so as to optimize the criterion,

$$J = \lim_{t\to\infty} E\left[\tilde{x}'(t)Q\tilde{x}(t)\right] \tag{1.5}$$

for some symmetric nonnegative definite matrix Q. As examples show, J does not necessarily possess a minimum when R is singular, but J always has an infimal value which we shall show can be approximated as closely as desired by an estimate of the form (1.3) if all unstable modes of A are either observable at the output or unaffected by additive noise at the input.

2. A geometric viewpoint when $R > 0$. Here we suppose that R is strictly positive definite. For convenience of notation let $\mathcal{X}$ denote the state space, so that $\mathcal{X} = \mathcal{R}^n$. Let $\mathcal{B}$ denote the range of B, let $\langle A|\mathcal{B}\rangle$ denote the subspace of controllable modes of the pair (A,B), and define $\eta = \bigcap_{i=1}^{n} \mathrm{Ker}(DA^{i-1})$, the subspace of unobservable modes of the pair (D,A). For details see chapters 1 and 3 of [6]. Let $\alpha(\lambda)$ denote the minimal polynomial of A and let $\alpha_+(\lambda)$ denote that factor of α which is associated with the unstable (closed right half plane) zeros. Define $\mathcal{X}^+ = \mathrm{Ker}\,\alpha_+(A)$. Then $\mathcal{X}^+$ is the subspace of unstable modes of A. We shall suppose that

$$\mathcal{X}^+ \subset \langle A|\mathcal{B}\rangle + \eta. \tag{2.1}$$

As applied to the system (1.1)-(1.2), relation (2.1) says that the unstable modes of A' are either observable at the output or are unaffected by the input noise. We shall suppose that k = dim η.

Let $\bar{\mathcal{X}}$ denote the space $\mathcal{X}$ reduced mod η and let T denote the canonical projection, $T: \mathcal{X} \to \bar{\mathcal{X}}$. Let $\bar{x} = Tx$, let $\bar{A}$ be the map induced in $\bar{\mathcal{X}}$ by A, let $\bar{B} = TB$, and let $\bar{D}$ be the unique map defined by $D = \bar{D}T$. It was shown in [6] that $(\bar{A},\bar{B})$ is stabilizable and $(\bar{D},\bar{A})$ is observable in $\bar{\mathcal{X}}$. It was shown in [7] that if (2.1) holds and R is positive definite then for the steady state regulator problem defined by,

$$\dot{x} = Ax + Bu, \qquad x(0) = \xi, \qquad J(\xi,u) = \int_0^\infty (x'D'Dx + u'Ru)dt\ , \tag{2.2}$$

(2.2)

the cost functional J has a minimum value $\xi' P\xi$ in the class of piecewise continuous control functions satisfying $\lim_{t\to\infty} Dx(t) = 0$ and $\lim_{t\to\infty} u(t) = 0$. The matrix P is given by

$$P = T'\bar{P}T, \tag{2.3}$$

where $\bar{P}$ is the unique, symmetric, positive definite solution of the algebraic Riccati equation,

$$\bar{A}'\bar{P} + \bar{P}\bar{A} + \bar{D}'\bar{D} - \bar{P}\bar{B}'R^{-1}\bar{B}\bar{P} = 0, \tag{2.4}$$

and the optimal feedback control is given by, $u = -R^{-1}B'Px$.

Here we shall prove the following theorem.

<u>Theorem 2.1</u>. If (2.1) holds, the steady state optimal estimation problem defined by (1.1)-(1.5) has a unique solution $\hat{x}$ with

$$K' = PBR^{-1}. \tag{2.5}$$

The matrix P is given by

$$P = T'\bar{P}T,$$

and $\bar{P}$ is the unique symmetric positive definite solution of (2.4). Furthermore,

$$\lim_{t\to\infty}(\exp(A' - K'B')t)\ K'y(t) = 0\ . \tag{2.7}$$

proof. From (1.1) we see that if $x(0) = 0$ then

$$x(t) = \int_0^t \exp(A'(t-s))D'w(s)ds,$$

and $x(\cdot)$ lives in a subspace $\mathcal{S}$ of $\mathcal{X}$ which is isomorphic to $\bar{\mathcal{X}}$. We shall show that we need to allow into the competition for the optimization only those estimates $\hat{x}(t)$ which also live in $\mathcal{S}$. In a coordinate system compatible with the decomposition $\mathcal{X} = \mathcal{S} \oplus \eta$ we have,

$$x' = (x_1', x_2'), D = (D_1, 0), K' = (K_1, K_2)', B' = (B_1', B_2'), \text{ and}$$

$$A = \begin{bmatrix} A_{11} & 0 \\ A_{21} & A_{22} \end{bmatrix}.$$

From (1.4) we obtain,

$$\dot{\tilde{x}}_2 = A_{22}' \tilde{x}_2 - K_2'(y - B'\hat{x}).$$

If we choose K_2 to be a null matrix we obtain $\tilde{x}_2(t) \equiv 0$ as well as $\hat{x}_2(t) \equiv 0$. Then from (1.3) we obtain

$$\dot{\hat{x}}_1 = A_{11}'\hat{x}_1 + K_1'(y - B_1'\hat{x}_1), \tag{2.8}$$

which describes a standard Kalman-Bucy filter in the reduced state space. If we define $\bar{K}' \triangleq TK'$ we see that our optimal filter problem is reduced to finding $\bar{K}$ when,

$$\dot{\bar{x}} = \bar{A}\,\bar{x} + \bar{K}'(y - \bar{B}'\bar{x})$$

to minimize $\lim_{t\to\infty} E\left[\bar{x}'(t)\bar{Q}\,\bar{x}(t)\right]$, when $\bar{Q} = TQT'$. The state and measurement equations are

$$\dot{\bar{x}} = \bar{A}'\bar{x} + \bar{D}'w,$$
$$y = \bar{B}'x + v.$$

The standard solution of this problem is given by $\bar{K} = \bar{P}\,\bar{B}\,R^{-1}$, where $\bar{P}$ satisfies (2.4). Premultiplying (2.4) by T', postmultiplying by T and defining P as in (2.3) we find that P satisfies the algebraic Riccati equation,

$$A'P + PA + D'D - PBR^{-1}B'P = 0. \tag{2.9}$$

Furthermore,

$$\lim_{t\to\infty} E\left[\tilde{x}'(t)Q\tilde{x}(t)\right] = \mathrm{tr}QP, \; P = \lim_{t\to\infty} E\left[\tilde{x}(t)\tilde{x}'(t)\right],$$

and for $\xi \varepsilon \mathcal{X}$,

$$\lim_{t\to\infty} E\left[\left|\xi'\tilde{x}(t)\right|^2\right] = \xi'P\xi = \bar{\xi}'\bar{P}\,\bar{\xi}.$$

The matrix $\bar{A} - \bar{B}\,\bar{K}$ is stable, and since K_2 is null then (2.7) holds and this proves the theorem. The problem defined by (1.1)-(1.5) is thus seen to be dual to the problem defined by (2.2).

3. <u>The case for singular R</u>. Hypothesis (2.1) as well as the assumption of positive definiteness on R will now be relaxed. Let $R^{\#}$ denote the Moore-Penrose pseudoinverse of R, let N denote a full rank matrix whose columns span Ker R, and for each positive integer μ define $R_\mu \triangleq R + \frac{1}{\mu}NN'$. Let B_0 denote the restriction of B to Ker R and let $\mathcal{B}_0$ denote $\mathrm{Im}B_0$. We consider the set $\underline{\mathcal{J}}$ of all subspaces $\mathcal{V}$ of Ker D satisfying

$$A\mathcal{V} \subset \mathcal{V} + \mathcal{B}_0 .$$

It is shown in Chapter 4 of [6] that $\underline{\mathcal{J}}$ is not empty, that $\underline{\mathcal{J}}$ has a supremal element $\mathcal{V}^*$, and that for every element $\mathcal{V}$ of $\underline{\mathcal{J}}$ there exists a map F satisfying,

$$(A - B_0F)\,\mathcal{V} \subset \mathcal{V}. \tag{3.1}$$

The set of all maps F satisfying (3.1) is denoted by $\underline{\mathcal{F}}(\mathcal{V})$. Let $F_0 \,\varepsilon\, \underline{\mathcal{F}}(\mathcal{V}^*)$ and define $A_0 \triangleq A - B_0F_0$. It is not hard to see that $\mathcal{V}^* = \bigcap_{i=1}^{n} \mathrm{Ker}(DA_0^{i-1})$, or $\mathcal{V}^*$ is the unobservable subspace of the pair (D, A_0). We now replace hypothesis (2.1) by

$$\mathcal{X}^+ \subset \langle A|\mathcal{B}\rangle + \mathcal{V}^*. \tag{3.2}$$

and note that since $\eta \subset \mathcal{V}^*$, (3.2) is weaker than (2.1).

Let T now denote the canonical projection of $\mathcal{X}$ onto the space $\bar{\mathcal{X}} = \mathcal{X}/\mathcal{V}^*$. We shall prove the following theorem.

<u>Theorem 3.1</u>. When (3.2) holds the steady state optimal estimation problem described by (1.1)-(1.5) is the dual of the partially singular steady state optimal regulator

problem described by (2.2). Furthermore,

$$\inf_{K} \lim_{t\to\infty} E\left[\tilde{x}'(t)Q\,\tilde{x}(t)\right] = \operatorname{tr} QP,$$

$$\lim_{t\to\infty} E\left[\tilde{x}'(t)\,\tilde{x}(t)\right] = P,$$

and

$$\inf_{K} \lim_{t\to\infty} E\left[\left|\xi'\tilde{x}(t)\right|^2\right] = \xi' P\xi.$$

The matrix P satisfies (2.3),

$$\overline{P} = \lim_{\mu\to\infty} \overline{P}_\mu, \tag{3.3}$$

$\overline{P}_\mu$ is the unique symmetric, positive definite solution of (2.4) when R is replaced by R_μ, and the covariance P satisfies the linear matrix inequality,

$$\begin{bmatrix} A'P + PA + D'D & PB \\ B'P & R \end{bmatrix} \geq 0. \tag{3.4}$$

In order to prove this result we shall first show that if (3.2) holds the singular estimation problem is equivalent to one in the reduced space, $\overline{\mathcal{X}}$. Let $F = NF_0$ and subtract $F'B' = F_0'B_0'$ from both sides of (1.1). Noting that $N'(y - B'x)$ is null, we obtain,

$$\dot{x} = A_0'x + D'w + F'y. \tag{3.5}$$

Letting $K' = F' + L'$ in (1.3) we find that,

$$\dot{\hat{x}} = A_0'\hat{x} + L'(y - B'\hat{x}) + F'y. \tag{3.6}$$

Subtracting (3.6) from (3.5) we find,

$$\dot{\tilde{x}} = A_0'\tilde{x} + D'w - L'(y - B'\hat{x}). \tag{3.7}$$

In a coordinate system compatible with the decomposition, $\mathcal{X} = \mathcal{S} \oplus \mathcal{V}^*$, we have

$x' = (x_1', x_2')$, $F = (F_1, F_2)$, $B' = (B_1', B_2')$, $K = (K_1, K_2)$, $L = (L_1, L_2)$, and

$$A_0 = \begin{bmatrix} A_{011} & 0 \\ A_{021} & A_{022} \end{bmatrix}.$$

Now $\mathcal{V}^*$ is A_0-invariant, and so there is a unique map $\bar{A}_0$ induced in $\bar{\mathcal{X}}$ by A_0, and $\bar{A}_0 = A_{011}$. Then,

$$\dot{\tilde{x}}_2 = A_{022}' \tilde{x}_2 - L_2'(y - B'\hat{x}). \tag{3.8}$$

If we let L_2 be null then again, $\tilde{x}_2(t) \equiv 0$. Again the estimation problem reduces to one in $\bar{\mathcal{X}}$, since if $\tilde{x}_2 \equiv 0$, we obtain from (3.7) and (1.2),

$$\dot{\tilde{\bar{x}}} = \bar{A}_0' \tilde{\bar{x}} + \bar{D}' w - \bar{L}' (y - \bar{B}'\bar{x}), \tag{3.9}$$

where $D = \bar{D}T$, $\bar{B} = TB$, $\bar{L}' = T\tilde{L}'$. This is again a standard problem in $\bar{\mathcal{X}}$ when R is positive definite, since the pair $(\bar{D}, \bar{A}_0)$ is observable and the pair $(\bar{A}_0, \bar{B})$ is stabilizable.

Now replace R by R_μ and solve the resulting steady state estimation problem in $\bar{\mathcal{X}}$. Denoting the unique optimal gain matrix by $\bar{L}_\mu$ and the optimal error covariance by $\bar{P}_\mu$, we obtain,

$$\bar{L}_\mu' = \bar{P}_\mu \bar{B} R_\mu^{-1} \tag{3.10}$$

where,

$$\bar{A}_0' \bar{P}_\mu + \bar{P}_\mu \bar{A}_0 + \bar{D}' \bar{D} - \bar{P}_\mu \bar{B} R_\mu^{-1} \bar{B} \bar{P}_\mu = 0, \tag{3.11}$$

$$\min_{\bar{L}} \lim_{t \to \infty} E\left[\bar{x}_\mu(t)\, \bar{Q}\, \bar{x}_\mu(t)\right] = \operatorname{tr} \bar{Q} \bar{P}_\mu,$$

and

$$\min_{\bar{L}} \lim_{t \to \infty} E\left[\left|\bar{\xi}' \bar{x}(t)\right|^2\right] = \bar{\xi}' \bar{P}_\mu \bar{\xi}, \tag{3.12}$$

for arbitrary $\bar{\xi} \in \bar{\mathcal{X}}$. Premultiplying (3.11) by T' and postmultiplying by T we find that

$$A_0'P_\mu + P_\mu A_0 + D'D - P_\mu BR_\mu^{-1}B'P_\mu = 0, \tag{3.13}$$

where $P_\mu = T'\overline{P}_\mu T$. Equation (3.13) may be rewritten as

$$A_0'P_\mu + P_\mu A_0 + D'D - P_\mu BR^{\#}B'P_\mu = nP_\mu BNN'B'P_\mu \ . \tag{3.14}$$

Since $\overline{P}_\mu$ is non-increasing, $\lim_{\mu\to\infty} \overline{P}_\mu = \overline{P}_\infty$ exists and $P_\infty = T' \ \overline{P}_\infty \ T$ satisfies

$$A_0'P_\infty + P_\infty A_0 + D'D - P_\infty BR^{\#}B'P_\infty \geq 0 \tag{3.15}$$

and

$$P_\infty BN = 0. \tag{3.16}$$

Substitution of (3.16) into (3.15) yields,

$$A'P_\infty + P_\infty A + D'D - P_\infty BR^{\#}B'P_\infty \geq 0. \tag{3.17}$$

It was shown in [7] that (3.16)-(3.17) are equivalent to (3.4) with P replaced by P_∞.

It must still be shown that

$$\inf_L \ \lim_{t\to\infty} \ E\left[\tilde{x}'(t)Q\tilde{x}(t)\right] = \mathrm{tr}QP_\infty.$$

The assumption,

$$\mathrm{tr}QP_\infty > \inf_L \ \lim_{t\to\infty} E\left[\tilde{x}'(t)Q\tilde{x}(t)\right]$$

can be shown to lead to a contradiction by an argument similar to one presented in [7], and thus proves the theorem.

It was shown in [7] that if (3.2) holds then the sequence $u_\mu = -R_\mu^{-1}B'P_\mu x$ of controls for the problem defined by (2.2) in infimizing when R is singular, and that $\inf J(\xi,u) = \xi'P_\infty\xi$. The infimization is taken over the class $\mathcal{U}$ of piecewise continuous functions $u(\cdot)$ satisfying,

$$\lim_{t\to\infty} \ D\,x(t) = 0, \ \ \int_0^\infty u'(t)Ru(t)dt < \infty,$$

where $x(\cdot)$ represents the trajectory corresponding to the input function $u(\cdot)$.

Since R is singular, $\lim_{t\to\infty} u(t)$ is not necessarily null. Thus we see that under the hypothesis (3.2) the problem described by (1.1)-(1.5) is again dual to the problem described by (2.2).

We may also suppose that R is only semidefinite when hypothesis (2.1) is invoked. The limiting process for $R_\mu = R + \frac{1}{\mu} NN'$ as $\mu \to \infty$ is the same. However, now the approximating optimal gains, K_μ, satisfy

$$\lim_{t\to\infty} (\exp(A' - K_\mu' B')t)K_\mu' = 0, \tag{3.18}$$

a conclusion that cannot be drawn under the weaker hypothesis (3.2).

References

1. A. E. Bryson and D. E. Johansen, "Linear Filtering for Time Varying Systems Using Measurements Containing Colored Noise, I.E.E.E. Trans. Automat. Contr., Vol. AC-10, pp. 4-10, 1965.

2. E. Tse and M. Athans, "Optimal Minimal Observer-Estimator for Discrete Time-Varying Systems," I.E.E.E. Trans. Automat. Contr., Vol. AC-15, pp. 416-426, 1970.

3. F. W. Fairman, "Optimal Observers for a Class of Continuous Linear Time-Varying Stochastic Systems," I.E.E.E. Trans. Automat. Contr., Vol. AC-22, pp. 136-137, 1977.

4. E. Fogel and Y. F. Huang, "Reduced Order Optimal State Estimator for Linear Systems with Partially Noise-Corrupted Measurements," I.E.E.E. Trans. Automat. Contr., Vol. AC-25, pp. 994-996, 1980.

5. Y. T. Ju and V. B. Haas, "A Duality Principle for State Estimation with Partially Noise-Corrupted Measurements," Proc. 1981 I.E.E.E. Conference on Decision and Control, San Diego, CA, Dec., 1981.

6. W. M. Wonham, Linear Multivariate Control, Springer-Verlag, New York, 1979.

7. V. B. Haas, "The Singular Steady State Linear Regulator," SIAM J. Contr. and Opt., Vol. 20, No. 2, 1982.

8. A. Albert, "Conditions for Positive and Nonnegative Definiteness in Terms of Pseudoinverses," SIAMJ Appl. Math, 17, pp. 434-440, 1969.

On a General Method for Solving Time-Optimal Linear Control Problems

by
O. Hájek
Department of Mathematics and Statistics
Case Western Reserve University, Cleveland,
Ohio 44106, USA
and
W. Krabs
Fachbereich Mathematik
Technische Hochschule Darmstadt
6100 Darmstadt, West Germany

Abstract

In this paper a class of methods for solving time-optimal linear control problems in an abstract setting is presented. Two convergent versions of this class, termed as first and second implementation of a basic algorithm, generalize the main two convergent algorithms that have been developed for linear systems governed by ordinary differential equations.

1. Introduction.

This article is an abbreviated version of [7] where a unified approach is given to algorithms for the computation of the minimal time and time-minimal controls for steering an abstract linear system into a time-independent target state by a family of admissible controls.

The general algorithm, termed as basic algorithm, which is described in Subsection 3.1 is based on a duality statement (see Theorem 2.3) which characterizes the minimal time by a maximum property, if a condition is met which generalizes the concept of properness in the sense of Hermes-LaSalle [8]. This duality statement generalizes a result of Neustadt in [12], who seems to have been the first to develop an algorithm for solving time-optimal control problems. Neustadt uses the maximum property of the minimal time in order to establish a differential equation from which time-minimal controls can be computed, if the system is normal. Normality, in general, is a stronger property than properness and guarantees uniqueness of time-minimal controls. In [5] Eaton gives a procedure for solving normal time-minimal control problems with time-dependent targets which, for fixed targets, can be considered as a special case of the basic algorithm developed in Subsection 3.1. He was, however, unable to prove convergence. This is also pointed out by Boltjanski who in [3] gives a unified representation of

Neustadt's and Eaton's results. In general, it is not possible to prove convergence for the basic algorithm of Subsection 3.1. By Theorem 3.1, however, a wide class of algorithms is admitted for which convergence can be proved. Among these there are two algorithms, termed as first and second implementation which generalize the main two classes of convergent algorithms developed for linear control problems governed by ordinary differential equations in [2],[4],[9], and [6].

Due to the limited space for this publication, applications to linear control problems with ordinary or partial differential equations cannot be presented. The interested reader is referred to [7].

2. Controllability and Time-Minimal Controllability.

We consider the following abstract version of a linear control problem: Let X be the dual space Z^* of a separable Banach space Z, let $\{S_t | t\in[o,T]\}$, for some T>o, be a family of continuous linear mappings from X into a finite-dimensional normed linear space Y such that S_o maps X into the origin of Y, and let $\hat{y}\in Y$ be a fixed element with $\hat{y}\neq o$. Further let, for some constant M>o,

$$U_M = \{u\in X \,|\, \|u\| \leq M\}. \tag{2.1}$$

Each element $u\in X$ is considered as a control of a physical system whose states are given by the elements of Y. The development of the system under a fixed control $u\in X$ with respect to the time is assumed to be described by the mapping $t\rightarrow S_t(u), t\in[o,T]$. The controls which lie in U_M (2.1) are called admissible. The state $\hat{y}\in Y$ is considered as a desired target state.

The problem of controllability then reads as follows:
Does there exist, for a given time $t\in(o,T]$, an admissible control u such that

$$S_t(u)=\hat{y}, \tag{2.2}$$

i.e., is it possible to reach the target state $\hat{y}$ by an admissible control within the time interval [o,t]? A necessary and sufficient condition for controllability which was derived by Antosiewicz in [1] for linear systems governed by ordinary differential equations (see also [11]) is the content of

<u>Theorem 2.1:</u> For each $t\in(o,T]$ we assume the mapping $S_t:X\rightarrow Y$ to be continuous with respect to the weak* convergence in X. Then, for some $t\in(o,T]$ there exists an admissible control u with (2.2) if, and only if

$$y^*(\hat{y}) \leq M\|S_t^*(y^*)\| \text{ for all } y^*\in Y^* \tag{2.3}$$

where Y^* denotes the dual space of Y and $S_t^*: Y^* \to X^*$ is the operator adjoint to S_t.

The proof (which will not be given here) is based on the fact that (2.2) holds for some $u \in U_M$ if, and only if $\hat{y}$ belongs to the reachable set

$$R_t = \{S_t(u) \mid u \in U_M\} \tag{2.4}$$

which is convex and closed, since U_M is weak* sequentially compact.

We define the infimal time by

$$t^* = \inf \{t \in (o,T] \mid S_t(u) = \hat{y} \text{ for some } u \in U_M\} \tag{2.5}$$

where t^* is put to $+\infty$, if controllability does not hold for any $t \in (o,T]$.

<u>Theorem 2.2:</u> In addition to the assumption of Theorem 2.1 let the mapping $t \to S_t$ from $[o,T]$ into $B(X,Y)$ be continuous with respect to the norm-topology of $B(X,Y)$.

If t^* defined by (2.5) is finite, then

a) there is some $u^* \in U_M$ such that $S_{t^*}(u^*) = \hat{y}$ which implies that $t^* \in (o,T]$, i.e. time-minimal controllability holds,

b) there is some $y^* \in Y^*$ with $y^* \neq o$ such that

$$y^*(\hat{y}) = M \| S_{t^*}^*(y^*) \|. \tag{2.6}$$

Assertion a) follows from [1o], Theorem 4.1, and assertion b) can be proved as in the case of linear systems governed by ordinary differential equations (see[1] and [11]). The general method for the computation of the infimal time t^* given by (2.5) to be described in Subsection 3.1 is based on

<u>Theorem 2.3:</u> In addition to the assumptions of Theorem 2.2 let the function $t \to \| S_t^*(y^*) \|$, for every fixed $y^* \in Y^*$ with $y^* \neq o$, be strictly increasing in $[o,T]$. If t^* defined by (2.5) is finite then

$$t^* = \max \{t \in (o,T] \mid (2.6) \text{ holds for some } y^* \in Y^*, y^* \neq o\}. \tag{2.7}$$

For the proof see [7].

The first implementation of the general method for the computation of t^* (2.5) to be described in Subsection 3.2 is based on

<u>Theorem 2.4:</u> Under the assumptions of Theorem 2.3, for each $t \in (o,T]$, the following equivalence holds true:

$$M(t) = \inf\{\| S_t^*(y^*) \| \mid y^* \in H\} \{\gtreqless\} M \iff t \{\lesseqgtr\} t^* \tag{2.8}$$

where

$$H = \{y^* \in Y^* \mid y^*(\hat{y}) = 1\} \tag{2.9}$$

and t^* is the infimal time defined by (2.5).

For the proof see [7].

3. Methods for the Computation of Optimal Time and Controls.

Throughout this Section we assume the assumptions of Theorem 2.3 to hold and t^*, given by (2.5), to be finite.

3.1. The Basic Algorithm.

Two sequences (t_k) in $[o,T]$ and (y_k^*) in Y^*, respectively, are constructed by putting $t_o=o$ and then, for each $k=o,1...$, performing the following two steps.

Step 1: For $t_k \in [o,T]$ given find $y_k^* \in Y^*$ such that

$$y_k^*(\hat{y}) > M \| S_{t_k}^*(y_k^*) \| . \tag{3.1}$$

Step 2: For $(t_k, y_k^*) \in [o,T] \times Y^*$ given with (3.1) determine the unique $t_{k+1} \in (t_k,T]$ such that

$$y_k^*(\hat{y}) = M \| S_{t_{k+1}}^*(y_k^*) \| , \tag{3.2}$$

replace t_k by t_{k+1} and go to step 1.

For $k=o$ there is always some $y_o^* \in Y^*$ with $y_o^*(\hat{y}) > o$ so that $t_1 \in (o,T]$ with $y_o^*(\hat{y}) = M \| S_{t_1}^*(y_o^*) \|$ is then uniquely defined. In general, the sequence (t_k) obtained by this algorithm has the property $t_k < t_{k+1} \leq t^*$ for all t_k for which there exists some $y_k^* \in Y^*$ with (3.1). This is a consequence of the determination of t_{k+1} and Theorem 2.3. If, for some t_k with $k \geq 1$, there is no $y_k^* \in Y^*$ with (3.1), then, by Theorem 2.1, we conclude that $t_k \geq t^*$ and hence $t_k = t^*$ because from the previous step 2 we know that $t_k \leq t^*$.

We assume that for each t_k there is some $y_k^* \in Y^*$ with (3.1) so that the algorithm never stops. Then $\hat{t} = \lim_{k\to\infty} t_k$ exists and $o < \hat{t} \leq t^*$.

Without specification of the choice of y_k^* in step 2 it is impossible to show that $\hat{t} = t^*$. This is, however, true under fairly general conditions to be formulated in

Theorem 3.1: If in step 1 of the above algorithm $y_k^* \in Y^*$ for $k \geq 1$ is chosen such that, for some fixed $\varepsilon > o$, an element $u_k \in U_M$ can be found with

$$y_k^*(\hat{y}) - M \| S_{t_k^*}^*(y_k^*) \| \geq \varepsilon \| \hat{y} - S_{t_k}(u_k) \| , \tag{3.3}$$

and if the sequence (y_k^*) is uniformly bounded, then the sequence (t_k) generated by the algorithm converges to the infimal time t^* given by (2.5). Further each weak* cluster point $u^* \in U_M$ of the sequence (u_k) (and there is at least one) is a time-minimal control (see assertion a) of Theorem 2.2).

For the proof see [7].

3.2. A First Implementation.

Starting with $t_o = o$ the two steps of the basic algorithm are performed to generate $y_o^* \in Y^*$ and $t_1 \in (o,T]$ with (3.2) for k=o. For each $k \geq 1$ we then continue as follows.

Step 1: Determine $y_{t_k}^* \in H$ (2.9) such that

$$\| S_{t_k}^* (y_{t_k}^*) \| = \frac{1}{M(t_k)} \text{ (see (2.8))}. \tag{3.4}$$

This is possible by routine arguments in approximation theory.

If $M(t_k) = M$, then put $t^* = t_k$ and stop.

If $M(t_k) > M$, then proceed to

Step 2: Determine the unique $t_{k+1} \in (t_k, T]$ such that

$$\| S_{t_{k+1}}^* (y_{t_k}^*) \| = \frac{1}{M}, \tag{3.5}$$

replace t_k by t_{k+1} and go to step 1.

Remarks: The case $M(t_k) < M$ cannot occur because by the previous step 2 we have

$$\| S_{t_k} (y_{t_{k-1}}^*) \| = \frac{1}{M} \text{ for some } y_{t_{k-1}}^* \in H \text{(2.9)}$$

which implies $\frac{1}{M(t_k)} \leq \frac{1}{M}$, hence $M \leq M(t_k)$.

If $M(t_k) = M$, $t^* = t_k$ follows from Theorem 2.4. Step 2 is exactly the same as in the basic algorithm.

Further it can be shown (see[7]) that the requirement (3.3) of Theorem 3.1 is met for each $k \geq 1$, if $\varepsilon = \|\hat{y}\|^{-1}$ and $u_k = (M/M(t_k)) u_{t_k}$ for some $u_{t_k} \in X$ with (2.2) for $u = u_{t_k}$ and $\| u_{t_k} \| = M(t_k)$, the existence being ensured. Since (y_k^*) is uniformly bounded (see also [7]), the convergence statements of Theorem 3.1 hold true.

3.3. A Second Implementation.

The beginning for k=o is the same as in the first implementation. For each $k\geq 1$ we then continue as follows.

Step 1: Determine $z_k \in R(t_k)$ (see (2.4)) such that

$$\|\hat{y}-z_k\| = \min\{\|\hat{y}-z\| \mid z \in R(t_k)\}. \tag{3.6}$$

The existence of z_k follows from routine arguments in approximation theory, since Y is finite-dimensional and $R(t_k)$ is closed (see Section 2).

If $z_k=\hat{y}$, then put $t^*=t_k$, $u^*=u_k$ where $z_k=S_{t_k}(u_k)$, $u_k \in U_M$ and stop.
If $z_k \neq \hat{y}$, then determine $y_k^* \in Y^*$ such that

$$\begin{aligned} &\|y_k^*\|=1,\ y_k^*(\hat{y}-z_k)=\|\hat{y}-z_k\|, \\ &y_k^*(z_k)=\max\{y_k^*(z) \mid z \in R(t_k)\} \end{aligned} \tag{3.7}$$

which is possible by a well-known duality theorem in convex approximation theory.

Step 2: Determine the unique $t_{k+1} \in (t_k, T]$ which satisfies (3.2), replace t_k by t_{k+1}, and go to step 1.

Remarks: If, in step 1, $z_k=S_{t_k}(u_k)=\hat{y}, u_k \in U_M$, then $\hat{y} \in R(t_k)$, hence $t_k \geq t^*$.

But also $t_k \leq t^*$, as a consequence of the previous step 2. Therefore $t_k=t^*$ and u_k is a time-minimal control.

If, in particular, Y is a Hilbert space, then $Y^*=Y$, there is exactly one $z_k \in R(t_k)$ with (3.6), and y_k^* with (3.7) is uniquely defined by

$$y_k^* = (\|\hat{y}-z_k\|)^{-1}(\hat{y}-z_k).$$

As a consequence of (3.7) we obtain, in general,

$$y_k^*(\hat{y})-M\|S_{t_k}^*(y_k^*)\| = y_k^*(\hat{y})-y_k^*(z_1) = \|\hat{y}-z_k\|.$$

Therefore (3.3) holds as an equality for $\varepsilon=1$. Since $\|y_k^*\|=1$ for all k, Theorem 3.1 is also applicable to the second implementation.

4. References

[1] Antosiewicz, H.A.: Linear Control Systems. Arch. Rat. Mech. Anal. 12 (1963), pp. 313-324.

[2] Babunashvili, T.G.: The Synthesis of Linear Optimal Systems. SIAM J. Control 2 (1965), 261-265.

[3] Boltjanski, W.G.: Mathematische Methoden der optimalen Steuerung. München: Carl Hanser Verlag 1972.

[4] Butkovskiy,A.G.: Distributed Parameter Systems. New York-London-Amsterdam: Elsevier 1969.

[5] Eaton,J.H.: An Iterative Solution to Time-Optimal Control. J.Math.Anal.Appl.5(1962),329-344.

[6] Fujisawa,T. and Y.Yasuda: An Iterative Procedure for Solving the Time-Optimal Regulator Problem. SIAM J.Control 5(1967),501-412.

[7] Hájek,O. and W.Krabs: On a General Method for Solving Time-Optimal Linear Control Problems. Preprint Nr.579 des Fachbereichs Mathematik der TH Darmstadt, Jan.81.

[8] Hermes,H. and J.P-LaSalle: Functional Analysis and Time Optimal Control. New York-London: Academic Press 1969.

[9] Krabs,W.: Einführung in die Kontrolltheorie. Darmstadt: Wissenschaftliche Buchgesellschaft 1978.

[10] Krabs,W.: Convex Optimization and Approximation. Preprint Nr.504 des Fachbereichs Mathematik der TH Darmstadt,Nov.1979.

[11] Marzollo,A.: Controllability and Optimization.International Centre for Mechanical Sciences, Courses and Lectures No.17, Wien-New York: Springer-Verlag 1969.

[12] Neustadt,L.: Synthesizing Time-Optimal Control Systems. J.Math. Anal.Appl.1(196o),484-493.

PERIODIC SOLUTIONS OF DISCRETE MATRIX RICCATI EQUATIONS WITH CONSTANT COEFFICIENT MATRICES

H. KANO* and T. NISHIMURA**

* the International Institute for Advanced Study of Social Information Science, Fujitsu Ltd., 410-03 Shizuoka, Japan
** the Institute of Space and Astronautical Science, Tokyo, Japan

ABSTRACT The well-known discrete matrix Riccati equations arising in the theory of optimal estimation and control problems are considered.

Existence conditions are established of real symmetric periodic solutions as well as of real symmetric nonnegative-definite periodic solutions. Furthermore, an algorithm is developed to derive such periodic solutions.

1. Introduction

As is well known, discrete matrix Riccati equations arise in the Kalman filter as well as in optimal regulator problem for time-invariant discrete systems. Especially, the steady-state solutions of Riccati equations play an important role in the design of fixed-gain Kalman filter and optimal regulators, and have been studied extensively.

Kalman [1] first showed that Riccati equation had positive-definite steady-state solution, if the system is completely controllable and observable. This condition was then weakened by Kucera [2], who showed that the stabilizability and detectability introduced by Wonham [3] were the necessary and sufficient condition for the existence of nonnegative-definite solution for continuous-time systems. The same result was shown to hold also for discrete-time systems [4].

For periodically time-varying systems, Kano and Nishimura [5] proved that the necessary and sufficient condition for the existence of nonnegative-definite periodic solution is that the system is stabilizable and detectable.

On the other hand, the existence of periodic solution is noticed by Hayase [6] for time-invariant systems, and the authors have given complete analyses for the existence of periodic solutions for continuous-time systems [7].

This paper is its discrete-time counterpart, which naturally is more suited for the implementation on digital computer. It should be noted that although the results of this paper are stated using the terminology in filtering problems, they are directly applicable to control problems by duality principle.

2. Problem Statement and Mathematical Preliminaries

The following discrete matrix Riccati equation is considered:

$$P(k+1) = \Phi\{P(k) - P(k)H^T[HP(k)H^T+I]^{-1}HP(k)\}\Phi^T + GG^T \tag{1}$$

in which Φ, G and H are respectively $n\times n$, $n\times r$ and $m\times n$ constant matrices.

The main purpose of this paper is to establish the existence conditions of periodic solutions in eq. (1) and to derive a computational procedure of such solutions.

The derivation of the results is based on the steady-state(constant) solution P of eq. (1), i.e., solution of the following algebraic Riccati equation:

$$P = \Phi\{P - PH^T[HPH^T + I]^{-1}HP\}\Phi^T + GG^T \tag{2}$$

It is known (e.g. see [8]) that the solution of eq. (2) can be obtained by means of canonical decomposition of certain $2n\times 2n$ matrix A defined by

$$A = \begin{bmatrix} \Phi + GG^T\Phi^{-T}H^TH & GG^T\Phi^{-T} \\ \Phi^{-T}H^TH & \Phi^{-T} \end{bmatrix} \tag{3}$$

A is then the so-called symplectic matrix. For the sake of simplicity, the matrix A is assumed to have distinct eigenvalues $\lambda_1, \ldots, \lambda_{2n}$. Then A can be transformed into its diagonal form as

$$AT = T\Lambda \tag{4}$$

where the matrix T is composed of eigenvectors of A and $\Lambda = \mathrm{diag}\{\lambda_1, \ldots, \lambda_{2n}\}$. Now partition T into four $n\times n$ matrices

$$T = \begin{bmatrix} Y & V \\ X & U \end{bmatrix} \tag{5}$$

then the solution P of eq. (2) is given, as long as $|X|\neq 0$, by

$$P = YX^{-1} \tag{6}$$

In addition, the solution P(k) of eq. (1) with an arbitrary initial condition P(0) is known to be given by

$$P(k) = Y(k)X(k)^{-1} \tag{7}$$

where Y(k) and X(k) are $n\times n$ matrices computed from

$$\begin{bmatrix} Y(k) \\ X(k) \end{bmatrix} = A \begin{bmatrix} Y(k-1) \\ X(k-1) \end{bmatrix} = A^k \begin{bmatrix} Y(0) \\ X(0) \end{bmatrix} \tag{8}$$

and Y(0) and X(0) are any matrices such that $Y(0)X(0)^{-1} = P(0)$.

We now consider to sample eq. (1) with period p. Let a matrix B be $B = A^p$. Then, it can be shown that B is of the following form

$$B = \begin{bmatrix} \Phi_p + G_pG_p^T\Phi_p^{-T}H_p^TH_p & G_pG_p^T\Phi_p^{-T} \\ \Phi_p^{-T}H_p^TH_p & \Phi_p^{-T} \end{bmatrix} \triangleq \begin{bmatrix} B_{11} & B_{12} \\ B_{21} & B_{22} \end{bmatrix} \tag{9}$$

in which matrices Φ_p, G_p and H_p can be derived recursively from Φ, G and H. This shows that B is also a symplectic matrix. The steady-state solution P_s of sampled equation is then the solution of

$$P_s = \Phi_p\{P_s - P_sH_p^T[H_pP_sH_p^T + I]^{-1}H_pP_s\}\Phi_p^T + G_pG_p^T \tag{10}$$

Similarly to eqs.(4)-(6), B matrix is decomposed by the eigenvector matrix T_s, and T_s is further partitioned as in eq. (5).

$$BT_s = T_sM, \qquad T_s = \begin{bmatrix} Y_s & V_s \\ X_s & U_s \end{bmatrix} \tag{11}$$

where $M = \mathrm{diag}\{\mu_1, \dots, \mu_{2n}\}$, $\mu_i = \lambda_i^p$. Then P_s is given by $P_s = Y_sX_s^{-1}$, if $|X_s| \neq 0$.

Notice that eq. (10) includes all solutions of eq. (2) and if, in addition, (1) has periodic solution of period p, then it appears as a solution of sampled equation (10). In terms of symplectic matrices A and B, if B has all distinct eigenvalues, then T_s=T up to a scale factor of each eigenvector. The solution formulas show that eqs. (2) and (10) have the same solutions and hence no periodic solution exists in this case. If, however, B has multiple eigenvalues, then T_s needs not coincide with T. As will be shown in Sec. 3, this leads to periodic solutions of eq. (1).

Several well-known facts [4] that are to be used henceforth are summarized below. Here ω_{ij} and $\bar{\omega}_{ij}$ denote ij-th elements of Ω and $\bar{\Omega}$ defined by

$$\Omega = X^*Y, \quad \bar{\Omega} = X^TY \tag{12}$$

<u>Lemma 1.</u> Let λ be an eigenvalue of A and $\begin{bmatrix} y \\ x \end{bmatrix}$ be the corresponding eigenvector. Then, λ^{-1} is an eigenvalue of A^T and $\begin{bmatrix} -x \\ y \end{bmatrix}$ is the corresponding eigenvector.

<u>Lemma 2.</u> If $\lambda_i^* \neq \lambda_j^{-1}$ $(\lambda_i \neq \lambda_j^{-1})$ for all i, j, $1 \leq i, j \leq n$, then $\omega_{ij} = \omega_{ji}^*$ $(\bar{\omega}_{ij} = \bar{\omega}_{ji})$.

<u>Lemma 3.</u> if $|\lambda_i| > 1$ $(|\lambda_i| < 1)$ for all i, $1 \leq i \leq n$, then $\Omega \geq 0$ $(\Omega \leq 0)$.

<u>Lemma 4.</u> Assume that the pair (H,Φ) is detectable and no eigenvalue of A lies on the unit circle. Then eq. (2) has a unique nonnegative-definite solution P iff the pair (Φ,G) is stabilizable.

<u>Lemma 5.</u> Eq. (2) posseses a unique nonegative-definite solution and the corresponding closed-loop system is asymptotically stable iff (Φ,G) is stabilizable and (H,Φ) is detectable.

Note that Lemmas 1-3 also hold for sampled system with the obvious change of symbols.

There are two special solutions in eq. (2) usually denoted as P^+ and P^- such that P^+ and P^- are derived from n eigenvalues of A that lie, respectively, outside and inside the unit circle. It is known that $P^+ \geq 0$ and $P^- \leq 0$. Furthermore, the solution described in Lemmas 4 and 5 is P^+.

3. Periodic Solutions

Suppose that eq. (2) has a solution P, then the sampled equation (10) introduced in Sec. 2 clearly possesses such a solution P_s that $P_s = P$. For instance, if the system (Φ, G, H) is detectable and stabilizable, eq. (2) has unique nonnegative-definite solution P^+ (Lemma 5) and hence the sampled eq. (10) has P_s^+ $(=P^+)$. If the system is detectable but not stabilizable, then eq. (2) has another solution $P^{(i)}$ in addition to P^+ (Lemma 4) and accordingly the sampled equation has $P_s^{(i)}$ $(=P^{(i)})$. Evidently such arguments also apply to general solutions including nonpositive and indefinite ones. In any case, denoting by $\{\lambda_1, \ldots, \lambda_n\}$ the set of eigenvalues of A to derive P, the solution $P_s(=P)$ in the sampled system can be obtained from the corresponding set $\{\mu_1, \ldots, \mu_n\}$, $\mu_i = \lambda_i^p$ of eigenvalues of B.

On the other hand, it is possible that eq. (10) has solutions P_s that are not solutions of eq. (2). Namely, if eq. (1) has a periodic solution of period p, then it appears as a constant solution in the sampled equation. Such a case arises when B has multiple eigenvalues as will be explained below (note that distinct eigenvalues are assumed in matrix A).

We will first derive an existence condition of general real symmetric periodic solution of eq. (1) including nonnegative-, nonpositive-definite, and indefinite periodic solutions.

The following conditions are introduced.

(c.1) A has complex eigenvalues $\lambda_{1,2} = re^{\pm j\phi\pi}$ $(r>1,\ \phi \in (0,1)$ is rational)

(c.2) A has no eigenvalue λ such that $|\lambda| = r$ except for $\lambda_{1,2}$

(c.3a) All the eigenvalues λ of Φ such that $|\lambda|>1$ are observable.

(c.3b) All the eigenvalues λ of Φ such that $|\lambda|<1$ are observable.

First, the conditions (c.1)-(c.3a) are assumed. Notice that A has no eigenvalues on the unit circle, and (c.3a) implies that the pair (H, Φ) is detectable. Now, (c.1) and Lemma 1 imply that $\lambda_{3,4} = r^{-1}e^{\pm j\phi\pi}$ are also eigenvalues of A. Let z_i be the eigenvector associated with λ_i, $i=1,\ldots, 4$, and let $q/p(=\phi)$ be the irreducible fraction of ϕ. Then as long as $k \neq mp$ with m a natural number, $B=A^k$ has distinct eigenvalues μ_1 and $\mu_2 (\neq \mu_1)$. In this case, eigenvectors of B corresponding to μ_1 and μ_2

are z_1 and z_2, respectively. However when $k=p$, μ_1 and μ_2 coincide, namely $\mu_1 = (-1)^q r^p = \mu_2 (\triangleq \mu)$. Then, the vector

$$\eta = \alpha_1 z_1 + \alpha_2 z_2 = \alpha_1 \begin{bmatrix} y_1 \\ x_1 \end{bmatrix} + \alpha_2 \begin{bmatrix} y_2 \\ x_2 \end{bmatrix} \triangleq \begin{bmatrix} v_1 \\ u_1 \end{bmatrix} \tag{13}$$

becomes the eigenvector of B associated with μ for arbitrary (complex) scalars α_1 and α_2, i.e., $B\eta=\mu\eta$. Similarly, in this case, μ_3 and μ_4 coincide, i.e., $\mu_3 = \mu_4 = \mu^{-1}$, hence for arbitraly (complex) scalars β_1 and β_2,

$$\xi = \beta_1 z_3 + \beta_2 z_4 = \beta_1 \begin{bmatrix} y_3 \\ x_3 \end{bmatrix} + \beta_2 \begin{bmatrix} y_4 \\ x_4 \end{bmatrix} \triangleq \begin{bmatrix} v_2 \\ u_2 \end{bmatrix} \tag{14}$$

is the eigenvector of B associated with μ^{-1}, namely $B\xi = \mu^{-1}\xi$.

The above eigenvectors η and ξ are to be employed to constitute a periodic solution. Although this choice of eigenvalues violates the condition of Lemma 2($\mu_1=\mu_3^{-1}$), real-symmetricity of solution can still be guaranteed by a proper choice of α's and β's. Let ω_{sij} and $\bar{\omega}_{sij}$ be the ij-th element of $\Omega_s = X_s^* Y_s$ and $\bar{\Omega}_s = X_s^T Y_s$, then Hermiteness of Ω_s and summetricity of $\bar{\Omega}_s$ require

$$\text{Hermiteness:} \quad \omega_{s12} - \omega_{s21}^* = u_1^* v_2 - v_1^* u_2 = 0 \tag{15}$$

$$\text{symmetricity:} \quad \bar{\omega}_{s12} - \bar{\omega}_{s21} = u_1^T v_2 - v_1^T u_2 = 0 \tag{16}$$

Such parameters can finally be found as, for $s = x_1^* y_3 - y_1^* x_3$,

$$\alpha_1 = e^{j\theta}, \quad \alpha_2 = e^{-j\theta}, \quad \beta_1 = js^* e^{j\theta}, \quad \beta_2 = -jse^{-j\theta} \tag{17}$$

where θ is arbitrary as far as $0 \leq \theta < \pi$.

In the sequel, it will be shown that the vectors η and ξ in eqs. (13) and (14) with α's and β's determined in this manner lead to real symmetric solution P_s of sampled system. Let

$$Y_{s\theta} = [v_{2\theta} \; v_{1\theta} \; Y_2] \triangleq [v_{2\theta} \; Y_{1\theta}] \tag{18}$$

$$X_{s\theta} = [u_{2\theta} \; u_{1\theta} \; X_2] \triangleq [u_{2\theta} \; X_{1\theta}] \tag{19}$$

in which the parameter θ is employed to denote the explicit dependence of vectors on θ, and Y_2 and X_2 are derived from the eigenvectors of A corresponding to other (n-2) eigenvalues that lie outside the unit circle. The proof that $P_{s\theta}$ given by

$$P_{s\theta} = Y_{s\theta} X_{s\theta}^{-1} \tag{20}$$

is a real symmetric solution of sampled system consists of two steps: The first step is to prove that $\Omega_{s\theta} = X_{s\theta}{}^* Y_{s\theta}$ is Hermitian and $\bar{\Omega}_{s\theta} = X_{s\theta}^T Y_{s\theta}$ is symmetric (hence $P_{s\theta}$ is real symmetric if it exists), and the second step is to prove that $X_{s\theta}$ is nonsingular for any θ, $0 \leq \theta < \pi$.

For the first part, $\Omega_{s\theta}$ can be written as

$$\Omega_{s\theta} = \begin{bmatrix} u_{2\theta}^* v_{2\theta} & u_{2\theta}^* Y_{1\theta} \\ X_{1\theta}^* v_{2\theta} & X_{1\theta}^* Y_{1\theta} \end{bmatrix} \tag{21}$$

Here, we have

$$u_{2\theta}^* Y_{1\theta} - v_{2\theta}^* X_{1\theta} = [u_{2\theta}^* v_{1\theta} - v_{2\theta}^* u_{1\theta}, \; u_{2\theta}^* Y_2 - v_{2\theta}^* X_2] = 0 \tag{22}$$

where eq. (15) and Lemma 2 are used. Furthermore since $X_{1\theta}^{*}Y_{1\theta} = Y_{1\theta}^{*}X_{1\theta}$ from Lemma 2, $\Omega_{s\theta}$ is Hermitian. Similar aragaments apply to the symmetricity of $\bar{\Omega}_{s\theta}$, but its proof is omitted.

As for the nonsingularity of $X_{s\theta}$, assuming that $X_{s\theta}$ is singular for some θ, $0 \leq \theta < \pi$, we will show that it leads to a contradiction. Under this assumption, there exists an (n-1)-dimensional vector $b \neq 0$ such that

$$u_2 = X_1 b \qquad (23)$$

because rank X_1 = n-1 owing to the detectability of (H,Φ). Note that here, and hereafter, the parameter θ is dropped for the sake of notational simplicity. Using eq. (23), we have

$$(b^{*}Y_1^{*} - v_2^{*})X_1 = b^{*}X_1^{*}Y_1 - v_2^{*}X_1 = u_2^{*}Y_1 - v_2^{*}X_1 = 0 \qquad (24)$$

On the other hand, we have from eqs. (11), (18) and (19)

$$B_{21}v_2 + B_{22}u_2 = \mu^{-1}u_2, \quad B_{21}Y_1 + B_{22}X_1 = X_1M_1 \qquad (25)$$

where $M_1 = \mathrm{diag}\{\mu, \mu_3, \ldots, \mu_n\}$, and from which it can be derived that

$$B_{21}(Y_1M_1^{-1}b - \mu v_2) + B_{22}(X_1M_1^{-1}b - \mu u_2) = 0 \qquad (26)$$

Since the inverse of B can be obtained by Lemma 1 as

$$B^{-1} = \begin{bmatrix} B_{22}^T & -B_{12}^T \\ -B_{21}^T & B_{11}^T \end{bmatrix} \qquad (27)$$

analogous procedure yields the following equation:

$$B_{22}^T(Y_1b - v_2) = Y_1M_1^{-1}b - \mu v_2 \qquad (28)$$

Using eqs. (23) and (26) in (28), we obtain

$$(Y_1b - v_2)^{*}B_{21}B_{22}^T(Y_1b - v_2) + b^{*}(M_1^{-1} - \mu I)^{*}Y_1^{*}X_1(M_1^{-1} - \mu I)b = 0 \qquad (29)$$

Since $B_{21}B_{22}^T = \Phi_p^{-T}H_p^TH_p\Phi_p^{-1} \geq 0$ from eq. (9) and $X_1^{*}Y_1 \geq 0$ from Lemma 3, the left hand side of above equation is the sum of nonnegative terms. Hence each term must be zero, yielding to

$$B_{21}B_{22}^T(Y_1b - v_2) = 0 \qquad (30)$$

Eqs. (30), (28) and (26) now lead to

$$X_1(M_1^{-1} - \mu I)b \triangleq X_1b_1 = 0 \qquad (31)$$

Since $|\lambda(M_1)| > 1$ and $|\mu| > 1$ imply $b_1 \neq 0$, this shows that X_1 is of rank less than (n-1), a contradiction. Hence $X_{s\theta}$ must be nonsingular for all θ, $0 \leq \theta < \pi$.

In the above, it has been proved that $P_{s\theta}$ in eq. (20) is a solution of sampled Riccati equation (10). Note that it is not a solution of (2), and therefore it derives a periodic solution of eq. (1) as shown below.

Suppose that the initial condition P(0) of eq. (1) is $P(0) = P_{s\theta} = Y_{s\theta}X_{s\theta}^{-1}$. Then the corresponding solution $P_{s\theta}(k)$ of eq. (1) is, as shown in eqs. (7) and (8), given by

$$P_{s\theta}(k) = Y_{s\theta}(k)\, X_{s\theta}(k)^{-1} \qquad (32)$$

where

$$\begin{bmatrix} Y_{s\theta}(k) \\ X_{s\theta}(k) \end{bmatrix} = A^k \begin{bmatrix} Y_{s\theta}(0) \\ X_{s\theta}(0) \end{bmatrix} = A^k \begin{bmatrix} Y_{s\theta} \\ X_{s\theta} \end{bmatrix} \tag{33}$$

Computations of matrices in this equation can further be simplified, as will be presented later in the algorithm. Clearly, $P_{s\theta}(k)$ is a periodic solution of eq. (1).

So far, the periodic solutions are derived from the solution P^+ of eq. (2). More specifically, they are constructed from P^+ by replacing eigenvectors corresponding to λ_1 and λ_2 by those of μ and μ^{-1}. If the condition (c.3b) is assumed, then another periodic solutions can also be derived from P^- in the similar way. The proof of existence of such periodic solution is mostly repetitive and it will be omitted.

Above all, we have the existence theorem of periodic solution.

Theorem 1 The conditions (c.1), (c.2) and, (c.3a) and/or (c.3b) are assumed. Then, there exists a real symmetric periodic solution $P_{s\theta}(k)$ of period p in eq. (1).

The above discussion is summarized to yield a computational algorithm of periodic solutions.

Algorithm

1) Compute eigenvectors $\begin{bmatrix} y_i \\ x_i \end{bmatrix}$, i=1,2,3,4, of A corresponding to complex eigenvalues $\lambda_{1,2} = re^{\pm j\phi\pi}$ and $\lambda_{3,4} = r^{-1}e^{\pm j\phi\pi}$, and find the period p.
2) Compute s by $s = x_1^* y_3 - y_1^* x_3$.
3) Compute the following vectors;

$$\begin{bmatrix} v_{1\theta}(k) \\ u_{1\theta}(k) \end{bmatrix} = \mathrm{Re}\left[e^{j(\theta+k\phi\pi)} \begin{pmatrix} y_1 \\ x_1 \end{pmatrix} \right], \quad \begin{bmatrix} v_{2\theta}(k) \\ u_{2\theta}(k) \end{bmatrix} = \mathrm{Re}\left[(js^*)e^{j(\theta+k\phi\pi)} \begin{pmatrix} y_3 \\ x_3 \end{pmatrix} \right]$$

4) Derive the eigenvector matrix $\begin{bmatrix} Y_2 \\ X_2 \end{bmatrix}$ corresponding to (n-2) eigenvalues of A that lie outside or inside the unit circle excluding $\lambda_{1,2}$ depending on whether (c.3a) or (c.3b) is satisfied.
5) Compute the periodic solution $P_{s\theta}(k)$, k=0,..., p-1 by
$P_{s\theta}(k) = [v_{2\theta}(k)\ v_{1\theta}(k)\ Y_2][u_{2\theta}(k)\ u_{1\theta}(k)\ X_2]^{-1}$

Remark Under the assumptions in Theorem 1, the sampled equation (10) has infinite number of solutions $P_{s\theta}$ where θ ranging over $[0,\pi)$, and correspondingly, eq. (1) has a set of periodic solutions $P_{s\theta}(k)$.

In particular, since nonnegative solutions are of practical interest in optimal estimation and control problems, we now state existence condition of such periodic solutions.

Theorem 2 Assume that the conditions (c.1)-(c.3a) are satisfied. If the complex eigenvalues $\lambda_{1,2}$ in (c.1) are uncontrollable, then eq. (1) has real symmetric nonnegative-definite periodic solutions $P_{s\theta}(k)$ of period p. (proof omitted)

It should be noted that the conditions in Theorem 2 imply that the system is unstabilizable.

4. Illustrative Example

Consider the following second order system:

$$\Phi = \begin{bmatrix} 0 & 1 \\ -2 & 2 \end{bmatrix} \qquad G = \begin{bmatrix} 0 & 0 \\ 0 & 0 \end{bmatrix} \qquad H = \begin{bmatrix} 0 & 1 \\ 0 & 0 \end{bmatrix} \tag{34}$$

The eigenvalues of A matrix are found to be $\lambda_1 = 1+j$, $\lambda_2 = \lambda_1^*$, $\lambda_3 = \frac{1}{2}(1+j)$, $\lambda_4 = \lambda_3^*$. The system is detectable but not stabilizable and we can easily verify that the conditions in Theorem 2 are satisfied. Hence eq. (1) has a nonnegative-definite periodic solution with period p=4 since $\lambda_{1,2} = 1\pm j = \sqrt{2}\, e^{\pm j\frac{\pi}{4}}$. This is computed via the algorithm presented in Sec. 3 as

$$P_{s\theta}(k) = \frac{5}{2\gamma_\theta(k)} \begin{bmatrix} \cos(2\theta+\frac{\pi}{2}k)+1 & \cos(2\theta+\frac{\pi}{2}k)-\sin(2\theta+\frac{\pi}{2}k)+1 \\ \cos(2\theta+\frac{\pi}{2}k)-\sin(2\theta+\frac{\pi}{2}k)+1 & -2\sin(2\theta+\frac{\pi}{2}k)+2 \end{bmatrix} \tag{35}$$

where, $\gamma_\theta(k) = \sqrt{5}\cos(2\theta + \frac{\pi}{2}k-\phi)+5\neq 0$ and $\phi=\tan^{-1}\frac{1}{2}$. The above $P_{s\theta}(k)$ is periodic with period p=4 and $P_{s\theta}(k) \geq 0$ as is guaranteed by Theorem 2. In particular, $P_{s\theta}(0) = P_{s\theta}$ is the solution of sampled equation (10). Moreover, the system has two steady-state solutions

$$P^+ = \begin{bmatrix} 3/4 & 1 \\ 1 & 3 \end{bmatrix} \qquad P^- = \begin{bmatrix} 0 & 0 \\ 0 & 0 \end{bmatrix} \tag{36}$$

and it can be easily verified that $P^- \leq P_{s\theta}(k) \leq P^+$.

In Fig.1, $P_{s\theta}$ and $P_{s\theta}(k)$ are shown in three dimensional space p_{11}, $p_{12}(=p_{21})$, p_{22} together with P^+ and P^-. $P_{s\theta}$ for various values of $\theta(0 \leq \theta < \pi)$ forms an ellipse on the singular cone ∂P^+, which describes the surface such that $p_{11}p_{22}-p_{12}^2=0$ and its inside corresponds to $P > 0$. As we see easily, $P_{s\theta}$ is the contour of periodic solutions $P_{s\theta}(k)$, k=0,..., p-1 for various θ. Furthermore, it can be shown that if the initial condition P(0) of eq.(1) lies on the cone ∂P^+, then the corresponding solution P(k) converges to a periodic solution $P_{s\theta}(k)$ for some θ, $0 \leq \theta < \pi$.

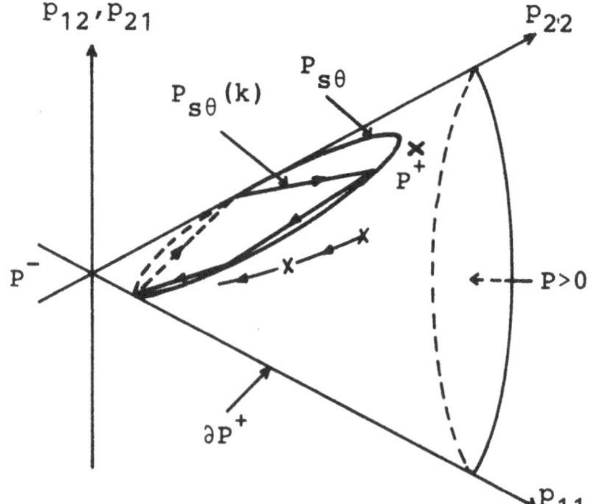

Fig.1 Singular cone and periodic solutions in three dimensional space.

5. Concluding Remarks

Periodic solutions are investigated of discrete matrix Riccati equations arising in optimal estimation as well as control problems of discrete time-invariant systems. The results have been derived by introducing the "sampled" Riccati equation, and by means of eigenvalue-eigenvector decomposition of the so-called symplectic matrix.

Existence Theorems 1 and 2 are established of periodic solutions in Riccati equation (1), and in particular we see that an unstabilizable complex mode derives nonnegative-definite periodic solutions. Notice that ϕ in the complex eigenvalues $\lambda_{1,2}$ in condition (c.1) has to be a rational number as differed from the continuous-time case. Quasi-periodic solutions are expected when ϕ is irrational. Also, a set of periodic solutions is derived from a pair of complex eigenvalues, whereas such eigenvalues yielded a periodic solution in the continuous-time case.

Furthermore, an algorithm is developed to determine such periodic solutions, which requires only a simple additional computation to deriving the steady-state (constant) solutions.

The results are exemplified in Sec. 4 and the convergence study of this example shows that certain nonnegative-definite initial condition P(0) renders the solution P(k) converge to a periodic solution.

Thus, possible application of this results may be for the examination of unstable phenomenon in Riccati equations, which are encountered, e.g., in the application of Kalman filter.

References

[1] R.E. Kalman: New methods and results in linear prediction and estimation theory, RIAS Rep. 61-1, Baltimore, MD, 1961.

[2] V. Kucera: A contribution to matrix quadratic equations, IEEE Trans. Automat. Contr., vol.AC-17, pp.344-347, June 1972.

[3] W.M. Wonham: On a matrix Riccati equation of stochastic control, SIAM J. Contr., vol.6, no.4, pp. 681-697, 1968.

[4] T. Nishimura: On the algebraic solution of matrix Riccati equations in discrete systems, presented at the 6th Symp. on Control Theory, Soc. Instrument and Contr. Eng. of Japan, Tokyo, May 1977.

[5] H. Kano and T. Nishimura: Periodic solutions of matrix Riccati equations with detectability and stabilizability, Int. J. Contr., vol.29, pp.471-487, Mar. 1979.

[6] M. Hayase: On the periodic solutions of Riccati equations (in Japanese), presented at the 6th Symp. on Control Theory, Soc. Instrument and Contr. Eng. of Japan, Tokyo, May 1977.

[7] T. Nishimura and H. Kano: Periodic oscillations of matrix Riccati equations in time-invariant systems, IEEE Trans. Automat. Contr., vol.AC-25, pp.749-755, Aug. 1980.

[8] D.R. Vaughan: A nonrecursive algebraic solution for the discrete Riccati equation, IEEE Trans. Automat, Contr., vol.AC-15, no.5, Oct. 1970.

A NEW SCHEME FOR DISCRETE IMPLICIT OBSERVER AND CONTROLLER

Myoung S. Ko and Uk Y. Huh
Dept. of Instrumentation and Control Eng.
Seoul National University
Seoul 151, KOREA

SUMMARY--Many different schemes of the adaptive observer and controller have been developed for both continuous and discrete system. In this paper we present a new scheme of the adaptive observer for the discrete linear system. The adaptation algorithm is derived based on the exponentially weighted least square method. The adaptive model following control system is also constructed according to the proposed observer scheme. The proposed observer and controller are simple structure and have fast convergence characteristic in adaptation algorithm. The effectiveness of the algorithm and structure are illustrated by the computer simulation of a third order system.

I. INTRODUCTION

The adaptive observer is a model reference adaptive scheme generating the inaccessible state variables of the unknown plant with only input and output measurement. And also the adaptive observer is essential to the adaptive model following indirect control. Since the discretization of the continuous algorithm is not suitable to the digital computer implementation, it is desirable to develop a simple and fast convergence adaptive observer for a adaptive system design.

Significant contributions to the adaptive continuous and discrete observer have been made by many authors [1]-[5]. Recently Kreisselmeier [6] proposed the parameterized adaptive observer and Suzuki, Nakamula and Koga [7] proposed its discretized form. Nuyan and Carroll [8] showed that the implicit observer is not necessary any auxiliary signals.

In case of adaptive control system, several schemes have been reported by many authors [9]-[11].

In this paper we propose a new scheme for the discrete implicit reduced order observer. Since the proposed scheme has fast convergence characteristics, it will be able to use for design of an indirect discrete adaptive control system design.

II. THE DISCRETE REDUCED ORDER OBSERVER AND CONTROLLER

Consider the single input single output linear discrete time invariant system described by

$$x_p(k+1)=A_p x_p(k)+b_p u(k)$$
$$y_p(k)=c^t x_p(k), \qquad x_p(0)=x_o \tag{1}$$

where $x_p(k)$ is the nth order state vector of the unknown plant, u(k) is a scalar input of the plant and A_p, b_p and c are nxn, nx1 and nx1 matrices respectively. Then we may assume, without loss of generality, that plant (1) is the following observable canonical form,

$$A_p=\left[\begin{array}{c|c} -a_p & \begin{array}{c} I_{n-1} \\ \hline 0 \end{array} \end{array}\right], \quad a_p=\begin{bmatrix} a_{p1} \\ a_{p2} \\ \vdots \\ a_{pn} \end{bmatrix}, \quad b_p=\begin{bmatrix} b_{p1} \\ b_{p2} \\ \vdots \\ b_{pn} \end{bmatrix}, \quad c=\begin{bmatrix} 1 \\ 0 \\ \vdots \\ 0 \end{bmatrix}$$

where vector a_p, b_p and x_o are unknown and I_{n-1} is a (n-1)st order unit matrix. Since the plant (1) is a time invariant observable system, its states can be estimated asymtotically by means of Luenburger's reduced order observer [12] such that

$$v(k+1)=Fv(k)+gy(k)+hu(k)$$
$$\hat{x}_1(k)=y_p(k) \tag{2}$$
$$\hat{x}_p(k)=v(k)+l_o y_p(k)$$

where F is an asymtotically stable matrix whose eigenvalues are located in the unit circle and different from those of A_p. The basic principle of the adaptive observer is to adjust g and h adaptively in real time to cause $\lim_{k\to\infty}\hat{x}_p(k)=x_p(k)$.

Taking the z-transformation of (1), we have the following output equation

$$(z^n+a_p^t d(z))Y_p(z)=b_p^t d(z)U(z)+zd^t(z)x_o \tag{3}$$

where $d(z)=[z^{n-1}\ z^{n-2}\dots\ z\ 1]$,

In order to investigate the relationship between (3) and state variable filter(S.V.F.) we introduce a nth order polynomial Q(z) and (n-1)st order polynomial M(z)

$$Q(z)=z^n+q^t d(z)$$
$$M(z)=z^{n-1}+m^t\bar{d}(z)$$
$$Q(z)=(z-\alpha_1)M(z)$$

where $q=[q_1\ q_2..q_n]^t$, $m=[m_1\ m_2..m_{n-1}]^t$ and $\bar{d}(z)=[z^{n-2}..z\ 1]^t$.

The elements of vector q and m will be chosen so that the roots lie within the unit circle and α_1 is a real root. (3) can be written by

$$(z-\alpha_1)Y_p(z)=(q_1-a_{p1})Y_p(z)+p_1^t \frac{\bar{d}(z)}{M(z)} Y_p(z)+b_{p1}U(z) + p_2^t \frac{\bar{d}(z)}{M(z)} U(z)+ \Theta^t(z)\bar{x}_o \qquad (4)$$

where $p_1=\bar{q}-\bar{a}_p-(q_1-a_{p1})m$, $p_2=\bar{b}_p-mb_{p1}$, $\Theta(z)=\mathcal{Z}[\theta(k)]=\frac{zd(z)}{M(z)}$, $\bar{a}_p=[a_{p2}\cdots a_{pn}]^t$, $\bar{b}_p=[b_{p2}\cdots b_{pn}]^t$, and $\bar{q}=[q_2\cdots q_n]^t$.

Since the elements of $\bar{d}(z)$ are linearly independent $\frac{d(z)}{M(z)}Y_p(z)$ and $\frac{d(z)}{M(z)}U(z)$ can be generated from (n-1)st order stable and controllable state variable filters with characteristic polynomial M(z). Then (n-1)-st controllable state variable filters are given by

$$\begin{aligned} r_1(k+1)&=M_1r_1(k)+l_1y_p(k) \\ r_2(k+1)&=M_2r_1(k)+l_2u(k) \end{aligned} \qquad (5)$$

where M_1, M_2 : (n-1)X(n-1) matrices

l_1, l_2 : (n-1) vector

there exist (n-1)st order nonsingular transformation matrices T_1, T_2 if the filters are controllable. Using these facts, the plant output can be written as

$$y_p(k+1)=\alpha_1y_p(k)+p^tr(k)+\theta(k)^t\bar{x}_o \qquad (6)$$

where $p^t=[q_1-a_{p1} \vdots p_1^tT_1 \vdots b_{p1} \vdots p_2^tT_2]$, $r^t=[y_p \vdots r_1^t \vdots u \vdots r_2^t]$

If we choose $l_o=m$, $g=\hat{p}_1(k)$, $h=\hat{p}_2(k)$ and det $[zI-F]=M(z)$ in (2), the reduced order discrete observer can be constructed.

Now the observer state v(k) can be constructed from the algebraic transformation of the state r(k) and we assume that their operators will be given by $H_1(\hat{p}_1)$, $H_2(\hat{p}_2)$ then the state vector of the observer is written as

$$v(k)=H_1(\hat{p}_1)r_1(k)+H_2(\hat{p}_2)r_2(k) \qquad (7)$$

Using (2) and (7), H_1 and H_2 can be derived as following

$$H_1=[(T_1^t)^{-1}\hat{p}_1 \vdots \cdots \vdots F^{n-2}(T_1^t)^{-1}\hat{p}_1][l_1 \vdots M_1l_1 \vdots \cdots \vdots M_1^{n-2}l_1]^{-1}$$

$$H_2=[(T_2^t)^{-1}\hat{p}_2 \vdots \cdots \vdots F^{n-2}(T_2^t)^{-1}\hat{p}_2][l_2 \vdots M_2l_2 \vdots \cdots \vdots M_2^{n-2}l_2]^{-1}$$

In addition to these results, the discrete reduced order adaptive observer can be formulated by the correct identification of the parameter p. We describe an indirect control scheme for discrete model following adaptive control using the proposed adaptive observer scheme. Now we assume that reference model system is described by

$$x_m(k+1)=A_m x_m(k)+b_m u_m(k)$$
$$y_m(k)=c^t x_m(k) \tag{8}$$

where $y_m(k)$, $x_m(k)$ and $u_m(k)$ are the model output, model state vector and reference input respectively and coefficient matrices of the model A_m, b_m and c are also observable canonical form.

At first we express the output error in terms of p and r(k) as observer scheme and then pick the control input u(k) so that output error $e_1(k)$ may approach to zero as $\hat{p}(k)\to p$. Finally we expect that the plant output will follow to the model's one. The output error $E_1(z)$ can be rearranged as previous scheme.

$$(z-\alpha_1)E_1(z)=(z-\alpha_1)Y_m(z)-(q_1-a_{p1})Y_p(z)-p_1^t\frac{\bar{d}(z)}{M(z)}Y_p(z)-b_{p1}U(z)-p_2^t\frac{\bar{d}(z)}{M(z)}U(z) \tag{9}$$

From above equation we pick the control input u(k) for model following control as follows:

$$u(k)=\frac{1}{b_{p1}}(y_m(k+1)-\alpha_1 y_m(k)-(q_1-a_{p1})y_p(k)-\hat{p}_1^t T_1 r_1(k)-\hat{p}_2^t T_2 r_2(k)) \tag{10}$$

In order to investigate the stability of the prescribed adaptive model following control system, we rearrange the (9) and (10) into as follows:

$$(z-\alpha_1)E_1(z)=\mathcal{Z}((\hat{p}(k)-p)^t r(k)) \tag{11}$$

If $\hat{p}(k)$ approaches to p as k increases, the output error $E_1(z)$ approach to zero for α_1 lies in unit circle of z-plane. In next section we propose such a stable algorithm. Fig. 1 and Fig. 2 show a structure of discrete reduced order adaptive observer and a structure of an adaptive controller respectively.

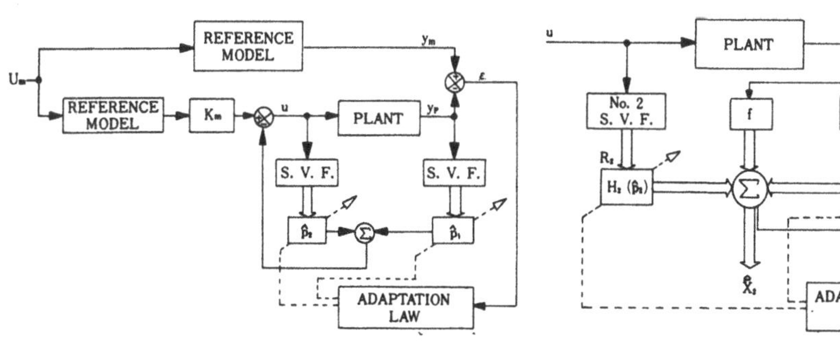

Fig.1 Block diagram of reduced order adaptive observer.

Fig.2 Block diagram of adaptive model following control system.

III. ADAPTATION ALGORITHM AND SIMULATION

We use the exponentially weighted least square method which is good convergence characteristic. If we denote $\hat{p}(k)$ as the estimate of parameter p at kth-iteration then from (6) the estimated plant output is written as

$$\hat{y}_{pk}(j+1)=\alpha_1 y_p(j)+\hat{p}^t(k)r(j)+cF^{j+1}v_0, \quad (j=0,1,\dots,k.\ v(0)=v_0) \tag{12}$$

We introduce the following criterion function to get the algorithm for error minimization.

$$J(k)=\sum_{j=0}^{k}\beta\lambda^{k-j}(\hat{y}_{pk}(j+1)-y_p(j+1))^2, \quad 0<\lambda<1, \quad \beta=1-\lambda \tag{13}$$

By taking gradient of J(k) with respect to $\hat{p}(k)$ and making zero, J(k) becomes minimum at each k. Therefore the following equations can be derived

$$\hat{p}(k)=[R^t(k)W(k)R(k)]^{-1}R^t(k)W(k)\Omega(k)=\Gamma(k)R^t(k)W(k)\Omega(k) \tag{14}$$

where $R(k)=[r(0)\,|\,r(1)\,|\,\dots\,|\,r(k)]^t$, $\quad \Omega(k)=[y_p(1),\ y_p(2),\dots,y_p(k+1)]^t$

$$W(k)=\begin{bmatrix}\beta\lambda^k & 0 & \cdots & 0\\ 0 & \beta\lambda^{k-1} & & \\ \vdots & & \ddots & \\ 0 & & & \beta\end{bmatrix}$$

From these equations, we can derive the following recursive equations.

$$\begin{aligned}\hat{p}(k+1)&=\hat{p}(k)-L(k+1)(\hat{y}_{pk}(k+1)-y_p(k+1))\\ \Gamma(k+1)&=\frac{1}{\lambda}[I-L(k+1)r^t(k+1)]\Gamma(k)\end{aligned} \tag{15}$$

where $L(k+1)=\frac{\Gamma(k)}{\lambda}r(k+1)\left(\frac{1}{\beta}+r^t(k+1)\frac{\Gamma(k)}{\lambda}r(k+1)\right)^{-1}$

The following lemma and theorem show the convergence of the proposed recursive algorithm.

Lemma: $\Gamma(k)$ defined by $\Gamma(k)=[R^t(k)W(k)R(k)]^{-1}$ is positive definite if W(k) is positive definite.

Proof: We claim that $\Gamma(k)$ is positive definite if the matrix $R^t(k)W(k)\cdot R(k)$ is positive definite. The elements of R(k) have the following properties

$$y_p(k)=\sum_{i=0}^{k-1} c^tA_p^{k-j}b_pu(i)$$

$$r_1(k)=\sum_{i=0}^{k-1} M_1^{k-i}l_1\sum_{j=0}^{i-1} c^tA_p^{i-j}b_pu(j)\triangleq\sum_{j=0}^{k-2}S_ju(j)$$

$$r_2(k)=\sum_{i=0}^{k-1} M_2^{k-i}l_2u(i)$$

Then R(k) is expressed by

$$R^t(k)=\begin{bmatrix}0 & c^tA_pb_p & c^tA_p^2b_p & \cdots & c^tA_p^kb_p\\ \underline{0} & \underline{0} & S_{k-2} & \cdots & S_0\\ 1 & 0 & 0 & \cdots & 0\\ \underline{0} & M_2l_2 & M_2l_2 & \cdots & M_2^kl_2\end{bmatrix}\cdot\begin{bmatrix}u(0) & u(1) & \cdots & u(k)\\ 0 & u(0) & & u(k-1)\\ \vdots & & \ddots & \vdots\\ 0 & & & u(0)\end{bmatrix}$$

$\triangleq S \times U$ (16)

where S is $2n\times(k+1)$ matrix and U is $(k+1)\times(k+1)$ matrix. Since S.V.F. and the plant are controllable system, the rank of S is 2n (for $k\geq 2n-1$) and the rank of U is k+1. So we can state that rank of R(k) is 2n. If W(k) is positive definite, $R^t(k)W(k)R(k)$ is positive definite. Therefore $[R^t(k)W(k)R(k)]^{-1}$ is positive definite also. Q.E.D.

Theorem: The value of $\Gamma(k)$ calculated by (15) is equal to $[R^t(k)W(k)\cdot R(k)]^{-1}$ for any initial value, and $\hat{p}(k)$ converges to p asymptotically.

Proof: $\Gamma(k)$ calculated by (15) can be described by

$$\Gamma(k)=[\beta\lambda^k\Gamma_o^{-1}+\sum_{j=0}^{k}\beta\lambda^{k-j}r(j)r^t(j)]^{-1} \quad (17)$$

Since $0<\lambda<1$, the effect of initial value Γ_o can be negligible as k approaches to infinite. Therefore $\Gamma(k)$ will be expressed by $[R^t(k)W(k)\cdot R(k)]^{-1}$ and positive definite, so $\hat{p}(k)$ converges to p asymptotically. Q.E.D.

In order to show the effectiveness of the observer and controller's structure and adaptation algorithm, the following third order plant are considered for simulation. The plant is

$$x_p(k)=\begin{bmatrix}-1 & 1 & 0\\ -0.31 & 0 & 1\\ -0.03 & 0 & 0\end{bmatrix}x_p(k) + \begin{bmatrix}1\\1\\1\end{bmatrix}u(k), \quad y_p(k)=[1\ 0\ 0]\,x_p(k) \quad (18)$$

and the model is

$$x_m(k)=\begin{bmatrix}-0.08 & 1 & 0\\ -0.104 & 0 & 1\\ -0.0096 & 0 & 0\end{bmatrix}x_m(k) + \begin{bmatrix}0.8\\1.0\\1.2\end{bmatrix}u_m(k), \quad y_m(k)=[1\ 0\ 0]x_m(k) \quad (19)$$

The reference input is stair-case waveform. In the adaptation algorithm λ is given by 0.95. In order to evaluate the effect of the S.V.F. characteristic polynomial, we take the following three cases for computer simulation.

First: The poles of filter are located at the origin of z-plane.

Second: The poles of the filter and that of the model are located at the same place: (0.2, 0.12, -0.4).

Third: The poles of the filter are located near inside of unit circle: (0.9, 0.7, -0.8).

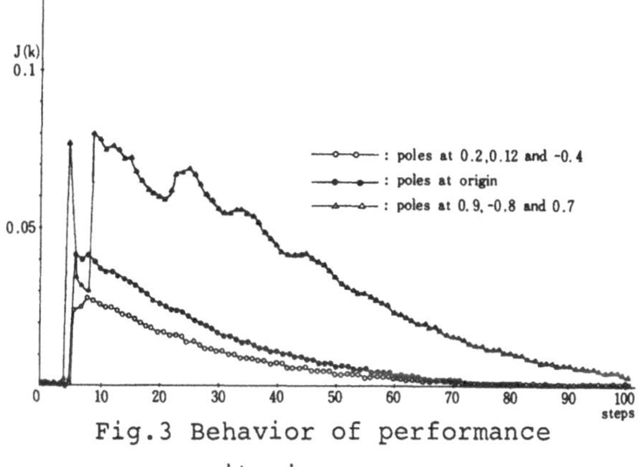

Fig.3 Behavior of performance criterion

Fig.3 shows the effectiveness of the proposed algorithm in view of performance criterion. The fact that the value of J(k) are very small in initial 6 steps means that the number of output sequences are less than that of coefficients to be adapted. In adaptive observer simulation, we investigate the state error distance(σ), which is defined by $\left(\sum_{j=1}^{n}(\hat{x}_j-x_j)^2\right)^{\frac{1}{2}}$ to show the convergence characteristics.

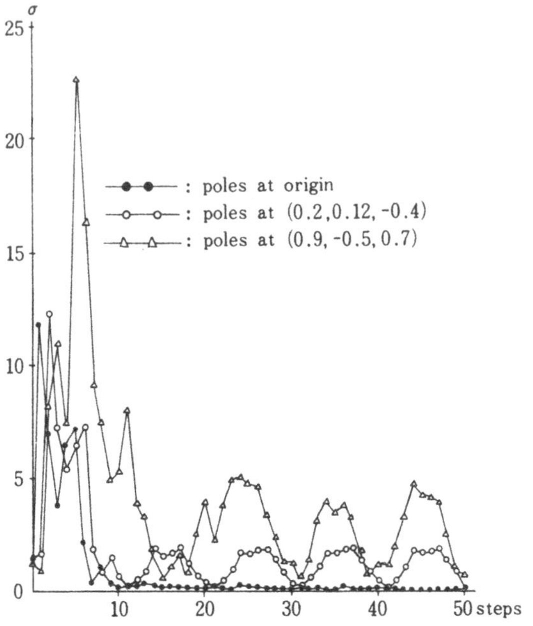

Fig.4 State error distance characteristics with λ=0.95.

Fig.4 shows the state error distance plot of three cases. The first case has the best convergence charateristics such that the state error distance converges to 5% of initial distance in 40 steps approximately. The convergence of the other two cases are not apparent. Maybe it is due to the relative position between the poles of plant and the poles of S.V.F..

For the adaptive model following control system, the output of the reference model and that of plant are shown in Fig.5. The second case has the best convergence characteristics such that the output error reaches within 5% of initial error in 40 steps. Fig.6 shows the comparisons of the state error distance

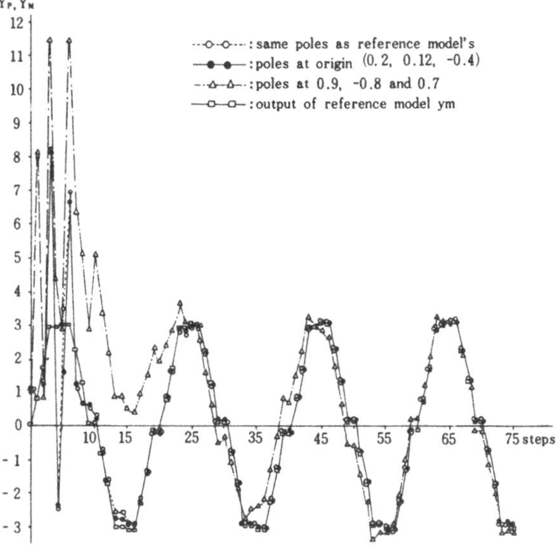

Fig.5 Plant and reference model output y_p and y_m with λ=0.95.

variation between the reduced order observer and the full order observer due to the value of λ in the first case. The convergence characteristics are similar to each other.

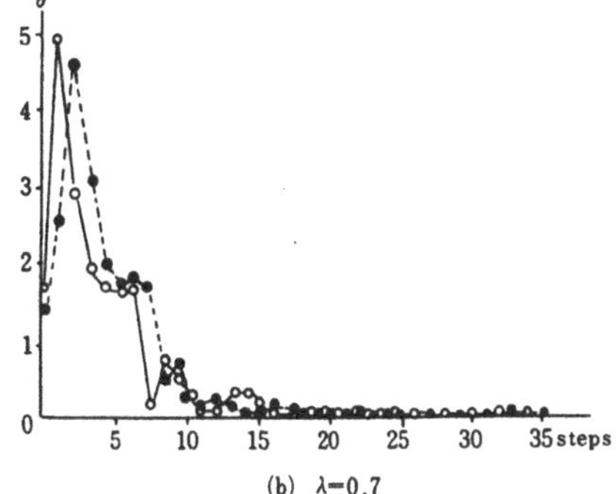

(b) $\lambda=0.7$

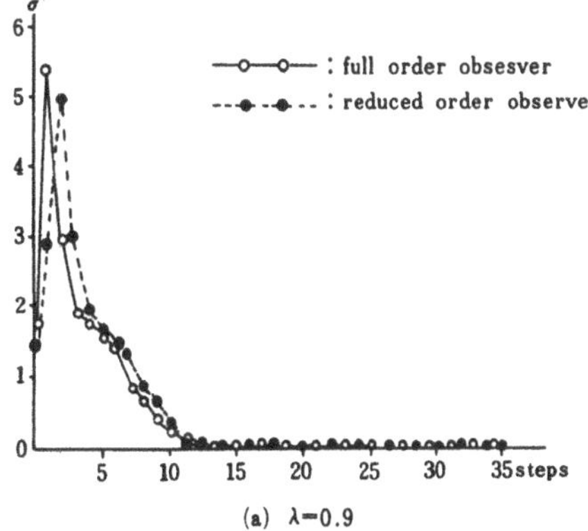

(a) $\lambda=0.9$

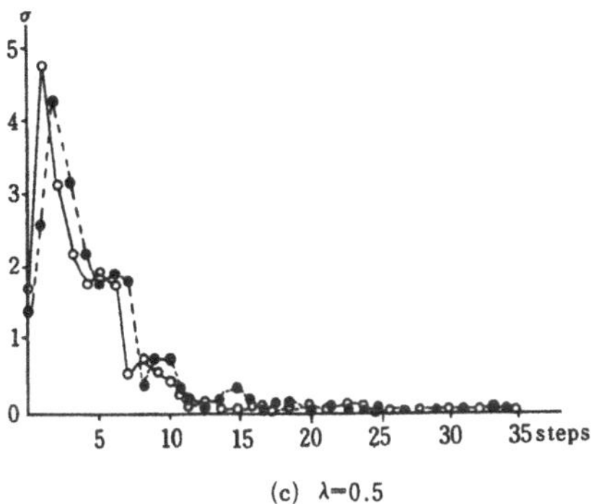

(c) $\lambda=0.5$

Fig.6 Comparison of the state error distance variation between the reduced and the full order observer.

IV. CONCLUSION

In this paper we have presented the discrete version of an adaptive reduced order observer without auxiliary signals. The adaptation law is derived based on the exponentially weighted least square method which has good convergence characteristic in deterministic system. And we derive the adaptive model following control system which is a similar structure to the reduced order adaptive observer.
From these results we can state that the reduced order discrete adaptive implicit observer and adaptive model following controller have the simple structure and fast convergence characteristics. Therefore it is desirable to apply these algorithms to the practical design problems. And these schemes can be expanded to the multi-input case easily. The optimal choice of M(z) and the biasing problem in adaptation algorithm are the future problem to be considered.

REFERENCES

[1] Carroll R.L. and Lindorff D.P., "An Adaptive Observer for Single-Input Single-Output Linear Systems", IEEE Trans. Automat. Contr., vol. AC-18, pp. 428-434 Oct. 1973.

[2] Lüders G. and Narendra K.S., "An Adaptive Observer and Identifier for a Linear System", IEEE Trans. Automat. Contr., vol. AC-18, pp. 496-499, Oct. 1973.

[3] Kudva P. and Narendra K.S., "Synthesis of an Adaptive Observer Using Lyapunov's direct method", Int. J. Contr., vol. 18, pp. 1201-1210, Dec. 1973.

[4] Kudva P. and Narendra K.S., "The Discrete Adaptive Observer", Proc. 1974 Conference on Decision and Control, Phoenix AZ, Nov. 20-22, pp. 307-312.

[5] Suzuki T. and Andoh M., "Design of a Discrete Adaptive Observer based on Hyperstability Theorem", Int. J. Contr., vol. 26, No. 4, pp. 643-653, 1977.

[6] Kreisselmeier G., "Adaptive Observers with Exponential Rate of Convergence", IEEE Trans. Automat. Contr.,vol. AC-22, pp. 2-8, Feb. 1977.

[7] Suzuki. T., Nakamura T and Koga M., "Discrete Adaptive Observer with fast Convergence", Int. J. Contr., vol. 31, No. 6, pp. 1107-1119, 1980.

[8] Nuyan S. and Carroll R.L., "Minimal Order Arbitrarily Fast Adaptive Observers and Identifiers", IEEE Trans. Automat. Contr. vol. AC-24, Apr. 1979.

[9] Narendra K.S. and Valavani L.S., "Stable Adaptive Controller Design-Direct Control", IEEE Trans. Automat. Contr., vol. AC-23, No. 4, Aug. 1978.

[10] Monopoli R.V., "Model Reference Adaptive Control with an Augmented Error", IEEE Trans. Automat. Contr., vol. AC-19, No. 5, Oct. 1974.

[11] Landau I.D., "Adaptive Control-The Model Reference Approach", New York: Dekker, 1979.

[12] Luenberger D.G. "An Introduction to Observers", IEEE Trans. on Automat. Contr., vol. AC-16, No. 6, Dec. 1971.

REACHABLE SETS AND GENERALIZED BANG-BANG PRINCIPLE FOR LINEAR CONTROL SYSTEMS

A.V. Levitin
Department of Mathematics
University of Kentucky
Lexington, Kentucky 40506

The notion of attainable set of the control system is one of the principal concepts of the optimal control theory. It has a clear geometric meaning and its properties are crucial for the analytic investigation of the system. The fundamental principle of the theory of linear systems - the famous bang-bang principle - can be naturally expressed in terms of attainable sets. Namely (see [1], p.46), the attainable set $X_t(U)$ at a moment t will not change if the set of admissible controls U is expanded to the set U_c by replacing each set of admissible control values by its convex hull, i.e.

$$X_t(U) = X_t(U_c) \ . \tag{1}$$

The equality (1) implies immediately that the attainable sets are convex and compact even if the sets of admissible control values at each moment t are not necessarily convex. This result was first proved for ordinary differential systems in [2] and since then was established for other types of linear systems without phase constraints (see the bibliography in [3]). In fact, the equality (1) holds for attainable sets of the following abstract system defined by a set of admissible controls

$$U = \{u \in L_p(T_1, E^r) : u(t) \in D(t) \quad \text{for a.e.} \quad t \in T_1\} \tag{2}$$

and an affine operator

$$A: u \in U \to x \in C(T_2, E^m) \tag{3}$$

if the latter is continuous when U is endowed with the weak-L_p (*-weak for $p=\infty$) topology. This result was first proved by Hermes [4] for $p=\infty$. It was generalized by Artstein [3] who also demonstrated that the bang-bang principle can be established for periodic controls.

We will consider the system (2),(3) in the presence of the following two types of phase constraints:

$$x(t_j) \in B_j \quad , \quad j=1,\dots,n \ , \tag{4}$$

where $B_1,\dots,B_n$ are closed convex subsets of E^m given at some fixed points $t_1,\dots,t_n$ of T_2, and

$$x \in R \tag{5}$$

where R is a given closed convex subset of $C(T_2, E^m)$.

Of course, the presence of restrictions (4) does not change anything (Theorem 3.2(a)). Nevertheless, this generalization, though technically trivial, is convenient since the constraints (4) incorporate, in particular, restrictions if any on initial and final states of the system as well as restrictions on its states at a finite number of intermediate values of t.

The presence of the phase constraints (5) does complicate the matter. As Example 4.2 shows, the equality (1) for such systems may fail. However, Theorem 3.6 asserts that the generalized version of the bang-bang principle

$$\text{cl } X_t(U) = X_t(U_c) \tag{6}$$

still holds provided the system has at least one "internal" trajectory. Of course, an arbitrary set with a convex closure can be of quite complicated structure. Fortunately, Theorem 3.6 also asserts that it is not the case for attainable sets which have convex relative interior. Consequently, though attainable sets of systems with phase constraints can be nonconvex ([5], Example 2), such systems can still be studied by the well-developed methods of convex analysis (see [6]).

Note that the generalized bang-bang equality is proved to hold not only for traditional attainable sets but for any reachable set of the system as it is defined below. This result and the fact that attainable sets and the set of all system trajectories are special cases of this definition may justify its introduction.

Definition. Let X(U), the set of admissible trajectories of a control system, be a set of functions continuous on a Hausdorff compact K. For any nonempty fixed subset T of K we define $X_T(U)$ the reachable set of the system on T as the restriction of X(U) on T, i.e.

$$X_T(U) = \{x(\cdot,u)|_T,\ u \in U\} .$$

So, in particular, $X_T(U) = X(U)$ is the set of admissible trajectories of the system if T=K and $X_T(U) = X_t(U)$ is the traditional attainable set at t if $T = \{t\}$.

2. SOME TOPOLOGICAL LEMMAS

Lemma 2.1. Let F,G be two subsets of a finite dimensional Euclidean space E^m such that G is convex, $G \subset F$ and cl F = cl G. Then $F \subset \text{aff } G$ and ri F = ri G.

Proof. Since $\text{cl } G \subset \text{aff } G$ ([6], p.44), we immediately have

$$F \subset \text{cl } F = \text{cl } G \subset \text{aff } G .$$

Since $G \subset F$ and aff G = aff F, $\text{ri } G \subset \text{ri } F$. On the other hand, $F \subset \text{cl } F$ implies $\text{ri } F \subset \text{ri (cl } F)$ and using Theorem 6.3 from [6] we obtain the inclusion

$$\text{ri } F \subset \text{ri (cl } F) = \text{ri (cl } G) = \text{ri } G$$

which completes the proof.

Let K be a Hausdorff compact and $C(K) = C(K,E^n)$ be the space of continuous vector functions $f\colon K \to E^n$ with the uniform topology induced by the norm $\|f\| = \sup_K |f(t)|$. Let $F \subset C(K)$ and let $T \subset K$. We will denote F_T the set of restrictions of all the functions from F on the set T, i.e.

$$F_T = \{f_T = f(\cdot)|_T,\ f \in F\} .$$

The following lemmas deal with some elementary properties of such sets and their proofs are straightforward.

Lemma 2.2. Let F,G be two subsets of C(K) such that $F \subset \mathrm{cl}\ G$. Then $F_T \subset \mathrm{cl}\ (G_T)$ for any $T \subset K$.

Lemma 2.3. Let $F,G \subset C(K)$. Then $\mathrm{cl}\ F = \mathrm{cl}\ G$ if and only if $\mathrm{cl}\ (F_T) = \mathrm{cl}\ (G_T)$ for any $T \subset K$.

Corollary 2.4. cl F is convex if and only if $\mathrm{cl}\ (F_T)$ is convex for any $T \subset K$.

Lemma 2.5. Let F,G be two subsets of C(K) such that $F \cup G$ is equicontinuous. Then the following conditions are equivalent:

i) cl F = cl G;

ii) $\mathrm{cl}\ (F_T) = \mathrm{cl}\ (G_T)$ for any finite $T \subset K$;

iii) $\mathrm{cl}\ (F_T) = \mathrm{cl}\ (G_T)$ for any $T \subset K$.

Corollary 2.6. Let $F \subset C(K)$ be a set of equicontinuous functions. Then the following properties are equivalent:

i) cl F is convex;

ii) $\mathrm{cl}\ (F_T)$ is convex for any finite $T \subset K$;

iii) $\mathrm{cl}\ (F_T)$ is convex for any $T \subset K$.

3. PROPERTIES OF REACHABLE SETS

We will study the system (2)-(5) under the traditional assumptions regarding (2) (see for example [3], Section 3) and the assumptions about (3), (4), (5) listed above in the introduction.

Consider first the reachable sets $X_T(U)$ of the system (2)-(3) along with the corresponding sets $X_T(U_c)$.

Theorem 3.1. (a) $X_T(U) = X_T(U_c)$ for any finite T and this set is nonempty, convex and compact;

(b) $\mathrm{cl}\ (X_T(U)) = X_T(U_c)$ for any T and this set is nonempty, convex and compact.

Proof. (a) Let T consist of n distinct points of K. Consider $X_T(U) = \tilde{A}U$ where $\tilde{A} = \mathrm{Res}_T \circ A$, A is the affine operator (3) and

$$\mathrm{Res}_T\colon x \in C(K) \to x_T \in C(T) .$$

For the case being considered $\tilde{A}$ maps U into $E^{m\times n}$ and it is affine and continuous when U is endowed with the weak-L_p (*-weak if $p=\infty$) topology. Theorem 3.2 and Corollary 3.3 from [3] (see also [4]) assert the statement.

(b) Let T be an arbitrary subset of K. The sets

$$X(U_c) = AU_c, \quad X_T(U_c) = (\mathrm{Res}_T \circ A)U_c$$

are convex and compact as affine continuous images of the set U_c which is convex and compact in the topology being considered. The compactness of $X(U_c)$ implies that $X(U_c) = X(U_c) \cup X(U)$ is equicontinuous (the Arzela-Ascoli Theorem). Then, since

$$\mathrm{cl}\ (X_T(U)) = \mathrm{cl}\ (X_T(U_c)) = X_T(U_c)$$

for any finite T as was proved in the first part of Theorem 3.1 the same equalities hold for any infinite T as well (Lemma 2.5).

Consider reachable sets of the system (2)-(4):

$$Y_T(U) = \mathrm{Res}_T(X(U) \cap B) \text{ where } B = \{f \in C(K):\ f(t_j) \in B_j,\ j=1,\dots,n\},$$

$$Y_T(U_c) = \mathrm{Res}_T(X(U_c) \cap B)\ .$$

<u>Theorem 3.2.</u> (a) $Y_T(U) = Y_T(U_c)$ for any finite T and this set is convex and compact;

(b) cl $(Y_T(U)) = Y_T(U_c)$ for any T and this set is convex and compact.

<u>Proof.</u> (a) Let T be a finite subset of K. It is enough to show that $Y_T(U_c) \subset Y_T(U)$ since the reverse inclusion is obvious. Let $x_T^c \in Y_T(U_c)$. Consider

$$T' = \left(\bigcup_{j=1}^{n} t_j\right) \cup T\ .$$

Since $X_{T'}(U) = X_{T'}(U_c)$ by Theorem 3.1(a), there exists $x_{T'} \in X_{T'}(U)$ such that $x_{T'} = x_{T'}^c$. But that implies

$$x(t_j) = x^c(t_j) \in B_j \quad \text{for} \quad j=1,\dots,n\ ,$$

i.e. $x_T \in Y_T(U)$. So, $x_T^c = x_T \in Y_T(U)$ which proves the equality in question. Finally, B is, obviously, convex and closed. Therefore $Y(U_c) = X(U_c) \cap B$ and, consequently, $Y_T(U_c)$ for any T are convex and compact.

(b) The assertion follows immediately from the first part of this theorem, Lemma 2.5 and Corollary 2.6.

Now we proceed to the reachable sets of the system (2)-(5):

$$Z_T(U) = \mathrm{Res}_T(X(U) \cap B \cap R)\ .$$

In addition, we consider the set

$$Z^\circ(U) = X(U) \cap B \cap (\mathrm{int}\ R)$$

where int R denotes the interior of R. (It is natural to refer to $Z^\circ(U)$ as the set of internal trajectories of the system (2)-(5).)

Lemma 3.3. (a) $\mathrm{cl}\ Z_T^\circ(U) = \mathrm{cl}\ Z_T^\circ(U_c)$ for any T;
(b) If T is finite then $Z_T^\circ(U) = Z_T^\circ(U_c)$.

Proof. (a) It is enough to prove the equality for T=K (see Lemma 2.3). By Theorem 3.2, $\mathrm{cl}\ Y(U) = \mathrm{cl}\ Y(U_c)$. Besides, int R is open. Therefore, making use of the formula from [7], p.45, we have

$$\mathrm{cl}\ Z^\circ(U) = \mathrm{cl}\ (Y(U) \cap \mathrm{int}\ R) = \mathrm{cl}\ (\mathrm{cl}\ Y(U) \cap \mathrm{int}\ R) =$$

$$= \mathrm{cl}\ (\mathrm{cl}\ Y(U_c) \cap \mathrm{int}\ R) = \mathrm{cl}\ (Y(U_c) \cap \mathrm{int}\ R) = \mathrm{cl}\ Z^\circ(U_c)\ .$$

(b) Let T be a finite subset of K. We shall prove that

$$Z_T^\circ(U_c) \subset Z_T^\circ(U)\ .$$

Let $x_T(\cdot,v) \in Z_T^\circ(U_c)$. Define

$$\tilde{T} = \left(\bigcup_{j=1}^{n} t_j\right) \cup T, \quad \tilde{B}(t) = \{x(t,v)\} \quad \text{for every} \quad t \in \tilde{T}\ .$$

Consider

$$\tilde{Y}(U) = \{x \in X(U):\ x(t) \in \tilde{B}(t) \ \text{for}\ t \in \tilde{T}\}, \quad \tilde{Z}^\circ(U) = \tilde{Y}(U) \cap \mathrm{int}\ R\ .$$

By the first part of this lemma proved above,

$$\mathrm{cl}\ \tilde{Z}_{\tilde{T}}^\circ(U) = \mathrm{cl}\ \tilde{Z}_{\tilde{T}}^\circ(U_c) = \{x_{\tilde{T}}(\cdot,v)\}\ .$$

Since this set is a singleton, there exists $x_{\tilde{T}}(\cdot,u) \in \tilde{Z}_{\tilde{T}}^\circ(U)$ such that $x_{\tilde{T}}(\cdot,u) = x_{\tilde{T}}(\cdot,v)$ which implies $x_T(\cdot,v) \in Z_T^\circ(U)$.

Corollary 3.4. (a) $\mathrm{cl}\ Z_T^\circ(U)$ is convex for any T;
(b) $Z_T^\circ(U)$ is convex for any finite T.

Lemma 3.5. Let $Z^\circ(U)$ be nonempty. Then $Z_T(U) \subset \mathrm{cl}\ Z_T^\circ(U)$ for any T.

Proof. According to Lemma 2.2 it is enough to prove this inclusion for T=K. Let

$$x \in Z(U) = Y(U) \cap R, \quad x^\circ \in Z^\circ(U) = Y(U) \cap \mathrm{int}\ R\ .$$

Obviously, $x \in \mathrm{cl}\ I$ where

$$I = \{f \in C(K):\ f = \lambda x + (1-\lambda)x^\circ,\ 0 \le \lambda < 1\}\ .$$

Since $x, x^\circ \in Y(U)$ and $\mathrm{cl}\ Y(U)$ is convex (Corollary 3.4), $I \subset \mathrm{cl}\ Y(U)$. In addition, $I \subset \mathrm{int}\ R$ ([8], p.110). Hence,

$$I \subset \mathrm{cl}\ Y(U) \cap \mathrm{int}\ R \subset \mathrm{cl}\ (\mathrm{cl}\ Y(U) \cap \mathrm{int}\ R) = \mathrm{cl}\ (Y(U) \cap \mathrm{int}\ R)\ ,$$

and

$$x \in \mathrm{cl}\ I \subset \mathrm{cl}\ (Y(U) \cap \mathrm{int}\ R) = \mathrm{cl}\ Z^\circ(U)\ .$$

Theorem 3.6. Let the system (2)-(5) have a trajectory which is interior for the constraint (5), i.e.

$$Z^\circ(U) = X(U) \cap B \cap \mathrm{int}\ R \neq \emptyset\ .$$

Then for any T

$$\mathrm{cl}\ (Z_T(U)) = Z_T(U_c) \tag{7}$$

and this set is nonempty, convex and compact. Moreover, if T is finite,

$$\operatorname{ri}\ (Z_T(U)) = \operatorname{ri}\ (Z_T(U_c)) \tag{8}$$

which is a nonempty convex set.

Proof. Making use of Lemma 3.5, we have for any T the inclusion

$$Z_T^\circ(U) \subset Z_T(U) \subset \operatorname{cl}\ Z_T^\circ(U)\ .$$

That implies

$$\operatorname{cl}\ Z_T(U) = \operatorname{cl}\ Z_T^\circ(U)\ . \tag{9}$$

By the same token,

$$\operatorname{cl}\ Z_T(U_c) = \operatorname{cl}\ Z_T^\circ(U_c)\ .$$

Since, by Lemma 3.3, the right hand sides of the last two equalites coincide, we have

$$\operatorname{cl}\ Z_T(U) = \operatorname{cl}\ Z_T(U_c) = Z_T(U_c)\ , \tag{10}$$

and the latter is obviously convex, compact and, of course, nonempty.

To prove the second assertion of the theorem we will apply Lemma 2.1 taking into account that E=C(T) is finite dimensional if and only if T is finite ([9], IV.13.15). So, if T is finite, $Z_T(U)$ contains the convex subset $Z_T^\circ(U)$ with the same closure (Corollary 3.4 and equality (9)). So, Lemma 2.1 asserts (8). Finally, ri $Z_T(U_c)$ is nonempty and convex since $Z_T(U_c)$ itself is a nonempty convex set ([6], Theorem 6.2).

4. EXAMPLES

The examples of this section are to demonstrate that the previous results cannot be strengthened to a considerable extent. All the systems considered below are described by linear ordinary differential equations and therefore (see [3]) are special cases of the general system (2)-(5).

Example 4.1. Consider the simplest control system

$$\dot{x} = u,\quad x(0) = 0\ ;$$
$$u(t) = 1 \text{ or } -1 \quad \text{for almost every } 0 \le t \le 1\ .$$

Consider the reachable set of the system on [0,1], i.e. the set X(U) of all its trajectories. Obviously, $x_1(t)=t$, $x_2(t)=-t$ belong to X(U) while $(x_1(t)+x_2(t))/2$ does not. Therefore X(U) is not convex and, of course, $X(U) \neq X(U_c)$. So, even for the trivial system with no phase constraints the reachable set on an infinite subset T may be nonconvex and the traditional bang-bang equality may fail.

Example 4.2. Consider the attainable set at t=2, i.e. $Z_{\{2\}}$, of the system

$$\dot{x}_1 = x_2,\quad x_1(0) = 0\ ,$$
$$\dot{x}_2 = u,\quad x_2(0) = 1\ ;$$
$$u(t) = 1 \text{ or } -1 \quad \text{for almost every } 0 \le t \le 2\ ;$$
$$x_2(t) \ge 0 \quad \text{for every } 0 \le t \le 2\ .$$

Using the integral representations

$$x_2 = 1 + \int_0^2 u(t)dt, \quad x_1 = \int_0^2 x_2(t)dt ,$$

one can easily show that a point (x_1,x_2) is reachable at t=2 if and only if $0 \le x_2 \le 3$ and

$$\frac{1}{2} + \frac{1}{2} x_2^2 < x_1 \le 4 - (3-x_2)^2/4 \quad \text{if} \quad 0 \le x_2 < 1 ,$$

$$4 - 2(3-x_2) + (3-x_2)^2/4 \le x_1 \le 4 - (3-x_2)^2/4 \quad \text{if} \quad 1 \le x_2 \le 3 .$$

The reachable set $Z_{\{2\}}$ is given at Fig. 1. It is not closed and does not coincide with $Z_{\{2\}}(U_c)$ which is its closure. Note that $Z_{\{2\}}(U)$ and $Z_{\{2\}}(U_c)$ have the same interior.

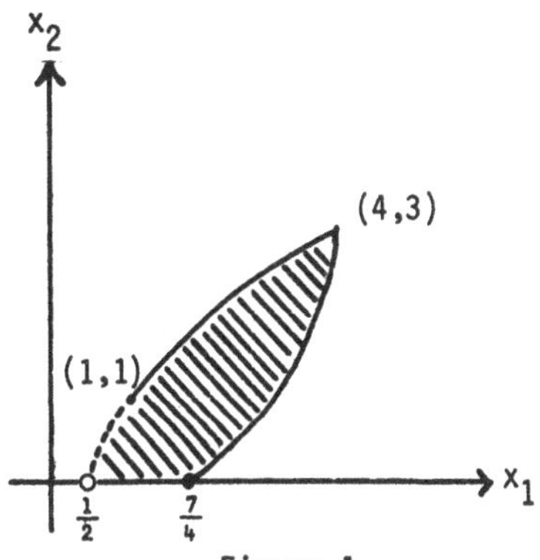

Figure 1

Example 4.3. Consider the attainable set at t=1 of the system

$$\dot{x}_1 = u_1, \quad x_1(0) = 0 ,$$
$$\dot{x}_2 = u_2, \quad x_2(0) = 0 ;$$

$$u(t) \in D(t) \quad \text{for a.e.} \quad 0 \le t \le 1 \quad \text{where} \quad D(t) = \left\{ \begin{pmatrix} 1 \\ t \end{pmatrix}, \begin{pmatrix} 0 \\ 0 \end{pmatrix} \right\} ;$$

$$R = \{x = (x_1,x_2) \in C(0,1) : tx_1(t) - 2x_2(t) = 0 \quad \text{for every} \quad 0 \le t \le 1\} .$$

It is obvious that int R is empty and, as it is shown in [5], the system has only two admissible trajectories. Consequently, $Z_{\{1\}}(U)$ has only two points, namely $(1,\frac{1}{2})$ and $(0,0)$, while $Z_{\{1\}}(U_c)$ is the whole interval with these end-points.

REFERENCES

1. H. Hermes and J.P. LaSalle, Functional Analysis and Optimal Control, Academic Press, New York and London, 1969.
2. J.P. LaSalle, The time optimal control problem, in Theory of Nonlinear Oscillations, vol. 5, pp. 1-24, Princeton Univ. Press, Princeton, New Jersey, 1959.
3. Z. Artstein, Discrete and continuous bang-bang and facial spaces or: look for the extreme points, SIAM Review 22(1980), pp. 172-185.
4. H. Hermes, On the structure of attainable sets for generalized differential equations and control systems, J. Diff. Eqs. 9(1971), pp. 141-154.
5. A.V. Levitin, The problem of optimal control with restrictions on the phase coordinates, Moscow Univ. Math. Bull. 26(1971), pp. 45-52 (translated from Russian).
6. R.T. Rockafellar, Convex Analysis, Princeton Univ. Press, Princeton, New Jersey, 1970.
7. K. Kuratowski, Topology, vol. I, Academic Press, New York and London, 1966.
8. J.L. Kelley and I. Namioka, Linear Topological Spaces, D. Van Nostrand Company, Inc. Princeton, New Jersey, 1963.
9. N. Dunford and J.T. Schwartz, Linear Operators Part I: General Theory, Interscience Publishers, Inc., New York.

PERIODICITY IN OPTIMAL CONTROL AND DIFFERENTIAL GAMES

Emilio O. Roxin
University of Rhode Island
Kingston, R.I.02881

Lynnell E. Stern
Simmons College
Boston, MA 02115

INTRODUCTION

Questions about optimality in the infinite time interval, and in particular periodic optimal solutions of control systems, have been studied extensively in the references given at the end of this paper, and in particular in the recent dissertation of Stern [7], where further references can be found. In the context of differential games (at least for two-player, zero sum games) very similar results can be established but the formulation of the problem is more complicated than in the control theory context. Indeed, in differential games, one player will not in general be able to force periodicity of the solution against the will of the opponent. Therefore it will be assumed that both players play optimally, and conditions will be established under which the resulting solution is periodic.

The objective functional will be defined for a sequence of terminal times $T_m = mT$, $m=1,2,3,\cdots$ where T is the period of the system. The existence of a periodic solution implies that each player has a certain optimal strategy which optimizes the objective functional for all these terminal times. (Left out of consideration is the possibility of each player optimizing at a different terminal time, since this would lead to a non-zero-sum game.)

Although the concepts introduced here are applicable to general non-linear games, the specific results given in this paper apply only to the case of linear games with terminal objective functional, where the optimal controls can be computed without integrating the state equations.

STATEMENT OF THE PROBLEM

The type of games which will be considered is:

$$\dot{x} = A(t)\ x + B(t)\ u + C(t)\ v \tag{1}$$

where $x(t) \in R^n$, $t \in [t_0,\infty)$, $u(t) \in U \subset R^{r_1}$, $v(t) \in V \subset R^{r_2}$, with U and V given constant, compact sets and A(t), B(t), C(t) are continuous and periodic of period T. The initial condition $x(t_0)$ is assumed to be given.

The objective functional considered here is of the type

$$J = \eta \cdot x(T_m) = \eta \cdot x(mT) \tag{2}$$

where m is any positive integer and η is a given constant vector. Player u should minimize and player v should maximize J.

Without loss of generality it may be assumed that $t_0 = 0$. The optimization of J is assumed to be achieved by both players in a periodic way, i.e. for all end-times mT, m=1,2,3,..., or for a period which is a multiple of T (substituting this by kmT, m=1,2,3,..., k any positive integer). More in particular, we are looking for solutions with periodic optimal controls and, more restrictively, with periodic optimal controls and periodic trajectories x(t).

It is well known [4] that the solution to the game (1), (2) is obtained by introducing the adjoint vector p(t) and the Hamiltonian

$$H(p,t,x,u,v) = p\cdot(A(t)\ x + B(t)\ u + C(t)\ v) \tag{3}$$

The adjoint vector p(t) should then satisfy

$$\dot{p} = -\frac{\partial H}{\partial x} = -\ p\ A(t) \tag{4}$$

The optimality of the solution requires that the optimal controls $u^*(t)$ maximize and $v^*(t)$ minimize the Hamiltonian $H(p(t),x^*(t),u,v)$ for almost every t (here $x^*(t)$ is the corresponding optimal trajectory). The boundary condition for p(t) is that p(mT) be parallel and opposite to η for all possible end-times mT.

The main problem now consists in determining for which values of η and x(0) the optimal controls $u^*(t)$ and $v^*(t)$ and possibly the trajectory $x^*(t)$ are periodic.

The particularly simple structure of (1) and therefore of (4) permits solving (4) independently of the integration of (1). The solution of (4) is

$$p(t) = p(0)\ \Phi(0,t) \tag{5}$$

where $\Phi(t,0)$ is the fundamental matrix of the homogeneous equation associated with (1). The solution of (1) can be expressed as

$$x(t) = \Phi(t,0)\ x(0) + \int_0^t \Phi(t,s)[B(s)\ u(s) + C(s)\ v(s)]\ ds \tag{6}$$

and this expression is very useful in analyzing the optimality and related properties of the solution.

Assume that $u^*(t)$, $v^*(t)$ are optimal controls for the game (1), (2), where J is to be optimized for all $m=1,2,3,\cdots$. Then, according to the maximum principle, for each $m=1,2,3,\cdots$, there exists an adjoint vector function $p_m(t)$ such that for almost every $t \in [0,mT]$,

$$\begin{aligned} p_m(t)\ B(t)\ u^*(t) &= \max_{u \in U}\ p_m(t)\ B(t)\ u \\ p_m(t)\ B(t)\ v^*(t) &= \min_{v \in V}\ p_m(t)\ B(t)\ v \end{aligned} \tag{7}$$

and

$$p_m(mT) = -\ k_m\ \eta \tag{8}$$

where k_m is a positive scalar.

The case considered in this paper assumes that all vector functions $p_m(t)$ satisfying (8) coincide, and therefore the subindex m may be deleted. Hence (8) becomes

$$p(mT) = -\ k_m\ \eta \tag{9}$$

where each k_m is a positive scalar. This is certainly not a necessary condition in order for (7) and (8) to be satisfied, but it is obviously sufficient to ensure periodic optimization. If $p(t)$ satisfies (9), it will be called "direction periodic", since the direction of all $p(mT)$ is then the same for $m=1,2,3,\cdots$.

LEMMA 1. If η is an eigenvector of $\Phi(T,0)$ and the corresponding eigenvalue is $\lambda > 0$, then the adjoint vector, solution of (4) with $p(0) = -\eta$, satisfies the periodic optimization condition (9) for $m=1,2,3,\cdots$, and the optimal controls $u^*(t)$, $v^*(t)$ satisfying (7) can be chosen to be T-periodic.

Proof: $p(mT) = p(0)\ \Phi(0,mT) = \lambda^{-m}\ p(0) = -\ \lambda^{-m}\ \eta$. This proves that $p(t)$ satisfies the periodic optimization condition (9). On the other hand,
$p(mT+\tau) = p(0)\ \Phi(0,mT+\tau) = p(0)\ \Phi(0,mT)\ \Phi(mT,mT+\tau) = \lambda^{-m}\ p(0)\ \Phi(0,\tau) = \lambda^{-m}\ p(\tau)$
and therefore from (7), $u^*(mT+\tau) = u^*(\tau)$ and $v^*(mT+\tau) = v^*(\tau)$ may be chosen.

LEMMA 2. If η is an eigenvector of $\Phi(T,0)$ and the corresponding eigenvalue is $\lambda < 0$, then the adjoint vector, solution of (4) with $p(0) = -\eta$, satisfies the periodic optimization condition (9) for $m=2,4,6,\cdots$, and the optimal controls $u^*(t)$, $v^*(t)$ satisfying (7) can be chosen to be 2T-periodic.

Proof: In this case, $p(0) = -\eta$ is an eigenvector of $\Phi(2T,0)$ with eigenvalue $\lambda^2 > 0$; the rest follows as above (lemma 1).

LEMMA 3. If, in addition to the assumptions in lemma 1, the matrix $I - \Phi(T,0)$ is nonsingular, then the periodic optimization problem stated in lemma 1 has a unique T-periodic solution $x_p(t)$.

Proof: According to lemma 1, $u^*(t)$ and $v^*(t)$ may be assumed to be T-periodic. Therefore the right side of the differential equation (1), considered as function of x and t, is T-periodic in t, hence the condition for the solution to be T-periodic is that $x(T) = x(0)$. Using (6) to express $x(T)$, this condition is equivalent to

$$(I - \Phi(T,0))\, x_p(0) = \int_0^T \Phi(T,s)[B(s)\, u^*(s) + C(s)\, v^*(s)]\, ds \tag{10}$$

and therefore

$$x_p(0) = (I - \Phi(T,0))^{-1} \int_0^T \Phi(T,s)[B(s)\, u^*(s) + C(s)\, v^*(s)]\, ds \tag{11}$$

determines $x_p(0)$ uniquely. A similar relation was given by Spyker [6] for periodic control systems.

LEMMA 4. If, in addition to the assumptions in lemma 2, the matrix $I - \Phi(2T,0)$ is nonsingular, then the periodic optimization problem stated in lemma 2 has a unique 2T-periodic optimal solution $x_p(t)$.

Proof: Similar to lemma 3.

REMARK 1. The condition that the matrix $I - \Phi(T,0)$ be nonsingular, is equivalent to the fact that the matrix $\Phi(T,0)$ does not have an eigenvalue equal to 1. Indeed, if $I - \Phi(T,0)$ is singular, then there exists a vector $p \neq 0$ such that $p\,(I - \Phi(T,0)) = 0$, hence $p\, I = p = p\, \Phi(T,0)$. Similarly for the matrix $I - \Phi(2T,0)$ in lemma 4.

THEOREM 1. If each of the eigenvalues of $\Phi(T,0)$ is different from 1, then to each eigenvector η with positive eigenvalue, there corresponds a T-periodic optimization problem and a unique T-periodic solution $x_p(t)$. The eigenvectors η and $k\,\eta$ with $k > 0$, lead to the same periodic optimization problem. The eigenvectors η and $k\,\eta$ with $k < 0$, lead to periodic optimization problems where η has been replaced by $-\eta$ which is equivalent to interchange the minimizer and the maximizer.

Proof: The first statement of this theorem follows from lemmas 1 and 3 and remark 1. For $k > 0$, the adjoint vectors $p(t)$ and $k\,p(t)$ determine the same optimal controls $u^*(t)$, $v^*(t)$ by (7). If, instead, $k < 0$, the substitution of $p(t)$ by $k\,p(t)$ substitutes η by a positive multiple of $-\eta$ in the objective functional (2). This is equivalent to keep the old η and having $u(t)$ maximizing and $v(t)$ minimizing J.

REMARK 2. It may certainly happen that different eigenvectors $p(0)$ and their corresponding $p(t)$, determine by (7) the same optimal controls and periodic solutions. Indeed, the determination of $u^*(t)$ and $v^*(t)$ depends on the control sets U and V, on which no other assumptions have been placed. This is clearly seen by the trivial case in which U and V reduce to single points.

THEOREM 2. If some of the eigenvalues of $\Phi(T,0)$ are equal to 1, then to each eigenvector $p(0)$ with positive eigenvalue there corresponds a T-periodic optimization problem and T-periodic optimal controls $u^*(t)$, $v^*(t)$, but there is a T-periodic optimal solution $x_p(t)$ if and only if $\int_0^T \Phi(T,s)[B(s)\, u^*(s) + C(s)\, v^*(s)]\, ds$ belongs to the range of the linear operator defined by $I - \Phi(T,0)$. If such periodic solutions corresponding to an eigenvalue equal to 1 exist, they are not unique.

Proof: Under the assumptions of this theorem, $I - \Phi(T,0)$ is singular according to remark 1. The rest follows from the possible existence of vectors $x_p(0)$ satisfying the periodicity condition (10).

REMARK 3. If $p(0) = -\eta$ is an eigenvector of $\Phi(T,0)$ and the corresponding eigenvalue is $\lambda < 0$, then it is also an eigenvector of $\Phi(2T,0)$ with eigenvalue $\lambda^2 > 0$, as seen in lemma 2. In this case the existence or non-existence of periodic solutions $x_p(t)$ of period 2T can be ascertained by considerations similar to those of theorems 1 and 2, replacing T by 2T.

REMARK 4. If $\lambda_{1,2} = e^{\sigma \pm i\omega}$ are complex eigenvalues of $\Phi(T,0)$, then there are real solutions of the adjoint equation (4) of the form

$p(t) = p(0)\, e^{\sigma t} \times$ (periodic trigonometric functions of period $2\pi/\omega$).

Assume that for the relatively prime positive integers m, c, the relation $\omega m T = 2\pi c$ holds, then p(0) is an eigenvector with eigenvalue $e^{\sigma m T} > 0$. In this case, $\eta = -p(0)$ defines a periodic optimization problem of period mT, and considerations similar to theorems 1 and 2 can again be made.

BEHAVIOR OF NON-PERIODIC SOLUTIONS

After having analyzed the problem of existence of periodic optimal controls $u^*(t)$, $v^*(t)$ and periodic optimal solutions $x_p(t)$ of a periodic optimization problem, consideration will now be given to the other (non-periodic) solutions x(t). Of particular interest is the case when the homogeneous equation associated to (1) is asymptotically stable.

THEOREM 3. For the periodic differential game (1), (2), if $u^*(t)$, $v^*(t)$ are periodic optimal controls and if the associated homogeneous system $\dot{x} = A(t)\,x$ is asymptotically stable, then there exists a unique optimal solution $x_p(t)$ and all other optimal solutions x(t) (corresponding to the same $u^*(t)$, $v^*(t)$) tend asymptotically to the periodic one..

Proof: The asymptotic stability implies that for all eigenvalues λ of $\Phi(T,0)$, $|\lambda| < 1$. This ensures that $I - \Phi(T,0)$ is nonsingular (lemma 3) and therefore there is a unique periodic optimal solution $x_p(t)$. For any other solution x(t) corresponding to the same $u^*(t)$, $v^*(t)$, the difference $x(t) - x_p(t)$ tends asymptotically to zero.

REMARK 5. From the practical point of view, theorem 3 means that all optimal solutions x(t), implemented over a sufficiently long time interval, tend to coincide with the periodic one $x_p(t)$.

REMARK 6. If the matrix $\Phi(T,0)$ has no eigenvalues λ such that $|\lambda| = 1$, then, in general, the solutions of $\dot{x} = A(t)\,x$ separate into families of stable and unstable behavior, and this determines the stability properties of the corresponding periodic solutions of (1) (see theorem 3).

APPLICATION TO CONSTANT COEFFICIENT SYSTEMS IN R^2

The method used in the following will be to search first for direction periodic adjoint vector functions p(t). Then, for any such p(t), $\eta = -p(0)$ defines a periodic optimization problem with the objective functional (2). This guarantees that the optimal controls u*(t), v*(t) will be periodic. Finally, the existence and uniqueness of the periodic solution $x_p(t)$ will be considered.

The system under consideration is $\dot{x} = A x + B u + C v$, where $x \in R^2$, A, B, C are constant matrices and u(t) and v(t) may be of dimensions 1 or 2, with corresponding compact control sets U and V.

<u>CASE 1:</u> Assume that the eigenvalues of A are real. In this case the only direction periodic solutions p(t) of the adjoint equation are vectors of constant direction and correspond to the eigenvectors of the matrix A. If the eigenvalues of the matrix A are distinct, then the only direction periodic p(t) are of the type $p(t) = p(0)\, e^{\lambda_i t}$, i=1,2; p(0) must be an eigenvector corresponding to the eigenvalue λ_i. It can be seen that the same is true in the case $\lambda_1 = \lambda_2$, even if A is non-diagonal. Therefore for each choice of η being an eigenvector of A, the optimization problem (2) leads to a direction periodic p(t) and to constant optimal controls u*(t), v*(t). The existence and uniqueness of periodic solutions $x_p(t)$ depends on the non-singularity of the matrix $I - \Phi(0,T)$. This matrix will be singular if and only if at least one of the eigenvalues of $\Phi(0,T)$ is 1.

<u>CASE 2.</u> Assume that the eigenvalues of A are complex. Then the real canonical form of the solutions of the adjoint equation is

$$\Phi(0,t) = e^{-\sigma t} \begin{pmatrix} \cos \omega t & -\sin \omega t \\ \sin \omega t & \cos \omega t \end{pmatrix}$$

The only real eigenvectors appear for t = T, where $\omega T = 2\pi$.

The corresponding solutions are direction periodic, of period T. Choosing η to be such an eigenvector, hence $p(0) = -\eta$, the corresponding optimal controls u*, v* will be T-periodic. The existence and uniqueness of a T-periodic solution $x_p(t)$ depends on the non-singularity of $I - \Phi(0,T)$. If $\sigma \neq 0$, then this matrix will be non-singular. For $\sigma = 0$, $I - \Phi(T,0)$ will be the zero matrix, hence the condition similar to (10) becomes

$$0 = \int_0^T \Phi(T,s)(B\, u^*(s) + C\, v^*(s))\, ds$$

If this condition is satisfied, then every $x_p(0)$ will lead to a periodic optimal solution $x_p(t)$.

REFERENCES

1. Gilbert, E. G. "Optimal Periodic Control: A General Theory of Necessary Conditions", SIAM J. Control Optim. 15 (1977), 717-746.

2. Guardabassi, G., Locatelli, A., and Rinaldi, S. "Status of Periodic Optimization of Dynamical Systems", J. Optim. Theory Appl. 14 (1974), 1-20.

3. Halanay, A. "Optimal Control of Periodic Solutions", Rev. Roum. Mat. Pure et Appl. 19 (1974), 3-16.

4. Isaacs, R. Differential Games, The SIAM Series in Appl. Math., John Wiley & Sons, Inc., N.Y., London, Sydney, 1965.

5. Lee, E. B. and Spyker, D. A. "On Linear Periodic Control Problems", IEEE Trans. Automat. Control (Feb. 1973), 39-40.

6. Spyker, D. A. "Application of Optimal Control Theory to Cardio-Circulatory Assist Devices", University of Michigan, Ph.D. Thesis, 1969.

7. Stern, L. E. "The Infinite Horizon Optimal Control Problem", University of Rhode Island, Ph.D. Thesis, 1980.

REDUCED APPROXIMATIONS IN PARAMETER IDENTIFICATION OF HEREDITARY SYSTEMS

E. M. Cliff* and J. A. Burns*
Virginia Polytechnic Institute and State University
Blacksburg, Virginia 24061-4097

I. INTRODUCTION

In recent studies of identification for certain aero-elastic systems, the dynamic models encountered were of the form

$$\dot{w}(t) = A(\gamma)w(t) + B(\gamma)w(t-r) + C(\gamma)u(t) \tag{1.1}$$

with output

$$v(t) = F(\gamma)w(t) + G(\gamma)u(t) . \tag{1.2}$$

Here the time delay r and the parameter γ are to be identified from input-output data. For the particular hereditary system studied, it was noted tht the "state" w(t) could be partitioned into $w(t) = col(x(t),y(t))$ where $x\in\mathbf{R}^p$, $y\in\mathbf{R}^q$ and the first p columns of $B(\gamma)$ were zero. It is clear that the histories of the first p components of w(t) play no part in the evolution of the system (1.1) - (1.2). The significance of this observation is that considerable economy can be realized in certain approximations that form the basis of the computational methods (see [3], [4]) used in the identification. Although such special structure has been observed before in controllability studies of hereditary systems, it has not been exploited conputationally.

In this paper we briefly discuss the aero-elastic system that was the motivation for the analysis and introduce a reduced approximation scheme. A numerical example is presented to illustrate the ideas.

II. A HEREDITARY MODEL FOR THEODORSEN'S PROBLEM

Consider the two-dimensional airfoil shown in Figure 2.1. Let h(t) denote the plunge and $\alpha(t)$ the pitch of the airfoil at time t. The equations of motion can be written (see [6], [10]) in the form

$$M_s\ddot{\theta}(t) + K_s\theta(t) = - f(t) , \tag{2.1}$$

where $\theta(t) = col(h(t),\alpha(t))$ and f(t) contains the aerodynamic loads on the airfoil. If L(t) and $M_\alpha(t)$ are the aerodynamic loads corresponding to total wing lift per unit depth and total moment about the 1/4 chord per unit depth, respectively, then

$$f(t) = col(L(t), M_\alpha(t)) . \tag{2.2}$$

* This research was supported in part by the National Science Foundation under grant ECS-8109245.

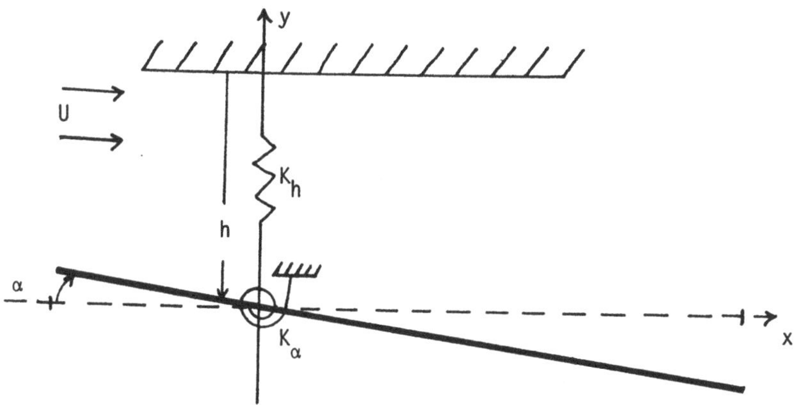

FIGURE 2.1

In order to obtain a state space model that is suitable for identification and control, one must provide a useful representation for the aerodynamic loads L(t) and $M_\alpha(t)$. It can be shown that L and M_α are linear functions of $\ddot{\theta}(t)$,$\dot{\theta}(t)$ and the output to a two-dimensional hereditary system (see [1], [4]). By extending the Jones type approximation of the Wagner function (see [6], [9]) to include a simple delay, this hereditary system is approximated by the two-dimensional delay equation (for $y = col(y_1, y_2)$)

$$\dot{y}(t) = Dy(t) + Ey(t-r) . \tag{2.3}$$

Using the linear relationship between the aerodynamic loads and $\ddot{\theta}(t)$, $\dot{\theta}(t)$ and y(t), Equation (2.1) is augmented by (2.3) to yield the model

$$\dot{w}(t) = Aw(t) + Bw(t-r) ,$$

where $w(t) = col(\dot{h}(t), \dot{\alpha}(t), h(t), \alpha(t), y_1(t), y_2(t))$. The 6×6 matrix B will have only zeros in the first 4 columns. Consequently, w(t) may be decomposed into $w(t) = col(x(t), y(t))$ where $x \in \mathbb{R}^4$, $y \in \mathbb{R}^2$ and the history of the system is carried by x(t) and $y(\cdot)$. Observe that this is precisely the type of system described in Section I.

In order to obtain a complete model, one must estimate some of the parameters in the matrices A and B and the delay r. We shall use spline approximations (see [2], [3], [4]) and a reduced state model to obtain a numerical algorithm for parameter estimation.

III. STATE SPACE MODELS

Let p, q be positive integers with $n = p + q$ and $r > 0$. Let L_2^μ denote the Lebesque space $L_2([-r,0];\mathbb{R}^\mu)$ for any integer $\mu > 0$. If $w:[-r,+\infty) \to \mathbb{R}^n$, then $w_t:[-r,0] \to \mathbb{R}^n$ is defined by $w_t(s) = w(t+s)$. In order to simplify notation we shall not distinguish between column and row vectors and simply write (η,ϕ) for $col(\eta,\phi)$.

Consider the delay differential equation

$$\dot{w}(t) = A(\gamma)w(t) + B(\gamma)w(t-r) + C(\gamma)u(t) \tag{3.1}$$

with initial data

$$w(0) = \eta \ , \qquad w_0(s) = \phi(s) \ , \qquad -r \le s < 0 \tag{3.2}$$

and output

$$v(t) = F(\gamma)w(t) + G(\gamma)u(t) \ . \tag{3.3}$$

Here, $w = (x,y)\varepsilon\mathbb{R}^p \times \mathbb{R}^q$, $\eta = (\eta^p,\eta^q)\varepsilon\mathbb{R}^p \times \mathbb{R}^q$ and $\phi = (\phi^p,\phi^q)\varepsilon L_2^p \times L_2^q$. Let Z denote the product space $\mathbb{R}^n \times L_2^n$ and define the operator $\mathcal{A}(\gamma)$ on

$$\mathcal{D}(\mathcal{A}(\gamma)) = \{(\eta,\phi)\varepsilon Z \mid \phi\varepsilon W^{1,2},\ \phi(0) = \eta\} \tag{3.4}$$

by

$$[\mathcal{A}(\gamma)](\eta,\phi) = (A(\gamma)\eta + B(\gamma)\phi(-r),\ \dot{\phi}) \ . \tag{3.5}$$

It is well known (see [4], [8]) that $\mathcal{A}(\gamma)$ generates a C_0-semigroup $\{S(\gamma,t)\}_{t\le 0}$ on Z. For $(\eta,\phi)\varepsilon Z$ and control function u, we define $z:[0,+\infty) \to Z$ by

$$z(t) = S(\gamma,t)(\eta,\phi) + \int_0^t S(\gamma,t-s)(C(\gamma)u(s),\theta^n)ds, \tag{3.6}$$

where θ^n denotes the zero function in L_2^n. It has been shown (see [4]) that $z(t) = (w(t),w_t)$, where $w(t)$ is the solution to (3.1) - (3.2). The function $z(t)$ defined by (3.6) is a mild solution of the abstract control system

$$\dot{z}(t) = \mathcal{A}(\gamma)z(t) + (C(\gamma)u(t),\ \theta^n) \tag{3.7}$$

$$z(0) = (\eta,\phi) \tag{3.8}$$

$$v(t) = \mathcal{F}(\gamma)z(t) + G(\gamma)u(t) \ , \tag{3.9}$$

where $\mathcal{F}(\gamma)$ is defined by $\mathcal{F}(\gamma)(\eta,\phi) = F(\gamma)\eta$.

The realization (3.7) - (3.9) is not "minimal" and therefore is not the most efficient model. In order to reduce the model we shall use the state space $Z_R = \mathbb{R}^p \times \mathbb{R}^q \times L_2^q$. Let $\pi:Z \to Z_R$ denote the natural projection

$$\pi(\eta^p,\eta^q,\phi^p,\phi^q) = (\eta^p,\eta^q,\phi^q) \ ,$$

and $I:Z_R \to Z$ the injection defined by

$$I(\eta^p,\eta^q,\phi^q) = (\eta^p,\eta^q,\hat{\eta}^p,\phi^q) ,$$

where $\hat{\eta}^p$ is the constant function $\hat{\eta}^p(s) \equiv \eta^p$. Define the operator in Z_R on

$$\mathcal{D}(A_R(\gamma)) = \{(\eta^p,\eta^q,\phi^q)\varepsilon Z_R | \phi^q \varepsilon W^{1,2}, \ \phi^q(0) = \eta^q\} \tag{3.10}$$

by

$$A_R(\gamma) = \pi \circ A(\gamma) \circ I \ . \tag{3.11}$$

We now make use of the special structure of (3.1) and assume that the first p columns of the $B(\gamma)$ matrix in (3.1) are zero so that $B(\gamma)$ can be partitioned as

$$B(\gamma) = \begin{bmatrix} 0 & B^p(\gamma) \\ & \\ 0 & B^q(\gamma) \end{bmatrix} , \tag{3.12}$$

where $B^p(\gamma)$ is $p \times q$ and $B^q(\gamma)$ is $q \times q$.

Theorem 3.1. Let $A_R(\gamma)$ be defined by (3.10) - (3.11). If $B(\gamma)$ is of the form (3.12), then $A_R(\gamma)$ generates a C_0-semigroup $\{S_R(\gamma,t)\}_{t\geq 0}$ on Z_R and $S_R(\gamma,t) = \pi \circ S(\gamma,t) \circ I$.

For $(\eta^p,\eta^q,\phi^q)\varepsilon Z_R$ and control function u, we define $z_R:[0,+\infty) \to Z_R$ by

$$z_R(t) = S_R(\gamma,t)(\eta^p,\eta^q,\phi^q) + \int_0^t S_R(\gamma,t-s)(C(\gamma)u(s),\theta^q)ds ,$$

where θ^q denotes the zero function in L_2^q. The following result is a direct consequence of the previous result.

Theorem 3.2. Let $(\eta^p,\eta^q,\phi^q) = \pi(\eta,\phi)$. If $B(\gamma)$ has the form (3.12) and u is locally integrable, then $z_R(t) = (w(t),y_t(\cdot))$ where $w(t) = (x(t), y(t))$ is the solution to (3.1) - (3.2).

In view of the previous two theorems we may construct a realization for the delay system (3.1) - (3.2) in Z_R. In particular, we consider the model

$$\dot{z}_R(t) = A_R(\gamma) z_R(t) + (C(\gamma)u(t), \theta^q) \tag{3.13}$$

$$z_R(0) = \pi(\eta,\phi) \tag{3.14}$$

$$v(t) = F_R(\gamma)z_R(t) + G(\gamma)u(t) , \tag{3.15}$$

where $F_R(\gamma)(\eta^p,\eta^q,\phi^q) = F(\gamma)(\eta^p,\eta^q)$. System (3.13) - (3.15) has the same input-output operator as systems (3.7) - (3.9) and (3.1) - (3.3).

The approximation schemes developed in [2], [3], [4], and [5] are based on the abstract model (3.7) - (3.9). In the next section we show that the model (3.13) - (3.15) can be used to reduce the dimension of the approximating systems.

IV. SPLINE APPROXIMATIONS

We construct approximation to the two systems (3.7) - (3.9) and (3.13) - (3.15) as follows. Let $\{t_j^N\}$ be the partition of $[-r,0]$ defined by $t_j^N = (jr/N)$, $j = 0, 1, \cdots, N$ and let L_N^μ denote the set of all $\psi:[-r,0] \to \mathbb{R}^\mu$ such that ψ is continuous on $[-r,0]$ and linear on each of the intervals $[t_j^N, t_{j-1}^N]$, $j = 1,2,\cdots,N$ (i.e. ψ is an $\mathbb{R}^\mu$-valued linear spline). Define Z^N and Z_R^N by $Z^N = \{(\psi(0),\psi)/\psi\varepsilon L_N^n\}$ and $Z_R^N = \mathbb{R}^p \times \{(\psi(0),\psi)/\psi\varepsilon L_N^q\}$. The subspaces $Z^N \subseteq Z$ and $Z_R^N \subseteq Z_R$ are closed and finite dimensional and we let $P^N:Z \to Z^N$ and $P_R^N:Z_R \to Z_R^N$ denote the orthogonal projections onto these subspaces. Define $A^N(\gamma):Z \to Z^N$ by $A^N(\gamma) = P^N A(\gamma) P^N$ and $A_R^N(\gamma):Z_R \to Z_R^N$ by $A_R^N(\gamma) = P_R^N A_R(\gamma) P_R^N$.

We consider the approximating systems

$$\dot{z}^N(t) = A^N(\gamma)z^N(t) + P^N(C(\gamma)u(t),\theta^n) \tag{4.1}$$

$$z^N(0) = P^N(\eta,\phi) \tag{4.2}$$

$$v^N(t) = F(\gamma)z^N(t) + G(\gamma)u(t) \tag{4.3}$$

and

$$\dot{z}_R^N(t) = A_R^N(\gamma)z_R^N(t) + P_R^N(C(\gamma)u(t),\theta^q) \tag{4.4}$$

$$z_R^N(0) = P_R^N\pi(\eta,\phi) \tag{4.5}$$

$$v_R^N(t) = F_R(\gamma)z_R^N(t) + G(\gamma)u(t)\ . \tag{4.6}$$

In [4] it is shown that the system (4.1) - (4.3) approximates the hereditary system (3.1) - (3.2) in the sense that $z^N(t) \to z(t)$ and the convergence is uniform for t and γ in compact sets. In order to establish the convergence of the reduced approximating system one can use a form of the Trotter-Kato Theorem [4] to establish that $z_R^N(t) \to z_R(t)$. A proof of this result will appear elsewhere. Consequently, both approximating systems yield approximations to the original hereditary system (3.1) - (3.2) and can be used for identification (see [2], [4]).

The important point is that the dimension of the reduced system (4.4) - (4.5) can be considerably less than the dimension of the system (4.1) - (4.2). In particular, the dimension of (4.1) - (4.2) is given by $DIM(N,n) = (N+1)n$ while the dimension of the reduced system (4.4) - (4.5) is given by $DIM_R(N,p,q) = (N+1)q + p$

A comparision of the dimensions of the two approximating systems is presented in Table 4.1 for the simple aero-elastic model described above. For this model, $n = 6$, $p = 4$ and $q = 2$. The system (4.1) - (4.2) is almost 3 times the size of the reduced system (4.4) - (4.5).

N	DIM(N,6)	$DIM_R(N,4,2)$
2	18	8
4	30	12
8	54	20
16	114	36
32	198	68
64	390	132

TABLE 4.1

Clearly there is an advantage in using the reduced approximating system provided that the rates of convergence of $z_R^N(t) \to z_R(t)$ and of $z^N(t) \to z(t)$ are comparable. Numerical simulations show that this is the case and we are currently working on obtaining theoretical error estimates for the reduced approximation. This analysis has been completed for the "full" spline approximations (4.1) - (4.3) (see [4], [5]). However, error analysis for the reduced approximating system (4.4) - (4.5) is more involved.

V. AN IDENTIFICATION ALGORITHM

We now illustrate how the approximation schemes can be used in the identification of parameters in the delay system. The problem may be described as follows. Given a fixed input u, initial data $(\eta,\phi) \in Z$ and observations $\bar{v}_i$ at times $\bar{t}_i$, $0 \le \bar{t}_1 < \bar{t}_2 < \cdots < \bar{t}_M = T$, find a parameter γ^* and delay r^* such that the output error

$$E(\gamma,r) = \sum_{i=1}^{M} \|v(\bar{t}_i;\gamma,r) - \bar{v}_i\|_Q^2$$

is minimized. Here $Q > 0$ is a weighting matrix and $v(t;\gamma,r)$ is the output to the system (3.1) - (3.3).

For each $N \ge 1$ we formulate a sequence of identification problems using the approximating systems (4.1) - (4.3) and (4.4) - (4.6). In particular, we minimize the errors

$$E^N(\gamma,r) = \sum_{i=1}^{M} \|v^N(\bar{t}_i;\gamma,r) - \bar{v}_i\|_Q^2$$

and

$$E_R^N(\gamma,r) = \sum_{i=1}^{M} \|v_R^N(\bar{t}_i;\gamma,r) - \bar{v}_i\|_Q^2$$

where v^N and V_R^N are the outputs to the systems (4.1) - (4.3) and (4.4) - (4.5), respectively. One can show that under reasonable assumptions the minimizers of E^N and E_R^N will provide approximations to (γ^*,r^*). This is the basis for the alogrithm.

The following numerical example is typical of the results obtained in a number of simulations. It illustrates the potential savings in CPU time that can be achieved using the reduced approximations.

Example 5.1.

The test model is described by the 2 dimensional delay equation

$$\dot{w}(t) = A(\gamma)w(t) + B(\gamma)w(t-r) + \begin{bmatrix}0\\1\end{bmatrix} u_{.1}(t) ;$$

with initial data

$$w(s) \equiv \begin{bmatrix}1\\0\end{bmatrix}, \quad s \le 0$$

and output

$$v(t) = w(t) ,$$

where $u_{.1}(t)$ is the unit step at $t = 1$ and

$$A(\gamma) = \begin{bmatrix}0 & 1\\ -\gamma_1^2 & 0\end{bmatrix}, \qquad B(\gamma) = \begin{bmatrix}0 & 0\\ 0 & -\gamma_2\end{bmatrix}.$$

Here $\gamma = (\gamma_1,\gamma_2)$ and the time delay r will be estimated. Data for the simulation was obtained on the interval [0,2] by selecting the true parameters $\gamma^* = (4,10)$ and $r^* = 1$, using the method of steps to solve the equation exactly and evaluating $v(\bar{t}_i)$ at 101 equally spaced values $\bar{t}_i$ on [0,2].

For various values of $N = 2,4,\cdots$, a maximum likelihood algorithm (see [7]) was used to estimate the parameters γ_1,γ_2 and r in the two sytems defined by (4.1) - (4.3) and (4.4) - (4.5). Startup values were taken to be $\gamma_1 = \sqrt{15}$, $\gamma_2 = 8.0$ and $r = .8$. Tables 5.1 and 5.2 contain a summary of the numerical results for this example. As indicated above, the reduced approximations required less CPU time to converge and generally produced better parameter estimates than the standard approximation (4.1) - (4.4). These computations were performed on an IBM 370/158 running under CMS.

PARAMETER ESTIMATES USING SYSTEM (4.1)-(4.3)				
N	γ_1	γ_2	r	CPU
2	3.4340	8.1849	.8725	31 sec
4	3.8313	9.0540	.8476	28 sec
8	3.9949	9.9439	.9989	60 sec
16	3.9986	9.9775	1.0002	241 sec
True	4.0000	10.0000	1.0000	

TABLE 5.1

PARAMETER ESTIMATES USING THE REDUCE SYSTEM (4.4)-(4.5)				
N	γ_1	γ_2	r	CPU
2	3.5183	9.0515	.9445	15 sec
4	3.8092	9.3691	.9457	19 sec
8	3.9919	10.0448	.9991	39 sec
16	3.9977	10.0002	1.0002	104 sec
32	3.9979	9.9862	1.0001	436 sec
True	4.0000	10.0000	1.0000	

TABLE 5.1

VI. CONCLUSION

In closing we comment that the practical observation: given the special structure of $B(\gamma)$ in (1.1) the "full" spline approximation will include superfluous variables; is not startling. However it is interesting to note that "equivalence" of the reduced and full systems is established via an abstract semi-group formulation. A direct comparision of the respective ordinary-differential equation models (4.1) - (4.2) and (4.4) - (4.5) is not a useful approach.

REFERENCES

[1] A. V. Balakrishnan, Active control of airfoils in unsteady aerodynamics, *Appl. Math. and Opt.*, *4* (1978), 171-195.

[2] H. T. Banks, J. A. Burns and E. M. Cliff, Spline-based approximation methods for control and identification of hereditary systems, in *International Symposium on Systems Optimization and Analysis*, A. Bensoussan and J. L. Lions, eds., Lecture

Notes in Control and Info. Sci., Vol. 14, Springer, Heidelberg, 1979, pp. 314-320.

[3] H. T. Banks, J. A. Burns and E. M. Cliff, A comparison of numerical methods for identification and optimization problems involving control systems with delays, *Brown University LCDS Tech. Rep.* 79-7, 1979, Providence RI.

[4] H. T. Banks, J. A. Burns and E. M. Cliff, Parameter estimation and identification for control systems with delays, *SIAM J. Cont. Opt., 19* (1981), 791-829.

[5] H. T. Banks and F. Kappel, Spline approximations for functional differential equations, *J. Differential Eqs., 34* (1979), 496-522.

[6] J. A. Burns and E. M. Cliff, Parameter identification for hereditary systems: final technical report on grant AFOSR-77-3221A, September 1979. *AFWAL TECHNICAL MEMORANDUM* 80-10-FLGC, Wright-Patterson AFB, Ohio, 1980.

[7] J. A. Burns and E. M. Cliff, Hereditary models for airfoils in unsteady aerodynamics, numerical approximations and parameter estimation, *AFWAL Technical Report*, Wright-Patterson AFB, Ohio, 1981, to appear.

[8] M. C. Delfour and S. K. Mitter, Hereditary differential systems with constant delays I. General Case, *J. Differential Eqs., 12* (1972), 213-235.

[9] R. T. Jones, Operational treatment of the nonuniform lift theory to airplane dynamics, *NACA TN 667*, 1938.

[10] D. L. York, Analysis of flutter and flutter suppression via an energy method, MS Thesis, Aerospace and Ocean Engineering Department, Virginia Polytechnic Institute and State University, Blacksburg, VA, June, 1980.

THE SMOOTHING PROBLEM - A STATE SPACE RECURSIVE COMPUTATIONAL APPROACH: APPLICATIONS TO ECONOMETRIC TIME SERIES WITH TRENDS AND SEASONALITIES.

Will Gersch and Tom Brotherton

Department of Information and Computer Sciences

University of Hawaii, Honolulu 96822

ABSTRACT

Let $y(n) = f(n) + \varepsilon(n)$, $n=1,\ldots,N$ with the $\varepsilon(n)$ i.i.d. from $\mathcal{N}(0,\sigma^2)$, σ^2 unknown and $f(\cdot)$ an unknown "smooth" function. The problem is to estimate $f(n)$, $n=1,\ldots,N$ in a statistically satisfactory manner. This problem was proposed by Whittaker, 1919. Wahba and Wold, 1975, is an $O(N^3)$ cubic spline solution; Akaike, 1979, is an $O(N^2)$ marginal likelihood "smoothness priors" solution.

In our approach alternative candidate smoothness constraint models are imbedded into dynamic state space forms. The recursive computational Kalman smoother procedure is invoked to achieve an $O(N)$ computation. Akaike's AIC statistical decision criterion is employed to determine the best of the alternative constraint Kalman filter-predictor modeled data. The Kalman smoother solution corresponding the AIC best Kalman filter, is then the best fixed interval smooth solution of the data.

Examples are shown including one in which the smoothing problem is generalized to the smoothing of econometric time series with trends and seasonalities.

I. INTRODUCTION

A "smoothing" problem and an approach to its solution, attributed to Whittaker 1919, is as follows: Let

$$y(n) = f(n) + \varepsilon(n) \qquad n=1,\ldots N \tag{1}$$

denote a sequence of observations. $f(\cdot)$ is an unknown "smooth" function, $\varepsilon(n)$, $n=1,\ldots N$ are independent normal identically distributed random variables with zero mean and unknown variance σ^2. The problem is to estimate $f(n)$ $n=1,\ldots N$ from the observations $y(1),\ldots y(N)$.

Whittaker suggested that the solution $f(n)$, $n=1,\ldots N$ balance a tradeoff between (in)fidelity to the data and (in)fidelity to a k-th order difference equation constraint. For a fixed value of λ, the solution satisfies

$$\min_{f,k}\left[\sum_{n=1}^{N}(y(n)-f(n))^2 + \lambda^2\sum_{n=1}^{N}(\nabla^{(k)}f(n))^2\right] \tag{2}$$

The first term in the brackets in equation (2) is the infidelity to the data measure, the second is the infidelity to the constraint measure and λ is the smoothness tradeoff parameter. Whittaker left the choice of k, the order of the difference equation constraint and the choice of λ, the smoothness trade off parameter, to the investigator.

Reinsch, 1967, demonstrated that the integral of the square of the m-th derivative smoothness constraint of the continuous function $f(\cdot)$ over the unit interval, led to an m-spline solution for $f(\cdot)$. Wahba, 1975, explored and exploited the convergence properties of the solution in a cubic spline solution-cross validation method of determinig the smoothness trade off parameter. Craven and Wahba, 1979, generalized the smoothing problem to determine the degree of the spline function fitted as well as the fidelity to the smoothness constaint trade off parameter. Schiller, 1973, and Akaike, 1979, invoked Bayesian approaches to the smoothing problem. In effect, Schiller interpreted the difference equation constraint to the solution of the smoothing problem as a Baysian or "smoothness prior" assumption on $f(n)$, $n=1,\ldots N$. Akaike exploited a marginal likelihood computation of that Bayesian model to obtain a two parameter constrained least squares solution to the original Whittaker problem.

In this paper, a Kalman smoother-Akaike AIC criterion solution is proposed for the smoothing problem. Constraints on the solution are expressed in a dynamical state space model of the observations. The role of λ, the fidelity to the data-fidelity to the smoothness constraint trade off parameter, is taken by the reciprocal of the state input noise to the observation noise. The Akaike AIC criterion, a statistic for determining the statiscically best of parametric models fitted to the data, allows us to automatically determine the best of the alternative difference order constraint models for smoothing. Examples of the smoothing problem including an application to the smoothing of nonstationary econometric time series, with trend and seasonal components are shown.

The recursive Kalman smoother solution to the smoothing parameter, are $O(N)$. In contrast, Wahba's generalized cross validation solution is $O(N^3)$ and Akaike's constrained least squares solution is $O(N^2)$.

2. A BAYESIAN SOLUTION TO THE SMOOTHING PROBLEM

A description of a Bayesian solution to the smoothing problem obtained by Akaike in 1979 that explicitly solves the problem articulated by Whittaker in 1919 follows: The minimization indicated in equation (2) with respect to $f(n)$, $n=1,\ldots N$ and k is equivalent to the maximization of

$$\ell(f) = \exp\left[\frac{-1}{2\sigma^2}\sum_{n=1}^{N}(y(n)-f(n))^2 \cdot \exp\frac{-\lambda^2}{2\sigma^2}\sum_{n=1}^{N}(\nabla^{(k)}f(n))^2\right] \tag{3}$$

For convenience, σ^2 is assumed known. Under the asumption of normality, equation (3) yields the Baysian posterior distribution interpretation

$$\ell(f) \propto p(y|\sigma^2,f)\ \pi(f|\lambda,k) \tag{4}$$

with $p(y|\sigma^2,f)$ the conditional data distribution, and $(f|\lambda,k)$ the smoothness prior distribution of f. Now, also consider λ to be unknown. Then, the marginal likelihood for λ is

$$L(\lambda,k) = \int p(y|\sigma^2,f)\ \pi(f|\lambda,k)\ df. \tag{5}$$

This "type II maximum likelihood method" of analysis was suggested by I.J. Good, 1965. In Bayesian parlance, λ is a hyperparameter. (See also Good and Gaskins, 1980.)

Directly integrating equation (5) and taking minus two times the logarithm of the likelihood yields an explicit closed form expression for $-2 \ln L(\lambda,k)$. Maximization of equation (3) is equivalent to minimization of $-2 \ln L(\lambda,k)$. Thus, the Bayesian optimal smoothness solution $f(n)$, $n=1,\ldots N$ corresponds to the minimum of $-2 \ln L(\lambda,k)$ that is achieved by a two parameter search over the parameters λ and k.

3. A KALMAN SMOOTHER-AKAIKE AIC CRITERION SOLUTION

A class of dynamic Gauss-Markov state space models of the observed data $y(n)$ $n=1,\ldots N$ (1) incorporates a difference equation constraint on the smoothing problem solution in the form,

$$\begin{aligned} x(n+1) &= F\, x(n) + G\, w(n+1) \\ y(n) &= H\, x((n) + v(n) \end{aligned} \tag{6a}$$

where $w(n)$ and $v(n)$ are i.i.d. random variables with

$$\begin{pmatrix} w(n) \\ v(n) \end{pmatrix} \sim \mathcal{N}\left[\begin{pmatrix} 0 \\ 0 \end{pmatrix} \begin{pmatrix} q^2 & 0 \\ 0 & r^2 \end{pmatrix}\right] . \tag{6b}$$

For the k=1,2,3 order stochastic difference eqution constraints respecively, the particular state space matrices F,G,H and the state in (6) are:

$$k=1:\; F = [1],\; G = [1],\; H = [1],\; x(nn) + [f(n)], \tag{7a}$$

$$k=2:\quad F = \begin{bmatrix} 2 & -1 \\ 1 & 0 \end{bmatrix} \qquad G = \begin{bmatrix} 1 \\ 0 \end{bmatrix} = H' \qquad x(n) = \begin{bmatrix} f(n) \\ f(n-1) \end{bmatrix} \tag{7b}$$

$$k=3:\quad F = \begin{bmatrix} 3 & -3 & 1 \\ 1 & 0 & 0 \\ 0 & 1 & 0 \end{bmatrix} \qquad G = \begin{bmatrix} 1 \\ 0 \\ 0 \end{bmatrix} = H' \qquad x(n) = \begin{bmatrix} f(n) \\ f(n-1) \\ f(n-2) \end{bmatrix} \tag{7c}$$

Higher order difference state space constraint models can be found similarly. With q^2 and r^2 known, equation (6) is an iterative formula for the generation of $y(n)$, $n=1,\ldots N$. It is initiated by assuming some initial condition on $x(0)$, say $x(0) \sim \mathcal{N}(x(0), P(0))$ where $P(k) = \operatorname{cov}(x(k))$. Also in equation (6) the quantity q^2/r^2 the ratio of the variances in the state process noise and observation noise takes the role of λ^2 the smoothness constraint trade off parameter.

The best one-step ahead predictor model of the $y(n)$, $n=1, \ldots, N$ data can be determined by a two parameter search minimization over k and q^2/r^2 via an interpretation of Akaike's, 1974 AIC criterion. That criterion is,

$$AIC(k,q) = -2 \ln \text{(maximized likelihood of the fitted model)} + 2 \text{(number of constraints in the fitted model)}. \quad (8)$$

The number of constraints on the solution is k+1, the number of states plus the trade off parameter. Following R.H. Jones, 1980 for example, for the Kalman filter/predictor and a given k, the likelihood of the parameter q^2 denoted $L(q^2)$, satisfies

$$L(q^2) = \prod_{i=1}^{N} (2\pi s(n))^{-\frac{1}{2}} \exp[-e^2(n)/2s(n)] \quad (9)$$

where

$$\begin{aligned} e(n) &= y(n) - H\hat{x}(n|n-1) \\ s(n) &= HP(n|n-1)H' + r^2 \\ x(\hat{n}|n) &= \text{Kalman filter/predictor estimate of } s(n) \text{ given } y(k),\ k=1,\ldots,n-1 \\ P(n|n-1) &= \text{cov}(\hat{x}(n|n-1)). \end{aligned} \quad (10)$$

For a given k the likelihood in (9) can be readily maximized with respect to q. The best one-step-ahead model of the data is that combination of k and q which gives rise to the minimum AIC (k,q). This model is then used with the conventional Kalman smoother. (For example see Meditch, 1969 for details of the smoothing computation.)

Other recent notable state space approaches to the smoothing problem are Weinart, 1979, Wecker and Ansley, 1978, and Kitagawa, 1981.

4. TWO EXAMPLES

In this section examples of smoothing problem and its Kalman smoother-Akaike AIC criterion solution are shown. The first example corresponds to the model situation discussed in section 3. Some illustrative results are shown in figures 1A-1D. Figure 1A shows simulated data $y(n) = f(n) + \varepsilon(n)$, $n = 1,2,\ldots,60$ with $\varepsilon(n)$ an i.i.d. sequence distributed $\mathcal{N}(0,1)$ and f(n) a truncated normal. Figures 1B-1D show the theoretical f(n) and the Kalman smoothed solutions for $q^2 = 0.001$, 0.005 and 0.1 respectively each computed for the k=2 state space constraint model. The solution of figure 1B is "too smooth" corresponding to excessive fidelity to the smoothing constraint. The solution shown in figure 1D is "too bumpy". This corresponds to excessive fidelity to the data. The corresponding AIC values are 77.6, 68.5 and 94.1 (the value q^2=0.005 was found using function minimization). The smoothed solution for q^2=0.005 is the best of the AIC criterion constraint models considered.

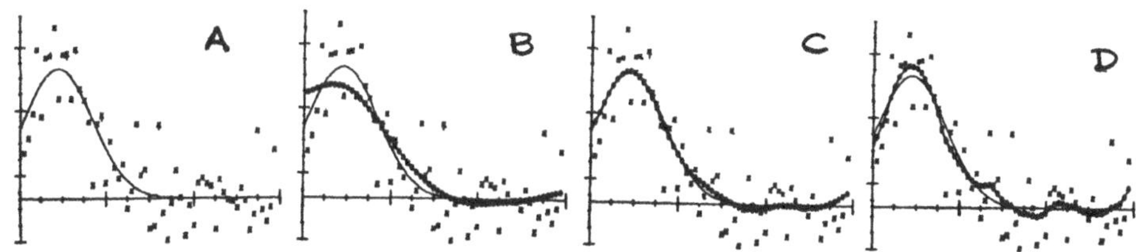

Figure 1: Theoretical smooth function ———, sample data xxx and smoothed estimates parametric in the smoothness parameter q^2 B: $q^2 = 0.001$, AIC = 77.6; C: $q^2 = 0.005$, AIC = 68.5; D: $q^2 = 0.1$, AIC = 94.1.

A nonstationary time series with a trend and seasonal mean value function can be modeled by the formula

$$y(n) = f(n) + s(n) + \varepsilon(n), \quad n = 1,\ldots,N. \tag{11}$$

Here $f(n)$ and $\varepsilon(n)$ are as before, $s(n)$ is a seasonal component of the mean value function of the time series $y(n)$.

A class of stochastic seasonal fixed difference equation constant models is given by

$$F = \left[\begin{array}{cccc|cccc} c_1 & \cdots & & c_k & & & & \\ 1 & & 0 & & & & 0 & \\ & \ddots & & & & & & \\ & & 1 & 0 & & & & \\ \hline & & & & 0 & \cdots & & 1 \\ & 0 & & & 1 & & 0 & \\ & & & & & \ddots & & \\ & & & & & & 1 & 0 \end{array}\right] \quad G = \begin{bmatrix} 1 & 0 \\ 0 & 0 \\ \vdots & \vdots \\ 0 & 0 \\ 0 & 1 \\ 0 & 0 \\ \vdots & \vdots \\ 0 & 0 \end{bmatrix}, \quad H' = \begin{bmatrix} 1 \\ 0 \\ \vdots \\ 0 \\ 1 \\ 0 \\ \vdots \\ 0 \end{bmatrix}, \quad x(n) = \begin{bmatrix} f(n) \\ f(n-1) \\ \vdots \\ f(n-k+1) \\ s(n) \\ s(n-1) \\ \vdots \\ s(n-p+1) \end{bmatrix} \tag{12}$$

$$\begin{bmatrix} w(n) \\ v(n) \end{bmatrix} \sim \mathcal{N}\left[\begin{pmatrix} 0 \\ 0 \\ 0 \end{pmatrix} \begin{bmatrix} q_1^2 & 0 & 0 \\ 0 & q_2^2 & 0 \\ 0 & 0 & r^2 \end{bmatrix}\right].$$

The $(c_1 \ldots c_k)$ are as in section (3), however $w(n)$, $n = 1,\ldots,N$ is now a two component i.i.d. vector random variable. The parameter p indicates the duration of the seasonal component; $p=4$ and $p=12$ for quarterly and monthly seasonal components respectively. There are $k + p + 2$ constraints in this model; k is the degree of the difference equation constraint on the trend, p is the order of the seasonality, q_1^2 and q_2^2 are the trade off trend and seasonal state input noise variances respectively. Now there are three terms in the trade off. They involve fidelity to the smoothness constraint, fidelity to the seasonal constraint and fidelity to the data.

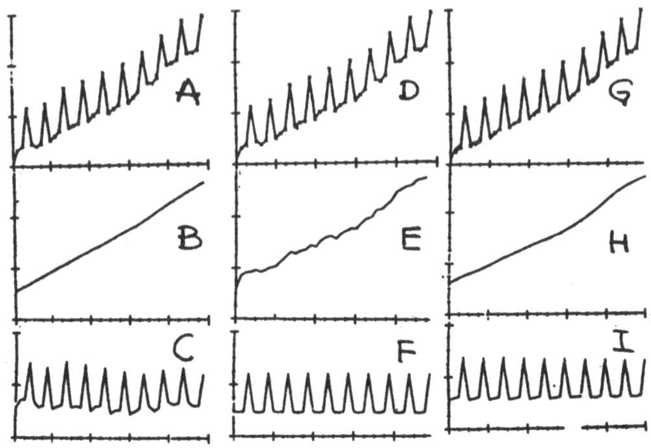

Figure 2: Trend plus Seasonality Component Estimates on the Japan GNP Data. [A,B,D]: $q_1^2 = 0.02$, $q_2^2 = 20$, AIC = -111.9 [DEF]: $q_1^2 = 20$, $q_2^2 = 0.02$, AIC = -69.7, [GHI]= $q_1^2 = 0.0218$, $q_2^2 = 0.0677$, AIC = -194.8

[A,D,G] = Original data ————, Trend plus seasonal component —xxx—,
[B,E,H] = Trend component estimate; G,F,I = Seasonal component estimate.

Figure 2 shows smoothing of a Japan GNP quarterly time series. The solution seen in Figures 2A and 2D enjoy almost perfect fidelity to the data. The first situation corresponds to excessive fidelity to the trend constraint the second corresponds to a excessive fidelity to the seasonal constraint. The best AIC smoothing solution shown in Figure 2 (G,H,I) in better agreement with our intuition about the appearance of the trend and seasonal components.

5. SUMMERY

We have shown a recursive computational approach to the classical smoothing problem and have illustrated an econometric time series smoothing problem variation. In our approach, candidate "smooth" constraint models are imbedded in dynamic state space forms. Akaike's AIC criterion is use to determine the best of the alternative Kalman filter-predictor modeled versions of the data. The Kalman smoother solution corresponding to the AIC best Kalman predictor model is then the best fixed interval smooth solution of the data.

The Kalman smoother iterative computational procedure yields an $O(N)$ computational solution to the smoothing problem. Heretofore $O(N^2)$ and $O(N^3)$ solutions were realized.

Potentially the Kalman smoother-Akaike AIC criterion recursive computational smoother for time series with trends and seasonalities can be applied to achieve an economic and objective alternative to the extensively employed X-11 seasonal adjustment procedures. Substantial Bayesian analyses of seasonal adjustment procedures have been done by Akaike, 1981 using the BAYSEA program. Those computations are $O(N^2)$. The methods shown here can realize O(N) alternatives to those procedures.

6. REFERENCES

[1] Akaike, H. 1974, 'A new look at the statistical model identification', IEEE Trans. on Auto. Control, AC-19, 716-723, 1974.

[2] Akaike, H. 1979 'Likelihood and the Bayes procedure', J.M. Bernardo, M.H. DeGrout, D.V. Lindley and A.F.M. Smith eds. Bayesian Statistics University Press, Vzlenciz, I.J. Good, Spain, (1080) pp. 143-166.

[3] Akaike, H. 1981, Seasonal adjustment by a Bayesian modeling, Journal of Time Series Analysis 1, 1-13.

[4] Craven, P. and Wahba, G. 1979, 'Smoothing noisy data with spline functions: Estimating the correct degree of smoothing by the method of generalized cross validation', Numer. Math. 31, 377-403, 1979.

[5] Good, I.J. 1965, The Estimation of Probabilities, MIT. Press, Cambridge, Mass.

[6] Good, I.J. and Gaskins, R.A. 1980, Density estimation and bump hunting on the penalized likelihood method exemplified by scattering and meteorite data, Journal of the Amer. Stat. Soc., 75, 42-73.

[7] Jones, R.H. 1980, Maximum likelihood fitting of ARMA models to time series, Technometrics, 22, 389-395.

[8] Kitagawa, G. 1981, "A nonstationary time series model and its fitting by a recursive technique", Journal & Time Series Analysis, 2, 103-110.

[9] Meditch, J.S. 1969, Stochastic Optimal Linear Estimation and Control, McGraw-Hill, 1969.

[10] Reinsch, C.H. 1967, 'Smoothing by spline functions', Numer. Math. 10, 177-183, 1967.

[11] Schiller, R. 1973, 'A distributed lag estimator derived from smoothness priors', Econometrica 41, 775-778, 1973.

[12] Wahba, G. and Wold, S. 1975, 'A completely automatic french curve: Fitting spline functions by cross validation', Comm. Statist. 4, 1975.

[13] Wecker, W.E. and Ansley, C.F. 1980, 'Linear and nonlinear regression viewed as a signal extraction problem', presented at the American Statistical Assoc. conf., 1980.

[14] Weinert, H.L. 1979, 'Statistical methods in optimal curve fitting', Comm. Statist., Vol. B7, 525-536, 1979

ON OBSERVABILITY AND UNBIASED ESTIMATION OF NONLINEAR SYSTEMS*

T. S. Lee
K. P. Dunn
C. B. Chang

Massachusetts Institute of Technology
Lincoln Laboratory
Lexington, Massachusetts 02173

1. Introduction

Motivated by estimation applications, we study the theory of observability and unbiased estimation for nonlinear systems. The purpose of this paper is two-fold. First, we study the nonlinear observability theory and present some new results. Second, we establish relationships between observability of nonlinear systems and the associated estimation problems through the use of the Fisher information matrix.

In solving a nonlinear state estimation problem, the theory of observability is extremely important. A representative discussion along this line is given in [1]. Other references can also be found in [6]. Practicing engineers have long been using system and measurement Jacobian matrices in place of linear observability conditions to derive a region of the state space where a practical nonlinear estimator can be designed, e.g., [2]. Furthermore, similar to linear systems, it has been observed that the Fisher information matrix of a nonlinear system shares the same form as the observability Gramian using Jacobian matrices, [3]. These practical considerations suggest that fruitful results established for linear systems can be easily extended to nonlinear applications. On the contrary, theoretical investigations, often motivated by solving nonlinear estimation problems, arrive at more complicated observability conditions than the linear counterparts. In this paper we address the similarity and the difference between linear and nonlinear systems related to observability and estimation problems.

The organization of this paper is described as follows. In Section 2, we establish observability conditions for nonlinear systems and then compare these results with linear system theories. In Section 3, we first point out that for linear systems minimum variance, unbiased estimates exist if and only if the systems are observable. For a nonlinear system, we find that the existence of bounded variance, unbiased estimates implies observability while the converse is not true in general. We finally give a condition which implies asymptotic observability for nonlinear systems. This same condition also assures asymptotic convergence in both the almost sure (a. s.) and the mean square (m. s.) sense, of the maximum likelihood estimate to the true state vector.

---

*This work was supported by the Department of the Army.

2. Observability

We consider an n-dimensional nonlinear state equation and an m-dimensional nonlinear observation equation given by

$$\dot{\underline{x}}(t) = \underline{f}(\underline{x}(t),t); \qquad \underline{x}(t_o) = \underline{x}_o \tag{2.1a}$$

$$\underline{y}(t) = \underline{h}(\underline{x}(t),t) \tag{2.1b}$$

respectively for an observation interval $[t_o,t_1]$ where $\underline{f}(\cdot)$ and $\underline{h}(\cdot)$ are differentiable with respect to $\underline{x}$ and t. The nonlinear differential equation (2.1a) is assumed to have a unique solution in $[t_o,t_1]$, denoted by $\underline{x}(t;\underline{x}_o)$.

Observability questions relate to the problem of determining the state trajectory from the observations obtained over some interval of time. Since we assume that the nonlinear state equation has a unique solution, knowing the initial state is sufficient for obtaining the entire state vector trajectory. We give the following definition for observability in a subset S of initial states in R^n.

Definition 2.1. The nonlinear system (2.1) is completely observable in S, if for any $\underline{x}_o$, $\underline{x}_1 \varepsilon S$, $\underline{h}(\underline{x}(t;\underline{x}_o),t) = \underline{h}(\underline{x}(t;\underline{x}_1),t)$ for all $t \varepsilon [t_o,t_1]$ then $\underline{x}_1 = \underline{x}_o$.
For linear systems, observability in S implies complete observability in R^n because the initial state can be constructed from the observation analytically, independent of the trajectory of the state vector, [4]. Therefore, the nonlinear observability problem is much more difficult than the linear counterpart. In this section, we shall study observability conditions in terms of properties of $\underline{f}$ and $\underline{h}$ in (2.1).

We first introduce the following notations. Let the Jacobian matrix F(t) be defined by

$$F(t) \triangleq \left. \frac{\partial \underline{f}(\underline{x},t)}{\partial \underline{x}} \right|_{\underline{x} = \underline{x}(t;\underline{x}_o)} \tag{2.2}$$

Likewise, H(t) is the Jacobian matrix of $\underline{h}(\underline{x},t)$. The transition matrix of F(t) is denoted by $\phi(t,\tau)$. For simplicity, the dependence of F(t), H(t), and $\phi(t,\tau)$ on the initial state $\underline{x}_o$ is not explicitly shown. Furthermore, we define

$$M(\underline{x}_o;t_o,t_1) \triangleq \int_{t_o}^{t_1} \phi^T(\tau,t_o)H^T(\tau)H(\tau)\phi(\tau,t_o)d\tau \tag{2.3}$$

where superscript "T" denotes matrix transpose. Again, for simplicity we may use $M(\underline{x}_o)$ to denote $M(\underline{x}_o;t_o,t_1)$.

For linear systems with system and measurement matrices given by $F(t)$ and $H(t)$, respectively, $M(\underline{x}_o;t_o,t_1)$ becomes independent of $\underline{x}_o$, which is commonly denoted by $M(t_o,t_1)$, and is the observability Gramian [4]. The linear system is completely observable in R^n if and only if $M(t_o,t_1) > 0$. We do not obtain exactly the same result for the nonlinear system (2.1) as illustrated by the following theorem.

Theorem 2.1. Let S be a set of initial states. System (2.1) is not completely observable in S if and only if for each pair of points $\underline{c}$, $\underline{d}$ in S, and $\delta\underline{c} \triangleq \underline{d}-\underline{c} \neq \underline{0}$, there exists a real number θ, $0 < \theta < 1$, such that for all t in $[t_o,t_1]$

$$\text{(i)}\quad M(\underline{e})\delta\underline{c} = \underline{0}$$

$$\text{(ii)}\quad \delta\underline{h} \triangleq \underline{h}(\underline{x}(t;\underline{d}),t)-\underline{h}(\underline{x}(t;\underline{c}),t) = \left.\frac{\partial\underline{h}(\underline{x}(t;\underline{x}_o),t)}{\partial\underline{x}_o}\right|_{\underline{x}_o=\underline{e}} \delta\underline{c} \qquad (2.4)$$

are true, where $\underline{e} = \underline{c} + \theta\delta\underline{c}$.

Proof. (a) Necessity

Suppose that there is a θ such that (2.4) holds. By (2.2) and (2.3), the Frechét derivative of $\underline{x}(t;\underline{x}_o)$ with respect to $\underline{x}_o$ is [1]

$$\frac{\partial\underline{x}(t;\underline{x}_o)}{\partial\underline{x}_o} = \phi(t,t_o) \qquad (2.5)$$

We can use (2.4) to obtain

$$g = \int_{t_o}^{t_1} ||\delta\underline{h}||^2 d\tau = (\delta\underline{c})^T M(\underline{e})(\delta\underline{c}) = 0 \qquad (2.6)$$

where $||\cdot||$ denotes the Euclidean norm. Therefore, $\underline{d}$ and $\underline{c}$ are not distinguishable from observation over $[t_o,t_1]$ and hence (2.1) is not completely observable in S.

(b) Sufficiency

Suppose that System (2.1) is not completely observable in S. By the mean-value theorem, we know that there exists a real number θ, $0 < \theta < 1$, such that (2.4) holds. Q.E.D.

Corollary 2.1.1. If $M(\underline{x}_o)$ is positive definite for all $\underline{x}_o$ in a convex set S then System (2.1) is completely observable in S.

Proof. Because of the convexity of S, we have $\underline{e} = \underline{c} + \theta\delta\underline{c}\varepsilon S$ for all θ, $0 < \theta < 1$. Condition (i) of Theorem 2.1 fails to be true due to the hypothesis that $M(\underline{x}_o) > 0$ for all $\underline{x}_o$ in S. Q.E.D.

The converse statement of Corollary 2.1.1 is not true in general. Examples are given in [6]. Note also that $M(\underline{x}_o)$ is a function of the trajectory $\underline{x}(t)$, therefore, using $M(\underline{x}_o)$ directly to examine the observability condition is not convenient. The remainder of this section is devoted to deriving an observability condition based on properties of $\underline{f}$ and $\underline{h}$. We first define a partial ordering among linear systems.

Definition 2.2. A linear system defined in R^n

$$\dot{\underline{x}}(t) = A(t)\underline{x}(t)$$
$$\underline{y}(t) = C(t)\underline{x}(t) \tag{2.7}$$

which is denoted by $(A(t),C(t))_n$ is said to be less or equally observable than another linear system $(F(t),H(t))_n$, if the observability Gramian $M(t_o,t_1)$ and $\hat{M}(t_o,t_1)$ of $(A(t),C(t))_n$ and $(F(t),H(t))_n$, respectively, have the partial order that $\hat{M}(t_o,t_1)-M(t_o,t_1)$ is semi-positive definite.

Note that the definition implies that the observable subspace associated with $(F(t),H(t))_n$ contains the observable subspace associated with $(A(t),C(t))_n$ [4]. Consequently, if $(A(t),C(t))$ is completely observable in R^n, it implies that $(F(t),H(t))_n$ is completely observable in R^n. The following theorem provides a sufficient condition for a linear system $(A(t),C(t))_n$ being less or equally observable than a linear system $(F(t),H(t))_n$:

Theorem 2.2. A linear system $(A(t),C(t))_n$ is less or equally observable than $(F(t),H(t))_n$, if

(i) $F(t)-A(t) = \lambda(t)I$ where $\lambda(t) \geq 0$, and (2.8)

(ii) $H^T(t)H(t) \geq C^T(t)C(t)$ (2.9)

for all t in $[t_o,t_1]$.

The proof of this theorem follows by applying the method used in [5] and the result of Lemma 2.1 established below.

Lemma 2.1. Let A and B be two nxn complex-valued matrices. The following inequality

$$AB + BA^* \geq 0 \tag{2.10}$$

holds for all semi-positive definite Hermitian matrix B, if and only if

$$A = \lambda I \quad \text{and} \quad Re(\lambda) \geq 0$$

where the superscript "*" denotes the conjugate and transpose of a complex-valued matrix and Re(•) denotes the real part of the enclosed complex variable.

The lemma presents a non-trivial result in the linear operator theory namely that the linear operator $L_A(B) \triangleq AB + BA^*$ is invariant in the subspace of semi-positive definite Hermitian matrices if and only if $A=\lambda I$ and $Re(\lambda) \geq 0$. The proof of the lemma follows from two facts that (i) B can be represented by a linear combination of the outer product of its eigenvectors, (ii) the inequality (2.10) is equivalent to the case when A is replaced by its Jordan form. Detailed proof of this lemma can be found in [6].

The application of Theorem 2.2 to nonlinear systems is summarized by the following corollary.

Corollary 2.2.1. (Global Observability Condition)

If there exists a linear observable system $(A(t),C(t))_n$ such that $\underline{f}$ and $\underline{h}$ of (2.1) satisfy the following conditions:

$$\text{(i)}\quad \underline{f}(\underline{x},t) = A(t)\underline{x} + \underline{g}(\underline{x},t), \text{ and} \tag{2.11}$$

$$\frac{\partial \underline{g}}{\partial \underline{x}} = \lambda(\underline{x},t)I \quad \text{where } \lambda(\underline{x},t) \geq 0;$$

$$\text{(ii)}\quad \left(\frac{\partial \underline{h}}{\partial \underline{x}}\right)^T \left(\frac{\partial \underline{h}}{\partial \underline{x}}\right) \geq C^T(t)C(t) \tag{2.12}$$

for all t in $[t_o,t_1]$ and $\underline{x}$ in R^n, then System (2.1) is completely observable in R^n.

The condition (2.11) restricts (2.1a) to be linear except when $\underline{f}$ is scalar. There are nonlinear differential equations [7] which can be transformed into linear differential equations. Corollary 2.2.1 is potentially applicable to this class of nonlinear systems.

3. Observability and Unbiased Estimation

We consider the problem of estimating the initial state $\underline{x}_o$ from noisy observations given by

$$\underline{z}(t) = \underline{h}(\underline{x}(t;\underline{x}_o),t) + \underline{n}(t) \tag{3.1}$$

The noise process $\underline{n}(t)$ is a zero-mean white Gaussian noise process with unity matrix spectral density function.

It is intuitively true that bounded variance, unbiased estimates of $\underline{x}_o$ based on noisy observations cannot be obtained if System (2.1) is not observable. The converse statement is also true for linear systems. For a discrete linear system, it is shown in [8] that the least-square, unbiased estimate of $\underline{x}_o$ exists if and only if the corresponding, static, stacked system matrix is of full rank. It is easy to show that the above condition is equivalent to the system being observable. If the corresponding Fisher information matrix $\mathcal{F}$ for a continuous linear system $(F(t),H(t))_n$ is invertible, the maximum likelihood estimate (MLE) of $\underline{x}_o$ is given by

$$\hat{\underline{x}} = \mathcal{F}^{-1} \int_{t_o}^{t_1} \phi^T(\tau,t_o)H^T(\tau)z(\tau)d\tau$$

where ϕ is the transition matrix of $F(t)$. The MLE is the minimum variance unbiased estimate. Furthermore, following the treatment in [9], we know that the Fisher information matrix is the observability Gramian.

In this section we shall study what we can expect for nonlinear systems using the Fisher information matrix. It can be shown that the observability Gramian $M(\underline{x}_o)$ defined by (2.3) is the Fisher information matrix. The following theorem assures that observability is necessary for the existence of unbiased estimates.

Theorem 3.1. Let S be a convex set of initial states. If there is an unbiased estimate for each $\underline{x}_o$ in S based on $\underline{z}(t)$, $t_o \leq t \leq t_1$, with a finite error covariance matrix P, then System (2.1) is completely observable in S.

Proof. Following exactly the same treatment described on page 80 of [10], we have, for each $\underline{x}_o \varepsilon S$,

$$\begin{bmatrix} P & I \\ I & M(\underline{x}_o) \end{bmatrix} \geq 0 \tag{3.2}$$

Let a be a constant such that $R \triangleq aI + P > 0$ and the inequality (3.2) preserves if P is replaced by R. Therefore, for any $\underline{x}_1$ and $\underline{x}_2$ with appropriate dimensions, we have

$$\underline{x}_1^T R \underline{x}_1 + \underline{x}_2^T \underline{x}_1 + \underline{x}_1^T \underline{x}_2 + \underline{x}_2^T M(\underline{x}_o)\underline{x}_2 \geq 0 \tag{3.3}$$

Letting $\underline{x}_1 = -R^{-1}\underline{x}_2$, it is seen that $M(\underline{x}_o) \geq R^{-1} > 0$ for all $\underline{x}_o$ in S. Corollary 2.1.1 implies that System (2.1) is completely observable in S. Q.E.D.

A necessary and sufficient condition for the existence of unbiased estimates with finite central moments was studied by Barankin in [11]. Although the Barankin's condition is not developed in the context of observability, it is useful to construct the following example to demonstrate that the converse statement of Theorem 3.1 is not true in general. For this purpose, let $S = [1,\infty)$, $\dot{x}=0$, and $z(t) = y(t)+n(t)$, where $y(t) = \sqrt{\ell n(x(t))/2}$. The deterministic system is completely observable in S because y(t) is monotone in x(t). However, using Theorem 2 of [11], it can be shown that there exists no unbiased estimate of x_o at 1 with finite variance.

It is clear that conditions in addition to observability are required for the existence of unbiased estimates for nonlinear systems. The remainder of this section is devoted to seeking for these extra conditions. For this purpose, we shall study the asymptotic property of the MLE of $\underline{x}_o$.

For mathematical rigor, (3.1) is redefined by

$$d\underline{m}(t) = \underline{h}(\underline{x}(t;\underline{x}_o),t)dt + d\underline{w}(t) \tag{3.4}$$

where $\underline{w}(t)$ is the standard Brownian motion with independent components and $\underline{h}$ is square-integrable. The MLE of $\underline{x}_o$ is derived from the likelihood ratio defined for (3.1). The likelihood ratio is the Radon-Nikodym derivative of the probability measure P_m induced by the $\underline{m}$-process with respect to the Wiener measure P_B induced by a Brownian motion process B(t), which is given by [12]

$$\frac{dP_m}{dP_B} = \exp(-\frac{1}{2}\int_{t_o}^{t_1} ||\underline{h}(\underline{x}(t;\underline{c}),t)||^2 dt + \int_{t_o}^{t_1} \underline{h}^T(\underline{x}(t:\underline{c}),t)d\underline{m}(t)) \tag{3.5}$$

for $\underline{c}$ in S. Let $L(\underline{c};t_1)$ be given by

$$L(\underline{c};t_1) = \frac{-1}{t_1}\ \ell n\ (\frac{dP_m}{dP_B}) \tag{3.6}$$

The MLE of $\underline{x}_o$, $\hat{\underline{x}}_o(t_1)$, is defined to be the nonlinear function of $\underline{m}(t)$, $t_o \leq t \leq t_1$, which minimizes $L(\underline{c};t_1)$ over S. We give the following theorem.

Theorem 3.2. Let S be a compact subset of R^n containing $\underline{x}_o$. If $\frac{1}{t_1} M(\underline{c};t_o,t_1)$ becomes positive definite asymptotically with respect to t_1 for all $\underline{c}$ in S then $\hat{\underline{x}}_o(t_1)$ converges a.s. and m.s. to $\underline{x}_o$ asymptotically.

Proof. Substituting (3.4) into $L(\underline{c};t_1)$ yields

$$L(\underline{c};t_1) = \frac{-1}{t_1}(-\frac{1}{2}\int_{t_o}^{t_1} ||\underline{h}(\underline{x}(t;\underline{c}),t)||^2 dt \tag{3.7}$$

$$+ \int_{t_o}^{t_1} \underline{h}^T(\underline{x}(t;\underline{c}),t)\underline{h}(\underline{x}(t;\underline{x}_o),t)dt + \int_{t_o}^{t_1} h^T(\underline{x}(t;\underline{c}),t)d\underline{w}(t))$$

Applying Khazminskii's Lemma (see [13] or [14]) to the last term of (3.7), we obtain that

$$|L(\underline{c};t_1)-Q(\underline{c};t_1)| \xrightarrow{t_1\to\infty} 0 \quad \text{a.s.} \tag{3.8}$$

and uniformly in S, where $Q(\underline{c};t_1)$ is defined by the first two terms of (3.7). Straightforward algebraic manipulation shows that

$$Q(\underline{c};t_1)-Q(\underline{x}_o;t_1) = \frac{1}{t_1}\int_{t_o}^{t_1} ||\underline{h}(\underline{x}(t;\underline{c}),t)-\underline{h}(\underline{x}(t;\underline{x}_o),t)||^2 \, dt \geq 0 \quad (3.9)$$

Using similar arguments in the proof of Theorem 2.1 and the hypothesis that $\frac{1}{t_1} M(\underline{c};t_o,t_1)$ is asymptotically positive definite for all $\underline{c}$ in S, we conclude that equality of (3.9) holds asymptotically iff $\underline{c} = \underline{x}_o$. The theorem is proved by the result reported in [15]. Q.E.D.

By using Hölder inequality in the original proof of Khazminskii in [13], Theorem 3.2 still holds if $\frac{1}{t_1^{\varepsilon}} M(\underline{x}_o;t_o,t_1) > 0$ for all $\underline{x}_o \varepsilon S$ asymptotically for any $\varepsilon > 1/2$.

Finally, we remark that the hypothesis of Theorem 3.2, implies asymptotic observability of (2.1) in S.

4. Summary and Conclusion

In this paper, we have studied the nonlinear observability theory and its relationship with nonlinear estimation problems. We first presented a necessary and sufficient condition for local observability of nonlinear systems. Based on the sufficient condition for local observability, we developed a global observability condition for a class of nonlinear systems. We have found that the observability of a nonlinear system is necessary for the existence of an unbiased, bounded-variance state estimate. However, different from the linear counterpart, the converse is not true in general. A sufficient condition for asymptotic observability for nonlinear systems is given. This condition also assures the asymptotic convergence, in both the almost sure and the mean square sense, of the maximum likelihood estimate to the true state vector.

References

[1] E. B. Lee and L. Markus, "Foundations of Optimal Control Theory", John Wiley and Sons, Inc., New York (1967).

[2] L. B. Weiner, and T. L. Homsley, "Application of Angle-Only Track to Ballistic Missile Defense," Proceeding of the 1976 Conference on Decision and Control, Clearwater, Florida, pp 579-584, (December 1976).

[3] C. B. Chang, "Ballistic Trajectory Estimation with Angle-Only Measurements," IEEE Trans. on Automatic Control, Vol. AC-25, pp 474-480, (June 1980).

[4] R. W. Brockett, "Finite Dimensional Linear Systems", John Wiley and Sons, Inc., New York, (1970).

[5] A. S. Gilman and I. B. Rhodes, "Cone-bounded Nonlinearities and Mean Square Bounds-Estimation Upper Bounds", IEEE Trans. on Automatic Control, 18, pp 260-265, (June 1973).

[6] K. P. Dunn, T. S. Lee and C. B. Chang, "On Observability and Unbiased Estimation of Discrete Nonlinear Systems", Submitted to IEEE Trans. on Automatic Control.

[7] J. M. Thomas, "Equations Equivalent to a Linear Differential Equation", Proc. AMS 3, pp 899-903, (1952).

[8] B. W. Rust and W. R. Burrus, "Mathematical Programming and the Numerical Solution of Linear Equations", American Elsevier Publishing Company, Inc., New York (1972).

[9] A. H. Jazwinski, "Stochastic Process and Filtering Theory", Academic Press, New York, (1970).

[10] H. L. Van Trees, "Detection, Estimation and Modulation Theory, Part I", John Wiley and Sons, Inc., New York, (1968).

[11] E. W. Barankin, "Locally Best Unbiased Estimates," Ann. Math. Stat. 20, pp 477-501, (1949).

[12] T. E. Duncan, "Evaluation of Likelihood Function," Information and Control 13, pp 62-74, (1968).

[13] R. Z. Khazminskii, "Stability of Systems of Differential Equations Under Random Disturbances of Their Parameters", (in Russian), Nauka, Moscow, (1969).

[14] T. S. Lee and F. Kozin, "Almost Sure Asymptotic Likelihood Theory for Diffusion Processes," J. Appl. Prob. 14, pp 527-537, (1977).

[15] F. Nakazima and F. Kozin, "A Characterization of Consistent Estimators," IEEE Trans. on Automatic Control, Vol. AC-24, pp 758-764, (October 1979).

AN ACCELERATED EXPERIMENTAL DESIGN ALGORITHM

T.G. Robertazzi* and S.C. Schwartz
Department of Electrical Engineering
and Computer Science
Princeton University
Princeton, NJ 08544

ABSTRACT

An efficient algorithm is presented for generating D-optimal designs. The usual sequential D-optimal design algorithm embodies the principle of the greedy algorithm of combinatorial optimization. It is shown that a sufficient condition for applying the accelerated greedy algorithm of M. Minoux to the design problem is satisfied. The actual implementation of the accelerated sequential design algorithm is based on a more general sufficient condition. This allows the evaluation of quadratic forms to replace determinant evaluations. A heap type data structure provides additional efficiency. While the standard sequential design algorithm requires a number of basis function evaluations proportional to the number of iterations, the accelerated design algorithm computation is proportional to a much smaller sum of coefficients.

I. INTRODUCTION

As optimal experimental design is some allocation, within a sampling region, of a limited number of replicated measurements which maximizes the statistical information obtained concerning some unknown parameters. Let the sampling region be χ. In the continuous formulations of the design problem, developed originally by Kiefer and Wolfowitz [1], one seeks an optimal "design measure", ξ, satisfying:

$$\xi(x)\geqslant 0 \quad x\varepsilon\chi \tag{1.1}$$

$$\int_{\chi}\xi(x)dx=1 \tag{1.2}$$

A D-optimal design, ξ^*, maximizes the determinant of the corresponding Fisher information matrix. A complete exposition of D-optimal design theory for static models appears in Federov [2] and more concise surveys are also available [3]. Mehra [4] and Titterington [5] discuss D-optimal design theory for dynamic systems. The accelerated sequential design algorithm described in this paper is applicable to static models; conditions under which it is economical for dynamic systems can be found in [14].

The static regression model is:

$$\underline{\theta}_k= \Phi, \; k=1,2,...M \tag{1.3}$$

*Now with Bell Telephone Laboratories, West Long Branch, New Jersey

$$y_k=\underline{H}_k\underline{\theta}_k+v_k \tag{1.4}$$

where $\underline{H}_k$ is the 1xN vector of basis functions at the kth measurement location. It is assumed that there is a grid of M such locations in the sampling region, χ. The Nx1 vector $\underline{\theta}_k$ represents the location (time) invariant parameters. Replicated measurements are allowed so that the measurement equation, (1.4), may be used several times for a particular location. The noise term, v_k, is assumed to be an independent and identically distributed, zero mean random variable with known variance, σ^2.

D-optimal experimental designs are asymptotically produced [2,6] by an iterative procedure which will be referred to as the "standard sequential design algorithm". The algorithm for un-normalized designs (1.2 does not hold) will be considered.

STANDARD SEQUENTIAL DESIGN ALGORITHM

1) Begin with some non-singular allocation of measurements (design) within the sampling region χ.
2) Find the measurement location (from those in the grid) where adding a measurement to the previous design would maximize the increase in the determinant of the Fisher information matrix. Update the old design by including an extra measurement at this location. Continue step 2 until some test for convergence of the design is satisfied.

Let the error covariance matrix be $\underline{P}_i=\sigma^2\underline{I}_i$. Since [6]:

$$|\underline{I}_i+\underline{H}^T\underline{H}|-|\underline{I}_i|=|\underline{I}_i|\,(\frac{\underline{H}\,\underline{P}_{i+1}\underline{H}^T}{\sigma^2}) \tag{1.5}$$

the many determinant evaluations that step 2 seems to require can be replaced by evaluations of the predicted measurement variance, $\underline{H}_k\underline{P}_{i+1}\underline{H}_k^T$. Moreover, a simple recursion may be used:

$$\underline{H}_k\underline{P}_{i+1}\underline{H}_k=\underline{H}_k\underline{P}_i\underline{H}_k - \frac{(\underline{H}_k\underline{P}_i\underline{H}_i^T)}{\underline{H}_i\underline{P}_i\underline{H}_i^T+\sigma^2} \tag{1.6}$$

Here i is the iteration number. The dominant computational cost in the above recursion is the evaluation of the basis functions at the kth location. Forming the product $\underline{H}_k\underline{P}_i\underline{H}_i^T$ also requires N multiplications per location.

In the above design procedure, the measurements which is appended to the old design at the ith iteration represents approximately $\frac{1}{i}$ of the new design. This corresponds to Wynn's [6] implementation of the standard sequential design algorithm. One could instead solve for the optimal weightings [2].

An alternative to the standard sequential design algorithm is Mitchell's DETMAX procedure [7,8]. The basic idea is to perform a series of "excursions" from a current solution. These consist of a sequence of additions and removals of several measurements. This procedure is more suited for the "discrete" design problem where the integral of (1.2) is replaced by a summation over a small number of replicated measurements. It is not very efficient when dealing with the large number of candidate grid points necessary to approximate a continuous sampling region.

2.0 GREEDY ALGORITHMS

The standard sequential design algorithm embodies the principle of the "standard" greedy algorithm [9] of combinatorial optimization. It can be viewed as an attempt to solve the optimal subset problem. Given a set E, of M elements, and a function defined on the power set of E, P(E), the optimal subset problem is to find the subset S* such that:

$$f(S^*)=\max_{S\subseteq E} f(s) \tag{2.1}$$

The standardgreedy algorithm may be written as:

STANDARD GREEDY ALGORITHM

1) Begin with the empty set S_o.
2) At each iteration, from the set of unused elements, $E-S_i$, find the element, e_j, which when appended to S_i, maximizes the incremental improvement in f.

For problems which are equivalent to finding the maximum weight basis of a matroid, the above type of procedure (with a provision for avoiding dependencies) is optimal [13]. More generally the greedy strategy is often used to produce approximate solutions. Recall that the standard sequential design algorithm asymptotically produces optimal designs.

The significant difference between the standard sequential design algorithm and the standard greedy algorithm is that the design algorithm may select any element of E, the set of measurement locations on the grid, at any iteration. That is since replicated measurements are allowed, any measurement location may be selected at any iteration.

The "accelerated" greedy algorithm of M. Minoux [9,10,11] depends on a particular set function property, submodularity, being inherent in problems it is applied to, or at least serving as a good approximation.

Let the incremental improvement in the objective function, that is obtained by appending an element e_j, to a set A ($e_j \notin A$), be defined as:

$$\delta(A,e_j)=f(A+\{e_j\})-f(A) \tag{2.2}$$

A set function is submodular if and only if ($B\subseteq A\subseteq E$ and $e_j \notin A$),

$$\delta(A,e_j)\leqslant\delta(B,e_j) \tag{2.3}$$

Submodularity can be thought of as a combinatorial version of the law of diminishing returns. The earlier that e_j is appended to A, the larger is the incremental improvement. This is suggestive of the situation in estimating statistical parameters. Generally the incremental improvement in the accuracy decreases, as the number of measurements already processed, increases.

In the standard greedy algorithm at the ith iteration one evaluates $\delta(S_i, e_j)$ for all e_j which are not contained in S_i. The element which provides the maximum δ is chosen. If all the elements of E are to be eventually selected, this requires $M^2/2$ evaluations. Minoux's accelerated greedy algorithm exploits submodularity so as to avoid performing many of these evaluations.

During the first iteration of the accelerated greedy algorithm, the incremental differences are calculated. The element associated with the maximum difference is selected and the remaining incremental differences are stored for use during the next iteration. These subsequent iterations involve a number of sub-iterations. In each sub-iteration the maximum incremental difference in storage is sought. This may be an old incremental difference. It is updated. This requires one function evaluation. If the new, updated value is larger than all the values in storage, it must be the true maximum. This is a consequence of (2.3). None of the values in storage that could possibly be updated may increase.

Suppose that an updated value is not larger than all the elements in storage. A flag, corresponding to the update value, is set to indicate that the updating has been performed and that value is returned to storage. The largest value in storage is found and another sub-iteration occurs. Eventually a sub-iteration produces a maximum value or the largest value found in storage has been updated previously. This element is appended to S_i to produce S_{i+1}, the flags are reset, and another iteration occurs.

A schematic outline of this algorithm appears in [9].

3.0 ACCELERATED SEQUENTIAL DESIGN ALGORITHM

If the accelerated greedy algorithm is to be applied to the experimental design problem, then a modification of the previous definition of submodularity is needed. Let a "replication function", $r(e_j)$, be defined over the individual elements of E. This is the number of times each element of E has been previously sampled by a greedy type algorithm. Thus associated with each measurement location, e_j, belonging to a design, A, is $r_A(e_j)$, the number of times a measurement at that location is made. Then $B \subseteq A$ if and only if $r_B(e_j) \leq r_A(e_j) \forall_j$. From this point on, a set function will be considered to be submodular if and only if for $B \subseteq A \subseteq E$ and $e_j \varepsilon A$, (2.3) holds.

The function f(S) for the un-normalized design problem is indeed submodular. The incremental improvement that results from appending a single measurement to the current estimate is:

$$|\underline{P}_{i+1}| - |\underline{P}_i| = \tag{3.1}$$

$$|(\underline{I} - \frac{\underline{P}_i \underline{H}_k{}^T \underline{H}_k}{\underline{H}_k \underline{P}_i \underline{H}_k + \sigma_i^2}) \quad (\underline{P}_i)| - |\underline{P}_i| \tag{3.2}$$

$$= \left(\left| \underline{I} - \frac{\underline{P}_i \underline{H}_k^T \underline{H}_k}{\underline{H}_k \underline{P}_i \underline{H}_k^T + \sigma_i^2} \right| - 1 \right) |\underline{P}_i|$$

$$= \frac{-\underline{H}_k \underline{P}_i \underline{H}_k^T}{\underline{H}_k \underline{P}_i \underline{H}_k^T + \sigma_i^2} |\underline{P}_i| \tag{3.3}$$

The incremental improvement is the product of two terms. The term at the right of (3.3), $|\underline{P}_i|$, is nonincreasing in the number of measurements that have been processed (see 1.5). This is also true of the term at the left. To see this, note that the term is monotonic in the predicted measurement variance, $\underline{H}_k \underline{P}_i \underline{H}_k^T$. This quantity is nonincreasing in the number of measurements processed (set k=i in equation 1.6). The incremental improvement, then, is non-increasing in the number of measurements that already have been processed. This means that (2.3) is satisfied and the function is submodular.

The covariance matrix determinant is not submodular if one considers normalized designs. Of course any unnormalized design can be scaled into a normalized design.

In section 1.0 it was pointed out that the determinant evaluations required by the standard sequential design algorithm could be replaced by evaluations of the predicted measurement variance. This is also true for the accelerated sequential design algorithm.

The greedy approach has so far been formulated in terms of incremental differences. What one actually wants is that $\{e_j\}$ which results in the largest value of the objective function f(S). The function actually examined by the accelerated greedy algorithm will be called f_{algo}. So far $f_{algo}(A + \{e_j\}) = \delta(A, \{e_j\})$. The underlying heuristic of the accelerated greedy algorithm is actually applicable so long as

$$f_{algo}(S + \{e_j\}) \leq f_{algo}(S) \tag{3.4}$$

for all $S \subseteq E$ and all $\{e_j\}$. The above inequality describes a class of problems which includes submodular set functions. It is also satisfied by the predicted measurement variance for the design problem. This follows from (1.6).

The design algorithm which exploits (3.4) for the predicted measurement variance can be written as:

ACCELERATED SEQUENTIAL (UN-NORMALIZED) DESIGN ALGORITHM

1) Begin with some initial allocation of measurements. This is in terms of a set of locations, S_0, and an associated replication function, $r(e_k)$. Compute the corresponding $\underline{P}_0$.
2) For every measurement location, $e_k \in E$, compute and store:

$$\underline{H}_k \underline{P}_0 \underline{H}_k^T$$

3) At iteration i:

3a) Find $\{e_k^{max}\}$ the measurement location with the largest corresponding value of predicted measurement variance in storage.

If i=0 or if $\{e_k^{max}\}$ has already been selected during this iteration, go to (3c).

3b) For $\{e_k^{max}\}$, update the predicted measurement variance:

$$\underline{H}_k{}^{max}\underline{P}_{i+1}\underline{H}_k{}^{T\ max}$$

Here $\underline{H}_k{}^{max}$ is the vector of basis functions associated with $\{e_k^{max}\}$.
If this updated predicted measurement variance is less than some value in storage, return to (3a). Otherwise go to (3c).

3c) The maximum element is appended to the current solution by updating S_i and $r(e_k{}^{max})$. The $i+1^{st}$ iteration begins. If the termination criterion is not satisfied, go to (3a).

The total number of function evaluations performed by the accelerated sequential design algorithm is:

$$\sum_{i=1}^{i_{total}} c_{bf}(i)M \tag{3.5}$$

Here M is the number of measurement locations, i_{total} is the number of iterations the algorithm performs and $c_{bf}(i)$ is the fraction of the total number of measurement locations that are evaluated at iteration i. This varies with the basis functions ("bf") in question.

The number of evaluations performed by the accelerated sequential design algorithm tends to decrease as the iterations progress. This is because the change in the predicted measurement variance that results from updating tends to decrease. As a consequence, there are fewer locations whose predicted measurement variance is larger than that of an updated location. This leads to fewer sub-iterations and fewer evaluations.

As $M\to\infty$, the c_{bf}'s represent the fraction of area in the sampling region whose grid points are updated during an iteration. As M increases, the increase in the number of function evaluations is simply proportional to M.

By way of comparison, the total number of evaluations performed by the standard greedy algorithm is:

$$i_{total}M \tag{3.6}$$

It is important to realize the cost of a standard sequential design algorithm evaluation is somewhat less than the cost of an accelerated algorithm evaluation. The standard algorithm can take advantage of the N multiplications per evaluation of (1.6) rather than the N^2 multiplications involved in forming the product $\underline{H}_k\underline{P}_i\underline{H}_k{}^T$. To acheive practical savings the accelerated algorithm must perform a small enough number of evaluations to compensate for this. As the cost of calculating the basis function values, $\underline{H}_k$, increases, this difference becomes less important.

The accelerated sequential design algorithm requires finding the maximum of a set of numbers (step 3a) many more times than the standard algorithm does. A naive implementation of this step can result in a total number of comparisons being made which is proportional to M^2. This can easily become the dominant computational cost. By storing the predicted measurement variances (and corresponding location numbers) in a heap type data structure [12], this problem can be eliminated. The total number of comparisons will now be bounded from above by

$$\sum_{i=1}^{i_{total}} c_{bf}(i) M \log_2 M \tag{3.7}$$

4.0 AN EXAMPLE

As an example of the use of the accelerated design algorithm, the following two dimensional basis functions on the unit square ($x \in [-1,1]$; $y \in [-1,1]$) will be used:

$$1,\ x,\ y,\ xy,\ x^2,\ y^2 \tag{4.1}$$

For the simulations which are summarized in the following two tables, the number of algorithm iterations was set equal to 1681. Equi-spaced grids of size 31x31, 41x41 and 49x49 were used. The number of iterations performed by either algorithm was the listed value, less the number of locations in the initial design. The initial design in all these simulations consisted of the D-optimal allocation of nine measurements.* It may seem odd to initialize a design algorithm with an optimal design but the rate of change in the computation required as a function of M would not be any different for any other choice of initial design. Arbitrary choices of initial designs provided results similar to those tabulated below. For the accelerated sequential design algorithm, the growth in the number of function evaluations that occurs as M is increased, can indeed be seen to be approximately linear. For the standard sequential design algorithm the growth is exactly linear, though the number of evaluations is much greater.

The CPU time in seconds that both algorithms required is tabulated below. To some extent this is machine (IBM 370), language (Fortran) and program dependent. The following figures are nonetheless informative. The increase in time required by the accelerated greedy algorithm is consistent with the increase in the number of function evaluations in the previous table. Even though an accelerated design algorithm evaluation is more expensive than a standard algorithm evaluation, the accelerated algorithm required significantly less time.

5.0 CONCLUSIONS

An efficient sequential algorithm has been presented for numerically producing D-optimal designs. It should be particularly useful for situations involving a very large number of candidate measurement location grid points.

ACKNOWLEDGEMENTS

This research has been supported by the Office of Naval Research under contracts N00014-80-C-0530 and N00014-81-K-0146.

*See Fig. 1

REFERENCES

[1]. J. Kiefer, "Optimum Experimental Designs," J. of the Royal Stat. Soc. B21, 1959, pp. 272-319.

[2]. V.V. Federov, "Theory of Optimal Experiments," Academic Press, 1972.

[3]. R.C. St. John and N.R. Draper, "D-Optimality for Regression Designs: A Review," Technometrics, Vol. 17, No. 1, Feb. 1975, pp. 15-23.

[4]. R.K. Mehra, "Optimization of Measurement Schedules and Sensor Designs for Linear Dynamic Systems," IEEE-AC-21, No. 1, Feb. 1976, pp. 55-64.

[5]. D.M. Titterington, "Aspects of Optimal Design in Dynamic Systems," Technometrics, Vol. 22, No. 3, August 1980.

[6]. H.P. Wynn, "The Sequential Generaton of D-Optimal Experimental Designs " Ann. of Math. Stat., Vol. 41, 1970, pp. 1655-1664.

[7]. T.J. Mitchell, "An Algorithm for the Construction of 'D-Optimal' Experimental Designs," Technometrics, Vol. 16, No. 2, May 1974, pg. 203.

[8]. Z. Galil and J. Kiefer, "Time and Space Saving Methods, Related to Mitchell's DETMAX, for Finding D-Optimal Designs," Technometrics, Vol. 22, No. 3, Aug. 1980.

[9]. M. Minoux, "Accelerated Greedy Algorithms for Maximizing Submodular Set Functions," 8th IFIP Conference, Wurzburg, 1977, Springer-Verlag, Part 2, pp. 234-243,

[10]. M.Minoux, "Minimisation de Fonctions D'Ensemble Surmodulaires et Sousmodulaires," Proc. IFORS 1981, Hamburg, W. Germany.

[11]. S. Biesel-Guitonneau, "Algorithmes Gloutons et Fonctions Surmodulaires: Theorie et Application a un Probleme de Securite dans les Reseaux de Telecommunications," Ann. Telecomm., 35, No. 7-8, 1980.

[12]. A.V. Aho, J.E. Hopcroft anf J.D. Ullman, "The Design and Analysis of Computer Algorithms," Addison-Wesley Publishing Co., 1976.

[13]. E.L. Lawler, "Combinatorial Optimization: Networks and Matroids," Holt, Rhinehart and Winston, 1976.

[14]. T.G. Robertazzi, "Measurement Processing Order and Recursive Statistical Estmation," Ph.D. Thesis, Dept. of Electrical Engineering and Computer Science, Princeton University, Sept. 1981.

Table 1

Number of Function Evaluations

$M(i_{total} = 1681)$	961	1681	2401
Standard Accelerated	1,615,441 32,281	2,825,761 51,261	4,036,081 69,965
Linear Increase in M Act. Incr. for Accel.		1.75 1.59	1.43 1.36

Table 2

Execution Time (seconds)

$M(i_{total} = 1681)$	961	1681	2401
Standard Accelerated	79. 14.	131. 21.	189. 29.
Linear Increase in M Actual Increase for Accel.		1.75 1.5	1.43 1.4

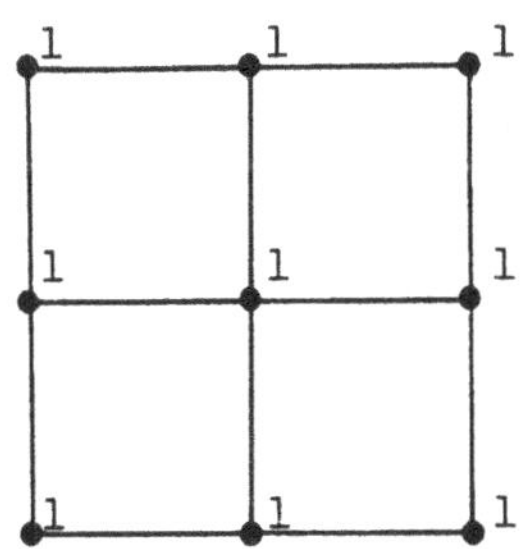

Figure 1 - D-Optimal Allocation

A ROBUSTIZED MAXIMUM ENTROPY APPROACH TO SYSTEM IDENTIFICATION

Chee Tsai and Ludwik Kurz
Department of Electrical Engineering and Computer Science
Polytechnic Institute of New York, Brooklyn, NY 11201

ABSTRACT

In this paper, a robust identification method of the system model is proposed. The robustizing process is in the form of a Robbins-Monro stochastic approximation (RMSA) algorithm and is based on the m-interval polynomial approximation (MIPA) method. The resulting algorithm, which estimates the coefficients of the system model, represents a recursive robustized version of the well-known maximum entropy method (MEM) for spectral estimation introduced by Burg, or of the popular Widrow least-mean-square (LMS) adaptive filter adopted widely in engineering. Furthermore, the MIPA robustizing algorithm leads naturally to a robustized Akaike's information criterion (AIC) to determine the order of the system model. The simplicity of implementation and flexibility make applications of the MIPA identification algorithms attractive in practice. The robustness of performance is confirmed by Monte Carlo simulations.

INTRODUCTION

The last decade was marked by a rapid progress in the area of identification or estimation of process parameters [1]. In spite of recent developments, however, it seems that the effect of some assumptions resulting from mathematical expediency about process statistics is not fully studied. For example, in many practical systems, the knowledge of process statistics being neither exact nor available, a Gaussian assumption is usually made. The resultant identification procedures, such as least square variants, may be sensitive to deviations from the Gaussian law and would degrade as measured by the convergence rate or may even diverge for some situations [2].

Though enormous literature on robust estimation of location parameters is available, it has been limited in its application to the problem of robust identification. One of the popular estimators is the min-max type developed by Huber [3] and used by many others [4,5]. As it was pointed out in [14], this method critically depends on the assumption about the "contaminated class" to which the process belongs; this information is not available and can not be estimated from the data. In addition, some nonparametric methods [6,7] have also been suggested. It was shown in [15] that poor efficiency and excessive computation associated with R-estimators preclude their real-time applications.

In this paper, a natural extension of the MIPA method introduced in [14] to the robust identification of a system model is considered. In particular, a recursive algorithm in the form of maximum likelihood estimators (MLE) robustized by the MIPA method is presented. The algorithm represents a robust version of the well-known MEM for spectral estimation [11], or of the Widrow least-mean-square (LMS) adaptive filter [10]. Moreover, the MIPA procedure leads to a robustized Akaike's information criterion (AIC) [13] to determine the order of the system model. Some properties about these algorithms are discussed and confirmed by Monte Carlo simulations.

SYSTEM MODEL

Consider the discrete-time system model [8]

$$x_{k+1} = A_k\, x_k + w_{k+1} \tag{1}$$

$$y_{k+1} = \theta^T x_k + v_{k+1} \tag{2}$$

where y_k is a scalar measurable output sequence, x_k is a p-vector measurable process, v_k is an independently, identically distributed (i.i.d.) nonmeasurable ergodic noise sequence while the (possibly stochastic) input sequence w_k could be partially or completely measurable to allow for known external inputs. The parameter vector θ is a constant and unknown, while A_k is possibly a function of θ.

Some special examples of interest are the autoregressive (AR) process model with

$$x_k^T = [y_k \;\; y_{k-1} \cdots y_{k-p}],\; A^{AR}(\theta) = \begin{bmatrix} \theta^T \\ I_{n-1} \;\; 0 \end{bmatrix},\; w_k^T = [1\;0]\; v_k,$$

a restricted class of moving average (MA) models with

$$x_k^T = [u_k \;\; u_{k-1} \cdots u_{k-p}],\; A^{MA} = \begin{bmatrix} 0 & 0 \\ I_{n-1} & 0 \end{bmatrix},\; w_k = u_k \text{ a measurable}$$

input sequence, which is independent of $v_{k+1}, v_{k+2}, \ldots$ and a restricted class of autoregressive moving average (ARMA) models with

$$x_k^T = [y_k \cdots y_{k-p}\; u_k \cdots u_{k-p}],\; A^{ARMA} = \left[\begin{array}{cc} A^{AR}(\theta_1) & [\theta_2 \;\; 0]^T \\ \hline 0 & A^{MA} \end{array}\right]$$

$$w_k^T = [v_k\; 0 \ldots 0\; u_k \ldots 0] \qquad y_{k+1} = [\hat{\theta}_1^T\; \hat{\theta}_2^T]\; x_k + v_{k+1}$$

ESTIMATION OF THE COEFFICIENTS θ

The properties of estimates which are valid when the error of estimation is small are generally referred to as asymptotic. One procedure for developing them formally is to study the behavior of the estimate as the number of data n approaches infinity. Since the MLE $\hat{\theta}_n$ is asymptotically consistent and efficient, then

$$\sqrt{n}\,(\hat{\theta}_n-\theta) \sim N(0,\, I_\theta^{-1}) \tag{3}$$

where the p x p square matrix I_θ is

$$I_\theta = E(\{\nabla_\theta[\ln f(v)]\}\{\nabla_\theta[\ln f(v)]\}) = -\,E(\nabla_\theta\{\nabla_\theta[\ln f(v)]\})$$

$$= \text{Fisher Information of } v \text{ for estimating } \theta$$

and I_θ^{-1} corresponds to the Cramér-Rao bound.

From the system model of Eq. (2), we have the likelihood function

$$L_n(\theta) = \ln f(y_1, y_2, \ldots, y_n|\theta) = \sum_{k-1}^{n} \ln f(y_k|\, x_{k-1}, \theta) = \sum_{k-1}^{n} \ln f(y_k - \theta^T x_{k-1}) \tag{4}$$

However, a recursive procedure is of particular interest in engineering applications.

Using standard procedures of finding MLE from $L_n(\theta)$ yields

$$\sum_{k=1}^{n} x_{k-1}(f'/f)(y_k - \hat{\theta}_n^T x_{k-1}) = 0 \tag{5}$$

It follows from the strong law of large numbers that, wp1,

$$\lim_{n\to\infty} (1/n) \sum_{k=1}^{n} x_{k-1}(f'/f)(y_k - \hat{\theta}_n^T x_{k-1}) = E_\theta[x_{k-1}\,(f'/f)(y_k - \hat{\theta}^T x_{k-1})]$$

and Eq. (5) implies

$$E_\theta\,[x_{k-1}(f'/f)(y_k - \hat{\theta}^T x_{k-1})] \begin{cases} = 0 & \hat{\theta} = \theta \\ \neq 0 & \text{otherwise.} \end{cases}$$

Therefore, asymptotically the MLE of Eq. (4) is equivalent to finding a root of the regression function $E_\theta[x_{k-1}(f'/f)(y_k - \hat{\theta}^T x_{k-1})]$ by using the RMSA algorithm [14]

$$\hat{\theta}_{n+1} = \hat{\theta}_n + \frac{1}{n\, I_\theta}\; x_n\,(f'/f)(y_{n+1} - \hat{\theta}_n^T x_n) \tag{6}$$

Thus,
$$\hat{\theta}_{n+1} = \hat{\theta}_n + \frac{x_n\, g(y_{n+1} - \hat{\theta}_n^T x_n)}{\sum_{k=1}^{n} x_k\, g(y_{k+1}-\hat{\theta}_k^T x_k)\, g(y_{k+1}-\hat{\theta}_k^T x_k)\, x_k^T} \tag{7}$$

or,
$$\hat{\theta}_{n+1} = \hat{\theta}_n + \frac{x_n\, g(y_{n+1} - \hat{\theta}_n^T x_n)}{\sum_{k=1}^{n} x_k\, g'(y_{k+1}- \hat{\theta}_k^T x_k)\, x_k^T} \tag{8}$$

where $g(\cdot) = -(f'/f)(\cdot)$.

As $v_k \sim N(0,1)$, $g(\cdot) = (\cdot)$, $g'(\cdot) = 1$ and the algorithm of Eq. (8) becomes

$$\hat{\theta}_{n+1} = \hat{\theta}_n + \frac{x_n (y_{n+1} - \hat{\theta}_n^T x_n)}{\sum_{k=1}^{n} x_k x_k^T} \tag{9}$$

which corresponds to the least square (LS) estimator [9].

To eliminate the calculation of the coefficient $1/(\sum_k^n x_k x_k^T)$, Widrow [10] applied the gradient-search technique to obtain the LMS adaptive filter

$$\hat{\theta}_{n+1} = \hat{\theta}_n + r_n x_n (y_{n+1} - \hat{\theta}_n^T x_n) \tag{10}$$

where r_n is a decreasing sequence satisfying $\sum_n^\infty r_n = \infty$, $\sum_n^\infty r_n^2 < \infty$.

The algorithm Eq. (7) or Eq. (8) involves searching recursively the estimate $\hat{\theta}$ of θ which satisfies

$$E_\theta[\ln f(y_k - \hat{\theta}^T x_{k-1})] = \text{Max !}$$

As noted by Akaike [13], it means equivalently that $\hat{\theta}$ is sought to maximize the Boltzman's entropy

$$S(f;\theta) = -\int \left(\frac{f(y_k - \theta^T x_{k-1})}{f(y_k - \hat{\theta}^T x_{k-1})} \right) \ln \left(\frac{f(y_k - \theta^T x_{k-1})}{f(y_k - \hat{\theta}^T x_{k-1})} \right) f(y_k - \hat{\theta}^T x_{k-1})\, dv$$

On the other hand, the popular Burg's MEM [11] for spectral estimation depends on the Gaussian assumption about the random process. As a consequence of duality between the MEM and the AR analysis [12], the algorithm of Eq. (7) turns out to be a robust version of the maximum entropy identification procedure.

M-INTERVAL POLYNOMIAL APPROXIMATION (MIPA) ALGORITHM

Though the algorithm of Eq. (7) is optimum in the sense that it achieves recursively the Cramér-Rao bound I_θ^{-1}, it requires the knowledge of process statistics, which is either inexact or unavailable in most system. It was shown in [14] that the degradation of any estimator from the Cramér-Rao bound is closely related to the mean-square-error

E_r between the nonlinear part of the estimator and the optimum nonlinearity g of Eq. (7): the smaller E_r is, the closer the asymptotic variance v_θ of estimates $\hat{\theta}$ approaches I_θ^{-1}. To achieve an adaptive approximation to the optimum nonlinearity with asymptotically negligible E_r, the m-interval polynomial approximation (MIPA) algorithm is introduced to estimate g.

Consider the nth order MIPA algorithm specified by

$$\psi_n(v) = \sum_{j=1}^{m} (c_{n\,j}\, v^n + c_{n-1\,j}\, v^{n-1} + \dots + c_{1\,j}\, v + c_{o\,j})\, I_{A_j}(v) = \sum_{j=1}^{m} (\sum_{i=1}^{m} c_{i\,j}\, v^i)\, I_{A_j}(v)$$

$$A_j = (a_{j-1}, a_j],\ \int_{A_j} f(v)dv = \lambda_j,\ \sum_{j=1}^{m} \lambda_j = 1,\ I_{A_j}(v) = \begin{cases} 1 & v \in A_j \\ 0 & \text{otherwise} \end{cases} \tag{11}$$

for which the optimum scores $C_j^T = [c_{n\,j} \dots c_{o\,j}]$ minimizing E_r are the solution of

$$U_j\, C_j = P_j \tag{12}$$

where $P_j^T = [P_{nj} \dots P_{oj}]$, $P_{nj} = n\, u_{n-1\,j} + a_{j-1}^n\, f(a_{j-1}) - a_j^n\, f(a_j)$ (13)

$$U_j = \begin{bmatrix} u_{2n\,j} & u_{2n-1\,j} & \dots & u_{n\,j} \\ u_{2n-1\,j} & u_{2n-2\,j} & \dots & u_{n-1\,j} \\ \dots & \dots & & \dots \\ u_{nj} & u_{n-1\,j} & \dots & u_{o\,j} \end{bmatrix},\quad u_{i\,j} = \int_{A_j} v^i\, f(v)\, dv \tag{14}$$

Thus,

$$v_\theta = 1/(I_\theta - E_r) \tag{15}$$

$$= 1/\sum_{j=1}^{m} \sum_{i=1}^{m} c_{i\,j}\, P_{ij}$$

In practice, n does not exceed 2 and m is chosen as 10.

The parameters a_j, $f(a_j)$ and $u_{i\,j}$ can be recursively estimated from the followings RMSA algorithms:

$$\omega_{1j}(k+1) = \omega_{1j}(k) + \frac{1}{k\, \omega_{2j}(k)}\, (\lambda_j - \Delta(\omega_{1j}(k) - v_k)) \tag{17}$$

$$\omega_{2j}(k+1) = \omega_{2j}(k) + \frac{1}{k}\, \left(\frac{\Delta(\omega_{1j}(k)+\delta_k-v_k) - \Delta(\omega_{1j}(k)-v_k)}{\delta_k} \right) - (\omega_{2j}(k)) \tag{18}$$

$$\bar{\omega}_{2j}(k+1) = \bar{\omega}_{ij}(k) + \frac{1}{k}\left([v_k]_j^i - \bar{\omega}_{ij}(k)\right) \tag{19}$$

where $\Delta(x) = \begin{cases} 1 & x \geq 0 \\ 0 & \text{otherwise} \end{cases}$ $\quad [v]_j = \begin{cases} v & v \varepsilon A \\ 0 & \text{otherwise} \end{cases}$ and $\{v_k\}$ is a positive sequence

satisfying $\sum_k^{\infty} 1/(k\delta_k)^2 < \infty$.

DETERMINATION OF THE ORDER p

As indicated above, the estimation of the coefficients θ of the system model requires that the order p of the system model be known. The correct choice of p is vital in obtaining a meaningful estimate of the system model. The estimating procedure for proposed by Akaike [13] gives reasonable results. However, the implementation of this procedure relies also on the Gaussian assumption on the data. Thus, in order to keep the estimator of p perform well independently of the underlying distributions, the algorithm based on the Akaike's information criterion (AIC) needs be robustized via the MIPA algorithm. This information criterion is defined by

AIC = - (maximum log likelihood) + (number of independently adjusted parameters)

When several competing models are being fitted by the method of maximum likelihood, the one with the smallest value of AIC is chosen as the best model. This procedure is called minimum AIC procedure and the final estimate is called the minimum AIC estimate (MAICE). The definition of MAICE gives a clear formulation of the principle of parsimony in statistical model building. When two or more models fit equally well to a given set of data in terms of the likelihood, the MAICE is the one with the smallest number of free parameters. The minimum final prediction error (FPE) procedure is a special case of the minimum AIC procedure.

The AIC for the system model can be expressed explicity as

$$AIC = - \sum_{k=1}^{n} \ln f(y_k - \sum_{i=1}^{M} \theta_i x_{k-i}) + M \tag{20}$$

with the upper limit L on the order of the system model (smallest number of free parameters) set large as enough not to exclude the efficient model.

It is obvious that a good approximation of the log function ln f by the MIPA method will generate a robustized minimum AIC estimator of p. This approximation is simply a integration of the negative of Eq. (11). The value M which yields the minimum AIC within M = 0, 1, ..., L is the order p of our model.

PERFORMANCE ANALYSIS AND SIMULATION RESULTS

Since the density which minimizes the Fisher Information subject to a variance constraint is Gaussian, improvements through the MIPA algorithm can be achieved for any nongaussian density. Asymptotically, $\sqrt{n}\,(\theta_n - \theta) \sim N(0,V_\theta)$ for both algorithms of Eqs. (7) and (9). Two heavy-tailed mixture distributions $f(v) = 0.9N(0,1) + 0.1N(0,3^2)$ and $f_2(v) = 0.9N(0,1) + 0.1\,(10/\pi)/(100/x^2)$ are studied in this section. Their respective values v_θ of Eqs. (7) and (9) are listed in Table 1. The dynamic range of all systems is assumed to be [-500,500] and a 2nd MIPA with m = 10 is used.

Monte Carlo simulations of 50 runs for the 1st and 2nd order AR case of $y_{k+1} = 0.50\ y_k + v_{k+1}$ and $y_{k+1} = 0.75\ y_k - 0.50\ y_{k-1} + v_{k+1}$ were undertaken. Some discrepancies were observed, however. For example, the convergence rate of Eq. (9) in Fig. 1 is exceptionally good. This can be explained as follows. In general, the random data generated by the IBM computer may be represented as a stream $(v_1, v_2, \ldots, v_n, v_{n+1})$ with a single high amplitude v_n. Thus, from Eq. (9)

$$\hat{\theta}_{n+1} = \hat{\theta}_n + \frac{y_n(y_{n+1} - \hat{\theta}_n\ y_n)}{\sum_{k=1}^{n} y_n\ y_n} \simeq \hat{\theta}_n + \frac{\theta\ y_n\ y_n - \hat{\theta}_n\ y_n\ y_n + y_n\ v_{n+1}}{y_n\ y_n} \simeq \theta$$

In real situations a high-variance "bust" noise will affect not only v_n but also v_{n+1}, which will lead to the breakdown of the algorithm in the sense that $\hat{\theta}_{n+1} \simeq \hat{\theta}_n + (y_n\ v_{n+1})/(y_n\ y_n) \neq \theta$. This phenomenon is verified by the simulations of Figs. 2 and 3 in which $v_{n+1} = v_n$ is used. On the other hand, as it is seen in Figs. 1,2 and 3, the algorithm of Eq. (7) retains uniformly good performance while the algorithm of Eq. (10) ($r_n = 1/n$ is used) has the least immunity to the burst noise.

Since the MEM was originally introduced in spectrum analysis, it is instructive to plot the corresponding power spectra of Fig. 3 (see Fig. 4). A simulation of the order determination by the AIC, corresponding the case of Fig. 2, is presented in Fig. 5. It demonstrates that a Gaussian assumption may not give rise to a minimum at the correct order p in nongaussian situations and the robustized version of the algorithm will yield the correct estimate p.

CONCLUSIONS

The effect of the Gaussian assumption in system identification is studied. Robustized algorithms based on the MIPA method to estimate the coefficients and order of a system model are introduced and verified by simulations. Some comparative analysis of robustized and unrobustized algorithms is also given.

REFERENCES

[1] G.E.P. Box and G.M. Jenkins, "Time Series Analysis, Forcasting and Control," Holden-Day, California, 1976.

[2] P.J. Huber, "Robust Regression: Asymptotics, Conjectures and Monte Carlo," Ann. of Statist., Vol. 1, pp. 799-821, 1973.

[3] P.J. Huber, "Robust Estimation of a Location Parameter," Ann. Math. Statist, Vol. 35, pp. 73-101, 1964.

[4] R.D. Martin and C.J. Masteliez, "Robust Estimation via Stochastic Approximation," IEEE Trans. Inform. Theory, Vol. IT-21, pp. 263-271, 1975.

[5] B.T. Poljak and Ya. Z. Tsypkin, "Robust Identification," Automatica , Vol. 16, pp. 53-63, 1980.

[6] L.A. Jackel, "Estimating Regression Coefficients by Minimizing the Dispersion of the Residuals," Ann. Math. Statist. Vol. 43, pp. 1449-1458, 1972.

[7] J. Jureckova, "Nonparametric Estimates of Regression Coefficients," Ann. Math. Statist. Vol. 42, pp. 1328-1338, 1971.

[8] J.B. Moore, "On Strong Consistence of Least Square Identification Algorithm," Automatica, Vol. 14, pp. 505-509, 1978.

[9] V. Strejc, "Least Square Parameter Estimation," Automatica, Vol. 16, pp. 535-550, 1980.

[10] B. Widrow, P.E. Mantey, L.J. Griffiths and B.B. Good, "Adaptive Antenna Systems," Proc. IEEE, Vol. 55, pp. 2143-2159, 1967.

[11] J.P. Burg, "A New Analysis Technique for time Series Data," Presented at the 37th Annual Meeting Soc. Explor. Geophr., Oklahoma City, Okla., 1967.

[12] A. van den Bos, "Alternative Intepretation of Maximum Entropy Spectral Analysis," IEEE Trans. Inform. Theory, Vol. IT-17, pp. 493-494, 1971.

[13] H. Akaike, "A New Look at the Statistical Model Identification," IEEE Trans. Automat. Contr., AC-19, pp. 716-723, 1974.

[14] C. Tsai and L. Kurz, "A Practical Approach to Robust Recusive Estimation/Detection and Its Application to Spread Spectrum Communications," Submitted to IEEE Trans. Inform. Theory.

[15] C. Tsai, "A Contribution to Robust Detection and Estimation," Ph.D. Thesis, Polytechnic Institute of New York.

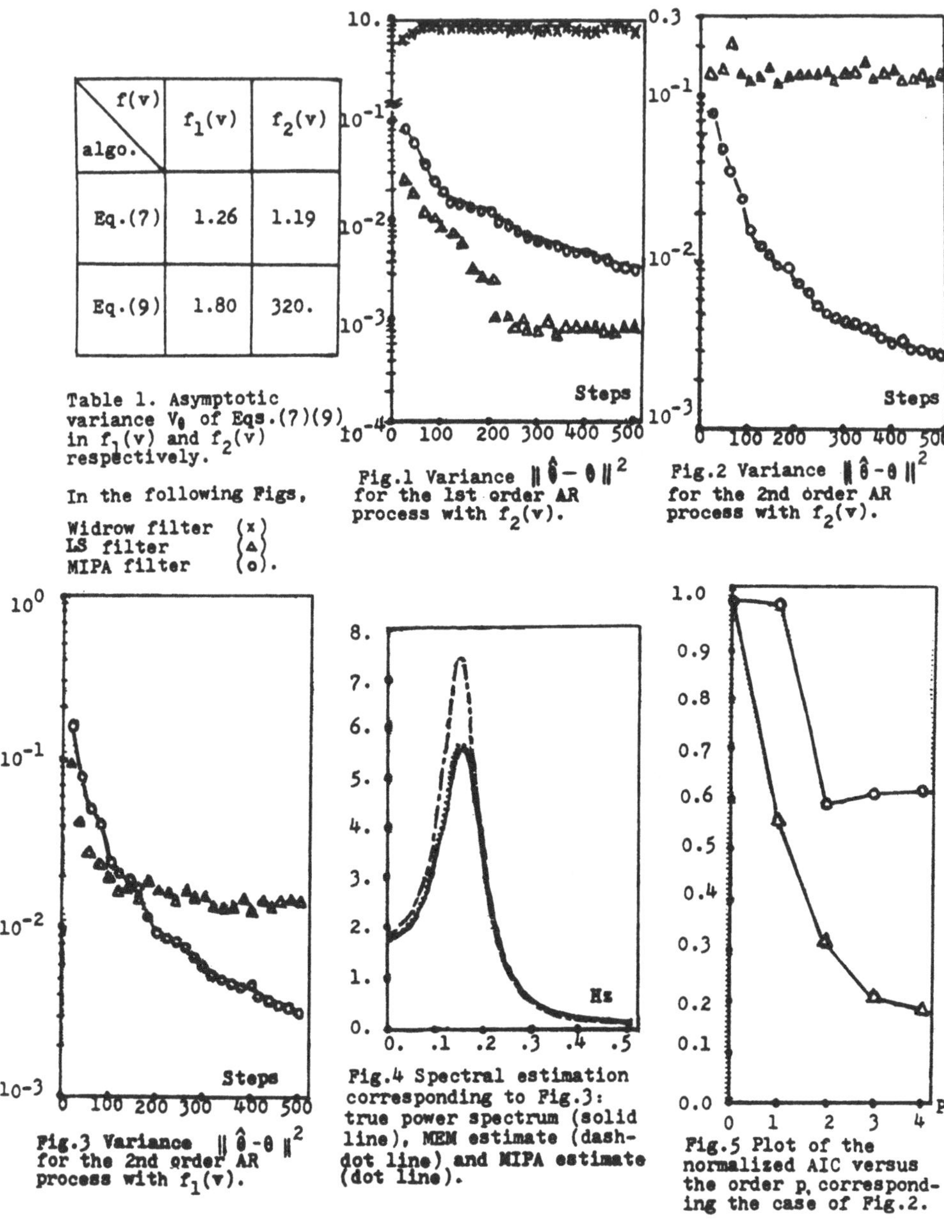

f(v) / algo.	$f_1(v)$	$f_2(v)$
Eq.(7)	1.26	1.19
Eq.(9)	1.80	320.

Table 1. Asymptotic variance V_θ of Eqs.(7)(9) in $f_1(v)$ and $f_2(v)$ respectively.

In the following Figs,

Widrow filter (x)
LS filter (Δ)
MIPA filter (o).

Fig.1 Variance $\|\hat{\theta}-\theta\|^2$ for the 1st order AR process with $f_2(v)$.

Fig.2 Variance $\|\hat{\theta}-\theta\|^2$ for the 2nd order AR process with $f_2(v)$.

Fig.3 Variance $\|\hat{\theta}-\theta\|^2$ for the 2nd order AR process with $f_1(v)$.

Fig.4 Spectral estimation corresponding to Fig.3: true power spectrum (solid line), MEM estimate (dash-dot line) and MIPA estimate (dot line).

Fig.5 Plot of the normalized AIC versus the order p, corresponding the case of Fig.2.

A Problem of Bilinear Control in Nonlinear Coupled Distributed Parameter Reactor Systems

Y. Kuroda, A. Makino and T. Ishibashi

Department of Nuclear Engineering
Tokai University, Tokyo, Japan

Introduction

The aim of this paper is to formulate some optimal control problems in distributed parameter nuclear reactor systems with the Xenon poisoning feedback effect. This effect, which appears in reactor power distributions as oscillations after long run operation, is particularly significant in case of start-up and shut-down. For instance, when the reactor power level is lowered, the Xenon distribution does not follow the fall of the power distribution, but once rises up and then drops. Therefore, some control rods are needed to be pull out to cancel this poisoning effect. This means that a certain control rods should be always inserted in the reactor core even in full power operation.

Optimal control problems on the spatial Xenon feedback effect so far have been dealt with linearization and modal expantion approaches in nuclear engineering (1). In this paper we formulate some aspects of these problems under weaker conditions, using techniques similar to those performed in biological systems (2) and other control and identification problems (3) (4) (5).

1. Reactor Model with Xenon Feedback

Let Ω be a bounded open domain (reactor core) in n-dimensional Euclidean space R^n ($n \leq 3$) with sufficiently smooth extrapolation boundary Γ. Let t be time in $]0,T[$ ($T < \infty$), $Q = \Omega \times]0,T[$ and $\Sigma = \Gamma \times]0,T[$. $H_0^1(\Omega)$ denotes the closure of $\mathcal{D}(\Omega)$ in $H^1(\Omega)$ which is the Sobolev space of order 1, $L^2(0,T;H_0^1(\Omega))$ denotes the class of functions, strongly measurable on $[0,T]$ with the range in $H_0^1(\Omega)$, (6).

The system model concerned is governed by two group diffusion equations in terms of fast neutron flux Y_1 and slow neutron flux Y_2, and two balance equations in terms of Xenon and Iodine distributions X and I, given by the system of distributed parameter systems (1.1), (2).

$$
\left.
\begin{aligned}
&\frac{1}{v_1}\frac{\partial Y_1}{\partial t} = \sum_{i,j=1}^{n}\frac{\partial}{\partial x_i}D_1(x,t)\frac{\partial Y_1}{\partial x_j} - \Sigma_{a1}Y_1 - \Sigma_R Y_1 - X\sigma_1^X Y_1 + \\
&\qquad\qquad + [\nu_1\Sigma_{f1}Y_1 + \nu_2\Sigma_{f2}Y_2] \\
&\frac{1}{v_2}\frac{\partial Y_2}{\partial t} = \sum_{i,j=1}^{n}\frac{\partial}{\partial x_i}D_2(x,t)\frac{\partial\ 2}{\partial x_j} - \Sigma_{a2}Y_2 + \Sigma_R Y_1 - X\sigma_2^X Y_2 + \Sigma_c Y_2 \\
&\frac{\partial I}{\partial t} = \lambda_I I + y_I\left[\Sigma_{f1}Y_1 + \Sigma_{f2}Y_2\right] \\
&\frac{X}{t} = -\lambda_X X + \lambda_I I + y_X\left[\Sigma_{f1}Y_1 + \Sigma_{f2}Y_2\right] - X\left[\sigma_1^X Y_1 + \sigma_2^X Y_2\right] \\
&Y_1(x,t) = 0\ ,\ Y_2(x,t) = 0 \qquad\qquad \text{on } \Sigma \\
&Y_1(x,0) = Y_{10}\ ,\ Y_2(x,0) = Y_{20}\ ,\ I(x,0) = I_0\ ,\ X(x,0) = X_0
\end{aligned}
\right] \qquad (1.1)
$$

where control is introduced bilinearly by macroscopic absorption cross section of the control rods $\Sigma_c(x,t)$ and observation is made by slow neutron flux $Y_2(x,t)$. $D_i(x,t)$, $\Sigma_{ai}(x,t)$, $\Sigma_R(x,t)$ and $\Sigma_{fi}(x,t)$, which are diffusion constant, absorption -, removal - and fission macroscopic cross-section respectively, are space and time dependent parameters belonging to $L^\infty(Q)$, and all parameters are positive. Note that nonlinear coupling between state variables appear in the first, the second and the fourth equations of (1.1).

The system (1.1) admits a unique solution in $(L^2(0,T;H_0^1(\Omega)))^2 \times (L^2(Q))^2 \cap (L^\infty(Q))^4$. The proof can be performed, based on the similar manner to that in biological systems (2), (8).

2. Optimal Flux Distribution Problem

2.1 Optimal Control Policy

Admissible control space is given by

$$
U_{ad} = \left[\begin{array}{l} \text{a bounded closed convex subset of the control space} \\ U = L^\infty(Q),\ \text{and } 0 < b_1 \leqq \Sigma_c \leqq b_2 < \infty\ . \end{array}\right. \qquad (2.1)
$$

Cost functional J is given by

$$
J(\Sigma_c) = \int_Q \left|Y_2(x,t) - Y_d\right|^2 dxdt
$$

The problem is to find an optimal control Σ_{co}, which is defined by

$$
J(\Sigma_{co}) = \inf J(\Sigma_c) \qquad \forall \Sigma_{co} \in U_{ad}. \qquad (2.3)
$$

2.2 Existence of Optimal Control

Let Σ_{cn} be a minimizing sequence, which converges in weak star topology of $L^\infty(Q)$, i.e. $J(\Sigma_{cn}) \longrightarrow \inf J(\Sigma_c)$, $\Sigma_{cn} \in U_{ad}$, and let $Y_{1n} = Y_1(\Sigma_{cn})$, $Y_{2n} = Y_2(\Sigma_{cn})$, $I_n = I(\Sigma_{cn})$ and $X_n = X(\Sigma_{cn})$.

I_n and X_n are bounded in $L^\infty(Q)$.

Since the admissible control space is bounded in $L^\infty(Q)$, from Eq. (1.1) we have

$$\frac{1}{v_1}(\frac{dY_{1n}}{dt}, Y_{1n}) + a_1(t; Y_{1n}, Y_{1n}) + ((\Sigma_{a1} + \Sigma_R - \nu_1\Sigma_{f1})Y_{1n}, Y_{1n})$$
$$= -(X_n\sigma_1^X Y_{1n}, Y_{1n}) + (\nu_2\Sigma_{f2}Y_2, Y_{1n}),$$

$$\frac{1}{v_2}(\frac{dY_{2n}}{dt}, Y_{2n}) + a_2(t; Y_{2n}, Y_{2n}) + (\Sigma_{a2}Y_{2n}, Y_{2n})$$
$$= (\Sigma_R Y_{1n}, Y_{2n}) - (X_n\sigma_2^X Y_{2n}, Y_{2n}) + (\Sigma_{cn}Y_{2n}, Y_{2n}).$$

Integrate over $[0,T]$, and using the condition of parameters, we have

$$\frac{1}{2v_1}|Y_{1n}(t)|^2 + \alpha_1\int_0^t \|Y_{1n}(\sigma)\|^2 d\sigma + c_1\int_0^t |Y_{1n}(\sigma)|^2 d\sigma$$
$$\leqq c_2\int_0^t (|X_n(\sigma)|\,|Y_{2n}|^2 + |Y_{2n}(\sigma)|\,|Y_{1n}(\sigma)|)d\sigma$$

$$\frac{1}{2v_2}|Y_{2n}(t)|^2 + \alpha_2\int_0^t \|Y_{2n}(\sigma)\|^2 d\sigma + c_3\int_0^t |Y_{2n}(\sigma)|^2 d\sigma$$
$$\leqq c_4\int_0^t (|Y_{1n}\;|Y_{2n}| + |X_n|\,|Y_{2n}|^2 + |Y_{2n}|^2)d\sigma.$$

Then

$$|Y_{1n}(t)|^2 + \int_0^t \|Y_{1n}(\sigma)\|^2 d\sigma \leqq C_1\int_0^t |Y_{1n}(\sigma)|^2 d\sigma + C_2\int_0^t |Y_{2n}(\sigma)|^2 d\sigma + C_3,$$

$$|Y_{2n}(t)|^2 + \int_0^t \|Y_{2n}(\sigma)\|^2 d\sigma \leqq C_4\int_0^t |Y_{2n}(\sigma)|^2 d\sigma + C_5\int_0^t |Y_{1n}(\sigma)|^2 d\sigma + C_6.$$

where bilinear forms are satisfied

$$\sum_{ij=1}^{n} a_k(t; \xi_i, \xi_j) \geqq \alpha_k|\xi_i|^2, \qquad (k = 1, 2, \ \xi_i \in R),$$

and α_k, c_h, C_h are positive constants,

and $\|\cdot\|$ (resp. $|\cdot|$) denotes the norm of $H_0^1(\Omega)$ (resp. $L^2(\Omega)$).
By using the inequality of Gromwall,

Y_{1n} and Y_{2n} are bounded in $L^2(0,T;H_0^1(\Omega))$.

Also.

$\frac{\partial Y_{1n}}{\partial t}$ and $\frac{\partial Y_{2n}}{\partial t}$ are bounded in $L^2(0,T;H^{-1}(\Omega))$,

$\frac{\partial I_n}{\partial t}$ and $\frac{\partial X_n}{\partial t}$ are bounded in $L^2(Q)$.

The injection from $H_0^1(\Omega)$ into $L^2(\Omega)$ is compact, so that we can take subsequances, which will be still denoted as Y_{1n}, Y_{2n}. Then we have

$$Y_{1n} \longrightarrow Y_1 \text{ weakly in } L^2(0,T;H_0^1(\Omega)),$$
$$Y_{2n} \longrightarrow Y_2 \text{ weakly in } L^2(0,T;H_0^1(\Omega)).$$

Also,

$$\frac{\partial Y_{1n}}{\partial t} \longrightarrow \frac{\partial Y_1}{\partial t} \text{ weakly in } L^2(0,T;H^{-1}(\Omega)),$$
$$\frac{\partial Y_{2n}}{\partial t} \longrightarrow \frac{\partial Y_2}{\partial t} \text{ weakly in } L^2(0,T;H^{-1}(\Omega)).$$

Therefore

$$Y_{1n} \longrightarrow Y_1 \text{ strongly in } L^2(Q),$$
$$Y_{2n} \longrightarrow Y_2 \text{ strongly in } L^2(Q).$$

Moreover

$$I_n \longrightarrow I \text{ weak star in } L^\infty(Q),$$
$$X_n \longrightarrow X \text{ weak star in } L^\infty(Q).$$

Also,

$$\frac{\partial I_n}{\partial t} \longrightarrow \frac{\partial I}{\partial t} \text{ weakly in } L^2(Q),$$
$$\frac{\partial X_n}{\partial t} \longrightarrow \frac{\partial X}{\partial t} \text{ weakly in } L^2(Q).$$

Therefore $\Sigma_{cn} Y_{2n} \longrightarrow \Sigma_c Y_2$.

Then we obtain that the mapping from control Σ_c to the state $\{Y_1, Y_2, I, X\}$ is continuous. It immediately follows that

there exist an optimal control Σ_{co},
i.e. $\liminf J(\Sigma_{cn}) \geq J(\Sigma_{co})$.

However, we can not admit the uniqueness of an optimal control.

2.3 Optimality Condition

We can show the necessary condition for Σ_c to be the optimal control by Eq. (2.7).

$$(J'(\Sigma_{co}),\ \widetilde{\Sigma}_c) \geqq 0 \qquad \forall \Sigma_c \in U_{ad} \tag{2.7}$$

We take derivatives of the state $\{Y_1, Y_2, I, X\}$ with respect to the control. In this case, differentiablity is admitted by applying the implicit function theorem (2), (4).

Put $\frac{d}{d\xi} Y_i(\Sigma_c + \xi\widetilde{\Sigma}_c)\Big|_{\xi=0} = \dot{Y}_i$, $\frac{d}{d\xi} I(\Sigma_c + \xi\widetilde{\Sigma}_c)\Big|_{\xi=0} = \dot{I}$,

$\frac{d}{d\xi} X(\Sigma_c + \xi\widetilde{\Sigma}_c)\Big|_{\xi=0} = \dot{X}$, where $i=1,2$, and $\widetilde{\Sigma}_c = \Sigma_c - \Sigma_{co}$.

Then the system (2.8) is introduced from the system (1.1),

$$\left.\begin{aligned}
\frac{1}{v_1}\frac{\partial \dot{Y}_1}{\partial t} &= \sum_{i,j=1}^{n}\frac{\partial}{\partial x_i}D_1(x,t)\frac{\partial \dot{Y}_1}{\partial x_j} - \Sigma_{a1}\dot{Y}_1 - \Sigma_R\dot{Y}_1 + (\nu_1\Sigma_{f1}\dot{Y}_1 + \\
&\quad + \nu_2\Sigma_{f2}\dot{Y}_2) - X\sigma_1^X\dot{Y}_1 - \dot{X}\sigma_1^X Y_1 \\
\frac{1}{v_2}\frac{\partial \dot{Y}_2}{\partial t} &= \sum_{i,j=1}^{n}\frac{\partial}{\partial x_i}D_2(x,t)\frac{\partial \dot{Y}_2}{\partial x_j} - \Sigma_{a2}\dot{Y}_2 + \Sigma_R\dot{Y}_1 - X\sigma_2^X\dot{Y}_2 - \\
&\quad - \dot{X}\sigma_2^X Y_2 - \Sigma_c\dot{Y}_2 - \widetilde{\Sigma}_c Y_2 \\
\frac{\partial \dot{I}}{\partial t} &= \lambda_I\dot{I} - y_I(\Sigma_{f1}\dot{Y}_1 + \Sigma_{f2}\dot{Y}_2) \\
\frac{\partial \dot{X}}{\partial t} &= -\lambda_X\dot{X} + \lambda_I\dot{I} + y_X(\Sigma_{f1}\dot{Y}_1 + \Sigma_{f2}\dot{Y}_2) - (\sigma_1^X\dot{Y}_1 + \sigma_2^X\dot{Y}_2)X - \\
&\quad - (\sigma_1^X Y_1 + \sigma_2^X Y_2)\dot{X} \\
&\dot{Y}_1(x,t) = 0\ ,\ \dot{Y}_2(x,t) = 0 \qquad \text{on } \Sigma \\
&\dot{Y}_1(x,0) = 0\ ,\ \dot{Y}_2(x,0) = 0\ ,\ \dot{I}(x,0) = 0\ ,\ \dot{X}(x,0) = 0
\end{aligned}\right] \tag{2.8}$$

Then Eq. (2.7) becomes

$$\int_Q \dot{Y}_2(Y_2 - Y_d)\ dxdt \geqq 0 \qquad \forall \Sigma_c \in U_{ad}. \tag{2.9}$$

Next we define adjoint states P_1, P_2, V and Z by the system (2.10).

$$\begin{aligned}
-\frac{1}{v_1}\frac{\partial P_1}{\partial t} &= \sum_{i,j=1}^{n}\frac{\partial}{\partial x_j}D_1\frac{\partial P_1}{\partial x_i} - \Sigma_{a1}P_1 - \Sigma_R(P_1 - P_2) + \nu_1\Sigma_{f1}P_1 + \\
&\quad + (y_I V + y_X Z)\Sigma_{f1} - X\sigma_1^X(P_1 + Z) \\
-\frac{1}{v_2}\frac{\partial P_2}{\partial t} &= \sum_{i,j=1}^{n}\frac{\partial}{\partial x_j}D_2\frac{\partial P_2}{\partial x_i} - \Sigma_{a2}P_2 - X\sigma_2^X(P_2 + Z) + \nu_2\Sigma_{f2}P_2 +
\end{aligned}$$

$$\left.\begin{array}{l} \qquad\qquad + (y_I V + y_X Z)\Sigma_{f2} - (Y_2 - Y_d) - \Sigma_c P_2 \\ -\frac{\partial V}{\partial t} = \lambda_I (V + Z) \\ -\frac{\partial Z}{\partial t} = -\lambda_X Z - (\sigma_1^X Y_1 + \sigma_2^X Y_2) Z + P_1 \sigma_1^X Y_1 + P_2 \sigma_2^X Y_2 \\ P_1(x,t) = 0 \ , \ P_2(x,t) = 0 \qquad\qquad \text{on } \Sigma \\ P_1(x,T) = 0 \ , \ P_2(x,T) = 0 \ , \ V(x,T) = 0 \ , \ Z(x,T) = 0 \end{array}\right| \quad (2.10)$$

In Eq. (2.10), multiply the first eq. by Y_1, the second eq. by Y_2, the third eq. by I, the fourth eq. by X, and integrate over Q, then Eq. (2.9) becomes Eq. (2.11).

$$\int_Q P_2 \cdot Y_2 (\Sigma_c - \Sigma_{co})\ dxdt \geq 0 \qquad \forall \Sigma_c \in U_{ad}. \qquad (2.11)$$

Therefore, the optimality system is given by Eq. (1.1), Eq. (2.10) (where $\Sigma_c = \Sigma_{co}$) and Eq. (2.11).

3. Optimal Shut Down Problem

As mentioned previously, after the shut-down of a reactor, the Xenon distribution once rises up and then drops. Therefore, in order to make restart-up feasible, it is required to obtain an operation program necessary for minimizing the Xenon poisoning effect. To do this, an optimal shut-down policy is presented which holds the Xenon distribution at the terminal of shut-down operation, t = T, below a specified level, given the Xenon distribution at the bigining of shut-down operation, t = 0.

Cost functional J is given by

$$J(\Sigma_c) = \int_\Omega |X(x,T) - X_d|^2\ dx \qquad \forall \Sigma_c \in U_{ad}. \qquad (3.1)$$

Mapping $\Sigma_c \longmapsto X(\Sigma_c)$ is continuous $L^\infty(Q) \longmapsto L^\infty(Q)$.

Then $X(x,T;\Sigma_c) \in L^\infty(\Omega)$, $\frac{\partial X}{\partial t} \in L^2(Q)$, i.e. $X(x,T;\Sigma_c) \in L^2(\Omega)$.

Therefore, an optimal control Σ_{co} is given by $J(\Sigma_{co}) = \inf J(\Sigma_c)$, and optimality condition is given by

$$\int_\Omega \dot{X}(x,T)(Xx,T - X_d)\ dx \geq 0$$

Next define adjoint states P_1, P_2, V and Z of Y_1, Y_2, I and X by Eq. (3.4),

$$-\frac{1}{v_1}\frac{\partial P_1}{\partial t} = \sum_{i,j=1}^{n}\frac{\partial}{\partial x_j}D_1\frac{\partial P_1}{\partial x_i} - \Sigma_{a1}P_1 - \Sigma_R(P_1 - P_2) + \nu_1\Sigma_{f1}P_1 +$$
$$+ (y_I V + y_X Z)\Sigma_{f1} - X\sigma_1^X(P_1 + Z)$$
$$-\frac{1}{v_2}\frac{\partial P_2}{\partial t} = \sum_{i,j=1}^{n}\frac{\partial}{\partial x_j}D_2\frac{\partial P_2}{\partial x_i} - \Sigma_{a2}P_2 - X\sigma_2^X(P_2 + Z) + \nu_2\Sigma_{f2}P_2 +$$
$$+ (y_I V + y_X Z)\Sigma_{f2} - \Sigma_c P_2 \qquad (3.4)$$
$$-\frac{\partial V}{\partial t} = \lambda_I(V + Z)$$
$$-\frac{\partial Z}{\partial t} = -\lambda_X Z - (\sigma_1^X Y_1 - \sigma_2^X Y_2)Z + P_1\sigma_1^X Y_1 + P_2\sigma_2^X Y_2$$
$$P_1(x,t) = 0 \; , \; P_2(x,t) = 0 \qquad \text{on}$$
$$P_1(x,T) = 0 \; , \; P_2(x,T) = 0 \; , \; V(x,T) = 0 \; , \; Z(x,T) = X(x,T) - X_d$$

The optimal control is characterized in the same way as the previous problem.

$$\int_Q P_2 \cdot Y_2(\Sigma_c - \Sigma_{co}) \, dxdt \geqq 0 \qquad \forall \Sigma_c \in U_{ad}. \qquad (3.5)$$

The optimality system is given by Eq. (1.1), Eq. (3.4) (where $\Sigma_c = \Sigma_{co}$) and Eq. (3.5).

4. Concideration and Conclusion

Concider the system of linearized perturbed equations corresponding to Eq. (1.1) by putting $\Sigma_c = \bar{\Sigma}_c + \delta\Sigma_c$, $Y_1 = \bar{Y}_1 + \delta Y_1$, $Y_2 = \bar{Y}_2 + \delta Y_2$, $I = \bar{I} + \delta I$ and $X = \bar{X} + \delta X$.

In this case the optimality system is given by this linearized system together with its adjoint system and Eq. (4.1),

$$\int_Q p_2 \cdot \bar{Y}_2(\Sigma_c - \Sigma_{co}) \, dxdt \geqq 0 \qquad \forall \Sigma_c \in U_{ad}. \qquad (4.1)$$

where p_2 is the adjoint state for perturbed term of δY_2.

Compare this optimality condition with that for the original problem, we observe : the linearized optimal problem is more advantageous concerning the numerical analysis, but is restricted only in the neighborhood of the steady state, becoming unsatisfactory when Xenon oscillations are increased locally.

In this paper we have solved some bilinear optimal control problems of non-linear coupled systems which stem from the nuclear reactor dynamics and so far have been treated with linearization. To obtain the optimality conditions, we have introduced the gradient of the state variable with respect to the control. This method is useful to solve control problems of non-linear control and/or non-linear distributed parameter systems.

References

(1) A. M. Christie and C. G. Poncelet ; On the control of spatial Xenon oscillations. Nuc. Sci. Eng. 51 10-24 1973.

(2) C. M. Brauner and P. Penel ; Un problêm de contrôle optimal non linéaire en biomathématique. Annali de l'Univ. Ferrera, XVIII 1-44 1972.

(3) G. Chavent ; Identification of distributed parameters. Proç. 3rd IFAC Symp. on Identification, The Hauge, 1973

(4) F. Mignot ; Contrôle dans les inéquations variationelles elliptiques. Jour. of Functional Analysis, 22 130-185 1976.

(5) F. Mignot, C. Saguez and J. P. Van de Wiele ; Contrôle optimal de systèmes gouvernés par des problèmes aux valeurs propres. App. Math. Opti. Vol. 3, No. 4, 291-320 1977.

(6)J. L. Lions and E. Magenes ; Non-homogeneous boundary value problems, Vol. I, II, Springer, 1972.

(7) J. L. Lions ; Contrôle optimal de systèmes gouvernès par des équations aux dérivées partielles. DUNOD 1968.

(8) Y. Konishi ; Sur un systèm dégenéré des équations paraboliques semi-linéaires avec les conditions aux limites non linéaires. Jour. Fac. Sci. Univ. Tokyo, Sec. IA. 19 353-361 1972.

(9) J. P. Kernevez and D. Thomas ; Numerical analysis and control of some biochemical systems. App. Math. Opt. Vol. 1, No. 3, 222-285 1975.

(10) G. Chavent, M. Dupuy and P. Lemonnier ; History matching by use of optimal theory. Soc. Petroleum Engineers Jour. 15(1) 74-86 1975.

OPTIMUM MANEUVERS OF A SUPERCRUISER

Ching-Fang Lin
Applied Dynamics International
Ann Arbor, MI 48104/USA

Nguyen X. Vinh
The University of Michigan
Ann Arbor, MI 48109/USA

Introduction

The analysis of the optimum maneuvers of high performance aircraft with control characteristics depending on the current state variables, and in particular on the Mach number, is a complex problem. In general, for any specified vehicle characteristics, direct numerical optimization technique has to be used. But this has the drawback that the structure of the optimum control, in terms of the aerodynamic and thrust modulation, is not characteristically displayed and the general behavior of the optimum trajectory is not clearly understood except for the particular example considered.

In this paper, we use the maximum principle to analyze the turning performance of a typical light-weight fighter aircraft. In addition, the following techniques have been used concurrently to alleviate the difficulties encountered:

a) The aerodynamics and engine characteristics are modelled as continuous functions of the Mach number. This allows a smooth application of the maximum principle and at the same time delimits the flight envelope in the phase space.

b) A set of dimensionless variables are introduced. This leads to general results for a whole class of vehicles having similar physical characteristics.

c) The optimal control is obtained by geometrical method through the use of the domain of maneuverability. This makes explicit the switching characteristics of the optimum control, and in particular the case where singular or chattering control is involved.

d) A backward integration coupled with a rotation of coordinates leads to an efficient evaluation of the unknown parameters and at the same time removes certain ambiguities in the selection of the optimum sequence for the control.

e) Through the use of the integrals of the motion, the totality of the optimum trajectories can be obtained as a family of curves depending on a certain number of arbitrary constants.

Equations of Motion

In horizontal flight, using standard notation with the dot denoting the time derivative, we have the equations

$$\dot{X} = V\cos\psi \qquad \dot{V} = (T-D)/m$$
$$\dot{Y} = V\sin\psi \qquad \dot{\psi} = L\sin\phi/mV \tag{1}$$
$$\dot{m} = -cT/g \qquad L\cos\phi = mg = W$$

The lift and drag forces have the form

$$L = \tfrac{1}{2}\rho S V^2 C_L , \qquad D = \tfrac{1}{2}\rho S V^2 C_D \tag{2}$$

while the lift and the drag coefficients are related by the parabolic relation

$$C_D = C_{D_0}(M) + K(M)C_L{}^2 \tag{3}$$

where the zero-lift drag coefficient, C_{D_0}, and the induced drag coefficient, K, are functions of the Mach number. For turning flight, during a relatively short time interval, we can neglect the mass flow equation and consider the weight as practically constant. Then, through the use of the dimensionless variables and parameter

$$x = gX/a^2,\ y = gY/a^2,\ M = V/a,\ \theta = gt/a,\ \omega = 2W/kpS \tag{4}$$

where a is the speed of sound which is related to the density ρ, pressure p, and ratio of specific heat k by the Hugoniot relation

$$a^2 = kp/\rho \tag{5}$$

we obtain the dimensionless equations

$$x' = M\cos\psi , \qquad M' = \tau - \frac{M^2}{\omega}\left(C_{D_0} + \frac{K\omega^2}{M^4\cos^2\phi}\right)$$
$$y' = M\sin\psi , \qquad \psi' = \frac{\tan\phi}{M} \tag{6}$$

where the prime denotes the derivative taken with respect to the dimensionless time θ, and τ is the thrust-to-weight ratio, $\tau = T/W$, subject to the constraint

$$0 \leq \tau \leq \tau_{max}(M, h) \tag{7}$$

It is clear that, besides the thrust control τ, the aerodynamic control is the bank angle ϕ. In turning flight, we consider the load factor

$$n = \frac{L}{W} = \frac{M^2}{\omega}C_L = \frac{1}{\cos\phi} \tag{8}$$

Because of physiological/structural constraint, n is bounded by an upper value n_S. On the other hand, the lift coefficient is bounded by an upper value $C_{L_{max}}(M)$, function of the Mach number. Hence we have the bounds on the bank angle

$$|\phi| \leq \inf.[\cos^{-1}(\frac{1}{n_S}) ,\ \cos^{-1}(\frac{\omega}{n^2 C_{L_{max}}(M)})] \tag{9}$$

In the equations, ω is a constant parameter which can be used to denote the flight altitude, and for a given altitude its variation provides the change in the wing loading. Hence, if numerical performance is generated for two values of ω, we can deduce the comparative performance for the same aircraft at two different altitudes, or keeping the altitude fixed we can compare the performance of two aircraft with same aerodynamics and engine characteristics but having different wing loadings. It is assumed that, $C_{D_0}(M)$, $K(M)$, $\tau_{max}(M, h)$ and $C_{L_{max}}(M)$ are known functions of the Mach number. For numerical computation, we used data for a supercruiser assembled in [1], but the same procedure applies to any other set of data.

Optimal Control

Using the maximum principle, we introduce the adjoint vector $\vec{p}$ to form the Hamiltonian

$$H = p_x M\cos\psi + p_y M\sin\psi - \frac{p_M M^2}{\omega}\left(C_{D_0} + \frac{K\omega^2}{M^4\cos^2\phi}\right) + p_\psi \frac{\tan\phi}{M} + p_M \tau \tag{10}$$

It is known that the problem has the integrals [2]

$$H = C_0,\ p_x = C_1,\ p_y = C_2,\ p_\psi = C_1 y - C_2 x + C_3 \tag{11}$$

From this , we deduce the adjoint p_M, and the set of adjoint equations are completely integrable. Regarding the thrust control, we consider the adjoint p_M, called the switching function. Then to maximize the Hamiltonian, if

$$\begin{aligned} &p_M > 0, \text{ we use } \tau = \tau_{max} \\ &p_M < 0, \text{ we use } \tau = 0 \\ &p_M \equiv 0 \quad \text{for } \theta \in [\theta_1, \theta_2], \text{ we use } \tau = \text{variable}. \end{aligned} \tag{12}$$

The optimum trajectory is a combination of boost arc (B), coast arc (C) and sustained arc (S). For a junction between subarcs, a B-C sequence is optimum if at the junction $p_M' < 0$. For a reverse condition, a C-B sequence is optimum [3].

Concerning the bank control, we consider the portion of the Hamiltonian containing ϕ

$$\overline{H} = P_1\Omega_1 + P_2\Omega_2 = \vec{P}\cdot\vec{\Omega} \tag{13}$$

where

$$\vec{P} = \left(\frac{p_\psi}{M},\ -\frac{K\omega p_M}{M^2}\right), \quad \vec{\Omega} = \left(\tan\phi,\ \frac{1}{\cos^2\phi}\right) \tag{14}$$

It is proposed to select the vector bank control $\vec{\Omega}$ to maximize the dot product $\vec{P}\cdot\vec{\Omega}$. The domain of maneuverability described by the terminus of the vector $\vec{\Omega}$ for all possible values of ϕ is the parabola $\Omega_2 = 1 + \Omega_1^2$ truncated at $\pm\phi_{max}$ (Fig. 1).

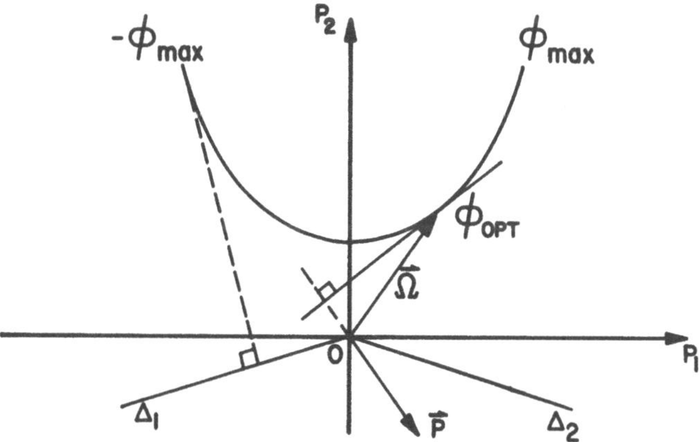

Fig. 1 Domain of maneuverability for the bank angle.

Then, if the vector $\vec{P}$ is outside the angle $\Delta_1 O \Delta_2$, we use $\phi = \pm\phi_{max}$ with the sign being the sign of p_ψ. If the vector $\vec{P}$ is inside the angle $\Delta_1 O \Delta_2$, the bank angle is an interior bank angle such that the tangent to the parabola at the terminus of the vector $\vec{\Omega}$ is orthogonal to $\vec{P}$. This leads to

$$\tan\phi = Mp_\psi/2\omega Kp_M \quad (15)$$

From the figure, it is clear that when $P_2 \geq 0$, that is when $p_M \leq 0$, we use maximum bank angle. This, of course, is on a coast or sustained arc. On the other hand, a boost arc, $\tau = \tau_{max}$, can be flown with either interior or boundary bank angle. In the special case where $P_1 \equiv 0$ and $P_2 > 0$, that is $p_\psi \equiv 0$, $p_M < 0$, we can have a chattering bank control in which the bank switches rapidly between $+\phi_{max}$ and $-\phi_{max}$. By the integral (11), we have $C_1y - C_2x + C_3 = 0$. Hence, the trajectory generated is a straight line with maximum deceleration. For interior bank control, using the optimum relation (15), with the aid of the integrals (11), we have a quadratic equation for evaluating $\Delta = \tan\phi$

$$p_\psi\Delta^2 - 2M[C_0 - M(C_1\cos\psi + C_2\sin\psi)]\Delta + \frac{M^2p_\psi}{\omega K}[\tau_{max} - \frac{M^2}{\omega}(C_{D_0} + \frac{K\omega^2}{M^4})] = 0 \quad (16)$$

Since p_ψ is linear in the constants of integration, we can divide the equation by any one of the constants C_i, $i = 0, .., 3$, and for the most general problem in turning flight the totality of the optimum trajectories depends on 3 arbitrary parameters. The difficulty in solving any particular problem depends on the discussion of the optimum thrust sequence, and the adjustment of the constants involved to satisfy the final and transversality conditions.

For the particular case of sustained arc, we have constantly $p_M = 0$, and hence also $p_M' = 0$. We have seen that it occurs at maximum bank angle, either with $n = n_s$ or with $C_L = C_{L_{max}}$. Expliciting the equation, and considering ϕ on its bound as given by condition (9), we have

$$p_M' = -\frac{\partial H}{\partial M} = p_\psi\left(\frac{\tan\phi}{M^2} - \frac{1}{M\cos^2\phi}\frac{d\phi}{dM}\right) - (C_1\cos\psi + C_2\sin\psi) = 0 \quad (17)$$

Since ϕ and its Mach derivative are functions of the Mach number, this equation gives a relation among the state variables along a sustained arc. The variable thrust appears upon taking the derivative of this equation.

Examples of Optimum Turns

The equations we have derived are sufficient to solve any turning maneuver problem. We shall consider some typical problems in performance assessment.

Turning with Minimum Radius. At the initial time, we have

$$\theta = 0, \quad x = 0, \quad y = 0, \quad M = M_0, \quad \psi = 0 \quad (18)$$

Since the instantaneous radius may vary during the turn, a realistic assessment of this performance can be considered using the following final conditions. It is either (Fig. 2)

$$\theta_f = \text{free}, \ x_f = \text{min.}, \ y_f = \text{free}, \ M = M_f, \ \psi_f = 90° \quad (19)$$

or

$$\theta_f = \text{free}, \ x_f = \text{free}, \ y_f = \text{min.}, \ M = M_f, \ \psi_f = 180° \quad (20)$$

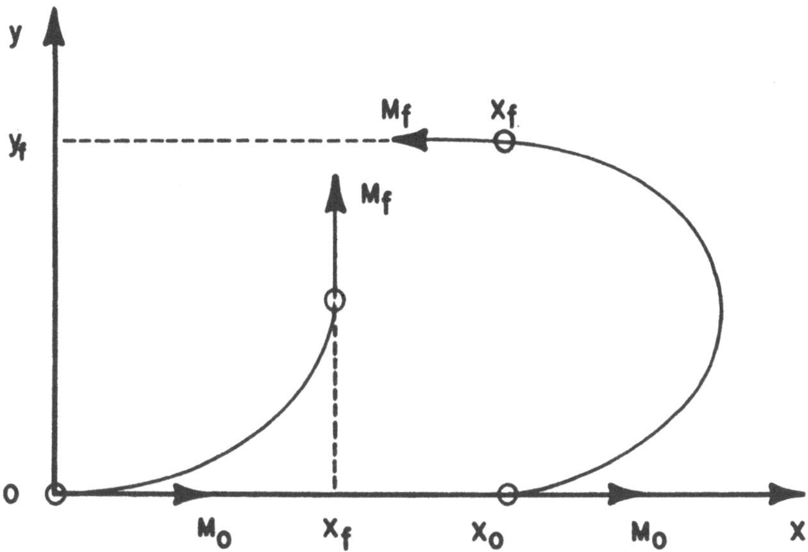

Fig. 2 Turn with minimum radius.

For this free time problem, $C_0 = 0$, and Eq.(16) is reduced to

$$(\frac{\Delta}{M})^2 + \frac{2M(k_1\cos\psi + k_2\sin\psi)}{(1 + k_1y - k_2x)}(\frac{\Delta}{M}) + \frac{1}{\omega K}[\tau_{max} - \frac{M^2}{\omega}(C_{D_0} + \frac{K\omega^2}{M^4})] = 0 \qquad (21)$$

where $k_1 = C_1/C_3$ and $k_2 = C_2/C_3$ are two arbitrary constants to be found. The equation can be solved for Δ/M which can be readily seen as the turning rate.

Along a sustained arc, $p_M = 0$, using the Hamiltonian integral with $C_0 = 0$ in the singular relation (17), we have after some manipulation

$$\frac{d}{dM}(\frac{\Delta}{M^2}) = 0 \qquad (22)$$

On the other hand, if r is the instantaneous radius of curvature

$$m\frac{V^2}{r} = L\sin\phi \qquad (23)$$

Since $L\cos\phi = mg$, we have

$$\frac{a^2}{gr} = \frac{\Delta}{M^2} \qquad (24)$$

Then, Eq.(22) shows that the instantaneous turning radius is minimized along a sustained arc. Since $\Delta = \tan\phi$ is on its maximum boundary, when Δ is evaluated with $C_{L_{max}}(M)$, Eq.(22), upon solving gives the best constant Mach number M_s for turning with thrust modulation. On the other hand, when Δ is on the bound $n = n_s$, it is constant, and the quantity Δ/M^2 is maximized, for minimum r, at the intersection

$$n_s\omega = M^2C_{L_{max}}(M) \qquad (25)$$

After solving, we have a Mach number M_c, called the corner Mach number to be used for turning. In both cases, the Mach number is function of ω, that is it depends on the altitude and the wing loading.

We now consider the case of 90° turn with minimum x_f as given by condition (19). We observe that the condition of prescribed final Mach number, $M = M_f$, can be disregarded since it suffices, at the end of the turn, to continue the flight with $x = x_f$ while readjusting the Mach number to its final value. Hence, with M_f = free, $p_M(\theta_f) = 0$. First, depending on the altitude and wing loading, through ω, the prescribed maximum load factor n_s, and the characteristic function $C_{L_{max}}(M)$, Eq.(25) is solved for the corner Mach number M_c. If a solution exists, M_c is the best Mach number for turning with minimum radius, provided that $M_c < M_s$, while using maximum bank with the corresponding variable thrust to maintain $M = M_c$. If $M_c > M_s$ or if no solution for M_c exists, then the best turning Mach number is $M = M_s$ as given by Eq.(22) while the maximum bank is used with $C_{L_{max}}$ with the variable thrust sufficient to maintain the required optimum Mach number. We call the singular Mach number M_* which is either M_s or M_c. Then, if $M_0 < M_*$, the trajectory starts with a B-arc until $\psi = 90°$ or until $M = M_*$ whichever occurs first. In the case where M_* is encountered, the turning Mach number is maintained at M_* until the end of the turn. The optimum trajectory is either a pure B-arc or a combination of BS in this order. For a B-arc, we use either $\phi = \phi_{max}$ or a variable ϕ as given by Eq.(21) with $k_2 = 0$ since y_f is free. The only unknown parameter is k_1, which should be selected such that the condition $p_M(\theta_f) = 0$ is satisfied. In general, for a high thrust propulsion system, Eq.(21) has no real root so that maximum bank angle is used throughout the turn. For the case where $M_0 > M_*$, the trajectory starts with a C-arc, hence with maximum bank angle, until $\psi = 90°$ or until $M = M_*$ whichever occurs first. In the case where M_* is encountered, the turning Mach number is maintained at M_* until the end of the turn. The optimum trajectory is either a pure C-arc or a combination of CS in this order.

For the problem of minimum y_f in a 180° turn as specified in condition (20), the solution is simple for a high thrust propulsion system which can deliver the required best turning Mach number M_* at maximum bank angle. It suffices to move first along the x-axis with $\phi = 0$, with a B-arc, or a C-arc, or as a matter of fact using any thrust program to bring the Mach number to the ideal value M_*. Then, the turn is made at this Mach number. The final Mach number is similarly adjusted after the completion of the turn.

When the thrust is not sufficiently high to maintain M_*, the problem consists of finding the optimum values M_0 and M_f at the beginning and at the end of the turn with adjustment along two straight lines parallel to the x-axis as necessary. Since x_f is free, $C_1 = 0$ and for the same reason as explained above for a 90° turn we can consider M_0 and M_f as free, that is $p_M(\theta_0) = p_M(\theta_f) = 0$. From the Hamiltonian integral, this leads to $p_\psi(\theta_0) = p_\psi(\theta_f) = 0$, that is $C_3 - C_2x_0 = C_3 - C_2x_f = 0$. This leads to $x_0 = x_f$, and if the turn starts at $x_0 = 0$, $C_3 = 0$. The only parameter of the problem is the initial Mach number selected such that $M_0 > M_*$. In general, the trajectory consists of a B-arc using maximum bank angle. The condition $\psi_f = 180°$ when $x_f = 0$ is used to find the correct initial value M_0.

Minimum Time Turn to a Point. The initial condition is as given in Eq.(18) but now with the final condition (Fig. 3)

$$\theta_f = \text{min.},\ x = x_f,\ y = y_f,\ M = M_f,\ \psi_f = \text{free} \tag{26}$$

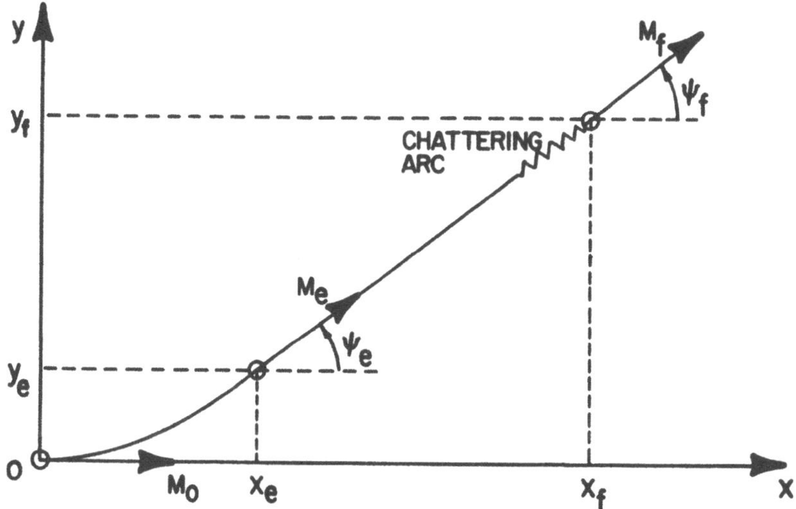

Fig. 3 Minimum time turn to a point.

Because of the condition of free ψ_f, we have at the final time

$$p_\psi(\theta_f) = C_1 y_f - C_2 x_f + C_3 = 0 \tag{27}$$

Hence, although the final position is prescribed, $C_1 \neq 0$, $C_2 \neq 0$, and the final time is minimized $C_0 > 0$, this condition, with the normalizing of the constants C_i, leads to a two-parameter problem. Using this in Eq.(16), we have the condition for interior bank angle along a B-arc

$$\left(\frac{\Delta}{M}\right)^2 - \frac{2[1 - M(k_1\cos\psi + k_2\sin\psi)]}{[k_1(y - y_f) - k_2(x - x_f)]}\left(\frac{\Delta}{M}\right) + \frac{1}{\omega K}\left[\tau_{max} - \frac{M^2}{\omega}\left(C_{D_0} + \frac{K\omega^2}{M^4}\right)\right] = 0 \tag{28}$$

where now $k_1 = C_1/C_0$ and $k_2 = C_2/C_0$ are two arbitrary constants to be selected such that when $M = M_f$, we have $x = x_f$, $y = y_f$.

If the given point is at a large distance, the last portion of the trajectory is a straight line, with $\tau = \tau_{max}$, $\phi = 0$. Along this line a limiting Mach number is ultimately reached and it is obvious that any prescribed final Mach number M_f must be less than this maximum Mach number. The final arc is therefore a decelerating C-arc, $p_M < 0$. Along this arc, we continue to have $\psi = \psi_f$, that is $p_\psi = 0$ and this is the condition of chattering bank control as discussed above. To compute this final chattering arc, it suffices to integrate backward, with $\tau = 0$, from $M = M_f$, using $\phi = \phi_{max}$ for the equation in M until the maximum Mach number is reached. Of course, if the final Mach number is not prescribed, the chattering arc disappears.

The problem is then to turn to the final rectilinear arc. Let subscript e denote the condition at entering of the straight line. At this point

$$\psi_e = \psi_f, \quad \tan\psi_e = \frac{y_f - y_e}{x_f - x_e} = \frac{k_2}{k_1} \tag{29}$$

The last equality is due to the condition that $\Delta = 0$ in Eq.(28). Since it is possible to remove the condition of prescribed M_f by subtracting the chattering arc, we can take $p_{M_f} = 0$ for the purpose of finding the final straight line. Then from the Hamiltonian integral, at the final time

$$k_1\cos\psi_f + k_2\sin\psi_f = \frac{1}{M_f} \tag{30}$$

At the point of entering the straight line, $\phi = 0$, for continuity of the bank angle the arc before must be a B-arc with interior bank. If it is a C-arc, we must have at that point $p_M = 0$ for a switch to B-arc along the straight line and hence, at the entering point

$$k_1\cos\psi_e + k_2\sin\psi_e = \frac{1}{M_e} \tag{31}$$

Since $\psi_e = \psi_f$ and $M_e < M_f$, the two equations above are not compatible and we should look for a pure B-arc. The procedure for numerical integration is as follows. A set of parameters (k_1, k_2) is used to evaluate the interior bank angle according to Eq.(28). At the point where $\tan\psi = k_2/k_1$, the condition $\Delta = 0$ is used to adjust the ratio k_2/k_1. This will insure that the final arc passes through the prescribed final position. The integration continues with $\Delta = 0$ until $x = x_f$. Then condition (30) is used to adjust the second parameter.

The physical data of a supercruiser as assembled in [1] have been used to compute the minimum time turn to a point at two different altitudes and the results are presented in Fig. 4 as isochronous lines from a turn starting at $M_0 = 1$. The lines can be seen as delimiting the reachable domain for turn with a prescribed time. The points of maximum lateral range which can be attained is 125 seconds in these altitudes are shown with a cross.

To find the reachable domain, we first for a prescribed θ_f, find the point of maximum lateral range with free x_f, hence $k_1 = 0$ in Eq.(28). Then, to use the equation for interior bank, initial guessed values for k_2 and x_f must be used. The first iteration is performed at the point of entering the final straight line which is parallel to the y-axis. Hence, at the point

$$\psi_e = \psi_f = 90°, \quad x_e = x_f \tag{32}$$

This will insure that $\Delta = 0$ at that point. The second iteration is performed at the prescribed final time, $\theta = \theta_f$, using the condition (30) which now becomes

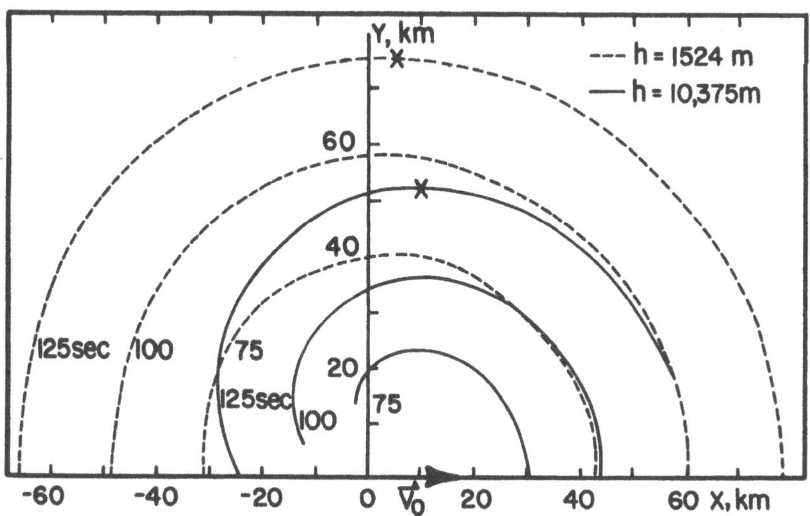

Fig. 4 Reachable domain for turn in prescribed time.

$$k_2 = \frac{1}{M_f} \tag{33}$$

To obtain the other points on the contour θ_f = constant we simply rotate the original system Oxy through an angle ψ_0 to the new position $O\bar{x}\bar{y}$ and again solve the problem of maximum lateral range but with the initial heading $\Psi(0) = -\psi_0$. In other words, ψ_0 is used as a scanning parameter to describe the contour.

References

1. Lin, Ching-Fang, "Optimum Maneuvers of Supersonic Aircraft," Ph.D. dissertation, The University of Michigan, 1980.
2. Vinh, Nguyen X., "Optimal Trajectories in Atmospheric Flight," Elsevier Scientific Publishing Company, Amsterdam, 1980.
3. Vinh, Nguyen X., "On Contensou Switching Theory" (in French), La Recherche Aerospatiale, No. 2, 1977. English Translation No. TT408 by European Space Agency.

APPLICATION OF CONSTRAINED CONSTANT OPTIMAL OUTPUT FEEDBACK TO MODERN FLIGHT CONTROL SYNTHESIS

by

E. Y. Shapiro and D. A. Fredricks

Lockheed-California Company
Burbank, California 91520

Abstract

This paper describes an applications oriented approach to the generation of optimal output feedback gains for linear time-invariant systems which is dependent of the open loop stability.

The standard requirements for the provision of initial stabilizing output feedback gains for priming the computational process is circumvented. In lieu of initial stabilizing gains, the proposed algorithm employs the full state feedback solution which guarantees stability under the mild condition of stabilizability.

In this paper the generation of sub-optimal output feedback problem is cast in the setting of a constrained parameter optimization problem. The solution of this constrained optimization employs Hestenes' method of multipliers with some modifications. A primal-dual problem is considered where the primal minimization employs a Davidon-Fletcher-Powell method, and the dual maximization is accomplished via a quasi-Newton procedure.

This approach provides the designer with a means of suppressing to zero selected gains corresponding to accessible output, either for the purpose of simplifying the controller structure or because prior knowledge indicates that certain gains are "nonproductive". In addition, the designer can easily incorporate certain linear constraints, which the feedback gains will satisfy, into the proposed procedure.

Detailed algorithm description and computational results for a realistic flight control design problem are provided.

I. Introduction

In practice, the designer of control systems rarely enjoys the luxury of having the freedom to feed back the system's state in its entirety. Lack of measurements of the entire state usually reflects economic or technical considerations which cause one to employ a limited sensor complement for control. In such a situation the well established theory dealing the quadratic optimization under full state availability motivates the designer to employ observer based compensator procedures in the deterministic setting, or a Kalman filter in the stochastic setting. These state reconstructors, while allowing the use of the well-developed theory of quadratic optimization with full state feedback, tend to complicate the structure of the feedback compensator.

The problem is compounded when dealing with the control of non-linear systems such as flight control systems for piloted vehicles where several linearized small perturbation models are used to describe the system through its entire operational regime. It is then necessary to schedule the optimal controllers relative to each operating point. Consequently, when dynamic compensators are introduced, the scheduling problem can become cumbersome and costly.

An important problem from a practical standpoint is the determination of constant output feedback gains for the sub-optimal control of systems with inaccessible states. This problem has recently been studied from the theoretical and numerical viewpoint by several researchers [1, 2, 3, 4, 5, 6]. Most of the existing numerical algorithms for the solution of the optimal output feedback problem call for initializing the computation procedure with stabilizing output gains for systems which are open loop unstable. This is a severe handicap, since the problem of system stabilization by output feedback is nontrivial and could be quite costly. Further, optimal output feedback, where each output is fed to each input, could still be complex. The designer is interested in simplifying the controller by eliminating some of the gains corresponding to accessible outputs.

It is well known that feedback gains do not contribute uniformly to improve total system performance. While many feedback gains can be termed "non-productive", as they have minimal effect on system performance, a small number of such gains carry the bulk of the effect of improving performance. Consequently, a significant reduction of controller complexity with minimal performance sacrifice becomes feasible. In addition, in order to achieve desired performance characteristics, e.g., handling qualities of piloted aircraft, it is sometimes necessary to force feedback gains to satisfy certain linear constraints. This method provides the designer with a means of achieving this, and, in addition, gives him a needed tool which enables him to assess cost effectiveness by weighing the available trade-offs between performance and controller structure complexity.

It is the purpose of this paper to outline the theoretical and computational procedures which resolve the problem of providing sub-optimal output feedback applicable to open loop unstable systems while allowing the designer to eliminate some preselected gains corresponding to accessible outputs and to force some of the feedback gains to satisfy specified linear constraints.

II. Problem Formulation

Consider the linear time invariant system

$$\dot{x} = Ax + Bu \qquad x(t=0) = x_0 \tag{1}$$

$$y = \tilde{C}x \tag{2}$$

where $x \in R_n$, $u \in R_m$, $y \in R_r$, and A, B, $\tilde{C}$ are real constant matrices of compatible order. Also let $\tilde{C}$ be of full rank, that is

$$\text{rank}\ \tilde{C} = r \tag{3}$$

With no loss of generality, we can assume that $\tilde{C}$ has the following partitioned form

$$\tilde{C} = \left[I_r \mid 0 \right] \tag{4}$$

where I_r is the rxt identity matrix. If $\tilde{C}$ does not have the form of (4), it is always possible to find a similarity transformation which will yield $\tilde{C}$ of the form of (4) as long as (3) is satisfied.

For the system (1) and (2) subject to (3), find a control law of the form

$$u = \bar{F}y = -\bar{F}\tilde{C}x \tag{5}$$

which minimizes the quadratic cost given by

$$J(\bar{F}) = E\left[\int_0^\infty (x'Qx + u'Ru)\, dt \right] \tag{6}$$

where $\bar{F}$ is an mxr constant matrix. Note that according to (4), y represents the first r components of the state x; that is

$$y = \begin{bmatrix} x_1 \\ x_2 \\ \cdot \\ \cdot \\ \cdot \\ x_r \end{bmatrix} \tag{7}$$

For convenience, we introduce the matrix $\bar{C}$, where

$$\bar{C} = \left[0 \mid I_{(n-r)} \right] \tag{8}$$

and augment $\tilde{C}$ with $\bar{C}$ to form the matrix C, given by the partitioned form

$$C = \begin{bmatrix} \tilde{C} \\ \hline \bar{C} \end{bmatrix} \tag{9}$$

Note that C is the nxn identity matrix, and we may now discuss a full state feedback problem with a particular feedback structure.

The feedback matrix F corresponding to the matrix C of (9) is an mxn matrix which we desire to have a special structure reflecting the threefold objective consisting of

a. Output feedback
b. Elimination of gains corresponding to accessible outputs
c. Forcing some gains to satisfy certain linear constraints

However, rather than forcing a certain structure on F, we may force certain conditions on the f_{ij}'s-the entries of the matix F. To see that more clearly, let $S(\cdot)$ be a column stacking operator. When $S(\cdot)$ operates on a pxq matrix V, it yields a pq vector which is made of the columns of V. Proceed by defining the mn vector f according to

$$f = S(F) \tag{10}$$

and consider the three constant matrices W_1, W_2, W_3 with dimensions $q_1$xmn, $q_2$xmn and $q_3$xmn respectively. We can form the qxmn matrix W according to

$$W = \begin{bmatrix} W_1 \\ \hline W_2 \\ \hline W_3 \end{bmatrix} \tag{11}$$

where q is given by

$$q = q_1 + q_2 + q_3 \tag{12}$$

Our three objectives, which were outlined above, can be accommodated by incorporating the set of constraints

$$Wf = d \tag{13}$$

where d is a vector in R_q, into the optimization problem (6).

Note that if w_i' is the i-th row of W, then the i-th equation implied by (13) is

$$w_i' f = d_i \tag{14}$$

Thus, if w_i' is of the form

$$w_i' = (0\,0\,0\dots0\,1\,0\dots0) \tag{15}$$

and the corresponding $d_i = 0$, (14) implies that a certain f_{ij} should be set to zero. Consequently, (13) can be viewed as a structural constraint on F. To accomplish our objectives, the matrix W_1 will enforce an output feedback by having rows of the form of (15) with

$$d_i = 0 \qquad 1 \leq i \leq q_1 \tag{16}$$

The matrix W_2 will allow the designer to eliminate gains corresponding to <u>accessible</u> outputs by having rows of the form of (15) with the single non-zero entry of "1" in a position corresponding to the f_{ij} which is to be eliminated, with

$$d_i = 0 \qquad q_1 + 1 \leq i \leq q_1 + q_2 \tag{17}$$

The third component of W, namely W_3, will enable the designer to force some of the feeback gains to satisfy certain linear constraints; therefore the rows of W_3 and the respective d_i's have no special structure. Consequently, our optimization problem can now be stated as: Given the linear system (1) with the output equation

$$y = Cx \tag{18}$$

find the matrix F of

$$u = -Fy \tag{19}$$

which minimizes (6) subject to the constraint

$$Wf = d \tag{20}$$

Before we proceed, it will be beneficial to illustrate the formation of the matrix W for a specific example. Thus we consider the system (1), (2), and (4) with dimensions n = 4, m = 2 and r = 3. The feedback gain matrix F is of dimension 2x4. Note that the state x_4 is not accessible; further assume that it is desired to eliminate the feedback gain f_{12}, and it is also required to maintain the following linear relationship between f_{11} and f_{23}

$$f_{11} - 3f_{23} = 0 \tag{21}$$

The column stacking of the matrix F according to (10) yields the vector f, where

$$f = (f_{11}, f_{21}, f_{12}, f_{22}, f_{13}, f_{23}, f_{14}, f_{24})' \tag{22}$$

from which we readily determine the structure of W, that is

$$W = \begin{bmatrix} W_1 \\ --- \\ W_2 \\ --- \\ W_3 \end{bmatrix} \quad \begin{bmatrix} 0\,0\,0\,0\,0\,0\,1\,0 \\ 0\,0\,0\,0\,0\,0\,0\,1 \\ ---------- \\ 0\,0\,1\,0\,0\,0\,0\,0 \\ ---------- \\ 1\,0\,0\,0\,0\,{-3}\,0\,0 \end{bmatrix} \begin{matrix} \}W_1 \\ \\ \}W_2 \\ \}W_3 \end{matrix} \tag{23}$$

and the vector d for this case if given by

$$d = 0 \tag{24}$$

Note that W_1 will cause the elimination of gains corresponding to inaccessible states, W_2 will cause the elimination of gains corresponding to accessible states, and W_3 takes care of the linear constraint (21).

III. Problem Development

The necessary conditions for the optimality of F have previously been derived [1, 5] by converting the dynamic optimation to a static optimization in the parameter space of the f_{ij}'s. It was shown that

$$J(F) = trP \tag{25}$$

where P satisfies the Lyapunov equation

$$(A-BFC)'P + P(A-BFC) = -(Q + C'F'RFC) \tag{26}$$

also, the gradient of J(f) with respect to F is given by [5]

$$\frac{\partial}{\partial F}\left[J(F)\right] = 2(RFCLC' - B'PLC) \tag{27}$$

where the matrix L satisfies the Lyapunov equation

$$(A-BFC)L + L(A-BFC)' = -I \tag{28}$$

Usually what one does is to employ (27) in a gradient search technique [6] and minimize J(F), provided an F_o which stabilizes $A-BF_oC$ can be found. However, in the present discussion, the matrix C is the identity matrix, and what is needed to initialize the gradient search is a full state feedback gain matrix, which stabilizes $A-BF_o$. To obtain such an F_o, one simply solves the Riccati equation corresponding to the quadrauple A, B, Q, R

$$A'K + KA - KBR^{-1}B'K + Q = 0 \tag{29}$$

for the unique positive definite matrix K, and the respective F_o is given by

$$F_o = R^{-1}B'K \tag{30}$$

A solution for K can be readily obtained provided the pair $\{A, B\}$ is stabilizable, independent of the stability of the A matrix [7].

The required solution of a Riccati equation does not represent additional computation effort, since the designer invariably will solve the full state feedback problem, thus establishing a data base for system performance. Consequently, the designer can always compare the performance of a system using output feedback to the performance of the ideal situation of full state feedback and exhibit the performance variation.

In view of the previous discussion, the problem addressed in this paper can be described by the following constrained minimization, that is:

$$\min_F J(F) = \min_F \operatorname{tr} P \tag{31}$$

subject to constraint

$$\xi = Wf - d = 0 \tag{32}$$

For convenience, we can replace (32) by an identical constraint given by

$$\|\xi\|^2 = \xi'\xi = 0 \tag{33}$$

The above minimization problem is equivalent to the following minimization problem [8]

$$\underset{F}{\text{minimize}} \ \{J(F) + \mu\, \xi'\xi\} \tag{34}$$

subject to

$$\xi'\xi = 0 \tag{35}$$

for any $\mu > 0$, in particular, for reasons of convenience, let $\mu = 1/2\gamma$ where $\gamma > 0$.

From a different point of view, one can adjoin the constraint (32) to the objective function (31) via a qxl vector Lagrange multiplier λ , and obtain an unconstrained minimization problem, that is,

$$\underset{F}{\text{minimize}} \quad \left\{J(F) + \lambda'\xi\right\} \tag{36}$$

However, since minimization of J(F) is equivalent to the minimization of $\left\{J(F) + 1/2\,\gamma\,\xi'\xi\right\}$ subject to the same constraint, we can combine (34) with (36) and obtain a single unconstrained minimization problem of the form

$$\underset{F}{\text{minimize}} \left\{J(F) + \lambda'\xi + 1/2\,\gamma\,\xi'\xi\right\} \tag{37}$$

Note that minimization of (37) w.r.t. the matrix F simultaneously provides a sub-optimal output feedback gain matrix having the prespecified structure as dictated by the matrix W.

IV. Problem Solution

The dual method is applied to the problem by defining the dual function

$$\Phi\gamma(\lambda) = \min_{F} \left\{J(F) + \lambda'\xi + 1/2\,\gamma\,\xi'\xi\right\} \tag{38}$$

If F* is the matrix minimizing the right hand side of (38), then

$$\frac{\partial}{\partial\lambda}\left[\Phi\gamma\right] = \xi^* = Wf^* - d \tag{39}$$

and it is possible to develop a modified Newton's method for solving the dual problem, namely maximization of (38) w.r.t. λ[8] . That is, using $\frac{1}{\gamma} I_{mr}$ as an approximate Hessian and ξ^* as a gradient, the following iterative scheme is applicable.

$$\lambda(k+1) = \lambda(k) + \gamma\xi^* \tag{40}$$

Now, as noted in [9, p527] , each $\Phi\gamma$ has the same local maximum λ^*. Therefore, it is possible to update the γ in some fashion, if this is advantageous. When the γ are updated in some way, the process becomes the method of multipliers.

The present algorithm employs the scheme suggested by Powell [10] and Buys [11] .

The inner loop unconstrained minimization is solved as in [6] by a Davidon-Fletcher-Powell type algorithm with an appropriately modified gradient formulation, that is, we let H(F) be defined as:

$$\begin{aligned} H(F) &= J(F) + \lambda'\xi + 1/2\,\gamma\,\xi'\xi = \\ &= J(F) + \lambda'(Wf-d) + 1/2\gamma\,(Wf-d)'(Wf-d) \end{aligned} \tag{41}$$

It is convenient, before the development of an expression for the gradient of H(F), to define the unstacking operator $S^{-1}{}_m(\cdot)$. This operator maps mn column vectors into the space of mxn matrices by taking m components at a time and using them as columns of the mxn matrix. Thus,

$$S^{-1}{}_m(f) = F \tag{42}$$

Using the notion of the unstacking operator, it can be shown that the gradient of H(F) with respect to F is given by

$$\frac{\partial}{\partial F}\left[H(F)\right] = \frac{\partial}{\partial F}\left[J(F)\right] + \sum_{j=i}^{q} \lambda_j S^{-1}{}_m(w_j) + \gamma S^{-1}{}_m(W'Wf - W'd) \tag{43}$$

where λ_j is the j-th component of the Lagrange multiplier λ, and w_j' is the j-th row of the matrix W.

The gradient of J(F) with respect to F in (43) is given by (27), but since in our case the matrix C is the identity matrix, we obtain

$$\frac{\partial}{\partial F}\left[J(F)\right] = 2\,(RFL - B'PL) \tag{44}$$

Thus, (43) together with (44) provide closed form information relative to the gradient needed in the Davidon-Fletcher-Powell algorithm.

A word about the existence of a solution to our problem is in order. Unfortunately, a compact analytical test to verify the existence of a solution is presently unavailable. However, it is possible to provide a geometrical necessary condition to this problem. This is, a solution exists if and only if a non-empty intersection exists between the intersection of the hyperplanes $w_i f = d_i$ and at least one of the regions of stability in the parameter space of the $\{f_{ij}\}$. At present, to the best knowledge of the authors, the question of hyperplane intersection with stability regions of moderate to large order systems is yet unanswered.

V. Algorithm Description

Based on the above discussion, an iterative algorithm for the determination of optimal output feedback for stable or unstable plants will now be briefly described. In what follows, recall that the matrix C is the n-th order identity matrix, and that output feedback, with the possibility of elimination of gains corresponding to accessible outputs and/or forcing some feedback gains to satisfy some linear constraints, is obtained by the special structure of the matrix W.

Step 1: Input the matrices: A(nxn), B(nxm), Q(nxn), R(mxm), W(qxmn). The qxl vector d, and the scalars β, ϵ, M.

Step 2: Initialize counter i = 0. Set the initial Lagrange Multiplier Vector $\lambda(0)$ to the qxl null vector, and select $\gamma(0)$ to be some "large" positive scalar.

Step 3: Use a Riccati solver to solve (29) and (30) to obtain the initial mxn feedback gain matrix F(0).

Step 4: Form

$$\xi(i) = Wf(i) - d \tag{45}$$

Step 5: If $\xi'(i)\,\xi(i) < \epsilon$ STOP

Else go on.

Step 6: Perform the inner loop minimization with respect to F, using the Davidon-Fletcher-Powell algorithm described in [6]. The cost function, recalling that $C = I_n$, is given by

$$H[F(i)] = J[F(i)] + \lambda'(i)\,\xi(i) + 1/2\,\gamma\,\xi'(i)\,\xi(i) \tag{46}$$

where

$$J[(F(i)] = \mathrm{tr}\,P(i) \tag{47}$$

and P(i) is a solution to

$$[A-BF(i)]'\,P(i) + P(i)\,[A-BF(i)] = -[Q + F'(i)RF(i)] \tag{48}$$

and the gradient of H $[F(i)]$ with respect to F is given by

$$\frac{\partial}{\partial F}\{H[F(i)]\} = \frac{\partial}{\partial F}\{J[F(i)]\} + \sum_{j=1}^{q}\lambda_j(i)\,S^{-1}{}_m[w_j] + \gamma(i)\,S^{-1}{}_m[W'\,\xi(i)] \tag{49}$$

and

$$\frac{\partial}{\partial F}\{J[F(i)]\} = 2\,[RF(i)\,L(i) - B'P(i)\,L(i)] \tag{50}$$

where L(i) is a solution of

$$[A-BF(i)]\,L(i) + L(i)\,[A - BF(i)]' = -I \tag{51}$$

Increment the counter for i by setting i ⟶ i+1, and denote the solution to the inner loop minimization by F(i).

Step 7: Update the Lagrande multiplier vector according to

$$\lambda(i+1) = \lambda(i) + \gamma(i)\,\xi(i) \tag{52}$$

Step 8: If

$$\frac{\|\xi(i-1)\|^2}{\|\xi(i)\|^2} > M$$

Then

$$\gamma(i) = \gamma(i-1)$$

Else

$$\gamma(i) = \beta\gamma(i-1)$$

Step 9: Go to Step 4.

VI. Numerical Example

The method presented will be illustrated through its application to a simplified flight control problem.

The inner loop lateral axis design problem, taken from [12], will be used to demonstrate how unnecessary or undesirable gains can be eliminated and the controller structure simplified without adverse effect on system performance. Progressively more restrictive controller structures, ranging from the full state feedback solution, Table A - Case I, to one which employs only six (6) of the twelve (12) gains available, Table A - Case VI, are utilized in this demonstration.

Table A. Optimal Feedback for Several Controller Structures

	OPTIMAL FEEDBACK MATRIX F							CLOSED LOOP EIGENVALUES (A-BF)				
		p	γ	β	δ_a	δ_r	ϕ	Dutch Roll	Roll Mode	Spiral Mode	Rudder Actuator	Aileron Actuator
I	δ_r	-1.28	-6.26	16.47	-.0044	9.56	.179	-.635±j1.406	-4.00	-.0503	-200.25	-100.03
	δ_a	5.64	1.83	-24.50	9.84	.0022	.298					
II	δ_r	-1.34	-6.07	16.51	.054	9.59	0	-.591±j1.433	-3.788	-.0093	-201.01	-102.81
	δ_a	5.32	2.89	-24.28	10.10	.250	0					
III	δ_r	-1.29	-6.15	16.52	0	9.53	0	-.616±j1.439	-3.83	-.0082	-199.82	-105.65
	δ_a	5.57	2.40	-24.23	10.38	0	0					
IV	δ_r	0	-5.97	16.67	0	9.42	0	-.619±j1.446	-4.19	-.0089	-197.34	-103.13
	δ_a	5.92	2.87	-24.32	10.15	0	0					
V	δ_r	0	-7.01	16.81	0	8.76	0	-.767±j1.456	-5.71	-.0058	-183.96	-124.19
	δ_a	10.04	0	-23.46	12.41	0	0					
VI	δ_r	0	-5.25	12.9	0	6.70	0	-.749±j1.460	-4.06	-.0065	-142.70	-135.80
	δ_a	7.33	0	-25.8	13.4	0	0					

The aircraft dynamics for this problem are described by

$$\dot{x} = Ax + Bu \tag{53}$$

$$y = \tilde{C}x \tag{54}$$

with

$$x = \begin{bmatrix} p_s \\ r_s \\ \beta \\ \delta_a \\ \delta_r \\ \Phi \end{bmatrix} \quad \begin{matrix} \text{stability axis roll rate} \\ \text{stability axis yaw rate} \\ \text{angle of sideslip} \\ \text{aileron deflection} \\ \text{rudder deflection} \\ \text{bank angle} \end{matrix}$$

$$u = \begin{bmatrix} \delta_{r_c} \\ \delta_{a_c} \end{bmatrix} \quad \begin{matrix} \text{rudder command} \\ \text{aileron command} \end{matrix}$$

and where A and B are given by

$$A = \begin{bmatrix} -.746 & .387 & -12.9 & 6.05 & .952 & 0 \\ .024 & -.174 & .4 & -.416 & -1.76 & 0 \\ .006 & -.999 & -.058 & -.0012 & .0092 & .0369 \\ 0 & 0 & 0 & -5.0 & 0 & 0 \\ 0 & 0 & 0 & 0 & -10.0 & 0 \\ 1.0 & 0 & 0 & 0 & 0 & 0 \end{bmatrix} \qquad (55)$$

$$B = \begin{bmatrix} 0 & 0 \\ 0 & 0 \\ 0 & 0 \\ 0 & 10.0 \\ 20.0 & 0 \\ 0 & 0 \end{bmatrix} \qquad (56)$$

Here we have destabilized the matrix A from [12] to exhibit the generation of optimal output feedback gains of open loop unstable systems. The change of A corresponds to an aft motion of the center of gravity of the aircraft. The open loop eigenvalues corresponding to A are given by

$$\left.\begin{array}{ll} \lambda_1 & = -.200 \text{ (Spiral Mode)} \\ \lambda_2 & = -.778 \text{ (Roll Subsidence Mode)} \\ \lambda_{3,4} & = .00013 \pm j\,.704 \text{ (Dutch Roll Mode)} \\ \lambda_5 & = -5.0 \text{ (Actuator)} \\ \lambda_6 & = -10.0 \text{ (Actuator)} \end{array}\right\}$$

Low actuator bandwidths have been chosen deliberately to illustrate the selection of an appropriate bandwidth by means of feedback.

Since we consider the inner loop design problem, the bank angle is an inaccessible state. Consequently,

$$\tilde{C} = (I_5 \mid 0) \qquad (57)$$

The solution of the full state feedback problem, given the weighting matrices from [12]

$$Q = H'_o \, H_o \qquad (58)$$

where

$$H_o = \begin{bmatrix} -.131 & -.612 & 1.64 & 0 & 1.0 & .0175 \\ .567 & .160 & -2.39 & 1.0 & 0 & .0303 \end{bmatrix} \qquad (59)$$

and

$$R = \begin{bmatrix} .01 & 0 \\ 0 & .01 \end{bmatrix} \qquad (60)$$

is shown as Case I in Table A.

For comparison, using the same weighting matrices (58, 60), the algorithm described in Section V was exercised to produce the various output feedback solutions shown in Table A, Cases II through VI.

Each case will be described below. Noting that each simplification prevails for all subsequent cases.

Recall that the bank angle, Φ, is an inaccessible state. Thus, gains, $f_{1,6}$ and $f_{2,6}$ are suppressed in Case II. Case III incorporates the elimination of the actuator coupling gains $f_{1,4}$ and $f_{2,5}$. Feedback of roll rate to rudder and yaw rate to aileron are known to be nonproductive, so $f_{1,1}$ and $f_{2,2}$ are eliminated in Cases IV and V, respectively, further simplifying the controller structure. Finally, the constraint

$$f_{23} = -2f_{13} \qquad (61)$$

is imposed.

The resulting controller structure for this case (VI) is given as

$$F = \begin{bmatrix} 0 & f_{12} & f_{13} & 0 & f_{15} & 0 \\ f_{21} & 0 & -2f_{13} & f_{24} & 0 & 0 \end{bmatrix} \tag{62}$$

Inspection of Table A, Case VI, reveals that the closed loop eigenvalues are in good agreement with the full state feedback solution, Case I, while the controller structure has been simplified by 50%. Also shown is the cumulative effect on system performance of systematic elimination of feedback gains.

VII. Conclusions

An approach for generating optimal constant output feedback gains which is dependent of open loop stability has been presented. This approach allows the designer to simplify further the controller structure by eliminating gains corresponding to accessible outputs while maintaining optimality of the remaining ones. Further, the designer is able to incorporate certain linear constraints on the feedback gains to accommodate desired performance characteristics. Computational procedures and a numerical example have been presented.

References

[1] W. S. Levine and M. Athans, "Determination of the Optimal Constant Feedback Gains for Linear Multivariable Systems," IEEE Transactions on Automatic Control, Volume AC-15, No. 1, February 1970, pp. 44-48.

[2] S. S. Choi and H. R. Sirisena, "Computation of Optimal Output Feedback Gains for Linear Multivariable Systems", IEEE Transactions on Automatic Control, Vol. AC-19, No. 3, June 1974, pp. 257-258

[3] H. P. Horisberger and P. R. Belanger, "Solution of the Optimal Constant Output Feedback Problem by Conjugate Gradients", IEEE Transactions on Automatic Control, Vol. AC-19, August 1974, pp. 434-435.

[4] C. H. Knapp and S. Basuthakur, "On Optimal Output Feedback", IEEE Transaction on Automatic Control, Vol. AC-17, No. 6, December 1972, pp. 823-825.

[5] J. Mendel, "A Concise Derivation of Optimal Constant Limited State Feedback Gains," IEE Transactions on Automatic Control, Vol. AC-19, No. 4, August 1974, pp. 447-448.

[6] E. Y. Shapiro and D. A. Fredricks, "An Algorithm for the Generation of Optimal Output Feedback Gains for Linear Time Invariant Systems", Proceedings of the Conference on Information Sciences and Systems, The John Hopkins University, Baltimore, Maryland, March 29-31,1978, pp. 467-472.

[7] H. Kwakernaak and R. Sivan, Linear Optimal Control Systems, Wiley-Interscience, 1972.

[8] D. G. Luenberger, Introduction to Linear and Nonlinear Programming, Addison-Wesley, 1973.

[9] D. P. Bertsekas, "Combined Primal-Dual and Penalty Method for Constrained Minimization", Siam Journal on Control, Vol. 13, May 1975, pp. 521-542.

[10] M.J.D. Powell, "A Method for Nonlinear Constraints in Minimization Problems", Optimization, R. Fletcher, ed., Academic Press, New York, 1969, pp. 283-298.

[11] J. D. Buys, "Dual Algorithms for Constrained Optimization", Phd. Thesis, Rijksuniversiteit de Leiden, the Netherlands, 1972.

[12] C. A. Harvey and G. Stein, "Quadratic Weights for Asymptotic Regulator Properties", IEEE Transactions on Automatic Control, Volume AC-23, No. 3, June 1978, pp. 378-387.

ADVANCED CONTROL LAWS FOR EXPERIMENTS IN FAST ROLLOUT AND TURNOFF OF THE B737-100 AIRCRAFT

A. S. C. Sinha
Purdue University
School of Engineering and Technology
1201 E. 38th Street
Indianapolis, Indiana 46205

Abstract

This paper presents a new approach in quasilinearization techniques for optimal control problems where bounds on control exist. The framework considered here is probably the most natural departure from the usual nonlinear system problem and is applicable to a large class of nonautonomous nonlinear cases. For this reason our results are extension to the existing qualilinearization techniques. The convergence of the algorithm is proven. Specifically, the application of quasilinearization techniques in solving the landing approach, rollout and turnoff of the B737-100 aircraft is presented.

INTRODUCTION

Research being conducted under the Terminal Configured Vehicle Program (TCV) at the Langley Research Center covers most aspects of automated landing of the B737-100 aircraft, including fast rollout and turnoff. One of the objectives of this program is to investigate and promote advanced control laws for use in fast rollout and turnoff experimentation. The advanced digital control schemes under consideration have much greater flexibility and logic capability than do comparable analog systems. However, because of their increased complexity, digital control systems often require careful theoretical analysis before design and development work can be implemented. The TCV Program is directed toward increasing traffic flow of commercial aircraft at congested airports by providing fast rollout and turnoff. The research program extends the concept of automated flight and landing to the rollout and turnoff phases in order to reduce runway occupancy during normal and adverse (CAT III) landing conditions. In order to achieve acceptable control and tracking performance in both dry and adverse weather conditions, a magnetic cable buried in the runway and three magnetic coil pickups mounted in the aircraft have been investigated [1,2].

The specific objective of this paper is to analytically design advanced control laws for fast rollout and turnoff experimentation. These methods will be applicable, in general, to the development of optimal feedback control of flight control systems of future aircraft.

The State Equations. The first and perhaps the most important step in analysis of the problem is to describe mathematically the essential features of the system, either in terms of system dynamic equations or aerodynamic coefficients having well-defined mathematical representations. Once the mathematical model of the automated landing system is known, the tools of analysis and computation can be used to carry out an optimization analysis. But, unless one has a realistic model of the system and basic analytic derivation of its control laws, numerical techniques alone will not yield meaningful answers. In an attempt to find a simple, yet realistic, model of automated landing, we discovered a mathematical model of the form

$$\dot{v}(t) = \Lambda\, v(t) + \zeta\, u\,(t) + \Gamma\, v(t)\, u\,(t) + \gamma \tag{1}$$

where the state equation $v(t)$ is given by

$$v = (\dot{x},\dot{y},\dot{z};\ p,q,r;\ \phi,\psi,\sigma;\ z_{1D},z_{2D},z_{3D})^T$$

and the matrices Λ and Γ are 12 x 12. The vectors ζ and γ have many negligible components. In particular, the equation of motions for the fast rollout and turnoff problem, where the pitch angle (θ) and the roll angle (ϕ) are considered very small, yield the following set of equations:

The equations of motion for the B737-100 aircraft in the runway frame are given by

$$\begin{bmatrix}\ddot{x}\\ \ddot{y}\\ \ddot{z}\end{bmatrix} = \begin{bmatrix}\Lambda_R\end{bmatrix}\begin{bmatrix}\phi\\ \theta\\ \sigma\end{bmatrix} + \begin{bmatrix}\Gamma_R\end{bmatrix}\begin{bmatrix}\phi\\ \theta\\ \sigma\end{bmatrix} u + \begin{bmatrix}\Lambda_R^{C}\gamma\end{bmatrix}\begin{bmatrix}p\\ q\\ r\end{bmatrix} + \zeta_R u + \gamma_R + O(\sigma) \tag{2}$$

where $(\dot{x}, \dot{y}, \dot{z})$, (p, q, r), (ϕ, θ, σ) and (z_{1D}, z_{2D}, z_{3D}) are aircraft inertial velocity, body rates, Euler angles and strut deflection respectively. The matrices Λ_R, Γ_R, $\Lambda_R^{C}\gamma$ and vectors ζ_R, γ_R and control input u are defined in terms of aerodynamic coefficients.These are not defined herein for the sake of brevity. $O(\sigma)$ is a perturbing term.

The kinematic equations relating the Euler angles and the body rates for fast rollout and turnoff conditions yield

$$\begin{bmatrix}\dot{p}\\ \dot{q}\\ \dot{r}\end{bmatrix} = - \begin{bmatrix}I_B\end{bmatrix}^{-1}\begin{bmatrix}\Lambda_W\end{bmatrix}\begin{bmatrix}\dot{p}\\ \dot{q}\\ \dot{r}\end{bmatrix} - \begin{bmatrix}I_B\end{bmatrix}^{-1}\begin{bmatrix}\Lambda_W^7\end{bmatrix}\begin{bmatrix}\phi\\ \theta\\ \sigma\end{bmatrix} + \zeta_W u + \begin{bmatrix}\Gamma_W\end{bmatrix}\begin{bmatrix}\phi\\ \theta\\ \sigma\end{bmatrix} u + \gamma_W \tag{3}$$

where again the matrices I_B, Λ_W, Λ_W^7, Γ_W are all 3 x 3 and are defined in relation to the aerodynamic coefficients as are the vectors ζ_W, γ_W.

The differential equations for the angular accelerations in the velocity frame and the strut deflection rate during fast rollout and turnoff are, respectively

$$\begin{bmatrix}\dot{\phi}\\ \dot{\psi}\\ \dot{\sigma}\end{bmatrix} = \begin{bmatrix}I_\phi\end{bmatrix}\begin{bmatrix}p\\ q\\ r\end{bmatrix} \quad \text{and} \quad \begin{bmatrix}\dot{z}_{1D}\\ \dot{z}_{2D}\\ \dot{z}_{3D}\end{bmatrix} = \begin{bmatrix}\Lambda_z\end{bmatrix}\begin{bmatrix}p\\ q\\ r\end{bmatrix} \tag{4}$$

where I_ϕ and Λ_z are 3 x 3 matrices. Combining the state equations from (2)-(4) results in

$$\dot{v}(t) = \Lambda\, v(t) + \zeta\, u(t) + \Gamma\, v(t)u(t) + \gamma \tag{5}$$

where the control input u represents the nose-wheel steering angle (γ_3). The rudder deflection (δ_R) is also a control input, but the nose-wheel steering (γ_3) and rudder deflection (δ_R) are related by $\delta_R = a\gamma_3$; ($a = -\frac{26}{7}$). The detailed derivation of equations is given in [2,3] and is omitted here for the sake of brevity.

OPTIMAL CONTROL OF BILINEAR SYSTEMS

Let $v(t_o) = v_o$ be the initial condition and let c^o be the path of fast rollout and turnoff. The problem is to minimize the functional defined on space C by the penalty function

$$J = \int_o^T \langle P[v(t,u,v_o) - c^o], [v(t,u,v_o - c^o]\rangle\, dt$$

$$+ \int_o^T \langle N\, u(t), U(t)\rangle\, dt \tag{6}$$

Here $v(t,u,v_o)$ is the solution of the dynamic equation representing the fast rollout and turnoff condition of B737-100 aircraft given by equation (1). Application of quasilinearization techniques as described by Leondes and Paine [4] has been used to show that the algorithm converges uniformly to the optimum solution.

THEOREM. There exists a unique control w^k given by

$$w^{k+1} = [N]^{-1} \{ \langle [\, I + f_w(w^k, v^k)\,], \psi^{k+1} \rangle \} \tag{7}$$

such that

$$w^{k+1} \leq w^k \text{ and } \psi^{k+1} \leq \psi^k \tag{8}$$

where $w = \zeta\, u + \gamma$, $f_w(w,v) = \Gamma\, v\, (\zeta^T\zeta)^{-1}\zeta^T$, and ψ^k is the solution of adjoint-system equations defined later.

Proof of Theorem 1. Consider equation (1) and define a control vector $\zeta\, u = \tilde{u}$. Here u is scalar representing the nose-wheel steering angle (γ_3). A simple manipulation gives

$$\dot{v} = \Lambda\, v + \tilde{u} + f\,(v,\tilde{u}) + \gamma \tag{9}$$

where $f(v,\tilde{u}) = \Gamma\, v\, (\zeta^T\zeta)^{-1}\, \zeta^T\tilde{u}$. Redefine another variable w such that $w = \tilde{u} + \gamma$ then the transformed equation becomes

$$\dot{v} = (\Lambda - \mu\, \Gamma)\, v + w + f(v.w) \tag{10}$$

where

$$\mu = (\zeta^T\zeta)^{-1}\, \zeta^T\, \gamma\, ;\; \zeta\, u = (w - \gamma)$$

$$f(v,w) = \Gamma\, v\, (\zeta^T\zeta)^{-1}\, \zeta^T\, w$$

In most practical problems, the magnitude of the control variable is usually constrained as $|w| \leq w_{max}$. In the quasilinearization form, it is possible to handle the problem with an additional term, and the recurrence relationship can be written as [4]

$$\dot{v}^{k+1} = (\Lambda - \mu \Gamma) v^k + w + f(v^k, w) + J(v^k)(v^{k+1} - v^k) \tag{11}$$

where

$$J_{ij}(v^k) = \left[\frac{\partial f_i}{\partial v_j} + \frac{\partial f_i}{\partial w} \cdot \frac{\partial w}{\partial v_j}\right] \Bigg|_{v^k} = [J_{ij}]$$

$i,j = 1,2,\ldots,n$; $\frac{\partial w}{\partial v_j} \mid_{v^n} = o$ on the boundary, and $w = \pm w_{max}$, and, off the boundaries w and its derivatives can be determined by maximizing H^o with respect to control variable w. We shall use the formalism of the maximum principle of Pontryagin [5] rather than the calculus of variations to attack this problem because it handles the degeneracy which is built into the problem in a natural way. Pontryagin's maximum principle asserts that if $w(\cdot)$ is an optimizing control then there exists a matrix ψ^{k+1} such that

$$\dot{\psi}^{k+1} = - J(v^k)\psi^{k+1} - P(v^{k+1} - C^o) - \alpha^k \tag{12}$$

and H^o, defined by

$$\begin{aligned} H^o(\cdot) = &< \psi^{k+1}, [(\Lambda - \mu\Gamma) v^k + w + f(v^k, w) + J(v^k)(v^{k+1} - v^k)] > \\ &+ 1/2 <(v^{k+1} - c^o), P(v^{k+1} - c^o) > + 1/2 < w, Nw > + \alpha^k(v^{k+1} - v^k) \end{aligned} \tag{13}$$

is minimized to w by the optimal control. Thus we have the optimal control given by

$$w_o^{k+1} = < - \psi^{k+1}, N^{-1} [I + \Gamma v^k (\zeta^T\zeta)^{-1} \zeta^T] > \tag{14}$$

This choice of w_o^{k+1} gives a pair of differential equations with the split boundary condition

$$\frac{d}{dt}\begin{bmatrix} v^{k+1} \\ \psi^{k+1} \end{bmatrix} = \begin{bmatrix} J(v^k) & -F(v^k) \\ -P & -J^T(v^k) \end{bmatrix} \begin{bmatrix} v^{k+1} \\ \psi^{k+1} \end{bmatrix} + \begin{bmatrix} (\Lambda - \mu\Gamma - J(v^k) & 0 \\ 0 & 0 \end{bmatrix} \begin{bmatrix} v^k \\ \psi^k \end{bmatrix} + \begin{bmatrix} f(v^k, w_o^k) \\ Pc^o - \alpha^k \end{bmatrix} \tag{15}$$

where

$$F(v^k) = N^{-1} [I + \Gamma v^k (\zeta^T\zeta)^{-1} \zeta^T] .$$

Denote the second matrix in equation (15) with a matrix [G]. The solution of equation (15) can be written as

$$\begin{bmatrix} v^{k+1}(t) \\ \psi^{k+1}(t) \end{bmatrix} = \phi(t,o) \begin{bmatrix} v^{k+1}(o) \\ \psi^{k+1}(o) \end{bmatrix} + \int_o^t \phi(t,\tau)\, G \begin{bmatrix} v^k(\tau) \\ w^k(\tau) \end{bmatrix} d\tau$$

$$+ \int_o^t \phi(t,\tau) \begin{bmatrix} f(v^k,w^k) \\ \\ Pc^o - \alpha^k \end{bmatrix} d\tau \tag{16}$$

Here $\phi = \begin{bmatrix} \phi_{11} & \phi_{12} \\ \phi_{21} & \phi_{22} \end{bmatrix}$ is the transition matrix associated with the first matrix in equation (15). From the transversality condition, $\psi^{k+1}(T) = 0$. Also, without loss of generality, it is possible to set $v^{k+1}(o) = 0$ [See 6]. Then solving for $\psi^{k+1}(o)$ from the transversality condition, we have

$$\psi^{k+1}(o) = - \phi_{22}^{-1}(T,o) \{\int_o^T [\phi_{21}(T,\tau) \quad \phi_{22}(T,\tau)]$$

$$\cdot G \begin{bmatrix} v^k \\ \\ w^k \end{bmatrix} d\tau + \int_o^T [\phi_{21}(T,\tau) \quad \phi_{22}(T,\tau)] \begin{bmatrix} f(\cdot) \\ \\ pc^o - \alpha^k \end{bmatrix} d\tau \tag{17}$$

Using the boundary conditions (17) with equation (16), we have,

$$\begin{bmatrix} v^{k+1}(t) \\ \\ \psi^{k+1}(t) \end{bmatrix} = \int_o^T \tilde{\phi}(t,\tau)\, G \begin{bmatrix} v^k(\tau) \\ \\ w^k(\tau) \end{bmatrix} d\tau + \int_o^T \tilde{\phi}(t,\tau) \begin{bmatrix} f(\cdot) \\ \\ Pc^o - \alpha^k \end{bmatrix} \tag{18}$$

where

$$\tilde{\phi}(t,\tau) = \phi(t,\tau) - \begin{bmatrix} \phi_{12}(t,o) \\ \\ \phi_{22}(t,o) \end{bmatrix} \phi_{22}^{-1}(T,o)[\phi_{21}(T,\tau) \quad \phi_{22}(T,\tau)]$$

In order to prove the convergence of equation (18) it is necessary to show that (v^k, ψ^k) converges uniformly to the optimum solution as $k\to\infty$. To show the convergence of (v^k,ψ^k) we utilize the conditions that $\frac{\partial H^o}{\partial \alpha^k} = 0$, and together with equation (14) forms the recursive relation as

$$\begin{bmatrix} v_o^{k+1} \\ \\ w_o^{k+1} \end{bmatrix} = \begin{bmatrix} I & o \\ \\ o & F^k(w^k, \psi^k) \end{bmatrix} \begin{bmatrix} v^k \\ \\ \psi^k \end{bmatrix} \tag{19}$$

Substituting for $(v^k, \psi^k)^T$ from equation (16), we have

$$\begin{bmatrix} v_o^{k+1} \\ \\ w_o^{k+1} \end{bmatrix} = \int_o^T \begin{bmatrix} I & o \\ \\ o & F^k(\cdot) \end{bmatrix} \tilde{\phi}(t,\tau)\, G \begin{bmatrix} v^k(\tau) \\ \\ w^k(\tau) \end{bmatrix} d\tau$$

$$+ \int_o^T \begin{bmatrix} T & o \\ \\ o & F^k(\cdot) \end{bmatrix} \tilde{\phi}(t,\tau) \begin{bmatrix} f^k(\cdot) \\ \\ Pc^o - \alpha^k \end{bmatrix} d\tau \tag{20}$$

Denote $L^k = (v^k,w^k)^T$. If A is the shift operator such that $A\,L^k = (v^{k+1},w^{k+1})$, then equation (20) becomes

$$A\,L^k = \int_0^T M^k(\cdot)\,\phi(t,\tau)\,G\,L^k(\tau)d\tau \;+\; \int_0^T M^k(\cdot)\,\phi(t,\tau)\begin{bmatrix} f^k(\cdot) \\ Pc^o - \alpha^k \end{bmatrix} d\tau \qquad (21)$$

where $M^k = \begin{bmatrix} I & o \\ o & F^k(\cdot) \end{bmatrix}$. Next, define a space C as follows:

Let $L_i(t)$ be continuous functions such that $L_i(o) = L_{oi}$, $L_i(T) = L_f$, and for some constant K, $\max_{i,t} |\; L_i(t) - L_{Ti}(t)| < K$. Then it is an elementary exercise to show the following:

Lemma 1. (C, ρ) is a complete metric space with the metric defined by

$$\rho(L_1{}^k, L_2{}^k) = \sum_{i=1}^{n} \max_t \| L_i^k(t) - L_{2i}(t) \|$$

If we define the shift operator A on C by equation (21) for $L^k \in C$ then we have the following:

Lemma 2. A is a contraction mapping on C if $R^n \subset D$ and in addition

$$\alpha \overset{def}{=} \{\mu^k + \nu^k \left[\frac{\max_t \left\| \begin{bmatrix} f^k(\cdot) \\ pc^o - \alpha^k \end{bmatrix} \right\|}{\max_t \| L^k(t) \|}\right] T\} < 1$$

where μ^k is the largest eigen-value of the matrix $M^k(\cdot)\,\phi(t,\tau)\,G$, and ν^k of $M^k(\cdot)\,\tilde{\phi}(t,\tau)$.

Proof of Lemma 2.

If $L^k \in C$, $A\,L^k$ is continuous; moreover, since by (21)

$$\|A\,L^k\| \leq \int_0^T \max_t \| M^k(\cdot)\,\phi(t,\tau)\,G\| \cdot \|L^k(\tau)\|\,d\tau$$

$$+ \int_0^T [\max_t \| M^k(\cdot)\,\tilde{\phi}(t,\tau) \begin{bmatrix} f^k(\cdot) \\ pc^o - \alpha^k \end{bmatrix} \| \,d\tau$$

$$\leq \mu^k\, T\, [\max_t \| L^k(t) \|]$$

$$+ \;\nu^k\, T \left[\frac{\max_t \left\| \begin{bmatrix} f^k(\cdot) \\ pc^o - \alpha^k \end{bmatrix} \right\|}{\max_t \| L^k(t) \|}\right] \cdot \;[\max_t \; \|L^k(t)\;\|] \qquad (22)$$

where μ^k is the largest eigen-value of the matrix $M^k(\cdot)\,\phi(t,\tau)\,G$; and ν^k of $\|M^k(\cdot)\,\tilde{\phi}(t,\tau)\|$. Therefore,

$$\|A\,L^k\| \leq \{\mu^k + \nu^k \left[\frac{\max_t \left\| \begin{bmatrix} f^k(\cdot) \\ pc^o - \alpha^k \end{bmatrix} \right\|}{\max_t \|L^k(t)\|}\right]\} \;T. \left[\max_t \|L^k(t)\|\right] \qquad (23)$$

we then have $A\,L^k \in C$

To prove that A is a contraction mapping, let L_1^k, $L_2^k \in C$ then by (22),

$$||A\,L_1^k, A\,L_2^k|| \leq \alpha\,\rho\,(L_1^k, L_2^k)$$

It should be noted that it is possible to choose $\alpha < 1$ since the matrix P and T can be chosen such that $\alpha < 1$. Thus, there is an open interval (o,T) where the algorithm converges uniformly to the optimum solution.

This completes the Proof of Lemma 2. Proof of Theorem 1 follows Lemmas 1 and 2.

ACKNOWLEDGEMENTS

The author wishes to thank Dr. J. S. Creedon and Dr. A. A. Nadkarni, Langley Research Center, NASA for their interest during the course of this work which was done partly at Langley Research Center, NASA, Hampton, VA.

REFERENCES

1. S. Pines and R. M. Hueschew, "Guidance and Navigation for Automatic Landing, Rollout, and Turnoff Using MLS and Magnetic Cable Sensors," 1978.

2. S. Pines, S. F. Schmidt, and F. Mann, "Automated Landing, Rollout, and Turnoff Using MLS and Magnetic Cable Sensors," NASA CR-2907, October 1977.

3. W. B. Horn, and T. J. W. Leland, "Influence of Tire Tread Pattern and Runway Surface Conditions on Braking Friction and Rolling Resistance of a Modern Aircraft Tire," NASA TND-1376, 1962.

4. C. T. Leondes and G. Paine, "Extensions in Quasilinearization Techniques for Optional Control," J. of Optimization Theory and Applications: Vol. 2, No. 5, pp. 316-330, 1968.

5. L. S. Pontryagin, V. Boltyanskii, R. Gamkrelidze and E. Mischenko, "The Mathematical Theory of Optimal Processes, Interscience," New York, 1962.

6. J. D. Pearson, "Dynamic Decomposition Techniques in Optimization Methods for Large Systems," McGraw-Hill, New York, 1971.

NUMERICAL SIMULATION OF AN ALLOY SOLIDIFICATION PROBLEM

A. BERMUDEZ* and C. SAGUEZ**

ABSTRACT

In this paper, the numerical simulation of the solidification of a binary alloy is studied. The mathematical model, we consider, takes into account the existence of a mushy region. By using the technique of quasi-variational inequalities, we prove the existence of a solution for a problem semi-discretized in time and we deduce a numerical algorithm, the convergence of which is demonstrated. Numerical results in two dimensional case, for the alloy Fe-C, are given.

INTRODUCTION

The problem of alloy solidification appears, for example, in steel industry to take into account the different components of the steel (Iron, Carbon, ...). This fact explains physically, the existence of phenomena, as dendrites, mushy region....

Here we study the numerical simulation of a binary alloy. Such problem differs from a classical tub phase. Stefan problem by the following points :

- we have a coupled system, temperature-concentration
- the temperature of solidification is unknown and depends on the concentration
- the concentration is discontinuous along the interface liquid-solid

In a first part, a general mathematical formulation is introduced. In a second part, for a problem semi-discretized in time, the existence of a solution is proved and a numerical algorithm is proposed. In a third part, numerical examples are presented.

1. - MATHEMATICAL FORMULATION

1.1. - Case without mushy region

Let $\Omega_1(t)$ be the liquid zone, $\Omega_2(t)$ the solid zone and S(t) the interface liquid-solid. We denote in $\Omega_i(t)$ by $\theta_i(x,t)$ temperature and by $c_i(x,t)$ the concentration of one of both components. Then we have the following equations :

* University of Santiago de Compostela Spain.
** INRIA, Domaine de Voluceau, 78153 Le chesnay Cédex, France

$$\rho\, k \frac{\partial \theta_i}{\partial t} - \mathrm{div}\,(\alpha_i \text{ grad } \theta_i) = 0 \quad \text{in } \Omega_i(t) \tag{1.1}$$

$$\frac{\partial c_i}{\partial t} - \mathrm{div}\,(\beta_i \text{ grad } c_i) = 0 \quad \text{in } \Omega_i(t) \tag{1.2}$$

Along the free-boundary :

$$(\alpha_1 \overrightarrow{\text{grad } \theta_1} - \alpha_2 \overrightarrow{\text{grad } \theta_2}).\vec{n} = - \rho\, L\, \vec{V}.\vec{n} \text{ on } S(t) \tag{1.3}$$

$$(\beta_1 \overrightarrow{\text{grad } c_1} - \beta_2 \overrightarrow{\text{grad } c_2}).\vec{n} = - (c_1 - c_2)\big|_{S(t)}\, \vec{V}.\vec{n} \text{ on } S(t) \tag{1.4}$$

$$\frac{c_1}{\sigma_1} = \frac{c_2}{\sigma_2} = \theta_1 = \theta_2 \text{ on } S(t) \tag{1.5}$$

Boundary conditions (for example)

$$\alpha \frac{\partial \theta}{\partial n}\Big|_\Sigma = q \quad ; \frac{\partial c}{\partial n}\Big|_\Sigma = 0 \tag{1.6}$$

Initial conditions :

$$\theta(x,o) = \theta_0(x) \quad ; \quad c(x,o) = c_0(x). \tag{1.7}$$

As in A.B. CROWLEY - J.R.OCKENDON [2], G.J. FIX [3], we introduce the new variable W defined by :

$$W = \frac{c_i}{\sigma_i} \quad \text{in } \Omega_i(t)$$

If we denote, by θ the temperature in $\Omega = \Omega_1 \cup \Omega_2$, we obtain :

$$\left\{ \begin{array}{ll} \dfrac{\partial u}{\partial t} - \mathrm{div}\,(\alpha(\theta,W) \text{ grad } \theta) = 0 \quad ; & u \in H_W(\theta) \quad (1.8) \\ \dfrac{\partial v}{\partial t} - \mathrm{div}\,(\gamma(\theta,W) \text{ grad } W) = 0 \quad ; & v \in G_\theta(W) \quad (1.9) \end{array} \right.$$

where :

$$\alpha = \alpha_i \quad \text{in } \Omega_i(t)$$

$$\gamma = - \sigma_i\, \beta_i \text{ in } \Omega_i(t)$$

and

$$H_W(\theta) = \{u \in L^2(\Omega) \mid u(x) \in \mathcal{H}_{W(x)}\, (\theta(x))\}$$

$$G_\theta(W) = \{v \in L^2(\Omega) \mid v(x) \in \mathcal{G}_{\theta(x)}\, (W(x))\}$$

with $\mathcal{H}_r(s)$, $\mathcal{G}_r(s)$ the following maximal monotone graphs :

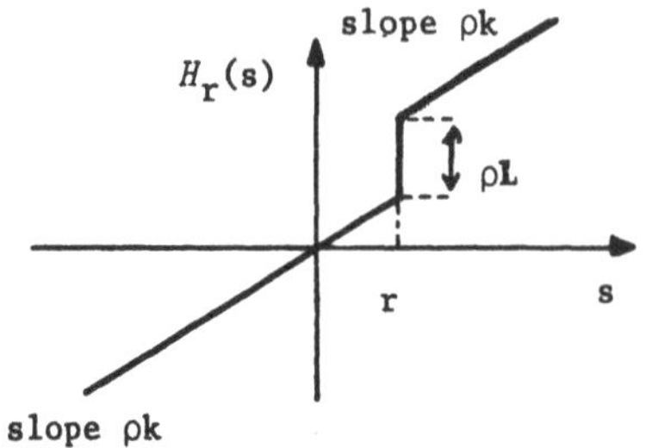

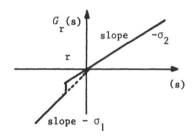

Figure 1.1

Then the liquid domain is defined as :

$$\Omega_1(t) = \{x \mid \theta(x,t) > W(x,t)\}$$

and the solid domain as :

$$\Omega_2(t) = \{x \mid \theta(x,t) < W(x,t)\}$$

1.2. - Case with a mushy region

When a mushy region exists, i.e., when $\mathcal{M}(\{(x,t) \mid \theta(x,t) = W(x,t)\}) > 0)$, the above formulation can be extend with :

$$\gamma = -f\ \beta_1\ \sigma_1 - (1-f)\ \beta_2\ \sigma_2$$

$$\alpha = f\ \alpha_1 + (1-f)\ \alpha_2$$

where f, the liquid fraction is defined by $f = \dfrac{u - \rho k\theta}{\rho L}$.

This formulation is detailed in A. BERMUDEZ, C. SAGUEZ [1].

Remark 1.1. The formulation and the results presented in this paper, can be extend for :

- ρ and k functions of θ,
- general phase-change diagram (in particular non linear).

Remark 1.2. The problem (1.8), (1.9) is a special case of the general system :

$$\begin{cases} \dfrac{\partial u}{\partial t} + A(\theta,W) = h \ ; \ u \in H_W(\theta) \\ \dfrac{\partial v}{\partial t} + B(W,\theta) = \theta \ ; \ v \in G_\theta(W) \\ u(x,o) = u_0 \in H_{W_0}(\theta_0) \ ; \ v(x,o) = v_0(x) \in G_{\theta_0}(W_0) \end{cases}$$

with H_r (s) ; $G_r(s)$ maximal monotone graphs and $A(\theta,W)$, $B(W,\theta)$ non linear operators.

Such systeme is studied in A. BERMUDEZ-C. SAGUEZ [1].

2. - STUDY OF A SEMI-DISCRETIZED PROBLEM

At each step of time,we have to solve the problem :

To find $\{u^{n+1}, v^{n+1}, \theta^{n+1}, W^{n+1}\}$ such that :

$$\left\{\begin{array}{ll} \dfrac{u^{n+1}-u^n}{\Delta t} + A^n.\theta^{n+1} = h^{n+1} \quad ; \quad u^{n+1} \in H_{W^{n+1}}(\theta^{n+1}) & (2.1) \\ \dfrac{v^{n+1}-v^n}{\Delta t} + B^n.W^{n+1} = 0 \quad ; \quad v^{n+1} \in G_{\theta^{n+1}}(W^{n+1}) & (2.2) \end{array}\right.$$

with :

$$(A^n\theta,z)_{V',V} = \int_\Omega \alpha^n \text{ grad } \theta \text{ grad } z d\Omega \quad (V = H^1(\Omega))$$

$$(B^n W,z)_{V',V} = \int_\Omega \gamma^n \text{ grad } W \text{ grad } z d\Omega$$

$$(h^{n+1},z)_{V',V} = \frac{1}{\Delta t} \int_{\partial\Omega} \left(\int_{n\Delta t}^{(n+1)\Delta t} q(\tau)\, d\tau \right) z\, d\Gamma$$

and

$$\alpha^n = f^h \alpha_1 + (1-f^n)\, \alpha_2$$

$$\gamma^n = -f^n \beta_1 \sigma_1 - (1-f^n)\, \beta_2 \sigma_2$$

$$f^n = \frac{u^n - \rho k \theta^n}{\rho L}$$

2.1. - Existence of a solution

Proposition 2.1. : The problem (2.1), (2.2) has a maximal solution and a minimal solution.

Demonstration : We use the techniques of quasi-variational inequalities (L. TARTAR [6], J.L. LIONS [4]).

1) If W^{n+1} is fixed in (2.1), (u^{n+1},θ^{n+1}) is solution of a classical variational inequality and the application $T_1 : W^{n+1} \mapsto \theta^{n+1}$ is non decreasing (i.e. : $W_1 \leq W_2 \Rightarrow T_1 W_1 \leq T_1 W_2$).

Similarly, if θ^{n+1} is fixed in (2.2), we have the existence of a solution (u^{n+1},W^{n+1}) and the application $T_2 : \theta^{n+1} \mapsto W^{n+1}$ is non decreasing.

2) There exist $\underline{W}$ and $\bar{W}$ such that :

$$\forall \theta \in L^2(\Omega) \quad \underline{W} \leq T_2\theta \leq \bar{W} \quad \text{a.e.}$$

3) We define the sequences $\{W_j\}$ and $\{\theta_j\}$ as follows :

$$W_0 = \underline{W} \quad ; \quad \theta_0 = T_1 \underline{W}$$

$$W_{j+1} = T_2\theta_j \ ; \ \theta_{j+1} = T_1 W_j$$

Then by 1) and 2) $\{W_j\}$ and $\{\theta_j\}$ are bounded increasing sequences.

4) By using classical a priori estimates for (2.1), (2.2) we obtain :

$$||\theta_j||^2_{H^1(\Omega)} \leq c \quad ; \ ||W_j||_{H^1(\Omega)} \leq c$$

Then by 3), we deduce that there exists $(\tilde{\theta},\tilde{W})$ such that :

$$\theta_j \rightharpoonup \tilde{\theta} \text{ in } H^1(\Omega) \text{ weakly}$$

$$W_j \rightharpoonup \tilde{W} \text{ in } H^1(\Omega) \text{ weakly.}$$

5) At the limit, we obtain that $(\tilde{\theta},\tilde{W})$ is solution of (2.1), (2.2). Because of the initial value W_0, if (θ,W) denote another solution of (2.1), (2.2), we have :

$$W_j \leq W \quad ; \quad \theta_j \leq \theta \quad \forall j$$

Then $(\tilde{\theta},\tilde{W})$ is a minimal solution.

6) To prove the existence of a maximal solution, we consider the same method, with $W_0 = \bar{W}$, by using a dual problem associated with (2.1), (2.2).

2.2. - Numerical algorithm

From the proposition (2.1), we deduce the following algorithm to compute a minimal solution (θ^{n+1},W^{n+1}) (simular algorithm can be defined for a maximal solution).

1) To compute :

$$\begin{cases} f^n = \chi\left(\dfrac{u^n - \rho k\theta^n}{\rho L}\right) \text{ with } \chi(\nu) = \begin{cases} 0 & \text{if } \nu \leq 0 \\ \nu & \text{if } 0 \leq \nu \leq 1 \\ 1 & \text{if } \nu \leq 1 \end{cases} \\ \alpha^n,\ \gamma^n \end{cases}$$

2) To compute $\underline{W}^{n+1}$ solution of the P.D.E:

$$\frac{\sigma 2}{\Delta t}\underline{W}^{n+1} + B^n\,\underline{W}^{n+1} = \frac{v^n}{\Delta t}$$

3) $j = 0$; $W_0^{n+1} = \underline{W}^{n+1}$.

4) To compute θ_j^{n+1} solution of the variational inequality :

$$\frac{u_j^{n+1}}{\Delta t} + A^n \theta_j^{n+1} = h^{n+1} + \frac{u^n}{\Delta t} \quad ; \quad u_j^{n+1} \in H_{W_j^{n+1}} (\theta_j^{n+1})$$

5) To compute W_{j+1}^{n+1} solution of the variational inequality :

$$\frac{v_{j+1}^{n+1}}{\Delta t} + B^n W_{j+1}^{n+1} = \frac{v^n}{\Delta t} \qquad ; \quad u_{j+1}^{n+1} \in G_{\theta_j^{n+1}} (W_{j+1}^{n+1})$$

6) Test of convergence :

if verified, n = n+1

if not , j = j+1 ; go to 4).

By similar demonstration as in proposition 2.1, we obtain.

Proposition 2.2. : We have a monotone convergence of the above algorithm.

Remark 2.1. : At point 4) and 5), we have to solve a variational inequality. For this, an algorithm, based on the equivalence :

$$u \in H(\theta) \iff u = H^\lambda (\theta+\lambda u) \ \forall\lambda > 0$$

(H maximal monotone operator, H^λ Yosida approximation of H), is used (A. BERMUDEZ, C. SAGUEZ [1], C. SAGUEZ [5]).

3. - NUMERICAL RESULTS

In two-dimensional case, we consider the solidification of the binary alloy Fe-C for a billet (K.H. TASKE, A. GRILL, H. MIGAZAWA, K. SCHWERDTFEGER [7]). The problem is discretizedby finite-elements P_1. The data are the following :

- Domain of integration :

  $\Omega =]0,10[\times]0,10[$, discretized in 200 triangles.

- Physical data :

  $\alpha_1 = \alpha_2 = 0,816$

  $\beta_1 = 0,0001$; $\beta_2 = 0,0000001$

  $\sigma_1 = -0,0128$; $\sigma_2 = -0,00448$

  $\rho = 7,4$; $k = 0,166$; $L = 65,28$;

  $\theta_S = 1536°C$ (temperature of solidification when the concentration of Carbon is equal to zero).

- Boundary conditions :

  The flux q(t) is given by :

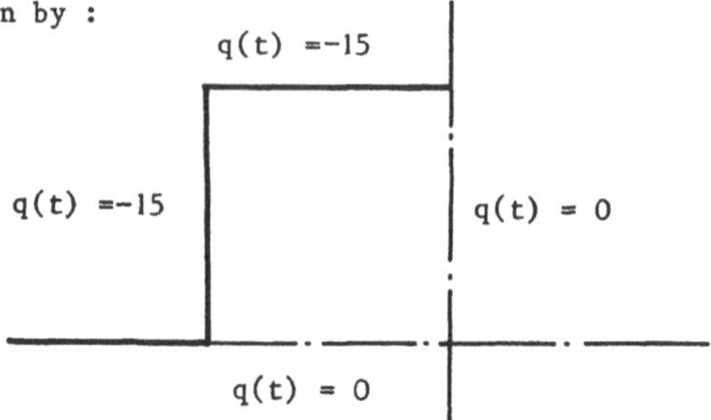

- Initial data :

  $\theta_0(x) = 1536°C$; $c_0(x) = 0,6$.

- Numerical parameter.

  $\Delta t = 10$ s.

Figures (3.1), (3.2) computed profiles of concentration and temperature are presented. The differendes between two curves of temperature is 50°C, and of concentration 0,001. In this case, we obtain the existence of a mushy region. Other results are given in A. BERMUDEZ, C. SAGUEZ [1].

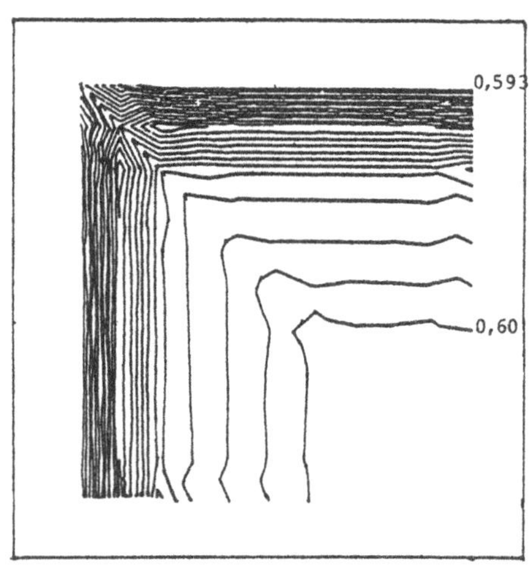

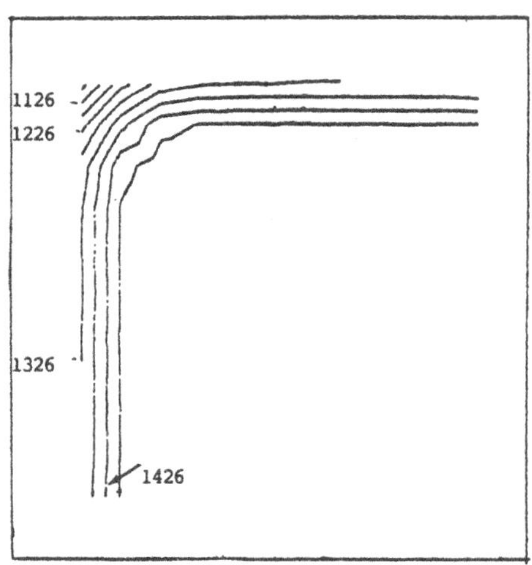

Concentration **Figure 3-1** Temperature

t = 50 s

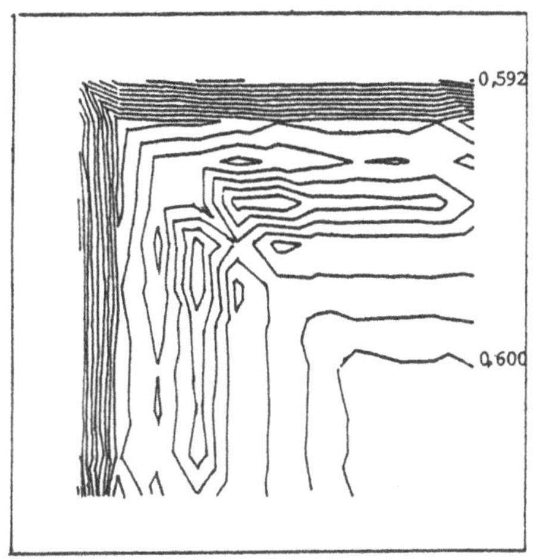

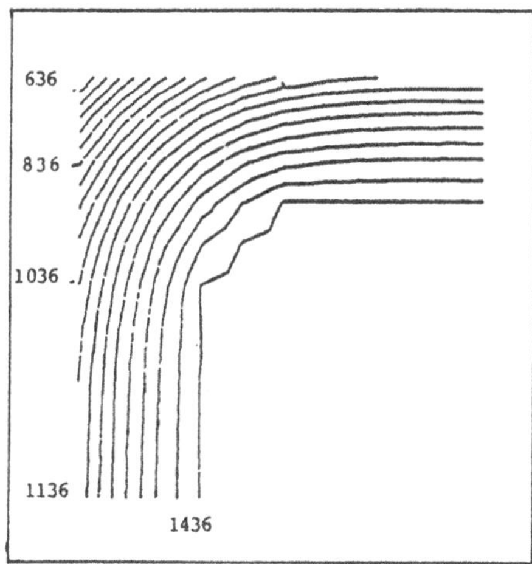

Concentration Figure 3-2 Temperature

t = 150 s

REFERENCES

[1] A. BERMUDEZ, C. SAGUEZ. Etude Numérique d'un problème de solidification d'un alliage (INRIA report).

[2] A.B. CROWLEY, J.H. OCKENDON. On the numerical solution of an alloy solidification problem (Int. J. Heat Mass. Transfer , Vol. 22, pp. 941-947, 1979).

[3] G.J. FIX : Numerical methods for alloy solidification problem (Moving boundary problems. D. Wilson, A.D. Salomon, P.S. Boggs, Ed., Academic Press, 1978).

[4] J.L. LIONS. Cours Collège de France (1974).

[5] C. SAGUEZ. Contrôle optimal de systèmes à frontière libre (Thesis U.T.C 1980).

[6] L. TARTAR. Inequations quasi-variationnelles abstraites (C.R.A.S. Paris 278, pp. 1193-1196, 1974).

[7] K.H. TASKE, A. GRILL, K. MIGAZAWA, K. SCHWERDTFEGER. Macrosegration in strand cast steel —Computation of concentration profiles with a diffusion model. (Max Planck Institut paper 1407, 1981).

THE THRESHOLD PROBLEM FOR A FITZHUGH-NAGUMO SYSTEM

C. Corduneanu
Department of Mathematics
The University of Texas at Arlington
Arlington, Texas 76019

A simplified mathematical model for the conduction of the nerve impulses along the axon is provided by the FitzHugh-Nagumo system

(FN) $$v_t = v_{xx} + f(v) - \langle c,u \rangle, \quad u_t = \sigma v - \gamma u ,$$

where $c = (c_1,c_2,c_3)$, $c_i > 0$, $i = 1,2,3$, $\sigma = \mathrm{col}(\sigma_1,\sigma_2,\sigma_3)$, $\sigma_i > 0$, $i = 1,2,3$, $\gamma = \mathrm{diag}(\gamma_1,\gamma_2,\gamma_3)$, $\gamma_i > 0$, $i = 1,2,3$, and $\langle .,. \rangle$ denotes the scalar product in R^3. The maps $v: [0,a] \times R_+ \longrightarrow R$ and $u: [0,a] \times R_+ \longrightarrow R^3$ are the unknown quantities in (FN), while $f: R \longrightarrow R$ is a given continuous function satisfying the sign condition

(1) $$v\, f(v) < 0 \quad \text{for} \quad v \neq 0 ,$$

in a neighborhood of the origin.

The following initial conditions are usually associated to (FN):

(IC) $$v(x,0) = \phi(x), \; u(x,0) = \psi(x), \; x \in [0,a] ,$$

where ϕ and ψ satisfy certain restrictions to be specified below (at least, continuity).

We will also assume that one of the following boundary value conditions is associated with (FN) and (IC):

$(BVC)_1$ $$v(0,t) = 0, \; v(a,t) = 0, \; t > 0 ,$$

$(BVC)_2$ $$v(0,t) = 0, \; v_x(a,t) + k\, v(a,t) = 0, \; (k > 0), \; t > 0$$

$(BVC)_3$ $$v_x(0,t) - hv(0,t) = 0, \; v_x(a,t) + kv(a,t) = 0, \; t > 0 ,$$

with $h > 0$, $k > 0$ in $(BVC)_3$.

Though we do not impose any boundary value condition on u, from the second equation of (FN), (IC), and $(BVC)_i$, $i = 1,2,3$, one can easily find what kind of conditions u satisfies.

The history of equations (FN) starts with the paper of Hodgkin and Huxley [4], and continues with numerous contributions that finally led to the above simplified form. In [3], Hastings gives a conspicuous account on this topic, and discusses the physical meaning of the FitzHugh-Nagumo equations in neurobiology. Without appealing to details, we mention that $v = v(x,t)$ stands for the electric potential, while $u = u(x,t)$ is a vector measuring the ionic permeability of the nerve membrane. As pointed out in

[9], where simplified versions of the original equations found by Hodgkin and Huxley are called "caricatures," such equations do provide a satisfactory description of the real phenomena involved in the propagation of the nerve impulses along the axon.

Recently, Rauch and Smoller [6], and Schonbek [7-8] dealt systematically with the qualitative investigation of FitzHugh-Nagumo equations, under various assumptions on f and the data (mainly on initial data, due to the fact they consider the equations on the whole real line R, or on the semi-axis $R_+ = [0,\infty)$).

In this paper, our interest is concentrated on the "threshold problem" for the system (FN), with initial data (IC), and one of the boundary value conditions $(BVC)_i$, $i = 1,2,3$. Unlike the authors quoted above [6-8], we assume the nerve to be of arbitrary finite length, which seems to be a more realistic assumption. Using only classical results, including an idea due to Popov [5] in regard to the stability of automatic control systems, a positive answer to the "threshold problem" as formulated below will be provided, under milder assumptions than those encountered in previous papers on this subject.

According to the physical evidence, a nerve is not really triggered, unless the level of excitation reaches a certain "threshold". In mathematical terms, this fact could be described by requiring that any solution $(v(x,t), u(x,t))$ of (FN), for which the initial data ϕ and ψ are small enough, tend to zero as $t \longrightarrow \infty$:

$$\lim_{t \longrightarrow \infty} (|v(x,t)| + \|u(x,t)\|) = 0, \quad \text{uniformly in } x \in [0,a]. \tag{2}$$

In other words, the problem under consideration is to prove the asymptotic stability of the solution $v = 0$, $u = 0$ of (FN), under boundary value conditions of the form listed above.

Let us point out that the condition $(BVC)_i$, $i = 1,2,3$, guarantee the fact that the maximum of any solution of the homogeneous equation $v_t = v_{xx}$ is attained at $t = 0$.

The positive answer to this problem will be obtained in this paper, based on "Liapunov's function technique", as well as on the variation of constants formula for parabolic equations.

A candidate for a "Liapunov's function" is the "energy functional"

$$W(v,u) = \frac{1}{2}\int_0^a (v^2 + \sum_{i=1}^{3} c_i\sigma_i^{-1}u_i^2)dx , \tag{3}$$

for which one obtains easily

$$\frac{d}{dt} W(v,u) = \int_0^a vv_{xx}dx + \int_0^a vf(v)dx - \sum_{i=1}^{3} c_i\gamma_i\sigma_i^{-1} \int_0^a u_i^2dx . \tag{4}$$

Since

$$\int_0^a vv_{xx}dx = vv_x \Big|_0^a - \int_0^a v_x^2dx , \tag{5}$$

taking into account the boundary value conditions $(BVC)_i$, $i = 1,2,3$, one can write

$$\int_0^a vv_{xx}dx = -kv^2(a,t) - hv^2(0,t) - \int_0^a v_x^2 dx , \tag{6}$$

which obviously holds true for any $(BVC)_i$, $i = 1,2,3$.

Taking into account condition (1), and assuming $v(x,t)\, f(v(x,t)) \leq 0$ on $[0,a]\times R_+$, one obtains from (4) by integrating in t:

$$\frac{1}{2}\int_0^a (v^2 + \sum_{i=1} c_i\sigma_i^{-1}u_i^2)dx +$$

$$\int_0^t\left\{\int_0^a (|vf(v)| + v_x^2 + \sum_{i=1}^3 c_i\gamma_i\sigma_i^{-1}u_i^2)dx + hv^2(0,t) + kv^2(a,t)\right\}dt \tag{7}$$

$$\leq \frac{1}{2}\int_0^a (\phi^2 + \sum_{i=1}^3 c_i\sigma_i^{-1}\psi_i^2)dx, \ t \geq 0 .$$

The inequality (7) has several implications, among which we mention first

$$\int_0^\infty\int_0^a (v^2 + \|u\|^2)dxdt < +\infty . \tag{8}$$

Indeed, the right hand side of (7) is constant, while the inequality holds true for any $t \geq 0$. This obviously implies the square integrability of u on $[0,a]\times R_+$. As far as v is concerned, let us remark first that one can write

$$v(x,t) = v(0,t) + \int_0^x v_x(\xi,t)d\xi, \ 0 \leq x \leq a, \ t \geq 0 ,$$

from which one derives

$$v^2(x,t) \leq 2v^2(0,t) + 2a\int_0^a v_x^2(\xi,t)d\xi . \tag{9}$$

On behalf of $(BVC)_i$, $i = 1,2,3$, and (7), one concludes that v is also square integrable in $[0,a]\times R_+$, hence (8). Inequality (8) is basic in establishing the uniform convergence of the series resulting from the variation of constants formula for the first equation (FN).

Let us remark that the solution of (FN), (IC), $(BVC)_i$, for any $i = 1,2,3$, can be represented as

$$v(x,t) = v_0(x,t) + \sum_{k=1}^\infty \left(\int_0^t e^{-\lambda_k(t-s)}f_k(s)ds\right)v_k(x) , \tag{10}$$

where λ_k, $v_k(x)$ are the eigenvalues and eigenfunctions (normalized) of the Sturm-Liouville problem associated to the second order equation $v'' + \lambda v = 0$ in $[0,a]$, with boundary value conditions generated by $(BVC)_i$, $i = 1,2,3$, respectively. For instance, in case $i = 1$ one has $\lambda_k = k^2\pi^2 a^{-2}$, $k = 1,2,\ldots$ with $v_k(x) = \sin k\pi a^{-1}x$, $k = 1,2,\ldots$

Further, the functions $f_k(t)$ are given by

(11) $$f_k(t) = \int_0^a [f(v) - \langle c,u\rangle]v_k(x)dx,\ k = 1,2,\ldots$$

while $v_0(x,t)$ denotes the solution of the homogeneous equation $v_t - v_{xx} = 0$, with initial conditions (IC), and one of the boundary value conditions $(BVC)_i$, $i = 1,2,3$ (of course, $v_0(x,t)$ is different for different i's, $i = 1,2,3$).

Since the maximum principle gives

(12) $$|v_0(x,t)| \leq \sup |\phi(x)|,\ (x,t) \in [0,a]\times R_+ ,$$

and the series expansion

(13) $$v_0(x,t) = \sum_{k=1}^{\infty} b_k e^{-\lambda_k t} v_k(x)$$

converges uniformly in $[0,a] \times [\varepsilon,\infty]$, for every $\varepsilon > 0$, one obtains from (13)

(14) $$\lim_{t \to \infty} v_0(x,t) = 0,\ \text{uniformly in } x \in [0,a] .$$

It will suffice to get similar properties for the series appearing in the right hand side of (10).

In order to obtain the uniform convergence of the series in the right hand side of (10), in $[0,a]\times R_+$, it will suffice to assume that f is a map verifying an inequality of the form

(15) $$|f(v)| \leq K|v|,\ K > 0,$$

in a neighborhood of $v = 0$. Then we can modify f to be defined on the whole real axis, such that (15) takes place, even when the given f does not verify such an inequality outside that neighborhood. As far as "small" solutions are concerned, this procedure of modifying f outside the neighborhood of $v = 0$ does not change the local behavior around $v = 0$, $u = 0$. For instance, for one of the most often encountered choices $f(v) = - v(v-a)(v-b)$, with $0 < a < b$, one can always choose $K > ab$, which will guarantee the validity of (15) in a fairly large neighborhood of the origin.

Let us remark now that

(16) $$\sum_{k=1}^{\infty} f_k^2(t) = \int_0^a [f(v) - \langle c,u\rangle]^2 dx,\ t \in R_+,$$

due to the fact the system of eigenvalues $\{v_k(x);\ k = 1,2,..\}$ is complete in $L^2(0,a)$. But

(17) $$|f(v) - \langle c,u\rangle |^2 \leq A(|f(v)|^2 + \|u\|^2) ,$$

where $A = \max(2,\|c\|^2)$, and from (8), (15) there results the right hand side of (16) is integrable on R_+. This leads to

$$(18)\qquad \sum_{k=1}^{\infty}\int_0^{\infty} f_k^2(t)dt = \int_0^{\infty}\left\{\int_0^a [f(v) - \langle c,u\rangle]^2 dx\right\}dt .$$

Consider now the general term of the series in the right hand side of (10), and apply Schwartz inequality to the integral, after taking into account the elementary inequality $|ab| \le 2^{-1}(a^2 + b^2)$. One obtains for every $k = 1,2,3,\ldots$

$$(19)\qquad \left|\int_0^t e^{-\lambda_k(t-s)} f_k(s)ds\right| \left|v_k(x)\right| \le \frac{1}{4}\int_0^{\infty} f_k^2(t)dt + \frac{1}{2}\frac{\phi_k^2(x)}{\lambda_k} ,$$

which proves the uniform (and absolute) convergence of the series under consideration in $[0,a]\times R_+$, as it is easily seen from (18), (19), and the Mercer theorem [2]. In general, the convergence is proven in $[0,a]\times[0,T]$, for any $T > 0$ (see [10]).

We can now prove that

$$(20)\qquad \lim_{t\to\infty}\sum_{k=1}^{\infty}\left(\int_0^t e^{-\lambda_k(t-s)} f_k(s)ds\right)v_k(x) = 0 ,$$

uniformly with respect to $x \in [0,a]$. Indeed, from $f_k \in L^2(0,\infty)$ one derives (see [1])

$$(21)\qquad \lim_{t\to\infty}\int_0^t e^{-\lambda_k(t-s)} f_k(s)ds = 0,\ k = 1,2,\ldots ,$$

and since the series is uniformly convergent in $[0,a]\times R_+$, (20) follows from (21).

From (13), (14), and (21) one finds out that $v(x,t)$, given by (10), satisfies $v(x,t) \to 0$ as $t \to \infty$, uniformly in $x \in [0,a]$. Finally, from the second equation (FN) and (IC) one obtains

$$(22)\qquad u(x,t) = e^{-\gamma t}\psi(x) + \int_0^t e^{-\gamma(t-s)}\sigma v(x,s)ds ,$$

which implies (see [1]) $u(x,t) \to 0$ as $t \to \infty$, uniformly in $x \in [0,a]$.

Property (2) is thus established.

There remains to prove

$$(23)\qquad \sup_{x\in[0,a],\ t\ge 0}(|v(x,t)| + \|u(x,t)\|) \to 0, \text{ as } \sup_{x\in[0,a]}(|\phi| + \|\psi\|) \to 0 .$$

Taking (22) into account, it suffices to show that

$$(24)\qquad \sup|v(x,t)| \to 0, \text{ as } \sup(|\phi| + \|\psi\|) \to 0 ,$$

which can be easily obtained using (10). Indeed, $v_0(x,t)$ verifies (24), according to (12). As far as the sum of the series in the right hand side of (10) is concerned, we notice that

$$(25)\qquad \int_0^{\infty}|f_k(t)|^2 dt \le A\int_0^{\infty}\left\{\int_0^a (K^2v^2 + \|u\|^2)dx\right\}dt \le A\int_0^a (\phi^2 + \|\psi\|^2)dx ,$$

if we take into account (15) - (18). Therefore, all terms in the (uniformly convergent!) series tend to zero, as $\sup(|\phi| + \|\psi\|) \to 0$. Hence, (24) holds true.

The above discussion leads to the following result, which provides the answer to the "threshold problem" formulated at the beginning of this paper.

Theorem Any classical solution of the system (FN), (IC), and $(BVC)_i$ $(i = 1,2,3)$, corresponding to small initial data, satisfies conditions (2) and (23), provided $f: R \longrightarrow R$ verifies (1) and (15) in a neighborhood of the origin. In particular, (1) and (15) are verified by any f, such that $f(0) = 0$, $f'(0) < 0$.

Remark An inspection of the proof conducted above shows that (FN) can be replaced by the more general system

$$(FN)_1 \qquad v_t = (p(x)v_x)_x - q(x)v + f(v) - \langle c,u \rangle , \; u_t = \sigma v - \gamma u ,$$

where $p(x) > 0$ on $[0,a]$ is continuously differentiable, and $q(x) \geq 0$ is continuous. The meaning of other quantities involved in $(FN)_1$ remains the same as above.

The existence problem of solutions whose behavior is investigated in this paper requires a lengthier discussion, and will be dealt with in another paper.

REFERENCES

1. C. Corduneanu, Integral Equations and Stability of Feedback Systems. Academic Press, New York, 1973.

2. C. Corduneanu, Principles of Differential and Integral Equations. Chelsea Publ. Co., New York, 1977.

3. S. Hastings, Some mathematical problems from neurobiology, Am. Math. Monthly, 82 (1975), 881-895.

4. A. L. Hodgkin, A. F. Huxley, A quantitative description of membrane current and its application to conduction and excitation in nerves. J. Physiology, 117 (1952), 500-544.

5. V. M. Popov, Noi criterii de stabilitate pentru sistemele automate neliniare, Stud. Cerc. Energ. 10 (1960), 159-171.

6. J. Rauch, J. Smoller, Qualitative theory of the FitzHugh-Nagumo equations. Advances in Math., 27 (1978), 12-14.

7. M. E. Schonbek, Boundary value problems for the FitzHugh-Nagumo Equations, Journal Diff. Equations, 30 (1978), 119-147.

8. M. E. Schonbek, A priori estimates of higher order derivatives of solutions of the FitzHugh-Nagumo equations, JMAA, 82 (1981), 553-565.

9. L. A. Segel (Editor), Mathematical models in molecular and cellular biology (Ch. 6). Cambridge Univ. Press, Cambridge, 1979.

10. H. F. Weinberger, Partial Differential Equations (A first course). Blaisdell, New York, 1965.

OPTIMAL DESIGN OF A THERMAL DIFFUSER WITH MINIMUM WEIGHT*

M.C.Delfour
Centre de recherche de mathématiques appliquées
Université de Montréal
C.P.6128, Succ.A
Montréal, Québec
Canada H3C 3J7

Guy Payre
Département de Génie Chimique
Université de Sherbrooke
Shebrooke, Québec
Canada J1K 2R1

J.-P. Zolésio
Département de Mathématiques
Université de Nice
06034 - Nice - Cedex
France

1. INTRODUCTION. This paper is concerned with the design of a thermal diffuser of minimum weight with a priori specifications on the input and output thermal power fluxes.This problem arises in connection with the use of high-power solid state devices (HPSSD's) in future communication satellites. The specifications for this diffuser came from the Center for Research in Communications (CRC) in Canada[1].

"An HPSSD dissipates a large amount of thermal power (typ. > 50W) over a relatively small mounting surface (typ. 1.25cm^2). Yet its junction temperature is required to be kept moderately low (typ. 110°C). The thermal resistance from the junction to the mounting surface is known for any particular HPSSD (typ. 1°C/W), so that the mounting surface is required to be kept at a lower temperature than the junction (typ. 60°C).

In a space application the thermal power must ultimately be dissipated to the environment by the mechanism of radiation. However to radiate large amounts of thermal power at moderately low temperatures, correspondingly large radiating areas are required. Thus we have the requirement to efficiently spread the high thermal power flux (TPF) at the HPSSD source (typ. 40W/cm^2) to a low TPF at the radiator (typ..04W/cm^2) so that the source temperature is maintained at an acceptably low level (typ. < 60°C at mounting surface). The efficient spreading task is best accomplished using heatpipes, but the snag in the scheme is that heatpipes can accept only a limited maximum TPF from a source (typ. max. 4W/cm^2).

Hence we are led to the requirement for a thermal diffuser. This device is inserted between the HPSSD and the heatpipes, and reduces the TPF at the source (typ. > 40W/cm^2) to a level acceptable to the heatpipes (typ. max. 4W/cm^2). The heatpipes then sufficiently spread the heat over large space radiators, reducing the TPF from a level at the diffuser (typ. max. 4W/cm^2) to that at the radiator (typ. .04W/cm^2). This scheme of heat spreading is depicted in Figure 1.

It is the design of the thermal diffuser which is the problem at hand. We may assume that the HPSSD presents a uniform thermal power flux to the diffuser at the HPSSD/diffuser interface. Heatpipes are essentially isothermalizing devices, and we may assume that the diffuser/heatpipes interface is indeed isothermal. Any other

* This research was supported by Canada Natural Sciences and Engineering Research Council Strategic Grants G-0573 and G-0654 (Communications) and FCAC Grant EQ-252 from the "Ministère de l'Education du Québec".

1 The statement of the problem and Figure 1 have been graciously provided by Dr. V.A.Wehrle of CRC.

surfaces of the diffuser may be treated as adiabatic."

Some early results were presented by Ph. Destuynder [1] with the requirement that the temperature at every point of the diffuser be less than a specified critical temperature. More details can be found in Delfour-Payre-Zolesio [2].

2. STATEMENT OF THE PROBLEM. We assume that the thermal diffuser is a volume Ω symmetrical about the z-axis (cf. Figure 2) whose boundary surface is made up of three regular pieces: the mounting surface Σ_1 (a disk perpendicular to the z-axis with center in $(r,z) = (0,0)$), the lateral adiabatic surface Σ_2 and the interface Σ_3 between the diffuser and the heatpipes saddle (a disk perpendicular to the z-axis with center in $(r,z) = (0,L)$).

The temperature distribution T over this volume Ω is the solution of

$$\Delta T = 0 \text{ in } \Omega, \quad k\frac{\partial T}{\partial n} = q_{in} \text{ on } \Sigma_1, \quad k\frac{\partial T}{\partial n} = 0 \text{ on } \Sigma_2, \quad T = T_3 \text{ on } \Sigma_3, \tag{2.1}$$

where ΔT is the Laplacian of T, n is the outward normal to the boundary surface Σ and

$$\frac{\partial T}{\partial n} = \nabla T \cdot n \quad (\nabla T = \text{gradient of } T) \tag{2.2}$$

is the normal derivative on Σ. The parameter k is the thermal conductivity (typ. 1.8W/cm×°C); the inward thermal power flux on Σ_1, the temperature on Σ_3 and the radius of the mounting surface Σ_1 are assumed to be constant and equal to $q_{in} \geq 0$, T_3 and R_0, respectively.

The diffuser is assumed to be solid without interior hollows or cutouts. The class of shapes for the diffuser is characterized by the design parameter $L \geq 0$ and the positive function $R(z)$, $0 \leq z \leq L$, with $R(0) = R_0 > 0$. They are volumes of revolution Ω about the z-axis generated by the surface A between the z-axis and the function $R(z)$ (cf. Figure 2a), that is

$$\Omega = \{(x,y,z) \mid 0 \leq z \leq L, \ x^2+y^2 \leq R^2(z)\}. \tag{2.3}$$

So the shape of Ω is completely specified by the length L and the function R on the interval $[0,L]$.

Assuming that the diffuser is made up of a homogeneous material of uniform density (no hollows) the design objective is to minimize the volume subject to a uniform constraint on the outward thermal power flux at the interface Σ_3 between the diffuser and the heatpipes saddle:

$$J(\Omega) = \pi\int_0^L R^2(z)dz, \quad \sup_{p \in \Sigma_3} -k\frac{\partial T}{\partial z}(p) \leq q_{out}, \tag{2.4}$$

where q_{out} is a specified positive constant.

3. REFORMULATION OF THE PROBLEM. Introduce the following changes of variables

$$\zeta = z/L, \quad 0 \leq \zeta \leq 1, \quad \xi_1 = x_1/R_0, \quad \xi_2 = x_2/R_0 \tag{3.1}$$

and the scaled shape parameter and shape function

$$\tilde{L} = L/R_0, \quad \tilde{\rho}(\zeta) = R(L\zeta)/R_0, \quad 0 \leq \zeta \leq 1. \tag{3.2}$$

Under the above change of variables the domain Ω is transformed into a ζ-axisymmetrical domain

(3.3) $$\tilde{\Omega} = \{(\xi_1,\xi_2,\zeta)\,|\,0 < \zeta < 1,\ \xi_1^2+\xi_2^2 < \tilde{\rho}(\zeta)^2\}$$

which is generated by the revolution of the surface

(3.4) $$D = \{(\rho,\zeta)\,|\,0 < \zeta < 1,\ 0 < \rho < \tilde{\rho}(\zeta)\}$$

about the ζ-axis (cf. Figure 2b). The domain $\tilde{\Omega}$ is completely specified by the new <u>shape function</u> $\tilde{\rho}$ subject to the conditions $\tilde{\rho}(0)=1$ and $\tilde{\rho}(\zeta)\geq 0$. We shall denote by $\tilde{\Sigma}$, $\tilde{\Sigma}_1$, $\tilde{\Sigma}_2$ and $\tilde{\Sigma}_3$ the transformed surfaces Σ, Σ_1, Σ_2 and Σ_3.

3.1. <u>Equations for the scaled temperature</u>. Define the scaled temperature

(3.5) $$y(\xi_1,\xi_2,\zeta) = T(R_0\xi_1,R_0\xi_2,L\zeta)k/(Lq_{in})$$

in the new variables (ξ_1,ξ_2,ζ) or in cylindrical coordinates

(3.5a) $$y(\rho,\zeta) = T(R_0\rho,L\zeta)k/(Lq_{in}),\quad \rho = \sqrt{\xi_1^2+\xi_2^2}.$$

It is readily seen that y is the solution of the variational equation

(3.6) $$y \in H_0(\tilde{\Omega}),\quad \int_{\tilde{\Omega}}[\tilde{L}^2(\frac{\partial y}{\partial \xi_1}\frac{\partial v}{\partial \xi_1} + \frac{\partial y}{\partial \xi_2}\frac{\partial v}{\partial \xi_2}) + \frac{\partial y}{\partial \zeta}\frac{\partial v}{\partial \zeta}]d\tilde{\Omega} = \int_{\tilde{\Sigma}_1} v d\tilde{\Sigma}.$$

for all v in the closed linear subspace $H_0(\tilde{\Omega})$ of $H^1(\tilde{\Omega})$:

(3.7) $$H_0(\tilde{\Omega}) = \{v \in H^1(\tilde{\Omega})\,|\,v|_{\tilde{\Sigma}_3} = 0\}.$$

It is readily seen that the solution of (3.6) is the solution of the following boundary value problem:

(3.8) $$A(y) = -[\tilde{L}^2(\frac{\partial^2 y}{\partial \xi_1^2} + \frac{\partial^2 y}{\partial \xi_2^2}) + \frac{\partial^2 y}{\partial \zeta^2}] = 0 \text{ in } \tilde{\Omega},\ \frac{\partial y}{\partial \nu_A}|_{\tilde{\Sigma}_1} = 1,\ \frac{\partial y}{\partial \nu_A}|_{\tilde{\Sigma}_2} = 0,\ y|_{\tilde{\Sigma}_3} = 0,$$

where the conormal derivative of y on the boundary $\tilde{\Sigma}$ associated with the operator A and the unit outward vector $\nu = (\nu_1,\nu_2,\nu_\zeta)$ is defined as

(3.9) $$\frac{\partial y}{\partial \nu_A} = \tilde{L}^2(\nu_1\frac{\partial y}{\partial \xi_1} + \nu_2\frac{\partial y}{\partial \xi_2}) + \nu_\zeta\frac{\partial y}{\partial \zeta}.$$

Notice that for $0 \leq r \leq R(L)$ and $\rho = r/R_0$

(3.10) $$\frac{k}{q_{in}}\frac{\partial T}{\partial n}(r,L) = \frac{k}{q_{in}}\frac{\partial T}{\partial z}(r,L) = \frac{\partial y}{\partial \zeta}(\rho,1) = \frac{\partial y}{\partial \nu_A}(\rho,1).$$

Hence the constraint takes the following form

(3.11) $$\mathrm{Sup}\{-\frac{\partial y(\sigma)}{\partial \nu_A}|\sigma \in \tilde{\Sigma}_3\} = \mathrm{Sup}\{-\frac{\partial y(\sigma)}{\partial \zeta}|\sigma \in \tilde{\Sigma}_3\} \leq q = q_{out}/q_{in}$$

where q is the <u>dimensionless flux ratio</u>. Finally the cost function can be rewritten

(3.12) $$J(\Omega) = R_0^3 J(\tilde{L},\tilde{\rho}),\quad J(\tilde{L},\tilde{\rho}) = \tilde{L}\pi\int_0^1\tilde{\rho}(\zeta)^2 d\zeta.$$

3.2. <u>The constrained minimization problem (P)</u>. Denote by (P) the constrained minimization problem which consists in minimizing $J(\tilde{L},\tilde{\rho})$ in (3.12) with respect to the dimensionless <u>design parameter</u> $\tilde{L}$ and <u>design function</u> $\tilde{\rho}$ subject to the constraint

(3.13) $$\sup\{-\frac{\partial y}{\partial \zeta}(\rho,1)\,|\,0 \leq \rho \leq \tilde{\rho}(1)\} \leq q,$$

where y is the solution of the variational problem (3.6).

Notice that the optimal design is only a function of the ratio q. The parameter

R_0 only appears as a scaling parameter:

(3.14) $L^* = R_0\tilde{L}^*$, $R^*(z) = R_0\tilde{\rho}^*(z/R_0\tilde{L}^*)$, $x_1 = R_0\xi_1$, $x_2 = R_0\xi_2$, $z = R_0\tilde{L}^*\zeta$.

4. APPROXIMATION OF THE SOLUTION TO THE CONSTRAINED MINIMIZATION PROBLEM (P). In the absence of existence and uniqueness results, we shall assume the existence of at least one solution to problem (P) and concentrate on the approximation of its solution. Constraint (3.11) is completely equivalent to the new constraint

(4.1) $$f(\tilde{L},\tilde{\rho}) \overset{\text{def.}}{=} \int_{\tilde{\Sigma}_3} [\frac{\partial y}{\partial \zeta} + q]^- d\tilde{\Sigma} = 2\pi \int_0^{\tilde{\rho}(1)} [\frac{\partial y}{\partial \zeta}(\rho,1)+q]^- \rho d\rho = 0,$$

where $u^- = \sup\{-u,0\}$.

Associate with an arbitrary family $\{\varepsilon : \varepsilon > 0\}$ of small positive numbers the penalized cost function

(4.3) $$J_\varepsilon(\tilde{L},\tilde{\rho}) = J(\tilde{L},\tilde{\rho}) + \frac{1}{\varepsilon} f(\tilde{L},\tilde{\rho}).$$

Replace the original constrained minimization problem (P) specified in section 3.2 by the following family of ε-indexed unconstrained minimization problems $(P_\varepsilon)_{\varepsilon>0}$: to find $(\tilde{L}_\varepsilon,\tilde{\rho}_\varepsilon)$ such that

(4.4) $J_\varepsilon(\tilde{L}_\varepsilon,\tilde{\rho}_\varepsilon) \le J_\varepsilon(\tilde{L},\tilde{\rho})$, $\forall \tilde{L},\tilde{\rho}$ such that $\tilde{L}\ge 0$ and $\forall\zeta$, $\tilde{\rho}(\zeta) \ge 0$ with $\tilde{\rho}(0)=1$.

It is readily seen that any limit point $\tilde{L},\tilde{\rho}$ of a sequence $(\tilde{L}_{\varepsilon_n},\tilde{\rho}_{\varepsilon_n})$ as ε_n goes to zero is a global minimum solution to problem (P). This is a consequence of the fact that the function f is non-negative.

5. DERIVATIVES WITH RESPECT TO THE SHAPE FUNCTION $\tilde{\rho}$. We make use of the techniques introduced by J. Céa [1], [2] (cf. also J.-P.Zolésio [3]). This is also known as the "speed method". The initial problem is equivalent to finding the diffusion coefficient $\tilde{L} \ge 0$ and the scaled ζ-axisymmetrical domain $\tilde{\Omega}$

(5.1) $$\tilde{\Omega} = \{(\xi_1,\xi_2,\zeta) \,|\, \zeta_1=x_1/R_0,\ \zeta_2=x_2/R_0,\ \zeta=z/L,\ (x_1,x_2,z) \in \Omega\}$$

which is completely specified by its boundary $\tilde{\Sigma}_2$.

When $\tilde{\Omega}$ is "deformed" into $\tilde{\Omega}_t$ by the speed V, the boundary $\tilde{\Sigma}_2$ must remain a ζ-axisymmetrical graph. So we shall only consider speeds V of the following special form:

(5.2) $$V(\rho,\zeta) = (\omega(\rho,\zeta),0),\quad \rho = \sqrt{\xi_1^2+\xi_2^2}$$

In the deformation, the ζ-axis and the boundary $\tilde{\Sigma}_1$ both remain fixed, that is

(5.3) $$\forall\zeta \in [0,1],\ \omega(0,\zeta) = 0,\quad \forall\rho \in [0,1],\ \omega(\rho,0) = 0.$$

Any point X of $\tilde{\Omega}$ is transformed into a point x(t,X) solution of the differential equation

(5.4) $$\frac{dx}{dt}(t,X) = V(x(t,X)),\quad x(0,X) = X,\quad t \ge 0.$$

This defines a transformation $T_t(V):X \to x(t;X)$ which changes the domain $\tilde{\Omega}$ into a new domain

(5.5) $$\tilde{\Omega}_t = T_t(V)(\tilde{\Omega}) = \{T_t(V)(x) \,|\, x \in \tilde{\Omega}\},\quad t \ge 0.$$

5.1 <u>Shape derivative of the volume</u>. Following J.-P.Zolésio [1], [3], we use the concept of <u>Eulerian derivative</u> (or shape derivative) at $\tilde{\Omega}$ in the direction of the field V. We obtain (cf. Delfour-Payre-Zolésio [2])

$$dJ(\tilde{L},\tilde{\rho};V) = \tilde{L}\int_0^1 2\pi\tilde{\rho}(\zeta)\omega(\tilde{\rho}(\zeta),\zeta)d\zeta. \tag{5.6}$$

5.2 <u>Shape derivative of the constraint functional f</u>. Here we avoid the case where the constraint is saturated everywhere on the boundary $\tilde{\Sigma}_3$ since in that case

$$f(\tilde{L},\tilde{\rho}) = \pi(1-q\tilde{\rho}(1)^2) \Rightarrow df(\tilde{L},\tilde{\rho};V) = -q2\pi\tilde{\rho}(1)\omega(\tilde{\rho}(1),1). \tag{5.7}$$

However we know that

$$-\frac{\partial y}{\partial\nu_A}(\rho,1) \leq q, \quad \forall 0 \leq \rho \leq \tilde{\rho}(1) \Rightarrow \tilde{\rho}(1) \geq \sqrt{1/q}. \tag{5.8}$$

This condition on $\tilde{\rho}(1)$ is necessary in order to satisfy the constraint on the boundary $\tilde{\Sigma}_3$.

<u>Hypothesis 5.1</u>. The shape function satisfies the conditions

$$\tilde{\rho}(0) = 1 \quad \text{and} \quad \tilde{\rho}(1) \geq \sqrt{1/q}. \quad \square \tag{5.9}$$

<u>Hypothesis 5.2</u>. For $0 < q < 1$, the function

$$\rho \to -\frac{\partial y}{\partial\zeta}(\rho,1): [0,\tilde{\rho}(1)] \to R \tag{5.10}$$

is monotone strictly increasing. □

Under Hypothesis 5.2 and for a flux ratio q, $0 < q < 1$, the subset of all ρ, $0 \leq \rho \leq \tilde{\rho}(1)$, such that

$$-\frac{\partial y}{\partial\zeta}(\rho,1) = q \tag{5.11}$$

has zero measure. Under Hypotheses 5.1 and 5.2, it can be shown that

$$df(\tilde{\Omega};V) = B(y,p) = -\int_{\tilde{\Sigma}_2}[\tilde{L}^2(\frac{\partial y}{\partial\xi_1}\frac{\partial p}{\partial\xi_1} + \frac{\partial y}{\partial\xi_2}\frac{\partial p}{\partial\xi_2}) + \frac{\partial y}{\partial\zeta}\frac{\partial p}{\partial\zeta}]\langle V,\nu\rangle d\sigma, \tag{5.12}$$

where p is the solution of the following boundary value problem

$$A_p = -[\tilde{L}^2(\frac{\partial^2 p}{\partial\xi_1^2} + \frac{\partial^2 p}{\partial\xi_2^2}) + \frac{\partial^2 p}{\partial\zeta^2}] = 0 \text{ in } \tilde{\Omega},\ p = \chi_+ \text{ on } \tilde{\Sigma}_3,\ \frac{\partial p}{\partial\nu_A} = 0 \text{ on } \tilde{\Sigma}_1\cup\tilde{\Sigma}_2, \tag{5.13}$$

χ_+ is the characteristic function of the set $\{\sigma \in \tilde{\Sigma}_3 : -\frac{\partial y}{\partial\zeta}(\sigma) > q\}$ (a disk centered in r=0 on the surface $\tilde{\Sigma}_3$). So we only have $\chi_+ \in H^{(1/2)-\varepsilon}(\tilde{\Sigma}_3)$, $\forall\varepsilon > 0$ and p is harmonic and belongs to $H^{1-\varepsilon}(\tilde{\Omega})$. However expression (5.12) requires that p belongs to $C^1(V_2)$ in some neighborhood V_2 of the boundary $\tilde{\Sigma}_2$. But this is always true under Hypotheses 5.1 and 5.2. Either $\chi_+ = 1$ on $\tilde{\Sigma}_3$ and p is equal to 1 on the closure of $\tilde{\Omega}$, or there exists a neighborhood N of the curve $C_3 = \tilde{\Sigma}_2\cap\tilde{\Sigma}_3$ such that

$$[\frac{\partial y}{\partial\zeta}(\sigma)+q]^- = 0, \quad \forall\sigma \in \tilde{\Sigma}_3\cap N, \tag{5.14}$$

in which case there exists a neighborhood V_2 of the boundary $\tilde{\Sigma}_2$, where p belongs to $C^1(V_2)$.

5.3 <u>Derivative of the penalized cost</u>. Under Hypotheses 5.1 and 5.2

$$dJ_\varepsilon(\tilde{L},\tilde{\rho};V) = \int_0^1 f^0(\zeta)\omega(\tilde{\rho}(\zeta),\zeta)d\zeta, \tag{5.15}$$

where (in cylindrical coordinates)

(5.16) $$f^0(\zeta) = \{\tilde{L}-[\tilde{L}\,\frac{\partial y}{\partial\rho}\,\frac{\partial p}{\partial\rho} + \frac{\partial y}{\partial\zeta}\,\frac{\partial p}{\partial\zeta}]/\varepsilon\}2\pi\tilde{\rho}(\zeta),$$

where p is the solution of (5.13) and y is the solution of (3.6).

6. SOLUTION OF THE PENALIZED PROBLEM. In this section we describe the method which will be used to compute the solution of the minimization problem

(6.1) $$\mathrm{Inf}\{J_\varepsilon(\tilde{L},\tilde{\rho}):\ \tilde{L} \ge 0,\ \tilde{\rho}\text{ satisfying (6.2)}\},$$

where

(6.2) $$\tilde{\rho}(\zeta) \ge 0,\quad 0 \le \zeta \le 1,\quad \tilde{\rho}(0) = 1,\quad \tilde{\rho}(1) \ge \sqrt{1/q}.$$

Problem (6.1) is first rewritten in the equivalent form

(6.3) $$\mathrm{Inf}\{\mathrm{Inf}\{J_\varepsilon(\tilde{L},\tilde{\rho}):\ \tilde{\rho}\text{ satisfying (6.2)}\}:\ \tilde{L} \ge 0\}.$$

So our original problem (6.1) can be split into the following two subproblems:

I) given $\tilde{L}$, find $\tilde{\rho}_\varepsilon(\tilde{L})$ satisfying (6.2) such that

(6.4) $$\forall\tilde{\rho}\text{ satisfying (6.2)},\ J_\varepsilon(\tilde{L},\tilde{\rho}_\varepsilon(L)) \le J_\varepsilon(\tilde{L},\tilde{\rho}),$$

II) find $\tilde{L}_\varepsilon \ge 0$ such that

(6.5) $$\forall\tilde{L} \ge 0,\ J_\varepsilon(\tilde{L}_\varepsilon,\tilde{\rho}_\varepsilon(\tilde{L}_\varepsilon)) \le J_\varepsilon(\tilde{L},\tilde{\rho}_\varepsilon(\tilde{L})).$$

Problem II can be solved by a one-dimensional search. So we only concentrate on problem I for a fixed $\tilde{L}$.

6.1 Approximation of y and p. We use a finite element method to approximate the solution y of equation (5.29) and p of (5.26). Both solutions are defined on the same domain $\tilde{\Omega}$. So we shall use the same triangulation of the domain D which generates $\tilde{\Omega}$ by revolution about the ζ-axis. The shape function $\tilde{\rho}$ which defines the boundary S_2 of D is first approximated by a continuous piecewise linear shape function $\tilde{\rho}_k$ in

(6.6) $$P_1^1 = \{\tilde{\rho} \in C^0([0,1])\,|\,\tilde{\rho} \ge 0,\ \tilde{\rho}(1) = 1,\ \tilde{\rho}\text{ linear on }[\zeta_j,\zeta_{j+1}],\ j=1,\dots,M\}$$

where $0 = \zeta_1 < \zeta_2 < \dots < \zeta_{M+1} = 1$ is a given partition of the interval [0,1] on the ζ-axis. This defines a polynomial domain D_k which is an approximation of the original domain D. The triangulation of D_k is obtained by "stretching" the triangulation of the fixed unit square D_0. Define the finite dimensional subspace $V_0^h(D_k)$ of $V_0(D_k)$

(6.7) $$V_0^h(D_k) = \left\{v_h \in C^0(\bar{D}_k)\,\middle|\,\begin{array}{l}v_h\text{ linear on each element}\\ v_h(\rho,1)=0,\ 0 \le \rho \le \tilde{\rho}_k(1)\end{array}\right\}.$$

The approximation $y_h \in V_0^h$ is the solution of

(6.8) $$\forall v_h \in V_0^h(D_k),\ \int_{D_k}(\tilde{L}^2\,\frac{\partial y_h}{\partial\rho}\,\frac{\partial v_h}{\partial\rho} + \frac{\partial y_h}{\partial\zeta}\,\frac{\partial v_h}{\partial\zeta})\rho\,d\rho\,d\zeta = \int_0^1 v_h(0,\rho)\rho\,d\rho.$$

Define

(6.9) $$\begin{cases}\chi_+^h(\rho) = 1,\ \text{ if } (\frac{\partial y_h}{\partial\zeta}(\rho,1)+q) < 0,\ \text{ and 0 otherwise}\\[2ex] V_\chi^h(D_k) = \left\{v_h \in C^0(\bar{D}_k)\,\middle|\,\begin{array}{l}v_h\text{ linear on each element}\\ v_h(\rho,1) = \chi_+^h(\rho),\ 0 \le \rho \le \tilde{\rho}_k(1)\end{array}\right\}.\end{cases}$$

The approximation $p_h \in V_\chi^h(D_k)$ of p is the solution of

$$(6.10) \qquad \forall v_h \in V_0^h(D_k), \quad \int_{D_k} (\tilde{L}^2 \frac{\partial p_h}{\partial \rho}\frac{\partial v_h}{\partial \rho} + \frac{\partial p_h}{\partial \zeta}\frac{\partial v_h}{\partial \zeta}) \rho d\rho d\zeta = 0.$$

6.2 Choice of the speed. For fixed D_k, $\tilde{\rho}_k$, y_h and p_h, the approximation to the gradient of the penalized cost in the direction v is of the form

$$(6.11) \quad j(v) = \int_0^1 f_h^0(\zeta) v(\zeta) d\zeta, \quad f_h^0(\zeta) = \left\{\tilde{L} - \frac{1}{\varepsilon}[\tilde{L}^2 \frac{\partial y_h}{\partial \rho}\frac{\partial p_h}{\partial \rho} + \frac{\partial y_h}{\partial \zeta}\frac{\partial p_h}{\partial \zeta}]_{\tilde{\Sigma}_2}\right\} 2\pi \tilde{\rho}_k(\zeta).$$

In general f_h^0 is discontinuous at the discretization points and $f_h^0(0) \neq 0$. To obtain an admissible direction of steepest descent we use the $L^2(0,1)$-projection $\mathcal{P}f_h^0$ of f_h^0 onto the subspace of continuous piecewise linear functions which are zero in zero

$$(6.12) \qquad P_0^1 = \{v \in C^0([0,1]) \mid v(0)=0, \ v \text{ linear on } [\zeta_j,\zeta_{j+1}], \ j=1,\ldots,M\}.$$

Then we use a method of steepest descent.

REFERENCES

J.Céa [1], Une méthode numérique pour la recherche d'un domaine optimal, Publication IMAN, Université de Nice, 1976.

[2], Optimization; theory and algorithms, Tata Institute, Springer-Verlag, 1978.

M.Delfour, G.Payre and J.-P.Zolésio [1], Design of a mass-optimized thermal diffuser, in "Optimization of Distributed Parameter Structures", J.Céa and E.Haug, eds., pp. 1265-1283, Sijthoff and Nordhoff, Alphen aan den Rijn, 1981.

[2], Optimal design of a minimum weight thermal diffuser with constraint on the output thermal power flux, CRMA report-1025, Université de Montréal, April 1981.

Ph.Destuynder [1], Etude théorique et numérique d'un algorithme d'optimisation de structures, Thèse de docteur ingénieur, Université de Paris VI, 1976.

J.-P.Zolésio [1], The material derivative, in Proceedings of the NATO-ASI on Optimization of Distributed Parameter Structural Systems, J.Céa and E.Haug, eds., pp. 1101-1207, Sijthoff and Nordhoff, Alphen aan den Rijn, 1981.

[2], Semi-derivative of repeated eigenvalues, in "Optimization of Distributed Parameter Structures", J.Céa and E.Haug, eds., pp. 1475-1491, Sijthoff and Nordhoff, Alphen aan den Rijn, 1981.

[3], Identification de domaines par déformations, Thèse de doctorat d'état, Université de Nice, 1979.

[4], An optimal design procedure for optimal control support, in "Proceedings of a conference held at Murat le Quaire" (1976), A.Auslender, ed., pp.200-230, Lecture Notes in Economical and Mathematical Systems, no.14, Springer-Verlag, New York, 1977.

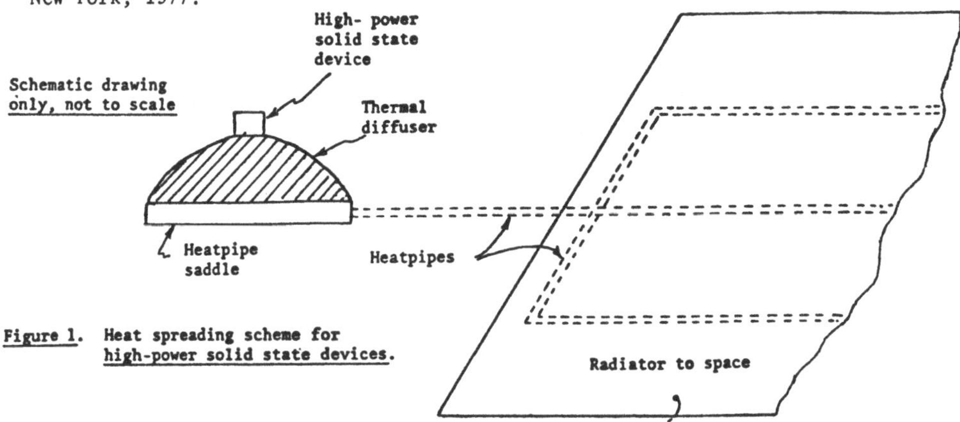

Figure 1. Heat spreading scheme for high-power solid state devices.

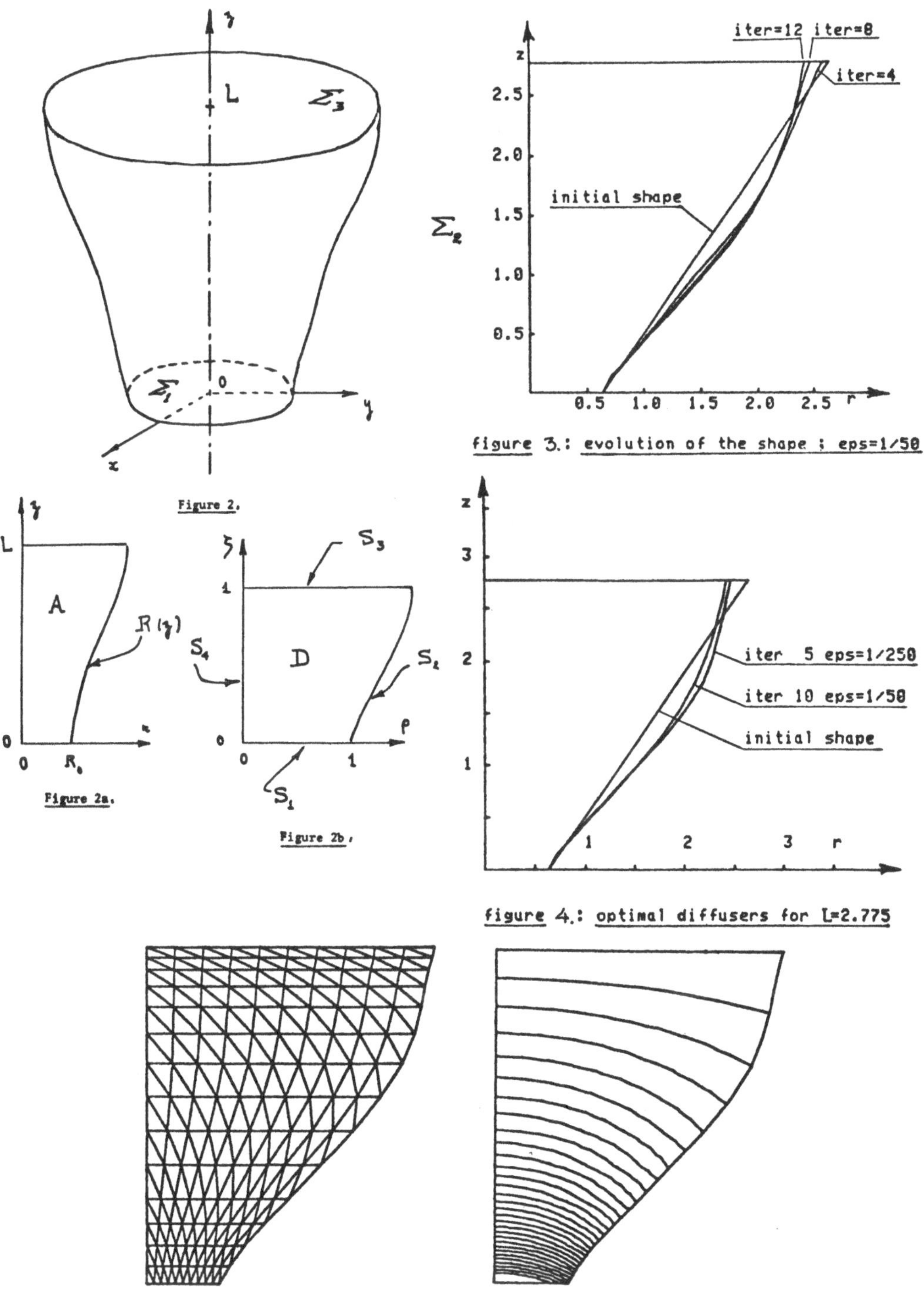

figure 3.: evolution of the shape ; eps=1/50

figure 4.: optimal diffusers for L=2.775

figure 5. : finite element grid and isotherms

DIFFERENTIAL INCLUSIONS
WITH MULTIVALUED BOUNDARY CONDITIONS

F. J. DI GUGLIELMO

Centre Universitaire d'Avignon
33, rue Louis Pasteur
84000 AVIGNON

1. Introduction

This paper gives existence and approximation results for differential inclusions with multivalued boundary conditions of the form :

$$(P_1)\quad \begin{cases} \frac{dx}{dt} \in -S(x(t)) & \text{a.e. in }]0,T[\\ x(0) = x_\circ \\ x(T) \in K \end{cases}$$

where S is a continuous multivalued mapping from $\mathbb{R}^\ell$ to $\mathbb{R}^\ell$ with convex compact images, K is a closed subset and $x_\circ$ a given point of $\mathbb{R}^\ell$.

We assume that the data $x_\circ$, S and K satisfy a consistency condition of the following form :

(L_1) For any $x \in K$ there exists an absolutely continuous function $\phi \in C([0,T] ; \mathbb{R}^\ell)$ such that $||\dot{\phi}||_\infty \leqslant M$, $\phi(T) = x \in K$, $\phi(t) - \phi(\tau) \in -\int_\tau^t S(\phi(s))ds$ for any $t,\tau \in [0,T]$ and satisfying moreover :

$$D_+V(\phi(T))(x_\circ - \phi(0)) + ||x_\circ - \phi(0)||^2 \leqslant 0$$

where V denotes a non negative lower semi continuous and lower semi compact function defined on K.

We associate with problem (P_1) a discrete implicit problem of the following form :

$$(P^{\varepsilon}_{1,m}) \begin{cases} \dfrac{x^{j+1} - x^j}{h} \in - S(x^{j+1}) + \varepsilon B \text{ for } j = 0,1,\dots m-1 \\ x^\circ = x_\circ,\ x^m \in K \end{cases}$$

with $h = \frac{T}{m}$ and $\varepsilon > 0$ and we prove the existence of a stationary point $x^m \in K$ of the mapping $F(x) = x_\circ - [I + h(S + \varepsilon B)]^m(x)$ by using a fixed point theorem due to AUBIN-EKELAND [1] from which we deduce the existence of a solution of the discrete problem $(P^{\varepsilon}_{1,m})$ and we prove by a compactness argument the existence of a subsequence of these approximate solutions converging to a solution of (P).

In this approach solutions of (P) which are described as trajectories of the differential inclusion $\frac{dx}{dt} \in -S(x(t))$ arriving at time T in the set K and having $x_\circ$ as an initial data are obtained by minimizing in a generalized sense the non negative function V on the closed set K.

The method can be extended to the case of two multivalued boundary conditions and also to the case where the boundary condition $x(T) \in K$ is replaced by a more general one of the form :

there exists $\bar{t} \in]0,T]$ such that $x(\bar{t}) \in K$.

2. Existence of approximate solutions of problem (P_1).

We shall prove the existence of solutions of the discrete problem $(P^{\varepsilon}_{1,m})$ by using the following fixed point theorem :

Theorem 1 (AUBIN-EKELAND)

Let K be a closed subset of a Hilbert space X, $F : K \to X$ be a set valued map and $V : K \to R_+$ be a lower semi continuous and *lower semi compact function satisfying :*

$\forall x \in K \quad \exists v \in F(x)$ *such that* $D_+V(x)(v) + ||v||^2 \leqslant 0$

Then there exists a stationary point $\bar{x} \in K$ *of* F.

We have the following result :

Proposition 1

(M) Let us assume that K is a compact subset of $\mathbb{R}^\ell$, that S is a continuous set valued map from $\mathbb{R}^\ell$ to $\mathbb{R}^\ell$, x_0 a given point of $\mathbb{R}^\ell$ and that the consistency condition (L_1) is satisfied. Then for any $\varepsilon>0$ and $m \geqslant m_\varepsilon$ the discrete problem $(P^\varepsilon_{1,m})$ has a solution. Moreover if the mapping S is such that :

$S(x) \subset [\alpha+\beta||\ x\ ||]B$ for any $x \in \mathbb{R}^\ell$,

then for any $\varepsilon \leqslant \varepsilon_0$ and $m \geqslant m_\varepsilon$ these solutions are contained in a bounded subset of $C([0,T] ; \mathbb{R}^\ell)$ which is independent of m and ε.

Proof

Let us denote by ϕ an arbitrary function satisfying condition (L_1). As $||\dot{\phi}||_\infty \leqslant M$ it satisfies $||\phi(t)-\phi(T)|| \leqslant M\ |t-T|$ for any $t \in [0,T]$ and since $\phi(T) \in K$ we have $\phi(t) \in K + MTB$. As the set valued map S is continuous, it is uniformly continuous on the ball with radius $|K| + MT$ with $|K| = \sup_{x\in K} ||x||$. Hence

(1) $\quad \forall\varepsilon>0 \quad \exists\ \delta>0$ such that $||x-y|| \leqslant \delta$ implies $S(x) \subset S(y) + \varepsilon B$.

Let us consider a net with steplength $h = \frac{T}{m}$ on the interval $[0,T]$. If $[jh,(j+1)h]$ is any subinterval of $[0,T]$, then for any $s \in [jh,(j+1)h]$ we have :

(2) $\quad ||\phi(s)-\phi([j+1]h)|| \leqslant M\ |s-(j+1)h| \leqslant Mh.$

For $h \leqslant \frac{\delta}{M}$ we deduce from (1) and (2) that :

$S(\phi(s)) \subset S(\phi([j+1]h)) + \varepsilon B$ for any $s \in [jh,(j+1)h]$

and hence by setting $t_j = jh$:

$$\int_{t_j}^{t_{j+1}} S(\phi(s))ds \subset \int_{t_j}^{t_{j+1}} [S(\phi(t_{j+1}))+\varepsilon B]ds = h[S(\phi(t_{j+1}))+\varepsilon B] .$$

As ϕ satisfies (L) we have for any $j \in \{0,1, \dots m-1\}$:

$$\phi(t_{j+1}) - \phi(t_j) \in -\int_{t_j}^{t_{j+1}} S(\phi(s))ds \subset -h[S(\phi(t_{j+1})) + \varepsilon B]$$

and hence :

(3) $\phi(t_j) \in [I+hS^{\varepsilon}]\phi(t_{j+1})$ for $j = 0,1, \dots, m-1$

where : $S^{\varepsilon}(\ .\ ) = S(\ .\ ) + \varepsilon B$.

We deduce from (3) that :

(4) $x_0 - \phi(0) \in x_0 - [I+hS^{\varepsilon}]^m(\phi(T))$ with $\phi(T) \in K$

and by applying the above cited result to the map

$F(.) = x_0 - [I+hS^{\varepsilon}]^m(.)$

we deduce from (4) the existence of $x_{\varepsilon}^m \in K$ satisfying :

$x_0 \in [I + hS^{\varepsilon}]^m(x_{\varepsilon}^m)$.

This implies the existence of $x_{\varepsilon}^1, x_{\varepsilon}^2, \dots x_{\varepsilon}^{m-1}$ such that

(5) $\dfrac{x_{\varepsilon}^{j+1} - x_{\varepsilon}^j}{h} \in -[S(x_{\varepsilon}^{j+1}) + \varepsilon B]$ with $h = \dfrac{T}{m}$.

Moreover we deduce from (5) and from condition (M) that for $m \geqslant \beta T$:

$$||x^{j+1}|| \leqslant \frac{1}{1-h\beta}[h(\alpha+\varepsilon) + ||x_{\varepsilon}^j||]$$

and hence by induction :

(6) $||x_{\varepsilon}^j|| \leqslant \dfrac{\alpha+\varepsilon}{\beta}\left[\dfrac{1}{(1-h\beta)^j} - 1\right] + \dfrac{||x_0||}{(1-h\beta)^j}$ for $j=0,1,\dots m$

Since $\lim_{m\to\infty} (1 - \frac{\beta T}{m})^m = e^{-\beta T}$ and $(1-h\beta)^{-j} \leqslant (1-h\beta)^{-m}$

for $j \leqslant m$, we deduce that the $(x_m^j)_{j=0,1,\dots m}$ are contained in a bounded subset of $\mathbb{R}^{\ell}$ which is independent of m.

3. Convergence of the approximate solutions

We will now study the convergence of these approximate solutions to an exact solution of Problem (P_1) and we will prove the following result :

Theorem 2

Under the assumptions of Proposition 1 Problem (P_1) *has a solution* $x \in \mathcal{C}([0,T];\, \mathbb{R}^{\ell})$ *absolutely continuous and such that* $\frac{dx}{dt} \in L^1(0,T;\, \mathbb{R}^{\ell})$.

Proof

We introduce the functions x_m^{ε} defined by :

$$(7)\quad x_m^{\varepsilon}(t) = x_m^j + (t-jh_m)\,\frac{x_m^{j+1} - x_m^j}{h_m} \quad \text{for } t \in [(j-1)h_m, jh_m]$$

which linearly interpolate the points x_m^j on $[0,T]$.
As the x_m^j are contained in a bounded subset independent of m and ε and the set valued map S is continuous with convex compact values. We deduce that the functions x_m^{ε} are contained in a compact subset of $\mathcal{C}([0,T];\, \mathbb{R}^{\ell})$ and that their derivatives $\dot{x}_m^{\varepsilon}$ are contained in a bounded subset of $L^{\infty}(0,T;\, \mathbb{R}^{\ell})$. We can then by extracting an appropriate subsequence of these approximate solutions and bu using Mazur's Lemma prove the existence of a function u obtained as a limit of this sequence which is a solution of problem (P_1) (see [2]). ∎

Problems with two multivalued boundary conditions of the form :

$$(P_2)\begin{cases} \frac{dx}{dt} \in - S(x(t)) & \text{a.e. in } [0,T] \\ x(0) \in K_o & \\ x(T) \in K_T & \end{cases}$$

can also be studied by the same method.
We have for example the following result :

Theorem 3.

Assume that K_o and K_T are compact subsets of $\mathbb{R}^\ell$, that S is a continuous set valued map from $\mathbb{R}^\ell$ to $\mathbb{R}^\ell$ with closed convex values verifying condition (M) and that these data satisfy the following consistency condition :

(L_2) *for any $x \in K_T$ there exists $\phi \in \mathbf{C}([0,T];\mathbb{R}^\ell)$ absolutely continuous such that $||\phi||_\infty \leq M$, $\phi(T)=x \in K_T$, $\phi(t)-\phi(\tau) \in -\int_\tau^t S(\phi(s))ds$ for any $t,\tau \in [0,T]$ and moreover :*
$\exists x_\phi \in K_o$ such that $D_+V(\phi(T))(x_\phi-\phi(0)) + ||x_\phi-\phi(0)||^2 \leq 0$

where V denotes a non negative lower semi continuous and lower semi compact function defined on K_T. Then problem (P) has a solution $x \in \mathbf{C}([0,T]; \mathbb{R}^\ell)$ absolutely continuous and such that $\frac{dx}{dt} \in L^1(0,T;\mathbb{R}^\ell)$.

4. Problems with a boundary condition of the "capture" type :

One can study by a similar method problems with boundary conditions of the following form :

there exists $\bar{t} \in]0,T]$ such that $x(\bar{t}) \in K$.

We have in this case the following result :

Theorem 4.

Assume that K is a compact subset of $\mathbb{R}^\ell$, that S is a continuous set valued map from $\mathbb{R}^\ell$ to $\mathbb{R}^\ell$ with closed convex images verifying condition (M), that $x_o \notin K$ is a given point of $\mathbb{R}^\ell$ and that these data satisfy the following consistency condition where V denotes a lower semi continuous and lower semi compact function from K to $[0,+\infty[$:

for any $x \in K$ there exists $t \in]0,T]$ and $\phi \in \mathbf{C}([0,T];\mathbb{R}^\ell)$ absolutely continuous such that $\phi(t)=x$, $||\phi||_\infty \leq M$,

(L_C) $\phi(\tau_2) - \phi(\tau_1) \varepsilon - \int_{\tau_1}^{\tau_2} S(\phi(s))ds$ *for any* $\tau_1, \tau_2 \in [0,2T]$

and moreover :

$D_+V(\phi(t))(x_o - \phi(0)) + ||x_o - \phi(0)||^2 \leqslant 0$.

Then the problem :

(P_C)
$$\begin{cases} \frac{dx}{dt} \in - S(x(t)) & a.e. \text{ in } [0,T] \\ x(0) = x_o \\ \exists \bar{t} \in [0,T] & \text{such that } x(\bar{t}) \in K \end{cases}$$

has a solution $x \in C([0,T]; \mathbb{R}^\ell)$ *absolutely continuous and such that* $\frac{dx}{dt} \in L^1(0,T ; \mathbb{R}^\ell)$.

Proof

Let us consider $\bar{x} \in K$. Then by (L_C) there exists $\bar{t}>0$ and a trajectory ϕ such that $\bar{x} = \phi(\bar{t})$ and since $\bar{t} \in [t_k, t_{k+1}[$ for some $k \in \{0,1,\dots m-1\}$ we have :

$$||\phi(t) - \phi(t_k)|| \leqslant M|t-t_k| \leqslant Mh$$

If we assume that $\delta > 0$ is associated with $\varepsilon > 0$ by (1), then for $h \leqslant \frac{\delta}{M}$ we have $\phi(t_k) \in \phi(t) + \delta B$ for any $t \in [t_k, t_{k+1}]$ and hence :

$$[I+hS^{\varepsilon}]\phi(t_k) = \phi(t_k)+hS(\phi(t_k))+h\varepsilon B \subset \phi(\bar{t})+\delta B+h[S(\phi(\bar{t}))+\varepsilon B]+h\varepsilon B$$
$$\subset [I+hS^{2\varepsilon}]\phi(\bar{t})+\delta B$$

which implies by induction for any integer $q > 0$:

(8) $[I+hS^{\varepsilon}]^q\phi(t_k) \subset [I+hS^{2\varepsilon}]^q\phi(\bar{t}) +\delta B.$

Since $\phi(t_o) \in [I+hS^{\varepsilon}]^k(\phi(t_k))$ for any $k \in \{1,2, \dots m\}$ it results from (8) that :

$$x_o - \Phi(t_o) \in x_o - [I+hS^{\varepsilon}]^k(\phi(t_k)) \subset x_o - [I+hS^{2\varepsilon}]^k\phi(\bar{t}) +\delta B$$

and this implies that :

(9) $x_o - \phi(t_o) \in F(\phi(\bar{t}))$

where $\phi(\bar{t}) = \bar{x}$ and F denotes the following set valued map :

$$F(x) = \bigcup_{k=1}^{m} \{x_o - [I+hS^{2\varepsilon}]^k (x)+ \delta B\}.$$

From (L_C) and (9) we deduce that :

$$\begin{cases} \text{for any } \bar{x} \in K \text{ there exists a vector } v \in F(\bar{x}) \text{ such that:} \\ D_+V(\bar{x})(v) + \|v\|^2 \leqslant 0 \end{cases}$$

and by Theorem 1 there exists $x^* \in K$ such that :
$0 \in F(x^*)$ or in other words there exists $j_\circ \in \{1,2,\dots m\}$ such that :

$$x_\circ \in [I+hS^{2\varepsilon}]^{j_\circ}(x^*) + \delta B \, .$$

We can now define the sequence (x^j) for all $j \geqslant j_\circ$ by taking a function $\tilde{\phi}$ satisfying (L_C) with $\tilde{\phi}(t_\circ) = x^* \in K$ and setting $x^j = \tilde{\phi}(t_j + t_\circ - t_{j_\circ})$ for $j \geqslant j_\circ$. The points x^j defined in this way satisfy :

$$\begin{cases} \dfrac{x^{j+1}-x^j}{h} \in -S(x^{j+1}) + 2\varepsilon B \quad \text{for } j = 0,1, \dots m-1 \\ \text{with } \|x^\circ - x_\circ\| \leqslant \delta \quad \text{and } x^{j_\circ} \in K. \end{cases}$$

By using condition (M) and the fact that $\|\dot{\phi}\|_\infty \leqslant M$ we deduce the existence of a constant $\rho_T > 0$ independent of m such that:

$\|x^j\| \leqslant \rho_T$ for any $j \in \{0,1,2, \dots m\}$ and $m \geqslant \beta T$.

If we introduce the functions (x_m^ε) which linearly interpolate the points x_m^j for $j = 0,1, \dots m$ then by the same argument as for Theorem 2 we can extract a subsequence (x_ν^ε) converging for $\nu \to +\infty$ to a limit x^ε in $C([0,T];\mathbb{R}^\ell)$ and such that $(\dot{x}_\nu^\varepsilon)$ converges to $\dot{x}^\varepsilon$ in $L^\infty(0,T;\mathbb{R}^\ell)$ weak.$*$
For every function x_ν^ε there exists a value $t_{j\nu} = j_\nu h$ such that $x_\nu^\varepsilon(t_{j\nu}) \in K$. Since the $t_{j\nu}$ belong to the compact interval $[0,T]$ a subsequence of $(t_{j\nu})$ converges to a limit $t_\varepsilon \in [0,T]$. Therefore on can assume that $x_\nu^\varepsilon(t_{j\nu}) \in K$ with $\lim_{\nu\to+\infty} t_{j\nu} = t_\varepsilon$.

Since K is closed and the x_ν^ε are equicontinuous we have finally $\lim_{\nu\to+\infty} x_\nu^\varepsilon(t_{j\nu}) = x^\varepsilon(t_\varepsilon)$ and hence $x^\varepsilon(t_\varepsilon) \in K$.
The proof can be concluded by letting ε tend to zero . ∎

5. Conclusion

The results given in this paper can be used to prove the existence of admissible controls for differential equations with a control parameter of the following form:

$$(P_{\mathcal{U}})\begin{cases}\frac{dx}{dt} = \varphi(x,u) & \text{a.e. in } [0,T]\ ,\ u \in \mathcal{U}\\ x(0) = x_o & \\ x(T) \in K. & \end{cases}$$

where $\mathcal{U}$ denotes a compact subset of $\mathbb{R}$ and (x,u) satisfies a Lipschitz condition with respect to x uniformly on $\mathcal{U}$. By applying theorem 2 to the set valued map $S(x) = \varphi(x,\mathcal{U})$ and using Filippov's measurable selection theorem we obtain the existence of an admissible trajectory and of the associated optimal control $u \in \mathcal{U}$. Existence and approximation results for problem (P_1) can also be proved when the set valued map S has non convex values. Details will be given elsewhere.

REFERENCES

[1] J.P. AUBIN
Contingent derivatives of set valued maps and existence of solutions to non linear inclusions and differential inclusions.
M.R.C. Technical Summary Report #2044 (1979)

[2] J.P. AUBIN, A. CELLINA, J. NOHEL
Monotone Trajectories of Multivalued Dynamical Systems.
Annali di Matematica Pure Appl.115(1977),99-117.

[3] J.P. AUBIN, F.H. CLARKE
Monotone invariant solutions to differential inclusions.
J. London Mathematical Society 16(1977),357-366.

ON THE OPTIMAL VALUE FUNCTION OF OPTIMIZATION PROBLEMS

Bernhard Gollan
Mathematisches Institut der Universität
8700 Würzburg, West Germany

1. Introduction

We consider the following family of optimization problems, parameterized by $p \in R^{n+m}$:

$$V(p) = \inf \{ f_0(x) \mid x \in S(p) \},$$
$$S(p) = \{ x \in X \mid f_i(x) \le p_i ,\ 1 \le i \le n ;\ f_j(x) = p_j ,\ n+1 \le j \le n+m \} . \quad (1)$$

Here X is a real Banach space, the functions $f_0, \dots, f_n$ are locally Lipschitz, and $f_{n+1}, \dots, f_{n+m}$ are continuously differentiable. As shown in [6], most results in this paper remain true when a further constraint $x \in C$ is present, with a closed subset C of X. We omit this detail here. As usual, we set $V(p) = +\infty$, if $S(p) = \emptyset$. Of course, the case $V(p) = -\infty$ may occur as well. Thus V maps R^{n+m} into the extended reals.

As indicated above, we are interested in the optimal value function or marginal function V(.). We take p = 0 as reference parameter and consider characterizations of V of the following type:

(a) Is $S(p) \neq \emptyset$ for small p, i.e., do perturbed problems have feasible solutions?

Upper bounds are given for the terms in (b) and (c):

(b) $V_+(0;p) = \liminf_{(\tau \downarrow 0, p' \to p)} (V(\tau p') - V(0))/\tau$,

$V^+(0;p) = \limsup_{(\tau \downarrow 0, p' \to p)} (V(\tau p') - V(0))/\tau$,

the lower resp. upper directional Hadamard derivatives.

(c) $V^{\uparrow}(0;p) = \lim_{\epsilon \downarrow 0} (\limsup_{(\eta \underset{V}{\to} 0, \tau \downarrow 0)} (\inf_{|p'-p| \le \epsilon} (V(\tau p' + \eta) - V(\eta))/\tau))$,

the generalized directional derivative of V (where $\eta \underset{V}{\to} 0$ iff $\eta \to 0$ and $V(\eta) \to V(0)$).

This formula applies to functions V which are strictly lower semicontinuous at p = 0, cf. [7,8].

(d) Set inclusions of the form $\partial V(0) \subseteq B$ are given, with certain sets B. Here $\partial V(0)$ denotes the generalized gradient of V at p = 0, which is defined as the subgradient of the convex function $p \to V^{\uparrow}(0;p)$ at p = 0.

Answers to questions (a), (b) can be found in Theorem 1, to questions

(c) and (d) in Theorem 2 and 3. These three theorems represent particular cases of more general results which are given in detail in [6]. The present paper serves as a preliminary report thereof. The respective proofs are extensions of the approach taken in [5] and rely mainly on separability properties of convex cones.
Problems (b), (c) and (d) are also considered by R.T.Rockafellar in [8] under slightly different assumptions. However, he uses quite different tools in proving his results. Other related work can be found in [1,3].

The answers to the above questions mainly involve the knowledge of multiplier sets which are obtained from necessary optimality conditions. Throughout this paper we use the set of all multipliers satisfying F. Clarke's multiplier rule, cf. [2]. If $\bar{x}$ is some optimal solution of problem (1) with $p=0$, this results in

$$\Lambda(\bar{x}) = \{ (\lambda_0,\lambda) \in \mathbb{R}^{1+n+m} \mid 0 \in \sum_{i=0}^{n+m} \lambda_i \partial f_i(\bar{x}) ;\ \lambda_i \geq 0,\ 0 \leq i \leq n ;\ \lambda_i f_i(\bar{x}) = 0,\ 1 \leq i \leq n \} .$$

As usual, we normalize λ_0 so that $\lambda_0 = 0$ or $\lambda_0 = 1$. With this distinction we derive from Λ the two multiplier sets

$$\Lambda_0(\bar{x}) = \{ \lambda \mid (0,\lambda) \in \Lambda(\bar{x}) \} , \quad \Lambda_1(\bar{x}) = \{ \lambda \mid (1,\lambda) \in \Lambda(\bar{x}) \} . \tag{2}$$

Since we assume the equality constraints to be continuously differentiable, the set Λ_1 is a closed convex set, and Λ, Λ_0 are closed convex cones. Furthermore we have

$$\Lambda_1(\bar{x}) = \Lambda_1(\bar{x}) + \Lambda_0(\bar{x}) ,$$

and $\Lambda_0(\bar{x})$ is the recession cone of $\Lambda_1(\bar{x})$.
Finally we define the support functions:

$$s_j(p,\bar{x}) = \sup \{ -\lambda p \mid \lambda \in \Lambda_j(\bar{x}) \} , \quad j=0,1. \tag{3}$$

Note that either $s_0(p,\bar{x}) = 0$ or $s_0(p,\bar{x}) = +\infty$, since $\Lambda_0(\bar{x})$ is a cone. The terms defined in (3) are crucial in order to answer questions (a) - (d).
Rockafellar, [8], uses slightly different multiplier sets, namely the following:

$$\tilde{\Lambda}(\bar{x}) = \{ (\lambda_0,\lambda) \mid 0 \in \partial (\sum_{i=0}^{n+m} \lambda_i f_i)(\bar{x}) ;\ \lambda_i \geq 0,\ 0 \leq i \leq n ;\ \lambda_i f_i(\bar{x}) = 0,\ 1 \leq i \leq n \} .$$

In general, $\tilde{\Lambda}(\bar{x}) \subseteq \Lambda(\bar{x})$.

2. Problems with inequality constraints

We treat this case separately for the following reason: Here perturbation results can be given without any regularity assumption about the reference solution and also without any tameness assumption (cf. [6,8]) about the parameter. Therefore, as a special case of (1), in this section let

$S(p) = \{x \in X \mid f_i(x) \leq p_i , 1 \leq i \leq n\}$, (4)

and let V be defined as in (1). We obtain the following result.

Theorem 1

Let $\bar{x}$ be a minimum of f_0 subject to (4), and $p \in R^n$.

(a) If $\Lambda_1(\bar{x}) \neq \emptyset$ and $\lambda p < 0$ for all $\lambda \in \Lambda_0(\bar{x})$, $\lambda \neq 0$, then there exist $\epsilon_1 , \epsilon_2 > 0$ such that

(i) $S(\epsilon p') \neq \emptyset$ for all $0 < \epsilon \leq \epsilon_2$, $|p' - p| \leq \epsilon_1$,

(ii) $V^+(0;p) \leq s_1(p,\bar{x}) < \infty$.

(b) If $\Lambda_1(\bar{x}) \neq \emptyset$ and $s_0(p,\bar{x}) = 0$, then for any $\epsilon_1 > 0$ there exist p' with $|p' - p| \leq \epsilon_1$, and $\epsilon_2 > 0$ such that

(i) $S(\epsilon p') \neq \emptyset$ for all $0 < \epsilon \leq \epsilon_2$,

(ii) $V_+(0;p) \leq s_1(p,\bar{x}) < \infty$.

(c) If $\Lambda_1(\bar{x}) = \emptyset$ and $s_0(p,\bar{x}) = 0$, then a statement as in (b) holds, but now with $V_+(0;p) = -\infty$.

Remarks

(A) This theorem is a simplified version of [6,Theorem 3.1]. There even sharper upper bounds than $s_1(p,\bar{x})$ in parts (a) and (b) of the preceding theorem are given.

(B) In the cases (b), (c), the results are sharp in the following sense: Examples can be given where

in case (b): $s_1(p,\bar{x}) = V_+(0;p) < V^+(0;p)$,

in case (c): $-\infty = V_+(0;p) < V^+(0;p)$.

3. Problems with equality and inequality constraints

In this section we consider the general problem (1). We make a weak regularity assumption about the parameter p, called tameness, and give upper bounds for the generalized directional derivative $V^{\uparrow}(0;p)$ as well as an estimate for the generalized gradient $\partial V(0)$. These two terms are defined as in [7,8]. The notion 'tameness' is taken from [8]. For our purposes it can be stated as follows.

Definition Problem (1) is called tame for $p = 0$, if there exist $\delta_1 > 0$, $\delta_2 > V(0)$, such that for any sequence $p_k \to 0$ with $|p_k| < \delta_1$ and $V(p_k) < \delta_2$, there is some $x_k \in S(p_k)$ with $V(p_k) = f_0(x_k)$ such that $\{x_k\}$ contains a convergent subsequence.

Note that tameness does not require that problem (1) has feasible or optimal solutions for all small p.

Below let M denote the set of all optimal solutions of problem (1) with $p=0$.

Theorem 2

Let problem (1) be tame for $p=0$.

(i) If $\Lambda_1(x) \neq \emptyset$ for some $x \in M$, then for any $p \in R^{n+m}$

$$V^{\uparrow}(0;p) \leq \sup_{x \in M} \sup \{ -\lambda p \mid \lambda \in \Lambda_1(x) + \Lambda_0(x) \} .$$

(ii) If $\Lambda_1(x) = \emptyset$ and $s_0(p,x) = 0$ for all $x \in M$, then $V^{\uparrow}(0;p) = -\infty$.

(iii) $-\partial V(0) \subseteq \text{cl conv} \left(\bigcup_{x \in M} (\Lambda_1(x) + \Lambda_0(x)) \right)$.

These results are very close to those in [8]. There $X = R^k$, but on the other hand also the equality constraints are assumed only locally Lipschitz. Furthermore, in [8] the multiplier set $\tilde{\Lambda}(x)$ is used, which in particular cases leads to better estimates than above.

However, under somewhat stronger assumptions than above, the results of Theorem 2 can be improved, as shown below. The crucial condition is the following:

All functions $f_0, \ldots, f_{n+m}$ are k-times continuously differentiable. (5)

With this condition, higher order necessary conditions for problem (1) are given in [4]. They depend on certain critical directions d, cf. [4]. Let $D^{(k)}(x)$ denote the set of all such critical directions d. Then for each $d \in D^{(k)}(x)$, define

$\Lambda^{(k)}(x,d) = \{ (\lambda_0, \lambda) \mid (\lambda_0, \lambda)$ satisfies the k-th order necessary conditions of [4, Theorem 2.5] $\}$.

Similar to (2), define $\Lambda_0^{(k)}(x,d)$ and $\Lambda_1^{(k)}(x,d)$. Obviously,

$$\Lambda^{(k)}(x,d) \subseteq \Lambda^{(1)}(x,d) = \Lambda(x) = \tilde{\Lambda}(x) \quad \text{for any } k,d .$$

As a refinement of Theorem 2 we obtain:

Theorem 3

Let problem (1) be tame for $p=0$ and condition (5) hold.

(i) If $\Lambda_1^{(k)}(x,d) \neq \emptyset$ for some $x \in M$, $d \in D^{(k)}(x)$, then for any $p \in R^{n+m}$

$$V^{\uparrow}(0;p) \leq \sup_{x \in M} \inf_{d \in D^{(k)}(x)} \sup \{ -\lambda p \mid \lambda \in \Lambda_1^{(k)}(x,d) + \Lambda_0^{(k)}(x,d) \} .$$

(ii) If $\Lambda_1^{(k)}(x,d) = \emptyset$ for all $x \in M$, $d \in D^{(k)}(x)$, and $\sup \{ -\lambda p \mid \lambda \in \Lambda_0^{(k)}(x,d) \} = 0$, then $V^{\uparrow}(0;p) = -\infty$.

(iii) For any choice $d = d(x) \in D^{(k)}(x)$, for each $x \in M$,

$$- \partial V(0) \subseteq \operatorname{cl} \operatorname{conv} \Big(\bigcup_{x \in M} [\Lambda_1^{(k)}(x,d(x)) + \Lambda_0^{(k)}(x,d(x))] \Big) .$$

In [6], an example is given where M contains one element $\bar{x}$, and for some critical direction $\bar{d}$ (with $k=3$)

$$- \partial V(0) = \Lambda_1^{(3)}(\bar{x},\bar{d}) = \{ (1/2, 1/2) \} \subset \Lambda_1(\bar{x}) = \{ (\lambda_1,\lambda_2) \mid \lambda_1, \lambda_2 \geq 0, \lambda_1 + \lambda_2 = 1 \} .$$

Thus here a third order condition leads to the 'right' multiplier set.

References

[1] Auslender,A.(1979). Differential Stability in Non Convex and Non Differentiable Programming. Math. Programming Study 10, 29-41.

[2] Clarke,F.H.(1976). A New Approach to Lagrange Multipliers. Math. Oper. Res. 1, 165-174.

[3] Gauvin,J.(1979). The Generalized Gradient of a Marginal Function in Mathematical Programming. Math. Oper. Res. 4, 458-463.

[4] Gollan,B.(1981). Higher Order Necessary Conditions for an Abstract Optimization Problem. Math. Programming Study 14, 169-176.

[5] Gollan,B.(1981). Perturbation Theory for Abstract Optimization Problems. J. Optimization Theory and Appl. 35, 317-341.

[6] Gollan,B. On the Marginal Function in Nonlinear Programming. Preprint No. 69, Mathematisches Institut, Univ. Würzburg, 1981.

[7] Rockafellar,R.T.(1980). Generalized Directional Derivatives and Subgradients of Nonconvex Functions. Canad. J. Math. 32, 257-280.

[8] Rockafellar,R.T. Lagrange Multipliers and Subderivatives of Optimal Value Functions in Nonlinear Programming. Math. Programming Study, to appear.

In the statement of Theorem 3 some further assumption has to be added. One possible choice is:

Let V be subdifferentially regular at $p = 0$, i.e.,

$$V_+(0;p) = V^{\uparrow}(0;p) \quad \text{for all } p .$$

It is not clear whether the statements of Theorem 3 remain true without such an additional assumption.

FINITE ELEMENT APPROXIMATION OF TIME OPTIMAL CONTROL PROBLEMS FOR PARABOLIC EQUATIONS WITH DIRICHELT BOUNDARY CONDITIONS

Irena Lasiecka
Mathematics Department
University of Florida
Gainesville, Fla. 32611

Abstract

Finite element approximation of the time optimal control problem with Dirichlet boundary conditions is considered. Convergence of optimal controls as well as rate of convergence is discussed. An approximation using subspaces which are not required to satisfy zero boundary conditions is also considered.

Introduction

Let $-A(x,\partial)$ be a second order uniformly strongly elliptic operator with real, smooth coefficients defined on Ω, Ω bounded open domain in R^n with smooth boundary Γ.
Consider the following parabolic equation:

$$\frac{\partial y(t)}{\partial t} = A(x,\partial)y(t) \quad ; \; x \in \Omega \; ; \; t \geq 0$$

(1.1) $$y(0) = y_0 \in L_2(\Omega)$$

$$y|\Gamma = u$$

Suppose that $y_1 \in L_2(\Omega)$ be an approximately controllable state in the following sense: There exists $T > 0$ and $\bar{u} \in U$ where

(1.2) $$U = \{u \in L_2[0T \times \Gamma] \; ; \; |u(x,t)| \leq 1 \; ; \; x, t \in \Gamma \times 0T\}$$

such that

$$\| y(\bar{u},T) - y_1 \|_{L_2(\Omega)} < \delta \text{ for some preassigned } \delta > 0.$$

Now we are in a position to formulate the following time-optimal control problem.
Minimize $\{T; \; T > 0; \; \exists \, u \in U; \; |y(u,T) - y_1| \leq \delta\}$

It is a well known fact ([F.1],[S.1] that the above problem has a unique solution, say u^0, T^0 with the bang-bang property:

(1.3) $|u^o(x,t)| = 1 \quad x,t \in \Gamma \times [0,T^o]$

(1.4) $\| y(u,T) - y_1 \|_{L_2(\Omega)} = \delta$

The major goal of the present paper is to introduce <u>finite element approximation</u> of the above control problem, to prove <u>convergence</u> of the approximation scheme and to estimate the rate of convergence. Let $h \to 0$ be a parameter of discretization, let V_h and U_h be finite dimensional subspaces approximating respectively the space of states and controls (spaces of splines).

Let $y_h(u_h)$ be an approximate solution of (1.1) corresponding to $u_h \in U_h$. Let u_h^o, $y_h^o = y_h(u_h^o)$, T_h^o stand for the optimal control, trajectory and time of the corresponding discrete problem (to be defined below). The main results of the paper are as follows:

Theorem 1

$$\| y(u,t) - y_h(u_h,t) \|_{L_2(\Omega)} \le C\, e^{wT} [h^{1/2-\varepsilon} + \| y_o - R_h y_o \|_{L_2(\Omega)} + \| u - u_h \|_{L_\infty[0T;H^{-1/2+\varepsilon}(\Gamma)]}]$$

where C, w do not depend on h, $\varepsilon > 0$ arbitrarily constant, and R_h is projection on V_h.

If we apply the above result with $u_h = P_h u$ where P_h is a projection on U_h we obtain:

<u>Corollary</u>

$$\| y(u) - y_h(P_h u) \|_{C[0T;L_2(\Omega)]} \le C h^{1/2-\varepsilon} \left[\| u \|_{L_\infty[0T;L_2(\Gamma)]} + \| y_o \|_{H^{1/2-\varepsilon}(\Omega)} \right]$$

Theorem 2

For y_o, y_1 in $H^{1/2-\varepsilon}(\Omega)$ we have the following convergence results:

(1.5) $\| y(u_h^o, T_h^o) - y_1 \|_{C[0T;L_2(\Omega)]} \le \delta + O(h^{1/2-\varepsilon})$

(1.6) $u_h^o \to u^o$ in $L_2[0T \times \Gamma]$

(1.7) $T_h^o \to T^o$

<u>Remarks</u>: The above results were <u>obtained using cubic splines.</u> It should be pointed out that the estimate of Theorem (1) while needed to prove convergence in Theorem 2, is also of independent interest in its own right. The author is not aware of any work providing the error estimate between exact and approximate trejectory in pointwise $C[0T;L_2(\Omega)]$-norm for boundary (Dirichlet

type) inputs being in $L_\infty[0T;L_2(\Gamma)]$. Estimates in the $L_2[0T;L_2(\Omega)]$-norm are available in [L.1], however it appears that the technique proposed there can not be applied to the present situation, which therefore requires an ad hoc new treatment.

We shall study boundary input parabolic equations through a semi-group approach. It is well known [B.1],[W.1], as abstract model for (1.1) we can take:

$$(1.8)\qquad y(t) = S(t)\, y_o + (Lu)(t)$$

$$(Lu)(t) = -A \int_0^t S(t-z)Du(z)dz$$

where A is the generator corresponding to $A(x,\partial)$ and zero boundary conditions, which generates an analytic semigroup S(t) and D is the Dirichlet map defined by:

$$A(x,d)Du = 0$$

$$Du|_\Gamma = u$$

The following regularity results will be extensively used in the sequel

$$(1.9)\qquad R(D) \subset D(A^{1/4-\varepsilon}) \qquad \varepsilon > 0$$

$$(1.10)\qquad D \in \mathcal{L}(H^s(\Gamma) \to H^{s+1/2}(\Gamma)) \qquad \text{for all real } s$$

$$(1.11)\qquad \|A^\alpha S(t)\| \le \frac{C}{t^\alpha} \qquad t > 0$$

$$(1.12)\qquad D(A^\alpha) = H_o^{2\alpha}(\Omega)\ ; \qquad \alpha < 3/4\ ; \qquad \alpha \neq 1/4$$

(see [F.2], [L.2]).

By using (1.9),(1.11) and (1.12) we obtain:

$$\|y(u,t)\|_{H^{1/2-\varepsilon}(\Omega)} \le C \quad e^{WT}\Big[\|y_o\|_{H^{1/2-\varepsilon}(\Omega)} +$$

$$+ \|u\|_{L_\infty[0T;L_2(\Gamma)]} \int_0^t \frac{1}{(t-z)^{1-\varepsilon}} dz \quad \le$$

$$\le C\, e^{WT}[\,\|y_o\|_{H^{1/2-\varepsilon}(\Omega)} + \|u\|_{L_\infty[0T;L_2(\Gamma)}\Big]$$

which proves that

$$(1.13)\qquad y \in C[0T;\ H^{1/2-\varepsilon}(\Omega)] \text{ for } u \in L_\infty[0T;L_2(\Gamma)].$$

REMARK

By applying a regularization procedure, Seidman in [S.2] treats the time optimal control problem with $\delta = 0$, which limit is a linut $\delta \to 0$ of regularized problems, i.e. $\|y(t) - y_1\| \le \delta$.

Therefore by combining his results with the one presented in this paper, we can claim that the proposed approximation scheme converges (when $h \to 0$, $\delta \to 0$) to the optimal control problem with $\delta=0$.

2. Approximation of control problem

We start with an approximation of parabolic equation (1.1), more precisely with an approximation of its semigroup version (1.8). We will be using the following finite-dimensional spaces!

$V_h \to N(h)$ dimensional approximation of $H'(\Omega)$

(ecs cubic splines).

$V_h{}^o \to \bar{N}(h)$ dimensional approximation of

$H_o{}'(\Omega)$

$U_h \to M(h)$ dimensional subspaces of $L_2(\Gamma)$ consisting of piecewise continuous function with the following approximation property:

$$(2.1) \quad \|y-R_h y\|_{H^\alpha(\Omega)} \le Ch^{\beta-\alpha}\|y\|_{H^\beta(\Omega)} \quad \text{for } 0 \le \alpha \le 1, \quad \beta-\alpha \le 2$$

Similar property we assume for $V_h{}^o$ with $R_h{}^o$ being projection on $V_h{}^o$, and U_h with P_h projection from $L_2(\Gamma)$ on U_h.
The spaces with above properties are well known in literature (for example splines, Hermite's polynomials, etc. see [B.1], [V.1]) Let $\xi^o_{h,j}$ $j = 1\ldots N(h)$ be a local basis in $V_h{}^o$.
Set:

$K_h = [\xi_{hj}{}^o, \xi_{hi}{}^o)]$ $i,j = 1\ldots\bar{N}(h)$

$M_h \stackrel{\Delta}{=} [a(\xi_{hj}{}^o, \xi_{hi}{}^o)]$ $i,j = 1\ldots\bar{N}(h)$

where $a(u,v)$ is a bilinear form **associated** with an elliptic operator $A(x,\partial)$.
Now we define approximation of all operators appearing in the semigroup formula (1.8). To this end, set:

$$(2.2) \quad A_h = -K_h^{-1} M_h$$

$$(2.3) \quad S_h(t) = e^{A_h t}$$

and D_h: $U_h \to V_h$ defined by (see [B.3])

$$(2.4) \quad (-A(x,\partial)D_h u_h, A(x,\partial)\xi_h)_{L_2(\Omega)} + h^{-3}\langle u_h - D_h u_h, \xi_h\rangle_{L_2(\Gamma)} = 0 \quad \text{for all } \xi_h \in V_h$$

It was proved in [B.3] that:

$$(2.5) \quad \|(D_h-D)u_h\|_{H^{1/2-\varepsilon}(\Omega)} \le C\,\|u_h\|_{L_2(\Gamma)}$$

(2.6) $\| D_h u_h \|_{H^{1/2-\varepsilon}(\Omega)} \leq C \| u_h \|_{L_2(\Gamma)}$

Now we are in a position to define a semidiscrete version of (1.8), namely

(2.7) $y_h(t) = S_h(t) R_h y_o + (L_h u_h)(t)$

$$L_h u_h(t) = -A_h \int_0^t S_h(t-z) D_h u_h(z) dz$$

Consequently we define semidiscrete approximation of control problem as follows:

Minimize $\{T_h ; \exists\, u_h \in L_\infty[0T; U_h] \cap U$ such that

$$\| y_h(u_h, T_h) - R_h y_1 \| \leq \delta \}$$

where $y_h(u_h, T_h)$ is defined by (2.7).

The above problem represents classial time optimal problem for ordinary differential equations systems and can be solved numerically using standard techniques. It can be shown that the state $R_h y_1$, is approximately controllable and consequently by standard arguments the above discrete control problem has the unique bang-bang solution. In the proof of Theorem 1 the following relation will play a crucial role

(2.8) $R_h^o S_h(t) - S_h(t) = A_h \int_0^t S_h(t-z)(R_h^1 - R_h^o) S(z) dz$

Using the above one can prove the following analogues of analytic estimates:

Lemma 2

(2.9) $\| A_h S_h(t) x \|_{L_2(\Omega)} \leq \frac{C}{t^{1-\varepsilon}} \| x \|_{H^\varepsilon(\Omega)}$

(2.10) $\| (A\, S(t) - A_h S_h(t)) x \|_{L_2(\Omega)} \leq \frac{Ch^\alpha}{t^{1-\varepsilon}} \| x \|_{H^{\alpha+\varepsilon}(\Omega)}$

To prove Theorem 1 one writes:

$$y(u,t) - y_h(u_h, t) = S(t) y_o - S_h(t) R_h y_o + $$

$$+ (Lu)(t) - (L_h u_h)(t)$$

On the other hand

$$\| (Lu)(t) - (L_h u_h)(t) \| \leq \| \int_0^t [A\, S(t-z) - A_h S_h(t-z)] Du(z) dz \|$$

(2.11) $+ \| \int_0^t A_h S_h(t-z)(D - D_h)\, u(z)\, dz \| + \| \int_0^t A_h(t-z) D_h (u_h - u)(z) \|$

To complete the proof one has to estimate each term on R H S of (2.11) which can be accomplished by using Lemma 2 and (2.5),(2.6).

3. Approximation of control problem using subspaces without boundary conditions.

Although the general question of approximating the optimal control problem is completely solved in sections 2-3, the proposed numerical techniques require however that certain finite dimensional subspaces satisfy zero boundary conditions. This requirement is in general not easy to accomplish (unless we work with polygonal domains). It is done in practice by the use of curvilinear elements [Z1,Z2]. Therefore a desiriable feature of the discretation scheme is to approximate among function which are not required to vanish on the boundary. In order to accomplish this, we shall use some results from elliptic theory given in [N.1] [B.2]. In order to obtain the same as before optimal (see Theorem 2) rate of convergence those methods will require using more restrictive splines. To be more specif, let $V_h \subset H'(\Omega)$ be finite dimensional subspace with the usual approximating properties and also satisfying the following requirements

(3.1) $V_h \,|_\Gamma \subset H'(\Gamma)$;

(3.2) $\left|\frac{\partial V_h}{\partial n}\right|_{L_2(\Gamma)} \leq Ch^{-1/2} |V_h|_{H'(\Omega)}$

Note V_h is not required to satisfy zero boundary conditions (as oppose to V_h^o) Let $A_h : U_h \to V_h$ be defined as follows [N1]:

(3.3) $(A_h u_h, V_h) \triangleq a(u_h, V_h) - (u_h, \frac{\partial}{\partial \eta} V_h)_{L_2(\Gamma)}$

$$-\left(\frac{\partial u_h}{\partial \eta}, V_h\right)_{L_2(\Gamma)} + \beta h^{-1} (u_h, V_h)_{L_2(\Gamma)} \qquad \forall V_h$$

It can be shown that the estimates (2.9),(2.10) are still valid. To approximate Dirichlet map using spaces without boundary conditions we choose Reyleigh-Ritz method (see [B.2] To elaborate, let $U_h \triangleq$ space of piecewise constant functions on $\Gamma \subset L_2(\Gamma)$.
Let V_h' be a space of cubic splines. Define: (see [B.2]).

$D_h : U_h \to V_h$; by the following formula

(3.4) $(-A(x,\partial) D_h u_h,\ A(x,\partial)\ \xi_h)_{L_2(\Omega)} +$

$$+\ h^{-3} (u_n - D_n u_n, \xi_n)_{L_2(\Gamma)} = 0 \ \forall\ \xi_h \in V_h$$

It is shown in [B.2] that the estimate (2.6), (2.5) are valid with D_h defined by (3.4) and with V_h being a space of cubic splines.

Remark

Note that V_h space is not required to satisfy zero boundary conditions and the alghorithm (3.1) can be applied to any function $u_h \in L_2(\Gamma)$. On a negative side in order to obtain the optimal rate of convergence (the one claimed by Theorem 2) we have to use more restrictive subspaces (see (3.1)(3.2)).

Similarly as before we define an approximation of a parabolic equation (1.8) by the formula (2.7), with D_h, A_h defined by (3.2), (3.1) respectively. For the above approximation (with cubic splines) finally we can prove the convergence results claimed in Theorem 1 and Theorem 2 (for details see [L.3]).

REFERENCES

[B.1] I. Babuska and A. Aziz, "The Mathematics Foundations of the Finite Element Method with Applications to Partial Differential Equations", Academic Press, New York, 1972.

[B.2] A.V. Babakrishnan, "Applied Functional Analysis", Springer-Verlag, Berlin, 1976.

[B.3] J. Bramble, A. Schatz Reyleigh-Ritz Galerkin Methods for Dirichlet Problem using subspaces without Bound Cond. Comm on Pure and Applied Math. Vol XXIII 653-675 (1970).

[F.1] H. Fattorini, The Time Optimal Problem for Boundary Control of the Heat Equation Calculus of Variations and Control Theory. D.L. Russell, ed. Academic Press, New York. pp. 305-320, 1976.

[F.2] D. Fujiwara, Concrete Characterization of the domains of fractional powers of some elliptic differential operators of the second order. Proc. Japan Acad. 43 (1967), 82-86.

[L.1] I. Lasiecka. Boundary Control of Parabolic Systems: Finite element approximation. Appl. Math. Optim. 6, 31-62, (1980).

[L.2] I. Lasiecka. Unified theory for abstract parabolic boundary problems. Applied Math. Optim. 6, 287-333. (1980).

[N.1] J. Nitsche, Uberin Variationsprinzip zur Losung von Dirichlet-Problems: App. Math. Sem. Univ. Hamburg 36 (1971).

[S.1] G. Schmidt, N. Weck. On the Behavior of solutions to elliptic and parabolic equations-with applications to boundary control for parabolic equations. SIAM J. Control and Optim. Vol. 16, 4, 493-538, 1978.

[S.2] T. Seidman. Approximation methods for distribed systems. Mathematics Program at UMBC, Research Report, 79-18.

[V.1] R.S. Varga. Functional Analysis and Approximation Theory in Numerical Analysis. Rep. Conf. Ser. Appl. Math. publ. by SIAM, Philadelphia, 1971.

[W.1] D. Washburn. A bound on the boundary input map for parabolic equations with applications to time optimal control. SIAM J. Control and Optim. Vol. 17, No.5, 1979.

[Z.1] M. Zlamal. "Curved elements in the finite element method I", SIAM J. Number. Anal. 10, 229-240, 1973.

[L.3] I. Lasiecka, Ritz Salerkin Approximation of time-optimal control problem for parabolic systems with Dirichlet boundary conditions (submitted to SIAM J. Control).

Dirichlet boundary control problems for parabolic equations with quadratic cost: Analyticity and Riccati's feedback synthesis

Irena Lasiecka and Roberto Triggiani
Mathematics Department, University of Florida
Gainesville, Fla. 32611

Abstract

Riccati type feedback synthesis of optimal controls for Dirichlet boundary parabolic equations is considered. The functional cost penalizes the L_2 -energy over $[0,T]$ of state and control action u and also final state $y(T)$ at $t = T$. This latter fact, makes the functional cost discontinuous on the space of admissible controls: $L_2(\Sigma)$; $\Sigma = [0T] \times \Gamma$. After overcoming some technical difficulties related to the above mentioned discontinuity, we prove that the optimal control u^o can be written in the desired feedback form:

$$u^o(t) = -CP(t)\, y^o(t) \qquad \text{for all } 0 \le t < T$$

2. Introduction and statement of the main results

We consider a parabolic equation in y, defined on a bounded domain Ω with boundary Γ and with control function u acting in the Dirichlet boundary conditions:

$$\frac{\partial y(t,\xi)}{\partial t} = -A(\xi,\partial)\, y(t,\xi) \text{ in } (0,T) \times \Omega \equiv Q$$

$$(1.1) \qquad y(0,\xi) = y_o(\xi) \qquad \xi \in \Omega$$

$$y(t,\sigma) = u(t,\sigma) \quad (0,T] \times \Gamma \equiv \Sigma$$

where $A(\xi, \partial)$ is a uniformly strongly elliptic operator of order two in Ω. We next study the optimal boundary control problem: Minimize the performance index:

$$J(u,y(u)) \equiv |u|_\Sigma^2 + |y|_Q^2 + \alpha |y(T)|_\Omega^2$$

over all $u \in L_2(\Sigma)$, subject to the dynamics (1.1). Here α denotes either 1, or else 0.

Remark 1 It is known [1] that the response y to an $L_2(\Sigma)$- control may not have a well defined final point $y(T)$ in $L_2(\Omega)$. In this case the value of y is $y(u,u(y)) = \infty$ for $\alpha = 1$. This is a pathology that will have to be treated. □

A main goal of the present paper is to establish a pointwise (in t) feedback synthesis of the optimal control u^o in terms of the corresponding optimal solution y^o. Our major result is the following feedback synthesis of Riccati type for the case $\alpha = 1$: The optimal control u^o can be written as:

$$u^o(t) = -D^*A^* \; P(t) \; y^o(t) \quad 0 \le t < T. \tag{1.2}$$

Here A is the differential operator $-A(\xi,\partial)$ with zero Dirichlet boundary conditions, D is the "Dirichlet map" associated with the corresponding elliptic problem: $v = Dg$, where $A(\xi,\partial)\, v = 0$ in Ω and $v = g$, on Γ and P(t) is the Riccati operator which satisfies the following Riccati type equation:

$$[\dot{P}(t)\, x,y] = -[x,y] - [P(t)\, x, Ay] - [P(t)Ax,y] + [D^*A^*P(t)\, x, D^*A^*P(t)y] \quad 0 \le t < T \quad \text{for } x,y \in \mathcal{D}(A^{\varepsilon}) \tag{1.3}$$

and

$$\lim_{t \to T} [P(t)x,y] = [x,y] \quad \text{for } x,y \in \mathcal{D}(A)^{1-\varepsilon} \tag{1.4}$$

Note that D^*A^* is an unbounded, unclosable operator (it represents the normal derivative on the boundary)

<u>Remark 2</u> If $\alpha = 0$ then Riccati equation is satisfied in a stronger sense, i.e. for all $x,y \in L_2(\Omega)$. Moreover, it can be shown that $P(t) \subset \mathcal{D}(A)$, that the terminal value $P(T)$ is well defined in $L_2(\Omega)$ and equal to zero. By contrast, the situation with final state penalization (i.e. $\alpha=1$) is regular only up to the final point T, excluding T, (see (1.3) and Remark 1).

3. The large literature on quadratic control problems and Riccati equations shrinks however to only a few references, where it comes to boundary control problems, see [2.3] in particular as to the case where the control function acts in the Dirichlet boundary conditions we can quote onlv Balaknishnan's work [2] which only the less technical case $\alpha = 0$. His approach however is <u>indirect</u>. In contrast, our approach is <u>direct</u>; i.e. the operator P(t) is first defined by an explicit formula in terms of the system's date, and only subsequently shown to satisfy a Riccati-type operator

equation. This way the problem of existence of a solution to the Riccati equation for $\alpha = 1$ is automatically guaranteed. Instead, the need to prove existence in the indirect approach or through other techniques meets serious technical difficulties. Our procedure develops alongthe following steps:

1) Step 1: By using Lagrange formalism, we prove that the optimal control can be expressed as:

$$(2.1) \qquad u^o = -[I + L_o^* L_o + L_{T_o}^* L_{T_o}]^{-1}[L_o^* S(\cdot)y_o + L_{T_o}^* S(T)y_o]$$

where

$$(L_s u)(t) \triangleq -A \int_s^t S(t-z)Du(z)dz$$

$$L_{sT}u \triangleq -A \int_s^T S(T-z)Du(z)dz$$

with S(t) the analytic semigroup generated by A.

Notice that by virtue of Remark 1, the operator L_T is an unbounded operator from $L_2(\Sigma)$into $L_2(\Omega)$. However, due to analyticity of $S(\cdot)$ and closedness,$(I+L_o^* L_o + L_T^* L_T]^x$ is invertible on $L_2(\Sigma)$. Thus the formula (2.1) defines an element in $L_2(\Sigma)$. By analizing farther the structure of right hand side at (2.1), we can prove that u^o is in fact an analytic $L_2(\Gamma)$- function on (0,T)and continuous at t = 0 the corresponding trajectory y^o is a continuous $L_2(\Omega)$-function on [0,T] and analytic on (0,T). A subset of the above regularity results, play a crucial role in the development.

2. Step 2: Define, the operator $\phi(t,s)$ by:

$$(2.2) \qquad \phi(t,s)x = S(t-s)x + [L_s[I+L_s^* L_s+L_{sT}^* L]^{-1}]\cdot$$

$$\cdot[L_s^* S(-s)x + L_{sT}^* S(T-s)x](t)$$

As before, since $I + L_s^* L_s + L_{sT}^* L_{sT}$ is invertible on $L_2(\Sigma)$, $\phi(t,s)$ is well defined in $L_2(\Omega)$ for all $s \leq t < T$ the formationed regularity results of the optimal solution (particularity), continuity of $y^o(t)$) yield that $\phi(t,s)$ is, in fact, an evalution operator, strongly continuous in $t \in [s,T]$ (actually analytic in (s,T)) and also strongly continuous in $s \in [0,t]$, $t < T$. (T excluded). As to $t = T$, the operator $\phi(T,s)$ is strongly continuous in $s \in [0,T]$.

Using these results, we also prove that the following integral:

$$\int_t^T S^*(z-t)\, \phi\, (z,t)\, x\, dz$$

is well defined in $L_2(\Omega)$ for $x \in L_2(\Omega)$.

3. Step 3: Next, we can define constructively an operator P(t) by:

$$P(t)x = \int_t^T S^*(z-t)\ \phi\ (z,t)x\ dz + S^*(T-t)\ \phi\ (T,t)x \tag{2.3}$$

for all $0 \le t < T$ and $x \in L_2(\Omega)$

4. Step 4: The regularity results established in Step 1. yield that $\phi(t,s)$ satisfies the following evaluation equation:

$$\frac{\partial\phi(t,s)x}{\partial t} = A\ [I-DD^*A^*\ P(t)]\ \phi\ (t,s)x \tag{2.4}$$

for all $x \in L_2(\Omega)$ and $0 \le s < t < T$ (excluding T).

The above equation and the fact that Range $P(t) \subset \mathcal{D}(A^{1-\varepsilon})$ for $t < T$ finally allow us to verify that the operator P(t) defined by formula (2.3), satisfies in fact Riccati equation.

5. Step 5: By means of formulas (2.1), (2.2) and (2.3) we finally arrive at the synthesis relation:

$u^o(t) = D^*A^*\ P(t)\ y^o(t)$ for $0 \le t < T$ (T excluding) as derived.

References:

[1] J.L. Lions, Optimal Control of Systems Governed by Partial Differential Equations, Springer Verlag 1971.

[2] R. Curtain - A. Pritchard "An Abstract theory for unbounded control action for distributed parameter systems" SIAM J. Control and Optimization 15 (1977), 566-611.

[3] A.V. Balaknishnan "Boundary Control of Parabolic Equations: L-Q-R Theory" Proc. V. Int. Sümmer Schl. Central Inst. Mel Mech. Acad. Sci. GDR Berlin, 1977.

OPTIMIZATION IN BANACH SPACES OF SYSTEMS
INVOLVING CONVEX PROCESSES

by
J.Ch. POMEROL

Université P. et M. Curie
Paris

1 - Introduction

The aim of this paper is to give the main Lagrange multiplier existence theorems in convex programs involving convex multiapplications. Namely we consider the program (P_y) :

$$(P_y) \begin{cases} \text{minimize } (f(x) - \langle x,y \rangle) \\ \text{subject to } x \in C \text{ and } Ax \cap T \neq \emptyset \end{cases} .$$

where X and U are two L.C.T.V.S., f is a convex, lower semicontinuous (l.s.c.) functional from X into $\overline{\mathbb{R}}$ (the extended real line), C is a closed convex subset of X, T (the target set) is a closed convex subset of U. The multiapplication $A : X \to 2^U$ is supposed to be convex and closed which means that $\mathrm{gr}\, A = \{(x,u)/u \in Ax\}$ is a closed convex subset of X x U. We denote by Y (resp. V) the topological dual of X (resp. U) endowed with the weak* topology, while $\langle .,. \rangle$ denotes the inner products. This kind of programs was already studied by Pham Hũu Sách [16, 17] and Oettli [15] .

We shall study (P_y) by means of the bifunctions as developed by Rockafellar [24, 26] . The first section is devoted to the primal results. In the second ones, where we only deal with convex processes in Banach spaces, we give some properties related to the conjugate of A.

2 - Primal results

Let us introduce the bifunction

$$F(x,u) = \begin{cases} f(x) \text{ if } \exists\, x \in C \text{ such that } (Ax + u) \cap T \neq \emptyset \\ + \infty \text{ otherwise.} \end{cases}$$

It is easy to check that F is a l.s.c. convex functional.

The conjugate of F is equal to $G(y,v) = \sup_{x,u} (\langle x,y \rangle + \langle u,v \rangle - F(x,u))$.

By a straightforward calculation we get

$$G(y,v) = \sup_{t \in T} \langle t,v \rangle + \sup_{x \in C} (\langle x,y \rangle - f(x) - \langle Ax,v \rangle)$$

with $\langle Ax, v \rangle = \inf_{z \in Ax} \langle z,v \rangle$.

A Lagrange multiplier for the program (P_y) is a vector v_o such that :

(1) $$h_y(u) = \inf_x (F(x,u) - \langle x,y\rangle) = -\inf_v G(y,v) = -G(y, v_o) .$$

Noticing that for every $v \in V$

$$\inf \{f(x) - \langle x,y\rangle + \langle Ax - t,v\rangle \,/\, (x,t) \in C \times T\} \leqslant$$

(2) $$\inf \{f(x) - \langle x,y\rangle \,/\, (x,t) \in C \times T \text{ and } \exists\, z \in Ax \text{ such that } \langle z-t,v\rangle = 0\}$$

$$\leqslant \inf \{f(x) - \langle x,y\rangle \,/\, x \in C \text{ and } Ax \cap T \neq \emptyset\}$$

we conclude that (1) is equivalent to (2) where the inequalies are replaced by equalities and v by v_o.

The equality of the primal and dual values is equivalent to the equalities (q_y).

(q_y) $$\inf \{f(x) - \langle x,y\rangle \,/\, x \in C \text{ and } Ax \cap T \neq \emptyset\} = \sup_v \inf\{ f(x) - \langle x,y\rangle \,/\, (x,t) \in C \times T \text{ and } \exists\, z \in Ax \text{ such that } \langle z-t,v\rangle = 0\} = \sup_v \inf \{f(x) - \langle x,y\rangle + \langle Ax - t,v\rangle / (x,t) \in C \times T\} .$$

As a direct application of [18, prop. V.3.5] we get the following result, where $A(D) = \bigcup_{x \in D} Ax$ for any subset D.

Proposition 1 - Assume that there exists $x_o \in C \cap \operatorname{dom} f$ satisfying $Ax_o \cap T \neq \emptyset$. If for any $\varepsilon > 0$ there exists a 0-neighborhood N_ε in U such that $N_\varepsilon \cap \overline{[T - A(C \cap \{x/f(x) \leqslant r\})]}$ is contained in $N_\varepsilon \cap [T - A(C \cap \{x/f(x) \leqslant r + \varepsilon\})]$ for every $r < f(x_o) + 1$ then (q_o) holds.

Remark 1 - The second assumption of Proposition 1 is satisfied if there exists a 0-neighborhood N in U such that $N \cap [T - A(C \cap \{x/f(x) \leqslant r\})]$ is closed for every $r < f(x_o) + 1$. Thus the proposition 1 generalizes the results of Rolewicz [27], [28, th. 2.1 and 5.1] [29, th. 1], Dolecki [6, th.III 6], and Singer [36, th. 2.1 et seq.] , all results which are concerned with linear systems.

In order to obtain the closedness of $T - A(C \cap \{x/f(x) \leqslant r\})$ it suffices that either T or $A(C \cap \{x/f(x) \leqslant r\})$ be compact, when A is upper semicontinuous it suffices that either C or $\{x/f(x) \leqslant r\}$ be compact [1, Ch. V., th.3] which generalizes a result of Pham Hũu Sách [16, th.4.2]. Some other closedness conditions in the line of Dieudonné's theorem can be founded in [5] and [10] .

Proposition 2 - Assume that X and U are normed T.V.S. then (P_0) has a Lagrange multiplier iff there exist $d \geq 0$ and a 0-neighborhood N in U such that :

$\forall \varepsilon > 0$ $\forall (x,u) \in (C \cap \text{dom } f) \times N$ satisfying $(Ax + u) \cap T \neq \emptyset$
there exists $x_\varepsilon \in C$ satisfying $Ax_\varepsilon \cap T \neq \emptyset$ and $f(x_\varepsilon) \leq f(x) + d\|u\| + \varepsilon$.

This condition is related to Gale duality theorem [9] (see [19]).

Corollary 1 - Assume that A is an injective, continuous linear operator and $T = \{t_0\}$. Then (P_0) has a Lagrange multiplier if f has a subgradient at $x_0 \in C$, with $Ax_0 = t_0$, and A is an homomorphism [34, p. 75].

It is well known that the existence of a Lagrange multiplier for (P_0) is equivalent to the existence of a subgradient of $h_0(u)$ at 0 [26, th. 16]. Thus the most popular sufficient conditions for the existence of a Lagrange multiplier are those implying the continuity of $h_0(u)$ at 0 (equivalently that h_0 is bounded from above on a 0-neighborhood).

Proposition 3 - (P_0) has a Lagrange multiplier if

(i) $\exists\, x_0 \in (\text{int } C) \cap \text{dom } f$ such that $t_0 \in Ax_0 \cap T$ and f is continuous at x_0,

(ii) A is open at (x_0, t_0) [18, XII. 2.5].

Proof - There exists a neighborhood W of x_0 such that f is bounded on W. For u belonging to the 0-neighborhood T - A(W), h_0 is bounded from above by the same real than f. Q.E.D.

Corollary 2 - Assume that A is a continuous linear operator and that (i) of Proposition 3 holds. Then (P_0) has a Lagrange multiplier if A is an homomorphism and $T \subset A(X)$.

Remark 2 - The previous results appear under various form in the literature. When $T = \{0\}$ and $C = X$ we can find them in Borwein [2, 3, 4]. The same vein is exploited by Dolecki [7], see also [8], where $T = \{0\}$ and A replaced by A^{-1}, so that the openness is replaced by the upper semi-continuity of A implying the openness of A^{-1}. For a linear operator see [36].

The existence of a Lagrange multiplier for every program (P_y) is asserted by the following result [12, prop. 3.2].

Proposition 4 - If there exists $x_o \in C \cap \mathrm{dom}\, f$ such that $Ax_o \cap T \neq \emptyset$ then (P_y) has a Lagrange multiplier for every $y \in Y$ iff $K = \{(y,t) / \exists\, v$ such that $G(y,v) \leq t\}$ is closed.

In Banach spaces, a condition implying that K is closed is the generalized Slater condition [12].

Proposition 5 - Assume that X and U are two Banach spaces then (P_y) has a Lagrange multiplier for every $y \in Y$ if

$$0 \in \mathrm{core}\ (T - A(C \cap \mathrm{dom}\ f))$$

Remark 3 - The above Slater condition is satisfied if there exists $x_o \in C$ such that either $Ax_o \cap \mathrm{int}\, T \neq \emptyset$ or $T \cap \mathrm{int}(Ax_o) \neq \emptyset$, conditions given by Phan Hũu Sắch [16, Th.4.1]. It appears in full generality in Oettli [15, th. 2 and 3] and for $T = \{0\}$ and $C = X$ in Borwein [2,4]. It also implies some boundedness properties of the Lagrange multiplier set [20].

3 - Dual conditions

To obtain the existence of a Lagrange multiplier, we shall prove the closedness of the set K defined in Proposition 4. The set K is closed when the following condition (C) is fulfilled [12, th. 4.2].

(C) X is a Banach space or Y is normed for a topology compatible with the pairing and $\forall\, k > 0\ \exists\, D_k$ a weak* compact subset of V such that for every (y, θ) satisfying $G(y,v) \leq \theta$, $\|y\| \leq k$, $|\theta| \leq k$ one can find $v' \in D_k$ satisfying $G(y,v') \leq \theta$.

Proposition 6 - Assume that A is a continuous linear operator. Then (P_y) has a Lagrange multiplier for every $y \in Y$ if

(i) $\exists\, x_o \in (\mathrm{int}\, C) \cap \mathrm{dom}\, f$, with $Ax_o \cap T \neq \emptyset$, and such that f is continuous at x_o

and (ii) X is a Banach space, $T \subset \overline{A(X)}$ and A is an homomorphism.

Proof - We have $G(y, v) = \sup_{t \in T} \langle t,v \rangle + (f + \psi_C)^*(y - A^*v)$ where ψ_C is the indicator function of the set C. Using the fact that the level sets of $(f + \psi_C)^*(.) - \langle x_o, . \rangle$ are equicontinuous and the properties of the homomorphism [18, VII.3.1] we can show that (C) is fulfilled, see [18,XI.3.5] .

Remark 4 - The assumption (ii) is satisfied when X and U are Banach spaces and A(X) is closed [34, IV.7.7]. Thus Proposition 6 generalizes various results of Singer [35, cor.1], [36, th.4.4] and Rolewicz [28,th.3.1], [29, th.2].

For simplicity sake we assume now that X and U are Banach spaces and that A is a closed convex process (i.e. the graph of A is a closed convex cone with apex 0). The basic facts on convex process are in [4, 13, 22, 23, 24, 25, 30, 31, 32, 33].

To follow the same way as in the linear case, we have to define $< x, A^{*}v >$.

We have already defined $< Ax, v> = \inf_{z\in Ax} < x,v >$ which means that A is a min-oriented process [24]. In that case we introduce $A^{*}v = \{y/(v,-y) \in (gr\ A)^{\circ}\}$ where $(gr\ A)^{\circ}$ is the polar set of gr A. Thus A^{*} is a max-oriented process [24]and $< x, A^{*}x> = \sup_{y\in A^{*}v} < x, y >$.

In a Banach space X we use the notation $B_X(k) =\{ x/ \|x\| \leq k\}$ and $B_X(\bar{x}, k) = \bar{x} + B_X(k)$.

Proposition 7 - Assume that either $C \subset \text{dom } A$ or $\text{dom } A^{*} = V$ then $\forall x \in C \ \ \forall v \in V <Ax, v > = < x, A^{*}v >$ provided that one of the following assumption (ai) is satisfied.

(a1) $\forall x \in C \ \forall v \in V \ \ w_v(x) = \inf_{u\in Ax} <u,v >$ is l.s.c. (see [21, lemma 1 and th.2], [32,th.1] and [18, Ch. XII, § 1]).

(a2) $\forall k > 0 \ \ A(C \cap B_X(k))$ is compact (boundedness condition, see [31 and 32]).

(a3) $\forall x \in C \ \exists k > 0$ such that $A(B_X(x,k))$ is compact

(a4) $\forall x \in C$ A is upper semicontinuous at x

(a5) $C \subset \text{core dom } A$ (a6) dom A = X (a7) $\text{dom } A^{*} = V$

(a8) $\exists k > 0$ such that $A^{*}(B_V(k)) \subset B_Y(1)$.

Assuming the equality $< Ax, v > = < x, A^{*}v >$ we get

$$G(y,v) = \sup_{t\in T} <t,v > + \sup_{x\in C} \inf_{z\in A^{*}v} (<x,y > - <x,z > - f(x)).$$

To inverse the supremum and the infinum in the above expression we need a new assumption.

Proposition 8 - If one of the following assumptions (bi) is satisfied one has

$$\forall v \in V \quad \sup_{x \in C} \inf_{z \in A^* v} (\langle x, y-z \rangle - f(x)) = \inf_{z \in A^* v} \sup_{x \in C} (\langle x, y-z \rangle - f(x))$$

(b1) $\exists\, k > \inf_{x \in X} f(x)$ such that $\{x / f(x) \leqslant k\} \cap C$ is compact [18,XI.1.8]

(b2) $\forall v \quad \forall y \; \exists x_o \in C \; \exists k < (\langle x,y \rangle - f(x_o) - \langle x_o, A^* v \rangle)$ such that $\{z / (\langle x_o, y \rangle - f(x_o) - \langle x_o, z \rangle) \leqslant k\} \cap A^* v$ is weak*-compact [14] .

(b3) $\forall v \in V \quad A^* v$ is weak*-compact

(b4) f is continuous at 0 with $0 \in \operatorname{int} C$ [11, prop. 8] .

Now with (ai) and (bi) we have $G(y,v) = \sup_{t \in T} \langle t,v \rangle + \inf_{z \in A^* v} (f + \psi_C)^*(y-z)$, and we can give two conditions implying that (C) is fulfilled.

Theorem - Assume that X and U are two Banach spaces, A is a closed convex process, either $C \subset \operatorname{dom} A$ or $\operatorname{dom} A^* = V$, one of the (ai) and one of the (bi) hold. Then (P_y) has a Lagrange multiplier for every $y \in Y$ if :

(i) $\exists x_o \in (\operatorname{int} C) \cap \operatorname{dom} f$ such that $A x_o \cap T \neq \emptyset$ and f is continuous at x_o ,

and either (ii) $\lim_{\|v\| \to +\infty} \inf_{y \in A^* v} \|y\| = +\infty$

or (iii) $\forall t \in T \quad \forall w \in \{w / \exists (v,v') \in V^2$ such that $w = v - v'$ and $A^* v \cap A^* v' \neq \emptyset\}$ one has $\langle t,w \rangle = 0$ and A is soft-open at zero [18, XII.2.1] .

Proof - In the case (i) + (ii) the proof is similar to that of Proposition 6. In the second alternative we use the fact that for A being soft-open if there exists k such that $\|z\| \leqslant k$ and $z \in A^* v$ then there exists k' such that $z \in A^* v'$ with $\|v'\| \leqslant k'$ [18, XII.3.1] .

Remark 5 - The condition (iii) is obviously satisfied if $T = \{0\}$ and $\operatorname{Im} A = U$. Actually, gr A being a cone, A is soft-open at 0 is equivalent to [18, XV.1.5] :

$$\exists\, k_o > 0 \text{ such that } \quad (\operatorname{gr} A + B_{X \times U}(k_o)) \cap (X \times \{0\}) \subset (A^{-1}(0) + B_X(1)) \times \{0\} .$$

REFERENCES

[1] Berge C., 1966 : Espaces topologiques, fonctions multivoques, Dunod, Paris.

[2] Borwein J., 1977 : Multivalued convexity and optimization : a unified approach to inequality and equality constraints, Math. Progr. 11, 183-199.

[3] ———— , 1980 : Convex relations in analysis and optimization, NATO Proceedings in generalized concavity, to appear.

[4] ———— , 1981 : Adjoint process duality, Carnegie-Mellon University D.P.

[5] Dedieu J.P., 1978 : Critères de fermeture pour l'image d'un fermé non convexe par une multiapplication, C.R. Acad. Sci. Paris 187, 941-943.

[6] Dolecki S., 1977 : Bounded controlling sequences, lower stability and certain penalty procedures, Applied Math. and Opt. 4, 15-26.

[7] ————, 1978 : Semicontinuity in constraints optimization I and II, Control Cyber. 7, 5-26 and 51-68.

[8] ———— and S. Rolewicz, 1979 : Exact penalties for local minima, SIAM J. Control 17, 596-606.

[9] Gale D., 1967 : A geometric duality theorem with economic applications, Rev. Econ. Studies 34, 19-24.

[10] Gwinner J., 1977 : Closed images of convex multivalued mappings in linear topological spaces with applications, J. Math. Anal. Appl. 60, 75-86.

[11] Lévine P. and J.Ch. Pomerol, 1978 : Quelques extensions des théorèmes "inf-sup", C.R. Acad. Sci. Paris 187, p. 565-567.

[12] ———————————— , 1979 : Sufficient conditions for Kuhn-Tucker vectors in convex programming, SIAM J. Control 17, 689-699. Erratum, same Journal 19, 1981, 431-432.

[13] Makarov V.L. and A.M. Rubinov, 1970 : Superlinear point-set maps and models of economic dynamics, Russ. Math. Surveys 25, N° 5, 125-170.

[14] Moreau J.J., 1964 : Théorèmes "inf-sup", C.R. Acad. Sci. Paris 258, 2720-2722.

[15] Oettli W., 1980 : Optimality conditions for programming problems involving multivalued mappings, to appear in Proc. Summer School on Opt. and Oper. Research, Bad Honnef 1979 (North Holland).

[16] Pham Hữu Sách, 1976 : Theory of control of processes with multivalued operators, Cybernetics, N° 2, 285-295.

[17] ———— , 1979 : Optimisation vectorielle des systèmes multivalués, Thông Báo Nghiên Cứu D.P. 1, Hànội, in Russian.

[18] Pomerol J.Ch., 1980 : Contribution à la programmation mathématique : existence de multiplicateurs de Lagrange et stabilité. P. and M. Curie University Thesis.

[19] Pomerol J.Ch., 1981 : Application of Gale's duality theorem to programming, Operations Research Verfahren 40, 137-140.

[20] ———————— , 1981 : The boundedness of the Lagrange multipliers set and duality in mathematical programming, Zeitschrift für Operations Research 7, in press.

[21] Pshenichnyi B.N. and I.B. Medvedovskii,1976 : A general result on convex analysis, Cybernetics, N° 1, 64-69;

[22] Robinson S.M., 1972 : Normed convex processes, Trans. Amer. Math. Soc. 174, 127-140.

[23] ———————— , 1976 : Regularity and stability for convex multivalued functions, Math. Oper. Res. 1, 130-143.

[24] Rockafellar R.T., 1970 : Convex analysis, Princeton University Press, Princeton.

[25] ———————— , 1973 : Convex algebra and duality in dynamic models of production, Math. Models in Economics, J. Łoś and M.W. Łoś eds; 351-378, North Holland, Amsterdam.

[26] ———————— , 1974 : Conjugate duality and optimization, Regional conf. series in Applied Mathematics 16, SIAM, Philadelphia.

[27] Rolewicz S., 1968 : On a problem of moments, Stud. Math. 30, 183-191.

[28] ————————, 1976 : On general theory of linear systems, Beiträge zur Analysis 8, 119-127.

[29] ———————— , 1976 : Linear systems in Banach spaces, Calculus of variations and control theory, D.L. Russell ed., 245-255, Academic Press, New York.

[30] Rubinov A.M., 1967 : A mathematical production model, Soviet. Math. Dokl.8, p. 681-683.

[31] ————————, 1968 : Duals models of productions, Soviet. Math. Dokl. 9, p. 691-694.

[32] ————————, 1977 : Sublinear operators and their applications, Russian Math. Surveys 36, N° 4, 115-175.

[33] Ruys P.H.M., 1974 : Public goods and decentralization, Tilburg University Press, the Netherlands.

[34] Schaefer H.H., 1971 : Topological vector spaces, Springer, New York.

[35] Singer I., 1973 : On a problem of moments of S. Rolewicz, Stud. Math. 48, 95-98.

[36] ————————, 1980 : Duality theorems for linear systems and convex systems, J. Math. Anal. Appl. 76, 339-368.

Address : Laboratoire Econométrie, 4 Place Jussieu, 75230 PARIS CEDEX 5, FRANCE.

A DECOMPOSITION ALGORITHM FOR A SECOND ORDER ELLIPTIC OPERATOR USING ASYMPTOTIC EXPANSIONS

H. Salhi and D. P. Looze
Coordinated Science Laboratory
University of Illinois
1101 W. Springfield Ave.
Urbana, Illinois 61801, U.S.A.

Introduction

The analysis of lumped and distributed systems whose models depend on a small parameter ε has received much attention over the last two decades [1]-[3]. The introduction of ε can have physical significance or may be completely artificial. For example, systems having the "time scales" or "space scales" properties fall into the first category. However, in the second category the small parameter ε is introduced artifically in order to obtain relevant information (such as regularity of solutions) about the problem at hand from the penalized problem, because the latter is simpler in some sense [4].

In this paper, we consider a distributed system of the first category, whose model is the following formal second order elliptic operator:

$$A_\varepsilon := \begin{bmatrix} -\sum_{i=1}^{p} \frac{\partial^2}{\partial x_i^2} & 0 \\ 0 & -\varepsilon \sum_{i=1}^{p} \frac{\partial^2}{\partial x_i^2} \end{bmatrix} := \begin{bmatrix} -\Delta & 0 \\ 0 & -\varepsilon\Delta \end{bmatrix}$$

$0 < \varepsilon << 1.$

Many physical problems can be described by models containing A_ε, e.g.,

1) Heat conduction in media with very different diffusivities [5].

2) Electromagnetic wave propagation in waveguides made of materials having different permittivities [5].

3) Small vibrations of elastic interfaced thin membranes with very different material densities [5].

4) Some stochastic problems [5].

This paper is organized as follows. In Section 1, the problem formulation is presented. In Section 2, we study the spectrum of A_ε. It is shown that the eigenvectors of A_ε are not analytic in ε. A counterexample is given, illustrating the

*This work was supported in part by the Joint Services Electronics Program under Contract N00014-79-C-0424 and in part by the National Science Foundation under Grant ENG-79-08778.

incorrectness of the formal asymptotic expansions. However, using the zeroth term of the incorrect expansions, we construct a basis for $L^2(\Omega)$. In Section 3, we solve a nonhomogeneous boundary value problem using the basis derived in Section 2 and compare the results obtained with those of [6]. A one-dimensional heat conduction example is solved to illustrate the various ideas. In Section 4, we employ a discrete approximation of the aforementioned example to plot some eigenvectors for some values of ε.

Finally, due to lack of space all theorem proofs and the plots of the eigenvectors shall be omitted. However, the interested reader should consult [5] for these omissions and generalizations of the results presented here.

. Preliminaries and Problem Formulation

Let Ω be an open subset of $\mathbb{R}^p$ with boundary Γ. We denote[(1)] by $H^m(\Omega)$ the Sobolev space of order m, i.e.

$$H^m(\Omega) = \{\varphi : D^k\varphi \in L^2(\Omega), \qquad \forall k \quad |k| \leq m\}$$

where

$$k = \{k_1, k_2, \dots, k_p\}$$

$$|k| = k_1 + k_2 + \dots + k_p$$

$$k_\ell \in \mathbb{N}, \qquad \ell = 1,2,\dots,p$$

$$D^k = \frac{\partial^k}{\partial x_1^{k_1} \partial x_2^{k_2} \dots \partial x_p^{k_p}}$$

where differentiation is interpreted in the distribution sense. Let

$$H^m(\Omega;\Gamma_o) = \{\varphi : \varphi \in H^m(\Omega), \quad D^k\varphi = 0 \text{ on } \Gamma_o \subseteq \Gamma, \quad |k| < m-1\}.$$

If $\Gamma_o = \Gamma$, then $H^m(\Omega;\Gamma) = H^m_o(\Omega)$. It is well known that $H^m(\Omega)$ is a Hilbert space for the following norm [7],[9]:

$$\|\varphi\|_m = \Big(\sum_{|k|<m} \int |D^k\varphi|^2 dx\Big)^{1/2}.$$

It is also worthy to mention that $H^m_o(\Omega)$ is a dense subspace of $L^2(\Omega)$.

Now let $\Omega := \Omega_o \cup \Omega_1$ with boundary $\Gamma := \Gamma_o \cup \Gamma_1$ and let S be the interface boundary as indicated below:

---

(1) See [4],[6].

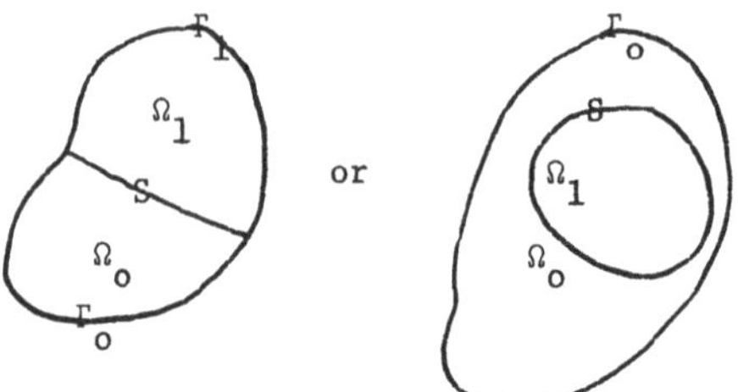

In the sequel, we shall assume:

1) Ω is bounded and satisfies the cone condition (A1)

2) Γ and S are sufficiently smooth (A2).

Let

$$W = \{(\varphi_o,\varphi_1) : \varphi_o \in H^2(\Omega_o),\ \varphi_1 \in H^2(\Omega_1)\} \cap H_o^1(\Omega)$$

be the domain of A_ε. With this definition, it can be easily verified that A_ε is self-adjoint. Because of assumptions A1-A2 the canonical injection of W into $L^2(\Omega)$ is compact and hence the spectrum of A_ε is a subset of $\mathbb{R}$, consisting only of the point spectrum. This eigenvalue problem is formulated as follows:

$$A_\varepsilon\varphi_\varepsilon = \lambda_\varepsilon\varphi_\varepsilon\ , \qquad \varphi_\varepsilon \in W^{(2)} \tag{1.1}$$

i.e.

$$-\Delta\varphi_o = \lambda\varphi_o \quad \text{in} \quad \Omega_o \quad , \quad \varphi_o|_{\Gamma_o} = 0 \tag{1.2}$$

$$-\varepsilon\Delta\varphi_1 = \lambda\varphi_1 \quad \text{in} \quad \Omega_1 \quad , \quad \varphi_1|_{\Gamma_1} = 0$$

$$\left.\begin{aligned} &\varphi_o = \varphi_1 \\ &\frac{\partial\varphi_o}{\partial\nu} + \varepsilon\frac{\partial\varphi_1}{\partial\nu} = 0 \end{aligned}\right\} \quad \text{on S} \tag{1.4}$$

where ν is the outward unit normal.

The weak formulation of this problem is as follows: Let $a(\varphi,\psi)$ be the bilinear form associated with A_ε, i.e.

$$\begin{aligned} a(\varphi,\psi) :&= (A_\varepsilon\varphi,\psi)_{L^2(\Omega)} \\ &= \sum_{i=1}^{p}\int_{\Omega_o}\frac{\partial\varphi}{\partial x_i}\frac{\partial\psi}{\partial x_i}dx + \varepsilon\sum_{i=1}\int_{\Omega}\frac{\partial\varphi}{\partial x_i}\frac{\partial\psi}{\partial x_i}dx \\ :&= a_o(\varphi,\psi) + \varepsilon a_1(\varphi,\psi) \qquad \forall\varphi,\psi \in H_o^1(\Omega). \end{aligned} \tag{1.5}$$

The weak formulation is then to seek $\varphi \in H_o^1(\Omega), \lambda \in \mathbb{R}$ such that

---

(2) We shall not indicate explicitly the dependence of φ, λ on ε for convenience.

$$a(\varphi,\psi) = \lambda(\varphi,\psi) \qquad \forall\psi\in H_o^1(\Omega). \tag{1.6}$$

We now summarize some well-known facts [7],[8] about the problem at hand:

Proposition 1.1: There exist unique sequences $\{\varphi_\ell\}_{\ell=1}^{\infty}\in H_o^1(\Omega)$, $\{\lambda_\ell\}_{\ell=1}^{\infty}\in\mathbb{R}$ such that (1.6) is satisfied. Furthermore,

1) $0<\lambda_1\le\lambda_2\le,\ldots,\ \lim\limits_{\ell\to\infty}\lambda_\ell=+\infty$.

2) The multiplicity of each eigenvalue is finite.

3) $\{\varphi_\ell\}_{\ell=1}^{\infty}$ is a complete orthogonal set of $L^2(\Omega)$.

4) The eigenvalues and the corresponding eigenvectors can be characterized using Raleigh quotients, i.e.

$$\lambda_\ell = \min_{V\subset H_o^1(\Omega),\ \dim V=\ell}\ \max_{\varphi\in V,\ \|\varphi\|_o=1} a(\varphi,\varphi). \tag{1.7}$$

2. Spectral Analysis of A_ε

Because of the structure of A_ε, let us assume that

$$\lambda = \lambda^o + \varepsilon\lambda^1 + \varepsilon^2\lambda^2 + \cdots \tag{2.1}$$

$$\varphi = \varphi^o + \varepsilon\varphi^1 + \varepsilon^2\varphi^2 + \cdots \tag{2.2}$$

Let $\{\bar\lambda_\ell\}_{\ell=1}^{\infty}$, $\{\underline{\lambda}_\ell\}_{\ell=1}^{\infty}$ be the ordered eigenvalues of $\bar A=-\Delta$, $\underline{A}=-\varepsilon\Delta$, respectively, with domain $\mathcal{D}(\bar A)=\mathcal{D}(\underline{A})=\mathcal{D}(A_\varepsilon)$. It can be easily shown using Raleigh quotients (see (1.7)) that $\{\lambda_\ell\}_{\ell=1}^{\infty}$ satisfy

$$\underline{\lambda}_\ell \le \lambda_\ell \le \bar\lambda_\ell .$$

With this fact in mind, we shall distinguish two cases:

Case 1: $\lambda^o\neq 0$

Putting (2.1),(2.2) in (1.2)-(1.4) and identifying formally equal powers of ε, we get

Lemma 2.1: The sequences $\{\lambda^k\}_{k=0}^{\infty}$, $\{\varphi^k\}_{k=0}^{\infty}$ satisfy

$$-\Delta\varphi_o^k = \sum_{i=0}^{k}\lambda^i\varphi_o^{k-i} \quad \text{in } \Omega_o, \quad \varphi_o^k\big|_{\Gamma_o}=0, \quad \frac{\partial\varphi_o^k}{\partial\nu}\Big|_S = 0 \tag{2.3}$$

$$\varphi_1^k = 0 \qquad \text{in } \Omega_1$$

$$k=0,1,2,\ldots$$

Lemma 2.2:

1) The sequence $\{\lambda^k\}_{k=1}^{\infty}$ is a null sequence.

2) The sequence $\{\varphi^k\}_{k=1}^{\infty}$ is constant, equal to $\varphi^o = (\varphi_o^o, 0)$.

Remark 2.1: Since $\varphi_o|_S \neq \varphi_1|_S$, $\varphi \notin H_o^1(\Omega)$ and hence it is not an eigenvector of A_ε. However, it can be easily seen that $\{\varphi_{o,\ell}^o\}_{\ell=1}^{\infty}$ is dense in $L^2(\Omega_o)$.

Case 2: $\lambda^o = 0$

As in case 1, by formal identification of equal powers of ε, we get:

Lemma 2.3: The sequences $\{\lambda^k\}_{k=1}^{\infty}$, $\{\varphi^k\}_{k=0}^{\infty}$ satisfy:

$$\varphi_o^o = 0 \quad \text{in } \Omega_o \tag{2.5}$$

$$-\Delta\varphi_1^o = \lambda^1\varphi_1^o \quad \text{in } \Omega_1 \,, \quad \varphi_1^o|_{\Gamma_1} = 0 \,, \quad \varphi_1^o|_S = 0 \tag{2.6}$$

$$-\Delta\varphi_o^k = \sum_{i=0}^{k-1}\lambda^{i+1}\varphi_o^{k-i-1} \quad \text{in } \Omega_o \,, \quad \varphi_o^k|_{\Gamma_o} = 0 \tag{2.7}$$

$$-\Delta\varphi_1^k = \sum_{i=0}^{k-1}\lambda^{i+1}\varphi_1^{k-i} \quad \text{in } \Omega_1 \,, \quad \varphi_1^k|_{\Gamma_1} = 0 \tag{2.8}$$

$$\left.\begin{aligned} &\varphi_1^k = \varphi_o^k \\ &\frac{\partial\varphi_o^k}{\partial\nu} + \frac{\partial\varphi_1^{k-1}}{\partial\nu} = 0 \end{aligned}\right\} \text{ on } S \tag{2.9}$$

$k = 1,2,3,\ldots$

Now we give a counterexample to show that the functions obtained through the processes described by (2.3) or (2.5)-(2.9) are not eigenvectors of A_ε.

Example 2.1: Let $\Omega_o = (-1,0)$, $\Omega_1 = (0,1)$, $\Gamma_o = \{-1\}$, $\Gamma_1 = \{1\}$, $S = \{0\}$, then (2.1)-(1.6) become

$$-\frac{d^2\varphi_o}{dx^2} = \lambda\varphi_o, \qquad \varphi_o(-1) = 0 \tag{1.2}$$

$$-\frac{d^2\varphi_1}{dx^2} = \lambda\varphi_1, \qquad \varphi_1(1) = 0 \tag{1.3}$$

$$\left.\begin{aligned} \varphi_o(0) &= \varphi_1(0) \\ \frac{d\varphi_o}{dx}(0) &= \varepsilon \frac{d\varphi_1}{dx}(0) \end{aligned}\right\} \qquad (1.4)$$

By straightforward calculation, we conclude that λ must satisfy

$$\cos\sqrt{\lambda}\ \sin\sqrt{\frac{\lambda}{\varepsilon}} + \sqrt{\varepsilon}\ \sin\sqrt{\lambda}\ \cos\sqrt{\frac{\lambda}{\varepsilon}} = 0. \qquad (2.10)$$

Clearly this is a transcendental equation in λ and therefore we cannot explicitly solve for λ as a function of ε. The process of case 1 gives:

$$\mu^o = ((2\ell-1)\frac{\pi}{2})^2 \qquad (2.11)$$

$$\begin{cases} \psi_{o,\ell} = \cos(2\ell-1)\frac{\pi}{2}\,x \\ \psi_{1,\ell} = 0 \end{cases} \qquad (2.12)$$

The process of case 2 gives:

$$\lambda_\ell = (\ell\pi)^2\varepsilon - 2(\ell\pi)^2\varepsilon^2 + (3(\ell\pi)^2 - \frac{2}{3}(\ell\pi)^4)\varepsilon^3 + 0(\varepsilon^4) \qquad (2.13)$$

$$\begin{cases} \varphi_{o,\ell} = \ell\pi(1+x)\varepsilon + (-\frac{(\ell\pi)^3}{6}(1+x)^3 + (\frac{(\ell\pi)^3}{2} - \ell\pi)(1+x))\varepsilon^2 + 0(\varepsilon^3) & (2.14) \\ \varphi_{1,\ell} = \sin \ell\pi x + \ell\pi(1-x)\cos \ell\pi x\ \varepsilon + 0(\varepsilon^2) & (2.15) \end{cases}$$

$\ell = 1,2,\ldots$

<u>Remark 2.2</u>: It is worthy to mention that $\varphi_\ell \to +\infty$ as $\ell \to \infty$ for any positive value of ε. This is not a peculiarity of this example. This can be predicted (see (5)).

<u>Remark 2.3</u>: Despite the transcendental nature of (2.20), it can be solved for some values of ε, e.g., for $\varepsilon = \frac{1}{25}$, we can see that

$$\begin{cases} \mu_1^* = (\frac{\pi}{2})^2 \\ \psi_1^* = (\cos\frac{\pi}{2}\,x,\ \cos\frac{\pi}{2\sqrt{\varepsilon}}x) \end{cases} \qquad (2.16)$$

$$\begin{cases} \lambda_5^* = \varepsilon(5\pi)^2 \\ \varphi_5^* = (\sqrt{\varepsilon}\sin\sqrt{\varepsilon}\,5\pi x,\ \sin 5\pi x) \end{cases} \qquad (2.17)$$

are two <u>exact</u> eigenvalue-eigenvector pairs of A_ε. Furthermore, we can easily see that

$$\|\psi_1-\psi_1^*\|_{L^2(-1,1)} = 0(1)$$

$$\|\varphi_5-\varphi_5^*\|_{L^2(-1,1)} = 0(\sqrt{\varepsilon}).$$

The last estimate is true regardless of how many terms φ_5 has. Consequently, one asks what is wrong with what we have undertaken? The answer suggests itself by examining A_ε and Eqs. (2.16),(2.17) and the remark made in [3, p.11], i.e.,

1) Zero is the limit of some eigenvalues of A_ε.

2) The eigenvectors of A_ε are not analytic in ε and hence we cannot expand then as in (2.2). Furthermore, we can always find eigenvectors of this type for ε as small as we wish.

Remark 2.4: We conjecture that the eigenvalues of A_ε are analytic in ε and are given by $\mu_\ell = ((2\ell-1)\frac{\pi}{2})^2$, $\lambda_\ell = (\ell\pi)^2\varepsilon$, $\ell = 1,2,\ldots$ This claim can be substantiated using Raleigh quotients with a "comparable" space, i.e., $V = \bar{V}xH_o^1(\Omega)$ [(3)] instead of $H_o^1(\Omega)$ in (1.9). (See numerical results in Sec. 4).

Remark 2.5: It seems that $\{\mu_\ell\}_{\ell=1}^\infty$, $\{\frac{\lambda_\ell}{\varepsilon}\}_{\ell=1}^\infty$ are the eigenvalues of A_1 (i.e., A_ε with $\varepsilon=1$) and the zeroth terms of the expansions obtained in case 1 (case 2) can be derived from the eigenvectors of A_1 by multiplying them by the characteristic function of Ω_o (Ω_1), i.e., in Example 2.1, the eigenvalue-eigenvector pairs of A_1 are given by:

$$\begin{cases} \alpha_\ell = ((2\ell-1)\frac{\pi}{2})^2 \\ \gamma_\ell = \cos(2\ell-1)\frac{\pi}{2}\, x \end{cases} \qquad x\in(-1,1)$$

$$\begin{cases} \beta_\ell = (\ell\pi)^2 \\ \nu_\ell = \sin \ell\pi x \end{cases} \qquad x\in(-1,1)$$

$$\ell = 1,2,\ldots$$

It is clear that

$$\begin{cases} \mu_\ell = \alpha_\ell \\ \psi_\ell^o = \gamma_\ell \chi_{(-1,0)} \end{cases}$$

$$\begin{cases} \lambda_\ell = \beta_\ell \varepsilon \\ \varphi_\ell^o = \nu_\ell \chi_{(0,1)} \end{cases}$$

Now we turn our attention to finding a basis for $L^2(\Omega)$ with elements of $H_o^1(\Omega)$. If we take the zeroth order terms of the expansions obtained in case 1 and case 2, i.e.,

---

(3) $\bar{V} = \{\varphi : \varphi \in H'(\Omega_o;\Gamma_o),\ \frac{\partial\varphi}{\partial\nu}\Big|_S = 0\}$.

$$\left.\begin{aligned} \psi_\ell &= (\psi^o_{o,\ell},0) \\ \varphi_\ell &= (0,\varphi^o_{o,\ell}) \\ \ell &= 1,2,\ldots \end{aligned}\right\} \qquad (2.18)$$

It can easily be seen from the way they are derived (see previous remarks) that:

1) $\{\psi^o_{o,\ell}\}^\infty_{\ell=1}$ is a basis of $L^2(\Omega_o)$.

2) $\{\varphi^o_{o,\ell}\}^\infty_{\ell=1}$ is a basis of $L^2(\Omega_1)$.

3) ψ_ℓ, φ_ℓ are trivially orthogonal in $L^2(\Omega)$.

Since $L^2(\Omega_o)\times L^2(\Omega_1)$ can be identified with $L^2(\Omega)$, the set of functions given by (2.18) form an orthogonal basis of $L^2(\Omega)$.

It is worthy to note that $\{\psi^o_\ell\}^\infty_{\ell=1}$ violate the part equation of (1.4), while $\{\varphi_\ell\}^\infty_{\ell=1}$ violate the second one. In the sequel, we modify these functions by adding appropriately selected functions so that (1.4) is satisfied. Let us start with $\{\varphi^o_\ell\}^\infty_{\ell=1}$.

Case 2: Let us add a function $\theta_\ell(\varepsilon)=\varepsilon\theta'_\ell+\varepsilon^2\theta^2_\ell+\cdots\varepsilon W^{(4)}$ to φ_ℓ which can be computed iteratively as follows:

$$\begin{aligned} -\Delta\theta^k_{o,\ell} &= 0 \quad \text{in } \Omega_o\,, \quad \theta^k_{o,\ell}\big|_{\Gamma_o}=0 \\ -\Delta\theta^k_{1,\ell} &= 0 \quad \text{in } \Omega_1\,, \quad \theta^k_{1,\ell}\big|_{\Gamma_1}=0 \end{aligned}$$

$$\left.\begin{aligned} \theta^k_{o,\ell} &= \theta^k_{1,\ell} \\ \frac{\partial\theta^k_{o,\ell}}{\partial\nu}+\frac{\partial\theta^{k-1}_{1,\ell}}{\partial\nu} &= 0 \end{aligned}\right\}\ \text{on } S \qquad (2.19)$$

$$k=1,2,3,\ldots$$

with $$\theta^o_{1,\ell}=\varphi^o_{1,\ell}$$

$$\ell=1,2,3,\ldots$$

Case 1: Let us add a function $\nu_\ell(\varepsilon)=\nu^o_\ell+\varepsilon\nu'_\ell+\cdots$ to ψ^o_ℓ which can be computed as

$$\begin{aligned} \nu^o_{o,\ell} &= 0 \\ -\Delta\nu^o_{1,\ell} &= 0 \quad \text{in } \Omega_1\,, \quad \nu^o_{1,\ell}\big|_{\Gamma_1}=0,\ \nu^o_{1,\ell}\big|_S=\psi^o_{o,\ell}\big|_S \end{aligned} \qquad (2.20)$$

Now the function $\psi^o_\ell+\nu^o_\ell\in H^1_o(\Omega)$ and hence $\{\nu^k_\ell\}^\infty_{\ell,k=1}$ are computed as in case 2.

---

(4) See p. 2.

3. Nonhomogeneous Boundary Value Problems

After the analysis carried out in Section 2, we consider the following boundary value problem

$$A_\varepsilon u = f, \quad u|_\Gamma = 0 \tag{3.1}$$

where $\quad f = (f_o, f_1) \in L^2(\Omega)$.

Theorem 3.1: The solution of (3.1) is given by

$$u = \sum_{\ell=1}^{\infty} \frac{c_\ell}{\mu_\ell} (\psi_\ell + \nu_\ell(\varepsilon)) + \sum_{\ell=1}^{\infty} \frac{d_\ell}{\varepsilon\lambda_\ell} (\varphi_\ell + \theta_\ell(\varepsilon)) \tag{3.2}$$

where

$$d_\ell = (f,\psi_\ell) = (f_o,\psi^o_{o,\ell}); \qquad d_\ell = (f,\varphi_\ell) = (f_1,\varphi^o_{1,\ell})$$

$\{\psi_\ell\}_{\ell=1}^\infty$, $\{\varphi_\ell\}_{\ell=1}^\infty$ are given in (2.28), $\{\mu_\ell\}_{\ell=1}^\infty$, $\{\lambda_\ell\}_{\ell=1}^\infty$ are the corresponding eigenvalues, $\{\nu_\ell(\varepsilon)\}_{\ell=1}^\infty$, $\{\theta_\ell(\varepsilon)\}_{\ell=1}^\infty$ are given by (2.30), (2.29) respectively.

Theorem 3.2: The solution u of (3.1) given by (3.2) is identical to the one obtained in [7].

We now give a simple example.

Example 3.1: (Example 2.1 continued)

$$-\frac{d^2u_o}{dx^2} = 1 \quad \text{in } (-1,0), \qquad u_o(-1) = 0$$

$$-\varepsilon\frac{d^2u_1}{dx^2} = 1 \quad \text{in } (0,1), \qquad u_1(1) = 0$$

$$u_o(0) = u_1(0)$$

$$\frac{du_o}{dx}(0) = \varepsilon\frac{du_1}{dx}(0)$$

The exact solution is

$$u_o = -\frac{x^2}{2}+\frac{1}{2}\,\frac{1-\varepsilon}{1+\varepsilon}x+\frac{1}{1+\varepsilon}\,; \qquad u_1 = -\frac{x^2}{2\varepsilon}+\frac{1}{2}\,\frac{1-\varepsilon}{\varepsilon(1+\varepsilon)}x+\frac{1}{1+\varepsilon} \tag{3.3}$$

The iterative process of [6] gives:

$$u_o = \left(-\frac{x^2}{2}+\frac{x}{2}+1\right)+\sum_{k=1}^{\infty}(-\varepsilon)^k(x+1); \quad u_1 = \frac{\left(-\frac{x^2}{2}+\frac{x}{2}\right)}{\varepsilon}+\sum_{k=1}^{\infty}(-\varepsilon)^{k-1}(1-x). \tag{3.4}$$

The solution as given by (3.5) is

$$\left.\begin{aligned} u_o &= \left(\sum_{\ell=1}^{\infty}\frac{2(-1)^{\ell-1}}{((2\ell-1)\frac{\pi}{2})^3}\cos(2\ell-1)\frac{\pi}{2}x+\frac{1}{2}(x+1)\right)+\sum_{k=1}^{\infty}(-\varepsilon)^k(1+x)\\ u_1 &= \frac{\sum_{\ell=1}^{\infty} 2\,\frac{1-(-1)^\ell}{(\ell\pi)^3}\sin\ell\pi x}{\varepsilon}+\sum_{k=1}^{\infty}(-\varepsilon)^{k-1}(1-x)\end{aligned}\right\} \tag{3.5}$$

Clearly if we expand $-\frac{x^2}{2}+\frac{x}{2}+1$, $-\frac{x^2}{2}+\frac{x}{2}$ in terms of $\{\psi^o_{o,\ell}\}_{\ell=1}^\infty$, $\{\varphi^o_{o,\ell}\}_{\ell=1}^\infty$ respectively, then (3.5) is equal to (3.4).

Remark 3.1: In this simple example, we obtain the exact solution.

4. Numerical Results

We divided $\Omega = (-1,1)$ into N equal interval of length $h = 2/N$. We selected the roof functions [9] as a basis of the finite dimensional approximation of $H_o^1(\Omega)$. The finite dimensional approximation of A_ε is

$$A_h = M_h^{-1} K_h$$

where

$$(M_h)_{i,j} = \int_\Omega \varphi_h^i \varphi_h^j dx, \qquad (K_h)_{i,j} = \int_\Omega a(x) \frac{d\varphi_h^i}{dx} \frac{d\varphi_h^j}{dx} dx \quad i,j = 1,2,\dots,N-1$$

where

$$a(x) = \begin{cases} 1 & x \in (-1,0) \\ \varepsilon & x \in (0,1) \end{cases}.$$

In this example, M_h, K_h can be computed explicitly, i.e. (for N=2)

$$M_h = \frac{h}{6}\begin{bmatrix} 4 & 1 & 0 & & \\ 1 & 4 & 1 & 0 & \\ 0 & 1 & 4 & \ddots & \\ & & & 4 & 1 \\ 0 & & & 1 & 4 \end{bmatrix} \quad K_h = \frac{1}{h}\begin{bmatrix} 2 & -1 & 0 & & & & \\ -1 & 2 & -1 & & & 0 & \\ 0 & -1 & 2 & \ddots & & & \\ & & & -1 & 1+\varepsilon & -\varepsilon & 0 \\ & & & & -\varepsilon & 2\varepsilon & -\varepsilon \\ & 0 & & & 0 & & \ddots \end{bmatrix} \leftarrow \frac{N}{2}$$

For all computer runs, we slected N=50. Table 1 contains λ_o, λ_1, μ_o computed for some values of ε.

ε	λ_o	λ_1	μ_o
.1	.77213	4.5741	2.3462
.04	.36170	1.3714	2.4778
.01	.09681	.38784	2.2461
.001	.009863	.039606	2.3667

Table 1

Remark 3.2: Note that $\lambda_o \approx \varepsilon\pi^2$, $\lambda_1 \approx \varepsilon(2\pi)^2$, $\mu_o \approx \frac{\pi^2}{4}$ except the first row.

Remark 3.3: The plots of the eigenvectors can be found in [5].

References

[1] P. V. Kokotovic, R. E. O'Malley, P. Sannuti, "Singular Perturbations and Order Reduction in Control Theory - An Overview," Automatica, Vol. 12, 123-132, 1976.

[2] W. Eckhaus, Asymptotic Analysis of Singular Perturbations, North-Holland Publ. Co., New York, 1979.

[3] P. P. N. de Groen, "Singular Perturbations of Spectra," in Asymptotic Analysis, F. Verhulst, ed., Lecture Notes in Mathematics, Springer-Verlag, New York, 1979.

[4] J. L. Lions, Optimal Control of Systems Governed by Partial Differential Equations, Springer-Verlag, New York, 1971.

[5] H. Salhi, Ph.D. thesis, University of Illinois, Urbana, Illinois, 1982.

[6] J. L. Lions, Perturbations Singulieres dans les Problemes aux Limites et en Controle Optimal, Lecture Notes in Mathematics, Springer-Verlag, New York, 1973.

[7] S. Kesavan, "Homogeneization of Elliptic Eigenvalue Problems: Part I," Appl. Math. Optim. 5, 153-167, 1979.

[8] J. Weidman, Linear Operators in Hilbert Spaces, Springer-Verlag, New York, 1980.

[9] G. Strang, G. J. Fix, An Analysis of the Finite Element Method, Prentice-Hall, Inc., Englewood Cliffs, N.J., 1971.

ON THE SEMI GROUP APPROACH FOR ERGODIC PROBLEMS OF OPTIMAL STOPPING

Maurice ROBIN - INRIA - BP 105 - 78153 LE CHESNAY CEDEX - FRANCE

INTRODUCTION

It is known that the maximum solution of :

$$(*) \qquad u \leq e^{-\alpha t}\,\phi(t)u + \int_0^t e^{-\alpha s}\,\phi(s)f\,ds \quad , \quad u \leq \psi ,$$

is the optimal cost function of a stopping problem, when $\phi(t)$ is a Markov semi-group. We refer to [1], [3], [9] for the study of that kind of system when $\alpha > 0$. When $f = 0$ and $\alpha = 0$, such a study can be found in [11].

This paper is devoted to the study of $(*)$ when $\alpha = 0$ and when the semi-group $\phi(t)$ has a "good" ergodic behaviour, namely, when :

$$||\phi(t)f - \int f\,d\mu|| \leq b.e^{-\gamma t}\,||f||, \forall f,$$

for an invariant measure μ of $\phi(t)$.

That kind of assumption was already used in [10] with probabilistic arguments to study the stopping time problem corresponding to $(*)$ with $\alpha = 0$. Here, we will use an analytic approach, starting with $(*)$ for $\alpha > 0$ and considering the asymptotic behaviour when $\alpha \to 0$.

Such a method, using partial differential equations techniques, for the case of diffusion processes was used in Bensoussan-Lions [4] and Lasry [7].

On the other hand, we will study the non linear semi-group associated to the finite horizon problem :

$$z(t) \leq \phi(t-s)\;z(s) + \int_0^{t-s} \phi(\sigma)f\,d\sigma,\; o \leq s \leq t,$$

$$z(t) \leq \psi$$

$$z(o) = \bar{u} \leq \psi$$

as defined in Bensoussan-Lions [3] and we consider the behaviour when $t \to \infty$ of $S(t)\bar{u}$, and the characterization of the maximum solution of $(*)$ (with $\alpha = 0$) as the solution of the equation $S(t)u = u$.

1. - NOTATIONS, ASSUMPTIONS AND STATEMENT OF THE PROBLEM

Let E be a compact metric space endowed with the Borel σ-algebra B(E). Let B be the space of Borel bounded functions on E, and C the space of continuous functions on E. We consider a Markov semi-group $\phi(t)$ on B, that is :

$$\begin{cases} \phi(t) \in \mathcal{L}(B,B),\; \phi(o) = I,\; ||\phi(t)|| = 1,\; \phi(t)\phi(s) = \phi(t+s) \\ \phi(t)\varphi \geq 0 \text{ if } \varphi \geq 0. \end{cases} \qquad (1.1)$$

Moreover, we will assume that $\phi(t)$ has the Feller property :

$$\begin{cases} \forall f \in C, \ \phi(t)f \in C, \ (\forall t \geq 0). \\ \lim\limits_{t \downarrow 0} \phi(t) \ f(x) = f(x) \quad \forall f \in C. \end{cases} \tag{1.2}$$

Since E is compact, $\phi(t)$ has at least one invariant probability measure (cf. [10], for instance). We state the additionnal hypothesis, there exist β, $\gamma > 0$ such that $\forall \Gamma \in B(E)$:

$$||\phi(t) \ \chi_\Gamma - \mu(\Gamma)|| \leq \beta \ e^{-\gamma t}, \tag{1.3}$$

where χ_Γ is the indicator function of the set Γ.L et A the generator of Φ (t).

The problems studied in the following sections are related to the set of solutions of the system :

$$u \in B \quad ; \quad u \leq \phi(t)u + \int_0^t \phi(s)f \ ds \quad ; \quad u \leq \psi \quad , \tag{1.4}$$

$$f, \ \psi \in C \text{ are given functions .} \tag{1.5}$$

2. - PRELIMINARY REMARKS AND TRANSFORMATION OF THE PROBLEM

Let us consider :

$$Av = f, \ f \in C, \ \int f \ d\mu(x) = 0 \ . \tag{2.1}$$

It follows from (1.1), (1.2), (1.3) and known results on Fredholm alternative (see the annex of [10] e.g.) that any solution of (2.1) can be written :

$$v = \int_0^\infty \phi(t) \ f \ dt + \text{constant}. \tag{2.3}$$

We can now replace the system (1.4) by another one.

Let $\gamma = \inf \psi$, and v^o the unique solution of :

$$\begin{cases} - Av^o = f - \bar{f}, \ v^o \in D_A, \ \text{(the domain of A in C)} \\ \max v^o = 0. \end{cases} \tag{2.4}$$

If w is any solution (assuming that it exists) of (1.4) then : $u=w-\gamma-v^o$ satisfies :

$$\begin{cases} u \in B \quad ; \quad u \leq g \quad ; \quad u \leq \phi(t)u + \bar{f}.t \\ \phi(t)v^o = v^o + \int_0^t \phi(s) \ Av^o \ ds, \ \forall t \end{cases} \tag{2.5}$$

where : $g = \psi - \gamma$,

$$g \geq 0, \ g \in C \ (\text{and } \min g = 0). \tag{2.6}$$

In the following, we will investigate the three cases : $\bar{f} > 0$, $\bar{f} = 0$, $\bar{f} < 0$.

3. - ANALYTIC STUDY OF THE PROBLEM

3.1. - The cas $\bar{f} > 0$

Let us first introduce the discounted problem. Let $\alpha > 0$ and consider :

$$w \le e^{-\alpha t}\,\phi(t)w + \int_0^t e^{-\alpha s}\,\phi(s)\,\bar{f}\,ds\ ;\ w \le g\ ;\ w \in C\,. \tag{3.1}$$

This problems is studied in A. Bensoussan [1] using the penalized problem (for $\varepsilon>0$) :

$$u_\alpha^\varepsilon = \int_0^\infty e^{-\alpha t}\,\phi(t)[\bar{f} - \frac{1}{\varepsilon}(u_\alpha^\varepsilon - g)^+]dt. \tag{3.2}$$

From [1], [9] one can take the following result.

Lemma 3.1. : under the assumptions (1.1), (1.2), (2.6).

(i) the set of function w satisfying (3.1) is non empty and has a maximum element u_α

(ii) (3.2) has a unique solution in C

(iii) $u_\alpha^\varepsilon \le u_\alpha^{\varepsilon'}$ if $\varepsilon \le \varepsilon'$

(iv) if $g \in D_A$, one has :

$$||u_\alpha^\varepsilon - u_\alpha|| \le \varepsilon\,||\bar{f} + Ag - \alpha g|| \tag{3.3}$$

more generally (i.e. $g \in C$), $u_\alpha^\varepsilon \to u_\alpha$ in C

(v) $0 \le u_\alpha^\varepsilon \le ||g|| + \varepsilon\,\bar{f}$. ■

Lemma 3.2. : under the assumptions (1.1), (1.2), (2.6)

$u_\alpha^\varepsilon \le u_{\alpha'}^\varepsilon$, if $\alpha \ge \alpha'$ and

$u^\varepsilon(x) = \lim_{\alpha\downarrow 0} u_\alpha^\varepsilon(x)$ is the unique solution, in B, of the equation

$$u^\varepsilon = \phi(t)u^\varepsilon + \int_0^t \phi(s)\,[\bar{f} - \frac{1}{\varepsilon}(u^\varepsilon - g)^+]\,ds,\ \forall t \ge 0 \tag{3.4}$$

Proof : From [1], we get that u_α^ε is the unique fixed point of the contraction mapping (in C) :

$$T_\varepsilon\,w = \int_0^\infty e^{-(\alpha+\frac{1}{\varepsilon})t}\,\phi(t)\,[\bar{f} + \frac{w\wedge g}{\varepsilon}]ds.$$

It is then clear that $u_\alpha^\varepsilon(x)$ is increasing when α decreases.

From (v) of lemma 3.1, u_α^ε is bounded, uniformly w.r.t α.

Therefore $u_\alpha^\varepsilon(x)$ has a limit $u^\varepsilon(x)$ when $\alpha \downarrow 0$ (which defines a lower semi-continuous fonction) and one can go to the limit in

$$u_\alpha^\varepsilon = e^{-\alpha t}\,\phi(t)u_\alpha^\varepsilon + \int_0^t e^{-\alpha s}\,\phi(s)\,[\bar{f} - \frac{1}{\varepsilon}(u_\alpha^\varepsilon - g)^+]ds$$

(which is equivalent to (3.2)), using the weak convergence of B. It remains to prove the uniqueness.

We first notice that (3.4) is equivalent to (see [1], [9], chap. 1):

$$u^{\varepsilon} = e^{-t/\varepsilon}\phi(t)u^{\varepsilon} + \int_{o}^{t} e^{-s/\varepsilon}\phi(s)\,[\bar{f} + \frac{u^{\varepsilon}\wedge g}{\varepsilon}]ds, \ \forall t \geq 0 \qquad (3.5)$$

Now let z_1, z_2 be two solutions of (3.4) and let γ the maximum number in $[0,1]$ such that : $\gamma\, z_1 \leq z_2$.

Let us show that $\gamma = 1$. For that, let us assume $\gamma < 1$. We are going to show that one can find β, $\gamma < \beta \leq 1$ such that $\beta\, z_1 \leq z_2$, contradicting the definition of γ. We have :

$$\beta z_1 = e^{-t/\varepsilon}\,\phi(t)\,\beta z_1 + \int_{o}^{t} e^{-s/\varepsilon}\phi(s)\,[\beta\bar{f} + \beta\,\frac{z_1\wedge g}{\varepsilon}]ds, \ \forall t. \qquad (3.6)$$

To obtain $\beta z_1 \leq z_2$, it is enough to have β such that :

$$\beta\bar{f} + \beta\,(\frac{z_1\wedge g}{\varepsilon}) \leq \bar{f} + \frac{z_2\wedge g}{\varepsilon} \qquad (3.7)$$

since the equation : $z = e^{-t/\varepsilon}\,\phi(t)z + \int_{o}^{t} e^{-s/\varepsilon}\,\phi(s)L\,ds$ has the unique solution :

$z = \int_{o}^{\infty} e^{-t/\varepsilon}\,\phi(t)L\,dt.$

Using $\gamma z_1 \leq z_2$, we see that (3.7) will be satisfied if :

$$\beta\bar{f} + \beta\,(\frac{z_1\wedge g}{\varepsilon}) \leq \bar{f} + \frac{(\gamma z_1)\wedge g}{\varepsilon}$$

and, a fortiori, if :

$$\beta\bar{f} = \beta\,(\frac{z_1\wedge g}{\varepsilon}) \leq \bar{f} + \gamma\,(\frac{z_1\wedge g}{\varepsilon}),$$

which will be obtained with $\beta = \frac{\rho + \nu}{1 + \rho}$, $\rho = \frac{\bar{f}}{C}$, $C \geq \frac{||z_1\wedge g||}{\varepsilon}$, $C > 0$.

Using the same argument in exchanging the roles of z_1 and z_2 we obtain $z_1 = z_2$. ∎

<u>Lemma 3.3.</u> : $u^{\varepsilon} \in C$

<u>Proof</u> : Let us consider :

$$\left| \begin{array}{l} u^{o} = \int_{o}^{\infty} e^{-t/\varepsilon}\,\phi(t)\,[\bar{f} + \frac{1}{\varepsilon}\,g]\,dt \\ u^{n} = \int_{o}^{\infty} e^{-t/\varepsilon}\,\phi(t)\,[\bar{f} + \frac{1}{\varepsilon}\,u^{n-1}\wedge g]dt, \quad n \geq 1. \end{array} \right. \qquad (3.8)$$

We have : $0 \leq u^{n} \leq u^{n-1} \leq u^{o} \leq \varepsilon\,\bar{f} + ||g||$, $\forall n$, $u^{n} \in C$.

And u^{n} converges weakly in B to a function w^{ε}. Going to the limit in (3.8), we obtain that w^{ε} satisfies (3.4). Hence $w^{\varepsilon} = u^{\varepsilon}$ which is therefore u.s.c and with Lemma 3.2 we get the result. ∎

<u>Lemma 3.4</u> : <u>if</u> w <u>is any solution of</u> (2.5), $u^{\varepsilon} \geq w$.

Proof : it is identical to similar result for $\alpha > 0$ as in [1], [3]. ∎

Lemma 3.5. :

(i) $u^{\varepsilon} \leq u^{\varepsilon'}$ if $\varepsilon \leq \varepsilon'$

(ii) $u_\alpha \leq u_{\alpha'}$, if $\alpha \leq \alpha'$ and if $u(x) = \lim_{\alpha \downarrow 0} u_\alpha(x)$, then

(iii) $u^{\varepsilon} \to u$ in C.

Proof : From (iii) of Lemma 3.4 : $u_\alpha^{\varepsilon} \leq u_\alpha^{\varepsilon'}$ if $\varepsilon \leq \varepsilon'$.

Therefore, when $\alpha \downarrow 0$, we obtain (i). (ii) is an easy consequence of Lemma 3.2 and Lemma 3.1.-(iv). Then since $0 \leq u_\alpha \leq g$, $u(x) = \lim_{\alpha \downarrow o} u_\alpha(x)$ exists.

Now assume $g \in D_A$ from Lemma 3.1 (iv), we have :

$$||u_\alpha^{\varepsilon} - u_\alpha|| \leq \varepsilon(||\bar{f} + Ag|| + ||g||) \text{ uniformly w.r.t.}\alpha\text{, if } \alpha \leq 1.$$

Therefore, when $g \in D_A$, we get :

$$||u^{\varepsilon} - u|| \leq \varepsilon(||\bar{f} + Ag|| + ||g||).$$

Then using the density of D_A in C, we complete the proof as in [1],[9]chap.1)(where $\alpha > 0$). ∎

Theorem 3.1 : Under the assumptions (1.1), (1.2), (1.3)

(i) the set of function satisfying (2.5) is non empty and has a maximum element $u \in C$.

(ii) $u_\alpha \nearrow u$ in C when $\alpha \downarrow 0$.

Proof : From Lemma 3.5. $u^{\varepsilon} \to u$ in C and, since $u^{\varepsilon} \in C$, $u \in C$ and u is also equal to $\lim_{\alpha \downarrow o} u_\alpha$. Since u_α is continuous, and increasing when $\alpha \downarrow 0$, $u_\alpha \to u$ in C.

Now from (3.1), going to the limit when $\alpha \downarrow 0$, u is clearly solution of (2.5).

But, by Lemma 3.4. if w is any solution of (2.5) $u^{\varepsilon} \geq w$.

Therefore when $\varepsilon \searrow 0$, we have $u \geq w$ and u is the maximum solution. ∎

Let now v_α be the maximum solution of the α-discounted problem corresponding to (1.4). We have :

Corollary 3.1. : under the assumptions of theorem 3.1

(i) $v = u + u_o + \gamma$ is the maximum solution of (1.4).

(ii) $v_\alpha \to v$ in C.

Proof : see [13] for details. ∎

We now investigate the convergence of a the discrete analogue of (2.5), namely we consider, for $h > 0$,

$$u_h = \min (g, \phi(h)u_h + \bar{f}.h). \tag{3.9}$$

We can state the following :

Theorem 3.2. : under the assumptions of theorem 3.1.

(i) (3.9) has a unique solution $u_h \in C$

(ii) $u_n \to u$ (the maximum solution of (2.5)) in C when $n \uparrow \infty$, if $u_n = u_{h.2^{-n}}$

Proof : This is an adaptation of one result included in [10]. We refer to [13]for details.

3.2. - The case $\bar{f} = 0$

Theorem 3.3. : under the assumptions (1.1), (1.2), (1.3), and $\bar{f} = 0$, the system (2.5) has a maximum solution $u \in B$ (which is upper semi-continuous).

Proof : Let β_n a decreasing sequence of positive numbers such that $\beta_n \searrow 0$ when $n \uparrow \infty$. Let us consider u_n the maximum solution of :

$$w \le g, \quad w \le \phi(t)\, w + \beta_n . t.$$

Since u_n is the limit of u_{nh}, solution of :

$$u_{nh} = \min\,(g, \beta_n . h + \phi(h) u_{nh})$$

it is clear that : $0 \le u_{n+1} \le u_n \le g$.

Therefore $u(x) = \lim\limits_{n \to o} u_n(x)$ is an upper semi-continuous function satisfying : $u \le g$, $u \le \phi(t) u$.

If w is another solution, we have $w \le g$, $w \le \phi(t) w + \beta_n t$, $\forall n$.

Therefore $w \le u_n$ and then $w \le u$. ∎

Unfortunately, in that case,we do not have the proof of the convergence of u_α to u.

3.3. $\bar{f} < 0$

This case is obvious and it is clear that (2.5) cannot have a bounded solution.

4. - STOCHASTIC INTERPRETATION

We will deal essentially with the case $\bar{f} > 0$. For the case $\bar{f} = 0$, the results are included in those of Shiryaev [11].

We denote by $\Omega = D(0,\infty;E)$ the space of right continuous, left limited functions from $\mathbb{R}^+$ into E, $x_t(w) = w(t)$ for $w \in \Omega$, $F_t = \sigma(x_s, s \le t)$, $F = F_\infty$, and we will denote by $X = (\Omega, x_t, \theta_t, P_x)$ a realization of the semi-group $\phi(t)$ (θ_t being the translation operator on Ω).

Since $\phi(t) 1 = 1$, the process is non terminating and since $\phi(t)$ is Feller and E compact, X is quasi-left-continuous-namely, if $\tau^n \uparrow \tau$ are F_t-stopping times, then :

$$x_{\tau^n} \to x_\tau \quad P_x \text{ a.s on } \{\tau < \infty\} \tag{4.1}$$

We then have :

Theorem 4.1. : under the assumptions of theorem 3.1 and if $\bar{f} > 0$, then, if v is the maximum solution of (1.4).

$$v(x) = \inf_{\tau} J_x(\tau), \text{ where} \tag{4.2}$$

$$J_x(\tau) = E_x \int_0^{\tau} f(x_t)dt + \psi(x_\tau)\chi_{\tau<\infty} \tag{4.3}$$

Moreover, the stopping time $\hat{\tau} = \inf (t \geq 0,\ v(x_t) = \psi(x_t))$ is optimal and satisfies $E_x\ \hat{\tau} \leq K$ for some constant, independant from x.

Proof : Let us begin with u, maximum solution of (2.5) and u_ε the solution of the corresponding penalized problem (3.4). It can be proved (as in [2], [9]), that (3.4) implies, with the Markov property :

$$z(t) = u_\varepsilon(x_t) + \int_0^t (\bar{f} - \frac{1}{\varepsilon}(u_\varepsilon - g)^+ (x_s)ds$$

is a martingale and therefore, for any stopping time τ,

$$u_\varepsilon(x) = E_x \int_0^{\tau\wedge t} (\bar{f} - \frac{1}{\varepsilon}(u_\varepsilon - g)^+ (x_s))ds + u_\varepsilon(x_{\tau\wedge t}). \tag{4.4}$$

Let τ_ε be the following stopping time :

$$\tau_\varepsilon = \text{Inf } (t \geq 0,\ u_\varepsilon(x_t) \geq \psi(x_t)),$$

we then have, using (4.4), $E_x\ \tau_\varepsilon \leq \frac{K}{\bar{f}}$. From now on, the proof is similar to the one [1], [9] chap 1 (when $\alpha > 0$). ■

5. - NON LINEAR SEMI-GROUP

Following A. Bensoussan, J.L. Lions [3], we consider for $\alpha \geq 0$, the set of functions $w \in C(0,T;C)$ satisfying :

$$w(t) \leq e^{-\alpha(t-s)}\phi(t-s)w(s) + \int_0^{t-s} e^{-\alpha\sigma}\phi(\sigma)fd\sigma,\ w(t) \leq \psi, w(o)=\bar{u}. \tag{5.1}$$

where $\bar{u} \in \mathcal{C} = \{\bar{u} \in C,\ \bar{u} \leq \psi\}$.

Then,we have the following :

Theorem 5.1. : (A. Bensoussan, J.L. Lions [3]) : under the assumptions (1.1), (1.2), (1.3) :

(i) the set of functions $w \in C(0,T,C)$ satisfying (5.1) is non-empty and has a maximum element w_α ;

(ii) $w_\alpha(t) = S_\alpha(t)\ \bar{u}$ defines a semi-group $S_\alpha(t)$ on $\mathcal{C}$ which is continuous (i.e. $\lim_{t\to 0} S(t)\bar{u} = \bar{u}$)

(iii) when $\alpha > 0$, $\lim_{t\to\infty} S_\alpha(t)\ \bar{u} = u_\alpha$, the maximum solution of :

$$w \leq e^{-\alpha t}\phi(t)w + \int_0^t e^{-\alpha s}\phi(s)f\ ds,\quad w \leq \psi,\quad w \in C. \quad ■ \tag{5.2}$$

For $\alpha = 0$, we will denote by S(t) the semi-group obtained from Theorem 5.1.

In this section, we will restrict ourselves to the case $\bar{f} > 0$.

Theroreme 5.2. : Under the assumptions of theoreme 3.1. ($\bar{f} > o$, u as in theoreme 3.1),
we have :

(i) $\lim_{t \to \infty} S(t)\bar{u} = u$, $\forall \bar{u} \in \mathscr{C}$

(ii) u is the unique solution of the equation

$S(t)\, u = u$, $u \in \mathscr{C}$.

Proof : The main steps are to prove, first, that, for $\alpha > u$, u_α is the unique solution of $S_\alpha(t)\, u_\alpha = u_\alpha$; then, we prove that $S_\alpha(t)\, \bar{u} \to S(t)\, \bar{u}$ in G, $\forall \bar{u} \in \mathscr{C}$, when $\alpha \to o$. We then deduce that $S(t)\, u = u$ and show the uniqueness by the property of maximaly of u. Because of lack of space, we refer to [13] for a detailed proof. ∎

The problem of the determination of the infinitesimal generator of S(t) (as for $S_\alpha(t)$, $\alpha > 0$) is an open question. If we were able to define this generator $\mathscr{A}$, the equation $S(t)u=u$ would mean $\mathscr{A}u = 0$ which should give some sort or variational inequality.

Actually, one can obtain a variational inequality in the case where $u \in D_A$ domain of the infinitesimal generator of $\phi(t)$:

Theorem 5.3. : Let us assume that $u \in D_A$, then u is the unique solution of variational inequality :

$$-Au \leq f \quad ; \quad u \leq \psi \quad ; \quad (Au + f)(u - \psi) = 0. \tag{5.16}$$

Proof : $u \leq \psi$ is already satisfied. Since $u \in D_A$, we have from (2.5) :

$$E_x \int_0^t (Au+f)\,(x_s)ds \geq 0$$

and since $\lim_{t \downarrow o} \frac{1}{t} \int_0^t \phi(s)(Au+f)ds = Au + f$, we obtain $-Au \leq f$.

An the other hand, we know that if : $\mathscr{O} = \{x, u(x) < \psi(x)\}$, $\mathscr{O}$ is an open subset of E and $\hat{\tau} = \inf\,(s \geq 0,\ u(x_s) = \psi(x_s))$ is optimal for u, namely :

$$u(x) = E_x\left(\int_0^{\hat{\tau}} f(x_s)ds + u(x_{\hat{\tau}})\right).$$

Since $E_x\tau < +\infty$, we can use the Dynkin's formula to obtain : $E_x \int_0^{+\infty} \varphi \cdot \chi_{\mathscr{O}}(x_{s \wedge \hat{\tau}})ds = 0$
where $\varphi = Au + f$.

In other words :

$$\int_0^\infty \tilde{\phi}(t)\tilde{\varphi}\ dt = 0, \tag{5.17}$$

where $\tilde{\varphi} = \varphi . \chi_{\mathcal{O}}$ and, $\tilde{\phi}(t)$ is the semi-group of the process stopped at $\hat{\tau}$.

Let us prove now that :

$$\lim_{t \downarrow o} \tilde{\phi}(t)\ \tilde{\varphi}(x) = \tilde{\varphi}(x). \tag{5.18}$$

In fact, if $x \in \mathcal{O}^c$, then $\tilde{\phi}(t)\ \tilde{\varphi}(x) = 0$, $\forall t$, since $\hat{\tau} = 0$, P_x a.s.

If $x \in \mathcal{O}$, then $\hat{\tau} > 0$ P_x.a.s (see [6]) and :

$$\tilde{\phi}(t)\ \tilde{\varphi}(x) = E_x \varphi(x_{t \wedge \hat{\tau}})\chi_{t<\hat{\tau}}$$

Therefore :

$$|\tilde{\phi}(t)\ \tilde{\varphi}(x) - \tilde{\varphi}(x)| \leq ||\varphi||.E_x(1-\chi_{t<\hat{\tau}}) + |E_x\varphi(x_t) - \varphi(x)| = I + II$$

I decreases to 0 when $t \downarrow 0$ since $P_x(\hat{\tau}>0) = 1$, and II goes to 0 when $t \downarrow 0$, by the Feller property of $\phi(t)$.

Now, since $\tilde{\varphi} \geq 0$, we have

$$0 \leq \int_0^t \tilde{\phi}(s)\ \tilde{\varphi}\ ds \leq \int_0^\infty \tilde{\phi}(s)\ \tilde{\varphi}\ ds = 0, \forall t.$$

Therefore, $\forall t$, $\frac{1}{t}\int_0^t \tilde{\phi}(s)\tilde{\varphi}\ ds = 0$ and when $t \downarrow 0$, we get, using (5.18), $\tilde{\varphi} = 0$.

Hence $Au + f = 0$, $\forall x \in \mathcal{O}$, and since $u = \psi$ on $\mathcal{O}^c$, we obtain (5.16).

Uniqueness can be proved from the stochastic interpretation or by a proof similar to the one use for u_ε in § 3.1 (see Bensoussan-Lions [2] for that kind of techniques for equations or variational inequalities). ∎

REFERENCE

[1] A. Bensoussan. On the semi-group approach to variational and quasi variational inequalities ; Proceedings of the 1st French-South East Asian Conference on Mathematical Sciences.

[2] A. Bensoussan, J.L. Lions. Application des Inéquations Variationnelles en Contrôle Stochastique. Dunod - Paris 1978.

[3] A. Bensoussan, J.L. Lions. Contrôle Impulsionnel et Inéquations quasi-variationnelles. to be published.

[4] A. Bensoussan, J.L. Lions. On the asymptotic behaviour of the solution of variational inequalities. Summer Scholl on Theory of non linear operator. Akademic Verlag - Berlin 1978.

[5] R.M. Blumenthal, R.K. Getoor. Markov processes and potential theory. Academic Press - 1968.

[6] E.B. Dynkin. Markov processes - Tome 1 and 2 - Springer Verlag-1965.

[7] J.M. Lasry. Control stochastique ergodique. Doct. Thesis. Paris IX Univ., 1974.

[8] M. NISIO. On non linear semi-group for Markov processes associated with optimal stopping - Applied Math and Optimization 4, (1978), pp. 143-169

[9] M. ROBIN. Contrôle impulsionnel des processus de Markov. Doctoral Thesis - Paris IX University - 1978.

[10] M. Robin. On some impulse control problems with long run average cost. SIAM J. Control and Optimization - May 1981.

[11] A.N. Shiryaev. Optimal stopping rules - Springer Verlag - Berlin 1978.

[12] J. Zabczyk. Semi-group methods in stochastic control theory-CRM-821- Montreal University - 1978.

[13] M. Robin. Semi-group approach for ergodic problem of optimal stopping. Internal report - INRIA - 1981.

THE PRINCIPAL EIGENVALUE OF A TRANSPORT OPERATOR - AN ASYMPTOTIC EXPANSION -

Rémi SENTIS

I.N.R.I.A
B.P. 105 78153 Le Chesnay (France)

1 - INTRODUCTION - MOTIVATION

For $\varepsilon > 0$, let us consider the following transport operator defined by

(1)
$$f = f(x,v) \to A^{\varepsilon} f(x,v) = -\frac{1}{\varepsilon} \sum_i v_i \frac{\partial f}{\partial x_i} + \frac{1}{\varepsilon^2} Q_x f$$

$$D(A^{\varepsilon}) = \{f \in L^2(\Omega \times V) \ / \ \sum_i v_i \frac{\partial f}{\partial x_i} \in L^2(\Omega \times V) \qquad f|_{\Gamma^-} = 0\}$$

with Ω a bounded connected open set, with smooth boundary $\partial\Omega$

V a compact set of $\mathbb{R}^N$, symetric with respect to 0, provided by a probability measure.

$$Q_x g = K_x g - \sigma g \qquad \forall g \in L^2(\Omega) \text{ where } K_x g(v) = \int_V \sigma_1 \ (x,v,w) g(w) dw$$

$$\sigma(x,v) = K_x 1(v)$$

$$\Gamma^- = \{(x,v) \in \partial\Omega . V \qquad n_x . v \le 0\} \qquad n_x \text{ is the outward normal in } x \text{ to } \partial\Omega.$$

Let us consider the solution $u = u(t,x,v)$ of the transport equation

$$\begin{cases} \dfrac{du}{dt} = A^{\varepsilon} u + \alpha u & \alpha \in \mathbb{R} \\ u(t,.,.)|_{\Gamma^-} = 0 & \\ u(t{=}0,x,v) = f & \text{where } f \text{ is given and positive.} \end{cases}$$

The physical meaning of u is the following : $u_\varepsilon(t,x,v)$ is the density of particules (neutrons) at time t in the position x and the velocity $v/_\varepsilon$ in a reactor Ω, when the initial density is $f(x,v)$.

The particules moves linearly between collisions which are described by the operator $(\frac{1}{\varepsilon^2} Q_x + \alpha)$. [The mean free path between two collisions is $\frac{\varepsilon}{\sigma(x,v)}$ and the mean number of secondary particules after collisions is $\frac{\sigma(x,v)}{\sigma(x,v) - \varepsilon^2\alpha} \simeq 1 + \frac{\varepsilon^2\alpha}{\sigma(x,v)}$].

The criticality problem for the domain Ω is to find α such that there exist u positive satisfying

$$A^\varepsilon u + \alpha u = 0 \qquad\qquad u \in D(A^\varepsilon)$$

The subject of this paper is to find an asymptotic expansion of α with respect to ε.

2 - RECALL OF RESULTS ON THE APPROXIMATION OF TRANSPORT OPERATORS.

Let us denote $C(\Omega \times V)$ or $C(V)$ the spaces of continuous functions from $\Omega \times V$ or V into $\mathbb{R}$. And $<.,.>$, $(.,.)_V$ or $(.,.)_\Omega$ the scalar product in $L^2(\Omega \times V)$, $L^2(V)$ or $L^2(\Omega)$ and $\|.\|$ the norm in $L^2(\Omega \times V)$.
We assume that $\sigma_1 \in C(\Omega \times V \times V)$ and that σ_1 is smooth with respect to x and bounded from below by σ_1 $(\sigma_1 > 0)$. Then we know (see for example Blankenship-Papanicolaou [1]) that there exist an unique eigenfunction π of Q' (Q' is the adjoint of Q) associated to 0 with $(\pi,1)_V = 1$, and we assume that π is independant of x and that :

$$(2) \qquad (\pi, v_i)_V = 0 \qquad\qquad \forall i = 1,2,\ldots..N$$

(If $\sigma_1(x,v,w) = \sigma_1(x,w,v)$ then $\pi = 1$ and (2) is satisfied).

Let us denote $\xi_i = \xi_i^x$ the function of $C(V)$ satisfying

$$Q_x \xi_i^x - v_i = 0 \qquad\qquad (\xi_i^x, \pi)_V = 0$$

$$a_{ij}(x) = (\pi, -v_i \xi_j^x)_V$$

We assume that V is not contained in any hyperplan of $\mathbb{R}^N$, then according to Sentis [2] we know that the matrix $(a_{ij})_{ij}$ is uniformely strictly positive definite, and depends smoothly on x.

Let us consider the unbounded operator A on $L^2(\Omega)$

$$A = \sum_{ij} \frac{\partial}{\partial_{xi}} (a_{ij} \frac{\partial}{\partial_{xj}}) \qquad D(A) = H^2(\Omega)_n H_0^1(\Omega).$$

Let us denote T_t^ε and T_t the semigroups on $L^2(\Omega \times V)$ and $L^2(\Omega)$ generated by A^ε and A. We know (see for example [1] in the case $\Omega = \mathbb{R}^N$, or [3]) that we have for any $t > 0$, when ε goes to 0 :

$$T_t^\varepsilon f \to T_t f \qquad \text{in } C(\Omega \times V) \quad \forall f \in C^\infty(\Omega) \quad f|_{\partial\Omega} = 0$$

$$T_t^\varepsilon f \to T_t \Pi f \qquad \text{in } C(\Omega \times V) \quad \forall f \in C(\Omega \times V) \text{ s.t. } f|_{\partial\Omega} = 0 \text{ , } f(o,v) \in C^\infty(\Omega).$$

with $\Pi f = (\pi, f)_V \quad \forall f \in L^2(\Omega \times V)$.

On the other hand we can show that the semigroup T_t^ε is bounded, uniformely with respect to t and ε, on $L^2(\Omega \times V)$ (indeed we have $(Qf,\pi f)_V \leq 0$ and $<A^\varepsilon f,\pi f> \leq 0$) and also on $C(\Omega \times V)$ (indeed we have $T_t^\varepsilon 1 \leq 1$). Thus according to the previous convergences we can see that for any $t > 0$:

(3) $$T_t^\varepsilon f \to T_t f \qquad \text{in } C(\Omega \times V) \quad \forall f \in C(\Omega) \text{ s.t. } f|_{\partial\Omega} = 0$$

(4) $$T_t^\varepsilon f \to T_t \Pi f \qquad \text{in } L^2(\Omega \times V) \ \forall f \in L^2(\Omega \times V)$$

Hence when $\varepsilon \to 0$, $(A^\varepsilon - \alpha)^{-1} \to (A - \alpha)^{-1}\Pi$ for the strong convergence of the operators on $L^2(\Omega \times V)$.

Let us denote ω_ε and ω the types of the semigroup T_t^ε and T_t (on the spaces L^2). These two types are non positive. We shall now prove that these types are also simple eigenvalue of A^ε and A, and that $\omega_\varepsilon \to \omega$ when ε goes to 0, before giving the final result. (The detailled proofs of the following results can be found in [4]).

3 - SPECTRAL PROPERTIES OF A^ε AND A.

Let us denote $\tilde{A}$ the operator A considered as an operator on $C(\Omega)$. Since any eigenfunction of A is smooth (indeed $f \in D(A^n) \subset H^{2n}\ \forall n \in N$) the eigenfunction of A and $\tilde{A}$ are the same and according to [5] we conclude that there exists $\omega_o \in \mathbb{R}$ greater than any other eigenvalue of A such that :

$$(5) \qquad \exists ! \phi \in D(A) \cap L^2_+(\Omega) \qquad A\phi = \omega_o \phi \qquad |\phi|_{L^2(\Omega)} = 1$$

$$(6) \qquad \inf (\phi(x)/x \in \Omega') > 0 \qquad \text{for any compact set } \Omega' \ (\Omega' \subset \Omega)$$

Proposition 1

We have $\omega = \omega_o$ and there exists $\hat{\phi}$ in $C(\overline{\Omega})$, positive such that

$$(7) \qquad \begin{array}{lll} e^{-\omega t}\, T_t f \to \phi(f,\hat{\phi}) & \text{in } C(\overline{\Omega}) & \text{when } t \to \infty \\ \forall f \text{ in } C(\Omega) \ \text{ such that } \alpha > o & & -\alpha\phi \le f \le \alpha\phi \end{array}$$

Proposition 2

$$(8) \qquad \exists ! \phi_\varepsilon \in D(A^\varepsilon) \cap L^2_+(\Omega \times V) \qquad A^\varepsilon \phi_\varepsilon = \omega_\varepsilon \phi_\varepsilon \qquad \| \phi_\varepsilon \| = 1$$

This result is a consequence of the compactness of the operator T^ε_t from $L^2(\Omega \times V)$ into $L^2(\Omega \times V)$ for t large enough (see Jorgens [6]) and of the following Lemma (which comes from Krein Rutman [7]).

Lemma Let B a Banach space and B_+ a closed cone of B (with $B = (\overline{B_+ - B_+})$). Let L an infinitesimal generator of a semigroup T_t of class $C°$ on B such that for t large enough :

T_t is compact

$T_t(B_+) \subset B_+$

Then the type ω of $\{T_t\}$ is an eigenvalue of L and

$$\exists\phi \in B_+ \cap D(L) \qquad L\phi = \omega\phi \qquad \phi \neq 0$$

4 - ASYMPTOTIC EXPANSIONS OF ω_ε.

Proposition 3 when ε goes to 0, we have

$$\underline{\lim}_{\varepsilon}\ \omega_\varepsilon \geq \omega$$

For the proof we need only the definition of the type, the positivity of T_t^ε and T_t ; and the properties (3) (6) (7).

Proposition 4 when $\varepsilon \to 0$, we have (for ϕ_ε satisfying (8)) :

i) $\phi_\varepsilon \to \phi$ in $L^2(\Omega \times V)$ (where ϕ is defined in (5))

ii) $\omega_\varepsilon \to \omega$

For the proof we remark first that, according to prop 3, we have

$$0 \leq - \langle Q\phi_\varepsilon, \pi\phi_\varepsilon \rangle \leq \varepsilon^2 |\omega_\varepsilon| C^{te} \to 0$$

Then we use the following implication (for a sequence f_n in $L^2(V)$) :

$$(Qf_n, \pi f_n)_V \to 0 \qquad \overline{\lim_n}\ f_n{}_{L^2} < +\infty \Rightarrow f_n \to (\lim_n |f_n|_{L^2}) \quad \text{in } L^2(V)$$

for showing that ϕ_ε converges to a function Ψ of $L^2(\Omega)$ in $L^2(\Omega \times V)$ and , last, it is sufficient to show that $\Psi = \phi$.

Now, with classicals tools of homogenization theory we can show that there exist an operator of the third degree (with respect to x) :

$$\mathcal{L} = \frac{\partial}{\partial x_i} \left(\beta_{ijk} \frac{\partial^2 \cdot}{\partial x_j \partial x_k} + \mu_{ij} \frac{\partial \cdot}{\partial x_j}\right) \quad \text{(using summation convention)}$$

such that for any f in $C^\infty(\Omega)$ if we denote (for $\alpha \geq 0$)

$$u^\varepsilon = (A^\varepsilon - \alpha)^{-1} f \qquad u = (A - \alpha)^{-1} f$$

Then we have

$$(9) \qquad \|u^{\varepsilon} - u - \varepsilon(\xi_i \frac{\partial u}{\partial x_i} + w_u)\| \leq \varepsilon^2 C^{te}$$

where w_u is the solution of :

$$(10) \quad \begin{cases} - \alpha w + \dfrac{\partial}{\partial x_i}(a_{ij} w) + \mathcal{L} u = 0 \\ w|_{\Gamma} = \gamma_i \dfrac{\partial u}{\partial x_i} \end{cases}$$

<u>Theorem</u> When $\varepsilon \to 0$, we have $\omega_{\varepsilon} = \omega + \varepsilon\omega_1 + O(\varepsilon^2)$, with $\omega_1 = (\phi, \mathcal{L}\phi)$

For proving this result we apply (9) (10) with $z_{\varepsilon} = (A^{\varepsilon})^{-1}\omega\phi$ and we write (with Ψ_{ε} solution of $A^{\varepsilon'}\Psi_{\varepsilon} = \omega_{\varepsilon}\Psi_{\varepsilon}$ $\qquad \|\Psi_{\varepsilon}\| = 1$) :

$$\frac{1}{\varepsilon} \frac{\omega_{\varepsilon} - \omega}{\omega_{\varepsilon}} <\phi,\Psi_{\varepsilon}> = \frac{1}{\varepsilon\omega_{\varepsilon}} (\omega_{\varepsilon}<\phi,\Psi_{\varepsilon}> - <z_{\varepsilon},A^{\varepsilon'}\Psi_{\varepsilon}>) = <w,\Psi_{\varepsilon}> + o(\varepsilon)$$

5 - <u>APPLICATION</u>.

We model a monogroup nuclear reactor by a homogeneous medium with periodic holes (which are control rods where neutrons are allmost all absorbed.

When we can use the previous presentation with $\Omega \subset [0,1[^2$

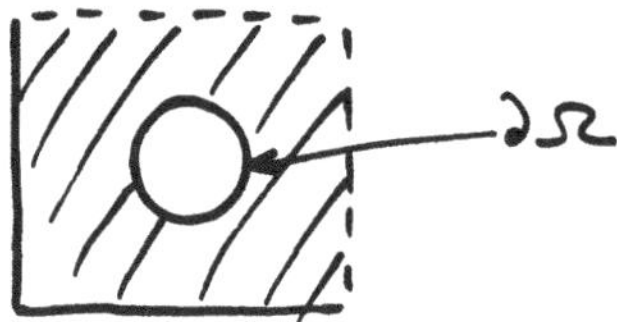

and with $\sigma_1(x,v,w) = 1 \qquad \forall x,v,w \qquad (\sigma \equiv 1$ also).

Then the medium will be critical for the α such that :

$$(*) \qquad A_{\varepsilon}u_{\varepsilon} + \alpha u_{\varepsilon} = 0 \qquad u_{\varepsilon}|_{\Gamma^-} = 0 \qquad u_{\varepsilon} \geq 0$$

According to the previous results we have

$$\alpha = |\omega| + O(\varepsilon^2) \qquad (\omega \text{ is the eigenvalue of } A)$$

(since in this case we can show that $\omega_1 = 0$)

The equation (*) can be an acceptable model for a problem in which the number of secondary particules is $(1 + \varepsilon^2|\omega|)$ which can be significantly larger than 1 (indeed in concrete case, we can take ε of the order of 0.2 or 0.25).

REFERENCES

[1] G. Blankenship - G. Papanicolaou."Stability and control of stochastic systems ..." (I) SIAM.J.Appl.Math. 34 (1978) p. 437-476.

[2] R. Sentis. "Homogénéization et approximation d'un processus de transport". Note C.R. Académie Sciences, Paris, 289 (1979) p. 567-570.

[3] R. Sentis. "Analyse asymptotique d'équations de transport dans un milieu périodique. Thèse d'état, partie A §3. Paris (1981).

[4] R. Sentis. "Study of the corrector of the eigenvalue of a transport operator".To appear

[5] Amman. Non linear operator ... in non linear operator proceedings. Bruxelles 1975. Lectures notes in Math. n° 543 Springer (1976).

[6] K. Jorgens. "An asymptotic expansion ..."Comm Pure and Appl. Math. 11 (1958) p. 219-242.

[7] M.G. Krein - M.A. Rutman. Linear operators leaving ... Transl. A.M.S. n° 26 (1950).

SHAPE SENSITIVITY ANALYSIS

FOR VARIATIONAL INEQUALITIES

J. SOKOLOWSKI
Systems Research Institute
Polish Academy of Sciences
01-447 WARSZAWA,
ul. Newelska 6, POLAND

J.P. ZOLESIO
Département de Mathématiques
Université de Nice
06034 - NICE-CEDEX, FRANCE

Let $y(\Omega)$ be the solution of a classical variational inequality posed in the Sobolev space $H_0^1(\Omega)$. We perturbe the domain Ω into Ω_t using a regular mapping T_t and we characterize the material derivative (see [3]) $\dot{y}(\Omega) = \left(\frac{d}{dt} y(\Omega_t) \circ T_t\right)_{t=0}$, element of $H_0^1(\Omega)$, as the solution of a variational inequality. We also characterize the domain derivative $y' = \nabla\dot{y}(\Omega) - \nabla y(\Omega).V$ as being the solution of an other variational inequality, see theorems 1 and 2 .

Using examples developped in [3] one can get the Eulerian semi-derivative $dJ(\Omega;V)$ for the classical differentiable or non differentiable domain functional written as $J(\Omega) = h(\Omega, y(\Omega))$. Even when h is a smooth function the cost function is not differentiable (at the domain Ω, in the direction V), that is $V \longrightarrow dJ(\Omega;V)$ is not linear, since $V \longrightarrow Y'$ is not linear. In other non differentiable situations, such as that of repeated eigenvalues (see [3]), the mapping $V \longrightarrow dJ(\Omega;V)$ is convex or concave (or could be written as sum of such mappings). The present one is, as far as we know, the first non differentiable shape problem arising without this property.

1 . The variational inequality problem
2 . The deformation of the domain, material and shape derivatives
3 . The variational inequality for which $\dot{y}$ is the solution
4 . Conical derivative of the projection
5 . The variational inequality for which y' is the solution

1 . The variational inequality problem.

Ω is a smooth bounded open set in $\mathbb{R}^n$, $K(\Omega)$ is the closed convex set defined by $K(\Omega) = \{\varphi \in H_0^1(\Omega) \mid \varphi \geq 0 \text{ a.e.}\}$, given f in $L^p(\mathbb{R}^n)$, $y = y(\Omega)$ is the solution of the variational inequality

$$(1) \qquad y \in K(\Omega) , \quad \forall\varphi \in K(\Omega) , \quad \int_\Omega \nabla y.\nabla(\varphi - y)dx \geq \int_\Omega f(\varphi - y)dx$$

We know from Stampacchia-Brezis [1] that y belongs to $W^{2,p}(\Omega)$, and for $p > n$, y is in $C^{1,\alpha}(\Omega)$ (for some positive α) .

2 . The deformations of the domain material and shape derivatives (see J.P. ZOLESIO [2], [3].

We study the continuous evolution of this configuration depending on a supplementary parameter t , which may be considered as time, any point X of Ω being transformed in $x = x(t,X)$. In the Eulerian viewpoint, the velocities of displaced particle x is, at time t , $V(t,x) = \frac{d}{dt} x(t,X)$.
Actually the velocity field V is considered as the direction of this continuous evolution, and for a given smooth field V in $C^0([0,\varepsilon[, C^1(\mathbb{R}^n, \mathbb{R}^n))$ (ε is an arbitrary positive number) we consider, at any t little enough, the mapping $T_t : X \mapsto x = x(t,X)$ where x is the solution of the (non autonomous) ordinary differential equation $\frac{d}{dt} x(t,x) = V(t,x(t,X))$ with initial condition $x(0,X) = X$.
This mapping is one to one, and T_t and T_t^{-1} belongs to $C^1(\mathbb{R}^n, \mathbb{R}^n)$.

We shall use some simple properties of the mapping T_t :

$V = \frac{\partial}{\partial t} T_t$ is the speed of deformation, $\frac{d}{dt}(\det DT_t)_{t=0} = \operatorname{div} V(0)$

$\frac{d}{dt}(DT_t) = DV(0)$, these differentiations being taken in $C^0(\mathbb{R}^n)$ and $C^0(\mathbb{R}^n, \mathbb{R}^n)$ topologies.

If $y_t = y(\Omega_t) \in K(\Omega_t)$ is the solution of the problem (1) on the perturbated domain $\Omega_t = T_t(\Omega)$, $y^t = y(\Omega_t) \circ T_t$ is an element of $H_0^1(\Omega)$.

The MATERIAL DERIVATIVE is the element (if it exists) of $H_0^1(\Omega)$ given by $\dot{y} = \dot{y}(\Omega) = \frac{d}{dt}\left(y(\Omega_t) \circ T_t\right)_{t=0}$ (this derivative being taken in $H_0^1(\Omega)$ norm).

If V is a field such that $V(t,X).n(X) = 0$ for $t \geq 0$ and $X \in \Gamma = \partial\Omega$ (n is the normal field on Γ) then $\Omega_t = \Omega$ (the continuous evolution just concerns the interior of the domain but not its shape Γ) and $y(\Omega_t) = y(\Omega) = y$ for any t , while the material derivative is not zero : $\dot{y} = \frac{d}{dt}(y \circ T_t)_{t=0} = \nabla y . V(0)$, then $\dot{y}(\Omega)$ cannot be considered as the shape derivative of $y(\Omega)$ (at Ω , in the direction V) ; therefore we consider $y' = y'(\Omega) = \dot{y}(\Omega) - \nabla y(\Omega) . V(0)$ to be the shape derivative of $y(\Omega)$.

If f belongs to $L^p(\mathbb{R}^n)$ with $p \geq 2$ then $y(\Omega)$ belongs to $H^2(\Omega)$ and $y'(\Omega)$ is an element of $H^1(\Omega)$. In général cases y' just depends on the germ on Γ of the field V(0) (under assumption that the mapping $V \longrightarrow \dot{y}$ is continuous, from $C^0([0,\varepsilon[, C^1(\mathbb{R}^n, \mathbb{R}^n))$ into $H_0^1(\Omega)$, see J.P. ZOLESIO [1]) .

Let $A(t)$ be the symmetric matrix, $A(t) = J_t(DT_t)^{-1} \cdot {}^*(DT_t)^{-1}$ (where $J_t = \det(DT_t)$ and ${}^*(DT_t)^{-1}$ is the transposed of the inverse of the Jacobian matrix). Using the change of variables formula in formula (1) (written for the domain Ω_t and the function $y(\Omega_t)$ and the property $(\nabla y_t) \circ T_t = {}^*(DT_t)^{-1} \cdot \nabla y^t$ (where $y^t = y_t \circ T_t$) we find that y^t is the solution of the problem

$$(2) \qquad y^t \in K(\Omega) \quad , \quad \forall \varphi \in K(\Omega) \ , \int_\Omega \langle A(t) \cdot \nabla y^t, \nabla(\varphi - y^t) \rangle dx \geq \int_\Omega f^t(\varphi - y^t) dx$$

where $f^t = J_t \ f \circ T_t$ and $\langle \ , \ \rangle$ denotes the scalar product in $\mathbb{R}^n$.

This problem (2) may be written as

$$(3) \qquad \int \nabla y^t \ \nabla(\varphi - y^t) dx \geq \int \langle (I_d - A(t)) \ \nabla y^t \ , \nabla(\varphi - y^t) \rangle \ dx + \int f^t(\varphi - y^t) dx \ .$$

Introduce F_t as the solution of the problem

$$(4) \qquad F_t \in H_o^1(\Omega) \quad , \quad \forall \varphi \in H_o^1(\Omega) \ , \int_\Omega \nabla F_t \cdot \nabla\varphi dx = \int_\Omega \left\{ \langle (I_d - A(t)) \cdot \nabla y^t, \nabla\varphi \rangle + f^t \right\} dx$$

Using (3) and (4) we see that y^t solves the problem :

$$(5) \qquad y^t \in K(\Omega) \ , \quad \forall \varphi \in K(\Omega) \ , \int_\Omega \nabla y^t \cdot \nabla(\varphi - y^t) dx \geq \int_\Omega \nabla F_t \cdot \nabla(\varphi - y^t) dx$$

Consider the Hilbert space $H_o^1(\Omega)$ equiped with the scalar product $\int_\Omega \nabla\varphi \cdot \nabla\psi \ dx$, then (5) means that y^t is the projection of F_r on the closed convex set $K(\Omega)$:

$$(6) \qquad y^t = P_{K(\Omega)}(F_t)$$

here we must notice that F_r defined by (41) depends on y^t, nevertheless the derivative (at $t = 0$) of the term $(I_d - A(t)) \cdot \nabla y^t$ which occurs in (4) don't depend on $\dot{y} = \left(\frac{d}{dt} y^t\right)_{t=0}$, for $(I_d - A(t))_{t=0} = 0$, but just depends on $A' = \left(\frac{d}{dt} A(t)\right)_{t=0}$ and $y(\Omega)$, and via (4) the dérivative $F' = \left(\frac{d}{dt} F_t\right)_{t=0}$ will not depnd on $\dot{y}(\Omega)$ too .

3 . Conical Derivative of the Projection on $K(\Omega)$

We now use the differentiability properties of $P_{K(\Omega)}$, (the fundamental result of F. Mignot [5]) following the approach of J. SOKOLOWSKI [6] : Consider

$CK(\Omega) = \left\{ \varphi \in H_o^1(\Omega) \mid \exists t > 0 \ , \ Y + t\varphi \geq 0 \ \text{ a.e.} \right\}$ and $\overline{CK}(\Omega)$ its closure in $H_o^1(\Omega)$ and

$$(7) \qquad S(\Omega) = CK(\Omega) \cap [\, P_{K(\Omega)}(F_o) - F_o \,]^{\perp}$$

$K(\Omega)$ is the positive cone of the Dirichlet space $H_o^1(\Omega)$ (mainly if φ belongs to $H_o^1(\Omega)$ then $|\varphi|$ too, see [5]) then it is shown in [5] we have

(8) $\overline{CK}(\Omega) = \{ \varphi \in H_o^1(\Omega) \mid \varphi \geq 0$ a.e. on $y^{-1}(0)\}$ where the level set $y^{-1}(0) = \{ x \in \Omega \mid y(\Omega)(x) = 0 \}$ is a closed subset while we suppose $p \geq 2$, then $y(\Omega) \in C^o(\overline{\Omega})$ for 2 or 3 dimensional domains.

With (7) and (8) one get directly

$$(9) \quad S(\Omega) = \{ \varphi \in H_o^1(\Omega) \mid \varphi \geq 0 \text{ a.e. on } y^{-1}(0) \text{ and } \int_\Omega (\Delta y+f)\,dx = 0 \}$$

from the density indicated above we get the expansion (see [5]) :

$$P_{K(\Omega)}(F_t) = P_{K(\Omega)}(F_o) + t\, P_{S(\Omega)}(\omega(t)) + O(t, \omega(t))$$

where $\omega(t) - (F_t - F_o)/t$ and $\frac{1}{t} \| O(t,\omega) \|_{H_o^1(\Omega)} \longrightarrow 0$, $t \longrightarrow 0$, uniformly in ω if remains in a compact subset of $H_o^1(\Omega)$. That is

$y^t = y(\Omega) + t\theta(t) + O(t, \omega(t))$ where $\theta(t)$ is the solution of the problem

$$(10) \quad \theta(t) \in S(\Omega), \quad \forall\varphi \in S(\Omega) , \quad \int_\Omega \nabla\theta(t).\nabla(\varphi - \theta(t))\,dx \geq \int_\Omega \nabla\omega(t).\nabla(\varphi - \theta(t))\,dx .$$

4 . <u>The variational inequality for which $\dot{y}$ is the solution.</u> Looking the definition of $\dot{y}$ we consider the limit, as t goes to zero, of the term $(y^t - y(\Omega))/t = \theta(t) + \frac{1}{t} O(t, \omega(t))$. The limit, if it exists (in $H_o^1(\Omega)$ norm), is the material derivative $\dot{y}$. Two problems arise : the limits of the two terms $\theta(t)$ and $\frac{1}{t} O(t, \omega(t))$. It is easily shown that if $\omega(t)$ possesses a limit in $H_o^1(\Omega)$, say $\omega(0)$, when t goes to zero, then the set $\{ \omega(t) \mid 0 \leq t \leq 1 \}$ is a vompact subset of $H_o^1(\Omega)$; by the above remarks about $\frac{1}{t} O(t, \omega(t))$, this term goes to zero in $H_o^1(\Omega)$. It follows easily by (10) that the term $\theta(t)$ possesses a limit (related to $\omega(0)$) which is $\dot{y}$.

4.1. The limit of $\omega(t)$ as t goes to zero.

From (4) we have $\omega(t) \in H_o^1(\Omega)$, $\forall\, \varphi \in H_o^1(\Omega)$,

$$(11) \quad \int_\Omega \nabla\omega(t).\nabla\varphi\,dx = \int_\Omega \{ < \frac{1}{t}(I_d - A(t)).\nabla y^t, \nabla\varphi > + (f^t - f)/t \}\,dx$$

For any fields $V \in C^o(\mathbb{R}^n, \mathbb{R}^n))$ we have the convergence facts (see [2], [3]) : as $t \to 0$,

(12) $$-\frac{1}{t}(I_d - A(t)) \longrightarrow A' = \operatorname{div} V(0)\, I_d - (DV(0) + {}^*DV(0))$$

in $C^o(\mathbb{R}^n, \mathbb{R}^n)^n$ topology, and as f belongs to $L^2(\Omega)$:

(13) $$(f^t - f)/t \longrightarrow \operatorname{div}(fV(0)) \quad , \text{ in } H^{-1}(\Omega) \text{ norm.}$$

Taking $\varphi = 0$ in (3) we get

$$\|y^t\|^2_{H^1_o(\Omega)} \leq \|I_d - A(t)\|_{L^\infty(\Omega)^n} \|y^t\|^2_{H^1_o} + C\,\|f^t\|_{L^2(\Omega)} \|y^t\|_{H^1_o(\Omega)}$$

using (12) we see that y^t remains in a bounded subset of $H^1_o(\Omega)$.

Taking $\varphi = y$ in (3) , $\varphi = y^t$ in (1) and adding the two inequalities we get $\|y^t - y\|_{H^1_o} \leq \|I_d - A(t)\|_{L^\infty} + \|f^t - f\|_{H^{-1}}$, that is

(14) $$y^t \longrightarrow y = y(\Omega) \quad , \quad t \to 0 \quad \text{in } H^1_o(\Omega) \text{ norm} \quad .$$

Using (12) , (13) and (14) in (11) we now find $\omega(t) \to \omega(0)$ as $t \to 0$, in $H^1_o(\Omega)$ norm ; where $\omega(0)$ is defined by the problem

(15) $$\omega(0) \in H^1_o(\Omega) , \ \forall\, \varphi \in H^1_o(\Omega), \int_\Omega \nabla\omega(0).\nabla\varphi\, dx = -\int_\Omega < A'.\nabla y + fV(0), \nabla\varphi > dx$$

5.2. $\dot{y}(\Omega)$ as the limit of $\theta(t)$.

We return to the limit problem for $\theta(t)$: the projection $P_{S(\Omega)}$ being lipschitzian (it can be seen in (10) as $\theta(t) = P_{S(\Omega)}(\omega(t))$) we get $\theta(t) \to \dot{y}(\Omega) = \theta(0)$ in $H^1_o(\Omega)$ norm. We have proven

<u>THEOREM 1</u> . Let $V \in C^o([0, \varepsilon[, C^1(\mathbb{R}^n, \mathbb{R}^n))$; $f \in L^p(\Omega)$, $p \geq 2$, Ω_t the perturbed domain, $\Omega_t = T_t(V)(\Omega)$, $y(\Omega)$ the solution of the variational problem (1) . Then $y = y(\Omega)$ possesses a material derivative in $H^1_o(\Omega)$,

$$\dot{y}(\Omega) = \left(\frac{d}{dt} y(\Omega_t) \circ T_t(V)\right)_{t=0} \quad \text{defined by the problem}$$

(16) $$\dot{y} \in S(\Omega), \ \forall\, \varphi \in S(\Omega), \int_\Omega \nabla\dot{y}\, \nabla(\varphi - \dot{y})\, dx \geq -\int_\Omega < A'.\nabla y + fV(0), \nabla(\varphi - \dot{y}) > dx$$

$S(\Omega)$ being defined by (9)

5 . <u>The variational inequality for which y' is the solution.</u> With $v = V.n$ on Γ consider

(17) $S_v(\Omega) = \{ \psi \in H_0^1(\Omega) \mid \psi \geq 0$ a.e. in $y^{-1}(0)$, $\int_\Omega (\Delta y + f)\psi \, dx = 0$, $\psi|_\Gamma = - \frac{\partial y}{\partial n} v \}$

and $S(\Omega) = S_0(\Omega)$. We first prove that y' belongs to $S_v(\Omega)$; we have two conditions to study. We know from STAMPACCHIA [7] that

(18) $\nabla y(x) = 0$ a.e.$x \in y^{-1}(s)$, now taking $p > n$, y belongs to $C^1(\bar{\Omega})$ and this equality holds for every x in $y^{-1}(0)$. (18) is sufficient to get y' in $S_v(\Omega)$: $\nabla y.V(0)$ is eqqual to zero on $y^{-1}(0)$, $\int_\Omega (\Delta y + f)\nabla y.V(0)dx = 0$ for it is well known that $\Delta y + f$ has its support included in $y^{-1}(0)$, we have proven the

LEMMA 1. If ψ belongs to $S_v(\Omega)$, then $\varphi = \psi + \nabla y.V(0)$ belongs to $S(\Omega)$.
we now consider, in (16), $\varphi - \dot{y} = \psi - (\dot{y} - \nabla y. V((0)) = \psi - y'$, and

(19) $y' \in S_v(\Omega)$, $\forall \psi \in S_v(\Omega)$, $\int_\Omega \nabla y'.\nabla(\psi - y')dx \geq B(\psi - y', V(0))$ where B is the bilinear form on $H_0^1(\Omega) \times C^1(\mathbb{R}^n, \mathbb{R}^n)$ defined by

$$B(\varphi, V) = - \int_\Omega < fV + A'.\nabla y + \nabla(\nabla y.V), \nabla\varphi > dx$$

where we recall that $A' = \operatorname{div} V(0) \, I_d - (DV(0) + {}^*DV(0))$.

If $v(t) = V(t).n$ is equal to zero on the boundary Γ then $\Omega_t = \Omega$ and $y' = 0$; then for such a field (19) is reduced to $0 \geq B(\psi, V)$, $\forall\psi \in S_v(\Omega)$; and taking V and $-V$, $B(\psi, V) = 0$ then using the Green's formula (and using $B(\psi, V) = 0$ for any $V \in \mathcal{D}(\Omega; \mathbb{R}^n)$) we get the

LEMMA 2. $\forall \varphi \in S(\Omega) \cap H^2(\Omega)$, $\forall V \in C^0([0, \varepsilon[, C^1(\mathbb{R}^n, \mathbb{R}^n))$, $B(\varphi, V) = 0$.
But $S(\Omega)$ is a closed convex set in $H_0^1(\Omega)$ and $H^2(\Omega) \cap S(\Omega)$ is dense in $S(\Omega)$, then this equality is true for any φ in $S(\Omega)$.

In the problem (16) ψ and y' belong to $S_v(\Omega)$ then $\psi - y' \in S(\Omega)$ and then $B(\psi - y', V(0)) = 0$. Hence

THEOREM 2. Let V belong to $C^0([0, \varepsilon[, C^1(\mathbb{R}^n, \mathbb{R}^n))$, f to $L^p(\Omega)$, $p > n$, Ω_t the perturbated domain, $y(\Omega)$ the solution of the variational problem (1) ; the shape derivative $y' = \dot{y}(\Omega).\nabla y(\Omega), V(0)$ is the elmeent of $H^1(\Omega)$ defined by the following problem

(20) $y' \in S_v(\Omega)$, $\forall \psi \in S_v(\Omega)$, $\int_\Omega \nabla y'.\nabla(\psi - y')dx \geq 0$.

[1] STAMPACCHIA - BREZIS, Inéquations elliptiques, Bull. Soc. Math. de France ; t. 86, 1968.

[2] ZOLESIO J.P., Identification de Domaines par déformation. Thèse de Doctorat d'Etat, Université de Nice, 1979

[3] ZOLESIO J.P. The material derivative in shapes optimization ... " optimization of Distributed parameters structures " J. Céa and Ed. Haug, eds, p. 1089-1194, 1447-1473, Série E, n° 50, Sijthoff and Hoordhof, Rackville, Mariland, 1981.

[4] DELFOUR - PAYRE - ZOLESIO, Theses proceedings

[5] F. MIGNOT, Contrôle dans les inéquations variationnelles. J. functional Analysis 22 (1976) p. 13

[6] J. SOKOLOWSKI Sensitivity analysis for a class of variational inequality in " optimization of Distributed parameters structures ", see [3]

[7] STAMPACCHIA Equation elliptique du second ordre à coefficients discontinus, Presses de l'Université de Montréal, 1966.

A CAUTIOUS TIME-OPTIMAL CONTROL ALGORITHM FOR STOCHASTIC CONTROL SYSTEMS WITH ADDITIONAL BOUNDARY CONSTRAINTS

J.H. de Vlieger

Department of Electrical Engineering, Control laboratory, Delft University of Technology, Mekelweg 4, 2628 CD Delft, The Netherlands.

Abstract.

The concept of the time-optimal control algorithm [10] for linear discrete time systems with additional boundary constraints is extended to stochastic systems with parameter and state uncertainty. The Deterministic Time-Optimal Control problem has been solved by means of linear programming techniques. However, the optimal control strategy for stochastic systems yields control decisions which are cautious of the uncertainty in the system and which probe it for estimation purposes. Mathematical difficulties make it is hard to derive optimal (dual) control algorithms. Several suboptimal algorithms which have ignored the probing property are referred to as cautious control algorithms. In this paper a Cautious Time-Optimal Control algorithm will be derived.

1. Introduction.

The major aim of this paper is to apply the concept of Deterministic Time-Optimal Control (D.T.O.C) [10], to stochastic control systems with parameter and state uncertainty. The D.T.O.C. algorithm can be applied in-line, since linear programming allows an efficient computation of the control sequence. In [10] the certainty equivalence property has been used heuristically to apply the D.T.O.C. algorithm to stochastic control problems. However, enforcement of this property does not guarantee the desired performance. Especially in the case of time-optimal control one has to be cautious of large values of the control variables, which have been based upon incorrect a priori knowledge of the parameters. The optimal control strategy yields control decisions which are cautious of the uncertainty in the system but also probe the system for estimation purposes. The reducible uncertainty, e.g. the uncertainty of deterministic but unknown system parameters, is thereby decreased The optimal control for this kind of stochastic systems has been called "Dual Control" [3] (dual: estimation versus control). Mainly because of mathematical difficulties it is hardly possible to derive optimal (dual) control algorithms. Several suboptimal control algorithms, such as Open-Loop Feedback Optimal [4], ignore the probing property and are referred to as cautious control algorithms [11]. The open-loop form of the D.T.O.C algorithm allows the application of linear programming to calculate the control sequence with the additional capability of adding constraints on the input and state variables. By making some assumptions a "Cautious Time-Optimal Control" (C.T.O.C.) algorithm quite similar to the D.T.O.C. algorithm, although without the disadvantage of the deterministic design of the D.T.O.C. algorithm, will be derived.

2. Notations.

The nxn identity matrix and the mxn null matrix are denoted by I and 0 . The Kronecker matrix product between two matrices is denoted by $[.] \otimes [.]$ (See further [6], [5] and [2]). Some useful lemmas which have been introduced in [8] can be found in Appendix A. The column string operation $\underline{cs}\{[.]\}$ is defined by

$\underline{cs}\{X\} = [\underline{x}'_{.1}\ \underline{x}'_{.2}\ \dots\dots\ \underline{x}'_{.n}]'$ where $\underline{x}_{.i}$ is the ith column of matrix X.

3. Problem formulation.

The following n^{th}-order linear stochastic state model will be used:

$$\begin{aligned}\underline{x}(k+1) &= A\underline{x}(k+1) + \underline{b}u(k) + \underline{\xi}(k)\\ \underline{y}(k) &= C\underline{x}(k) + \underline{\eta}(k)\ .\end{aligned} \tag{1}$$

It is assumed that $\underline{x}(0)$, $\underline{\xi}(k)$ and $\underline{\eta}(k)$ are realizations of stochastic variables with normal distributions:

$$\underline{\xi}(k) \sim N\{\underline{0},\ P_{\xi}(k)\}\ ,\quad \underline{\eta}(k) \sim N\{\underline{0},\ P_{\eta}(k)\}\ ,\quad \underline{x}(0) \sim N\{\underline{\bar{x}}(0),\ P_{x}(0)\}\ , \tag{2}$$

where $N\{\underline{\bar{x}}(0),\ P_x(0)\}$ defines a normal distribution function with mean value $\underline{x}(0)$ and covariance matrix $P_x(0)$. Furthermore, only the parameters of the control vector $\underline{b}$ are assumed to be unknown and vector $\underline{b}$ is assumed to be selected from a normal distribution $\underline{b} \sim N\{\underline{\bar{b}},\ P_b\}$.
The following boundary constraints can be imposed on the control variables:

$$\underline{u}^- \leqslant \underline{u}(k) \leqslant \underline{u}^+\ , \tag{3}$$

where $\underline{u}^+$ and $\underline{u}^-$ define the upper and lower bounds of the control variables. Given the cost function

$$C = \sum_{k=0}^{N} C_k(\underline{x}(k),\ u(k)) \tag{4}$$

we shall minimize the mathematical expectation of the future costs $J = E\{\ C\ \}$. In the theory of dynamic programming [1] it is stated that a cost function of this form can be minimized by successive minimizations:

$$\min_{u(k)} J_k = J_k^{\star} = \min_{k(k)} E\{C_k + J_{k+1}^{\star}\ |I^k\ \}\ , \tag{5}$$

where $J_k^{\star}$ is the <u>optimal cost-to-go</u> for sampling instant kT_s with $J_N^{\star} = C\ (\underline{x}(N))$. The general approach to solving the dynamic programming equation is to generate a sequence of control laws (functions) which minimize the cost function. The laws u(i) can be calculated from the expected future cost based on the future information set at sampling instant kT_s and the optimal control laws $u^{\star}(I^k)$, $k=i+1,i+2,\dots,N-1$. Feedback is a natural outcome of this approach. However, a second way to minimize the costs is to determine the control decisions u(i) from the open-loop solutions of the minimization problem at the successive sampling instants iT_s (open-loop feedback [9]). This approach is only optimal for deterministic control processes and some special cases of stochastic control processes. Generally, these two approaches lead to two different mathematical forms of the control algorithms: the feedback and the open-loop forms.

Consider a terminal guidance problem with the quadratic costs

$$C\ = \underline{x}'(N)\underline{x}(N)\ . \tag{6}$$

Note that in the deterministic case these costs will be zero for a certain value of N. In stochastic control problems the horizon N is usually assumed to be fixed. However, in section 4 we shall introduce a criterion for the minimum value of N.
To evaluate the expectation of future costs we shall start by determining of the cost

J_{N-1} at instant N-1:

$$J_{N-1} = E\{ \underline{x}'(N)\underline{x}(N) | I^{N-1}\} = E\{\underline{x}'(N-1)A'A\underline{x}(N-1) + \underline{b}'\underline{b}\, u^2(N-1) + 2\underline{x}'(N-1)A'\underline{b}u(N-1) + \underline{\xi}'(N-1)\underline{\xi}(N-1) | I^{N-1}\} . \quad (7)$$

Introducing the trace operation results in

$$J_{N-1} = \text{trace}[\, A'A\, E\{\underline{x}(N-1)\underline{x}'(N-1)|I^{N-1}\}\,] + 2\,\text{trace}[\, A'E\{\underline{bx}'(N-1)|I^{N-1}\}\,]\, u(N-1) + \text{trace}[\, E\{\underline{bb}'|I^{N-1}\}\,]u^2(N-1) + \text{trace}[\, P_\xi(N-1)\,] . \quad (8)$$

The deterministic matrix A and control variable u(N-1) have been separated from the expectation operator. Although it is not difficult to derive the optimal control law for the unconstrained control problem, the optimization problem for instant $(N-2)T_s$ is more difficult, since minimizing the expectation of the cost-to-go

$$J^\star_{N-2} = \min_{u(N-2)} E\{\, C_{N-2} + J^\star_{N-1}\, | I^{N-2}\,\} \quad (9)$$

with respect to the information which will be gathered at future sampling instants cannot be performed analytically. Moreover, the information set will increase with time ("the curse of dimensionality" [1]). Note that C_{N-2} is equal to zero for the terminal guidance problem. The terms $E\{\underline{bb}\, | I^{N-1}\}$ and $E\{\underline{bx}\,(N-1)|I^{N-1}\}$ are involved in the optimization, since they depend on u(N-2). In most cases this optimization problem cannot be solved and the optimal cost-to-go $J^\star_{N-2}$ must be approximated. At this point the probing property often disappears because the control law $u(I^{N-2})$ does not take into account the future system uncertainty. In the following we shall avoid this problem by making two assumptions and using the open-loop form to develop the C.T.O.C. algorithm which allows us to easily add constraints.
Proceeding with the evaluation of the costs J_{N-2}, it is assumed that

$$E\{\underline{bx}'(N-1)|I^{N-1}\} = E\{\underline{bx}'(N-2)|I^{N-2}\}A' + E\{\underline{bb}'|I^{N-2}\}u(N-1) \quad (10)$$

and

$$\text{trace}\{A'A\, E\{\underline{x}(N-1)\underline{x}'(N-1)|I^{N-1}\} = E\{\underline{x}'(N-1)A'A\underline{x}(N-1)|I^{N-2}\}$$
$$= \text{trace}[\, (A^2)'A^2\, E\{\underline{x}(N-2)\underline{x}'(N-2)|I^{N-2}\} + \text{trace}[A'A\, E\{\underline{bb}'|I^{N-2}\}]u^2(N-2) + \text{trace}[A'A\, E\{\underline{\xi}(N-2)\underline{\xi}'(N-2)|I^{N-2}\} + 2\,\text{trace}[\, (A^2)'A\, E\{\underline{bx}'(N-2)|I^{N-2}\}]u(N-2) . \quad (11)$$

Substitution of (10) and (11) in (9) results in

$$J_{N-2} = \text{trace}[\, E\{\underline{bb}'|I^{N-2}\}\,]u^2(N-1) + \text{trace}[\, A'A\, E\{\underline{bb}'|I^{N-2}\}u^2(N-2) + 2\,\text{trace}[\, (A^2)'E\{\underline{bx}'(N-2)|I^{N-2}\}\,]u(N-1) + 2\,\text{trace}[\, A'E\{\underline{bb}'|I^{N-2}\}\,]u(N-1)u(N-2) + \text{trace}[\, (A^2)'A^2\, E\{\underline{x}(N-2)\underline{x}'(N-2)|I^{N-2}\}\,] + 2\,\text{trace}[\, A'AP_\xi(N-2) + P_\xi(N-1)\,] . \quad (12)$$

The following calculation rule has been used:

$$\text{trace}[A'\underline{bx}'A'] = \underline{x}'A'A'\underline{b} = \text{trace}[(A^2)'\underline{bx}'] \quad (13)$$

The evaluation of J_0 can be performed similarly by deriving J_1 , J_2 through J_{N-1}. The general expression of J becomes:

$$J_0 = \sum_{i=1}^{N} \text{trace}[(A^{i-1})'A^{j-1}M_{bb}]u(N-i)u(N-j) + 2\sum_{i=1}^{N} \text{trace}[(A^N)'A^{i-1}M_{bx}(0)]u(N-i)$$
$$+ \text{trace}[(A^N)'A^N M_{xx}(0)] + \text{trace}[\sum_{j=1}^{N} (A^{i-1})'A^{j-1}P_\xi(N-i)] , \quad (14)$$

where M_{bb}, M_{bx} and M_{xx} are defined by:

$$M_{bb} = E\{\underline{bb}'|I^0\} , \quad M_{bx}(0) = E\{\underline{bx}'(0)|I^0\} , \quad M_{xx}(0) = E\{\underline{x}(0)\underline{x}'(0)|I^0\} . \quad (15)$$

The minimization of J_0 can be performed by:

$$\frac{\partial J_0}{\partial u(N-i)} = 2\sum_{j=1}^{N-1} \text{trace}[(A^{i-1})'A^{j-1}M_{bb}]u(N-j) + 2\,\text{trace}[(A^N)'A^{i-1}M_{bx}(0)] = 0 , \quad i=1,\ldots\ldots,N . \quad (16)$$

By transposing the trace and addition operations we get

$$\text{trace}[A^{i-1}\{\sum_{j=1}^{N} M_{bb}(A')^{j-1}u(N-j) + M_{bx}(0)\,(A')^N\}] = 0 . \quad (17)$$

Using the column string operation results in (see Appendix A):

$$[\underline{cs}\{(A')^{i-1}\}]'[\sum_{j=1}^{N} \underline{cs}\{M_{bb}(A')^{j-1}\}u(N-j) + \underline{cs}\{M_{bx}(0)(A')^N\}] = 0 , \quad i=1,\ldots\ldots,N. \quad (18)$$

The open-loop formulation results from conditions (18) and (3). Condition (18) implies N matrix equations which only differ from each other by the matrix A^{i-1}. However, one could reduce (18) to n equations if N is "sufficiently" large, as will be considered in section 4.

4. Determination of the final-time horizon.

The final-time horizon for D.T.O.C. problems is defined by the smallest integer N for which $\underline{x}(N)=\underline{0}$. Considering quadratic costs $C=\underline{x}'(N)\underline{x}(N)$ the minimum value of N is found for C=0. The following cases can be distinguished: 1) $N < \sigma$, 2) $N = \sigma$, and 3) $N > \sigma$, where σ is the controllability index of the deterministic system. Cases 1) and 2) are referred to as minimum-time dead-beat control. If $N<\sigma$ the state x(0) is already a part of a particular optimal trajectory and fewer steps are required to reach the terminal state. When $N>\sigma$ the controls are subjected to additional constraints.
In the case of the stochastic system (1) the controllability index is not defined properly. However, it is possible to use an equivalent definition of the final-time horizon for the stochastic case by considering the quadratic costs with an undefined horizon. For this reason we consider the terminal guidance problem, which can generally be evaluated to the following form:

$$J_0 = \sum_{i=1}^{N}\sum_{j=1}^{N} \underline{\alpha}_i'\,\Lambda\underline{\alpha}_j\; u(N-i)u(N-j) + 2\sum_{i=1}^{N} \underline{\alpha}_i'\underline{\beta}\; u(N-j) + \underline{\gamma} , \quad (19)$$

where $\underline{\alpha}_i$, $\underline{\beta}$ and $\underline{\gamma}$ are assumed to be vectors which will have different definitions, specified below, for the deterministic and the stochastic case. In order to minimize J_0 by u(0) through u(N-1), the following conditions have to be satisfied:

$$\frac{\partial J_0}{\partial u(N-i)} = 2\underline{\alpha}_i'[\sum_{j=1}^{N} \Lambda\underline{\alpha}_j u(N-j) + \underline{\beta}] = 0 , \qquad i=1,2,\ldots,N . \quad (20)$$

Let us now consider the deterministic case. The vectors $\underline{\alpha}_i$ and $\underline{\beta}$ are equal to

$$\underline{\alpha}_i = A^i\underline{b} , \qquad \underline{\beta} = A^N\underline{x}(0) , \quad (21)$$

and the matrix Λ is equal to the identity matrix. A dead-beat solution is found if

$$N = \sigma = \min\{ k \mid \operatorname{rank}[\, \underline{b}\ A\underline{b}\ A^2\underline{b}\ \ldots\ A^{k-1}\underline{b}] = n \} , \tag{22}$$

where $[\underline{b}\ A\underline{b}\ \ldots\ A^{\sigma-1}\underline{b}]$ is the controllability matrix [7]. If additional boundary constraints are imposed the horizon N is equal to σ or is sufficiently larger. When the stochastic system (1) is considered the vectors $\underline{\alpha}_i$ and $\underline{\beta}$ will be equal to

$$\underline{\alpha}_i = \underline{cs}\{(A^{j-1})'\} , \qquad \underline{\beta} = \underline{cs}\{ M_{bx}(0)(A^N)'\} , \tag{23}$$

according to expression (18), and the matrix Λ can be determined by

$$\Lambda = I_n \otimes M_{bb} . \tag{24}$$

From Appendix A we know that $\underline{cs}\{M_{bb}(A')^{j-1}\} = (I_n \otimes M_{bb})\underline{cs}\{(A')^{j-1}\} = \Lambda\underline{\alpha}_j$. In both cases we will define the final-time horizon by

$$N = \min\{ k \mid \operatorname{rank}[\underline{\alpha}_1\ \underline{\alpha}_2\ \ldots\ \underline{\alpha}_k] = n \} . \tag{25}$$

The above definition of the final-time horizon implies both the deterministic and the stochastic case. Note that in the deterministic case condition (18) would be reduced to

$$(A^{i-1}\underline{b})'\{ \sum_{j=1}^{N} A^{j-1}\underline{b}u(N-j) + A^N\underline{x}(0)\} = 0, \qquad i = 1,2,\ldots,N . \tag{26}$$

Condition (26) can be reduced to n instead of N equations, since the rank of the matrix $[\underline{b}\ A\underline{b}\ \ldots A^{N-1}\underline{b}]'$ is equal to n if $N \geq n$. We shall now define the matrix by

$$C_k = [\ \underline{cs}\{I_n\}\ \underline{cs}\{A'\}\ \ldots\ \underline{cs}\{(A')^{k-1}\}\ ] . \tag{27}$$

Equation (18) can be rewritten into:

$$C_N' \quad [\sum_{j=1}^{N} \underline{cs}\{M_{bb}(A')^{j-1}\}u(N-j) + \underline{cs}\{M_{bx}(0)(A')^N\}] = 0 . \tag{28}$$

Appendix B proves that matrix C_N has a rank at most equal to n when $N \geq n$. Matrix C_N in (28) can now be replaced by C_n , yielding a set of n instead of N equality conditions:

$$C_n' \quad [\sum_{j=1}^{N} \underline{cs}\{M_{bb}(A')^{j-1}\}u(N-j) + \underline{cs}\{M_{bx}(0)(A')^N\}] = \underline{0} . \tag{29}$$

The open-loop form has been formulated by condition (29) and boundary conditions (3). The control actions u(j) can be calculated by means of linear programming in a way similar to that described in [10].

5. State and parameter estimation.

The advantage of the proposed model appears to be the linearity of the state equations in the stochastic variables. Because of the assumption of a known system matrix and an unknown input matrix a linear augmented state equation can be introduced :

$$\underline{z}(k+1) = \begin{bmatrix} A & I_n u(k) \\ 0 & I_n \end{bmatrix} \underline{z}(k) + \underline{v}(k) , \tag{30}$$

$$\underline{y}(k) = H\underline{z}(k) + \underline{\eta}(k) ,$$

where

$$\underline{z}'(k) = [\ \underline{x}'(k)\ \underline{b}'\ ] , \qquad \underline{v}'(k) = [\ \underline{\eta}'(k)\ \underline{0}'\ ] . \tag{31}$$

The augmented state $\underline{z}(k)$ can be determined by means of a Kalman filter. The previously defined matrices M_{bb}, M_{bx} and M_{xx} can be calculated from

$$\begin{bmatrix} M_{xx}(k) & M_{xb}(k) \\ M_{bx}(k) & M_{bb}(k) \end{bmatrix} = \hat{\underline{z}}(k)\hat{\underline{z}}'(k) + P_{zz}(k|k) , \tag{32}$$

where $\hat{\underline{z}}(k) = E\{\underline{z}(k)|I^k\}$, $P_{zz}(k|k) = E\{(\underline{z}(k)-\hat{\underline{z}}(k))(\underline{z}(k)-\hat{\underline{z}}(k))|I^k\}$. (33)

6. Experimental results.

The computational effort required by the proposed C.T.O.C. algorithm is equivalent to the D.T.O.C. case. To compare the two algorithms the stable third-order system of [8] is used. The model of the system and the a priori knowledge are:

$$A = \begin{bmatrix} 1 & 0.2 & 0 \\ 0 & 1.0 & 0.2 \\ -1 & -1.4 & 0.4 \end{bmatrix} , \quad C = [\ 1\ 0\ 0\ ] , \qquad P_\eta = 0.09 ,$$

$$P_\xi = \begin{bmatrix} 0.04 & 0.08 & 0.12 \\ 0.08 & 0.16 & 0.24 \\ 0.12 & 0.24 & 0.36 \end{bmatrix} \times 0.01 , \quad \hat{\underline{x}}(0) = [\ 1\ 1\ 1\ ] , \tag{34}$$

$$P_x = P_{xx}(0|0);\underline{x}(0)\} = 4I_n , \quad \hat{\underline{b}}(0) = [\ 0\ 0\ -0.4\ ] , \ P_b = P_{bb}(0|0) = I_n ,$$

where A has eigenvalues 0.8 and 0.8 + 0.4j. Monte Carlo simulation runs have been performed in order to compare the proposed cautious control algorithm to the D.T.O.C. algorithm [10]. For each run a new initial state and new values of the $\underline{b}$ parameters have been selected randomly by using $N(\hat{\underline{b}}(0),P_b)$ and $N(\hat{\underline{x}}(0),P_x)$. The algorithms have been compared under the same conditions by using the performance index $\underline{x}'(k)\underline{x}(k)$. For all cases we shall consider the mean costs and the instantaneous mean costs over 20 runs and 25 stages (sampling instants). Fig. 1 shows the instantaneous mean cost, for the dead-beat control case. The costs of the D.T.O.C. are much higher than the C.T.O.C. Figs. 2 and 3 show the convergence of the parameter b and the instanteneous costs of a typical run. The high costs in the D.T.O.C. case are related to the passive learning of this CE control law. The parameter estimates rapidly converge to the real values at high costs. Although the cautious control does lead to a slower convergence of the parameter estimates, a better performance results.
The same simulations have been performed for the bounded-control variable case. The constraint $-2 \leqslant u(k) \leqslant 2$ has been added. The performance of the D.T.O.C. algorithm is better than the previous case, but remains inferior to that of the cautious control (Fig.4). The sensitivity to different control bounds and levels of initial uncertainty has been investigated. Fig. 5 shows the performance as a function of the control bounds and Fig. 6 that of the initial uncertainty. For small control bounds and initial uncertainties the mean cost of the C.T.O.C. law tends to that of the D.T.O.C.

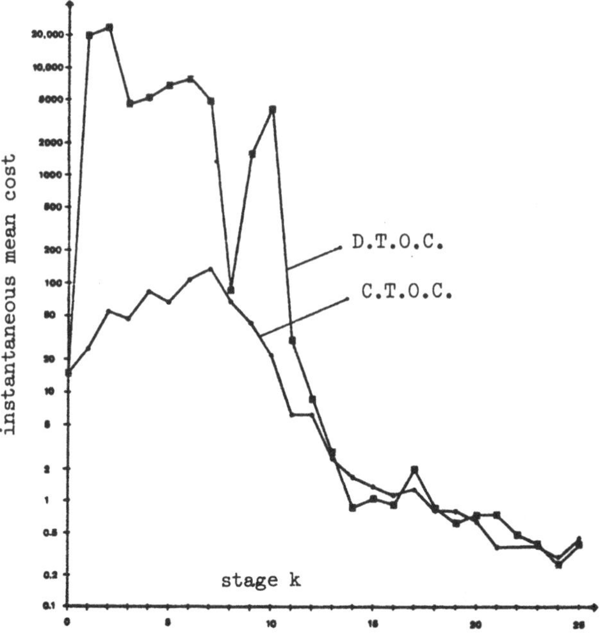

Fig. 1 Time history of the instantaneous mean cost for the two control algorithms (dead-beat case)

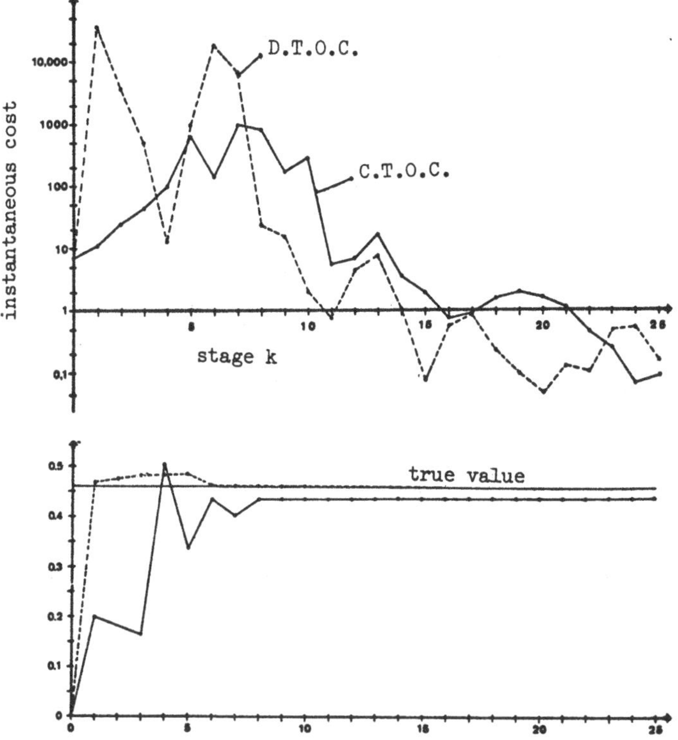

Fig. 2 Time history of the instantaneous cost and the b_1 estimate of a typical run

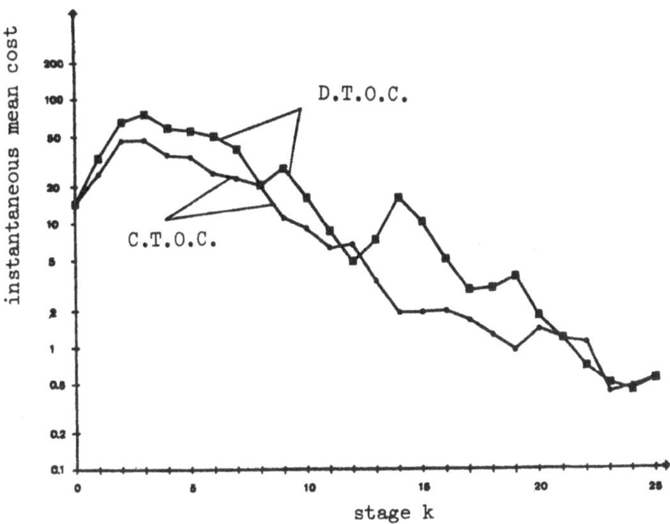

Fig. 3 Time history of the instantaneous mean cost for the two algorithms (bounded control)

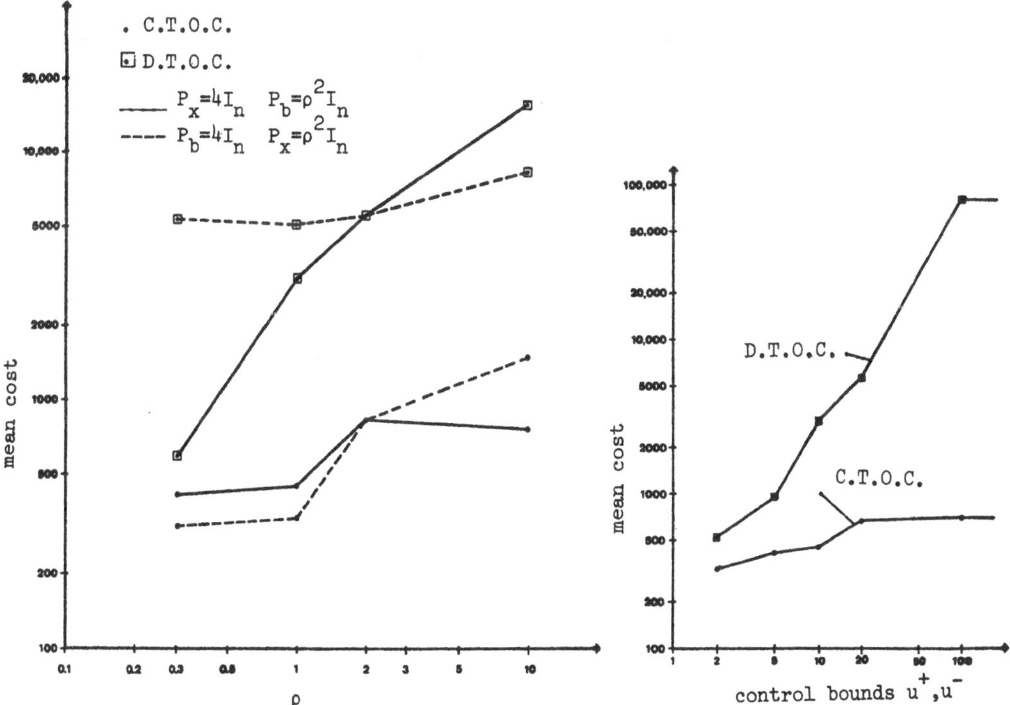

Fig. 4 Dependence of the performance on the initial uncertainties ρ

Fig. 5 Dependence of the performance on the magnitude of the control bounds

7. Conclusions.

A cautious time-optimal control algorithm (C.T.O.C.) has been derived for linear stochastic systems with parameter and state uncertainty. The algorithm has been compared to the deterministic time-optimal control algorithm [10] (D.T.O.C.) by means of Monte Carlo simulation runs for various cases. In all cases the cautious control algorithm resulted in a better performance. Extension of the algorithm to a wider class of stochastic control problems is a subject of current research.

Appendix A.

Consider the following lemmas [8]:

Lemma A1: $$\underline{cs}(ABC) = (C' \otimes A)\underline{cs}(B) \; . \tag{A1}$$

Lemma A2: $$\text{trace}\{AB\} = [\underline{cs}(A')]'\underline{cs}(B) \; , \quad \text{trace}\{AC'BC\} = [\underline{cs}(C)]'(A \otimes B')\underline{cs}(C) \; . \tag{A2}$$

Appendix B.

The following statement is proved: $\quad \text{rank} [\, C_N] \leqslant n \, , \qquad N \geqslant n \; .$ (B1)

The N x n dimensional matrix C_N is defined by

$$C_N = [\, \underline{cs}(I_n), \, \underline{cs}(A'), \ldots\ldots, \, \underline{cs}((A')^{N-1}) \,] \; . \tag{B2}$$

Consider the characteristic matrix $\lambda I - A$ and the characteristic equation $c(\lambda) = |\lambda I - A| = 0$. By the Caylay-Hamilton theorem A^n can be expressed in a linear combination $A^n = -\sum_{i=1}^{n} c_i A^{i-1}$ The column string operation performed on $(A')^n$ yields

$$\underline{cs}\{(A')^n\} = -\sum_{i=1}^{n} c_i \underline{cs}\{(A')^i\} \; . \tag{B3}$$

Since the same result can be obtained for matrices $A^{n+1}, \ldots A^{N-1}$, the matrix contains at most n linear independent columns if $N \geqslant n$.

REFERENCES.

[1] Bellman, R., "Adaptive Processes-A Guided Tour", Princeton Univ. Press, Princeton, N.J., 1961.

[2] Bellman, R., "Introduction to matrix analysis", McGraw-Hill, New York, 1970.

[3] Feld'baum, A., "Optimal Control Systems", Academic Press, New York, 1965.

[4] Ku, R., M. Athans, "On the adaptive control for a class of linear systems using the Open-loop Feedback Optimal approach", I.E.E.E. Trans on AC-18, 1973, p. 489.

[5] Neudecker, H., "Some theorems on matrix differentiation with special reference to Kronecker matrix products", J. Amer. Statist. Assoc. 64, 1969, p. 953.

[6] Nissen, D.H., "A note on the variance of a matrix", Econometrica 36, 1968, p. 603.

[7] O'Reilly, J.,"The discrete linear time invariant time-optimal control problem - an overview", Automatica, 17, no. 2, 1981, p.363.

[8] Toda, M., R.V. Patel, "Algorithms for adaptive control for a class of linear systems", J. Math. Anal & Appl, 72, 1979, p. 122.

[9] Tse, E., Y. Bar-Shalom, L. Meier, "Wide-sense adaptive dual control of stochastic nonlinear systems", I.E.E.E. Trans on AC-18, 1973, p. 98.

[10] Vlieger, J.H. de, H.B. Verbruggen, P.M. Bruijn, "An in-line time-optimal algorithm for digital computer control", Proceedings of the 2nd IFAC/IFIP Symposium on Software in Computer Control, Prague, Czechoslovakia, 1979.

[11] Wittenmark B., "Stochastic adaptive control: a survey", Int. J. Control, 21, 1975, p. 705.

ON LINEAR-QUADRATIC-GAUSSIAN CONTROL OF SYSTEMS WITH UNCERTAIN STATISTICS

D. P. Looze, H. V. Poor, K. S. Vastola, and J. C. Darragh
Department of Electrical Engineering and
Coordinated Science Laboratory
University of Illinois
1101 W. Springfield Ave.
Urbana, Illinois 61801

Abstract

The problem of linear-quadratic Gaussian control of multivariable linear stochastic systems with uncertain second-order statistical properties is considered. Uncertainty is modeled by allowing process and observation noise spectral density matrices to vary arbitrarily within given classes, and a minimax control formulation is applied to the quadratic objective functional. General theorems proving the existence and characterization of saddle-point solutions to this problem are presented, and the relationship of these results to earlier results on minimax state estimation are discussed.

1. Introduction

The design of optimum decision and control procedures for a linear stochastic system requires an accurate description of the statistical behavior of the system. However, because of nonideal effects such as nonstationarity, nonlinearity, and other modeling inaccuracies, there is always a degree of uncertainty in such statistical descriptions. A useful approach to design in the presence of small modeling inaccuracies is to use a game-theoretic formulation in which one optimizes worst-case performance, and this approach has been applied successfully to many aspects of decision and control system design (see, for example, Huber [1] and Mintz [2]). In a recent paper [3], two of the authors have applied this approach in considering the problem of designing linear minimax-mean-square-error state estimators for linear systems observed in and driven by noise processes with uncertain second-order statistics. In particular, it is shown in [3] that, for two general formulations, such estimators can often be designed by designing linear minimum-mean-square-error filters for least-favorable pairs of noise spectra or covariance matrices. Related minimax state estimation results are found in a paper by Morris [4].

In this paper, we consider the analogous problem of minimax linear-quadratic-Gaussian control (LQG) of systems with uncertain second-order statistics. In particular, we consider the control of linear multivariable systems with white Gaussian

This research was supported in part by the Joint Services Electronics Program under Contract N00014-79-C-0424 and in part by the Office of Naval Research under Contract N00014-81-K-0014.

process and observation noises with uncertain spectral density matrices. It is shown here that, within mild conditions, this problem can be solved by designing an optimal control for a least-favorable model, although the model which is least-favorable for control may not be the same as that which is least-favorable for state estimation for the same type of noise uncertainty. However, it is also shown that, for uncertainty in either the process or observation noise only, a given minimax linear-quadratic-Gaussian control problem does have the same least-favorable model as does a particular minimax state estimation problem with a weighted-mean-square-error criterion. Thus, as might be expected, a limited duality exists between these two problems. Another phenomenon which is shown to be associated with minimax control is that the separation principle which separates the problems of optimal control and optimal state estimation is not necessarily valid for minimax control and minimax state estimation. In particular, it is shown that, although the minimax control law is independent of the minimax state estimator, the reverse is not true.

2. Preliminaries

Consider the linear time invariant stochastic system

$$\dot{x}_t = Ax_t + Bu_t + \xi_t \qquad , t \geq 0 \tag{1}$$

$$y_t = Cx_t + \theta_t \qquad , t \geq 0 \tag{2}$$

where x_t and ξ_t are in $\mathbb{R}^n$, u_t is in $\mathbb{R}^m$, and y_t and θ_t are in $\mathbb{R}^p$ for each t. The matrices A, B and C are assumed to have compatible dimensions (as required by (1)-(2)) with the pairs (A,B) and (A,C) stabilizable and detectable respectively. The noise processes ξ_t and θ_t are assumed to be zero-mean white Gaussian processes with second order statistics

$$\begin{aligned} &E\{\xi_t \theta_s^T\} = 0 \\ &E\{\xi_t \xi_s^T\} = \Xi \, \delta(t-s) \\ &E\{\theta_t \theta_s^T\} = \Theta \, \delta(t-s) \end{aligned} \tag{3}$$

where δ is the Dirac impulse. It is assumed that $(A,\sqrt{\Xi})$ is stabilizable and that $\Theta > 0$. The objective of the problem is to choose u_t to minimize the time-averaged quadratic cost

$$J = \lim_{T\to\infty} \frac{1}{T} \int_0^T (x_t^T Q x_t + u_t^T R u_t)\, dt \tag{4}$$

where $Q \geq 0$ with $(A,\sqrt{Q})$ detectable and $R > 0$.

When Ξ and Θ are known, the solution to the stochastic regulator problem (1)-(4) is given by the feedback system:

$$u_t = -G\hat{x}_t \qquad , t \geq 0 \tag{5}$$

$$\dot{\hat{x}}_t = A\hat{x}_t + Bu_t + H(y_t - C\hat{x}_t) \qquad , t \geq 0 \tag{6}$$

where

$$G = R^{-1}B^TK, \tag{7}$$

$$A^TK + KA + Q - KBR^{-1}B^TK = 0, \tag{8}$$

$$H = PC^T\Theta^{-1}, \tag{9}$$

and

$$AP + PA^T + \Xi - PC^T\Theta^{-1}CP = 0. \tag{10}$$

The matrices K and P are the unique positive semi-definite stabilizing solutions to (8) and (10) respectively.

As discussed in Section 1, the second order statistics for the processes ξ_t and θ_t are often not known precisely. A common representation of this type of uncertainty is to assume that Ξ and Θ are contained in compact sets $\mathcal{X}$ and $\mathcal{N}$ respectively. The objective is then to choose u_t to minimize the worst possible performance (4) given $(\Xi,\Theta) \in \mathcal{X} \times \mathcal{N}$. We will restrict our consideration to controls generated by causal, appropriately measurable[1] functions of the measurement. Denote this class of operators as $\mathcal{L}_s^+$. The problem can then be stated as the minimax problem:

$$\min_{L \in \mathcal{L}_s^+} \quad \max_{(\Xi,\Theta) \in \mathcal{X} \times \mathcal{N}} J(L,\Xi,\Theta) \tag{11}$$

where the dependence of J defined by (1)-(4) on L, Ξ, and Θ has been explicitly noted. Note that the optimal linear feedback law defined by (5)-(10) is a member of $\mathcal{L}_s^+$.

3. Existence and Characterization of a Saddlepoint

Two important results concerning solutions to the minimax problem formulated in Section 2 are presented in this section. The first result establishes an equivalence between a saddlepoint solution to (11) and an optimal stochastic regulator solution (5)-(10) corresponding to a particular (Ξ,Θ) pair. The second result establishes the existence of a saddlepoint when the sets $\mathcal{X}$ and $\mathcal{N}$ are convex.

To obtain these results, we will need the following well-known theorem (cf. [5]) which establishes the fact that the existence of a saddlepoint is a necessary and sufficient condition for the minimax problem (11) to be equivalent to the corresponding maximin problem

$$\max_{(\Xi,\Theta) \in \mathcal{X} \times \mathcal{N}} \quad \min_{L \in \mathcal{L}_s^+} J(L,\Xi,\Theta). \tag{12}$$

Theorem 1: There exists a triplet $(L_o,\Xi_o,\Theta_o) \in \mathcal{L}_s^+ \times \mathcal{X} \times \mathcal{N}$ satisfying the saddlepoint condition

---

[1]See, for example, Chapter 16 of [6] for the explicit measurability conditions.

$$J(L_o,\Xi,\Theta) \leq J(L_o,\Xi_o,\Theta_o) \leq J(L,\Xi_o,\Theta_o) \tag{13}$$
$$\forall\, L \in \mathcal{L}_s^+, \Xi \in \mathcal{X}, \Theta \in \mathcal{N}$$

if and only if the values of (11) and (12) are equal.

We will also require the following lemma which expresses the cost for any Ξ and Θ when the control is generated by (5)-(8) with H being any matrix such that (A-HC) is asymptotically stable.

Lemma 1: Assume that the control u_t is generated by the system (5)-(6) with feedback gain G determined by (7)-(8), and that H is any matrix such that all eigenvalues of (A-HC) have negative real parts. Then the cost J defined by (1)-(8) is:

$$J = tr(\Xi K) + tr(\Xi + H\Theta H^T)X \tag{14}$$

where K is given by (8) and X is the unique positive semi-definite solution of

$$(A-HC)^T X + X(A-HC) + G^T RG = 0. \tag{15}$$

Proof: Straightforward.

Theorem 2 provides the desired characterization of a saddlepoint.

Theorem 2: Assume there exists $\Xi_o \in \mathcal{X}$ and $\Theta_o \in \mathcal{N}$ which satisfy

$$tr\{\Xi Y\} \leq tr\{\Xi_o Y\} \quad \forall\, \Xi \in \mathcal{X} \tag{16}$$

and

$$tr\{\Theta H_o X H_o^T\} \leq tr\{\Theta_o H_o X H_o^T\} \quad \forall\, \Theta \in \mathcal{N}, \tag{17}$$

where H_o is the Kalman filter gain corresponding to Ξ_o and Θ_o (given by (9)-(10)), X is given by (15), Y is the solution to

$$(A-H_oC)^T Y + Y(A-H_oC) + Q + KH_oC + C^T H_o^T K = 0, \tag{18}$$

and G and K are given by (7)-(8). Let L_o be the operator representing the optimal stochastic regulator (5)-(6) corresponding to Ξ_o and Θ_o. Then (L_o,Ξ_o,Θ_o) is a saddlepoint solution to (11).

Conversely, assume that (L_o,Ξ_o,Θ_o) is a saddlepoint for (11). Then L_o is the LQG regulator (5)-(10) and (Ξ_o,Θ_o) satisfy (15)-(18).

Proof: (Sufficiency) Consider the maximin problem (12). Let Ξ_o and Θ_o satisfy (15)-(18) and let L_o be the corresponding optimal stochastic regulator. Let H_o be the Kalman gain for Ξ_o and Θ_o given by (9)-(10). Then, by Lemma 1,

$$J(L_o,\Xi,\Theta) = tr\{\Xi(X + K)\} + tr\{H\Theta H^T X\} \tag{19}$$

for every $(\Xi,\Theta) \in \mathcal{X} \times \mathcal{N}$. Adding (15) and (8) gives:

$$(A-H_oC)^T(X + K) + (X + K)(A-H_oC) + Q + KH_oC + C^T H_o^T K = 0. \tag{20}$$

Hence

$$Y = X + K. \tag{21}$$

Also, by (16)

$$tr\{\Xi(X + K)\} \leq tr\{\Xi_o(X + K)\}. \tag{22}$$

Adding (22) and (17), and using (19) gives the lower inequality of (13)

$$J(L_o,\Xi,\Theta) \leq J(L_o,\Xi_o,\Theta_o) \quad \forall \,\Xi \in \mathcal{X},\ \Theta \in \mathcal{H}. \tag{23}$$

The upper inequality of the saddlepoint condition (13) follows trivially from the fact that L_o is the optimal stochastic regulator. Thus, (L_o,Ξ_o,Θ_o) is a saddlepoint for (11).

(Necessity) Suppose (L_o,Ξ_o,Θ_o) satisfies (13). The upper inequality of (13) implies that L_o is the optimal stochastic regulator (for which one realization is (5)-(10)). Hence Lemma 1 can be used to express the cost. The lower inequality and Lemma 1 imply:

$$tr\{\Xi\, K\} + tr\{(\Xi + H_o\Theta H_o^T)X\} \leq tr\{\Xi_o\, K\} + tr\{\Xi_o + H_o\Theta_o H_o^T)X\} \tag{24}$$

for every $\Xi \in \mathcal{X}$ and $\Theta \in \mathcal{H}$. By (21) this can be written as

$$tr\{(\Xi - \Xi_o)Y\} + tr\{(\Theta - \Theta_o)H_o^T X H_o\} \leq 0 \quad \forall \,\Xi \in \mathcal{X}, \Theta \in \mathcal{H}. \tag{25}$$

In particular, $\Theta = \Theta_o$ gives (16) and $\Xi = \Xi_o$ gives (17). □

Thus, we see that conditions (15)-(18) are equivalent to the existence of a saddlepoint. If such a saddlepoint exists, then the minimax controller is simply the optimal stochastic regulator designed for the particular (Ξ_o,Θ_o) pair which satisfies (15)-(18). This result can be used to establish the existence of a saddlepoint.

Theorem 3: Assume $\mathcal{X}$ and $\mathcal{H}$ are convex, compact sets such that if $\Xi \in \mathcal{X}$ then $\Xi \geq 0$ and $(A,\sqrt{\Xi})$ is stabilizable and if $\Theta \in \mathcal{H}$ then $\Theta > 0$. Then a saddlepoint solution for the minimax problem (11) exists.

Proof: The proof shows that a solution to the maximin problem (12) exists and satisfies conditions (15)-(18) of Theorem 2.

By Lemma 1, and equations (7)-(10) and (15),

$$\min_{L \in \mathcal{L}_s^+} J(L,\Xi,\Theta) = tr\,\Xi\, K + tr(\Xi + \bar{H}\Theta\bar{H}^T)X \triangleq M(\Xi,\Theta) \tag{26}$$

is continuous in Ξ and Θ (with $\bar{H}$ given by (9)-(10) for each Ξ and Θ). Since $\mathcal{X}$ and $\mathcal{H}$ are compact, a solution to (12) exists. Let (L_o,Ξ_o,Θ_o) be such a solution. Then the Fréchet differential of (26) with respect to Ξ and Θ at (Ξ_o,Θ_o) must be nonpositive in every direction into the set $\mathcal{X} \times \mathcal{H}$. The Fréchet differential of (26) is given by:

$$\begin{aligned}\delta M(\Xi,\Theta;\Delta\Xi,\Delta\Theta) = {} & tr\{\Delta\Xi(K+X)\} + tr\{\Delta\Theta\bar{H}^T X\bar{H}\} \\ & + tr\{\Theta(\delta\bar{H}^T X\bar{H} + \bar{H}^T X\delta\bar{H})\} + tr\{(\Xi + \bar{H}\Theta\bar{H}^T)\delta X\}.\end{aligned} \tag{27}$$

In (27), $\delta\bar{H}$ and δX represent the Frechet differentials of $\bar{H}$ and X with respect to Ξ and Θ. From (15), δX can be computed as the solution of

$$(A-\bar{H}C)^T\delta X + \delta X(A-\bar{H}C) - C^T\delta\bar{H}^TX - X\delta\bar{H}C = 0. \tag{28}$$

Thus, δX is given by:

$$\delta X = -\int_0^\infty e^{(A-\bar{H}C)^Tt}[C^T\delta\bar{H}^TX + X\delta\bar{H}C]e^{(A-\bar{H}C)t}\,dt. \tag{29}$$

Substituting (29) into (27) and using a few trace manipulations gives

$$\begin{aligned}\delta M(\Xi,\Theta;\Delta\Xi,\Delta\Theta) &= tr\{\Delta\Xi(X+K)\} + tr\{\Delta\Theta\bar{H}^TX\bar{H}\} \\ &-tr\{\int_0^\infty e^{(A-\bar{H}C)t}(\Xi+\bar{H}\Theta\bar{H}^T)e^{(A-\bar{H}C)^Tt}dt[C^T\delta\bar{H}^TX+X\delta\bar{H}C]\} \\ &+tr\{PC^T\delta\bar{H}^TX + PX\delta\bar{H}C\}.\end{aligned} \tag{30}$$

But the integral in the third term of (30) is the solution to (10); i.e., P. Hence

$$\delta M(\Xi,\Theta;\Delta\Xi,\Delta\Theta) = tr\{\Delta\Xi(X+K)\} + tr\{\Delta\Theta\,\bar{H}^TX\bar{H}\}. \tag{31}$$

Consider an arbitrary point $(\Xi,\Theta) \in \mathcal{X}\times\mathcal{N}$. Since $\mathcal{X}$ and $\mathcal{N}$ are convex, the line segment joining (Ξ_o,Θ_o) and (Ξ,Θ) is in $\mathcal{X}\times\mathcal{N}$ and hence

$$(\Delta\Xi,\Delta\Theta) = (\Xi - \Xi_o,\Theta - \Theta_o) \tag{32}$$

is a direction into $\mathcal{X}\times\mathcal{N}$. Substituting (32) into (31), requiring (31) to be non-positive and using (21) gives:

$$tr\{\Xi - \Xi_o)Y\} + tr\{\Theta - \Theta_o)H_o^TXH_o\} \leq 0. \tag{33}$$

The choice $(\Xi,\Theta) = (\Xi,\Theta_o)$ in (33) gives (16) while the choice $(\Xi,\Theta) = (\Xi_o,\Theta)$ in (33) gives (17). Thus, by Theorem 2, (Ξ_o,Θ_o) is a saddlepoint for (28). □

This section has provided two major results. First, every saddlepoint solution to the minimax problem formulated in Section 2 has been characterized by the conditions of Theorem 2. In addition to providing a means of identifying a particular solution, these conditions can be used to characterize the set of possible solutions. Theorem 3 provides the second important result by demonstrating the existence of a saddlepoint solution to the minimax problem where the sets $\mathcal{X}$ and $\mathcal{N}$ are convex and compact.

4. Discussion

There are several interesting observations which can be made concerning the results of the previous section. First we note that, since (6), (9), and (10) give the linear least-squares state estimator for a fixed (Θ,Ξ) pair, the optimal linear regulator problem for fixed (Θ,Ξ) is solved by feeding back optimal state estimates through the gain G (which does not depend on (Θ,Ξ)). Thus, as is well known, there is a separation between the estimator and regulator design problems in the case of fixed (Θ,Ξ). However, it follows from Theorem 2, (16), and (17) that such a separation does not generally exist in the minimax problem. In particular we see from Theorem 2 that, although the feedback gain does not depend on (Θ,Ξ), the state estimates used for minimax control are not generally the minimax-mean-square-error state estimates. This follows because the equations determining the least-favorable pair for control depend directly on the cost matrices Q and R, which of course have no effect on which pair is least-favorable for state estimation (as in [3]).

The above observation also implies that the (Θ,Ξ) pair which is least favorable for control is not necessarily the same as that which is least favorable for state estimation. However, the conditions that Theorem 2 requires for minimax control are similar in structure to conditions required by Theorem 5 of [3] for minimax state estimation. Using the similarity it follows that, for fixed Ξ, the Kalman filter corresponding to (Θ_o,Ξ) where Θ_o is from (17) also solves the minimax state estimation problem

$$\min_{\hat{x}_t} \max_{\Theta \in \mathcal{N}} E\{(x_t - \hat{x}_t)^T G_o^T R G_o (x_t - \hat{x}_t)\} \tag{34}$$

where G_o is the regulator feedback gain from (7). A similar statement applies if Θ is fixed and Ξ is unknown; however if both Ξ and Θ are unknown, there generally is not a single minimax-mean-square-error state estimation problem which has the same least favorable pair as (5).

References

1. P.J. Huber, Robust Statistical Procedures. SIAM: Philadelphia, 1977.

2. M. Mintz, "A Kalman filter as a minimax estimator," J. Opt. Theory Appl., vol. 9, pp. 99-111, Feb. 1972.

3. H.V. Poor and D.P. Looze, "Minimax state estimation for linear stochastic systems with noise uncertainty," IEEE Trans. Autom. Control, vol. AC-26, pp. 902-906, August 1981.

4. J.M. Morris, "The Kalman filter: a robust estimator for some classes of linear quadratic problems," IEEE Trans. Inform. Theory, vol. IT-22, pp. 526-534, Sept. 1976.

5. V. Barbu and Th. Precupanu, Convexity and Optimization in Banach Spaces. Editura Academiei: Bucharest, 1978.

6. R.S. Lipster and A. N. Shiryayev, Statistics of Random Processes II, Springer-Verlag, New York, 1978.

A LIAPUNOV-LIKE CRITERION AND A FIRST PASSAGE-TIME PROBLEMS IN NON-LINEAR STOCHASTIC SYSTEMS

Sueo Sugimoto
Department of Applied Physics
Osaka University
Suita, Osaka 565 Japan

I. Introduction

We investigate the finite time stochastic stability as well as the exponential type stochastic stability of the systems govered by the following Ito equation,

$$d\underline{x}(t) = \underline{f}(\underline{x})dt + G(\underline{x})d\underline{B}(t) \quad ; \quad \underline{x}(0) = \underline{x}_0 \quad , \tag{1-1}$$

where $\underline{B}(t) \triangleq [\, B_1(t),..,B_m(t)\,]^T$ and $B_i(t)$, $i = 1,..,m$, are mutually independent standard Wiener processes. The non-linear vector function $\underline{f}(\underline{x})$ and matrix function $G(\underline{x}) \triangleq [\, \underline{g}_1(\underline{x}),..,\underline{g}_m(\underline{x})\,]$ are satisfying the regularity conditions for the existence and uniqueness of the solution to Eq.(1-1). Also assume the condition of having the zero-equilibrium solution of Eq.(1-1), namely, $\underline{f}(\underline{0}) = \underline{g}_i(\underline{0}) = \underline{0}$.

The main purpose in this paper is to evaluate an upper bound of the probability,

$$P[\sup_{t\varepsilon[T_1,T_2]} V(\underline{x}(t)) \geq \delta\,] \quad ; \quad 0 \leq T_1 \leq T_2 \quad , \tag{1-2}$$

where $V(\underline{x})$ is some suitable function related to the generalized energy or the envelope of the solution process to Eq.(1-1).

It is well known that the powerful tool in the stability study for non-linear stochastic systems has been estabilished by Bucy[1] and Kushner[2], that is so-called stochastic Liapunov method (also see [3, 4]). The method relies heavily upon the property of the positive supermartingale for the diffudion Markov process.

In this paper, we attempt to present another Liapunov-like criterion so that we may evaluate a bound of the probability of (1-2) via directly applying Ito's differential rule to the non-linear function of $\ln V(\underline{x})$, instead of the supermartingale inequality applied by Bucy and Kushner.

It will be shown in the sequal that the two conditions for a Liapunov-like function $V(\underline{x})$ quarantee interrelation between the objective probability and the first passage time probability. The evaluation of an upper bound of (1-2) is reduced to the problem of the level crossing or the first passage time for the Wiener process with a drift.

II. Preliminary Discussions

Before Showing the main result(Theorem 3.1), we will preliminarily dicsuss the stability problem for a first order linear Ito system that will subsequently guide us to importance of considering the first passage time or the so-called level crossing problem for the diffusion Markov process[5,6].

2.1 Linear Ito Systems and First Passage Time Problems

Consider a scalar linear Ito system,

$$dx(t) = ax(t) + \sigma x(t)dB(t) \quad ; \quad x(0) = x_0 \tag{2-1}$$

where a and σ are constants and B(t) is a scalar standard Wiener process. Then it is well known that asymptotical behavior of the solution process to Eq.(2-1) is almost surely stable, namely,

$$P[\lim_{t\uparrow\infty} |x(t)| = 0] = 1 , \tag{2-2}$$

if and only if

$$a - \frac{1}{2}\sigma^2 < 0 \quad . \tag{2-3}$$

However we are now interesting in the behavior of the process in a finite time period from the aspect of the practical engineering problems such as reliability and many types of tracking for stochastic dynamical systems. Since we know that some sample solutions of the stable system (i.e., the system whose coefficients satisfy the conditions (2-3)) behave like as unstable in a finite time period via simulation result[7,8], we may have a question about the finite time behavor of the process x(t). This question motivates us to evaluate the probability in (1-2) for stochastic systems.

For such purpose let us obtain the exact sample solutions to Eq.(2-1). Applying Ito's differential rule to a non-linear function lnx(t) for Eq.(2-1), we have

$$d(\frac{1}{2}\ln x^2) = L(\frac{1}{2}\ln x^2)dt + \frac{1}{2}\frac{\partial \ln x^2}{\partial x}\sigma x dB(t) \tag{2-4}$$

where L is a differential operator generally defined by

$$L(\cdot) = \underline{f}^T(\underline{x})\frac{\partial}{\partial \underline{x}}(\cdot) + \frac{1}{2}\mathrm{tr.}[G(\underline{x})G^T(\underline{x})\frac{\partial^2}{\partial \underline{x}^2}(\cdot)] \tag{2-5}$$

for Eq.(1-1)[9]. Integrating both sides in Eq.(2-4) with respect to t from 0 to t, we have

$$\ln x(t) - \ln x_0 = (a - \tfrac{1}{2}\sigma^2)t + \sigma^2 B(t),$$

namely,

$$x(t) = x_0 \exp[(a - \tfrac{1}{2}\sigma^2)t + \sigma^2 B(t)], \tag{2-6}$$

where we assume $x_0 > 0$ without loss of generality. Therefore, our objective probability in Eq.(1-2) with setting $V(\underline{x}) \triangleq |x|$, $T_1 \triangleq 0$, and $T_2 \triangleq T$, may be written as

$$P[\sup_{t\in[0,T]} |x(t)| \geq \delta] = P[\sup_{t\in[0,T]} (a-\tfrac{1}{2}\sigma^2)t + \sigma^2 B(t) \geq \ln\frac{\delta}{x_0}], \tag{2-7}$$

where the supremum of the stochastic process appears.

It should be here remarked that the distribution of the supremum of the stochastic process over finite time interval is closely related to the first passage time density. Denote by $q_L(t+t_0|y(t_0)=y_0)dt$ the probability that the process $y(t)$ first crosses the level L between time $t+t_0$ and $t+dt+t_0$, given that it is assumed $y(t_0)=y_0$ at time t_0. Then it is well known[6] that the following relation holds, i.e.,

$$P[\sup_{t\in[0,T]} y(t) \leq L \mid y(0)=y_0] = 1 - \int_0^T q_L(s|y_0)ds \ . \tag{2-8}$$

Therefore our problem of the evaluation of the probability in (2-7) may turn to consider the first passage time problem for the Wiener process with a drift, namely, the process $y(t)$ formed by $y(t) = ct + B(t)$, where c is a constant.

<u>Lemma 2.1 [10,5]</u>

For the process $y(t) = -\alpha t + B(t)$ with $E[B(t)B(s)] = \sigma^2\min(t,s)$,

$$q_L(t_0+t|y(t_0)=y_0) = \frac{L-y_0}{\sqrt{2\pi\sigma^2 t^3}} \exp[-\frac{(L+\alpha t-y_0)^2}{2\sigma^2 t}] \tag{2-9}$$

<u>Lemma 2.2</u>

Let $B(t)$ be a Wiener process with a covariance parameter σ^2, then the probability of the maximum of the Wiener process with a drift $-\alpha t$, crossing over a barrier level L during the time interval between T_1 and T_2 may be evaluated by the following equation,

$$P[\sup_{t\in[T_1,T_2]} -\alpha t + B(t) \geq L] = \frac{1}{2}(1 - \mathrm{erf}[\frac{L+\alpha T_1}{\sqrt{2\sigma^2 T_1}}])$$

$$+ \int_{-\infty}^{L+\alpha T_1} dr\{\frac{1}{\sqrt{2\pi\sigma^2 T_1}} \exp[-\frac{r^2}{2\sigma^2 T_1}] \frac{1}{2} \{1 - \text{erf}[\frac{L+\alpha T_2-r}{\sqrt{2\sigma^2(T_2-T_1)}}]$$

$$+ \exp[-\frac{2\alpha(L+\alpha T_1-r)}{\sigma^2}](1 - \text{erf}[\frac{L+2\alpha T_1-\alpha T_2-r}{\sqrt{2\sigma^2(T_2-T_1)}}])\}\} , \qquad (2\text{-}10)$$

for $0 \le T_1 \le T_2$, where $\text{erf}[z] \triangleq \frac{2}{\sqrt{\pi}} \int_0^z \exp(-s^2) ds$

Proof:

Remark the following identities,

$$P[\sup_{t\varepsilon[T_1,T_2]} -\alpha t + B(t) \ge L \mid B(0) = 0]$$

$$= P[\sup_{t\varepsilon[0,T_2-T_1]} -\alpha t + B(t+T_1) \ge L + \alpha T_1 \mid B(0) = 0]$$

$$= P[B(T_1) \ge L+\alpha T_1 | B(0)=0] + \int_{-\infty}^{L+\alpha T_1} P[\sup_{t\varepsilon[0,T_2-T_1]} -\alpha t+B(t+T_1) \ge L+\alpha T_1$$

$$|B(T_1)=r] \; p[B(T_1)=r|B(0)=0]dr, \qquad (2\text{-}11)$$

where $p[.|.]$ is the transition probability density function of the Wiener process. Then the first term of the last equation can be calculated by using the distribution function of the Wiener process and the quantity inside of the integral in the second term can be evaluated by applying the relation in Eq.(2-8) and Lemma 2.1, that is

$$P[\sup_{t\varepsilon[0,T_2-T_1]} -\alpha t+B(t+T_1) \ge L+\alpha T_1 | B(T_1)=r] = \int_0^{T_2-T_1} q_{L+\alpha T_1}(s|y(0)=r)ds$$

$$= \int_0^{T_2-T_1} \frac{L+\alpha T_1-r}{\sqrt{2\pi\sigma^2 s^3}} \exp[-\frac{(L+\alpha T_1+\alpha s-r)^2}{2\sigma^2 s}]ds .$$

After elementary manipulation of the exponents, we have the formula in (2-10).

As special cases of Lemma 2.2, we easily show the following result.

Corollary 2.1

i) $P[\sup_{t\varepsilon[0,T_2]} -\alpha t + B(t) \ge L]$

$$= \frac{1}{2} \{1 - \text{erf}[\frac{L+\alpha T_2}{\sqrt{2\sigma^2 T_2}}] + \exp[-\frac{2\alpha L}{\sigma^2}](1 - \text{erf}[\frac{L-\alpha T_2}{\sqrt{2\sigma^2 T_2}}])\}$$

ii) $$P[\sup_{t\in[T_1,\infty)} -\alpha t + B(t) \geq L]$$

$$= \frac{1}{2}\{1 - \mathrm{erf}[\frac{L+\alpha T_1}{\sqrt{2\sigma^2 T_1}}] + \exp[-\frac{2\alpha L}{\sigma^2}](1 + \mathrm{erf}[\frac{L-\alpha T_1}{\sqrt{2\sigma^2 T_1}}])\}$$

iii) $$P[\sup_{t\in[0,\infty)} -\alpha t + B(t) \geq L] = \begin{cases} \exp[-\frac{2\alpha L}{\sigma^2}], & \text{if } \alpha > 0 \\ 1, & \text{if } \alpha \leq 0 \end{cases}$$

iv) $$P[-\alpha t + B(t) \geq L] = \frac{1}{2}(1 - \mathrm{erf}[\frac{L+\alpha T}{\sqrt{\sigma^2 T}}])$$

Therefore Eq.(2-7) can be completely evaluated by Corollary 2.1.

In what follows, we will present the auxiliary results for the stochastic integral which will play important role in the proof of Theorem 3.1 for finding an upper bound of the probability (1-2) for non-linear systems (1-1).

2.2 Auxiliary Results for the Stochastic Integral

We define the stochastic integral,

$$\int_0^T h(t,\omega)dB(t,\omega) \tag{2-12}$$

with respect to the Wiener process $B(t,\omega)$.

For all $t \in [0,T]$, let the σ-algebra of the event F_t defined so as to possess the properties, a) for $t_1 < t_2$, $F_{t_1} \subset F_{t_2}$, b) $B(t,\omega)$ is F_t-measurable, c) the process $B(t+s,\omega) - B(t,\omega)$ does not depend on the σ-algebra F_t. Let the symbol $H_2[0,T]$ be the space of random function $h(t,\omega)$ defined for $t \in [0,T]$ and F_t-measurable for each t, and for which $\int_0^T h^2(t,\omega)dt$ is finite w.p.l.

Lemma 2.3[11, p.31]

Let $h(t,\omega)$ be defined for $t \geq 0$ and for each $T > 0$, let $h(t,\omega) \in H_2[0,T]$. We assume that $\int_0^\infty h(t,\omega)dt = \infty$ w.p.l and set

$$\tau_t \triangleq \inf\{ s: \int_0^s h^2(u,\omega)du \geq t \} \tag{2-13}$$

Then the process defined by

$$\zeta(t,\omega) \triangleq \int_0^{\tau_t} h(s,\omega)dB(s,\omega) \tag{2-14}$$

is a standard Wiener process.

Lemma 2.4 (comparison theorem)

Let $h(t,\omega) \in H_2[0,T]$ and if $|h(t,\omega)| \leq k$, w.p.l, then

$$P[\sup_{t\varepsilon[0,T]} \int_0^t h(s,\omega)dB(s,\omega) \geq \delta]$$

$$\leq P[\sup_{t\varepsilon[0,\,k^2T]} B(t,\omega) \geq \delta] = P[\sup_{t\varepsilon[0,T]} \bar{B}(t) \geq \delta] \tag{2-15}$$

where $B(t,\omega)$ and $\bar{B}(t,\omega)$ are Wiener processes with $E[B(t,\omega)B(s,\omega)] = 1$ $\min(t,s)$ and $E[\bar{B}(t,\omega)\bar{B}(s,\omega)] = k^2\min(t,s)$ respectively.

Proof:

Define the random variable $\xi_T(\omega)$ as

$$\xi_T(\omega) \triangleq \int_0^T h^2(s,\omega)ds \tag{2-16}$$

then for $\zeta(t,\omega)$ defined in Eq.(2-14), we have the following identities with applying Lemma 2.4,

$$P[\sup_{t\varepsilon[0,\xi_T]} \zeta(t,\omega) \geq \delta] = P[\sup_{t\varepsilon[0,\xi_T]} \int_0^{\tau_t} h(s,\omega)dB(s,\omega) \geq \delta] \tag{2-17}$$

$$= P[\sup_{t\varepsilon[0,T]} \int_0^t h(s,\omega)dB(s,\omega) \geq \delta]$$

where the second equality holds by the relation of $\tau_0 = 0$, and $\tau_{\xi_T} = T$ In addition, we have

$$\xi_T(\omega) = \int_0^T h^2(s,\omega)ds \leq \int_0^T k^2ds = k^2T, \text{ w.p.1},$$

or

$$[\,0,\ \xi_T(\omega)\,] \subset [\,0,\ k^2T\,], \quad \text{w.p.1} \quad . \tag{2-18}$$

Therefore, we have

$$P[\sup_{t\varepsilon[0,\xi_T]} \zeta(t,\omega) \geq \delta] \leq P[\sup_{t\varepsilon[0,k^2T]} \zeta(t,\omega) \geq \delta] \tag{2-19}$$

Combining relations in Eqs.(2-17) and (2-19), we have this lemma.

III. A Liapunov-Like Criterion

We turn to discuss the finite time stability problem and obtain an upper bound of the probability (1-2) for the Ito systems Eq.(1-1).

Theorem 3.1

For the Ito system in Eq.(1-1), if $V(\underline{x})$ is a positive definite and twice continuously differential function which satisfies the following conditions,

$$\text{c.1)} \quad LV(\underline{x}) - \frac{1}{2V(\underline{x})} V_{\underline{x}}^T(\underline{x})G(\underline{x})G^T(\underline{x})V_{\underline{x}}(\underline{x}) \leq -\alpha V(\underline{x})$$

and

$$\text{c.2)} \quad V_{\underline{x}}^T(\underline{x})G(\underline{x})G^T(\underline{x})V_{\underline{x}}(\underline{x}) \leq k^2V^2(\underline{x})$$

then $P[\sup_{t\varepsilon[T_1,T_2]} V(\underline{X}(t)) \geq \delta] \leq P[\sup_{t\varepsilon[T_1,T_2]} -\alpha t + k\bar{B}(t) \geq \ln \frac{\delta}{V(\underline{x}_0)}]$ (3-1)

where $\bar{B}(t)$ is a scalar valued Wiener process with $E[\bar{B}(t)\bar{B}(s)] = 1.\min(t,s)$, and L is the differential operator defined by

$$LV(\underline{x}) = V_{\underline{x}}^T(\underline{x})\underline{f}(\underline{x}) + \frac{1}{2}\,\mathrm{tr.}(V_{\underline{xx}}(\underline{x})G(\underline{x})G^T(\underline{x})) \quad (3\text{-}2)$$

with $V_{\underline{x}}$ and $V_{\underline{xx}}$ denoting a gradient vector and Hessian matrix of V respectively (see Eq.(2-5)).

Proof:

The proof is quite similar to Khas'minskii's derivation for obtaining the necessary and sufficient condition of almost surely sample stability to the linear Ito system (see [12,13]).

Let us apply Ito's differential rule to the non-linear function $\ln V(\underline{x})$ for Eq.(1-1) instead of $\ln\|\underline{x}\|$ employed in [12], then we have

$$d\ \ln V(\underline{x}) = Q(\underline{x}(t))dt + R(\underline{x}(t))d\underline{B}(t) \quad (3\text{-}3)$$

where

$$Q(\underline{x}) = \frac{1}{V(\underline{x})}\left[LV(\underline{x}) - \frac{1}{2V(\underline{x})}V_{\underline{x}}^T(\underline{x})G(\underline{x})G^T(\underline{x})V_{\underline{x}}(\underline{x}) \right] \quad (3\text{-}4)$$

and

$$R(\underline{x}) = \frac{1}{V(\underline{x})} V_{\underline{x}}^T(\underline{x})G(\underline{x}) \quad . \quad (3\text{-}5)$$

Therefore, we have relation

$$P[\sup_{t\varepsilon[T_1,T_2]} V(\underline{x}(t)) \geq \delta]$$

$$= P[\sup_{t\varepsilon[T_1,T_2]} V(\underline{x}_0)\exp\{\textstyle\int_0^t Q(\underline{x}(s))ds + \int_0^t R(\underline{x}(s))d\underline{B}(s)\} \geq \delta]$$

$$= P[\sup_{t\varepsilon[T_1,T_2]} \textstyle\int_0^t Q(\underline{x}(s))ds + \int_0^t R(\underline{x}(s))d\underline{B}(s) \geq \ln \frac{\delta}{V(\underline{x}_0)}] \ .$$

Now applying Lemma 2.4 with the conditions c.1) and c.2), we have the inequality (3.1).

Example:

Let us show an illustrative example for applying Theorem 3.1. Consider the system described by

$$dx(t) = (ax - f(x))dt + g(x)dB(t), \quad x(0) = x_0 \quad (3\text{-}6)$$

with assuming

i) $f(0) = g(0) = 0$

ii) $sf(s) \geq 0$ for all s

iii) $g_1 s^2 \leq sg(s) \leq g_2 s^2, \quad 0 \leq g_1 \leq g_2$: sector condition.

Then letting $V(\underline{x}) = x^2(t)$, we have

$$Q(\underline{x}) = \frac{1}{x^2}\left[2x(ax-f(x)) + g^2(x) - \frac{1}{2x^2}(2x)^2 g^2(x)\right]$$

$$= 2a + \frac{1}{x^2}\left[-2xf(x) - g^2(x)\right] \le 2a - g_1^2$$

$$R(\underline{x}) = \frac{1}{x^2}\left[2xg(x)\right] \le 2g_2 \quad .$$

Therefore, we have

$$P\left[\sup_{t\varepsilon[T_1,T_2]} x^2 \ge \delta\right]$$

$$\le P\left[\sup_{t\varepsilon[T_1,T_2]} (2a - g_1^2)t + 2g_2B(t) \ge \ln\frac{\delta}{x_0^2}\right] \quad . \qquad (3\text{-}7)$$

It is also clear that the sufficient condition of almost sure sample stability for Eq.(3-6) is $2a - g_1^2 \le 0$.

IV. Concluding Remarks

We have develop a Liapunov-like method to evaluate an upper bound of the probability (1-2) via directly applying Ito's differential rule to the non-linear function $\ln V(\underline{x})$ with the conditions c.1) and c.2) in Theorem 3.1. The results obtained previously in [14] for the first order linear Ito system with $V(\underline{x}) = x^2$, can be treated as a special case of Theorem 3.1. Also it should be remarked that the condition c.1) is similar to the condition appeared in [2, Theorem 4] for the study of the exponential type stochstic stability, c.3) $LV(\underline{x}) \le -\alpha V(\underline{x})$. It is clearly shown that the function satisfying c.3) is alway satisfied by the condition c.2), because of $V_{\underline{x}}^T(\underline{x})G(\underline{x})G^T(\underline{x})V_{\underline{x}}(\underline{x})$ being a non-negative definite function.

Acknowledgement:

A part of this paper contains the results [14,15] obtained during the author's reserach activity, supervised by Professor F. Kozin, Polytechnic Institute of New York. The author would like to thank him. Also the author wishes to thank Professor Y. Sunahara, Kyoto Institute of Technology, for his continuous suggestions and encouragement.

References

[1] R.S. Bucy: Stability and Positive Supermartingales, J. Diff. Eqs., vol.1, pp.151-155, (1965).

[2] H.J. Kushner: Stochastic Stability and Control, Academic, N.Y. (1967).

[3] F. Kozin: A Survey of Stability of Stochastic Systems, Automatica, vol.5, pp.95-112, Pergamon Press, (1969).

[4] R.Z. Has'minskii: Stochastic Stability of Differential Equations, Sijthoff & Noordhoff, Maryland, (1980).

[5] I.F. Blake and W.C. Lindsey: Level-Crossing Problems for Random Processes, IEEE Trans. Inf. Th., vol.IT-19, no.3, pp.295-315, (1973).

[6] D.A. Darling and A.J.F. Siegert: The First Passage Problems for a Continuous Markov Process, Ann. Math Stat., vol.24, pp.624-639, (1953).

[7] F. Kozin and S. Sugimoto: Determining Stability Properties of Stochastic Systems from Sample Observation, Proc. 1977 JACC, pp.1049 -1055, (1977).

[8] F. Kozin and S. Sugimoto: Decision Criteria for Stability of Stochastic Systems, Stochastic Problems in Dynamics, ed. by B.L. Clarkson, pp.8-35, Pitman, London, (1977).

[9] Y. Sunahara: Stochastic Systems Theory, Inst. Electr. and Com. Eng. of Japan, Tokyo, in Japanese, (1979).

[10] C.B. Mehr and J.A. Mcfadden: Certain Properties of Gaussian Processes and their First Passage Times, J. Roy. Statist. Soc.(B), pp.505-522, (1965).

[11] I.I Gihman and A.V. Skorohod: Stochastic Differential Equations, Springer, N.Y., (1972).

[12] R.Z. Khas'minskii: Necessary and Sufficient Conditions for the Asymptotic Stability of Linear Stochastic Systems, Th. Prob. Appls., vol.1. pp.144-147, (1967).

[13] F. Kozin and S. Prodromou: Necessary and Sufficient Conditions for the Almost Sure Sample Stability of Linear Ito Equations, SIAM J. Appl. Math., vol.21, no.3, pp.413-424, (1971).

[14] S. Sugimoto and F. Kozin: Some Properties on First Order Ito Systems and its Application to Finite Time Stochastic Stability Studies, Proc. 5th SICE Symp. on Contr. Th., pp.167-170, Hachioji, Tokyo, (1976).

[15] S. Sugimoto: Relation between Sample and Moment Stability and Related Topics, Ph. D. Thesis, Dept. of Electr. Eng., Polytech. Inst. of New York, (1974).

STABILITY ANALTSIS FOR LARGE SCALE STOCHASTIC SYSTEMS

E.E.Zakzouk(Ph.D.)
Military Technical College,
Cairo, Egypt.

S.A. Hassan (Ph.D.), and
M.A.Bisher (M.Sc)
Faculty of Engineering and Tech.,Shebin El-Kom, Egypt.

Abstract.

Stability properties of stochastic systems with unknown parameters and are subjected to optimum controllers, have been the matter of several reports [1-2].However, the need to reach at least the stable sub-optimal controller in a finite number of iteration steps (on-line systems) is a challanging aim.

In this work, stability analysis (based on Lyapunov's $2^{\underline{nd}}$ method) is carried out for the class of stochastic systems that can be identified by linear multi-dimensional regression models, and are subjected to optimum controllers that minimize the conditional mean of a quadratic cost function (Minimum variance control).

1. Introduction.

The design of an optimum controller for stochastic control process with unknown parameters is not an easy task. Several techniques havebeen derived to handle such problem [1] . One of the interesting techniques was established by V. Peterka [2]. However, stability analysis of the derived controller was not obvious. That is why reformulation of the given technique is carried out in order to discuss the stability of the over all system in a rather easy way.

Regression model was choosen to be an adequate model for the given stochastic process. The model parameters was indentified(real time updating) through effective routines [2] . The model equation is transfered to state-space representation [4] and the calculation of the optimum controller is obtained using dynamic programming. Stability of the over all system is tested through liapunov 2nd method.

The present paper amplifies and extends the work given in [4] .

2. Statement of the problem.

Let us now investigate the stability properties for the multidimensional regres. model given by its equivelent state-space equation (1) ,

$$\mathbf{Y}(t) = \mathbf{A}\,\mathbf{Y}(t-1) + \mathbf{B}\,\mathbf{U}(t) + \mathbf{E}(t) \qquad (1)$$

where

$$\underset{nr \times nr}{\mathbf{A}} = \begin{bmatrix} A_1 & A_2 & \cdots\cdots & A_n \\ 1 & 0 & \cdots\cdots & 0 \\ 0 & 1 & \cdots\cdots & 0 \\ \cdot & & & \\ \cdot & & & \\ 0 & 0 & \cdots\cdots & 0 \end{bmatrix} \qquad \underset{nr\times 1}{\mathbf{E}(t)} = \begin{bmatrix} E(t) \\ 0 \\ \cdot \\ \cdot \\ \cdot \\ 0 \end{bmatrix}$$

$$\underset{nr\times(n+1)p}{\mathbb{B}} = \begin{bmatrix} B_o & B_1 & \cdots\cdots & B_n \\ 0 & 0 & \cdots\cdots & 0 \\ \cdot & \cdot & & \cdot \\ \cdot & \cdot & & \cdot \\ 0 & 0 & \cdots\cdots & 0 \end{bmatrix}$$

E(t) : Vector of r elements (sequence of Gaussion random noise with zero mean).
A_i : (r x r) matrix coefficient of the output at time(t-i).
B_i : (r x p) matrix coefficient of the controller at time (t-i).

$$\underset{nr\times 1}{\mathbb{Y}(t)} = \begin{bmatrix} Y(t) \\ Y(t-1) \\ \cdot \\ \cdot \\ Y(t-n+1) \end{bmatrix} \quad ; \quad \underset{(n+1)p\times 1}{\mathbb{U}(t)} = \begin{bmatrix} U(t) \\ U(t-1) \\ \cdot \\ \cdot \\ U(t-n) \end{bmatrix}$$

Y(t) : The output vector at time t.
U(t) : The control vector at time t.
Equation (1) is the state space reformulation of the known multidimensional regression model equation,

$$Y(t) = \sum_{i=1}^{n} A_i \; Y(t-i) + \sum_{i=0}^{n} B_i \; U(t-i) \; + E(t) \cdots\cdots \qquad (2)$$

The control vector $\mathbb{U}(t)$ is introduced to minimize the cost function,

$$\Psi = -\frac{1}{N}\,\xi\left[\sum_{t=t_o}^{N+t_o} \mathbb{Y}^T(t)\,\mathbb{Q}_Y\,\mathbb{Y}(t) + \sum_{t=t_o+N-n_i}^{N+t_o} \mathbb{U}^T(t)\,\mathbb{Q}_U\,\mathbb{U}(t) \Big/ \mathbb{Y}(t_o),\ \mathbb{U}(t_o)\right] \cdots\cdots \quad (3)$$

where,

$$\underset{nr\times nr}{\mathbb{Q}_Y} = {}^{r}\begin{bmatrix} Q_Y & 0 \\ 0 & 0 \end{bmatrix} , \quad \underset{(n+1)p\times(n+1)p}{\mathbb{Q}_U} = {}^{p}\begin{bmatrix} Q_U & 0 \\ 0 & 0 \end{bmatrix}$$

Q_Y , Q_U : are positive definite matrices.
ξ : the conditional mean.
n_i : the minimum number sufficient to bring the optimum controller to a stable region of operation.
N : length of the interval needed to reach the stable suboptimum controller.

Equation (3) is minimized for $\mathbb{U}(t_o), \mathbb{U}(t_o + 1), \ldots , \mathbb{U}(t_o + N)$.

The procedure of determing the controller $\mathbb{U}(t)$ is based on dynamic programming technique. The controller $\mathbb{U}(t)$ is found to be [5] .

$$\grave{\mathbb{U}}(t) = \mathbb{S}(t)\;\grave{\mathbb{U}}(t-1) + \mathbb{L}(t)\;\mathbb{Y}(t) \cdots\cdots\cdots\cdots \qquad (4)$$

where

$$\mathbb{U}(t) = \begin{bmatrix} U(t) \\ \grave{\mathbb{U}}(t-1) \end{bmatrix} \begin{matrix} p \\ np \end{matrix} \quad ; \quad \underset{np\times nr}{\mathbb{L}(t)} = \begin{bmatrix} -\lambda(t) \;\; L^T(t) \\ 0 \end{bmatrix}$$

$$\underset{np \times np}{S(t)} = \begin{bmatrix} -\lambda(t)\ S^T(t) & & & \\ 1 & 0 & 0 \dots & 0 \\ 0 & 1 & 0 \dots & 0 \\ \cdot & & & \\ \cdot & & & \\ 0 & \dots\dots\dots & 1 & 0 \end{bmatrix}$$

$$\lambda(t) = \left[N_{11}(t) + B_o^T M_1(t) B_o + 2K_{11}(t) B_o\right]^{-1}$$

$$L^T(t) = K'^T(t)A + B_o^T M'^T(t)A$$

$$S^T(t) = \frac{1}{2}(N'^T(t) + N''^T(t)) + B_o^T M_1(t) B'^T + K_{11}(t) B'^T + B_o^T K''^T(t)$$

$$M(t) = Q_Y + A^T M(t-1)A - L(t-1)\lambda^T(t-1) L^T(t-1)$$

$$K(t) = K''^T(t-1)A + B''^T M(t-1) A - \frac{1}{2} S(t-1)(\lambda^T(t-1)L^T(t-1)+\lambda(t-1). L^T(t-1))$$

$$N(t) = Q_U + N''(t-1) + B''^T M(t-1) B'' - S(t-1)\lambda^T(t-1) S^T(t-1)+2K''^T(t-1)B''$$

The matrices $M(t)$, $K(t)$ and $N(t)$ appear during the process of minimizing the cost function. Noting that the equation for $N(t)$ is valid only for the condition $(t \geq t_o + N - n_i)$, other wise for $(t < t_o+N-n_i)$,

$$N(t) = N''(t-1) + B''^T M(t-1) B'' + 2K''^T(t-1) B'' - S(t-1)\lambda^T(t-1)S^T(t-1)$$

$$M'(t) = \begin{bmatrix} M_1 \\ M_2 \\ \cdot \\ \cdot \\ M_n \end{bmatrix} \quad \text{(The } 1^{st}\ r \text{ column of matrix } M(t))$$

$$\tilde{K}(t) = \begin{matrix} np \\ \\ p \end{matrix}\overset{nr}{\begin{bmatrix} K(t) \\ \\ 0 \end{bmatrix}}$$

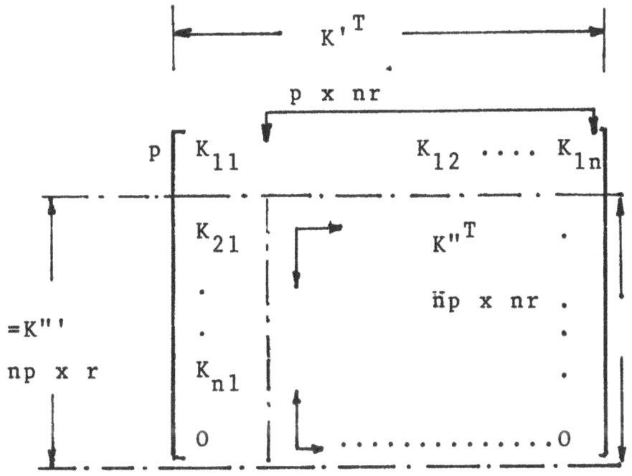

K'^T: $p \times nr$

$$\begin{bmatrix} K_{11} & K_{12} & \dots & K_{1n} \\ K_{21} & & K''^T & \cdot \\ \cdot & & np \times nr & \cdot \\ \cdot & & & \cdot \\ K_{n1} & & & \cdot \\ 0 & \dots & \dots & 0 \end{bmatrix}$$

$=K''$: $np \times r$

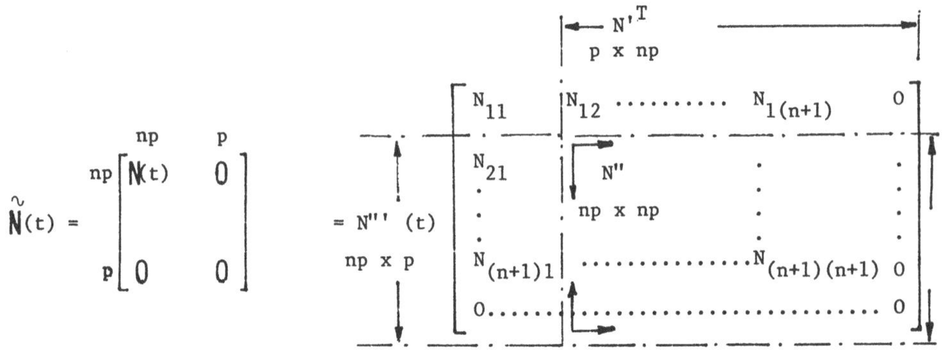

From equantion (1) and (4), we construct the state-space representation for the closed loop system,

$$X(t) = \beta(t)\, X(t) + \grave{E}(t) \quad \ldots\ldots (5)$$

where

$$\underset{n(r+p)x1}{X(t)} = \begin{bmatrix} Y(t) \\ \cdot \\ \cdot \\ Y(t-n+1) \\ U(t) \\ \cdot \\ \cdot \\ U(t-n+1) \end{bmatrix} \quad ; \quad \underset{n(r+p)xn(r+p)}{\beta(t)} = \begin{bmatrix} A + B''_o L(t) & B'' + B''_o S(t) \\ & \\ L(t) & S(t) \end{bmatrix}$$

and

$$\underset{nr \times np}{B''_o} = \begin{bmatrix} B_o & 0 \\ & \\ 0 & 0 \end{bmatrix} \quad ; \quad \underset{nr \times np}{B''} = \begin{bmatrix} B_1 & B_2 & \cdots & B_n \\ 0 & 0 & & 0 \\ \cdot & \cdot & & \\ \cdot & \cdot & & \\ 0 & 0 & \cdots & 0 \end{bmatrix}$$

$$\grave{E}(t) = \begin{bmatrix} E(t) \\ 0 \\ \cdot \\ \cdot \\ \cdot \\ 0 \\ 0 \end{bmatrix} \qquad (nr + np)\ x1$$

Equation (5) represents the closed loop system behaviour from which, it is possible to build the necessary and sufficient stability conditions.

As $t \to \infty$, then $\beta(t) \to \beta = \beta$ (constant matrix), i.e. as the nº of iteration increases we can say that the matrix $\beta(t)$ converges to its steady value. Using theories(8-16) ; (9-17) and (8-19) given in reference [3] , it results that, the stability condition (necessary and sufficient) Lies in the fact that the absolute value of each of the eigen values of matrix β must be less than unity.

3. Computer implementation.

Since obtaining a closed formula for the derived stable optimum controller is quite impossible, computer is implemented to find the corresponding controller. A computer algorithm is stabilished through the flow chart drown in Fig. 1. It is important to said here that, the fast convergence to a stable region of system operation is also our aim. Special interest is devoted to the weighting matrices M, N and K. Computer simulation showes that the initial norm choice of that matrices rapids the convergence to a stable region of operation.

4. Illustrative Example.

One example is chosen here to present all ideas that had been depicted through the analysis. Mainly, how we transfer the closed loop system to stable region of operation (through the choise of the no n_i) and the effect of the initial norms of the matrices M, N and K on the speed of convergence to optimum stable controller.

Refering to equation (1) , Let

$$A = \begin{bmatrix} 1.65 & 0.0 & 0.665 & 0.0 \\ 0.0 & 0.95 & 0.0 & 0.0 \\ 1.0 & 0.0 & 0.0 & 0.0 \\ 0.0 & 1.0 & 0.0 & 0.0 \end{bmatrix} ;$$

$$B = \begin{bmatrix} 0.13 & 0.0 & 0.045 & 0.0 & 0.165 & 0.0 \\ 0.0 & -2.0 & 0.0 & 2.05 & 0.0 & 0.0 \\ 0.0 & 0.0 & 0.0 & 0.0 & 0.0 & 0.0 \\ 0.0 & 0.0 & 0.0 & 0.0 & 0.0 & 0.0 \end{bmatrix}$$

The results are presented in table 1.

5. Conclusion.

Stability properties of stochastic control system (identified by linear regression model) has been easly discussed through transfering the model equation to state space repressentation given by equation (1) . The effect of penalizing terms (2 nd term in the cost function) on stability of the closed loop system is quite clear. It is important to say that, the initail norms of the Matrices M, N and K affect the convergence to the stable optimum controller, however they must be choosen carefully.

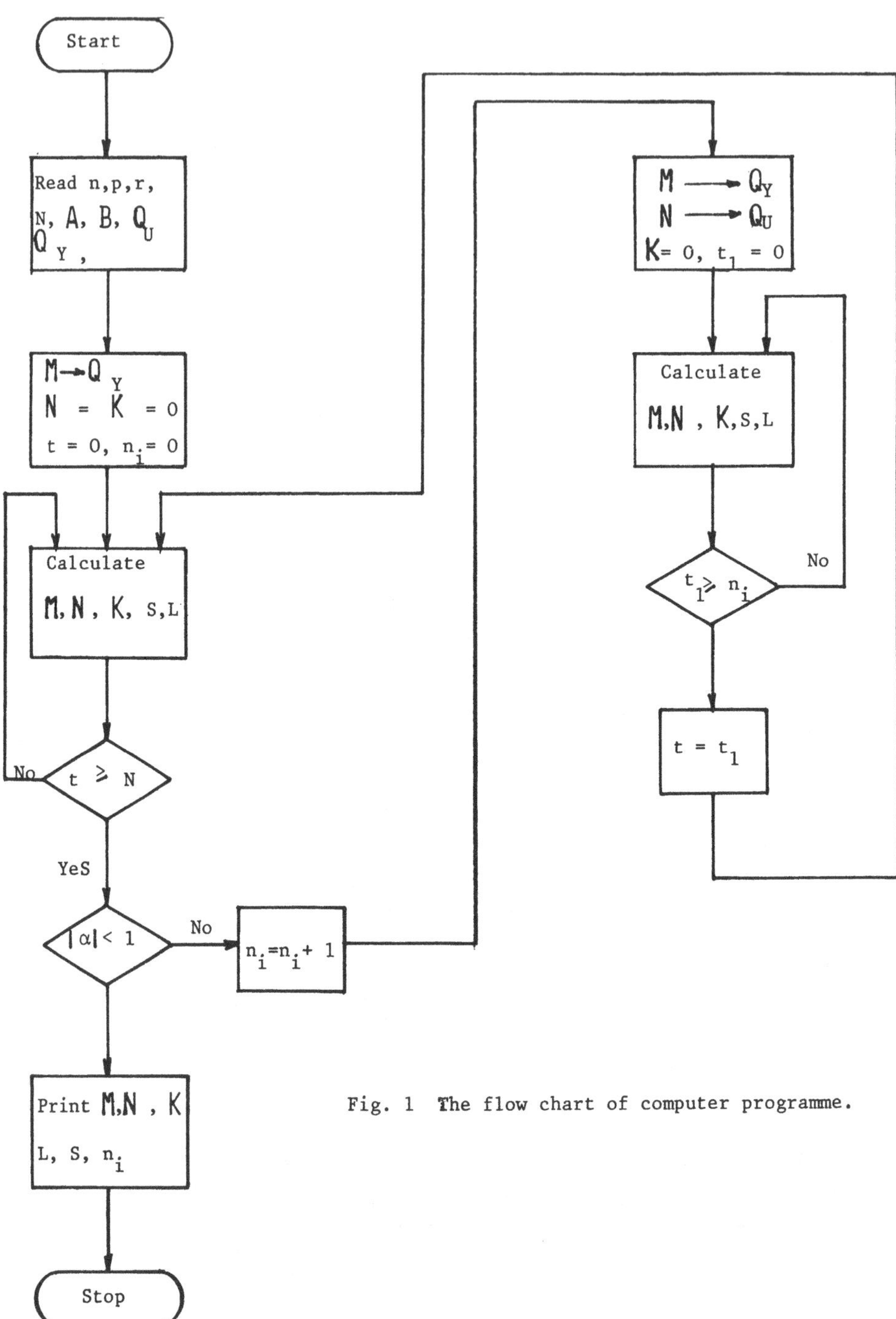

Fig. 1 The flow chart of computer programme.

Table 1 .

n_i	No of iterat. steps	Initial $Q_U(N(0))$	Initial $Q_Y(M(0))$	$Q_U(N(t))$	$Q_Y(M(t))$	Dominant eigen value	Stability	Note
0	200	---------	$\begin{bmatrix}1 & 0\\0 & 1\end{bmatrix}$	---------	$\begin{bmatrix}1 & 0\\0 & 1\end{bmatrix}$	1.0237	X	The optimum dominat Eigen Value = 0.9765
1	100	$\begin{bmatrix}1 & 0\\0 & 1\end{bmatrix}$	$\begin{bmatrix}1 & 0\\0 & 1\end{bmatrix}$	---------	$\begin{bmatrix}1 & 0\\0 & 1\end{bmatrix}$	0.9827	✓	
2	2	$\begin{bmatrix}100 & 0\\0 & 100\end{bmatrix}$	$\begin{bmatrix}100 & 0\\0 & 100\end{bmatrix}$	$\begin{bmatrix}1 & 0\\0 & 1\end{bmatrix}$	$\begin{bmatrix}1 & 0\\0 & 1\end{bmatrix}$	0.9213	✓	
2	5	$\begin{bmatrix}100 & 0\\0 & 100\end{bmatrix}$	$\begin{bmatrix}100 & 0\\0 & 100\end{bmatrix}$	$\begin{bmatrix}1 & 0\\0 & 1\end{bmatrix}$	$\begin{bmatrix}1 & 0\\0 & 1\end{bmatrix}$	0.9349	✓	
2	10	$\begin{bmatrix}100 & 0\\0 & 100\end{bmatrix}$	$\begin{bmatrix}100 & 0\\0 & 100\end{bmatrix}$	$\begin{bmatrix}1 & 0\\0 & 1\end{bmatrix}$	$\begin{bmatrix}1 & 0\\0 & 1\end{bmatrix}$	0.9487	✓	
2	5	$\begin{bmatrix}1000 & 0\\0 & 10000\end{bmatrix}$	$\begin{bmatrix}1000 & 0\\0 & 10000\end{bmatrix}$	$\begin{bmatrix}1 & 0\\0 & 1\end{bmatrix}$	$\begin{bmatrix}1 & 0\\0 & 1\end{bmatrix}$	0.7549	✓	

RERERENCES

1. Feldbaum A.A., "Optimal control systems", Academic Press, 1965.

2. Peterka V., "Digital control process with random noise and unknown parameters", D.Sc. Thesis, Institute of information control Prague, 1975.

3. Ogata K., "State space analysis of control systems" Prentice-Hall, N.Y., 1967.

4. Zakzouk E.E., "Design of stable suboptimal controller using regression models", Irria Conference, Paris, 1979.

5. Hassan S.A., Zakzouk E.E., . and Bisher M., "Optimum controller for multidimensional linear regression model", Proceeding of the 7 th annual operations research conference 1981, Zagazig, Egypt.

Algorithms for Some Simple Infinite Dimensional Linear Programs

E.J.Anderson and A. B. Philpott
University Engineering Department
Cambridge, England

1. Introduction

The following problem was introduced by Bellman[5]:

$$\begin{aligned} &\text{Minimize} && \int_0^T c(t)^T x(t)\,dt \\ &\text{subject to} && Bx(t) + \int_0^t Kx(\tau)\,d\tau = b(t), \\ &&& x(t) \geq 0,\ t \in [0,T]. \end{aligned}$$

He called this a bottleneck problem. It can be regarded as a linear program over the function space within which x(t) lies and is now generally called a Continuous Linear Program (CLP). It can be thought of as an optimal control problem with linear dynamics and objective function, and linear constraints on the states and controls.

The problem CLP has been studied by a number of authors, including Levinson[13], Tyndall[18] and Grinold[8]. Their work has been primarily concerned with establishing duality theorems when CLP and an appropriate dual problem are both posed in L_∞. At the same time Kretschmer[10], Nakamura and Yamasaki[15], and others have worked on the duality theory of abstract linear programs, generally posed over a linear topological space.

Lehmann[11], and Drews, Hartberger and Segers[6] have investigated the possibility of a solution algorithm for CLP. This would amount to a generalization of the simplex method to a function space setting. Perold[16] has dealt in detail with the problems involved in the construction of such an algorithm. There is no doubt that the construction of an implementable algorithm for this problem will be extremely difficult. In this paper, we review some recent work on algorithms for certain linear programs which, while being infinite dimensional, are not as complex as CLP.

2. Continuous Network Flow Problems

We define a network Ω by a set of nodes, numbered 1 through n, where node 1 is the source and node n is the sink. The classical network flow problem is that of maximizing the total flow between the source and the sink subject to capacity constraints on the arcs. It can be formulated as the following linear program.

$$\text{NP} : \quad \text{Maximize} \sum_{k=1}^{n} [x_{kn} - x_{nk}]$$

$$\text{subject to} \sum_{k=1}^{n} [x_{jk} - x_{kj}] = 0, \; j = 2,3,\ldots,n-1,$$

$$0 \leq x_{jk} \leq b_{jk}, \; j,k = 1,2,\ldots,n.$$

Here x_{jk} is the flow in the arc (having capacity b_{jk}) joining nodes j and k. The problem NP was first solved by Ford and Fulkerson whose elegant and famous Maximum Flow-Minimum Cut theorem formed the basis of their labelling algorithm (see [7]).

A straightforward extension of NP is to the case where a choice of flows has to be made at successive times, and the arc capacities vary with time. If storage is allowed at the nodes of the network, the problem becomes more complicated. Letting $b_{jk}(t)$ be the capacity of the arc (j,k) at time t, and $a_j(t)$ the storage capacity of node j at time t, a continuous-time version of this problem, called the Time-Continuous Network Flow Problem (TCNP), can be formulated as follows.

$$\text{TCNP} : \quad \text{Maximize} \int_0^T \sum_{k=1}^{n} [x_{kn}(\tau) - x_{nk}(\tau)]\, d\tau$$

$$\text{subject to} \int_0^t \sum_{k=1}^{n} [x_{kj}(\tau) - x_{jk}(\tau)]\, d\tau \leq a_j(t), \; j = 2,3,\ldots,n-1, \quad (1)$$

$$\int_0^t \sum_{k=1}^{n} [x_{jk}(\tau) - x_{kj}(\tau)]\, d\tau \leq 0, \quad j = 2,3,\ldots,n-1, \quad (2)$$

$$0 \leq x_{jk}(t) \leq b_{jk}(t), \; j,k = 1,2,\ldots,n, \; t \in [0,T]. \quad (3)$$

The time-continuous analogue of the Ford-Fulkerson cut is a straightforward extension of the classical case. This generalized cut is a set valued function of time C, defined on [0,T] with C(t) a subset of S such that for every $t \in [0,T]$, $1 \in C(t)$, $n \in \bar{C}(t)$ and for each j in S, the set $\Gamma_j = \{t : j \in C(t)\} \cap (0,T)$ is open.

Since Γ_j is open we can express it as a countable union of disjoint open intervals, unique up to order, whence

$$\Gamma_j = \bigcup_{i=1}^{\infty} (\alpha_{ij},\beta_{ij}).$$

Now defining R_j as $\{\beta_{ij}: i = 1,2,...\} \cap [0,T)$, we define the value V of the cut C as

$$V = \sum_{j=1}^{n} \sum_{k=1}^{n} \int_{\Gamma_j \cap \bar{\Gamma}_k} b_{jk}(\tau)d\tau + \sum_{j=2}^{n-1} \sum_{t \in R_j} a_j(t).$$

If v is the value of any flow which is feasible for TCNP and V is the value of any generalized cut C in Ω, then it is not hard to show that $v \leq V$. Furthermore, it is possible using a labelling procedure to construct a feasible flow and a generalized cut having equal values, thus solving TCNP. For a feasible flow in the network Ω defined by x_{jk}, the labelled nodes of Ω are defined by the finite application of the following recursive rules.

(i) Node 1 is labelled for every time $t \in [0,T]$.

(ii) If node j is labelled at time t and there is some $\delta > 0$ such that

$$\underset{t-\delta<t'<t+\delta}{\text{ess inf}} \{b_{jk}(t') - x_{jk}(t')\} > 0$$

then k is labelled at time t.

(iii) If node j is labelled at time t and there is some $\delta > 0$ such that

$$\underset{t-\delta<t'<t+\delta}{\text{ess inf}} x_{kj}(t') > 0$$

then k is labelled at time t.

(iv) If node j is labelled at time t_1 and for every $t \in [t_1,t_2)$, $y_j(t) < a_j(t)$, then j is labelled at all times $t \in (t_1,t_2)$.

(v) If node j is labelled at time t_1 and for every $t \in (t_2,t_1]$, $y_j(t) > 0$, then j is labelled at all times $t \in (t_2,t_1)$.

(vi) If $y_j(T) > 0$, then node j is labelled from node 1 at time T.

These ensure that if node n is labelled at any time $t \in (0,T)$, then it is possible to construct a chain of nodes and arcs connecting the source and sink through which a feasible increase in flow can be made. This improvement can be made on some set of non-zero measure and thus the augmented flow has a value strictly greater than that of the original flow. As in the classical network flow problem, the optimal generalized cut is specified by the set of nodes which are labelled when the continuous labelling algorithm fails to label node n at any time. This makes it straightforward to prove the continuous analogue of the Max Flow-Min Cut theorem, which asserts that the maximum value of the flow in the network

Ω is equal to the minimum value of all generalized cuts in Ω. The reader is referred to [2] for a detailed discussion of these results.

The above analysis leads naturally to the specification of a continuous labelling algorithm which, given a trial flow in Ω, labels the nodes in specified subintervals of (0,T). If there is some interval where node n is labelled, breakthrough occurs and we can increase the flow through Ω to obtain an improved trial flow. This process continues until a succession of passes fails to label node n for any time interval, and we have an optimal flow by virtue of the theorem. This algorithm has been programmed for networks having constant storage capacities, and arc capacities which are piecewise linear functions of time. This leads to piecewise linear trial flows and quadratically varying storage. In practice, for simple networks, the algorithm converges in a finite number of iterations. Further work is needed to establish theoretical convergence results, which seem to rely on the arc capacities being sufficiently well behaved.

An alternative generalization of NP is to what we have called the <u>Space-Continuous Network Flow Problem</u> (SCNP). Here the network is a region Ω in space and we wish to maximize the flow of material from one subset of the boundary $\partial\Omega$ of Ω to another, where the flow is incompressible and subject to capacity constraints throughout the region. This problem can be posed, for example in $\mathbb{R}^2$, as the following mathematical program.

SCNP : Maximize λ

subject to $\nabla . \sigma = 0$ in Ω,

$\sigma . n = \lambda f$ on $\partial\Omega$,

$|\sigma| \leq c$ in Ω.

Here the flow is given by a vector field $\sigma = (\sigma_1(x,y), \sigma_2(x,y))$ with $|\sigma(x,y)| = \sqrt{(\sigma_1^2 + \sigma_2{}^2)} \leq c(x,y)$ at every (x,y) in Ω. The function f gives the distribution of sources and sinks on $\partial\Omega$ and must satisfy

$$\int_{\partial\Omega} f ds = 0.$$

We can define a continuous cut as a subset C of Ω; this has a value defined as

$$V = \int_B c ds \Big/ \int_A f ds,$$

where $A = \partial C \cap \partial\Omega$ and $B = \partial C \setminus A$. Although it can be shown that the maximum value of λ in SCNP equals the infimum of V over all cuts C in Ω, the specification of a space-continuous labelling algorithm appears less straightforward than in the time-continuous case and awaits further work. The above formulation of SCNP and the Max Flow-Min Cut result are due to Strang[17] who gives a a comprehensive discussion of this class of problems.

3. Continuous Transportation Problems

The classical transportation problem (TP) is that of satisfying a fixed demand for some commodity at n destinations from a fixed supply at m sources at least transportation cost. It can be posed as the following linear program.

$$\text{TP : Minimize} \sum_{i=1}^{m} \sum_{j=1}^{n} c_{ij}x_{ij}$$

$$\text{subject to} \quad \sum_{j=1}^{n} x_{ij} = a_i,\ i = 1,2,\ldots,m, \tag{4}$$

$$\sum_{i=1}^{m} x_{ij} = b_j,\ j = 1,2,\ldots,n, \tag{5}$$

$$x_{ij} \geq 0,\ i = 1,2,\ldots,m,\ j = 1,2,\ldots,n.$$

A possible generalization of TP, which we have called the Space-Continuous Transportation Problem (SCTP), is to the case where there are an infinite number of sources distributed in some space X, and an infinite number of destinations distributed in some space Y. To formulate SCTP we require the following definitions. For continuous functions f_1 on X and f_2 on Y define $\hat{f}_1$ and $\hat{f}_2$ on X×Y by $\hat{f}_1(x,y) = f_1(x)$, $\hat{f}_2(x,y) = f_2(y)$. We may now formulate SCTP as

$$\text{SCTP : Minimize} \int_{X\times Y} c(x,y)d\rho(x,y)$$

$$\text{subject to} \quad \int_{X\times Y} \hat{f}_1(x,y)d\rho(x,y) = \int_X f_1(x)d\mu_1(x), \tag{6}$$

for all continuous functions f_1 on X,

$$\int_{X\times Y} \hat{f}_2(x,y)d\rho(x,y) = \int_Y f_2(y)d\mu_2(y), \tag{7}$$

for all continuous functions f_2 on Y.

Here ρ, μ_1 and μ_2 are non-negative Radon measures and c is a continuous function. X and Y are compact spaces with $\mu_1(X)=\mu_2(Y)$. The constraints

(6) and (7) are the continuous versions of (4) and (5); they amount to asking that the projections of ρ onto the two coordinate axes are equal to the given measures μ_1 and μ_2. The problem SCTP was first recognized in 1781 by Monge[14] who posed it as a mass transfer problem, and has since been considered by Appell[4], Kantorovitch[9] and Levin and Milyutin[12].

For simplicity we shall take X and Y to be the closed line interval [0,1], and we let μ_1 and μ_2 both equal the Lebesgue measure on the line. We also assume that c has continuous first partial derivatives. To specify an algorithm to solve SCTP under these conditions, it is sufficient to restrict attention to a subclass of measures which we have called assignments.

A non-negative measure ρ satisfying (6) and (7) is called an assignment if the support of ρ is within the closure of the graph of a 1-1 function. This amounts to saying that the pattern of mass transfer between the two line segments is given by a 1-1 measure preserving function $f:[0,1]\to[0,1]$, with mass from point x being transferred only to f(x). For convenience we also refer to the function f as an assignment. We may now formulate the <u>Space-Continuous Assignment Problem</u> (SCAP) as follows.

$$\text{SCAP : Minimize } \int_0^1 c(x,f(x))dx$$

$$\text{subject to } \mu(f(S)) = \mu(S) \text{ for all measurable sets } S,$$

$$f:[0,1]\to[0,1],\ \text{1-1 and onto.}$$

This problem is equivalent to SCTP with the extra condition that the solution be an assignment. In many cases the optimal solution will not be an assignment; however it can be shown[3] that the algorithm described below for SCAP converges to an optimal solution to SCTP.

If the solution to SCAP is piecewise continuous, then it will consist of a finite number of line segments, which will be at 45 degrees to the horizontal. We will describe an algorithm which, given such a solution, will find an improved solution whenever it is not optimal.

The first step is to specify functions r and s. These are used to test for optimality, and in the event that the current solution is not optimal, they indicate how to improve it. Put r(0) = 0, and when f is continuous at x with f(x) = y, set

$$\frac{d}{dx} r(x) = \frac{\partial}{\partial x} c(x,y).$$

This determines a unique continuous function r and we define s by setting $s(f(x)) = c(x,f(x)) - r(x)$ for $x \in [0,1]$.

If $r(x) + s(y) \leq c(x,y)$ everywhere then it can be shown that the current solution is optimal for SCTP. Otherwise if $r(x^*) + s(y^*) > c(x^*,y^*)$ for some x^* and y^* then we can construct an improved solution as follows. Let

$z = f^{-1}(y^*)$ and suppose first that $z > x^*$. As c is continuous and r, s and f are piecewise continuous we can choose x^* so that f is continuous at x^* and z. Define f' from f as follows.

$$\begin{aligned} f'(x) &= f(z + x^* - x), \quad x \in [x^*, x^* + \varepsilon), \\ &= f(x - \varepsilon), \quad x \in [x^* + \varepsilon, z], \\ &= f(x), \quad \text{otherwise}. \end{aligned}$$

Then f' is also a 1-1 measure preserving function and is a new feasible solution for SCAP. It can be shown that when ε is small enough f' is an improvement on f. A similar f' can be defined for the case $z < x^*$.

With minor additions this algorithm has been succesfully implemented on a minicomputer at Cambridge for cost functions which are quartics.

A second generalization of TP, which we have called the Time-Continuous Transportation Problem (TCTP), is to the case where demand, availability and transportation costs vary with time and storage is allowed at both the sources and destinations. This problem can be formulated as

$$\text{TCTP : Minimize } \int_0^T \Big[\sum_{i=1}^{m}\sum_{j=1}^{n} c_{ij}(\tau)x_{ij}(\tau)\Big]\, d\tau$$

$$\text{subject to } \int_0^t \Big[\sum_{j=1}^{n} x_{ij}(\tau)\Big]\, d\tau \le \int_0^t a_i(\tau)d\tau, \quad i = 1,2,\dots,m,$$

$$\int_0^t \Big[\sum_{i=1}^{m} x_{ij}(\tau)\Big]\, d\tau \ge \int_0^t b_j(\tau)d\tau, \quad j = 1,2,\dots,n,$$

$$x_{ij}(t) \ge 0, \ i = 1,2,\dots,m, \ j = 1,2,\dots,n, \ t \in [0,T].$$

The time continuous transportation problem is an example of a more general type of problem called a Separated Continuous Linear Program (SCLP). The form of SCLP is

$$\text{SCLP : Minimize } \int_0^T c(t)^T x(t)dt$$

$$\text{subject to } \int_0^t Gx(s)ds + y(t) = a(t), \tag{8}$$

$$Hx(t) + z(t) = b(t), \tag{9}$$

$$x(t), y(t), z(t) \ge 0, \ t \in [0,T].$$

Here x, z, b and c are bounded measurable functions and y and a are absolutely continuous functions. The description 'separated' refers to the fact that the constraints are in two sets, the integral constraints

(8) and the instantaneous constraints (9). This class of problems is treated by Anderson[1] who discusses a simplex like algorithm for SCLP.

4. Conclusions

In this paper we have discussed a number of infinite dimensional linear programs and algorithms for their solution. It is clear that there are many interesting problems which can be formulated in this way. Moreover it seems that algorithms for the solution of these problems can often be specified quite easily.

These algorithms have a common structure which makes them similar to the simplex algorithm for finite LP. Essentially they consist of two parts. Firstly there is a check for the optimality of a given solution. This optimality check is generally related to the duality structure of the problem, though space has precluded our demonstrating this here. The optimality check is often only applicable to a restricted set of solutions, the extreme points of the feasible set.

Secondly there is an improvement mechanism. This is a procedure for producing from a given non-optimal solution another solution with an improved value for the objective function. Successive applications of the optimality check and the improvement mechanism constitute an algorithm for the problem. However this does not guarantee that the optimum is reached, or indeed that the solutions converge to the optimal solution. Further work is being carried out on the convergence behaviour of these types of algorithm. The actual behaviour observed when such algorithms are implemented is extremely good. In fact these algorithms will generally reach the exact optimum in a finite number of steps if the optimal solution has a form which makes this possible.

References

[1] E.J.ANDERSON, Basic solutions and a simplex method for a class of continuous linear programs, Optimization techniques, 9th IFIP Conference, Springer-Verlag, Berlin, 1980.

[2] E.J.ANDERSON, P.NASH & A.B.PHILPOTT, A class of continuous network flow problems, internal technical report, CUED/F-CAMS/TR 214, 1981.

[3] E.J.ANDERSON & A.B.PHILPOTT, Duality and an algorithm for a class of continuous transportation problems, in preparation.

[4] P.APPELL, Le problème géométrique des déblais et remblais, Memorial

des Sciènces Mathématiques, 1928.

[5] R.BELLMAN, Dynamic Programming, Princeton University Press, 1957.

[6] W.P.DREWS, R.J.HARTBERGER and R.G.SEGERS, A simplex-like algorithm for continuous-time linear optimal control problems, Optimisation methods in resource allocation, R.W.Cottle and J.Kraup, ed., Crane Russak and Co Inc., New York, 1974.

[7] L.FORD & D.R.FULKERSON, Flows in Networks, Princeton University Press, 1962.

[8] R.GRINOLD, Symmetric duality for a class of continuous linear programming problems, SIAM J. Appl. Math., 18 (1970), 84-97.

[9] L.V.KANTOROVITCH, On the translocation of masses, Doklady Akad. Nauk.SSSR, 37 (1942), 199-201.

[10] K.S.KRETSCHMER, Programmes in paired spaces, Canadian J. Math., 13 (1961), 323-334.

[11] R.S.LEHMANN, On the continuous simplex method, RM-1386, Rand Corporation, 1954.

[12] V.L.LEVIN & A.A.MILYUTIN, The problem of mass transfer with a discontinuous cost function, Russian Math. Surveys, 34 (1978), 1-78.

[13] N.LEVINSON, A class of continuous linear programming problems, J. Math. Anal. and Appl.,16 (1966), 73-83.

[14] G.MONGE, Mémoires de l'Académie des Sciènces, 1781.

[15] T.NAKAMURA and M.YAMASAKI, Sufficient conditions for duality theorems in infinite linear programming problems, Hiroshima Math. J., 9 (1979), 323-334.

[16] A.F.PEROLD, Fundamentals of a continuous time simplex method, Stanford University technical report, SOL 78-26, 1978.

[17] G.STRANG, Maximal flow through a domain, to appear.

[18] W.F.TYNDALL, An extended duality theory for continuous linear programming problems, SIAM J. Appl. Maths., 15 (1967), 1294-1298.

ENTROPY OPTIMIZATION VIA ENTROPY PROJECTIONS

Yair Censor
Department of Mathematics
University of Haifa
Mt. Carmel, Haifa 31999, Israel

1. INTRODUCTION

Linearly constrained entropy optimization problems arise in various fields of applications, e.g., (i) transportation planning (the gravity model), (ii) statistics (adjustment of contingency tables, estimation of transition probabilities of doubly stochastic Markov chains), (iii) linear numerical analysis (preconditioning of a matrix prior to calculation of eigenvalues and eigenvectors, (iv) chemistry (the chemical equilibrium problem), (v) image reconstruction from projections (radioastronomy, X-ray tomography), (vi) geometric programming (the dual problem). Some details and further references may be found in [1,6,7,8,9].

The use of entropy is rigorously founded in several areas (see, e.g., [11]) while in other situations entropy optimization is used on a more empirical basis. In image reconstruction from projections (where our own motivation to study entropy optimization comes from) arguments in favour of maximum entropy imaging are given (e.g., [5]) but often there is merely the conviction that the maximum entropy approach yields the most probable solution consistent with the available data, i.e., a solution which is most objective or maximally uncommitted with respect to missing information.

In this note we touch upon several entropy optimization algorithms which are based on or are related to Bregman's general method for convex programming [2,3,4].

2. ENTROPY PROJECTIONS ONTO HYPERPLANES

We are interested in entropy maximization problems of the form $\text{Max}\{-f(x)\}$ where $f: \mathbb{R}^n_+ \longrightarrow \mathbb{R}$ is given by $f(x) = \sum_{j=1}^n h(x_j)$ with $h(y) = y \ln y$ if $y > 0$, and $h(0) = 0$, which are linearly constrained by equality constraints $Ax = b$, inequality constraints $Ax \leq b$ or interval constraints $c \leq Ax \leq b$ with possibly added box constraints $u \leq x \leq v$. ($\mathbb{R}^n_+$ stands for the nonnegative orthant of the n-dimensional Euclidean space).

The algorithms presented here make use of entropy projections onto hyperplanes which are defined as follows.

Definition 1. Let $H = \{x \in \mathbb{R}^n \mid \langle a,x \rangle = b\}$ be a given hyperplane and $y \in \mathbb{R}^n_+$ then the system

$$\begin{cases} x^*_j = y_j \exp(\lambda a), & j = 1,2,\dots,n, \\ \langle a,x^* \rangle = b, \end{cases}$$

determines a point $x^* \in \mathbb{R}^n$ and a real λ.

It follows from the general theory [4, Lemma 3.1] that these x^* and λ are uniquely determined and therefore we call them the entropy projection of y onto H, and the entropy projection parameter, respectively. The entropy projection onto a hyperplane is a concrete realization of Bregman's notion of a D-projection, see, e.g., [4], which generalizes the orthogonal projection in a manner that entails several interesting optimization algorithms.

3. ALGORITHMS

We describe three entropy optimization algorithms which employ entropy projections.

Algorithm 1.

Initialization: $z^0 = 0,\ x^0 = e^{-1}\underset{\sim}{1}$;

Typical Step: $\begin{cases} x_j^{k+1} = x_j^k \exp(B^k a_{ij}), & j = 1,2,\dots,n. \\ z^{k+1} = z^k - B^k e_i \ ; \end{cases}$

Control: $i \equiv i_k,\ \{i_k\}_{k=0}^{\infty}$ is cyclic on $M = \{1,2,\dots,m\}$.

This algorithm produces a sequence $\{x^k\}_{k=0}^{\infty}$ which converges to the solution of the problem

$$\text{Min} \sum_{j=1}^{n} x_j \ln x_j, \text{ s.t. } Ax = b \text{ and } x \geq 0 ,$$

where the following conventions are made: $\underset{\sim}{1}$ stands for the n-dimensional vector of ones, e is the natural logarithms base, a_{ij} is the i'th row j'th column entry of the $m \times n$ matrix A, $e_i \in \mathbb{R}^m$ is the i'th unit vector with one in its i'th coordinate and zeros elsewhere. B^k is the entropy projection parameter associated with entropy projecting x^k onto the hyperplane described by the i'th equation of $Ax = b$. Therefore, B^k is such that $\langle a_i, x^{k+1} \rangle = b_i$ is ensured where b_i is the i'th coordinate of $b \in \mathbb{R}^m$ and a_i^T is the i'th row of A. Cyclic control means $i_k = k(\text{mod } m) + 1$.

Algorithm 2.

Initialization: $z^0 = 0,\ x^0 = e^{-1}\underset{\sim}{1}$;

Typical Step:
$$\begin{cases} x_j^{k+1} = x_j^k \exp(c^k a_{ij}) \ ; \quad j = 1,2,\ldots,n, \\ z^{k+1} = z^k - c^k e_i \ ; \end{cases}$$

with $c^k = \min\{z_i^k, B^k\}$.

Control: $i \equiv i_k$, $\{i_k\}_{k=0}^{\infty}$ is cyclic on M.

This algorithm produces a sequence $\{x^k\}_{k=0}^{\infty}$ which converges to the solution of the problem:

$$\text{Min} \sum_{j=1}^{n} x_j \ln x_j \ , \text{ s.t. } Ax \leq b \text{ and } x \geq 0.$$

Here, x^{k+1} is the entropy projection of x^k onto the bounding hyperplane of the halfspace represented by the i'th inequality of $Ax \leq b$ provided that x^k violates this inequality or lies on the bounding hyperplane. If $\langle a_i, x^k \rangle < b_i$ then a move towards the bounding hyperplane is made which depends on the value of the i'th coordinate of the current dual vector z^k.

Algorithm 3.

Initialization: $z^0 = 0,\ x^0 = e^{-1}\underset{\sim}{1}$;

Typical Step:
$$\begin{cases} x_j^{k+1} = x_j^k \exp(d^k a_{ij}), \quad j = 1,2,\ldots,n, \\ z^{k+1} = z^k - d^k e_i \ ; \end{cases}$$

with $d^k = \text{Mid}\{z_i^k, \Delta^k, \Gamma^k\}$

Control: $i \equiv i_k$, $\{i_k\}_{k=0}^{\infty}$ is cyclic on M.

Here Δ^k and Γ^k stand for the entropy projection parameters associated with the entropy projection of x^k onto the bounding hyperplanes of the hyperslab obtained from the i'th row interval inequality of $c \leqslant Ax \leqslant d$, respectively. Mid denotes the median of the latter and the i_k'th coordinate of the current dual vector z^k. This algorithm produces a sequence $\{x^k\}_{k=0}^{\infty}$ which coverges to the solution of the problem:

$$\text{Min} \sum_{j=1}^{n} x_j \ln x_j, \text{ s.t. } c \leqslant Ax \leqslant b , \quad x \geqslant 0.$$

Algorithm 3 is derived from the general scheme for interval convex programming given in [4, Algorithm 5.1]. Its advantages over a direct application of Algorithm 2 to the interval constrained problem are explained there.

4. MART vs. BREGMAN

Another entropy optimization algorithm is MART (≡ Multiplicative Algebraic Reconstruction Technique), which was studied in [10].

MART

Initialization: $x^0 = e^{-1} \underset{\sim}{1}$;

Typical Step: $x_j^{k+1} = x_j^k \left(\frac{b_i}{\langle a_i, x^k \rangle}\right)^{\lambda_k a_{ij}}$, $j = 1,2,\ldots,n$;

Control: $i \equiv i_k$, $\{i_k\}_{k=0}^{\infty}$ cyclic on M.

Relaxation: $\{\lambda_k\}_{k=0}^{\infty}$ are such that $0 < \varepsilon \leqslant \lambda_k \leqslant 1$.

Designed to solve the above mentioned equality constrained entropy maximization problem this algorithm was thought not to be a special case of Bregman's method (unless the elements a_{ij} are all 0 or 1), see [9, p. 248]. A recent analysis of this question (Censor, Lent and Kuo - in preparation) reveals the intimate connection between MART and Bregman's algorithm. It shows that under some additional conditions MART may indeed be viewed as a particular underrelaxed version of Algorithm 1 above. These results will appear elsewhere.

ACKNOWLEDGEMENTS

Preparation of this report was partially supported by NIH Grant No. HL 28438 while the author was with the Medical Imaging Section, Department of Radiology, Hospital of the University of Pennsylvania, Philadelphia, PA., during the summer of 1981. Thanks are due to Mrs. Ana Burcat for typing the manuscript.

REFERENCES

[1] Altschuler, M.D., Censor, Y., Herman, G.T., Lent, A., Lewitt, R.M., Srihari, S.N., Tuy, H., and Udupa, J.K., Mathematical aspects of image reconstruction from projections, Progress in Pattern Recognition (L.N. Kanal and A. Rosenfeld, Editors), North-Holland, Amsterdam, v.1 (1981), In press.

[2] Bregman, L.M., The relaxation method of finding the common point of convex sets and its application to the solution of problems in convex programming. USSR Comp. Math. and Math. Phys., v.7 (3), (1967), pp. 200-217.

[3] Censor, Y., Row-action methods for huge and sparse systems and their applications. SIAM Review, v.23 (1981), pp.444-466.

[4] Censor, Y., and Lent, A., An iterative row-action method for interval convex programming. J. Optimization Theory and Applications, v.34 (1981), pp. 321-353.

[5] D'Addario, L.R., Maximum entropy imaging: theory and philosophy. In: Image Analysis and Evaluation,(R. Shaw, Editor), Society of Photographic Scientists and Engineers, Washington, D.C., 1977.

[6] Elfving, T., On some methods for entropy maximization and matrix scaling. Linear Algebra and Its Applications, v. 34 (1980), pp. 321-339.

[7] Erlander, S., Entropy in linear programs. Mathematical Programming, v. 21 (1981), pp. 137-151.

[8] Frieden, B.R., Statistical models for the image restoration problem. Computer Graphics and Image Processing, v. 12 (1980), pp. 40-59.

[9] Lamond, B., and Stewart, N.F., Bregman's balancing method. Transportation Research, v. 15B (1981), pp. 239-248.

[10] Lent, A., A convergent algorithm for maximum entropy image restoration, with a medical X-ray application. In: Image Analysis and Evaluation, (R. Shaw, Editor), SPSE, Washington, D.C., 1977, pp. 249-257.

[11] Levine, R.D., and Tribus, M., (Editors), The Maximum Entropy Formalism, The MIT Press, Cambridge, MA., 1978.

RESOLUTION OF A QUADRATIC COMBINATORIAL PROBLEM BY DYNAMIC PROGRAMMING

J.C. Hennet
Laboratoire d'Automatique et d'Analyse des Systèmes du C.N.R.S.
7, avenue du Colonel Roche
31400 - Toulouse - France

ABSTRACT

The problem of arranging heliostats on the collector field of a solar central receiver system can give rise to the following quadratic combinatorial formulation :

Maximize $X^t P X$ subject to $X^t X \leq M$

With $X = (X_1, \ldots X_i, \ldots X_n)^t$ and $i \in (1, \ldots N)$; $X_i = 0$ or 1.

We compute average intrinsic efficiencies of each location (p_{ii}) and average interaction rates due to shadow effects for each pair or heliostats ($-p_{ij}$). Then we want to optimally choose a maximum of M heliostat locations among N possible ones.

For finding the optimal vector, X^*, we propose an original approach based on dynamic programming.

INTRODUCTION

Combinatorial programming provides clear representations of decision problems where many elementary decisions can occur and interact. Each elementary decision is represented by a variable x_I which can take the value 0 or 1. Interrelations between variables can be implicitely contained in the set of constraints or directly expressed by cross products $\prod_{J \in Q} x_J$ in the objective function. The problem presented in this paper includes the two types of interactions in the simplest form, that is a quadratic criterion and a budget constraint in which all the variables have the same weight.

The most current approaches used for solving quadratic combinatorial programs are implicit enumeration (Hansen 1972, Mc Bride and Yormark 1980) and conversion into 0-1 linear programs by addition of variables and constraints (Glover and Wolsey 1974, Granot, Granot and Kallberg, 1979). In both methods the set of optimal elementary decisions can be constructed through a "branch and bound" algorithm : different chains of decisions are successively tested and the best one is selected. In some cases (Nemhauser and Ullman, 1969) it is possible to build the set of optimal decisions by constructing the optimal sequence of a discrete dynamic programming problem. We have found that dynamic programming could be applied to our problem with considerable savings in computation time.

PRESENTATION OF THE OPTIMIZATION PROBLEM

Solar Energy is currently considered as one of the major energy sources of the future. However, the costs and benefits which would result from industrial mass production of solar plants are still controversial and hard to determine. For example, the accuracy of performance evaluation of a "power tower" thermodynamic system depends on how many simulation points can be analyzed by a computer at a reasonable cost. In order to decrease computation times, we have built analytical models of flux density calculations and shadow effets (Hennet 1980 , Hennet and Abatut 1981). For any project of solar central receiver system, we can determine the "performance matrix" of the field of heliostats. Diagonal terms of the matrix represent intrinsic efficiencies of heliostats. Out of diagonal terms are average rates of energy losses due to shadow effects (Hennet 1982).

The concept of performance matrix has been created for evaluating existing fields of heliostats. But its generalization to dummy heliostat locations is straightforward. As an example,we can define a set of possible locations at the nodes of a uniform radial network (fig. 1) projected onto the topography of the site.

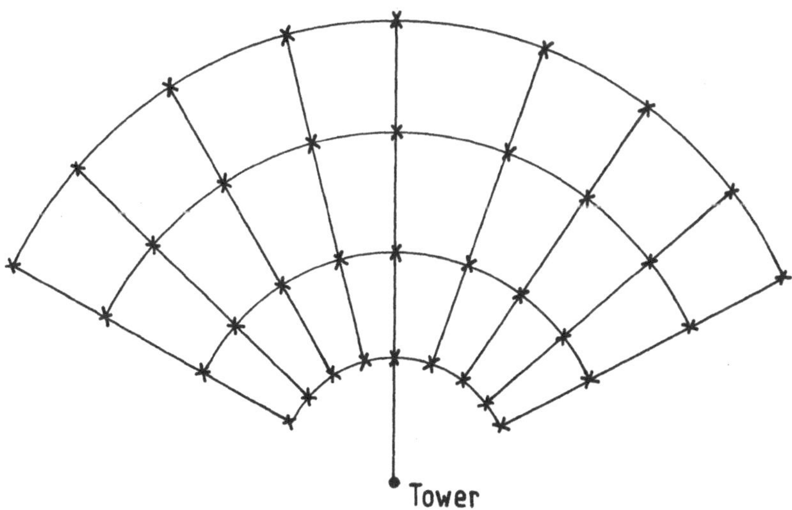

Fig. 1. A radial network

Among possible locations, we want to select the best ones relatively to a performance criterion under a budget constraint :

$$(I)\begin{cases} \text{Maximize } \sum_{I=1}^{N} (p_{II}\, x_I + \sum_{\substack{J=1\\ J\neq I}}^{N} p_{IJ}\, x_J\, x_I) & (1)\\ \text{Subject to } \sum_{I=1}^{N} x_I \leqslant M & (2)\\ \text{and for any } I \in (1,\ldots,N),\ \ x_I=0 \text{ or } 1 & (3)\end{cases}$$

If heliostat I is selected, its efficiency in the field is :

$$\rho_I = p_{II} + \sum_{\substack{J=1\\ J\neq I}}^{N} p_{IJ}\, x_J$$

with $0 < p_{II} < 1$

$- p_{II} \leqslant p_{IJ} \leqslant 0$

$0 \leqslant \rho_I \leqslant p_{II}$

Because all x_I are 0 or 1, the optimization problem can also be written:

Maximize $Q(X) = X^T P X$

Subject to $X^T X \leqslant M$

and for $I \in (1,\ldots,N)$, $x_I = 0$ or 1

(I) with $X = (x_1, x_2, \ldots x_N)^T$

and $P = \begin{bmatrix} p_{11} & & p_{1N} \\ & p_{II} \quad p_{IJ} & \\ p_{N1} & & p_{NN} \end{bmatrix}$ (Performance matrix of the field consisting of all possible locations)

Analytical models for flux density and shadow effects calculations are too complicated to be integrated in a continuous optimization problem. This is the reason why we have defined a discrete optimization problem (I) with N "a-priori" locations. If the mean spacing between possible positions is sufficiently small, the optimal solution of (I) can be considered as the optimal arrangement of a maximum of M heliostats on the site.

This is just an approximation since N is finite. In practice, matrix P can only be computed up to $N \simeq 200$. But it is often possible to decompose the location problem into subproblems (Hennet 1982). Possible locations are defined without considering mechanical interference between heliostats : if relative values of M and N are reasonable, p_{IJ} terms play the role of penalties tending to exclude unsufficiently spaced locations.

CONSTRUCTION OF THE SET OF FEASIBLE SOLUTIONS

Let us consider a feasible solution of problem (I) :

$$\hat{X} = (\hat{x}_1, \dots, \hat{x}_N)$$

for any $I \in (1, \dots, N)$, $\hat{x}_I = 0$ or 1.

We define :

$$\mathcal{L}(\hat{X}) = \{ I ; \; x_I = 1 \}$$

We assume that

$$0 < M < N \; ; \quad M \text{ and } N \text{ are integer}$$

For solution $\hat{X}$ let us assume that inequality (2) is strict :

$$\text{Card}\,(\mathcal{L}(\hat{X})) = \sum_{I=1}^{N} \hat{x}_I < M$$

The value of the criterion for $\hat{X}$ is:

$$Q(\hat{X}) = \sum_{I \in \mathcal{L}(\hat{X})} \Big(p_{II} + \sum_{\substack{J \in \mathcal{L}(\hat{X}) \\ J \neq I}} p_{IJ}\Big) \tag{4}$$

From $\hat{X}$, it is possible to construct other admissible solutions denoted $\hat{X}^K$ such that :

$$\hat{X}^K = (\hat{x}_1^K, \dots, \hat{x}_N^K)$$

For all $\quad I \in \mathcal{L}(\hat{X}), \quad \hat{x}_I^K = 1$

For all $\quad \begin{cases} I \notin \mathcal{L}(\hat{X}), \\ I \neq K \end{cases} \quad \hat{x}_I^K = 0$

and with $\quad K \notin \mathcal{L}(\hat{X}), \quad \hat{x}_K^K = 1$

If $0 < M < N$, it is always possible to find an index K for which $\hat{x}_K = 0$.
The number of solutions $\hat{X}^K$ is : **M** - Card $(\mathcal{L}(\hat{X}))$. $\hat{X}^K$ is feasible since

$$\sum_{I=1}^{N} \hat{x}_I^K = \sum_{I=1}^{N} \hat{x}_I + 1 \leq M$$

The value of the criterion for $\hat{X}^K$ is :

$$Q(\hat{X}^K) = Q(\hat{X}) + p_{KK} + \sum_{J \in \mathcal{L}(\hat{X})} (p_{JK} + p_{KJ}) \tag{5}$$

In order to find the best solution constructed from $\hat{X}$ we have to compare quantities $Q(\hat{X}^K)$ for all the possible values of K.
The order of solution $\hat{X}$ is Card$(\mathcal{L}(\hat{X}))$
The order of solution $\hat{X}^K$ is Card$(\mathcal{L}(\hat{X})) + 1$.
Reciprocally, if $\tilde{X}$ is a solution of order j with $1 \leq j \leq M$, it is always possible to construct a feasible solution ${}^K\tilde{X}$ such that :

for all $\quad I \notin \mathcal{L}(\hat{X}), \quad {}^K\tilde{x}_I = 0$

for all $\quad \begin{cases} I \in \mathcal{L}(X), \\ I \neq K \end{cases} \quad {}^K\tilde{x}_I = 1$

and with $\quad K \in \mathcal{L}(\tilde{X}), \quad {}^K\tilde{x}_K = 0$

Then, $\tilde{X}$ is a solution of type $\hat{X}^K$ for $\hat{X} \equiv {}^K\tilde{X}$

Consequently, all the admissible solutions of problem (I) can be constructed step by step from the solution of order 0 : (0,0,...,0).

All the feasible solutions of order j can be constructed from all the feasible solutions of order j - 1.

DETERMINATION OF A SUB-OPTIMAL SOLUTION BY DYNAMIC PROGRAMMING

The best solution of order 1 is directly obtained by applying relation (5) to the initial solution $\hat{X} = (0,0,\ldots,0)$:

$$X^* = \hat{X}^{K^*}$$

with $Q(\hat{X}^{K^*}) = \underset{K=1,\ldots,N}{\text{Max}} \; Q(\hat{X}^K) = \underset{K=1,\ldots,N}{\text{Max}} \; p_{KK} = p_{K^*K^*}$

In order to apply dynamic programming, we assume a Markov-type property between the sets of optimal solutions of orders $j = 2,\ldots,M$.

- Markov-type property : The best solution of order j associated with the decision "$x_I = 1$ at stage j" can be constructed from one of the best solutions of order j-1.

When the Markov-type property applies, all the possible states of the decision process at stage j can be simply characterized by the last decision : "x_I=1 at stage j", the previous decisions corresponding to an optimal policy.

The set $\mathcal{L}(X)$ associated with each state can be simply denoted $L_j(I)$.

The objective function Q (X) is now denoted $f_j(I)$ and calculated by the following relations :

for any I $\quad f_1(I) = p_{II}$

and for $2 \leq j \leq m \quad f_j(I) = p_{II} + \underset{K}{\text{Max}} \left(f_{j-1}(K) + \sum_{L \in L_{j-1}(K)} (p_{IL} + p_{LI})\right)$

The choice of K^* such that :

$$f_j(I) = p_{II} + f_{j-1}(K^*) + \sum_{L \in L_{j-1}(K^*)} (p_{IL} + p_{LI})$$

is restricted to the set of nodes K for which $I \notin L_{j-1}(K)$.

Finally, information on the best vector X^* is contained in $L_{j^*}(I^*)$, the pair (I^*, j^*) being such that

$$f_{j^*}(I^*) = \underset{\substack{I = 1,\ldots,N \\ j = 1,\ldots,M}}{\text{Max}} f_j(I)$$

EVALUATION OF THE METHOD

If the Markov-type property is not verified, the method still works, but optimality of the solution is notguaranteed. Then we wish to know how far the solution is from the optimal one. Since we have no theoretical method for evaluating that distance, we can only compare the solution provided by the D.P. algorithm with the optimal solution in small-size examples.

Due to the structure of problem (I), it is very easy to solve small-size problems by a Branch and Bound algorithm. We use the general procedure described by Geoffrion and Marsten (1972). At each step of the process, we solve a relaxed problem which is linear of the knapsack type (Nauss 1976). The solution of the candidate problem is evaluated by finding a lower bound and on upper bound from the solution of the relaxed problem.

Comparison of the two methods (fig. 2 and fig. 3) shows that the branch and bound algorithm is sufficiently fast up to $N \simeq 55$, but the exponential growth of its computing time does not allow for solving large size problems. For large values of N, the D.P. algorithm is very efficient in terms of computing time. The difference between the optimal solution and the sub-optimal one is usually less than 5 ‰.

The choice of the D.P. algorithm for large size problems carries back the limits in computing time on calculating matrix P (as shown on fig. 2 and fig. 3).

	Matrix P	Time B.B.	Time D.P.	Sol B.B.	Sol D.P.	Difference
Problem 1 (5 among 10)	10 sec	0,96sec	0,88sec	4,033	4,023	0,010
Problem 2 (20 among 45)	58 sec	38,55sec	2,10sec	16,643	16,643	0
Problem 3 (20 among 55)	1mn 16 sec	3mn 16sec	2,61sec	16,713	16,655	0,057
Problem 4 (25 among 91)	4 mn	Not finished in 25 mn	12,71sec	20,814	20,784	0,030

Figure 2

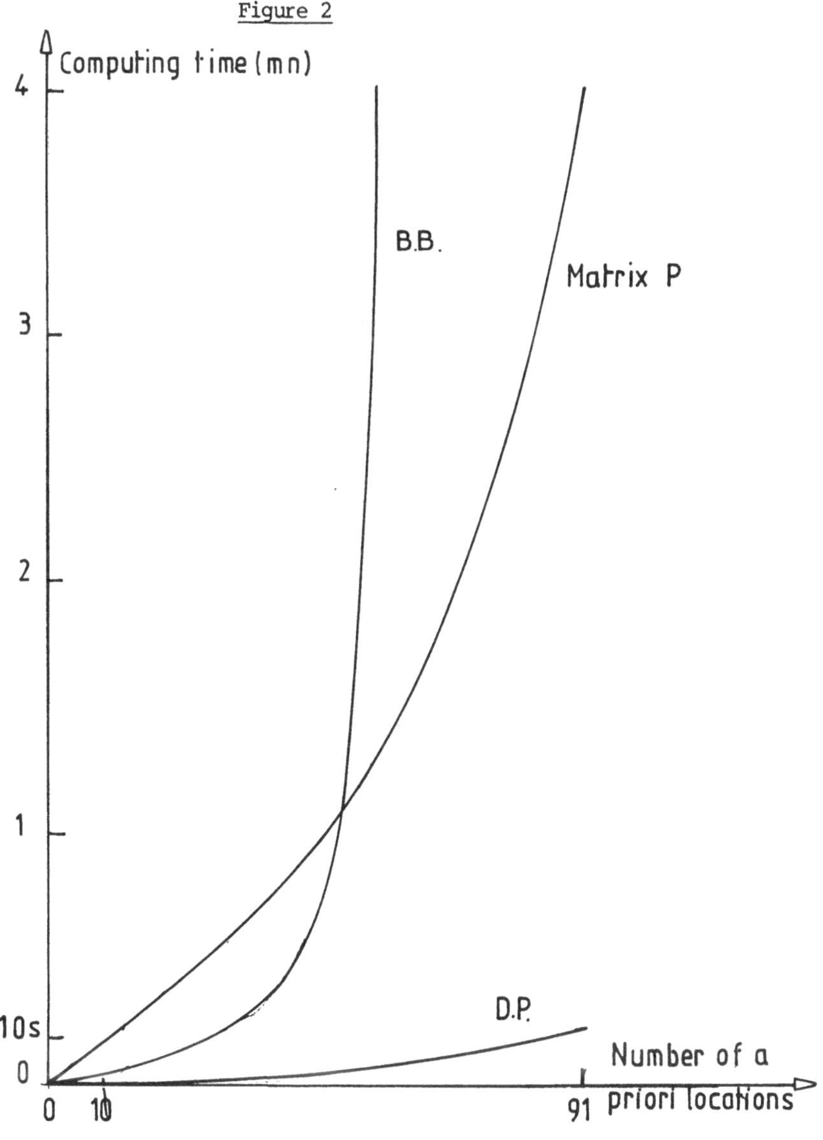

Figure 3

CONCLUSION

The problem of choosing M mutually interacting elements among N possible ones can be expressed as a quadratic combinatorial program. This program can be solved by an implicit enumeration approach up to $M \times N \simeq 1100$ (Problem 3). For large size problems we must use a sub-optimal method. We can decrease the computing time by fathoming only a part of the set of feasible solutions, with the condition that this part should contain satisfactory solutions. We propose a solution construction process which allows for verifying such a condition. It is an algorithm basec on dynamic programming. We have tested this algorithm and shown that it makes possible the resolution of large size problems without moving away too far from the optimal solution.

REFERENCES

R. BELLMAN, Dynamic Programming, Princeton University Press, Prindeton, N.J., 1957.

A.M. GEOFFRION, R.E. MARSTEN, Integer Programming Algorithms : A Framework and a State-of-the-Art Survey, Management Science, Vol. 18, no. 9, pp. 465-491.

F. GLOVER, R.E. WOLSEY, Converting the 0-1 Polynomial Programming Problem to a 0-1 Linear Program, Operations Research, Vol. 22, 1974, pp. 180-182.

D. GRANOT, F. GRANOT, J. KALLBERG, Covering Relaxation for Positive 0-1 Polynomial Programs, Management Science, Vol. 25, no. 3, March 1979, pp. 264-273.

P. HANSEN, Quadratic 0-1 Programming by Implicit Enumeration, in Numerical Methods for Non Linear Optimization, Lootsma (ed.), Academic Press, 1972, pp. 282-296

J.C. HENNET, Etude des Effets d'Ombre entre Héliostats d'une Centrale Solaire, Note interne LAAS-ASE, Nov. 1980.

J.C. HENNET, J.L. ABATUT, An Analytical Method for Reflected Flux Density Calculations - Solar World Forum, Brighton 1981.

J.C. HENNET, Méthodologie d'Evaluation et de Conception de Champs d'Héliostats Focalisants. Thèse de Doctorat d'Etat, Toulouse 1982 (to appear)

R.D. Mc BRIDE, J.S. YORMARK, An Implicit Enumeration Algorithm for Quadratic Integer Programming, Management Science, vol. 26, no. 3, March 1980, pp. 282-296

R.M. NAUSS, An Efficient Algorithm for the 0-1 Knapsack Problem, Management Science, Vol. 23, no. 1, Sept. 1976

G.L. NEMHAUSER, Z. ULLMAN, Discrete Dynamic Programming and Capital Allocation, Management Science, vol. 15 (1969), pp. 494-505.

THE STRUCTURE AND COMPUTATION OF SOLUTIONS TO CONTINUOUS LINEAR PROGRAMS

J.Jasiulek
Department of Mathematics
Simon Fraser University
Burnaby, BC V5A 1S6, Canada

ABSTRACT

Basic solutions for continuous linear programs are defined in analogy with the finite dimensional case. Not every feasible extreme point solution is necessarily basic; however, under appropriate regularity conditions it can be shown that extreme points of the feasibility set are indeed basic. Duality theory and parametric linear programming techniques constitute the primary tools of this analysis. Feasible trajectories through the state space are studied as they pass through the boundaries of the elementary cones into which the state space is partitioned. Switches from one basis to another are classified as primally or dually induced depending on how the trajectory passes through the boundaries of the elementary cones. A simple example illustrates how these ideas can be used to construct optimal solutions and interpret them geometrically.

Introduction

The continuous linear programming problem is described as

$$\text{maximize} \int_0^T c(t)x(t)\,dt$$

subject to the constraints

$$A(t)x(t) \leq b(t) + \int_0^T K(t,s)x(s)ds$$

$$x(t) \geq 0,$$

where $x(t)$ is an n-dimensional vector of (possibly) generalized functions, $b(t)$ is an m-dimensional vector, $c(t)$ is an n dimensional row vector, and $A(t)$, $K(t,s)$ are m x n matrices such that the integrals above are well defined. The ordering is to be understood componentwise. In the case of Lebesgue integrable functions, the inequalities hold almost everywhere, in the case of Schwartz distributions $x \geq y$ if and only if for all nonnegative C^∞ - functions $z(t)$ with compact support $(z,x) \geq (z,y)$.

Problems of this type arise in investment and planning models as well as in engineering applications. Associated with this primal program is the following dual program

$$\text{minimize} \int_0^T y(t)b(t)dt$$

subject to

$$y(t)A(t) \geq c(t) + \int_t^T y(s)K(s,t)ds$$

$$y(t) \geq 0.$$

The existence of optimal solutions can depend critically on the proper choice of the space of admissable functions as the following example shows [1]:

$$\text{minimize} \int_0^3 [y_1(t) + 3y_2(t)]dt$$

subject to

$$y_2(t) + \int_t^3 y_1(s)ds \geq 1$$

$$y(t) \geq 0.$$

It can be shown that the minimum is not achieved for any integrable function, but that $y_1(t) = 0$, $y_2(t) = \delta(3-t)$, where $\delta(t)$ denotes the Dirac δ- distribution, solves this problem optimally.

It turns out that the notion of state is useful here. We call

$$B(t) = b(t) + \int_0^t K(t,s)x(s)ds \qquad \text{and} \qquad C(t) = c(t) + \int_t^T y(s)K(s,t)ds$$

the <u>primal</u> and <u>dual</u> state, respectively. The pair $(B(t),C(t))$ is referred to as a <u>total trajectory</u>. We also associate the following <u>instantaneous linear program</u> at time t with the given continuous linear program

$$\text{maximize} \qquad C^T(t)u$$

subject to

$$A(t)u \leq B(t)$$

$$u \geq 0 , \qquad u \in R^n.$$

In this terminology the weak maximum principle [1] can be stated. Let $(x(t),y(t))$ be optimal solutions of a continuous linear program and its dual. Then $u = x(t)$ solves the corresponding instantaneous linear program optimally almost everywhere.

Basic Solutions and Extreme Points

The simplex method for standard linear programs uses the notion of basic solution successfully to arrive at an optimal extreme point. Thus it seems reasonable to extend this notion. A solution $x(t)$ is called basic if and only if there exists a partition of the interval $[0,T]$ such that within each subinterval no more than m components of $x(t)$ are positive. Unfortunately not all extreme points of the convex set of constraints are basic in this sense, not even when $A(t)$, $K(t,s)$ are restricted to be constant matrices.

Counterexample: Let E be a closed, nowhere dense set of positive measure contained in $[0,T]$, and denote its characteristic function by $\chi(t)$. Let F be an open interval contained in the complement of E. Then $(x_1(t),x_2(t)) = (\chi(t), \chi(t))$ is an optimal extreme point solution of the problem

$$\text{maximize} \int_0^1 (\chi(t) - 1)x(t)dt$$

subject to

$$-x_1(t) + \int_0^t x_1(s)ds + x_2(t) = \int_0^t \chi(s)ds$$

$$x(t) \geq 0.$$

Clearly the optimal value is zero. The extreme point property follows from the fact that $e(t)$ is an extreme point solution of a linear equation $Le(t) = b$, if and only if the homogeneous equation $Lu(t) = 0$ has only the trivial solution if we require $-e(t) \leq u(t) \leq e(t)$ (see [2]). This shows that some regularity conditions are required.

Theorem: Let $x(t)$ be a piecewise continuous extreme point of the set of state constraints, and suppose that $A(t)$ and $K(t,s) = K(s)$ are continuous. Then $x(t)$ is basic.

A proof of this result is contained in [2], and it should be noted that it need not hold when the kernel $K(t,s)$ depends on s and t [4].

Continuous linear programs allow nice geometric representations. As an illustration Gale's problem #1 is chosen [5], which has the parameters $T = 8$, $m = 2$, $n = 2$,

$$A = \begin{bmatrix} 1 & 2 \\ 2 & 1 \end{bmatrix} \quad K = \begin{bmatrix} 1 & 0 \\ 0 & 1 \end{bmatrix} \quad b = \begin{bmatrix} 1 \\ 4 \end{bmatrix} \quad \text{and } c = (1 \quad 0).$$

Assuming the total trajectory $(B(t),C(t))$ is known we can graph the instantaneous problems for different values of t (see figure 1). The optimal solution of each of these instantaneous problems is obvious. Moreover we can see that the primal

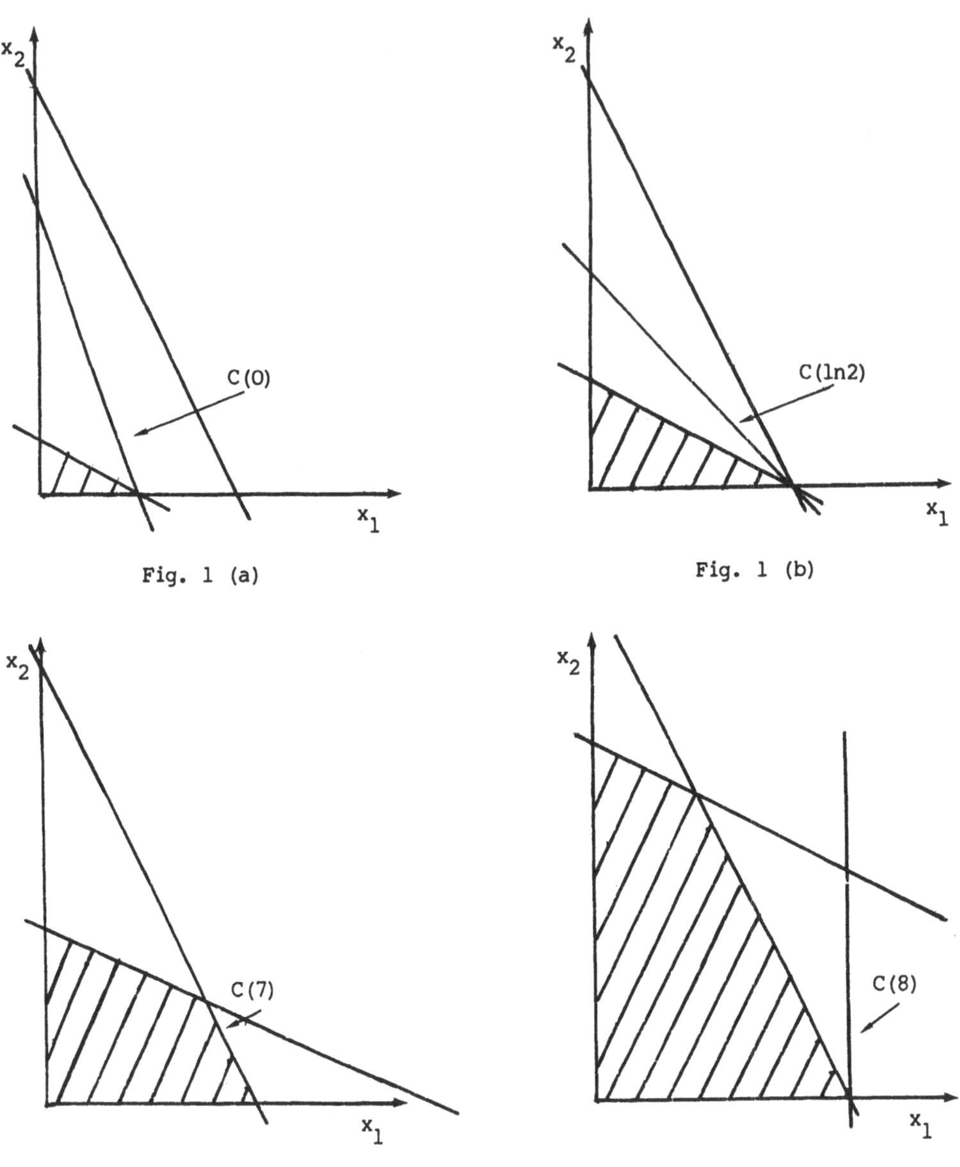

Fig. 1 Representation of Instantaneous Linear Programs associated with Gale's Problem # 1 at

(a) $t = 0$ (b) $t = \ln 2$

(c) $t = 7$ (d) $t = 8$

constraints represent moving hyperplanes and that C(t) is a moving supporting hyperplane. The optimal extreme point is determined locally by different intersections of these hyperplanes at different times. This suggests that parametric programming techniques might provide some insight into the local structure of extreme points of continuous linear programs and lead to their construction.

Let A denote the set of pairs $(d,f) \in R^{n+m}$ for which the standard linear program

$$\text{maximize} \quad d^T u$$

subject to

$$A\,u \leq f$$
$$u \geq 0,\ u \in R^n$$

has a finite solution. This set of parameters has the following known structure[3]:

<u>Theorem</u> A can be decomposed into finitely many polyhedral cones A_ℓ such that

(i) $A_\ell = B_\ell \times C_\ell$, where B_ℓ and C_ℓ are polyhedral cones of dimension m and n
(ii) a unique set of basic indices corresponds to the interior of each A_ℓ.

This result also provides more insight in the geometry of continuous linear programs. The total trajectory (B(t),C(t)) represents a curve in the parameter space A. Clearly, whenever the total trajectory lies in the interior of some elementary cone A_ℓ the basic indices of the optimal solution are uniquely determined, and they change whenever a boundary face is crossed. This means that the corresponding instantaneous linear program is degenerate when the total trajectory crosses a boundary face. A change of basic indices over a subinterval of [0,T] is referred to as a switch. We can distinguish between primally and dually induced switches, depending on whether the total trajectory crosses a boundary face of some elementary cone B_ℓ or C_ℓ.

Complementarity relations can be established to characterize optimal pairs further. An optimal dual pair (x(t),y(t)) is called complementary nondegenerate, if the corresponding instantaneous linear program is never degenerate in both the primal and dual variables at the same time. This means that the total trajectory avoids the simultaneous crossing of a primal and dual boundary face. This generic condition of avoiding a manifold of co-dimension two facilitates the analysis. The case when switches are both primally and dually induced plays the role corresponding to degeneracy in standard linear programs, and is not considered here. It turns out that complementary nondegeneracy implies a complementary smoothness condition, i.e. for any t either x(t) or y(t) is discontinuous, but not both. This can be seen as follows. Suppose that x(t) is discontinuous at a given switch. This implies that the dual solution of the corresponding instantaneous problem is degenerate. The assumption of nondegeneracy implies that the dual solution is unique. Consequently the primal state B(t) is differentiable at a primally induced switch, provided the right hand side b(t) is also differentiable at that point. This implies that the left and right limits of y(t) at that switch coincide as t approaches the switching time. A similar argument applies when y(t) is discontinuous. These complementarity conditions are helpful in determining into which elementary cone the total trajectory proceeds at primally or dually induced switches.

Construction of Simple Solutions

The ideas presented so far will now be applied to the construction of optimal solutions. For simplicity assume that the optimal solution of a given continuous linear program does not have any switches of bases. This means that the total trajectory is contained in exactly one elementary cone $\mathcal{A}_\ell = \mathcal{B}_\ell \times \mathcal{C}_\ell$. Suppose the dual state C(T) lies in the interior of $\mathcal{C}_\ell$. Consider the standard linear program

$$\text{maximize } d^T u$$
$$\text{subject to } Au \leq f,\ u \geq 0,\ u \in R^n.$$

For d = C(0) and f = B(0) this is the instanteneous linear program at t = 0. Its basic solution is unchanged when C(0) is replaced with any other d in the interior of $\mathcal{C}_\ell$ by the decomposition theorem. In particular, we can choose the known vector C(T) = c(T). Consequently, by solving the problem with d = c(T) the value x(0) = u will be determined uniquely. Hence the optimal solution satisfies the equation

$$A_{ba}(t)x_{ba}(t) = b(t) + \int_0^t K_{ba}(t,s)x_{ba}(s)ds$$

where the subscript ba refers to the basic components. In the case of constant coefficients this becomes a trivial initial value problem. In case that C(T) lies on the boundary the resulting linear program will have more than one optimal basic solution, while the continuous linear program may still have a unique solution. If B(0) is an interior point of the elementary cone the dual solution will be unique. Hence we can compute C(t), and if it lies in the interior for some t = r we can solve the standard linear program again with f = B(0) and d = C(r) to determine the basic indices of the solution for the continuous linear program uniquely. In more general cases a tie breaking mechanism may be required. This is analogous to degeneracy in standard linear programming.

As soon as there are switches, the situation can become complicated in general. Suppose there is exactly one primally induced switch, and that the total trajecory stays within the same two elementary cones when the duration T is allowed to vary between 0 and the value specified. This means that the problem is considered as imbedded into a class parametrized by the length of the interval [0,T]. In this case we proceed just as discussed above. Only now the resulting basic solution x_{ba} will not stay nonnegative for all $t \in [0,T]$. Thus we find the switching time t_s as root of the equation $x_i = 0$, where i denotes the index of the component which becomes negative. If we proceed ε units beyond $t = t_s$ using x_{ba} the primal state B(t) will enter the desired elementary cone since B(t) is differentiable at the switching time t_s. Now one dual step in the primal-dual simplex method applied to the standard linear program with the cost $d = C(T)$ and the right hand side $f = B(t_s + \varepsilon)$ will identify the new basic indices. Then the new dynamic basis can be determined over $[t_s,T]$ and we arrive at the optimal solution. A similar construction applies when there is exactly one dually induced switch and the imbedded problem parametrized by T has a total trajectory contained in the same two elementary cones for all durations between 0 and the specified value for the given problem. The conditions stated here are quite restrictive, but the construction has been carried out for more general cases in [2] and will be developed further in a future paper.

Example

To illustrate this construction of optimal solutions the discussion of Gale's problem # 1 is continued. We introduce slack variables $x_3(t)$ and $x_4(t)$ to convert the inequalities into equations. Thus the problem becomes

$$\text{maximize} \int_0^8 x_1(t)dt$$

subject to

$$x_1(t) + 2\, x_2(t) - \int_0^t x_1(s)ds + x_3(t) = 1$$

$$2\, x_1(t) + x_2(t) - \int_0^t x_2(s)ds + x_4(t) = 4$$

$$x_i(t) \geq 0, \; i = 1,2,3,4$$

Then the optimal solution structure is uniquely described by specifying the optimal partition of [0,8] and the basic indices for each subinterval:

interval	basic variables
[0,ln 2]	$x_1(t)$, $x_4(t)$
[ln 2,7]	$x_1(t)$, $x_2(t)$
[7, 8]	$x_1(t)$, $x_3(t)$

This solution can be constructed as follows (see fig. 1):

The known cost functional C(8) = (1 0) gives rise to the same instantaneous solution at t = 0 as the unknown cost functional C(0). So we attempt to use the basis $(x_1(t),x_4(t))$ which is the locally optimal solution corresponding to C(8) until it becomes infeasible, i.e. $x_4(t)$ becomes negative beyond t = ln2, which corresponds to a primally induced switch. One dual step in the primal-dual simplex method indicates switching to the basis $(x_1(t),x_3(t))$, which stays feasible to the endpoint at t = 8. Now the dual problem is solved backwards in time. Beyond t = 7 the dual solution becomes infeasible, and a primal simplex step suggests using the local basis $(x_1(t),x_2(t))$ for $\ln 2 \leq t \leq 7$. A final feasibility check for the primal and dual solution indicates that it is indeed the optimal solution for this problem.

In general, more forward and backward feasibility checks are needed before the optimum is reached. This heuristic technique works when the structure of the solution is sufficiently simple.

References

[1] R.C. Grinold, Continuous Linear Programming, Part One: Linear Objectives, Journal of Mathematical Analysis and Applications, vol. 28, 1969, pp 32-51.

[2] J. Jasiulek, Continuous Linear Programming; Theory and Computation, Ph.D. thesis, University of California, Davis.

[3] F. Nozicka et al., Theorie der linearen parametrischen Optimierung, Akademie Verlag, Berlin 1974.

[4] A. Perold, Extreme Points and Basic Feasible Solutions in Continuous Linear Programming, SIAM Journal on Control and Optimization, vol. 19, no. 1, 1981.

[5] R.G. Segers, A General Function Setting For Dynamic Optimal Control Problems, NATO Conference on Applications of Optimization Methods for Large Scale Resource Allocation Problems, Elsinore, Denmark, July 1971.

THE BAYESIAN APPROACH TO GLOBAL OPTIMIZATION

J.Mockus

Institute of Mathematics and Cybernetics
Lithuanian SSR Academy of Sciences
Vilnius, USSR

1. Introduction

There exists some split between the theory and practice in the field of global optimization.

Mathematical methods are usually of the "worst case" type. Efficient practical methods often have no mathematical justification [1]. To make the split as small as possible the so-called Bayesian approach was developed [2], [3]. The procedures of search which minimize not maximal (as in the worst case analysis) but the average deviation from the global minimum are called the Bayesian methods of optimization. The deviation is defined as

$$\delta(f) = f(x_{N+1}) - f(x_0)$$

where $f(x_0) = \min_{x \in A} f(x)$

A is a compact subset of R^m ,

f is a continuous function,

x_{N+1} is the final decision made by the given algorithm of optimization,

N is the fixed number of observations (calculations of at fixed points $x_i \in A$, $i = 1, \dots, N$).

The average deviation is defined as an integral

$$\int_C \delta(f) P(df) \qquad (1)$$

where C is the set of all continuous functions and P is some measure which represents the a priori information about the likelihood and importance of different subsets of C .

The Bayesian methods should minimize the average deviation (1) which depends on P . Linearity of δ as a function of $f(x_{N+1}) - f(x_0)$ follows from the usual assumptions of utility theory.

The proper choise of an apriori measure P is a difficult problem of the Bayesian approach. In order to consider this problem axiomatically it was supposed that a subjective likelihood relation $\succcurlyeq$ can be defined on the set of events, where the event B_i means that the vector $(f(x_1), \dots, f(x_l)), x_j \in A, j = \overline{1,l}, l = 1,2,\dots$ belongs to the l-dimensional interval.

A pair of intervals B_i, B_j will be called an s-pair if $a_k^i = a_k^j$ and $b_k^i = b_k^j$, when $k \neq s$,

$a_k^i = a_k^j$ or $b_k^i = b_k^j$, when $k = s$.

A pair will be called a lower (upper) s-pair if $a_s^i = a_s^j$ ($b_s^i = b_s^j$). It is obvious that the union of an s-pair is an interval.

It was assumed that the relation $\succcurlyeq$ satisfies seven conditions:

1. It is a complete ordering on a set of all intervals B_i and an empty set ϕ.
2. If $B_1 \cap B_2 = B_3 \cap B_4 = \phi$, where B_1, B_2 and B_3, B_4 are s-pairs for some $s = 1, \ldots, l$, then from $B_1 \succcurlyeq B_3$, $B_2 \succcurlyeq B_4$ it follows that $B_1 \cup B_2 \succcurlyeq B_3 \cup B_4$. If one of the first two relations is strict $\succ$ then the last relation should also be strict $\succ$.
3. $B \succ \phi$ for all intervals B.
4. If $B_1 \succ B_2 \succ \phi$, then there exist B_3, B_4 such that $B_3 \subset B_1, B_4 \subset B_1, B_2 \sim B_3 \sim B_4$, where B_3, B_1 is a lower s-pair and B_4, B_1 is an upper s-pair for any $s = 1, \ldots, l$. Here $\sim$ denoted the equivalence relation.
5. If $B_i \sim C_i$, then $\bigcup_{i=1}^{n} B_i \sim \bigcup_{i=1}^{n} C_i$.
6. $B^1 \times \cdots \times B^n \times B_\infty^{n+1} \times \cdots \times B_\infty^l \sim B^1 \times \cdots \times B^n$ here $B_\infty^l = (-\infty, \infty)$.
7. $B^1 \times \cdots \times B^n \sim B^{i_1} \times \cdots \times B^{i_n}$, $\{i_k\} = \{1, \ldots, n\}$

Condition 1 appears rather general and natural. Condition 2 can be regarded as independence from irrelevant events, conditions 3,4 mean a sort of continuity of relation $\succcurlyeq$, condition 5 defines the extension of the relation $\succcurlyeq$ from intervals to an algebra of intervals, conditions 6,7 define the consistency of the relation $\succcurlyeq$.

It was proved [4] that under conditions 1÷7 there exists a unique consistent family of l-dimensional probability density functions which correspond to the relation $\succcurlyeq$ in the sense that the event B_i is no less probable than the event B_j if and only if B_i is no less likely than B_j, namely

$$P(B_i) \geqslant P(B_j) \text{ if and only if } B_i \succcurlyeq B_j$$

Here the probability corresponds to the l-dimensional distribution function which defines the stochastic function $f(x)$. This means that any function f can be regarded as a sample path of some stochastic function.

The conditions were defined [5] when the methods which are optimal in the average sense (1) converge to a global minimum of any continuous function. The most restrictive conditions are

$$\lim_{N\to\infty} \sigma^2 = \begin{cases} 0, & \text{if } d \to 0 \\ \alpha > 0, & \text{if } d \to \beta > 0, \end{cases} \quad \lim_{N\to\infty} \mu = f(x)$$

Here is a conditional variation of $f(x)$, $d = \|x - x_i\|$, x_i is the point of the nearest observation and μ is a conditional expectation.

It was also proved [6], [7] that fairly reasonable assumptions of homogeneity, independence of the m-th differences and continuity of sample paths are satisfied when P corresponds to the Gaussian stochastic function with a constant mean μ_0 and the covariance S_{jk} between the values of function f at the points $x_j, x_{ij} \in [-1,1]^m$

$$S_{jk} = \sigma_0^2 \prod_{i=1}^{m} \left(1 - \frac{|x_j^i - x_k^i|}{2}\right) \tag{2}$$

where σ_0^2 is a variation. This stochastic function can be regarded as a generalization of the Wiener field. Actually it is a sum of 2^m Wiener fields with zero points on the vertexes of the hyper cube $[-1,1]^m$ plus the constant μ_0.

The parameters of this stochastic function (expectation μ_0 and variation σ_0^2 are usually unknown and ought to be evaluated on the base of available observations. It means some updating of an a priori measure using the observed results. It can be considered as some generalization of classical Bayesian techniques where the a priori distribution remained unchanged.

2. Standard one-step Bayesian models

The minimization of (1) is time consuming. Therefore one-step Bayesian methods have been introduced: each time a new point has to be identified the algorithm acts as if the new point is the last one. The formula for the one-step Bayesian approach is

$$x_{n+1} \in \arg\min_{x \in A} (1/\sigma) \int_{-\infty}^{\infty} \min(y,c) \exp(-1/2)((y-\mu)/\sigma)^2 dy \tag{3}$$

Here μ is the conditional expectation and σ is the conditional standard variation of $f(x)$ when the observed values are

$$y_1 = f(x_1), \ldots, y_n = f(x_n)$$

and

$$c = \min_{x \in A} \mu - \varepsilon \quad , \quad \varepsilon > 0 \tag{4}$$

Here ε approximately takes into account the influence of subsequent observations. When ε is large the method becomes a nearly uniform search. When ε approaches zero the method is strictly one-step.

Usually

$$\min_{x \in A} \mu = \min_{1 \le i \le n} \{f(x_i)\},$$

and

$$\varepsilon = \sigma_0 (\sqrt{2 \ln (N+1)} - \sqrt{2 \ln n} \quad .$$

Relation (3) can be expressed as

$$x_{n+1} \in \arg\max_{x \in A} (1/\sigma) \int_{-\infty}^{c} (c-y) \exp(-(1/2)((y-\mu)/\sigma)^2) dy \tag{5}$$

Denote

$$u = (y-\mu)/\sigma \tag{6}$$

and

$$a = (c-\mu)/\sigma \tag{7}$$

Parameter $a>0$ because $c<\mu$ what follows from (4). Substituting u and a from (6),(7) into (5) we can write

$$x_{n+1} \in \arg\max_{x\in A} \varphi(a) \tag{8}$$

where

$$\varphi(a) = \sigma\int_{-\infty}^{a} (a-u)\exp(-1/2)u^2\,du \tag{9}$$

The differential of φ

$$\frac{d\varphi}{da} = \frac{\partial\varphi}{\partial a} + \frac{\partial\varphi}{\partial\sigma}\frac{d\sigma}{da} \tag{10}$$

where

$$\frac{\partial\varphi}{\partial a} = \frac{\partial}{\partial a}\left(\sigma a\int_{-\infty}^{a}\exp((-1/2)u^2)\,du - \sigma\int_{-\infty}^{a} u\exp((-1/2)u^2)\,du = \right.$$

$$= \sigma\int_{-\infty}^{a}\exp((-1/2)u^2)\,du \geqslant 0, \quad \text{because} \quad \sigma \geqslant 0 \tag{11}$$

and

$$\frac{\partial\varphi}{\partial\sigma} = \int_{-\infty}^{a}(a-u)\exp((-1/2)u^2)\,du > 0, \text{ because } a-u>0 \text{ for all } u\in(-\infty,a) \tag{12}$$

and

$$\frac{d\sigma}{da} = \frac{d}{da}\left(\frac{c-\mu}{a}\right) = \frac{\sigma^2}{\mu-c}\left(1+\frac{\sigma}{(\mu-c)\sigma'/\mu'-\sigma}\right) \geqslant 0 \tag{13}$$

In the neighbourhood of any local maximum of (5) if N-n is large, here $\sigma' = d\sigma/dx$, $\mu' = d\mu/dx$

It follows from (10)(11)(12)(13) that

$$\frac{d\varphi}{da} \geqslant 0 \tag{14}$$

because

$$\frac{\partial\varphi}{\partial a} \geqslant 0, \quad \frac{\partial\varphi}{\partial\sigma} \geqslant 0 \quad \text{and} \quad \frac{d\sigma}{da} \geqslant 0 \tag{15}$$

Inequality (14) means that φ is an increasing function of a and since its maximum corresponds to the maximal value of a , if N-n is large,

$$x_{n+1} \in \arg\max_{x\in A} a = \arg\max_{x\in A}(-1/a) = \arg\max_{x\in A}\psi(x) \tag{16}$$

Here

$$\psi(x) = \sigma/(\mu-c) \tag{17}$$

The solution of (5)(16) happens to be rather time consuming in a multidimensional case. No more than 100-200 observations can be handled when the expressions for μ and σ correspond to the usual multidimensional distribution, a special case of which is the stochastic model (2).

It appears that no further substantial simplifications can be made if one wishes to satisfy the Kolmogorov consistency conditions. These consistency conditions provide that the same sample path of a stochastic function can be considered during the process of optimization. Some additional requirements [7] allow to consider only the continuous sample paths.

It is easy to notice that the consistency conditions make the stochastic model less adaptive. For example, updating the parameters μ_0 and σ_0 of a stochastic function on the base of the results of observations we are violating the consistency conditions. However, it usually improves the results of optimization because it adapts the stochastic model to the observed data.

Since by updating the unknown parameters we are violating the consistency conditions, any way it is reasonable to omit those conditions at all if doing so we will get some computational advantage.

3. Adaptive models.

There exist many ways to simplify the stochastic model (3) if the consistency and continuity conditions are dropped. Let us consider the situation when the conditions of consistency and continuity of sample paths are omitted but there remain the conditions of convergence[5] and continuity of the function $\psi(X)$ the maximum of which defines the next point of observation in accordance with (16).

The adaptive model will be defined in the following way:

$$f(x)=f_i(x);\ x\in A_i;\ \cup A_i=A;\ A_i\cap A_j=\psi;\ i\neq j;\ i,j=1,\dots,n \qquad (18)$$

where each A_i contains one observation x_i and $f_i(X)$ is Gaussian. Its conditional expectation μ_i is equal to the observed value y_i and its conditional variation is an increasing function Δ of the distance $d_i=\|x-x_i\|$ from the point of observation x_i, namely

$$\mu_i=y_i \quad \text{and} \quad \sigma_i^2=\Delta(d_i) \qquad (19)$$

It follows from (6)(8) that

$$\psi_i(x)=\sigma_i/(\mu_i-c) \qquad (20)$$

and

$$\psi(X)=\psi_i(X),\ X\in A_i \qquad (21)$$

Since the $\psi_i(X)$ are fixed by (20) the postulated continuity of

$\psi(x)$ can be provided only by the proper choise of A_i, namely, when

$$A_i=\{x: \psi_i(x) \le \psi_j(x),\ j=1,\dots,n\} \tag{22}$$

From (21)(22)

$$\psi(x)=\min_{1\le i\le n} \psi_i(x) \tag{23}$$

Hence and from (16)

$$x_{n+1} \in \arg\max_{x\in A} \min_{1\le i\le n} \psi_i(x) \tag{24}$$

or taking into consideration (20)

$$x_{n+1} \in \arg\max_{x\in A} \min_{1\le i\le n} \sigma_i/(\mu_i-c) \tag{25}$$

Most of the calculations so far were made using the conditional variation

$$\sigma_i^2=\sigma_0^2\|x-x_i\|^2 \tag{26}$$

The method (24)(25) satisfies the conditions given in [5] sufficient for the convergence of sequence(24) to a global minimum of any continuous function. It means that the use of the adaptive model(24), developed without the conditions of consistency and continuity of sample paths, shows the same asymptotic results as the standard one-step stochastic model (2)(3) which is consistent and continuous but more complicated. The results of calculations do not show any substantial differences between the standard and adaptive models so far.

4. A case of noisy observations

In some cases the exact values of $f(x)$ cannot be defined because of errors in calculations or physical experimentation. E.g., errors of calculations usually arise when $f(x)$ is obtained by numerical integration of some differential equations. The errors of physical experimentations often happen in the optimal experimental design.

The problems with errors can be naturally considered by the Bayesian approach. In such a case the formulas in (3) for conditional expectation and variation should be changed respectively.

5. The results of computer simulation

Different algorithms were investigated using the family of two-dimensional functions

$$f(x)=f(x^{(1)},x^{(2)})=\Big\{\Big[\sum_{i,j=1}^{J}(a_{ij}\sin(\pi i x^{(1)})\sin(\pi i x^{(2)})+ b_{ij}\cos(\pi i x^{(1)})\cos(\pi i x^{(2)}))\Big]^2+\Big[\sum_{i,j=1}^{J}(c_{ij}\sin(\pi i x^{(1)})\sin(\pi i x^{(2)})- d_{ij}\cos(\pi i x^{(1)})\cos(\pi i x^{(2)}))\Big]^2\Big\}^{1/2} \tag{27}$$

where $J=7$.

It is some approximation of the stress function in the elastic square plate under the cross section load[8].

50 sample paths corresponding to the random uniformly distributed parameters $a_{ij}, b_{ij}, c_{ij}, d_{ij} \in (0,1)$ were considered.

A local optimization was carried out using about 40 observations of f by the well known simplex method of Nelder and Mead from the best point of global optimization.

The relation between the percentage of successful cases (when the global minimum was found) and the total number of observations $N_t = N + N_L$ (where N_L is the number of observations for the local search) is represented in Table 1.

Index 1 corresponds to the adaptive Bayesian algorithm (24), index 2 to the standard one-step Bayesian algorithm (2)(3), index 3 to Strongin's algorithm[9] , index 4 to uniform random search, index 5 to uniform deterministic search, index 6 to the search procedure performed by the expert[10].

Table 1

The relation between the percentage of successful cases and the total number of observations

N_t \ Index	1	2	3	4	5	6
60						48
80	46	46		30	26	
90	60					
100		56		38		
105		62	56		44	
110						81
125	80	72				
135						92
140	88	86	68	44	68	
200			82	52		
240	96				84	
340					92	
370			94			
400			100	78	94	

The results of simulations of the adaptive Bayesian algorithm (24)(25) using the family (27) are even better than those of the standard one-step Bayesian methods (3). However, preliminary results of simulation using another set of functions in[11]are more favourable for the standard method (3).

Programming and calculations were made by J.Valevičiene using the package of FORTRAN programs for Bayesian optimization[12].

6. Possible applications

The standard one-step Bayesian methods were successfully applied [6] to a number of multiextremal problems of optimal design and the planning of experiments. However, the use of such methods was restricted by considerable amount of auxiliary computations. The adaptive Bayesian methods avoid this disadvantage.

The use of Bayesian methods is most natural in the case of "noisy" observations when the regression function to be minimized is multiextremal.

The noise is usually present when the values of f are obtained either by physical experimentation or computer simulation. The multiextremal functions f often occur when the relations between f and x are complicated and the domain A is large.

7. References

1. J.B.Mockus. Multiextremal problems of design. Nauka, Moscow, 1967 (in Russian).
2. J.B.Mockus. On Bayesian methods for seeking the extremum. Automation and Computers, No.3,1972, p.53 (in Russian).
3. J.B.Mockus. On Bayesian methods of optimization. Towards Global Optimization, edited by L.C.W.Dixon and G.P.Szegö, North-Holland, Amsterdam, 1975, p.166.
4. A.Žilinskas and A.Katkauskaite. Development of the stochastic models of complicated functions under uncertainty. Proceedings of the VIIth Conference on the theory of coding and information transmission. Moscow-Vilnius, 1978, p.70 (in Russian).
5. J.B.Mockus. Sufficient conditions for the convergence of the Bayesian methods to the global minimum of any continuous function. The Optimal Decision Theory, Vilnius, vol.4, 1978, p.67 (in Russian).
6. J.B.Mockus. On Bayesian methods for seeking the extremum and their applications. Information Processing 77, North-Holland, 1977, p.195.
7. A.Katkauskaite. Random fields with independent differences, Lithuanian Mathematical Transactions, Vilnius, XII, No.4, 1972, p.75 (in Russian).
8. V.A.Grishagin. Operative characteristics of some algorithms of global search. Problems of Random Search, Riga, vol.7, 1978, p.198 (in Russian).
9. R.G.Strongin. Numerical methods in multiextremal problems, Nau-

ka, Moscow, 1978 (in Russian).

10. V.R.Shaltenis. The analysis of problems in the interactive systems of optimization. Proceedings of the Conference on the Application of Random Search Methods in C.A.D., Tallin, 1979 (in Russian).
11. J.B.Mockus, V.Tiešis, A.Žilinskas. The application of Bayesian method for seeking the extremum. Towards Global Optimization 2, edited by L.C.W.Dixon and G.P.Szegö, North-Holland, Amsterdam, 1978, p.117.
12. V.A.Tiešis. The package for nonlinear programming. Proceedings of the Conference on Computers. Kaunas, 1979 (in Russian).

AN OPTIMIZATION MODEL FOR ENERGY SAVING IN THE HEATING OF BUILDINGS (*)

M.L. Nitti - M.G. Speranza
Istituto di Matematica - Università di Milano
via L. Cicognara, 7
20129 Milano (ITALY)

Abstract: The use of optimization techniques has been recently advocated in order to improve the energy performance of the design of buildings. The objective function in our model accounts both for the heating cost and the cost of insulation materials so that its minimization, with respect to a meaningful set of technological and architectural variables, yields a sequence of designs of decreasing "cost", converging to that design which ensures, for the weather conditions of the site of the buildings, the optimal balance between the cost of additional insulation and the related energy saving.

1. Introduction.

The increasing cost of energy has made imperative to investigate the thermal behaviour of buildings, in order to improve their structure and reduce both construction cost and future energy consumption.

The most accepted index of the energy performance of a building is the heating load q_j, defined as the heat quantity supplied by the heating plant at the time j to keep the inside air temperature between some prefixed values.

Obviously it is easy to reduce the energy consumption of the building investing much money in its structure; the problem is to find a balanced solution in some sense optimal between the cost of construction and the cost of operation. For this reason the "performance index" of our model accounts for both the heating cost and the cost of insulation materials and depends on the weather data of the site of the building and a number of technological and architectural parameters.

(*) This research has been developed in the framework of the "Progetto Finalizzato Energetica" of the Consiglio Nazionale delle Ricerche.

An accurate scanning of the feasible domain of such parameters to recognize effective designs is often too much burdensome in terms of computer time. Thus, in order to identify optimal building designs, it is expedient to select some parameters as control variables and to carry out on these variables the optimization of the "performance index". Optimization techniques for this purpose have been already developed by various authors [1], [4], [5].

The cost of the evaluation of the objective function is rather high, due to the many factors which have to be accounted for in a realistic model of buildings; in view of this fact an important feature of the optimization model is a mathematical procedure, outlined in 4., which gives the value of the gradient of the objective function.

This procedure results in a substantial reduction in computer time.

2. The structure of the optimization model.

In our model the building is assumed to be a uni-modular structure, from now on called "room", that is, we disregard the internal partition walls and we assume that the roof is always an even surface.

The energy consumption $E_{i,j}$ at the hour j of the day i depends on $q_{i,j}$ by the formula:

$$E_{i,j} = q_{i,j} / \rho_{i,j}$$

where $q_{i,j}$ is the heating load and $\rho_{i,j}$ are the values derived experimentally and subsequently tabulated of a nonlinear function $\rho(q(t))$, expressing the efficiency of the heating plant. The yearly consumption, on a prefixed year period of N days, which must coincide with the heating period, is given by the relation:

$$E = \sum_{i=1}^{N} \sum_{j=1}^{24} E_{i,j}$$

In order to compute $q_{i,j}$ to evaluate the contribution due to the heat conduction, the model calculates the "response factors" for every wall [6]; the contributions due to radiation and convection are accounted for by empirical relations. Then, the model calculates the outside temperatures and the outside heating loss of the walls at any hour ; the system

$$A * \underline{T}_j = \underline{B}_j$$

expresses the thermal balance conditions for every wall and, if required, for the air; A and $\underline{B}_j$ depend on the geometry of the room and the different heat contributions. In fact, being NS the number of surfaces considered in the building, the system is formed by NS equations when the air temperature has been fixed assuming the existence of an ideal thermostat keeping such a temperature always constant; the system is formed instead by NS+1 equations when to such a temperature has been assigned a variability range, assuming the heating plant to be switched on at its lower bound and switched off at its upper bound. The solution $\underline{T}_j$ of the system gives the inner temperature $T_{j,l}$ for every surface l and, if required, the inside air temperature TA_j.

Finally, we can calculate the heating load $q_{i,j}$ at the hour j of the day i:

$$q_{i,j} = \sum_{l=1}^{NS} A_l H_l \{T_{j,l} - TA_j\} + K_j$$

where A_l = area of the l-th surface,

H_l = inside surface convection heat transfer coefficient for A_l,

K_j = heat gain to domestic equipment, lights, occupants and loss due to air leakage.

In the experiments we have been performing we have usually assumed as control variables the ratio between the South length and the West length of the room, the ratio between the area of the South glass window and the area of the South wall, the same ratio for the North wall, the thickness of a structural layer in common among some walls (concrete layer or brick layer, for instance), the thickness of insulation layer in various external walls.

If C_r is the present construction cost, the cost of the building over M years, assumed is its life span, is:

$$CM = C_r \frac{M(1+\alpha)^M}{(1+\alpha)^M - 1}$$

where α is the index of yearly interest [7].

Assuming the yearly energy consumption E constant, the cost of M years of heating is

$$CH = E * C_0 * \frac{1 - \exp((M+1)*\varepsilon)}{1 - \exp \varepsilon}$$

where C_0 is the starting average cost of the energy and ε is the yearly increase percentage of the energetic cost [7].

The objective function takes into account both the heating cost CH and the construction cost CM of the building, obviously with regard to the structure components which have been assumed as variables, according to the following formula:

$$C = CM + CH$$

Then the optimization problem is minimizing C with regard to the variables X_i, i = 1,...,NV, selected among the admissible variables, with the constraints:

$$L_i \leq X_i \leq U_i \qquad i = 1,\dots,NV \qquad L_i, U_i \geq 0$$
$$X_4 + X_{4+i} \leq b_{4+i} \qquad i = 1,\dots,NV-4$$

where the variables from X_4 to X_{NV} represent the thickness of the walls. The method chosen to solve this nonlinear constrained optimization problem is the REQP (Recursive Equality Quadratic Programming) [2] using the program OPRQP of the OPTIMA package, developed at the Numerical Optimization Centre of the Hatfield Polytechnic.

3. The gradient of the objective function: observations.

In order to apply the REQP method the gradient of the objective function is required. The calculation of the gradient by a suitable finite-difference formula is affected by two sources of error: if the step h is too small, rounding errors can became a major influence in the computations, if h is too large, the difference could be a poor approximation to the gradient; in both cases a remarkable slowdown of convergence has been experienced [3]. Then we have the problem of the choice of an optimal step, indeed, in our case, since it is a matter of the derivation of a NV-variable function, of the choice of an optimal NV-dimensional vector.

The Stepleman-Winarsky algorithm [8] gives a method to choose such a step in the case of a one-variable function; however, such a method requires an average number of 20 function evaluations.

Extending this algorithm to our case, we should have on average 20*NV objective function evaluations, for the calculation of h, to which we should add 2*NV evaluations for the computation of the numerical gradient, what would be extremely burdensome from the point of view of computer time. To give an example, for NV=8 the calculation time of the objective function is about 4.5", then the total time for the calculation of the gradient in one point would be 13'12", using a computer UNIVAC 1108.

For these reasons we felt that the better accuracy and the reduction in time requirements allowed by the computation of the analytical gradient were well worthy the complex implementation of the procedure required to compute the analytical gradient by the theorem of the implicit function.

For NV=8 the calculation time of the analytical gradient has been about 47' on the same computer.

4. The gradient of the objective function: calculation.

The greater difficulty in the calculation of the gradient of the objective function is given by the term connected to the heating load.

For such a calculation we need the roots β_k and respective derivatives $\frac{\partial \beta_k}{\partial x_i}$ of a particular function named B(p), which is typical of a multi-lavel wall, defined by the relation:

$$\begin{bmatrix} A(p) & B(p) \\ C(p) & D(p) \end{bmatrix} = \prod_{i=1}^{N} \begin{bmatrix} \cosh\left(\sqrt{\frac{p}{\alpha_i}} \cdot x_i\right) & \frac{1}{\lambda_i}\sqrt{\frac{\alpha_i}{p}} \sinh\left(\sqrt{\frac{p}{\alpha_i}} . x_i\right) \\ \lambda_i \sqrt{\frac{p}{\alpha_i}} \sinh\left(\sqrt{\frac{p}{\alpha_i}} . x_i\right) & \cosh\left(\sqrt{\frac{p}{\alpha_i}} . x_i\right) \end{bmatrix}$$

where α_i = thermal diffusivity of the layer

λ_i = thermal conductivity

N = number of layers of the wall

x_i = thickness of the layer

The roots of B(p) can be computed numerically by an interative method. As far as the derivatives of the β_k functions are concerned, such functions are the solutions to the equation $B(p,x)=0$, and this defines implicitly:

$$\beta_k = \beta_k(x)$$

In view of the above definition of $\beta_k(x)$ the derivatives $\frac{\partial \beta_k}{\partial x_i}$ can be computed applying the local theorem for implicit functions, whose applicability can be easily checked.

5. A case - study.

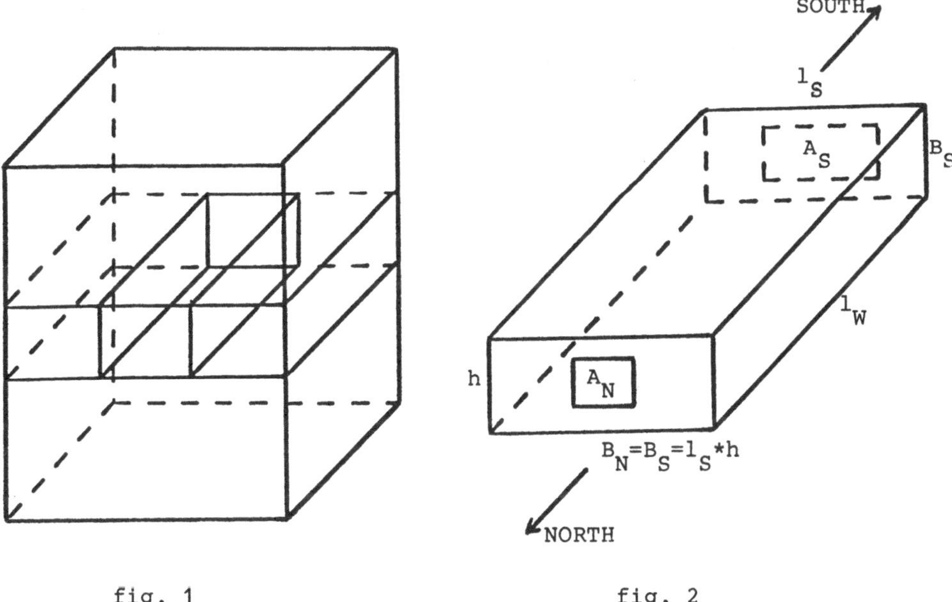

fig. 1 fig. 2

We report in this section the results of the optimization of one design of a room in which the following parameters has been selected as control variables (for the symbols, fig.2):

$w(1) = l_S / l_W$

$w(2) = A_S / B_S$

$w(3) = A_N / B_N$

w(4) = thickness of brick layer in North and South wall

w(5) = thickness of insulation layer in the South wall

w(6) = thickness of insulation layer in the North wall

The constraints are:

$0.20 \leq w(1) \leq 3.00$

$0.04 \leq w(2) \leq 0.80$

$0.04 \leq w(3) \leq 0.80$

$0.3937 \leq w(4) \leq 0.8202$ (ft)

$0.1312 \leq w(5) \leq 0.2297$ (ft)

$0.1312 \leq w(6) \leq 0.3281$ (ft)

The structure (number and thickness of layers) is fixed for the roof, the floor and the East and West walls; the North and South walls are the only external ones (fig. 1) and have four layers, two variables and two fixed.

Assuming a life time of a building of 20 years, being α the yearly interest of the money and ε the yearly percentage increase in the cost of energy, we studied three cases. For all cases the variables w(3), w(4), w(5) have been set by the optimization routine to their lower bound, and w(2) to its upper bound.

Case 1) $\alpha = 20$ $\varepsilon = 16$

The optimal values are:

$w^*(1) = 0.87$ $w^*(6) = 0.27$

Case 2) $\alpha = 20$ $\varepsilon = 19$

The optimal values are:

$w^*(1) = 0.81$ $w^*(6) =$ upper bound

Case 3) $\alpha = 10$ $\varepsilon = 16$

The optimal values are:

$w^*(1) = 0.72$ $w^*(6) =$ upper bound

We report some interesting computer times:

for function evaluation	4.5"
for analytical gradient evaluation	42"
for numerical gradient evaluation (fixed step)	54"
for optimization in case 3)	3'48"

In the same case 3) optimization required 5 function evaluations and 5 gradient evaluations.

References

[1] ARCHETTI,F. - VERCELLIS,C. *An Application of nonlinear programming techniques to the energy - economic optimization of building design* Proc. of the 9th IFIP, Warsaw 1979.

[2] BIGGS,M.C. *Constrained minimization using recursive equality quadratic programming* in "Numerical Methods for Nonlinear Optimization", Lootsma ed., Academic Press, 1971.

[3] DUMONTET,J. - VIGNES,J. *Détermination du pas optimal dans le calcul des dérivées sur ordinateur* RAIRO, Numerical Analysis, vol.11,1, 1977, p. 13-25.

[4] JUROVICS,S.A. *Optimization applied to the design of Energy Efficient Building* IBM Journal of Res. and Develop., vol.22,4, 1978.

[5] JUROVICS,S.A. *An investigation of the minimization of building energy load through optimization techniques* Los Angeles Scient. Center IBM Corporation - Los Angeles, Calif., 1979.

[6] KUSUDA,T. *Thermal Response Factors for multi-layer structures of various heat conduction systems* Paper N. 2108, ASHRAE Semi-annual Meeting, 1969.

[7] SILVESTRINI *Il clima come elemento di progetto* Liguori, Napoli, 1978.

[8] STEPLEMAN,R.S. - WINARSKY,N.D. *Adaptive numerical differentiation* Mathem. of comput., vol.33, n. 148, oct. 1979, p.1257-1264.

AN OPTIMAL DISPATCHING STRATEGY FOR VEHICLES IN A TRANSPORTATION SYSTEM

A. Schornagel

Mathematisch Centrum
Kruislaan 413
1098 SJ Amsterdam
The Netherlands

1. INTRODUCTION

We consider a transportation system with N vehicles which are either in a depot or on route. From the depot a vehicle takes passengers to various destinations and returns after a random trip time at the depot where it stays until departure for a next trip. The trip times are independent and identically distributed random variables. Passengers arrive at the depot according to a Poisson process with a known and constant arrival rate λ. However, we cannot observe the arrivals and consequently the number of passengers waiting at the depot is not known. If a vehicle is dispatched from the depot it picks up all the passengers waiting at that time at the depot. The waiting costs for the passengers are $c > 0$ per passenger per unit time. A fixed cost of $K \geq 0$ is incurred each time that a vehicle is dispatched from the depot. One can control the system by dispatching from time to time a vehicle from the depot. The objective is to find a dispatching strategy in order to minimize the long run average expected cost per unit time.

This problem was first considered by ASGHARZADEH and NEWELL [1] and OSUNA and NEWELL [4] for the case $c = 1$ and $K = 0$. If $K = 0$ it can easily be seen that the objective is equivalent to minimizing the average expected waiting time per customer. It was shown in [4] that if $\{H_i\}$ is a sequence of subsequent intervals between dispatch times under a stationary strategy, then the average wait per customer is

$$(1.1) \qquad E(W) = E(H^2)/2E(H)$$

where $E(H) = \lim_{n\to\infty} \Sigma_{i=1}^{n} H_i/n$ and $E(H^2) = \lim_{n\to\infty} \Sigma_{i=1}^{n} H_i^2/n$. This formula is independent of the distribution of the trip times. In both [1] and [4] one is concerned with the problem how to express E(W) in terms of a given dispatching strategy using (1.1). OSUNA and NEWELL [4] focus on the case of small N and give a detailed analysis for N = 1 and N = 2 whereas ASGHARZADEH and NEWELL [1] give most attention to the limiting case $N \to \infty$ where they approximate the number of vehicles in the depot by a continuum. For finite N and under the assumption that the trip times have an exponential distribution they give an algorithm to compute an optimal dispatching strategy.

Also in this paper we emphasize on the case of finite N and exponentially distributed trip times. Using Little's formula $L = \lambda W$ and (1.1) it can easily be shown that for general c and K the average expected costs are equal to

$$(1.2) \qquad \frac{K}{E(H)} + \frac{\lambda c\ E(H^2)}{2\ E(H)}$$

Thus the algorithm in [1] can also be applied to the case K > 0. However we shall derive a more efficient algorithm by modelling the dispatching problem as a continuously controlled Markov drift process.

In the following sections we show how the theory of generalized Markov decision processes [3] can be used. We propose an algorithm to compute an optimal dispatching strategy and give some results on the structure of an optimal dispatching strategy. We conclude with some numerical results.

2. THE MODEL FORMULATED AS A GENERALIZED MARKOV DECISION PROCESS.

For reasons of space, we refer to [3] for the theoretical background of the theory of generalized Markov decision processes and we use without further explanation the concepts and notations given in [3]. A more detailed analysis of the dispatching problem and proofs can be found in the technical report [6].

By assumption, trip times of the vehicles are independent random variables having a common exponential probability distribution function $F(t) = 1 - e^{-\mu t}$, $t \geq 0$. This implies that at any time the number of vehicles in the depot gives sufficient information to describe the situation with respect to the vehicles. Since it is assumed that the number of passengers waiting at the depot cannot be observed, we have as only information about the customers the time elapsed since the last dispatch. Hence at any time we can represent the state of the system by the pair (i,s) where i is the number of vehicles in the depot and s is the time elapsed since the last dispatch. For each stationary dispatching strategy the future behaviour of the system depends only on the present state. Since a vehicle in the depot may be dispatched at each time, we have a continuously controlled Markov drift process with state space $\{(i,s) \mid i = 0,1,\dots,N;\ s \geq 0\}$.

We now proceed to verify the elements 1 - 7 as described in [3]. The notations are the same as in [3] and [6]. The state space has already been defined and is denoted by $X = \{(i,s) \mid i = 0,1,\dots,N;\ s \geq 0\}$. The natural process and the feasible decisions in each state must be chosen in such a way that for each strategy the corresponding decision process is a result of the superposition of the natural process and the interventions prescribed by the strategy. We choose the natural process such that no vehicles will be dispatched in the natural process whereas an intervention is the decision to dispatch a vehicle. By a_0 we denote the null-decision not to disturb the natural process and by a_1 we denote an intervention. An intervention in state (i,s) causes an instantaneous transition to the state (i-1,0) (i = 1,2,...,N). Clearly the natural process eventually reaches the set $\{(N,s) \mid s \geq 0\}$ with probability 1.

In order to define the set A_0 of all states in which we always have to disturb the natural process (i.e. to dispatch a vehicle), we fix some sufficiently large number L. Then for each state (i,s) the set D(i,s) of feasible decisions is chosen as

$$D(i,s) = \begin{cases} \{a_0, a_1\} & i = 1,2,\ldots,N;\ 0 \le s < L \\ \{a_1\} & i = 1,2,\ldots,N;\ s \ge L \\ \{a_0\} & i = 0; \qquad s \ge 0. \end{cases}$$

Hence we always dispatch a vehicle when the time since the last dispatch reaches L and at least one vehicle is available and so $A_0 = \{(i,s) \mid i = 1,2,\ldots,N;\ s \ge L\}$.

Although other choices of the natural process and feasible decisions are possible, the above choice results in a very simple calculation of the k- and t-functions as defined in element 6 in section 2 of [3]. Since we have only one intervention in the model we write k(i,s) and t(i,s) instead of $k((i,s);a_1)$ and $t((i,s);a_1)$ respectively. From elementary analysis it follows that (see [6] for details)

$$\text{(2.1a)} \qquad k(i,s) = \begin{cases} K + \dfrac{c\lambda s^2}{2}, & i = 2,3,\ldots,N;\ 0 \le s < L \\ K + \dfrac{c\lambda s^2}{2} + \dfrac{c\lambda}{\mu} e^{-\mu L}(L + \dfrac{1}{\mu}), & i = 1; \qquad s \ge 0 \end{cases}$$

$$\text{(2.1b)} \qquad t(i,s) = \begin{cases} s, & i = 2,3,\ldots,N;\ 0 \le s < L \\ s + \dfrac{1}{\mu} e^{-\mu L}, & i = 1; \qquad s \ge 0. \end{cases}$$

As has been mentioned in [1] it is intuitively obvious that an optimal strategy belongs to the class of structured stationary strategies which can be described by an N-vector $z = (a_1, a_2, \ldots, a_N)$ such that a vehicle is dispatched in state (i,s) only if $s \ge a_i$ $(i = 1,2,\ldots,N)$. Such a strategy is called connected. Also it follows from intuition that an optimal connected dispatching strategy is monotone non-increasing, i.e. $a_1 \ge a_2 \ge \ldots \ge a_N$. Clearly $a_1 \le L$. Thus we consider the set of strategies given by $Z = \{z = (a_1, a_2, \ldots, a_N) \mid L \ge a_1 \ge a_2 \ge \ldots \ge a_N\}$. For a given strategy $z = (a_1, \ldots, a_N) \in Z$ the set of intervention states is given by $A_z = \{(i,s) \mid i = 1,2,\ldots,N;\ s \ge a_i\}$. It will be clear that L should be chosen large enough such that $a_1^* < L$ for an optimal strategy $z^* = (a_1^*, \ldots, a_N^*)$. However since A_z does not depend on L for $z \in Z$, it will appear that the computational effort in our algorithm is independent of L. For reasons of notation it appears efficient to define $a_0 = \infty$ and $a_{N+1} = 0$.

Figure 1 shows the state space and some elements described above. Also it shows two typical realizations of the decision process and the natural process. The figure is given at the end of this paper.

Now we are able to specify for each strategy $z \in Z$ the set of functional equations for the average cost g and the relative values $v(i,s)$ (see (8) in [3]). This set of functional equations is embedded on the set of intervention states A_z which contains a continuum of states in our model. From figure 1 it can be seen that the embedded Markov chain $\{I_n\}$ on A_z as defined in section 3 in [3] only assumes states in the set $\tilde{A}_z = \{(i,s) \mid 1 \le i \le N;\ a_i \le s < a_{i-1}\}$. However, in each intervention state (i,s), $a_i \le s < a_{i-1}$, the intervention causes an instantaneous transition to the same state $(i-1,0)$, $i = 1,2,\dots,N$. Hence the set of functional equations can be embedded on the finite set $\{(i,0) \mid i = 0,1,\dots,N\}$, cf. Remark 2 in [3]. If we denote by $S[(i,s);A]$ the first state in the set $A \subset X$ taken on by the natural process starting from state (i,s), then we have

$$\text{(2.2a)} \qquad v(i,0) = Ev(S[(i,0);A_z]), \qquad i = 0,1,\dots,N$$

$$\text{(2.2b)} \qquad v(i,s) = k(i,s) - gt(i,s) + v(i-1,0), \qquad i = 1,2,\dots,N;\ a_i \le s < a_{i-1}.$$

The relations (2.2) together with an additional normalizing equation, e.g. $v(i,0) = 0$ for some i, determine uniquely the $N+2$ unknowns g, $v(0,0), v(1,0),\dots,v(N,0)$ for each $z \in Z$ where $g = g_z$ is the average cost under strategy z. For each state (i,s) where z prescribes the null-decision we have

$$\text{(2.3)} \qquad v(i,s) = Ev(S[(i,s);A_z]), \qquad i = 0,1,\dots,N;\ 0 \le s < a_i.$$

To make (2.2a) tractable we shall now express $Ev(S[(i,0);A_z])$ in terms of the $v(j,0)$. From figure 1 it follows that starting outside A_z, there are two typical possibilities for the natural process to enter $\tilde{A}_z$; either at an arrival of a vehicle in the depot, or by reaching one of the boundary points (i,a_i). If the entrance in $\tilde{A}_z$ is not at an arrival, then, starting from that given state $(i,0)$, the first entrance state in $\tilde{A}_z$ is one of (j,a_j), $j \ge i$. Because of the non-increaseing character of the dispatching strategies the probability of the event $\{S[(i,0);A_z] = (j,a_j)\}$ is given by the probability that out of $N-i$ vehicles on route $j-i$ arrive in the depot during a period of length a_j. Since the residual trip time of each vehicle on route at any time has an exponential distribution with mean $1/\mu$, the above probability is from a binomial distribution:

$$\text{(2.4)} \qquad \Pr\{S[(i,0);\tilde{A}_z] = (j,a_j) \mid \text{no arrival in state } (j-1,a_j)\}$$

$$= P^{N-i}_{j-i}(a_j) = \binom{N-i}{j-i}(1 - e^{-\mu a_j})^{j-i} e^{-(N-j)\mu a_j},$$

$$i = 0,1,\dots,N;\ j = i,i+1,\dots,N.$$

If from state (i,0) the first entrance in $\tilde{A}_z$ is at an arrival then $S[(i,0);\tilde{A}_z] \in \{(j,s) \mid j = i+1,i+2,\dots,N;\ a_j \le s < a_{j-1}\}$. Now the possible events form a continuum and we find as probability density function

$$(2.5) \qquad f\{S[(i,0);\tilde{A}_z] = (j,s) \mid \text{arrival at state } (j-1,s)\}$$

$$= f^{N-i}_{(j-i)}(s) = (N-i)\binom{N-i-1}{N-j-1}(1-e^{-\mu s})^{N-j-1}e^{-(j-i+1)\mu s},$$

$$i = 0,1,\dots,N-1; \quad j = i+1,i+2,\dots,N; \quad a_j \le s < a_{j-1},$$

which is equal to the probability density of the j-ith order statistic out of a sample of N-i independent random variables with a common exponential distribution with mean $1/\mu$, ([2], page 21). From (2.4), (2.5) and the fact that for different $(j,s) \in \tilde{A}_z$ the events $\{S[(i,0);\tilde{A}_z] = (j,s)\}$ are mutually disjoint, it follows that

$$(2.6) \qquad Ev(S[(i,0);\tilde{A}_z] = \sum_{j=i}^{N} P^{N-i}_{j-i}(a_j)v(j,a_j) + \sum_{j=i+1}^{N} \int_{a_j}^{a_{j-1}} f^{N-i}_{(j-i)}(s)v(j,s)ds,$$

$$i = 0,1,\dots,N-1.$$

Substitution of (2.6) and (2.2b) in (2.2a) gives

$$(2.7a) \qquad v(i,0) = \sum_{j=i}^{N} q_{i,j}v(j-1,0) - a_i g + b_i, \qquad i = 1,2,\dots,N$$

$$(2.7b) \qquad v(0,0) = \sum_{j=1}^{N} q_{0,j}v(j-1,0) - a_0 g + b_0, \qquad i = 0$$

where

$$q_{i,j} = P^{N-i}_{j-i}(a_j) + \int_{a_j}^{a_{j-1}} f^{N-i}_{(j-i)}(s)ds$$

$$a_i = \sum_{j=i}^{N} P^{N-i}_{j-i}(a_j)t(j,a_j) + \sum_{j=i+1}^{N} \int_{a_j}^{a_{j-1}} f^{N-i}_{(j-i)}(s)t(j,s)ds$$

$$b_i = \sum_{j=i}^{N} P^{N-i}_{j-i}(a_j)k(j,a_j) + \sum_{j=i+1}^{N} \int_{a_j}^{a_{j-1}} f^{N-i}_{(j-i)}(s)k(j,s)ds$$

For i = 0 and i = N a_i and b_i need obvious modifications. Observe the simple form of the system (2.7) which can be solved recursively by a proper normalization (see [6]).

By (2.7) we are able to compute the average cost g_z for each strategy $z \in Z$. We now show how a strategy can be improved. The following is based on the modified policy iteration algorithm as described in [3], section 5. This algorithm contains two procedures to improve a strategy. We first discuss the so-called policy improvement operations in which the current strategy is improved by replacing null-decisions by interventions. To describe this operation we consider a fixed strategy $z = (a_1,\dots,a_N) \in Z$ and a bounded solution $\{g_z, v_z(0,0),\dots,v_z(N,0)\}$ to (2.7). In [6] it is pointed out that the policy improvement operation can be performed for each set $\{(i,s) \mid s \geq 0\}$ separately, $i = 1,2,\dots,N$. Therefore we fix some j, $1 \leq j \leq N$. Since an improved strategy must be non-increasing the only states where the null-decision may be replaced by an intervention are in the set $\{(j,s) \mid a_{j+1} \leq s < a_j\}$. Moreover, from the fact that each strategy must be connected it follows that adding an intervention in state (j,u) is only feasible if for all states (j,s) with $s > u$ an intervention is prescribed. Now suppose that by applying the policy improvement operation an intervention is added to the states (j,s) with $u < s < a_j$. Let $v_z(a_1.(i,s)) = k(j,s) - g_z t(j,s) + v_z(j-1,0)$ for $s \geq 0$ and let $\tilde{v}_z(i,s) = v_z(i,s)$ for all states $(i,s) \in X$ except for the states (j,s) with $u < s < a_j$ where $\tilde{v}_z(j,s) = v_z(a_1.(j,s))$. Now, for small $\Delta > 0$, we compare $v_z(j,u-\Delta)$ with the expected $\tilde{v}_z$-value of the first state in $A_z \cup \{(j,s) \mid s > u\}$ taken on by the natural process starting from state $(j,u-\Delta)$ if strategy z is used. For small Δ the expected $\tilde{v}_z$-value can be written as

$$(2.8) \qquad (N-j)\mu\Delta\tilde{v}_z(j+1,u) + (1-(N-i)\mu\Delta)\tilde{v}_z(i,u) + o(\Delta)$$

$$= (N-j)\mu\Delta\{k(j+1,u) - g_z t(j+1,u) + v_z(j,0)\} +$$

$$+ (1-(N-j)\mu\Delta)\{k(j,u) - g_z t(j,u) + v_z(j-1,0)\} + o(\Delta),$$

using that the residual trip times of the N-j vehicles on route are exponentially distributed with mean $1/\mu$. By substituting (2.1) in (2.8), subtracting (2.8) from $v_z(a_1.(j,u))$, dividing the result by Δ and letting $\Delta \to 0$ we find that assigning an intervention to the state (j,u) improves the strategy if

$$(2.9) \qquad \lambda u - g_z - (N-j)\mu\{v_z(j-1,0) - v_z(j,0)\} \geq 0.$$

The zero of the left-hand side of (2.9) is equal to

$$(2.10) \qquad a_j^0 = \frac{g_z + (N-j)\mu\{v_z(j-1,0) - v_z(j,0)\}}{\lambda}.$$

Further we define

$$(2.11) \qquad a_j^! = \min\{\max\{a_{j+1}, a_j^0\}, a_j\}.$$

Then it follows from the fact that the left side of (2.9) is increasing in u, that each strategy $(a_1,\dots,a_{j-1},w,a_{j+1},\dots,a_N)$ with $a_j^! \leq w \leq a_j$ is a better strategy than z. It appears to be reasonable (see [6]) to take $w = a_j^!$. Without solving (2.7) with

$z' = (a_1,\dots,a_{j-1},a_j',a_{j+1},\dots,a_N)$ one may improve z' as above with z replaced by z' and using the solution to (2.7) for the previous strategy z. This can be repeated for each i, $1 \le i \le N$.

In the above described policy improvement operation only null-decisions may be replaced by interventions. We shall now discuss the cutting operation in which interventions may be replaced by null-decisions. Again, let $z \in Z$ be some fixed strategy and let $\{g_z, v_z(0,0),\dots,v_z(N,0)\}$ be a bounded solution to (2.7) for this strategy. It turns out (see [6]) that also the cutting operation can be performed for each set $\{(i,s) \mid s \ge 0\}$ separately, $i = 1,2,\dots,N$. Therefore we fix some j $(1 \le j \le N)$ and show how the cutting operation works out for the set $\{(j,s) \mid s \ge 0\}$. For the same reason as in the policy improvement operation the set of states for which the intervention is replaced by the null-decision must be of the form $\{(j,s) \mid a_j \le s \le u\}$ where u is some number with $a_j \le u < a_{j-1}$. Let us denote strategy $f_u = (a_1,\dots,a_{j-1},u, a_{j+1},\dots,a_N)$ with $a_j \le u < a_{j-1}$. Consider now a stopping problem for the natural process where a cost $v_z(i,w)$ is incurred when the natural process is stopped in state (i,w). Then strategy f_u is better than the current strategy z if for this stopping problem the set A_{f_u} is as stopping set at least as good as the set A_z for each initial state (j,s) with $s \ge a_j$, cf. Theorem 4 in [3]. For this stopping problem it can be shown [6] that the infinitesimal look ahead rule as described in [5] can be used. In our model this means that for some small $\Delta \ge 0$ and u with $a_j \le u < a_{j-1}$, the value $v_z(j,u-\Delta)$ is compared with the expected v-value of the state taken on by the natural process after a time Δ when it starts in $(j,u-\Delta)$. This expected v-value is equal to (2.8) with $\tilde{v}$ replaced by v. Thus, using the same analysis as in the policy improvement operation, f_u is a better strategy than z if

$$(2.12) \qquad \lambda u - g_z - (N-j)\mu\{v_z(j-1,0) - v_z(j,0)\} \le 0.$$

With a_j^0 as in (2.10) we define

$$(2.13) \qquad a_j'' = \max\{\min\{a_{j-1},a_j^0\},a_j\}.$$

Because the left side of (2.12) is increasing in u, each strategy $(a_1,\dots,a_{j-1},w, a_{j+1},\dots,a_N)$ with $a_j \le w < a_j''$ is a better strategy than z. Again it appears to be reasonable to take $w = a_j''$. As with the policy improvement operation, the cutting operation may be repeatedly applied for each i $(1 \le i \le N)$ without solving (2.7) each time.

3. A POLICY ITERATION ALGORITHM

With the above described techniques to improve a strategy, several algorithms to compute an optimal non-increasing dispatching strategy can be composed (see [6]). However the formulas (2.10), (2.11) and (2.13) suggest the following algorithm where the policy improvement operation and the cutting operation are combined to a so-called generalized policy improvement operation. Let $z_k = (a_1^k,\dots,a_N^k)$ be the strategy obtained at the end of the $k-1^{th}$ iteration, then the kth iteration is as follows:

Step (1) *Value determination operation.* Determine a bounded solution $\{g_{z_k}, v_{z_k}(0,0),\dots,v_{z_k}(N,0)\}$ to (2.7) with $z = z_k$.

Step (2) *Generalized policy improvement operation.* Take $z_{k+1} = (a_1^{k+1},\dots,a_N^{k+1})$ according to

$$(3.1) \qquad a_i^{k+1} = \frac{g_{z_k} + (N-i)\mu\{v_{z_k}(i-1,0) - v_{z_k}(i,0)\}}{\lambda}, \qquad i = 1,2,\dots,N.$$

The choice of a_i^{k+1} in (3.1) is in fact a combined policy improvement operation and cutting operation as long as a_i^{k+1} is non-increasing in i. In [6] the following conjecture is partially proved.

CONJECTURE. *If the non-increasing strategy* $z_1 = (a_1,\dots,a_N)$ *is given by* $a_i = [g_{z_0} + (N-i)\mu\{v_{z_0}(i-1,0) - v_{z_0}(i,0)\}]/\lambda$ *for some non-increasing strategy* $z_0 \in Z$, *then any bounded solution* $\{g_{z_1}, v_{z_1}(0,0),\dots,v_{z_1}(N,0)\}$ *of* (2.7) *with* $z = z_1$ *has the property that* $v_{z_1}(i,0)$ *is convex and decreasing in* i.

It is easy to see from (3.1) that if $v_{z_k}(i)$ is convex decreasing in i, then a_i^{k+1} is decreasing in i.

In [6] we prove the following:

LEMMA 1. *For each strategy* $z_1 = (b,\dots,b)$ *with* $b > 0$ *each bounded solution* $\{g_{z_1}, v_{z_1}(0,0),\dots,v_{z_1}(N,0)\}$ *to* (2.7) *with* $z = z_1$ *is such that* $v_{z_1}(i,0)$ *is convex and decreasing in* i.

If we start the above algorithm with $z_0 = (b,\dots,b)$ with $b > 0$ then it follows with Lemma 1 and the conjecture that each z_k in the algorithm is non-increasing. This means that in each iteration of the algorithm step (2) can indeed be seen as a combination of the policy improvement operation and the cutting operation. Thus it follows from the theory in [3] that $g_{z_{k+1}} \le g_{z_k}$ for all k and moreover that the algorithm converges to an optimal dispatching strategy. With $i = N$ in (3.1) we have $a_N^{k+1} = g_{z_k}/\lambda$ and consequently a_N^{k+1} is non-increasing in k. This is not generally true for a_i^k $(1 \le i < N)$ as appeared from our numerical experiments.

For an optimal strategy $z^* = (a_1^*,\dots,a_N^*)$ application of the policy improvement operation and the cutting operation will not lead to a different strategy. Hence

$$(3.2) \qquad a_i^* = \frac{g_{z^*} + (N-i)\mu\{v_{z^*}(i-1,0) - v_{z^*}(i,0)\}}{\lambda}, \qquad i = 1,2,\dots,N.$$

With $i = N$ in (3.2) we have $a_N^* = g_{z^*}/\lambda$. This relation for the optimal dispatching strategy was also found in [1] and [4] in a completely different way.

4. NUMERICAL RESULTS

For our numerical experiments we used the algorithm as described in Section 3. In each iteration the integrals in the coefficients a_i and b_i were computed by numerical procedures. For $N \le 50$ and a required accuracy of 0.001 percent in the average cost and the numbers a_i, the algorithm needed between 3 and 5 iterations. The computation time never exceeded 1 minute on a CDC 170 computer.

In Table 1 the parameters are the same as in Table 1 in [1]. The results in [1] are shown in the last column of Table 1. The Tables 2 and 3 show the effects on different K. Observe that in all examples a_N and g decrease with each iteration. This is not necessarily true for a_i with $i < N$, e.g. a_2, a_3 and a_4 in Table 3. Further, cf. (3.2), $a_N^* = g^*/\lambda$.

	Iteration number			
	0	1	2	AαN [1]
a_2	1.00000	0.83997	0.83521	0.838
a_1	1.00000	1.12023	1.11923	1.125
g=E(W)	0.83997	0.83521	0.83521	0.835

TABLE 1. $N=2$, $\lambda=1$, $\mu=1$, $c=1$, $K=0$

	Iteration number			
	0	1	2	3
a_5	1.00000	0.75737	0.73762	0.73737
a_4	1.00000	0.82024	0.80104	0.80078
a_3	1.00000	0.91711	0.89176	0.89153
a_2	1.00000	1.07863	1.03993	1.03997
a_1	1.00000	1.37644	1.36198	1.36211
g=E(W)	1.54474	1.47525	1.47473	1.47473

TABLE 2. $N=5$, $\lambda=2$, $\mu=1$, $c=1$, $K=0$

	Iteration number			
	0	1	2	3
a_5	2.00000	1.26006	1.11666	1.11547
a_4	2.00000	1.26607	1.14564	1.14610
a_3	2.00000	1.28089	1.19557	1.19704
a_2	2.00000	1.32870	1.29439	1.29584
a_1	2.00000	1.54521	1.55437	1.55434
g	2.52013	2.23331	2.23095	2.23094

TABLE 3. $N=5$, $\lambda=2$, $\mu=1$, $c=1$, $K=1$

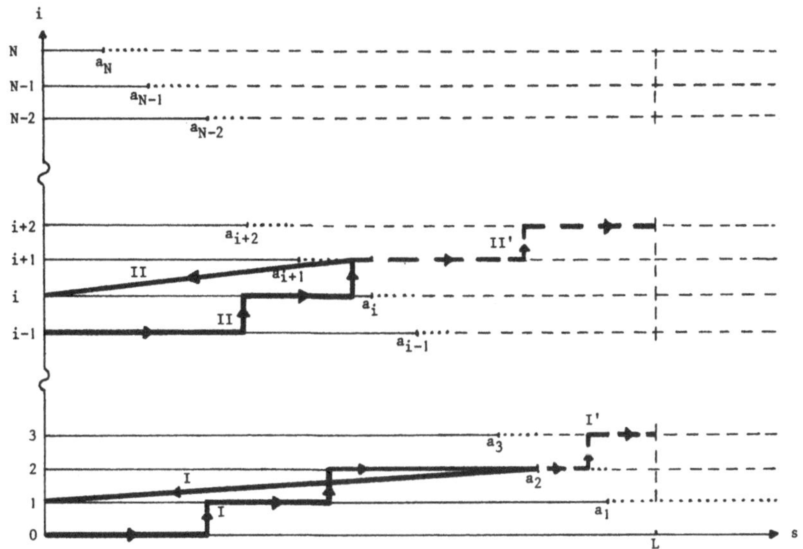

FIGURE 1. The state space. The light solid lines $[0,a_i)$ are the set of states where no interventions are prescribed by the strategy $\underset{\sim}{z}$, i.e. no vehicles are dispatched. The dotted lines $[a_i,a_{i-1})$ form the set $\tilde{A}_z$, i.e. the set of states, reachable by the decision process where an intervention is prescribed. The dotted lines together with the light dashed lines $[a_i,\infty)$ form the set of intervention states A_z. The set A_0 consists of the light dashed lines $[L,\infty)$. Two possible realizations of the decision process I and II are drawn in dark solid lines. In I a vehicle is dispatched in state $(2,a_2)$ and in II a vehicle is dispatched at an arrival of a vehicle in the depot. The two dark dashed lines I' and II' are realizations of the natural process until this reaches A_0.

ACKNOWLEDGEMENT

I wish to thank Prof. Henk Tijms for many fruitful discussions and for his helpful comments on the original manuscript of this paper.

REFERENCES

[1] K. ASGHARZADEH, G.F. NEWELL (1978), "Optimal Dispatching Strategies for Vehicles Having Exponential Distributed Trip Times", *Naval Res. Log. Quart.* 25, 489-509.

[2] W. FELLER (1966), *An Introduction to Probability Theory and its Applications*, Vol. II, Wiley, New York.

[3] G. DE LEVE, A. FEDERGRUEN, H.C. TIJMS (1977), "A General Markov Decision Method I: Model and Techniques", *Adv. Appl. Prob.* 9, 296-315.

[4] E.E. OSUNA, G.F. NEWELL (1972), "Control Strategies for an Idealized Public Transportation System", *Transp. Science* 6, 52-72.

[5] S.M. ROSS (1971), "Infinitesimal Look-Ahead Stopping Rules", *Ann. Math. Statist.* 42, 297-303.

[6] A. SCHORNAGEL (1982), *"An Optimal Dispatching Strategy for Vehicles in a Transportation System"*, Technical Report, Mathematisch Centrum (in preparation).

ASYNCHRONOUS PARALLEL SEARCH IN GLOBAL OPTIMIZATION PROBLEMS

F. Archetti - F. Schoen
Istituto di Matematica - Università di Milano
via L. Cicognara, 7
20129 Milano (ITALY)

Abstract: A class of asynchronous parallel search methods is proposed in this paper in order to solve the global optimization problem on a multiprocessor system, consisting of several processors which can communicate through a set of global variables contained in a memory shared by all processors. The speed-up ratio and memory contention effects are experimentally analyzed for some algorithms of this class.

1. Introduction.

Many approaches have been recently suggested in order to solve the global optimization problem (Archetti & Szego (1980)). A main distinction can be drawn between probabilistic techniques, which perform a statistical inference about the global optimum, and deterministic ones, mainly represented by "space covering" techniques, which can be regarded as a way of searching implicitly the feasible set. These latter techniques are usually more costly than probabilistic ones but can provide, under wide conditions, an approximation of prefixed accuracy to the global optimum.

Anyway, the global optimization also of not very complex problems is still rather costly, and none of the methods as yet suggested has established itself as a computational tool of general use.

In this situation it is only too natural for the designer of global optimization algorithms to cast an interested look to the latest developments in computer technology, in the hope that new system architectures and the related computational paradigms can provide new and more effective solutions.

In this paper we propose a class of parallel search methods derived in the framework of the "space covering" approach, bearing in mind Multiple Instruction Multiple Data (M.I.M.D.) machines. Our computer model consists in K independent processors P_j, $j=1,\ldots,K$, each with its own CPU and private memory, which communicate through a set of global variables contained in a shared memory to which all processors are connected.

How such a multiprocessor architecture can be exploited in parallel processing is a matter of increasing research, about the operating systems, the programming languages and, more generally, the analysis of concurrent processes.

In this paper we shall not be concerned with these aspects, even if it is clear that their development will have a major influence on the very field of parallel algorithms, but only with investigating whether the simultaneous availability of several processors could result in an increased flexibility in the design of algorithms and in speeding up the computations.

2. Strategies, coverings and optimality criteria.

In this paper we are concerned with the problem of finding $f^* = \underset{x \in B}{\mathrm{Max}}\, f(x)$ where B is a compact in R^N and $f \in L = \{ \phi : B \to R^N : \forall x,y \in B \; | \phi(x) - \phi(y) | \leq k\rho(x,y) \}$ where ρ is some continuous distance function and k is a known constant.

Definition 1:

Let n be a positive integer. A n-point strategy S_n on L is an n-tuple of mappings $y_i : L x B^{i-1} \longrightarrow B$, i=1,...,n such that

$$\forall f \in L, \quad \forall (x_1, x_2, \ldots, x_{i-1}) \in B^{i-1}$$

(1) $$x_i = y_i(x_1, f(x_1), x_2, f(x_2), \ldots, x_{i-1}, f(x_{i-1})) \in B$$

In the following we shall use the symbol S_n to denote the class of n-point strategies. We shall also use the term "passive" or "a priori" to indicate the constant strategies in L.

Definition 2:

The accuracy of a strategy $S_n \in S_n$ is the quantity:

(2) $$A(S_n;f) = f^* - \underset{i=1,n}{\mathrm{Max}}\, f(x_i)$$

The accuracy over the whole class L can be characterized as follows:

Definition 3:

The guaranted accuracy of a strategy $S_n \varepsilon \mathcal{S}_n$ on the class L is the quantity:

(3) $$A(S_n) = \underset{\phi \varepsilon L}{\text{Sup}} \{ \phi^* - \underset{i=1,n}{\text{Max}} \phi(x_i) \}$$

The following optimality criteria can be given:

Definition 4:

Let $\varepsilon > 0$; a strategy S_{n^*} such that $A(S_{n^*}) \leq \varepsilon$ is said to be n-optimal if:

(4) $$n^* = \min \{ n: \exists\, S_n \varepsilon \mathcal{S}_n,\ A(S_n) \leq \varepsilon \}$$

Sub-classes of the class of optimal strategies have been defined in Sukharev (1972) in order to exploit the sequential nature of the scheme (1).

Let "choice" functions $C_m: B^m \longrightarrow B$ be given such that:

$C(z_1, z_2, \ldots, z_m) = z_{m_1}$, where $z_j \varepsilon B$, $j=1,m$; $m_1 \varepsilon \{1,2,\ldots,m\}$ and let

$L_{|i}(f) = \{ \phi \varepsilon L : \phi(x_j) = f(x_j),\ j=1,i \}$, where $f \varepsilon L$

Definition 5:

An n-optimal strategy $S_{\bar{n}}$ is said to be best-n-optimal in L if $\forall f \varepsilon L$

$$S_{\bar{n}} = (y_1, y_2, \ldots, y_{\bar{n}}) \quad \text{where:} \quad y_{i+1} = C_{n_i} \circ P_{(m_i|i)} \quad i=0,\bar{n}-1$$

$n_i = m_i - i$; $P_{(m_i|i)}$ is an n-optimal a-priori strategy in $L_{|i}(f)$.

The following theorem, whose proof can be found in Sukharev(1972), relates optimal strategies to optimal coverings:

Theorem 1:

Let $x_{i+1}, \ldots, x_m \varepsilon B$ be such that $\forall x'_{i+1}, \ldots, x'_{m-1} \varepsilon B$

$$\{ \bigcup_{j=1,i} S(x_j; (f_i^* - f_j + \varepsilon)/k \quad \bigcup_{j=i+1,m} S(x_j; \varepsilon/k) \} \supset B, \quad \text{while}$$

$$\{ \bigcup_{j=1,i} S(x_j; (f_i^* - f_j + \varepsilon)/k \quad \bigcup_{j=i+1,m-1} S(x'_j; \varepsilon/k) \} \not\supset B \quad \text{where:}$$

$f_j = f(x_j)$; $f_i^* = \underset{j=1,i}{\text{Max}}\, f_j$; $S(c;r) = \{x \varepsilon B : \rho(x,c) \leq r\}$

Then the points $x_{i+1},\dots,x_m$ constitute a passive n-optimal strategy in $L_{|i}(f)$.

Finding a best-optimal strategy can be very costly if $N > 2$. A best-n-optimal 1-dimensional scheme is given in Schoen(1981).

Now we introduce an optimal scheme of implicit search which can be shown to be an approximation to a best-n-optimal strategy and lends itself naturally to the design of a class of parallel algorithms which will be presented in section 3.

Let S_n be an n-optimal a-priori strategy in B guaranteeing in L a prefixed accuracy ε.

Once the objective function has been computed in $x_1,x_2,\dots,x_i$, the implicit search choses the next evaluation point x_{i+1} in the discrete set:

(6) $$D_i = \{ x_l \varepsilon S_n : \forall j=1,i \quad \rho(x_l,x_j) > (f_i^* - f_j)/k \}$$

where an improvement over f_i^* could be observed.

A sensible criterion for chosing x_{i+1} in D_i is given by the following rules:

We consider in S_n a family of total orderings defined as follows: let $\delta=(d_1,d_2,\dots,d_N)$ be a permutation of the first N integers and let $S \varepsilon\{-1;1\}^N$ Then $x_l \prec x_m$ if $\exists c: 1 \le c \le N$ such that:

$$x_l^{(d_j)} = x_m^{(d_j)} \qquad 1 \le j < c$$

$$(x_l^{(c)} - x_m^{(c)}) S_c < 0$$

We also denote by $\prec$min and $\prec$Max the min and Max operators with respect to the ordering $\prec$.

The point x_{i+1} is then chosen as

(7) $$x_{i+1} = \prec\min \{ x_l : x_l \varepsilon D_i \}$$

This choice can be implemented by the following procedure SEARCH:

```
x̂:= ≺min { x_l ε S_n }; x̂_n:= ≺Max { x_l ε S_n }
while x̂ ≼ x̂_n do
begin
    if x̂ ε D_i then x_{i+1}:= x̂; i:= i+1
               else find l̃: ρ(x̂,x_l̃) ≤ ( f*_i - f(x_l̃))/k
```

$\tilde{x} := \arg \langle \min\{x_l \varepsilon S_n : x_l \rangle \tilde{x}, \rho(x_l, x_{\tilde{l}}) > (f_i^* - f(x_{\tilde{l}}))/k\}$

end

end

It is worth remarking that also in this scheme the cardinality of D_{i+1} can usually be more quickly reduced choosing, in the earlier stages, x_{i+1} at random in D_{i+1}. Therefore a more effective implementation of SEARCH performs the first choices of x_{i+1} from a discrete uniform distribution in D_{i+1}.

This search scheme can be regarded as an approximation to a best-n-optimal one as it is shown below:

After the i-th step, a best-n-optimal strategy would be obtained choosing a centre of an n-optimal covering of the set:

$$U_i = \{ x \varepsilon B: \forall l: 1 \le l \le i \quad \rho(x, x_l) > (f_i^* - f(x_l) + \varepsilon)/k \}$$

with radius ε/k.

The above search scheme instead choses one of the centres of an n-optimal covering of the set:

$$\tilde{U}_i = \{ \bigcup_{j \varepsilon J} S(x_j; \varepsilon/k) \}^C \qquad \text{where:}$$

$$J = \{ j: x_j \varepsilon S_n, \forall l: 1 \le l \le i \quad \rho(x_j, x_l) \le (f_i^* - f(x_l))/k \}$$

Clearly, the points in $S_n \cap \tilde{U}_i$ are the centres of an n-optimal covering of $\tilde{U}_i$, due to the optimality of S_n.

The sequence generated by the implicit search is only an approximation to a best-n-optimal one because $\tilde{U}_i \supseteq U_i$. Indeed, if $x \notin \tilde{U}_i$ then there exists $j \varepsilon J$ such that $\rho(x, x_j) \le \varepsilon/k$ and, by definition of J, there exists $l \le i$ such that $\rho(x_l, x_j) \le (f_i^* - f(x_l))/k$ from which follows $x \notin U_i$.

3. A class of asynchronous parallel algorithms.

For a thoroughful discussion of the concept of an asynchronous parallel algorithm as it is meant in this paper, we feel more appropriate to address the reader to Baudet(1978) and Kung(1976).

A major step in the design of parallel algorithms is establishing the level of parallelism: in this paper we choose the highest possible level - also referred to as the "holistic" approach in Mc Keown (1980). After this approach, the whole set of data and the whole algorithm are entrusted to each processor and inherent to each of the computational tasks into which the execution of the algorithm is decomposed.

Thus the implicit search method outlined in section 2 can be exploited to design different parallel algorithms which differ by (i) the choice of the global variables, (ii) the processor-memory communication and (iii) the policy of scheduling the computational tasks.

Five asynchronous holistic parallel algorithms of this class are now outlined.

A1) (i) a scalar variable F^*.

(ii) each processor sets $\min(f_i^*;F^*)$ to $\mathrm{Max}(f_i^*;F^*)$

(iii) all processors perform SEARCH with the same δ and S

A2) (i) F^* and a vector $X^* \varepsilon R^N$ such that $f(X^*)=F^*$.

(ii) as in A1 for F^*; $(x_i^*;X^*)$ are consequently updated

(iii) as in A1

A3) (i) F^*,X^* and a vector $\tilde{X} \varepsilon R^N$.

(ii) as in A2 for $(X^*;F^*)$; $\measuredangle\min(\hat{x};\tilde{X}) := \measuredangle\mathrm{Max}(\hat{x};\tilde{X})$

(iii) as in A1

A4) (i) F^*,X^*.

(ii) as in A2

(iii) each processor performs SEARCH with its own choice of and S; thus the processors "sweep" in different directions

A5) (i) F^*,X^* and a vector γ in R^K.

(ii) as in A2 for (X^*, F^*). The components of γ, together with δ and S, define that part of B still to be searched: they are used to exclude from the search the half-space $\{ x: (x^{(d_N)} - \gamma_N)\, S^{(N)} \geq 0 \}$ implicitly exhaustively searched.

4. Experimental results.

The computer program ASPASIA, acronym for Asynchronous Parallel

Simulation of Algorithms, has been written in order to simulate, on a serial machine, the parallel execution of an algorithm.

The following class of test functions was used in the simulation:

$$f(x) = \underset{i=1,m}{\text{Max}} \ (\rho\ (x,x_i)\ \alpha_i + z_i) \text{ where } \rho\ (x,x_i) = \underset{j=1,N}{\text{Max}} \ | x^{(j)} - x_i^{(j)} |,$$

B is the unit N-dimensional hypercube, x_i, α_i, z_i are uniformly distributed in B, $[-L,0]$, $(0,F)$, where $L > k$ and $F > 0$.

The computations reported in this section are for N=2, k=70, L=100 and F=100.

The simulated clock of each processor has been set to the CPU time of a UNIVAC 1100/80, the machine used in the simulation.

The diagram plotted in the following figure depicts the speed-up $\bar{\tau} = \bar{\tau}_1 / \bar{\tau}_K$, where $\bar{\tau}_1$ and $\bar{\tau}_K$ are respectively the average time required by 1 and by K processors, computed over 20 runs.

The symbols are associated to the algorithms in the following way:
A1 ⟷ △ A2 ⟷ + A3 ⟷ ✱ A4 ⟷ ⊞ A5 ⟷ ○

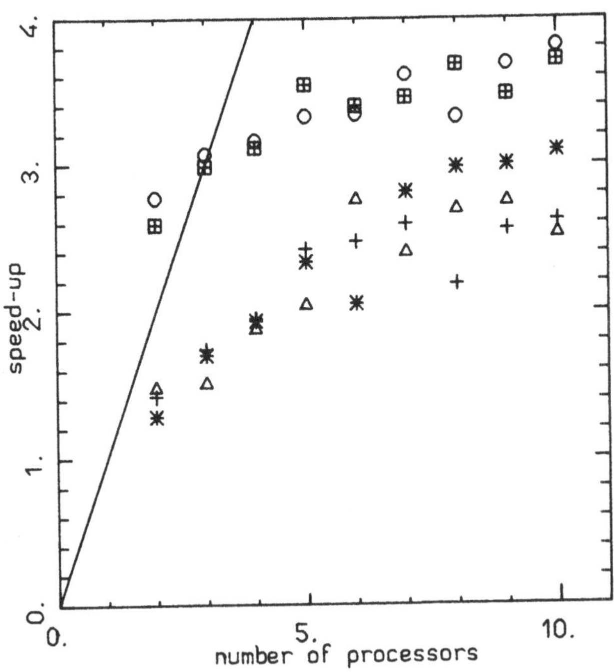

The sequential search expressed by (7) is optimal in L in a minimax sense, and this explains why better than linear speed-up, as displayed by A4 and A5 for K=2, can be observed for a particular function.

A4 and A5 display a consistently better behaviour than the other algorithms: problems of higher dimension could reasonably push to larger values of K the flattening of the speed-up curve. A similar pattern to that of figure 1 has been observed for $\eta = k\bar{\eta}_1/\bar{\eta}_K$, where $\bar{\eta}_1$ and $\bar{\eta}_K$ are the average number of function evaluations performed respectively by 1 and by K processors.

References.

(1) F. Archetti, G.P. Szego (1980): *Global Optimization Algorithms*, in "Nonlinear Optimization: Theory and Algorithms", L.C.W. Dixon, E. Spedicato and G.P. Szego eds., Birkhauser, Boston.

(2) G.M. Baudet (1978): *The Design and Analysis of Algorithms for Asynchronous Multiprocessors*, Ph.D. Thesis, Carnegie-Mellon University, Dept. of Computer Science, Pittsburgh.

(3) H.T. Kung (1976): *Synchronized and Asynchronous Parallel Algorithms for Multiprocessors*, in "Algorithms and Complexity: New Directions and Recent Results", J.F. Traub ed., Academic Press, New York.

(4) J.J. McKeown (1980): *Aspects of Parallel Computation in Numerical Optimization*, in "Numerical Techniques for Stochastic Systems", F. Archetti and M. Cugiani eds., North-Holland, Amsterdam.

(5) F. Schoen (1981): *On a Sequential Search Strategy in Global Optimization Problems*, to appear in "Calcolo"

(6) A.K. Sukharev (1972): *Best Sequential Strategies for Finding an Extremum*, Zh. vychisl. Mat. mat.Fiz., vol. 18, n.1

A CLASS OF METHODS FOR THE SOLUTION OF OPTIMIZATION PROBLEMS WITH INEQUALITIES

G.Di Pillo*, L.Grippo**, F.Lampariello**

* Istituto di Automatica, Università di Roma, Via Eudossiana, 18 - 00184 Roma, Italy.

** Istituto di Analisi dei Sistemi ed Informatica-CNR Via Buonarroti, 12 - 00185 Roma, Italy.

ABSTRACT

The purpose of this paper is that of describing and evaluating a new class of Newton-type and Quasi-Newton algorithms for the solution of nonlinear optimization problems with inequality constraints. These algorithms are based on the use of a continuously differentiable exact augmented Lagrangian function which allows to obtain the solution of the constrained problem by employing unconstrained minimization techniques. The algorithms considered are evaluated by means of a set of standard test problems.

1. INTRODUCTION

It has been recognized in recent years that the most effective approach for solving optimization problems with nonlinear constraints is based on methods which employ estimates of the Lagrange or Kuhn-Tucker multipliers [1].

For the equality constrained case it has been shown [2] that most of these methods can be considered in a unified framework, as Newton-type or Quasi-Newton algorithms for the search of a saddle point of the Lagrangian function. Interesting special cases are the augmented Lagrangian method proposed in [3][4], the exact penalty function methods described in [5][6] and the recursive quadratic programming approach [7] [8].

For the inequality constrained case, extensions of these methods have been proposed either by resorting to squared slack variables [9] [10] or by employing suitable estimates of the active set [8] [11] or by approximating the original problem with a sequence of inequality constrained quadratic programs [12].

The main difficulty encountered by these methods appears to be that of conciliating an ultimate superlinear convergence rate with global convergence towards constrained minima from poor starting approximations. In particular, the exact penalty function approach requires the minimization of a non differentiable line search function.

The aim of this paper is that of describing and evaluating a new class of computational algorithms for the solution of inequality constrained problems. These algorithms are based on the use of a continuously differentiable exact augmented Lagrangian function which allows to obtain the solution of constrained problems by employing unconstrained minimization techniques.

This approach was first introduced in [13] [14] for the case of equality constraints and was extended in [15] to the inequality constrained case. Further results were obtained in [16] [17] [18].

The relevant feature of the proposed approach is that the search for a saddle point of the Lagrangian function is replaced with the search for an unconstrained minimum of the exact augmented Lagrangian in the extended space of the problem variables and the associated multipliers. In this way we get a continuously differentiable line search function on the extended space, which can be used to enforce global convergence of unconstrained descent methods towards local solutions of the constrained problem. On this basis, Newton-type and Quasi-Newton methods can be devised in which global convergence can be ensured, while retaining the ultimate quadratic or superlinear convergence rate.

The paper is organized as follows. Section 2 contains the problem formulation and the definition of the exact augmented Lagrangian function. In section 3 the properties of the augmented Lagrangian are summarized. Section 4 is devoted to Newton-type algorithms based on consistent approximations of the Newton's direction for the augmented Lagrangian. In section 5 some Quasi-Newton algorithms are described and it is shown that, under suitable assumptions, the corresponding search directions are descent directions. In section 6 a scheme for the automatic selection of the penalty coefficient is described and convergence results are established. Finally, in section 7 the results obtained for a set of standard test problems are reported.

In the sequel all the proofs are omitted because of the space limitations. The interested reader is referred to [19].

2. PROBLEM FORMULATION

The problem under consideration is the nonlinear programming problem:

$$\text{minimize } f(x), \quad \text{subject to } g(x) \leq 0 \tag{1}$$

where the functions $f : R^n \to R^1$ and $g : R^n \to R^m$ are three times continuously differentiable on R^n.

We denote by $L(x,\lambda) \triangleq f(x) + \lambda' g(x)$ the Lagrangian function of problem (1). $\nabla_x L(x,\lambda)$ and $\nabla_x^2 L(x,\lambda)$ will be, respectively, the gradient and the Hessian matrix of $L(x,\lambda)$ with respect to x. Moreover the Jacobian matrix of $g(x)$ will be denoted by $A(x)$.

The algorithms considered here for the solution of problem (1) are based on the unconstrained minimization, with respect to both x and λ, of the following augmented Lagrangian function:

$$\begin{aligned} T(x,\lambda) \triangleq\; & f(x) + \lambda'[g(x) + Y(x,\lambda)y(x,\lambda)] + \frac{1}{\varepsilon}\|g(x) + Y(x,\lambda)y(x,\lambda)\|^2 \\ & + \eta\|A(x)\nabla_x L(x,\lambda) + 4Y^2(x,\lambda)\lambda\|^2 + \tau\|\Lambda[g(x) + Y(x,\lambda)y(x,\lambda)]\|^2 \end{aligned} \tag{2}$$

where

$$y_i^2(x,\lambda) \triangleq -\frac{\min[0, w_i(x,\lambda)]}{2(1+16\varepsilon\eta\lambda_i^2+\varepsilon\tau\lambda_i^2)}, \qquad i = 1,\dots,m \tag{3}$$

with

$$w_i(x,\lambda) \triangleq 2g_i(x)+\varepsilon\lambda_i+2\varepsilon\tau\lambda_i^2 g_i(x)+8\varepsilon\eta\lambda_i \frac{\partial g_i(x)}{\partial x}\nabla_x L(x,\lambda), \qquad i=1,\dots,m \tag{4}$$

and

$$Y(x,\lambda) \triangleq \operatorname{diag}[y_i(x,\lambda)], \qquad \Lambda \triangleq \operatorname{diag}[\lambda_i]$$

for given values of the coefficients $\varepsilon > 0$, $\eta > 0$ and $\tau \geq 0$.

It can be easily verified that function (2) is continuously differentiable with respect to x and λ, and that its gradient is given by

$$\begin{aligned} \nabla_x T(x,\lambda) &= \nabla_x L(x,\lambda) + \frac{2}{\varepsilon}A'(x)\varphi(x,\lambda) + 2\eta N(x,\lambda)\psi(x,\lambda) + 2\tau A'(x)\Lambda^2\varphi(x,\lambda) \\ \nabla_\lambda T(x,\lambda) &= \varphi(x,\lambda) + 2\eta K(x,\lambda)\psi(x,\lambda) + 2\tau[G(x) + Y^2(x,\lambda)]\Lambda\varphi(x,\lambda) \end{aligned} \tag{5}$$

where

$$\varphi(x,\lambda) \triangleq g(x) + Y(x,\lambda)y(x,\lambda), \qquad \psi(x,\lambda) \triangleq A(x)\nabla_x L(x,\lambda) + 4Y^2(x,\lambda)\lambda,$$

$$K(x,\lambda) \triangleq A(x)A'(x) + 4Y^2(x,\lambda), \qquad G(x) \triangleq \operatorname{diag}[g_i(x)]$$

$$N(x,\lambda) \triangleq \nabla_x^2 L(x,\lambda)A'(x) + \sum_{i=1}^m \nabla_x^2 g_i(x)\nabla_x L(x,\lambda)e_i',$$

being e_i in $N(x,\lambda)$ the i-th column of the $m \times m$ identity matrix.

We will refer to any point $(\bar{x},\bar{\lambda}) \in R^n \times R^m$ satisfying the Kuhn-Tucker necessary conditions: $g(\bar{x}) \leq 0$; $\bar{\lambda}'g(\bar{x}) = 0$; $\bar{\lambda} \geq 0$; $\nabla_x L(\bar{x},\bar{\lambda}) = 0$ as a *K-T point* for problem (1), and to any point $(\bar{x},\bar{\lambda})$ such that $\nabla_x T(\bar{x},\bar{\lambda}) = 0$, $\nabla_\lambda T(\bar{x},\bar{\lambda}) = 0$ as a *stationary point* of the function $T(x,\lambda)$.

In the sequel we shall make use of the following index sets

$$I_\pi(x,\lambda) \triangleq \{i: w_i(x,\lambda) > 0\}; \qquad I_\mu(x,\lambda) \triangleq \{i: w_i(x,\lambda) \leq 0\};$$

accordingly, given a matrix R with rows r_i, $i = 1,\dots,m$, we define the submatrices:

$$R_\pi = [r_i],\ i \in I_\pi(x,\lambda); \qquad R_\mu = [r_i],\ i \in I_\mu(x,\lambda)$$

whose rows appear in the same order as in R.

3. PROPERTIES OF THE AUGMENTED LAGRANGIAN

The main results concerning the relationships between the solutions of problem (1) and the unconstrained minima of T are summarized in this section. It can be shown that the augmented Lagrangian (2) introduced here enjoys, in essence, the same properties established in [15] for the case $\tau = 0$. However the presence of the extra term $\tau\|\Lambda\varphi\|^2$ allows to ensure the boundedness from below of T in $R^n \times R^m$, provided that $\inf f(x) > -\infty$, and this can be computationally advantageous with respect to the case $\tau = 0$.

THEOREM 1. *Let* $(\bar{x},\bar{\lambda})$ *be a* K-T *point for problem (1). Then for any* $\varepsilon > 0$, $\eta > 0$ *and* $\tau \geq 0$, $(\bar{x},\bar{\lambda})$ *is a stationary point of* T *and* $T(\bar{x},\bar{\lambda}) = f(\bar{x})$.

THEOREM 2. *Let* $X \times \Omega$ *be a compact subset of* $R^n \times R^m$ *and assume that at every feasible point in* X, *the gradients of the active constraints are linearly independent.*

Then, there exists a scalar $\bar{\tau} > 0$ *and, for each* $\tau \in [0,\bar{\tau}]$, *a scalar* $\bar{\varepsilon}(\tau) > 0$ *such that for all* τ, ε *with* $\tau \in [0,\bar{\tau}]$, $\varepsilon \in (0,\bar{\varepsilon}(\tau)]$ *if* $(\bar{x},\bar{\lambda}) \in X \times \Omega$ *is a stationary point of* T, $(\bar{x},\bar{\lambda})$ *is a* K-T *point for problem (1).*

THEOREM 3. *Let* $(\bar{x},\bar{\lambda})$ *be a* K-T *point for problem (1) and assume that* $\bar{x}$ *is an isolated local minimum point of problem (1), and that at* $\bar{x}$ *the gradients of the active constraints are linearly independent.*

Then, there exists a scalar $\bar{\tau} > 0$ *and, for each* $\tau \in [0,\bar{\tau}]$, *a scalar* $\bar{\varepsilon}(\tau) > 0$ *such that for all* τ, ε *with* $\tau \in [0,\bar{\tau}]$, $\varepsilon \in (0,\bar{\varepsilon}(\tau)]$ *the point* $(\bar{x},\bar{\lambda})$ *is an isolated local minimum point of* T.

THEOREM 4. *Let* $X \times \Omega$ *be a compact subset of* $R^n \times R^m$ *and assume that at every feasible point in* X *the gradients of the active constraints are linearly independent, and that strict complementarity holds at any* K-T *point* $(\bar{x},\bar{\lambda})$ *in* $X \times \Omega$.

Then, there exists a scalar $\bar{\tau} > 0$ *and, for each* $\tau \in [0,\bar{\tau}]$, *a scalar* $\bar{\varepsilon}(\tau) > 0$ *such that for all* τ, ε *with* $\tau \in [0,\bar{\tau}]$, $\varepsilon \in (0,\bar{\varepsilon}(\tau)]$ *if* $(\bar{x},\bar{\lambda}) \in \text{int}(X\times\Omega)$ *is a local unconstrained minimum point of* T, $\bar{x}$ *is a local minimum point for problem (1) and* $\bar{\lambda}$ *is the corresponding* K-T *multiplier.*

4. NEWTON-TYPE METHODS FOR THE MINIMIZATION OF T

In this section it will be shown how the augmented Lagrangian function (2) can be used to obtain Newton-type methods for the solution of problem (1).

We observe first that the function T is twice continuously differentiable at any point (x,λ) where $w_i(x,\lambda) \neq 0$ for all i. In fact, as suming $w_i(x,\lambda) \neq 0$ for all i, and making use of (3) and (4) it can be easily verified that (2) reduces to:

$$T = f + \frac{1}{\varepsilon}\|g\|^2 + \lambda' g + \eta\|A\nabla_x L\|^2 + \tau\|\Lambda g\|^2 - \sum_{i\in I_\mu} \frac{w_i^2}{4\varepsilon(1+\varepsilon\tau\lambda_i^2+16\varepsilon\eta\lambda_i^2)} \tag{6}$$

where, in this case, $I_\mu = \{i: w_i < 0\}$. Since, by continuity, the index set I_μ does not change in a neighbourhood of (x,λ), it is easily seen from (6) that T is twice continuously differentiable in this neighbourhood. In particular, if $(\bar{x},\bar{\lambda})$ is a K-T point and strict complementary holds at $(\bar{x},\bar{\lambda})$, the function T is twice continuously differentiable in a neighbourhood of $(\bar{x},\bar{\lambda})$.

Then we can state a local optimality result based on second order sufficiency conditions.

THEOREM 5. *Let $(\bar{x},\bar{\lambda})$ be a K-T point for problem (1) and assume that: (i) strict complementarity holds at $(\bar{x},\bar{\lambda})$; (ii) $\bar{x}$ is an isolated local minimum point of problem (1) satisfying the second order sufficiency condition: $x'\nabla_x^2 L(\bar{x},\bar{\lambda})x > 0$ for all x such that $\frac{\partial g_i}{\partial x}(\bar{x})x = 0$, $i \in \{i: g_i(\bar{x}) = 0\}$, $x \neq 0$; (iii) at $\bar{x}$ the gradients of the active constraints are linearly independent.*

Then, for every $\tau \geq 0$, there exists an $\bar{\varepsilon}(\tau) > 0$ such that, for all $\varepsilon \in (0,\bar{\varepsilon}(\tau)]$ the Hessian matrix $\nabla^2 T(\bar{x},\bar{\lambda})$ is positive definite and $(\bar{x},\bar{\lambda})$ is an isolated local minimum point of T.

The Hessian matrix $\nabla^2 T$ contains, in general, third order derivatives of the problem functions f and g. However, it can be easily verified that, under the strict complementarity assumption, the terms containing third order derivatives vanish at the K-T points for problem (1). On this basis we can define Newton-type algorithms on the extended space of the problem variables and of the associated multipliers, by resorting to consistent approximations of the Hessian matrix of T which are defined everywhere and do not require the evaluation of third order derivatives.

In particular consider the index sets $I_\pi(x,\lambda)$ and $I_\mu(x,\lambda)$ and define the following symmetric matrix:

$$H(x,\lambda) = (H_{ij}) \qquad i,j = 1,2,3$$

where

$$\begin{aligned} H_{11} &= \nabla_x^2 L + \frac{2}{\varepsilon} A_\pi' A_\pi + 2\eta\nabla_x^2 L A' A \nabla_x^2 L + 2\tau A_\pi' \Lambda_\pi^2 A \\ H_{22} &= 2\eta A_\pi A' A A_\pi' \\ H_{33} &= 2\eta[A_\mu A' A A_\mu' + 16G_\mu^2 - 4G_\mu A_\mu A_\mu' - 4A_\mu A_\mu' G_\mu] - \frac{\varepsilon}{2}E_\mu \\ H_{12} &= A_\pi' + 2\eta\nabla_x^2 L A' A A_\pi' \\ H_{13} &= 2\eta\nabla_x^2 L[A' A A_\mu' - 4A_\mu' G_\mu] \\ H_{23} &= 2\eta A_\pi[A' A A_\mu' - 4A_\mu' G_\mu] \end{aligned} \tag{7}$$

being E_μ the appropriate identity matrix.

Then, the matrix H enjoys the properties summarized as follows.

PROPOSITION 1. *Let $(\bar{x},\bar{\lambda})$ be a K-T point for problem (1). Then, under the assumptions of Theorem 5, it results $H(\bar{x},\bar{\lambda}) = \nabla^2 T(\bar{x},\bar{\lambda})$. Moreover, for every $\tau \geq 0$, there exists an $\bar{\varepsilon}(\tau) > 0$ such that, for all $\varepsilon \in (0,\bar{\varepsilon}(\tau)]$ the matrix $H(x,\lambda)$ is positive definite in a neighbourhood of $(\bar{x},\bar{\lambda})$.*

By using the approximation H of the Hessian matrix we can define the following iterative procedure:

ALGORITHM 1.

$$\begin{bmatrix} \hat{x} \\ \hat{\lambda} \end{bmatrix} = \begin{bmatrix} x \\ \lambda \end{bmatrix} + \alpha d$$

$$H(x,\lambda)d = -\nabla T(x,\lambda)$$

where $(x' \quad \lambda')'$, $(\hat{x}' \quad \hat{\lambda}')'$ are respectively the present and the next iterate, d is the search direction and α is the stepsize.

By Proposition 1, the search direction d in Algorithm 1 is a consistent approximation of the Newton's direction; thus superlinear convergence rate follows from the standard convergence theory for Newton's method. Of course, suitable precautions must be taken for the case in which $H(x,\lambda)$ cannot be guaranteed to be positive definite. In particular, global convergence results can be established by employing the search direction $-\nabla T(x,\lambda)$, whenever some standard angle condition is not satisfied (see, for instance, [17]). However, an important point which must be remarked is that, in a neighbourhood of any K-T point satisfying the assumptions of theorem 5, the search direction d is a descent direction for T.

The main drawback of Algorithm 1 is the large amount of calculations required, in the solution of high dimensional problems, for constructing the matrix H and for computing the search direction d. However, following the procedure suggested in [17], it can be shown that a consistent approximation of the Newton's direction can often be obtained by solving a much simpler system.

In fact, consider the following system

$$\begin{bmatrix} \nabla_x^2 L & A'_\pi & A'_\mu \\ A_\pi & 0 & 0 \\ 0 & 0 & I \end{bmatrix} \begin{bmatrix} d_x \\ d_\pi \\ d_\mu \end{bmatrix} = - \begin{bmatrix} \nabla_x L \\ g_\pi \\ \lambda_\mu \end{bmatrix} \tag{8}$$

and assume that the above matrix is non singular. Then we can state the following proposition.

PROPOSITION 2. *Let* $d \triangleq [d_x' \; d_\pi' \; d_\mu']'$ *be the solution of system (8) and let* $(\bar{x},\bar{\lambda})$ *be a* K-T *point for problem (1). Then, under the assumptions of theorem 5, the vector* d *is a consistent approximation of the Newton's direction for* T. *Furthermore, for every* $\tau \geq 0$ *there exists an* $\bar{\varepsilon}(\tau) > 0$ *such that, for all* $\varepsilon \in (0,\bar{\varepsilon}(\tau)]$, d *is a descent direction for* T *in a neighbourhood of* $(\bar{x},\bar{\lambda})$.

Proposition 2 allows to define a Newton-type algorithm based on the consistent approximation of the Newton's direction obtained by solving system (8):

ALGORITHM 2.

$$\begin{bmatrix} \hat{x} \\ \hat{\lambda} \end{bmatrix} = \begin{bmatrix} x \\ \lambda_\pi \\ \lambda_\mu \end{bmatrix} + \alpha \begin{bmatrix} d_x \\ d_\pi \\ d_\mu \end{bmatrix}$$

$$\begin{bmatrix} \nabla_x^2 L & A'_\pi \\ A_\pi & 0 \end{bmatrix} \begin{bmatrix} d_x \\ d_\pi \end{bmatrix} = - \begin{bmatrix} \nabla_x f + A'_\pi \lambda_\pi \\ g_\pi \end{bmatrix}$$

$$d_\mu = - \lambda_\mu .$$

The main advantage of Algorithm 2 is that the dimension of the system to be solved at each iteration is $n + m_\pi$ where m_π is the number of constraints which are perceived as active.

5. QUASI-NEWTON METHODS

The minimization of $T(x,\lambda)$ by means of Quasi-Newton methods would require the evaluation of second-order derivatives of $f(x)$ and $g(x)$ which appear in the gradient $\nabla T(x,\lambda)$. However, as already pointed out for the case of equality constraints [14], also in this case it is possible to define algorithms which make use only of first order information on the problem functions, by employing a suitable finite-difference approximation. In fact, considering the expression of $\nabla_x T$, we note that second-order derivatives need only be evaluated along specific directions.

Let $u \triangleq A'\psi$, $z_i \triangleq \nabla_x L\psi_i$, $i=1,2,\dots m$; then, for a sufficiently small value of a scalar $t>0$, it results

$$N\psi \simeq \frac{1}{t}[\nabla_x L(x+tu,\lambda)-\nabla_x L(x,\lambda)] + \frac{1}{t}\sum_{i=1}^{m}[\nabla_x g_i(x+tz_i)-\nabla_x g_i(x)] \triangleq \tilde{N}\psi.$$

Therefore in the algorithms described in this section the gradient ∇T can be replaced with the approximation obtained by substituting $\tilde{N}\psi$ for $N\psi$ into (5).

The most straightforward way to define a Quasi-Newton algorithm for the minimization of T is that of producing an approximation of $\nabla^2 T$ by means of some standard Quasi-Newton update.

Thus we can define an iterative procedure of the following type:

ALGORITHM 3.

$$\begin{bmatrix}\hat{x}\\ \hat{\lambda}\end{bmatrix} = \begin{bmatrix}x\\ \lambda\end{bmatrix} + \alpha d$$

$$Bd = -\nabla T$$

$$\hat{B} = B + \nabla B$$

where ∇B is a standard updating matrix such that $\hat{B}$ satisfies the Quasi-Newton equation:

$$\hat{B}\left[\begin{bmatrix}\hat{x}\\ \hat{\lambda}\end{bmatrix} - \begin{bmatrix}x\\ \lambda\end{bmatrix}\right] = \begin{bmatrix}\nabla_x T(\hat{x},\hat{\lambda}) - \nabla_x T(x,\lambda)\\ \nabla_\lambda T(\hat{x},\hat{\lambda}) - \nabla_\lambda T(x,\lambda)\end{bmatrix}.$$

Structured Quasi-Newton algorithms of the type considered in [2] can also be defined by approximating only the term $\nabla_x^2 L(x,\lambda)$ in connection with Algorithms 1 and 2 described in the preceding section.

More specifically we can define the following

ALGORITHM 4.

$$\begin{bmatrix}\hat{x}\\ \hat{\lambda}\end{bmatrix} = \begin{bmatrix}x\\ \lambda\end{bmatrix} + \alpha d \qquad (9)$$

$$Cd = -\nabla T \qquad (9)$$

where $C = (C_{ij})$, $i,j=1,2,3$ is a symmetric matrix in which $C_{22}=H_{22}$, $C_{23}=H_{23}$, $C_{33}=H_{33}$ and C_{11}, C_{12} and C_{13} are obtained from H_{11}, H_{12} and H_{13} respectively by replacing $\nabla_x^2 L$ with an approximation D defined by an updating process $\hat{D}=D+\Delta D$, such that $\hat{D}[\hat{x}-x] = \nabla_x L(\hat{x},\hat{\lambda}) - \nabla_x L(x,\hat{\lambda})$.

An appropriate updating formula for D could be one which ensures that D remains positive definite. In this case it can be shown that, under suitable assumptions, the matrix C is positive definite, for sufficiently small values of ε.

THEOREM 6. *Let* D *be a compact set of symmetric positive definite* $n\times n$ *matrices; let* $X\times\Omega$ *be a compact subset of* $R^n\times R^m$ *and assume that at every point* $x\in X$ *the matrix* $A(x)A'(x)-4G(x)$ *is non singular. Then, there exists a scalar* $\bar{\varepsilon}>0$ *such that, for all* $\varepsilon\in(0,\bar{\varepsilon}]$ *the matrix* C *of system (9) is positive definite for all* $D\in\mathcal{D}$ *and all* $(x,\lambda)\in X\times\Omega$.

ALGORITHM 5.

$$\begin{bmatrix}\hat{x}\\ \hat{\lambda}\end{bmatrix}=\begin{bmatrix}x\\ \lambda_\pi\\ \lambda_\mu\end{bmatrix}+\alpha\begin{bmatrix}d_x\\ d_\pi\\ d_\mu\end{bmatrix}$$

where:

$$\begin{aligned}Dd_x+A_\pi'd_\pi&=-[\nabla_x f+A_\pi'\lambda_\pi]\\ A_\pi d_x&=-g_\pi\\ d_\mu&=-\lambda_\mu\end{aligned}\tag{10}$$

and D is a positive definite approximation of $\nabla_x^2 L$, defined by an updating process $\hat{D}=D+\Delta D$, such that $\hat{D}[\hat{x}-x]=\nabla_x L(\hat{x},\hat{\lambda})-\nabla_x L(x,\hat{\lambda})$.

THEOREM 7. *Let* $\mathcal{D}$ *be a compact set of symmetric positive definite* $n\times n$ *matrices; let* $X\times\Omega$ *be a compact subset of* $R^n\times R^m$ *and assume that at every point* $x\in X$ *the vectors* $\nabla_x g_i(x)$, $i\in\{i:\ g_i(x)\geq 0\}$ *are linearly independent.*

Then, there exists a scalar $\bar{\eta}>0$ *and, for each* $\eta\in(0,\bar{\eta}]$ *a scalar* $\bar{\varepsilon}(\eta)>0$ *such that for all* η, ε *with* $\eta\in(0,\bar{\eta}]$, $\varepsilon\in(0,\bar{\varepsilon}(\eta))$ *if at* $(D,x,\lambda)\in\mathcal{D}\times X\times\Omega$ *system (10) admits a solution* $d=(d_x'd_\pi'd_\mu')'$ *and* $\nabla T\neq 0$, *it results* $d'\nabla T<0$.

We note that Algorithm 5 can be considered as an extension to the inequality constrained case of the Quasi-Newton algorithm proposed in [18] for equality constrained problems.

6. AUTOMATIC SELECTION OF THE PENALTY COEFFICIENT

In this section we describe a procedure for the automatic selection of the penalty coefficient, which is based on the algorithm model given in [20] and extends to the inequality constrained case a similar scheme given in [14] for equality constrained problems.

We assume that it is available an unconstrained minimization algorithm, defined by an iteration map $M: R^n\times R^m\to 2^{(R^n\times R^m)}$ which for given values of the parameters ε, η, τ converges to stationary points of T (in the sense that limit points of the sequence produced by M are stationary points of T).

The automatic scheme for the adjustment of the penalty coefficient described below has the objective of avoiding convergence to stationary points of T which are not K-T points for problem (1).

ALGORITHM EPS

DATA: $0<\rho<1$, $0<\sigma<1$, $\delta>0$, ε°, τ°, $z^\circ=(x^\circ,\lambda^\circ)$.

STEP 0 : Set $j=0$.

STEP 1 : Set $i=0$ and set $z^j=(x^\circ,\lambda^\circ)$.

STEP 2 : If $\nabla_x T\neq 0$ or $\nabla_\lambda T\neq 0$ or $\varphi\neq 0$ continue; else stop.

STEP 3 : If $\min_h|1+2\tau^j\lambda_h\varphi_h|\geq\delta$ continue, else set $\tau^j=\rho\tau^j$ and go to step 6.

STEP 4 : If $\|\nabla_\lambda T\|^2 + \nabla_x T' A' K\varphi \geq \|K\varphi\|^2$ continue; else set $\varepsilon^j = \sigma\varepsilon^j$ and go to step 6.

STEP 5 : Compute $(x^{i+1}, \lambda^{i+1}) \in M[(x^i, \lambda^i)]$, set $i = i+1$ and go to step 2.

STEP 6 : Set $z^{j+1} = (x^i, \lambda^i)$, $j = j+1$ and go to step 1.

The convergence properties of Algorithm EPS are given in the following theorem.

THEOREM 8. *Assume that the gradients of the active constraints are linearly independent at any feasible point* $x \in R^n$ *and that for every* $\varepsilon > 0$, $\eta > 0$, $\tau \geq 0$ *and* $(x^\circ, \lambda^\circ) \in R^n \times R^m$, *any limit point of the sequence* $\{(x^i, \lambda^i)\}$ *generated by the iteration map* M, *is a stationary point of* T. *Then, if the algorithm constructs a finite sequence* $\{(x^i, \lambda^i)\}_{i=0}^{\nu}$ *and stops,* (x^ν, λ^ν) *is a K-T point for problem (1); if the sequence* $\{z^j\}$ *is finite and the algorithm constructs an infinite sequence* $\{(x^i, \lambda^i)\}$, *any limit point* $(\bar{x}, \bar{\lambda})$ *is a K-T point for problem (1); if the algorithm constructs an infinite sequence* $\{z^j\}$, *this sequence has no limit point.*

7. NUMERICAL RESULTS

In order to evaluate the performance of the algorithms described in this paper a set of five standard test problems was considered.

TEST PROBLEM 1. (Rosen and Suzuki)

Minimize $f(x) = -5(x_1 + x_2) + 7(x_4 - 3x_3) + x_1^2 + x_2^2 + 2x_3^2 + x_4^2$

subject to:

$$\left(\sum_{i=1}^{4} x_i^2\right) + x_1 - x_2 + x_3 - x_4 - 8 \leq 0$$
$$x_1^2 + 2x_2^2 + x_3^2 + 2x_4^2 - x_1 - x_4 - 10 \leq 0$$
$$2x_1^2 + x_2^2 + x_3^2 + 2x_1 - x_2 - x_4 - 5 \leq 0.$$

Solution: $x^* = (0,1,2,-1)'$ with $f(x^*) = -44$.
Starting point: $x^\circ = 0$, $\lambda^\circ = 0$.

TEST PROBLEM 2. (Beale)

Minimize $f(x) = -8x_1 - 6x_2 - 4x_3 + 2x_1^2 + 2x_2^2 + x_3^2 + 2x_1x_2 + 2x_1x_3 + 9$

subject to:

$$x_1 + x_2 + 2x_3 - 3 \leq 0$$
$$x_i \geq 0, \quad i = 1,2,3.$$

Solution: $x^* = (4/3,\ 7/9,\ 4/9)'$ with $f(x^*) = 1/9$
Starting point: $x^\circ = (1/2,\ 1/2,\ 1/2)'$, $\lambda^\circ = 0$.

TEST PROBLEM 3. (Wong)

Minimize $f(x) = (x_1 - 10)^2 + 5(x_2 - 12)^2 + x_3^4 + 3(x_4 - 11)^2 + 10x_5^6 + 7x_6^2 + x_7^4 - 4x_6x_7 - 10x_6 - 8x_7$

subject to:

$$2x_1^2 + 3x_2^4 + x_3 + 4x_4^2 + 5x_5 - 127 \leq 0$$
$$7x_1 + 3x_2 + 10x_3^2 + x_4 - x_5 - 282 \leq 0$$
$$23x_1 + x_2^2 + 6x_6^2 - 8x_7 - 196 \leq 0$$
$$4x_1^2 + x_2^2 - 3x_1x_2 + 2x_3^2 + 5x_6 - 11x_7 \leq 0.$$

Solution: $x^* = (2.33050,\ 1.95137,\ -0.47754,\ 4.36573,\ -0.62448,\ 1.03813,\ 1.594)'$ with $f(x^*) = 680.630$.
Starting point: $x^\circ = (1,2,0,4,0,1,1)'$, $\lambda^\circ = 0$.

TEST PROBLEM 4. (Transformer design problem)

$$\text{Minimize } f(x) = .0204x_1x_4(x_1+x_2+x_3) + .0187x_2x_3(x_1+1.57x_2+x_4) + .0607x_1x_4x_5^2(x_1+x_2+x_3) + .0437x_2x_3x_6^2(x_1+1.57x_2+x_4)$$

subject to:

$$x_i \geq 0,\quad i=1,\ldots,6$$

$$2.07 - .001x_1x_2x_3x_4x_5x_6 \leq 0$$

$$0.00058x_2x_3x_6^2(x_1+1.57x_2+x_4) + 0.00062x_1x_4x_5^2 - 1 \leq 0.$$

Solution: $x^* = (5.33,\ 4.66,\ 10.43,\ 12.08,\ .752,\ .878)'$ with $f(x^*) = 135.07596$.
Starting point: $x^\circ = (5.54,\ 4.4,\ 12.02,\ 11.82,\ .702,\ .852)'$, $\lambda^\circ = 0$.

TEST PROBLEM 5. (Shell Dual)

$$\text{Minimize } f(x) = \sum_{j=1}^{5} e_jx_j + \sum_{j=1}^{5}\sum_{i=1}^{5} c_{ij}x_ix_j + \sum_{j=1}^{5} d_jx_j^3$$

subject to:

$$b_i - \sum_{j=1}^{5} a_{ij}x_j \leq 0,\quad i=1,\ldots,10$$

$$x_j \geq 0,\quad j=1,\ldots,5$$

where the coefficients are given in [21]

Solution: $x^* = (.3000,\ .3335,\ .4000,\ .4285,\ .224)$ with $f(x^*) = -32.349$.
Starting point: $x^\circ = (0,0,0,0,1)$, $\lambda^\circ = 0$.

The numerical experiments were performed by employing Algorithms 1-5 with the same line search procedure. Algorithms 3, 4 and 5 were implemented by using the BFS updating formulas.

A first group of numerical experiments was performed by assuming $\tau = 0$ and fixed values of the parameters ε and η. The results obtained are collected in Table 1. For each case we report the number LS of line searches and the corresponding number NT of function evaluation needed to attain a specified accuracy δx on the solution. In particular, the condition employed was: $|x_i - x_i^*| \leq \delta x$, $i = 1,\ldots,n$ where δx is a given number.

For all problems a maximum number of 200 line searches was allowed.

The unsuccessful cases correspond either to divergence of the sequence $\{T(x^{(k)}, \lambda^{(k)}, \varepsilon)\}$, or to a failure in attaining the required accuracy within 200 line searches. These cases are indicated with F.

From the results given in Table 1 it appears that in most of cases Algorithms 2 and 5 perform better than the remaining algorithms. All algorithms exhibit a similar behaviour with respect to the parameters η and ε. More specifically, for each problem there is a preferential range of values for the parameters and, for a given η, too large values of ε give rise to divergence, whereas too small values of ε produce slow convergence.

A second group of experiments was worked out in order to evaluate the effect of non zero values of the parameter τ. In particular, it results that for Test Problem 1, the failures shown in Table 1 are due to the unboundedness of T with respect to λ. We report in Table 2 the results obtained with the same values of ε and η and with $\tau > 0$. We note that non zero values of τ prevent divergence; however, in some cases, the required accuracy was no more achieved. Moreover it can be noted that the numbers LS and NT increase with τ.

Finally, a third group of numerical experiments was performed by employing the Algorithm EPS described in section 6 for the automatic selection of the penalty coefficient ε.

The algorithm was implemented with $\tau^\circ = 0$, $\varepsilon^\circ = 1$ and $\sigma = .5$.

In Table 3 are shown, in particular, the results obtained for Test Problem 1 by employing Algorithms 1-5 and in Table 4 are reported the results obtained for all Test Problems by using Algorithm 5. By comparing Tables 3 and 4 with Table 1 it can be seen that in most cases the automatic selection procedure allows to improve the convergence rate while retaining the same accuracy on the solution point.

7. CONCLUSIONS

On the basis of the theoretical and computational results reported in this paper it appears that the class of methods based on the exact augmented Lagrangian function T is quite promising. In particular, Algorithms 2 and 5 performed better than the remaining algorithms.

Although no comparison was attempted with alternative approaches, the algorithms considered here seem to be competitive with the most effective techniques presently available, at least on higly nonlinear problems. On the other hand, it is to be expected that problems with a large number of linear constraints are more conveniently handled by the recursive quadratic programming approach of [12].

An additional effort would be required in order to develop more effective computational codes. In particular, the selection rules for the parameters ε, η, τ and the expedients for maintaining the search trajectory within a compact set deserve a further attention.

REFERENCES

[1] M.J.D.POWELL: *Optimization Algorithms in 1979*, in Optimization Techniques, K.Iracki, K.Malanowski, S.Walukiewicz eds., Springer Verlag, 1980, Part 1, pp. 83-98.

[2] R.A.TAPIA: *Quasi-Newton Methods for Equality Constrained Optimization: Equivalence of Existing Methods and a New Implementation* - in Nonlinear Programming 3, O.L.Mangasarian, R.R.Meyer, S.M.Robison eds., Academic Press, 1978, pp. 124-164.

[3] M.R.HESTENES: *Multipliers and Gradient Methods* - in Computing Methods in Optimization Problems, vol. 2, L.A.Zadeh, L.W.Neustadt, A.V.Balakrishnan eds. Academic Press, New York, 1969, pp. 143-163.

[4] M.J.D.POWELL: *A Method for Nonlinear Constraints in Minimization Problems* - in Optimization, R.Fletcher ed., Academic Press, New York, 1969, pp. 283-298.

[5] R.FLETCHER: *A Class of Methods for Nonlinear Programming with Termination and Convergence Properties* - in Integer and Nonlinear Programming, J.Abadie ed., North Holland, Amsterdam, The Netherlands, 1970.

[6] H. MUKAI, E.POLAK: *A Quadratically Convergent Primal-Dual Algorithm with Global Convergence Properties for Solving Optimization Problems with Equality Constraints* - Mathematical Programming, Vol. 9, 1975, pp. 336-349.

[7] M.C.BIGGS: *Constrained Minimization Using Recursive Equality Quadratic Programming* - in Numerical Methods for Nonlinear Optimization, F.A.Lootsma ed., Academic Press, London 1972.

[8] M.C.BIGGS: *The Recursive Quadratic Programming Approach to Constrained Optimization*, 2nd IFAC Workshop on Control Applications of Nonlinear Programming, Oberpfaffenhofen, FRG., 1980.

[9] D.P.BERTSEKAS: *Multiplier Methods: a Survey*, Automatica, Vol. 12, 1976, pp. 133-145.

[10] R.A.TAPIA: *Diagonalized Multiplier Methods and Quasi-Newton Methods for Constrained Optimization*, Journal of Optimization Theory and Applications, Vol. 22, 1977, pp. 135-194.

[11] R.FLETCHER: *An Exact Penalty Function for Nonlinear Programming with Inequalities*, Math. Programming, Vol. 5, 1973, pp. 129-150.

[12] M.J.D.POWELL: *A Fast Algorithm for Nonlinear Constrained Optimization Calculations* - in Numerical Analysis, Dundee 1977, G.A.Watson ed., Lecture Notes in Mathematics No. 630, Springer Verlag, Berlin 1978.

[13] G.DI PILLO, L.GRIPPO: *A New Class of Augmented Lagrangians in Nonlinear Programming*, SIAM Journal on Control and Optimization, Vol. 17, 1979, pp. 618-628.

[14] G.DI PILLO, L.GRIPPO, F.LAMPARIELLO: *A Method for Solving Equality Constrained Optimization Problems by Unconstrained Minimization* - in Optimization Technique, 9th IFIP Conference, K.Iracki, K.Malanowski and S.Walukiewicz eds., Springer Verlag, Berlin, FRG, 1980.

[15] G.DI PILLO, L.GRIPPO: *A New Augmented Lagrangian Function for Inequality Constraints in Nonlinear Programming Problems*, (to appear in Journal of Optimization Theory and Applications).

[16] L.C.W.DIXON: *Exact Penalty Function Methods in Nonlinear Programming*, The Hatfield Polytechnic, Numerical Optimization Centre, Technical Report No. 103, 1979.

[17] D.P.BERTSEKAS: *Enlarging the Region of Convergence of Newton's Method for Constrained Optimization*, M.I.T., Laboratory for Information and Decision Systems, Report No. 985, 1980.

[18] D.P.BERTSEKAS: *Variable Metric Methods for Constrained Optimization Using Differentiable Exact Penalty Functions*, Proc. Eighteenth Annual Allerton Conference on Communication, Control and Computing, Allerton Park, Ill., Oct. 1980.

[19] G.DI PILLO, L.GRIPPO, F.LAMPARIELLO: *A Class of Methods for the Solution of Optimization Problems with Inequalities*, IASI-CNR, Rep. n. 18, September 1981.

[20] E.POLAK: *On the Stabilization of Locally Convergent Algorithms for Optimization and Root Finding*, Automatica, Vol. 12, 1976, pp. 337-342.

[21] A.R.COLVILLE: *A Comparative Study of Nonlinear Programming Codes*, I.B.M. New York Scientific Centre, Techn. Rep. 320-2949, 1968.

Table 1.

	η	$\frac{1}{\varepsilon}$	ALG.1 LS	ALG.1 NT	ALG.2 LS	ALG.2 NT	ALG.3 LS	ALG.3 NT	ALG.4 LS	ALG.4 NT	ALG.5 LS	ALG.5 NT
P1 (δx= .001)	10^{-4}	10^2	17	36	16	28	38	78	24	50	24	48
		10^3	F		33	54	56	118	F		42	62
	10^{-3}	10^2	F		21	48	F		F		19	41
		10^3	28	60	26	45	51	111	39	91	27	46
P2 (δx= .001)	10^{-1}	10	3	4	3	4	9	13	7	11	5	7
		10^2	5	13	5	13	18	27	7	14	4	9
	1	10	2	3	2	3	15	22	20	34	6	8
		10^2	3	4	3	4	19	33	15	23	6	8
P3 (δx= .001)	10^{-6}	10	13	27	15	44	28	68	23	44	37	77
		10^2	25	59	20	57	52	110	40	92	44	86
	10^{-5}	10	28	53	32	70	66	124	41	78	93	183
		10^2	58	129	148	269	112	222	65	124	F	
P4 (δx= .01)	10^{-4}	10^3	19	44	12	25	F		14	20	188	404
		10^4	39	91	F		F		39	73	F	
	10^{-3}	10^3	13	27	F		F		25	43	16	21
		10^4	F		27	52	F		31	50	17	33
P5 (δx= .0001)	10^{-4}	10^3	F		30	85	F		F		14	34
		10^4	51	137	36	103	93	200	F		31	98
	10^{-3}	10^3	F		20	55	65	104	41	100	20	59
		10^4	60	152	43	117	99	220	F		39	99

Table 2.

η	$\frac{1}{\varepsilon}$	τ	ALG.1 LS	ALG.1 NT	ALG.1 δx	ALG.3 LS	ALG.3 NT	ALG.3 δx	ALG.4 LS	ALG.4 NT	ALG.4 δx
10^{-4}	10^3	10	71	122	.005				75	119	.005
		10^2	174	223	.001				107	196	.001
10^{-3}	10^2	10^2	55	93	.005	51	95	.005	84	100	.005
		10^3	97	129	.005	105	178	.005	187	276	.001

Table 3.

η	ALG.1 LS	ALG.1 NT	ALG.2 LS	ALG.2 NT	ALG.3 LS	ALG.3 NT	ALG.4 LS	ALG.4 NT	ALG.5 LS	ALG.5 NT
10^{-4}	15	45	13	23	57	119	20	48	11	28
10^{-3}	12	23	10	22	49	103	13	29	14	32

Table 4.

	η	LS	NT
P1	10^{-4}	11	28
	10^{-3}	14	32
P2	10^{-1}	6	9
	1	5	9
P3	10^{-6}	38	78
	10^{-5}	163	340
P4	10^{-4}	30	59
	10^{-3}	16	33
P5	10^{-4}	8	19
	10^{-3}	15	43

A PROBABILISTIC APPROACH TO THE MINIMIZATION OF STOCHASTIC FUNCTIONS BY SEQUENTIAL, NEAR-CONJUGATE SAMPLING

W. Jarisch
Scientific Systems, Inc.
Cambridge, MA 02140
Harvard University
Cambridge, MA 02138

ABSTRACT

We consider the problem of minimizing the convex function $f(\underline{x})$ when $z=f(x)+e$ is observed. Deterministic nonlinear parameter estimation schemes are often hampered by the observation noise e, and stochastic approximation methods often suffer from inefficiency.

In order to achieve an efficient algorithm for dealing with "high dimensional" problems a stochastic modeling approach is taken. We make use of a hybrid of ideas of conjugate gradient algorithms [6] and stochastic function minimization [4]. First, the orthogonal projection idea of conjugate gradient methods is adapted to a particular sequential sampling and regression procedure. Second, convergence requirements are relaxed, e.g. we strive only for convergence with probability P ($0 \ll P < 1$) into a region R around the minimum of $f(\underline{x})$. By means of a stochastic modeling approach the adequacy of the hypothesis of such convergence is tested.

1.0 INTRODUCTION

Efficiency, speed, robustness and the capability to negotiate "high dimensional" problems are of major concern in modern minimization problems. Examples of such problems, requiring minimization with respect to many variables, arise in finite element analysis, systems parameter estimation, process control, and Monte Carlo studies.

In order to cope with problem settings such as above, several rather distinct techniques have been developed in the past. Among these techniques are conjugate gradient and stochastic function minimization methods. Conjugate gradient methods are popular in many multivariable problems. Much work has been devoted to studying the convergence properties of these algorithms when $f(\underline{x})$ is twice continuously differentiable; see for example the basic texts by Bard [3], and Avriel [2], and the extensive literature in the Journal of Optimization Theory.

With regard to stochastic approximation see the text by Kushner & Clark [8], which reviews constrained and unconstrained systems and and gives some new approaches to these methods. Nevelson & Hasminskii [11] address the inefficiency of certain stochastic approximation methods and present an "optimal" design of finite differences for a scalar Robbins-Monro type algorithm [12]. Along the line of adaptive design, see Lai & Robbins [9] who discuss such an algorithm with accelerated convergence properties for a general scalar regression problem. Common to all these approaches is that some critical information is taken from prior assumptions of the problem at hand. However, making the regression adaptive is an important step in the direction of relaxing the need for prior information.

For our purposes we assume a local quadratic approximation to the function $f(\underline{x})$, as is common practice in the minimization of deterministic functions. In view of the uncertainty in a finite number of measurements z_i, where

$$z_i = f(\underline{x}_i) + e_i \tag{1.1}$$

we suggest to test the adequacy of a local quadratic approximation. In fact, by formulating a testing problem for a finite number of sample points one can specify a region for which the quadratic approximation is useful.

Let us further discuss this region. Assume for the moment that the function $f(\underline{x})$ is truly quadratic with a unique minimum. Clearly, for a fixed distribution with finite variance for error e, the error in estimating derivatives (or similarly coefficients of a quadratic approximation) in a given point increases, say in a mean square sense, with decreasing distance between sample points. This observation suggests that it is necessary to choose large finite differences in order to make estimation error (arbitrarily) small.

Consider next a "near-quadratic" function $f(\underline{x})$, strictly convex and twice continuously differentiable. Assume this function can be observed without any error. To find a quadratic approximation at a point $\underline{x}$, it is obviously desired to choose sample points "close" together: sample points "far apart" will usually lead to "strongly" biased estimates of derivatives at the point of interest.

For our problem consider now a combination of the above properties of the observations: let $f(\underline{x})$ be "near-quadratic" and let the measurements z, each drawn at some cost, be corrupted by noise e. In order to keep sampling risk low, while achieving a (sufficiently) good quadratic approximation, a compromise in the choice of finite differences is necessary. How can one choose such differences ?

For the scalar case, Bard [3 , p. 119] points at the need to know higher derivatives of $f(x)$ in order to pick optimal finite differences. Generally, it seems some additional information about the surface is necessary to pick at least good values. We believe that for many practical situations such information is not available a priori.

As an alternative, we suggest acquiring the information for the choice of finite differences by a suitable number and distribution of test sample points. In this problem formulation, a model fitting and hypothesis testing approach can be taken.

For our purposes we make use of a hybrid of ideas from orthogonal projections, statistical modeling, sampling, and testing theories. A by-product of the proposed approach is an apparently new understanding of "high dimensional" problems. A number of different schemes can be constructed by using this hybridization. Depending, among other factors, on tradeoffs between sampling costs, computational cost and measurement error, one will form different hybrids of above methods. In this paper we will give an outline of a basic version of the method.

The organization of the paper is as follows. In section 2.1 we give a rough outline of the procedure; in section 2.2 we develop methods for computing on-line error bounds. Next, in section 2.3 we discuss the choice of finite differences, while section 2.4 addresses aspects of sequential sampling via orthogonal projection. Section 2.5 outlines termination criteria and associated diagnostics. Finally, section 3 discusses the anticipated benefits of the proposed approach.

2.0 OUTLINE OF PROCEDURES

2.1 The Basic Algorithm

Estimates of conjugate directions are constructed, locally, in an iterative fashion. Two different types of statistical testing problems now arise in a subspace of X, say X_n of dimension n. They concern in an alternating way: (i) the choice of the size of finite differences along the most recently obtained estimate of the n'th conjugate direction; and (ii) the need to iterate sampling along refined estimates of conjugate directions generated in X_n, starting with an a priori estimate of that direction.

By solving these two testing problems and expanding X_n (when both tests have been successful) until $X_n = X$ we can construct an estimate of the Hessian H, which is more robust against noise e than a Hessian obtained from mutually orthogonal sample directions in X, based on arbitrary finite differences. For purposes of discussion we consider here a linear transform of the space X, say Y.

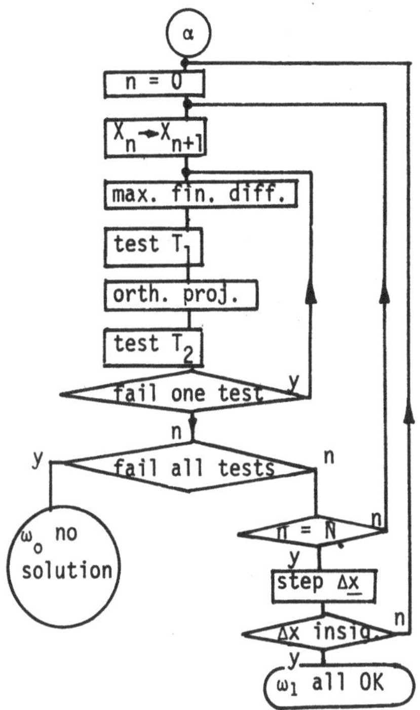

Figure 2.1: Flow chart

Following figure 2.1, we assume initially that Y = X. Then, sequentially X_n is expanded whenever a good estimate of the n'th conjugate direction has been found (when both tests have been successful). Corresponding to these estimates of conjugate directions, estimates of orthogonal directions in Y are specified.

Assume we are now at a location $\underline{x}$. At this location a stage n, $0<n\leq N-1$ of the sequential sampling scheme is defined: assume the first n estimates of an orthogonal basis in Y_n have been found and the (n+1)st estimate is desired. For this purpose draw the necessary samples (of the order of n points) from $X_{(n+1)}$ based on the estimates of orthogonal directions in Y_n (equivalent to the estimates of conjugate directions in X_n) and the (n+1)st basis vector in X (not conjugate). Now, by means of orthogonal projection, find a preliminary estimate of the (n+1)st orthogonal basis vector in $Y_{(n+1)}$. Based on two statistical tests this preliminary estimate is accepted or rejected. When the estimate is acceptable the estimate of the Hessians is updated and the sampling

scheme goes on to stage n+1. When the estimate is rejected a new set of samples may be drawn from the same subspace $Y_{(n+1)}$ based on the current (n+1) estimates of conjugate directions. Drawing of new sample points depends on the status of the two tests. If one test is failed the subspace Y(n+1) is resampled. When both tests are failed the procedure terminates and informs the user of insufficient accuracy of measurements to insure the requested performance.

The two tests involve the following: first, on every sampling along a particular direction the sampling points are chosen as far as possible from the current location $\underline{x}$. The directions are specified by the vertices forming the prism spanned by preliminary estimates of orthogonal directions and sampling (end-) points in the space $Y_{(n+1)}$. The maximal distance for (end-) points from $\underline{x}$ is determined by means of a quadratic regression line: lack of fit indicates the maximal permissible distance has been exceeded. In this way "maximal" finite differences "minimize" the influence of noise on finding good estimates for first and second derivatives along any particular direction. Now the test T1 checks whether the estimated confidence interval (for some suitably chosen probability level $P < 1$) of derivatives is small enough to provide sufficient information to keep the size of the error in an eventual step $\Delta\underline{y}$, along an orthogonal direction in Y_n, bounded at a suitable magnitude.

Second, one performs the test T2, checking whether the new, say (n + 1)st, estimate of the conjugate direction was of sufficient accuracy. How this may be done is described in section 2.3. When only this test is failed a refined orthogonal direction in Y_{n+1} is estimated by orthogonal projection

Then, when the stage n = N is reached after passing all statistical tests, the algorithm performs a step of the type $\Delta\underline{y} = H_y^{-1}$ grad $[f(\underline{y})]$, where H_y is the Hessian in the space Y. Finally, when this step is considered statistically insignificant, the procedure terminates.

The above rough outline of the basic procedure permits us to turn to the discussion of some technical details.

2.2 Error Bounds

Aside from the effects of the usually non-quadratic nature of $f(\underline{x})$, the error in steps $\Delta\underline{x}$ arises from the measurement noise e and leads to uncertainty when estimating the gradient and the Hessian. For a step

$$\Delta\underline{x} = H^{-1} \text{ grad } [\, f(\underline{x}) \,] \tag{2.1}$$

the error can be bounded in the following way by resorting to the space Y: let V be the error in estimating the Hessian H_y, and let $\underline{g}$ be the error in estimating the gradient. Associated with the measurement error e (and dependent on the particular sampling procedure) V and $\underline{g}$ are characterized by particular (usually mutually dependent) distributions. By truncating the tails of these distributions, such that they contain only a suitably small probability mass, the error in the step $\Delta\underline{y}$ corresponding to equation (2.1) can be bounded with probability $P < 1$, relative to some metric by

$$\max // \Delta \underline{y} // = \max // B \; \text{grad} [f(\underline{y})] + H_y^{-1} \underline{g} + B \underline{g} //$$

$$< \max // B \; \text{grad} [f(\underline{y})] // + \max // H_y^{-1} \underline{g} //$$

$$+ \max // B \; \underline{g} // \qquad (2.2)$$

where B is related to V and reflects the uncertainty in the inverse Hessian (after truncation of the tails of the distribution of e). In practice, of course, all the above quantities have either to be replaced by estimates or by estimated bounds. Estimating bounds associated with B requires the most effort, and hence we shortly describe this process. The estimation of bounds for $\underline{g}$ can be done along similar lines, while estimates of H_y and grad [$f(\underline{y})$] are obtained from the main body of the basic algorithm.

At first, a bound W for V can be found in the following way. From the particular known finite differences the distribution of the individual second differences can be specified for a given distribution of e . One may now truncate the distributions of the second finite differences (on both sides) at levels $\pm w(i,j)$. If these resulting confidence intervals are small enough in the sense of

$$|w(i,j)| < \sqrt{h_{yii} \, h_{yjj}} \; / \; N \qquad (2.3)$$

then the infinite sum over $(H_y^{-1} W)^i$, $i = 0, \ldots,$ converges. Instead of $(H_y + V)$ one can now use bounds for equation (2.2) by computing bounds on products of the type

$$//(H_y + V)^{-1} // = // H_y^{-1} + B // < \sum_i //(H_y^{-1} W)^i H_y^{-1} // \; (2.4)$$

When the elements of W are very small in the sense of inequality (2.3), rapidly converging sums for evaluating bounds on the individual components in inequality (2.2) may be computed easily. Note that the algorithm is designed to provide small $w(i,j)$'s.

A remark should be made on the computation of the probabilities that all of the bounds are satisfied. The individual quantities involved may be mutually dependent (depending on one's sampling scheme). Thus it is useful to consider the following bounds on probabilities: Assume there are N statements s_i , each true with probability $P = 1 - \varepsilon_i$. Then, for all statements to be true simultaneously the probability is bounded by

$$1 - \max(\varepsilon_i) > P(\bigcap_i s_i = \text{true}) > 1 - \sum_i \varepsilon_i \qquad (2.5)$$

This bound may, for example, be applied to W. In this case one can find by Tchebycheffs inequality with var[e] = σ^2

$$P[\; |w(i,j)| \; / \; \sqrt{h_{yii} \, h_{yjj}} < \rho \; \forall \; i,j \;]$$

$$> 1 - \sum_i \sum_j (c_{ij}/ \rho^2) \; \sigma^2/(\Delta_i^2 f)(\Delta_j^2 f) \qquad (2.6)$$

where

$$\Delta_k^2 f = f(\underline{y} + \delta \underline{y}_k) - 2 f(\underline{y}) + f(\underline{y} - \delta \underline{y}_k) \qquad (2.7)$$

Here c_{ij} is a numerical factor which depends on the particular sampling scheme. For finite differences similar to equation (2.7), it is between 4 and 6. Observe the signal-to-noise ratio in equation (2.6) formed by the second finite differences and σ^2; note the strong dependence of this bound on the dimensionality N of the problem.

In summary, the sampling in conjugate directions does not only lead to a "minimization" of error in estimating derivatives, but it also allows, by use of simple algorithms, evaluating "on-line" estimates of bounds on the performance of the minimization method. The "minimization" of the effect of noise is further enhanced by a suitable choice of finite differences.

2.3 Choice Of Finite Differences

As explained earlier, the choice of finite differences is a tradeoff between the expected bias in estimates of derivatives when "large" values are used and the spread of estimates when "small" values are used. Although one can test the goodness-of-fit of a quadratic regression line along any particular direction (e.g. by the analysis of residuals, or by testing the order of a regression model with Akaike's AIC [1]) there is a problem with picking initial values of finite differences. It seems that initial guesses have to be provided based on one's understanding of the underlying (physical) nature of the problem. Once such initial guesses have been provided (possibly also including a description of uncertainty in these parameters) doubling or bisection type methods seem to be a suitable approach to refining initial estimates of finite differences until "large" finite differences have been found which satisfy the goodness-of-fit tests.

2.4 Orthogonal Projection And Sequential Sampling

Assume a prior estimate of the (n + 1)st direction, conjugate to the subspace X_n is given, and the "maximal" finite differences have been determined. One may now compute, say in $Y_{(n+1)}$ the Hessian $H_{y(n+1)}$ by the finite differences along orthogonal directions, say $\delta \underline{y}_{(n+1)}$ and $\delta \underline{y}_i$, $1 \leq i \leq (n+1)$, where $\delta \underline{y}_k$ has only its k'th component not equal to zero. The resultant Hessian $H_{y(n+1)}$ will generally be of the form

$$H_{y(n+1)} = \begin{bmatrix} D_n & \underline{h} \\ \underline{h}' & c \end{bmatrix} \qquad (2.9)$$

which can be expressed, using standard matrix manipulation, as

$$H_{y(n+1)} = U' \; D_{(n+1)} \; U \qquad (2.10)$$

where D indicates a diagonal matrix. The last column of the matrix U can directly be used for the last column in S to express

$$H_{x(n+1)} = S' \; D_{(n+1)} \; S \qquad (2.11)$$

while redefining orthogonal directions in $Y_{(n+1)}$ by an updated estimate of $H_{y(n+1)} = D_{(n+1)}$. The finite difference $\delta \underline{y}_{(n+1)}$ may now again be maximized. This procedure will be repeated when the h_i, $i \neq n+1$, in equation (2.9) are considered too large to be accounted for by the confidence interval w(i,n+1).

The main value of this procedure is to sample along potential ridges of the object function $f(\underline{x})$. If this is not done, sample points in the larger space X would be wasted, as they cannot give much information on the direction and the curvature of the ridge. When "very high" dimensional problems are considered, multidimensional ridges may arise, e.g. when the eigenvalues of the Hessian H_x are spread over many orders of magnitude. Trying to minimize a function under those conditions will be in vain unless "informative" samples are drawn from the response surface by distributing them well.

In retrospect the outlined orthogonal projection and resampling procedure is designed to solve the problem of allocating informative samples in the space X. When a problem can still not be solved one is informed about the nature of the problem. For example, one finds out which variable is associated with a problem. Some possible cures for this situation are discussed later. The remaining question is how the termination of the procedure is specified in a probabilistic way after all of the above tests have been passed.

2.5 Termination

The uncertainty in measurements can be used to define a natural termination criterion. From equation (2.2) the uncertainty in a step can be found in full generality. However some simplification is possible by observing that the error due to the second term $t_2 = H_y^{-1} \underline{g}$ will be the dominant part near termination. If one assumes measurement error e to have a Gaussian distribution, this term t_2 has a χ^2 distribution (the degrees of freedom depend on the particular sampling scheme, however). Thus one may simply test, based on a χ^2 distribution, when a step $\Delta\underline{x}$ satisfies

$$P[\ // \ \Delta\underline{x} \ // < r\ ] > 1 - \varepsilon \qquad (2.12)$$

with a suitable value ε ; the quantity r specifies an estimate of a corresponding confidence region R as a centered hyperellipsoid. In case of "high dimensional" problems and for non-Gaussian distributions of e one may still use a suitable distribution by invoking limiting arguments.

3.0 DISCUSSION

Some of the issues concerning the evaluation of anticipated performance, potential to overcome a number of difficulties (e.g. associated with non-convex functions), and the limitations built into the outlined method, deserve some discussion:

First, note the difficulty of estimating a quadratic approximation in an n-dimensional space: when standard regression is used, n(n+1)/2 coefficients describing this quadratic have to be estimated. The system of equations becomes unsolvable (by todays standards) when n reaches a value of about 50. As an alternative to such an approach, we propose to sample the function $f(\underline{x})$ somewhat more often. This allows, by retaining only points along estimated conjugate directions (which are fitted by quadratic lines), the reduction of the regression problem to a reasonable size.

Next, we remark that the bound in equation (2.6) appears to be rather loose. This can be seen from the following consideration. When the estimation of all finite differences is based on mutually independent samples (in the sense that the errors are independent) then one can invoke limiting arguments for the sum of absolute values of the elements in columns of W. This allows to find a fairly tight bound on the Hessian: such a bound would involve only a single sum (when, by Taylor series expansion, an approximation of the product of probabilities associated with the individual columns in W is determined), instead of the double sum in equation (2.6). Furthermore, the argument inside the sum would tend to be a slow function of the dimensionality of the problem. This might suggest that difficulties with the proposed scheme grow only slightly superlinearly with respect to the dimensionality of the problem at hand. Note, that for estimation of independent derivatives only about twice as many sample points are necessary compared to an approach which tries to minimize the number of necessary sample points by utilizing them repeatedly for the various derivatives.

Several benefits can be drawn from the sequential expansion of subspaces. For example, when a hyperbolic point on $f(\underline{x})$ is encountered, one can, by use of Green's method [as cited by 3], negotiate the problem by limiting minimization to this subspace until one has come out of such a region. Since hyperbolic regions (which may be viewed as being due to higher terms of a Taylor series expansion at the minimum of the function) are not well treated with quadratic approximations in minimization methods, sampling effort may be reduced by avoiding uninformative sampling in "bad" regions. Yet another advantage of the method may be realized in certain situations: consider the case where measurment error decreases near the minimum of the function. In case of having difficulties in minimizing the N-dimensional problem, one restricts minimization, in the beginning, to a subspace. In this event, measurement error decreases until it is possible to handle the full problem.

Finally, we mention the interest to also evaluate the bias in estimates of derivatives. One approach would be to fit fourth-order polynomials to the directions of interest. The cubic and the quartic coefficients allow to estimate the bias in the estimate of the first and the second derivative of the surface. For estimating these higher order coefficients one may include the sampling points which had not been acceptable for fitting a quadratic regression line along conjugate directions.

In summary, an algorithm has been presented for nonlinear parameter estimation which accounts explicitly for measurement error. Measurement error is often a cause for unsuccessful attempts to minimize functions of many variables. By means of bounds the concept of "high dimensionality" is related to estimates of one's ability to converge with a desired probability to a desired accuracy. Beside this new formulation of the problem, the bounds will allow in many cases to provide on-line evaluation of the performance of the minimization scheme. The bounds and tests in the procedure also permit diagnosis of specific types of problems in a systematic fashion. These possibilities may lead to very powerful optimization schemes.

REFERENCES

[1] Akaike, H., "A New Look at the Statistical Model Identification," IEEE Trans. Auto. Contr., Vol. AC-19, No. 6, pp716-723, 1974.

[2] Avriel, M., Nonlinear Programming Analysis and Methods, Prentice-Hall, Inc., Englewood Cliffs, NJ, 1976.

[3] Bard, Y., Nonlinear Parameter Estimation, Academic Press, New York, San Franzisco, London, 1974.

[4] Dvoretzky A., "On Stochastic Approximation Methods," Proc. 3rd Berkeley Symp. Math. and Prob. (J. Neyman ed.), Univ. of Calif. Press, Berkeley, Calif., pp39-55, 1956.

[5] Fletcher, R., Freeman, T. L., "A Modified Newton Method for Minimization," J. of Optim. Theory and Applications, Vol. 23, No. 3, pp357-372.

[6] Fletcher R., Powell M.J.D., "A Rapidly Convergent Descent Method for Minimization," The Computer J., Vol. 6, p136, 1963.

[7] Gupta, N. K., Mehra, R. K., "Computational Aspects of Maximum Likelihood Estimation and Reduction in Sensitivity Function Calculations," IEEE Trans. on Autom. Control, Vol. AC-19, No. 6, pp774-783, 1974.

[8] Kushner, H. J., Clark, D. S., Stochastic Approximation Methods for Constrained and Unconstrained Systems, Springer Verlag, New York, Heidelberg, Berlin, 1978.

[9] Lai, T., Robbins, H., "Adaptive Design and Stochastic Approximation," The Ann. of Statist., Vol. 7, No. 6, pp1196-1221, 1979.

[10] Nazareth, L., "A Conjugate Direction Algorithm Without Line Searches," J. of Optim. Theory and Applications, Vol. 23, No. 3, pp373-387.

[11] Nevelson, M. B., Hasminskii, R. Z., Stochastic Approximation and Recursive Estimation, Providence, RI: American Mathematical Society, 1973.

[12] Robbins, H., Monro, S., "A Stochastic Approximation Method," Ann. Math. Statist., Vol. 22, pp400-407, 1951.

EXTENSIONS TO SUBROUTINE VF02AD

by

M.J.D. Powell

Department of Applied Mathematics and Theoretical Physics,
University of Cambridge,
Silver Street, Cambridge CB3 9EW, England.

1. Algorithms for constrained optimization

Many optimization calculations are of the form: minimize $F(\underline{x})$, $\underline{x} \in \mathbb{R}^n$, subject to the constraints

$$\left.\begin{array}{ll} c_i(\underline{x}) = 0, & i = 1,2,\ldots,m', \\ c_i(\underline{x}) \geq 0, & i = m'+1,\ldots,m. \end{array}\right\} \qquad (1.1)$$

We assume that the functions F and $\{c_i;\ i = 1,2,\ldots,m\}$ are all real-valued, are all differentiable, and that their values and first derivatives can be calculated for any $\underline{x}$. Numerical results show that the algorithm described by Powell (1978), which is implemented as Fortran subroutine VF02AD in the Harwell Library, is a particularly promising technique for solving this minimization problem. However, because the algorithm has some deficiencies, the purpose of this paper is to describe some extensions to the subroutine that overcome a few of them.

Ample justification for the statement that VF02AD is particularly promising is given in the comparative study of Hock and Schittkowski (1981). They describe many test examples for nonlinear programming codes and, using them, they compare six algorithms for constrained optimization that are available as computer programs. We quote some results that are obtained by four of these algorithms, which apply the following four main approaches to constrained optimization:

Penalty Function	FMIN
Reduced Gradient	GRGA
Augmented Lagrangian	VF01A
Successive Quadratic Programming	VF02AD,

where the column on the right lists the names of the algorithms. The results that we quote are for the following four test problems:

#43	Rosen-Suzuki	n=4	m=3
#86	Colville 1	n=5	m=10

#108	Himmelblau 16	n=9	m=13
#117	Colville 2	n=15	m=5.

Here the left hand column gives reference numbers that are used by Hock and Schittkowski (1981), and the right hand columns give numbers of variables and numbers of constraints. The constraints of the Colville 1 problem are all linear, but, except for some non-negativity conditions on variables which are not counted, all other constraints are nonlinear.

Table 1 gives pairs of figures of the form a/b, where a is the number of times the objective function $F(\underline{x})$ is evaluated during an optimization calculation, and where b is the number of evaluations of the gradient vector $\underline{\nabla}F(\underline{x})$. Of course the numbers of constraint function evaluations are important also, but they are not given because they do not provide information that alters the main conclusions. Table 2 gives computer times in seconds on a Telefunken TR440 computer for the solutions of the four problems by the four algorithms.

Table 1 Number of evaluations of F and $\underline{\nabla}F$

	FMIN	GRGA	VF01A	VF02AD
#43	366/92	426/101	66/66	12/12
#86	1190/112	246/15	57/57	9/9
#108	984/210	678/176	201/201	9/9
#117	1324/206	1142/197	201/201	17/17

Table 2 Execution times

	FMIN	GRGA	VF01A	VF02AD
#43	5.84	2.95	0.42	1.47
#86	49.97	1.81	0.55	1.26
#108	46.47	21.85	3.58	14.17
#117	57.34	17.62	6.15	59.89

The figures for the four test problems are typical of the figures of Hock and Schittkowski (1981) for 119 test problems. The first table suggests clearly that the successive quadratic programming approach is more economic than the other three methods in terms of function and gradient evaluations. Partly this is because FMIN and VF01A have to solve a sequence of unconstrained calculations in n variables for each of the test problems, while GRGA has to follow constraint boundaries closely. Thus there are good reasons for studying further the properties of VFO2AD, even though sometimes, as shown in Table 2, the execution times of VFO2AD are quite high. Another successive quadratic programming routine, namely OPRQP (Biggs, 1975), is usually faster than VFO2AD on the 119 test problems, but it tends to require more function and gradient evaluations; in particular it solves problem #117 in only 9.00 seconds, using about 40 function and gradient evaluations.

However, there are situations where VFO2AD is inefficient, which are not shown by Hock and Schittkowski (1981). Two of them are mentioned, and are illustrated by numerical examples, in this paper. One is due to the "Maratos effect", which is considered in Section 3, and the other is due to the adjustment of penalty parameters of a line search objective function, which is considered in Section 4. Suitable modifications to VFO2AD are proposed, and they are discussed briefly in Section 5. They are included in a new Fortran subroutine, called VMCWD, which is available from the author, except that the user has to provide a procedure for quadratic programming. Some numerical results that have been obtained by VMCWD are given in Sections 3, 4 and 5.

2. Some details of VFO2AD

Each iteration of VFO2AD begins with an estimate, $\underline{x}_k$ say, of the required vector of variables, an $n\times n$ positive definite matrix, B_k say (which can be regarded as an approximation to the second derivative matrix of the Lagrangian function of the minimization problem), and a set $\{\mu_i;\ i = 1,2,\ldots,m\}$ of non-negative parameters that are used after modification in a line search. On most iterations the search direction, $\underline{d}_k$ say, is the vector $\underline{d}$ that minimizes the quadratic function

$$Q(\underline{d}) = F(\underline{x}_k) + \underline{d}^T\underline{\nabla}F(\underline{x}_k) + \tfrac{1}{2}\underline{d}^T B_k\underline{d}\ , \tag{2.1}$$

subject to the linear constraints

$$\left.\begin{aligned} c_i(\underline{x}_k) + \underline{d}^T\underline{\nabla}c_i(\underline{x}_k) &= 0\ , \quad i = 1,2,\ldots,m',\\ c_i(\underline{x}_k) + \underline{d}^T\underline{\nabla}c_i(\underline{x}_k) &\geq 0\ , \quad i = m'+1,\ldots,m. \end{aligned}\right\} \tag{2.2}$$

However, $\underline{d}_k$ is modified (Powell, 1978) if the constraints (2.2) are inconsistent. Having calculated $\underline{d}_k$, a positive multiplier α_k is chosen, and $\underline{x}_{k+1}$ is given the value

$$\underline{x}_{k+1} = \underline{x}_k + \alpha_k \underline{d}_k \ . \tag{2.3}$$

Finally, B_{k+1} is calculated from B_k and from a change in gradient of an estimate of the Lagrangian function.

We give particular attention to the choice of α_k. It depends on the parameters $\{\mu_i;\ i = 1,2,\ldots,m\}$, but they may differ from the values that were given at the start of the iteration, because on some iterations they are revised at the end of the quadratic programming calculation that determines $\underline{d}_k$. The step-length α_k has to satisfy the condition

$$W_k(\underline{x}_k + \alpha_k \underline{d}_k) < W_k(\underline{x}_k) \ , \tag{2.4}$$

where W_k is the function

$$W_k(\underline{x}) = F(\underline{x}) + \sum_{i=1}^{m'} \mu_i |c_i(\underline{x})| + \sum_{i=m'+1}^{m} \mu_i \max[0,-c_i(\underline{x})] . \tag{2.5}$$

One reason for revising $\underline{\mu}$ is to ensure that condition (2.4) can be obtained by choosing α_k to be sufficiently small and positive.

In the line search a step-length of one is tried initially, but it is reduced if a condition that is a little stronger than inequality (2.4) is not obtained. Reductions are made in the trial values of α_k, each reduction being at most a factor of ten, until the step-length is acceptable, except that there is an error return from VF02AD if five reductions are insufficient. Another error return from VF02AD, which is mentioned in Section 5, occurs if the constraints (2.2) are not only inconsistent, but also there is no value of $\underline{d}$ that can reduce all the constraint violations that are present when $\underline{d} = 0$. An error return of this kind is necessary, in case the nonlinear constraints (1.1) are inconsistent.

3. The Maratos effect and the watchdog technique

Suppose that an iteration of VF02AD is started at a point $\underline{x}_k$ that satisfies all the constraints and that is close to the required solution $\underline{x}^*$. Then, using the data of the quadratic programming calculation that determines $\underline{d}_k$, one can predict the reduction that occurs in the line search objective function (2.5) when $\underline{x}$ is changed from $\underline{x}_k$ to $\underline{x}_k + \underline{d}_k$. This predicted reduction is of order $\|\underline{d}_k\|^2$. However, since only linear approximations to the constraints are satisfied at $\underline{x}_k + \underline{d}_k$, it is possible for $|c_i(\underline{x}_k+\underline{d}_k)|$ to also be of order

$\|\underline{d}_k\|^2$. Thus inequality (2.4) need not be satisfied when $\alpha_k = 1$, even if $\underline{x}_k$ is very close to $\underline{x}^*$ and B_k is an excellent approximation to the Lagrangian function at the solution. In this case the step-length is reduced, which usually makes the rate of convergence of the iteration only linear; this phenomenon is called the "Maratos effect".

For example, consider the "circular constraint problem":

$$\left.\begin{array}{ll} \text{minimize} & F(\underline{x}) = -x_1 + 10(x_1^2 + x_2^2 - 1) , \\ \text{subject to} & x_1^2 + x_2^2 - 1 = 0 , \\ \text{starting at} & (x_1,x_2) = (0.8,0.6) . \end{array}\right\} \qquad (3.1)$$

Because of the Maratos effect, it is highly inefficient to solve it by subroutine VFO2AD. Specifically, Table 3 gives the calculated components of $\underline{x}_{k+1}$ for $k = 33$, 34 and 35; the 35 iterations required 100 function and gradient evaluations. The table suggests correctly that the Maratos effect can persist for many iterations, and that it causes very slow convergence.

Table 3 Subroutine VFO2AD applied to problem (3.1)

Iteration	x_1	x_2
33	0.9817	0.1994
34	0.9856	0.1793
35	0.9887	0.1613

Remedies have been proposed by Maratos (1978) and by Chamberlain, Lemaréchal, Pedersen and Powell (1980), the latter one being called the "watchdog technique". This technique is based on the observation that, if $\underline{x}_k + \underline{d}_k$ is so infeasible that condition (2.4) fails when $\alpha_k = 1$, then a normal iteration from the point $\underline{x}_{k+1} = \underline{x}_k + \underline{d}_k$ may give the reduction

$$W_k(\underline{x}_{k+1} + \underline{d}_{k+1}) < W(\underline{x}_k) . \qquad (3.2)$$

Therefore on some iterations the watchdog technique allows $\alpha_k = 1$, even though condition (2.4) may not be obtained. If a pre-set number of further iterations fail to achieve a new least value of $W_k(\underline{x})$, then "back-tracking" occurs, i.e. the next iteration is started at the vector of variables that has given the least calculated value of $W_k(\underline{x})$, and on the new iteration the reduction (2.4) is mandatory.

The watchdog technique is included in the extended version of VF02AD, namely subroutine VMCWD, and one refinement has been added to the description of Chamberlain et al. (1980). It is that, if back-tracking occurs, then the reduction (2.4) is mandatory for the next ten iterations. Problem (3.1) is solved very efficiently by subroutine VMCWD, the vector $\underline{x}_{k+1}$ being correct to six decimal places after only 5 iterations (9 function and gradient evaluations).

4. The choice and adjustment of $\underline{\mu}$

A usual way of forcing convergence of an algorithm from a poor starting point is to force each iteration to reduce a function of the variables. However, due to changes in $\underline{\mu}$ on each iteration, the line search objective function (2.5) may not be suitable. Indeed, Chamberlain (1979) gives an example where, even though each iteration of VF02AD makes the reduction $W_k(\underline{x}_{k+1}) < W_k(\underline{x}_k)$, the numbers $\{W_k(\underline{x}_k);\ k = 1,2,3,...\}$ are all the same; in this example a cycle occurs such that $\underline{x}_{k+2} = \underline{x}_k$ for all k, and such that the vector $\underline{\mu}$ is also repeated every two iterations.

Therefore, in VMCWD no component of $\underline{\mu}$ is ever reduced, and, when an increase occurs, at least one component of $\underline{\mu}$ is increased by at least a pre-set relative amount. It follows that, provided $\underline{\mu}$ remains bounded, it also remains constant after a finite number of iterations. Therefore it is hoped that, for most calculations, the algorithm will correct automatically an initial value of $\underline{\mu}$ that is too small, and that the algorithm will have the global convergence properties that can be obtained for a constant vector $\underline{\mu}$.

A danger of this approach, however, is that an unnecessarily large value of $\underline{\mu}$ may penalize constraint violations so heavily that inefficiencies occur due to following curved constraint boundaries closely. The use of the watchdog technique lessens this danger, but the following example shows that disastrous behaviour can occur.

The example is the "poorly scaled banana constraint problem":

$$\left.\begin{array}{ll} \text{minimize} & F(\underline{x}) = 10^{-3}(x_1-1)^2 , \\ \text{subject to} & 10^{-3}(x_1^2-x_2) = 0 , \\ \text{starting from} & (x_1,x_2) = (-0.8,1.0) . \end{array}\right\} \qquad (4.1)$$

In both VFO2AD and VMCWD the matrix B_k is set to the unit matrix for the first iteration. Therefore, because of the scaling factors 10^{-3} that occur in the problem, and because of the change that has to be made to the initial $\underline{x}$ to satisfy a linear approximation to the constraint, it follows that, in the initial quadratic programming calculation that determines the search direction, the Lagrange parameter has a rather large value, namely 102.74. This number is the initial value of μ_1 in VFO2AD, but, during the first 14 iterations, μ_1 is reduced automatically to 0.0446, and at this stage the solution of problem (4.1) has been found to six decimals accuracy. Thus the calculation of VFO2AD takes only 15 function and gradient evaluations.

However, if VMCWD is applied to this problem, and if $\mu_1 = 102.74$ on the first iteration (which is not the choice that would be made automatically), then the need to keep close to the constraint boundary causes the matrix B_k to become nearly singular. Thus, after only seven iterations, the search direction $\underline{d}_k = (445.09,-819.07)^T$ is calculated, which is so long that five reductions to the initial value $\alpha_k = 1$ do not provide a suitable step-length. Therefore, as mentioned in the last paragraph of Section 2, there is an error return from VMCWD.

The example suggests that it is more important than before to choose the initial value of $\underline{\mu}$ carefully. The main requirement on $\underline{\mu}$ is that $\underline{d}_k$ is a descent direction at $\underline{x}_k$ for the line search objective function (2.5). In particular, in the case of problem (4.1), it is sufficient if the initial value of μ_1 satisfies the condition

$$\mu_1 |c_1(\underline{x}_1)| > |\underline{d}_1^T \nabla F(\underline{x}_1)| \,. \tag{4.2}$$

The actual choice of μ_1 is based on this inequality, and normally VMCWD sets $\mu_1 = 3.216$, which allows problem (4.1) to be solved in only 13 iterations (14 function evaluations).

In the general case VMCWD applies the following rules for choosing and adjusting $\underline{\mu}$. Before the first iteration all components of $\underline{\mu}$ are given tiny positive values, in case F is a constant function. If the search direction of an iteration satisfies the condition

$$\underline{d}_k^T \nabla F(\underline{x}_k) + \tfrac{1}{2}\underline{d}_k^T B_k \underline{d}_k \leqslant 0 \,, \tag{4.3}$$

then no change is made to $\underline{\mu}$. Otherwise we require the inequality

$$\sum_{i=1}^{m'} \mu_i |c_i(\underline{x}_k)| + \sum_{i=m'+1}^{m} \mu_i \max[0,-c_i(\underline{x}_k)] - \sum_{i=1}^{m'} \mu_i |c_i(\underline{x}_k) + \underline{d}_k^T \nabla c_i(\underline{x}_k)|$$

$$-\sum_{i=m'+1}^{m} \mu_i \max[0,-c_i(\underline{x}_k)-\underline{d}_k^T\underline{\nabla} c_i(\underline{x}_k)] \geqslant \beta_k |\underline{d}_k^T\underline{\nabla} F(\underline{x}_k)| \tag{4.4}$$

to hold, where β_k is a positive constant. If the $\underline{\mu}$ that is set at the beginning of the iteration satisfies this condition when $\beta_k = 1.5$, then $\underline{\mu}$ is not altered. Otherwise the following procedure is used to obtain a value of $\underline{\mu}$ such that inequality (4.4) holds for $\beta_k = 2$. The increase in β_k ensures that at least one component of $\underline{\mu}$ is multiplied by a number that exceeds 4/3.

For $i = 1,2,\ldots,m$ we let λ_i be the Lagrange multiplier of the i-th of the constraints (2.2) at the solution of the quadratic programming problem that determines $\underline{d}_k$. For each i we leave μ_i unchanged if increasing its value would not increase the left hand side of expression (4.4). Otherwise μ_i is given the value

$$\mu_i(\text{new}) = \max[\mu_i(\text{old}),\sigma|\lambda_i|] , \tag{4.5}$$

where the positive parameter σ, which is independent of i, is determined by the condition that expression (4.4) is satisfied as an equation when $\beta_k = 2$.

It is straightforward to implement this procedure. The presence of λ_i in expression (4.5) causes the line search objective function to be independent of constraint scaling, except for the tiny contribution from the tiny components of $\underline{\mu}$ that are set initially. The reason for condition (4.3) is that, if it holds, then the quadratic programming problem that determines $\underline{d}_k$ suggests that a substantial reduction would be obtained in the line search objective function even if $\underline{\mu}$ were zero; therefore there is no need to increase $\underline{\mu}$. It can be proved that the given choice of $\underline{\mu}$ makes the search direction $\underline{d}_k$ a direction of descent for the line search objective function (2.5).

5. Discussion of the modifications in VMCWD

We give one more example to guide our discussion, namely the "sine-cosine wriggle problem":

$$\left.\begin{array}{ll} \text{minimize} & F(\underline{x}) = x_1 , \\ \text{subject to} & x_2 - x_1\sin(10\pi x_1) \geqslant 0 \\ \text{and} & [2+\cos(10\pi x_1)]x_1 - x_2 \geqslant 0 , \\ \text{starting from} & (x_1,x_2) = (1.0,1.0) . \end{array}\right\} \tag{5.1}$$

The solution to this example is at the origin, there are no feasible points in the left half plane $x_1 < 0$, and the oscillations of the constraint functions imply that linear approximations to the constraints can be very misleading.

When VF02AD and VMCWD are applied to this problem, there are very large excursions into the infeasible region from which there is no recovery. The reason is that the automatic choice of $\underline{\mu}$ is so small that the line search objective function (2.5) does not indicate that moves into the infeasible region should be treated with suspicion. Therefore, subroutine VMCWD includes an option that allows the user to specify the initial value of $\underline{\mu}$. Thus $\mu_1 = 102.74$ was assigned for one of the numerical experiments on problem (4.1) that are mentioned in Section 4.

When this option gives the initial values $\mu_1=\mu_2=2$, then subroutine VMCWD solves the problem (5.1) in 43 iterations (67 function evaluations). A similar option in VF02AD would be useful only if some of the automatic reductions in $\underline{\mu}$ were suppressed.

Another option in VMCWD that is sometimes helpful is that the user may also specify simple upper bounds on the moduli of the components of the search directions. For example, if the algorithm is applied to problem (4.1), if $\mu_1=102.74$ initially, and if the bounds $|d_1| \leq 0.2$ and $|d_2| \leq 0.2$ are added to the constraints (2.2) in order to prevent the error return that is noted in Section 4, then the solution to the problem is found in 106 iterations (201 function evaluations). In this case the watchdog technique is a hinderance until the last five iterations, which is the reason for the refinement that is given in the last paragraph of Section 3.

If VMCWD is applied to problem (5.1), and if the bounds $|d_1| \leq 0.1$ and $|d_2| \leq 0.1$ and the initial values $\mu_1=\mu_2=2$ are set, then the solution is obtained in 47 iterations (52 function evaluations), so in this example the bounds reduce the number of function evaluations that are taken. However, if the same bounds are set but the initial value of $\underline{\mu}$ is assigned automatically, then x_1 converges to -0.12149 and x_2 has a similar value, so neither constraint is satisfied. Because the gradients $\underline{\nabla} c_1(\underline{x})$ and $\underline{\nabla} c_2(\underline{x})$ are parallel and have opposite signs when $x_1=-0.12149$, there is an error return from VMCWD, as mentioned at the end of Section 2, because it seems that the constraints are inconsistent, even though the starting point $(x_1,x_2) = (1.0,1.0)$ was feasible!

These examples show that, for problems with highly nonlinear constraints, it may be necessary for the user to assign bounds on the search directions and/or the initial value of $\underline{\mu}$. The availability of these options and the watchdog technique are the main features that are in VMCWD but not in VF02AD.

The arguments of VMCWD are the same as those of VF02AD, except that the values -101, -110 and -111 of the input parameter INFO are special. In VF02AD this parameter has to be set to any negative integer, and in all cases $\underline{\mu}$ is initialized automatically and there are no bounds on the search directions. However, the special values indicate to VMCWD that bounds on $\underline{d}$ and/or the initial value of $\underline{\mu}$ are provided in a working space array. Therefore, unless a current user of VF02AD has made the unfortunate choice INFO = -101, or -110, or -111, he may switch to subroutine VMCWD by changing only the name of the constrained optimization subroutine in his computer program. Further, it is planned that, in the Harwell Library, VMCWD will replace VF02AD and will inherit its name. Thus users' programs will be switched to the new routine automatically.

Further research needs to be done on several parts of the calculation, including the interface and method of the quadratic programming subroutine that calculates the search directions, the initial choice of the variable metric matrix B , and the estimation of Lagrange multipliers.

References

M.C. Biggs (1975), "Some improvements to the OPTIMA subroutines", Technical Report No. 69, Numerical Optimization Centre, The Hatfield Polytechnic.

R.M. Chamberlain (1979), "Some examples of cycling in variable metric algorithms for constrained minimization", Math. Programming, Vol. 16, pp.378-383.

R.M. Chamberlain, C. Lemaréchal, H.C. Pedersen and M.J.D. Powell (1980), "The watchdog technique for forcing convergence in algorithms for constrained optimization", Report DAMTP 80/NA9, University of Cambridge (to be published in Math. Programming Stud.).

W. Hock and K. Schittkowski (1981), Test Examples for Nonlinear Programming Codes, Lecture Notes in Economics and Mathematical Systems 187, Springer-Verlag (Berlin).

N. Maratos (1978), "Exact penalty function algorithms for finite dimensional and control optimization problems", Ph.D. thesis, Imperial College, University of London.

M.J.D. Powell (1978), "A fast algorithm for nonlinearly constrained optimization calculations", in Numerical Analysis, Dundee 1977, Lecture Notes in Mathematics 630, ed. G.A. Watson, Springer-Verlag (Berlin), pp.144-157.

ON GLOBALLY STABALIZED QUASI-NEWTON METHODS FOR INEQUALITY CONSTRAINED OPTIMIZATION PROBLEMS

E. Polak and A. L. Tits

Department of Electrical Engineering and Computer Sciences
and the Electronics Research Laboratory
University of California, Berkeley, California 94720

1. INTRODUCTION

Over the last several years there have been a number of successful attempts to construct superlinearly converging algorithms for the solution of constrained optimization problems. A common starting point in the construction of these new methods is the use of Newton's method, in some form, for solving the Kuhn-Tucker first order optimality condition equations and inequalities. These methods can be grouped into two categories: those traceable to R W. Wilson's successive quadratic programming method (SQP) [15], and those which emanate from the ordinary Newton method for the solution of equations.

Wilson's method is a form of Newton's method which solves a quadratic program with equality and inequality constraints at each iteration. For optimization problems of the form $\min\{f(x) \mid h(x) = 0\}$, it yields exactly the same iterates (x_i,λ_i) as the ordinary Newton method does when applied to the optimality equations $h(x) = 0$, $\nabla f(x) + (\partial h(x)/\partial x)^T\lambda = 0$; for optimization problems of the form $\min\{f(x) \mid g(x) \leq 0\}$, it yields iterates which differ only by a second order term from those constructed by the extended Newton method, developed by Robinson [13], when applied to the Kuhn-Tucker optimality equations and inequalities, viz. $\mu^j g^j(x) = 0$, $\nabla f(x) + (\partial g(x)/\partial x)^T\mu = 0$, $g(x) \leq 0$, $\mu \geq 0$. It was shown by Robinson [12] that when intialized sufficiently closely to a "strong" Kuhn-Tucker pair (x,μ), the SQP method was quadratically convergent. SQP was extended to a quasi-Newton version by Han [4,5,6]. Han also globalized the local method, i.e., extended its domain of convergence as well as eliminated the possibility of convergence to a local maximum instead of to a local minimum, by using an exact penalty function for step size determination: a technique subsequently refined and improved upon by Powell [11] and Mayne and Polak [8]. The main drawback of successive quadratic programming is that it is difficult to find reliable quadratic programming codes, capable of solving non-positive-semidefinite problems, that find a solution of smallest norm, as required by Robinson's theory [12] for superlinear convergence.

The extended Newton method was never tried for solving the Kuhn-Tucker relations of general optimization problems because of a persisting erroneous belief that it would fail because the relations did not satisfy the Robinson LI conditions [13] and because it was not clear how it could be globalized. However, it was considered for problems of the form $\min\{f(x) : h(x) = 0\}$ by Tapia [14] and by Bertsekas [1].

Furthermore, Bertsekas was able to globalize Newton's method by using an exact differentiable penalty function, proposed by DiPillo and Grippo [2], as a descent function in step size determination. He showed that Newton's method yields a direction which, asymptotically, approaches the Newton direction for the DiPillo and Grippo penalty function. For problems with both equality and inequality constraints, Bertsekas has proposed an "active set" strategy, as a means of removing the need to solve inequalities as well as equations. The obvious advantage of the ordinary Newton method over successive quadratic programming is that it only needs to solve a linear equation at each iteration.

In the present paper, we show that when a sufficiently good initial proximation to a "strong" Kuhn-Tucker triplet is available, optimization problems with both equality and inequality constraints can be solved <u>without</u> using an active set strategy, by applying Newton's method, or a quasi-Newton method, only to the equations part of the Kuhn-Tucker conditions. The resulting local method is superlinearly convergent. For problems with inequality constraints only, we show that globally convergent methods with excellent overall properties can be obtained by combining quasi-Newton methods with a phase I - phase II method of feasible directions.

2. Local Methods

Consider the problem

$$\min\{f(x) \mid g(x) \le 0,\ h(x) = 0\} \tag{1}$$

where $f : \mathbb{R}^n \to \mathbb{R}$, $g : \mathbb{R}^n \to \mathbb{R}^m$ and $h : \mathbb{R}^n \to \mathbb{R}^\ell$ are all twice continuously differentiable. Let x^* be a local minimizer for (1) such that the triplet $z^* = (x^*,\mu^*,\lambda^*)$ satisfies the Kuhn-Tucker first order conditions:

$$\nabla_x L(x,\mu,\lambda) = 0 \ ; \tag{2a}$$

$$h(x) = 0 \ ; \tag{2b}$$

$$\mu^j g^j(x) = 0, \quad j \in \underline{m} \ ; \tag{2c}$$

$$g(x) \leqq 0 \ ; \tag{2d}$$

$$\mu \geqq 0 \ ; \tag{2e}$$

where $L(x,\mu,\lambda) = f(x) + \langle \mu, g(x)\rangle + \langle \lambda, h(x)\rangle$ and $\underline{m} = \{1,2,\ldots,m\}$.

<u>Assumption 1</u>: With $J^* \triangleq \{j \in \underline{m} \mid g^j(x^*) = 0\}$, we assume

$$\langle y, \frac{\partial^2 L(x^*,\mu^*,\lambda^*)}{\partial x^2} y\rangle > 0 \quad \forall y \in \{y' | \frac{\partial h(x^*)^T}{\partial x} y' = 0; \langle \nabla g^j(x^*), y'\rangle = 0 \quad \forall j \in J^*\} \tag{3}$$

(ii) that $\mu^{*j} > 0$ for all $j \in J^*$, and

(iii) that the vectors $\nabla h^k(x^*)$, $k \in \underline{\ell}$, $\nabla g^j(x^*)$, $j \in J^*$, are linearly independent.

¤

Now consider the equalities part of the Kuhn-Tucker conditions (2), viz:

$$\nabla_x L(x,\mu,\lambda) = 0; \tag{4a}$$

$$h(x) = 0 ; \tag{4b}$$

$$\mu^j g^j(x) = 0 \quad \forall j \in \underline{m}. \tag{4c}$$

We define our <u>local</u> algorithm as a quasi-Newton method applied to (4), viz., given $z_i \triangleq (x_i,\mu_i,\lambda_i)$,

$$z_{i+1} = z_i + \Delta z_i , \tag{5}$$

where $\Delta z_i = (\Delta x_i,\Delta\mu_i,\Delta\lambda_i)$ is a solution of the linear system

$$\nabla_x L(x_i,\mu_i,\lambda_i) + G(z_i)\Delta x_i + \frac{\partial g^T}{\partial x}(x_i)\Delta\mu_i + \frac{\partial h^T}{\partial x}(x_i)\Delta\lambda_i = 0 ; \tag{6a}$$

$$h(x_i) + \frac{\partial h}{\partial x}(x_i)\Delta x_i = 0; \tag{6b}$$

$$\mu_i^j g^j(x_i) + \mu_i^j \frac{\partial g}{\partial x}(x_i)\Delta x_i + \Delta\mu_i^j g^j(x_i) = 0, \quad \forall j \in \underline{m}. \tag{6c}$$

Clearly, when $G(z_i) = \frac{\partial^2 L}{\partial x^2}(z_i)$, (5-6) defines the ordinary Newton method for solving (4).

The Jacobian of the system (6) is given by

$$J(z,G) = \begin{pmatrix} G(z) & \frac{\partial g^T}{\partial x}(x) & \frac{\partial h^T}{\partial x}(x) \\ \frac{\partial h}{\partial x}(x) & 0 & 0 \\ \mu^1 \frac{\partial g^1}{\partial x}(x) & g^1(x) & 0 \\ \vdots & \vdots & \vdots \\ \mu^m \frac{\partial g^m}{\partial x}(x) & g^m(x) & 0 \end{pmatrix} \tag{7}$$

For $G = \frac{\partial^2 L}{\partial x^2}$, it was shown by McCormick [7] that under Assumption 1, $J(z^*, \frac{\partial^2 L}{\partial x^2})$ is nonsingular.

We define the norm $\|\cdot\|$ on $\mathbb{R}^{n+m+\ell}$ by

$$\|z\|^2 = \|x\|^2 + \|\mu\|^2 + \|\lambda\|^2 , \tag{8}$$

so that $\|(x,0_m,0_\ell)^T\| = \|x\|$. Then, using induced norms for matrices, we get

$$\|J(z,G_1) - J(z,G_2)\| = \|G_1(z) - G_2(z)\| \tag{9}$$

Theorem 1 (Local convergence): Suppose that for all i,

$$\|G(z_i) - \frac{\partial^2 L}{\partial x^2}(z_i)\| < \frac{1}{2\|J^*(z^*)\|}, \tag{10}$$

where $J^*(z^*) = J(z^*, \frac{\partial^2 L}{\partial x^2})$. Then there exists a $\delta > 0$ such that if $z_0 \in B(z^*,\delta)$ then

(i) The sequence $\{z_i\}$ constructed according to (5), (6) is well defined;

(ii) $z_i \to z^*$ R-linearly in the norm $\|\cdot\|$.

(iii) If, in addition,

$$\|[G(z_{i-1}) - \frac{\partial^2 L}{\partial x^2}(z_{i-1})] \frac{(x_i - x_{i-1})}{\|z_i - z_{i-1}\|}\| \to 0 \text{ as } i \to \infty, \tag{11}$$

then $z_i \to z^*$ superlinearly.

(iv) If for some $k > 0$ and $i = 0,1,2,\ldots$,

$$\|[G(z_{i-1}) - \frac{\partial^2 L}{\partial x^2}(z_{i-1})](x_i - x_{i-1})\| < k\|z_i - z_{i-1}\|^2 \tag{12}$$

then $z_i \to z^*$ quadratically.

Proof: This theorem follows directly from theorems A1 and A2 in the Appendix of [10] and (9). ¤

3. Stabilization of the Local Method

In this section we shall restrict ourselves to the important subclass of problems of the form (1) which have inequality constraints only, viz. to problems of the form

$$\min\{f(x) \mid g(x) \leq 0\} \tag{13}$$

Newton's method is particularly attractive for such problems because, assuming that at least some inequalitites are active, the optimality conditions for a local minimum are quite distinct from those for a local maximum, so that Newton's method cannot, inadvertently, produce a local maximum rather than a local minimum when simple precautions are used.

Obviously, we can use any globally convergent first order method on problem (13) to obtain an approximation $\tilde{z}$ to z^*, a local minimizer satisfying Assumption 1. The difficulty is in determining whether $\tilde{z}$ is in the domain of convergence of the Newton method (5), (6). We propose to do this adaptively, by monitoring whether $\tilde{\mu}$ is sufficiently "positive", $g(\tilde{z})$ sufficiently "negative" and whether Newton's method is giving signs of at least linear convergence. We shall use the phase I - phase II algorithm described in [9] for stabilization. This algorithm requires the following quantities:

$$\psi(x) \triangleq \max_{j \in \underline{m}} g^j(x) \;, \tag{14}$$

$$\psi(x)_+ \triangleq \max\{0,\psi(x)\}. \tag{15}$$

For $\varepsilon > 0$, $x \in \mathbb{R}^n$ given,

$$I_\varepsilon(x) \triangleq \{j \in \underline{m} | g^j(x) \geq \psi(x)_+ - \varepsilon\}. \tag{16}$$

For $\varepsilon > 0$, $\delta > 0$ and $x \in \mathbb{R}^n$ given,

$$\theta_\varepsilon(x) \triangleq \min_{\bar{\mu}}\{\bar{\mu}^0\gamma\psi(x)_+ + \tfrac{1}{2}\|\bar{\mu}^0\nabla f(x) + \sum_{j\in I_\varepsilon(x)} \bar{\mu}^j\nabla g^j(x)\|^2 \Big| \bar{\mu} \geq 0,\ \Sigma\bar{\mu}^j = 1\} \tag{17}$$

For $\varepsilon_0 > 0$, $\nu \in (0,1)$ given,

$$E \triangleq \{0,\varepsilon_0,\nu\varepsilon_0,\nu^2\varepsilon_0,\ldots\}, \tag{18}$$

$$\varepsilon(x) \triangleq \max\{\varepsilon \in E | \theta_\varepsilon(x) \geq \varepsilon\} \;, \tag{19}$$

$$h(x) \triangleq -[\bar{\mu}^0_{\varepsilon(x)}\nabla f^0(x) + \sum_{j\in I_{\varepsilon(x)}(x)} \bar{\mu}^j_{\varepsilon(x)}\nabla g^j(x)] \;, \tag{20}$$

where $\bar{\mu}^k_{\varepsilon(x)}$, $k = 0,1,\ldots,m$, are the solutions of (17) for $\varepsilon = \varepsilon(x)$.

We assume that the matrices G_i in the algorithm below will be constructed by one of the quasi-Newton formulas or set equal to $\frac{\partial^2 L}{\partial x^2}(z_i)$. In addition, we need the follwoing standard hypothesis:

<u>Assumption 2</u>: For all $x \in \mathbb{R}^n$ such that $\psi(x) > 0$ $\quad 0 \notin co\{\nabla g^j(x) | j \in I_0(x)\}$, where co denotes convex hull. ¤

<u>Algorithm 1</u>:

<u>Parameters</u>: ε_0, K_g, K_μ, $K_z > 0$; α, β, $\gamma \in (0,1)$.

<u>Data</u>: $x_0 \in \mathbb{R}^n$, $\bar{x}_0 = x_0$, $k = 0$, $s = 0$.

<u>Step 0</u>: Compute $\mu_0 \in \mathbb{R}^m$ by solving

$$\mu_0 = \arg\min_{\mu\geq 0}\{\sum_{j=1}^m \mu^j g^j(x_0) + \frac{1}{2}\|\nabla f(x_0) + \sum_{j=1}^m \mu^j g^j(x_0)\|^2\} \tag{21}$$

and set $i = 0$.

<u>Step 1</u>: If $\min_{j\in\underline{m}} \mu_i^j < -K_\mu\gamma^k$ or $\max_{j\in\underline{m}} g^j(x_i) > K_g\gamma^k$ go to step 3. Else, compute $\Delta z_i = (\Delta x_i, \Delta\mu_i)$ by solving the linear system of equations

$$\nabla_x L(x_i,\mu_i) + G_i\Delta x_i + \frac{\partial g^T}{\partial x}(x_i)\Delta\mu_i = 0 \tag{22}$$

$$\mu_i^j g^j(x_i) + \mu_i^j \frac{\partial g}{\partial x}(x_i)\Delta x_i + \Delta\mu_i^j g^j(x_i) = 0, \quad \forall j \in \underline{m} \tag{23}$$

Step 2: If $\|\Delta z_i\| \leq K_z \gamma^k$, set $x_{i+1} = x_i + \Delta x_i$, $\mu_{i+1} = \mu_i + \Delta\mu_i$, $i = i + 1$, $k = k + 1$ and go to step 1. Else, set $i = 0$, $k = k + 1$ and go to step 3.

Step 3: Compute $\varepsilon(\bar{x}_s)$, $h(\bar{x}_s)$ according to (18) and (19).

Step 4: If $\varepsilon(\bar{x}_s) \leq \varepsilon_0 \nu^k$, set $x_0 = \bar{x}_s$ and go to step 0. Else, if $\psi(x_s)_+ > 0$ compute largest $t_s \in \{1, \beta, \beta^2, \ldots\}$ such that

$$\psi(\bar{x}_s + t_s h(\bar{x}_s)) - \psi(\bar{x}_s) \leq -\alpha t_s \varepsilon(\bar{x}_s) \tag{24}$$

if $\psi(\bar{x}_s) \leq 0$, compute largest $t_s \in \{1, \beta, \beta^2, \ldots\}$ such that

$$\psi(\bar{x}_s + t_s h(\bar{x}_s)) \leq 0$$

$$f(\bar{x}_s + t_s h(\bar{x}_s)) - f(\bar{x}_s) \leq -\alpha t_s \varepsilon(\bar{x}_s) \tag{25}$$

set $\bar{x}_{s+1} = x_s + t_s h(\bar{x}_s)$, set $s = s + 1$ and go to step 3. ¤

Theorem 2: Suppose that (10) is satisfied for all i and that the sequence $\{\bar{x}_s\}$ is bounded. (i) If $\{\bar{x}_s\}$ is infinite then, (a) every accumulation point x^* of $\{\bar{x}_s\}$ satisfies $g(\bar{x}_s) \leq 0$ and the F. John first order conditions of optimality; (b) let $\{\bar{x}_s\}_K$ be the subsequence of $\{\bar{x}_s\}$ at which a transfer to step 0 takes place (i.e. $x_0 = \bar{x}_s$), then no accumulation point of $\{x_s\}_{s \in K}$ satisfies Assumption 1. (ii) If $\{\bar{x}_s\}$ is finite, then $z_i \to \hat{z}$ as $i \to \infty$, with $\hat{z} = (\hat{x}, \hat{\mu})$ a Kuhn-Tucker pair. Furthermore, if $\hat{z}$ satisfies Assumption 1, then Theorem 1 gives rate of convergence, provided its assumptions are satisfied.

Proof: (i) (a) If $\{\bar{x}_s\}$ is infinite, then every accumulation point of $\{\bar{x}_s\}$ is a feasible F. John point by [9]. Furthermore, $\varepsilon(\bar{x}_s) \to 0$ as $s \to \infty$. (i) (b) Suppose that $\bar{x}_s \overset{K'}{\to} \bar{x}^*$ with $K' \subset K$ and that $\bar{x}^*$, together with the corresponding multiplier μ^* satisfy Assumption 1. We note that because of Assumption 1, μ^* is a unique Kuhn-Tucker multiplier for $\bar{x}^*$. Now, let $\{\mu_{0,s}\}_{s \in K'}$ be the multipliers μ_0 computed in Step 0 for $x_0 = \bar{x}_s$, $s \in K'$. Then, because μ^* is unique and the solutions $\mu_{0,s}$ are u.s.c. in $\bar{x}_s$, it follows that $\mu_{0,s} \overset{K'}{\to} \mu^*$ as $s \to \infty$. Consequently, there must exist an $s' \in K'$ such that the local algorithm converges superlinearly from $\mu_0 = \mu_{0,s'}$, $x_0 = \bar{x}_{s'}$ and satisfies the tests in step 1 and step 2 for all $i \geq 0$. Thus we get a contradiction that $\{\bar{x}_s\}$ is infinite.

(ii) If $\{z_i\}$ is infinite, then, since we must have that $k = i + k_0$, for some k_0, it follows that $g^j(x_i) \leq K_g \gamma^{k_0+i}$, $\forall j \in \underline{m}$ and $\mu_i^j \geq -K_\mu \gamma^{k_0+i}$ $\forall j \in \underline{m}$, for all i, so that $\overline{\lim}\, g^j(x_i) \leq 0$, and $\underline{\lim}\, \mu_i^j \geq 0$, $j \in \underline{m}$. Since $\|\Delta z_i\| \leq K_z^{k_0+i}$ for all i, it follows that $\{z_i\}$ is Cauchy and hence that $z_i \to \hat{z}$ as $i \to \infty$. It follows then from (14a,b), that $\hat{z} = (\hat{x}, \hat{\mu})$ is a Kuhn-Tucker pair. The rate of convergence result follows from Theorem 1. ¤

ACKNOWLEDGEMENT

Research sponsored by the National Science Foundation Grants ECS-79-13148 and ENV76-04264 also by the Air Force Office of Scientific Research (AFSC) United States Air Force under Contract No. F49620-79-C-0178.

REFERENCES

[1] Bertsekas, D. P., "Enlarging the region of convergence of Newton's method for constrained optimization," to appear in JOTA

[2] Dipillo, G. and Grippo, L., "A new class of augmented lagrangians in nonlinear programming," SIAM J. Control, Vol. 17, No. 5, pp. 618-628, 1979.

[3] Garcia-Palomares, U. M. and Mangassarian, O. L., "Superlinear convergent quasi-Newton algorithms for nonlinearly constrained optimization problems," Mathematical Programming, Vol. 11, No. 1, pp. 1-13, 1976.

[4] Han, S. P., "A globally convergent method for nonlinear programming," JOTA, Vol. 22, 1977.

[5] Han, S. P., "Superlinearly convergent variable metric algorithms for general nonlinear programming problems," Math. Programming, Vol. 11, 1976.

[6] Han, S. P., "A hybrid method for nonlinear programming," in Nonlinear Programming 3, R. R. Meyer and S. M. Robinson eds., Academic Press, pp. 65-95, 1978.

[7] McCormick, G. P., "Penalty function versus nonpenalty function methods for constrained nonlinear programming," Mathematical Programming, Vol. 1, pp. 217-238, 1971.

[8] Mayne, D. Q., and Polak, E., "A superlinearly convergent algorithm for constrained optimization problems," to appear in Mathematical Programming Study on Constrained Minimization. (See also publication No. 78/52, revised 15/1/1980, Department of Computing and Control, Imperial College, London).

[9] Polak, E., Trahan, R., and Mayne, D. Q., "Combined phase I - phase II methods of feasible directions," Mathematical Programming, Vol. 17, No. 1, pp. 32-61, 1979.

[10] Polak, E. and Tits, A. L., "On globally stabilized Newton methods for inequality constrained optimization problems," University of California, Berkeley, ERL Memo No. M81/87, December 7, 1981.

[11] Powell, M. J. D., "A fast algorithm for nonlinearly constrained optimization calculations," The Dundee Conference on Numerical Analysis, 1977.

[12] Robinson, S. M., "Perturbed Kuhn-Tucker points and rates of convergence for for a class of nonlinear programming algorithms," Math. Programming, Vol. 7, No. 1, pp. 1-16, 1974.

[13] Robinson, S. M., "Extension of Newton's method to mixed systems of nonlinear equations and inequalities,' Technical Summary Report No.1161, Mathematics Research Center, University of Wisconsin, 1971.

[14] Tapia, R., "Diagonalized multiplier methods and quasi-Newton methods for constrained optimization," JOTA, Vol. 22, No. 2, pp. 135-194, 1977.

[15] Wilson, R. B., "A simplified algorithm for concave programming," Ph.D. Dissertation, Grad. School of Bs. Ad., Harvard Univ., Cambridge, 1963.

APPENDIX

The following results are similar, but somewhat stronger to the ones in the open literature, cf. [3]. Consider the equation

$$f(x) = 0 \tag{A.1}$$

when $f : \mathbb{R}^n \to \mathbb{R}^n$ is continuously differentiable. A quasi-Newton method is defined by

$$G(x_k)(x_{k+1} - x_k) + f(x_k) = 0 \ . \tag{A.2}$$

We use the notation

$$F(x) \triangleq \frac{\partial f}{\partial x}(x) \ . \tag{A.3}$$

Let x^* be a solution of (A.1).

Assumption A1:

(i) $F(x^*)$ is nonsingular.

(ii) There exists an $\varepsilon > 0$ such that

$$\|G(x) - F(x)\| < \frac{1}{2\|F(x^*)^{-1}\|} \quad \forall x \in \bar{B}(x^*,\varepsilon) \tag{A.4}$$

The following result is obvious.

Lemma A1. There exist $\rho \in (0,\varepsilon)$, $\mu > 0$, $\beta > 0$, $\alpha < 1/2\beta$ such that $\forall x, x' \in B(x^*,\rho)$, $F(x)$ is nonsingular and

$$\|F(x)^{-1}\| < \beta \tag{A.5}$$

$$\|F(x) - G(x)\| < \alpha \tag{A.6}$$

$$\|f(x') - f(x) + G(x)(x' - x)\| \leq M\|x' - x\|^2 + \|(F(x) - G(x))(x' - x)\| \tag{A.7}$$

Furthermore, x^* is the unique solution to (A.1) in $\bar{B}(x^*,\rho)$. ¤

Lemma A2: Let ρ, α, β be as in Lemma A1. Suppose that $\hat{x} \in B(x^*,\rho)$. Then $G(\hat{x})$ is nonsingular and the solution v of

$$G(\hat{x})\, v + f(\hat{x}) = 0 \ . \tag{A.8}$$

satisfies

$$\|v\| \leq 2\beta\|f(\hat{x})\| \tag{A.9}$$

Theorem A1. There exists a $\delta > 0$ such that, if $x_0 \in B(x^*,\delta)$, then

(i) The sequence $\{x_i\}$ constructed by (A.2) is well defined and remains in $B(x^*,\rho)$;

(ii) $\{x_i\}$ converges R-linearly to x^* in the norm $\|\cdot\|$;

(iii) for i = 1,2,...

$$\|x_{i+1} - x_i\| \leq 2\beta[M\|x_i - x_{i-1}\|^2 + \|(F(x_{i-1} - G(x_{i-1}))(x_i - x_{i-1})\|] \tag{A.10}$$

QUASI-NEWTON METHODS FOR A CLASS OF NONSMOOTH CONSTRAINED OPTIMIZATION PROBLEMS

Ekkehard Sachs
Technische Universität Berlin

In this paper we consider a class of problems of the following type: Let G be a nonlinear Fréchet-differentiable map of $\mathbb{R}^n$ into some finite-dimensional linear space Y. Furthermore, let ϕ be a locally Lipschitz continuous function from Y into $\mathbb{R}$. For a given convex subset $X \subset \mathbb{R}^n$ we minimize $\phi \circ G$ on X, i. e. find an $\hat{x} \in X$ such that

$$\phi(G\hat{x}) \leq \phi(Gx) \quad \text{for all } x \in X. \tag{1}$$

Problems of this type with $X = \mathbb{R}^n$ have been considered by Bertsekas [1] and Poljak [2] by using augmented Lagrangeans for the equivalent problem of minimizing $\phi(y)$ subject to $y = Gx$ over $(x,y) \in \mathbb{R}^n \times Y$. Here we extend an approach being used for discrete Chebyshev approximation by Osborne and Watson [3] and Ishizaki and Watanabe [4]. This algorithm has been altered by Madsen [5] using a certain Broyden-update instead of derivatives of G. He showed under a certain step-size procedure that superlinear rate of convergence is obtained. In this paper we give for locally Lipschitz continuous functions ϕ general conditions on the step-sizes and the update formulas which allow to prove the superlinear convergence rate, thus extending results by Gruver and Sachs [6]. A certain growth condition for the problem, in Chebyshev approximation known under the name of strong uniqueness condition, is imposed.

<u>Assumption:</u> For each optimal point $\hat{x} \in X$ there exist a neighborhood N of $\hat{x}$ and a constant $\gamma > 0$ such that for all $x \in N \cap X$

$$\phi(Gx) \geq \phi(G\hat{x}) + \gamma \|x - \hat{x}\| \quad . \tag{2}$$

It is known that a weakening of this condition can yield no convergence at all for certain examples.

Let $L(\mathbb{R}^n, Y)$ denote the space of all linear continuous mappings from $\mathbb{R}^n$ into Y.

Algorithm: Let $\{\lambda_i\}_{\mathbb{N}} \subset [0,1]$, $\{B_i\}_{\mathbb{N}} \subset L(\mathbb{R}^n, Y)$, $x_o \in X$.

Find $z_i \in X$ such that for all $x \in X$

$$\phi(Gx_i + B_i(z_i - x_i) \leq \phi(Gx_i + B_i(x - x_i)). \tag{3}$$

Define

$$x_{i+1} = x_i + \lambda_i(z_i - x_i). \tag{4}$$

It is clear from step (3) that ϕ and X have to be of such a structure that the minimization in (3) can be carried out with a reasonable effort.

The theorem on superlinear convergence in its general form is as follows:

Theorem 1: Let ϕ be locally Lipschitz continuous, G Fréchet-differentiable and let (2) hold. For given $\{\lambda_i\}_{\mathbb{N}} \subset [0,1]$, $\{B_i\}_{\mathbb{N}} \subset L(\mathbb{R}^n, Y)$ we construct sequences $\{x_i\}_{\mathbb{N}}$, $\{z_i\}_{\mathbb{N}} \subset X$ as in (3) and (4) described. Suppose that

$$\lim_{i\to\infty} \lambda_i = 1, \tag{5}$$

$$\lim_{i\to\infty} x_i = \hat{x}, \tag{6}$$

$$\lim_{i\to\infty} \frac{\|(B_i - G'_{\hat{x}})(x_{i+1} - x_i)\|}{\|x_{i+1} - x_i\|} = 0. \tag{7}$$

Then there exist $\varepsilon > 0$, $i_o \in \mathbb{N}$ such that if

$$\|B_i - G'_{\hat{x}}\| \leq \varepsilon \qquad \text{for } i \geq i_o \;, \tag{8}$$

then

$$\lim_{i\to\infty} \frac{\|x_{i+1} - \hat{x}\|}{\|x_i - \hat{x}\|} = 0. \tag{9}$$

Proof: (4), (5) and (6) imply

$$\lim_{i\to\infty} z_i = \hat{x}. \tag{10}$$

(2) and (3) yield

$$\begin{aligned}\gamma \|z_i - \hat{x}\| &\le \phi(Gz_i) - \phi(G\hat{x}) \\ &\le \phi(G\hat{x} + G'_{\hat{x}}(z_i - \hat{x})) - \phi(Gx_i + B_i(z_i - x_i)) \\ &+ \phi(Gx_i + B_i(\hat{x} - x_i)) - \phi(G\hat{x}) \\ &+ \phi(Gz_i) - \phi(G\hat{x} + G'_{\hat{x}}(z_i - \hat{x})). \end{aligned} \tag{11}$$

Because of (6), (8) and (10), all arguments of ϕ in (11) lie in a neighborhood of $G\hat{x}$ for i large enough and we can use the local Lipschitz continuity of ϕ and the Fréchet-differentiability of G to obtain from (11) for some $\kappa > 0$

$$\begin{aligned}\gamma \|z_i - \hat{x}\| &\le \kappa(\|(G'_{\hat{x}} - B_i)(z_i - x_i)\| + \alpha_i \|x_i - \hat{x}\| \\ &+ \|(G'_{\hat{x}} - B_i)(x_i - \hat{x})\| + \alpha_i \|x_i - \hat{x}\| \\ &+ \beta_i \|z_i - \hat{x}\|) \\ &\le \kappa(2 \|(G'_{\hat{x}} - B_i)(z_i - x_i)\| + 2\alpha_i \|x_i - \hat{x}\| \\ &+ \|(G'_{\hat{x}} - B_i)(z_i - \hat{x})\| + \beta_i \|z_i - \hat{x}\|) \end{aligned} \tag{12}$$

where

$$\lim_{i\to\infty} \alpha_i = \lim_{i\to\infty} \beta_i = 0 . \tag{13}$$

Because of (4), (7), and (13) there is $i_o \in \mathbb{N}$ such that for all $i \ge i_o$.

$$\beta_i \le \frac{\gamma}{6\kappa}, \quad \varepsilon \le \frac{\gamma}{6\kappa}, \quad \frac{\|(B_i - G'_{\hat{x}})(z_i - x_i)\|}{\|z_i - x_i\|} \le \frac{\gamma}{12\kappa} .$$

Then (12) yields

$$\begin{aligned}\frac{\gamma}{2}\|z_i - \hat{x}\| &\le (\gamma - \kappa(\beta_i + \varepsilon + 2\frac{\|(B_i-G'_{\hat{x}})(z_i-x_i)\|}{\|z_i - x_i\|}))\|z_i - \hat{x}\| \\ &\le 2\kappa \left(\frac{\|(G'_{\hat{x}}-B_i)(z_i-x_i)\|}{\|z_i - x_i\|}\|x_i - \hat{x}\| + \alpha_i \|x_i-\hat{x}\|\right) \end{aligned} \tag{14}$$

From (4) and (14) we obtain

$$\|x_{i+1} - \hat{x}\| \leq (1 - \lambda_i)\|x_i - \hat{x}\| + \lambda_i \|z_i - \hat{x}\|$$

$$\leq (1 - \lambda_i + 4\lambda_i \frac{\kappa}{\gamma}(\frac{\|(B_i - G'_{\hat{x}})(z_i - x_i)\|}{\|z_i - x_i\|} + \alpha_i))\|x_i - \hat{x}\| .$$

Hence, (5), (7), and (13) yield (9).

In the Quasi-Newton methods for differentiable functions, (7) has been shown to be a necessary and sufficient condition for superlinear convergence, see Dennis and Moré [7]. The next theorem deals with the local convergence. It shows that the mapping B_i needs to approximate $G'_{\hat{x}}$ only within a certain bound. Step (3) of the algorithm is of a somewhat global nature and in proving a local convergence result the determination of z_i has to be changed slightly. We confine the minimization onto a ball

$$K_\rho(\hat{x}) = \{x \in \mathbb{R}^n : \|x - \hat{x}\| \leq \rho\} \qquad , \quad \rho > 0.$$

Then (3) is substituted by:

Find $z_i \in X \cap K_\rho(\hat{x})$ such that for all $x \in X \cap K_\rho(\hat{x})$

$$\phi(Gx_i + B_i(z_i - x_i)) \leq \phi(Gx_i + B_i(x - x_i)) \quad . \tag{15}$$

Obviously, Theorem 1 also holds with (3) replaced by (15) because its assumptions imply (10), i.e. $z_i \in X \cap K_\rho(\hat{x})$ for i large enough.

<u>Theorem 2:</u> Let ϕ be locally Lipschitz continuous, G Fréchet-differentiable and let (2) hold. Given $\{\lambda_i\}_{\mathbb{N}} \subset [0,1]$, $\{B_i\}_{\mathbb{N}} \subset L(\mathbb{R}^n, Y)$, $x_o \in X$ with

$$\liminf_{i\to\infty} \lambda_i = \lambda^* > 0 . \tag{16}$$

There exist $\varepsilon > 0$, $\rho > 0$ such that if $\{x_i\}_{\mathbb{N}}$, $\{z_i\}_{\mathbb{N}} \subset X$ are constructed according to the rules (15) and (4) and if

$$||x_o - \hat{x}|| \leq \varepsilon \quad \text{and} \tag{17}$$

$$||B_i - G'_{\hat{x}}|| \leq \varepsilon \quad \text{for i large enough} \tag{18}$$

then

$$\lim_{i\to\infty} x_i = \hat{x} .$$

Proof: By assumption, ϕ is Lipschitz continuous on bounded sets. Take any $x \in X$, $B \in L(\mathbb{R}^n, Y)$ with

$$||x - \hat{x}|| \leq \varepsilon \quad \text{and} \quad ||B - G'_{\hat{x}}|| \leq \varepsilon . \tag{19}$$

Then (11) holds omitting indices, if ε is small enough. All arguments of ϕ in (11) lie in a sufficiently large ball around $G\hat{x}$ as can be checked easily. Then there exists a Lipschitz constant κ such that a similar estimate as in (12) holds, namely

$$\begin{aligned}\gamma \, ||z - \hat{x}|| \leq \kappa(2\varepsilon \, ||z - x|| &+ (\varepsilon + \beta(||z - \hat{x}||) \, ||z - \hat{x}|| \\ &+ 2\alpha(||x - \hat{x}||) \, ||x - \hat{x}||) ,\end{aligned} \tag{20}$$

where

$$\lim_{\eta\to\infty} \alpha(\eta) = \lim_{\eta\to\infty} \beta(\eta) = 0 .$$

Select ρ and ε so small such that

$$\varepsilon \leq \frac{\gamma}{12\kappa} \quad \text{and} \quad \beta(\eta) \leq \frac{\gamma}{6\kappa} \quad \text{for} \quad 0 < \eta \leq \rho . \tag{21}$$

Then (20) implies

$$\frac{\gamma}{2} \, ||z - \hat{x}|| \leq 2\kappa \, (\varepsilon + \alpha(|| x - \hat{x}||)) \, ||x - \hat{x}|| .$$

Suppose furthermore that ε is chosen small enough such that besides (21) also

$$\varepsilon \leq \frac{\gamma}{16\kappa} \quad \text{and} \quad \alpha(\eta) \leq \frac{\gamma}{16\kappa} \quad \text{if} \quad \eta \leq \varepsilon$$

holds. Then

$$||z - \hat{x}|| \leq \frac{1}{2} \, ||x - \hat{x}|| , \tag{22}$$

if (19) is satisfied. Therefore, from (17) and (18) we obtain with (4) and (22)

$$\|x_1 - \hat{x}\| \leq (1-\lambda_o)\|x_o - \hat{x}\| + \lambda_o \|z_o - \hat{x}\|$$
$$\leq (1-\tfrac{1}{2}\lambda_o)\|x_o - \hat{x}\| \quad .$$

Similarly, if for any $i \in \mathbb{N}$ the inequalities (19) hold, we have with (16)

$$\|x_{i+1} - \hat{x}\| \leq (1-\tfrac{1}{2}\lambda_i)\|x_i - \hat{x}\| \leq (1-\tfrac{1}{2}\lambda^*)\|x_i - \hat{x}\|$$

and consecutively

$$\|x_{i+1} - \hat{x}\| \leq (1-\tfrac{1}{2}\lambda^*)^i \|x_o - \hat{x}\| , \tag{23}$$

which proves the convergence.

As it can be seen from the proof, under the assumptions of Theorem 2 we do obtain a linear rate of convergence.

In Theorem 2 we have considered more closely condition (6) of Theorem 1. Let us now investigate the Broyden update formula for $\{B_i\}_{\mathbb{N}}$ and the convergence assumptions (7) and (8). This update formula is given by

$$B_{i+1} = B_i - \frac{]B_i p_i - y_i, p_i[}{\|p_i\|^2} \tag{24}$$

with

$$p_i = x_{i+1} - x_i \qquad \text{and} \qquad y_i = Gx_{i+1} - Gx_i$$

and for $z \in \mathbb{R}^n$, $y \in Y$ the mapping $]y,z[\; : \mathbb{R}^n \to Y$

$$]y,z[(x) = \langle z,x\rangle \, y \qquad \text{for all } x \in \mathbb{R}^n ,$$

$\langle \; , \; \rangle$ denoting the scalar product in $\mathbb{R}^n$.

In Gruver and Sachs [6] more general classes of updates have been considered, however for this particular case (24) we have the following estimate.

Let $G'_{( )}$ be Lipschitz continuous. Then for some $\delta > 0$

$$\|B_{i+1} - G'_{\hat{x}}\| \leq \|B_i - G'_{\hat{x}}\| \quad \|I - \frac{]p_i, p_i[}{\|p_i\|^2}\|$$

$$+ \frac{\| \,]y_i - G'_{\hat{x}}(p_i), \, p_i[\, \|}{\|p_i\|^2}$$

$$\leq \|B_i - G'_{\hat{x}}\| + \delta \max\{ \|x_{i+1} - \hat{x}\| , \|x_i - \hat{x}\| \}. \quad (25)$$

As in [6,pp.97] , (25) can be incorporated in the proof of linear convergence rate using the Broyden update where the step sizes are uniformly bounded away from zero.

<u>Theorem 3:</u> Let ϕ be locally Lipschitz continuous, G Fréchet-differentiable with Lipschitz continuous derivative and let (2) hold. Given $\{\lambda_i\}_{\mathbb{N}} \subset [0,1]$ with (16). There exist $\varepsilon > 0$, $\delta > 0$ such that if $\{x_i\}_{\mathbb{N}}$, $\{z_i\}_{\mathbb{N}} \subset X$, $\{B_i\}_{\mathbb{N}} \subset L(\mathbb{R}^n, Y)$ are constructed according to (15), (4), and (24), respectively, and if

$$\|x_o - \hat{x}\| \leq \varepsilon \quad \text{and} \quad \|B_o - G'_{\hat{x}}\| \leq \varepsilon \ ,$$

then

$$\|B_i - G'_{\hat{x}}\| \leq \varepsilon \quad \text{for all } i \in \mathbb{N}$$

and

$$\lim_{i \to \infty} x_i = \hat{x}$$

at a linear rate.

For the superlinear convergence and its crucial condition (7) we have the following lemma.

Lemma 4 [6,pp.99]: Let $\{x_i\}_{\mathbb{N}} \subset X$ such that

$$\sum_{i=1}^{\infty} \|x_i - \hat{x}\| < \infty \tag{26}$$

for some $\hat{x} \in X$. Then

$$\lim_{i\to\infty} \frac{\|(B_i - G'_{\hat{x}})(x_{i+1} - x_i)\|}{\|x_{i+1} - x_i\|} = 0 .$$

Therefore the linear convergence rate is sufficient to show (26) and hence condition (7) of the theorem on superlinear convergence rate is fulfilled.

It should be noted that we have investigated in this paper questions concerning rates of convergence for locally Lipschitz continuous functions and inexact step sizes. It remains to analyze global convergence properties as well as descent properties for step (3). Partially, this has been done for the Broyden update in [5].

References:

[1] D. P. Bertsekas, Approximation procedure based on the method of multipliers, J. Optim. Theory Appl. 23 (1977), 487-510.

[2] B. T. Poljak, On the Bertsekas method for minimzation of composite functions, in International on System Optimization and Analysis, Rocquencourt 1978, eds. A. Bensoussan, J. L. Lions, Springer, Berlin- Heidelberg-New York, pp. 179-186, 1979.

[3] M. R. Osborne and G. A. Watson, An algorithm for minimax approximation in the nonlinear case, Computer J. 12 (1969), 64-69.

[4] Y. Ishizaki and H. Watanabe, An iterative Chebyshev approximation method for network design, IEEE Trans. Circuit Theory 15 (1968), 326-336.

[5] K. Madsen, Minimax solutions of non-linear equations without calculating derivatives, Math. Progr. Study 3 (1975), 110-126.

[6] W. A. Gruver and E. Sachs, Algorithmic Methods in Optimal Control, Pitman, London, 1981.

[7] J. E. Dennis and J. J. Moré, A characterization of superlinear convergence and its application to quasi-Newton methods, Math. Comp. 28 (1974), 549-560.

Current address:

Department of Mathematics
North Carolina State University
Raleigh, N. C. 27650
U. S. A.

PROBABILISTIC ANALYSIS OF THE SOLUTION OF THE KNAPSACK PROBLEM

G. Ausiello, A. Marchetti - Spaccamela
Istituto di Automatica, University of Rome
M. Protasi
Istituto di Matematica, University of L'Aquila

ABSTRACT

In this paper the value of the optimal solution of a random instance of the Knapsack problem is analyzed. With respect to this value, the performance of a simple greedy heuristic for the solution of this problem is evaluated. The results are compared with the performance of other greedy heuristics.

1. INTRODUCTION

The problem considered in this paper is the 0-1 Knapsack problem namely the problem of finding z* such that

$$z^* = \max \sum_{i=1}^{n} p_i x_i$$

$$\text{subject to } \sum_{i=1}^{n} a_i x_i \leq b$$

where $x_i \in \{0,1\}$ and p_i (profit of item i) and a_i (occupancy of item i) are positive integers, for $i = 1,\dots,n$.
This problem is very interesting both for its practical relevance, and also for its peculiarity with respect to the computational complexity. In fact the Knapsack problem has been shown to be a pseudopolynomial and fully approximable NP-complete problem [GJ.79, IK75]. This means that even if a polynomial algorithm that solves the Knapsack problem exactly does not exist unless P=NP, for any $\varepsilon > 0$ it is possible to find an approximate algorithm which, in time bounded by a polynomial in n and $1/\varepsilon$, gives a solution z such that

$$\left| \frac{z - z^*}{z^*} \right| < \varepsilon$$

In this paper we are interested in the solution of the Knapsack problem by means of simple approximate algorithms based on greedy heuristics. The reason for being interested in such algorithms is twofold. First of all this kind of heuristics give rise to programs which are simpler in nature and easier to implement than, for example those in [L 78]; second, there is a strong theoretical motivation in determining the performance of such algorithms when applied to random instances of the Knapsack problem and to compare it with the performance that the same kind of heuristics have in the case of other NP-complete problems. In particular in [AMP 81] the behaviour of two simple greedy beuristics applied to one strong NP-complete problem, such as "clique", and to one pseudopolynomial problem, such as "Knapsack", was analyzed from a probabilistic point of view. Indeed while a detailed probabilistic analysis was developped for the problem clique, in the case of Knapsack the behaviour was only determined for particular classes of instances. The difference between the two heuristics is the following: in one case (blind heuristics) the optimal solution is achieved by adding new items to a partial solution at random, provided that the new solution is still feasible. Instead, in the second case (short sighted heuristics) the items are selected according to some quality criterion, n to the "density" (profit/ occupancy) of the items.

In [D'A 79] the second type of heuristics has been extensively, analyzed and it has been shown that the ratio between the value of the optimal solution z* and the value z_{SG} obtained by such algorithm is such that

$$\lim_{n \to \infty} \frac{z^*}{z_{SG}} = 1 \text{ almost surely}$$

In this paper we are interested in analyzing what is the behaviour of the first kind of algorithms, in order to determine whether the same good behaviour is achieved or whether we lose in approximation when we do not exploit the information on the density of the items.

2. ANALYSIS OF THE EXACT SOLUTION

In order to perform, the probabilistic analysis of the exact solution of the Knapsack problem and, subsequently, of the proposed greedy algorithms we will assume that the coefficients a_i, p_i are uniformly distributed over the interval $\{1,2,\dots,c\}$ for some c.

Since in this model no hypothesis is made on the value of b, we will consider b as a parameter. To perform a meaningful analysis we will limit ourselves to consider b in the following range:

$$\frac{n}{c^2} < b \leq \frac{n(c+1)}{2}$$

In fact for any ε, if we choose $b < n/c^2(1-\varepsilon)$ it is not difficult to prove (by Chernoff's inequality) that the optimal solution is only given by those elements which have maximum profit c and minimum occupancy 1 almost everywhere. We will not consider these cases anymore because the optimal solution can be trivially found almost everywhere.

On the other hand, for $b > \frac{n(c+1)}{2}$, we have the following result

THEOREM 2.1. *If* $b > \frac{n(c+1)}{2}$ *then any algorithm which puts all items in the Knapsack will almost surely achieve the optimal solution.*

PROOF. Let us first observe that the expected value of the total profit and the total occupancy of all items is $n\frac{(c+1)}{2}$. Hence if b is larger than such value all items can be expected to be put in the knapsack. Therefore, the thesis follows from the law of large numbers.

Q.E.D.

In order to evaluate the value of the exact solution in this meaningful range, we could apply an enumeration technique calculating, for each fixed b, the values of the exact solution by a counting argument.

Instead of carrying on such a tedious calculation for every value of b and c we use the following approximate argument.

Let us represent the space of items of an instance of knapsack as a set of points in the two dimensions, occupancy (a_i) and profit (p_i); we obtain the following diagram

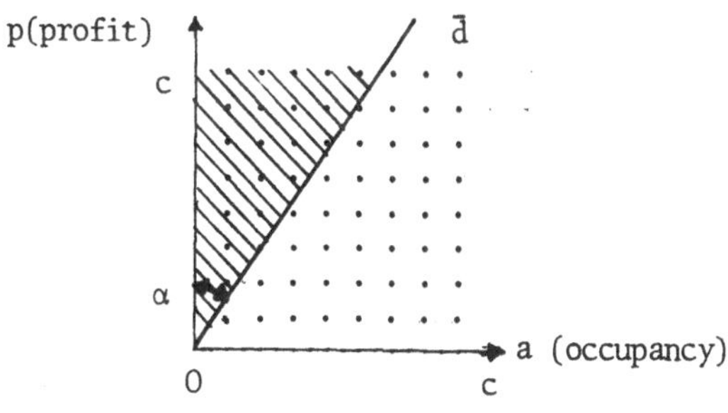

Fig. 1

For any sufficiently large instance and for each pair $\langle \bar{a},\bar{p} \rangle$ we may expect that n/c^2 items with such values of occupancy and profit occur in the instance.

Hence, for a given value of density (the ratio between profit and occupancy) $\bar{d}$ corresponding to the line $p = \bar{d}a$ where $\bar{d} = 1/\mathrm{tg}\,\alpha$, a good extimate of z* may be obtained by computing the integral of the shaded area. The results which may be obtained are stated in the following:

THEOREM 2.2. *If* $\alpha \leq 45°$ *the expected optimal value* z* *is bounded as follows*

$$R' - ERR < z^* < R' + ERR$$

where $R' = \frac{c^3}{3}\,\mathrm{tg}\alpha + \frac{c^2}{4}$

and $ERR = \frac{c^2}{2} + \frac{c}{2} + \frac{1}{6}$.

PROOF. In order to establish an upper bound to the value of z* we may calculate the contributions of all items which are indicated in Fig. 2 (a blow-up of Fig. 1)

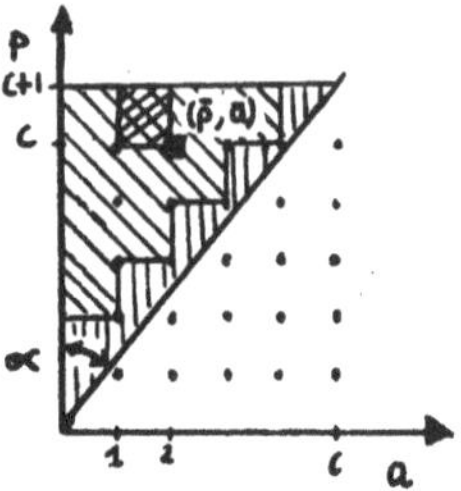

Fig. 2

That is, not only the items contained in the area with oblique shading but also the area with vertical shading. Hence we have the following condition

$$z^* \leq \int_0^{c+1} \left[\int_0^{p\,tg\alpha} (p - \tfrac{1}{2})\,da \right] dp$$

Note that the integral is calculated for $p - \frac{1}{2}$ in such a way that

$$\int_{\bar{p}}^{\bar{p}+1} \int_{\bar{a}-1}^{\bar{a}} (p - \tfrac{1}{2})\,da\,dp = \bar{p}$$

as it may be seen in the doubly shaded area.

By evaluating the integral we obtain that, since $tg\alpha < 1$

$$z^* \leq (\frac{c^3}{3} + \frac{3}{4}c^2 + \frac{c}{2} + \frac{1}{12})tg\alpha \leq \frac{c^3}{3}tg\alpha + \frac{c^2}{4} + \frac{c^2}{2} + \frac{c}{2} + \frac{1}{6} = R' + ERR$$

In order to compute the lower bound to the value of z* we underestimate the contributions of the items as shown in Fig. 3.

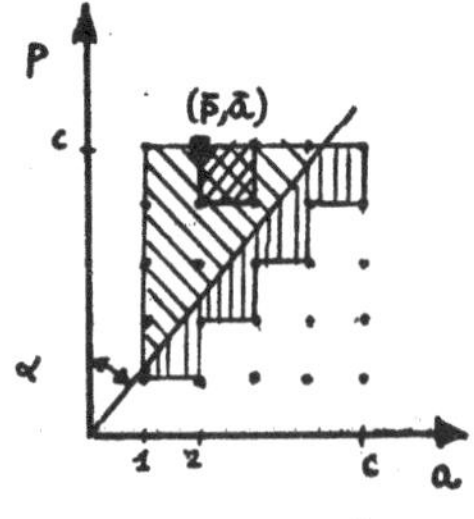

Fig. 3

To determine the exact value of z* we should consider all the shaded, (oblique and vertical) area; instead we limit ourselves to compute the oblique shaded area.

Note that in this case the integral is calculated for $p + \frac{1}{2}$ in such a way that

$$\int_{\bar{p}-1}^{\bar{p}} \int_{\bar{a}}^{\bar{a}+1} (p+ \tfrac{1}{2})da\ dp = \bar{p}$$

so we obtain

$$z^* \geq \int_{1/p\mathrm{tg}\alpha}^{c} [\int_{0}^{p\mathrm{tg}\alpha} (p + \tfrac{1}{2})da]\,dp = \frac{c^3}{3}\,\mathrm{tg}\alpha + \frac{c^2}{4}\,(\tfrac{1}{2}\mathrm{tg}\alpha - 1) - \frac{c}{2} - \frac{1}{3\mathrm{tg}^2\alpha} -$$

$$- (\tfrac{1}{2}\mathrm{tg}\alpha - 1)\frac{1}{2\mathrm{tg}^2\alpha} + \frac{1}{2\mathrm{tg}\alpha} \geq \frac{c^3}{3}\mathrm{tg}\alpha + \frac{c^2}{4} - \frac{c^2}{2} - \frac{c}{2} - \frac{1}{6} = R' - ERR$$

Q.E.D.

A theorem corresponding to the preceding one, in the case in which $\alpha > 45°$, may be obtained in a similar way by computing the global profit of all items and subtracting the non-shaded area

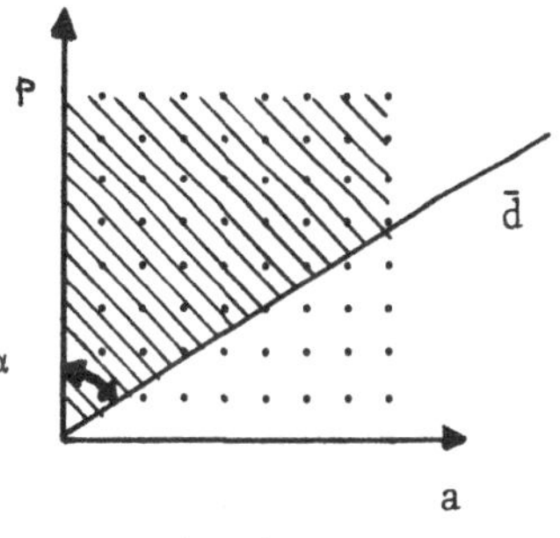

Fig. 4

THEOREM 2.3. <u>If</u> $\alpha > 45°$ <u>the expected optimal value</u> z^* <u>is bounded as follows</u>

$$R'' - ERR < z^* < R'' + ERR$$

where $R'' = \frac{c^3}{2} - \frac{c^3}{6}\,\mathrm{cotg}^2\alpha + \frac{c^2}{4}$.

<u>PROOF</u>. It is sufficient to compute the global profit $T = \frac{c^3}{2} + \frac{c^2}{2}$ and subtracting

$$\int_{0}^{c+1} [\int_{0}^{a\mathrm{cotg}\alpha} (p + \tfrac{1}{2})\,dp]\,da$$

for the upper bound function and

$$\int_{1/\mathrm{cotg}\alpha}^{c} [\int_{1}^{a\mathrm{cotg}\alpha} (p - \tfrac{1}{2})\,dp]\,da$$

for the lower bound function.

Q.E.D.

3. ANALYSIS OF A BLIND GREEDY ALGORITHM

In [D'A 79] the probabilistic behaviour of an approximate algorithm for the solution of the knapsack problem has been studied. The algorithm is essentially based on a shortsighted greedy strategy where the quality of the items is given by the ratio between profit (p_i) and occupancy (a_i). For this reason the list of items is assumed to be sorted in decreasing order, that is

$$\frac{p_i}{a_i} > \frac{p_j}{a_j} \text{ implies } i < j;$$

If $\frac{p_i}{a_i} = \frac{p_j}{a_j}$ and $a_i > a_j$ then $i < j$.

ALGORITHM SG:
T:=∅; S:=sorted list of items;
<u>while</u> S≠∅ <u>do</u>

1. pick up smallest item k in S; S:=S-{k};
2. if $\sum_{i \in T} a_i + a_k \leq b$ <u>then</u> T:=T ∪ {k}

<u>end</u>;
<u>output</u> T.

The result of the analysis is the following:

THEOREM 3.1. [D'A 79]. <u>Algorithm</u> SG <u>provides an optimal solution of the knapsack problem almost everywhere</u>.

As it has been announced in §1, we are interested in comparing the behaviour of various greedy strategies applied in the solution of the knapsack problem. Hence our further step is to consider the behaviour of an algorithm (BG) based on a blind strategy, obtained by eliminating the preprocessing phase of sorting the items. This is equivalent to assuming that the items are chosen at random. The assumption on the distribution of the items is the same as for algorithm SG.

The study of the value of the solution achieved by algorithm BG can be carried on, following the approach introduced in §2.

THEOREM 3.2. <u>If</u> $\alpha \leq 45°$ <u>the expected value of the occupancy</u> b <u>is bounded as follows</u>

$$S' - ERR < b < S' + ERR$$

<u>where</u> $S' = \frac{c^3}{6} tg^2\alpha + \frac{c^2}{4} tg\alpha + \frac{c}{2} tg\alpha$ <u>and the error is the same as in theorem 2.2.</u>

<u>PROOF</u>. We can calculate the upper and lower bounds of occupancy b analougously to theorem 2.2.

$$b < \int_0^{c+1} \left[\int_0^{p\,tg\alpha} (a + \tfrac{1}{2}) da \right] dp = \frac{(c+1)^3}{6} tg^2\alpha + \frac{(c+1)^2}{4} tg\alpha \leq S' + ERR$$

$$b > \int_{1/tg\alpha}^{c} [\int_{1}^{p\,tg\alpha} (a+\tfrac{1}{2})da]\,dp = \frac{c^3 tg^2\alpha}{6} - \frac{c^2}{4} tg\alpha + \frac{1}{12 tg\alpha} > S' - ERR$$

QED

THEOREM 3.3. If $\alpha > 45°$ the expected value of the occupancy b is bounded

$$S'' - ERR < b < S'' + ERR$$

where $S'' = \frac{c^3}{2} - \frac{c^3}{3}\cot g\alpha + \frac{c^2}{2} - \frac{c^2}{4}\cot g\alpha.$

PROOF. As for Theorem 2.3 by computing the global occupancy $T = \frac{c^3}{2} + \frac{c^2}{2}$ and subtracting

$$\int_{1/\cot g\alpha}^{c} [\int_{1}^{a \cot g\,\alpha} (a + \tfrac{1}{2})dp]\,da$$

for the upper bound and

$$\int_{0}^{c+1} [\int_{0}^{a \cot g\,\alpha} (a - \tfrac{1}{2})dp]\,da$$

for the lower bound

QED

Note that the preceding theorems of §2 and 3 could also be strenghtened using the almost everywhere convergence.

In order to evaluate the quality of the approximation obtained by the blind heuristic, given by the ratio $\frac{z^*}{z_{BG}}$ we should determine the expected value of the approximate solution z_{BG}. However we may notice that the preceding results, concerning the expected value of the occupancy $\sum_i a_i x_i$, may be also used to provide the value $z_{BG} = \sum_i p_i x_i$ because the profit variables p_i and the occupancy variables a_i are both uniformly distributed over $1 \div c$ and uncorrelated and therefore, by the law of large numbers, the relative difference

$$\left| \frac{\sum_i a_i x_i - \sum_i p_i x_i}{\sum_i a_i x_i} \right|$$

asymptotically vanishes.

The error introduced is not negligeable for small values of c and α, but, in this case, we can apply an enumeration argument to compute more precise bounds.

In every other case, if we evaluate the ratio $\frac{z^*}{z_{BG}}$, the obtained values are very close to those that can be found by specializing R'/S', R"/S". For example, let us consider those instances such that the optimal solution may be obtained by considering all items with:

1) profit larger than or equal to $\lceil \frac{c}{2} \rceil$ and occupancy 1, ($tg\alpha = \frac{2}{c}$),

2) profit larger than or equal to the occupancy ($tg\alpha = 1$),

3) profit greater than 1 or occupancy smaller than or equal to $\lfloor \frac{c}{2} \rfloor$ ($tg\alpha = \frac{c}{2}$).

THEOREM 3.4. Corresponding to the said instances we have

1') $b = \lfloor \frac{n}{c^2} \rfloor \cdot \lceil \frac{c}{2} \rceil \Rightarrow \frac{z^*}{z_{BG}} \to \frac{3}{4} c$ a.e.

2') $b = \lfloor \frac{n}{c^2} \rfloor \cdot (c(c+1) \frac{c+2}{6}) \Rightarrow \frac{z^*}{z_{BG}} \to \frac{2c+1}{c+2}$ a.e.

3') $b = \lfloor \frac{n}{c^2} \rfloor (\frac{(c-1)(c+1)c}{2} + \frac{\lfloor \frac{c}{2} \rfloor (\lfloor \frac{c}{2} \rfloor +1)}{2}) \Rightarrow \frac{z^*}{z_{BG}} \to \frac{c^2(c+1)-2\lfloor c/2 \rfloor}{c(c^2-1)+\lfloor c/2 \rfloor (\lfloor c/2 \rfloor +1)}$ a.e.

PROOF. By enumeration arguments. QED

In correspondance to the cases 1') 2') by evaluating R'/S' and to the case 3') by evaluating R"/S", we obtain respectively the following values

1") $\frac{11}{14} c$

2") $\frac{4c+3}{2c+3}$

3") $\frac{4c^3+2c^2-\frac{16}{3}c}{4c^3-\frac{4}{3}c^2-4c}$

We can see that 1"),2"),3") are very close to 1'),2'),3') and this substantiates the good quality of the approximation obtained in the theorems.

In conclusion both the results obtained by approximate analytic methods and the results of Theorem 3.4 obtained by counting arguments show that the blind greedy algorithm globally has a poor performance for the knapsack problem. In fact as it is summarized in figure 5, the approximation ratio becomes worse and worse as c (the range of values of the items) increases.

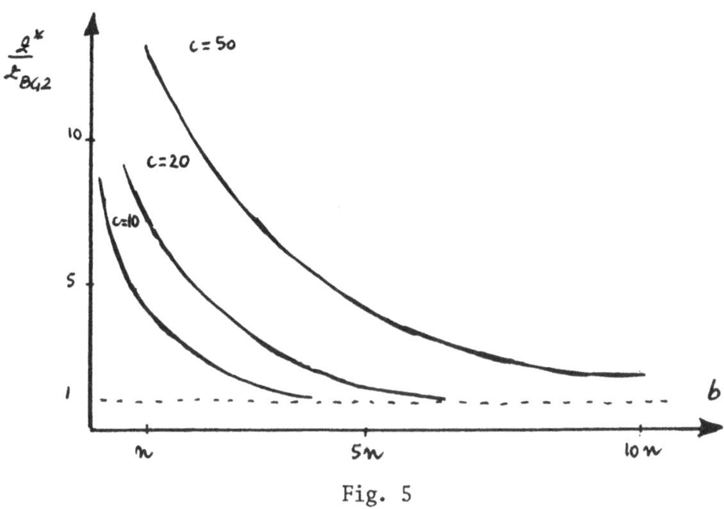

Fig. 5

REFERENCES

AMP 81 G. AUSIELLO, A. MARCHETTI-SPACCAMELA, M. PROTASI: "Probabilistic analysis of the performance of greedy strategies over different classes of combinatorial problems". Proc. FCT '81, Lecture Notes Comp. Sc. n. 117, Springer Verlag, 1981.

DA 79 G. D'ATRI: "Analyse probabiliste du problème du sac-a-dos". Thèse de 3 cycle, University Paris 6, 1979.

GJ 79 M.R. GAREY, D.S. JOHNSON: "Computers and intractability: a guide to the theory of NP-completeness" Freeman and Company, 1979.

IK 75 O.H. IBARRA, C.E. KIM: "Fast approximation algorithms for the knapsack and sum of subset problems" J. ACM vol. 22 n. 4, 1975.

L 77 E.L. LAWLER: "Fast approximation algorithms for knapsack problems". Proc. 18th FOCS, 1977.

A LINEAR TIME ALGORITHM TO MINIMIZE MAXIMUM LATENESS FOR THE TWO-MACHINE, UNIT-TIME, JOB-SHOP, SCHEDULING PROBLEM

P. Brucker
University of Osnabrück
4500 Osnabrück, FRG

Abstract: A linear time algorithm is given for the two-machine, job-shop scheduling problem with n unit-time tasks in which maximum lateness is to be minimized. This algorithm generalizes a linear time algorithm for the corresponding makespan problem given by HEFETZ and ADIRI and improves an O(n log n)-algorithm developed by the author.

1. Introduction

Consider a job-shop problem with r jobs i=1,...,r and two machines denoted by A and B. Each job i=1,...,r has n(i) tasks (i,j)(j=1,...,n(i)) each of length 1. For all i=1,...,r task (i,j+1) cannot be started before task (i,j) finishes and if (i,j) is processed on machine A resp. B task (i,j+1) must be processed on machine B resp. A (j=1,...,n(i)-1). Let us denote by m(i,j) the machine where task (i,j) is to be processed. Then a job i may be characterized by the number n(i) of its tasks and the first machine m(i,1) where i is to be processed. Finally let

$$n = \sum_{i=1}^{r} n(i)$$

be the total number of tasks.

We assume that time zero is the earliest time a task can be started. Furthermore let t_{max} be an upper bound for the largest start time of any job. For example we may choose $t_{max} = n$. Then a schedule may be defined by two arrays A(t) and B(t) with $t=0,\dots,t_{max}$ where A(t) = (i,j) if task j of job i is to be processed on machine A at time t and A(t) = λ if machine A is idle during the time period form t to t+1. In this connection we call λ an empty task. Furthermore for each task (i,j) to be processed on A there exists a time t with A(t) = (i,j). B(t) is defined similarly. A schedule is feasible if (i,j) = A(t) resp. (i,j) = B(t) with $1<j\leq n(i)$ implies that (i,j-1) = B(s) resp. (i,j-1) = A(s) for some $s < t$. A permutation of all tasks is called a list. Given a list L a feasible schedule can be constructed in the following way. Schedule the tasks in an order given by L where each task is scheduled as early as possible. Such a schedule is called the list schedule corresponding with L.

The finish time y(i) of job i in a feasible schedule Y = (A(t), B(t)) is given by

$$y(i) = \max \{t+1 \mid A(t) \text{ or } B(t) \text{ is a task of job } i\}.$$

Given a due date d(i) ≥ 0 associated with each job i lateness of job i is defined by

$$L(i) = y(i) - d(i) \text{ for } i=1,\dots,r.$$

In this paper we are interested in the problem of finding a schedule Y which minimizes maximum lateness

$$\max \{L(i) \mid i=1,\dots,r\}.$$

For the special case in which d(i) = 0 for i=1,...,r this problem was solved by HEFETZ and ADIRI [1979] by an O(n)-algorithm. The following algorithm 1 due to BRUCKER [1980] solves the general problem.

Algorithm 1

1. Associate with each task (i,j) the label l(i,j) = d(i) - n(i) + j.
2. Construct a list L of all tasks in which the tasks are ordered according to nondecreasing l(i,j)-values.
3 Find a list-schedule corresponding with L.

This algorithm can be implemented in O(n log n)-time. We will show in this paper that by using some type of hash techniques (compare MONMA [1981]) the complexity can be improved to O(n). In section 2 such an O(n)-algorithm is presented. A correctness proof is given in section 3.

In the following sections we assume that besides d(i)≥ o for all jobs i we have d(i) = o for at least one job i. Otherwise for each job i we could replace d(i) by d'(i) = d(i)-d where d = min{d(i)|i=1,...,r}. Algorithm 1 shows that the problem with these new d(i)-values has the same optimal schedule as the original one. Furthermore d'(i)≥o for all jobs i. Thus in schedule Y we have L(i) = y(i) - d(i) ≥ o for at least one job i.

2. An O(n)-algorithm

To get an O(n)-algorithm for the problem we sort the tasks according to nondecreasing l(i,j)-values using some kind of hash technique, i.e.

we create lists L(k) containing all tasks (i,j) with l(i,j) = d(i) - n(i) + j = k. The smallest possible l(i,j)-value is -n+1. Furthermore in each list schedule the largest finish time y(i) is bounded by n. Thus in any list schedule no job i with d(i) ≥ n is late and because L(i)≥o for at least one job i we can put the task of all these jobs in an arbitrary order at the end of L. If we add all tasks (i,j) of jobs i with d(i) ≥ n to a list L(n) all other tasks (i,j) can be inserted in some list L(k) with -n < k < n because for such a task (i,j) we have

$$l(i,j) = d(i) - n(i) + j \leq d(i) < n.$$

The list L(k) (k=1-n,...,n) are created by steps 1,2 and 3 of the following algorithm 2.

<u>Algorithm 2:</u>

```
 1. FOR k ← 1-n UNTIL n DO L(k) ← ϕ;
 2. FOR i ← 1 UNTIL r DO
 3.    IF d(i) < n THEN
       FOR j ← 1 UNTIL n(i) DO add (i,j) to L (d(i) - n(i) + j)
       ELSE
       FOR j ← 1 UNTIL n(i) DO add (i,j) to L(n);
 4. FOR i ← 1 UNTIL r DO LAST(i) ← o;
 5. T1 ← T2 ← o;
 6. FOR t ← 1 UNTIL n DO A(t) ← B(t) ← λ;
 7. FOR k ← 1-n UNTIL n DO
 8.    WHILE L(k) ≠ ϕ DO
           BEGIN
 9.        Choose a task (i,j) in L(k) and eliminate this task from L(k);
10.        IF m (i,j) = A THEN DO
11.            IF T1 ≤ LAST(i) THEN
                   BEGIN
12.                t ← LAST(i);
13.                A(t) ← (i,j)
                   END
               ELSE
                   BEGIN
14.                t ← T1;
15.                A(t) ← (i,j);
16.                WHILE A(t) ≠ λ DO T1 ← T1 + 1
                   END
           ELSE
```

```
17.         IF T2 ≤ LAST(i) THEN
                BEGIN
18.             t ← LAST(i);
19.             B(t) ← (i,j)
                END
            ELSE
                BEGIN
20.             t ← T2
21.             B(t) ← (i,j);
22.             WHILE B(t) ≠ λ DO T2 ← T2+1
                END
        END
```

Using these lists L(k) a corresponding list schedule is constructed in steps 4 to 22 of algorithm 2. In this algorithm T1 resp. T2 denotes the first time time period $t \geq o$ where machine A resp. B is idle. Furthermore LAST(i) is used to store the finish time of the last scheduled task of job i. It can be seen easily that the complexity of algorithm 2 is O(n).

3. Correctness of algorithm 2

First we will show that by algorithm 2 a feasible schedule is constructed. This is true if and only if before setting $A(t) \leftarrow (i,j)$ resp. $B(t) \leftarrow (i,j)$ in step 13 resp. step 19 of algorithm 2 we have $A(t) = \lambda$ resp. $B(t) = \lambda$. Otherwise two different tasks are scheduled at the same time on the same machine.

<u>Lemma</u>: Let $Y = (A(t),B(t))$ a list schedule with $A(t) = \lambda$ resp. $B(t)=\lambda$. Then for each $s > t$ if $A(s) = (i,j)$ resp. $B(s) = (i,j)$ then $B(s-1) = (i,j-1)$ resp. $A(s-1) = (i,j-1)$.

<u>Proof</u>: We show by induction on s that $B(s) = (i,j)$ with $s > t$ implies $A(s-1) = (i,j-1)$. This is certainly true for $s = t+1$ because if $B(t+1) = (i,j)$ and task A(t) does not belong to job i then (i,j) must be scheduled earlier by construction of the list schedule.
Let us assume that the Lemma is true for all v with $t < v < s$ and we have $B(s) = (i,j)$. Choose a maximal r with $t \leq r < s$ and $B(r) = \lambda$. By induction assumption A(v-1) and B(v) belong to the same job for each $v = r+1,\dots,s-1$. Now suppose that A(s-1) does not belong to job i. Then for each $v \in \{r,r+1,\dots,s-1\}$ task A(v) do not belong to job i because B(v+1) belongs to the same job as A(v) for $v=r,\dots,s-1$ (see Figure 1).

Thus (i,j) can be already processed at time r which is a contradiction

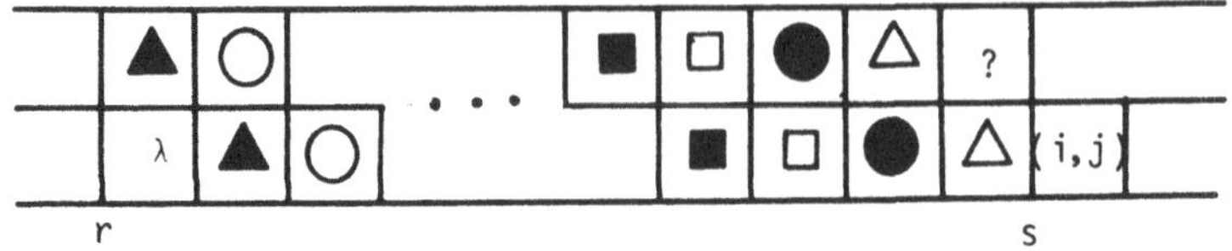

Figure 1

to the fact that Y is a list schedule.

Theorem 1: Let (i,j) be a task to be scheduled in step 13(resp. 19) of algorithm 2 and $t = LAST(i) > T1$ (resp. T2). Then $A(t) = \lambda$(resp. $B(t) = \lambda$).

Proof: Assume that $A(t) \neq \lambda$ (resp. $B(t) \neq \lambda$). We have $A(T1) = \lambda$ (resp. $B(T2) = \lambda$). Thus the previous lemma implies that A(t) and B(t-1) (resp. B(t) and A(t-1))are tasks of a same job $j \neq i$. This is a contradiction to $LAST(i) = t$.

Theorem 2: The list schedule constructed by algorithm 2 is optimal.

Proof: Let L' be the list of all tasks ordered according to nondecreasing l(i,j)-values and Y' be the corresponding list schedule. Furthermore let L = L1L2 be the list which corresponds to the order in which the tasks are scheduled by algorithm 2 where L1 is the partial list of all jobs in L(k) for k=1-n,...,n-1 and L2 is the partial list of all jobs in L(n). Let Y be the list schedule corresponding with L. For all jobs in L1 i.e. for all jobs i with $d(i) < n$ we have

$$y(i) \leq y'(i) \text{ and thus } L(i) = y(i) - d(i) \leq L'(i) = y'(i) - d(i).$$

Because $L(i) \leq o$, $L'(i) \leq o$ for all jobs i with $d(i) \geq n$ and because there exists at least one job i with $L(i) \geq o$ and $L'(i) \geq o$ we have

$$\max\{L(i) \mid i=1,\dots,r\} = \max\{L(i) \mid d(i)<n\} \leq \max\{L'(i) \mid d(i)<n\} = \\ = \max\{L'(i) \mid i=1,\dots,r\}.$$

Thus optimality of schedule Y' implies the optimality of schedule Y constructed by algorithm 2.

REFERENCES

P. BRUCKER [1980], Minimizing maximum lateness in a two-machine unit-time job-shop. To appear in Computing.

N. HEFETZ and I. ADIRI [1979], An efficient optimal algorithm for for two-machines, unit-time, job-shop, schedule-length, problem. Operations Research, Statistics and Economics Mimeograph Series No. 237, Technion-Israel Institute of Technology.

C.L. MONMA [1981], Linear time algorithm for scheduling equal length tasks with due dates on parallel processors subject to precedence constraints. To appear in Operations Research.

AN APPROXIMATE SOLUTION FOR THE PROBLEM OF OPTIMIZING THE PLOTTER PEN MOVEMENT

M. Iri, K. Murota and S. Matsui

Department of Mathematical Engineering and Instrumentation Physics
Faculty of Engineering, University of Tokyo
Hongo, Bunkyo-ku, Tokyo, Japan 113

Summary. An efficient way of drawing a figure by a mechanical plotter is proposed, where the figure is a (generally unconnected) graph with straight-line edges, the coordinates (abscissae and ordinates) of whose vertices are given. The objective is to minimize the plotting time T, which depends on the number N_1 of raising and lowering the pen, the number N_2 of unit pen movements (a unit pen movement being either a movement for drawing an edge with the pen on the paper or a movement from a vertex to another with the pen off) and the total length L of pen movement.

If the graph is connected, N_2 can be minimized by matching (with new edges) the odd-degree vertices in pairs, and N_1 by traversing the Eulerian path (with the pen off the paper when moving along new edges) [1],[2]. For those purposes there is an algorithm linear in space and in time.

To minimize L we need the minimum-weight perfect matching on the complete graph (whose vertices are the odd-degree vertices of the original graph and the weights of whose edges are the distances between their end vertices). The known exact algorithm running in time $O(n^3)$ (n: the number of relevant points) is too complex for large problems and even the known approximation algorithms [3] are of complexity at least $O(n \log n)$. We propose a family of linear-time approximation algorithms for this matching problem. The algorithms divide a square or rectangular region, within which the relevant points lie, into small square cells (or buckets) of equal size, scan those cells one after another according to a prescribed order, and match the points in pairs as soon as two unmatched points are found during the course of scanning. (There are several variants in scanning and matching.) The worst-case performance of the proposed algorithms is analyzed by means of linear programming approach. The average-case performance is also analyzed theoretically; the theoretical results show fairly good agreement with the experimental results for a large number of randomly generated patterns with up to

2048 points.

For an unconnected graph, a similar approximation algorithm is proposed.

Experiences of application of our approximation algorithm to a number of graphs (a small graph with 362 edges and 223 vertices to a large one with 25336 edges and 20726 vertices) were more than satisfactory. The plotting time was nearly halved in most cases compared with more primitive plotting strategies, whereas the extra computer time needed was negligible (e.g. 20ms for the small graph and 850ms for the large).

1. Motivation and Example

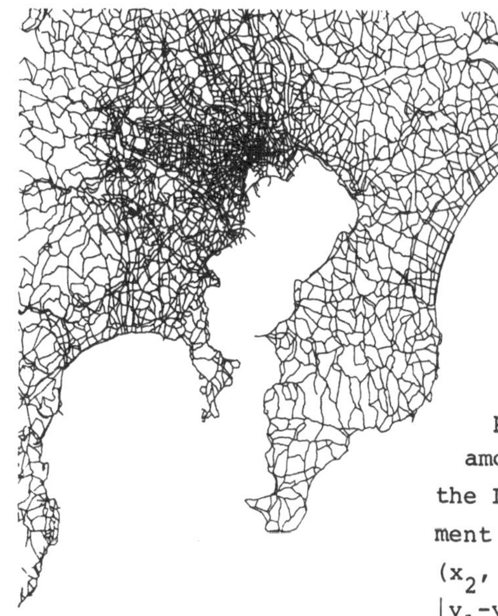

20726 nodes (including degree-2 nodes)
25336 links (straight line segments)
25 connected components

Total L_∞-length of lines (on 60cm×60cm sheet): 44.47m

Fig.1. Road map of Kanto district of Japan

Fig. 1 is the map of main roads in Kanto district (about 160km×160km) of Japan, which contains 20726 nodes (including degree-2 nodes) and 25336 links (of straight line segments). We want to draw the map on the 60cm×60cm square sheet with a mechanical plotter. Total length of lines to be drawn, i.e., movement of the pen holder with pen on the paper, amounts to 44.47m in L_∞-measure, where the L_∞-measure of a straight line segment connecting points (x_1, y_1) and (x_2, y_2) is defined by $\max(|x_1-x_2|, |y_1-y_2|)$.

Figs. 2-4 show the wasted pen movements involved in three different ways of plotting: (1) Input data shuffled (most stupid strategy), (2) Input data prepared by an intelligent person (from west to east and then from north to south), and (3) Input data preprocessed by our algorithm.

This example will illustrate the practical importance of the problem of optimizing the plotter pen movement.

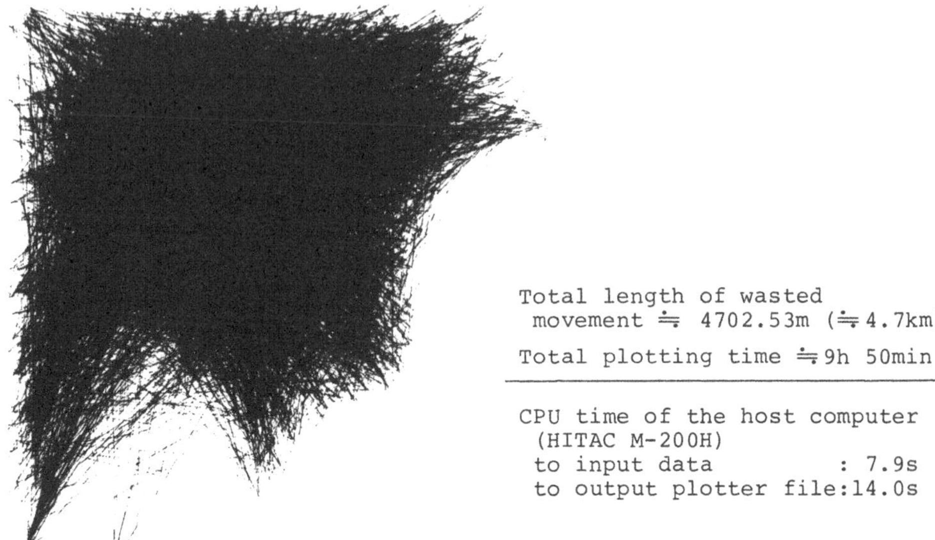

Total length of wasted
movement ≒ 4702.53m (≒4.7km)
Total plotting time ≒9h 50min
CPU time of the host computer
(HITAC M-200H)
to input data : 7.9s
to output plotter file:14.0s

Fig.2. Pen movement wasted —— Case (1)
Input data shuffled

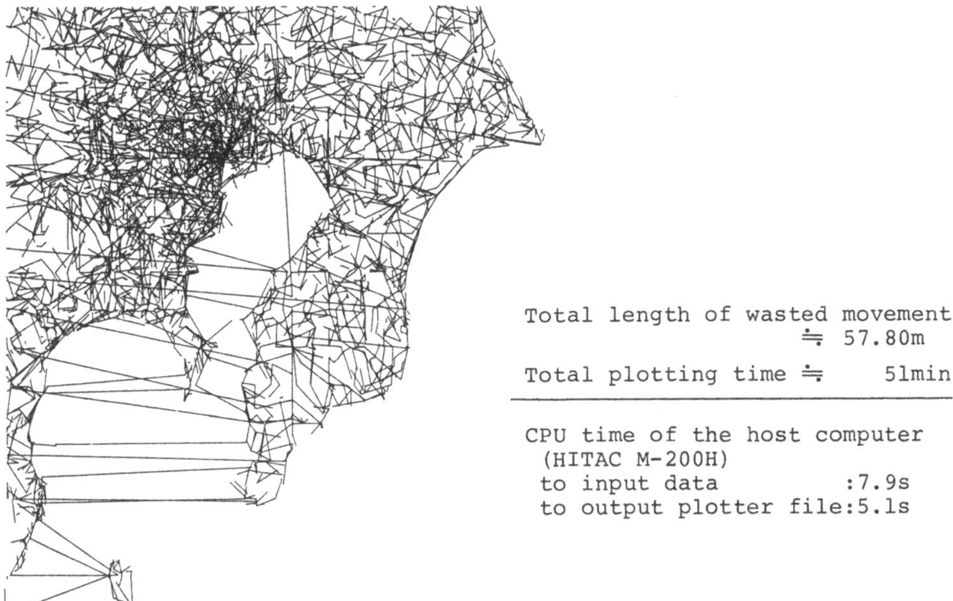

Total length of wasted movement
≒ 57.80m
Total plotting time ≒ 51min
CPU time of the host computer
(HITAC M-200H)
to input data :7.9s
to output plotter file:5.1s

Fig.3. Pen movement wasted —— Case (2)
Input data prepared by an intelligent person

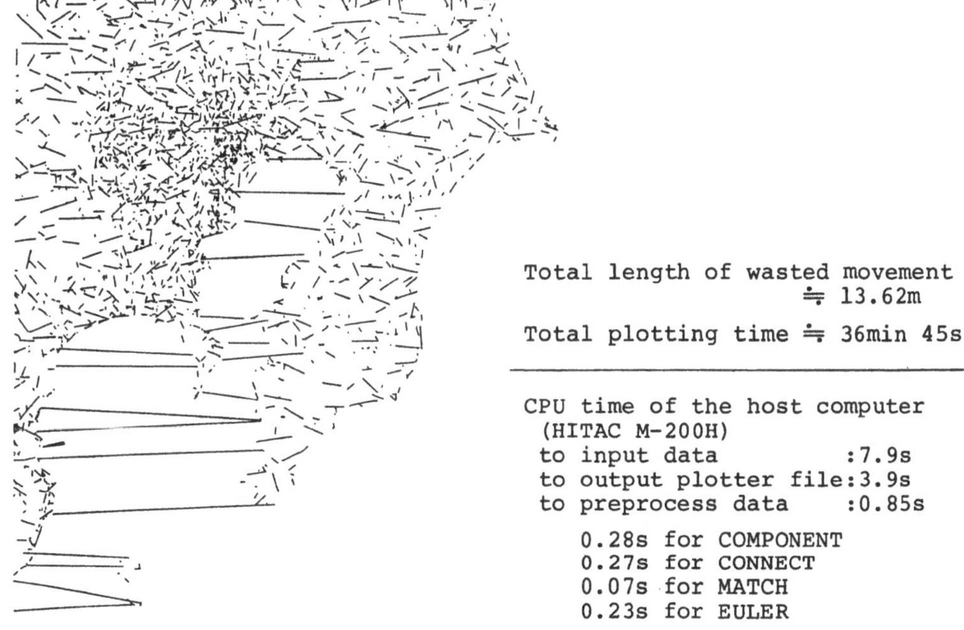

Total length of wasted movement ≒ 13.62m
Total plotting time ≒ 36min 45s
CPU time of the host computer (HITAC M-200H)
to input data :7.9s
to output plotter file:3.9s
to preprocess data :0.85s
0.28s for COMPONENT
0.27s for CONNECT
0.07s for MATCH
0.23s for EULER

Fig.4. Pen movement wasted —— Case (3)
Input data preprocessed by our algorithm

2. Mathematical Formulation of the Problem

We consider the problem of minimizing the plotting time T under the assumption that no line should be traced more than once with pen down on the paper, to get a high-quality drawing. The plotting time T may be estimated by the formula:

$$T = a_1N_1 + a_2N_2 + a_3L,$$

where

- N_1: number of times of raising and lowering pen,
- N_2: number of unit pen movements (i.e., number of real drawn line segments and blank line segments corresponding to movements of the pen off the paper),
- L : total length of pen movements (in L_∞-length).

For instance, we had a_1=0.01s, a_2=0.078s, a_3=0.07s/cm for the plotter used in our experiment. (We got a_1=0.033s, a_2=0s, a_3=0.10s/cm for another type).

Then, minimizing T is equivalent to minimizing N_1, N_2 and L. Here it should be noted that raising (lowering) pen immediately after lowering (raising) pen can be replaced by no operation, so that we can assume N_1 equals twice the number of pen movements off the paper. Thus we may concentrate on minimizing N_2 and L.

In case the graph to be drawn is Eulerian, we may use an algorithm for finding a Eulerian path and can traverse all the edges with pen kept on the paper. An exact algorithm, named EULER, for finding a Eulerian path is of complexity O(V, E) both in time and in space, where V is the number of nodes and E is the number of edges.

When the graph is connected but not Eulerian, we have to match the odd-degree vertices in pairs in such a way that the total distance between the matched pairs of vertices (corresponding to the wasted pen movements) may be as small as possible. Then N_2 as well as L is minimized. Known exact matching algorithms have time complexity at least $O(V_o^3)$ (V_o: number of odd-degree vertices), which is too complex for problems of practical size, so that we would propose an approximation algorithm MATCH of complexity $O(V_o)$ both in time and in space.

In the general case where the graph is not connected, we preliminarily add some blank edges (corresponding also to wasted pen movements), with the total length as small as possible, to make the graph connected. An approximation algorithm CONNECT of complexity O(V) is used. Given a graph structure, our algorithm first decomposes it into connected components by COMPONENT of complexity O(E) and makes it connected by CONNECT, then matches the odd-degree vertices by MATCH, and finally traverses all the edges unicursally by EULER. All these procedures are performed in O(V, E) time and space.

3. Approximation Algorithm for Matching —— MATCH

Suppose that n points (V_o odd-degree points in the plotter problem) are distributed in a unit square, where n is even. We propose a family of linear-time approximation algorithms, which we call straightforward algorithms, for finding the minimum-weight (with respect to L_∞-measure) perfect matching. The basic idea is as follows:

1. Divide the square into small k×k square cells and order them.
2. Determine to which cell each point belongs.
3. Scan all points, cell by cell, according to the prescribed cell order.
4. Match the points as soon as they are scanned.

An example of the cell order, called "spiral rack" order, is shown in

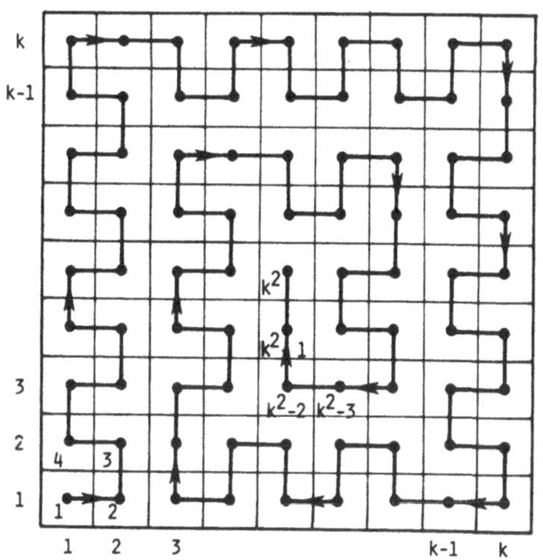

Fig.5. Spiral rack cell order ($k \sim 1.26\sqrt{n}$, odd)

Fig. 5, which is one of the orders to be recommended.

The following modifications improve the performance of the basic algorithms to some extent: (a) Preprocess: —— First, match points in one and the same cell and, then, scan the remaining points; (b) Tour: —— Compare two matchings, one starting from the first point and the other from the second (the last point is matched with the first in the latter case), and then adopt the better.

4. Analysis of the Performance of Straightforward Algorithms

We define the cost of a matching by the total distance between matched pairs of points and measure the performance of an approximation algorithm in terms of this absolute cost, not by the ratio to the exact minimum. Let $M_n(k)$ be the total distance between matched pairs of n points when the square is divided into k×k cells. The following notation is used:

$\bar{M}_n(k)$: supremum of $M_n(k)$ over all possible configurations of n points in the square

$$\bar{\mu}(\alpha) = \lim_{n\to\infty} \bar{M}_n(\alpha\sqrt{n})/\sqrt{n}$$

$$\bar{\mu}_o = \bar{\mu}(\bar{\alpha}_o) = \min_{\alpha} \bar{\mu}(\alpha)$$

$E(M_n(k))$: average of $M_n(k)$ for randomly distributed points

$$\mu(\alpha) = \lim_{n\to\infty} E(M_n(\alpha\sqrt{n}))/\sqrt{n}$$

$$\mu_o = \mu(\alpha_o) = \min_{\alpha} \mu(\alpha)$$

We will adopt $\bar{\mu}_o$ and μ_o for the performance measure for the worst and the average case, respectively.

The worst-case performance is analyzed theoretically by means of linear programming approach [4]. The spiral rack algorithm with preprocess and tour turned out to be the best among several linear-time algorithms considered, including the linear-time implementation of the four-square algorithm, which is claimed to be of complexity O(n logn) in [3]. The optimal value $\bar{\alpha}_o=\sqrt{3}$ yields $\bar{\mu}_o=\sqrt{3}/2=0.866$. The average-case performance is also analyzed theoretically by a stereotyped method. The spiral rack algorithm with preprocess and tour is again the best, but with a different optimal value of α; i.e., $\mu_o=0.443$ for $\alpha_o=1.07$. Considering the trade-off between the worst- and the average-case performance, we would recommend the value $\alpha=1.26$, for which we have $\bar{\mu}\leq 0.91$ and $\mu=0.449$.

The asymptotic behavior of the recommended algorithm, i.e., the spiral rack algorithm with preprocess and tour with $\alpha=1.26$, has been investigated by Monte-Carlo experiment for randomly distributed points. Fig. 6 illustrates that $M_n/\sqrt{n}$ approaches the theoretical limit. It is observed that the variance of $M_n/\sqrt{n}$ decreases as rapidly as $O(1/n)$ as n tends to infinity. The costs by the recommended algorithm have also been compared with those by the exact minimum solution for n=16 to 256. It is observed that the ratio clusters around 1.5 for n large.

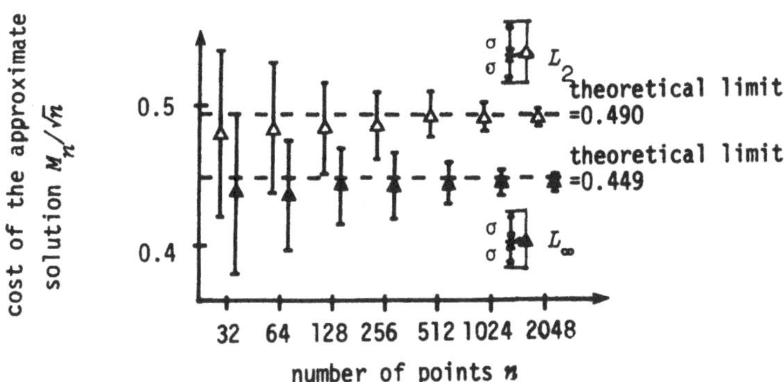

Fig.6. Asymptotic behavior of the recommended algorithm (200 repetitions for each instance)

5. Algorithm for Making the Graph Connected —— CONNECT

In case the graph is unconnected, we make it connected by adding some blank edges and, possibly, new vertices, so that the total length of the added edges may be as small as possible. Let c be the number of connected components. Obviously c-1 additional edges are necessary and sufficient to make it connected.

We could consider to introduce new vertices as well as new edges to make the graph connected. However, the problem of minimizing the total length of the additional edges with new vertices is a kind of Steiner problem and therefore hard to solve. Under the restriction that new vertices are not allowed, i.e., that new edges should connect only vertices of the original graph, the additional edges with the exact minimum total length could be found in a polynomial time, but not in linear time, with the aid of the Voronoi diagram in the L_∞-measure [5], [6]. However, we should like to adopt an approximation algorithm CONNECT to make our algorithm as a whole run in linear time.

The algorithm CONNECT is basically in the same vein as the algorithm MATCH for matching. Consider a graph (generally unconnected) with its vertices lying in the unit square. We assume that the graph is decomposed into c connected components. The idea is as follows:

1. Divide the square into small $k \times k$ square cells and order them.
2. Determine to which cell each vertex belongs.
3. Scan all vertices, cell by cell, according to the prescribed cell order.
4. If two consecutive vertices belong to distinct connected components, connect them with a new edge, and merge the two connected components into one.

Evidently the algorithm CONNECT adds c-1 new edges and can be implemented to run in O(V) time if $k=O(\sqrt{V})$, where V is the number of vertices of the graph.

The average-case performance of CONNECT is analyzed theoretically under the assumption that the vertices are uniformly distributed in the unit square, independently of the connectedness among them. The average length in the L_∞-measure of an additional edge is estimated theoretically and confirmed experimentally; for the spiral rack cell order, it attains the approximate minimum of $0.9/\sqrt{V}$ for $k \sim 1.2\sqrt{V}$.

Acknowledgements

The authors thank Mr. H. Imai for his nice cooperation in laborious computation and map drawing.

The geographical data of the road map of Kanto district have been compiled from the data bank "Digital National Land Information" prepared by the Geographical Survey Institute of the Ministry of Construction of Japan.

This work was supported in part by the Grant in Aid for Scientific Research of the Ministry of Education, Science and Culture of Japan

(Cooperative Research (A)00535012 (1981)) and in part by the Kajima Foundation's Research Grant.

References

[1] Reingold, E. M., and Tarjan, R. E.: On a Greedy Heuristic for Complete Matching. (1978)

[2] Iri, M., and Taguchi, A.: The Determination of the Pen-Movement of an XY-Plotter and Its Computational Complexity (in Japanese). Proc. Spring Conference of the Operations Research Society of Japan, 1980, P-8, pp.204-205.

[3] Supowit, K. J., Plaisted, D. A., and Reingold, E. M.: Heuristics for Weighted Perfect Matching. Proc. 12th Annual ACM Symposium on Theory of Computing, 1980, pp.398-414.

[4] Iri, M., Murota, K., and Matsui, S.: Linear-Time Approximation Algorithms for Finding the Minimum-Weight Perfect Matching on a Plane. Information Processing Letters, 1981, Vol.12, pp.206-209.

[5] Shamos, M. I., and Hoey, D.: Closest-Point Problems. Proc. 16th Annual Symposium on Foundations of Computer Science, 1975, pp.151-162.

[6] Lee, D. T., and Wong, C. K.: Voronoi Diagrams in L_1 (L_∞) Metrics with 2-Dimensional Storage Applications. SIAM J. Computing, 1980, Vol.9, pp.200-211.

OPTIMAL CONTROL WITH CONSTRAINED BINARY SEQUENCES

Marc Kaltenbach
Bishop's University
Lennoxville, Quebec
Canada J1N 1Z7

INTRODUCTION

It is fairly common in practice to encounter processes controlled by binary sequences in which there are constraints on the number of consecutive identical values in a sequence. Such processes are models for industrial production operations and a variety of waiting line systems, including as a particular case the control of traffic lights in urban street networks.

This paper originated within the context of a traffic control application [1]; however the approach it describes is general and can be adapted to other specific cases with minor changes.

The paper begins with a summary description of the control process and control action applied to it; then an open loop optimization approach is obtained via a series of refinements over a proposed Branch and Bound enumeration procedure. The paper terminates with numerical illustrations and a brief description of an application in a wider setting.

I SUMMARY DESCRIPTION OF THE CONTROL PROCESS

Assume that a discrete time model of the process under control is available [1] and of the form,

$$X_{t+1} = F(X_t, u_t) \qquad (1)$$

$$X_0 \text{ given}, \quad t = 0,1,2,...$$

where $X_t \in R^m$ and u_t is a control variable with values restricted to $\{0,1\}$.

A performance index of additive type,

$$PI[t] = \sum_{n=0}^{t} H(X_n, u_n), \quad t = 1,...,T \qquad (2)$$

is associated with a T time steps simulation of (1).

The problem under consideration is to find, within a set of admissible sequences to be described, a sequence that minimizes $PI[T]$.

II ADMISSIBLE SEQUENCES

The control space is defined as the set of sequences $\{u_t\}_{t=-\infty,\infty}$ such that $u_t \varepsilon \{0,1\}$ and at least L_1 and not more than L_2 consecutive terms of the sequence have the same value.

In the context of Traffic Control, a value $u_t = 1$ indicates a green light and $u_t = 0$ a red light for a specific direction at an intersection during time interval $[t\Delta, (t+1)\Delta)$, where Δ is a time discretization step size associated with (1). In a sequence $\{u_t\}_{t=-\infty,\infty}$ the successive strings of consecutive identical values represent the "phases", green or red, of classical Traffic Control Theory [2]. A minimum phase duration $L_1\Delta$ is specified to correspond to a safe intersection crossing duration by a vehicle or pedestrian. The maximum phase duration $L_2\Delta$ is also provided in order to prevent monopoly of the right-of-way by one particular direction; it may be set as the maximum wait time that may be tolerated by a vehicle at the intersection.

In order to further characterize admissible control sequences in the case of simulations with a finite number of time steps, the concept of a control state is required. Denote by ℓ_n the number of identical consecutive terms in the last phase of sequence $\{u_t\}_{t=-\infty,n}$; the pair (ℓ_n, u_n) is called the control state at step n. At step $n + 1$, the control state may be either

a) $(1 + \ell_n, u_n)$,

b) $(1, 1 - u_n)$, or

c) a free choice of either a) or b)

case a) applies if $\ell_n < L_1$; case b) if $\ell_n = L_2$ and case c) if $L_1 \le \ell_n < L_2$.

An actual simulation calls for a control sequence $\{u_t\}_{t=1..T}$ with given initial control state (ℓ_0, u_0). A phase initiated before the beginning of the simulation must satisfy the phase length requirements already mentioned. It follows that the first phase in the sequence, with number of steps denoted ph_1, must satisfy the additional requirement: if $\ell_0 < L_1$ then $u_1 = u_0$ and if $u_1 = u_0$ then $L_1 \le \ell_0 + ph_1 \le L_2$. Also, the last phase of the sequence may have less than L_1 steps.

The control space thus defined admits a number of sequences growing exponentially with n the number of terms in a sequence. More specifically if S_n is that number, then $S_n \sim 0(c^n)$ as $n \to \infty$, where c is the unique real root existing between 1 and 2 for the equation,

$$\frac{1 - x^{-L_2}}{1 - x^{-L_1}} = x \tag{3}$$

To show this establish that $S_n = S_{n-L_1} + \dots + S_{n-L_2}$ and study the convergence of $\frac{S_{n+1}}{S_n}$ as $n \to \infty$. For many applications it is desirable to develop an optimization approach adapted to these large control spaces.

III OPTIMIZATION WITH CONSTRAINED SEQUENCES

Let $\{u_t\}_{t=1..T}$ denote a control sequence and $\{\tilde{u}_t\}_{t=1..T}$ the best sequence known so far in the search for an optimal one. An enumeration algorithm for admissible sequences is proposed next.

1. Basic Enumeration Algorithm

An enumeration of all admissible sequences is obtained by performing the following steps.

Step 1. Start with any admissible sequence $\{u_t\}_{t=1..T}$

Step 2. Determine the corresponding trajectory of phase lengths $\{\ell_t\}_{t=1..T}$

Step 3. Construct sequence $\{d_t\}_{t=1..T}$ such that for $t = 1..T$,
$d_t = 0$ if $2 \le \ell_t \le L_1$ or if $\ell_t = 1$ and $\ell_{t-1} = L_2$,
and
$d_t = 1$ otherwise.
At the start of the algorithm, $d_k = 1$, where $k \in \{1,..,T\}$, means that the value of the last term in sequence $\{u_t\}_{t=1..k}$ may be changed without violating the admissibility of the new sequence thus obtained.

Step 4. The enumeration is performed through successive calls to two procedures. Procedure <u>Backward</u> selects the last term u_k in the sequence $\{u_t\}_{t=1..T}$ for which $d_k = 1$. If no such term exists, the enumeration is complete. If value u_k is changed, then d_k is set to zero and control is transferred to Procedure <u>Forward</u>. The purpose of this procedure is to complete the sequence $\{u_t\}_{t=1..k}$ with any sequence, provided the resulting sequence for t=1,..,T is admissible. In Procedure Forward, the new values for $\{\ell_t\}_{t=k+1..T}$ and $\{d_t\}_{t=k+1..T}$ are determined according to Steps 2 and 3.

2. Search For Optimal Sequences

Conceptually an optimal sequence is identified by computing the cost associated with each one of the enumerated sequences. It can be shown the enumeration procedure just described reduces the number of iterations of (1) to only a fraction of the number of iterations required by independent simulations of all admissible sequences of T terms; this fraction is $\frac{c}{n(c-1)}$ in the limit as $n \to \infty$. In exchange, it must be possible to store a system state trajectory for T simulation steps. Assuming this, a further major computational saving is obtained by terminating procedure Forward as soon as the partial cost associated with a sequence being simulated exceeds the lowest value $\tilde{PI}[T]$ found so far. This is called "implicit enumeration".

3. Methods of Successive Approximations

Implicit enumeration is now combined with (heuristic) rules inspired from gradient approaches to optimal control problems [1]. With no integer restriction on the control variables, it is observed that a gradient step updates simultaneously many terms in a control sequence and, at least close to an optimum, an updated control sequence $\{u''_t\}$ is close to the previous one $\{u'_t\}$ in the sense that the distance between two sequences, defined as $\sum_{t=1}^{T} | u''_t - u'_t |$, is a small number. The following are attempts to elicit a similar behavior in an optimization procedure for admissible $0 - 1$ sequences.

a) Optimal Completion of a Perturbed Sequence

Recall that procedure Backward selects a component u_k of an admissible sequence and changes its value. We call this operation a perturbation of nominal sequence $\{u_t\}_{t=1..T}$. Procedure Forward then generates a completion $\{u_t\}_{k+1..T}$ of sequence $\{\{u_t\}_{t=1..k-1}, 1 - u_k\}$ unless the perturbed sequence is found not to be optimal. With the definition of distance between two sequences already given, the problem is to complete the perturbed sequence in a way that minimizes the distance of the perturbed sequence to $\{\tilde{u}_t\}_{t=k+1..T}$. This is achieved by selecting among just two sequences. The first one is obtained by applying the following rule in Procedure Forward. If there is choice for the value of a term in the completion sequence, it should be selected as the value of the corresponding term in $\{\tilde{u}_t\}_{t=k+1..T}$. The second sequence is obtained similarly, except that the rule is reversed for the first phase of the completion sequence. As in practice the first sequence is selected most of the time, it may be chosen always without significant resulting loss in algorithm efficienty.

b) Toward simultaneity in Updating Sequence Components

Simultaneity in updating sequence components is approximated by restricting deviations from a nominal sequence to a subset of the set of admissible sequences. In the terminology of Chernousko [3] a "tube" is constructed "around" a nominal sequence and the search for a better sequence is restricted to sequences lying within the tube. This operation is repeated until no improvement is obtained within a newly created tube. At this point, the procedure is either terminated or continued with a larger tube. An optimal sequence is guaranteed as a final result only when the tube is the entire set of admissible sequences.

We found this idea very effective in speeding up convergence toward an optimal sequence when combined with implicit enumeration. The tubes may be constructed in dynamic association with the sequence enumeration process as shown in the following.

i) Limits on phase length variations: The 0-1 sequence enumeration process may be viewed as a succession of phase length expansions or contractions. A tube is obtained by specifying a limit NK, $NK \leq L_2 - L_1$, on the number of terms by which a phase may be expanded or contracted. The NK limit applies as well to the beginning as to the end of a phase upon entering procedure Backward. In the context of the sequence optimization problem, limit NK is not applied if the immediate previous sequence in the phase length expansion or contraction has resulted in an improved sequence. For this reason, limit NK is called a <u>maximum depth for unsuccessful variation</u>.

ii) Methods based on partial costs: Associate sequence $\{PI(t)\}_{t=1..T}$ with sequence $\{u_t\}_{t=1..T}$ and sequence $\{\tilde{PI}(t)\}_{t=1..T}$ with $\{\tilde{u}_t\}_{t=1..T}$, the best sequence known so far. Sequence $\{u_t\}_{t=1..T}$ is said to belong to the proposed tube if, for t=1,..,T, $PI(t) - \tilde{PI}(t) < C_{max}$ where C_{max} is a positive number. Since C_{max} is a bound on the value by which a partial cost may exceed the corresponding partial cost of the best sequence known so far, C_{max} is also called a <u>maximum depth for unsuccessful variation</u>.

These two approaches for tube construction can be applied simultaneously; furthermore the maximum depths for unsuccessful variation may be progressively increased to provide a sequence of tubes varying from very constraining to not constraining at all.

iii) Trade-off Between Short Term and Long-Term Optimization: The amount of computing required by the successive terms in the control sequence may be equalized further by specifying the depths for unsuccessful variation as decreasing functions of the index of the term of the control sequence to which they are applied.

IV NUMERICAL ILLUSTRATIONS

Consider (1) and (2) as representing the effect of traffic light control at one node within a street network, with PI denoting total delay to vehicles resulting from conflicts of right-of-way. If $L_1 = 6$, $L_2 = 12$ and $\ell_0 = 6$ then there are exactly 153 366 sequences for a 50 time step simulation, ($c \cong 1.255$). The number of simulation steps required by the enumeration algorithm in III is 753 828. The number of actual simulation steps required by an arbitrarily selected test problem was found as 50 150 when implicit enumeration alone was used; PI [50] then was reduced from an initial value 3 959 to 3 393. Using $C_{max} = 0$, the same reduction was obtained with only 140 simulation steps. When depths for unsuccessful variation are used, the quality of resulting optimized sequences is of course problem dependent. At present, it is claimed that their applications constitute worthwhile heuristics in the context of some important network flow problems.

V AN APPLICATION

The proposed algorithm has been applied to the simultaneous (parallel) control of nodes in large networks. The optimization at each node is carried over a subnetwork of dependent nodes so that coordination between nodes is achieved by the overlap of subnetworks. In addition, the whole procedure is repeated in the manner of rolling schedules so as to achieve real time on-line control. The overall performance has been found to compare favorably with standard traffic control approaches.

REFERENCES

M. KALTENBACH

[1] Modelling and control of traffic networks, Ph.D. Thesis, Electrical Engineering, University of Toronto, 1976.

TORONTO TRANSIT COMMISSION

[2] State of the Art, Vol. 1, 1974.

F.L. CHERNOUSKO

[3] A local variation method for numerical solution of variational problems, Zh. vȳchisl. Nat. Fiz. 5,4, 749-754, 1965.

Generalized Augmenting Paths for the Solution of Combinatorial Optimization Problems

Aaron Kershenbaum

Polytechnic Institute of New York

Abstract

Alternating chain procedures can be thought of as generalizations of the greedy algorithm in that instead of accepting the best remaining element, they seek to obtain a better augmentation by examining a wider range of alternatives. It is possible to generalize the notion of an augmenting sequence to include augmentations which are in effect trees as opposed to simply paths such that these augmentations are sufficient to guarantee optimality. Unfortunately, in the worst case, these trees are of exponential size. We examine the application of such generalized augmenting sequences to the solution of NP-complete problems and examine their effectiveness and efficiency.

Introduction

The theory of NP-completeness, which was first expounded by Cook [1], has led to a search for a unified treatment of combinatorial optimization problems. Cook was able to characterize a very large class of interesting and important problems as being equivalent in the sense that an efficient algorithm capable of finding an optimal solution to any one of these problems can be used to obtain optimal solutions to all of the others. Many papers by many authors and an excellent compendium [2] of problems in this class (as well as techniques for proving that a problem is in this class) have been published since Cook's seminal paper. Problems in this class are called NP-complete problems (or, more properly, NP-hard when they are optimization problems as opposed to decision problems).

Cook's results can be interpreted in several ways. One of these is to say that many clever people have spent many years trying and failing to find efficient algorithms for individual problems in this class. Surely one of them would have succeeded if, in fact, such algorithms existed. Hence, it is unlikely that such an algorithm will be found and it is tempting to stop looking for one. This leads to the development of heuristics for the solution of such problems [3] and to probabilistic methods [4].

An alternate interpretation is that this pessimistic view is justified only with respect to algorithms which guarantee optimal solutions and reasonable runtimes for all instances (input data sets) of a problem. In this paper we speak of an algorithm's runtime being reasonable if it grows polynomially rather than exponentially with the size of the problem. This does not preclude the existence of algorithms with guaranteed reasonable runtimes and which yield optimal (or near-optimal) solutions with high probability. Nor does it preclude the existence of algorithms which guarantee optimal solutions and which have reasonable runtimes with high probability. There are many examples of both types of algorithms which are used in practice to solve specific NP-complete problems. Most important, the theory of NP-completeness does not preclude or even lessen the likelihood of the existence of algorithms which solve specific (nontrivial) instances of a problem and guarantee both an optimal solution and reasonable runtime.

In this paper, we explore this second, more optimistic, point of view and present a family of algorithms for the solution of an NP-complete problem. Some algorithms in this family have guaranteed reasonable runtimes. Others guarantee optimal solutions. While the algorithms are presented for the solution of a specific problem, the technique can be extended to the solution of other problems as well.

Matroid Theory

A specific way of approaching the solution of many combinatorial optimization problems is via matroid theory. The excellent book by Lawler [5] gives a complete treatment of this. Here we outline the fundamentals of this theory which are necessary for the presentation which follows.

A matroid is a couple (E,F) where E is a finite set of m elements:

$$E = \{e_j \mid j = 1,2, \ldots M\}$$

and F is a family of independent subsets of E. The notion of independence is quite general. We require, however, that it satisfy two properties:

P1: Every subset of an independent set is independent, i.e., if

$I \in F$ and $J \subset I$ then $J \in F$

P2: If I_P and I_{P+1} are independent subsets of E containing P and P + 1 elements, respectively, then there exists an element, $e \in I_{P+1}$ $(e \notin I_P)$ such that $I_P \cup \{e\}$ is an independent set containing P + 1 elements.

Given two matroids, (E,F_1) and (E,F_2), defined on the same set of elements, but using two different notions of independence, we define an intersection of them to be any subset, $I \subset E$, such that $I \in F_1$, and $I \in F_2$. This definition can be extended to cover three or more matroids as well.

Many combinatorial optimization problems can be thought of as finding the best independent set in a matroid or the best intersection of two or more matroids. If weights, w_j, are associated with the elements, e_j, in E, then one can speak of the best set as being the one with largest total weight. The maximal (or minimal) spanning tree problem can be thought of as finding the maximum (or minimum) weight independent set in a matroid (E,F) where E is the set of edges in the graph and F is the family of forests. A forest is defined to be a set of 0 or more edges which do not contain a circuit. As another example, Lawler [5, p. 304] shows that the Traveling Salesman Problem can be thought of as finding the best intersection of three matroids. The problem of finding the maximum weight intersection of three matroids has been shown to be NP-complete [2]. Lawler shows [5, p. 364] that the problem of finding intersections of four or more matroids can be reduced to that of finding intersections of three. There are many other combinatorial optimization problems which can be naturally thought of as matroid intersection problems. The theory of NP-completeness assures us that all problems can be thought if in this way.

We will consider one of the simplest possible 3-Matroid Intersection Problems in the sequel for the sake of clarity. The problem considered is the Three Dimensional Assignment Problem (TDAP). In this problem, we are given N people, N jobs, and N days. There is a cost, C_{ijk} of having person i doing job j on day k. Each person is to do only one job, each job is to be done only once, and only one job is to be done on a day. Formally the problem is:

$$\text{Minimize} \quad Z = \sum_{i,j,k} C_{ijk} X_{ijk}$$

such that

$$\sum_{i,j} X_{ijk} = \sum_{i,k} X_{ijk} = \sum_{j,k} X_{ijk} = 1 \text{ for } i,j,k = 1,2, \ldots N \qquad X_{ijk} \in \{0,1\}$$

Thus, setting X_{ijk} to 1 corresponds to having person i do job j on day k. This problem can be viewed as an intersection of three partition matroids. Given a set of elements, E (in this case, the X_{ijk}), a partition matroid can be defined by a partition of E and a vector, A, constraining the number of elements of E which may be selected from any part of the partition. Formally, we have the partition of E into subsets E_j, j = 1, ... k, where

$$\bigcup_j E_j = E \qquad \text{and} \qquad E_i \cap E_j = \phi \text{ for } i \neq j$$

and an integer vector $A = \{a_j | \; j = 1, \ldots k\}$

A matroid (E,F) is then defined where F consists of all subsets, I, of E formed by selecting no more than a_j elements of E_j.

In the case of the TDAP, the first partition of the X_{ijk} is by person, i.e.,

$$E_i = \{X_{ijk} | j = 1, \ldots N; \; k = 1, \ldots N\}$$

and $a_i = 1$ for all i. The independent sets in this first matroid correspond to assigning each person at most one job. Similarly, two more matroids can be defined to constrain jobs and days. Intersections of these three matroids correspond to feasible partial assignments and intersections of maximum cardinality correspond to feasible complete assignments. If we define weights W_{ijk} associated with the X_{ijk}:

$$W_{ijk} = C - C_{ijk}$$

where C is larger than any C_{ijk}, then the maximum weight intersection corresponds to the optimal solution to the TDAP.

Augmenting Paths

We now define a family of algorithms for the solution of matroid intersection problems. These are generalizations of the basic procedure given in [6].

Given a matroid (E,F) (and hence a notion of independence) and a (not necessarily independent) subset S, of E, we define the span of S, denoted sp(S), as S together with all elements of E not independent of the elements in S, that is

$$sp(S) = \{ e \mid I \cup \{e\} \notin F \text{ where I is any independent subset of } S\}$$

If S is an independent set and $e \in sp(S)$ then e forms a unique cycle, which we denote by C(e), with S. A cycle is a dependent set which becomes independent if any element is removed from it.

If the matroid intersection problem only involves two matroids, we can obtain a maximum weight intersection by producing a sequence of intersections, $I^{(K)}$, containing K elements, for K = 1,2, ... m. Each $I^{(K)}$ is the maximum weight intersection containing K elements. The algorithm which produces the $I^{(K)}$ is called an augmenting path procedure because it augments $I^{(K)}$ to produce $I^{(K+1)}$ by finding the longest path in the graph $G^{(K)}$ defined below.

We define $G^{(K)}$ to be a bipartite graph with nodes corresponding to the elements, e_j, of E plus distinguished start and finish nodes, a and z. Directed arcs are defined as follows:

(a,i)	$i \in E - sp_1(I^{(K)})$	(i,j)	$i \in E - I^{(K)},\ j \in C^{(2)}(i)$
(i,z)	$i \in E - sp_2(I^{(K)})$	(j,i)	$i \in E - I^{(K)},\ j \in C^{(1)}(i)$

Paths from a to z correspond to augmentations of $I^{(K)}$, that is, to sets of elements to be added or deleted from $I^{(K)}$ to produce an intersection with K + 1 elements. Notice that all a to z paths go alternately through nodes not contained in $I^{(K)}$ (which are to be added to $I^{(K)}$) and nodes in $I^{(K)}$ (which are to be deleted). Note also that there is one more node of the former type than there is of the latter and hence an augmentation results. If we associate lengths with the arcs equal to the weights of the elements which the nodes correspond to (positive for elements to be added and negative for elements to be deleted), then the length of a path corresponds to the incremental weight of the augmentation. The longest path results in an optimal augmentation. Such a path can be found using a shortest path algorithm suitably modified to find longest paths. $G^{(K)}$ contains no positive cycles and so the algorithm converges.

These augmentations do, indeed, result in intersections. As one passes through nodes from a to z we see that an element is added preserving independence in the first matroid but not the second. An element is then deleted restoring independence in the second matroid and hence the intersection. A node is then added which, because of the deleted node, maintains independence in the first matroid. This process continues until the added element maintains independence in the second matroid as well as the first, thus completing the augmentation.

As an example, consider a two dimensional assignment problem (involving, say, only people and jobs.) The W_{ij}'s for this problem are given in Figure 1. $I^{(2)}$ is clearly 11,22, i.e., person 1 assigned to job 1, and person 2 assigned to job 2. $G^{(2)}$ is shown in Figure 2. The arc lengths are shown as are the lengths of the longest paths to each node from node a. The longest a to z path is a,11,12,22,23,z which corresponds to deleting 11 and 22 from the intersection and adding 31,12, and 23. The length of this path, 7, is the difference between the weight of $I^{(3)}$ and $I^{(2)}$. A complete description of this process and a proof of its validity is given in [5].

Generalized Augumenting Paths

In the graph shown in Figure 2, one can obtain an optimal augmentation (i.e., one which takes us from an optimal assignment of K elements to an optimal assignment of K + 1) because:

1. If the current intersection is not maximal then an augmenting path exists.
2. The labels given to the nodes during the longest path algorithm completely summarize the augmenting paths.

We now wish to generalize the notion of an augmenting path, and hence the entire procedure, to the problem of the intersection of three matroids. One way of doing this is to "freeze" one of the matroids and only consider alternating sequences within the other two. In this case the first node, s_1 in an augmenting path would be independent of $I^{(K)}$ in two of the three matroids (or in all three, in which case it is the only node in the augmenting path). Say s_1 is independent of $I^{(K)}$ in the first and third matroids. We could then freeze the third matroid and maintain the same span within the third matroid throughout the augmenting path. Thus, the deletion of s_i for i even reduces this span and the addition of s_i for i odd restores it. We thus reduce the search space to two matroids and the same polynomial bounded procedure will work. Note that, alternatively, we could have considered the first matroid frozen. Indeed, it is so frozen in the two matroid intersection algorithm. Thus, there are three types of augmenting paths, one for each matroid within which s_1 is dependent. Unfortunately, while this procedure is polynomial bounded, it does not guarantee optimal solutions as there are augmentations which have no such corresponding argumenting path.

In order to guarantee that all augmentations are explored, we must relax the definition of an augmenting path still further to include cases where independence is not necessarily restored by the deletion of s_i for i even. Thus, an augmenting path may start with any element, s_1, which is independent of $I^{(K)}$ in at least one of the matroids. Unlike the procedure given for two matroids, one may begin with independence in any matroid. Consider the graph shown in Figure 3 corresponding to two augmenting paths, Path 1 and Path 2, for the partial assignment 111,222,333 (i.e., person 1 to job 1 on day 1, etc.) in a TDAP. These paths are not strictly comparable in that they exclude different elements along the way. Thus in Figure 2, when node 22 is labeled using the path a,32,22 it is equivalent (in terms of how the path can continue, not necessarily in terms of the numerical value of the label) to being labeled using the path a,31,11,12,22. In Figure 3, however, when node 111 is labeled using the path a,411,111 it is different from labeling 111 using the path a,154,111 because different continuations of these paths are possible. Thus starting with a,411,111 we can continue to 152 but not 215 and, conversely, starting with a,154,111 we can continue with 215 but not 152. Thus, Path 1 and Path 2 are not comparable in terms of their lengths only.

Such paths must also be compared in terms of their spans. We note that if two paths from a to some node i result in sets having identical spans then the same continuations of both paths are possible. (This was the case for intersections of two matroids.) Indeed, it is possible for paths to have slightly different spans and still have the same set of possible continuations. In particular, if the only difference in the intersections of the spans of two paths are nodes outside the intersection of the spans of $I^{(K)}$, then the paths are comparable. We can thus generalize the augmenting path procedure to consider all undominated a to z paths where one path dominates another only if it has the same continuations and a larger length.

The notion of a path itself, however, must be generalized as well. In the case of 3 matroids, not all augmentations correspond to paths. We see an example of this for a TDAP. The augmentation [412,234,341,123] - [111,222,333] does not correspond to any path in the conventional sense. It is possible however, to extend the augmenting path procedure to include such augmentations by extending the notion of a path.

We define a generalized augmenting path with respect to an intersection $I^{(K)}$ to be a sequence of nodes $S = (s_1, s_2, \dots s_m)$ where $s_i \in E - I^{(K)}$ for odd i and $s_i \in I^{(K)}$ for even i. As before, $I^{(K)} + s_1 - s_2 + s_3 - \dots + s_m$ is an intersection. Also, the even s_i are deleted in order to remove dependencies created by the inclusion

Person \ Job	1	2	3
1	10	9	5
2	5	10	9
3	9	5	1

FIGURE 1 - Cost Matrix

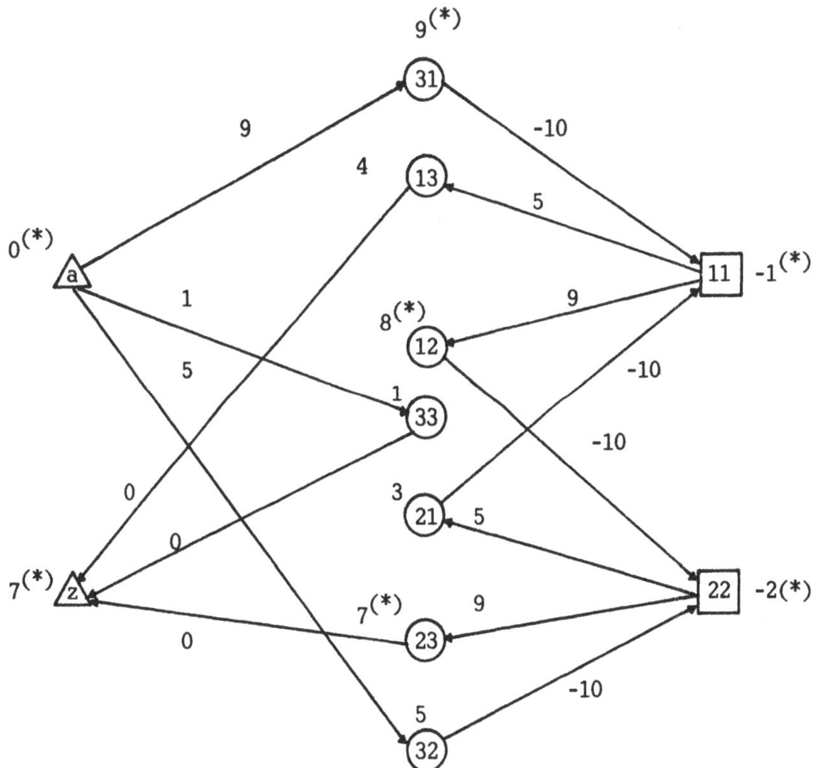

FIGURE 2 - Bipartite Graph $G^{(2)}$

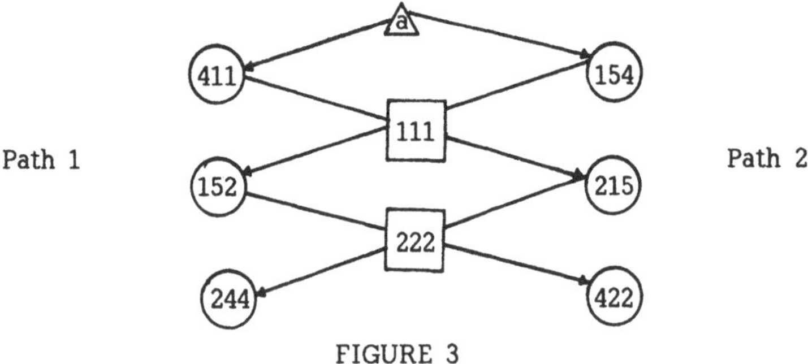

FIGURE 3

of the odd s_i. Now, however, the subsequences $I^{(K)} + s_1 - s_2 + \ldots - s_j$ for even j need not correspond to intersections.

One can thus guarantee an optimal intersection as in the case of two matroids. The number of generalized augmenting paths one may need to consider, however, may grow exponentially with K. In practice, however, the number of such paths can be controlled at the expense of optimality. First, the length of any path, (s_1, s_2, ..., s_j), can be reduced by a penalty to account for the nodes which still must be deleted to restore the intersection. In the case of arbitrary matroids, this may be complex to compute. In the case of the TDAP, however, where 3 partition matroids are involved, and all cycles contain 2 elements, it is easily computed.

In some cases the above may keep the computations reasonable. In others, it may be necessary to reduce the number of paths considered by relaxing the definition of dominance. This will also result in a heuristic rather than an optimal solution. In the case of the TDAP, one such relaxation is to ignore differences in the spans outside the intersection of the span of $I^{(K)}$. This is motivated by the fact that we consider deleting elements in $I^{(K)}$ in order to include elements blocked by them.

We can thus consider a hierarchy of generalized augmenting path procedures with increasingly stringent dominance criteria and increasing runtime. A tradeoff between optimality and runtime is then available. We are currently investigating this tradeoff using the TDAP as an example.

Acknowledgment

This research has been supported by U.S. Army, CORADCOM, under contract DAAK-80-80-K-0579 and the National Science Foundation under Grant ENG7908120.

References

1. Cook. S.A., "The Complexity of Theorem-Proving Procedures," Proc. of Third Ann. ACM Symposium on Theory of Computing, 1971, p 151-158.

2. Garey, M.R. and D.S. Johnson, Computers & Intractability, W.H. Freeman, 1979.

3. Sahni, S. and E. Horowitz, "Combinatorial Problems: Reducibility and Approximation," Operations Research 26(4), 1978.

4. Karp, R.M., "The Probabalistic Analysis of Some Combinatorial Search Algorithms," in Algorithms and Complexity, Academic Press 1976.

5. Lawler, E., Combinatorial Optimization: Networks and Matroids, Holt, Rinehart & Winston, 1976.

6. Edmonds, J., "Matroid Intersection," in Discrete Optimization I, North Holland Publishers Co. p. 39-49.

A VERSATILE INTERCONNECTION PATTERN LAID ON O(n) AREA

F. Luccio and L. Pagli
Istituto di Scienze dell'Informazione
Università di Pisa
56100 Pisa, Italy

Abstract

The problem of interconnecting n points by a planar line pattern is crucial in VLSI design.

The standard layout is a rectilinear tree, H-shaped and laid on O(n) area; it allows to reach all terminal nodes after $\log_2 n$ branchings.

In this note we introduce new H-type layouts with exactly the same area of the standard H layout. In addition to the original broadcast capabilities, the new layouts provide faster access to subsets of terminal nodes. In particular direct access is provided to $4\sqrt{n}-6$ subtrees of the original tree, by modifying the routing capabilities of branching nodes. Tree access is provided to $4\sqrt{n}-5$ subsets of terminal nodes, wich are not contained in subtrees of the original tree, by laying a new family of indipendent H-shaped trees penetrating in the empty space left by the main tree connection. Finally some tree reconfigurations are obtained by putting different branching nodes at the root, and by modifying the routing structure. In particular, the tree can be divided into a set of reconfigurable subtrees, accessed independently.

1. Introduction

An important interconnection problem is the one of reaching n points in a plane, by a line pattern that includes branching nodes. This is crucial in VLSI layout, where n elementary processors (terminal nodes) must be reached in parallel by a message, broadcast from a source.

The most suitable pattern is a binary tree, whose internal nodes have the function of duplicating the message, until n copies are generated after $\log_2 n$ branchings. See for example [1,4,5,7]. The tree is laid on the chip plane, with the arcs (wires) parallel to the coordinate axes.

Arbitrary trees of bounded degree have been shown to be enbeddable in O(n) area, without arc crossings [8]. For binary trees, two typical layouts have been proposed [2]. The "H" layout (fig. 1) is the most widely used, since its area is of O(n) [6]. In this case, only $O(\sqrt{n})$ terminal nodes lay on the boundary, and can therefore be connected to an external pin. The alternative layout [2] has a straightforward shape on O(nlogn) area, but has the advantage that all terminal nodes lay on the boundary. (In fact, to verify this condition, the minimum chip area if of order nlogn if the boundary is convex [2]).

All these studies refer to complete binary trees. Generation of a planar embedding of an arbitrary binary tree has been examined in [3].

In this note we propose new H-type layouts with exactly the same area of the standard H layout. In addition to the original broadcast capabilities, the new layouts provide faster access to proper subsets of terminal nodes, and allow tree reconfigurations by putting at the root different internal nodes. The study is carried out under the usual assumption that all nodes are squares of unit area connection wires have unit width; and a unit distance separates nodes and wires not directly connected. If, in particular techological implementations the node area must be increased over the unit square, the basic features of the proposed layout can be adapted for the new model without difficulties.

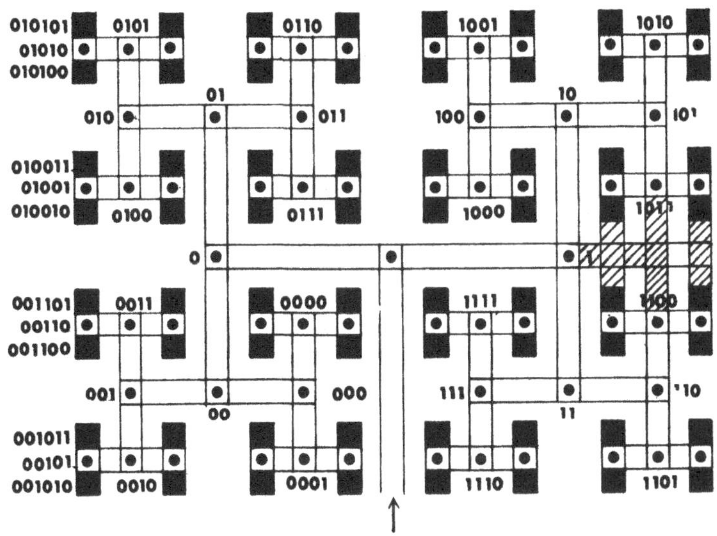

Fig. 1. The H layout for a tree of $n=2^6$ terminal nodes. ■ and ● represent processors and branching nodes, respectively. The arrow shows the external access to the tree. Shaded is a portion of unused space, which could accomodate additional wiring.

2. Node numbering

Our first concern is to enumerate the branching and terminal nodes in the H-layout.

Each node will be labelled by a binary number of d bits, where d is the level, i.e. distance from the root, of the node. Let the number of terminal nodes be $n=2^k$. A node Q is reached from the external access by a path containing a succession of d left or right turns, $0 \leqslant d \leqslant k$ (see fig. 1). By coding left and right turns with 0 and 1 respectively, the above succession is represented by an ordered sequence N of d bits:

$$N = b_1 \, b_2 \, \dots \, b_d .$$

N is interpreted as a binary number, and uniquely codes node Q. The root is at level d=0, and carries no number.

This node enumeration is shown in fig. 1, up to the nodes at level d=4. Sample codings for nodes at level 5 and 6 are shown in the left side of the figure.

The total area used for the layout is by definition, the minimum rectangular area containing the pattern. This assumption is not restrictive in VLSI technology, where almost all chips have a rectangular boundary. Note now that part of the rectangular area containing the H-layout is a actually unused, and could be utilized, for example, for laying additional wiring. Fig. 1 shows a portion of this area. In fact, while the pattern has empty spaces, no contraction aimed to pack the pattern more tightly would result in a smaller boundary which is still rectangular.

We will now show how to use the empty spaces, to increase access capabilities. The H-layout will be kept intact, while additional wiring will provide new access to defferent subsets of processors.

3. Direct access to subtrees

Taking into account the total unused area it is readily seen that several nodes can be directly accessed by laying new straight wires from the boundary. Referring to the node numbering of fig. 1, for example, branching nodes 0, 010, 001, 01010, 01001, 00110, 00101 can be accessed from the left side, if their structure is slightly modified to include two input ports.

We will now identify all branching nodes accessible in such a fashion (terminal nodes are not modified, i.e., no new access is provided for them). Obviously, this new provision allows fast access to subtrees of the original tree.

Branching nodes can be newly accessed "horizontally or vertically" (i.e., from the left and right sides, or from the upper and lower sides).

In the first case, they have odd level; in the second case they have even level.

It can be easily shown (see fig. 1) that each node at level d, horizontally accessed, is labelled by a binary sequence N of the form:

$$N = b_1 S_1 S_2 \dots S_{(d-1)/2}, \qquad d \text{ odd} \geqslant 1, \tag{1}$$

where $b_1 = 0,1$, for $d>1$;

$$S_i = \begin{cases} 01 \\ 10 \end{cases}, \qquad 1 \leqslant i \leqslant (d-1)/2.$$

Similarly, a node vertically accessed is labelled by a sequence:

$$N = S_1 S_2 \dots S_{d/2}, \qquad d \text{ even} \geqslant 2, \tag{2}$$

where S_1=00,01,10,11, and for $d>2$:

$$S_i = \begin{cases} 01 \\ 10 \end{cases}, \qquad 2 \leqslant i \leqslant d/2.$$

Note that the original tree root could be vertically accessed from the upper side also, however, this new wiring is not taken into account, since it would not provide a new access to a subtree).

Obviously, the subtree with root N includes 2^{k-d} terminal nodes (recall that $n=2^k$ is, the total number of terminal nodes) labelled by the sequence concatenation NS, where S is any sequence of k-d bits.

Let s_d indicate the number of subtrees with root at level d 1, directly accessible from the boundary. For d odd (horizontal accesses), we have from relation (1):

$$s_d = 2, \qquad d=1,$$

$$s_d = 2s_{d-2}, \qquad d \text{ odd} > 1,$$

that is immediately solved as:

$$s_d = 2^{(d+1)/2}, \quad d \text{ odd}, 1 \leqslant d < k. \quad (3)$$

For d even (vertical access) we have from relation (2):

$$s_d = 4, \quad d = 2,$$

$$s_d = 2s_{d-2}, \quad d \text{ even} > 2,$$

that yields:

$$s_d = 2^{d/2+1}, \quad d \text{ even}, 2 \leqslant d < k. \quad (4)$$

The total number s_{tot} of subtrees thus accessible, is readily computed from relations (3) and (4) as:

$$s_{tot} = \sum_{1}^{k-1}{}_d \, s_d = \begin{cases} 4\sqrt{n}-6, & k \text{ even}, \\ 3\sqrt{2n}-6, & k \text{ odd}. \end{cases}$$

4. Penetrating trees

In the previous section we have seen that some nodes at level d can be directly reached from the boundary. Let N be one such node. Insted of reaching N directly from outside we could use the empty space between N and the boundary, to build a new H-shaped tree, that reaches a subset of leaves belonging to the subtree with root N (see fig. 2). We will now show how these new penetrating trees can be built.

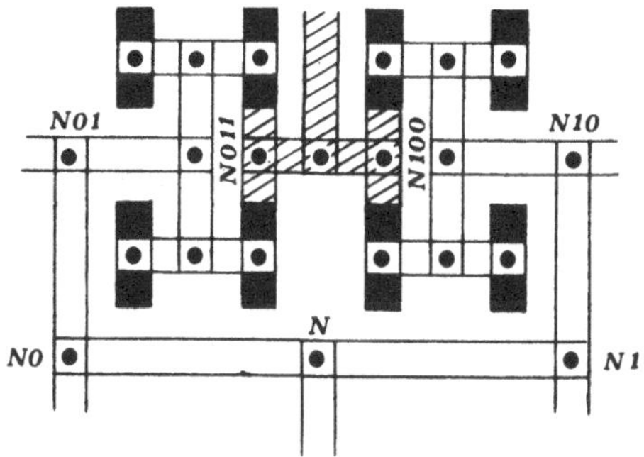

Fig. 2. A penetrating tree (shaded area) reaching 4 terminal nodes.

First we assume that two input port can be built in the terminal nodes, to access such nodes both from the original tree, and from a penetrating tree.

Each terminal node, wich is a descendant of N, is labelled by a sequence NS, where S is a sequence of (k-d) bits. In particular, it can be shown that the terminal nodes accessible from the penetrating trees are labelled by:

$$NS = NS_1 \ldots S_{h-1}S_h, \quad 1 \leqslant h \leqslant \lceil (k-d)/3 \rceil, \quad (5)$$

where

$$S_i = \begin{cases} 011 \\ 100 \end{cases}, \quad 1 \leqslant i \leqslant h-1,$$

and S_h is a prefix of 011 or 100, of proper length (1,2 or 3). These accessible nodes are then 2^h; however, not all of them can in general be accessed. We distinguish two cases.

1. k-d=3h.
   All the accessible nodes are reached: see fig. 2, where k-d=6, i.e. h=2, and the accessible nodes are:

$$NS_1S_2 = \begin{cases} N011100 \\ N011011 \\ N100100 \\ N100011 \end{cases}$$

2. k-d=3(h-1)+C, C=1,2.
   Only 2^{h-1} nodes can be reached, out of the 2^h accessible nodes.
   See fig. 3(a), where k-d=5, i.e. h=2 and c=2; and fig. 3(b), where k-d=4, i.e. h=2 and c=1.

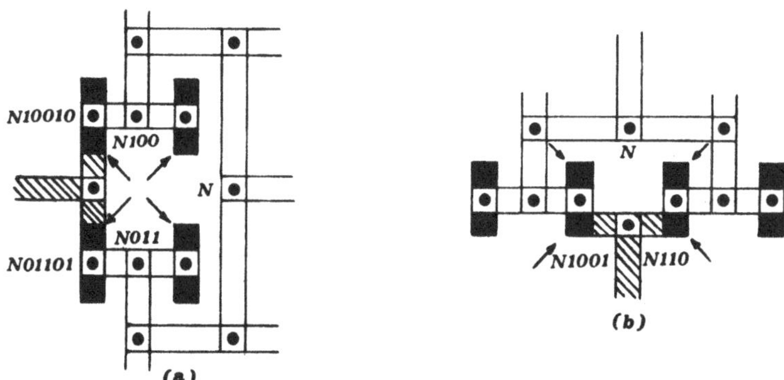

Fig. 3. Penetrating trees for (a): k-d=5, and (b): k-d=4 reaching two terminal nodes. The four accessible nodes are indicated by arrows.

Note that, in case 2 (fig. 2) one half of the accessible nodes can be reached, because there is no space for a penetrating tree to reach all such nodes. In fig. 2 we have assumed to lay the penetrating tree of minimal lenght. However, it can be easily seen that any subset of 2^{h-1} accessible nodes can be reached.

Note that a penetrating tree can be built in front of the root of the original tree (e.g., fig. 2 where N is such a root). In this case, the string labelling the root is empty, and N does not a appear in relation (5).

Finally, for K-d=1,2, i.e; h=1, the penetrating trees degenerate to direct accesses to single terminal nodes on the boundary.

Penetrating trees are built in front of the roots of the subtrees directly accessible, discussed in section 3. In addition, a penetrating tree is built in front of the root of the original tree. Hence, the number p_{tot} of penetrating trees is:

$$p_{tot} = \begin{cases} 4\sqrt{n}-5 \ , & k \text{ even,} \\ 3\sqrt{2n}-5 \ , & k \text{ odd.} \end{cases}$$

5. Tree reconfiguration

Let us now discuss how to change the tree configuration, by taking different internal nodes as root of the total tree, or by partitioning the tree in different sections, each of wich can be in turn reconfigured.

The usual properties are satisfied, namely: 1) the original node set, tree structure and boundary size are fully maintained,while additional connections tree arcs are inserted in the available space; 2) new input ports are added to the internal nodes, as in the case of direct access to subtrees (section 3), and tree branches are allowed to transmit information in both directions.

Several patterns may support reconfigurations. The ones proposed here satisfy the additional properties: 3) all the terminal nodes of the original tree are actually reacheable in the reconfigured pattern; 4) each original internal node traversed in a new pattern is actually used to duplicate the message, that is each new path traverses minimum number of nodes.

A reconfiguration of the whole tree that meets the above properties is shown, in fig. 4. The new root is a node originally at level two, and a very limited amount of new connections is needed. Patterns with roots taken from level four and even levels can be similarly built. It can be shown that the root cannot be taken from an odd level, whitout violating the stated properties.

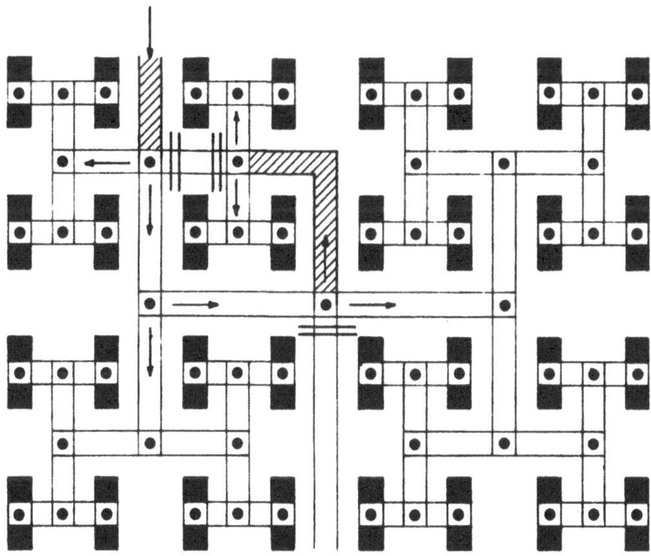

Fig. 4. The reconfiguration (shaded area) with the root taken from level 2. Arrows show the directions of transmission, and double bars indicate blocked node ports.

Let us now discuss a general pattern that allows several reconfigurations. (Note that two of the above discussed reconfigurations cannot be supported by the same pattern, since both require a common portion of empty space). Fig. 5 shows additional connections supporting the whole tree reconfiguration of fig.4, or a partition of the tree in two sutrees of n/2 leaves each, wich are in turn reconfigured by taking the root from their level two (i.e., from level three of the whole tree).

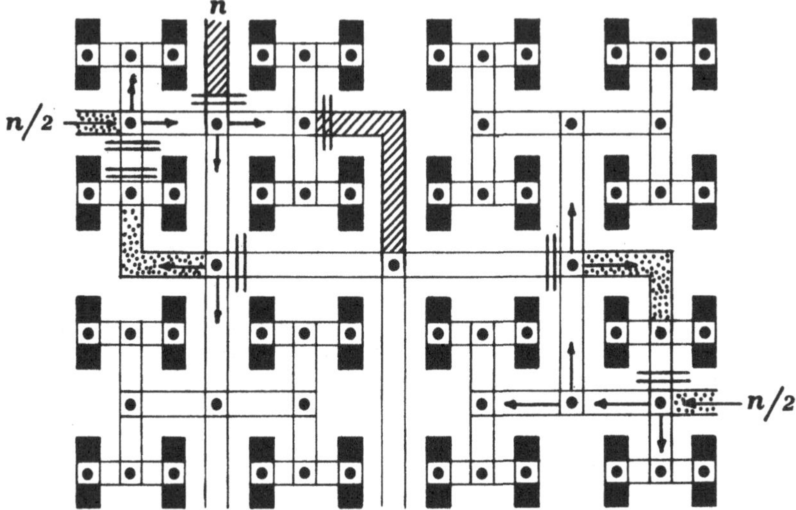

Fig. 5. Layout suppoting the n leaves tree reconfiguration of Fig. 4 (shaded area), and reconfiguration of two n/2 leaves subtrees (dotted area, arrows and bars).

This rule can be iterated on the tree with n/2 leaves whose area does not support the connections for the reconfigured n leaves tree (i.e., on the right subtree), by inserting new connections to partition the subtree in two trees of n/4 leaves with root taken from their level two (original level four). This provision is shown in fig. 6. Similarly, new connections can be added to divide the upper right subtree in two reconfigured trees of n/8 leaves, and so on.

According to this scheme we have the following possible reconfigurations to be chosen in alternative, that are all supported by the same layout:

1 reconf. tree of n leaves, root from level 2;
2 reconf. trees of n/2 leaves, roots from level 3;
3 reconf. trees of n/2, n/4, n/4 leaves, roots from levels 3,4,4;
4 reconf, trees of n/2, n/4, n/8, n/8 leaves, roots from levels 3,4,5,5;
....
k reconf, trees of $n/2^1, n/2^2, \ldots, n/2^{k-2}, n/2^{k-1}, n/2^{k-1}$, roots from levels 3,...,k,k+1,k+1.

Similar patterns may be built in connection with the n leaves tree reconfiguration with root taken from level four, or from any even level.

6. Concluding remarks

We have shown how a classical binary tree layout, particularly suitable for VLSI implementation, can be extended to allow complex accesses to substes of nodes, without altering the original connection capabilities and the boundary of the figure.

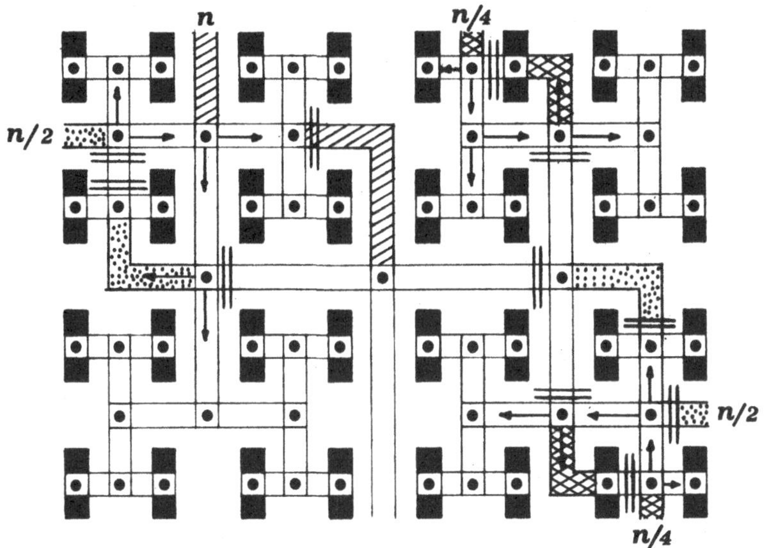

Fig. 6. The right subtree can be partitioned in two reconfigured trees of n/4 nodes each (crossed area, arrows and bars).

Three different aspects of this problem have been consideraded, namely access to subtrees of the original tree, construction of new connecting trees penetrating in the spaces unused in the original pattern, and reconfigurations of the tree by changes of root and partitioning. These aspects could be treated comtemporarily, by partial use of all the new possible connections, to give rise to a extremely rich family of patterns. For example, the tree could be used as if it were partitioned in two subtrees of n/2 leaves, one of which is directly accessed from its original root (as in section 3), and the other is reconfigured by taking a new root (as in section 5); while some penetrating trees provide faster access to proper subsets of their leaves (as in section 4). The only constraint is that all the new connections needed, are laid in different available portions of the figure.

Many variations to the proposed schemes are possible. This note shows a possible line of work.

Aknowledgement.

This work has been partially supported by Consiglio Nazionale delle Ricerche of Italy.

References

[1] J.L. Bentley and H.T. Kung, A tree machine for searching problems, Proc. IEEE 1979 International Conference on Parallel Processing (1979) 257-266.

[2] R.P. Brent and H.T. Kung, On the area of binary tree layouts, Information Processing Letters 11 (1) (1980) 46-48.

[3] E.Horowitz and A. Zorat, The binary tree as an Interconnection network: application to multiprocessor system and VLSI, IEEE Trans, Comp. 30 (4) (1981) 247-253.

[4] H.T. Kung, Let's design algorithms for VLSI systems, Proc. Conference on Very Large Scale Integration: Architecture, Design, Fabrication, California Institute of Technology (1979) 65-90.

[5] C.A. Mead and L.A. Conway, Introduction to VLSI Systems. Addison-Wesley, Reading, MA, 1980.

[6] C.A. Mead und M. Rem, Cost and performance of VLSI computing structures, IEEE J. Solid State Circuits 14 (2) (1979) 455-462.

[7] S.W. Song, A highly concurrent tree machine for database applications, Carnegie-Mellon University, Departement of Computer Science (1980).

[8] L.G. Valiant, Universatily considerations in VLSI circuits, IEEE Trans. Comp. 30 (2) (1981) 135-140.

RANK, CLIQUE AND CHROMATIC NUMBER OF A GRAPH

Cyriel VAN NUFFELEN
U.F.S.I.A.
B-2000 Antwerpen, Belgium

ABSTRACT

In this note we show that the rank of the adjacency matrix (vertex - vertex matrix) of a graph is aan upper bound for the clique number and the chromatic number of this graph.

0. INTRODUCTION

It is well-known that finding an algorithm for the clique number or the chromatic number is an NP-complete problem (see e.g. KARP [4]). Bounds for the clique number and the chromatic number are given in many papers. We propose an other upper bound which is not included in the known bounds.

We deal with a simple graph G, i.e. undirected, no multiple edges and no loops. The adjacency matrix $A(G) = [a_{ij}]$ of a labeled simple graph G with n vertices, is an n x n-matrix in which $a_{ij} = 1$ if the edge (i,j) is in G, otherwise $a_{ij} = 0$

The rank of the matrix A(G) calculated over the real number is called the rank of the graph G and denoted by r(G)

The clique number of a simple graph G is denoted by $\omega(G)$ and its chromatic number by $\gamma(G)$.

In a complete k-partite graph, the vertex set can be partitioned into k subsets such that vertices in the same subset are never adjacent and vertices in different subsets are always adjacent.

Other definitions and symbols we use in this note can be found in C. BERGE [2].

1. CLIQUE NUMBER

In [1], AMIN & HAKIMI showed that the rank $r(G)$ is an upper bound for the clique number $\omega(G)$. They proved this by means of the eigenvalues of the adjacency matrix. Through the eigenvalues, it is rather difficult to find when the equality holds for this bound; because HOFFMAN & HOWES proved in [3] that two cospectral graphs are not necessarily isomorphic.

We also propose the rank $r(G)$ as an upper bound for $\omega(G)$, but in addition, we give the conditions for the equality : $\omega(G) = r(G)$.

THEOREM 1

For a non trivial simple graph G we have :

$$\omega(G) \leq r(G)$$

and $\omega(G) = r(G) = k$, iff G is complete k-partite.

Proof

a) Let C be a maximum clique in G. Choosing in the adjacency matrix $A(G)$, this rows and columns which correspond to the vertices of C, we obtain a submatrix A_1 whose entries are 1 except for the main diagonal whose entries are 0. The order of A_1 is $\omega(G) = k$ and the determinant of A_1; $\det A_1 = (-1)^{k-1}(k-1)$. So the rank of A_1 is k if $k > 1$ and therefore $r(G) \geq \omega(G)$.

b) If G is a complete k-partite then $r(G) = \omega(G) = k$. Indeed. In that case the vertex set of G can be partitioned into k subsets so that vertices in the same subset are all adjacent and vertices in different subsets are never adjacent. If the vertices in the same subset are successively labeled, then the adjacency matrix $A(G)$ has the form :

$$A(G) = \left[\begin{array}{c:c:c:c} 0 & 1 & & 1 \\ \hdashline 1 & 0 & & 1 \\ \hdashline & & & \\ \hdashline 1 & 1 & 1 & 0 \end{array}\right]$$

In the blocks on the diagonal all entries are 0, and in the other blocks all entries are 1. Now, it is clear that the rank $r(G)$ of this matrix is $k = \omega(G)$.

c) If $r(G) = \omega(G) = k$, then G is a complete k-partite graph.
Indeed. In that case G contains a k-clique C. Consider the vertices of C and an arbitrary vertex $x_0 \notin C$.
The submatrix A_1 of A(G) which corresponds with the vertices of $C \cup \{x_0\}$ can then be written as :

$$A_1 = \begin{bmatrix} 0 & 1 & 1 & 1 & \cdots\cdots & \alpha_1 \\ 1 & 0 & 1 & 1 & \cdots\cdots & \alpha_2 \\ 1 & 1 & 0 & 1 & \cdots\cdots & \alpha_3 \\ & & \vdots & & & \\ & & & & & \alpha_k \\ \alpha_1 & \alpha_2 & \cdots\cdots & & \alpha_k & 0 \end{bmatrix}$$

in which $\alpha_i \in \{0,1\}$ for $i = 1,2,\ldots,k$.
Because the first rows of A_1 are linearly independent (they correspond to the k-clique), the last row must be a linear combination of the foregoing. This means :

$$\begin{array}{lr} a_2 + a_3 + \cdots\cdots = \alpha_1 & (1) \\ a_1 + \quad + a_3 + \cdots\cdots = \alpha_2 & (2) \\ a_1 + a_2 \quad + \cdots\cdots = \alpha_3 & (3) \\ \vdots & \\ \cdots\cdots\cdots\cdots = \alpha_k & (k) \\ \alpha_1 a_1 + \alpha_2 a_2 + \cdots\cdots + \alpha_k a_k = 0 & (k+1) \end{array}$$

in which $a_i \in \mathbb{R}$ for $i = 1,2,\ldots,k$.
This system of equations can also be written as :

$$a_2 + a_3 + \cdots\cdots + a_k = \alpha_1$$

$$a_1 - a_i = \alpha_i - \alpha_1 \qquad \text{for } i = 2,3,\ldots,k$$

$$\sum_{i=1}^{k} a_i \alpha_i = 0$$

Let us now make the following remark.
Assume there exists a linear combination in which the coefficient a_1 is different from zero. Then it is possible to multiply the equations so that we have : $a_1 = 1$. This means $a_1 \in \{0,1\}$. By considering now the equation :

$$a_1 - a_i = \alpha_i - \alpha_1 \text{ in which } \forall i : \alpha_i \in \{0,1\}$$

it follows that $a_i \in \{0,1\}$ for $i = 1,2,\dots,k$.

From the equations (1) to (k) it follows that at most one of the coefficients a_i may be equal to 1, and all others have to be 0.

So we have either : $\forall i \; a_i = 0$ and this gives the trivial zero-solution for the system of equations, or exactly one coefficient, say a_j, is such that $a_j = 1$ and $a_i = 0$ for $i \neq j$.
This is the unique non-zero-solution namely :

$$\begin{array}{l} a_j = 1 \text{ and } \forall i \neq j : a_i = 0 \\ \alpha_j = 0 \text{ and } \forall i \neq j : \alpha_i = 1 \end{array} \qquad (*)$$

Unther these conditions, A_1 is the adjacency matrix of a complete k-partite graph; which is formed by the vertices of the k-clique on one side and the vertex x_0 on the other side is adjacent to all vertices of the k-clique except for the vertex x_j.

The conditions (*) must hold for every vertex $x_0 \notin C$ and for every new k-clique of the form $(C - \{x_j\}) \cup \{x_0\}$ we can form. From this we deduce, G itself is complete k-partite. This completes the proof.

2. CHROMATIC NUMBER

Several bounds are known for the chromatic number. We conjecture that the rank $r(G)$ is an upper bound for $\gamma(G)$. We were able to prove this conjecture for several special classes of graphs, see [5]. The aim of the following is to construct a proof-technique so that we can give a proof for all classes of graphs. We shall explicitely show this proof for all graphs with chromatic number $\gamma(G) = 3$. To do this we have to consider 4 cases. If the chromatic number is either $\gamma(G) = k$, then the number of cases to be examinated can roughly be estimated at 2^{k^2}. So, this technique implicitely indicates why the computing of the chromatic number is NP-complete.

PROOF TECHNIQUE

(i) Assume $\gamma(G) = k \geqq 2$, then we start to colour the vertices of G with colour c_1 until it is no longer possible, after that we use colour c_2 and so on. It is clear that there exists an elementary chain of length k - 1, between k vertices in G with k different colours. Let us consider the subgraph G_1 of G induced by these k vertices.

(ii) The edges of G_1 are : these of the elementary chain, but there may or may not exist other edges in this subgraph G_1. We construct all possible such subgraphs G_{1i} and calculate their rank. If for some i we have $r(G_{1i}) \geqq k$, then we found a subgraph to confirm the conjecture.

(iii) If for a certain subgraph G_{1i} we have $r(G_{1i}) < k$, then it is easy to see that G_{1i} is different from a complete graph. Therefore, it is always possible to consider a new subgraph G_2 of G induced by the vertices of G_{1i} and one other vertex of G that we choose in the following way.
In G_{1i} the vertices are coloured with the colours $c_1, c_2, \ldots, c_k$ in that order, see (i). We take in G_{1i} the first vertex (say x_p) which is not adjacent to a vertex (in G_{1i}) with a previous colour (say c_d). Then x_p has to be adjacent to a vertex x_q not in G_{1i} and with colour c_d. Now we construct all possible such subgraphs G_2 of G induced by the vertices of G_{1i} and the vertex x_q.

(iv) Either for all subgraphs G_2 we have $r(G_2) \geqq k$ or we can repeat the construction (iii) by adding one new vertex to the subgraph G_2 (unless $r(G_2) < k$, which would disprove the conjecture).

THEOREM 2

If $\gamma(G) \leqq 5$, then $\gamma(G) \leqq r(G)$.

In the following theorem we give the construction for a graph with $\gamma(G) = 3$ but constructions are available for graphs with chromatic numbers up to $\gamma(G) = 5$.

Proof for $\gamma(G) = 3$

We construct the subgraphs G_{1i} induced by an elementary chain with the three colours, like we described before.

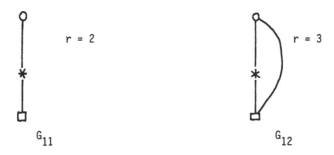

Because of G_{11} we also have to consider the following subgraphs of the form G_2.

These graphs satisfy the conditions, which in turn completes the proof.

REFERENCES

1. AMIN A.T. & HAKIMI S.L., Upper bounds on the order of a clique of a graph, SIAM J. Appl. Math. 22 (1972) 569-573.
2. BERGE C., Graphes et Hypergraphes, Dunod, Paris, 1970
3. HOFFMAN A.J. & HOWES L., On eigenvalues and colorings of graphs II, Ann. New York Acad. Sci. 175 (1970), 238-242.
4. KARP R.M., Reducibility among combinatorial problems, in R.E. Miller and J.W. Thatcher (eds), Complexity of Computer Problems, Plenum Press, New York, 1972, 85-103.

5. VAN NUFFELEN C., On the rank of the adjacency matrix, in Problèmes combinatoires et théorie des graphes, Editions du CNRS, Paris, 1978, 321-322.

EXTENSIONS OF SOME TWO-DIMENSIONAL BIN PACKING ALGORITHMS

P.Y. Wang
Department of Mathematics
University of Maryland-Baltimore County
Catonsville, MD 21228/USA

Introduction

The problem of packing a finite set of rectangles into a minimum number of bins is one which has widespread applications in computer science and operations research. Two-dimensional bin packing problems can be regarded as simplifications of the more general cutting stock problem which appears in numerous industries [2,9]. In computer science, these packing problems also have applications in the areas of memory allocation and resource scheduling [4-8].

We consider the following two-dimensional bin packing problem. Let $L = \{R_1, R_2, \ldots, R_n\}$ be a list of n rectangles where each rectangle $R_i = h_i \times w_i$ has height h_i and width w_i which are real numbers in $(0,1]$. Place the rectangles of L into a minimum number of 1×1 bins, N_1^*, so that the edges of the rectangles are parallel to the sides of the bins and no rectangles overlap or rotate in any bin.

Alternatively, by removing the rotation restriction, we obtain a second bin packing problem. Place the rectangles of L into a minimum number of 1×1 bins, N_2^*, so that the edges of the rectangles are parallel to the sides of the bins and no rectangles overlap in any bin. Thus, 90° rotations of the rectangles are permitted.

Each of these packing problems is NP-hard. Hence, we are interested in formulating and analyzing algorithms which obtain approximate solutions to the problems. In particular, if A is an approximation algorithm that packs N bins by successively allocating the rectangles of L to the bins, we seek asymptotic worst case performance bounds of the form

$$N \leq \alpha_1 N_1^* + \beta_1 \quad \text{or} \quad N \leq \alpha_2 N_2^* + \beta_2$$

for all lists L. α_1, α_2, β_1, and β_2 are constants.

Several popular two-dimensional packing algorithms have previously been developed in [1,3]. These algorithms were employed to pack the rectangles of L into a single bin of unit width and infinite height with the objective of minimizing the total height of the packing. Worst case error bounds were also proved. We will show how these techniques can be modified and extended to the problem of packing 1×1 bins.

Level-Oriented Packing Algorithms

The first approach we shall examine was discussed by Coffman et al in [3]. By their definition, a level-oriented packing of a bin is a packing in which the bin is filled as a sequence of levels. Each level is defined by a horizontal line drawn through the top of the first rectangle placed on the previous level. Within each level, the bottom edge of each rectangle is placed on the horizontal line as far to the left as possible. The bottom edge of the bin is considered to be the first level.

With the assumption that the rectangles of the list L are reordered by nonincreasing heights, i.e. $h_1 \geq h_2 \geq \cdots \geq h_n$, the following algorithms are outlined by the above authors. A single unit width, infinite height bin is to be packed.

(1) Next Fit Decreasing Height (NFDH)

Each rectangle of L is successively placed on the highest nonempty level in the bin if it fits on that level. If not, a new level is defined above the current level, and further packing continues with the new level.

(2) First Fit Decreasing Height (FFDH)

Each rectangle of L is successively placed on the lowest level on which it fits. If there is no such level, a new one is defined above the highest nonempty level.

For a given list L, let the heights of the packings produced by the Next Fit and First Fit Decreasing Height algorithms be denoted by h_{NFDH} and h_{FFDH}, respectively. Define h_{OPT} to be the minimum height of the bin within which all the rectangles of L can be packed. Figure 1 illustrates a sample NFDH and FFDH packing.

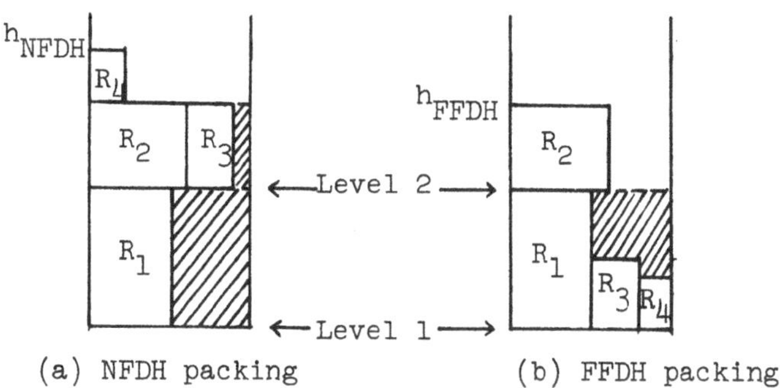

(a) NFDH packing (b) FFDH packing

Figure 1.

The following error bounds were proved in [3] for these algorithms.

Theorem 1. $h_{NFDH} < 2 \sum_{i=1}^{n} h_i w_i + 1$

Theorem 2. $h_{FFDH} \leq 1.7\, h_{OPT} + 1$

Theorem 3. $h_{FFDH} \leq (1 + \frac{1}{r}) \sum_{i=1}^{n} h_i w_i + 1$ where no rectangle in L has width exceeding $\frac{1}{r}$ for some $r \geq 2$.

We may extend the NFDH and FFDH algorithms to the problem of packing a minimum number of 1 x 1 bins with a given list L of rectangles. Assume that the 1 x 1 bins are indexed B_1, B_2, B_3,... and that the rectangles of L are again reordered by nonincreasing heights. The NFDH algorithm can be modified into the following

Algorithm A. Pack each successive rectangle of L on the highest level of the highest indexed nonempty bin, B_k, if it fits in that bin. If not, define a new level in bin B_k. If this is also not possible, the new level is defined in the next empty bin, B_{k+1}. Further packing continues in this bin.

Similarly, the First Fit Decreasing Height algorithm can be modified to obtain the following

Algorithm B. Pack each successive rectangle of L on the lowest level of the lowest indexed bin in which it fits. If there is no level in all of the nonempty bins B_1, B_2,..., B_k which will accomodate the rectangle, a new level is defined in bin B_k. If this is not possible, the new level is defined in the next empty bin, B_{k+1}.

Algorithms A and B can also be described as combinations of the NFDH and FFDH packing algorithms with the one-dimensional Next Fit packing algorithm described in [8]. That is, the strips resulting from the level-oriented packings of the infinite height bin can be subsequently packed into bins of finite height by using a Next Fit approach. The following results can be obtained for Algorithms A and B.

THEOREM. Let N denote the number of bins packed by Algorithm A or B for a given list L. Then $N < 2h + 1$ where $h = h_{NFDH}$ for Algorithm A and $h = h_{FFDH}$ for Algorithm B.

Proof sketch. Recalling the above remarks, we note that

$$N = \sum_{i=1}^{k} H_i + \sum_{j=1}^{N} \beta_j$$

where H_i is the height of level i in the infinite height packing, and β_j is the height of the unused space between the top of the highest rectangle packed in bin B_j and the top of the bin. See Figure 2. Using either algorithm, we have that β_j must not exceed the height of the first level of bin B_{j+1}. Otherwise, the first rectangle of this bin would have been used to define a new level in bin B_j.

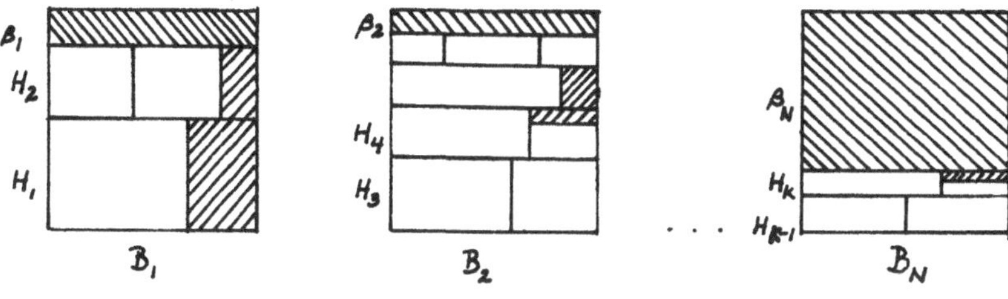

Figure 2.

Thus, $\sum_{j=1}^{N} \beta_j = \sum_{j=1}^{N-1} \beta_j + \beta_N < \sum_{i=1}^{k} H_i + 1$. But $\sum_{i=1}^{k} H_i =$ h_{NFDH} and h_{FFDH} for Algorithms A and B respectively, so we obtain the desired results.

Corollary 1. For a given list L, the number of bins packed by Algorithm A satisfies $N < 4 N_2^* + 3$.

Proof. Since $h = h_{NFDH} < 2 \sum_{i=1}^{n} h_i w_i + 1$ from Theorem 1,

$$N < 2 (2 \sum_{i=1}^{n} h_i w_i + 1) + 1 = 4 \sum_{i=1}^{n} h_i w_i + 3 .$$

The optimal number of bins, N_2^*, into which the rectangles of L can be packed with rotations must satisfy $N_2^* \geq \sum_{i=1}^{n} h_i w_i$. Hence, the desired bound is obtained.

Corollary 2. For a given list L, the number of bins packed by Algorithm B satisfies $N < 3.4 N_1^* + 3$.

Proof. Since $h = h_{FFDH} \leq 1.7 h_{OPT} + 1$ from Theorem 2,

$$N < 2 (1.7 h_{OPT} + 1) + 1 = 3.4 h_{OPT} + 3 .$$

The optimal number of bins, N_1^*, satisfies $N_1^* \geq h_{OPT}$ since the bins have unit height.

Corollary 3. For a given list L, the number of bins packed by Algorithm B satisfies $N < 2 (1 + \frac{1}{r}) N_2^* + 3$ where no rectangle of L has width exceeding $\frac{1}{r}$ for some fixed $r \geq 2$.

Proof sketch. Substitute the inequality of Theorem 3.

The results of our Theorem can be improved if we have additional information concerning the height of the tallest rectangle in L. We can show the following.

THEOREM. Let N denote the number of bins packed by Algorithm A or B for a given list L. If $m = \lfloor 1/h_1 \rfloor$, then $N < (1 + \frac{1}{m})h + 1$ where $h = h_{NFDH}$ for Algorithm A and $h = h_{FFDH}$ for Algorithm B.

Proof sketch. As before, we begin with $N = \sum_{i=1}^{k} H_i + \sum_{j=1}^{N} \beta_j$. It can be shown that $\beta_j < \frac{1}{m+1}$ for $j = 1,2,\ldots,N-1$, from which it follows that

$$N < \sum_{i=1}^{k} H_i + \frac{N-1}{m+1} + 1 .$$

Then,
$$(N - 1)(1 - \frac{1}{m+1}) < \sum_{i=1}^{k} H_i$$
or
$$N - 1 < \frac{m+1}{m} \sum_{i=1}^{k} H_i .$$

Corollary 4. For a given list L, the number of bins packed by Algorithm A satisfies $N < (1 + \frac{1}{m})\, 2\, N_2^* + 3$.

Corollary 5. For a given list L, the number of bins packed by Algorithm B satisfies $N < (1 + \frac{1}{m})\, 1.7\, N_1^* + 3$.

Corollary 6. For a given list L, the number of bins packed by Algorithm B satisfies $N < (1 + \frac{1}{m})(1 + \frac{1}{r})\, N_2^* + 3$ where no rectangle of L has width exceeding $\frac{1}{r}$ for some fixed $r \geq 2$.

Remark. For the same list L, the optimal number of bins N_2^* and N_1^* into which the rectangles can be packed with and without rotations, respectively, are related by $N_2^* \leq N_1^*$. Removing the rotation restriction increases the number of ways in which the rectangles can be allocated to the 1×1 bins. Thus, upper bounds for Algorithms A and B are easily obtained with N_1^* replacing N_2^* in Corollaries 1,3,4, and 6.

The multiplicative constants in the upper bounds for Algorithm A can also be shown to be as small as possible.

THEOREM. For any $\delta > 0$, there exists a list L for which the number of bins packed by Algorithm A satisfies $\frac{N}{N_2^*} > 4 - \delta$.

Proof. Let L_1 consist of 4k repetitions of the pair of rectangles

$$\langle (\tfrac{1}{2} + \varepsilon) \times (\tfrac{1}{2} - \varepsilon),\ (\tfrac{1}{2} + \varepsilon) \times 3\varepsilon \rangle.$$

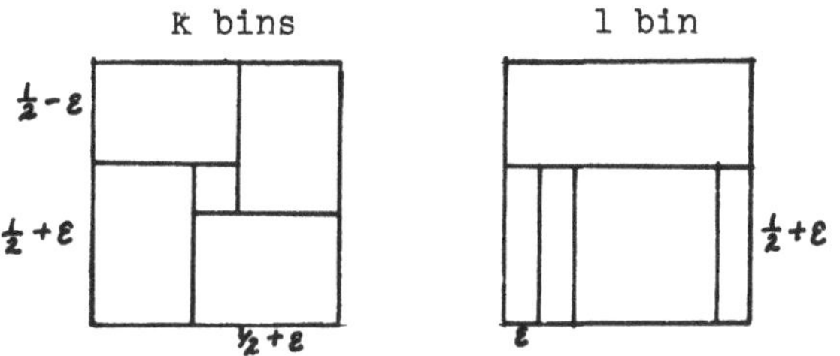

Figure 3. Optimal packing of L_1

Using Algorithm A, 4k bins are packed. However, $N_2^* = k + 1$ bins as illustrated in Figure 3. This implies that $\frac{N}{N_2^*} = \frac{4k}{k+1} > 4 - \delta$ when k is chosen so that $\frac{1}{k+1} < \delta$ and ε is sufficiently small.

THEOREM. For any $\delta > 0$, there exists a list L for which $m = \lfloor 1/h_1 \rfloor$ and the number of bins packed by Algorithm A satisfies

$$\frac{N}{N_2^*} > 2\left(1 + \frac{1}{m}\right) - \delta .$$

Proof sketch. Choose α and k to be positive integers where $\frac{4\alpha}{2\alpha-1} \frac{1}{k+1} < \delta$. Let L_2 be the list of $4\alpha k$ repetitions of the pair of rectangles

$$\left\langle \tfrac{1}{\alpha}(\tfrac{1}{2} + \varepsilon) \times (\tfrac{1}{2} - \varepsilon),\ \tfrac{1}{\alpha}(\tfrac{1}{2} + \varepsilon) \times 3\varepsilon \right\rangle$$

where ε is sufficiently small. Then $m = \lfloor 1/\tfrac{1}{\alpha}(\tfrac{1}{2}+\varepsilon) \rfloor = 2\alpha - 1$, $N = \lceil 4\alpha k/(2\alpha - 1) \rceil$ bins and $N_2^* = k + 1$ bins.

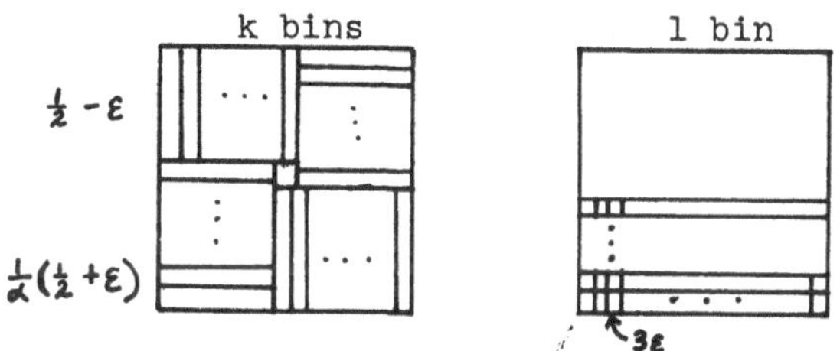

Figure 4. Optimal packing of L_2

$$\text{Thus, } \frac{N}{N_2^*} = \frac{\lceil 4\alpha k/(2\alpha-1) \rceil}{k+1} > 2\left(1 + \frac{1}{2\alpha-1}\right) - \delta = 2\left(1 + \frac{1}{m}\right) - \delta .$$

A Bottom-Up Left-Justified Packing Algorithm

An alternative approach to packing two-dimensional bins was proposed by Baker et al in [1]. In the bottom-left (BL) packing approach, a single unit width, infinite height bin is packed by allocating each successive rectangle of the list L to the bottom-most and left-most position at which it will fit. With the given list L arranged by nonincreasing widths of the rectangles, the authors proved the following result.

Theorem 4. The height of the BL packing satisfies $h_{BL} \leq 3h_{OPT}$ where h_{OPT} is the minimum height of the infinite bin into which all rectangles of L can be packed.

We can also extend this BL packing approach to the problem of packing a minimum number of 1 x 1 bins with a given list L of rectangles. As before, the bins to be packed are indexed $B_1, B_2, B_3, \ldots$. However, the rectangles of L are first reordered by nonincreasing widths. We formulate the following modified BL algorithm.

<u>Algorithm</u> <u>C</u>. Pack each successive rectangle of L into the bottom-most, left-most position of the highest indexed nonempty bin, B_k, if it will fit in that bin. If not, pack the rectangle into the next empty bin B_{k+1}. Further packing continues in this bin.

This algorithm also combines the BL approach of packing an infinite height bin with the one-dimensional Next Fit approach. The following error bounds can be obtained for Algorithm C.

<u>THEOREM</u>. For any positive integer M, $M > 1$, there exists a list of rectangles for which the number of bins, N, packed by Algorithm C satisfies $N > M N_2^*$ and $N > M N_1^*$.

<u>Proof</u>. Choose $k > M$ and let L_3 be the list consisting of $(M+1)k$ repetitions of the rectangles

$$\left\langle 1 \times \frac{1}{M+1}\ ,\ M \text{ repetitions of } \varepsilon \times \frac{1}{M+1} \right\rangle .$$

Then $N = (M+1)k$ bins while $N_2^* = N_1^* = k+1$ bins.

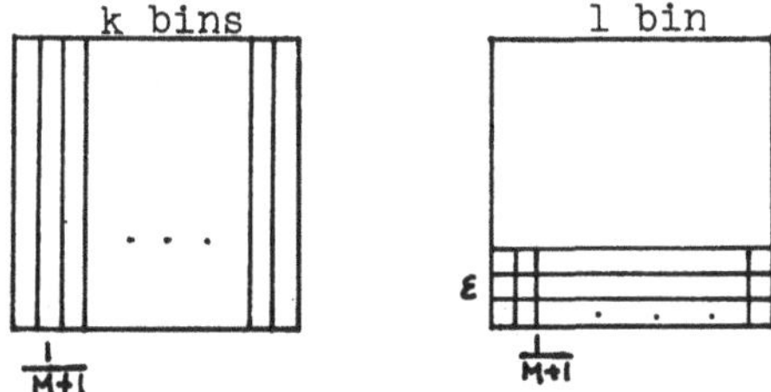

Figure 5. Optimal packing of L_3

$$\text{Therefore,}\quad \frac{N}{N_2^*} = \frac{N}{N_1^*} = \frac{(M+1)k}{k+1} > \frac{Mk+M}{k+1} = M .$$

<u>THEOREM</u>. Let N be the number of bins packed by Algorithm C for a given list L of rectangles whose heights do not exceed $\frac{1}{m}$ where $m \geq 2$. Then,

$$N < 2\,\frac{m}{m-1}\,N_2^* + 1 .$$

<u>Proof</u> <u>sketch</u>. For each bin B_i, define h_i^* to be the height of the bin at which the first rectangle in B_{i+1} would have been packed but for the resulting height violation. Using a result from [1], it can be shown that the region below h_i^* is at least half-occupied, i.e.

$$\sum_{R_j \in B_i} h_j w_j > \tfrac{1}{2} h_i^* \qquad \text{for } i=1,2,\ldots,N-1 .$$

Further, all $h_i^* > 1 - \frac{1}{m}$ so that

$$N_2^* \geq \sum_{j=1}^{n} h_j w_j > \tfrac{1}{2}(N-1)\left(1 - \tfrac{1}{m}\right) .$$

$$\text{Thus,}\qquad N < 2\,\frac{m}{m-1}\,N^* + 1 .$$

THEOREM. For any $\delta > 0$, there exists a list L of rectangles whose heights do not exceed $\frac{1}{m}$ where m is even, and the number of bins packed by Algorithm C satisfies $\frac{N}{N_2*} > 2\frac{m}{m-1} - \delta$.

Proof. Choose m and k to be positive integers where m is even and $2\frac{m}{m-1}\frac{1}{k+1} < \delta$. Define L_4 to be the list containing 2mk rectangles $D_1 = \frac{1}{m}(1-2\epsilon) \times (\frac{1}{2}+\epsilon)$ and 2mk rectangles $D_2 = 3\epsilon \times (\frac{1}{2}+\epsilon)$, where L_4 consists of $\lfloor 2mk/(m-1) \rfloor$ repetitions of

$$\langle \text{m-1 repetitions of } D_1, D_2 \rangle$$

followed by the remaining rectangles. Then $N = \lceil 2mk/(m-1) \rceil$ bins and $N* = k+1$ bins.

$$\frac{N}{N_2*} = \frac{\lceil 2mk/(m-1) \rceil}{k+1} > \frac{2m}{m-1}\frac{k}{k+1} > \frac{2m}{m-1} - \delta.$$

Conclusion

In this paper, we have presented some extensions of the level-oriented and bottom-up packing algorithms. These allocation procedures have previously been used to obtain approximate solutions to the problem of packing, with minimum total height, a set of rectangles into a single unit width, infinite height bin. We have modified these approaches and applied the algorithms to the problem of packing a given list of rectangles into a minimum number of 1 x 1 bins. In addition, error bounds for these methods were obtained by extending the results which were previously proved. In several cases, the multiplicative constants of the upper bounds for our algorithms were also shown to be as small as possible.

References

[1] Baker, B., E.G. Coffman Jr., and R.L. Rivest, "Orthogonal Packings in Two Dimensions," SIAM J. Computing, Vol. 9, No. 4, pp. 846-855 (1980).

[2] Brown, A.R., Optimum Packing and Depletion: The computer in space- and resource-usage problems, American Elsevier Inc., New York, 1971.

[3] Coffman, E.G., Jr., M.R. Garey, D.S. Johnson, and R.E. Tarjan, "Performance Bounds for Level-Oriented Two-Dimensional Packing Algorithms," SIAM J. Computing, Vol. 9, No. 4, pp. 808-826 (1980).

[4] Garey, M.R. and R.L. Graham, "Worst-Case Analysis of Memory Allocation Algorithms," Proceedings of the 4th Annual ACM Symposium on The Theory of Computing, Denver, No. 4, pp. 143-150 (1972).

[5] Garey, M.R., R.L. Graham, D.S. Johnson, A.C. Yao, "Resource Constrained Scheduling as Generalized Bin Packing," Journal of Combinatorial Theory, Vol. 21, pp. 257-298 (1976).

[6] Graham, R.L., "Bounds on the Performance of Scheduling Algorithms," Computer and Job-Shop Scheduling (E.G. Coffman, Jr., ed.), John Wiley and Sons, 1975.

[7] Johnson, D.S., A. Demers, J.D. Ullman, M.R. Garey, and R.L. Graham, "Worst-case Performance Bounds for Simple One-Dimensional Packing Algorithms," SIAM J. Computing, Vol. 3, No. 4, pp. 299-325 (1974).

[8] Johnson, D.S., "Fast Algorithms for Bin Packing," Journal of Computer and System Sciences, Vol. 8, pp. 272-314 (1974).

[9] Wang, P.Y., Computational Techniques for Two-Dimensional Rectangular Cutting Stock Problems, Ph.D Dissertation, University of Wisconsin-Milwaukee, August 1980.

OPTIMIZATION IN HIERARCHICAL SETTING OF A SIMPLE WORLD INDUSTRIALIZATION MODEL

A. Bagchi*
Department of System Science
University of California, Los Angeles, CA 90024

and

M. Moraal and G.J. Olsder
Department of Applied Mathematics
Twente University of Technology
7500AE Enschede, The Netherlands

1. INTRODUCTION

In the late sixties, in a U.N.I.D.O. conference held at Lima, it was agreed that by the year 2000 the industrial production of the less developed countries should be at least one-fourth the total industrial production of the world as a whole. As a follow-up to this declaration, a simple world industrialization model has been developed in [1] and the feasibility of "Lima target" has been studied within the production and trade possibilities. It was a dynamic, multisectoral, multiregional linear input-output model in which some simple linear programming exercises have been carried out. Our present study differs fundamentally from [1] in the role of the different regions in the optimization process. Furthermore, we assume that each region has its own optimizing criterion which may be in conflict with one another. Formulating the problem from this viewpoint leads to a multicriteria Stackelberg decision making problem in the LP-set up. For keeping the number of model constraints within reasonable bounds while retaining the basic features of the problem, we aggregate the model in [1] and divide the world into three regions of the developed market economies, the less developed market economies and the centrally planned economies. The centrally planned economies are assumed to enter the model exogeneously. The developed and less developed economies have different decision (instrument) variables and different optimizing criteria and the developed economies have the role of a leader in the optimization process. This leads to the Stackelberg game problem mentioned above. In section 2, we describe the aggregated world industrialization model. In section 3, we study the general Stackelberg LP-problem and develop a simplex-type algorithm for its solution. Finally, in section 4, we make some preliminary studies towards applying this algorithm to the world industrialization model and analyze the results.

2. A SIMPLE WORLD INDUSTRIALIZATION MODEL

The world is divided into 3 regions: (1)D.M.E.'s(developed market economies), (2)L.M.E.'s(less developed market economies) and (3)C.P.E.'s(centrally planned economies). Regions are denoted by suffix r. Each region is composed of 3 sectors: (1)Traditional (agriculture, mining and food), (2)Industry (chemicals, metals and

* On sabbatical leave from Twente University of Technology, Department of Applied Mathematics, P.O. Box 217, 7500 AE Enschede, The Netherlands

equipment) and (3)Services (transport-communication, construction and services). Suffix t will denote time and base year variables have time index zero.

Variable vectors are denoted by small Roman letters, scalar variables by capital Roman letters. Coefficients and vectors of coefficients are denoted by small Greek letters, matrices of coefficients by capital Roman letters with an upper bar and diagonal matrices of coefficients by small Greek letters with a circumflex.

We take the time steps to be 10 years with 1970 as the base. We describe the model in its barest essential. Details may be found in [2]. The balance equation is given by

$$x_{rt} \geq \bar{A}_r x_{rt} + c_{rt} + i_{rt} + e_{rt} - m_{rt}$$

where x stands for gross output, c is final consumption, i is total investments by sector of origin, e stands for exports and m for imports. Per capita consumption is specified by a simple Engel curve:

$$c_{rt} = \gamma_{or} P_{rt} + \gamma_{1r} C_{rt}$$

where P stands for total population and C, the total consumption expenditure is given by

$$C_{rt} = Y_{rt} - S_{rt} - \sigma_{or} P_{rt}.$$

Here Y is the gross domestic product satisfying

$$Y_{rt} = \alpha_{rt} x_{rt}$$

and S denotes the controllable part of the savings.

We now turn to investments and capital formations:

$$i_{rt} = i_{rt}^{repl.} + i_{rt}^{new}$$

where $i^{repl.}$ stands for replacement investments and i^{new} for new investments.

$$i_{rt}^{repl.} = \hat{\delta}_{or} k_{\overline{rt-1}} \quad \text{(k:capital stocks)}$$

$$k_{rt} = \bar{K}_r x_{rt}$$

$$i_{rt}^{new} = \hat{\omega}_1^{-1} \left(\frac{1}{\theta} \bar{D}_r h_{rt} - \hat{\omega}_o i_{\overline{rt-1}}^{new}\right)$$

$$h_{rt} = \hat{\kappa}_r (x_{rt} - x_{\overline{rt-1}})$$

h is the accumulated new investments over a period of θ years by sector of destination. θ is 10 years in the model. The basic model can be completed by specifying the trade equations. So far, we have been only interested in regions $r = 1$ and 2. The region $r = 3$ will enter the model through trade equations. m^w and e^w denote imports and exports of sectors producing world goods, while m^{wi} will denote imports of world goods imported from region i. With these conventions, we have

$$e_{rt} = B_2 e_{rt}^w + B_3 m_{rt}^w$$

$$m_{rt} = B_1 m_{rt}^w$$

$$e_{rt}^w = \sum_{i=1}^{3} m_{it}^{wr} \quad \text{and} \quad m_{rt}^w = \sum_{i=1}^{3} m_{rt}^{wi}$$

Let $R_r \stackrel{\Delta}{=} 1 - \bar{A}_r - \gamma_{1r}\alpha_r' - \hat{\omega}_1^{-1} \frac{1}{\theta} \bar{K}_r$. Then the above sets of equations give two basic constraints for the model:

$$(1)R_r x_{rt} \geq (\hat{\delta}_{or} - \hat{\omega}_1^{-1} \tfrac{1}{\theta})\bar{K}_r x_{\overline{rt-1}} - \hat{\omega}_1^{-1}\hat{\omega}_o i_{\overline{rt-1}}^{new} + (B_3-B_1)m_{rt}^{wi} - \gamma_{1r}S_{rt} + B_2 m_{it}^{wr} + (\gamma_{or}-\gamma_{1r}\sigma_{or})P_{rt}$$

and

$$(2)i_{rt}^{new} = \hat{\omega}_1^{-1} \tfrac{1}{\theta} \bar{K}_r x_{rt} - \hat{\omega}_1^{-1} \tfrac{1}{\theta} \bar{K}_r x_{\overline{rt-1}} - \hat{\omega}_1^{-1}\hat{\omega}_o i_{\overline{rt-1}}^{new} \quad ; \; r = 1,2; \; i = 1,2, \; i \neq r, \; t = 1,2,3.$$

To specify the model completely, we have to impose some additional constraints. We denote by x_{rtj} the j-th component of the vector x_{rt} and use similar notations for other vectors appearing in the model. Balance of trade restrictions, excluding c.i.f. margins (see [1] for explanation of c.i.f. margins) and with $m_{3t}^{wr} = m_{rt}^{w3}$(C.P.E.'s have perfect trade balance with other regions), give

$$(3)B_{rt}^{\ell} \leq [1,1](m_{it}^{wr} - m_{rt}^{wi}) \leq B_{rt}^{u}.$$

Import substitution constraint is

$$(4)m_{rtj}^{wi} \geq \{(1 - \varepsilon_{rj}^{i})^{t} m_{r0j}^{wi}/x_{r0j}\}x_{rtj}.$$

(In the original model [1], $\varepsilon_{11}^2 = \varepsilon_{12}^2 = \varepsilon_{21}^1 = \varepsilon_{22}^1 = 0.2$). The following constraint is imposed due to the export growth limitations for world goods by region of destination:

$$(5)m_{it}^{wr} \leq (I + \hat{\pi}_r^{i})^{t\theta} m_{i0}^{wr}.$$

We have also the savings constraint

$$(6)S_{rt} \leq 0.25Y_{rt} = 0.25\alpha_r' x_{rt}$$

and finally, the "Lima target" in our notation yields the constraint

$$(7)x_{232} \geq \alpha(x_{132} + x_{232})$$

$$\alpha = 0.25(x_{102} + x_{202} + x_{302})/(x_{102} + x_{202}); \quad [r = 1,2; \; i = 1,2; \; i \neq r; \; t = 1,2,3]$$

Equations (1) - (7) constitute the model constraints. There are altogether 91 inequalities in this model. Decision variables for regions r, r = 1,2, are x_{rt}, i_{rt}^{new}, m_{rt}^{wi}, S_{rt}, i = 1,2, i ≠ r, t = 1,2,3. Both DME's and LME's have altogether 27 decision (instrument) variables.

Specifying the optimizing criteria of D.M.E.'s and L.M.E.'s is complicated. We consider here discounted total consumption in each region as the criterion of that region. Thus, for r = 1,2, region r wants to minimize

$$J_r = -\sum_{t=1}^{3} C_{rt}/(1+\eta_r)^{\theta(t-1)} = \{\sigma_{or}P_{r1} + \tfrac{1}{2}\sigma_{or}P_{r2} + \tfrac{1}{4}\sigma_{or}P_{r3}\}$$
$$+ \{-\alpha_r' x_{r1} + S_{r1} - \tfrac{1}{2}\alpha_r' x_{r2} + \tfrac{1}{2}S_{r2} - \tfrac{1}{4}\alpha_r' x_{r3} + \tfrac{1}{4}S_{r3}\}.$$

Given decision variables of the D.M.E's, r = 1, the L.M.E.'s, r = 2, determine their decision variables so that J_2 is a minimum. Given those optimum decision variables for L.M.E.'s and substituted in the criterion J_1, D.M.E.'s determine their decision variables so that J_1 is a minimum. Now fixing the decision variables for D.M.E.'s, optimization problem for region 2 is a standard linear programming problem. The problem, however, becomes entirely nonclassical when we want to optimize at the higher level of region 1. In the next section, we give an abstract framework for this Stackelberg LP-problem and indicate a simplex-type algorithm for its solution. In the terminology of game theory, we have a two-player leader-follower (Stackelberg) game with the D.M.E.'s having the role of a leader and the L.M.E.'s having the role of a follower.

3. MATHEMATICAL SET-UP: STACKELBERG LP-PROBLEM

The abstract formulation of the problem has the following features:

(a) We have two decision makers who want to minimize their respective objectives;

(b) Decision of one (player) influences that of the other;

(c) One decision maker has more power than the other, he can impose his decision on the other.

Mathematically, the Stackelberg problem we are faced with can be described as follows: We have two players, the leader (1) and the follower (2). Decision variables for the leader, $u_1 \in \mathbb{R}^{m_1}$ and those for the follower, $u_2 \in \mathbb{R}^{m_2}$. The leader chooses u_1 and the follower chooses u_2 to minimize respectively

$$J_1(u_1,u_2) = c_{11}' u_1 + c_{12}' u_2$$

and

$$J_2(u_1,u_2) = c_{21}' u_1 + c_{22}' u_2; \; c_{i1} \in \mathbb{R}^{m_1}, \; c_{i2} \in \mathbb{R}^{m_2}; \; i = 1,2.$$

Decisions of both the players are restricted to a feasible set

$$FS = \{(u_1,u_2) \mid A_1u_1 + A_2u_2 \le b; \; u_1 \ge 0, \; u_2 \ge 0\}$$

with A_i being $m\times m_i$ matrix $(i = 1,2)$ and $b \in \mathbb{R}^m$.

The leader announces u_1 and the follower then minimizes $J_2(u_1,u_2)$. The leader chooses u_1 such that $(u_1,u_2) \in FS$. This leads to the follower's LP problem:

(P1) With announced u_1^o,

$$\text{minimize } c_{21}' u_1^o + c_{22}' u_2$$

$$\text{subject to } A_2u_2 \le b - A_1u_1^o; \; u_2 \ge 0.$$

Assume that, given $u_1 \in \mathbb{R}^{m_1}$, there exists a unique optimal $u_2^o(u_1)$ for problem (P_1). Then the leader's optimization problem is:

(P2) minimize $c_{11}' u_1 + c_{12}' u_2$

$$\text{subject to } A_1u_1 + A_2u_2 \le b; \; u_1 \ge 0, \; u_2 \ge 0$$

$$u_2 = u_2^o(u_1).$$

Let us define the reaction curve by

$$RC \triangleq \{(u_1,u_2) \mid u_1 \text{ is admissible and } u_2 = u_2^o(u_1)\}.$$

Then we can reformulate problem (P2) as

(P2') minimize $c_{11}' u_1 + c_{12}' u_2$

subject to $(u_1,u_2) \in RC$.

Solution of the problem (P2') is called a Stackelberg solution.

A Team solution for the leader is a minimum point for $J_1(u_1,u_2) = c_{11}' u_1 + c_{12}' u_2$ on FS.

The two following properties play a crucial role in developing a simplex-type algorithm for solving the Stackelberg game problem formulated above.

Property 1 If a Team solution (for the leader) exists, a Stackelberg solution also exists. Moreover, if (u_1^o, u_2^o) is a Team solution and $(u_1^o, u_2^o) \in RC$, then (u_1^o, u_2^o) is a Stackelberg solution.

Property 2 If a Stackelberg solution exists, there is a Stackelberg solution at an extreme point of FS.

Using these two properties we can think of the following procedure for determining a Stackelberg solution: start by computing a Team solution for the leader. If it is on RC we stop. Otherwise, move from one extreme point of FS to an adjacent extreme point, just as in the simplex algorithm. The actual algorithm is quite complex because of the nonconvexity of RC. Details of the algorithm, developed by one of the authors, may be found in [3].

4. SIMULATION STUDIES

In all the simulation runs performed for different α's, the Team solution of the leader turned out to be on RC and therefore, yielded already a Stackelberg solution. The following values were taken for the coefficients appearing in the model:

$$A_1 = \begin{bmatrix} 0.3429 & 0.0500 & 0.0358 \\ 0.0826 & 0.0351 & 0.1116 \\ 0.1208 & 0.1436 & 0.2067 \end{bmatrix}$$

$$A_2 = \begin{bmatrix} 0.1760 & 0.1044 & 0.0195 \\ 0.0710 & 0.3010 & 0.0755 \\ 0.1493 & 0.1866 & 0.1418 \end{bmatrix}$$

$$\gamma_{11}' = \gamma_{12}' = [0.1688 \quad 0.0868 \quad 0.7444]$$

$$\gamma_{01}' = \gamma_{02}' = [64.48 \quad -5.51 \quad -58.97]$$

$$\sigma_{01} = \sigma_{02} = 5.68$$

$$\alpha_1' = [0.4538 \quad 0.4713 \quad 0.6459]$$

$$\alpha_2' = [0.6037 \quad 0.4078 \quad 0.7632]$$

$$\hat{\delta}_{01} = \text{diag.}[0.0900 \quad 0.0900 \quad 0.0630]$$

$$\hat{\delta}_{02} = \text{diag.}[0.0900 \quad 0.0900 \quad 0.0662]$$

$$K_1 = \begin{bmatrix} 0.2087 & 0 & 0 \\ 0.4900 & 0.7660 & 0.5396 \\ 0.4654 & 0.3523 & 1.1571 \end{bmatrix}$$

$$K_2 = \begin{bmatrix} 0.2610 & 0 & 0 \\ 0.3867 & 1.5419 & 0.4399 \\ 0.7075 & 0.5737 & 0.6378 \end{bmatrix}$$

$\hat{\omega}_1 = 0.45 I_3 \qquad \hat{\omega}_2 = 0.55 I_3 \qquad \theta = 10$

$$B_1 = \begin{bmatrix} 1 & 0 \\ 0 & 1 \\ 0 & 0 \end{bmatrix} \quad B_2 = \begin{bmatrix} 1 & 0 \\ 0 & 1 \\ 0 & 0.01958 \end{bmatrix} \quad B_3 = \begin{bmatrix} 0 & 0 \\ 0 & 0 \\ 0.09964 & 0 \end{bmatrix}$$

$P_{11} = 784 \qquad P_{12} = 857 \qquad P_{13} = 928$

$P_{21} = 2187 \qquad P_{22} = 2745 \qquad P_{23} = 3412$

give the population in millions.

The initial conditions are, in millions of U.S. dollars,

$$x_{10} = \begin{bmatrix} 654371 \\ 996664 \\ 2068711 \end{bmatrix} \quad x_{20} = \begin{bmatrix} 253819 \\ 97018 \\ 249547 \end{bmatrix} \quad x_{30} = \begin{bmatrix} 441015 \\ 453719 \\ 566269 \end{bmatrix}$$

$$i_{10}^{new} = \begin{bmatrix} 5503 \\ 117230 \\ 154944 \end{bmatrix} \quad i_{20}^{new} = \begin{bmatrix} 2651 \\ 22239 \\ 20163 \end{bmatrix}$$

$B_{11}^u = -B_{11}^\ell = B_{21}^u = -B_{21}^\ell = 15000$

$B_{12}^u = -B_{12}^\ell = B_{22}^u = -B_{22}^\ell = 30000$

$B_{13}^u = -B_{13}^\ell = B_{23}^u = -B_{23}^\ell = 45000.$

$(m_{10}^{w2})' = [43852 \quad 11448]$ and $(m_{20}^{w1})' = [18093 \quad 35807]$

The choice of α is, of course, subjective. With the initial values, "Lima" goal would make α approximately equal to $0.25\ (996664 + 97018 + 453719)/(996664 + 97018) \approx 0.35$. But with the expected gradual increase in the share of industrial production of the L.M.E.'s a more reasonable choice of α is 0.3. Simulations were performed with several choices for α. Here we shall give results only for $\alpha = 0.3$ and $\alpha = 0.2$ and based on these figures, we draw some general conclusions.

Team solution and the same Stackelberg solution for $\alpha = 0.2$

In the following tables, the optimum values of the different variables are given (we take their 1970 values to be 1·000)and next to those quantities we give in parenthesis the corresponding yearly growth percentages (averaged over a period of 10 years).

TABLE 1

t	x_{1t1}	x_{1t2}	x_{1t3}	x_{2t1}	x_{2t2}	x_{2t3}
1	1.504 (4.16)	1.764 (5.84)	1.701 (5.46)	1.730 (5.64)	1.417 (3.55)	2.126 (7.83)
2	2.655 (5.85)	2.923 (5.18)	2.890 (5.44)	2.354 (3.13)	4.199 (11.48)	3.922 (6.31)
3	5.062 (6.66)	4.470 (4.34)	4.818 (5.24)	2.354 (0.00)	11.481 (10.58)	7.761 (7.06)
1970 values:	654371	996664	2068711	253819	97018	249547

t	i^{new}_{1t1}	i^{new}_{1t2}	i^{new}_{1t3}	i^{new}_{2t1}	i^{new}_{2t2}	i^{new}_{2t3}
1	1.555 (4.51)	1.673 (5.28)	1.789 (5.99)	2.834 (10.98)	1.352 (3.06)	2.454 (9.39)
2	4.450 (11.09)	2.847 (5.46)	2.980 (5.23)	0.000 (——)	5.088 (14.17)	3.092 (2.34)
3	7.834 (5.82)	4.302 (4.21)	4.807 (4.90)	0.000 (——)	8.877 (5.72)	7.422 (9.15)

1970 values: 5503 117230 154944 2651 22239 20163

t	m^{w2}_{1t1}	m^{w2}_{2t2}	m^{w1}_{2t1}	m^{w2}_{2t2}	S_{1t}	S_{2t}
1	2.594 (10.00)	1.411 (3.50)	1.384 (3.31)	2.509 (9.64)	1.840 (6.29)	2.066(7.52)
2	2.435 (-0.63)	2.362 (5.29)	1.507 (0.85)	2.688 (0.69)	3.123 (5.43)	3.624(5.78)
3	2.592 (0.62)	30.993 (29.36)	16.745 (27.23)	5.878 (8.14)	5.206 (5.24)	6.523(6.05)

1970 values: 43852 11448 18093 35807 482175 87880

Team solution and the same Stackelberg solution for $\alpha = 0.3$

In the following tables, the optimum values of the different variables are given (we take their 1970 values to be 1·0000) and next to those quantities we give in parenthesis the corresponding yearly growth percentages (averaged over a period of 10 years).

TABLE 2

t	x_{1t1}	x_{1t2}	x_{1t3}	x_{2t1}	x_{2t2}	x_{2t3}
1	1.504 (4.16)	1.764 (5.84)	1.701 (5.46)	1.730 (5.64)	1.417 (3.55)	2.126 (7.83)
2	2.575 (5.53)	2.913 (5.15)	2.814 (5.16)	2.613 (4.21)	4.720 (12.79)	4.885 (8.67)
3	4.490 (5.72)	3.446 (1.69)	4.027 (3.65)	2.928 (1.15)	15.170 (12.38)	10.424 (7.88)

1970 - values: 654371 996664 2068711 253819 97018 249547

t	i^{new}_{1t1}	i^{new}_{1t2}	i^{new}_{1t3}	i^{new}_{2t1}	i^{new}_{2t2}	i^{new}_{2t3}
1	1.555 (4.51)	1.673 (5.28)	1.789 (5.99)	2.834 (10.98)	1.352 (3.06)	2.454 (9.39)
2	4.010 (9.94)	2.626 (4.61)	2.682 (4.13)	1.435 (-6.58)	7.176 (18.16)	5.612 (8.62)
3	5.656 (3.50)	1.290 (-6.86)	1.989 (-2.94)	0.000 (——)	13.237 (6.31)	9.894 (5.84)

1970 - values: 5503 117230 154944 2651 22239 20163

t	m^{w2}_{1t1}	m^{w2}_{2t2}	m^{w1}_{2t1}	m^{w2}_{2t2}	S_{1t}	S_{2t}
1	2.594 (10.00)	1.411 (3.50)	1.384 (3.31)	2.509 (9.64)	1.840 (6.29)	2.066(7.52)
2	2.719 (0.47)	1.864 (2.83)	1.672 (1.91)	3.919 (4.56)	3.056 (5.21)	4.316(7.65)
3	2.299 (-1.66)	39.135 (35.58)	17.449 (26.43)	7.767 (7.08)	3.492 (1.34)	0.632(7.18)

1970 - values: 43852 11448 18093 35807 482175 87880

The figures given above are only preliminary in nature and are used merely to draw some general conclusions. We first conclude that so long as D.M.E.'s and L.M.E.'s are only interested in maximizing their own consumptions, Team solution of the leader for the 3-sector model always turns out to be a Stackelberg solution as well. Furthermore, comparison of the figures in Tables 1 and 2 leads to some general observations (which are also confirmed by other simulation runs with different α's):

① In all the simulation runs, figures for the year 1980 are identical. Both regions have maximum savings and no overproduction. The D.M.E.'s have maximum trade deficit possible and this is effected by maximum trade deficit in the traditional sector and a slight trade surplus in the industrial sector. This is not surprising since the value-added for the L.M.E.'s are higher for the traditional and service sectors than for the industrial sector while those sectors have lower marginal capital/output coefficients; for the D.M.E.'s on the other hand the value-added in the industrial sector is higher than in the traditional sector, while the capital/output coefficient for the industrial sector is lower than in the traditional sector.

② Difference in the figures for different α's start to appear in the year 1990. Savings are still maximum and no overproduction takes place. Trade deficit for D.M.E.'s is not maximum possible for $\alpha = 0.2$, but is again maximum for $\alpha = 0.3$. This change comes about because of increase in trade surplus for the D.M.E.'s in the industrial sector while the trade deficit in the traditional sector remains more or less the same. The effect of "Lima target" set for the year 2000 is already noticeable in the year 1990. The L.M.E.'s have strong tendency to increase their industrial production while the D.M.E.'s pay some more attention to their traditional sector (see the growth percentages in the tables for the year 1990).

③ In the year 2000, the effect of "Lima target" is clear. There is still no overproduction while savings for L.M.E.'s is maximum during all the simulation runs. This is, however, not the case anymore for the D.M.E.'s when $\alpha = 0.3$. Thus for $\alpha = 0.3$, D.M.E.'s do not use their maximum growth possibilities. In all the simulation runs, the L.M.E.'s in the year 2000 have maximum trade deficit possible, effected through maximum allowed trade deficit in the traditional sector and a slight trade surplus in the industrial sector, an exact replica of the trade pattern of the D.M.E.'s in the year 1980. This is clearly the effect of imposing the "Lima target."

5. CONCLUSION

Our study indicates a general pattern of optimum world economic growth in a simple industrialization model, when the decisions of the D.M.E.'s are binding on the L.M.E.'s ; C.P.E.'s appear in the model exogeneously and the "Lima target" of the minimum share of the L.M.E.'s in the total world industrial production in the year 2000 is taken as a model constraint.

REFERENCES

[1] H. Opdam and A. Ten Kate, "A Simple World Industrialization Model," Report, Erasmus University, Rotterdam, December 1978.

[2] M. Moraal, "State Space Representation and Simulation of a Simple World Industrial-

ization Model," TW-Memorandum, No.300, Twente University of Technology, Enschede, The Netherlands, March 1980.

[3] M. Moraal, "Stackelberg Solutions in Linear Programming Problems," Methods of Operations Research, Vol.44, (Proceedings of the VIth Symposium über Operations Research held at Augsburg, September 7-9),1981.

APPENDIX

Notations:

x: vector of gross output by sector

$\bar{A}$: technology matrix of input-output coefficients

c: vector of final consumption by sector

i: vector of total investments by sector of origin

e: vector of exports by sector

m: vector of imports by sector

Y: gross domestic product by region

α: vector of gross value added coefficients

P: total population by region

C: total consumption expenditures by region

γ_o: vector of coefficients with the sum of its elements equal to zero

S: controllable part of savings

σ_o: scalar coefficient giving autonomous part of savings per capita

$i^{repl.}$: vector of replacement investments by sector of origin

i^{new}: vector of new investments by sector of origin

h: vector of accumulated new investments over a period of θ years by sector of destination

$\hat{\kappa}$: diagonal matrix of marginal sectoral capital-output ratios

$\hat{\omega}_o, \hat{\omega}_1$: diagonal matrix of weights; $\hat{\omega}_o + \hat{\omega}_1 = I$

k: vector of capital stocks

$\hat{\delta}_o$: diagonal matrix of replacement ratios

γ_1: vector of coefficients with the sum of its elements equal to one

LONG TERM NUCLEAR SCHEDULING
IN THE FRENCH POWER SYSTEM

P. COLLETER, P. LEDERER, J. ORTMANS

ELECTRICITE DE FRANCE

Etudes Economiques Générales
2, rue Louis Murat
75384 PARIS Cedex 08 - France

1 - INTRODUCTION : THE FRENCH SYSTEM

The main features of the French electrical system are :

- a demand with pronounced variations between the seasons as well as during the day. Furthermore, this demand is subject to important random variations ;
- diversified thermal plants. This means the cost per produced kWh is rapidly increasing with the thermal demand ;
- an hydraulic system composed of about twenty large seasonal reservoirs. Most of these reservoirs receive the greatest part of their inflows outside of the high demand periods.

The main problem in the yearly cycle consists in managing the different storages of energy : reservoirs [1-6] and nuclear plants. The purpose of the RELAX model is to perform the optimal scheduling of the nuclear system according to the available controls (power and date of refueling) so as to minimize the global operating cost.

2 - THE PROBLEM

2.1 State, controls, constraints, criterion

Roughly speaking, a nuclear unit can be described by the remaining energy in the core : at time t, the next normal date of refueling is a function of this stock, and some flexibilities around this date (anticipation on stretch-out) are possible, as shown below.

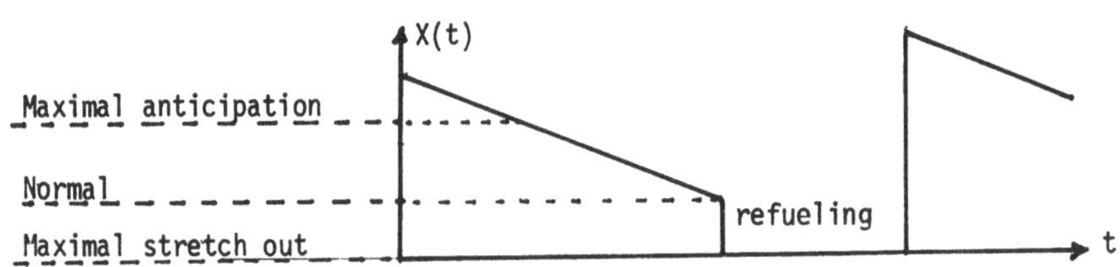

For sake of simplicity, we dropped here the influence of the number of the cycle and the flexibilities used during the preceding cycles, although they are taken into account in the RELAX model.

We are thus dealing with a one dimensional state driven by the operating power of the unit and the date of shut down.

If we refer to a base loaded unit, with normal availability, the average length of a cycle (operation + refueling) is about 14 months, with up to one month of stretch-out or two months of anticipation.

The global objective of the management will be the sum of the different operation costs, taken as a mathematical expectation with respect to inflows, demand and thermal availability.

The nuclear plants are supposed to have an average rate of availability.

In addition to the proper constraints of each unit (anticipation and stretch-out), two units on a same site cannot be refueled at the same time because of maintenance staff and equipment availability. Considering that many french sites are made up of four units, the feasible domain is considerably reduced.

2.2 A fitting formalization

The different features that we pointed out imply that a good choice for the model results in the use of :

- dynamic programming for feedback controls,
- a several years horizon because of transient states and connection between two cycles (caused by the flexibilities),
- a minimum discounted cost over this period.

The size of the nuclear french system (up to 60 units in 1990) makes it necessary to break down the problem into smaller ones by means of a relaxation : the nuclear system is then considered as a team where each unit tries to minimize the global criterion. This algorithm ensures a decreasing sequence of the cost. But unfortunately, the conditions for this algorithm to converge to the true optimum do not hold. It was then necessary to check the quality of the result by comparing it with a minorant that was obtained on a simplified problem : the difference between the two results was always less than 1 % (and most of time less than 0.5 %).

3 - THE "RELAX" MODEL

3.1 Problem formulation

3.11 State of the system and controls

For each reactor i (i = 1 to n), two kinds of controls are available :

a) -$de_i(t)$ is the first week after week t during which reactor i will be shut down for maintenance and refueling.

-$dd_i(t)$ is the week during which the reactor i will be put into operation again ($dd_i(t) = de_i(t) + d$, where d is the duration of maintenance and refueling).

b) -$m_i(t)$ is the fraction of the available power of reactor i used during week t.

$X_i(t)$ is the state variable of reactor i. It represents the energy which can be produced by the reactor i from week t till maximum stretch-out. This energy is measured in terms of weeks of operation at full available capacity so that $X_i(t)$ is homogenous to time. As the reactor cannot be simultaneously shut down for refueling and operating, the two possibilities have been gathered in the same state and $X_i(t) = -k$ means that the reactor i has been shut down for maintenance and refueling for k weeks.

The following two figures show the evolution of the state variable with the time and the corresponding power delivered by the reactor :

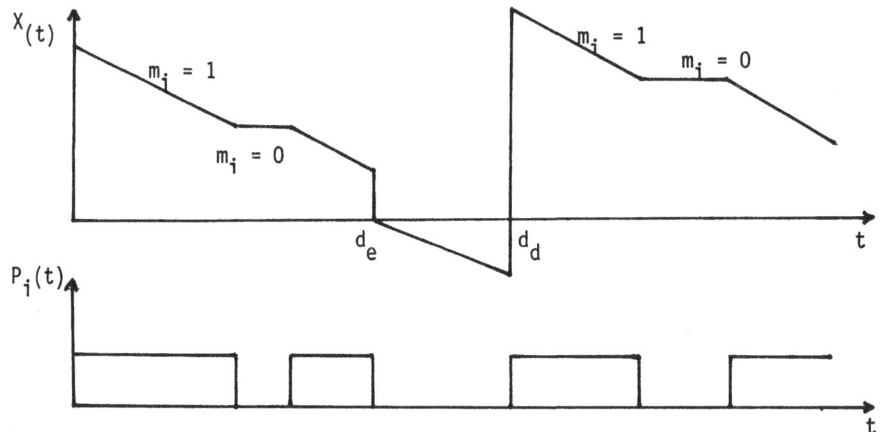

Let us introduce some more notations :

X_{max} is the state of a reactor just after refueling,

D_{max} = maximum number of weeks at full available capacity,

X_{min} is the limit of stretch out,

D_{min} = minimum number of weeks at full available capacity,

d = duration of maintenance and refueling,

W = $D_{max} - D_{min}$ (W is the duration of the "window" when shut down for maintenance has to be decided).

3.12 Optimal observation frequency

In the deterministic case, the maximal frequency for taking a decision of shut-down for maintenance and refueling is determined by $\Delta T = D_{min} + d$.

It is then sufficient to study the system at discrete times T_k with $T_{k+1} = T_k + \Delta T$. Thus, given $X_i(T_k)$ and $de_i(T_k)$, $m_i(t)$ for $T_k < t < T_{k+1}$ the trajectory of the state during the period $[T_k, T_{k+1}]$ can be computed as well as the power delivered by the reactor.

In this way, the necessary computations are significantly reduced as compared to a week by week optimization.

3.2 Model formulation

We are now able to introduce the formulation of the RELAX model. We just need a slight modification concerning the control $m_i(t)$.

Notation : let :

m^b_{ik} be the number of weeks t with $T_k < t < \min(de_i(T_k), T_{k+1})$ where the reactor i is shut down.

m^a_{ik} be the number of weeks t with $\dot{dd}_i < t < T_{k+1}$ where the reactor i is shut down.

With these control variables $(de_i(T_k), m^b_{ik}, m^a_{ik})$ we can write the motion equations of the state variable X_i.

$$X_i(T_{k+1}) = X_i(T_k) - \Delta T + m^b_{ik} \qquad \text{if} \qquad de_i(T_k) > T_{k+1}$$
$$X_i(T_k) > 0$$

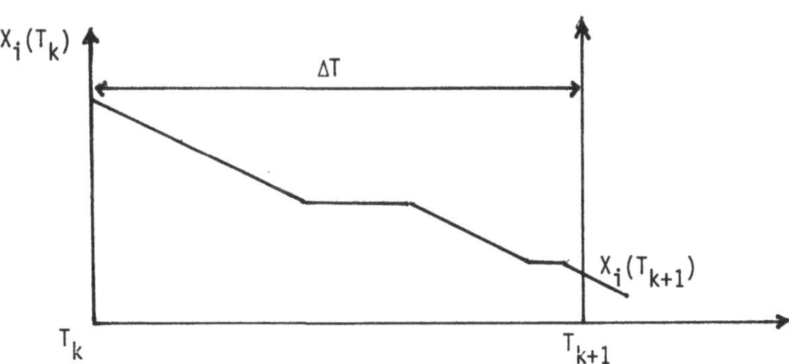

The other cases are straight forward to derive.

The problem can now be solved using dynamic programming and relaxation techniques :

Dynamic programming for one unit :

$$\text{MIN} \left[\sum_{t = T_k}^{T_{k+1}} C \left[(D(t) - P_i(t, de_i, m_{ik}^a, m_{ik}^b) \right] + V \left[X_i(T_{k+1}), T_{k+1} \right] \right]$$

$(de_i \; \varepsilon D \; (X_i(T_k), m_{ik}^b), m_{ik}^b, m_{ik}^a)$.

where : - $D \; (X_i(T_k), m_{ik}^b)$ is an interval in time which can be seen as a function of $X_i(T_k)$ and m_{ik}^b ;

- $V(X, T_k)$ is the Bellman function of the problem ;
- $D(t)$ is the residual demand to be satisfied by unit i and the fossil plants ;
- $C(D - P_i)$ is the mathematical expectation of the fuel costs with respect to demand, hydraulic inflows and fossil plants availability ;
- $P_i(t, de_i, m_{ik}^a, m_{ik}^b)$ is the power delivered by unit i. It can be seen that this function comes itself as the result of a sub-optimization : m_{ik}^b weeks have to be allocated to $(T_k, de_i(T_k))$, so as to minimize the cost on this period ; in the same way m_{ik}^a weeks have to be allocated to (dd_i, T_{k+1}).

As a result of this optimization, we get for each unit :

- the optimum number of refuelings for a given horizon,
- the optimal management of its stock between two refuelings.

The coordination of the different sub-problems derives then from a relaxation scheme using an auxiliary problem as formulated in [7] .

3.3 The relaxation step

For a given management of the other units, each unit faces a residual demand D(t) and minimizes the global criterion.

Although we can ensure the global criterion to decrease, we cannot prove that the global optimum will be reached. However, different experiences showed that the difference between the global optimum and the reached stationnary point is not great :

a) changing the initial solution changes the plannings, but the effect on the final criterion is small (generally less than 0.5 %) ;

b) the difference with a minorant of the criterion obtained on a simplified problem was never greater than 1 %, and most of time less than 0.5 %.

3.4 Computation of the criterion for a unit

Let us call J the global criterion for a unit :

$$j. = \Sigma_k j_k \quad \text{with} \quad J_n = \sum_{T_n}^{T_{n+1}} C(D^t - P^t).$$

Let us call cm the derivative of the fossil cost :

$$C(D^t - P^t) = \int_0^{D^t} cm(P)\, dP - \int_{D^t - P^t}^{D^t} cm(P)\, dP.$$

Minimizing J is then equivalent to maximize $J' = \sum_k J_k'$:

$$J_k' = \sum_t \int_{D^t - P^t}^{D^t} cm(P)\, dP = \sum_t P^t\, cm(\alpha^t),\ \alpha^t\ \varepsilon(D^t - P^t,\ D^t).$$

As P^t is small compared with D^t($\sim$ 2 %), we can find a good approximation for J' by choosing for example $\alpha^t = D^t - P^t$.

The relaxation becomes then quite similar to a price decomposition where the prediction of the price $cm(\alpha^t)$ is updated after the contribution of each unit.

4 - IMPROVING THE MODEL

4.1 Teething effect

As the system is in a renewing period, the teething effect is particularly important during the first cycles.

The following table gives the rate of availability, the average duration of operation, and the duration of the refueling and maintenance, according to the cycle number.

Cycle	1	2	3 and following
Availability	. 75	. 85	. 85
Weeks of operation	48	39	42
Refueling	17	8	8

Taking this into account implies the introduction of a new parameter and yields a two dimensional state :

(X, N) = (Stock, cycle number).

4.2 Heredity in flexibilities

In fact, the average duration of a cycle and the available flexibilities depend on the flexibilities that have been used during the preceding cycles.

Let us call :

D^k_{nor} = normal length of operation (weeks) at cycle k,

D^k_{max} = maximum length of operation (weeks) at cycle k,

D^k_{min} = minimum length of operation (weeks) at cycle k.

$F^k = X_{max} - X(d^k_e) - D^k_{nor}$ flexibility used at cycle k.

Then, as a first approximation of the heredities caused by the flexibilities, we use the following model :

$$D^k_{nor} = f(F^{k-1}) + f_o,$$

$$D^k_{max} - D^k_{nor} = g(F^{k-1}) + g_o,$$

$$D^k_{min} - D^k_{nor} = h(F^{k-1}) + h_o,$$

when f, g, h are linear functions.

We need now a three dimensional state to compute the possible window of refueling at each state.

$$\begin{bmatrix} X \\ N \\ F \end{bmatrix} = \begin{bmatrix} \text{Stock of energy} \\ \text{Number of the cycle} \\ \text{Flexibility of cycle } N - 1 \end{bmatrix}$$

where :

Maximal anticipation $< F <$ Maximal stretch out,

in practice $- 8 < F < + 4$ (weeks).

Thanks to the relatively small dimension of F and to the low frequency of observation, we are still able to carry out dynamic programming in the three dimensional grid (X, N, F).

4.3 Constraints on a site

Several sites are made up of 4 units. In the beginning it was taken for granted that two units could not be refueled at the same time because of a lack of maintenance staff and specialized tools. This constraint introduced a coupling between the dynamics of the units of a same site.

An ordinary relaxation scheme would have been untractable and we had to use a DPSA-like algorithm [8] :

each unit is limited to the controls which do not change the final state of the others (on the same site).

5 - NUMERICAL RESULTS

5.1 Levelizing of the residual demand for fossil energy

Figure (1) shows week by week for year 1990 the global demand for power, the available nuclear power and the thermal margin, and this for the peak demand (8 most loaded hours of the week).

The yearly period of the global available nuclear power appears clearly, and thus we can say that the refueling becomes a meaningful tool for levelizing of the demand for fossil energy, as well as the management of the hydroelectric reserves. The refueling periods are gathered in summer as much as possible by the use of :

- the flexibilities around the normal date of refueling,
- voluntary shut down, when the demand is low to shift the window of refueling towards periods where the marginal cost is lower, or to avoid overlapping with a unit of the same site.

5.2 Scheduling on a site

Looking at an example of scheduling (fig. 2) gives an idea of the weight of the no-overlapping constraint on a site, and confirms the importance of the voluntary shut down to find a feasible solution.

Let us briefly mention some particular cases on unit 1 :

1984 : voluntary shut down to shift the refueling period from the other unit,

1985 : voluntary shut down to shift the refueling out of the cold period.

5.3 Further developments

The RELAX model gives answers to the operating people as far as scheduling is concerned. Moreover, it can provide the long term manager with useful indications concerning :

- the optimal length of the nuclear cycle,
- the development of special uses consuming nuclear energy when available,
- the cost of the no-overlapping constraints on a site.

Several studies concerning theses questions have already been worked out.

They made it clear, that generally speaking a yearly period fits better than the actual one to the structure of the electrical system, and helped to decide between several possibilities of given costs.

The constraint on a site was proven to be very expensive and a special effort to lessen this constraint is being done.

The opportunity of selling nuclear energy to special uses (exportation, hydrogen production, ...) has been studied for different prices, and prospective studies have been fed with the corresponding results.

CONCLUSION

The RELAX model has already come to a point where it can help people to get acquainted with the management of the new nuclear power system.

It can provide the operating people with answers to problems which have merely been stated by now, and give prospective elements to the futur developments of the electrical power system.

THERMAL DEMAND AND AVAILABLE POWER

1990 - 8 most loaded hours in the week

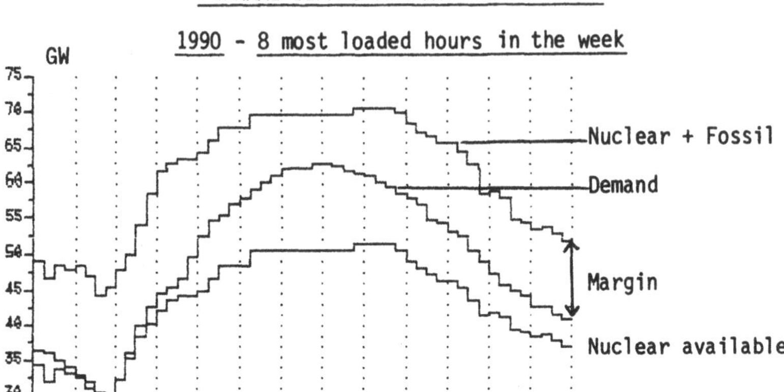

Figure 1

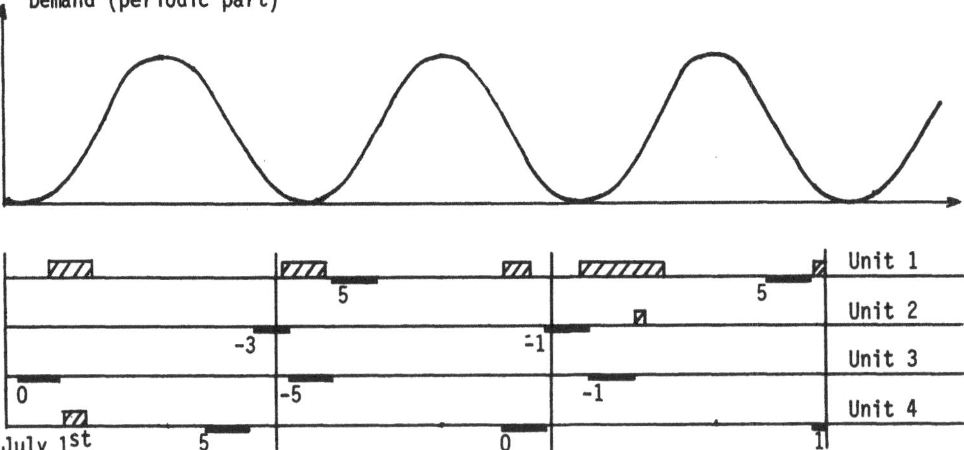

Figure 2

REFERENCES

[1] Contribution of stochastic control singular perturbation averaging and team theories to an example of large scale system : management of hydropower production.

DELEBECQUE, QUADRAT - IEEE A.C n° 28 pp 209.222, April 1978.

[2] Planification de la production énergétique au moyen de modèles à réservoirs multiples.

PRONOVOST - Colloque "Théorie des systèmes et Application à la gestion des Services Publics" - Montréal - Presses Universitaires Janvier 1975.

[3] Optimal operation of multireservoir power system with stochastic inflows.

TURGEON - Water Resources Research - vol. 16 - n° 2, April 1980.

[4] Gestion en stratégie d'un grand système hydroélectrique, le cas français.

FALGARONE, LEDERER - Communication au TIMS-ORSA - New York - Mai 1978.

[5] Applications of stochastic control methods in the management of energy production in New Caledonia.

COLLETER, DELEBECQUE, QUADRAT, FALGARONE - Applied stochastic control in econometrica and management science - North-Holland 1980.

[6] Optimal operation feedbacks for the french hydropower system.

COLLETER, LEDERER - CORS-TIMS-ORSA Meeting - Toronto - May 3-6 1981.

[7] Optimisation by decomposition and coordination : a unified approach.

COHEN - in IEEE Automatic Control - vol. AC 23 - April 1978.

[8] A dynamic programming successive approximations technique with convergence proofs.

LARSON, KORSAK - Automatica - vol. 6 - p. 245-252 - 1970.

PREDICTION OF SOCIO-ECONOMIC POLICY: INFORMATION GAIN BY INTERACTIVE DECISION ACTIVITIES

R. Fahrion
Department of Economics
University of Heidelberg,
Grabengasse 14, 69oo Heidelberg/F.R.G.

Abstract: The prediction of endogenous variables in dynamic econometric models is normally based on the concept of rational expectations. The ex ante expectation proxies are assumed to fulfil strong conditions on information availability which are generally not given. We suggest a process of interaction between model structure and policy decision making yielding a successive gain of information. In every stage of prediction the two 'players' are mutually confronted with the results of their actions. We formalize the decision behavior by using preference values with Weibull-distributed error residuals. The adaptation of control and endogenous variables and of the model specification is continued until a Pareto compromise is found.

1. Introduction

The error-learning principle for the prediction of endogenous variables in dynamic econometric models is normally based on the concept of rational expectations. The practical importance of rational expectations however, is restricted by the size of required information availability (Friedman (1979)). The applicability of rational expectations depends of the true but unknown model structure, and one usually has to cope with very strong statistical conditions on the expectation patterns: In order to converge the control path of endogenous variables to an equilibrium, orthogonality of errors and independence between ex ante predetermined variables and prediction-errors is supposed. A kind of well-behavior in the model structure is required such that the deviations of ex ante endogenous variables are independent from all predetermined variables. These deviations, however, are non-controllable and sometimes may cumulate. Hence, the classical process of extrapolation-predictions has to be modified into the facility of a successive alteration process in the model structure.

In most stability models quadratic utility functionals are used. The elements of the weighting matrix reflect the policy maker's preference structure. Sometimes it will be argued that roughly specified weights yield similar results than more refined ones. We will not follow this argument but suggest a process of gaining more experience in formulating and formalizing the policy maker's behavior, although it is often argued that an accurate specification of the planning authorities' preferences is not possible. Furthermore, we will not concentrate on the

fulfilment of some given optimality criterion concerning the regressions and expectations, but rather on a process of interaction between model and policy which successively yields an information gain.
In the following section the linear econometric model used within the interaction principle will be described. Then we will specify the intensity of assocation according to which the decision authority is willing to influence their current control variables by past experiences. In section 3 we define choice probabilities for the stochastic specification of the policy maker's preference values, and suggest the alteration principle between policy decision making and adaptive model specification. Thus, information exchange delivers information gain. In every stage of ex ante prediction the two 'players' are mutually confronted with their results. This interactive processing will be performed until Pareto optimality is attained, i.e. the level of relevant control variables combined with the correspondingly generated endogenous variables is accepted.

2. The linear econometric model and the definition of the policy maker's preference pattern

We consider a linear econometric model in the general form

$$(1)\qquad y_t = A_{ot}y_t + A_{1t}y_{t-1} + \ldots + A_{mt}y_{t-m} + C_{ot}x_t + \ldots + C_{nt}x_{t-n} + b_t + u_t,$$

$t=1,\ldots,T$, where y_t is the $p\times 1$ vector of endogenous variables at time t, x_t the $s\times 1$ vector of exogenous variables, $A_{jt}\in\mathbb{R}_{p,p},(j=o,\ldots,m)$, $C_{jt}\in\mathbb{R}_{p,s}$ $(j=o,\ldots,n)$ represent the matrices of structural parameters for the endogenous and predetermined variables, respectively.The stochastic properties of the $p\times 1$ vector of residuals u_t are assumed $E(u_t)=o, E(u_t u_t')=V$, V positive definite. From (1) follows the difference-equation of m-th order (endogenous) and n-th order (predetermined)

$$(2)\qquad y_t = B_{1t}y_{t-1} + \ldots + B_{mt}y_{t-m} + D_{ot}x_t + \ldots + D_{nt}x_{t-n} + d_t + v_t, \text{ with}$$

$B_{jt}:=(I-A_{ot})^{-1}A_{jt}$ $(j=1,\ldots,m)$, $D_{it}:=(I-A_{ot})^{-1}C_{it}$ $(i=o,\ldots,n)$, $d_t:=(I-A_{ot})^{-1}b_t$, $v_t:=(I-A_{ot})^{-1}u_t$. Suppose some of the components in x_t are used as control variables, which we denote with the $\ell\times 1$ vector $x_t^S:=(x_{s_1},\ldots x_{s_\ell})'$, $s_j\in\{1,\ldots,n\}$. Then (2) has the form

$$(3)\qquad y_t = \sum_{j=1}^{m} B_{jt}y_{t-j} + \sum_{i=o}^{n} D_{it}x_{t-i} + \sum_{i=o}^{n} \tilde{D}_{it}x_{t-i}^S + d_t + v_t.$$

In order to have the difference equation (3) only dependent of the current and predetermined endogenous variables and the control variables, we summarize the exogenous non-control variables and the vector d_t of absolute values in the vector $\tilde{d}_t$:

(4) $$y_t = \sum_{j=1}^{m} B_{jt} y_{t-j} + \sum_{i=o}^{n} \tilde{D}_{it} x_{t-i}^{S} + \tilde{d}_t + v_t, \quad t=1,\ldots,T \; .$$

According to Chow (1975), p.153, we write (4) as the system

(5) $$\eta_t = \mathfrak{A}_t \, \eta_{t-1} + \mathfrak{D}_t x_t^S + d_t + v_t$$

of first-order difference equations. η_t is a vector with mp+n components, $\mathfrak{D}_t$ has ℓ columns. Here, the control variables themselves are embedded in the vector of dependent variables, the advantage of such a transformation turns out in an easy algebraic handling within the framework of a specific control-mechanism. The dimension of the system grows very fast, but in practice difference equations of first and second order are mostly used, so that the size of the system remains acceptable.

As it will be demonstrated in the description of the step-by-step interactive forecasting processing, we use the binary incidence matrix J_T of all endogenous and control variables occurring in the structural equations. In reflecting the short-term global preference structure of the policy maker, we assume that for a short period it is sufficient to consider this incidence matrix, multiplied by a weighting factor. Therefore, the only relevant problem is how to determine this weighting factor. Let e_t^i (i=1,...,N) be a vector of control variables at time t, where the policy maker is assumed to have N alternatives representing his experience. Substituting the vector of control variables e_{T+1}^i into (5) results in a vector $a_{T+1}^i = \mathfrak{R}_T(e_{T+1}^i \mid \eta_T, d_T)$ which corresponds to η_{T+1}. Hereby $\mathfrak{R}_T$ denotes the model specification in (5). Note that we directly use the model to generate this vector a_{T+1}^i of 'state' variables (in the terminology of control theory) and not some expectation $a_{T+1}^i = E(\mathfrak{R}_T(e_{T+1}^i) \mid I_T))$ using all known information I_T in T. The reason is that we do not know precisely whether the policy maker uses his total information available to him at time T or only a subset of I_T (complete or partial rationality). Furthermore, because we are not able to specify a preference or utility function in a sufficient precise manner, we only consider the preference value

(6) $$W(a_{T+1}^i) = \alpha \sum_{j=1}^{pm+sn} \frac{1}{|a_{j,T+1}^i - \eta_{j,T}|} + \frac{\alpha}{\gamma(T)\,|P_T - P^*|}$$

as a crude measure corresponding to the i-th alternative. P_T denotes the popularity index of the government at time T, P^* some kind of ideal poularity index characterizing the theoretically possible degree of governmental popularity, and $\gamma(T)$ represents in the form of a concave

weight function the current distance of time between two elections. Without loss of generality we may assume $\alpha=1$, since in the framework of the interactive forecasting principle the range of the values $W(a^1_{T+1}),\ldots,$ $W(a^N_{T+1})$ is of secondary importance. Within the second term on the right side of (6) we implicitly suppose a popularity function with some measurable indicators such as unemployment rate, real per capita GNP (see for example the approach of Fair (1975) for the U.S.). Alternative approaches for the popularity function in the Federal Republic of Germany are due to Frey and Garbers (1972) and Kirchgässner (1976).
As to the government's primary target of winning the next elections, we formulate the ideal popularity index P^* as a minimum amount of agreement to the government's policy which is necessary to attain the outlined target. The closer the government approaches this value the more accentuated is its tendency to take unpopular measures. This is effected by the term $|P_T-P^*|$ in (6). The time-dependent factor $\gamma(t)$ implicitly reflects the assumption that the policy authority is more reluctant to take unpopular measures, the closer the next date of election is. Let t^*_j, t^*_{j+1} be two consecutive election dates, then we define $\gamma(t)$ in the form

$$(7)\qquad \gamma(t)=\frac{K}{1+\exp(-(t-t_j))}\;,\quad t^*_j < t \leq t^*_{j+1}.$$

Thus, $\gamma(t)$ is a periodic function over the whole time axis. The constant K has its effect in strengthening or dampening the time component in the popularity index.
Finally, let us consider

$$(8)\qquad U(a^i_{T+1})=W(a^i_{T+1})+\varepsilon_i,\quad i=1,\ldots,N\;,$$

where $W(a^i_{T+1})$ equals to the right side of (6). For each $i\in\{1,\ldots,N\}$ we determine the choice probability

$$(9)\qquad p_i:=\text{prob}(W(a^i_{T+1})+\varepsilon_i)\geq \max_{j\in\{1,\ldots,N\}}(W(a^j_{T+1})+\varepsilon_j)),\quad (j\neq i),$$

and define the weight matrix for the control loss function by

$$(10)\qquad K_T=p_{i^*}J_T,\quad p_{i^*}=\max_{i\in\{1,\ldots,N\}} p_i,$$

where J_T is the binary incidence matrix of all endogenous and control variables occurring in the structural equations. For practical computations we use the Weibull probabilities

$$(11)\qquad p_i=\frac{\exp(W(a^i_{T+1})-\beta_i)}{\exp(W(a^i_{T+1})-\beta_i)+\exp(-\log(\sum_{j\neq i}\exp(W(a^j_{T+1})-\beta_j)}$$

(see McFadden, Domencich (1975), p. 64).

3. Information gain by interactive one-step control

Let us consider the quadratic preference functional

$\mathcal{P}^{i^*}(\eta_{T+1})=(\eta_{T+1}-a^{i^*}_{T+1})'K_T(\eta_{T+1}-a^{i^*}_{T+1})$, where, according to (11), i^* denotes the most probable alternative of policy strategies. We are now interested in the compatability of the model with the policy maker's strategy $e^{i^*}_{T+1}$. In order to receive some insight into the control mechanism of this bipolar field of actions between policy and model, we assume the economy to react on $e^{i^*}_{T+1}$ in such a way that $e^{i^*}_{T+1}$ is attained best possible:

$$(12) \qquad \min_{x^S_{T+1}} E(\mathcal{P}^{i^*}(\eta_{T+1})) \;\Big|\; \eta_{T+1}=\mathcal{A}_T\,\eta_T+\mathcal{D}_T x^S_{T+1}+d_T+v_T.$$

Take into account that x^S_{T+1} is contained in η_{T+1}. The solution x^S_{T+1} has now to be compared with the components in $e^{i^*}_{T+1}$ in order to be able to give a statement about the compatability of model and policy.

Since we practically have no information on how to form expectations of $\mathcal{P}(\eta_{T+1})$, we suggest to use $\mathcal{P}(\eta_{T+1})$ deterministically, even though we have defined K_T by probabilities. As to the performance of the minimization process in (12), we have two alternatives:

1) We may solve (12) as a problem of optimal control over the known estimation horizon from 1 to T, according to the procedure of Chow (1975), p.157.

2) We perform only one backward and one forward step of the optimal control algorithm and receive a solution of control variables x^S_{T+1} by the linear feed-back rule between control variables and η_T:

$$(13) \qquad x^S_{T+1}=G_T\,\eta_T+g_T,\; G_T=-(\mathcal{D}_T'K_T\,\mathcal{D}_T)^{-1}\,\mathcal{D}_T'K_T\,\mathcal{A}_T,\; g_T=-(\mathcal{D}_T'K_T\,\mathcal{D}_T)^{-1}\cdot(K_T\,d_T-K_Ta^{i^*}_{T+1}).$$

We suggest to proceed as described in 2), since we are especially interested in x^S_{T+1}. This requires only the weak assumption that the structure of the model remains constant in the time interval $[T,T+1]$ only. Having generated x^S_{T+1}, we

a) compare the policy maker's a priori given strategies e^i_{T+1} $(i=1,\dots,N)$ with the computed x^S_{T+1},
b) generate all endogenous variables in T+1 using x^S_{T+1} in (13),
c) present both x^{T+1}_S and η_{T+1} to the decision authority (model reaction to policy decision),
d) evaluate and interprete the deviations between the control components in x^S_{T+1} and the 'corner variables' e^i_{T+1}, and between the generated η_{T+1} and η_T (deviation in model reaction).

This initiates an adaptation process between model builder and policy authority. Since we may consider η_{T+1} as a kind of pseudo-observation, we assume the ex ante developing process of the economic structure to 'forget', by omitting the first observation of our original estimation

horizon. By estimating the model in (5) over the ex post time domain $\{2,\ldots,T+1\}$ we produce new estimators for $\mathcal{O}\!\mathcal{l},\mathcal{D}$ and d which in most cases are slightly different from the original ones. The structural change from time date T to T+1 is explicitly given by the deviation of the estimated coefficient matrices in T and T+1. We continue this procedure in the same way in order to generate η_{T+2}, η_{T+3}, etc.
After having adapted some (or all) of the policy maker's control components in e^i_{T+1}, we may repeat the following steps:
1) Determine $a_{T+1}=\mathcal{R}_T(e^*_{T+1})$ for the revised strategy e^*_{T+1},
2) determine the maximal probability in (11),
3) define K_T according to (1o),
4) find a further solution within the control step in (12) and (13).

This repeated execution of steps 1 to 4 defines a principle of interaction issuing an increase in information due to alteration. Since it is totally unrealistic to assume optimality of the model specification and policy strategy a priori, an interactive procedure is the most promising possibility to attain new information. Needless to say, that a Pareto 'point' η^i_{T+1} , $i\in\{1,\ldots,N\}$, reflecting precisely the conception of the two players, is probably not attainable, but rather a certain Pareto 'domain' which constitutes a kind of satisfactory compromise. Without an exact delimitation of such a compromise, sufficient agreement on e_{T+1} and η_{T+1} is found respectively, when the amount of information gain - generated according to the procedure above - is sufficiently large.

4. References

Chow G.C. (1975): Analysis and Control of Dynamic Economic Systems. Wiley: New York, London, Sydney, Toronto.

Fair R.C. (1975): On controlling the economy to win elections. Cowles Foundation Disc.paper no. 397.

Frey B.S., Garbers H. (1972): Politometrics - On measurement in political economy. Political Studies 19, 316-32o.

Friedman B.M. (1979): Optimal expectations and the extreme information assumptions of 'rational expectations' macro-models. J. of Monetary Economics 5, 23-41.

Kirchgässner G. (1977): Wirtschaftslage und Wählerverhalten. Politische Vierteljahresschrift 18, 51o-536.

McFadden D., Domencich T.A. (1975): Urban Travel Demand. North-Holland American Elsevier, Inc., New York.

QUALITY ASSURANCE SPECIFICATIONS FOR TIME DEPENDENT AEROMETRIC DATA

Turkan K. Gardenier
George Washington University
Washington, D.C. 20052

I. INTRODUCTION

Monitoring environmental measurements to detect temporary and permanent trend changes poses special statistical queries. Most environmental data represent averages of successive data and are autocorrelated. The present paper introduces the parametric characteristics between averaging time, degree of autocorrelation and, in turn, the variance of the statistical distribution. An application to a set of environmental data is presented through the use of an autoregressive approach with heuristic modifications in order to classify data into temporary or permanent change indices.

II. THE IMPACT OF AUTOCORRELATION

Time dependence of successive values is important in modeling of aerometric data, particularly in continuous monitors. The shorter the time interval between successive measurements, the higher the time dependence or autocorrelation.

Figure 1 shows two versions of a hypothetical data set following a lognormal distribution: one where successive observations are independently and identically distributed, and another after impact of autocorrelation. The figure demonstrates what the value of a standard or threshold limit would be if determined solely by the probability that emissions exceed the standard. If we were to set the probability of exceedence at .1, the limit for sulfur dioxide (SO_2) would need to be raised, allowing for the sources to emit more SO_2. Thus, for similar averaging times of successive data, autocorrelation would increase the variance and thus demand a higher value for the standard. If the standard is defined in terms of pounds of SO_2 emitted into the air, this would mean more air pollution.

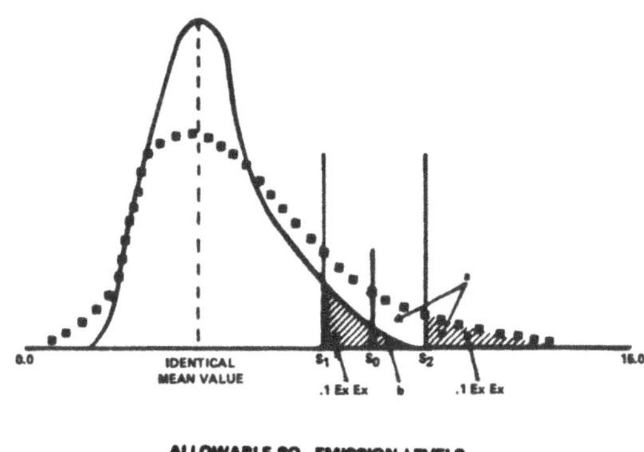

Figure 1. Example of Impact of Autocorrelation in Setting Standards

Note that both versions of the data set have the same mean value or average emissions. An expected exceedences approach to standard setting sets the probability of excess emissions at a specific level (such as .1). Therefore, the value of the standard fluctuates according to the value of the variance (geometric standard deviation). If, on the other hand, we set an absolute standard, there will be more expected violations because of the impact of autocorrelation.

This issue became of concern when a 30-day moving average was being considered as a standard for sulfur scrubber efficiency in flue gas desulfurization units. Scrubber efficiency is determined by the percent of sulfur removed. The Utility Air Regulatory Group, the Edison Electric Institute, and the National Rural Electric Cooperative Association debated whether the 90% scrubbing efficiency requirement for sulfur scrubbers for a 30-day moving average was consistent with the previous 24-hour averaging standard (1). Claiming that autocorrelation increases variance, they demanded a lower scrubber efficiency requirement. Data simulations and analytic solutions explored the impact of changes in standard deviation in the expected number of times a specific level is exceeded during a year.

Under a threshold-oriented definition of standard setting, an exceedence corresponds to those levels of emission which impact human health, or to the probability of exceeding the standard which has been established. This means extra pollution, incremental to the limit established in setting the standard. Regardless of whether the threshold level is chosen to protect the most sensitive subgroup (such as asthmatics) or the general public, each exceedence corresponds to the probability of either a risk-eliciting event or a noncompliance penalty. In setting regulatory standards, therefore, it is essential that our simulation models of exceedence define the statistical parameters which affect the expected number of exceedences.

III. AVERAGING TIME AND AUTOCORRELATION EFFECT UPON VARIANCE AND STANDARD DEVIATION

The longer the time interval for averaging, the smoother the function if plotted over time. Averaging time, in turn, impacts the coefficient of variation, i.e., the standard deviation relative to the mean (RSD). PEDCO Environmental, Inc. (5, 6) and Versar, Inc. (7, 8) have reported RSDs for sulfur variability in lbs/10^6 Btu for coal using averaging periods of 3 hours, 24 hours, 1 week, and 1 month. Both find approximately a 2:1 ratio in the reduction of RSD as the averaging time lengthens from one day to a month. Results are shown in Table 1.

Table 1. Averaging Time and Changes in Relative Standard Deviation (RSD) in Sulfur Variability

Averaging Period	Coal Burn (Tons)	Relative Standard Deviation: Percent By Weight	Relative Standard Deviation: Lbs/10^6 Btu
3 hour	375	.20	.21
24 hour	3,000	.17	.18
1 week	21,000	.14	.14
1 month	90,000	.09	.09
1 year	1 million	.03	.03

Kendall and Stuart (9) have presented the data shown in Table 2 for the percentage variance reduction with a moving average of extent k iterated Q times. That is, if we were to average three numbers, the variance would be 1/3 or .33 of the initial observations; if seven were averaged, the expected variance would shrink by 1/2 to .14. Reading across columns, we observe how much incremental reduction in variance may be expected as we iterate the moving average process over a number of sequences. For example, we find a further decrease by approximately 1/2 as we repeat the averaging four times.

Table 2. Percent Variance Reduction With a Moving Average of Extent K: K Iterated Q Times

		Q				
		1	2	3	4	5
k	3	0.33	0.23	0.19	0.17	0.15
	4	0.25	0.17	0.14	0.12	0.11
	5	0.20	0.14	0.11	0.10	0.09
	6	0.17	0.11	0.09	0.08	0.07
	7	0.14	0.10	0.08	0.07	0.06

In reality, the issues addressed in Sections II and III have opposing influences upon the variance and, in turn, expected exceedences. While averaging reduces variance, autocorrelation generated by averaging increases variance. Averaging over a longer period would make the standard more restrictive (a lower required emission level) if there were no autocorrelation. The presence of autocorrelation makes the standard less restrictive. Do the two factors balance each other? Analytical or simulation-oriented solutions are presented below to answer this query.

Switzer (10) has formulated the following parametric relationships between averaging time and autocorrelations. Assuming that C_1, C_2 ..., C_j are successive hourly threshold values and $D_1, D_2, \ldots, D_j$ denote the corresponding actual hourly emissions, the probability that a J-hour exceedence will occur is:

$$\frac{1}{J}\sum_{i=1}^{J}\frac{D_i}{C_i} > 1.$$

If we replace all the individual C_i values by their average $\bar{C}_j$, exceedences would be of the form $\bar{D}_j > \bar{C}_j$. We can then estimate exceedence probabilities and rates by the same method for single-hour exceedences.

However, while the mean of the distribution would remain unchanged, the variance would decrease with averaging time. If S^2 is the variance of one-hour emissions, the variance of the distribution of J-hour averages is:

$$S_J^2 = \frac{S^2}{J}\left[1 + (J-1)\bar{r}_J\right]$$

where $\bar{r}_J$ is the average autocorrelation between hourly emissions during an average period of J hours. The autocorrelations of the emission time series needs to be specified, in order to calculate exceedence probabilities.

As an example, consider a continuous-time concentration process with exponentially decaying autocorrelation function

$$\tilde{\rho}(\Delta) = e^{-\alpha\Delta}$$

If this process were observed at evenly spaced (hourly) time points, it would be a first-order autoregressive process or time series. Suppose for l g Δ = 1 hour, the autocorrelation of instantaneous values of concentration is

$$\rho(1) = 0.90 = e^{-\alpha}$$

Then the variance of one-hour averages is 97% of the variance of the instantaneous values, using formulations developed by Switzer (10). The variance of three-hour averages is 90% of the variance of the instantaneous values; the corresponding result for eight-hour averages is 80%.

Using the same lag one-hour autocorrelation of instantaneous values (0.90), the lag one-hour autocorrelation for hourly averages is 0.93, the lag one-hour correlation for three-hour (overlapping) averages is 0.76, the autocorrelation between successive nonoverlapping three-hour averages is 0.82, and the autocorrelation between successive nonoverlapping eight-hour averages is 0.46.

IV. AUTOREGRESSIVE APPROACH TO DETECTING EXCEEDENCE

We may use the concept of autocorrelation in devising special monitoring and control techniques for autoregressive processes in developing guidelines and specifications to detect changes in environmental conditions.

Stoodley and Mirnia (11) analyzed several time-indexed variables using Box-Jenkins methods. Four of the five series were found to conform to the ARIMA model.

$$z_t = z_{t-1} + a_t - \theta a_{t-1}$$

where a_t is white noise with variance σa^2, and 0 is a parameter. The forecasting for this model is done by simple exponential smoothing,

$$\hat{z}_{t+1} = \theta \hat{z}_t + (1-\theta)_{z_t}$$

or in terms of the error

$$e_t = z_t - \hat{z}_t$$
$$\hat{z}_{t+1} = \hat{z}_t + (1-\theta)e_t$$

If changes occur in level and slope, the ARIMA (0,2,2) model would be appropriate.

Greenberg (12) has constructed an example showing the application of the exponential smoothing model advocated by Stoodley and Mirnia to a set of environmental data. Results are shown in Table 3.

Table 3. Sample Calculations for ARIMA-Based Categorizations of Exceedences

X_t	X_t	e_t	e_t^2	D_t	F_t	X_t	X_t	e_t	e_t^2	$\ddot{X}_t$ Corr	e_t Corr	D_t	F_t
217				55	- 55	383	421.0	- 38.0	1444.00	319.8	- 63.2	11.8	- 138.2
207	217.0	- 10.0	100.00	85	- 65	327	386.8	- 59.8	3576.04	376.7	- 49.7	81.5	- 25.3
190	208.0	- 18.0	324.00	93	- 57	270	333.0	- 63.0	3966.48	332.0	- 62.0	137.0	- 13.0
175	191.8	- 16.8	282.24	91.8	- 58.2	270	276.3	- 6.3	39.66	276.2	- 6.2	81.2	- 68.8
170	176.7	- 6.7	44.62	81.7	- 68.3	270	270.6	- 0.6	0.40	271.2	- 1.2	76.2	- 73.8
175	170.7	4.3	18.77	70.7	- 79.3	270	270.1	- 0.1	0.00	270.1	- 0.1	75.1	- 74.9
176	174.6	1.4	2.05	73.6	- 76.4	225	270.0	- 45.0	2025.57	270.0	- 45.0	120.0	- 30.0
180	175.9	4.1	17.17	70.9	- 79.1	225	229.5	- 4.5	20.26	229.5	- 4.5	79.5	- 45.5
175	179.6	- 4.6	21.03	79.6	- 70.4	203	225.5	- 22.5	504.01	225.5	- 22.5	97.5	- 43.0
180	175.5	4.5	20.62	70.5	- 79.5	180	205.3	- 25.3	637.31	205.3	- 25.3	100.3	- 37.7
182	179.5	2.5	6.02	72.5	- 77.5	180	182.5	- 2.5	6.37	182.5	- 2.5	77.5	- 55.2
180	181.8	- 1.8	3.08	76.8	- 73.2	192	180.2	11.8	138.00	180.2	11.8	63.2	- 86.6
179	180.2	- 1.2	1.38	76.2	- 73.8	203	190.8	12.2	148.22	190.8	12.2	62.8	- 87.2
185	179.1	5.9	34.60	69.1	- 80.9		201.8						
186	184.4	1.6	2.52	73.4	- 76.6								
193	185.8	7.2	51.25	67.8	- 82.2			- 17.13	58634.77				
204	192.3	11.7	137.26	63.3	- 86.7								
206	202.8	3.2	10.06	71.8	- 78.2								
196	205.7	- 9.7	93.76	84.7	- 65.3								
196	197.0	- 1.0	0.94	76.0	- 74.0								
196	196.1	- 0.1	0.01	75.1	- 74.9								
187	196.0	- 9.0	81.17	84.0	- 66.0								
193	187.9	5.1	26.00	69.9	- 80.1								
196	192.5	3.5	12.32	71.5	- 78.5								
179	195.7	- 16.7	278.89	91.7	- 58.3								
173	180.7	- 7.7	58.83	82.7	- 67.3								
155	173.8	- 18.8	352.20	93.8	- 56.2								
162	156.9	5.1	26.25	69.9	- 80.1								
270	161.5	108.5	11774.93	- 33.5*	- 183.5								
493	259.2	179.9	32346.47	- 104 **	- 254								
(326.5)		(67.3)		(55)	(- 55)								

$n = 42 \quad \bar{e} = \frac{-17.13}{42} = -0.41$

$Var\ e_t = \frac{58634.77}{41} - \frac{42}{41}(-0.41)^2 = 1429.96$

$\sigma = \sqrt{1429.96} = 37.8$

$L_0 = 56 \qquad C = 20 \qquad n_s = 5 \qquad m = 37.8$

* NOT SLOPE CHANGE (ONLY 2 + ERRORS)
NOT SHIFT CHANGE (NEXT $> \delta_m$ BUT NEXT HAS OPPOSITE SIGN)
NOT TRANSIENT ($3\sigma = 113.4$): FALSE ALARM

** NOT SLOPE CHANGE
NOT SHIFT CHANGE
TRANSIENT $\hat{X}_t = (270+383)/2 + 326.5$

A control chart applied to the exponential smoothing model would be a chart of e_j with mean zero and standard deviation σ. Unfortunately, little can be said about the distribution of the e_j. The classical exponential smoothing model considers the e_j to be independent and identically distributed. In practice this is rarely so. A temporary effect which causes an increase in X_j in one period will continue to be reflected in Y_{j+1} for several future periods. Notice that the equal number of positive and negative erros in Table 3 array themselves in a series of runs, or successive similar signs in error. For this reason, cumulative sum (CUSUM) charts are almost always used with exponential smoothing models. CUSUM charts are less dependent upon distributional assumptions than are the other types of monitoring techniques. In constructing CUSUM charts, one cumulates the difference between the successive observations and the mean

$$S_n = \sum_{i=1}^{n} (X_i - \epsilon(X_i))$$

The Stoodley and Mirnia method of applying cumulative sum control techniques to an exponentially smoothed series is based on the concept that an out-of-control indication can be due to one of four causes, as shown in Figure 2:

- Change in slope
- Step change
- Transient change
- False alarm.

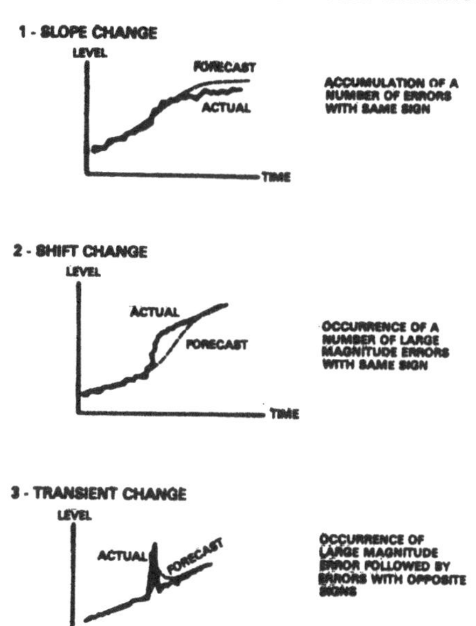

Figure 2. Three Modes of "Out of Control" Behavior Representation

Lack of control is signaled by either the quantity D_t or F_t , at t=0. For specified positive parameters $\underline{L_0}$ and $\underline{c}$,

$$D_t = \min\left(D_{t-1}, L_0\right) + c - e_t \,, \quad D_0 = L_0$$

$$F_t = \max\left(F_{t-1}, L_0\right) - c - e_t \,, \quad F_0 = -L_0$$

Lack of control is indicated if D_t shows negative or F_t shows positive. The values of L_0 and $\underline{c}$ are set by first determining the magnitude of a step change that it is important to detect, and then calculating the L_0 and $\underline{c}$ values that will detect this with near certainty within three or four time units of the shift.

Initially, the slope is taken to be zero, i.e., no long-term up or downward trend is assumed. If an out-of-control indication is the cumulation of a long sequence of residuals with the same sign, it is concluded that a slope is now present in the data. The author recommends that a sequence of five consecutive residuals with the same sign is sufficient to conclude the change in slope.

Once the change in slope has been signaled by the control chart, the smoothing equation must be corrected. For example, assume that the out-of-control condition is indicated following observation X_j, and there are c consecutive residuals with the same sign. The new slope is estimated from the last C+1th observation. The least squares estimate of the slope can be shown to be

$$B_j = \left(12 \sum_{i=0}^{c} X_{j-c+i} - 6c \sum_{i=0}^{c} X_{j-c+i}\right) / c(c+1)(c+2)$$

The new value of Y_j is

$$Y_j = Y_{j-c} + cB_j$$

The prediction for the next observation is $Y_{j+1} + B_{j+1}$ with $B_{j+1} = B_j$. Subsequently,

$$Y_{j+i} = AX_{j+i-1} + (1-A)(Y_{j+i-1}) + B_{j+1}$$

$$e_{j+i} = X_{j+i} - (Y_{j+i} + B_{j+i})$$

with $B_{j+1} = B_{j+i-1}$ until the next change of slope is indicated.

A step change is an abrupt shift in the process to a new level where it remains for a period of time. This is contrasted with a transient change, where the shift to the new level is only apparent for one or two observations before the process reverts to its former level. Both the step change and the transient change are indicated by an out-of-control condition in which the residual has a magnitude in excess of three standard deviations. Thus, following observation S_j the residual is

$$e_j = X_j - (Y_j + B_j)$$

If the CUSUM chart shows an out-of-control situation and no slope change is indicated, and if

$$|e_j| > 3\sqrt{\text{var}(e_j)}$$

either a step change or transient change is suspected. To decide which has occurred, two additional observations are taken. Their residuals are based on the Y_j values calculated prior to the out-of-control indication, that is

$$e_{j+1} = X_{j+1} - (Y_j + 2B_j)$$
$$e_{j+2} = X_{j+2} - (Y_j - 3B_j)$$

If both e_{j+1} and e_{j+2} have the same sign as e_j, and if their magnitudes both exceeded two standard deviations, then a step change is concluded to have occurred. The control procedure is continued with

$$Y_{j+2} = (1/3)(X_j + X_{j+1} + X_{j+2}) + B_j$$

If both ej+1 and ej+2 do not meet the conditions for the step change, the shift is concluded to be transient. The exponential smoothing is modified by replacing X_j with the average of the previous and following values (1/2)(Xj-1 + Xj+1), and recalculating the subsequent Y_{j+1}, e_{j+1}, and e_{j+2} values.

To illustrate, assume that out-of-control is signaled by observation z_t. A step change is concluded to have occurred if z_t, z_{t+1} and z_{t+2} generate forecast errors l_t, l_{t+1}, and l_{t+2} whose magnitudes exceed some prespecified $\delta \leq 3\sigma_\alpha$ in the same direction. This is similar to saying, "look for an assignable cause whenever three consecutive points fall outside the same one-sigma limit (for $\delta = \delta_\alpha$)." The L_0 and c values would be chosen to yield a high probability of this occurring if the shift magnitude was of a value important to detect.

A change in slope is indicated when the point going out of control is at the end of a sequence of N_s (or more) consecutive errors with the same sign. The author suggests $N_s = 5$ as giving good results. If this change in slope is detected, then the last k errors ($k \geq N_s$) will have the same sign. The new slope (b_t) is obtained from a least squares fit to their k z values.

If neither a step change nor a slope change is indicated, then the effect is concluded to be transient when the error term e_t has magnitude larger than $3\sigma_\alpha$. If neither a step change, slope change, nor transient change is indicated, then a false alarm is concluded. In all cases, t is set back to zero to calculate D_t and F_t. In addition, the following calculations are made:

- For slope change, calculate $\hat{b}_t$ as described earlier
- For step change, calculate a new m_t:

$$m_t = (1/3)(z_t + z_{t+1} + z_{t+2})$$

- For transient change, replace z_t by a value that will cause it to die out of the forecast more rapidly:

$$z_t = (1/2)(z_{t-1} + z_{t+1})$$

Figure 3 is constructed to show control charts and typological classification of exceedence using the data of Table 3 and the methods illustrated above for a series of daily sulfur dioxide emissions.

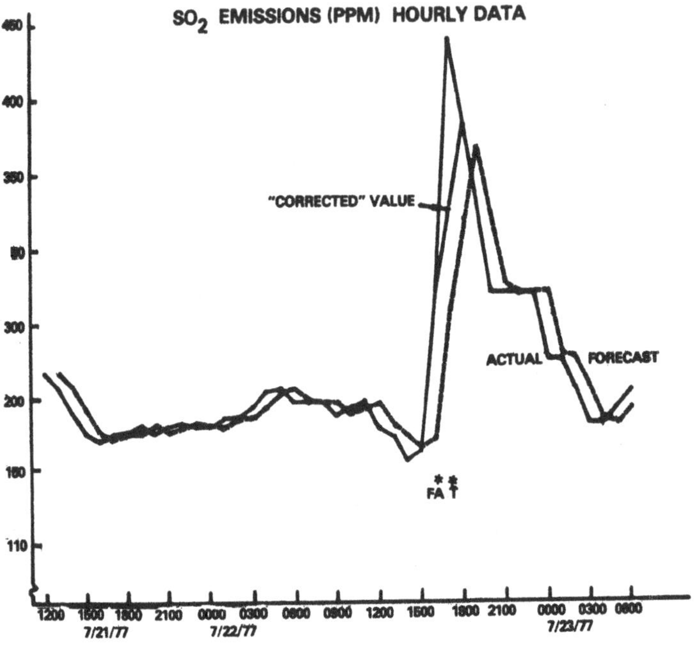

Figure 3. Schematic Classification and Sample Update of Actual/Forecast Values

REFERENCES

1. Entropy Environmentalists, Inc., A Statistical Evaluation of the EPA FDG System Data Base Included in the Subpart DA NSPS Docket. Prepared for Utility Air Regulatory Group and Edison Electric Institute, July, 1979.

2. Vector Research, Inc., Analysis of FGD System Efficiency Based Upon Existing Utility Boiler Data. EPA Report Number QAQPS-78-1, VI-B-13, November, 1979.

3. Foster Associates, Inc., A Statistical Study of Coal Sulfur Variability and Related Factors. EPA Contract No. 68-02-2592, July, 1979.

4. Larsen, R.I., "A New Mathematical Model of Air Pollutant Concentration Averaging Time and Frequency," J. Air Pollution Control Association, 1969, (18) 24.

5. PEDCO Environmental, Inc., Preliminary Evaluation of Sulfur Variability in Low-Sulfur Coals from Selected Mines. EPA Contract No. 68-02-1321, December 1979.

6. PEDCO Environmental, Inc., Statistical and Engineering Evaluation of Continuous Emissions Monitoring (CEM) NO_x and SO_2 Data from FGD-Controlled Electric Utility Steam Generating Units. EPA Contract No. 68-02-1321, December, 1979.

7. VERSAR, Inc., SO_2 Emission Reduction Data from Commercial Physical Coal Cleaning Plants and Analysis of Product Sulfur Variability. Prepared for U.S. Environmental Protection Agency, January, 1979.

8. VERSAR, Inc., Effect of Physical Coal Cleaning Upon Sulfur Variability. Prepared for U.S. Environmental Protection Agency, January, 1979.

9. Kendall, M.G., Stuart, A., The Advanced Theory of Statistics, Vol. 3, New York: Hafner Publishing Co., 1968, p. 3.

10. Switzer, P., Internal Memorandum to the U.S. Environmental Protection Agency, 1981.

11. Stoodley, K.D.C., and Mirnia, M., "The Automatic Detection of Transients, Step Changes and Slope Changes in the Monitoring of Medical Time Series," The Statistician: 1979 (20), 163-170.

12. Greenberg, I., "Statistical Quality Control of Data Bases." Unpublished internal document, U.S. Environmental Protection Agency, Washington, D.C., 1980.

LABOR MARKET IMPLICATIONS OF TECHNICAL CHANGE IN A MULTI-REGIONAL MULTI-SECTORAL SYSTEM

Agostino La Bella

Institute for System Analysis
and Computer Science of the Italian
National Research Council
Via Buonarroti 12 - 00185 Roma - Italy

ABSTRACT

This paper is devoted to the analysis of the effects of technical change on labor market dynamics, performed on the basis of a mathematical setting general enough to represent a wide variety of real situations, allowing also for a spatial and sectorial disaggregation. A conceptual frame for this work is provided both by the theory of dynamic equilibrium and by the mathematics of convex structures.

1. INTRODUCTION

This paper contributes to the analysis of the relationships between population and the economics by studying the effects of structural changes in the production process on labor market dynamics at a disaggregated spatial-sectoral level. The analysis is performed on the basis of a linear dynamic system of equations with time - varying parameters.

In a previous paper (Caravani and La Bella, 1980) a similar approach has been used to investigate equilibrium of labor market and production under different assumptions regarding structural dependencies and causal relationships prevailing among sub-systems. The main purpose of the study was that of assessing how the overall economic growth on a balanced path is affected by the individual behaviour of system components.

In this paper, production dynamics is assumed to drive labor market dynamics. Technology change is introduced in the form of a time - varying convex combination of a set of Leontief matrices, each representing a technology option arising at a given time t. Labor market, in its turn, is modeled as a linear dynamic system, whose equilibrium is governed both by prices and quantities in a flexible way.

The two models for production and labor market are introduced in the next two sections. Then, in section 4, an analysis of dynamic properties of production, is performed, showing that in presence of technical change the notion of balanced growth path can be generalised into that of balanced growth cone.

Labor market consequences of these structural changes in the production process are then investigated in section 5, proving some invariance properties of asymptotic behavior of labor market.

The main conclusions of this work are then summarized in section 6.

2. PRODUCTION AND TECHNOLOGY

In the framework of a Leontief system, consider the dynamic behavior of sectoral production under a scheme of varying technological coefficients and final demand.

Under the assumption of one period lag between expenditure and production, the Leontief equations take on the form:

$$y(t+1) = T(t)y(t) + u(t) \tag{1}$$

where y is an $N = M \times K$ vector for an M-sector, K-region economy, T is an $N \times N$ time varying matrix describing technology and investment pattern at time t, and u(t) is the vector of final demand. As it is well known, matrix T combines input-output and capital coefficients matrices: although these two matrices are usually kept separate in input-output models, no such distinction is needed for the present purposes (for a formal derivation of (1), see Livesey, 1973).

We assume here that a finite number of competing technologies are available at each period of time; some of them would be "mature" technologies, and some others would be new and, in different measures, more profitable than the old ones.

Since the process of adoption of a new technology depends on micro-economic behavior of individual firms, it is reasonable to assume that technological substitution will be a gradual process, and therefore that, at each period, more production technologies will co-exist. Hence the time path followed by T(t) is regarded here as an intertemporal decision process over a finite number of technological options $T_1, T_2, \ldots, T_n$, and the technology of production at time t is represented as the convex combination

$$T(t) = \sum_{i=1}^{M} \nu_i(t)\, T_i \tag{2}$$

$$\sum_{i=1}^{M} \nu_i(t) = 1 \tag{3}$$

$$\nu_i(t) \geq 0 \qquad \forall i,t \tag{4}$$

where the decision variables $\nu_i(t)$ represent the level at which the corresponding technology is being adopted.

3. LABOR MARKET DINAMICS

The variables of our labor market model are non-negative N-vectors indexed by sector and region. Let

$$w(t),\ l(t),\ e(t),\ y(t) \qquad t = 1,2,\ldots..$$

be wage, labor supply, employment and output vectors at time t.

The following linear system is assumed to describe the dynamics of the labor market:

$$w(t+1) = Aw(t) + Bl(t) + [Cy(t) - \beta Bl(t)] \tag{5}$$

$$l(t+1) = El(t) + \beta Dw(t) + \gamma Fe(t) \tag{6}$$

$$e(t+1) = Pe(t) + (1-\gamma)Hl(t) + \beta Gw(t) + Qy(t) \tag{7}$$

where A, B, C, D, E, F, G, H, P, Q are non negative square parameter matrices of order N, and the two parameters β and γ have been introduced in order to represent the prevailing direction of the coupling between each pair at variables (see Caravani and La Bella, 1980, for a full discussion of the model structure).

A compact representation of the labor market model is provided by the following linear dynamic system

$$x(t+1) = Sx(t) + Zy(t) \tag{8}$$

where

$$S = \begin{bmatrix} A & (1-\beta)B & 0 \\ \beta D & E & \gamma F \\ \beta G & (1-\gamma)H & P \end{bmatrix} \qquad z' = \left[C' \;\vdots\; 0' \;\vdots\; Q' \right] \tag{9}$$

4. ANALYSIS OF PRODUCTION

In this section we shall investigate the balanced growth properties of the linear dynamic system (1). We recall firstly that a sequence of vectors y (t) is said to trace out a balanced growth path if the ratios of its components remain unchanged over time. Therefore such a sequence moves on a certain ray emanating from the origin, and having non-negative direction cosines.

For a linear dynamic stationary system it can be proved that, under appropriate growth of the forcing term (see Nikaido, 1968 and 1972), the system state will converge to the balanced growth ray. The relevant aspects in studying this kind of dynamic equilibrium regard the normative significance that balanced growth assumes when a "turnpike proposition" holds (see Sanmelson 1960, 1965), i.e. when in moving from the initial state to a desired economic state it can be proved that the most efficient path lays close to a given balanced growth ray at intermediate periods on the way to the final goal.

Here we tackle the problem of estabishing some balanced groth properties for the system (1), where T is a time-varying matrix. This is, as opposed to the stationary case, a relatively unexplored field.

A first result in the investigation of dynamic equilibrium for a non-stationary linear system is provided by the following theorem.

THEOREM 1. *Consider system (1), with the specifications (2) - (4).*

Assume:

(i) $u(t) = \rho^t u$

(ii) $\rho > \max \lambda\, (T_i + T_i')/2$

(iii) there exist at least N + 1 *technology options, and a set* M *of indeces including at least* N + 1 *elements, such that, for some* ρ *satisfying (ii), all the vectors* y_i *generated by taking*

$$y_i = [\rho I - T_i]^{-1} .\ u \qquad \forall\ i \in M \tag{10}$$

are affinely independent.

Then, production grows at a constant rate ρ, *and its long-run trajectory belongs to the convex cone*

$$K(Y) = \{\ \sum \alpha_i Y_i;\ Y_i \in Y,\ \alpha \geq 0\} \tag{11}$$

where

$$Y = \{Y_i : Y_i = [\rho I - T_i]^{-1} .\ u, \qquad \forall\ i\}$$

PROOF. We first define the vector $\nu' = [\nu_1, \ldots, \nu_n]$ of technology mix coefficients and the set

$$\Gamma = \{\nu |\ \sum_i \nu_i = 1; \quad \nu_i \geq 0\}$$

For any technology mix $\nu \in \Gamma$ it is possible to identify a balanced growth path which will prevail in the long run, under a forcing term of the type:

$$u(t) = \rho^t .\ u \qquad\qquad u \geq 0,\ \rho > 0$$

Letting $y(t) = \rho^t y(\nu)$ be the balanced path associated with ν, and substituting in the state equation we get

$$\rho y(\nu) = \sum \nu_i T_i y(\nu) + u \tag{12}$$

Equations (12) have a non-negative solution $y(\nu)$ only for the values of ρ making $[\rho I - \sum_i \nu_i T_i]$ non-negatively invertible, i.e. for

$$\rho > \lambda (\sum_i \nu_i T_i) \tag{13}$$

From (Caravani and La Bella, 1980), we know that

$$\lambda(\sum_i \nu_i T_i) \leq \sum_i \nu_i \lambda(T_i + T_i')/2 \tag{14}$$

and we can ensure that (13) holds for any convex combination of the single technology options by making

$$\rho > \max_i \lambda(T_i + T_i')/2 \geq \sum_i \nu_i \lambda(T_i + T_i')/2 \geq \lambda (\sum_i \nu_i T_i)$$

Therefore condition (i) ensures that, for any $\nu \in \Gamma$, a balanced growth path exists, given by

$$y(\nu) = [\rho I - \sum_i \nu_i T_i]^{-1} . \ u$$

Then, the set of all feasible balanced growth path corresponding to the various technology mix can be represented as

$$\tilde{Y} = \{Y(\nu) | \ [\rho I - \sum_i \nu_i T_i] y(\nu) = u, \ \nu \in \Gamma\} \tag{15}$$

Let now $C(\tilde{Y})$ be the convex hull of $\tilde{Y}$. From Carathéodory's Theorem (Rockafellar, 1970) we know that any point of $C(\tilde{Y})$ can be expressed as a convex combination of N + 1 or fewer affinely independent points belonging to $C(\tilde{Y})$.

Since

$$\{y_i, \ i \in M\} \subset \tilde{Y} \subset C(\tilde{Y})$$

we get

$$C(\tilde{Y}) = \{y : y = \sum_i \alpha_i y_i, \ \alpha_i \geq 0, \ \ i \in M, \ \sum_{i \in M} \alpha_i = 1\}$$

and therefore

$$C(\tilde{Y}) \subset \{y : y = \sum_i \alpha_i y_i, \ \alpha_i \geq 0\} = K(Y)$$

which concludes the proof.

REMARK. The above discussion allows us to generalize the notion of balanced growth path into that of balanced growth cone. This generalization can be useful in studying the dynamic equilibrium properties of economic systems with time-varying parameters.

REMARK 2. Theorem 1 states that the long-run trajectory of production belongs to the convex cone spanned by the balanced growth rays associated with the single technological options. Up to a certain extent, the theorem therefore establishes a "property of separation" regarding the effect of technological change on production growth paths.

5. ANALYSIS OF LABOR MARKET

Very little theoretical work has been done on the consequences that technical changes in production may, ceteris paribus, have on the dynamics of labor market, even if this subject is of utmost importance for designing industrial policies. As a matter of fact labor demand and, more generally, the structure of labor market, depend not only on the levels, but also on the technology of production.

In this section we tackle the problem of relating labor market and production from the dynamic equilibrium perspective. We know that, when technology changes over time, y may range within the convex come $K(Y)$. In this case, the resulting properties for long-run behaviour of labor market are established by the following:

THEOREM 2. *Consider system (8), with*

$$y(t) = \rho^t \cdot y \quad , \qquad\qquad y \in K(Y)$$

and assume $\rho > \lambda(S)$. *Then labor market variables grow at a constant rate* ρ, *and their long-run trajectory belongs to the convex cone* $\tilde{K}$ *spanned by the column vectors of* $[\rho I - S]^{-1}$.

PROOF. Since $\rho > \lambda(S)$, matrix $[\rho I - S]$ is non-negatively invertible, and we can define the set X of long-run trajectories for labor market as the convex cone:

$$X = \{x : x = [\rho I - S]^{-1} \cdot y, \quad y \in K(Y)\}$$

Therefore, since

$$\tilde{K} = \{x : x = [\rho I - S]^{-1} \cdot y, \quad y \geq 0\}$$

it follows $X \subseteq \tilde{K}$, which concludes the proof.

The interesting result pointed out in theorem 2 is that, when technology changes over time, the long-run trajectories for labor market may vary within a convex cone, whose boundary depends on the set of technology options. However, whatever choice is made regarding the set of technology options, we can identify a larger cone, depending only on structural labor market parameters, which will always include those trajectories. This establishes a kind of "invariance property" of labor market with respect to technical changes.

6. CONCLUSIONS

In this paper a joint analysis of production and labor market has been presented, with particular emphasis on dynamic equilibrium properties. The discussion has been based on a model which allows structural changes in the production process, in the form of time-varying technological coefficients.

The main conclusions of the work, which opens the ground for further investigation, can be summarized as follows:

1. The notion of balanced growth path can be generalized into that of balanced growth cone, to account for economic systems with time varying parameters.
2. The effects of adapting a new technology option on the overall growth process can be identified and analyzed.
3. The labor market dynamics is conditioned by the technology of production; however, it can be proved that there exists a portion of the state space whose points are not reachable by means of changing levels or technology of production. Steering the system into that region of the state space requires some structural changes in the internal functionning of labor market also.

REFERENCES

N.D. CANON, G.D. CULLUM Jr., E. POLAK (1970): Theory of Optimal Control and Mathematical Programming. McGraw-Hill, New York.

P. CARAVANI, A. LA BELLA (1980): Labor - Production Equilibrium in a Multi-sectoral Multi-Regional System. Papers Regional Science, v. 47.

D.A. LIVESEY (1973): The Singularity Problem in the Dynamic Input-Output Model. Int. J. System Sci., v. 4, 437-440.

H. NIKAIDO (1968): Convex Structure and Economic Theory. Academic Press, New York.

H. NIKAIDO (1972): Introduction to Sets and Mappings in Modern Economics. North-Holland, Amsterdam.

R.T. ROCKAFELLAR (1970): Convex Analysis, Princeton University Press.

F.A. SAMUELSON (1960): Efficient Paths of Capital Accumulation in Terms of the Calculus of Variations.
In Mathematical Methods in the Social Sciences, K.J. Arrow, S. Karlin and P. Suppes eds, Stanford University Press.

P.A. SAMUELSON (1965): A Catenary Turnpike Theorem Involving Consumption and the Golden Rule, Amer. Ec. Rev., v. 55, N. 3.

FUNCTIONAL SENSITIVITY ANALYSIS OF MATHEMATICAL MODELS

D H Martin
National Research Institute for Mathematical Sciences
CSIR, P O Box 395, PRETORIA 0001, South Africa

ABSTRACT

When real dynamical systems or processes are modelled using ordinary differential equations, the question arises of gauging the sensitivity of predictions derived from the model to possible perturbations in various uncertain or assumed expressions or functions which appear in the differential equations. Often, for example, various terms will be assumed to be linear, and sensitivity analysis should then study not only the effects of perturbing the constant of proportionality (parameter sensitivity analysis) but also the effect of abandoning the assumption of linearity (functional sensitivity analysis).

In a recent paper the writer introduced a new and potentially useful framework for such functional sensitivity analysis, in which, however, the entire system of differential equations was considered as being subject to functional perturbations. It has subsequently been realised that this is not always the case, and the present paper extends the methodology to the case of constrained functional perturbations to the model.

1. FUNCTIONAL PERTURBATIONS OF MODELS

When 'real' dynamical systems or processes are modelled using ordinary differential equations

$$\frac{dx}{dt} = X(t,x), \quad t_0 \leqslant t \leqslant t_f, \quad x(t_0) = x_0, \tag{1}$$

where $x \in \mathbb{R}^n$, there is frequently considerable uncertainty regarding the most appropriate choice of the vector function X. It is not only that certain parameters upon which X depends may be poorly known, but also that the *functional form* of X may be arbitrary to some extent. For example, a modeller may assume that certain effects depend linearly, or depend exponentially, upon others and then select values for the corresponding parameters as best he can. At the later stage of sensitivity analysis (in the usual sense of parameter sensitivity analysis; see for example [1] [2] [3]), the sensitivity of predictions to possible errors in these parameters will be revealed, but the sensitivity of predictions to the underlying assumptions of linear or exponential dependence will not be revealed.

As a simple illustration of this we consider the motion of a mass m suspended on a spring. If we assume that the spring obeys Hooke's Law, i.e. that the restoring force is proportional to the extension, and if we ignore air-resistance, the differential equations of motion become

$$\frac{dx_1}{dt} = x_2$$
$$\frac{dx_2}{dt} = -\frac{k}{m} x_1 \tag{2}$$

where x_1 is the spring extension and k is the spring constant.

If we wished to examine to the sensitivity of predictions, not only to the choice of spring constant k, but to the assumption of Hooke's Law itself, we should have to compare predictions of the nominal model (2) with those of models of the form

$$\frac{dx_1}{dt} = x_2$$
$$\frac{dx_2}{dt} = -\frac{1}{m} u(x_1,x_2) \tag{3}$$

where $u(x_1,x_2)$ is an arbitrary continuous function representing a modified force law for the spring.

In the general situation, we consider a *nominal model*, represented by (1), in which however we recognise that certain expressions within some or all of the components of the vector-function X are uncertain, and are to be perturbed in a sensitivity analysis. Thus the nominal model can be represented in the form

$$\frac{dx}{dt} = X(t,x) \equiv F(t,x,\hat{u}(t,x)), \quad t_0 \leqslant t \leqslant t_f, \quad x(t_0) = x_0 \tag{4}$$

where $\hat{u}(t,x)$ is the vector of expressions which are to be perturbed, while $F(\cdot,\cdot,\cdot)$ is a fixed function. If $\hat{x}(\cdot)$ denotes the solution to (4), then $\hat{x}(t_f)$ is the *nominal prediction*. Perturbed models take the form

$$\frac{dx}{dt} = F(t,x,u(t,x)), \quad t_0 \leqslant t \leqslant t_f, \quad x(t_0) = x_0 \tag{5}$$

where $u(\cdot,\cdot)$ represents an arbitrary alternative set of expressions. If $x_u(\cdot)$ denotes the solution to (5), then $x_u(t_f)$ is the *perturbed prediction*. Note that we do not perturb the *data* t_0, t_f, x_0.

This paper describes a theory of the analysis of sensitivity to such functional perturbations. The theory is a direct extension of that presented in (4), where the case of entirely unrestricted perturbations ($F(t,x,u) \equiv u$) was presented. The end product of the theory is an easily computed matrix quantity, the insensitivity tensor, which contains all information about sensitivity to 'infinitesimal' functional perturbations.

2. SENSITIVITY ANALYSIS

In broad terms, sensitivity analysis signifies the comparison of perturbations in the model with corresponding perturbations in the prediction. Thus, if 'small' model perturbations can result in 'large' prediction perturbations, we should say that the model is sensitive. To effect such a comparison, we need *measures* of both prediction perturbations and model perturbations.

Let us suppose, for example, that our modeller is satisfied that the usual Euclidean distance

$$d(x_u(t_f),\ \hat{x}(t_f)) = \|x_u(t_f) - \hat{x}(t_f)\| \tag{6}$$

represents a reasonable measure of the perturbation in the prediction. How should he measure the model perturbation incurred on replacing (4) by (5)? First, we note that as the data t_0, t_f, x_0 are not perturbed, the measure may involve these quanti= ties. Second, the measure should directly assess the difference between the models - i.e. between the vector functions $F(t,x,\hat{u}(t,x)) = X(t,x)$ and $F(t,x,u(t,x))$. Third, the measure should have units of Euclidean distance, so as to render comparison with (6) possible. A model perturbation measure which meets these criteria is the measure

$$\begin{aligned} m[u] &= (t_f-t_0) \times \{\text{root-mean-square value of } \|F(t,x,u(t,x))-F(t,x,\hat{u}(t,x))\| \text{ along the perturbed trajectory } x_u(\cdot)\} \\ &= (t_f-t_0)\{\frac{1}{t_f-t_0}\int_{t_0}^{t_f}\|F(t,x_u(t),u(t,x_u(t))) - X(t,x_u(t))\|^2 dt\}^{\frac{1}{2}} \end{aligned} \tag{7}$$

More generally, as is described in [4], our modeller may select any Riemannian metric $d(\cdot,\cdot)$ on the state-space to replace (6), and then adopt (7) directly, with the under= standing that the norm involved in (7) is that induced on each tangent space by the Riemmanian metric.

It is not difficult to prove that $m[u] \geqslant 0$ with equality if and only if $x_u(\cdot) \equiv \hat{x}(\cdot)$, and in fact $x_u(t) \to \hat{x}(t)$ uniformly on $[t_0,t_f]$ as $m[u] \to 0$.

The required comparison of model and prediction perturbation measures can now be effected by asking: for each state vector p near the nominal prediction $\hat{x}(t_f)$, and considering all perturbed models (5) for which the prediction is the state p, what is the least value of m[u]? That is, what is the smallest model perturbation which produces a given prediction perturbation? This leads us to define, as in [4], the *insensitivity ratio at* p:

$$\sigma(p) = \frac{\inf_u\{m[u]\,|\,x_u(t_f)=p\}}{d(p,\hat{x}(t_f))}, \quad p \neq \hat{x}(t_f). \tag{8}$$

This gives a scalar function defined in a neighborhood around the nominal prediction.

Clearly a 'large' value of $\sigma(p)$ means that large model perturbations are needed to produce the point p as perturbed prediction - characteristic of insensitivity.

On the other hand, 'small' values of $\sigma(p)$ would signal sensitivity.

An immediate question, of course, is how large is large? The following two results, proved in [4], establish the value *unity* as the neutral value between sensitivity and insensitivity. Both theorems apply only to the case of unconstrained model per= turbations ($F(t,x,u) \equiv u$).

THEOREM 1 *As* $t_f \downarrow t_0$, $\sigma(p) \to 1$ *for each fixed state* $p \neq \hat{x}(t_f)$.

THEOREM 2 *If* $d(\cdot,\cdot)$ *is a flat metric and* X *is a time-independent parallel vector field, then* $\sigma(\cdot) \equiv 1$.

The prototype example of the situation envisaged in Theorem 2 is the case of a *constant* vector field X = v = constant, coupled with the usual Euclidean distance (6).

3. EVALUATION OF $\sigma(p)$

For any perturbed expressions $u(t,x)$, we may write

$$m^2[u] = (t_f - t_0) \int_{t_0}^{t_f} \|\dot{x} - X(t,x(t))\|^2 dt$$

where $x(t)$ satisfies

$$\dot{x}(t) = F(t,x(t),v(t))$$

with

$$v(t) = u(t,x(t)).$$

It follows that for a given state p, the numerator in (8) may be evaluated by solving the optimal control problem

$$J(v(\cdot)) \triangleq \int_{t_0}^{t_f} \|\dot{x} - X(t,x(t))\|^2 dt \to \min \tag{9}$$

subject to

$$\dot{x} = F(t,x,v(t)) \qquad t_0 \leqslant t \leqslant t_f$$

$$x(t_0) = x_0, \quad x(t_f) = p.$$

Note that for $p = \hat{x}(t_f)$, the nominal model gives a global minimum value of zero to $J(\cdot)$.

Of course, one may ask whether arbitrary points p in a neighborhood of the nominal prediction $\hat{x}(t_f)$ are reachable by this control system. A sufficient condition (see, for example, [5]) for this is that the *linearisation* of the system, around values

$$(x(t),v(t)) = (\hat{x}(t),\hat{u}(t,\hat{x}(t)))$$

corresponding to the nominal model, should be reachable. This linearisation is

$$\dot{z} = A(t)z + B(t)w \quad t_0 \leq t \leq t_f, \tag{10}$$

where

$$A(t) = \frac{\partial X}{\partial x}(t,\hat{x}(t)), \quad B(t) = \frac{\partial F}{\partial u}(t,\hat{x}(t),\hat{u}(t,\hat{x}(t))).$$

In practice one would seldom be inclined to solve the optimal control problem (9), and we turn to the potentially more useful characterisation of the asymptotic beha= viour of $\sigma(p)$ as $p \to \hat{x}(t_f)$.

4. ASYMPTOTIC BEHAVIOUR FOR SMALL PERTURBATIONS

The insensitivity ratio $\sigma(p)$ is not defined at the nominal prediction $p = \hat{x}(t_f)$, and in general it has a non-removable discontinuity there. In the previous section mention was made of the question of the reachability of the linear control system (10). In this section we assume a slightly stronger condition, under which the asymptotic behaviour of $\sigma(p)$ as $p \to \hat{x}(t_f)$ can be rather easily determined.

ASSUMPTION *The control system* (10) *is reachable over arbitrary intervals* $[t_0,t]$ *with* $t_0 < t \leq t_f$.

Note that this condition is automatically satisfied for the case of unconstrained perturbations considered in [4]. This condition guarantees that the nominal trajec= tory $\hat{x}(t)$, $t_0 \leq t \leq t_f$, can be imbedded in a central field of optimal trajectories, all passing through the 'pole' of the field at $x(t_0) = x_0$, but for each $t \in (t_0,t_f]$, covering a neighbourhood of $\hat{x}(t)$.

Let $\varepsilon \to c(\varepsilon)$, be any smooth curve in $\mathbb{R}^n$ passing through the nominal prediction - say $c(0) = \hat{x}(t_f)$, and let its tangent vector there be

$$\left.\frac{dc}{d\varepsilon}\right|_{\varepsilon=0} = \xi \neq 0.$$

For $|\varepsilon|$ sufficiently small, the point $c(\varepsilon)$ will lie within the neighborhood covered by the central field at $t = t_f$, and a standard variational calculation shows that

$$\inf_u \{m^2[u] \,|\, x_u(t_f) = c(\varepsilon)\} = (t_f-t_0)\varepsilon^2\xi^T P(t_f)\xi + o(\varepsilon^2), \tag{11}$$

where $P(t_f)$ is the value at $t = t_f$ of the symmetric solution $P(\cdot)$ of the matrix Riccati equation*

$$-\dot{P} = A(t)P + PA^T(t) + PB(t)[B^T(t)B(t)]^+B^T(t)P,$$

all eigenvalues of $P(t)$ tending to $+\infty$ as $t \downarrow t_0$. This initial condition reflects the pole of the central field, and guarantees the existence of the inverse matrix

*) We denote the Moore-Penrose psuedo-inverse of a matrix M by M^+.

$$K(t) = P^{-1}(t), \tag{12}$$

at least for t near t_0, with

$$K(t) \to 0 \quad \text{as} \quad t \downarrow t_0.$$

One verifies easily that $K(\cdot)$ satisfies the *linear* matrix differential equation

$$\dot{K} = KA(t) + A^T(t)K + B(t)(B^T(t)B(t))^+B^T(t) \tag{13}$$

with initial condition

$$K(t_0) = 0. \tag{14}$$

If $\Phi(\cdot,\cdot)$ denotes the fundamental matrix solution of the linear homogeneous system

$$\dot{z} = A^T(t)z \qquad t_0 \leqslant t \leqslant t_f,$$

we have

$$K(t) = \int_{t_0}^{t} \Phi(t,s)B(s)(B^T(s)B(s))^+B^T(s)\Phi^T(t,s)ds.$$

It follows easily from this and our arbitrary-interval reachability assumption that $K(t)$ is defined and *positive definite* over the whole interval $[t_0,t_f]$.

Turning to the denominator in (8), we have simply

$$d^2(c(\varepsilon),\hat{x}(t_f)) = \varepsilon^2\|\xi\|^2 + o(\varepsilon^2) \quad \text{as} \quad \varepsilon \to 0.$$

Combining this with (11), (12) and (13) we see that if we define the *insensitivity tensor* to be the positive definite symmetric matrix

$$M = (t_f - t_0)K^{-1}(t_f), \tag{15}$$

where $K(t_f)$ is derived by solution of the linear initial-value problem (13), (14), then we have the following result:

THEOREM 3 *Provided the above assumption holds, then as* $p \to \hat{x}(t_f)$ *in the direction of a non-zero vector* ξ, *so*

$$\sigma^2(p) \to \frac{\xi^T M \xi}{\|\xi\|^2} \quad . \tag{16}$$

Thus the easily computed insensitivity tensor M fully describes the asymptotic behaviour of the insensitivity ratio $\sigma(p)$ for $p \to \hat{x}(t_f)$. Values of the vector ξ which give extreme values to the ratio (16) identify directions of extreme sensiti= vity of the nominal model - i.e. directions from the nominal prediction $\hat{x}(t_f)$ in which the prediction is most/least easily perturbed by model perturbations. As is well-known, these directions will be given by eigenvectors of the matrix M, and the corresponding eigenvalues will give the squares of the extreme value s. Thus the least eigenvalue is associated with the most sensitive direction.

5. EXAMPLE - THE HARMONIC OSCILLATOR

Consider the nominal model (2) with k = m = 1, and perturbed models of the form (3), with $t_0 = 0$, $t_f = T$. Choose Euclidean distance

$$d(x,y) = [(x_1-y_1)^2 + (x_2-y_2)^2]^{\frac{1}{2}}$$

as prediction perturbation measure.

We have

$$X(t,x) \equiv \begin{bmatrix} x_2 \\ -x_1 \end{bmatrix}, \quad F(t,x,u) \equiv \begin{bmatrix} x_2 \\ -u \end{bmatrix},$$

and hence for any nominal trajectory,

$$A(t) = \begin{bmatrix} 0 & 1 \\ -1 & 0 \end{bmatrix} \quad B(t) = \begin{bmatrix} 0 \\ -1 \end{bmatrix} .$$

This pair satisfies the arbitrary-interval reachability assumption. By solving the initial-value problem (13), (14), we find

$$K(t) = \tfrac{1}{2} \begin{bmatrix} t - \sin t \cos t & \sin^2 t \\ \sin^2 t & t + \sin t \cos t \end{bmatrix},$$

and hence that

$$M = TK^{-1}(T) = \frac{2T}{T^2 - \sin^2 T} \begin{bmatrix} T+\sin T\cos T & -\sin^2 T \\ -\sin^2 T & T-\sin T\cos T \end{bmatrix} .$$

Eigenanalysis identifies the directions

$$\begin{bmatrix} \sin T \\ 1+\cos T \end{bmatrix}, \quad \begin{bmatrix} -\sin T \\ 1-\cos T \end{bmatrix}$$

with corresponding insensitivities (square roots of eigenvalues)

$$\sqrt{\frac{2}{1+\frac{\sin T}{T}}}, \quad \sqrt{\frac{2}{1-\frac{\sin T}{T}}} .$$

Both exceed unity, and tend to $\sqrt{2}$ for large T, indicating a degree of functional insensitivity. Note that as $T \downarrow 0$, one of these tends to unity while the other grows without bound.

6. CONCLUDING REMARKS

As noted earlier, a general Riemannian metric may be chosen instead of the Euclidean distance as a measure of prediction perturbation. The reader is referred to [4] for more details of this, from which the further modifications required to Section 4 above will be clear. The theory presented in terms of a general Riemannian metric has the attractive feature of being fully invariant under transformations of state variables.

Finally it should be pointed out that the nature and the amount of the work involved in computing the insensitivity tensor M is very similar to that required to carry out a standard parameter sensitivity analysis, in that the latter also requires the solu= tion of a linear system of differential equations with coefficients in this equation derived by differentiation and substitution from the nominal model and nominal tra= jectory.

REFERENCES

[1] R Tomavic and M Vukobratovic, *General Sensitivity Theory*, Elsevier, New York, 1972.

[2] J B Cruz, *System Sensitivity Analysis*, Dowden, Hutchinson and Ross, Stroudsberg, 1973.

[3] P M Frank, *Introduction to System Sensitivity Theory*, Academic Press, New York, 1978.

[4] D H Martin, Prediction sensitivity to functional perturbations in modelling with ordinary differential equations, *Appl. Math. Optim.*, Vol. 6, pp123-137, 1980.

[5] E B Lee and L Markus, *Foundations of Optimal Control Theory*, John Wiley and Sons, Inc., New York, 1967, Ch.6.

ON THE ROLE OF THE IMPULSE FIXED COST IN STOCHASTIC OPTIMAL CONTROL. AN APPLICATION TO THE MANAGEMENT OF ENERGY PRODUCTION

J.L. MENALDI (*), J.P. QUADRAT (**) AND E. ROFMAN (**)

INTRODUCTION

There are lot of practical problems in which the system is controled by two kind of variables :

- impulse control which brings on some jump on the state of the system,
- continuous control which modifies the trajectory of the system continuously.

Associated to each impulse there is an impulse cost, and to the continuous control there is an integral cost.

It is possible to optimize such system by dynamic programming methods. Unfortunately this kind of method can be applied only to systems of small dimension. Perturbation technique is a way to decrease the dimensionality of the system. In this paper we study the situation where the impulse cost is small, and there is a reduction of the dimension of the system when we neglect the impulse cost. We give a result of continuity which proves that by this way we do only a small error.

We give an important example the management of hydropower system on a yeartime scale in which the continuous control are the release through the turbines, and the impulse control are the starting decisions of the power plant. With this time scale the starting cost is small. When we neglect the starting cost there is a decomposition of the problem : we have to manage the dams with a cost function defined by the static optimisation of the thermal production. If we don't neglect the impulse cost we have to introduce a lot of state variables which define the state of the thermal production system. This decomposition is currently used in practise. This paper gives a justification of such heuristics.

The plan will be the following :

1) presentation of the hydropower management problem,
2) continuity result of the optimal cost when the impulse cost go to zero,
3) application to the hydropower system.

This work has been partly supported by D.O.E., Office of Electric Energy Systems, under contract 01 - 80RA-50154.

(*) University Paris IX (Dauphine) - 75775 PARIS (France)
(**) INRIA - 78153 LE CHESNAY CEDEX (France)

1. THE HYDROPOWER MANAGEMENT PROBLEM

The problem consists on minimizing the cost of satisfaction of the electricity demand by hydro and thermal means (cf. [1]).

The hydro system can be model by a set of valleys $i = 1,\dots,n$, and dams in each valley $j = 1,\dots,J_i$ where J_i is the number of dams in the valley i.

We denote by :

- x_j^i the input flow of water between dams j-1 and j.
- y_j^i the stock of water in the dam (i,j).
- u_j^i the water release through turbines of the dam (i,j).
- v_j^i the water release through spill way.
- $e_j^i(y_j^i,u_j^i)$ the power generated by the dam (i,j).

Then the evolution equation of the stocks of water are :

$$\dot{y}_j^i = x_j^i + u_{j-1}^i + v_{j-1}^i - u_j^i - v_j^i \text{ if } \hat{y}_j^i \le y_j^{i'} \le \check{y}_j^i \qquad (1.1)$$

with the constraints :

$$x_j^i + u_{j-1}^i + v_{j-1}^i - u_j^i - v_j^i \ge 0 \text{ if } y_j^i = \hat{y}_j^i \qquad (1.2)$$

$$x_j^i + u_{j-1}^i + v_{j-1}^i - u_j^i - v_j^i \le 0 \text{ if } y_j^i = \check{y}_j^i \qquad (1.3)$$

$$0 \le u_j^i \le \check{u}_j^i \qquad (1.4)$$

$$0 \le v_j^i \qquad (1.5)$$

with :

- $\check{y}_j^i$ the capacity of storage, $\hat{y}_j^i$ a lower bound on the storage level,
- $\check{u}_j^i$ the maximal releases through turbines.

Remarks

- It is clear that for the head dam $u_{-1}^i = 0$ and $v_{-1}^i = 0$
- (1.1) is valid only for valley in which the dams are in cascad. For a more general network of dams the corresponding equation is easily obtained.
- The input of water are stochastic processes.
- We have neglected the transport delay.

We have a demand of electricity $D(t)$ and thermal power plant denoted by the index $k = 1,\dots,K$. Each thermal plant is described by an admissible level of production $P_k \in \mathcal{P}_K = \{\{0\} ; [\hat{P}_k,\check{P}_k]\}$. $P_k = 0$ means that the power plan is stopped; $P_k \in [\hat{P}_k,\check{P}_k]$ means that when the power plant produces electricity there are limit on admissible modulation of the production level. To each power plant there are two costs :

- α_k the impulse cost (sometime is only the cost of heating the power plant before it produces any energy) ;
- β_k the marginal production cost when the power plant is hot.

$Z_k(t)$ shall denotes a boolean variable with the following meaning :

- $Z_k = 1$ the power plant k is hot,
- $Z_k = 0$ the power plant k is cold.

$N_k(t)$ is the number of starts in the $(0,t($ period.

We have the relation :

$$N_k(t) = \sum_{s<t} \Delta Z_k^+(s) \tag{1.6}$$

The optimization problem can be formulated :

$$V^\alpha = \min_{\substack{u\\v\\\Delta z\\p}} \mathbb{E} \int_0^T (\sum_k \beta_k P_k)dt + \sum_k \alpha_k N_k(T) \tag{1.7}$$

on the constraint of demand satisfaction :

$$(\sum_k P_k + \sum_{i,j} e_j^i(u_j^i,y_j^i))(t) \geq D(t) . \tag{1.8}$$

The purpose of this work is to show that :

$$\lim_{\alpha_k \to 0} V^\alpha = \min_{\substack{u\\v\\p}} E \int_0^T C(D - \sum_{i,j} e_j^i)dt \tag{1.9}$$

Where C is defined by :

$$C(R) = \min_P \sum \beta_k P_k \quad , \quad C(R) = 0 \quad \text{pour } R \leq 0 \tag{1.10}$$

on the constraints :

$$P_k \in \mathcal{P}_k$$
$$R = \Sigma P_k$$

The simplification which happens is that in the right hand side of (1.9) the state variables Z_k does not appear. This is an important simplification. It means that to solve the complete problem we have to describe the thermal system by 2^k situations, but for the simplified one we have to solve only, for each time t, the static optimization of the thermal system. In particular when $\hat{P}_k = 0$ the problem (1.10) can be solved analyticaly and so the time cost to compute C(R) is very small. We have to remark that in general the complete problem is yet insolvable (K is of order 200 in the french situation).

2. THE CONTINUITY OF THE OPTIMAL COST

2.1. Notations

For simplicity we shall consider $\alpha_k = \alpha, \forall k$ and we shall rewrite (1.7) as :

$$\min_{W_{ad}} (J^I(w) + \alpha\, N(w)) = \min_{W_{ad}} J_\alpha(w) \qquad (2.1)$$

with :

$$W_{ad} = \{w \;/\; w = (u,v,\Delta z,P) \;;\; u,v,\Delta z,P \text{ verifying the restrictions imposed by the model}\}$$

- $J^I(w)$ is the mathematical expectation of the integral cost associated to the continuous control contained in w ;
- N(w) is the mathematical expectation of the number of impulses contained in w ;
- $J_\alpha(w)$ is the mathematical expectation of the total cost related to the policy w, and α is the fixed cost of each impulse. Furthermore $\hat{w}_\alpha$ is the optimal policy when α is the impulse cost.

2.2. The problem

Our purpose is to give some results concerning the validity of :

$$\lim_{\alpha\to 0} \min_{W_{ad}} [J^I(w) + \alpha\, N(w)] = \inf_{W_{ad}} J_0(w), \qquad (2.2)$$

where $J_0(w)$ denotes the mathematical expectation of the cost related to w if the impulse cost α is neglected.

Remark 2.1. In our model, as it was said at the end of § 1, the assumption $\alpha = 0$ implies that the component z of the control w has not influence for computing $\inf_{W_{ad}} J_0(w)$. Nevertheless we will conserv the notation W_{ad} in the second member of (2.2).

Remark 2.2. The difficulty of our problem is that when $\alpha \to 0$, $N(\hat{w}_\alpha)$ a priori may increase and go to ∞. In general we have not an upperbound of $N(\hat{w}_\alpha)$ independent of α. Instead of that, we can give for each α a value $\tilde{N}_\alpha$ such that $N(\hat{w}_\alpha) \leq \tilde{N}_\alpha$ and, at most, it is known (cf. [2]) that the product $\alpha\, N(\hat{w}_\alpha)$ is bounded.

Remark 2.3. In the second member of (2.2) we have now $\inf\limits_{W_{ad}}$ instead of $\min\limits_{W_{ad}}$ because, in general, it is not possible to insure the existence in W_{ad} of a control giving the optimal cost. When the existence is proved it is obtained thanks to additional hypothesis. For example, in [3], it is shown -using a combination of analytic and probabilistic techniques related to impulse control theory and taking advantage of the main hypothesis : "impulses are lower bounded by a positive number"- that $N(\hat{w}_0) < \infty$ for different types of problems.

In what follows we will present two theorems concerning (2.2). Then, in § 3, we will return to the model presented in § 1 to justify the simplification introduced in (1.9), (1.10).

Theorem 2.1. Let us suppose known $\hat{w}_0$ (having a finite number of impulses $N(\hat{w}_0)$) optimal admissible policy when α is neglected, i.e. :

$$J_0(\hat{w}_0) = J^I(\hat{w}_0) = \inf_{w\in W_{ad}} J_0(w). \tag{2.3}$$

Then :

$$\lim_{\alpha\to 0}\ \min_{W_{ad}}\ J_\alpha(w) = J_0(\hat{w}_0) \tag{2.4}$$

the excess of cost introduced by using $\hat{w}_0$ instead of $\hat{w}_\alpha$ is less than $\alpha\, N(\hat{w}_0)$ (2.5).

Proof. After our assumptions the expected cost related to $\hat{w}_0$ is :

$$J^I(\hat{w}_0) + \alpha\, N(\hat{w}_0) \tag{2.6}$$

On the other hand, taking into account the cost α we have :

$$\min_{w\in W_{ad}} J_\alpha(w) = J_\alpha(\hat{w}_\alpha) = J^I(\hat{w}_\alpha) + \alpha\, N(\hat{w}_\alpha) \tag{2.7}$$

Obviously, from (2.6) and (2.7) we have :

$$J^I(\hat{w}_\alpha) + \alpha\, N(\hat{w}_\alpha) \leq J^I(\hat{w}_0) + \alpha\, N(\hat{w}_0) \tag{2.8}$$

But, from (2.3) we know that :

$$J^I(\hat{w}_0) \leq J^I(\hat{w}_\alpha) \tag{2.9}$$

so, after (2.9) we obtain from (2.8) an upperbound (independent of α) for the mathematical expectation of the number of impulses contained in the optimal policies obtained with the fixed cost α :

$$N(\hat{w}_\alpha) \le N(\hat{w}_0) \tag{2.10}$$

Then as $\lim_{\alpha\to 0} \alpha\, N(\hat{w}_\alpha) = 0$, we obtain easily (2.4). ■

To show (2.5) we have, from (2.8) and (2.9) :

$$J^I(\hat{w}_0) \le J^I(\hat{w}_\alpha) \le J_\alpha(\hat{w}_\alpha) \le J^I(\hat{w}_0) + \alpha\, N(\hat{w}_0) \; ; \tag{2.11}$$

so :

$$[J^I(\hat{w}_0) + \alpha\, N(\hat{w}_0)] - J_\alpha(\hat{w}_\alpha) \le \alpha\, N(\hat{w}_0) \quad ■ \tag{2.12}$$

<u>Theorem 2.2.</u> Let us consider as admissible controls those having finite mathematical expectation of the number of impulses, i.e. :

$$W_{ad} = \{w \;/\; N(w) < +\infty\} \tag{2.13}$$

We will introduce as parameters the initial conditions t,x ; the cost will be function of those parameters and we shall have as optimal cost :

$$V_\alpha(x,t) = \inf_{W_{ad}} J_\alpha(x,t;w) \; ; \tag{2.14}$$

in particular, when α is neglected, we put :

$$V_0(x,t) = \inf_{W_{ad}} J^I(x,t;w). \tag{2.15}$$

Then, under assumption (2.13) we have, as $\alpha \to 0$, a pointwise convergence of (2.14) to (2.15), i.e. :

$$\lim_{\alpha\to 0} V_\alpha(x,t) = V_0(\alpha,t) \tag{2.16}$$

<u>Proof</u> : From (2.15), $\forall\varepsilon > 0\ \exists w^\varepsilon \in W_{ad}$ such that :

$$J^I(x,t;w^\varepsilon) < V_0(x,t) + \varepsilon.$$

We shall denote $N_{xt}(w^\varepsilon)$ the mathematical expectation, variable with x,t of the number of impulses of w^ε. So, the expected cost of using w^ε is :

$$J_\alpha(x,t;w^\varepsilon) = J^I(x,t;w^\varepsilon) + \alpha\, N_{xt}(w^\varepsilon)^{(*)} \tag{2.18}$$

(*) In this case we cannot obtain, as we did in (2.10), an upperbound for $N_{xt}(\hat{w}_\alpha)$ independent of α.

From (2.14), (2.15) and (2.18) we obtain :

$$V_0(x,t) \le V_\alpha(x,t) \le J_\alpha(x,t;w^\varepsilon) \tag{2.19}$$

On the other hand, from (2.17) :

$$J^I(x,t;w^\varepsilon) + \alpha N_{xt}(w^\varepsilon) \le V_0(x,t) + \varepsilon + \alpha N_{x,t}(w^\varepsilon) ; \tag{2.20}$$

so we can also write :

$$0 \le V_\alpha(x,t) - V_0(x,t) \le \varepsilon + \alpha N_{xt}(w^\varepsilon) \tag{2.21}$$

Hence, as (2.21) holds for all $\varepsilon > 0$, we obtain (2.16). ∎

Remark 2.4. If we assume $V_0(x,t)$ and $V_\alpha(x,t)$ continuous the convergence just proved becomes uniform oves compact sets. In particular the continuity of $V_0(x,t)$ and $V_\alpha(x,t)$ can be established as a consequence of regularity properties of the function having intervention in the integral cost.

The uniform convergence in a small neighbourhood of a generic point (x_0,t_0) is also achieved if in (2.17) we can choose w^ε independent of (x,t) of that neighbourhood. Also in this case we obtain the uniform convergence over compact sets.

3. APPLICATIONS OF RESULTS OF § 2 TO THE HYDROPOWER SYSTEM

3.1. In § 1 we have said that sometimes the impulse cost α_k is only the cost of heating the power plant before it produce any energy. In this case, even if in the model we have considered α_k as an impulse fixed cost, the action of heating a thermal power plants needs a period of time $\tau > 0$. So, in $[0,T]$ we can obtain a finite upper-bound $\tilde{N}$ for $N(w), \forall w \in W_{ad}$, independent of α. Then (2.10) is satisfied with $\tilde{N} \ge N(\hat{w}_0)$ and the results of theorem 2.1 holds.

Furthermore we can roughly give a first evaluation of the error introduce by the simplification proposed at the end of § 1 ; in fact, after (2.12), we have :

$$\frac{J_\alpha(\hat{w}_0) - J_\alpha(\hat{w}_\alpha)}{J_\alpha(\hat{w}_\alpha)} \le \frac{\alpha N(\hat{w}_0)}{J^I(\hat{w}_0)} \le \frac{\alpha \tilde{N}}{J^I(\hat{w}_0)} . \tag{3.1}$$

3.2. To apply the results of theorem 2.2 to the model of § 1 we shall show now that we can restraint our initial class W_{ad} of admissible controls to the class W^*_{ad} of those w having $N(w)$ finite. Clearly, we will take advantage of some "density argument".

In order to simplify techniques let us suppose that :

$$\begin{cases} x_j^i(t),\ D(t) \text{ are adapted processes in } L^1 \\ e_j^i(u,y) \text{ is Borel measurable.} \end{cases} \tag{3.2}$$

We remark that a classical argument of measurability shows, if $w = (u,v,\Delta z,P)$, that:

$$\inf_{W_{ad}} E \{\int_0^T (\sum_k \beta_k P_k)dt\} = \inf_{(u,v)} E \int_0^T C\,(D(t) - \sum_{i,j} e_j^i\,(u_j^i,y_j^i))dt\ , \tag{3.3}$$

with the notations introduced in § 1, i.e. :

$$\begin{cases} C(R) = \min \{\sum_k \beta_k P_k \ /\ \hat{P}_k \le P_k \le \check{P}_k \text{ or } P_k = 0,\ \sum_k P_k = R\} \text{ if } R \ge 0\ ; \\ C(R) = 0 \qquad\qquad \text{if } R < 0. \end{cases} \tag{3.4}$$

We will assume :

$$\hat{P}_k = 0 \quad \forall\, k \tag{3.5}$$

in order to have a convex problem for (3.4) ; so, if we put $\beta_1 \le \beta_2 \le \dots \le \beta_k$, we will have :

$$C(R) = \sum_{k=1}^{K-1} \beta_k \check{P}_k + \beta_K P_K \tag{3.6}$$

where :

$$\begin{cases} K = \min\ \{\chi \ge 1 \ /\ \sum_{k=1}^{\chi} \check{P}_k \ge R\} \\ P_K = R - \sum_{k=1}^{K-1} \check{P}_k \end{cases} \tag{3.7}$$

We can now show the following

Theorem 3.1. Under assumptions (3.2) and (3.5) we have :

$$\inf \{J(w) \ /\ w \in W_{ad}\} = \inf \{J(w) \ /\ w \in W_{ad}^*\} \tag{3.8}$$

with $J(w)$ denotes the cost functional given by (3.3).

Proof. Let us define :

$$R(t) = [D(t) - \sum_{i,j} e_j^i(u_j^i(t),\ y_j^i(t))]^+ \tag{3.9}$$

After assumptions (3.2), $R(t)$ is an adapted L^1 process. Thus, there exists an adapted L^1 process, $R_n(t)$ such that :

$$\begin{cases} R_n(t) \to R(t) \text{ in } L^1, \text{ as } n \to \infty ; \\ R_n(t) \text{ is piecewise constant.} \end{cases} \tag{3.10}$$

Then, given any admissible control w, we have :

$$E \{ \int_0^T C(R_n(t))dt \} \to J(w) \quad \text{as } n \to \infty. \tag{3.11}$$

Since $R_n(t)$ is piecewise constant, the index $K = K(R_n)$ given by (3.7) have finite variation, i.e. :

$$\sum_{t\geq 0} |\Delta K(t)| < \infty \tag{3.12}$$

Finally, noticing that, in our model, $\Delta K(t) = \Delta Z(t)$, we conclude the theorem. ∎

FINAL COMMENTARY

After 3.1 and, more generally, with the use of theorem 2.2 as it was indicated in 3.2, we are able to insure the validity of (1.9). So, it is possible to simplify the numerical solution of the management problem, as it was proposed at the end of § 1.

BIBLIOGRAPHY

[1] F. DELEBECQUE, J.P. QUADRAT - Contribution of Stochastic control singular perturbation averaging and team theories to an example of large scale systems : Management of hydropower production - IEEE AC Avril 1978.

[2] J.L. MENALDI - "Sur le Problème de Contrôle Impulsionnel et l'Inéquation Quasi-Variationnelle Dégénérée Associée" : C.R. Acad. Sc. Paris, Tome 284, (1977), Série A, pp. 1499-1502.

[3] J.L. MENALDI, E. ROFMAN - "On stochastic control problems with impulse cost vanishing" : International Symposium on semi-infinite programming and applications - Univ. Texas at Austin - Sept. 8-10, 1981.

WATER DISTRIBUTION NETWORK SELF-TUNING CONTROL

R. Ortega & R. Canales-Ruíz
Instituto de Ingeniería, UNAM
Apdo. 70-472, Delegación Coyoacán
04510, México, D.F. MEXICO

1. INTRODUCCION

Most of the methods used in the design of controllers for multivariable systems use a model (generally linear) of the process and its environment. However in practical applications the process dynamics is non-linear and often unknown, hence some model building or identification technique must be used before the controller is designed. Using the certainty-equivalence concept a class of stochastic adaptive control schemes for unknow systems (self-tuning controllers, STC) which have gained enormous popularity in the last years (Fauvier/Guillemin, 1980) has been defined. Since the appearance of the seminal works of Aström/Wittenmark, (1973), and Peterka (1970), many significant practical contributions have been made. Clarke and Gawthrop (1975) improved the basic self-tuning regulator allowing control effort to be adjusted and set-points to be simply included. Recently, multivariable extensions to the STC have been reported (Koivo, 1980; Borisson, 1979), and several applications to process control proved succesfull (Unbehaven/Schmid, 1979).

Here, using a state-space setting, an extended minimum variance self-tuning controler (EMV/STC) for multivariable state measurable systems is introduced. Following the same reasoning of Clarke/Gawthrop (1975), and Koivo (1980), the controller structure is obtained by penalizing the control terms in the stochastic cost function. The control law turns out to be the solution of a deterministic ptim zation problem.

This strategy was developed for the control of a water distribution network (WDN) in which the boundary conditions are not readily known, because of the existence of domiciliary tanks. The tanks served from a node are modelled (Sanchez/Fuentes, 1979) as a single local tank of unknown time-varying geometry. The levels in these local tanks constitute the state variables which are estimated from local measurements. A set of valves distributed along the WDN provides the control action, whose aim is to distribute the water through the pipes in such a way that the state variables follow a reference with minimum variance. The daily water demand is represented by a known deterministic profile corrupted with noise. The control of the WDN will then consist of: 1) a recursive identification stage, in which a state-model is adjusted with the local tank levels and the valves position data; 2) the determination of the EMV/STC. This procedure is repeated at each sampling time.

The paper is organized as follows: in section 2, the multivariable EMV/STC theory is presented. The WDN control problem and the application of the EMV/STC to it are described in section 3 and finally in section 4, simulation results are presented.

2. SELF-TUNING CONTROL OF MULTIVARIABLE STATE-MEASURABLE SYSTEMS

The systems considered are discrete, multi-input state-measurable and randomly disturbed, described by:

$$\underline{X}_{t+1} = A\underline{X}_t + B\underline{U}_t + \underline{D} + C\underline{\xi}_t \tag{1}$$

where $\underline{X}_t \in R^n$ and $\underline{U}_t \in R^m$ are the state and input vector, $\underline{D} \in R^n$ is an offset term (to be explained later), $\underline{\xi}_t \in R^\ell$ and $\{\underline{\xi}_t\}$ is a sequence of independent equally distributed random vectors with zero mean value and A, B and C are unknown matrices of the proper dimensions.

This type of models arise from local linearizations, the signals are increments around nonzero levels $\bar{X}_i$, $\bar{U}_i$. These levels, in general, do not satisfy the incremental model, so an offset term ($\underline{D}$) must be added.

Consider now the problems of designing a closed-loop regulator for system (1) which minimizes the following criterion:

$$J = E\left\{||\underline{X}_{t+1} - \underline{X}_t^+||_R + ||\underline{U}_t||_M \;\; /t\right\} \tag{2}$$

where $\underline{x}_t^+$ is the known reference signal and R and M are polynominal matrices in the delay operator q^{-1} and $E\{(\cdot)\}$ represents de expected value of $(\cdot)$ given $\{\underline{X}_t, \underline{X}_{t-1},\ldots\}$ and $\{\underline{U}_t, \underline{U}_{t-1},\ldots\}$. Typical choices of R and M are:

$$\begin{aligned} R &= \text{diag}\,\{R_i\} \\ M &= \text{diag}\,\{M_i(1 - q^{-1})\} \end{aligned} \tag{3}$$

It is assumed that the control law at time t is a function of the observed states up to and including time t and of all past control signals, that is,

$$\underline{U}_t = \underline{g}(X^t, U^{t-1})$$

Substituting (1) in the criterion gives

$$J = E\left\{||A\underline{X}_t + B\underline{U}_t + \underline{D} + C\underline{\xi}_t - \underline{X}_t^+||_R + ||\underline{U}_t||_M \;/t\right\} \tag{4}$$

The last equation can be written as

$$J = \left\{||A\underline{X}_t + B\underline{U}_t + \underline{D} - \underline{X}_t^+||_R + ||\underline{U}_t||_M + E\;||C\underline{\xi}_t||_R/t\right\} \tag{5}$$

where the following relations have been used

$$E\left\{||\underline{U}_t||_M/t\right\} = ||\underline{U}_t||_M \; ; \; E\left\{||A\underline{X}_t + B\underline{U}_t + \underline{D} - \underline{X}_t^+||_R/t\right\} = ||A\underline{X}_t + B\underline{U}_t + \underline{D} - \underline{X}_t^+||_R$$

$$E\left\{D^T RC\underline{\xi}_t/t\right\} = 0 \; ; \; E\left\{\underline{X}_i \; \underline{\xi}_t^{\;T}/t\right\} = E\left\{\underline{U}_i \; \underline{\xi}_t^{\;T}/t\right\} = 0 \quad \forall i \leq t$$

As can be seen from (5) the stochastic problem has been reduced into a deterministic optimization. The minimizing $\underline{U}_t^{\;OPT}$ in (5) is found by setting to zero the gradiente of J with respect to $\underline{U}_t$, that is

$$\frac{1}{2}\frac{\partial J}{\partial \underline{U}_t} = (M + B^T RB)\underline{U}_t^{\;opt} + B^T R(A\underline{X}_t - \underline{X}_t^+ + \underline{D}) = 0 \tag{6}$$

For the known parameter case this equation defines the optimal control over all the admissible strategies (Eq. 4).

Given that the disturbing noise is assumed uncorrelated, the least squares estimation for the unknown parameter case, will be unbiased and the estimated parameters will asymptotically tend to the real values. Hence the substitution of the system parameter by its estimated values, assures that, after the parameters have converged, the control law coincides with the optimal.

A least-squares identification scheme may now be proposed to estimate recursively the parameters of the following model

$$\underline{X}_{t+1} = \hat{A}\underline{X}_t + \hat{B}\underline{U}_t + \hat{\underline{D}} + \underline{\varepsilon}_{t+1} \triangleq \hat{\underline{X}}_{t+1/t} + \underline{\varepsilon}_{t+1} \tag{7}$$

which can be written as

$$\underline{X}_{t+1} = \theta_t \; \psi_{t+1} + \underline{\varepsilon}_{t+1} \tag{8}$$

where

$$\theta_t \triangleq [\hat{A} : \hat{B} : \hat{\underline{D}}] \quad \text{and} \quad \underline{\psi}_{t+1}^T = \{\underline{X}_t^{\;T} : \underline{U}_t^{\;T} : 1\}$$

the least square estimation of θ is done by estimating one column $\underline{\theta}_i$ at a time. The weighted least squares criterion is

$$V_N(\underline{\theta}_i) = \frac{1}{N}\sum_{t=n}^{N+n} \rho_i^{\;N+n-t} \; \varepsilon_i^2(t) \quad i = 1, 2 \ldots n+m+1 \tag{9}$$

where the weighting (forgetting) factors are choosen to satisfy $0 < \rho_i < 1$ in order to allow the system to track parameter changes. The following recursive equations insure the minimum of V_N:

$$\begin{aligned} \underline{\theta}_i(t+1) &= \underline{\theta}_i(t) + \underline{\Gamma}_t \left[X_i(t+1) - \underline{\psi}_{t+1}^{\;T} \underline{\theta}_i(t)\right] \\ \underline{\Gamma}_t &= \{\rho_i + \psi_{t+1}^T P_t \underline{\psi}_{t+1}\}^{-1} \; P_T \underline{\psi}_{t+1} \\ P_{t+1} &= \frac{1}{\rho_i}\{I - \underline{\Gamma}_t \; \underline{\psi}_{t+1}^T\} P_t \end{aligned} \tag{10}$$

Significant saving in the computations is obtained by choosing the same initial value for P_t for every $\underline{\theta}_i$, them the corresponding gain vectors $\underline{\Gamma}_t$ will also be the same for all parameter vectors.

The EMV/STC must update the estimated parameters at each sampling time. Substituting these values in (6) the optimal control law can be found.

3. THE WATER DISTRIBUTION NETWORK FROM THE CONTROL POINT OF VIEW

Mexico City's WDN currently supplies water to about 13 million people. Mexico like most fast growing population centers, has suffered from shortages due to both distribution inadequacy and insufficient supply. In order to guarantee water disponibility during the day the user saves it at high pressure hours in domiciliary tanks. This particular situation makes the analysis of Mexico's WDN different from American and most of the Europeans WDN, given that boundary conditions are not readily known. The control objective significantly differs too, because due to high demand Mexico's WDN may not insure high pressure along the day, but the aim of the control must be to distribute the water in such a way that the domiciliary installed capacity is optimally use, for example, by making their level follow a reference signal with minimum variance.

3.1 Description of the process

Mexico City's primary WDN consists of 238 pipes and 209 nodes which supply water through the secondary WDN to the end users. Flow in pipes and pressure in nodes are related through the following equations (see fig 1):

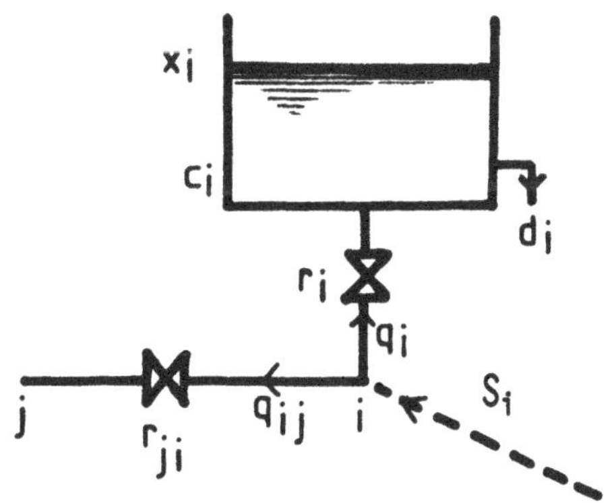

Fig. 1 Element of the water Distribution Network.

a) Continuity equations at the nodes

$$S_i - Q_i + \sum_{j\varepsilon\psi_i} Q_{ij} = 0 \; ; \; i\varepsilon NN = \text{set of nodes} \quad (11a)$$

where S_i is an external water supply flow, Q_i is the flow delivered to the secondary WDN from the i-th node, Q_{ij}, the water flow in the pipe joining the i-th and j-th nodes and ψ_i is the set of pipes conected to node i,

b) Pipe equations

$$h_i - h_j - R_{ij}\, Q_{ij}|Q_{ij}| = 0 \quad i\varepsilon NP = \text{set of pipes} \quad (11b)$$

where h_i is the pressure in the node i, R_{ij} is the restriction of the pipe (denoted by U_{ij} when it is variable restriction to be defined by the control algorithm).

From the nodes of the primary WDN, water is supplied along pipes of various diameters and lengths to the domiciliary tanks. A first approximation to model this so-called secondary WDN is to consider the domiciliary tanks served by a node as a single local tank of unknown time-varying geometry, conected to the node by means of a single pipe. The levels of the local tanks ($X_i(t)$, $i \in NN$) which are indicative of the neighboring domiciliary tank levels, define the dynamic part of the process by:

$$\dot{X}_i(t) = f_i(Q_i - d_i) \; ; \; i \in NN \quad (12)$$

where d_i is the user demand associated to the i-th node.

Generally the nature of the water demand is quiet homogeneous in the neighboring zone of each node, consequently d_i can be represented as a known deterministic profile corrupted with noise (whose variance σ_d^2, will increase with increasing uncertainty in the users habits) for each of the nodes, e.g. $d_i(t) = d_i^\circ(t) + \xi_i(t)$, $i \in NN$.

3.2 Control Problem

The control objective is posed as: impose a reference signal ($X_i^+(t)$, $i \in NN$) to the local tanks levels, such that they have enough water to satisfy at all times a demand of the form $d^\circ(t) + 2\underline{\sigma}_d$.

It is expected that the noise component of $d_i(t)$ will deteriorate the reference following and the control algorithm must take into account the stochastic nature of this signals to improve its performance. It is of particular interest to minimize the variance of the tracking error because in this way, first water supply tends to equal the demand and secondly the minimum level insuring (under certain confidence limits) water disponibility may be lowered.

With the controlobjective determined, measurable and controllable variables must be defined and a control strategy proposed. It is evident that we are dealing with a non-measurable-state system, but since there are adaptive schemes used to estimate the state from current measurements we will assume that the local tank levels are known at all times.

3.3 Self-tuning control

The implementation of the EMV/STC to the WDN will consist of:

1) Using the least-squares recursive estimator given in (10) calculate the matrices $\hat{A}$, $\hat{B}$ and the bias $\hat{\underline{D}}$ of the prediction model
2) Substitute this estimations in (7) and solve for $\underline{U}_t^{opt}$. This will give a set of values for the NC controllable valves.

   The process is repeated at each sampling time.

Fast changing signals must be avoided in valve actuators, otherwise undesirable transients arise. Choosing the weighting matriz M of the form (3) in the control criterion, allows us to reduce the variation of the control signal. The control signal in this case can be explicity written as:

$$\underline{U}_t^{opt} = \{M(0) + \hat{B}^T R\hat{B}\}^{-1} \{M(0)\underline{U}_{t-1} - B^T R(A\underline{X}_t - \underline{X}_t + \hat{\underline{D}})\} \tag{14}$$

the least-squares model employed is:

$$\underline{X}_{t+1} = A\underline{X}_t + \hat{B}\underline{U}_t + \hat{\underline{D}} + \underline{\varepsilon}_{t+1} \tag{15}$$

4. EXPERIMENTAL RESULTS.

A simple three-mode two-pipe network with two control valves (see fig.2) was used for the experiments.

A 100 step open-loop identification stage with $\underline{U}_t$ random functions was carried out first with

$$\rho_i = 0.98,\ \forall i\ ;\ P_0 = (1000)I,\ \theta_\circ = 0$$

the resulting values were used as initial values in the control.

Perturbations about a dynamic equilibrium point given by:

$$Q_i = D_i\ ,\ \forall_i$$

$$\underline{X}^\circ = \{10\ 20\ 10\}^T\ ;\ \underline{U}^\circ = \{11/3\quad 5/3\}\ ;\ \underline{d}^\circ = \{3\quad 4\quad 3\}$$

where introduced.

Step response tests (with Δt = 0.5 sec) showed an approximate fastest time constant of 20 sec, hence the sampling time was choosen as T_s = 2 sec.

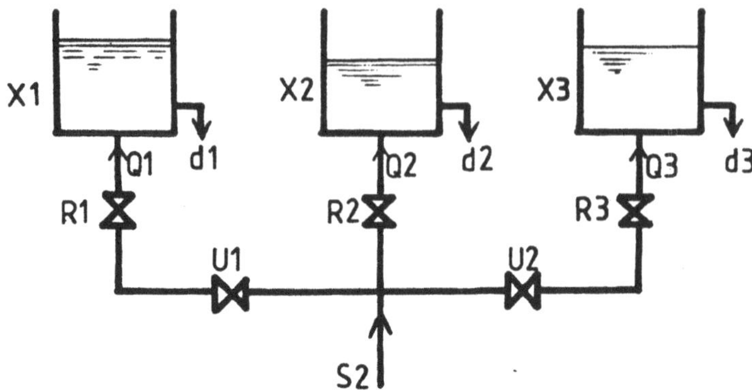

Fig. 2 Testing Network

A noise component generated from a pseudorandom binary sequence was added to the water demand, chosen such as to satisfy the following condition:

$$\sum_{i=1}^{3} d_i^{\xi}(t) = 0, \forall t$$

The noise sequence value had an amplitude of $\pm$ 0.5 and was changed every 10 sec.

The simulation results, shown in fig. 3, were obtained with:

$$\underline{X}_t^+ = \underline{X}^\circ \; ; \; R_{ii} = 0.1, \forall i; \; R_{ij} = 0, \forall i \neq j \; ; \; M = 0$$

A possible choice for the control criterion when there is no knowledge of the water demand is to keep the tank levels equal, that is

$$J = E\{\Sigma(x_i - x_j)^2/t\} \; \forall i \neq j$$

With these criterion the system was simulated for decorrelated noise disturbances in the demands and the results are shown in fig. 4.

5. CONCLUSIONS.

An extended minimum variance self-tuning controller for state measurable multivariable systems is presented. The properties of the algorithm have been tested with a simple simulated water distribution network in which the control action is exerted through valve-controlled pipes. The distrubance rejection of the closed loop system were shown. In spite of the poor controlability of the system, the control law considerably improves the behaviour of the system what gives confidence about the performance of the controller for a large WDN.

The preliminary experiments with a medium size WDN show that unless the state is con-

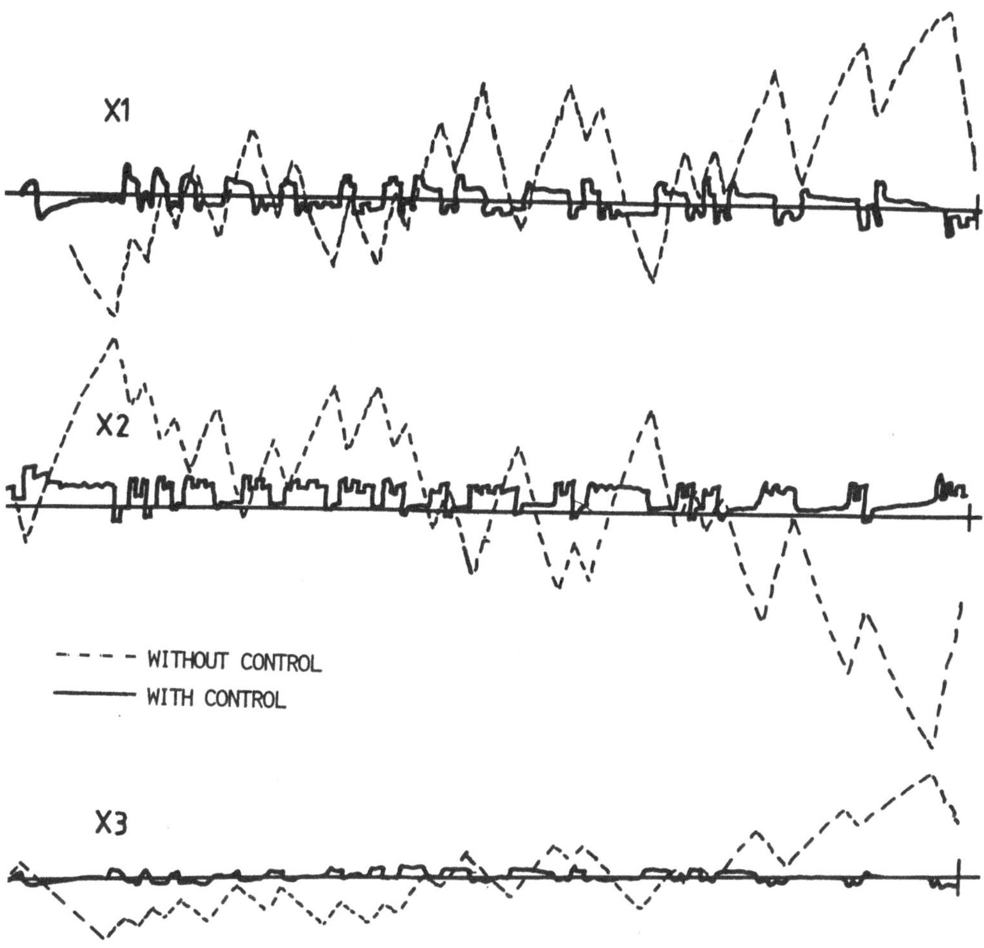

Fig. 3 Disturbance rejection results.

trollable in one step, as the described three tank example is, the reference following capability is not an acceptable criterion. A better water distribution scheme is obtained based on the criterion

$$J = E\Big\{\Big[X_i - \sum_{j=i} \alpha_j X_j\big)^2_{t+1} + ||\underset{\sim}{U}_t||_M\Big]/t\Big\}$$

where the α_i's are scalars functions of time that can be used to change the relative importance of the nodes.

Acknowledgement

Most of the simulations results were obtained by Augusto Hernández

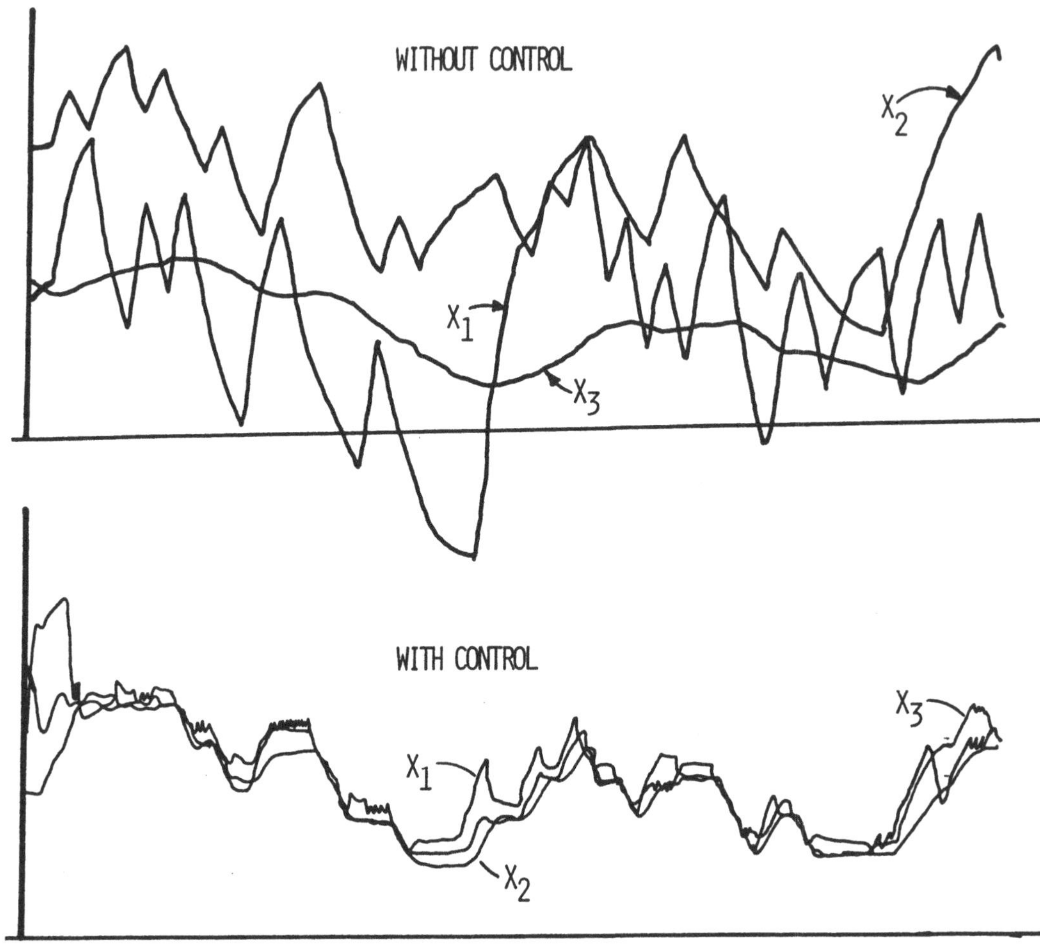

Fig. 4 Equal level criteria results

References

Astrom, K.J. and B. Witternmark (1979) "On self-tuning regulators" AUTOMATICA V. 9, 185-199.

Clarke, D.W., et.al. (1975) "Self-tuning controller" Proc. IEE., V. 122, 929-934.

Fauvier, G. And Guillermin, P. (1980) "A comparative study of self-tuning regulators" 3 rd IFAC Symp. on Syst. appr. for develop. Rabat, Morrocco, 27-45.

Unbehaven, H and Scmid Chr. (1979) "Application of adaptive systems in process control" Intl. Workshop on appl of adapt. control Aug. 23-25, Yale Univ. USA.

Koivo, H.N. (1984) "A multivariable self-tuning controller" AUTOMATICA V. 16, 351-366.

Sánchez, J. and Fuentes O. (1979) "Consideraciones sobre el cálculo de red de distribución de agua potable" Instituto de Ingeniería, No. 421, UNAM, México.

CONFLICTS OVER NORTH SEA OIL PROFITS AND MACRO-ECONOMIC POLICY

Rick van der Ploeg and Martin Weale
Department of Applied Economics
University of Cambridge
Sidgwick Avenue
Cambridge CB3 9DE, U.K.

1. Introduction

The main objective of this paper is to take advantage of a medium-term multi-sectoral model of the UK economy to formulate depletion policy and to analyse its long-term macro-economic implications. Thus the effects of low, high and optimal oil extraction policies on the British economy are examined within the framework of a large disaggregated econometric model.

The second objective of this paper is to demonstrate the usefulness and feasibility of applying the optimal control approach to a very large disaggregated econometric model. The paper illustrates that optimal control is a practical approach for regulating large non-linear econometric models.

2. Depletion Policy for Private Agents

2.1. Pricing and depletion of exhaustible resources

Modern extraction theory finds its roots in Hotelling's principle [11], which shows that if an individual holding a stock of an exhaustible resource is to be indifferent at the margin between selling an extra unit of that resource today or keeping it in the ground the price (p) of the resource must rise at the rate of return on other assets. Although in many aspects of economic theory results which are derived for a large number of small producers do not hold in cases of monopoly or oligopoly in this case provided the demand curve is isoelastic market power of producers makes no difference [5]. The extraction path which depletes at the rate $-\eta r = -\eta \frac{\dot{p}}{p}$, where η is the elasticity of demand, will be chosen by a planning board which wants to maximise the net present value of expected gross consumer surpluses.

However where resource stocks are owned by companies the pressure to pay dividends [19] may lead to a rate of extraction faster than that given by the above formula and close to the actual rate of discount, provided the resource is one such as oil with price elasticity of demand below one.

2.2. The Cambridge Growth Project model

The Cambridge Growth Project model was originally developed as a static Leontief input-output model suitable for analysing the equilibrium impact of structural change on Britain's industries [17]. It has subsequently been extended by introducing dynamics in the econometric relationships for the components of final demand, imports, employment and wages, to allow the year by year projection of transient paths [2]. The model does not ensure equilibrium in the labour market or in the market for foreign exchange. The size of the Cambridge Growth Project model, that is 3000 endogenous, 850 policy and more than 4000 other predetermined variables at any point of time, is due to its industrial nature and provides an ideal challenge to the optimal control procedures proposed in the appendix. It is used in the rest of this paper to examine and derive a number of oil extraction paths and their impact on the U.K. economy.

2.3. Extraction of North Sea oil and the British economy

Two extraction paths are studied in this section a "company" or high depletion case in which an estimate of the stock of oil reserves in the UK is depleted at the rate of return, say 5%, and a "world" or low depletion case in which UK oil stocks are run down at the same rate as those in the world as a whole, say 3%. Obviously such paths assume that there is no further oil to be discovered in the UK and that the feasibility of alternative sources of fuel will not change greatly. The price of oil is assumed to grow in real terms at 5% p.a. (the assumed rate of discount and roughly the rate of return on UK equities), which implies a price elasticity of -0.6. The exchange rate is adjusted with the optimal control algorithm to maintain balance of payments equilibrium. Both cases are compared with the expected steady path of extraction, which is a consequence of a government policy which places a high weight on net self-sufficiency in oil.

In the real world it can sensibly be argued that what matters is not only the discounted value of future income but also unemployment, public borrowing, the terminal capital stock (proxied by the present discounted value of gross fixed capital formation) and the remaining stock of unextracted oil. Table 1 and Figures 1-6 show that the low extraction case is superior to either the standard or the 5% case even before allowance is made for unemployment or terminal oil reserves, since it leads to 700000 fewer unemployed, and yields the country terminal reserves larger by 400m tonnes. The mechanism which makes this possible is the lower exchange rate allowing the country to remain competitive in export markets and produce more in industries which, unlike oil extraction, generate employment. These conclusions should be contrasted with the static analysis in [7]. The lower exchange rate of the 3% depletion case implies higher prices and lower real wages. A further consequence of the low depletion is lower tax revenues and thus higher public borrowing. However the higher level of income and employment generates extra savings to fund a borrowing requirement higher by £8000m over the period as a whole.

3. Depletion as a Policy Variable

3.1. Priorities

Other empirical studies of depletion paths or investigations of the consequences of oil [1,3,6,7,10,16,18] have paid particular attention to the effects of oil revenues on manufacturing industry, although not in a formal regulatory framework. This section instead considers the problem of an economy which wants to steer towards an optimal extraction path (exemplified by the 3% case) but is also concerned about the effects of the route towards this path. In particular it looks at the consequences

Table 1 The Extraction Paths

Oil Extraction Rate	Base View 80mn tonnes in 1980 then 100mn tonnes	5% p.a. 120mn tonnes in 1980 reduced by 5% p.a.	3% p.a. 72mn tonnes in 1980 reduced by 3% p.a.
Real Oil price	+ 5% p.a.	+ 5% p.a.	+ 5% p.a.
Terminal Oil stock	1320 m tonnes	1353 m tonnes	1713 m tonnes
P.D.V. of private consn. + GFCF (£m)	450643	449130	450850
Average Unemployment (thou.)	3136	3044	2353
PSBR (total 1980-90 £m)	189846	189126	197412
Closing Consumer Price Index (1970=1)	9.15	9.23	9.47

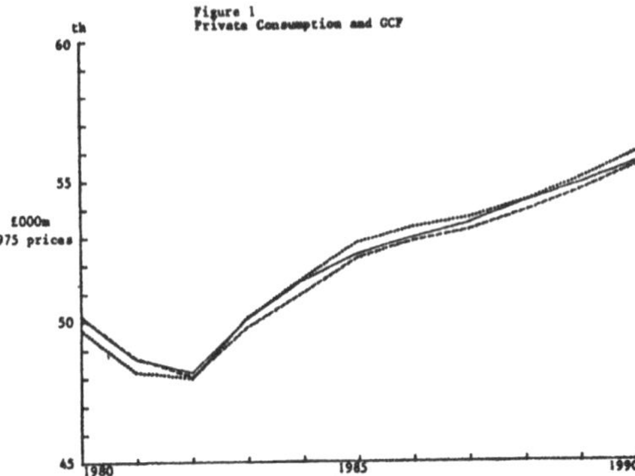

Figure 1
Private Consumption and GCF

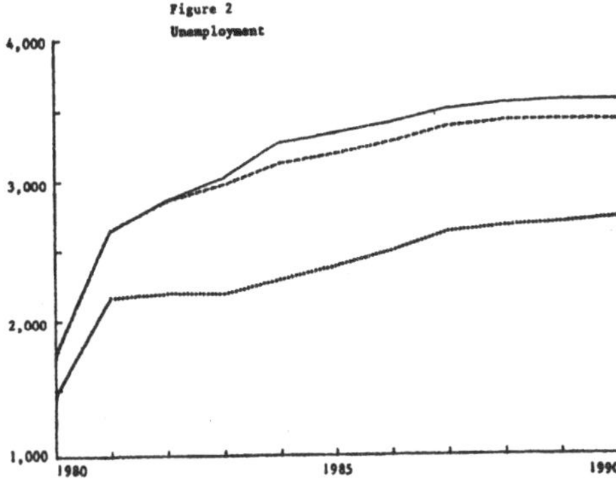

Figure 2
Unemployment

Key to Figures 1-6:

The extraction paths and their influence on the macro-economy

Key

..... 3% p.a. depletion of oil reserves

——— 5% p.a. depletion of oil reserves

——— Standard view (Government statement of depletion policy)

The initial stock is assumed to be 2400 m tonnes. This includes probable and some possible as well as proven reserves.

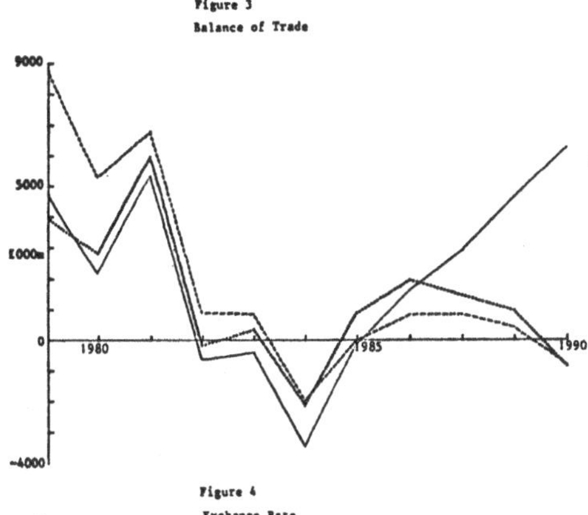

Figure 3
Balance of Trade

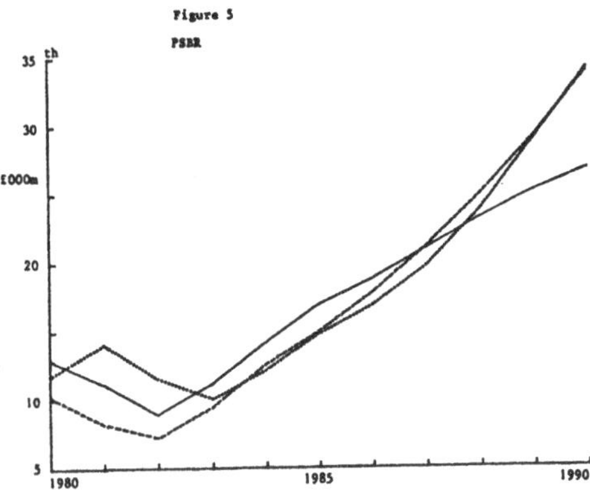

Figure 5
PSBR

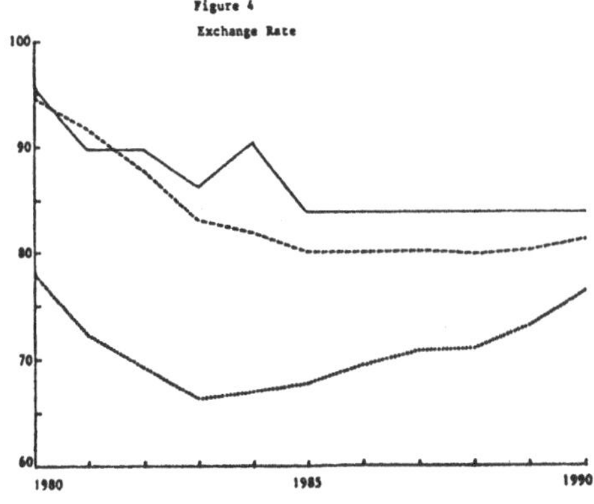

Figure 4
Exchange Rate

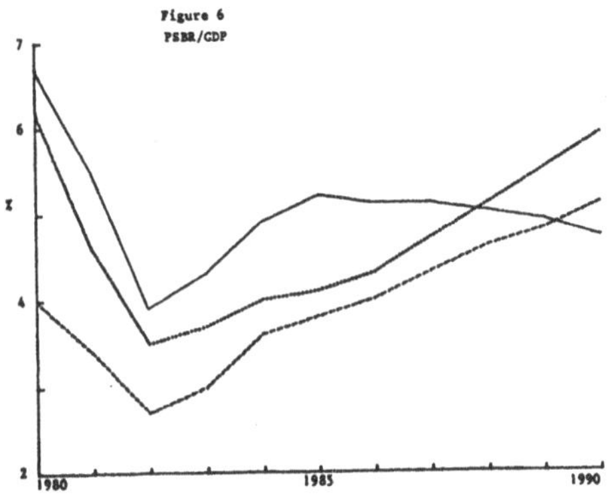

Figure 6
PSBR/GDP

of the paths for the PSBR and unemployment. Optimal disequilibrium paths are calculated for the exchange rate and the output of North Sea oil to ensure either a satisfactory level of public borrowing or of unemployment. Table 2 shows the target paths and some hypothetical standardised priorities. To avoid policy instability a penalty is given to exchange rate movements as well as the exchange rate itself.

3.2. Steering towards the 3% path

Table 3 and Figures 7-11 show the disequilibrium paths derived from the two preference criteria described above. The PSBR case requires a higher rate of oil extraction than the standard case. The extra oil output leads to higher oil company profits and thus to increased tax revenues. The fact that the 3% path is taken as a reference point leads to lower unemployment and a lower exchange rate than in the 5% extraction path discussed in Section 2.3.

The second disequilibrium path considered shows how lower unemployment can be achieved from lower oil output, because this leads to a lower exchange rate. This low exchange rate generates increased exports which raise investment through the accelerator mechanism and further add to output through Lord Kahn's multiplier process. It should be noted that this increased activity also generates extra tax revenue, so that after 1983 the PSBR remains below 4% of GDP. In the economy as represented by our model a given financial target may be consistent with many states of the real economy. Our analysis shows that rapid depletion of UK oil reserves may generate some additional tax revenues, but has unpleasant consequences for the labour force and the non-oil sector. The deflationary effects of rapid depletion may however be avoided if it is accompanied by appropriate measures, such as investment incentives to highly innovative industries or export subsidies to manufacturing [3].

4. Future work

The extraction rate in section 3 is regarded as a policy variable rather than the outcome of the response to tax rates and incentives provided by the Government, although [5] shows how taxes can be used to set the rate at which a profit-maximising company will extract its oil. Any extension of the analysis to endogenise the extraction rate is relatively straightforward following the construction of an appropriate model of the oil companies behaviour. Another useful extension would be to optimise the extraction path on the basis of a maximised present discounted value of private demand.

Appendix: Optimal control of large non-linear econometric models

The problem may be formulated as minimise $J = \|Y - Y^*\|_Q$ subject to the non-linear constraints of the econometric model $Z = F(Z,U)$ and the linear target restraints $Y = H\,Z + h$, where the vectors Y, U and Z denote respectively the vectors containing the target, control and endogenous variables for all periods of the planning horizon. The dimension of F is ± 4000 and dim(U) = 22 in the applications of this paper. The

Table 2: Preferences of two optimal depletion runs

Variable	Target	Priority PSBR case	Priority Unemployment case
Balance of Trade	0	1	1
Exchange Rate	Path followed in 3% extraction case	40	40
Change in exchange rate	0	40	40
Unemployment	1 million	0	40
PSBR/GDP	3%	200	0
Oil extraction	3% extraction path	40	40

Table 3: Outcome of two optimal depletion runs

	PSBR case	Unemployment case
Terminal Oil Stock (m. tonnes)	1517	1877
Average Unemployment (thousands)	2485	1713
P.D.V. of Private Consumption + GFCF (£m 1970 prices)	446203	447330

Figure 7: Exchange Rate

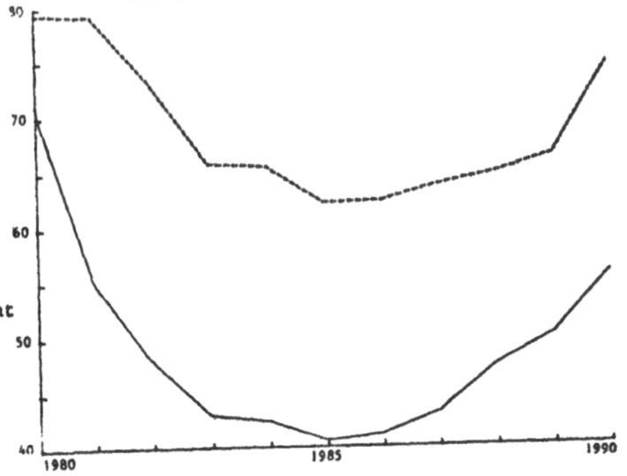

Figure 8: Oil extraction

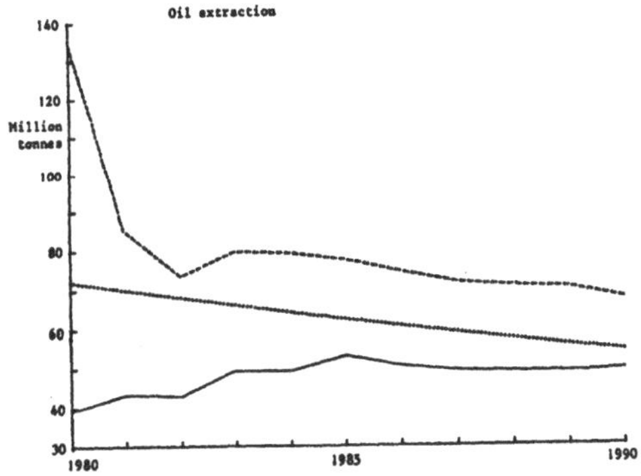

Figure 10: PSBR/GDP

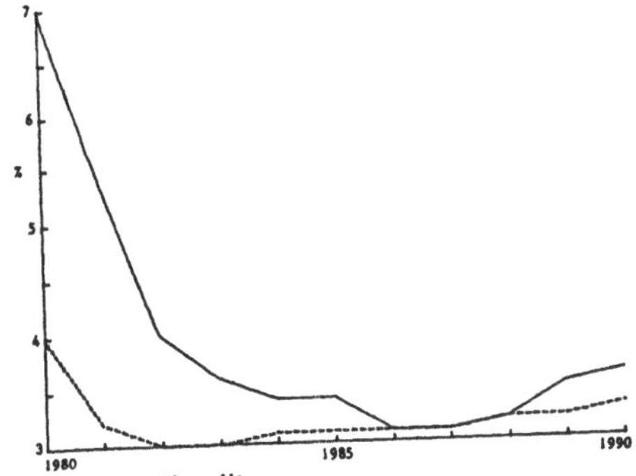

Figure 9: Unemployment

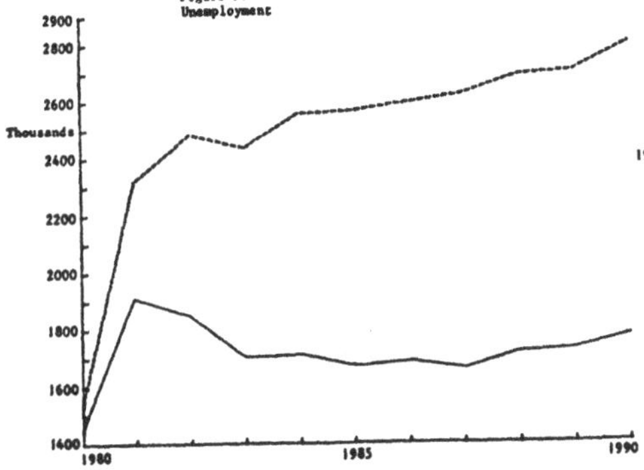

Figure 11: Private Consumption + GFCF

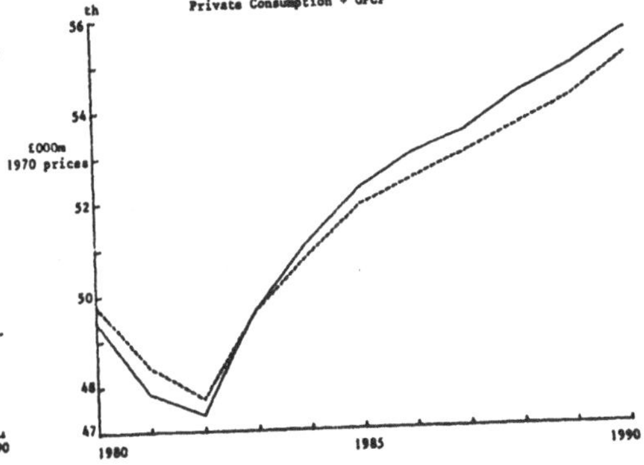

regulation criterion is the same form as in [12] and F(Z,U) represent the equations of the econometric model stacked for all periods of the planning horizon. If it is assumed that the mapping F is continuous and invertible one can appeal to the implicit function theorem which says that the model may be written in reduced form, $Z = F^r(U)$, as a function of the exogenous variables only. Usually is is not possible to do this analytically but the Gauss-Seidel or Newton-type solution program of the econometric model generates the endogenous variables Z once values for the exogenous variables U are known. Since Y is linearly related to Z the regulation criterion may be written as a function of U, J(U), only.

The optimal solution to the reduced problem should satisfy the normal equations of non-linear least squares $(\partial^2 J(U)/\partial U^2 U)\ \delta U + \nabla J(U) = 0$, where the reduced gradient and reduced Hessian are given by $\nabla J(U) = 2\ M'Q\ \{Y(U) - Y^*\}$ and $(\partial^2 J/\partial U^2) = 2\left[M'QM + \sum_{i=1}^{\dim(U)} (\partial M'/\partial U(i))Q\ \{Y(U) - Y^*\}\right]$. The dynamic multiplier matrix is given by $M = H(\partial F^r/\partial U)$ and for causal (without rational expectations) econometric models is lower block-triangular. The normal equations yield the Newton-Raphson iteration step defined by $\delta U = -\ (\partial^2 J(U)/\partial U^2)^{-1}\ \nabla J(U)$, whilst if one is prepared to ignore second order derivative information about the econometric model (which is typically very expensive to obtain) one could apply the Gauss-Newton correction $\delta U = -\ (M'QM)^{-1}\ M'Q\ \{Y(U) - Y^*\}$. The Gauss-Newton expression may be found with the aid of the techniques of [8]. The Gauss-Newton algorithm proceeds by adjusting U with the correction δU until convergence is achieved [10]. The Gauss-Newton method does not guarantee a global optimum, hence the iteration step is complemented with the line search $U^{k+1} = U^k + \alpha\ \delta U^k$, $(0 < \alpha < 1)$, where the step length parameter α is chosen such that U^{k+1} ensures a reduction in the regulation penalty [9, 14]. Overall convergence should be tested by checking that the norm of the vectors J^k, $(J^{k+1} - J^k)$ and $(U^{k+1} - U^k)$ are all sufficiently close to zero. The dynamic multiplier matrix M may be estimated by numerical perturbation, which requires $\{\frac{1}{2}\ T(T+1) \dim u_t\}$ model simulations per iteration, the adjoint variable technique [15] or a cheap sub-optimal approach of updating time-invariant dynamic multipliers. The latter technique updates M at each iteration with the aid of the variable metric formula for updating derivatives [4], which ensures that the errors in the estimate of the dynamic multipliers are reduced at each iteration. The new update of the dynamic multipliers should preserve the dynamic structure. In other words we wish the correction to be lower triangular, otherwise damage is done to the convergence properties of the least squares algorithm. Hence we make use of sparse variable metric updating formulae [15]. The extension to the case of a non-quadratic welfare criterion and inequality constraints may be found in [13].

References

1. Barker T.S. (1981). 'Depletion Policy and the De-industrialisation of the UK economy' - Energy Economics, 3, 2, 71-79.

2. Barker T.S., Borooah V.K., van der Ploeg, F., and Winters A.L. (1980). 'The Cambridge Multisectoral Dynamic Model: an Instrument for National Economic Policy Analysis' - Journal of Policy Modeling, 2, 319-344.

3. Barker T.S. and Brailovsky V. (1981). Oil or Industry? Academic Press, London.

4. Broyden C.G. (1965). 'A class of methods for solving nonlinear simultaneous equations' - Maths. of Comp., 21, 368-381.

5. Dasgupta P.S. and Heal G.M. (1979). Economic Theory and Exhaustible Resources - Cambridge Economic Handbooks, James Nisbet, Welwyn Garden City, UK.

6. Ellman M. (1977). 'Report from Holland: the economics of North Sea oil hydrocarbons' - Cambridge Journal of Economics, 1, 281-290.

7. Forsyth P.J. and Kay J.A. (1981). 'Oil Revenues and Manufacturing Output' - Fiscal Studies, 2, 2, 9-18.

8. Gill P.E. and Murray W. (1976). 'Algorithms for the Solution of the Nonlinear Least Squares Problem' - National Physical Laboratory Report NAC71 - Teddington, UK.

9. Golub G. (1965). 'Numerical Methods for Solving Linear Least Squares Problems' - Numerische Mathematik, 7, 206-217.

10. Holbrook R.S. (1975). 'Optimal Policy Choice Under a Nonlinear Constraint: an Iterative Application of Linear Techniques' - Journal of Money, Credit and Banking, 7, 1, 33-49.

11. Hotelling H. (1931). 'The Economics of Exhaustible Resources' - Journal of Political Economy, 131-178.

12. Ploeg F. van der (1981). 'The industrial implications of an optimal reflationary mix for the British economy' - forthcoming in Applied Economics.

13. Ploeg F. van der (1981). 'Medium-Term Planning with a Multisectoral Dynamic Model of the U.K. Economy' - forthcoming in the International Journal of Systems Science.

14. Rustem B. and Zarrop M.B. (1978). 'Newton Type Methods for the Optimisation and Control of Nonlinear Econometric Models' - P.R.E.M. Discussion Paper 25 - Imperial College, London.

15. Schubert L.K. (1970). 'Modifications of a quasi-Newton method for nonlinear equations with a sparse Jacobian' - Maths. of Comp., 24, 27-30.

16. Singh A. (1979). 'North Sea Oil and the Reconstruction of UK Industry' - pp. 202-224 in Blackaby (ed.) De-industrialisation, Heinemann, London.

17. Stone J.R.N. and Brown A. (1962). A Computable Model of Economic Growth - Chapman and Hall, London.

18. Worswick G.D.N. (1980). 'North Sea Oil and the Decline of Manufacturing' - National Institute Economic Review, November, 22-26.

19. Wood A. (1975). A Theory of Profits - Cambridge University Press.

THE GATHERING OF A COMMODITY THROUGH A PERIODIC MARKETING RING

A. H. Zemanian

State University of New York at Stony Brook
Stony Brook, N.Y. 11794

Abstract. A dynamic economic analysis of a single isolated periodic marketing ring is given. The analysis determines a trajectory for the vector of prices, commodity flows, and stored amounts from an appropriate set of initial conditions. Under certain conditions on the slopes of the various supply and demand functions in the system, the ring has a unique equilibrium state, and that state is asymptotically stable.

Introduction. We present herein some results concerning one form of a periodic marketing network. Such networks occur commonly in third-world countries, especially in the rural areas. The particular form examined herein is shown symbolically in Figure 1. It is an isolated ring of markets consisting of n-1 rural markets $\phi_1,\ldots,\phi_{n-1}$ and a single urban market ϕ_n. (In Figure 1, we have chosen n-5.) The trading activity in the ring is taken to be as follows. ϕ_1 is assumed to open only on the first day of each marketing week, ϕ_2 on the second day,..., ϕ_{n-1} on the penultimate day, and ϕ_n on the last day of the marketing week. A group of traders proceeding out of ϕ_n arrive at ϕ_1 on the first market day and individually buy various amounts of a certain commodity, which we may think of as an agricultural staple. The other agents in ϕ_1 are primarily local suppliers, perhaps farmers, but some of them may be local consumers. The traders proceed on to ϕ_2 and buy more of the same commodity. This process continues on through $\phi_3,\ldots,\phi_{n-1}$. Finally, the traders return to ϕ_n where they either sell or store the commodity. Thus, the other agents in ϕ_n are taken to be primarily buyers, possibly wholesalers. In short, the traders gather the commodity in ϕ_1 through ϕ_{n-1}, transport it to ϕ_n, and supply it to the urban community. They then proceed on to ϕ_1 again to repeat the process for the next marketing week.

Our objective in this paper is to create a dynamic economic model of this marketing ring and to state some conclusions that can be drawn from the model. Space limitations do not allow a complete mathematical exposition, and so we shall merely explain our ideas with the aid of graphs. This has the advantage of rendering the presentation more understandable as well as concise, at the expense of precision. A complete rigorous analysis of the system will appear elsewhere [1].

A Traders Excess-Supply Function in ϕ_k, k=1,...,n-1. The economic behavior of each trader in any one of the ϕ_k, where k=1,...,n-1, on day t is specified by an excess-supply function $S_k^i(p,t)$ having the typical shape shown in Figure 2. This curve is

This work was supported by the National Science Foundation under Grant MCS 80-20386.

derived by treating each trader as a profit-maximizing firm that supplies the service of transferring ownership of the commodity over space and time.

$C^i(t)$ is the amount of goods the ith trader brings into ϕ_k from ϕ_{k-1}. For ϕ_1 the value of $C^i(t)$ is always zero, for the trader never carries goods out of ϕ_n. $E^i(t)$ is the price the ith trader expects to receive in ϕ_n the next time he returns to ϕ_n. $E^i(t)$ is determined by that trader's prior experience and knowledge of prices in the system; it is a function of those prior prices. (That function was called a "memory function" in other works.) T^i is the minimum per-unit cost to the ith trader for transporting the commodity from ϕ_k to ϕ_n. The location of the trader's excess-supply curve $S_k^i(p,t)$ is determined by the value of $E^i(t)$ and $C^i(t)$. Then, the clearance price P(t) in ϕ_k at t determines the amount $Q^i(t)$ the ith trader buys. Thus, $C^i(t+1)=C^i(t)-Q^i(t)$ is the amount the trader carries on to ϕ_{k+1} at the end of day t.

Ordinarily, the various traders have similar excess-supply curves and face a negative aggregate excess-demand function from the farmers and local consumers in ϕ_k. As a result, a typical price in P(t) will lie below the ordinate cross-over point P_c^i, as is indicated in Figure 2, and the trader in fact buys the amount $-Q^i(t)>0$ in ϕ_k. However, this model is flexible in that it allows the trader to sell $Q^i(t)>0$ goods in ϕ_k if P(t) is larger than P_c.

A Trader's Excess-Supply Function in ϕ_n. We assume that the ith trader has a storage facility in ϕ_n and therefore is free to store goods as well as sell (or buy) goods in ϕ_n. Figure 3 illustrates his excess-supply function in ϕ_n on day t. It is determined by his weekly storage costs, the price $F^i(t)$ he expects to receive in ϕ_n one week hence, the amount $C^i((t)$ he has just transported into ϕ_n, and the amount $A^i(t-n)$ he has in storage from the preceding week. B^i is his maximum storage capacity. I^i is his marginal storage cost for a unit of the commodity when he is storing the amount B^i. P_c^i again denotes a cross-over price. When $P(t)>F^i(t)$, he sell everything he holds, both $C^i(t)$ and $A^i(t-n)$. For $P_e^i<P(t)<F^i(t)$, he sells $Q^i(t)$ and stores $A^i(t)$ for the next market day in ϕ_n. For $P(t)<P_c^i$, he buys some goods and adds it to his storage. This model allows $C^i(t)+A^i(t-n)-B^i$ to be positive, in which case there is no cross-over price - or equivalently we can set $P_c^i=0$; when this is so and when $P(t)<F^i(t)-I^i$, the trader stores the amount B^i and sells the goods he holds in excess of his storage capacity, namely, $C^i(t)+A^i(t-n)-B^i$.

Ordinarily, the various traders have similar excess-supply curves in ϕ_n, and they face a positive demand function. As a result, the condition $P(t)>P_c^i$ will hold for most, if not all, of them. This means that most of them will sell at least part of their goods.

Clearance in ϕ_k, k=1,...,n-1. Our model assumes that each market clears on its market day. Hence, in ϕ_k, where k=1,...,n-1, a price P(t) is achieved where supply equals demand. This is illustrated in Figure 4, where $S_k(p,t)$ is the aggregate excess-supply function of all the traders in ϕ_k and $D_k(p,t)$ is the aggregate excess-demand function they face. $P(t)$ is the clearance price and $Q(t) = \Sigma Q^i(t)$ is the total amount of goods exchanged. Figure 4 illustrates the usual case where $D_k(P(t),t)$ and $S_k(P(t),t)$ are negative. This means that the traders as a whole buy goods from the farmers.

Clearance in ϕ_n. This is illustrated in Figure 5. As before, $S_n(p,t)$ is the aggregate excess-supply function of all the traders and $D_n(p,t)$ is the aggregate excess-demand function they face. We assume in this case that $D_n(p,t)$ is positive for all p; this occurs when no agents other than the traders supply the commodity in ϕ_n. Thus, at the clearance price $P(t)$ the traders as a whole sell the amount $Q(t) = \Sigma Q^i(t)$ to the wholesalers in ϕ_n.

The Dynamic Behavior of the Marketing Ring. The above construction assumes that there is perfect competition in each market and that each market achieves its unique equilibrium state on its market days before any appreciable exchange of goods takes place. Observers of periodic markets have reported that at least in some cases these are not bad assumptions. (Another common situation occurs when groups of agents collide and form oligopolies or oligopsonies. Even monopolies and monopsonies can happen. (The latter case is discussed in another paper.) However, the occurrence of equilibrium from day to day in the various markets does not mean that the system as a whole is in equilibrium. It ordinarily will not be. Instead, disequilibrium is the rule, and the system follows a dynamic variation.

Such a dynamic trajectory is uniquely determined once an appropriate set of initial conditions is assumed. Recall that each trader's expected price $E^i(t)$ or $F^i(t)$ in a given ϕ_k is determined by a rule (unspecified in this paper but different in general for different traders) by which the trader prognosticates about $E^i(t)$ or $F^i(t)$ from various past prices. Assume that ϕ_n is open on $t=0$, ϕ_1 on $t=1$, ϕ_2 on $t=2$, and so forth for both positive and negative integer values of t. Assume also that enough prices are specified for $t=0,-1,-2,\ldots$ so that every trader's $E^i(T)$ or $F^i(T)$, where $T>0$, can be determined from the appropriate prognostication rule once the prices for $t=1,2,\ldots, T-1$ are also determined. Assume furthermore that the amount $A^i(0)$ every trader puts into storage at $t=0$ is also given. Finally, assume that the $D_k(p,t)$ are given for all $t\geq 1$ and for all k. Then, clearance in ϕ_1 at $t=1$ is established by using Figures 2 and 4. (In this model $C^i(1) = C^i(n+1) = C^i(2n+1) = \ldots = 0$, for the traders never carry goods from ϕ_n to ϕ_1.) This determines $P(1)$, $Q^i(1)$, and $C^i(2)$ for every i. (Recall that $C^i(2)$ is the amount the ith trader carries on toward ϕ_2 at the end of the market day $t=1$. Moreover, each expected price $E^i(2)$ is determined by one of the prognostication rules. Thus, the needed initial conditions are updated by one unit of t. So, by the same procedure $P(2)$, $Q^i(2)$, $C^i(3)$ can all be determined through Figures 2 and 4 again. Also, the $E^i(3)$ can be determined by the prognostication rules. Proceeding in this fashion, we can determine all the prices, quantity flows, and expected prices for $t=1,2,\ldots,n-1$.

At t=n, we have the $F^i(n)$ from the prognostication rules, the $C^i(n)$ from the equilibrium in ϕ_{n-1} at t=n-1, and the $A^i(0)$ from the assumed initial conditions. Then, Figures 3 and 5 can be used to determine the equilibrium conditions in ϕ_n at t=n and thereby P(n), Q(n), and all the $Q^i(n)$ and $A^i(n)$. Thus, all the initial conditions have been updated by one week. Moreover, the prognostication rules determine the $E^i(n+1)$. So, the analysis can be continued in this recursive fashion from week to week. In summary, under appropriate initial conditions, our model has a completely determined dynamic economic behavior.

Overall Equilibrium for the Entire Marketing Ring. An equilibrium state for the entire marketing ring can occur only when the aggregate excess-demand function $D_k(p,t)$ for all k=1,..., n are fixed with respect to t. In this circumstance an overall equilibrium is said to occur if, for each k, the price $P_k(t)$ does not vary over the sequence of weekly market days for market ϕ_k and, for each i, stored amount $A_n^i(t)$ does not vary from week to week. The set of constant prices, one for each market, in conjunction with the set of all stored amounts, one for each trader, is called an equilibrium state.

One can show that, if the supply and demand curves of Figure 2 through 5 are continuous and have the shapes indicated therein, then our model has a unique equilibrium state. Moreover, the stored amounts are all zero in an equilibrium state; this means that in an equilibrium state no trader stores a nonzero amount of goods indefinitely. It can also be shown that, under certain assumptions upon the magnitudes of the slopes of the supply and demand functions, the equilibrium state is asymptotically stable. For proofs of these assertions, see the reference cited below.

A Final Comment. We have discussed herein only one form of a periodic marketing system. However, this analysis can be - and has been - applied to several different types of periodic marketing systems. In fact, it has also been used for a dynamic economic analysis of certain daily (rather than periodic) marketing networks found in the industrial countries. Under various assumptions, the conclusions obtained above can also be drawn for these other kinds of marketing systems.

REFERENCE

[1] A.H. Zemanian, "Equilibrium and stability in a periodic marketing ring," SIAM Journal on Algebraic and Discrete Methods, in press.

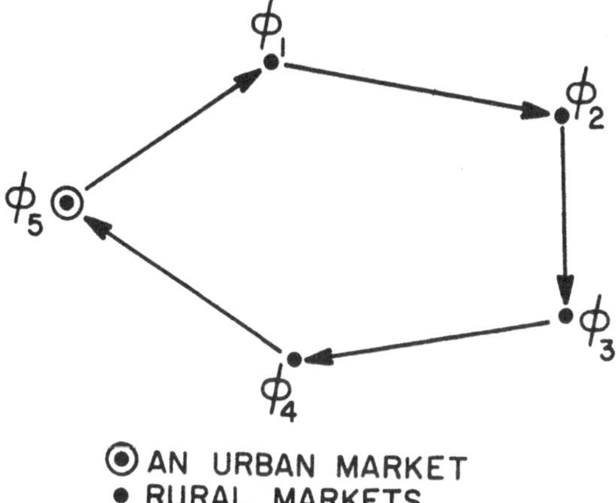

Figure 1. A periodic marketing ring.

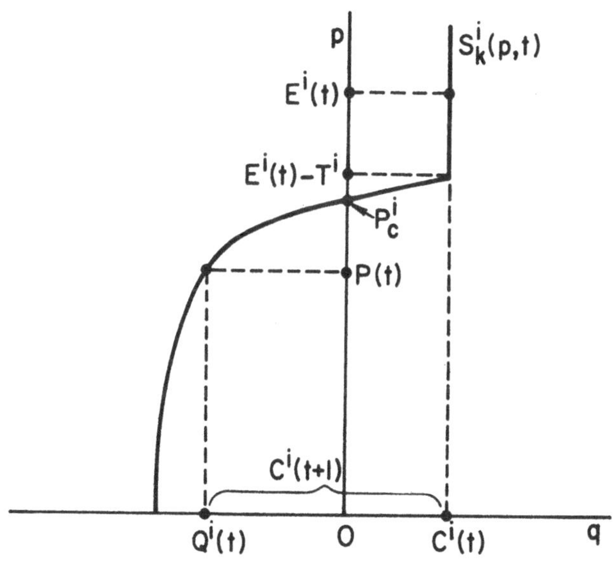

Figure 2. A trader's excess-supply function in Φ_k, k = 1, ... , n-1 on day t.

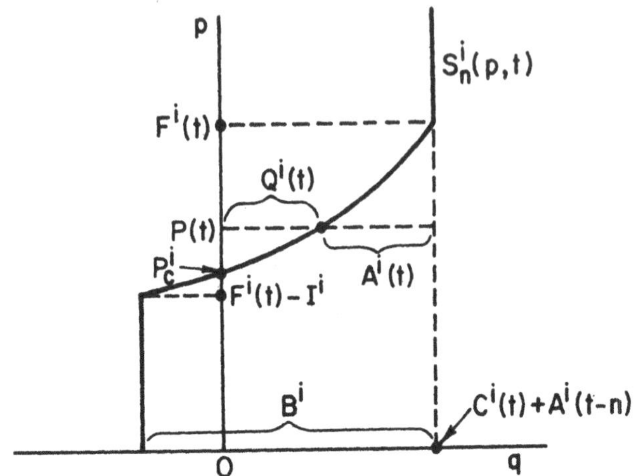

Figure 3. A trader's excess-supply function in ϕ_n on day t.

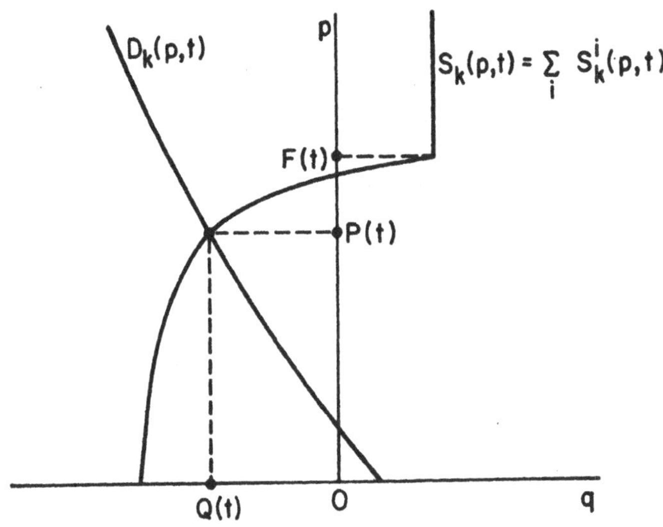

Figure 4. Clearance in ϕ_k, k=1, ... , n-1.

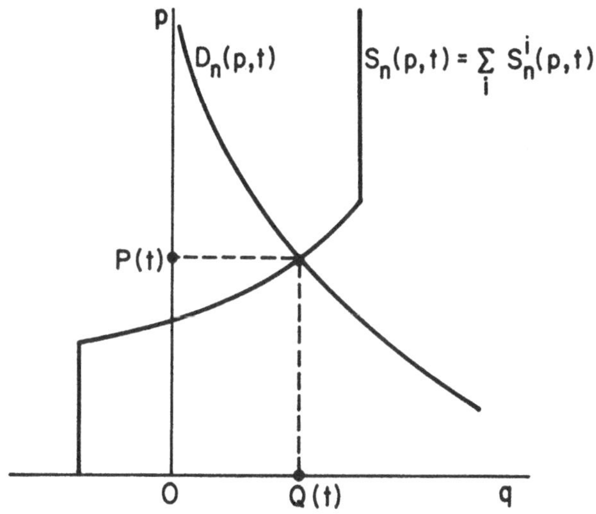

Figure 5. Clearance in Φ_n.

FORECASTING SECTORIAL OUTPUTS UNDER UNCERTAINTY

Paolo Caravani
Istituto di Analisi dei Sistemi ed Informatica del C.N.R.
Via Buonarroti, 12
00185 Roma

ABSTRACT

Information on technology, investment and consumption, as collected in traditional dynamic models of the Leontief type, is usually insufficient to determine future output. Indeterminateness can be handled mathematically by abstract theory of point-to-set mappings but the rôle of intersectoral flows, technical substitution and investment remains unexplained. When substitution dynamics prevails over growth-decline of total output sectoral predictions are possible on the basis of a bilinear convex model whose matrices retain the economic significance of the Leontief model. This approach, intermediate between traditional and abstract theory, permits to establish a certain long-run invariance of the production vector in presence of uncertainties affecting technology investment, and consumption pattern.

1. INTRODUCTION

Growth and distribution are inseparable aspects of multisector economies. Yet considerable efforts in separating the two were spent by economic theorists of the neoclassical tradition willing to pay, as usual, the gains in insight at the cost of abstraction. Their efforts were not addressed equally, greater attention being focused on growth, much less on distribution of sectoral output.

In economies experiencing fast growth, weak sectoral interaction and perfect competition balanced growth assumptions may prove a reasonable device to make predictions or suggest development policies. Even when more unbalanced conditions prevail the advantages of such a harmonious and ideal path have all-too-well been established in turnpike theory.

However, recent trends in all industrialized countries pushed turnpike time into the realm of science fiction, with yearly growth rates of aggregate output seldom exceeding 3 ÷ 4 percent in real terms, while sectoral quota under the swift upsurge of new technology and investment options suffer, or enjoy, shocks of up to 100 percent. A considerable toll for some, a modest pursuit for all.

If relative growth over average growth of the system as a whole became a crucial concern to sectorized societies, such awareness should surface theoretical thought by examining the consequences of new and more realistic assumptions. In what follows differences in production in different times and sectors will be looked at through normalizing lenses. The balanced growth abstraction will be replaced by the assumption that output vector is measured yearly so as to keep the sum of its components equal to one.

Sectoral output changes, relatively to one another, result from technological innovation which requires investment which depends on interest rates which relate to global economic growth. Assuming zero growth rate of aggregate output, as it is done here, no progress can be made at clarifying the nature of that link: sectoral dynamics will not be explained in terms of growth.

Sectoral dynamics will rather be described here as a function of technological choice. A set of rival technologies is assumed to exist in each sector, together with a pre-specified set of investment options and consumption possibilities. What will be known ex-post as the prevailing technology investment and consumption pattern is only known ex-ante as the cartesian product of those three sets. It is plausible to expect some of the uncertainty embodied in the technology, investment consumption sets to be transferred on output. By this transfer, the effects of technological improvement in sector A as they were expected assuming no interaction are offset partly by (uncertain) investment in sector B, partly by (uncertain) consumption in sector C and output of A, relative to B's or C's, may turn out lower than anticipated. The question addressed here is the nature and the consequence of that transfer mechanism.

After reviewing the Leontief dynamic scheme in sec. 2, a bilinear convex model is introduced in sec. 3 with the aim of retaining the original significance of the intersectoral matrices while leaving room for predictive uncertainty in a more convenient mathematical form.

The resulting dynamic model is studied in sec. 4 and its properties are illustrated in sec. 5 with the aid of a numerical example. The main conclusion of the paper is that a certain invariance of long-run sectoral outputs can be established despite imperfect knowledge of the intersectoral matrices.

2. A REVIEW OF DYNAMIC LEONTIEF MODEL

In 1951 Prof. Wassily W. Leontief [5] introduced his now classical input-output model

$$x(t) = Ax(t) + y(t) \tag{1}$$

($x \sim$ output; $y \sim$ final demand; $A \sim$ technology) on the basis of the assumptions

A1 the economy comprises n productive sectors each producing one homogeneous good, with no joint production

A2 all plants operate at full capacity, so x denotes indifferently produced quantity or installed capacity

A3 production is characterized by constant returns to scale.

The model, in this form, is static.

When a model of final demand (y) is required, dynamic aspects emerge. In a closed economy y comprises consumption and investment. The former is linked to the level of output via a propensity matrix C, the latter to the increase in capacity via a capital coefficient matrix B. Therefore

(2) $$y(t) = Cx(t) + B[x(t+1) - x(t)]$$

which substituted into (1), yields the model in dynamic closed form (Leontief [6], 1953)

(3) $$Bx(t+1) = [B + I - A - C]x(t)$$

Equation (3) may have different uses [9] depending on what subset of variables is known and what is not. When the use is that of predicting future outputs given x(0), the output vector at some base year, one should - in principle - solve (3) as a forward difference equation. This, however, strictly depends on two additional, awkward assumptions

A4 matrix B is invertible at all t > 0.

A5 A, B, C are known at all t > 0.

Assumption A4 fails when the economy includes a sector producing no capital goods, like agriculture [1], [3], [7], [8]. In that case, there are more capacity increases than capital goods so the latter do not explain the former uniquely. Matrix B is not full rank(*) and next year production lies somewhere within an uncertainty region. That region is the intersection of $\mathbb{R}^n_+$ with the set $\{x\colon x = \hat{x} + b,\ b \in \text{null}(B)\}$ where $\hat{x}$ is a solution of (3), a closed convex polyhedron. Denoting $M[x(t)]$ this region, eq. (3) must be viewed as a point-to-set mapping associating x(t) to

(4) $$x(t+1) \in M[x(t)]$$

Assumptions A5 fails because of natural uncertainty.

What is known is that the triple {A,B,C} prevailing at each future time is an element of the threefold cartesian product $\mathcal{A} \times \mathcal{B} \times \mathcal{C}$, a composite finite set collecting all possible options for technology, investment, and consumption. If the rule $t \to \{A(t), B(t), C(t)\}$ is left unspecified, eq. (3) should, again, be viewed as point-to-set, with the l.h.s. taking on as many values as there are elements in $\mathcal{A} \times \mathcal{B} \times \mathcal{C}$. Some form ought to be assumed for these sets and we will explicitly replace A4, A5 by

A6 (convexity of alternatives) If $\{A_i, B_i, C_i\}$, $i=1,2,\ldots,m$ are possible options, then $u_i \geq 0$ and $\sum_{i=1}^{m} u_i = 1$ imply that $\sum_{i=1}^{m} u_i\{A_i, B_i, C_i\}$ is also a possible option.

Thus, for instance, if sector i's unit production requires from sector j 10 units of input under technology A_1 and 20 under A_2, and neither A_1 or A_2 is known to prevail in all plants but some use A_1, and the remaining A_2; then i's unit requirements from j shall fall anywhere between 10 and 20. Under convexity of alternatives eq. (3) is a point-to-set mapping of convex polyhedral type, just like (4).

We finally remark that if A2 is dropped, output increase $x(t+1) - x(t)$ in the r.h.s. of (2) ought to be replaced by capacity increase $z(t+1) - z(t)$ with $\Delta z \geq \Delta x$, so that (3), written in terms of output, holds with $\leq$ replacing =. The mathematical consequence, again, is indeterminacy and (3) becomes a point-to-set mapping of type (4).

(*) b_{ij} is the quantity of investment goods produced by sector i per unit net capacity increase in sector j. If, for instance, the production process includes a sector k which produces no investment goods, typically agriculture, B will contain all zeros in row k and hence be singular.

Our conclusions are

i) Leontief model poses severe restrictions to its use as a predictive tool

ii) when those restrictions are violated a point-to-set mapping of convex polyhedral type appears as the most natural mathematical substitute.

3. A BILINEAR CONVEX MODEL OF MULTISECTORAL PRODUCTION

Motivated by the preceding discussion, we consider an abstract characterization of a production process

$$y \in M[x] \tag{5}$$

where x and y are sectoral production quota at time t and t+1. As we are concerned with relative dynamics rather than growth, we shall assume normalized outputs so that, at all times, x and y are elements of the n-dimensional unit simplex

$$S^{n-1} = \langle e_i \rangle_{i=1}^{n} \tag{6}$$

where e_i is the unit vector along the i-th axis of $\mathbb{R}^n$ and $\langle\cdot\rangle$ denotes convex combination. Several authors studied the case in which the graph of M

$$G(M) = \{(x,y):\ y \in M[x]\} \tag{7}$$

is a polyhedral convex cone [2],[10],[11],[12]. In our case, given the restraints on x and y, $G(M)$ is the convex hull of a finite number N of points in $S^{n-1} \times S^{n-1}$, or

$$G(M) = \langle (x_i,y_i):\ y_i \in M(x_i) \rangle_{i=1}^{N} \tag{8}$$

A necessary and sufficient characterization of M can be given in terms of $n \times n$ Markov matrices (*).

Given a vertix (x,y) of $G(M)$, say the i-th, consider the set A^i of all Markov matrices satisfying $y = Ax$. To study the nature of this set, note that the elements a_{kj} of A must satisfy

$$\begin{cases} Ax = y \\ \mathbf{1}^T A = \mathbf{1}^T \\ a_{kj} \geq 0;\ k,j \in [1,n] \end{cases} \tag{9}$$

So, the solution set is the intersection of 2n hyperplanes in $\mathbb{R}_+^{n^2}$ containing at least the point $a_{kj} = y_k$ $(k,j \in [1,n])$ as direct substitution shows. Therefore, A^i is a non-void convex polyhedron expressible in terms of no more than a finite number m_i of such matrices

$$A^i = \langle A^\ell \rangle_{\ell=1}^{m_i}$$

Collecting all these matrices for each vertix i, $i \in [1,N]$, let m be the cardinality of the resulting set A. Then

$$G(M) = \langle (x_i,y_i):\ y_i \in \langle A^\ell x_i \rangle_{\ell=1}^{m} \rangle_{i=1}^{N}$$

(*) By $n \times n$ Markov matrix it is meant here a square matrix whose columns belong to S^{n-1}.

and $M(\cdot)$ is sufficiently characterized by

(10) $$y \in \langle A^{\ell}x \rangle_{\ell=1}^{m}$$

This characterization is also necessary for otherwise there would be entries of A, say a_{ij}, either less than zero or greater than one and there would be vectors in S^{n-1} like e_j whose image under A would not be in S^{n-1} as assumed. Consequently, M-processes lead naturally to a production model of the form

(11) $$x(t+1) = [\sum_{i=1}^{m} A^i u_i(t)]x(t)$$

with, at any t

$$\begin{cases} x(t) \in S^{n-1} & \text{i.e., no growth assumption;} \\ u(t) \in S^{m-1} & \text{i.e., convexity of alternatives;} \end{cases}$$

in the sense that any sequence $\{x(1),x(2),\ldots\}$ such that all pairs $(x(t),x(t+1))$ are in the graph of M will be also generated by (11) by a proper choice of $\{u(1),u(2),\ldots\}$(*). Comparing (11) and (3) we get

(12) $$[\sum_{i=1}^{m} A^i u_i(t)] = B^{-1}(t)[I + B(t) - A(t) - C(t)]$$

and we may interpret the A^i matrices in (11) in terms of technology investment and consumption matrices of a nonsingular zero-growth Leontief model.

4. ANALYTICAL DEVELOPMENTS

Any trajectory of (3) coincides with some of the trajectories of (11) when u(t) is arbitrarily chosen in S^{m-1}, so properties common to all trajectories of (11) are also shared by (3). This motivates our interest in (11).

When the measure of output is taken to be its value (one) at constant prices, each column of A^i, say the j-th, represents the unit-sum vector of marginal values of the output with respect to the j-th input. Average marginal value of output k with respect to all inputs is

$$\frac{1}{n}\sum_{j=1}^{n} a_{kj}^{i}$$

while its average value is simply $\frac{1}{n}$.

As inputs to sector k are outputs of other sectors, the difference

$$\Delta_k^i = \frac{1}{n}\sum_{j=1}^{n} a_{kj}^{i} - \frac{1}{n}$$

measures the potential gain (in terms of value) of sector k as it is allowed by all other sectors under technology i. Notice that Δ_k^i can be positive, zero or negative. We will assume that the technology eventually prevailing in each sector is such that the marginal value of the output with respect to every input exceeds its potential gain. In other words, the criterion for acceptance in the A-family is that no candidate matrix should include intersectoral transactions yielding lower marginal value

(*) Formally, the structure of (11) is that of a bilinear control system with state variable x and control u. However, no control problem will be formulated at this stage, outside the mentioned choice of a control sequence for which (11) reproduces any given trajectory of (3).

than the allowed potential gain, for otherwise technological substitution would have no reason to take place.

Formally, this implies the restriction on the A^i matrices

$$r_k^i - n\, a_{km}^i < 1 \tag{13}$$

where

$$a_{km}^i = \min_{\ell,h,j} a_{hj}^\ell \qquad r_k^i = \sum_{j=1}^n a_{kj}^i$$

For subsequent developments, it is useful to introduce a family of closed convex polyhedra $\{S_\alpha\}$ defined by

$$S_\alpha \triangleq \langle \frac{1-\alpha}{n}\mathbf{1} + \alpha\, e_k \rangle_{k=1}^n, \quad \alpha \in [0,1]$$

with $\mathbf{1} \triangleq \{1\ 1\ 1\ \dots\ 1\}^T$.

Note that for $\alpha = 1$ $S_\alpha = S_1 = S^{n-1}$, the whole unit simplex; for $\alpha = 0$ $S_\alpha = S_o = \{\frac{\mathbf{1}}{n}\}$, the set containing just the unit simplex centroid.

The family $\{S_\alpha\}$ is directed by inclusion in the sense

$$\alpha < \beta \Rightarrow S_\alpha \subset S_\beta$$

Therefore $S_\alpha \cap S_\beta = S_{\min(\alpha,\beta)}$ $S_\alpha \cup S_\beta = S_{\max(\alpha,\beta)}$ and $\tau = \{S_\alpha, \emptyset\}$ costitutes a topology of sub-simplices in S^{n-1}. As it will be convenient to work with metric spaces, we equip τ with the metric $\rho = |\alpha-\beta|$ and denote this space by $S(\tau,\rho)$. It is then easy to prove that S is complete [4].

Next we turn to asymptotic properties of

$$x(t+1) = [\sum_{i=1}^m A^i u_i(t)]x(t) \qquad x \in S^{n-1} \qquad u \in S^{m-1} \tag{11}$$

and search for the smallest invariant in the $\{S_\alpha\}$ family under arbitrary control law $u(t) \in S^{m-1}$. More precisely, we wonder 1. whether there exists an S_α with $\alpha < 1$ such that $x(t) \in S_\alpha \Rightarrow x(t+1) \in S_\alpha$, 2. what is the minimum value $\hat{\alpha}$ of α for which this holds and 3. whether such an invariant is reachable from any initial state under arbitrary control.

Let the initial state be x. After one period, the set of states reachable by (11) is the image of x under a point-to-set mapping A_o: $S^{n-1} \to 2^{S^{n-1}}$

$$x \to A_o(x) \triangleq \langle A^i x \rangle_{i=1}^m$$

When x is let to vary in S_α, the reachable set is the image of S_α under the set-to-set mapping A_1: $\tau \to 2^{S^{n-1}}$

$$S_\alpha \to A_1(S_\alpha) \triangleq \bigcup_{u \in S_\alpha} \langle A^i x \rangle_{i=1}^m$$

This set is, in turn, contained in a minimal set S_β defined as

$$S_\beta \triangleq \{\bigcap_\gamma S_\gamma : S_\gamma \supset \bigcup_{x \in S_\alpha} \langle A^i x \rangle_{i=1}^m\}$$

and a second set-to-set mapping A_2: $2^{S^{n-1}} \to \tau$ is established

$$A_2(\bigcup_{x \in S_\alpha} \langle A^i x \rangle_{i=1}^m) \triangleq S_\beta$$

The composition $A = A_2 \cdot A_1$ yields finally the mapping A: $\tau \to \tau$

$$S_\alpha \to A(S_\alpha) = S_\beta$$

which can be regarded as point-to-point in τ.
Stated in words, S_β is the smallest element in the $\{S_\alpha\}$ family containing the set of all states reachable from S_α in one period.

We can now prove the following

THEOREM 1. A is a contraction mapping on $S(\tau,\rho)$.

Proof. Fix α. Using the fact that convex polyhedra remain such under a linear transformation, we get

$$\{\sum_{i=1}^{m} A^i u_i x: x\in S_\alpha;\ u\in S^{m-1}\} = \bigcup_{u\in S^{m-1}} \langle \frac{1-\alpha}{n}\sum_{i=1}^{m} r^i u_i + \alpha\sum_{i=1}^{m} a_k^i u_i\rangle_{k=1}^{n}$$

where $r^i = A^i\mathbb{1}$ and $a_k^i = a^i e_k$. Now we seek the smallest γ such that this union is contained in $\langle \frac{1-\gamma}{n}\mathbb{1} + \gamma e_j\rangle_{j=1}^{n}$. This yields the condition

$$\sum_{i=1}^{m} [\frac{r^i}{n}(1-\alpha) + \alpha a_k^i]u_i - \frac{1-\gamma}{n}\mathbb{1} \geq 0 \quad \forall\ k\in[1,n];\ \forall\ u\in S^{m-1} \tag{14}$$

As the bracketed quantity is the convex hull of two nonnegative vectors, condition (14) is violated for γ smaller than a limiting value β satisfying

$$\beta = 1 - \min_{\substack{k\in[1,n]\\ u\in S^{m-1}}} \text{component}\ \{\sum_{i=1}^{m}[r^i(1-\alpha) + n\alpha a_k^i]u_i\}$$

For fixed k, the j-th component of this vector is an element of a convex bounded set, in fact of a closed interval of the real line. As the minimal element in a collection of closed intervals is the smallest number in the collection, we have

$$\beta = 1 - \min_{ijk}(r_j^i(1-\alpha) + n a_{jk}^i\alpha) = 1 - \min_{ij}(r_j^i(1-\alpha) + n\min_k a_{jk}^i\alpha)$$

Rewrite this expression for $\alpha' \neq \alpha$ and let β' the corresponding β. Then

$$\rho(A(S_\alpha),A(S_{\alpha'})) = \rho(S_\beta,S_{\beta'}) = |\beta-\beta'| = |\min_{ij}(r_j^i(1-\alpha') + n\min_k a_{jk}^i\alpha') - \min_{ij}(r_j^i(1-\alpha) + n\min_k a_{jk}^i\alpha)|$$
$$\leq (r_m^\ell - n\, a_{mn}^\ell)\,|\alpha-\alpha'|$$

where $a_{mn}^\ell = \min_{ijk} a_{jk}^i$ and the inequality holds by virtue of Lemma 1 (see Appendix). As the quantity in bracket is contained in (0,1) (see condition (13)) the theorem follows.

A contraction mapping on a complete metric space has a unique fixed point. Thus there exists just one $S_{\hat\alpha}$ in the $\{S_\alpha\}$ family such that

$$A(S_{\hat\alpha}) = S_{\hat\alpha}$$

Furthermore, the contractive property ensures monotonic convergence to $S_{\hat\alpha}$. In terms of our problem, these results can be rephrased in the following

THEOREM 2. There exists just one set $S_{\hat\alpha}$ in the $\{S_\alpha\}$ family such that, for system (11)

i) No trajectory starting in $S_{\hat\alpha+\varepsilon}$ goes outside $S_{\hat\alpha+\varepsilon}$, $\varepsilon > 0$.

ii) All trajectories starting outside $S_{\hat\alpha}$ are eventually in $S_{\hat\alpha}$.

iii) All trajectories starting in $S_{\hat\alpha}$ stay in $S_{\hat\alpha}$.

Therefore $S_{\hat\alpha}$ contains all equilibrium points of (11). In particular, it contains the ergodic set, i.e. all equilibrium states under constant control. However, while all points outside $S_{\hat{\alpha}}$ are disequilibrium, only some points of $S_{\hat{\alpha}}$ are equilibrium (*).

Next we turn to the evaluation of $S_{\hat{\alpha}}$. The results are summarized in

THEOREM 3. The set $S_{\hat{\alpha}}$ in the $\{S_\alpha\}$ family is

$$S_{\hat{\alpha}} = \langle \frac{1-\hat{\alpha}}{n} \mathbf{1} + \hat{\alpha} e_k \rangle_{k=1}^{n}$$

with

i) $\hat{\alpha} = 0$ iff A^i are doubly Markov for all i

ii) $\hat{\alpha} = \dfrac{1}{1+n\hat{h}}$ otherwise

where $\hat{h}$ is computable finitely by the algorithm below

Proof. From Thm 2 $S_{\hat{\alpha}}$ is characterized by the minimum value α for which

$$A(S_\alpha) \subset S_\alpha$$

This leads to the minimization problem: find min α such that

$$\sum_{i=1}^{m} u_i [\frac{1-\alpha}{n} r^i + \alpha a_k^i] = \sum_{j=1}^{n} \lambda_j [\frac{1-\alpha}{n} \mathbf{1} + \alpha e_j]; \quad k \in [1,n]; \; u \in S^{m-1}; \quad \lambda \in S^{n-1}$$

or, find min α such that

$$\sum_{i=1}^{m} u_i [\frac{1-\alpha}{n} r^i + \alpha a_k^i] = \frac{1-\alpha}{n} \mathbf{1} + \alpha\lambda; \quad k \in [1,n]; \quad u \in S^{m-1}; \quad \lambda \geq 0$$

If all A^i are doubly Markov, all rows add up to one and $r^i = \mathbf{1}$ in which case $\sum_{i=1}^{m} u_i \alpha a_k^i = \alpha\lambda$; $u \in S^{m-1}$; $\lambda \geq 0$ is satisfied for all $\alpha \in [0,1]$ and $\hat{\alpha} = 0$. On the other hand, if at least one of the A^i is not doubly Markov, $\alpha = 0$ would imply

$$\sum_{i=1}^{m} u_i [\frac{r^i}{n} - \frac{\mathbf{1}}{n}] = 0, \quad u \in S^{m-1}$$

which is false for some u_i, since some components of r^i are less than one. This proves i).

If $\alpha \neq 0$, we have

$$\lambda = \sum_{i=1}^{n} u_i [\frac{1-\alpha}{n\alpha}(r^i - \mathbf{1}) + a_k^i] \geq 0, \quad k \in [1,n]; \quad u \in S^{m-1}$$

Thus, letting $h = \dfrac{1-\alpha}{n\alpha}$

$$h(r^i - \mathbf{1}) + a_k^i \geq 0, \quad k \in [1,n] \tag{15}$$

As some components of $(r^i - \mathbf{1})$ are negative, the largest in absolute value is upper binding for h. The upper bound $\hat{h}$ is computable as follows.

Let $I \triangleq \{(m,n)\}$ be the set of index-pairs where

$$\{r_\ell^i : i \in [1,m], \ \ell \in [i,n]\}$$

attains its minimum p. Clearly $(p-1)<0$ is the minimal component of the bracket in (15). Let q be the minimal element in the array

$$\{a_{jk}^i : (i,j) \in I, \ k \in [1,n]\}$$

Then (15) implies $h \leq q/(1-p) = \hat{h}$, or

$$\alpha \geq \frac{1}{1+n\hat{h}} = \hat{\alpha}$$

(*) A finer topology than τ may improve the situation in this respect.

This justifies the following

Algorithm (for the evaluation of $\hat{h}$)

1. Evaluate vector r^i by adding up rows of A^i
2. Find minvalue p in the array $\{r^i: i \in [1,m]\}$
3. Store index pairs where p is attained in set I
4. Find minvalue q in the array $\{a^i_{jk}: (i,j) \in I, k \in [1,n]\}$
5. Compute $\hat{h} = q/(1-p)$ and $\hat{\alpha} = 1/(1+n\hat{h})$.

5. INTERPRETATION OF THE RESULTS

The significance of the above results is that present uncertainty in the structural matrices of a multisector model can be related to the uncertainty affecting future output. More precisely, Thm 3 states that under a given set of technology investment and consumption options some output combinations are ruled out independently of the order in which those options may be adopted in time. No set of states outside $S_{\hat{\alpha}}$ is stably attained by the economy. Comparing $S_{\hat{\alpha}}$ with present sectoral composition of the output permits to judge which sectors are likely to enjoy a relatively stable situation, which ones are bound to suffer more or less drastic changes in their production quota.

As an illustrative example, assume a three sector economy with normalized output at period 1

$$x(1) = \{.147 \;\; .655 \;\; .198\}^T$$

From t = 2 onwards, assume total uncertainty over the following options

$$A^1 = \begin{bmatrix} .1 & .3 & .2 \\ .2 & .1 & .4 \\ .7 & .6 & .4 \end{bmatrix}; \qquad A^2 = \begin{bmatrix} .5 & .2 & .6 \\ .3 & .2 & .1 \\ .2 & .6 & .3 \end{bmatrix}; \qquad A^3 = \begin{bmatrix} .7 & .6 & .2 \\ .2 & .2 & .2 \\ .1 & .2 & .6 \end{bmatrix}$$

We are then led to study the asymptotic behaviour of (11) with $m = 3$ and arbitrary control. By Thm 3, x(t) will eventually enter $S_{\hat{\alpha}}$ with $\hat{\alpha} = 4/7 = .571$. The quota presently held by sector 2, $x_2(1) = .655$ is in a dis-equilibrium situation. Sector 2 will have to reduce its output by at least $.571 - .655 = -.084$, that is more than a 8.4% drop from present share of total output.

Howmuch more? That remains in the uncertainty margin: it depends on what the other sectors do within their respective growth margins which are $.571 - .147 = +.424$ for x_1 and $.571 - .198 = +.373$ for x_3. This, of course, depends on what technology investment and consumption pattern is actually going to prevail.

ACKNOWLEDGMENT

I whish to thank dr. L. Grippo and dr. C. Leporelli for useful discussion and suggestions. Rome, 15, Nov. 1981.

APPENDIX

In the proof of Thm 1, use is made of the following

LEMMA. Let A_i, a_i: $i = 1,2,\ldots,n$ be scalars satisfying $A_i > a_i$, $A_p - a_p = K$ where $a_p = \min_i a_i$. Then, for all α and β in [0,1]

$$|\min_i[A_i(1-\alpha)+\alpha a_i] - \min_i[A_i(1-\beta)+\beta a_i] \leq K|\alpha-\beta|$$

PROOF. For any i we have $a_p \leq a_i$ and let $A_q \leq A_i$. If $p = q$ we have the situation in fig. 1.

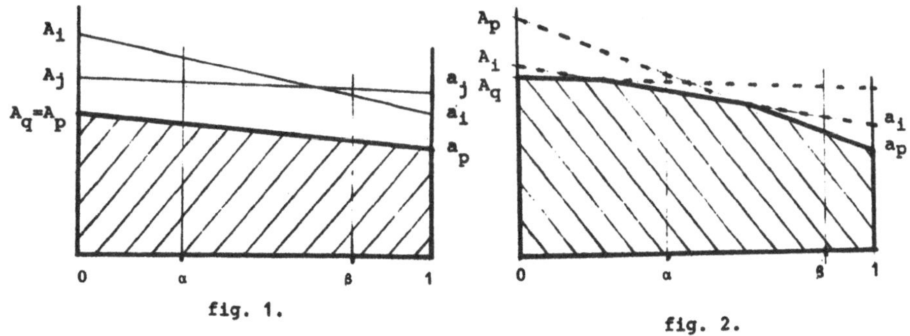

fig. 1.

fig. 2.

and

$$|\min_i[A_i(1-\alpha)+\alpha a_i] - \min_i[A_i(1-\beta)+\beta a_i]| = (A_p - a_p)|\alpha-\beta| = K|\alpha-\beta|$$

If $p \neq q$ we have the situation in Fig. 2, and

$$|\min_i[A_i(1-\alpha)+\alpha a_i] - \min_i[A_i(1-\beta)+\beta a_i] \leq (A_p - a_p)|\alpha-\beta| = K|\alpha-\beta|$$

REFERENCES

[1] S.L. CAMPBELL: Non regular Singular Dynamic Leontief Systems. Econometrica, Vol. 47, 1979, pp. 1565-68.

[2] J.J.M. EVERS: The Dynamics of Concave Input/Output Processes. Lect. Notes in Eco. and Math. Systems, Springer Verlag, Vol. 168, Ed. Kriens, 79, pp. 73-114.

[3] D. KENDRICK: On the Leontief Dynamic Inverse. Quanterly Journal of Economics, Vol. 86, 1972, pp. 693-696.

[4] A.N. KOLMOGOROV, S.V. FOMIN: Introductory Real Analysis. Dover Publ., 1975.

[5] W.W. LEONTIEF: The Structure of American Economy 1919-1939. 2nd ed. Oxford University Press, New York, 1951.

[6] W.W. LEONTIEF: Studies in the Structure of American Economy. Oxford University Press, New York, 1953.

[7] D.A. LIVESEY: The Singularity Problem in the Dynamic Input-Output Model. International Journal of System Science, 1973, p. 437-440.

[8] D.G. LUENBERGER, A. ARBEL: Singular Dynamic Leontief Systems. Econometrica, Vol. 45, 1977, pp. 991-995.

[9] A.S. MANNE: Multisector Models for Development Planning.
A survey in: Frontiers of Quantitative Economics, Vol. 11, M.D. Intriligator and D.A. Kendrick edts. North Holland 74, pp. 449-489.

[10] H. NIKAIDO: Introduction to Sets and Mappings in Modern Economics.
North Holland 1972.

[11] H. NIKAIDO: Convex Structures and Economic Theory.
Acad. Press 1968.

[12] R.T. ROCKAFELLAR: Convex Algebra and Duality in Dynamic Models of Production.
In Mathematical Models in Economics, Los editors, North Holland 1974, pp. 351-378.

ACCEPTABLE EQUILIBRIA IN DYNAMIC BARGAINING GAMES

A. Haurie
Ecole des Hautes Etudes Commerciales
5255 Decelles Ave., Montréal, Canada H3T 1V6

B. Tolwinski
Systems Research Institute,
Polish Academy of Sciences
ul. Newelska 6, 01-447 Warsaw, Poland

Abstract: The paper proposes a theory for two-player dynamic deterministic games where the players cannot make formal and unconditionally binding contracts, but can cooperate by the way of reaching informal agreements whose violation at any stage of the game is prevented by the threat of retaliations at subsequent stages. The theory is based on the introduction of equilibrium memory strategies with the embodied retaliation mechanism, and on the extension of the classical bargaining theory to the case where the set of acceptable outcomes of the game is a nonconvex and possibly disconnected subset of the set of all achievable pay-offs. The proposed approach permits the definition of a bargaining solution over which neither player will be tempted to cheat, and which is selected according to a set of rules that seem reasonable, even though they are to some extent arbitrary.

1. INTRODUCTION

In this paper, we introduce a theory for two-player dynamic deterministic bargaining games. This theory is based on a solution concept presenting both the characteristics of the arbitrated solutions of the classical static bargaining games, and of the equilibrium solutions in which no temptation exists for unilateral deviation from the solution by one player. This will be achieved by the consideration of memory strategies permitting the players to incorporate threat of retaliations in the strategies considered in the bargaining process.

Bargaining games have been the object of considerable study since the work of Zeuthen (1930) and Nash (1950,1953). Recent contributions to the subject have been those of Kalaï and Smorodinsky (1975), Rosenthal (1976) and Roth (1979). These theories are concerned with two-person games in normal form. A bargaining solution singles out a particular outcome which is "undominated" and which reflects the relative strength of the players in presence. In most models the ingredients of a bargaining game are the components of the triple (S, m, ψ) where S is a subset of $\mathbb{R}^2$ representing the feasible payoffs or outcomes, m is an element of S corresponding to the outcome in the case of no agreement[1] and $\psi: (s,m) \to \mathbb{R}^2$ is an "arbitration" function which associates a unique cooperative outcome to the data (S,m).

[1] This is not the case for Rosenthal's arbitration model (Rosenthal 1976) which, instead of using an outcome m corresponding to the case of no agreement, makes use of two points $m^1=(m_1^1,m_2^1)$, $m^2=(m_1^2,m_2^2)$ corresponding to the two possible outcomes depending on the player who commits himself first to use his threat. In that case the arbitration function is $\psi: (S,m^1,m^2) \to \mathbb{R}^2$.

Usually the mapping ψ used for the definition of the arbitrated solution is obtained through an axiomatic approach based on the definition of the game in the normal form. For games defined over a controlled dynamical system, the proper way to define bargaining strategies without temptation for cheating or breaking the agreement, is to consider memory strategies which can incorporate retaliation threats in their definition. (Basar 1974, Ho et al. 1980, Tolwinski 1980, 1981 a and b).

In section 2, we extend the classical bargaining models to the case where negociated outcomes are constrained to belong to a subset S^* of the set S of feasible pay-offs. In section 3 we formulate the dynamic deterministic game, define its equilibrium in the class of memory strategies, and introduce the concepts of retaliation threats, acceptable equilibrium memory strategies and the set of acceptable outcomes. In section 4, the results of section 2 and 3 serve to explore dynamic bargaining games where binding agreements are impossible. In particular, two approaches to the definition of threats are discussed, and a procedure producing a cooperative "no-temptation" solution of the game is presented.

2. A BARGAINING MODEL WHOSE OUTCOME HAS TO BE ACCEPTABLE

Let Γ_1, Γ_2 be the strategy sets of players 1 and 2 respectively and $J_1 : \Gamma_1 \times \Gamma_2 \to \mathbb{R}$, $J_2 : \Gamma_1 \times \Gamma_2 \to \mathbb{R}$, be the corresponding pay-off functions. Then

$$S \triangleq \{(y_1,y_2) : y_i = J_i\ (\gamma_1,\gamma_2),\ \gamma_1 \in \Gamma_1,\ \gamma_2 \in \Gamma_2,\ i=1,2\} \qquad (1)$$

is the set of feasible outcomes.

Let S^* be a subset of S called the set of <u>acceptable outcomes</u>. An outcome (y_1^*,y_2^*) in S^* can be achieved by using "acceptable strategies" (γ_1^*,γ_2^*), e.g. strategies which are "cheating-proof".

Our bargaining game will be thus defined by (S,m,ψ, S^*,φ), where m is an element of $\mathbb{R}^2$ called the "<u>threat point</u>* or "<u>status quo</u>", $\psi : (S,m) \to \mathbb{R}^2$ is a mapping which associates with a set of feasible outcomes S and a threat point m, a unique element of S called the "<u>ideal settlement</u>", S^* is the subset of S corresponding to the acceptable outcomes, and $\varphi : \left(m,\ \psi(S,m),\ S^*\right) \longrightarrow P(S^*)$ is a mapping which associates a subset of S^*, called the set of "<u>acceptable negociated outcomes</u>", with a threat point m, an ideal settlement point $\psi(S,m)$ and a set S^* of acceptable outcomes.

The basic idea is that a bargaining game played by "gentlemen" would lead to the ideal settlement $\psi(m,S)$. However each player is not sure of the "morality" of his opponent. So one has to look for a "second-best" negociated outcome in the subset S^*.

We will now propose a definition of this set of acceptable negociated outcomes, assuming S^* to be compact.

Let $m = (m_1,m_2)$ and $\hat{y} = (\hat{y}_1,\hat{y}_2) = \psi(S,m)$ correspond to the threat point and the ideal settlement point respectively. Define the slope

$$\alpha = (\hat{y}_2 - m_2)\ /\ (\hat{y}_1 - m_1) \qquad (2)$$

of the line passing by these two points in $\mathbb{R}^2$. Let $\varphi^1(m,\alpha,S^*)$ be the set of vectors (y_1^*,y_2^*) which are solution to

$$\left.\begin{array}{l} \text{Max } y_1 \\ \text{s.t.} \\ (y_1,y_2) \in S^* \\ y_2 - m_2 \geq \alpha(y_1 - m_1) \end{array}\right\} \quad (3)$$

Let $\varphi^2(m,\alpha,S^*)$ be the set of vectors (y_1^*,y_2^*) which are solution to

$$\left.\begin{array}{l} \text{Max } y_2 \\ \text{s.t.} \\ (y_1,y_2) \in S^* \\ y_2 - m_2 \leq \alpha(y_1 - m_1) \end{array}\right\} \quad (4)$$

Denote $P(m,\alpha,S^*)$ the subset of undominated (Pareto) points in the set $\varphi^1(m,\alpha,S^*) \cup \varphi^2(m,\alpha,S^*)$.

Assumption 1: <<The mapping φ is defined by $\varphi(m, \psi(S,m), S^*) = P(m,\alpha,S^*)$ where α is the slope given by (2) and $P(m,\alpha,S^*)$ is defined as above.>>

Figure 1 gives an illustration of the definition of φ.

This assumption can be justified by the following considerations:

a) If $\psi(S,m)$ is an element of S^* then
$$\varphi(m, \psi(S,m), S^*) = \psi(S,m). \quad (5)$$

b) An acceptable negociated outcome must be undominated in the set S^*.

c) Given the threat point m, any agreed outcome (y_1,y_2) must be selected in the orthant:
$$\vartheta = \{(y_1,y_2) : y_i - m_i \geq 0,\ i=1,2\}\ . \quad (6)$$

When the ideal settlement $\psi(S,m)$ is given, the line of slope α passing through these two points separates the orthant in two regions. The region I above the line is "more favorable" to player 2, the region II, below the line is "more favorable" to player 1. Therefore one leaves player 1 choose the acceptable outcome in region I and player 2 choose the acceptable outcome in region II. If one of these points dominate the other, it will be the only candidate for an acceptable negociated outcome. Notice that the number of candidates, i.e. of elements of $\varphi(m, \psi(S,m), S^*)$, will be 0, 1 or 2. It will be 0 if S^* is disjoint of the orthant ϑ. It will be 1 if S^* is convex, or if the solutions of (3) and (4) either coincide (they are then located on the line of slope α) or are such that one solution dominate the other. It will be 2 if no Pareto optimal point in S^* can be found on the line of slope α and if the two solutions of (3) and (4) are not comparable.

It is expected that in most cases $\varphi(m, \psi(S,m))$ will reduce to a singleton. This will define a solution of the game. If this set is empty, or if it has two elements, we cannot propose a theory for what will happen. In the first case the cooperative game is not playable with acceptable strategies. In the second case an arbitration scheme has to be proposed for the selection of an outcome between the two possible ones. A random choice may be one of the possible ways to effectuate a fair choice.

3. ACCEPTABLE OUTCOMES IN A DYNAMIC DETERMINISTIC GAME

The dynamical system controlled by the two players is described by the state equation

$$x(t+1) = f^t\big(x(t),\ u_1(t),\ u_2(t)\big),\ t \in \{0,1,\dots,T-1\} \tag{7}$$

where t is the time period of stage, T the time horizon, $x(t) \in X^t \subset \mathbb{R}^n$ is the state variable at t, and $u_i(t) \in U_i^t \subset \mathbb{R}^{m_i}$ is the control (decision) variable of player i at t. The functions $f^t : X^t \times U_1^t \times U_2^t \to X^{t+1}$ define the possible state transitions from stage t to t+1.

Given a state $x^t \in X^t$ at t and two control sequences $\tilde{u}_i^t \triangleq \big(u_i(t),u_i(t+1),\dots,u_i(T-1)\big)$, i=1,2, there is a unique trajectory $\tilde{x}^t \triangleq \big(x(t),x(t+1),\dots,x(T)\big)$ which is the solution of (7) generated by $\tilde{u}_1^t$ and $\tilde{u}_2^t$ and such that $x(t) = x^t$. Associated with the triple $(x^t,\tilde{u}_1^t,\tilde{u}_2^t)$ let us consider the pay-off functions

$$G_i(t,x^t,\tilde{u}_1^t,\tilde{u}_2^t) \triangleq \sum_{s=t}^{T-1} g_i^s\big(x(s),u_1(s),u_2(s)\big) + g_i^T\big(x(T)\big),\ \text{where } g_i^s{:}X^s{\times}U_1^s{\times}U_2^s\to\mathbb{R},g_i^T{:}X^T\to\mathbb{R} \tag{8}$$

are given functions.

We assume perfect information; each player at stage t knows the initial state x^0 at stage 0 and recalls all past decisions $u_i(s)$, for s=0,1,...,t-1, i=1,2. The strategies are thus sequences $\gamma_i = (\gamma_i^t)_{t\in\{0,1,\dots,t-1\}}$ of mappings

$$\gamma_i^t : \big(x^0;\ u_1(0),\ \dots,\ u_1(t-1);\ u_2(0),\ \dots,\ u_2(t-1)\big) \to U_i^t \quad \text{for } i=1,2. \tag{9}$$

Given the initial state x^0, the game can be reduced to its normal form by defining the pay-off functions

$$J_i\ (\gamma_1,\gamma_2) \triangleq G_i\ (0,x^0,\tilde{u}_1^0,\tilde{u}_2^0),\ i=1,2 \tag{10}$$

where $\tilde{u}_i^0$, i=1,2, are the control sequences obtained through the use of the strategies γ_1 and γ_2 respectively, i.e.

$$u_i(t) = \gamma_i^t\ \big(x^0;\ u_1(0),\ \dots,\ u_1(t-1);\ u_2(0),\ \dots,\ u_2(t-1)\big) \tag{11}$$

t=0,1,...,T-1, i=1,2

$$x(t+1) = f^t\ \big(x(t),\ u_1(t),\ u_2(t)\big),\ t=0,1,\dots,T-1 \tag{12}$$

$$x(0) = x^0 \tag{13}$$

<u>Definition 1</u>: <<A strategy pair (γ_1^*,γ_2^*) is an <u>equilibrium</u> at x^0 if for all stages t=0,1,...,T-1 and for i=1,2

$$G_i\ \big(t,x^*(t),\ \tilde{u}_1^{*t},\ \tilde{u}_2^{*t}\big) \geq \sup_{\tilde{u}_i^t} G_i\ \big(t,x^*(t),\ \tilde{u}_1^t,\ \tilde{u}_2^t\big) \tag{14}$$

where, for $j \neq i$

$$u_j(s) = \gamma_j^{*s}\ \big(x^0;\ u_1^*(0),\ \dots,\ u_1^*(t),\ u_1(t+1),\ \dots,\ u_1(s-1);$$
$$u_2^*(0),\ \dots,\ u_2^*(t),\ u_2(t+1),\ \dots,\ u_2(s-1)\big)\ , \tag{15}$$

while $\tilde{x}^{*0}$ is the trajectory and $\tilde{u}_i^{*t}$, i=1,2 are the control sequences generated by (γ_1^*,γ_2^*). The set of all equilibria at x^0 is denoted $E^*(x^0)$>>

This definition corresponds to a Nash equilibrium in the class of memory strategies for the game in normal form defined by (4). Moreover this Nash equilibrium satisfies the dynamic programming principle of optimality along the equilibrium trajectory $\tilde{x}^{*0}$, since for t=0,1,2,...,T-1, the restricted strategies $\left((\gamma_i^{*s})_{s=t}^{T-1}\right)_{i=1,2}$ *constitute a* Nash-equilibrium for the game starting at $\left(t,x^*(t)\right)$.

In this paper we shall define a threat from player i as a pure feedback strategy $d_i : \left(t,x(t)\right) \to U_i^t$. We will now show that we can associate with a given threat pair (d_1,d_2) a whole class of equilibria in the sense of Definition 1. This class contains the acceptable equilibrium memory strategies.

Let (d_1,d_2) be a pair of given feedback strategies. Let us consider the following algorithm.

1) Define the two auxiliary control systems

$$x(t+1) = f^t\left(x(t),\ u_1(t),\ d_2(t,x(t))\right),\quad t=0,1,\dots,T-1 \tag{16}$$

$$x(t+1) = f^t\left(x(t),\ d_1(t,x(t)),\ u_2(t)\right),\quad t=0,1,\dots,T-1 \tag{17}$$

with the respective pay-offs:

$$H_1(t,x^t,\tilde{u}_1^t) = \sum_{s=t}^{T-1} g_1^s\left(x(s),\ u_1(s),\ d_2(s,x(s))\right) + g_1^T\left(x(T)\right) \tag{18}$$

$$H_2(t,x^t,\tilde{u}_2^t) = \sum_{s=t}^{T-1} g_2^s\left(x(s),\ d_1(s,x(s)),\ u_2(s)\right) + g_2^T\left(x(T)\right) \tag{19}$$

associated with state $x^t \in X^t$ and control sequences $\tilde{u}_1^t$ and $\tilde{u}_2^t$ respectively.

2) For $t \in \{0,1,\dots,T-1\}$ and $x^t \in X^t$ define

$$D_i(t,x^t) \triangleq \sup_{\tilde{u}_i^t} H_i\ (t,x^t,\tilde{u}_i^t) \tag{20}$$

where the sup is subject to the constraints (16) or (17) depending on i=1 or 2.

3) For $t \in \{0,1,\dots,T-2\}$, $x^t \in X^t$, $u_i^t \in U_i^t$, i=1,2, define, for j=1,2, j≠i

$$M_i(t,x^t,u_j^t) = \sup_{u_i(t)} \left[g_i^t\left(x^t,\ (u_j^t,\ u_i(t))\right) + D_i\left(t+1,\ f^t\ (x^t,\ (u_j^t,\ u_i(t))\right)\right] \tag{21}$$

Here we have used the notation $\left(u_j^t,\ u_i(t)\right)$ for the pair $\left(u_1^t,\ u_2(t)\right)$ or $\left(u_1(t),\ u_2^t\right)$ depending on i=1 or 2.

4) End.

We can easily interpret the expression $D_i(t,x^t)$ given in (20) as the maximum pay-off that player i can secure for himself for the stages t to T given that the system is in state x^t at t and that player j will play according to his threat strategy d_j.

The expression $M_i(t,x^t,u_j^t)$ given in (21) is the maximum pay-off that player i may expect for the stages t to T if the system is in state x^t at t, if the opponent is bound to play u_j^t at t but will use his threat strategy from stage t+1 onward. This expression plays a fundamental role in the definition of temptations.

<u>Definition 2</u>: <<Let x^0 be a fixed initial state, (d_1,d_2) a fixed pair of threat strat-

egies and $(\tilde{u}_1^{*0}, \tilde{u}_2^{*0})$ a pair of control sequences generating the trajectory $\tilde{x}^{*0}$.

We say that there is <u>no temptation</u> associated with $(\tilde{u}_1^{*0}, \tilde{u}_2^{*0})$ at x^0 if, along the trajectory $\tilde{x}^{*0}$ the following holds:

a) $\forall t \in \{0,1,\ldots,T-2\}\ G_i\left(t,x^*(t), \tilde{u}_1^{*t}, \tilde{u}_2^{*t}\right) \geq M_i\left(t, x^*(t), u_j^*(t)\right),\ i,j=1,2,\ j\neq i$ (22)

b) $$G_i\left(T-1, x^*(t-1), (u_i^*(T-1), u_j^*(T-1))\right) = \sup_{u_i(T-1)} G_i\left(T-1, x^*(T-1), (u_i(T-1), (u_j^*(T-1))\right) \quad i=1,2 \quad j\neq i \qquad (23)$$>>

Notice that the absence of temptation at stage T-1 corresponds to a static Nash equilibrium. This is due to the fact that there is no possible retaliation after stage T-1. At stage T-2, T-3, ..., the absence of temptation is due to the threat of retaliation, the cheating being possible for one stage only.

<u>Proposition 1</u>: <<Let (d_1,d_2) be a fixed pair of threat strategies. Let $A(x^0)$ be the class of control sequences $(\tilde{u}_1^{*0}, \tilde{u}_2^{*0})$ at x^0 <u>without temptation</u>. Define the class $\Gamma^*(x^0)$ of memory strategy pairs (γ_1^*, γ_2^*) where:

$$\gamma_i^{*0}(x^0) = u_i^*(0) \qquad (24)$$

$$\gamma_i^{*t}\left(x^0; u_1(0), \ldots, u_2(t-1); u_2(0), \ldots, u_2(t-1)\right) = \begin{cases} u_i^*(t) & \text{if } u_j(s) = u_j^*(s) \text{ for } s=0,\ldots,t-1 \\ d_i\left(t,x(t)\right) & \text{otherwise} \end{cases} \qquad (25)$$

where $(\tilde{u}_1^{*0}, \tilde{u}_2^{*0})$ is an element of $A(x^0)$, and where $x(t)$ is the state resulting from x^0 and the control $u_i(s)$, $s=0,\ldots,t-1$, $i=1,2$.

Then every pair $(\gamma_1^*, \gamma_2^*) \in \Gamma^*(x^0)$ is an <u>equilibrium</u> at x^0 according to Definition 1.>>

<u>Proof</u>: We have to check that the conditions (8), (9) are satisfied. With the notations of (8) and (9), and according to (18), (19) and the property a) of Definition 2 we have:

$$\sup_{\tilde{u}_i^t} G_i\left(t, x^*(t), \tilde{u}_1^t, \tilde{u}_2^t\right) = M_i\left(t, x^*(t), u_j^*(t)\right) \leq G_i\left(t, x^*(t), \tilde{u}_1^{*t}, \tilde{u}_2^{*t}\right) \qquad (26)$$

for $i,j=1,2$, $j\neq i$, $t=0,1,\ldots,T-1$.

According to property b) of Defnition 2 the equilibrium property holds at $x^*(T-1)$ too. Therefore (γ_1^*, γ_2^*) is an equilibrium. ∎

<u>Definition 3</u>: <<For a given pair of threat strategies (d_1,d_2), the class $\Gamma^*(x^0)$ defined by (24), (25) is called the class of <u>acceptable equilibrium memory strategies at x^0</u>.>>

<u>Definition 4</u>: <<The set $S^*(x^0)$ of vectors (y_1^*,y_2^*) where

$$y_i^* = G_i\ (0, x^0, \tilde{u}_1^{*0}, \tilde{u}_2^{*0}), \quad i=1,2 \qquad (27)$$

for any pair of control sequences $(\tilde{u}_1^{*0}, \tilde{u}_2^{*0})$ in $A(x^0)$, is called <u>the set of acceptable outcomes at x^0</u>.>>

4. BARGAINING IN A DYNAMIC DETERMINISTIC GAME

The two players control the dynamical system (7). The set $S(x^0)$ is thus defined as the convex hull of the set

$$\mathcal{g}(x^0) \triangleq \{(y_1,y_2) : y_1 = G_i(0,x^0,\tilde{u}_1^0,\tilde{u}_2^0) \quad i=1,2,$$
$$\forall\, \tilde{u}_1^0,\ \tilde{u}_2^0 \text{ admissible control sequences}\} . \tag{28}$$

The definition of $S(x^0)$ as the convex hull of $[\mathcal{g}(x^0)]$ is justified since the players could ideally randomize their choice of cooperative control sequences. Furthermore, we assume the following:

Assumption 2: <<The set $\mathcal{g}(x^0)$ is compact.>>

The players have to select an element of $S(x^0)$ through a bargaining process and they will restrict their choice to acceptable outcomes. They will thus play a threat game in order to reach a favorable acceptable outcome. We suggest the following procedure as the rules of the bargaining game.

Phase 1: The two players know everything about the system they control and their relative pay-offs. They know the initial state x^0. Independently they formulate threats which determine a status-quo point $m = (m_1,m_2)$ in $\mathbb{R}^2$. An arbitrator, using a mapping ψ thus proposes an ideal settlement $\psi(S(x^0),\ m)$ which is Pareto optimal in $S(x^0)$. This phase I of the game corresponds to a classical static bargaining game.

Phase 2: The two players formulate now retaliation threats (d_1,d_2) which they will use in case of deviation from the agreed cooperative strategy in the course of the game. This permits them to compute the set $S^*(x^0)$ of acceptable outcomes at x^0. Using the mapping φ they obtain the set of acceptable negociated outcomes $\varphi(m,\psi(S(x^0),m\ ,S^*(x^0))$. If this set is a singleton this is the solution of the game. If the set contains two elements the arbitrator chooses randomly one of them. If the set is empty, the bargaining game has no solution.

In each of these two phases the players have to formulate threats. Also there must be a pre-specified arbitration mapping ψ. This could be done in Phase 1 as according to any of the proposed philosophies for static bargaining games. We discuss below two possibilities for the definition of retaliation threats in Phase 2.

4.1 Retaliation threats as Min Max strategies

The potentially most damaging threat that player i can raise against his opponent, player j, is to use a minmax feedback strategy defined for the pay-offs G_j. More precisely d_i is such that

$$\sup_{\tilde{u}_j^t} H_j(t,x^t,\tilde{u}_j^t) = \inf_{\tilde{u}_i^t} \sup_{\tilde{u}_j^t} G_j(t,x^t,\tilde{u}_1^t,\tilde{u}_2^t) \quad \forall t\in\{1,2,\dots,T-1\},\ \forall x^t\in X(t) \tag{29}$$

where H_j is defined as in (16), (19)

The threat of player 1 is thus obtained by considering a zero-sum multistage game defined by the state equation (7) and the "cost" G_2 as defined in (8). We will call this game the <u>associated duel</u> for player 1. Similarly defined, there is also an <u>associated duel</u> for player 2.

<u>Assumption 3</u>: <<Each of the two associated duels admits a saddle point in pure feedback strategies: (d_1^1, d_2^1) for the duel associated with player 1, (d_1^2, d_2^2) for the duel associated with player 2.>>

Under this last assumption the minmax threat pair is $(d_1^1,\ d_2^2)$. The following result shows that the set of equilibrium strategies can then have a nice characterization.

<u>Proposition 2</u>: <<Under Assumption 3 a strategy pair (γ_1^*, γ_2^*) is an equilibrium at x^0, according to Definition 1, only if the control sequences $(\tilde{u}_1^{*0},\ \tilde{u}_2^{*0})$ generated by (γ_1^*, γ_2^*) are without <u>temptation</u> with respect to the threat pair (d_1^1, d_2^2).>>

<u>Proof</u>: Suppose that $(\tilde{u}_1^{*0},\ \tilde{u}_2^{*0})$ is not an element of $A(x^0)$, the set of control pairs without temptations at x^0. The equilibrium conditions at stage T-1 for the strategy pair (γ_1^*, γ_2^*) imply that condition b) of Definition 2 holds. Therefore we must have

$$G_i\left(t,\ x^*(t),\ \tilde{u}_1^{*t}\ \tilde{u}_2^{*t}\right) < M_i\left(t,\ x^*(t),\ \tilde{u}_j^*(t)\right) \tag{30}$$

for some stage $t \in \{0,1,\ldots,T-2\}$ and some player $i \in \{1,2\}$. Here $\tilde{x}^*$ is the trajectory generated by (γ_1^*, γ_2^*) and $M_i\left(t, x^*(t),\ \tilde{u}_j^*(t)\right)$ is defined by the algorithm (16), (21).

Suppose, for example, that (30) holds for player 1, then, by the saddle point property of the associated duel for player 2, it implies the existence of $u_1(t) \in U_1^t$ such that

$$\begin{aligned} G_1\left(t, x^*(t), \tilde{u}_1^{*t}, \tilde{u}_2^{*t}\right) &< g_1^t\left(x^*(t), u_1(t), u_2^*(t)\right) \\ &\quad + \inf_{\tilde{u}_2^{t+1}} \sup_{\tilde{u}_1^{t+1}} G_1\left(t+1,\ f\left(x^*(t), u_1(t), u_2^*(t)\right),\ \tilde{u}_1^{t+1},\ \tilde{u}_2^{t+1}\right) \\ &= g_1^t\left(x(t), u_1(t), u_2^*(t)\right) + \inf_{\tilde{u}_2^{t+1}} G_1\left(t+1, f\left(x^*(t), u_1(t), u_2^*(t)\right),\ \tilde{v}_1^{t+1},\ \tilde{u}_2^{t+1}\right) \end{aligned} \tag{31}$$

where $\tilde{v}_1^{t+1}$ is the control sequence from state t to stage T-1 obtained for player 1, from the application of the feedback strategy d_1^2.

Player 1 has thus access to a control sequence $\left(u_1(t),\ \tilde{v}_1^{t+1}\right)$ which will be such that the resulting pay-off for him will be higher, when player 2 sticks to his strategy γ_2^*, than the pay-off obtained through $(\tilde{u}_1^{*t},\ \tilde{u}_2^{*t})$. Therefore, this contradicts (14) and the fact that (γ_1^*, γ_2^*) is an equilibrium. ∎

<u>Proposition 3</u>: (i) 'Under Assumption 3 the set $S^*(x^0)$ is nonempty.

(ii) Suppose that $T < \infty$; U_i^t, i=1,2, t=0,1,...,T-1 are compact, f^t, t=0,1,...,T-1 and g_i^t, i=1,2, t=0,1,...,T are continuous in all their arguments. Then $S^*(x^0)$ is compact.

<u>Proof</u>: (i) We have $(d_1^2, d_2^1) \in A(x^0)$, so $S^*(x^0)$ is non empty.

(ii) Under our assumptions $D_i\left(t,\ x(t)\right)$ is continuous in x(t), and in consequence

$M_1(t, x(t), u_j(t))$ is continous in $x(t)$ and $u_j(t)$ for $i=1,2$, $j\neq i$. Thus all functions entering the definition of $A(x^0)$ are continuous, so the set $S^*(x^0)$ is closed; it is also bounded hence compact. ∎

4.2 Credible retaliation threats

Introduce the notation $G_i(t,x^t,d_1,d_2)$ for the pay-off accrued to players i from stage t and state x^t onward, when the feedback strategies d_1 and d_2 are used, i.e.:

$$G_i(t,x^t,d_1,d_2) = \sum_{s=t}^{t-1} g_i^s\left(x(s),\ d_1(s,x(s)),\ d_2(s,x(s))\right) + g_i^T\left(x(T)\right) \qquad (32)$$

where $x(t) = x^t$

$$x(s+1) = f^s\left(x(s),\ d_1(s,x(s)),\ d_2(s,x(s))\right),\quad s=t,\dots,T-1 \qquad (33)$$

The definition of threats as minmax strategies of associated duels neglects the credibility aspect of the problem. The realization of such threats can be very damaging to the player who declared them, and what is especially disturbing, even more damaging to the player himself than to his opponent. If one accepts the reasoning that a threat is believable when its realization hurts the player who declared it no more than his opponent, then (d_1,d_2) should be defined as saddle point strategies of the difference of the players' pay-off functions, that is (d_1,d_2) are such that

$$\forall\ t \in \{1,2,\dots,T-1\},\quad \forall\ x^t \in X^t$$

$$G(t,x^t,d_1,d_2) = \sup_{\tilde{u}_1^t}\inf_{\tilde{u}_2^t} G(t,x^t,\tilde{u}_1^t,\tilde{u}_2^t) = \inf_{\tilde{u}_2^t}\sup_{\tilde{u}_1^t} G(t,x^t,\tilde{u}_1^t,\tilde{u}_2^t) \qquad (34)$$

$$\text{where}\quad G(t,x^t,\tilde{u}_1^t,\tilde{u}_2^t) = G_1(t,x^t,\tilde{u}_1^t,\tilde{u}_2^t) - G_2(t,x^t,\tilde{u}_1^t,\tilde{u}_2^t) \qquad (35)$$

For a more detailed discussion of this approach see Luce and Raïffa (1958) and Tolwinski (1981b).

Another concept of credible threat is based on the assumption that a threat is believable if it is the player's equilibrium decision. In other words (d_1,d_2) are defined as pure feedback Nash equilibria that is, with an obvious adaptation of the notations (8) and (32), (33).

$$G_1(t,x^t,\tilde{d}_1^t,\tilde{d}_2^t) = \sup_{\tilde{u}_1^t} G_1(t,x^t,\tilde{u}_1^t,\tilde{d}_2^t)\qquad \forall t\in\{1,2,\dots,T-1\},\ \forall x^t\in X(t) \qquad (36)$$

$$G_2(t,x^t,\tilde{d}_1^t,\tilde{d}_2^t) = \sup_{\tilde{u}_2^t} G_2(t,x^t,\tilde{d}_1^t,\tilde{u}_2^t) \qquad (37)$$

The threats obtained in this way are very believable but the resulting sets of no-temptation controls can be rather restricted. For example, in the case of finite time horizon T, $A(x^0)$ reduces to the singleton $\{(d_1,d_2)\}$ and there does not exist any acceptable solution other than no-memory Nash equilibrium.

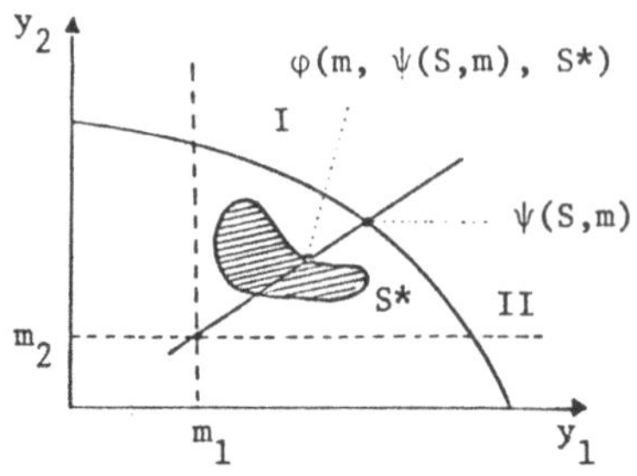

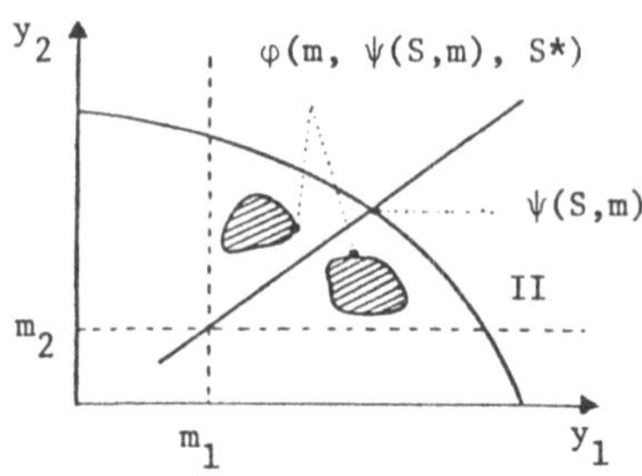

Figure 1: Construction of $\varphi\left(m,\psi(S,m),S^*\right)$ for two different possible sets S^*.

5. CONCLUSION

We have presented a theory for two-player dynamic deterministic games, where the players cannot make formal binding contracts but can cooperate by the way of reaching informal agreements whose violation at any stage of the game is prevented by the threat of retaliations at subsequent stages. The theory consists of two basic elements. The first one is the introduction of equilibrium memory strategies with the embodied retaliation mechanism. The second element is the extension of the classical bargaining theory developed in the abstract setting of the games in the normal form, to the case where the set of acceptable outcomes of the game is a non-convex subset of the set of all achievable pay-offs.

The proposed approach permits the definition of a bargaining solution in a dynamic deterministic game, with no temptation for cheating. There is still a dose of arbitrariness in the definition of the acceptable negociated outcomes, which seems to be inescapable.

It should be noted that, although this definition is valid for any choice of retaliation threats and bargaining model for the selection of the ideal settlement, it does not cover all situations which can possibly arise in dynamic bargaining games. It assumes in particular that the negociations between the players are held only once before the beginning of the game, even though repeated bargaining is more common in real-life situations (a first investigation of the case where the bargaining is repeated at every stage of the game can be found in Tolwinski (1981b)). Another restrictive assumption is that the retaliation for violating an agreement always lasts until the end of the game, ruling out any possibility of reconciliation and return to the cooperative mood of play. However, the extension of our theory to repeated bargaining and short term retaliation games is possible and will be explored in the near future.

REFERENCES

AUMANN, R.J., (1959), "Acceptable points in general cooperative n-person games", A.W. Tucker and R.D. Luce (eds.), Contributions to the Theory of Games, Vol. IV, pp. 287-324, Princeton, 1959).

BASAR, T., (1974), "A counter-example in linear-quadratic games: Existence of non-linear Nash solutions", J. Optimiz. Theory Appl., Vol. 14, No. 4, pp.425-430, 1974.

BLAQUIERE, A., F. GERARD, and G. LEITMANN, (1969), Quantitative and Qualitative Games, Acad. Press, 1969.

FRIEDMAN, J.W., (1971), "A non-cooperative equilibrium for supergames", Review of Economic Studies, Vol. 38, pp. 1-12, 1971.

FRIEDMAN, J.W., (1977), Oligopoly and the Theory of Games, North Holland, 1977.

HAURIE, A., (1975), "On some properties of the characteristic function and the core of a multistage game of coalitions", IEEE Trans. Automat. Contr., Vol. AC-20, pp. 238-241, 1975.

HAURIE, A., (1976), "A note on nonzero-sum differential games with bargaining solution", J. Optimiz. Theory Appl., Vol. 18, No. 1, pp. 31-39, 1976.

HO, Y.C., P.B. LUH and G.J. OLSDER, (1980), "A control-theoretic view on incentives", Proc. of the Fourth International Conference on Analysis and Optimization of Systems, Versailles, France, December 1980.

KALAI, E. and M. SMORODINSKY, (1975), "Other solutions to Nash's bargaining problem", Econometrica, Vol. 43, pp. 513-518, 1975.

LUCE, R.D. and H. RAIFFA, (1957), Games and Decisions: Introduction and Critical Survey, Wiley, 1957.

NASH, J., (1950), "The bargaining problem", Econometrica, Vol. 18, pp. 155-162, 1950.

NASH, J., (1953), "Two-person cooperative games", Econometrica, Vol. 21, pp. 128-140, 1953.

RAIFFA, H., (1953), "Arbitration schemes for generalized two-person games", H.W. Kuhn and A.W. Tucker (eds.), Contributions to the Theory of Games, Vol. II, pp. 361-387, Princeton, 1953.

RAY, A. and A. BLAQUIERE, (1981), "Sufficient conditions for optimality of threat strategies in a differential game", J. Optimiz. Theory Appl., Vol. 30, No. 1, pp. 99-109, 1981.

ROSENTHAL, R.W., (1976), "An arbitration model for normal form games", Mathematics of Operations Research, Vol. 1, No. 1, pp. 82-88, 1976.

ROTH, A.E. (1979), "Axiomatic Models of Bargaining", Springer-Verlag, 1979.

STARR, A.W. and Y.C. HO, (1969), "Nonzero-sum differential games", J. Optimiz. Theory Appl., Vol. 3, No. 3, 1969.

TOLWINSKI, B., (1980), "Stackelberg solution of dynamic games with constraints", Proc. of the 19th IEEE Conference on Decision and Control, Albuquerque, New Mexico, December 1980.

TOLWINSKI, B., (1981a), "Closed-loop Stackelberg solution to multistage linear-quadratic game", J. Optimiz. Theory Appl., Vol. 34, No. 4, 1981.

TOLWINSKI, B., (1981b), "A concept of cooperative equilibrium for dynamic games", Technical Report of Systems Research Institute, Warsaw (also submitted for publication).

ZEUTHEN, F., (1930), "Problems of Monopoly and Economic Warfare",Chap. 4, G. Routledge, London 1930.

Additional Reference:

HAURIE, A. and TOLWINSKI, B. (1981) "Acceptable Equilibria in Dynamic Bargaining Games", (unabridged version). Cahiers du GERAD - G81-13. GERAD, Ecole H.E.C., 5255 Decelles Montréal H3T 1V6 Que. Canada.

ACKNOWLEDGEMENT: Research supported by SSHRC - Canada, Grant 410-78-0603 and Ecole des H.E.C., Montréal (Fonds internes de recherche).

CYCLICAL TAXONOMY AND LARGE ECONOMETRIC MODELS

Ullrich Heilemann[+]
Rheinisch-Westfälisches Institut für Wirtschaftsforschung
D-43oo Essen 1, West Germany

The cyclical behavior of an econometric short-term model is an important criterion of its usefulness. Consequently, it has been a subject of econometric model research from its beginnings. In most cases, this interest focuses on the question, how well a model catches the (economic) turning points of important variables such as GNP or inflation rate. The results of such a turning point analysis are, however, seldom clearcut and it is rather difficult to get an undisputed picture of a model's cyclical performance because the model-simulations catch the different variables' turning points in varying degrees[1].

Recently, an approach intended to overcome this difficulty was suggested by Meyer and Weinberg in their report on the classification of the U.S. economy into four cycle stages by multivariate discriminant analysis[2]. They selected 2o variables[3], computed three discriminant functions and classified with their help - the stages or phases being recession, recovery, demand-pull (-inflation) and stagflation - each month from February 1947 to September 1973 into their scheme. To illustrate the working of this scheme, in Table 1 a simplified classification pattern, only based on two variables is shown. Unlike the similar traditional four-phase-scheme with upswing, upper turning point, downswing, lower turning point, however, the Meyer/Weinberg-scheme does not require every cycle to include all four phases[4].

The classificatory variables chosen by Meyer/Weinberg are endogenous in most of the current large econometric models of the U.S. economy. Therefore, the cyclical performance of these models in terms of the Meyer/Weinberg-scheme can be tested by comparing

---

[+]Financial support by the German Academic Exchange Service, Bonn, is gratefully acknowledged. I am indebted to Data Resources, Inc., Lexington, MA, USA, for the use of their computational facilities. For helpful comments on earlier drafts of this paper I have to thank J.R. Meyer, D.H. Weinberg, O. Eckstein, H. Spelter, G. Uebe and H.J. Münch.

[1]See e.g. V. Zarnowitz, Ch. Boschan, G.H. Moore, ass. by V. Su, Business Cycle Analysis of Econometric Model Simulations. In: B.G. Hickman (Edtr.), Econometric Models of Cyclical Behavior. (Studies in Income and Wealth, 36) New York and London 1972, p.311-533.

[2]J.R. Meyer, D.H. Weinberg, On the Classification of Economic Fluctuations. "Explorations in Economic Research", vol.2(1975), p.167-2o2. - For a description of the working of multivariate discriminant analysis see M. Kendall, Multivariate Analysis. London 1975, p.145f.

[3]The selection of these variables had been made under theoretical and practical considerations, special attention being given to variables "that had figured prominently in the development of formal econometric models of the U.S. economy" - contd n.p.

Simplified Classification Scheme of Meyer/Weinberg — Table 1

		Growth Rate of GNP	
		High	Low
Price Increases	High	Demand-Pull	Stagflation
	Low	Recovery	Recession

the classification of a certain period resulting from a model solution with the classification of this period established by Meyer/Weinberg. The score may serve as an indicator of a model's cyclical properties.

The present paper reports on such a test performed with the 1976 version of the Data Resources, Inc. (DRI) econometric model of the U.S. economy[5] to find out how well different stages of the business cycle are reflected by this version. The first section describes the modifications of the Meyer/Weinberg estimations which were necessary to compute appropriate discriminant functions. In section two the simulation results are compared with the modified Meyer/Weinberg classification. The final section summarizes the findings and draws some conclusions as to the usefulness of multivariate discriminant analysis for studying the cyclical properties of large econometric models.

1. The discriminant functions and the resulting classifications

The discriminant functions were computed[6] with the same variables employed by Meyer/Weinberg, with the exception of the "New York Stock Exchange Composite Price Index" which had to be replaced by the "Standard & Poors Index"; the price base of the variables was changed from 1958 to 1972. As the functions were to classify quarterly models, quarterly instead of monthly data were used. The beginning of the sample period had to be shifted for data reasons from September 1947 to the second quarter 1948.

The computation of discriminant functions requires an a priori classification of all sample period observations. This was made by assigning all observations from 1948:2 to 1973:2 to one of the four phases. In detail, the procedure was as follows: firstly

(J.R. Meyer, D.H. Weinberg, op.cit., p.176).

[4] Ibid, p.191. For a theoretical justification of a non-continuous cyclical movement see e.g. A.B. Adams, Economics of Business Cycles. New York 1925, p.2o6f.

[5] O. Eckstein, Model Description U.S./Macro. Lexington 1977 (Data Resources, Inc.).

[6] All computations were made with the discriminant analysis routine of the Statistical Package for the Social Sciences (SPSS). For a non-technical description of this routine see N.H. Nie, C.H. Hull, J.G. Jenkins, K. Steinbrenner, D.H. Bent, Statistical Package for the Social Sciences. Second Edition. New York 1975, chapter 23 - Discriminant Analysis, p.434f.

all quarters were classified into the same stage as most of their months had been classified by Meyer/Weinberg. In cases where a change of the stage had happened in midquarter, the period was assigned to the phase of the first month of this quarter. After the computation of the first set of discriminant functions (3), the classification of these border quarters was changed if a priori assignment and a posteriori classification differed. This helped much to reduce the misclassification of border periods, but nevertheless, six misclassifications remained: 1955:2, 1959:3, 1966:4, 1967:1, 1969:1 and 197o:3. It should be noted, however, that Meyer/Weinberg had failed their a priori (NBER -) classification in four out of these six periods (1959:3, 1966:4, 1967:1, 197o:3).

Classification of U.S. Business Cycles into a Four Stage Scheme
1948:2 to 1977:1 Table 2

Cycle	Classification			
	Recession	Recovery	Demand-Pull	Stagflation
1	-	-	-	1948:2 - 1948:4
2	1949:1 - 1949:4	195o:1 - 195o:2	195o:3 - 195o:4	1951:1 - 1953:3
3	1953:4 - 1954:2	1954:3 - 1955:1	1955:2 - 1957:3	-
4	1957:4 - 1958:2	1958:3 - 196o:2	-	-
5	196o:3 - 196o:4	1961:1 - 1965:1	1965:2 - 1967:4	1968:1 - 1969:4
6	197o:1 - 197o:4	1971:1 - 1972:4	1973:1 - 1974:3	-
7	1974:4 - 1975:1	1975:2 -(1977:1)		

As noted in the opening section, the Meyer/Weinberg study used here[7] covered only the period February 1947 to September 1973. The restriction of the model test on this sample would have been too severe and an extension of the classification period up to 1977:1 seemed to be necessary. This was made in a two step procedure. First, the period 1973:4 to 1977:1 was classified with the discriminant functions[8] computed from the (original) sample period 1948:2 to 1973:3. In a second step, new discriminant functions were computed for the now larger sample with the a posteriori classifications for 1973:4 to 1977:1 serving as a priori classification. The a priori classification was met by the discriminant functions fairly well and after consulting other statistical material, corrections proved necessary only at one border period. The final classification of the complete sample period is shown in Table 2[9].

[7] An extrapolation made by Meyer/Weinberg was confirmed by the quarterly classification. See J.R. Meyer, D.H. Weinberg, op.cit., p.18o. - An update of the classification by the two authors till September 1976 differs considerably from their extrapolation, but came too late to be used for the present study (J.R. Meyer, D.H. Weinberg, On the Classification of Economic Fluctuations: An Update. "Explorations in Economic Research", vol.3(1976), p.584f.).

[8] The difference between the discriminant functions and their derivatives - the classification functions - may be neglected in the present context.

[9] More detailed information than presented here is available from the author upon request.

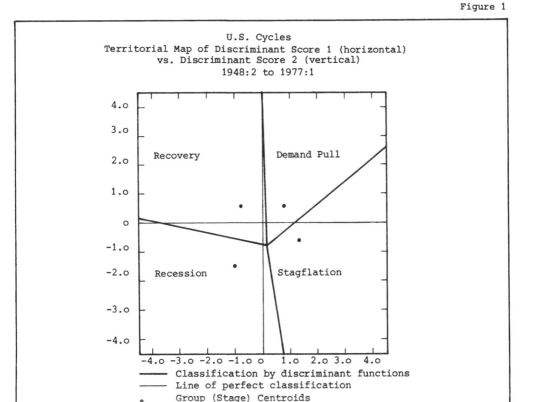

Variables, standardized and unstandardized parameters and explanatory powers of the discriminant functions are presented in the following Table. A territorial map of two discriminant scores of the observations is shown in Figure 1: The group centroids seem to be well separated and the discrimination between the stages (groups) is sharp, with the exception of demand-pull/stagflation. In general, all results are well in line with those of the monthly data-based classification, the differences being mainly due to different data bases.

An important criterion in judging discriminant functions is the explanatory rank of the variables used, measured by their F-value. As the results (not shown in this paper) indicate, Unemployment rate, Real GNP and Unit labor cost play an important role in the quarterly data-based functions - similar to their influence in the monthly data-based functions. Most of the other variables, however, differ in their explanatory ranks in both sets of functions, especially GNP deflator, Compensation per man hour and Prime rate. An analysis of the causes of this discrepancy would go beyond the scope of this paper, but a substitution of some variables by variables more appropriate to quarterly data seems to be promising[10].

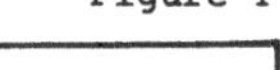

Figure 1

---

[10] Especially with respect to the monetary aggregates and the interest rates.

Coefficients of the Discriminant Functions
1948:2 to 1977:1

Table 3

Variable	Function[1]		
	First	Second	Third
Unemployment rate[2]	-0.691	0.148	0.166
	-0.971	0.209	0.233
Real GNP (1972 Dollars)[3]	-0.404	0.359	0.726
	-1.909	1.694	3.428
Unit labor cost[3]	0.059	-0.204	0.144
	0.277	-0.956	0.672
Govt.surplus as % of GNP	-0.142	0.108	-0.095
	-0.267	0.202	-0.178
GNP deflator[3]	-0.462	0.074	0.381
	-1.457	0.233	1.203
Prime rate[4]	0.003	0.007	0.001
	0.096	0.230	0.015
Gross govt.expenditures[3]	0.008	-0.008	0.011
	0.132	-0.122	0.172
Money supply M1[3]	-0.060	-0.257	0.270
	-0.173	-0.744	0.781
Money supply M2[3]	0.012	0.231	-0.214
	0.044	0.854	-0.792
Net exports as % of GNP	0.311	0.019	-0.361
	0.196	0.012	-0.228
Wholesale price index, industr.comm.only[3]	0.010	0.054	-0.071
	0.062	0.347	-0.454
Compensation per man-hour[3]	0.016	0.251	-0.048
	0.058	0.925	-0.177
Corporate bond rate[3]	0.007	0.017	-0.025
	0.099	0.228	0.338
Consumer price index[3]	-0.052	-0.270	0.449
	-0.181	-0.935	1.556
Consumer price index, food only[3]	0.028	0.086	-0.135
	0.172	0.526	0.826
Output per man-hour[3]	-0.004	-0.345	-0.001
	-0.016	-1.345	-0.004
Standard & Poors index[4]	0.008	0.007	-0.006
	0.185	0.157	-0.134
Consumer price index, all comm.exc.food[3]	0.098	0.156	-0.111
	0.291	0.462	-0.328
Money GNP[3]	0.449	-0.144	-0.534
	-0.173	-0.744	0.781
Wholesale price index[3]	0.004	-0.007	-0.065
	0.027	-0.049	-0.427
Constant	2.483	-1.540	-0.906
Explanatory Power[5]	65.7	22.5	11.7

[1]The first line shows the unstandardized, the second line the standardized (standard deviation for all phases) coefficients.-[2]Seasonally adjusted at annual rate.-[3]Per cent change, seasonally adjusted at annual rate.-[4]Per cent change.-[5]In % of the a priori classification.

For reasons explained below, discriminant functions were also computed for shorter periods. To avoid any bias as far as possible, these periods were chosen in such a way that they included "complete" cycles only. A shortening of the sample period beyond 196o:3 to 1974:3 did not yield reasonable results and is neglected here. Sample periods and prediction results can be learned from the synopsis below (Table 4). - If cycle classification with the method and variables presented in this paper is regarded as an adequate description of reality, these results could lead to the suggestion that the period 1957:4 to 1974:3 is most representative for the postwar U.S. short term development. In addition, it might be concluded that cycle patterns did not change very much, since there are no great differences in the prediction results of the various sets of functions.

Prediction Results of Discriminant Functions for Various Sample Periods

Table 4

Sample Period	Predicted Group Membership[1]				
	Recession	Recovery	Demand-Pull	Stagflation	All
1948:2-1977:1	94.4	97.8	9o.o	95.5	94.8
1948:2-1974:3	93.8	92.1	93.3	95.5	93.4
1949:1-1974:3	93.8	92.1	93.3	1oo.o	94.2
1953:4-1974:3	1oo.o	91.7	82.1	87.5	89.3
1957:4-1974:3	1oo.o	97.o	1oo.o	1oo.o	98.5
196o:3-1974:3	1oo.o	1oo.o	1oo.o	87.5	98.3

[1] In % of the a priori classification.

2. The classification of simulation results

With the discriminant functions at hand, it is possible to examine the cyclical performance of an econometric model. This is simply done by replacing the observed values in the classification procedure by a model's simulation results. In the present case these results were obtained from a dynamic simulation of a test version of the 1976 DRI-model of the U.S. economy over the period 1966:2 to 1976:4 ("Dynamic 1"). "Dynamic 1" fails to meet the a priori classification in 1o cases (1967:1, 1967:3, 1967:4, 1968:3, 1969:1, 1969:2, 1969:4, 197o:1, 1974:4, 1975:1). It should be noted again, that the 1967:1 and the 1969:1 misclassifications are already made by the monthly and quarterly data-based functions. The summary (Table 5) shows that the simulation reproduces the recovery and the demand-pull phases fairly well, while the record for the stagflation and the recession periods is remarkably worse. As more detailed results reveal, half of the misclassifications happen in border quarters; one cycle stage (the 1974/75 recession) is totally lacking in the simulation. With 77 % of all a priori classifications met, the overall record of "Dynamic 1" does not seem to be too bad.

Summary of the Classification of "Dynamic 1"
1966:2 to 1976:4

Table 5

Stage	Number of Classifications		
	a priori	a posteriori[1]	% met
Recession	6	3	5o
Recovery	15	15	1oo
Demand-Pull	14	11	79
Stagflation	8	4	5o
Total	43	33	77

[1]Without misclassifications.

A proper analysis of the cyclical qualities of the model would require a stepwise exogenizing of all discriminant variables, re-running the simulation and re-classifying the results. However, such extensive tests could not be carried out for the present study, so that the analysis has to be confined to a comparison of the mean absolute error (MAE) of each classification-variable with its specific errors in the 1o quarters of misclassification. The comparison indicates that only for one quarter (1967:1) of the widely-missed 1967-demand-pull-phase, above average errors at important variables (Real Growth, Unemployment) are responsible. The causes for missing the 1968/69 stagflation seem to be more complicated: estimations of the Unemployment rate as well as the Wholesale price index show comparatively large errors which are only partly offset by errors in the Real GNP. The same holds for the lack of the 197o recession, the only difference being the lack of offsetting effects by Real GNP errors. Finally, the lack of the 1974/75 recession must be assigned to nearly all variables, since each of them shows large above-average errors (offsetting effects neglected).

It could be argued, that these track-records might differ according to the period over which the discriminant functions have been fitted. To test the sample period sensitivity of the classification results, it seemed interesting to classify the simulation-results by the various discriminant functions. The results presented in Table 6 clearly confirm the conjecture, that the various sample periods lead to very different scores for "Dynamic 1": a maximum of 14 misclassifications is reached when the simulations are classified with the discriminant functions based on the 1948:2 - 1974:3-sample[11] and a minimum of only 6 misclassifications when the functions are computed over the sample 196o:3 to 1974:3. Furthermore, it should be noticed, that the range of the total explanation of the simulation results is twice as large as that of the classification of the observed values; the dispersion in the classifications of the simulation results is for all stages (groups) considerably higher than that for the observed values; third, the relation between group- or stage-wise explanation and total explanation seems to be roughly the same in the classification of observed and simu-

[11]This period was chosen to exclude the uncertainties linked with the extrapolation.

Summary of the Classification of "Dynamic 1"
by Discriminant Functions from various sample periods

Table 6

Sample Period	Predicted Group Membership[1]				
	Recession	Recovery	Demand Pull	Stagflation	All
1948:2-1977:1	5o.o	1oo.o	78.6	5o.o	76.7
1948:2-1974:3	83.3	73.3	64.3	5o.o	67.4
1949:1-1974:3	66.7	73.3	57.1	87.5	69.8
1953:4-1974:3	5o.o	66.7	64.3	1oo.o	69.8
1957:4-1974:3	33.3	86.7	78.6	87.5	76.7
196o:3-1974:3	33.3	1oo.o	92.9	87.7	86.o

[1] In % of the a priori classification.

lated values. These findings indicate that the classification is very sensitive as to the sample period of the discriminant functions used. Obviously, this sensitivity depends on the simulation quality or forecasting property of the model: if the model had met the observed development perfectly, the sensitivity would only have been as great as that of the discriminant functions.

3. Summary and conclusions

The cycle classification scheme of Meyer/Weinberg, originally based on monthly data, also yields comparatively good results when performed with quarterly data. The explanatory power of the (new) discriminant functions is sufficient, but might be improved by replacing some of the variables by more appropriate to quarterly data ones.

The comparison of the classification of observed and simulated values seems to be a good indicator of an econometric model's capability to trace the complete business-cycle as well as its various phases. The classifications of simulation results disclosed a comparatively high sensitivity towards the sample period of the discriminant functions. This sensitivity may serve as a further measure of a model's tracking quality, but it also suggests a careful selection of the sample period of the functions to be used for classifying. In the present case, the best results were obtained when the functions were based on the shortest sample period (196o:3 to 1974:4).

How accurate is this test? In the present study a final answer to this question is not possible, however two points should be remarked upon. First, most of the classificatory variables are growth rates, which makes the test to be an accurate and sharp one. Second, as the standardized coefficients of the functions clearly show, there are only a few variables of dominant influence, such as Unemployment rate, Real GNP, Unit labor cost and the various price indices. Hence, once these few variables - of which several show a high correlation, too - are simulated well, the "true" cycle stage can hardly be missed.

The applications of the classification scheme also reveal some of its limitations. One of these origninates from the missing (internal) determination or uniformity of the scheme used, as not every cycle has to include all stages. A second limitation is caused by the discrete character of the classification: nothing is said about the degree of meeting or missing the cycle stage in question. In short: cycle classification based on multivariate discriminant analysis cannot replace a detailed traditional error analysis of the simulation results of an econometric model, but will support it in an efficient manner.

ANTIPOLLUTION ACTIVITIES IN INPUT-OUTPUT ANALYSIS

Kwang-Soo Lee
Department of Management-Finance
Indiana State University
Terre Haute, Indiana 47809/USA

I. Introduction

Input-output analysis can be extended to an economic system with antipollution measures to describe interdependency among economic sectors and antipollution sectors in the economy. Such a model is formulated by Professor Leontief (3) as a system of equations extending his original input-output framework. Leontief introduces a set of tolerable pollution levels to maintain a certain desired environmental quality. Steenge (7) further modifies Leontief's extended model changing Leontief's pollution abatement policy. However, there seems to be an unresolved question concerning the existence of a unique nonnegative set of output levels for any given bill of goods and the pollution tolerable levels in these models.

This paper 1) examines the treatment of the antipollution activities in Leontief's extended model and Steenge's modification, and studies their implications; 2) presents an alternative approach that introduces an automatic mechanism in the input-output framework for initiating the antipollution activities once specified levels of tolerable pollution have been reached; and 3) discusses the conditions under which the proposed model provides a unique nonnegative solution of output levels for any given bill of goods and tolerable pollution.

The alternative approach proposed in this paper is first to modify the pollutant balance equations in Leontief's extended model to make pollutant flows independent of the tolerable levels, and then introduce the tolerable levels in conjunction with a mechanism to initiate the antipollution activities. Through a modified physical input-output table, complementarity relationships between the antipollution activities and "slack tolerance" levels can be derived mathematically. The resulting model is a complementarity problem with upper bounding conditions. The existence and uniqueness of the solution to the proposed model are guaranteed under a much weaker condition than Hawkins-Simon condition, and the problem is easily solvable under the condition.

II. Fixed Pollution Tolerance Level Model

Leontief suggests a set of prescribed pollution tolerable levels to ensure a desired environmental quality in an economic system. The economy requires an antipollution activity for a polluting substance, a by-product generated by production and consumption activity, not to exceed its pollution tolerable level. The anti-pollution activity demands various industrial outputs as its inputs and, in turn, the increased industrial outputs result in an increase in demand of the antipollution activity generating more pollution. Leontief describes the interrelationships among industrial and antipollution sectors with the following extended input-output model:

$$(I - A_{11})X_1 - A_{12}X_2 = \quad Y_1 \qquad (1)$$

$$- A_{21}X_1 + (I - A_{22})X_2 = DY_1 - Y_2 \qquad (2)$$

where

A_{11} = a matrix of unit commodity input requirements for industrial sectors
A_{12} = a matrix of unit commodity input requirements for antipollution sectors
A_{21} = a matrix of unit pollutant generation for industrial sectors
A_{22} = a matrix of unit pollutant generation for antipollution sectors
D = a matrix of unit pollutant generation for the final consumption sector
X_1 = a vector of the gross output levels of industrial sectors
X_2 = a vector of the gross output levels of antipollution sectors
Y_1 = a final demand vector of commodities
Y_2 = a pollution tolerance vector of pollutants

The extended input-output model provides a unique solution of gross output levels for each sector for a given bill of final consumption Y_1 and tolerable pollution Y_2. However, the solution may be unrealistic since, in (1) - (2), each pollution level should be the same as the prescribed tolerable level regardless of the amount of the pollution generated. Especially, when the amount of a certain pollutant generated is less than its tolerable level, the extended model forces the corresponding antipollution sector to increase the pollution up to the tolerable level. As a result, the model provides erroneous total output levels for all the sectors.

Steenge claims that this difficulty can be avoided by simply using Leontief's original model if the amount of pollution generated is less than the tolerable level. His claim invites two criticisms. First, for a given Y_1, the amount of pollution generated is not known without using either Leontief's original or extended model. In (1) - (2), the amount of pollution generated is: $A_{21}(I - A_{11})^{-1}Y_1 + DY_1$. In order to use the original model the following condition should be satisfied:

$$A_{21}(I - A_{11})^{-1}Y_1 + DY_1 \leq Y_2 \qquad (3)$$

Note that, in (3), $X_1 = (I - A_{11})^{-1}Y_1$ is the general solution of Leontief's original model. Second, when the condition, (3), is not satisfied, some antipollution sectors may remain idle unless there is only one pollutant identified in the economy. Leontief's example (3) deals with one pollutant. Hence, the required input-output model is either Leontief's original or extended model. However, when there are more than one pollutant, neither model provides any systematic way of determining active and idle antipollution sectors.

If there are r number of pollutants identified in an economic system, we can realize 2^r (two to the r power) different cases with different combinations of active and idle antipollution sectors while changing the bill of goods. In other words, we require 2^r number of different input-output models for analyzing the economy. The main problem embedded in the extended model is to identify the proper combination of active and idle antipollution sectors for every given bill of goods.

Moore (5) attributes the cause of the difficulty to the fixed pollution tolerance policy by saying that "the idea of a tolerated level of pollution is fine provided the total amount of each pollutant generated is greater than, or equal to, the tolerable level." He advocates Steenge's modification that employs a fixed percentage pollution abatement policy. Nevertheless, the pollution tolerable level should be independent of the amount of pollution generated or economic activity levels. It is an exogenous quantity. We will discuss what would happen once we endogenize it in the next section.

The cause of the inconsistency is not the fixed pollution tolerance policy but the structure of the pollutant balance relationships. As Lee (2) elaborated, if we set up Leontief's example in the form of (1) - (2), the structural matrix is Leontief; all the off-diagonal elements are nonpositive,and all the principal minors are positive. As a result, the inverse of the structural matrix is a positive matrix and the output levels of the industrial sectors and antipollution sectors should be expressed as positive sums of the right-hand-side(r-h-s) values. However, unlike Leontief's original model, some r-h-s's of (1) - (2) may take negative values. Flick's counter-example (1) assigns a large negative value to the r-h-s of the pollutant balance equation by increasing the tolerable pollution level, so that the extended model gives a negative output level for the antipollution sector.

III. Fixed Percentage Abatement Policy

Steenge suggests an alternative pollution abatement policy and provides a modification to Leontief's extended input-output model, (1) - (2). Steenge assumes that the tolerable pollution level is proportional to the output level of the corresponding antipollution sector. That is,

$$Y_2 = \alpha \overset{\vee}{X}_2 \qquad (4)$$

with a fixed α. Then, he expresses the tolerable pollution level as a fixed percentage, $100(1+\alpha)^{-1}\%$, of the total pollution generated. Substituting (4) into (1) - (2), the following system is obtained:

$$(I - A_{11})X_1 - A_{12}X_2 = Y_1 \qquad (5)$$
$$- A_{21}X_1 + (I^* - A_{22})X_2 = DY_1 \qquad (6)$$

with $I^* = (1+\alpha)I$. Using (6), Steenge eliminates X_2 term from (5) and claims that the following adjusted system can be used in place of (1) - (2) to analyze the economy:

$$(I - A^*)X_1 = D^*Y_1 \qquad (7)$$

with $A^* = A_{11} + A_{12}(I^* - A_{22})^{-1}A_{21}$, and $D^* = I + A_{12}(I^* - A_{22})^{-1}D$. For a given Y_1, the industrial output levels can be obtained from (7) as:

$$X_1 = (I - A^*)^{-1}D^*Y_1 \qquad (8)$$

and the antipollution output levels can be expressed as:

$$X_2 = (I^* - A_{22})^{-1}A_{21}(I - A^*)^{-1}D^*Y_1 \qquad (9)$$

Finally, the tolerable pollution level can be obtained from (4).

The fixed percentage abatement policy makes Leontief's example predict nonnegative output levels, but it creates more problems than cures. It is obvious to see that the tolerable pollution level is proportional to final demands. Substituting (9) into (4), we obtain:

$$Y_2 = BY_1 \qquad (10)$$

with $B = \alpha(I^* - A_{22})^{-1}A_{21}(I - A^*)^{-1}D^*$. For a fixed α, matrix B is fixed. Now, without loss of generality, let final demands (or Gross National Product) be increased by 100β %; that is,

$$Y_1 = (1 + \beta)\bar{Y}_1 \qquad (11)$$

where $\bar{Y}_1$ is the old final demand vector. Substituting (11) into (10), we obtain $Y_2 = (1 + \beta)\bar{Y}_2$ where $\bar{Y}_2$ is the old tolerable pollution level vector. As expected, the tolerable level is increased at the same rate as GNP growth rate. Since the tolerable level changes with respect to the final demand, the adjusted model does not have the desirable feature of incorporating the environmental quality standards into the system that the extended model possesses.

Since the adjusted model assumes that α is fixed, unnecessary pollution abatement may be required. For example, let $\alpha = 1$ in Leontief's example. Then, the adjusted model gives $x_1 = 44.06$, $x_2 = 33.04$, and $x_3 = 14.32$ for $y_1 = 20$ and $y_2 = 20$. Without the antipollution activity (using Leontief's original model), $x_1 = 42.38$, $x_2 = 29.48$, and the total pollution generated amounts to 27.09. The antipollution activity should be idle, instead of $x_3 = 14.32$, since the total amount of pollution generated is below the prescribed pollution tolerable level, 30.

It is still possible that the adjusted model predicts negative antipollution output levels. Leontief and Ford (4) indicate that "substances described as pollutants can sometimes be and often actually are utilized as inputs in the production of useful goods." If we assume that, in Leontief's example, the manufacturing sector uses 0.38 grams of the pollutant per unit production of its commodity, at $\alpha = 1$ the adjusted model predicts $x_1 = 67.23$, $x_2 = 101.06$, and $x_3 = -2.39$ for $y_1 = 10$, and $y_2 = 80$. Furthermore, since $y_3 = x_3$ at $\alpha = 1$, the tolerable pollution level should be -2.39. Both negative antipollution activity and tolerable levels are not acceptable.

The fixed percentage abatement policy may be imposed on individual industrial sectors as a part of their production function. Nevertheless, the policy is not suitable for setting the tolerable pollution levels. In order to maintain acceptable environmental quality in an economic system, we still need antipollution sectors separate from industrial pollution abatement activities, and the prescribed tolerable pollution levels that reflect the acceptable environmental quality.

IV. A Generalized Input-Output Model

The fixed pollution tolerance level policy is kept since it is a sound policy to maintain the environmental quality standards. In Section II, we noticed that the pollution balance equations in Leontief's extended model is not appropriate. An alternative approach is to modify the pollutant balance equations to make pollutant flows independent of the tolerable pollution levels.

Let us define:

x_{ij} = amount of commodity i purchased by sector j

x_{iq} = amount of commodity i purchased by antipollution sector q

x_i = the total output of commodity i

x_{pj} = amount of pollutant p generated by sector j

x_{pq} = amount of pollutant p generated by antipollution sector q

x_p = the total amount of pollutant p eliminated by the corresponding antipollution sector

t_p = amount of pollutant p discharged into environment (externality)

y_i = amount of commodity i delivered to the final consumption sector

c_p = amount of pollutant p generated by the final consumption sector

Within the traditional input-output framework, Leontief's physical input-output table may be modified by introducing an "environment" sector column separately from the final consumption sector as Table 1. The final demand column only represents the amount of each pollutant generated by households at each pollutant row, and the

TABLE 1.-- A PHYSICAL INPUT-OUTPUT TABLE

Input and Pollutant	Output Sectors: Industrial Sectors 1 . . . m	Output Sectors: Antipollution Sectors m+1 . . . m+n	Environment	Final Demand	Total Output
1 . . . m	x_{ij}	x_{iq}	0	y_i	x_i
m+1 . . . m+n	x_{pj}	x_{pq}	$-t_p$	c_p	x_p

environment sector column accounts for the amount of uneliminated pollutant with a negative sign. Each column of the table shows the input of each commodity and the amount of pollution generated for each sector. The total output column gives the output of each commodity and the amount of each pollutant eliminated, so that we obtain:

$$\sum_{j=1}^{m} x_{ij} + \sum_{q=m+1}^{m+n} x_{iq} + y_i = x_i \qquad i = 1, 2, . . ., m \qquad (12)$$

$$\sum_{j=1}^{m} x_{pj} + \sum_{q=m+1}^{m+n} x_{pq} + c_p = x_p + t_p \qquad p = m+1, . . ., m+n \qquad (13)$$

Note that the tolerable pollution level is not included in (13).

Extending Leontief's fixed-coefficient assumption to the antipollution sectors, we can express each input requirement and pollutant flow as:

$$a_{ij} = x_{ij}/x_j \qquad i, j = 1, . . ., m \qquad (14)$$

$$a_{pj} = x_{pj}/x_j \qquad p = m+1, \ldots, m+n,\ j = 1, \ldots, m \qquad (15)$$

$$a_{iq} = x_{iq}/x_q \qquad i = 1, \ldots, m,\ q = m+1, \ldots, m+n \qquad (16)$$

$$a_{pq} = x_{pq}/x_q \qquad p, q = m+1, \ldots, m+n \qquad (17)$$

Now we can derive technical coefficients of each output sector (including Anti-pollution) from Table 1. We divide the entries in j-th column of output sector by j-th element in the last column and obtain Table 2. Each entry in the table represents either unit input requirement or unit pollution load.

TABLE 2.--TECHNICAL COEFFICIENTS

Input and Pollutant	Output Sectors: Industrial Sectors 1 . . . m	Output Sectors: Antipollution Sectors m+1 . . . m+n
1 . . . m	a_{ij}	a_{iq}
m+1 . . . m+n	a_{pj}	a_{pq}

System (12) - (13) is not suitable for analyzing the structure of the economy. To convert the system into an analytical model, we utilize the technical coefficients defined by (14) - (17), and presented in Table 2. Also, we assume

$$c_p = \sum_{i=1}^{m} d_{pi} y_i$$

where d_{pi} is the amount of pollutant p generated by consumption sector consuming one unit of commodity i. Then (12) - (13) can be converted as the following system of equations:

$$(1 - a_{ii})x_i - \sum_{j \neq i} a_{ij} x_j - \sum_{q=m+1}^{m+n} a_{iq} x_q = y_i \qquad i = 1, \ldots, m \qquad (18)$$

$$(1 - a_{pp})x_p - \sum_{j=1}^{m} a_{pj} x_j - \sum_{q \neq p} a_{pq} x_q + t_p = \sum_{i=1}^{m} d_{pi} y_i$$

$$p = m+1, \ldots, m+n \qquad (19)$$

Unlike Leontief's extended model, the system of equations, (18) - (19), contains the amount of pollutant p discharged, t_p, as an extra variable in each pollutant balance equation. However, the variable, t_p, together with the prescribed tolerable pollution level, can be used to initiate the corresponding antipollution activity.

Let y_p be the tolerable pollution level of pollutant p. Then the economy discharges the pollutant into the environment as long as t_p does not exceed y_p ($y_p - t_p > 0$), and the antipollution sector is idle ($x_p = 0$). Once t_p reaches y_p ($y_p - t_p = 0$), the economy initiates the antipollution activity and starts eliminating pollutant p ($x_p > 0$). In other words, x_p and $(y_p - t_p)$ are complements of each other and; therefore, we may require the following complementarity conditions:

$$(y_p - t_p)x_p = 0 \qquad p = m+1, \ldots, m+n \qquad (20)$$

Furthermore,

$$t_p \leq y_p \qquad p = m+1, \ldots, m+n \qquad (21)$$

The quantity, $y_p - t_p$, can be interpreted as "slack tolerance level," so that

$$w_p = y_p - t_p \geq 0 \qquad p = m+1, \ldots, m+n \qquad (22)$$

Substituting (22) into (19), (20), and (21), we can obtain the following complementarity problem:

$$(I - A_{11})X_1 - A_{12}X_2 = Y_1 \qquad (23)$$

$$-A_{21}X_1 + (I - A_{22})X_2 - W_2 = DY_1 - Y_2 \qquad (24)$$

$$X_2'W_2 = 0;\ X_1,\ X_2,\ W_2 \geq 0 \qquad (25)$$

$$W_2 \leq Y_2 \qquad (26)$$

Note that Leontief's extended model, (1) - (2), is a special case of the proposed model, (23) - (26), with $W_2 = 0$.

V. Existence and Uniqueness of a Solution

As shown by Somelson et. al. (6), a necessary and sufficient condition for the complementarity problem, (23) - (25), to have a unique solution is that all principal minors of the structural matrix are positive. If none of the industrial sectors uses any pollutant as input, all the off-diagonal elements are nonpositive, and the amount of each pollutant discharged is nonnegative. As a result, $t_p \geq 0$, and $w_p \leq y_p$ in (22); the unique nonnegative solution to (23) - (25) satisfies the additional upper bounding constraint, (26).

If at least one industrial sector utilizes some pollutant as its input, some off-diagonal elements are positive and the unique solution may not satisfy (26). The amount of the pollutant discharged becomes negative in the case where the total amount of the pollutant required by the pollutant-consuming industry exceeds the total amount of the pollutant generated in the economy. Since all the principal minors of the structural matrix should be positive for the economy to be productive, it is appropriate to identify additional conditions on the structural matrix that makes the unique solution satisfy (26). If the pollutant-consuming industry also uses, as its input, outputs of other industries that generate the pollutant, the pollutant-consuming industry also generates "indirect" pollution through the input-providing industries. If the pollutant-consuming industry is a net polluting industry, the unique nonnegative solution to (23) - (25) also satisfies (26).

The unique nonnegative solution to the proposed model for any given Y_1 can be obtained by Lee's reduction algorithm (2). The algorithm first pivots on x_i's in the commodity balance equations. Then it selects a pollutant balance equation with a nonnegative r-h-s value one at a time, sets the slack tolerance level, w_p, equal to zero (the zero variable of the complementary pair), and pivots on x_p in the equation. The algorithm terminates arriving at either Case I: The principal pivot operation has been carried out to the last equation of the system, or Case II: All the unselected (remaining) pollutant balance equations have negative r-h-s values. For Case I, the required solution is immediately available since the reduced system without those zero variables, w_p's, is an identity system. For Case II, we set x_p's in the remaining balance equations equal to zero and multiply -1 to both sides of the remaining equations. The required solution is again immediately available since the reduced system becomes an identity system.

References

1. Flick, Warren A., "Environmental Repercussions and the Economic Structure: an Input-Output Approach: A Comment," The Review of Economics and Statistics 56 (Feb. 1974), 107-109.
2. Lee, Kwang-Soo, "A Generalized Input-Output Model of an Economy with Environmental Protection," The Review of Economics and Statistics (forthcoming).
3. Leontief, Wassily W., "Environmental Repercussions and the Economic Structure: an Input-Output Approach," The Review of Economics and Statistics 52 (Aug. 1970), 262-271.
4. ___________, and Daniel Ford, "Air Pollution and the Economic Structure: Empirical Results of Input-Output Computations," in Andrew Brody and Anne P, Carter (eds.), Input-Output Techniques (Amsterdam-London: North-Holland Publishing Co. 1972), 9-30.
5. Moore, Stuart A., "Environmental Repercussions and the Economic Structure: Some Further Comments," The Review of Economics and Statistics 63 (Feb. 1981) 139-142.
6. Samuelson, Jans, R. M. Thrall, and Oscar Wesler, "A Partition Theorem for Euclidean n Space," Proceedings of the American Mathematical Society, 9 (1958), 805-807.
7. Steenge, Albert E., "Environmental Repercussions and the Economic Structure: Further Comments," The Review of Economics and Statistics 60 (Aug. 1978), 482-486.

MODELLING AND CONTROL OF MARKET PENETRATION

Claudio Leporelli
Istituto di Automatica
Università di Roma

Abstract: In this paper different approaches to the modelling of the market penetration of new goods, processes and brands are briefly discussed. A warning is given on the use of purely descriptive models based on the common features (the s- shaped market penetration curve) of phenomena that widely differ in causes and possible controls; diffusion and capacity expansion models are compared to this end. An optimal pricing exercise is finally presented to clarify the importance of interpretative and control models in technological forecasting problems.

1. Introduction

Many of the recently developed application-oriented sectoral economic models (particularly in the energy field) have been motivated by the contingency forecasting and control needs of agencies involved in funding long range research and development projects and in promoting the adoption of innovative systems and practices /1/.
On the other hand, from a model - builder's point of view, the problem of market penetration is of deep interest for its challenging interpretation and specification issues:

- The dynamics of the phenomenon can be related to a number of different causes (the information flow and the changing attitudes in the social system, the economic attractiveness and its evolution, connected to learning and scale effects and dynamic pricing, behavioural and financial constraints in the capacity expansion process, etc.). Control models are however difficult to estimate because of multicollinearity.
- Uncertainties about the future performance and costs of the innovative system and on the growth of the demand for its services affect the decision process and can be modelled through alternative approaches: behavioural constraints and myopic optimization in a deterministic setting /2/ adaptative control /3/, multiperiod stochastic programming with recourse /4/; a choice between the neoclassical intertemporal optimization /5/ hypothesis and more "evolutionary" theories /6/ is also

implicitly made in this way.

- The realistic description of the technological options - a crucial problem in many economic models - is even more important in this context; a priori information coming from engineering knowledge (scale economies, construction times, inputs requirements, cost structure, operating conditions, etc.) should be used in the specification and estimation process. On the other hand many technologically oriented forecasting models lack of a proper analysis of the economic and financial setting of the phenomenon.

2. The common phenomenology and the different application contexts

Suppose:

(1) $\dot{A}(t) = a(t)A(t) \quad , \quad \dot{B}(t) = b(t)B(t) \quad , \quad a(t) - b(t) = r(t)$

$$f_A(t) = A(t) \;/\; (A(t) + B(t)) \qquad f_B(t) = 1 - f_A(t)$$

Then is easily seen that:

(2) $$\dot{f}_A(t) = r(t)f_A(t)f_B(t)$$

or

(3) $$f_A(t) = f_A(t_0) \;/\; \left(f_A(t_0) + f_B(t_0)\, e^{-\int_{t_0}^{t} r(t)dt} \right)$$

If $r(t) = r$ you obtain the Fisher-Pry logistic market penetration model that has been widely used / / / / for its good descriptive properties and because it is very simple to estimate through the

(4) $$\ln(f_A(t) \;/\; f_B(t)) = \ln(\frac{f_A(t_0) - rt_0}{f_B(t_0)}) + rt$$

Moreover the equation (2) suggests a prey-predator or diffusion interpretation of the s- shaped time evolution of (3). But, as we have seen, the simple hypothesis in (1) of growth rates differing for a constant r is sufficient to obtain the logistic curve (3) for the shares. This example is useful to clarify the purely descriptive character of the model: the constant and intercept term in (4) summarize the information relative respectively to the time and to the (mid-point) rate of adoption (a third parameter could be inserted, and estimated with a search procedure, to take into account asymptotic levels of the shares different from 1 and 0). Within this framework any interpretation is left to the comparison of the parameters obtained for different substitution processes: Mansfield /9/, for example, found that the rate constant r is positively correlated with the profitability of the new technology and negatively correlated with the capital investment needed to introduce it; Bungaard - Nielsen /10/ analysed the international diffusion of two major innovations in the steel industry and found that the diffusion rate in the late adopter countries is higher than in the early adopter countries.

The forecasting use of the model (2)-(4) implies the hypothesis of stability of the effect of the economic and institutional factors that influence the process /11/ and the renouncing to any possibility of controlling it. Both the hypothesis and the consequence are questionable, specially in a period of rapidly changing economic environment and shortened life cycles of innovations /12/. In /13/, for example, is shown that the market penetration of synthetics in the textile industry in Italy, that until 1973 had followed model (2)-(4) quite well, suddenly stopped from that year on because of the combined effect of changes in relative prices, new laws on consumer's information and the evolution in the role of the Italian textile industry in the international market.

However the building of interpretative models is complicated by the variety of application contexts in which the phenomenon takes place:

- The population of adopters may be formed by individuals or organizations, the adoption can concern consumption or production/investment activities and durable goods or not. The latter alternatives usually involve long range decisions in a much more formalized decision process.
- The population may be static or time-varying, homogeneous or not in terms of needs, preferences, potential uses of the new good, stages in the adoption process (unaware, aware, adopter, past-adopter /12/). The adoption may consist in a single long lasting event or in a repeat-purchase with possible discontinuation. A dishomogeneous and time varying situation requires several state variables and specific state - transition equations.
- The interest of the analyst (and the data he can use) may be directly related to the adoption phenomenon per se or to secondary effects of it (the consumption of production inputs in an industry adopting innovative processes /14/, the residential demand for energy in connection with the diffusion of energy-saving appliances /15/, etc.). The use of derived data make it more difficult specially in a dishomogeneous population case to disentangle the adoption effects from the capa-

city utilization and composition effects.

On the other hand, the approach to the analysis of market penetration phenomena also depends on the disciplinary setting of the studies:

- In the technological forecasting approach a well defined set of goods/processes is analysed, at a very disaggregated level; the economic competitiveness of the new process is assessed a priori (in terms of costs and not of market prices) and then postulated and generally not imbedded in the subsequent market penetration analysis. /16/ /17/

  The approach has been criticized /18/ as being purely descriptive, deterministic, rigid when theoretically well developed (in a diffusion framework) and resorting to ad hoc procedures without rigorous foundations in other cases.
- In economics the interest focused historically on the study of substitution among primary factors of production, on an aggregate basis, and on the analysis of consumer behaviour; the change in the demand for intermediate goods was mainly estimated comparing and empirically updating input-output matrices /19/; a few studies tried the so-called "engineering production function" approach /20/. In recent years the interest in disaggregated general equilibrium models and in sectoral models, in relation also with the increasing importance of forecasts on energy demand, motivated a number of studies on production inputs demand also at industry level /21/ /22/; many of them attempt to econometrically estimate the relative contribution of technical progress, scale economies, and, price induced substitution to the determination of production inputs demand /19/. An alternative approach involves the use of technological process models /23/ /24/, possibly integrated with behavioural hypotheses in the recursive programming approach /2/, or used to generate pseudo-data in the estimation of production functions /25/.

  Adjustment costs in the investment process are involked to, in the production function approach, to explain, in a neoclassical framework, the partial and lagged response of input demand to changes in relative prices /22/.

  The cost minimization hypothesis does not seem sufficient however, specially in a deterministic framework, to obtain reliable forecast of market penetration of new technologies: in /26/ a sensitivity analysis of the Brookhaven paper industry model /24/ shows that dramatically different structures for input demand and utilization of technologies are compatible with a cost level only 1% above the minimum; that means that room is left for secondary goals and behavioural constraints and that "cheap" industrial policies can be conceived to control the market penetration process.

A lot of work has also been done in econometrics on limited dependent variable models for quantal and discrete choice analysis /27/; it is interesting to note that these studies are usually static, use cross-section data (individual choices) and emphasize the probabilistic setting of the choice and the consequences of the use of specific functional forms /28/.

- In the quantitative marketing literature the emphasis of the analysis is on the short term and on the brand choice process; the effect of marketing policies is modelled with a view to the control problem of choosing a market penetration strategy at firm level; in particular in an oligopolistic setting, advertising technical assistance and other sales promotion activities increase their importance with respect to price as instruments of competition; so the informational aspects are a central issue of the problem and diffusion type models are widely used.

3. A comparison of the diffusion and capacity expansion approaches to market penetration

In the diffusion interpretation of model (2), a possibly time varying population of potential plus actual adopter P(t) progressively adopts the new practice because of the word of mouth or confidence diffusion effect of actual adopters A(t)

(5) $$\dot{A}(t) = r(t)\, A(t)\, (P(t) - A(t))$$

The parameter r can be considered proportional to the ease of communication but also to the strenght of the diffuser's arguments, i.e. to the reliability and economic attractiveness of the innovation; on the other hand also the value of potential users $P(t)$ depends on the economic factor, so both the rate and the asymptotic level of the substitution process can be influenced by the prices prevailing on the market. The population however can be partially formed by innovators that don't need the contact with adopters /29/

(6) $$\dot{A}(t) = r(t)\, (A(t) + a(t))\, (P(t) - A(t))$$

The coefficient $a(t)$ can be considered proportional to the "goodwill" accumulated through the advertising and thus a new control variable as added /30/; moreover "autonomous adopters" (for example government supported demonstration programs) could be added to $a(t)$ /31/.

Peterka /32/ obtains an equation very similar to (5) with the procedure used in equations (1)-(3) and with a completely different interpretation of the market penetration process: he supposes that the external capital extended to an industry is negligible in the long run so that the industry growth is financed by internally generated profits; if α_i is the amount of capital needed for a unit increase in production capacity of process i, supposing that any production $P_i(t)$ doesn't ever decrease at a rate larger than physical depreciation rate and that there are no distributed profits, you obtain:

$$\alpha_i \dot{P}_i(t) = P_i(t)\,(p_i(t) - C_i(t)) \qquad i=1,\dots n \tag{7}$$

that is the analogue of (1) for the n-good case. Using (7) for generic processes i and j you obtain

$$\frac{d}{dt}\left(\ln \frac{f_i(t)}{f_j(t)}\right) = \frac{d}{dt}\left(\ln \frac{P_i(t)}{P_j(t)}\right) = \frac{p_i(t) - C_i(t)}{\alpha_i} - \frac{p_j(t) - C_j(t)}{\alpha_j} \tag{8}$$

where $f_i(t) = P_i(t) / P(t)$ and $P(t) = \sum_i P_i(t)$.

Peterka focuses on products/processes that perfectly substitute so $p_i(t) = p(t)$ and, putting $\rho = \dot{P}(t) / P(t)$ solves for the price

$$p = \sum_i (\alpha_i \rho + C_i)\frac{f_i}{\alpha_i} \Big/ \sum_i \frac{f_i}{\alpha_i} \tag{9}$$

obtaining a weighed average of the long run marginal costs of the competing goods/processes. This is not compatible with any price theory for market economies, and conceals the effects of interactions between supply and demand, but is a direct consequence of the equally dubious hypotheses about financing, complete investment of profits, full production capacity utilization and, first of all, the possible simultaneous expansion of several production capacities /33/. A simultaneous investment in competing technologies is however a striking feature of market penetration processes in actual economies and could be optimal in a stochastic modelling setting; moreover Spence /34/ has shown that the financial constraint (perhaps in the form of a maximum debt-equity ratio) is crucial in optimal investment strategies for individual firms in a new market and that means that, in the early stages of growth, all the profits are invested. Similar arguments have been used by Eichner /35/ to explain the time evolution of the mark-up in the oligopolistic pricing.

4. Optimum pricing for a self-financing firm facing a price sensitive diffusion demand

In a two processes case with $\alpha_1 = \alpha_2$ and $p_1 = p_2$ the Peterka model implies

$$(10)\quad \dot{f}_1(t) = f_1(t)\ (1 - f_1(t))\ (C_2(t) - C_1(t))\ /\alpha$$

Without loss of generality we can assume that $C_1(t)$ and $C_2(t)$ coincide with the prices of two production inputs used respectively by the first and the second production processes; therefore, with a constant volume of total sales, the demand function for the (monopolistic) firm that produces the first production input is

$$(11)\quad S(t) = a\,f_1(t) = a\,f_1(t_o)\ /\ \left(f_1(t_o) + (1 - f_1(t_o))\ \exp - \int_{t_o}^{t} \frac{C_2(t) - C_1(t)}{\alpha}\, dt\right)$$

Suppose that firm has unit costs $C(t)$ and needs an amount of capital β for a unit increase in its production capacity; suppose further that the firms choose $C_1(t)$ to maximize the present value (at a rate δ) of earnings net of investment over a finite horizon for a given evolution of $C(t)$ and $C\ (t)$, with $C_2(t) > C(t)$

$$(12)\quad \max_{C_1(t)} J = \int_{t_o}^{T} \left[(C_1(t)-C(t))\,S(t) - \beta\,\dot{S}(t)\right] e^{-\delta(t-t_o)}\, dt$$

subject to

$$(13)\qquad C(t) \leqslant C_1(t) \leqslant C_2(t) \qquad\qquad t_o \leqslant t \leqslant T$$

$$(14)\qquad (C_1(t) - C(t))\ S(t) - \beta\,\dot{S}\ (t) \geqslant 0 \qquad\qquad t_o \leqslant t \leqslant T$$

In words: the price $C_1(t)$ of the production input must cover costs and be consistent with sales growth and self-financing.

These constraints are satisfied if, and only if

$$(15)\quad C_1(t) = C_2(t) - \lambda(t)\ (C_2(t) - C(t))\ /\ (1 + (1 - f(t))\ \beta/\alpha\ )\ ,\quad 0 \leqslant \lambda(t) \leqslant 1$$

Therefore the maximization problem reduces to

$$(16)\quad \max_{(t)} J = \int_{t_o}^{T} e^{-\ (t-t_o)} \alpha (C_2(t) - C(t))\ \ f(t)\ (1 - \lambda(t))\ dt$$

subject to

(17) $\dot{f}(t) = \lambda(t)\ (C_2(t) - C\ (t))f(t)\ (1 - f(t))\ /\ (\alpha + \beta(1 - f(t)))\qquad 0 \leqslant \lambda(t) \leqslant 1$

For a constant (expected) value of $C_2(t) - C(t)$, with a straightforward application of the maximum principle, we obtain:

(18)
$$\lambda(t) = 1 \quad \text{if} \quad \frac{1}{1 - f(t)} < \frac{(C_2 - C)}{\alpha\delta}\ (1 - e^{-\delta(T - t)}) - \frac{\beta}{\alpha}$$
$$\lambda(t) = 0 \quad \text{otherwise}$$

Therefore the price $C_1(t)$ is set to the decreasing minimum level that permits self-financing of the sales growth and all profits are invested in a first period if $f(t_o)$ is small enough to satisfy the first of (18); then, when the share reaches the level given in (18), the price $C_1(t)$ is set equal to the maximum level that permits to mantain a non decreasing share and all the profits are distributed. This simple example clarifies how economic goals and constraints can modify the smooth s- shaped market penetration pattern usually assumed by technological forecasting descriptive models and why interpretative and control models, although more difficult to build and use, are necessary for any realistic scenario analysis of market penetration phenomena.

References

/1/ CHARLES RIVER ASSOCIATES: "Review and Evaluation of Selected Large-Scale Energy Models: a State of the Art Study for the Electric Power Research Institute" CRA Report n. 231 Jan. 1977.

/2/ R.H. DAY and A. CIGNO: "Modelling Economic Change", North Holland 1978.

/3/ BAR-SHALOM Y. and TSE E.: Caution, Probing, and the Value of Information in the Control of Uncertain Systems" ; Annals of Economic and Social Measurement, 5/7, 1976

/4/ LOUVEAUX F.V.: "A Solution Method for Multistage Stochastic Programs with Recourse with Application to an Energy Investment Problem" Operations Research vol. 28, n. 2 1980.

/5/ CASS D. and SHELL K.: "The Hamiltonian Approach to Dynamic Economics" Academic Press, 1976.

/6/ NELSON R.R. and WINTER S.G.: "Dynamic Competition and Technical Progress" in NELSON R.R. and BALASSA B. eds. "Economic Progress Private Values and Public Policies" North Holland 1977.

/7/ FISHER J.C. and PRY R.H.: "A Simple Substitution Model of Technological Change" Report 70-C-215, General Electric Company, Research and Development Center, Schenectady, N.Y., Technical Information Series; 1970.

/8/ ROGERS E.M.: "New Product Adoption and Diffusion", J. Consumer Res. 2, 209-301, 1976.

/9/ MANSFIELD E.: "Technical Change and the Rate of Imitation", Econometrica vol. 29 October 1961, pp. 741-766.

/10/ BUNDGAARD-NIELSEN M.: "The International Diffusion of New Technology" Technol. Forecast. Soc. Change 8, 365-370, 1976.

/11/ MARCHETTI C. and NAKICENOVIC N.: "The Dynamics of Energy Systems and the Logistic Substitution Model" IIASA - RR - 79 - 13 Dec. 1979.

/12/ KALISH S. and LILIEN G.L.: "Models of New Product Diffusion Current Status and Research Agenda" MIT ENERGY LAB Work. PAP. n. MIT - EL - 79 - 054 WP, Oct. 1979.

/13/ ALESSANDRONI A., LEPORELLI C., REY G.M.: "A Disequilibrium Model at Industry Level" Res. Rep. 81 - 06 of Istituto di Automatica, Università di Roma.

/14/ ALMON C. et al.: "Interindustry Forecasts of the American Economy" Lexington: Lexington Books, D.C. Heath and Company, 1974.

/15/ HAUSMAN J.A.: "Individual Discount Rates and the Purchase and Utilization of Energy-Using Durables" Bell Jour. Econ. Vol. 10 pp. 33-54 1979.

/16/ HURTER A.P. and RUBENSTEIN A.H.: "Market Penetration by New Innovations: the Technological Literature" Technol. Forecast. Soc. Change 11, 197-221 1978

/17/ PHILIPSON L.L.: "Market Penetration Models for Energy Production Devices and Conservation Techniques" Technol. Forecast. Soc. Change 11, 223-236 1978.

/18/ WARREN E.H.: "Solar Energy Market Penetration Models: Science or Number Mysticism?" Technol. Forecast. Soc. Change 16, 105-118 1980.

/19/ SATO R. and RAMACHANDRAN R.: "Measuring the Impact of Technical Progress on the Demand for Intermediate Goods: a Survey" Journ of Econ. Liter. vol. 18 n. 3 1980.

/20/ CHENERY H.B.: "Engineering production functions" Quaterly Journal of Economics 63, pp. 507-531, Nov. 1949.

/21/ SHEININ Y.: "The Demand for Factor Inputs under a Three Level CES Four Factor Production Function" Ph.D. thesis Univ. of Pennsylvania, 1980.

/22/ DENNY M., FUSS M., WAVERMAN L.: "An Application of Optimal Control Theory to the Estimation of the Demand for Energy in Canadian Manufacturing Industries" Proc. of the Ninth IFIP Conf. on Optimization Techniques, Springer Verlag 1980.

/23/ ADAMS F.G. and GRIFFIN J.M.: "An Economic-Linear Programming Model of the U.S. Petroleum Refining Industry" Journal of the Americal Statistical Association, 67 pp. 542-551, 1972.

/24/ PILATI A. D. et al.: "A Process Model of the U.S. Pulp and Paper Industry" Brookhaven National Laboratory, New York, 1980.

/25/ GRIFFIN J.M.: "Joint Production Technology: the Case of Petrochemicals" Metroeconomica, Vol. 46, n. 2, March 1978.

/26/ PILATI A.D. et al.: "Industry Process Models: Applications for Market Penetration and Energy Use Projections" Brookhaven National Laboratory, New York 1980.

/27/ Annals of Economic and Social Measurement: Special Issue on Discrete Qualitative and limited Dependent Variables. Vol. 5 n. 4 1976.

/28/ TAE HOON OUM: "A Warning on the Use of Linear Logit Models in Transport Mode Choice Studies" Bell Jour. Econ. Vol. 10 pp. 374-388 1979.

/29/ BASS F.M.: "A New Product Growth Model for Consumer Durables" Man. Science Vol. 15 n. 5 pp. 215-227, 1969.

/30/ SETHI S.P.: "Dynamic Optimal Control Models in Advertising: A Survey" SIAM Rev. Vol. 19 1977.

/31/ LILIEN G.L.: "The Implications of Diffusion Models for Accelerating the Diffusion of Innovation" Technol. Forecast. Soc. Change 17, PP. 339-351 1980.

/32/ PETERKA V.: "Macrodynamics of Technological Change: Market Penetration by New Technologies" RR-72-22 IIASA Laxenburg Austria 1977.

/33/ SPINRAD B.I.: "Market Substitution Models and Economic Paremaeters" RR-80-28 IIASA Laxenburg Austria 1980.

/34/ SPENCE A.M.: "Investment Strategy and Growth in a New Market" Bell Jour. Econ. Vol. 10 pp. 3-18 1979.

/35/ EICHNER A.S.: "The Megacorp and Oligopoly" Cambridge University Press 1976.

EQUILIBRIUM ADVERTISING IN AN OLIGOPOLY WITH NERLOVE-ARROW ADVERTISING DYNAMICS: EXISTENCE AND STABILITY

Ram C. Rao
Krannert Graduate School of Management
Purdue University
W. Lafayette, IN 47907

1. Introduction

The purpose of this paper is to examine the equilibrium level of advertising in an oligopoly. In making their choices on the level of advertising expenditures firms in an oligopoly can be expected to take into account the actions of rivals. A second element in deciding on advertising outlay is the consideration that advertising in a given period will influence sales in that as well as subsequent periods. The optimal level of advertising for a firm therefore depends on the dynamic effects of advertising as well as on the choices of rival firms.

The industry studied in this paper is characterized by firms whose asymmetry derives from a locational choice made by each firm. The product marketed by each firm is assumed to be uniquely, and differently, located in the space of product attributes. Thus each firm's product is differentiated. Additionally, this paper takes these choices as given and unalterable. Demand for each firm's product, to the extent that it depends on its choice of location relative to others in the industry is stationary; also it is assumed to be a deterministic function of other variables. The explicit dynamics of concern is that of the effect of advertising. Advertising in a given period is assumed to affect demand in that period as well as that in future periods. In this sense, firms' choices must be made with an eye to the future even though demand is a stationary deterministic function. In addition to advertising, firms' choice of prices are also assumed to affect demand. It should be noted that many empirical studies of demand functions in oligopoly, in fact, concern themselves with industries characterized by the foregoing technology.

Firms are assumed to behave non-co-operatively to maximize discounted profits over an infinite horizon. An industry equilibrium is then defined to be a Nash solution to a set of maximizing problems faced by firms, each firm choosing its advertising level for all periods. The existence of industry equilibrium defined in this manner is the main focus of this paper. Additionally, stationary states along an equilibrium path are obtained as a function of exogenous variables. Finally, the stability of stationary states is briefly addressed.

Optimality of advertising policies for the monopolist was first studied by Nerlove and Arrow [5]. Gould [3] examined a variation of that, and Sethi [8] for a different specification of the sales advertising relation. The case of oligopoly has been examined by Schmalansee [7] where the interdependence of firms' choices is not explicitly modeled, and therefore no formal notion of equilibrium developed.

Lambin et. al [4] consider the behaviour of one firm in an oligopoly wherein it essentially acts as a monopolist taking into account behavioural responses of others. In contrast to the foregoing, this paper attempts analysis of a formal model of advertising rivalry in a dynamic setting.

The paper is organized as follows. Section 2 outlines the basic model of the industry and a variant of the Nerlove-Arrow model of advertising process. In section 3 industry equilibrium is defined and its existence proved. Conditions for the stability of stationary states are established. Section 4 ends with concluding comments.

2. Model of Industry

This section presents a model of the oligoplistic industry whose advertising decisions are the focus of interest. Demand functions are first specified, followed by a formal statement of the maximizing problems faced by firms.

2.1 Demand Functions and Short-Run Pricing Decisions

The industry under study consists of n firms, each of whom is assumed to produce (and market) strictly positive quantities of branded products. Brand (and, firm) i, i = 1, 2,..., n, is assumed to be characterized by a quality index s_i. In general s_i may be thought of as a vector that locates brand i in a space of product attributes, the location of brand i relative to others being a determinant of demand for i. For ease of exposition it is assumed here that the location can be summarized by a scalar index s_i. This paper will explicitly deal with certain dynamic phenomena, and it is useful to stress here that once a firm has chosen s_i, it is assumed to be fixed and invariant over time. Demand for brand i, at time t, q_{it}, is then assumed to be given by the (suitably) differentiable function

$$q_{it} = f_i(s_i, \hat{s}_i, x_{it}, \hat{x}_{it}, p_{it}, \hat{p}_{it}) \tag{1}$$

where

$$\hat{s}_i = \sum_{\substack{j=1 \\ j\neq i}}^{n} s_j \;, \quad \hat{x}_{it} = \sum_{\substack{j=1 \\ j\neq i}}^{n} x_{jt} \;, \quad \hat{p}_{it} = \sum_{\substack{j=1 \\ j\neq i}}^{n} p_{jt}$$

and x_{it} is the stock of advertising goodwill of firm i at time t, and p_{it} is the price chosen by firm i at time t. Assume that $\partial f_i/\partial p_{it} < 0$ and $\partial f_i/\partial x_{it} > 0$. Additional assumptions on f_i are made when necessary. The stock of goodwill x_{it} is assumed to be a known function of past advertising expenditures over the interval $\tau = t, t-1, \ldots$. The exact specification of this process is taken up in the next section. It can be observed here that the demand function for i is symmetric with respect to all i's rivals. Moreover, the effect of each competitive instrument (quality, price and advertising) chosen by rivals is summarized by that of a linear function of the rivals' choices of that instrument. More general specification of demand functions will not violate any of the results in the paper. The specification chosen here is to facilitate the presentation of ideas in general, and the stability result in particular. It should also be noted that empirical demand investigations frequently specify independent variables, such as price, in relative form, which is the specification implicitly chosen here.

Each firm at time t is assumed to first determine its goodwill x_{it} in a manner to be defined later. Given x_{it}, $i = 1, 2,\ldots, n$, each firm faces the problem of choosing price so as to maximize its profits in period t. Let c_i and p^*_{it}, $i = 1, 2,\ldots, n$ respectively be the (constant) marginal cost of production and chosen price of firm i. Then in period t firm i is assumed to solve the ith problem in the following set of maximization problems:

$$\begin{aligned} (2) \qquad \pi_i(x_{it}, \hat{x}_{it}) &= (p^*_{it} - c_i)\, f_i(s_i, \hat{s}_i, x_{it}, \hat{x}_{it}, p^*_{it}, \hat{p}^*_{it}) \\ &= \max_{p_{it}} \left[(p_{it} - c_i)\, f_i(s_i, \hat{s}_i, x_{it}, \hat{x}_{it}, p_{it}, \hat{p}^*_{it})\right], \\ & \qquad\qquad i = 1, 2,\ldots, n \end{aligned}$$

where π_i is the resulting profits to firm i given $(x_{it}, \hat{x}_{it})$ and p^*_{it} solve the n problems in (2). It is clear that (2) defines a non-co-operative equilibrium where p^*_{it} represent the Nash prices. We now make implicit assumptions on the underlying technology of (2) so that the following assumptions on p^*_{it} and π_i hold:

<u>A1</u> There exist unique Nash prices p^*_{it}, $i = 1, 2,\ldots, n$ which are strictly positive.

<u>A2</u> The profits π_{it} are continuous in $(x_{it}, \hat{x}_{it})$ and strictly concave in x_{it}.

It is of course possible to make explicit assumptions of f_i to assure A1 and A2. For example, if the price elasticity of demand is constant, then the prices p^*_{it} are a function of c_i, and the elasticity of demand alone. In that case A1 is automatically satisfied. And, since p^*_{it} is no longer a function of $(x_{it}, \hat{x}_{it})$, a sufficient condition for A2 to hold is that f_i be concave in x_{it}: given the assumption that $\partial f_i/\partial x_{it} > 0$, this implies $\partial^2 f_i/\partial x^2_{it} < 0$. Alternatively, prices can be assumed to be exogenous to the model. We now turn to the dynamics of advertising.

2.2 Dynamics of Advertising

The model of advertising goodwill is similar to that of Nerlove and Arrow [5]. Advertising expenditures act in the nature of investment whose stock depreciates when fresh investment is absent. Denote $a_{it} \geq 0$ to be the advertising goodwill acquired by firm i in period t, and postulate that

$$(3) \qquad x_{it} = \rho x_{it-1} + a_{it}, \qquad\qquad i = 1, 2,\ldots, n$$

where $0 < \rho < 1$ is the rate at which goodwill stock declines. The above implies a geometric decay of goodwill stock summarized by a first order process. Assume further that the cost of acquiring a goodwill of a_{it} is given by the convex function $u_i(a_{it})$ with

$\partial u_i / \partial a_{it} > 0$ $\qquad\qquad$ $\partial^2 u_i / \partial a_{it}^2 > 0$

This is a familiar investment specification in the presence of adjustment costs. Gould [1970], for example, has used it in modeling advertising.

2.3 Statement of the Advertising Problem

We are now in a position to state the problem of how firms invest in advertising. Assume that firms discount cash flows by a factor β, $0 < \beta < 1$. Denote V_i to be the value of firm i given by

$$(4) \qquad V_i = \sum_{t=1}^{\infty} \beta^t (\pi_i(x_{it}, \hat{x}_{it}) - u_i(a_{it})) \qquad\qquad i = 1, 2, \dots, n$$

where

$$(5) \qquad x_{it} = \rho x_{it-1} + a_{it} \qquad\qquad \begin{array}{l} i = 1, 2, \dots, n \\ t = 1, 2, \dots \end{array}$$

$$(6) \qquad a_{it} \in [0, a_{max}] \qquad\qquad \begin{array}{l} i = 1, 2, \dots, n \\ t = 1, 2, \dots \end{array}$$

$$(7) \qquad x_{i0} \text{ given} \qquad\qquad i = 1, 2, \dots, n$$

The restrictions (6) amount to upper bounding u_i, the amount spent on advertising in any given period. The strict concavity of π_i in x_{it}, convexity of u_i, and the geometric decay of x_{it} guarantee that V_i is upper bounded. We can therefore, without affecting the optimal solution, suitably bound u_i, and as a consequence, a_{it}. Denote the sequence $\{a_{it}\}$ by a_i. Condition (6) can then be rewritten as

$$(8) \qquad a_i \in A = \{a_{it}: \; a_{it} \in [0, a_{max}]\}$$

where A is a set defined on the (vector) space of bounded sequences. Firms are assumed to behave non-co-operatively to maximize their discounted net profits. The problem facing each firm i in the industry is then to choose $a_i \in A$ to maximize V_i given by (4), subject to (5) and (7). This constitutes a Nash equilibrium in a_i. Because the sequence a_i is chosen at time 0 the equilibrium represents an open loop Nash solution.

3. Equilibrium and Stability of Stationary States

First the problem defined by (4), (5), (7) and (8) in Section 2.3 is reformulated to prove the primary result of the paper: the existence of an equilibrium. Following that a simple condition is established which ensures the local stability of stationary states on an equilibrium path.

3.1 Existence of Equilibrium

First rewrite (5) as

$$(9) \qquad a_{it} = x_{it} - \rho x_{it-1}.$$

Corresponding to any $a_i \in A$, and given x_{i0}, there is a unique sequence $x_i = \{x_{it}\}$. Define a set X_i as follows:

$$x_i \in X_i = \{ x_{it}: \ x_{i0} \text{ given, } x_{it} \text{ satisfies } x_{it} = \rho x_{it-1} + a_{it}, \ a_i \in A\}$$

and make the assumption that $x_{i0} < a_{max}/(1-\rho)$. Then, it is obvious that X_i is a set from the space of bounded sequences, since $x_{it} \in [0, x_{max}]$ where $x_{max} = a_{max}/(1-\rho)$. Associated with x_i define the metric

$$d(x_i, x_j) = \sum_{t=1}^{\infty} \beta^t |x_{it} - x_{jt}|, \qquad x_i, x_j \in A$$

It can be seen that X_i is compact and convex. The problem in Section 2.3 can now be stated as

$$\text{(10)} \quad \max_{x_i} \ V_i(x_i, \hat{x}_i) = \sum_{t=1}^{\infty} \beta^t [\pi_i(x_{it}, \hat{x}_{it}) - u_i(x_{it} - \rho x_{it-1})] \qquad i = 1, 2,\dots, n$$

$$\text{(11)} \quad x_i \in X_i \qquad i = 1, 2,\dots, n$$

In the above the number of firms n is finite and X_i is a convex compact set. By assumption A2 of Section 2, and because $\partial^2 u_i/\partial a_{it}^2 > 0$, it is easy to see that V_i is concave in x_i. Finally assumption A2 also guarantees that V_i is continuous on $(x_i, \hat{x}_i)$. Therefore the following result obtains as a consequence of Friedman's [2] theorem 7.1:

Theorem 3.1 There exists an open loop Nash equilibrium to the problem defined by (4) - (7) and equivalently to the problem (10) - (11), i.e., there exists an industry equilibrium.

3.2 Stationary Equilibria

Stationary equilibria refer to stationary states resulting from equilibrium sequences x_i, $i = 1, 2,\dots, n$. Denote $\bar{x}_t$ to be the vector $\bar{x}_t = [x_{1t}, x_{2t},\dots, x_{nt}]$. The strict concavity of V_i in x_i implies that given other firms' goodwill stock sequences x_j, $j = 1, 2,\dots, n$, $j \neq i$, there is a unique sequence $x_i \in X_i$ which maximizes V_i. Moreover such a sequence must satisfy the first order condition obtained form differentiating (10). Dentoe w_{it} to be

$$w_{it} = \partial \pi_i(x_{it}, \hat{x}_{it})/\partial x_{it} - \partial u_i(x_{it} - \rho x_{it-1})/\partial a_{it} + \rho\beta\partial u_i(x_{it+1} - \rho x_{it})/\partial a_{it}$$

for each $i = 1, 2,\dots, n$. Then we have either

$$\text{(12)} \quad w_{it} = 0 \qquad \text{if } x_{it} > \rho x_{it-1}$$

or

$$\text{(13)} \quad w_{it}|_{x_{it} = \rho x_{it-1}} \leq 0 \qquad \text{if } x_{it} = \rho x_{it-1}$$

The relations (12) and (13) completely characterize the equilibrium path of x_i, $i = 1, 2,\dots, n$, and therefore of advertising investment in the industry.

Definition: Define $\bar{x}_s$ to be a steady state if for some t, the equilibrium sequences are such that

$$\bar{x}_\tau = \bar{x}_s = [x_s, \dots x_{ns}], \qquad \tau \geq t.$$

Clearly at steady state, $x_{it} > \rho x_{it-1}$, and therefore (12) must hold for each i.

Thus the solution to the n equations given by (12) yields the steady state by setting $x_{it} = x_{it-1} = x_{it+1} = x_{is}$, $i = 1, 2, \ldots, n$. Since, in general, there may be more than one solution to the system of non-linear equations in (12), there may be more than one steady state.

3.3 Stability of Stationary States

The analysis of local stability of the stationary state presented here closely follows Flaherty [1]. First introduce the following notation:

$$\pi_i^{11} = \partial^2\pi_i(x_{it}, \hat{x}_{it})/\partial x_{it}^2$$

$$\pi_i^{12} = \partial^2\pi_i(x_{it}, \hat{x}_{it})/\partial x_{it}\partial x_{jt}$$

$$u_i^{11} = \partial^2 u_i(x_{it} - \rho x_{it-1})/\partial x_{it}^2$$

$$u_i^{22} = \partial^2 u_i(x_{it} - \rho x_{it-1})/\partial x_{it-1}^2$$

$$u_i^{12} = \partial^2 u_i(x_{it} - \rho x_{it-1})/\partial x_{it}\partial x_{it-1}$$

for each $i = 1, 2, \ldots n$, $j = 1, 2, \ldots, n$ $j \neq i$. Also let

$$\bar{u}_i = u_i^{11}(x_{is} - \rho x_{is}) > 0$$

Then note that

$$u_i^{12}(x_{is} - \rho x_{is}) = -\rho\bar{u}_i$$

$$u_i^{22}(x_{is} - \rho x_{is}) = \rho^2\bar{u}_i$$

Suppose also the following assumption holds at a stationary state:

<u>A3</u> $\quad (1/\rho\beta + \rho) + (\pi_i^{12}(x_{is}, \hat{x}_{is}) - \pi_i^{11}(x_{is}, \hat{x}_{is}))/\rho\beta\bar{u}_i > 1 + 1/\beta,$

$i = 1, 2, \ldots, n$

Let $y_t = (\bar{x}_t, \bar{x}_{t-1})$. Let the quartet $(x_{it-1}, x_{it}, \hat{x}_{it}, x_{it+1})$ be sufficiently close to a steady state $(x_{is}, x_{is}, x_{is}, x_{is})$ so that w_{it} in equation (12) can be approximated by a Taylor series expansion around the steady state. Then we can write (12) as

$$\begin{aligned}(14)\qquad 0 = {} & \partial\pi_i(x_{is}, \hat{x}_{is})/\partial x_{it} - (1 - \rho\beta)\,\partial u_i(x_{is} - \rho x_{is})/\partial x_{it} \\ & - (x_{it-1} - x_{is})\,[u_i^{12}(x_{is}, x_{is})] \\ & + (x_{it} - x_{is})\,[\pi_i^{11}(x_{is}, \hat{x}_{is}) - u_i^{11}(x_{is}, x_{is}) - \beta u_i^{22}(x_{is}, x_{is})] \\ & + (x_{it+1} - x_{is})\,[-\beta u_i^{12}(x_{is}, x_{is})] \\ & + (\hat{x}_{it} - \hat{x}_{is})\,[\pi_i^{12}(x_{is}, \hat{x}_{is})]\end{aligned}$$

The above can be put in the form

$$y_{t+1} = Ry_t + \text{constant}$$

where

$$R = \begin{bmatrix} R_1 & -I_n \\ I_n & 0 \end{bmatrix} \quad \text{, a (2n x 2n) matrix}$$

and

$$R_1 = \begin{bmatrix} d_1 & e_1 \cdots & e_1 \\ e_2 & d_2 \cdots & e_2 \\ \cdot & & \\ \cdot & & \\ \cdot & & \\ e_n & e_n & d_n \end{bmatrix} \quad \text{, a (n x n) matrix}$$

(15) $d_i = [-u_i^{22} - u_i^{11}/\beta + \pi_i^{11}/\beta]/u_i^{12} \Big|_{\bar{x} = \bar{x}_s}$

(16) $e_i = \pi_i^{12}/\beta u_i^{12} \Big|_{\bar{x} = \bar{x}_s}$

Since this system is mathematically identical to the one studied by Flaherty [1] we invoke her Theorem 6.8 to prove the following result:

Theorem 3.2 A stationary state is locally stable if assumption A3 is satisfied.

Proof From Flaherty's Theorem 6.8, a stationary state is locally stable if d_i, e_i from (15), (16) satisfy

$$\min_i \{d_i - e_i\} > 1 + 1/\beta$$

Now, we have

$$d_i - e_i = (\pi_i^{11} - \pi_i^{12} - u_i^{11} - \beta u_i^{22})/\beta\, u_i^{12}$$

This can be simplified to obtain

$$d_i - e_i = (1/\rho\beta + \rho) + (\pi_i^{12} - \pi_i^{11})/\rho\beta\bar{u}$$

By assumption A3 this is satisfied for all i. Hence the desired result. Q.E.D.

It is useful to note here that a sufficient condition for A3 to hold is that

$$\pi_i^{12}(x_{is}, x_{is}) \geq \pi_i^{11}(x_{is}, \hat{x}_{is}) \text{ for all i.}$$

4. Conclusions

The equilibrium advertising levels chosen by firms in an oligopoly were modeled as a Nash solution to an appropriate set of maximization problems. It was found that for the Nerlove-Arrow advertising dynamics with adjustment costs, there exists an open loop Nash equilibrium in the oligopoly. Additionally, the local stability of stationary states along an equilibrium path was examined and a sufficient condition for the stability established.

There remain several directions for further research. One possibility is to numerically characterize equilibrium configurations in an n-firm oligopoly where an entrant, i.e., an (n+1)th firm would encounter negative profits thereby assuring the perpetuation of the n-firm oligopoly. Among other things, this will depend on the parameters governing the advertising dynamics as well as the adjustment costs. Another possibility is to examine if observed advertising expenditures in an industry correspond to a stable Nash solution as developed here. This would require estimation of the demand functions and the advertising dynamics. A further discussion of this is to be found in Rao [6].

References

1. Flaherty, M. Therese: Industry structure and cost reducing investment. Econometrica, vol. 48, pp. 1187-1209 (1980).

2. Friedman, James W.: Oligopoly and the theory of games. Amsterdam: North Holland (1977).

3. Gould, John P.: Diffusion processes and optimal advertising policy. in E.S. Phelps et. al. eds: Microeconomic foundations of employment and inflation theory. New York: W.W. Norton (1970).

4. Lambin, Jean J.; Naert, Philippe A.; Bultez, Alain.: Optimal Marketing behaviour in oligopoly. European Economic Review, vol. 6, pp. 105-128 (1975).

5. Nerlove, Marc.; Arrow, Kenneth J.: Optimal advertising policy under dynamic conditions. Economica, vol. 29, pp. 129-142 (1962).

6. Rao, Ram C.: Advertising decisions in oligopoly: an industry equilibrium analysis. Paper no. 752, Institute for Research in the Behavioural, Economic and Management Sciences, Purdue University, W. Lafayette, IN (1981).

7. Schmalansee, Richard: The economics of advertising. Amsterdam: North Holland (1972).

8. Sethi, Suresh P.: Optimal control of the Vidale-Wolfe advertising model. Op. Res., vol. 21, pp. 998-1013 (1973).

Applications of Advances in Nonlinear Sensitivity Analysis

Paul J. Werbos, U.S. Department of Energy
Forecast Analysis and Evaluation Team

The following paper summarizes the major properties and applications of a collection of algorithms involving differentiation and optimization at minimum cost. The areas of application include the sensitivity analysis of models, new work in statistical or econometric estimation, optimization, artificial intelligence and neuron modelling. The details, references and derivations can be obtained by requesting "Sensitivity Analysis Methods for Nonlinear Systems" from Forecast Analysis and Evaluation Team, Quality Assurance, OSS/EIA, Room 7413, Department of Energy, Washington, DC 20461.

Context of the Work

The Energy Information Administration (EIA) provides data and analysis on all aspects of energy supply and demand. It uses dozens of models, including econometric (statistical, empirical) models, linear programming models based on technological data, a nonlinear micro equilibrium model solving for thousands of variables simultaneously across a 50-year span, hybrids and combinations of these, etc.

Many users of EIA's analyses do not accept EIA's conclusions at face value, especially when reports from other sources disagree. Thus the Forecast Evaluation and Analysis Team of EIA and its predecessors have carried out a broad program to evaluate and explain the qualitative assumptions of EIA models and forecasts. This program includes the development of tools to characterize the properties of large models, studies of estimation methods which are robust against outliers or model misspecification (i.e., correlated errors), proofs of convergence and existence properties, and many other projects. The first part of this paper describes how a small part of this work - the minimum cost calculation of first and second order derivatives of nonlinear systems - makes an essential contribution to the rest. The second part elaborates on another application, a method for stochastic optimization which becomes feasible only with the help of low-cost derivatives. This method opens up a wholly new approach to the field of artifical intelligence and neuron modelling; it is especially efficient with the new generation of "parallel" computers.

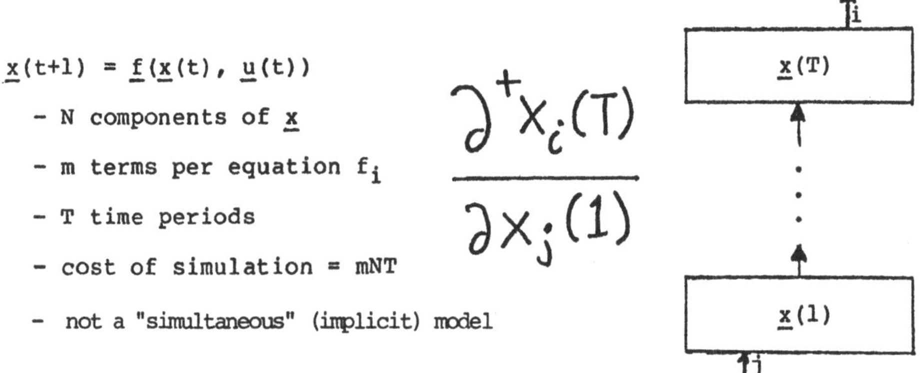

$\underline{x}(t+1) = \underline{f}(\underline{x}(t), \underline{u}(t))$
- N components of $\underline{x}$
- m terms per equation f_i
- T time periods
- cost of simulation = mNT
- not a "simultaneous" (implicit) model
$\frac{\partial^+ x_i(T)}{\partial x_j(1)}$
i
$\underline{x}(T)$
$\underline{x}(1)$
j

Figure 1: A Simple Example

Figure 1 shows a simple example of the kind of "derivative" we are trying to compute. Suppose that we have a nonlinear system, with a vector $\underline{x}$ of N endogenous variables and a vector $\underline{u}$ of exogenous variables. Suppose that the system is governed by the equation shown in Figure 1. The cost of simulating the model over the whole time range is mNT, because in each of the T time periods we compute a forecast for each of the N variables in $\underline{x}$, and each such forecast involves m terms. Please note that N is often much larger than m. Given a small change in the variable x_i in time period 1, we want to know how large the resulting change in x_i is in the final time period T.

The change in $x_i(T)$ per change in $x_j(1)$, holding the rest of $\underline{x}(1)$ constant, is a fundamental quantity of the system. It goes by many different names. In modelling, it is often called a "sensitivity coefficent." In economics, it is traditionally called an "impact multiplier." Electrical engineers often call it a "transient response," or "constrained derivative." Here we will call it an "ordered derivative," using the notation shown in Figure 1, for two reasons: (1) the notation is somewhat more explicit than what is usually used; and (2) the concept of ordered derivative is somewhat more general and rigorous, as will be seen.

Well-known applications which require the use of such first-order derivatives are sensitivity analysis, maximization of a system result (i.e., "deterministic optimization"), and statistical estimation. In the last two cases, one actually is concerned with the derivatives of a function of $\underline{x}(T)$ or of $\underline{x}(t \leq T)$ rather than the derivatives of $x_j(T)$ for some j, but it is easy to make this extension of the methods; for example, the function to be differentiated or a running total for it may be added to the list of system variables.

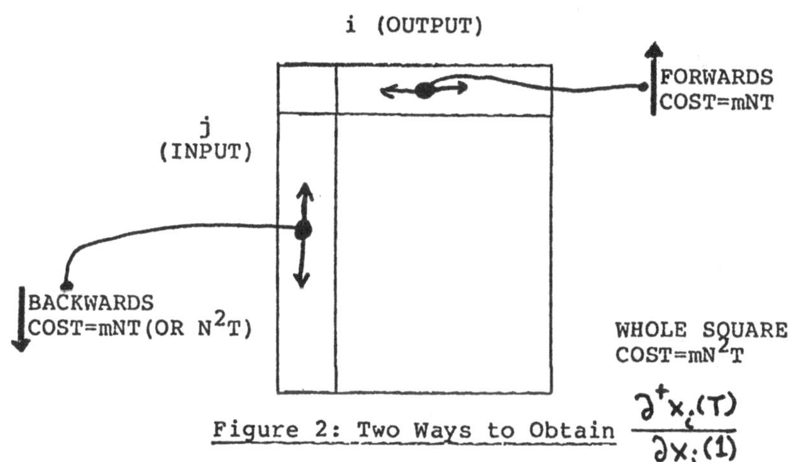

Figure 2: Two Ways to Obtain $\dfrac{\partial^+ x_i(T)}{\partial x_j(1)}$

Figure 2 describes two methods for computing ordered derivatives in the example above. The corresponding equations are:

$$\uparrow: \quad \underline{z}(t) = \frac{\partial^+ \underline{x}(t)}{\partial x_j(1)} \; ; \quad \underline{z}(t+1) = f'(t)\underline{z}(t)$$

$$\downarrow: \quad \underline{z}^*(t) = \frac{\partial^+ x_i(T)}{\partial \underline{x}(t+1)} \; ; \quad \underline{z}^*(t) = f'^T(t)\underline{z}^*(t+1) \quad \text{(or transpose)}$$

The large square in Figure 2 represents the entire matrix of ordered derivatives of all $x_i(t)$ with respect to all $x_j(1)$. The conventional or "forwards" method (indicated by an arrow pointing upwards) is based on perturbing one of the initials values $x_j(1)$, and observing the impact on all the final results, i.e., on the vector

$\underline{x}(T)$. Each time we apply this method, we perturb only one of the initial values; thus we obtain only one row of the matrix of ordered derivatives, as shown in Figure 2. This costs us mNT calculations, as shown. Often the initial value $x_j(1)$ is actually changed, and the model resimulated. (This costs mNT operations, as did the original run of the model.) However, this leads to problems with the numerical accuracy of the results, because one computes each derivative by subtracting two numbers very close to each other in size. MIT's Troll System uses the forwards closed-form Jacobian formula, shown at the bottom of Figure 2, which has the same cost but is more accurate.

The backwards method, shown with a downwards pointing arrow in Figure 2, computes an entire column of the matrix, using only mNT calculations. In engineering, this sort of method has been used for many years with "constrained derivatives," but has not been applied more generally.

The key point about these methods is that the forwards method is often used when the backwards method would be more appropriate. This can multiply costs (by a factor of N) to the point where it becomes infeasible to do what one wants to do. For example, it has long been known that economic data, like engineering measurements, are fraught with many errors, and that these errors invalidate conventional estimation methods. Statisticians like Hannan observed years ago that white noise converts a simple econometric model (like our example, but linear) into a "vector mixed autoregressive moving average process." In other words, one can account for such errors in data by estimating the corresponding vector ARMA process. However, because of the sheer cost of such estimation, it has rarely been done in economics. Instead, an approximation suggested by Hibbs has become popular of late: a conventional model is estimated by regression, and then simple unvariate ARMA ("Box-Jenkins") modeling is used on the residuals, and the process may be iterated. Yet in statistical estimation, one only needs a single column of the derivative matrix (i.e., the derivatives of error), not the whole matrix; using the backwards method, one can compute all the derivatives needed in an iteration at the cost of only mNT, which is what it takes to exercise the model. This method was applied to vector ARMA estimation in the early 1970's, but has yet to receive wide application in economics. It now appears that vector ARMA estimation (and thus Kalman filtering estimation, which is formally equivalent to it) may have less value in social science than other more robust methods based on a generalization of Hartley's simulation path approach; however, those methods, too, require a set of derivatives, as part of minimizing a complicated loss function.

Likewise, in sensitivity analysis, a user often wants to know the sensitivity of a few key results to all the initial values, or to be sure he knows the largest of these sensitivity coefficients. Again, only a few columns of the matrix are required; it is wasteful to pay for the whole matrix.

With large models or network systems, N may range from the hunreds to the millions or more. Thus cutting the cost of computing derivatives by a factor of N is often crucial to feasibility. One may be sure that the cost of exercising the system (mNT) is affordable, or the system would be of no interest; more than this, by a multiple of N, may be unacceptably expensive.

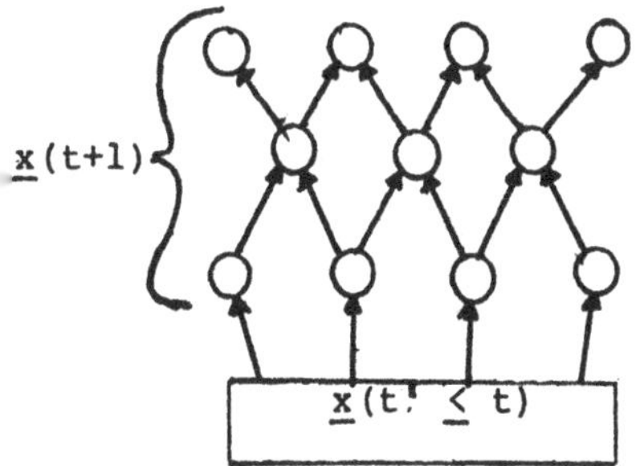

"CHAIN RULE"(DYNAMIC FEEDBACK):

$$\frac{\partial^+ x_i}{\partial x_j} = \sum_{k=j+1}^{i} \frac{\partial^+ x_i}{\partial x_k} \cdot \frac{\partial f_k}{\partial x_j} \qquad i>j$$

CONVENTIONAL PERTURBATION:

$$\frac{\partial^+ x_i}{\partial x_j} = \sum_{k=j}^{i-1} \frac{\partial f_i}{\partial x_k} \cdot \frac{\partial^+ x_k}{\partial x_j} \qquad i>j$$

Figure 3: A More General Example: $\underline{x}(t+1) = f(\underline{x}(\text{all } t' \leq t+1), \underline{u}(\text{all } t'))$

Now let us consider the more general model shown in Figure 3. All nonnegative lags - including zero - are permitted in the endogenous variables. However, we still assume here (as in the paper) that the model has been reduced to "explicit" form. (In economics, one would call this a recursive model; in mathematics, one calls it a nonrecursive system.) We assume that the functions f_i, which make up $\underline{f}$, can be ordered in such a way that we can use them one by one to calculate the vector $\underline{x}(t+1)$. The dynamic feedback "chain rule" has not been published before.

Figure 3 illustrates an example where $\underline{x}(t+1)$ has eleven components, each represented by a circle; the arrows flowing into a circle represent inputs required to compute that component of $\underline{x}$.

The forwards and backwards methods are generalized as shown in Figure 3. The subscripts here refer to an ordered index of all time/variable-number combinations; the formulas are given in more conventional form in the main paper. The key thing to note is that there are only m calculations per time/variable combination. Thus we still only need to make mNT calculations to get a complete row or column of ordered derivatives, as in our earlier example. This has not previously been published. With conventional matrix methods for constrained derivatives, based on our earlier example, one would have to use N by N matrices f', which would not usually be sparse; thus the generalization here makes it feasible to differentiate large network systems which would have been too expensive to differentiate with conventional methods.

The methods shown in Figure 3 remain efficient even if one uses "parallel" computers. Parallel computers - based on many processors operating in parallel rather than one CPU - are becoming increasingly common. With a conventional computer, it would take roughly 11 calculation times to compute $\underline{x}(t+1)$ in our example (1 for each component of $\underline{x}$). With a parallel computer, it need only take 3: in the first period, 4 processors would calculate the lower tier in parallel, since none of the 4 lower components depends on the others; in the second period, the middle tier would be calculated; etc. The backwards method shown here allows similar economies: one can calculate ordered derivatives of a model result with respect to the top tier in the first period of calculation, then to the middle tier, and then to the bottom tier. The forwards method is similiar.

Large scale models or systems typically can be represented as relatively sparse networks, as in this example. Actual physical networks, made up of units operating in parallel, have a similar structure. To optimize such a system (except in unusual special cases) it is essential to know the derivatives of the desired performance measure with respect to all parameters in the system; for this to be feasible, it is essential to use a method such as the generalized backwards method which does not multiply the cost of getting the derivatives far beyond the cost of exercising the system.

In this paper, we have discussed derivatives with respect to initial values of the variables only; however, the EIA report does consider parameters, and the case of exogenous variables is a trivial extension of the endogenous variable case. To avoid making a complicated discussion even more complicated, the EIA report only mentions our earlier example when discussing second derivatives; however, it is trivial to substitute the general formulas in Figure 3 for those in Figure 2, whenever they apply in the second derivative calculation, to arrive at more general methods. The section on stochastic optimization provides a partial example of the possibilities.

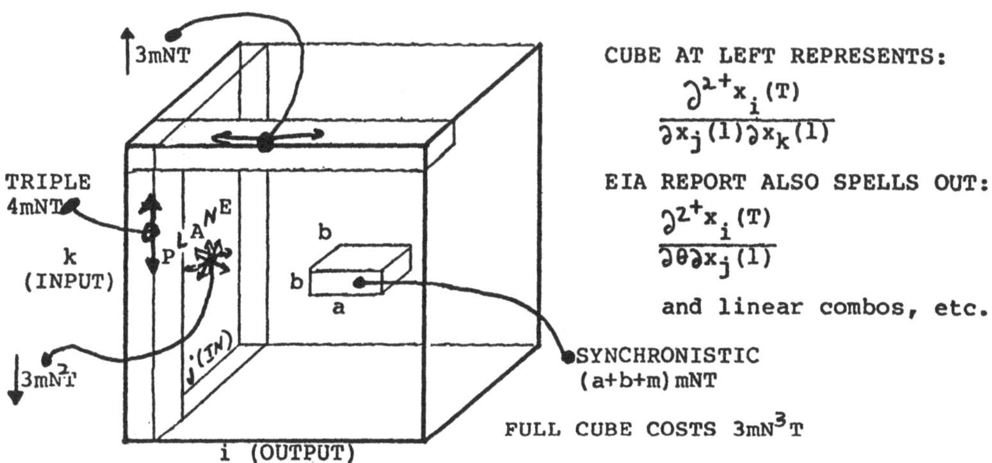

Figure 4: Costs of Obtaining Various Sets of Second Derivatives

Figure 4 provides a summary of the properties of the four variable-variable second derivative calculation methods provided in the EIA report. The set of ordered derivatives of $x_i(T)$ to $x_j(1)$ and $x_k(1)$ form an N by N by N cube, as shown; each method computes a subset of the cube, at approximate costs shown. Again, in practice, the key point is to compute only the subset required, and not pay for the entire cube. The five methods for computing variable-parameter second derivatives offer the same subsets (except that an upwards column and a row pointing backwards count as two separate cases) for the same rough costs.

The EIA report notes that variable-parameter second derivatives provide meaningful information about a model, essentially equivalent to what MIT provides for linear systems by looking at changes in eigenvalues. In effect, they tell us, for a change in a parameter of the system, how its dominant dynamics (revealed in the matrix of ordered derivatives to variables) change.

Among the possible applications is the use of Newton's method in estimation and optimization. It is straightforward to use the full backwards approach here for parameter-parameter derivatives; this allows computation of all the second derivatives one needs in order to use Newton's method, for a rough cost of only $3mN^2T$, about the same as what people have paid to get only first derivatives when using forwards methods.

Applications of Stochastic Optimization Over Time

The remainder of this paper will discuss a way to implement "GDHP," a previously published approach to stochastic optimization over time. The methods discussed above make GDHP cheap enough to be feasible with very large scale models or systems. This opens up many possible applications.

GDHP provides an approximate solution to Howard's problem:

GIVEN THAT $\underline{x}(t+1) = \underline{F}(\underline{x}(t),\underline{u}(t),\underline{e}(t))$
where $\underline{e}(t)$ is random, $\underline{u}(t)$ a control,
MAXIMIZE $\langle U(\underline{x})\rangle$, the expectation across all future times

Applications include: (1) use of large stochastic policy models in decision making; (2) devising distributed/decentralized systems to output optimal policy; (3) general adaptive artifical intelligence; and (4) modelling of learning in the brain ("neural plasticity").

The first of these applications is straightforward. Let "$\underline{F}$" be the policy model, and let "U" represent the values to be served by the decisions. How to formulate "U" is an old and unavoidable issue, beyond the scope of this paper.

The second application is an extension of the fields of micro-economics and hierarchical control theory. Economists have proven many theorems about how a large deterministic optimization problem (devising an efficient pattern of production) may be decentralized; however, these theorems typically assume no uncertainty or fundamental structural "change" (nonlinearity). If the actual problem is general and stochastic (beset by uncertainty), one needs a method of solution to the stochastic problem which can be implemented efficiently in a decentralized (i.e., parallel processor) system. The EIA report describes how GDHP may be implemented in this way.

In artificial intelligence, if one does not impose severe constraints on the range of models $\underline{F}$ or of action strategies to be considered (which would be inappropriate in an adaptive system), one must develop a generalized approach to optimization in order to find the optimal action strategy. However, numerical analysts have found, even with the much simpler task of maximizing a function of a few variables, that first derivatives at a minimum are essential to finding a maximum in a reliable, efficient manner. Until the failure of the Minsky-Selfridge "jitters" machine, there was widespread recognition of the fact that generalized optimization and adaptive intelligence are almost inseparable concepts. The "jitters" failure does not invalidate the original concept, however; it merely shows that one needs a full set of valid derivatives in each iteration (as is well known to numerical analysts) and that one cannot make do with a factor of N less information. (i.e., One derivative per time period, as in "jitters.") Likewise, a system needs to have an explicit model of its environment in order to properly optimize actions over time as in dynamic programming.

There is an additional reason why artificial intelligence research abandoned the idea of explicit optimization in the 1960's. Much of the work at that time was inspired by an early description of the neuron (brain cell) developed by McCulloch and Pitts. In that description, the variables x_i computed by the brain could take on only two values, 0 or 1 ("all" or "none"); thus the functions $\underline{F}$, J and $\underline{u}$ above could not be differentiable. However, more recent work has shown that neurons in the human brain use a "code" based on "volleys," such that the variables vary over a continuous range; thus optimization of a network of model neurons is not intrinsically different from conventional optimization based on a specified functional form.

Similar concepts apply to the field of actual neuron modelling. Like objects in solid state physics, neurons are very complicated, and will never be totally encompassed by theory; as with solid state physics, however, a general theory of neuron networks as intelligent systems may help increase the range of important phenomena which can be understood. This is a far cry from the ad hoc approach to brain modelling, in which a given set of equations is derived by appeals to simplification and common sense only. When the actual functional ability of a model to reproduce intelligence is considered, many otherwise credible models can be quickly ruled out; for example, many theorems have been proven about the Grossberg neuron model, but in statistical terms, this model uses simple correlation coefficients as the parameters of multivariate forecasting equations. Some neurologists have compared themselves to a man who studies a radio as follows: he pulls out a transistor, notices that the radio whines, and calls the transistor the "whine center"; a deeper understanding of brain functioning requires more consideration of the mathematical elements which are necessary in order to produce generalized intelligence, defined as the ability to learn and adapt to totally new problems.

The method discussed here - GDHP - does not address all aspects of stochastic optimization. The theory needs further development. However, there is a very close parallel between GDHP and conventional statistical estimation methods. As with statistics, specific examples of problems will be important to improving our understanding of stochastic optimization. However, as with statistics, examples alone will not be enough; the underlying theory needs explicit development at a generalized level, if we are ever to cope with very complex problems and improve the general methods.

General Approach Used in GDHP

The exact solution to Howard's problem, as described in Howard's book on dynamic programming, requires the calculation of a scalar function $J(\underline{x})$ and a vector function $\underline{u}(\underline{x})$. $\underline{u}(\underline{x})$ yields the optimal action (or motor output or policy variables) as a function of the state of the environment, $\underline{x}$. $J(\underline{x})$ is a measure of how good the results of the actions are in terms of their total long-term impact. Howard proves that one can normally converge on a choice for $\underline{u}(\underline{x})$ which maximizes the future expected value of utility ($U(\underline{x})$) by alternately: (1) finding the unique function J which solves a certain equation for the current guess for $\underline{u}(\underline{x})$; and (2) picking a new guess for $\underline{u}(\underline{x}(t))$ so as to maximize the expected value of J at time t+1.

In GDHP, one does not find the exact theoretical J and $\underline{u}$, because this tends to be infeasible with large systems. Instead, one assumes that the user or a higher-level system has proposed functional forms, J^* and $\underline{u}^*$, for J and $\underline{u}$. GDHP attempts to adjust the parameters of J^* and $\underline{u}^*$ to make them fit the conditions for optimality as closely as possible, over a finite (not necessarily fixed) set of possible scenarios, $\underline{x}$. This involves two steps, to be carried out alternately or in parallel until convergence:

- Maximize $\langle J^*(\underline{F}(\underline{x},\underline{e},\underline{u}^*(\underline{x},\underline{b}))\rangle$ over parameters $\underline{b}$, scenarios $\underline{x}$, random $\underline{e}$.

- Pick $\underline{a}$ in $J^*(\underline{x},\underline{a})$ to minimize the sum over scenarios $\underline{x}$ of:

$$E \triangleq \sum_i W_i \left(\frac{\partial}{\partial x_i} \langle J^*(\underline{F}(\underline{x},\underline{e})) - \underline{u}^*(\underline{x}) - J^*(\underline{x})\rangle\right)^2,$$

where $\underline{f}(\underline{x},\underline{e})$ is defined as $\underline{F}(\underline{x},\underline{e},\underline{u}(\underline{x}))$ and where the W_i are a set of weights. If J* can solve Howard's equation exactly with $\underline{u}$*, for some $\underline{a}$, then E will always equal zero for that $\underline{a}$.

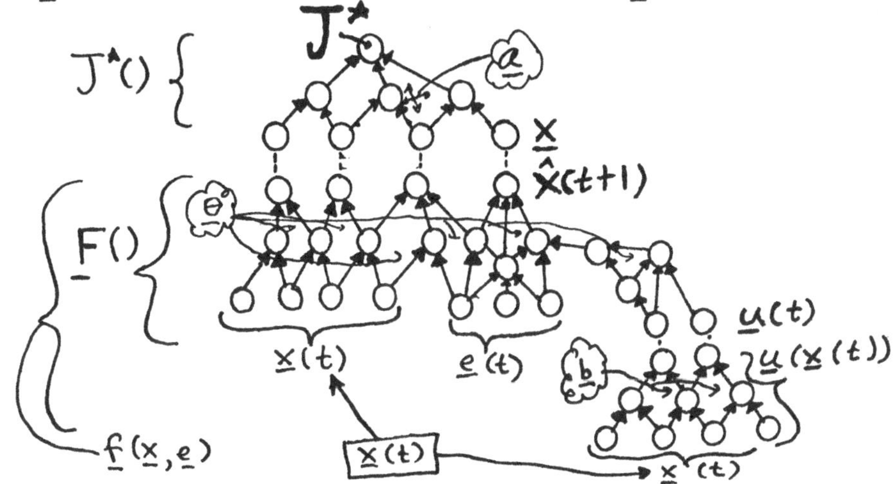

Figure 5 : Realization of GDHP As a Triple Network to Make Decisions

GDHP can be implemented as part of a procedure to optimize the choices of actions over time, in a situation where the dynamics of the environment and the optimal actions are both to be learned from empirical observation rather than specified a priori. This requires updating the parameters of three functions: J*, $\underline{u}$* and $\underline{F}$*. (Updating $\underline{F}$* is a well studied problem in statistics.) Such a system can be implemented as a three-level network as shown in Figure 5; each level contains a network realization of one of the three functions. GDHP and statistics would be used to adopt the parameters; the network needs no external guidance except for data input and functional forms.

Figure 5 has an interesting analogy to the mammalian brain, which is essentially made up of three interlocking networks of brain cells: (1) the "limbic" network, which, like J calculates system values ("values" as in "values we cherish") which are the basis for reinforcing actions; (2) the cortico-thalamic system which, like $\underline{F}$, embodies a model of the external environment; and (3) the brain stem, which, like $\underline{u}$, directly controls actions, in response to what is known about the external environment. There are further correspondences and features to improve performances of this system far beyond the scope of this paper.

Rationale of the Approximations Used in GDHP

The approximations used by GDHP require some explanation.

First, consider the use of specific functional forms instead of general functions J and $\underline{u}$. No realistic system can actually use an estimate of J which involves a functional form which exceeds the available storage space; thus it is necessary to limit explicit attention to specific realizable functional forms. We can first analyze the problem of parameter estimation for a fixed functional form, before studying how to compare and improve functional forms, exactly as in statistics. In artificial intelligence, successful game playing machines have required "static position evaluators," which correspond with the approximate J function used in GDHP to evaluate $\underline{u}(\underline{x})$.

There are several obvious problems in assuming a functional form for J*: (1) the true J which obeys Howard's equation will usually not be expressable exactly as $J^*(\underline{a})$ for any $\underline{a}$; (2) the need for a user or metaprogram to choose the functional form; (3) the problem of choosing weights W_i; (4) dangers of autocorrelation invalidating the adjustment process; and (5) the need to worry about robustness of the results, influential observations and unobserved variables. All these problems are precisely parallel to known problems in using statistical forecasting models $\underline{F}^*$ with fixed functional forms; as in statistics, these problems require analysis but do not invalidate the approach. Indeed, the methods of analysis needed to extend GHDP are very similar to those used in statistics.

Reasons for choosing E as a loss function for J* are discussed in the EIA report: (1) the derivatives of J are the "shadow prices" or "Lagrange multipliers"; (2) decisions ($\underline{u}$) are based on local comparisons, such that the accuracy of the derivatives of J determine the accuracy of the decisions; (3) this eliminates Howard's "$\overline{U}$" constant; (4) other reasons are mentioned in Policy Analysis and Information Systems (Elsevier), 6/79.

Implementation of GDHP

The major difficulty in implementing GDHP is the difficulty of performing the minimization and maximization described above. In the general case, where no strong special assumptions are made about J and $\underline{F}$ and $\underline{u}$, and where there are many parameters in these functions, this requires that we get the derivatives of the quantities to be minimized or maximized. Here, however, E already involves first derivatives; the derivatives of E involve second derivatives. The EIA report derives in detail the calculationswhich yield the required derivatives. These calculations only cost about three or four times what it costs to compute $J(\underline{F}(\underline{x},\underline{u}(\underline{x})))$ itself. If the functional forms for J, $\underline{F}$ and $\underline{u}$ are "realizable," this means that the cost of calculating $J(\underline{F}(\underline{x},\underline{u}(\underline{x})))$ is bearable; the cost of obtaining derivatives, then, is also bearable for all realizable J, $\underline{F}$ and $\underline{u}$. The EIA report explicitly spells out the details in the case where multiple processors can be used to reduce the cost of calculating J, $\underline{F}$ and $\underline{u}$; it demonstrates that the cost of getting derivatives can be kept in line with that of getting J, $\underline{F}$ and $\underline{u}$ even in that case.

Minimization and maximization are nontrivial problems, even with the derivatives available. "Sparse quasiNewtonian" methods are being developed, however, which show promise in handling large systems. EIA has a crude but adequate method which it now uses to solve 100,000 nonlinear simultaneous equations.

In a real-time system, one would want to carry out all the iteration processes above in parallel rather than wait, for example, to complete a minimization before getting new data. For analytical purposes, however, it is convenient for now to consider decision problems focused at one moment in time.

Implications of GDHP Feasibility

This approach makes it possible to develop generalized, adaptive artifical intelligence, capable of achieving results comparable to what is discussed in science fiction, by a rational development of statistics, optimization theory and numerical analysis in the directions indicated above. The implications for psychology and economics raise issues too complex to permit adequate discussion here.

FREEWAY INCIDENT DETECTION BASED ON STOCHASTIC DYNAMIC MODELS OF TRAFFIC VARIABLES

Samir A. Ahmed
School of Civil Engineering
Oklahoma State University
Stillwater, Oklahoma 74078, USA

INTRODUCTION

Improving the operational efficiency of the existing urban freeway network represents a challenging problem in the face of increasing traffic demands and growing pressures against building new freeways. Research efforts over the past two decades have led to the development of automated freeway surveillance and control systems, where a digital computer monitors the traffic on the freeway and determines on-line control decisions to optimize the corridor operations. Real-time observations of prevailing traffic conditions are typically gathered from electronic detectors buried in the pavement, generally at one-half mile (0.8 Km) intervals in each lane.

This paper outlines the development of an on-line computer algorithm for the automatic detection of freeway capacity-reducing incidents (accidents, vehicle breakdowns, spilled loads, etc.) using surveillance data obtained from detectors. Incident detection is not an end in itself, but is rather a part of an overall real-time traffic management system. This system involves coordinated use of ramp metering controls, variable message signs, and incident management services. The developed algorithm is based on discrete-time stochastic models which describe the dynamic behavior of freeway traffic. Automatic incident detection is achieved by utilizing real-time estimates of the variability in the state variables as detection thresholds. The performance of the proposed detection algorithm is evaluated in terms of the probabilities of false-alarms and correct detections, in addition to the mean time-lag to detection using actual data on 50 representative traffic incidents.

DISCRETE-TIME MODELS FOR FREEWAY TRAFFIC

Lane occupancy, the percentage of time a detector is occupied by vehicles, is the basic state variable used in the surveillance and control of freeway traffic. It is a surrogate measure for density that provides information about all the important aspects of the traffic stream. The computer scans each detector approximately once every 0.10 sec and aggregates the occupancy times over 30 to 60 sec intervals. Most surveillance and control applications require an average lane occupancy which is computed from the different lane occupancies at each detector station. These averages are continuously updated every 20, 30, or 60 sec depending on the system design. The resulting average lane occupancies represent aggregate traffic conditions over the different freeway sections. Figure 1 is a plot of representative average lane occupancies observed at a detector station on I-35 in Minneapolis during the afternoon peak period.

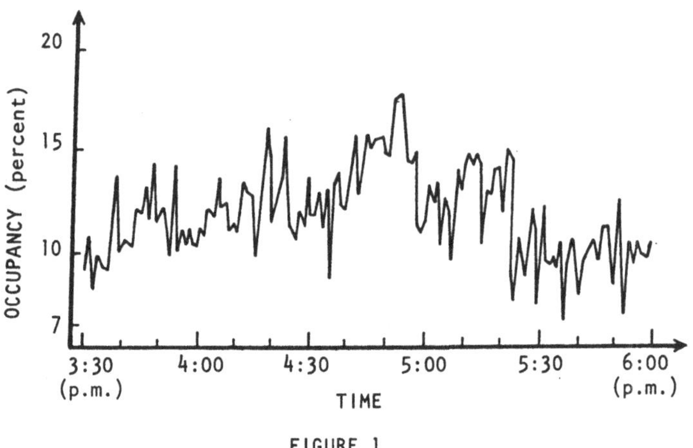

FIGURE 1

FREEWAY TRAFFIC OCCUPANCY SERIES

The stochastic behavior of occupancy observations sampled at uniformly spaced points in time can be regarded as a realization of a discrete time point process $\{X_t\}$. This process can be represented by one of the broad class of linear models

$$\Phi_p(B)\ \nabla^d\ (X_t - \mu) = \Theta_q(B)\ a_t \tag{1}$$

where p, d, and q are non-negative integers; μ is the mean of the series; $\Phi_p(B)$ is an autoregressive operator of order p; $\Theta_q(B)$ is a moving average operator of order q; ∇^d is a difference operator of order d; and a_t's are white noise variables independently distributed as $N(0, \sigma_a^2)$. The models in Equation 1 are usually referred to as ARIMA models of order (p,d,q), and they are formally discussed in many references including the well known textbook by Box and Jenkins [1].

ARIMA models are fitted to an observed series by a three-stage interative procedure: identification, estimation, and diagnostic checking. At the identification stage, the tentative model to fit to the data is determined by inspecting the autocorrelation and partial autocorrelation functions of the series and its differences, and comparing them with those of some basic stochastic processes. At the estimation stage, maximum likelihood estimates are obtained for the model parameters. The output observations are operated upon by the inverse of the filter that is computed to have produced the white noise sequence $\{a_t\}$. Finally, the fitted model is diagnosed to ensure that the estimated model residuals are white noise deviations, otherwise the model should be redesigned by repeating the three stages of model construction.

The described ARIMA model building procedure was applied to 84 different time series of traffic occupancies representing more than 13,500 minutes of observations on incident-free traffic conditions. These observations were recorded on the Los Angles, Minneapolis, and Detroit freeway systems during the afternoon peak periods. Experimentation with different values of p and q in the general class of models described by Equation 1 indicated that the stochastic process generating the data stream is ARIMA (0,1,3), i.e., the first differences of traffic occupancies can be represented by a third order moving average model of the form

$$\nabla X_t = (1 - \Theta_1 B - \Theta_2 B^2 - \Theta_3 B^3)\ a_t \tag{2}$$

Diagnostic checks applied to the residuals $\hat{a}_t$ indicate the adequacy of the form of the ARIMA (0,1,3) model. The residual autocorrelations are all quite small in magnitude and exhibit no significant structure remaining in the data.

The minimum mean-square error predictor of X_{t+1} given the history of the process up to time t, denoted $\hat{X}_t(1)$, is given by the conditional expectation

$$\hat{X}_t(1) = E\ [X_{t+1}/(X_t, X_{t-1}, \ . \ . \ . \ )] = X_t - \theta_1 a_t - \theta_2 a_{t-1} - \theta_3 a_{t-2} \tag{4}$$

and the forecast error made at time t would be

$$e_t(1) = \hat{X}_{t+1} - X_t(1) = a_{t+1} \tag{5}$$

In other words, the random distrubances $\{a_t\}$ which generate the process are the succession of the one-step ahead forecast errors. Hence, the forecast made at time t for X_{t+1} can be expressed as

$$\hat{X}_t(1) = X_t - \Theta_1 \cdot e_{t-1}(1) - \Theta_2 \cdot e_{t-2}(1) - \Theta_3 \cdot e_{t-3}(1) \tag{6}$$

and the approximate 95 percent confidence limits for X_{t+1} would be

$$X_{t+1}(\pm) = \hat{X}_t(1) \pm 2\ \hat{\sigma}_a \tag{7}$$

where $\hat{\sigma}_a$ is the estimate of the standard deviation of the white noise variables.

APPLICATION TO INCIDENT DETECTION

Several algorithms have been proposed for freeway incident detection. These algorithms can be categorized as pattern recognition algorithms [4, 5, 6], and smoothing algorithms [2, 3]. The first category of algorithms attempts to distinguish between incident and incident-free conditions by means of comparing the observed value of a key state variable with a preselected threshold level determined through calibration. The second category of algorithms employs short-term forecasts of the state variables and a set of calibrated thresholds to detect the sudden perturbations in traffic stream behavior generated by incidents. This is similar to the problem of signal detection in noisy communications systems. The appealing characteristic of this second category of algorithms is that past trends of the state variables can be used to predict recurrent congestion which takes some time to build up, whereas non-recurrent congestion due to incidents is unpredictable.

All previously developed algorithms share two major problems: high false-alarm rates and threshold calibration requirements. In fact, the two problems are strongly related, because the threshold levels for detection cannot allow for the multiplicity of factors causing variations in traffic flow (e.g., time of day, geometrics, pavement and environmental conditions, etc.). Intuitively, the use of real-time estimates of the variability in state variables in the setting of detection thresholds should lessen the false-alarm problem and should improve the overall performance of the detection algorithm.

Motivated by the preceding remarks, the ARIMA algorithm described here has been structured so that it includes real-time estimates of the variability in traffic occupancy. An incident is detected if the observed occupancy value lies outside the confidence limits constructed two standard deviations away from the corresponding point forecasts. This approach was applied to a total of 50 lane blockage incidents which took place on a 2-mile (3.2-km) section of the Lodge Freeway in Detroit. This section contained a closed network of ultrasonic vehicle presence detectors located at four surveillance stations in addition to fourteen television cameras. Historical records of 1-minute average occupancies were observed at the nearest upstream detector station for each of the 50 incidents. Each record commenced about 10 minutes preceding the television log time of incident occurrence and continued until about 10 minutes after the logged time of removal of the incident or until the dissipation of congestion. This provided a total of 1,692 minutes of observations on occupancy data associated with capacity-reducing incidents.

Figure 2 illustrates the performance of the ARIMA occupancy algorithm during one of the 50 incidents used in the analysis. The solid line represents the observed upstream occupancy while the broken lines are the 95 percent confidence limits of the point forecasts. The algorithm's response to this incident is indicated by the deviation of the observed occupancy value from the corresponding confidence limits at the onset of congestion.

The overall performance of the ARIMA occupancy algorithm is shown in Figure 3 by an operating characteristic curve relating the detection rate to the false-alarm rate. This curve was obtained by changing the width of the constructed confidence intervals from 3 to 5 standard errors in increments of one standard error. As noted, the algorithm detected all of the 50 incidents with 1.4 percent false-alarm rate. In addition, the average time-lag to detection was found to be 0.39 minute.

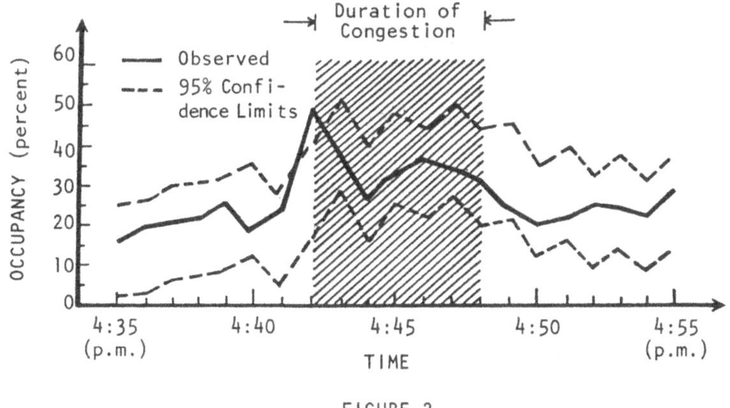

FIGURE 2

PERFORMANCE OF ARIMA OCCUPANCY ALGORITHM DURING AN INCIDENT

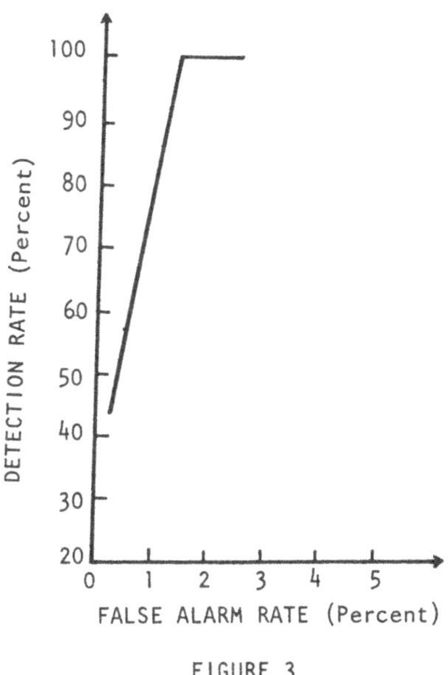

FIGURE 3

OPERATING CHARACTERISTIC CURVES FOR ARIMA INCIDENT DETECTION ALGORITHM

CONCLUSION

High false-alarm rates and threshold calibration requirements are two interrelated problems handicapping the operational performance of existing incident detection algorithms. In this paper, a new detection algorithm has been proposed based on a stochastic model that describes the dynamics of freeway traffic occupancies. This algorithm has three promising characteristics: accuracy, freedom from calibration, and computational efficiency.

REFERENCES

1. Box, G.E.P. and Jenkins, G.M. Time Series Analysis: Forecasting and Control. Holden-Day, San Francisco, 1976.
2. Cook, A. R. and Cleveland, D.E. The Detection of Freeway Capacity-Reducing Incidents by Traffic Stream Measurements. Transportation Research Record 495, 1974, pp. 1-11.
3. Dudek, C. L., Messer, C. J., and Nuckles, N. B. Incident Detection on Urban Freeways, Transportation Research Record 495, 1974, pp. 12-24.
4. Levin, M. and Krause, G. M. Incident Detection--A Bayesian Approach. Transportation Research Record 682, 1978, pp. 52-58.
5. Payne, H. J. and Tignor, S. C. Freeway Incident Detection Algorithms Based on Decision Trees with States. Transportation Research Record 682, 1978, pp. 30-37.
6. Willsky, A. S., Chow, E. Y., Gershwin, S. B., Greene, C. S., Houpt, P. K., and Kurkjian, A. L. Dynamic Model-Based Techniques for the Detection of Incidents on Freeways. IEEE Transactions on Automatic Control, Vol. AC-25, 1980, pp. 347-360.

IDENTIFICATION OF SYNCHRONOUS MACHINE PARAMETERS FOR STABILITY MODELS, USING SYNCHRONIZING TESTS

N. Jaleeli and W.J. Echeverria
Department of Electrical Engineering
Ohio University
Athens, Ohio 45701/USA

Abstract

Connecting a synchronous machine to a power system, when the magnitude and the angle of the machine terminal voltage are different from those of the system, makes a synchronizing test in which the transient behavior is affected by the parameters of both axes of the machine. The transient data of several synchronizing tests is obtained using a computer program, hereafter called the first program, to simulate the tests. Each set of transient data obtained from this computer program together with quasilinearization are used in another computer program, called hereafter the second program, to identify the machine reactances along the two axes. The values determined for x_{ℓ}, x_d, x_d', x_d'' and x_q from the second computer program are compared with those used in the first program to measure the goodness of quasilinearization and synchronizing tests in the identification of synchronous machine parameters.

Introduction

A reliable mathematical model of the synchronous machine is required to accurately assess the transient stability of power systems. The structure of the model used for the generator is not the only factor affecting the accuracy of the transient analysis of power systems. The parameters employed in these structures are equally important. The accurate determination of synchronous machine parameters have been the subject of considerable industry effort over the past several years [1-8]. Frequency [7,8] and time domain [2,6] response study of the synchronous machine are the two major bases for obtaining the parameters. The method proposed and investigated in this paper falls in the latter category. It removes the disadvantage of [2-4] in imposing at least three disturbances on the power system. It avoids severe disturbance caused by a three phase short circuit test required in [5]. Because of the detailed model used for the synchronous machine in this paper, subtransient parameters, to the contrary to [6], are also identified.

The method described in this paper requires the measurement of the instantaneous values of the stator currents, terminal voltages, rotor angle and speed during a synchronizing test in which the magnitude and the angle of the machine terminal voltage are different from those of the system. This transient data is obtained for several synchronizing tests using a computer simulation program. In this first program, $\dot{\lambda}_d$ and $\dot{\lambda}_q$ in the synchronous machine are neglected, and the damping effects are considered by including one short circuited coil along each axis of the machine. The same machine model is employed in the second computer program. In this program, the transient data obtained in every synchronizing test together with an identification algorithm developed in this paper are used to identify the reactances of the machine. The closeness of the values obtained for the reactances in the second computer program with those used in the first program gives a basis to assess the goodness of the proposed method for the identification of machine parameters.

A detailed synchronous machine model

The synchronous machine model used in this study is the E" model [9]. This model accounts for the transient and subtransient effects of the machine and its state equations are given below:

$$\dot{\lambda}_D = -\frac{1}{\tau''_{do}} \lambda_D + \frac{1}{\tau''_{do}} e'_q + \frac{1}{\tau''_{do}} (x'_d - x_\ell)\, i_d \tag{1}$$

$$\dot{e}''_d = -\frac{1}{\tau''_{qo}} e''_d - \frac{1}{\tau''_{qo}} (x_q - x''_q)\, i_q \tag{2}$$

$$\dot{e}'_q = \frac{K_d}{\tau'_{do}} \lambda_D - (1 + K_d) \frac{e'_q}{\tau'_{do}} + \frac{X_{xd}}{\tau'_{do}} i_d + \frac{e_F}{\tau'_{do}} \tag{3}$$

$$\dot{\omega} = \frac{T_m}{\tau_j} - \frac{K_2}{\tau_j} \lambda_D\, i_q - \frac{e''_d}{\tau_j} i_d - \frac{K_1}{\tau_j} e'_q\, i_q \tag{4}$$

$$\dot{\delta} = \omega - 1 \tag{5}$$

Where:

$$K_1 = \frac{x''_d - x_\ell}{x'_d - x_\ell} \quad , \quad K_d = \frac{(x_d - x'_d)(x'_d - x''_d)}{(x'_d - x_\ell)^2}$$

$$K_2 = 1 - K_1 \quad , \quad X_{xd} = \frac{(x_d - x'_d)(x''_d - x_\ell)}{x'_d - x_\ell}$$

Saturation effects are neglected in the work presented in this paper. However, this is not a limiting factor of the method and work for inclusion of saturation is in progress.

The following two equations, which gives the d-q components of the machine terminal voltage in terms of the state variables and the d-q components of the stator current, complete the model.

$$v_d = e''_d - r\, i_d - x''_q\, i_q \tag{6}$$

$$v_q = -\, r\, i_q + x''_d\, i_d + \frac{x''_d - x_\ell}{x'_d - x_\ell}\,(e'_q - \lambda_D) + \lambda_D \tag{7}$$

Synchronizing tests

Synchronizing tests are of special interest for the purpose of the identification of synchronous machine parameters. For proper choice of differences between the magnitude and the angle of the machine terminal voltage and those of the system, the test is safe and excites all the transient modes of the machine. In addition, for this particular test all the initial values of the state variables are independent from the parameters and are:

$$\lambda_D = V_t \;,\quad e''_d = 0 \;,\quad e'_q = V_t \;,\quad \omega = \omega_R \text{ and } \delta = 0.$$

Thus, the problem of identification is reduced to improve the initial estimate of the parameters while the initial values of the state variables remain fixed at their known values.

Instead of obtaining the transient data from synchronizing a real machine with a system, as a first step in evaluating the method presented here, a computer program is developed to simulate the synchronization of a 160 MVA machine to an infinite bus. In this program, the model presented in the previous section with the constants given in Table 1 are used to present the synchronous machine. In simulating the synchronization, the phase 'a' terminal voltage of the generator and that of the infinite

x_d per unit	x_q per unit	x'_d per unit	$x''_d = x''_q$ per unit	x_ℓ per unit	r per unit	τ''_{do} seconds	τ'_{do} seconds	τ''_{qo} seconds	τ_j seconds
1.7	1.64	0.245	0.185	0.15	0.00078	0.033	5.9	0.076	4.74

Table 1. Machine constants used in the first computer program

bus, before the synchronization, are taken as:

$$v_{ga} = \sqrt{2}\, V_t \cos \omega_R t$$

$$v_{\infty a} = \sqrt{2}\, \eta\, V_t \cos (\omega_R t + \alpha)$$

The voltages of phases 'b' and 'c' lag these voltages by 120° and 240° respectively. In these equations, α represents the phase difference between the machine terminal voltage and that of the infinite bus. η is the ratio of the magnitude of the two voltages.

Several synchronizations are made using different values of η and α. For each synchronization, the observable variables i_d, i_q, v_d, v_q and ω at equal intervals are recorded. Each set of transient data obtained is used in the second computer program to identify the machine reactances and hereby the proposed identification method is assessed.

An identification algorithm using quasilinearization and Euler's method

The machine state and observation equations given previously may be summarized in the following form:

$$\dot{\underline{x}} = \underline{f}(\underline{x}, \underline{u}, \underline{p}, t) \tag{8}$$

$$\underline{y} = \underline{h}(\underline{x}, \underline{u}, \underline{p}, t) \tag{9}$$

where:

$$\underline{x}^t = [\lambda_D , e''_d , e'_q , \omega, \delta]$$

$$\underline{x}^t(t_o) = \underline{x}_o^t = [V_t , 0, V_t, \omega_R, 0]$$

$$\underline{y}^t = [v_d, v_q, \omega]$$

$$\underline{p}^t = [P_1, P_2, P_3, P_4, P_5] = [x_d, x'_d , x''_d , x_\ell , x_q]$$

$$\underline{u}^t = [i_d , i_q]$$

The objective is to identify $\underline{P}$ using a set of measured quantities of $\underline{y}$, $\hat{\underline{y}}$, at Δt intervals over $t_o \leqslant t \leqslant t_f$. To identify $\underline{P}$, equations 8 and 9 together with $\underline{x}_o$ and $\underline{P}^o$, an estimated vector for $\underline{P}$, are used to get $\bar{\underline{x}}$ and $\bar{\underline{y}}$ at the same Δt intervals. A better estimate of $\underline{P}$ is then obtained by minimizing:

$$\rho = \int_{t_o}^{t_f} \sum_{k=1}^{3} (\bar{y}_k + \delta y_k - \hat{y}_k)^2 \, dt \tag{10}$$

where:

$$\delta y_k = y_k - \bar{y}_k$$

y_k is the nominal trajectory of the kth observable output which could be computed for any instant of time, if the real values of the parameters were available.

As $\bar{y}_k$ and $\hat{y}_k$ are only available at Δt intervals, ρ may be approximated by:

$$\rho \simeq \sum_{m=0}^{m_f} \sum_{k=1}^{3} (\bar{y}_k + \delta y_k - \hat{y}_k)^2 \, \Delta t \tag{11}$$

where:

$$m_f = \frac{t_f - t_o}{\Delta t} - 1$$

To minimize ρ, the partial derivative of it with respect to the change of each parameter is set to zero.

$$\frac{1}{2\Delta t} \frac{\partial \rho}{\partial \delta P_j} = \sum_{m=0}^{m_f} \sum_{k=1}^{3} (\bar{y}_k + \delta y_k - \hat{y}_k) \frac{\partial \delta y_k}{\partial \delta P_j} = 0 \tag{12}$$

Different values of j may be used in this equation to obtain a set of five simultaneous equations. These equations can then be solved for $\delta \underline{P}$, if δy_k is expressed in terms of $\delta \underline{P}$ at the beginning of every interval.

To express δy_k as a function of $\delta \underline{P}$, equations 8 and 9 are linearized about the nominal trajectory of $\underline{x}$ to give:

$$\delta \dot{x}_i = \sum_{j=1}^{5} a_{ij} \delta x_j + \sum_{\ell=1}^{5} b_{i\ell} \, \delta p_\ell \tag{13}$$

$$\delta y_k = \sum_{i=1}^{5} c_{ki} \, \delta x_i + \sum_{\ell=1}^{5} g_{k\ell} \, \delta P_\ell \tag{14}$$

where:

$$a_{ij} = \frac{\partial f_i}{\partial x_j} \quad , \quad b_{i\ell} = \frac{\partial f_i}{\partial p_\ell}$$

$$c_{ki} = \frac{\partial h_k}{\partial x_i} \quad , \quad g_{k\ell} = \frac{\partial h_k}{\partial p_\ell}$$

The values of a_{ij}, $b_{i\ell}$, c_{ki} and $g_{k\ell}$ may be computed at any instant of time using $\underline{p}^o$, $\bar{\underline{x}}$ and $\underline{u}$ at the same instant, as shown in the appendix. Applying Euler's integration algorithm on equation 13 gives:

$$\delta x_i \Big|_{m+1} = \delta x_i \Big|_m + \Delta t \, \delta \dot{x}_i \Big|_m = \delta x_i \Big|_m + \Delta t \, \Big[\sum_{j=1}^{5} a_{ij} \delta x_j + \sum_{\ell=1}^{5} b_{i\ell} \delta p_\ell \Big]_m \tag{15}$$

where $Z\big|_m$ denotes the value of Z at $t = t_o + m \, \Delta t$.

Since $\underline{x}(t_o)$ is known, $\delta\underline{x}(t_o) = \underline{0}$ and therefore:

$$\delta x_i\Big|_1 = \Delta t \sum_{\ell=1}^{5} b_{i\ell}\Big|_1 \;\delta p_\ell = \sum_{\ell=1}^{5} d_{i\ell}\Big|_1 \delta p_\ell \tag{16}$$

where $d_{i\ell}\Big|_1 = b_{i\ell}\Big|_1 \Delta t$

Using equation 15 for the second time yields:

$$\delta x_i\Big|_2 = \delta x_i\Big|_1 + \Delta t \left[\sum_{j=1}^{5} a_{ij}\,\delta x_j + \sum_{\ell=1}^{5} b_{i\ell}\,\delta p_\ell\right]\Big|_1 = \sum_{\ell=1}^{5} d_{i\ell}\Big|_2 \delta p_\ell \tag{17}$$

where $d_{i\ell}\Big|_2 = [d_{i\ell} + \Delta t(b_{i\ell} + \sum_{j=1}^{5} a_{ij}\, d_{j\ell})]\Big|_1$

Continuation of the above procedure yields:

$$\delta x_i\Big|_m = \sum_{\ell=1}^{5} d_{i\ell}\Big|_m \delta p_\ell \tag{18}$$

where:

$$d_{i\ell}\Big|_m = [d_{i\ell} + \Delta t(b_{i\ell} + \sum_{j=1}^{5} a_{ij}\, d_{j\ell})]\Big|_{m-1} \tag{19}$$

Substituting equation 18 into equation 14 gives the required expression for δy_k as:

$$\delta y_k = \sum_{\ell=1}^{5} v_{k\ell}\,\delta p_\ell \tag{20}$$

where:

$$v_{k\ell} = g_{k\ell} + \sum_{i=1}^{5} c_{ki}\, d_{i\ell} \tag{21}$$

Replacing δy_k in equation 12 by the expression obtained in 20 yields:

$$\sum_{m=0}^{m_f} \sum_{k=1}^{3} (\bar{y}_k - \hat{y}_k + \sum_{\ell=1}^{5} v_{k\ell}\,\delta p_\ell)\, v_{kj} = 0 \tag{22}$$

or:

$$\sum_{\ell=1}^{5} s_{j\ell}\delta p_\ell = e_j \tag{23}$$

where:

$$s_{j\ell} = \sum_{m=0}^{m_f} \sum_{k=1}^{3} v_{kj} \, v_{k\ell} \tag{24}$$

$$e_j = \sum_{m=0}^{m_f} \sum_{k=1}^{3} v_{kj} \, (\hat{y}_k - \bar{y}_k) \tag{25}$$

Equation 23 is obtained by taking partial derivative of ρ with respect to δp_j. A set of five linear equations is obtained, if the partial derivative is taken with respect to all five elements of $\delta \underline{p}$. These equations may then be written in the matrix form to give:

$$[S] \; \delta \underline{p} = \underline{e} \tag{26}$$

Starting from an estimated vector $\underline{p}^o$ for the parameters, and computing $\bar{\underline{x}}$ and $\bar{\underline{y}}$, every element of S, s_{ij}, is determined for each set of the transient data by using equations 19, 21, 24 and those given in the appendix. Also each element of $\underline{e}$, e_j, is similarly computed by using equation 25 instead of 24. Then, $\delta \underline{p}$ is determined by solving equation 26. The process is completely repeated using a new estimate for $\underline{p}$, $\underline{p}^{n+1} = \underline{p}^n + \delta \underline{p}^n$, until all elements in $\delta \underline{p}^n$ become very small. If the solution is converged in n iterations, elements of $\underline{p}^{n+1}$ are the identified values for the machine parameters.

Identification results

The transient behavior of the synchronous machine, whose parameters are given in Table 1, is simulated when it is connected to an infinite bus for various values of η and α. The set of values used for η and α together with the simulation period, T, and the sampling rate, Δt, are tabulated in Table 2.

Simulation Number	η	α (degrees)	T (seconds)	Δt (milliseconds)
1	1.1	10	5.0	5.0
2	1.02	2	5.0	5.0
3	1.1	0	5.0	5.0
4	1.01	0	5.0	5.0
5	1.0	5	5.0	5.0
6	1.0	2	5.0	5.0
7	1.0	2	2.0	1.0
8	1.0	0	--	--

Table 2. The condition of each synchronization, the period and interval of recording.

Based on the identification algorithm described in the previous section, a computer program is developed to identify parameters P_1 to P_5. The identification process is examined for three sets of initial estimate of P_1 to P_5 given in Table 3. Other than P_1 to P_5, the values used for the parameters are those given in Table 1.

Initial Estimate Set Number	$P_1 = x_d$	$P_2 = x'_d$	$P_3 = x''_d$	$P_4 = x_\ell$	$P_5 = x_q$
1	2.04	0.294	0.222	0.18	1.968
2	1.36	0.196	0.148	0.12	1.312
3	2.21	0.2205	0.148	0.15	1.148

Table 3. Different sets of values used as initial estimate of P_1 to P_5

In Set 1, all the estimates are 1.2 times of the true values.

In Set 2, all the estimates are 0.8 times of the true values.

In Set 3, the estimates are taken at random between 70% to 130% of the true values.

For all initial estimates of P_1 to P_5, when the transient data of simulations 1 to 4 are used, in a few iterations the parameters converge to within 1% of the values given in Table 1. Exceptions of this rule are x_ℓ and x_q, whose maximum differences are about 1%.

For all initial estimates of P_1 to P_5, when the transient data of simulations 5, 6 or 8 are used, the method is not able to identify the parameters. However, when the transient data of simulation number 7, which is the same as number 6 with shorter T and Δt, is used, the parameters converge to their expected values in a few iterations.

Conclusions

The identification algorithm based on quasilinearization and developed in this work is capable to identify the machine parameters, when the transient data is obtained from another computer program using the same machine model. This identification is very sensitive to the value of η used in the synchronization. If the difference between the magnitude of the terminal voltage and that of the system is chosen large enough, the synchronization test gives adequate information to identify the machine parameters.

The result obtained in this work is an encouragement for examining this and other identification algorithms using the transient data obtained from a higher order model and different from the one used in the identification process.

References

1. Task Force on Definitions , Supplementary Definitions and Associated Test Methods for Obtaining Parameters for Synchronous Machine Stability Study Simulations, IEEE Transactions on Power Apparatus and Systems, Vol. PAS 99, No. 4, 1980, pp. 1625-1633.
2. Power Technologies, Inc., Determination of Synchronous Machine Stability Study Constants, EPRI EL 1424, Research Project 997-3, Final Report, June 1980.
3. deMello, F. and Hannett, L., Validation of Synchronous Machine Models and Derivations of Model Parameters from Tests, IEEE PES Summer Meeting, Paper 80 SM 537-1, 1980.
4. deMello, F.P. and Ribeiro, J., Derivation of Synchronous Machine Parameters from Test, IEEE Transactions on Power Apparatus and Systems, 1977, PAS 96, pp.1211-1218.
5. Lee, C.C. and Tan, O., A Weighted Least-Squares Parameter Estimator for Synchronous Machines, IEEE Transactions on Power Apparatus and Systems, Vol. PAS 96,1, 1977, pp. 97-100.
6. Nishiwaki, N., Yokokawa, S. and Ohtsuka, K., Identification of Parameters for Power System Stability Analysis Using Kalman Filter, IEEE PES Winter Meeting, Paper 81 WM 012-4, 1981.
7. Dandeno, P.L., et al., Adaptation and Validation of Turbogenerator Model Parameters Through On-Line Frequency Response Measurements, IEEE PES Summer Meeting, Paper 80 SM 576-9, 1980.
8. Manchur, G., Lee, D., Griffin, J. and Watson, W., Generator Models Established by Frequency Response Test on a 555 M.V.A. Machine, IEEE Trans., 1972, PAS 91, pp. 2077-2084.
9. Anderson, P.M. and Fouad, A.A., Power System Control and Stability, Iowa State University Press, Ames, 1977.

Nomenclature

v_t	generator terminal voltage
i_d, i_q	d-q components of the stator currents
v_d, v_q	d-q components of the terminal voltages
λ_D	d axis damper winding flux linkage
e''_d	d component of emf produced in the stator by the subtransient flux linkage
e'_q	stator emf corresponding to the field flux linkage
e_F	stator emf corresponding to the field voltage
ω, ω_R; δ	instantaneous and rated angular frequency; rotor angle
x_ℓ, r	armature leakage reactance and resistance
x_d, x'_d, x''_d	d axis synchronous, transient and subtransient reactances
x_q , x''_q	q axis synchronous and subtransient reactances.
τ'_{do}, τ''_{do}	d axis transient and subtransient open circuit time constants
τ''_{qo}	q axis subtransient open circuit time constant

APPENDIX

Non zero elements of a_{ij} coefficients:

$$a_{11} = \frac{\partial f_1}{\partial x_1} = -\frac{1}{\tau''_{do}} \qquad a_{13} = \frac{\partial f_1}{\partial x_3} = \frac{1}{\tau''_{qo}}$$

$$a_{22} = \frac{\partial f_2}{\partial x_2} = -\frac{1}{\tau''_{qo}} \qquad a_{31} = \frac{\partial f_3}{\partial x_1} = \frac{(p_1 - p_2)(p_2 - p_3)}{(p_2 - p_4)^2\ \tau'_{do}}$$

$$a_{33} = \frac{\partial f_3}{\partial x_3} = -\left[1 + \frac{(p_1 - p_2)(p_2 - p_3)}{(p_2 - p_4)^2}\right]\frac{1}{\tau'_{do}}$$

$$a_{41} = \frac{\partial f_4}{\partial x_1} = -\left[1 - \frac{p_3 - p_4}{p_2 - p_4}\right]\frac{u_1}{\tau_j} \qquad a_{42} = \frac{\partial f_4}{\partial x_2} = -\frac{1}{\tau_j}u_1$$

$$a_{43} = \frac{\partial f_4}{\partial x_3} = -\left[\frac{p_3 - p_4}{p_2 - p_4}\right]\frac{u_2}{\tau_j} \qquad a_{54} = \frac{\partial f_5}{\partial x_4} = 1$$

Nonzero elements of $b_{i\ell}$ coefficients:

$$b_{12} = \frac{\partial f_1}{\partial p_2} = \frac{u_1}{\tau''_{do}} \qquad b_{14} = \frac{\partial f_1}{\partial p_4} = -\frac{u_1}{\tau''_{do}}$$

$$b_{23} = \frac{\partial f_2}{\partial p_3} = \frac{u_1}{\tau''_{qo}} \qquad b_{25} = \frac{\partial f_2}{\partial p_5} = -\frac{1}{\tau''_{qo}}u_2$$

$$b_{31} = \frac{\partial f_3}{\partial p_1} = \left[\frac{(p_2 - p_3)}{(p_2 - p_4)^2}(x_1 - x_3) + \frac{(p_3 - p_4)}{(p_2 - p_4)}u_1\right]\frac{1}{\tau'_{do}}$$

$$b_{32} = \frac{\partial f_3}{\partial p_2} = \left[\frac{p_3 + p_1 - 2p_2}{(p_2 - p_4)^2} - 2.\frac{(p_1 - p_2)(p_2 - p_3)}{(p_2 - p_4)^3}\right](x_1 - x_3)\frac{1}{\tau'_{do}} + \left[\frac{(p_3 - p_4)(p_4 - p_1)}{(p_2 - p_4)^2}\right]\frac{u_1}{\tau'_{do}}$$

$$b_{33} = \frac{\partial f_3}{\partial p_3} = \left[\frac{p_2 - p_1}{(p_2 - p_4)^2}\right]\frac{1}{\tau'_{do}}(x_1 - x_3) + \left[\frac{p_1 - p_2}{p_2 - p_4}\right]\frac{u_1}{\tau'_{do}}$$

$$b_{34} = \frac{\partial f_3}{\partial p_4} = 2\cdot\left[\frac{(p_1 - p_2)(p_2 - p_3)}{(p_2 - p_4)^3}\right]\frac{1}{\tau'_{do}}(x_1 - x_3) + \left[\frac{(p_1 - p_2)(p_3 - p_2)}{(p_2 - p_4)^2}\right]\frac{u_1}{\tau'_{do}}$$

$$b_{42} = \frac{\partial f_4}{\partial p_2} = \left[\frac{p_3 - p_4}{(p_2 - p_4)^2}\right]\frac{u_2}{\tau_j}(x_1 - x_3) \qquad b_{43} = \frac{\partial f_4}{\partial p_3} = \left[\frac{1}{p_2 - p_4}\right]\frac{u_2}{\tau_j}(x_1 - x_3)$$

$$b_{44} = \frac{\partial f_4}{\partial p_4} = \frac{(p_3 - p_2)}{(p_2 - p_4)^2}\ \frac{u_2}{\tau_j}(x_1 - x_3)$$

Nonzero elements of c_{ki} elements:

$$c_{12} = \frac{\partial h_1}{\partial x_2} = 1 \qquad c_{21} = \frac{\partial h_2}{\partial x_1} = \left[1 - \frac{p_3 - p_4}{p_2 - p_4}\right]$$

$$c_{23} = \frac{\partial h_2}{\partial x_3} = \frac{p_3 - p_4}{p_2 - p_4} \qquad c_{34} = \frac{\partial h_3}{\partial x_4} = 1$$

Nonzero elements of $g_{k\ell}$ elements:

$$g_{13} = \frac{\partial h_1}{\partial p_3} = -u_2 \qquad g_{22} = \frac{\partial h_2}{\partial p_2} = \left[\frac{p_4 - p_3}{(p_2 - p_4)^2}\right](x_1 - x_3)$$

$$g_{23} = \frac{\partial h_2}{\partial p_3} = \frac{-1}{p_2 - p_4}(x_1 - x_3) + u_1 \qquad g_{24} = \frac{\partial h_2}{\partial p_4} = -\frac{p_3 - p_2}{(p_2 - p_4)^2}(x_1 - x_3)$$

A CONTRIBUTION TO THE OPTIMAL GENERATION SCHEDULLING OF LARGE HYDROTHERMAL POWER SYSTEMS

*C. Lyra H. Tavares S. Soares
DEE/FEC/UNICAMP
13100 - Campinas-SP - Brasil

Abstract - In this paper an approach is proposed to optimize the operation of large energy production systems, over periods where the independent water inflows can be well known. A precise model is considered to represent the hydraulic system (with individualized reservoir in cascade, water transportation delays, nonlinear generation, ...) as well as the non - hydraulic system (thermal generations, interchanges, load sheddings, ...). Difficulties resulting from the large scale are overcomed by decomposition techniques. An example ilustrates the approach.

INTRODUCTION

A power system must attend the electric energy necessities of a community at the lowest possible cost. In Brazil, where the hydroelectric energy is responsible for 93% of the total electric energy production, an efficient operation of the system is important not only for economic reasons, but to prevent load shedding. In most systems, the way to reduce the cost is to operate efficiently the hydroelectric plants in order to avoid thermal generation, expansive non-contractual importation from other markets and undesirable load shedding. During periods, when there is some surplus of energy, secondary loads (non-contractual, exportations, interruptible, internal loads, etc) can be attended. The profits of these sales contribute to lower the total operational cost for the planning horizon.

Many approachs have been suggested for the problem of finding an optimizing policy for energy production in power systems. Variational calculus [1], dynamic programming [2-5] Pontryagin's maximum principle [6] and many other mathematical programming techniques have been used to solve the problem with different formulations. In [7] a decomposition and coordination method is adopted. The approach enables planning optimal operation of large hydrothermal power systems, considering random

* This research was partially supported by the Conselho Nacional de Desenvolvimento Científico e Tecnológico - CNPq.

load demand, many hydroelectric plants coupled in cascade, time delays, water head variations in reservoirs, importation and exportation of energy.

This paper treats again the problem with a more general and precise model. The accurate hydroelectric model proposed in [7] is adopted, but a better representation is used for the non-hydraulic generation, importation and exportation from/to other electrically coupled neighbours systems, and load sheddings. For instance, importation and exportation are limited and the unitary cost varies with the amount of energy changed. Random load demand and secondary markets are considered. The randomness of load is counterbalanced, not only with the help of other systems, but also by internal non-hydraulic generation, arranged to obtain the minimum operational cost. The difficulties imposed by the large dimension of the problem are overcomed by duality decomposition. A hierarchical calculation structure results from the manipulation.

MATHEMATICAL MODEL

Non-Hydraulic Generation Model

The non-hydraulic generation (g^m) comprises thermal generation plants, energy imports not stipulated by contracts and a dummy generation to take into account the load shedding. The thermal generation and secondary imports are bounded and have increasing marginal production cost. Load shedding is an undesirable contingency.

Each non-hydraulic generation unit is represented by its marginal production cost function $\mu_i(g_i^m)$. All the generation units have a bounded range of operation

(1) $\underline{g}_i \leqslant g_i^m \leqslant \overline{g}_i$, $i \in I$, $m \in M$, where I is the non-hydraulic generation index set and M is the time intervals index set.

The functions μ_i are combined in order to produce the minimum non-hydraulic marginal production cost (μ) as a function of the total non-hydraulic energy production g^m,

$$(2) \qquad \begin{array}{ll} g^m = \sum\limits_{i \in I} g_i^m \ , & \underline{g} \leqslant g^m \leqslant \overline{g} \\ \underline{g} = \sum\limits_{i \in I} \underline{g}_i = 0 \ , & \overline{g} = \sum\limits_{i \in I} \overline{g}_i \end{array} \ , \quad m \in M$$

The optimization necessary to find the minimum marginal production cost function can be done by one of the simple methods described by Stevenson

[8] and Elgerd [9]. The function μ is monotonically increasing. The optimum participation of each mean of non-hydraulic production as a function of the total non-hydraulic generation, $g_i^*(g^m)$, is also found during the proceeding of determining μ.

Energy load model

The system energy load is divided in primary and secondary load. The primary load (d^m) is determined by the internal non-interruptible loads plus the contracted energy exports. It must be served, unless load shedding is inevitable. This condition is expressed as

(3) $p^m \geqslant d^m \quad , \quad m \in M \quad ,$ where p^m is the total energy produced

by the system during interval $m(p^m = g^m + h^m$, h^m being the hydroelectric generation).

The primary load at each period $m(d^m)$ is a random variable defined in the interval $[0, \infty)$. Its continuous cumulative distribution function $F_m(d^m)$ at each time period m is assumed known and no correlation between different periods of time is supposed.

The secondary loads are comprised by export markets not covered by contract and internal interruptible loads. Each secondary load is represented by its marginal revenue function $\delta_l(e_l^m)$, which are assumed monotonically decreasing functions of the served load. The secondary loads are bounded

(4) $0 \leqslant e_l^m \leqslant \overline{e}_l \quad , \quad l \in L \, , \quad m \in M \quad ,$ where L is secondary load

index set.

The marginal revenue functions δ_l are optimally combined to form the maximum marginal revenue of secondary loads (δ) as a functions of e^m.

(5) $e^m = \sum_{l \in L} e_l^m \quad , \quad 0 \leqslant e^m \leqslant \overline{e} \quad , \quad \overline{e} = \sum_{l \in L} \overline{e}_l \quad , \quad m \in M$

The same method used to find the minimum marginal production cost function (μ) may be applied to obtain the maximum marginal revenue function (δ). The function δ is monotonically decreasing.

Whilst determining the function δ, we also proceed the optimum distribution of secondary load as a function of the total secondary load ($e_l^*(e^m)$).

Hydroelectric generation model

The hydroelectric system considered is composed of many plants, which can be run of the river or with reservoir, located in the same or in different river basins. Hydroelectric generation, is represented by h^m,

(6) $h^m = \sum_{j \in J} h_j^m \quad , \quad m \in M \quad ,$ where J is the hydroelectric plant index set and h_j^m the hydroelectric generation of plant j during interval m. The generation of a hydroelectric plant is a nonlinear function of water discharge (u_j^m) and storage (x_j^m)

(7) $h_j^m = \phi_j(u_j^m, x_j^m) \quad , \quad j \in J \quad , \quad m \in M$

The reservoir dynamics are described by a difference equation

(8) $x_j^{m+1} = x_j^m + y_j^m + z_j^m - u_j^m - v_j^m \quad , \quad j \in J \quad , \quad m \in M \quad ,$ where y_j^m is the independent water inflow and v_j^m the spill. The dependent water inflow z_j^m is the coupling variable connecting hydroelectric plants over the same hydraulic valley and is expressed by

(9) $z_j^m = \sum_{k \in S_j} (u_k^{m-t_{kj}} + v_k^{m-t_{kj}}) \quad , \quad j \in J \, , \quad m \in M \quad ,$ where S_j is the index set of immediate upstream neighbouring hydroelectric plants and t_{kj} is the time lag for water displacement.

There are also other local constraints

(10)
$$x_j^m \in X_j = \{x_j^m / \underline{x}_j \leqslant x_j^m \leqslant \overline{x}_j\}$$
$$u_j^m \in U_j = \{u_j^m / \underline{u}_j \leqslant u_j^m \leqslant \overline{u}_j\} \quad , \quad v_j^m \geqslant 0$$

The initial and final reservoir states are known.

(11) $x_j^o \, , \; x_j^T \quad (j \in J)$ are fixed

Otherwise, the final states can be considered free and values assigned to them.

PROBLEM FORMULATION AND STRATEGY OF RESOLUTION

The system should be operated in order to satisfy the load at the lowest possible cost. This problem can be formulated as follows: minimize the functional J, which describes the total net operational cost of production for the planning horizon.

(12) $$J = \sum_{m \in M} \{ \int_{h^m}^{h^m+g^m} \mu(\zeta - h^m) d\zeta - \int_{d^m}^{d^m+e^m} \delta(\zeta - d^m) d\zeta \}$$

The energy balance equation that couples the hydraulic and non-hydraulic systems is

(13) $p^m = h^m + g^m = d^m + e^m \quad , \quad m \in M$

The optimization can be proceeded in steps. In the first step the minimum production cost of the non-hydraulic system is found for a known hydroelectric generation and load demand - the problem is projected in the space of the hydroelectric generations. Next, the randomness of load is taken into account and, finally, the optimum generation trajectories are found.

Let θ be the function that represents the minimum operational cost as a function of the hydroelectric generation and load demand. Then

(14) $\theta(h^m, d^m) = \min_{p^m} \overline{\theta}\,(h^m, d^m, p^m) \quad , \quad m \in M$

(15) s.t. $\max(h^m, d^m) \leqslant p^m \leqslant \min(d^m + \overline{e}, h^m + \overline{g}) \quad , \quad m \in M$, where $\overline{\theta}$

represents the production cost as a function of the hydraulic generation, load demand, and total generation (see Fig. 1).

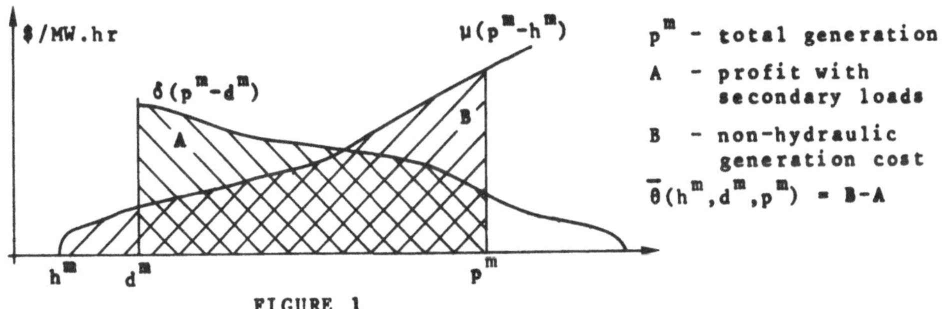

FIGURE 1

(16) $\overline{\theta}(h^m, d^m, p^m) = \int_{h^m}^{p^m} \mu(\zeta - h^m)\,d\zeta - \int_{d^m}^{p^m} \delta(\zeta - d^m)\,d\zeta \quad , \quad m \in M$

The following problem (P_1) determines the optimum trajectories of the hydroelectric generation:

(17) $P_1\{\min J' \quad , \quad$ s.t. (6)-(11) where $J' = \sum_{n \in M} \theta(h^m, d^m)$

Due to the randomness of load demand, it follows that J' is also a random quantity. So, the minimization cannot be proceeded without an additional manipulation. It was decided to work with an equivalent deterministic problem, taking the mathematical expectation of J' with the hypothesis that randomnes of load demand is compensated by non-hydraulic generation and secondary loads, combined in order to produce the minimun generation cost. In addition, to calculate J" (mathematical expectation of J') it is assumed no correlation between time intervals.

$$(18) \quad J'' = \sum_{m \in M} E_{d^m} \{\theta(h^m, d^m)\} = \sum_{m \in M} w^m(h^m)$$

Problem P_1 is replaced by P_2, obtained by changing J' for J"

$$(19) \quad P_2 \{\min \sum_{m \in M} w^m(h^m) \quad , \quad \text{s.t. } (6)-(11)$$

$w^m(h^m)$ is a convex decreasing function. A duality decomposition approach can be applied to find the solution of P_2 for large scale hydrothermal systems (see [10] for more details). The minimization can be carried out as two separated subproblems (see Fig. 2)

Other approachs without applications of decomposition can also be applied to solve problem P_2. In [11] an efficient technique that can be used to solve P_2 for large hydrothermal systems is discussed.

Solving problem P_2, the trajectories of hydraulic generations (h^m) are obtained. Then, the optimum non-hydraulic generation function $g^{*m}(d^m)$ and $e^{*m}(d^m)$, for secondary load, can be determined for all $m \in M$. These functions and the already known g_1^* and e_1^*, enable the operator to determine the generation of each non-hydraulic plant and how much energy must be served to each secondary load after the knowledge of the actual load demand.

Minimization of the expected cost energy

$$\text{P2-a} \{\min \sum_{m \in M} [w^m(h^m) + \lambda^m h^m]$$

Problem P2-a can be solved with independent optimizations, one for each time interval m. If the analytical expression of $w^m(h^m)$ ($m \in M$) is known (and differentiable), the solution is given by $\hat{h}^m$, where

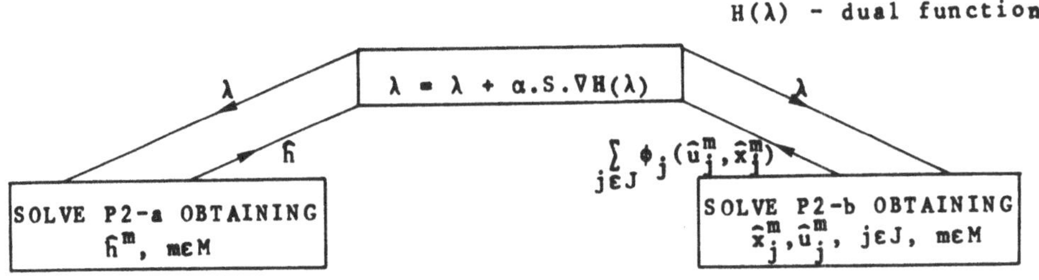

FIGURE 2

$$\left.\frac{dw^m(h^m)}{d\,h^m}\right|_{\hat{h}^m} + \lambda^m = 0 \quad , \quad m \in M$$

Minimization of the hydroelectric generation

$$\text{P2-b } \{\max \sum_{m \in M} \lambda^m \sum_{j \in J} h_j^m \qquad \text{s.t. } (7)-(11)$$

The great computation effort to solve problem P2 is due to the hydroelectric problem P2-b. A natural approach to problem P2-b is dynamic programming. To overcome the problem of dimensionality present in large systems with hydraulic coupling, special techniques, like differential dynamic programming (DDP) [4] and successive approximations dynamic programming (SADP) [3] must be used. However, these methods can be efficient only when no time lag is present.

Other techniques can be applied to solve P2-b, especially when considering time lags. First, there exist global procedures which can solve P2-b, as a whole. Some works [11-13] may be emphasized. Second, there are decomposition approachs which breaks P2-b, into smaller subproblems easier to solve [7], [10], [14].

EXAMPLE

The method is illustrated in an example with four coupled hydroelectric and two thermal plants, importation and exportation of energy (Fig. 3).

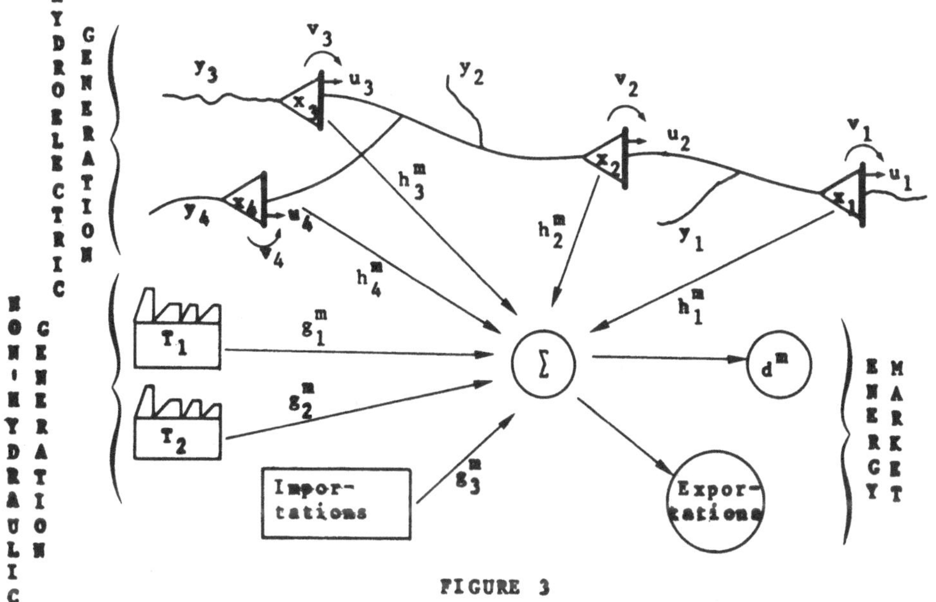

FIGURE 3

in a time horizon with twelve stages.

The non-hydraulic generation data are: $\mu_1 = 20.0 + 1.00\ g_1$; $\mu_2 = 13.2 + 1.66\ g_2$; $\mu_3 = 100.0$; $\bar{g}_1 = 70$; $\bar{g}_2 = 60$; $\underline{g}_1 = \underline{g}_2 = \underline{g}_3 = 0$.

The load demand at each stage has a normal distribution: variance = 5, mean varying from 190 to 250. The secondary load is such that: $\delta = 10$; $\underline{e} = 0$

The hydroelectric generation functions are:

$$\phi_1(x_1,u_1) = -.001x^2 - .1u^2 + .01xu + .38x + 3.8u - 30.;$$
$$\phi_2(x_2,u_2) = -.001x^2 - .1u^2 + .01xu + .30x + 3.0u - 30.;$$
$$\phi_3(x_3,u_3) = -.001x^2 - .1u^2 + .01xu + .38x + 3.5u - 30.;$$
$$\phi_4(x_4,u_4) = -.001x^2 - .1u^2 + .01xu + .40x + 4.0u - 30.$$

Table 1 gives other necessary hydraulic data.

PLANT	$\underline{x}$	$\bar{x}$	$\underline{u}$	$\bar{u}$	x^o	x^T	Inflows at each stage											
							1	2	3	4	5	6	7	8	9	10	11	12
1	70	160	13	25	120	140	19	19	19	16	0	0	0	0	0	0	0	0
2	100	240	10	30	170	170	0	0	2	2	3	4	3	2	1	1	1	2
3	60	100	6	15	80	70	8	8	9	9	8	7	6	7	8	9	9	8
4	80	150	5	15	100	120	10	9	8	7	6	7	8	9	10	11	12	10

The program, using SADP routine, required 17 minutes on a PDP-10 computer. With a DDP routine only 7 minutes were necessary.

The optimum control trajectories are

u_1 = {20,22,24,25,20,21,21,21,22,23,24,24}, u_2 = {16,15,14,13,20,21,21,21,20,19,18,16}, u_3 = {6,6,8,9,9,10,10,10,9,10,10,9}, u_4 = {6,6,7,8,8,10,9,8,7,6,7,5}

CONCLUSIONS

The method can be used to plan the production of a large hydrothermal power system. A precise model is used, dealing with features of real systems: coupling of cascade reservoirs, operative plant restrictions, nonlinear generation functions, energy interchanges, secondary markets, etc... It must be emphasized that a solution of the problem provides only the optimum trajectories for hydroelectric production. Instead of determining the sequence of thermal generations controls $\{g_i^m\}$, optimum decision curves $g_i^*(g^{*m}(d^m))$ and $e_1^*(e^{*m}(d^m))$ are found. This is a more flexible and precise approach to deal with randomness of load.

An example with four hydroelectric and two thermal plants, importation and exportation of energy, was solved. The computational effort to solve the problem in a PDP-10 computer was 7 minutes with the hydroelectric subproblem solved by differential dynamic programming.

The authors believe that faster convergence can be attained with better initial values and improved techniques to solve the hydroelectric sub-problem.

The model is better suited to plan the operation for short horizons, since the natural inflows in the rivers must be well predicted.

REFERENCES

[1] J.H. Drake, L.K. Kirchmayer, R.B. Mayall and W. Wood, "Optimum Operation of Variable Head Hydroelectric Plants". AIEE Transactions on Power Apparatus and Systems 80 (1962) 242-250.

[2] N.V. Arvanitidis and J. Rosing, "Optimal Operation of Multireservoir Systems using Composit Representation". IEEE Transactions on Power Apparatus and Systems 89 (1970) 327-335.

[3] F.J. Ress and R.E. Larson, "Computer - Aided Dispatching and Operation Planning for an Electric Utility with Multiple Types of Generations". IEEE Transactions on Power Apparatus and Systems 90 (1971) 891-899.

[4] M. Heidari, V.T. Chow, P.K. Kokotovic and D.D. Meredith, "The Discrete Differential Dynamic Programming Approach to the water Resources Systems Optimizations". Water Resources Res. 7 (1971) 273-282.

[5] R. Pronovost and J. Boulva, "Long - Range Operation Planning of a Hydro-Thermal System - Modelling and Optimization". Canadian Electrical Association Spring Meeting, 1978.

[6] I. Hano, Y. Tamura and S. Narita, "An Applications of the Maximum Principle to the Most Economical Operation of Power Systems" IEEE Transactions on Power Apparatus and Systems 85 (1966) 486-494.

[7] S. Soares, C. Lyra and H. Tavares, "Optimal Generations Scheduling of Hydro-thermal Power Systems". IEEE Transactions on Power Apparatus and Systems 99 (1980) 1107-1115.

[8] W.D. Stevenson, "Elements of Power Systems Analysis". McGraw Hill Book Company, New York, 1962.

[9] O.I. Elgerd, "Electric Energy Systems Theory: An Introducitons" McGraw Hill Book Company, New York, 1971.

[10] C. Lyra, "Otimização da Escala de Geração em Sistemas de Potência Hidrotérmicos". Master Thesis, Universidade Estadual de Campinas (UNICAMP), 1979.

[11] C. Lyra, A. Friedlander and J.C. Geromel, "Coordenação da Operação Energética no Médio São Francisco por um Método de Gradiente Reduzido". 4º Congresso Nacional de Matemática Aplicada e Computacional, Rio de Janeiro, 1981.

[12] J.C. Geromel and L.F.B. Baptistella, "A feasible direction Method for Large Scale Nonconvex Programs - A Decomposition Approach". Report Interne LASS, CNRS, Toulouse, 1978.

[13] J.L.D. Faco, "Application of the Greco Algorithms to the Optimal Generation Scheduling for Electric Power Systems". Tenth Int. Symp. on Math. Program., Montreal, 1979.

[14] T. Ohishi, "Aspectos da Otimização da Escala de Geração em Sistemas de Potência Hidrotérmicos", Master Thesis, Universidade Estadual de Campinas (UNICAMP), 1981.

MODELING OF FOURDRINIER PAPER MAKING MACHINES AND BASIS WEIGHT CONTROL

Masao Murata
Dept. of Electrical Engineering
Utsunomiya University
Utsunomiya-shi, Tochigi-ken, 321/Japan

Introduction

The basis weight computer control systems on paper making machines were originally designed mainly for the high-speed Fourdrinier paper making machines (the machine speed is 700 m/min grade). Recently, however, those systems have come to be applied to the lower-speed Fourdrinier paper making machines as well. This recent tendency may well have brought on the problem of whether those basis weight computer control systems could be applied to any kind of Fourdrinier paper making machine.
On the other hand, little has been known about the common basic measures for introducing basis weight control to Fourdrinier paper making machines of various kinds.
The complicated structure of paper making systems is perhaps the main source of difficulty in theoretically establishing those basic measures.
This paper consists of the following two parts. (1) The modeling of the paper making systems in basis weight control is discussed under the new approach, where attention is paid to the retention of solid materials on the wire. This investigation leads to classifying those paper making systems into type A,B,C and D. (2) The two basis weight control theorems for practical use, namely, Smith's method and the deadbeat performance, are investigated in terms of their application on each type of paper making system.

Description of basis weight control

Fig.1 shows a schematic diagram of basis weight computer control systems on the Fourdrinier paper making machine. The thickstock consistency shown in Fig.1 is ordinarily from 3 to 4 percent. The thickstock is diluted by backwater. The consistency of the thinstock in the head box is from 0.3 to 1.2 percent. In basis weight control, basis weight at the dry end is to be controlled at a constant value, by controlling the aperture of the thickstock valve in accordance with the basis weight variation detected by the BM guage at the dry end. The effect of thickstock flowrate change (thickstock valve aperture change) appears only on thinstock consistency, almost disappearing on thinstock flowrate. Accordingly, in basis weight control, the thinstock consistency is controlled by thickstock valve aperture.

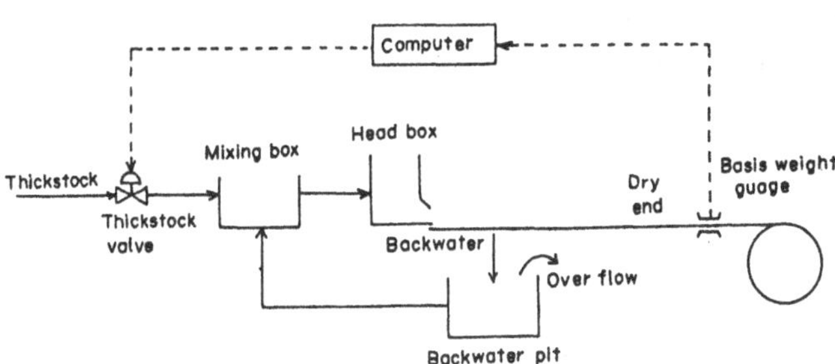

Fig.1 Schematic diagram of the basis weight computer control system.

Various and diversified factors are complicatedly combined to affect the basis weight variation. Although it is very difficult even at present to systematically identify all the affecting factors, roughly classifying the dominant factors influencing the basis weight variation, the basis weight variation W(t) could be expressed by the expression (1) in many cases:

$$W(t)= W_1(t) + W_2(t) + W_3(t) + W_0 \quad \cdots\cdots\cdots\cdots\cdots(1)$$

where W_0 is a constant value and the power spectral density of $W_1(t)$, $W_2(t)$and $W_3(t)$, respectively, equals zero in the period range of more than 1 min., of less than 1 min. and of more than 5 min., and of less than 5 min.

Earlier investigations have ascertained that the variation $W_1(t)$ in basis weight has a considerably high correlation with the total head variation in stockinlet, i.e., the thinstock flowrate variation; however, there is little correlation between $W_1(t)$ and the thickstock consistency variation, while the variation $W_2(t)$ and $W_3(t)$ in basis weight can be considered to be largely affected by the thickstock consistency variation. The main reason for this lies with the fact that the instruments to accurately measure both the thick and thin stock consistency have not been developed so far. Stock flowrate control is far more easier than stock consistency control and various flowrate control methods have been applied so far to the stockinlet and other various processes. On the other hand, accurately controlling the stock consistency is in general very difficult; that is, controlling the stock consistency is most necessary for keeping basis weight constant.

Mathematical modeling of paper making systems

In Fig.2 some typical examples of the retention of solid materials in the thinstock flowed out onto the wire are shown. As shown in Fig.2, the ratio of mixing chemical pulp (CP) with ground wood pulp (GP) is different in webs. For example, the numerical values of the ratio for higher grade webpapers are CP:GP= 10:0 or 8:2, while those for middle or lower grade webpapers, newsprints, etc. are CP:GP=3:7. The experimental results shown in Fig.2 are summerized in the following: (A) when CP:GP=10:0 or 8:2, backwater consistency is approximately proportional to headbox consistency,(B) when CP:GP=3:7, backwater consistency can be considered to be constant without any connection with variations in headbox consistency. These two remarkable results indicate that the basis weight response to a step of thickstock flowrate may be different in pulps, even if the same machine is used. On the other hand, in our experiments, the retention of solid materials in thinstock on the wire was estimated to be above 90 percent, when CP alone is used for stock. In comparing this fact with the results shown in Fig. 2, where the retention of stock on the wire is about 40 percent, the addition of filler such as clay is considered to be the cause of the lower retention as shown in Fig.2. Fig.2 leads to the conclusion that adequate information as to what kind of pulps are used and in which section clay is added in the paper making process are of prime importance for modeling paper making systems in order to improve the systems.

The mathematical modeling methods of paper making systems are roughly classified into the following two methods:

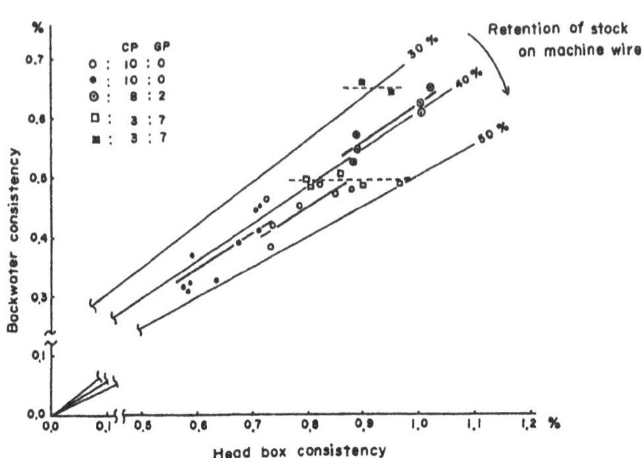

Fig.2 Relation between headbox consistency and backwater consistency.

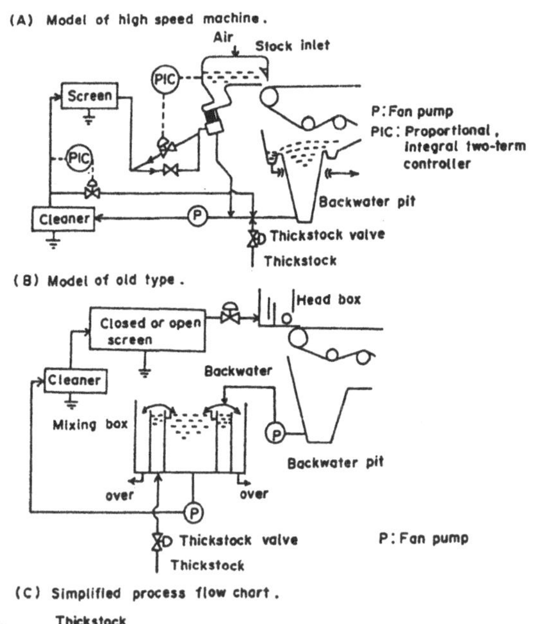

Fig.3 Principal modeling of paper making systems.

(I) black box modeling, where the system is regarded as a black box, the mathematical representations of which are calculated from any input (ordinarily, step or random disturbance) and the resultant output[1]; (II) breaking down the machine into separate parts which can be represented by simple equations, then linking up the equations on the assumption that backwater consistency is proportional to headbox consistency.[2] Comparing the two methods, it can be concluded that the latter method is much more useful than the former method in terms of improving each part of paper making systems, despite the fact that knowing all the constants necessary for modeling each part is difficult in many cases. In this paper, therefore, we try to improve method II so that models of the systems may be built when some of the constants are known and the basis weight response to a step of thickstock flowrate is given, with special attention being given to the fact the backwater consistency is not always proportional to headbox consistency.

Fig.3A shows the layout of a typical high speed Fourdrinier paper making system. In contrast with Fig.3A, many Fourdrinier paper making machines in common use today belong to the older type shown in Fig.3B. The outstanding difference between the two types is great; for example, in the newer type of machine, the headbox is replaced by stockinlet; furthermore, in each part of the newer type of machine, various flowrate control methods are applied to cope with high speed paper making. However, no significant difference is found between the newer and older type machines in terms of the fundamental paper making princeple. In both types of machine, thickstock is diluted by backwater, and the thinstock diluted by backwater is flowed out onto the wire to form paper sheet. Therefore, in mathematically modeling paper making systems, simplifying the systems using the following method is considered to be most suitable; the basic model is at first made on the basis of the older type machine, then the basic model is slightly modified in accordance with the characteristics of each machine.

In Fig.3C, the fundamental principle of this method is shown by block diagram, and Fig.4 shows the simulation model developed by using the method above. The thickstock valve and each part of (I) to (VI) in Fig.3C, indicate the most fundamental components of Fourdrinier paper making systems of various kinds. The mathematical model represented in Fig. 4 is formed on the assumption which is shown below, and represented by the

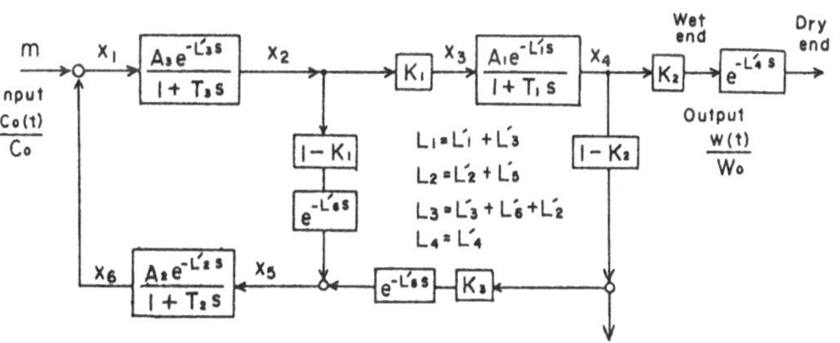

Fig.4 Block diagram of simulation model.

steady material balance equations in the wet end system; thickstock valve, (I), (III), (IV) and (VI) are regarded as consistency variable systems, while (II) and (V) are regarded as flowrate variable systems.

The amount of solid material in thinstock is variable. Focusing attention on the variable components of solid materials in thinstock, the value K_2, which is the retention of the solid material variable component on the wire, may be generally classified as follows.

(a) $K_2 > 0.9$: when major component of applied pulp consists of GP like newspaper, backwater consistency does not change even if thinstock consistency is changed. Therefore, $K_2=1$ may hold. On the other hand, when applied pulp consists only of CP like electrical insulating paper, $K_2 > 0.9$ may hold.

(b) $0.8 < K_2 < 0.9$: when major component of applied pulp consists of CP and a filler such as clay is added at the point (A) shown in Fig.3C, clay component in thinstock flowing out on the wire is not affected at all even if thickstock valve aperture is changed. Therefore, the effect of clay on K_2 may be completely ignored.

(c) $K_2 < 0.5$ (mainly $0.3 < K_2 < 0.5$): when adding point of clay is shifted from A to B in Fig.3C, clay component contained in thinstock changes depending on the aperture change of thickstock valve, even if applied pulp is the same as that of (b).

A large scale digital computer or a hybrid computer is in general needed for the simulation models of paper making systems. According to the model presented in this paper, however, a small digital computer (32K words grade) is enough for the simulating paper making systems. In addition, if the state space method is used for representing both inputs and outputs, Fourdrinier paper making systems can be simulated by Fortran language without special simulation language.

Classification of paper making systems

As shown in Fig.4, the constants necessary for modeling are time constant T_i, dead time L_k and constant K_j. Compared with the dead time L_k and constant K_j, the time constant T_i is merely given at a rough estimate in many cases; the main reason is that measuring the transfer functions of pipes connecting each machine chest is difficult, although the influence of the pipes on each chest is not negligible. Then, the time constants T_i (i=1,2,3) shown in Fig.5 are the estimate values, which minimize the root mean square of the difference between the actual response and the simulation response within the limits of $(1 - \alpha)\, T_i'$ to $(1 + \alpha)\, T_i'$ when T_i' is given as the time constant. Adding here a supplementary explanation to α, it is experimentally confirmed that the minimum value of the root mean square does not change even if α is extended beyond the limit estimated at $\alpha = 20\%$.

Concerning basis weight response to a step of thickstock flowrate, Fig.5 gives the comparison between the simulation response and the actal response. Fig.5 leads to the conclusion that the Fourdrinier paper making process may be fairly approximated by the simulation model shown in Fig.4. As shown in Fig.4, the basis weight responses are affected by the values of time constant T_i, dead time L_k and constant K_j. K_2 especially has great influences on the responses. Fig.6 shows that paper making systems can be classified into types A,B,C and D by K_2, which is concerned with the retention of solid materials in thickstock flowed out onto the machine wire. Type A is in the case of $K_2 > 0.9$, type B in the case of $0.8 < K_2 < 0.9$, and type C and D in the case of $0.3 < K_2 < 0.5$. The transfer functions of types A,B, and C, respectively, are considered to be represented approximately by the models with one time constant T and dead time L, where the ratios L:T for types A,B, and C, respectively, are $T < L$ (normally $\frac{1}{4}L < T < \frac{1}{3}L$), $T \fallingdotseq L$ and $T > L$ (normally $T > 3L$). Although K_2 of type C is the same as that of type D, namely $K_2 < 0.5$ (mainly $0.3 < K_2 < 0.5$), type C is limited only to the case where $T_3 > 3T_1$ (T_1: time constant of headbox, T_3: time con-

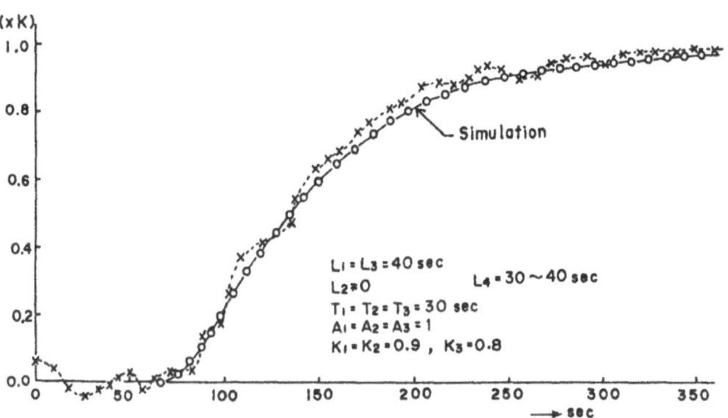

Fig.5 Basis weight response to a step of thickstock flow rate.

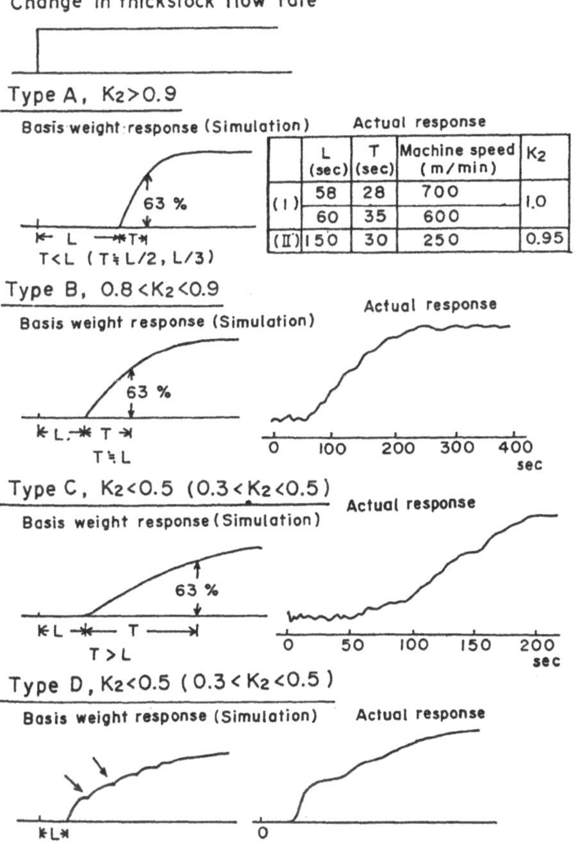

	L (sec)	T (sec)	Machine speed (m/min)	K2
(I)	58	28	700	1.0
	60	35	600	
(II)	150	30	250	0.95

Fig.6 Classification of paper making systems.

stant of mixing box) holds. Therefore, type C is regarded as a special case of type D.

The basis weight step response in type D changes stepwise as shown in Fig.6. Smoothing here the basis weight step response, the smoothed response could be approximately represented by the exponential function in the same manner as the basis weight step responses in types A,B and C. In fact, it is often observed in the actual basis weight step responses of types A,B and C that perturbations are superposed on the exponential function. Therefore, it is considered to be meaningless to discriminate particularly type D, as far as basis weight step response is concerned. However, discrimination between type D and the other types becomes important in regard to basis weight control application. The reason of this is because the perturbations observed in the actual basis weight step responses of types A,B and C are caused by factors other than thickstock valve aperture change. Such superposed perturbations could be eliminated if those factors are removed by machine tuning. On the contrary, the stepwise change of the basis weight step response in type D is caused by thickstock valve aperture change itself. Therefore, such conditions remains unchanged unless the paper making system is modified to either type A,B,or C.

The basis weight control theorems for practical use are mainly Smith's method and deadbeat performance. Both of the two theorems are based on the assumption that paper making systems can be represented approximately by the models with one time constant and dead time, and that the variations in thickstock consistency alone can be considered as the disturbance input in the basis weight control systems. Accordingly, it is concluded from the above description the discrimination between type C and D is absolutely necessary for the design of basis weight control.

Application limit of basis weight control

The main tendency toward basis weight control theorems seems to be based on Smith's method,[3] although in Japan many are based on deadbeat performance.[4] There exists a partial difference between the application of the two control theorems to basis weight control. The main reason for this appears to lie with the difference of requirements for the machine tuning prior to the application of basis weight control. In the basis weight control based on deadbeat performance, as shown in Fig.7, it is worthy of note that variations in the output with the control become larger than those in the output without the control, according to the wavelength of the input disturbances. Then, the relationship between the wavelength of the input disturbances and the control ratio R_1 is shown in Table 1, where models with one time constant T and dead time L are used as simulation models. The ratios L:T shown in Table 1 are the representative values of types A,B,and C, respectively. Table 1 emphasizes the necessity for taking away the input disturbances with the following wavelength ℓ by machine tuning prior to basis weight cintrol: (1)$\ell < 15L$ for type A, (2)$\ell < 10L$ for type B, (3) $\ell < 10L$ for type C.

To obtain deadbeat performance in type D, on the other hand, the procedure is much more complicated than that in types A, B, and C, even on the assumption that the basis weight control in type D is not discrete but continuous. In addition, in applying this basis weight control theorem to type D after representing type D by a model with one time constant and dead time, the basis weight control is apt to produce an unstable system. Hence, the paper making system of type D may have to be improved by taking one of the following procedures, prior to the basis weight control. (1) Remodeling type D into type A or B by taking away the influence of clay and other fillers added in the thickstock; for example, changing the section to add clay as Fig.3C. (2) Remodeling type D into type C by enlarging the time constant of mixing box. (3) Developing the new theorem as a basis weight control. (4) Taking away variations in the thickstock consistency in some way.

In general, short period basis weight variation $W_1(t)$ is eliminated by the procedure of data-processing, while medium period basis weight variation $W_2(t)$ must be removed by machine tuning. For example, machine chests are sometimes remodeled for the purpose of uniformly mixing thickstock, backwater, clay, etc.

In the case using deadbeat performance, the machine tuning for eliminating $W_2(t)$ (medium period basis weight variation) and the control accuracy of the thickstock valve are highly required, although long period basis weight variation $W_3(t)$ is controlled with high accuracy. In the high speed machines which belong to type A or B, the flowrate control system of various kinds are installed for the purpose of coping with the high speed paper making; as a result, short and medium period variations of basis weight are removed to a great extent. Therefore, the introduction of the basis weight control

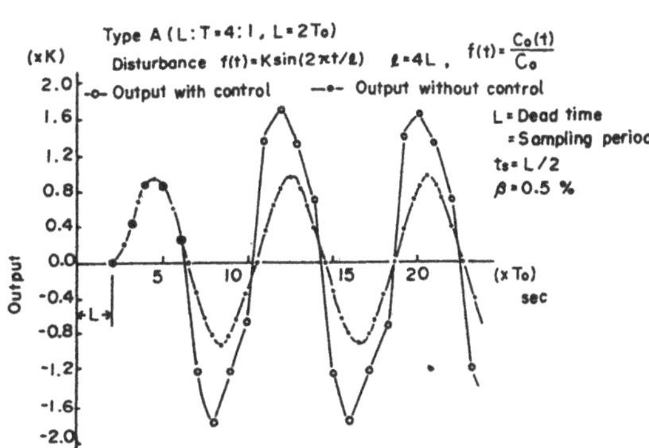

Fig.7 Example of output with control.

Table 1. Features of deadbeat performance.

ℓ	R_1 Type A (L:T=4:1)	Type B (L:T=1:1)	Type C (L:T=1:4)
0.4L	0.5 (β=0.5 %)	0.19 (β=0.5 %)	0.07 (β=0.1 %)
0.75L	0.82 (β=0.5 %)	0.3 (β=0.5 %)	0.1 (β=0.1 %)
L	0.5 * (β=1 %)	0.15 * (β=0.5 %)	0.04 * (β=0.5 %)
4L	1.74 (β=0.5 %)	1.07 (β=0.5 %)	0.5 (β=0.5 %)
10L	1.14 (β=5 %)	1.0 (β=5 %)	0.35 (β=5 %)
20L	0.61 (β=5 %)	0.45 (β=5 %)	0.19 (β=5 %)
30L	0.41 (β=5 %)	0.31 (β=2 %)	0.13 (β=1 %)
60L	0.21 (β=1 %)	0.16 (β=0.5 %)	0.064 (β=0.5 %)

Disturbance is $f(t)=\sin(2\pi t/\ell)$, $f(t)=\frac{C_0(t)}{C_0}$

L = Dead time (sec)

$R_1 = \frac{D_2 \text{ with control}}{D_1 \text{ of disturbance}}$

D_1 = (Maximum value of disturbance) − (Minimum value of disturbance)

D_2 = (Maximum value of output) − (Minimum value of output)

* Outputs of the case ℓ = L are shown in Table 2.

using deadbeat performance is relatively easier in these machines. The control accuracy of the case based on deadbeat performance is in general very high in actual cases as well; the standard deviations of the output variations are 0.25% to 1%.

In the basis weight control based on Smith's method, it seems to be no problem to moderate machine tuning for $W_2(t)$ to some degree, although the control accuracy for $W_3(t)$ is presumed to reduce sharply. Modifying Smith's method is necessary for raising the control accuracy. The applications of modified Smith's method are widely found in the small scale or older type machine as well as the newer type machine.[5] The standard deviations of the output variations are 2.5% to about 3% in the case of modified Smith's method in regard to the actual basis weight control. The basis weight control for type D is difficult unless type D is removed into types A,B, and C, even in the case of modified Smith's method.

Conclusion

In this paper, the modeling of Fourdrinier paper making machines in basis weight control is discussed under the new approach,where attention is paid to the retention of solid matherials on the wire.

This investigation leads to classifying those paper making systems into four types; thus, this classification has been proved to enable us to easily established the basic measures for introducing basis weight control to Fourdrinier paper making machines of various kinds.

References

(1) I.D. Landau: "Unbiased recursive identification using model reference adaptive techniques," IEEE Trans. on Automatic Control,Vol. AC-21, No.2, pp.194-202, April, 1976.

(2) A.L.Ramaz, S.Bauduin, and D.Marcé:"Automation of a papermachine and its ancilliaries," Papermaking Systems and Their Control, Trans. of the fourth international symposium organized by the Fundamental Research Committee of the Technical Section of the British Paper and Board Maker's Association, pp.443-457, September, 1969.

(3) O.J. Smith: "A controller to overcome dead time," ISA Journal, Vol.6, No.2, pp. 28-33, 1959.

(4) J.T.Tou: Modern Control Theory,pp.118-129, McGRAW-HILL, 1964.

(5) I.D.McFarlane: "Economics of computer control for the smaller mill," Paper Technology, pp.341-345, October, 1972.

PHASE FREQUENCY APPROXIMATION IN SYSTEM MODEL REDUCTION

Awad I.Saleh and Mamdouh I.Fouad
Department of Electrical Engineering
Faculty of Engineering
Assiut, Egypt.

Magda O.EL-Shenawee
Department of Mathematics
Faculty of Engineering
Giza, Egypt.

ABSTRACT

This paper proposes a phase frequency approximation. It is based on Thiele's continued fraction expansion to obtain a reduced order system whose phases interpolate the phase frequency response at a set of discrete frequencies that cover the workable frequency range.

1-INTRODUCTION

It is often desirable to represent a higher order mathematical model representing a complex system by an equivalent lower order model. System reduction is carried out either in the time domain or in the frequency domain. The two main methods used in the time domain are Davison's dominant eigenvalue approach[1,2,3] and the Schwarz canonical form[4]. The main methods in the frequency domain are the continued fraction expansion[5], the moment matching[6], the method due to Hsia[7], and the Routh approximation approach[8]. These methods implement Padé approximation about a point in the s-plane while Hsia's method is based on the Padé approximation of the square of the amplitude of the frequency transfer function about $\omega = 0$. Beside the above mentioned methods, optimization techniques are used to obtain a reduced order optimal model which has a minimum meaningful error either in the time or frequency domain[9,10,11,12,13].

This paper presents another frequency domain approach. It is based on Thiele's continued fraction to obtain a reduced order system whose phases interpolate the phase frequency response at a set of discrete frequencies. This set, covers mainly the workable frequency range. Interpolation of phases results in a recursion relation that leads to a reduced order transfer function. An algorithm for the reduction is proposed. It should be noted that the reduction approach, through phase approximation, is applied to minimum phase unreduced transfer function since there exist a unique relationship between the amplitude and phase functions[14]. A minimum phase reduced system can always be guaranteed.

2-REDUCTION PROCEDURE

Consider a single input-single output minimum phase unreduced system having the transfer function

$$H_0(s) = K \frac{1 + a_1 s + a_2 s^2 + \dots + a_m s^m}{1 + b_1 s + b_2 s^2 + \dots + b_n s^n} \quad , \quad m \leq n . \qquad (1)$$

The object of system reduction is to approximate $H_o(s)$ by a lower order transfer function H(s), where

$$H(s) = K\,\frac{1 + c_1 s + c_2 s^2 + \dots + c_p s^p}{1 + d_1 s + d_2 s^2 + \dots + d_q s^q}\,, \qquad p \le q < n. \tag{2}$$

For convenience, the transfer function H(s) can be written as

$$H(s) = \frac{Q(s)}{D(s)} = \frac{m_1(s) + n_1(s)}{m_2(s) + n_2(s)}\,, \tag{3}$$

where $\{m_1(s), m_2(s)\}$ and $\{n_1(s), n_2(s)\}$ are the even and odd parts of Q(s) and D(s) respectively. The criterion proposed for reduction is to make the phase function of the reduced system $\psi(s)$ interpolates the original phase function of the unreduced system $\psi_o(s)$ at a set of discrete ascending frequencies $\omega_1, \omega_2, \dots, \omega_{p+q}$ over the workable frequency range (say one and half the unreduced system bandwidth).
Equation (3) can be written as

$$H(s) = \frac{\{m_1(s).m_2(s)-n_1(s).n_2(s)\}+\{n_1(s).m_2(s)-m_1(s).n_2(s)\}}{m_2^2(s) - n_2^2(s)}. \tag{4}$$

So, the hyperbolic tangent is obtained as

$$\tanh\psi(s) = \frac{n_1(s).m_2(s)-m_1(s).n_2(s)}{m_1(s).m_2(s)-n_1(s).n_2(s)} = \frac{s\,M(s)}{N(s)}\,, \tag{5}$$

where M(s) and N(s) are even polynomials of the complex frequency variable s.
Using Thiele's continued fraction expansion[15], equations(5) can be written in the form

$$\tanh\psi(s) \simeq \cfrac{s}{\alpha_1 + \cfrac{(s^2+\omega_1^2)}{\alpha_2 + \cfrac{(s^2+\omega_2^2)}{\alpha_3 + \cfrac{\ddots}{\alpha_{p+q-1} + \cfrac{(s^2+\omega_{p+q-1}^2)}{\alpha_{p+q}}}}}}\,, \tag{6}$$

Equation(6) can be arranged in the form

$$\frac{s}{\tanh\psi(s)} = F(x) = \alpha_1 + \cfrac{(x_1 - x)}{\alpha_2 + \cfrac{(x_2 - x)}{\alpha_3 + \cfrac{\ddots}{\alpha_{p+q-1} + \cfrac{(x_{p+q-1} - x)}{\alpha_{p+q}}}}} \tag{7}$$

with $s^2 = -x$ and consequently $\omega_i^2 = x_i$ $(i = 1,2, \dots, p+q-1)$.

At this point it is advantageous to make the following remarks:

1. If F(x) is terminated after k terms (k=1,2,...,p+q), then its rational fraction expansion is given by

$$F(x) = \frac{N_k}{M_k} \quad ,$$

where N_k and M_k satisfy the following recurrence relation[16]:

$$N_{k+1} = \alpha_{k+1} N_k + (x_k - x) N_{k-1} \ , \tag{8}$$

and

$$M_{k+1} = \alpha_{k+1} M_k + (x_k - x) M_{k-1} \ ,$$

with the initial conditions

$$N_o = 1,\ N_1 = \alpha_1 \ ,\ M_o = 0 \text{ and } M_1 = 1.$$

2. A simplified algorithm can be proposed for the calculations of the α_i's. Invoking the properties of Thiele's continued fraction[15], substitute F(x) by

$$F(x) = \nu_1(x)$$

where

$$\nu_k(x) = \nu_k(x_k) + (x_k - x)/\nu_{k+1}(x) \ , \quad k = 1,2,\ldots,p+q \tag{9}$$

i.e.

$$F(x) = \nu_1(x) = \nu_1(x_1) + \cfrac{(x_1 - x)}{\nu_2(x_2) + \cfrac{(x_2 - x)}{\nu_3(x_3) + \cdots + \cfrac{(x_{p+q-1} - x)}{\nu_{p+q}(x_{p+q})}}} \tag{10}$$

Clearly

$$\alpha_k = \nu_k(x_k) \tag{11}$$

From equation (7)

$$\nu_1(x) = F(x) = \frac{s}{\tanh \psi(s)} = \frac{\omega}{\tan \psi(\omega)} \quad ,$$

i.e

$$\nu_1(x_i) = \frac{\omega_i}{\tan \psi(\omega_i)} = \frac{\omega_i}{\tan \psi_o(\omega_i)} \tag{12}$$

where $\psi(\omega_i) = \psi_o(\omega_i)$ at $i=1,2,\ldots,p+q$

also equation (9) yields for k=1

$$\nu_2(x) = \frac{x_1 - x}{\nu_1(x) - \nu_1(x_1)}$$

and consequently

$$\nu_2(x_i) = \frac{x_1 - x_i}{\nu_1(x_i) - \nu_1(x_1)} \quad ,$$

for k=2

$$\nu_3(x) = \frac{x_2 - x}{\nu_2(x) - \nu_2(x_2)} \quad , \text{ and so on.}$$

Using equation (9), the sequence $\nu_k(x_k)$ is obtained as shown in Table I.

Table I. The reciprocal difference table.

x	$\nu_1(x)$	$\nu_2(x)$	$\nu_3(x)$	$\nu_4(x)$	$\nu_5(x)$	...
x_1	$\nu_1(x_1)$					
x_2	$\nu_1(x_2)$	$\nu_2(x_2)$				
x_3	$\nu_1(x_3)$	$\nu_2(x_3)$	$\nu_3(x_3)$			
x_4	$\nu_1(x_4)$	$\nu_2(x_4)$	$\nu_3(x_4)$	$\nu_4(x_4)$		
:	:	:	:	:	...	...

In the above table it is clear that the diagonal elements are the desired α_i which appear in equation (6). After obtaining the α_i's of the reduced model,we use the recurrence relation given by equations (8) to determine the rational function $\tanh\psi(s)$.

Thus, the transfer function H(s) can be determined following the conventional methods of determining the transfer function from a known phase function[17]. Using equation (5), we have

$$s.M(s) + N(s) = \{m_1(s) + n_1(s)\}\ \{m_2(s)-n_2(s)\}$$

In fact, $\{s.M(s)+N(s)\}$ can be written in the form of a polynomial of order (p+q) which is a function of the complex frequency variable s. From equation (3) we observe that

$$s.M(s)+N(s) = Q\ (s)\ .\ D(-s) \tag{13}$$

Thus the zeros of the polynomial $\{s.M(s)+N(s)\}$ are those of Q (s) and D(-s). To get the zeros and poles of the reduced transfer function H(s) we can follow these steps:

1- Solve the polynomial $s.M(s)+N(s) = 0$ for its roots.

2- Assign the roots in the right half-plane to D(-s) and those in the left half-plane to Q (s).

3- If all the roots are in the right half-plane, then assign these roots to D(-s) to get an all pole reduced order system. Steps(2) and (3) yield a minimum phase reduced order system.

4- If in step (2) the number of zeros exceeds the number of poles (i.e. $p > q$), the resultant reduced order transfer function will be unbounded and this is physically unrealizable. So we have either to choose a different group of discrete frequencies over the workable frequency range, or to propose a reduced order model of different (p+q).

3-DETERMINATION OF OPTIMAL REDUCED ORDER TRANSFER FUNCTION

A more refined reduced transfer function can be obtained if the coefficients of the reduced order transfer function, given by equation(2), are selected to minimize the square error phase deviations of both the unreduced and reduced systems over the workable frequency range. The performance index

$$J = \int_0^A \{\psi_0(\omega) - \psi(\omega)\}^2 d\omega \tag{14}$$

where A is the upper frequency bound.

To determine the coefficients of the optimal reduced order transfer function, we can use the values of the coefficients obtained in section (2) using the proposed phase approximation method as an initial guess for the minimization procedure.

4-NUMERICAL EXAMPLES

Example 1: The transfer function to be reduced is

$$H_0(s) = \frac{1+1.773s+1.1844s^2+0.3746s^3+0.0564s^4+0.0031s^5}{1+2.45s+2.255s^2+1.021s^3+0.243s^4+0.029s^5+0.0014s^6}$$

The zeros of $H_0(s)$ are at

$$s = -1.554,\ -2.0689,\ -8.092,\ -3.24 \pm J1.378$$

and the poles are at

$$s = -1,\ -1.982,\ -3.245 \pm J0.355,\ -5.692 \pm J1.186$$

If we assume that the reduced order transfer function is:

a) $H(s) = \dfrac{1+c_1s+c_2s^2}{1+d_1s+d_2s^2+d_3s^3}$, i.e. p+q=5

The bandwidth of the unreduced system is 1.37 rad/sec.

The chosen set of discrete frequencies are

$\omega_1=0.5136$, $\omega_2=1.0273$, $\omega_3=1.37$, $\omega_4=1.54$ and $\omega_5=2.055$ rad/sec.

Using equation (10),(11),(12) and table I, we get

$\alpha_1=-1.5643$, $\alpha_2=3.776$, $\alpha_3=-1.56$, $\alpha_4=18.772$ and $\alpha_5=-0.655$

From equations(8) and (13), the reduced order system zeros are at

$$s = -1.5512, \quad -3.344$$

and poles are at

$$s = -1.001, \quad -2.2933, \quad -5.3755$$

The third order reduced model transfer function is

$$H_0(s) = \frac{1 + 0.944\ s + 0.193\ s^2}{1 + 1.6208\ s + 0.7024\ s^2 + 0.081\ s^3} \tag{15}$$

b) $H(s) = \dfrac{1 + c_1s}{1 + d_1s + d_2s^2}$, i.e. p+q=3.

The chosen frequencies are 1.027, 1.37 and 2.055 rad/sec.

The obtained α's are $\alpha_1=-1.774$, $\alpha_2=4.97$ and $\alpha_3=-1.543$

The second order reduced system zero is at $s=-1.946$, while the poles are at $s=-1.057$, -4.206.

The corresponding transfer function is

$$H(s) = \frac{1 + 0.5138\ s}{1 + 1.1887\ s + 0.2249\ s^2} \tag{16}$$

Comparison of the frequency responses and unit-step output responses of the un-reduced system and reduced order models are shown in Figs.(1)and (2).

The optimum reduced second order system obtained by minimizing the performance index J, equation (14), over the workable bandwidth (A=2.055 rad/sec.) is

$$H_o(s) = \frac{1 + 0.51428\ s}{1 + 1.1864\ s + 0.2249\ s^2} \tag{17}$$

Its zero is -1.944 and its poles are -1.053, -4.222.

The location of the zeros and poles of the transfer functions (16) and (17) are almost the same.

For the sake of comparison the following second order reduced models are obtained using other existing reduction methods.

(1) Davison's reduced model is

$$H_d(s) = \frac{1 + 0.75\ s}{1+1.504s+0.504s^2}$$

Its zero is at $s = -1.333$, while its poles are at $s = -1$, -1.982.

(2) Schwarz canonical form model-the method failed to give a reduced second order model.

(3) Continued fraction reduced model is

$$H_c(s) = \frac{1 + 0.568s}{1+1.245s+0.2557s^2}$$

Its zero is at $s = -1.7605$, while its poles are at $s = -1.0147$, -3.854.

(4) Routh reduced model is

$$H_r(s) = \frac{1 + 1.773s}{1+2.45s+1.839s^2}$$

Its zero is at $s = -0.564$, while its poles are at $s = -0.666 \pm J0.316$.

(5) Reddy's reduced model[9], is

$$H_{re}(s) = \frac{1+0.848\ s}{1+0.0876\ s+1.536\ s^2}$$

Its zero at $s = -1.179$ while its poles are at $s = -0.0285 \pm J0.806$

(6) Levy's reduced model[10] the method failed to give a stable reduced second order model.

For evaluating the performance of the proposed phase approximation method for reduction, Table II demonstrates as a measure the values of the integral square error (ISE) of the step response.

Table II

Method	Authors	Optimum Phase approximation	Davison	Continued Fraction	Routh	Reddy
ISE	1.341×10^{-6}	6.284×10^{-7}	4.67×10^{-4}	9.43×10^{-7}	3.34×10^{-3}	3.3349

Example 2: The transfer function to be reduced is, an all-pole one given by

$$H_0(s)=\frac{1}{1 + 2.45\,s + 2.255\,s^2 + 1.021\,s^3 + 0.243\,s^4 + 0.029\,s^5 + 0.0014\,s^6}$$

Its poles are the same as in example (1). The bandwidth is 0.731 rad/sec.

a) Assume that the reduced order model has a five zeros and poles (p+q=5).

The chosen set of frequencies are

0.274, 0.548, 0.731, 0.822 and 1.096 rad/sec.

In this case it happened that all the zeros of the polynomial {s M(s)+N(s)} are in the right - half plane which results in an all pole reduced order model of the fifth order (p=0, q=5) to get a minimum phase reduced model whose

$$H(s)=\frac{44.877}{(s+0.99)(s+1.97+J0.372)(s+1.97-J0.372)(s+2.62+J2.076)(s+2.62-J2.076)}$$

b) Assume that the reduced model has p+q=3.

The chosen discrete frequencies are 0.548, 0.731 and 1.096 rad/sec.

The resultant minimum phase reduced model is a third order one (p=0, q=3) of

$$H(s) = \frac{1.1826}{(s+0.734)(s+0.898+J0.895)(s+0.898-J0.895)}$$

Comparison of the frequency responses and unit step responses are shown in Figs.(3) and (4).

The values of ISE of the unit step output response in cases (a) and (b) are 2.0545×10^{-9} and 5.794×10^{-6} respectively.

5-CONCLUSION

The proposed phase approximation method gives very satisfactory results. The various cases considered, indicate that the obtained reduced model is always a minimum phase one, and consequently the resulting amplitude frequency response is in good agreement with that of the unreduced system over the workable frequency range. Again, results show that both the reduced and optimum phase approximated reduced models are virtually similar to each other, since there is no significant zero or pole displacements between the two models.

Results obtained, indicate how close is the agreement between both the frequency, and step responses of the reduced and optimum models to their counterparts of the unreduced models. Comparing with the already existing reduction methods, it is clear that the proposed method compares favourably with them without the need of lengthy minimization procedures.

REFERENCES

[1] E.J.Davison, "A Method of Simplifying Linear Dynamic Systems", IEEE Trans. Automatic Cont., vol. AC-11, pp. 93-101, 1966.

[2] M.R.Chidambara and E.J.Davison, "On A Method for Simplifying Linear Dynamic

Systems", IEEE Trans. Automatic Cont., vol. AC-12, pp. 119-121, 1967.

[3] A.Kuppurajulu and S.Elangovan, "System Analysis by Simplified Models" IEEE Trans. Automatic Cont. vol. AC-15, pp. 234-237, 1970

[4] M.Arumugam and M.Ramamoorty, "A Method of Simplifying Large Dynamic Systems", Int. J.Control, vol. 17, pp. 1129-1135, 1973.

[5] C.F.Chen and L.S.Shieh, "A Novel Approach To Linear Model Simplification", Int. J.Control, vol. 8, pp. 561-570, 1968.

[6] M.J.Boseley and F.P.Less, "A Survey of Simple Transfer Function Derivation from High Order State Variable Models", Automatica, vol. 8, pp. 765-775, 1972.

[7] T.C.Hsia, "On The Simplification of Linear Systems", IEEE Trans. Automatic Cont., vol. AC-17, pp. 372-374, 1972.

[8] M.F.Hutton and B.Friedland, "Routh Approximation for Reducing Order of Linear, Time-Invariant Systems", IEEE Trans. Automatic Cont., vol. AC-20, pp. 329-337, 1975.

[9] A.S.S.R.Reddy, "A Method for Frequency Domain Simplification of Transfer Functions", Int. J.Control, vol. 23, pp. 403-408, 1976.

[10] E.C.Levy, "Complex-Curve Fitting", IRE Trans. Automatic Cont., vol. AC-4, pp. 37-43, 1959.

[11] F.D.Galiana, "On The Approximation of Multiple Input-Multiple Output Constant Linear Systems", Int. J.Control, vol. 17, pp. 1313-1324, 1973.

[12] Gerd and Gerhard, "On Optimal Approximation of Higher Order Systems by Low Order Models", Int. J.Control, vol. 22, pp. 399-408, 1975.

[13] A.I.Saleh, S.S.Bedair and W.H.Abou-Shohoud, "A Complex Frequency Approach for Reduction of Higher Order Systems", 22 nd Midwest Symposium on Circuits and Systems, University of Pennsylvania-Philadelphia, pp. 253-256, June 1979.

[14] H.W.Bode, "Network Analysis and Feedback Amplifier Design", D.Van Nostrand, New York, Ch.14 and 15, 1945.

[15] F.B.Hilderbrand, "Introduction to Numerical Analysis", McGraw-Hill, New York, Ch.9, 1956.

[16] M.F.Fahmy, "Transfer Functions With Arbitrary Phase Characteristics", Circuit Theory and Applications, vol. 7, pp. 21-29, 1979.

[17] E.Guillmin,"Synthesis of Passive Network, J.Wiley, New York, Ch. 8, 1957.

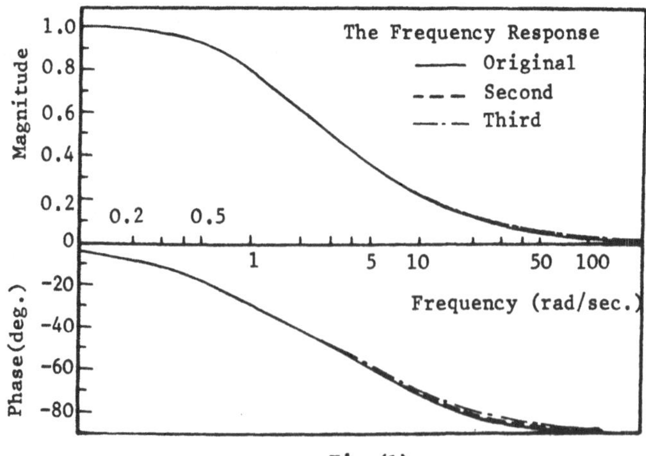

Fig.(1)

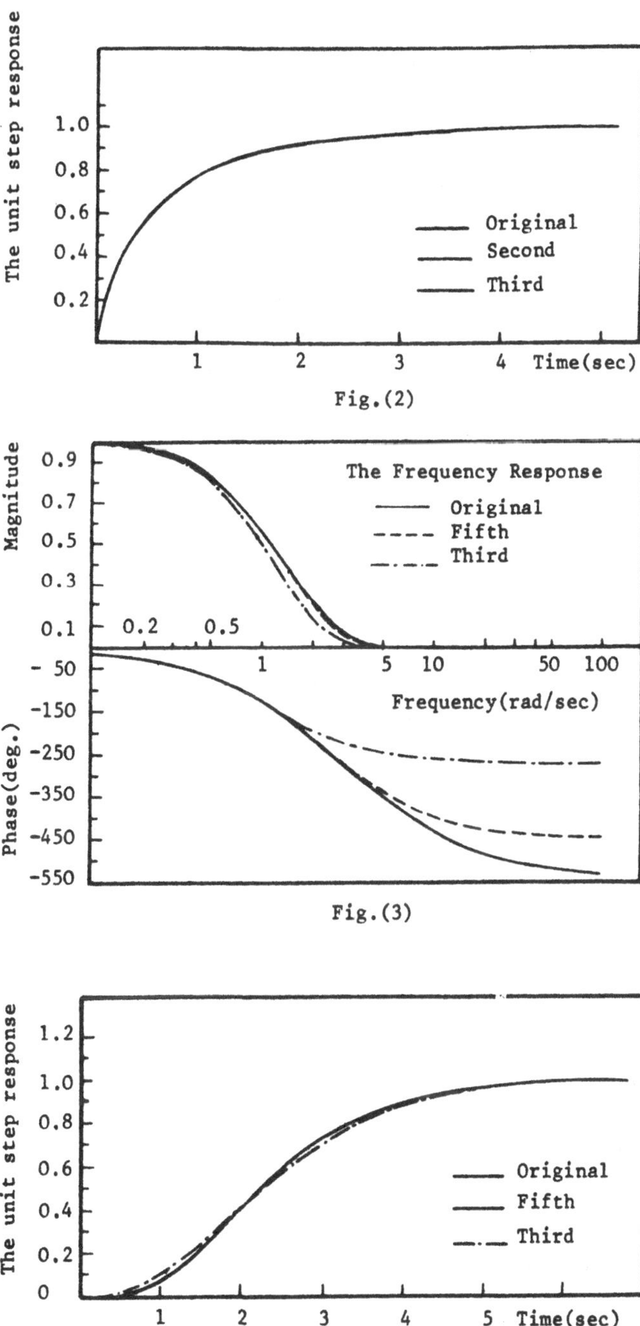

Fig.(2)

Fig.(3)

Fig.(4)

CONTROL APPLICATIONS IN ANALYZING TRANSPORTATION SYSTEM PERFORMANCE UNDER DYNAMIC CONSTRAINTS

Yorgos J. Stephanedes
Department of Civil and Mineral Engineering
University of Minnesota
Minneapolis, Minnesota 55455
U.S.A.

ABSTRACT

A feedback dynamic model of transportation systems is presented. The dynamic method used emphasizes the long-term, delayed causal mechanisms that determine transportation demand, transportation supply and transportation resources over time. The model uses demographic and socioeconomic data and transportation system characteristics as inputs and evaluates alternative transportation management policies by analyzing their impacts on transportation system performance through time. Control variables available to the management include transit fare and transit frequency, and are subject to fuel availability and price. Performance indeces considered minimize energy consumption and passenger waiting time, and maximize load factors and system revenue.

1. INTRODUCTION

Existing models of transportation demand and supply are aggregate in nature, are based on deterministic components or lead to static estimation. Most do not consider delayed feedback effects, have not been built on a priori causal relations, and are not consistent. As a result, such models can be used for explanatory analysis, but not for normative purposes.[1]

Recent work by two research teams has developed a purely dynamic representation for the transportation system. However, the first approach[2] deals almost exclusively with demand and simply assumes that transit supply tends to adjust to that demand. As a result, it cannot be used for transit scheduling determination. The second approach[3,4,5] has been applied to both rural and metropolitan areas, and models the demand-supply interactions in detail. In addition, it takes system finances and energy consumption into account. It is used to determine the expected impact of several management policies on transit performance with scheduling and pricing policies being only a subset of the scenarios analyzed. Owing to system complexity, simulation is used throughout. Under certain simplifying assumptions, however, the dynamic model structure can be described by a set of nonlinear differential equations, which may be solved analytically.

2. DYNAMIC MODEL STRUCTURE

The dynamic model structure is broken down into five sectors, each characterized by a set of differential equations: (a) Demand, (b) Service, (c) Capital Acquisition, (d) Financial, and (e) Energy. Each sector is segmented in terms of trip purpose, trip destination, travel mode availability and travel mode choice. Owing to space limitations, only the transit demand and supply sectors are described in greater detail in this paper.

Demand for transit depends on the transportation system characteristics, and the socioeconomic variables characterizing the population in the service area as described by a logit behavioral travel-demand estimation model[6,7] that has been tested and shown to be transferable to areas of differing characteristics.[8,9] From the estimated choice probabilities, weekly ridership and transit frequency are found using the following equations:

$$\dot{D} = -\frac{D}{k_1} + \frac{1}{k_3+k_4\exp(v)} \qquad (1)$$

$$\dot{F} = \frac{u_2}{k_2}\, F \qquad (2)$$

where:

$$v = -k_5u_1+k_6/F+k_7D/F \qquad (3)$$

$$u_2 = u_3-u_4\exp(-k_8u_5D/F) \qquad (4)$$

and:

D, weekly ridership

F, transit frequency in hr^{-1}

u_i, i=1 to 4, control variables

k_1, k_2, time delays

k_j, j=3 to 8, positive constants

3. PRELIMINARY RESULTS

System equations (1-4) were solved at equilibrium and the sensitivity of the state variables D and F to the system controls was examined. System controls include u_1, the transit fare and u_2, the transit frequency change multiplier. In turn, u_2 is a function of u_3, frequency change when demand is much larger than transit capacity; u_3-u_4, frequency change at zero demand; and u_5, management service constant. A preliminary analysis assumed u_3 and u_4 to be known and constant. From the equilibrium solution, u_5 was then determined.

Sensitivity of system performance to controls u_1 and u_5 was examined for transit systems in urban and rural areas. For example, it was found that, as u_5 increases, it becomes decreasingly effective. Similarly, transit fare u_1 does not influence modal choice once it reaches below a certain value. Finally, the transit system is not stable at high fares and low u_5 values.

In a different experiment, it was determined that control over system behavior improves dramatically when $u_2=u_2(D,\dot{D})$. In particular, it was found that this strategy contributes in keeping performance indicators within a narrow range. In addition, the strategy improved system performance. For example, it decreased non-capital net cost per vehicle mile by 30% while keeping service frequency 60% higher.

Current investigations are being conducted toward optimizing an index of interest to transportation decision makers under certain constraints. Performance indeces considered minimize energy consumption and passenger waiting time, and maximize load factors and system revenue. For example, maximizing efficient utilization of transit service can be expressed by maximizing the load factor:

$$\text{Maximize:} \quad \frac{D}{F}$$

with dynamic equations (1) and (2) and constraints:

$$\frac{u_1 D}{q_1 F + q_2} \geqq P \tag{5}$$

$$0 \leqq D \leqq c_1 \tag{6}$$

$$c_2 \leqq F \leqq c_3 \tag{7}$$

where:

P, minimum operating ratio allowed

$q_1(t)$, $q_2(t)$, energy and capital unit costs

c_1, c_2, c_3, positive constants

Constraints (6) and (7) are imposed by the total traveler population and vehicle characteristics respectively. Constraint (5) is imposed by system management or government agencies depending on whether the transit operation is private or public and requires a certain return on funds invested in the system. Functions $q_1(t)$ and $q_2(t)$ reflect equipment and operating costs and depend on the state of the economy and the cost of energy. Additional travel time and money budget constraints[10] may also be considered.

State equations (1) and (2) admit at most two solutions. With all constants known, and assuming capital and energy costs $q_1(t)$ and $q_2(t)$ to be constant, these solutions were found for various transportation system types. System controls are now sought such that orbits maximizing the given objective under the constraints can be chosen across time. To be sure, the complete dynamic model is characterized by a larger[4] set of equations describing all sectors of the transportation system. Solution to the complete set of equations and constraints is the goal of research currently conducted by this author.

4. ACKNOWLEDGEMENT

This work was partially supported by the U.S. Department of Transportation, OST Program of University Research contract DOT-OS-80006 to Dartmouth College and by the Urban Mass Transportation Administration contract DOT-MN-11-0004 to the University of Minnesota.

5. REFERENCES

1. Stephanedes, Y.J., "A Feedback Dynamic Model for Transportation Systems", G.J. Savage and P.H. Roe, eds., Large Engineering Systems, Waterloo, Ontario, Canada, Sanford Education Press, 1978.
2. Deneubourg, J.L. and A. de Palma, "Dynamic Models of Competition Between Transportation Modes", Envir. and Plann., vol. 11, pp. 665-673, 1979.
3. Stephanedes, Y.J., Performance Indicators and Policy Evaluation in Rural Transit, Ph.D. Dissertation, Thayer School of Engineering, Dartmouth College, Hanover, NH, 1978.
4. Adler, T.J., S.R. Stearns and Y.J. Stephanedes, "Techniques for Analyzing the Performance of Rural Transit Systems", Final Report, DOT-RSPA-DPB-50/80/23, U.S. Dept. of Transportation, Office of University Research, Washington, DC, 1980.
5. Stephanedes, Y.J., P.G. Michalopoulos, H. Hanna and D. Gabriel, "Development and Implementation of Dynamic Methodologies for Evaluating Energy Conservation Strategies", Draft Final Report, DOT-MN-11-0004, U.S. Dept. of Transportation, UMTA, Washington, DC, 1981.
6. McFadden, D., "Conditional Logit Analysis of Qualitative Choice Behavior", Zarembka, P., ed., Frontiers in Econometrics, Academic Press, New York, NY, 1973.
7. Domencich, T.A. and D. McFadden, Urban Travel Demand, North-Holland Publ. Co., Amsterdam, Netherlands, 1975.
8. Ben-Akiva, M. and T. Atherton, "Transferability and Updating of Disaggregate Travel Demand Models", Center for Transportation Studies Report #76-2, MIT,

Cambridge, MA, 1976.

9. Stephanedes, Y.J., "Improved Demand Estimation for Rural Work Trips", Transportation Research Record, Transportation Research Board, Washington, DC, forthcoming 1982.

10. Zahavi, Y., "The UMOT Project", Final Report, DOT-RSPA-DPB-20-79-3, U.S. Dept. of Transportation, Washington, DC, 1979.

TO THE TEMPORAL AGGREGATION IN DISCRETE DYNAMICAL SYSTEMS

H.-J. Werner
Institute for Econometrics and
Operations Research, Econometrics Unit
University of Bonn

D-53oo Bonn 1, W.Germany

1. Introduction.

The transformation from a (true) dynamic structural econometric model into a statistically observable system involves, in general, a temporary aggregation. In economics, many structural behavioural equations point to a time interval of 1 week, 1 month, or so. Nevertheless, observational data are often available only on a quarterly or on a yearly basis. Then a temporary aggregation becomes necessary.

In this paper we consider a linear dynamic system of n equations with n endogeneous variables in reduced form. The microsystem may thus be written in the form

$$(1) \qquad y(t) = \sum_{i=1}^{H} A_i y(t-i) + w(t), \quad t \in Z,$$

where the n×n matrices A_i contain the structural parameters, $w = (w(t))$, $t \in Z$, is an autonomous input process, H is the order of the autoregressive part (i.e. $A_H \neq 0$), and t measures time in some microperiods (e.g. three month periods). We further suppose that the available data are in some other - e.g. annual - form, so that the analyst prefers to replace the microsystem (1) by a macrosystem which contains only available parts of y on both sides of the equality sign. If $Y = (Y(\theta)) = (y(m\theta))$, $\theta \in Z$, with $m \in N$ fixed, is the observable (available) variable, then it is clear that by eliminating the nonobservable parts of y in (1) we can obtain without difficulty an exact macrosystem of the form

$$(2) \qquad \tilde{A}_0 Y(\theta) = \sum_{s=0}^{H} \tilde{A}_i Y(\theta-i) + W(\theta), \quad \theta \in Z,$$

where

$$(3) \qquad W(\theta) = \sum_{s=0}^{(m-1)H} B_s w(m\theta-s)$$

and the matrices B_i and $\tilde{A}_i$ are functions of the matrices A_j ($B_0 = \tilde{A}_0$).

The question we will treat in this paper is the following one:

(P1): Can we find, for each $n \in N$, an exact <u>nonsingular</u> macrosystem (not necessarily of order H of the autoregressive part)? The word <u>nonsingular</u> is used here to indicate that $\tilde{A}_0$ is nonsingular, so that such an associated macrosystem can be written in reduced form. (Without loss of generality we can hence require that $\widetilde{A}_o = I$).

Though the problem of temporal aggregation is known in the econometric literature for a long time (see e.g. Theil (1955), Mundlak (1961), Telser (1967), Sims (1971), Geweke (1978), Schönfeld (198o), to mention only a few) our above question (P1) has yet been studied only for the single autoregressive process (i.e. n=1) by Telser (1967), whereas an analysis for the above multivariate case (i.e. n>1) seems still to be lacking. Our treatise is purely algebraic, constructive and different from Telser's approach. In section 2 we deal with the solution of a matrix problem (P2) which is strongly related to our above problem (P1). Section 3 is to state precisely our solution of (P1). As a by-product we obtain a least upper bound for the order of the autoregressive part in the associated nonsingular macrosystem at the worst. In section 4 we conclude this paper with a program for computing an associated nonsingular macrosystem being of minimal order (with respect to the autoregressive part) in the class of all the associated nonsingular macrosystems.

2. The Matrix Problem.

Using the lag operator λ defined as

$$\lambda^k y(t) = y(t-k) , \qquad k = \dots,-1,o,1,\dots ,$$

one can write the reduced form (1) as

$$(1)' \qquad 0 = A(\lambda)y(t) + w(t)$$

with the $n \times n$ polynomial matrix (lambda matrix)

$$A(\lambda) := \sum_{i=o}^{H} A_i \lambda^i$$

where

$$A_o := -I .$$

If we now define

$$f = f(m,H) := \min \{ \nu \in N \mid \nu m \geq H \} ,$$
$$K = K(m,H) := fm - H ,$$
$$g = g(m,H) := H - f ,$$

and introduce for any $\ell \in N_o$ the classes

$$\Pi_\ell := \{ B(\lambda) = \sum_{i=o}^{K+\ell m} B_i \lambda^i \mid B_i \in R^{n\times n}, \; B_o = I \} ,$$

$$\Gamma_\ell := \{ C(\lambda) = \sum_{i=o}^{K+H+\ell m} C_i \lambda^i \mid C_i \in R^{n\times n},$$
$$\forall\, j \in \{o,1,\dots,K+H+\ell m\} \setminus \{o,m,\dots,(f+\ell)m\}: \; C_j = 0 \}$$

we get at once as simple propositions:

$$o \leq K \leq m-1 , \qquad g \geq o .$$

Further it is now evident that problem (P1) is equivalent to the following matrix problem:

(P2): Suppose that $A(\lambda)$ is any given lambda matrix of order $n\times n$ having degree H. Let $A_o := -I$. The question is: Can we find an $\ell \in N_o$ such that

$$B(\lambda)A(\lambda) \in \Gamma_\ell$$

holds for some $B(\lambda) \in \Pi_\ell$?

If the answer to this problem is in the affirmative a closely related question is the following one: What is the smallest degree of $B(\lambda)$ at the worst?

Before handling the first question it is convenient to introduce some further notation. We introduce:

$$\Omega := \{1,\ldots,K+\ell m\} \setminus \{m,2m,\ldots,\ell m\} \quad ,$$
$$W := \{1,2,\ldots,H\} \setminus \{m-K,2m-K,\ldots,fm-K=H\} \quad ,$$

and define for $(i,j,w) \in N_o \times \{o,1,\ldots,m-1\} \times \{1,\ldots,H\}$:

$$G(im+j,w) := \sum_{s=o}^{im+j} B_{im+j-s}A_{w+s} \quad . \tag{4}$$

Then we easily obtain as a first result:

<u>Lemma 1.</u> Let $\ell \in N_o$ be given. Then the following condition is necessary and sufficient for the relations

$$B(\lambda)A(\lambda) \in \Gamma_\ell , \quad B(\lambda) \in \Pi_\ell$$

to have a common solution:

The equations

$$B_i = \sum_{s=o}^{i-1} B_s A_{i-s} \qquad (i \in \Omega) , \tag{5}$$

$$0 = G(\ell m+K,w) \qquad (w \in W) \tag{6}$$

all simultaneously hold for some sequence $B_o=I, B_m, \ldots, B_{\ell m}$.

Observe that the conditions (5) can always be satisfied by defining the matrices on the left hand side by the corresponding terms on the right hand side of the equality sign. By doing so, we can choose freely only the matrices $B_m, \ldots, B_{\ell m}$ (recall that $B_o = I$) but we have to choose them, of course, such that (6) is fulfilled. Hence we now have to treat the following question: When considering the equations (5) as defining relations for B_i $(i \in \Omega)$, can we construct a series of matrices $B_m, \ldots, B_{\ell m}$ solving (6)? That is, we have to investigate the solvability of (6).

Before proceeding it seems useful to determine mathematical representations of the matrices B_i $(i \in \Omega)$ and of $G(im+j,w)$ in (4) only in terms of $B_o = I$ and the above mentioned "free input matrices" $B_m, B_{2m}, \dots$.

Because of page limitation it is not possible to present proofs of the following two results; the interested reader is refered to Werner (198o).

Lemma 2. If we define matrices $R_{i,j} \in R^{n\times n}$ by

$$R_{o,o} := I ,$$

$$R_{i,o} := 0 \quad \text{for} \quad i \in N ,$$

$$R_{i,j} := \sum_{r=1}^{j-1} R_{i,r}A_{j-r} + \sum_{v=o}^{i-1} \sum_{r=1}^{m-1} R_{v,r}A_{(i-v)m+j-r} + A_{im+j}$$

$$\text{for all} \quad (i,j) \in N_o \times \{1,\dots,m-1\} ,$$

then we get for all $(i,j) \in N_o \times \{1,\dots,m-1\}$:

$$B_{im+j} = \sum_{r=o}^{i} B_{(i-r)m}R_{r,j} . \tag{7}$$

Inserting (7) in (4) yields a similiar representation for the matrix $G(im+j,w)$. Precisely we obtain:

Lemma 3. If we define matrices $S_{i,j,w} \in R^{n\times n}$ by

$$S_{i,j,w} := \sum_{r=1}^{j} R_{i,r}A_{w+j-r} + \sum_{v=o}^{i-1} \sum_{r=1}^{m-1} R_{v,r}A_{w+(i-v)m+j-r} + A_{w+im+j} ,$$

then we get

$$G(im+j,w) = \sum_{r=o}^{i} B_{(i-r)m}S_{r,j,w} \tag{8}$$

for all $(i,j,w) \in N_o \times \{o,1,\dots,m-1\} \times \{1,\dots,H\}$.

The definitions of the functions $R_{i,j}$ and $S_{i,j,w}$ may (at a first sight) seem to be quite complicated but it should be recognized that the representations (7) and (8) then merely reflects some intuitively clear facts: So you may easily deduce from (5) that for instance B_o has the same influence on B_j as B_m has on B_{m+j}. And second, you will certainly feel that the pure effects of, for instance, B_o and B_m on B_{m+j} can be separated. Furthermore, it should be noted that although the definitions of $R_{i,j}$ and $S_{i,j,w}$ do not allow for simple hand computations, their recursive structure is ideally suited to machine calculation.

Lemma 3 enables us to give the following answers to our matrix problem (P2):

Theorem 4. Let $A(\lambda)$ be an n×n polynomial matrix of degree H with $A_o := -I$. Further define

$$S_{r,K} := (S_{r,K,w})_{w\in W}$$

and let $Z(S_{r,K})$ denote the row space of the block matrix $S_{r,K}$. Then we get: The relations

$$B(\lambda)A(\lambda) \in \Gamma_\ell , \quad B(\lambda) \in \Pi_\ell$$

are simultaneously consistent if and only if

$$S_{\ell,K} \equiv 0 \qquad \text{for} \quad \ell = o$$

or

$$Z(S_{\ell,K}) \subset \sum_{i=o}^{\ell-1} Z(S_{i,K}) \qquad \text{otherwise.}$$

Proof. Since by Lemma 3 (5) can be rewritten as

$$\sum_{r=o}^{\ell} B_{(\ell-r)m} S_{r,K,w} = 0 \qquad \text{for all} \quad w \in W$$

or as

$$(B_{\ell m} \mid \dots \mid B_m)(S_{r,K})_{r=o,1,\dots,\ell-1} = -S_{\ell,K} ,$$

the assertion should be obvious. □

Theorem 5. Let $A(\lambda)$ be any given n×n polynomial matrix of degree H and let $A_0 := -I$. Then we have:
There always exists for $\ell := ng$ a lambda matrix $B(\lambda) \in \Pi_\ell$ with $B(\lambda)A(\lambda) \in \Gamma_\ell$.

Proof. If

$$Z(S_{j,K}) \subset \sum_{i=o}^{j-1} Z(S_{i,K}) \qquad \text{is true for some} \quad j < \ell ,$$

then the assertion follows from Theorem 4. Otherwise we certainly obtain the following inequality chain

$$\dim \sum_{i=o}^{\ell-1} Z(S_{i,K}) \geq 1 + \dim \sum_{i=o}^{\ell-2} Z(S_{i,K}) \geq \dots \geq \ell-1 + \dim Z(S_{o,K}) \geq \ell .$$

Since $S_{i,K} \in R^{n\times\ell}$ the assertion now trivially holds for $j = \ell$. □

3. The solution of problem (P1).

Since it turned out that (P2) is equivalent to (P1) we now readily obtain the following solution to (P1):

Theorem 6. Given a microsystem in the form (1) composed of n equations. Let H be the order of this microsystem and suppose that data are only available to the macrotimes o, ± m, ±2m, ... ($m \in N$ fixed). Then there always exists an associated nonsingular macrosystem of an order $\leq H + g(n-1)$.

Proof. We directly get from Theorem 5 that the smallest degree is (f+gn) at the worst. Since g is defined as $g := H-f$ the proof is complete. □

It is worth noting that Theorem 6 reduces for n=1 to the known result of Telser (1967).

4. Program.

We conclude this paper by introducing (without proof) a program for computing an associated nonsingular macrosystem being of minimal order. For a more detailed discussion of the procedure given below we refer the reader again to Werner (1980).

For the rest of the paper let us define

$$S_i := (S_{r,K})_{r=o,\dots,i}$$

and

$$P_i := S_i^+ S_i$$

where S_i^+ denotes the Moore-Penrose Inverse of S_i (see e.g. Ben-Israel/Greville (1974) or Werner (1979)).

Then a program for evaluating the smallest order of all the associated nonsingular macrosystems is as follows:

STEP 0: Compute $S_{o,K}$.
IF $S_{o,K} = 0$, then STOP → f(m,H) is the smallest possible order.
OTHERWISE, GO TO STEP 1.

.

.

.

STEP i: Compute $S_{i,K}$ and P_{i-1} .
IF $S_{i,K}P_{i-1} = S_{i,K}$, then STOP → f(m,H)+i is the smallest possible order.
OTHERWISE, GO TO STEP i+1.

.

.

By Theorem 6, it is clear that the above program stops at least after STEP g(m,H)n. When the procedure stops in STEP i, then we know that all the associated nonsingular macrosystems have an order greater or equal to (f(m,H) + i). Further it can be shown that an associated nonsingular macrosystem being of order f+i (in the AR-part) is then given as

$$(9) \qquad Y(\theta) = \sum_{\tau=1}^{f+i} \tilde{A}_\tau Y(\theta-\tau) + W(\theta), \quad \theta \in Z,$$

with

$$(10) \qquad W(\theta) := \sum_{\tau=0}^{im+k} B_\tau w(m\theta-\tau), \quad \theta \in Z,$$

where

$$(11) \qquad (B_{im} \vdots \; . \; . \; . \; \vdots B_m) := -S_{i,K} S^+_{i-1} \quad ,$$

respectively

$$(12) \qquad B_\tau := \sum_{s=o}^{\tau-1} B_s A_{\tau-s} \qquad \text{otherwise,}$$

and

$$(13) \qquad \tilde{A}_\tau := \sum_{r=o}^{im+K} B_r A_{\tau m-r} \qquad \text{for} \quad \tau = 1,\dots,f+i \quad ,$$

or equivalently,

$$(14) \qquad \tilde{A}_\tau = \sum_{r=o}^{i} B_{rm} \{ \sum_{j=o}^{m-1} \sum_{v=r}^{i-1} R_{v-r,j} A_{(\tau-v)m-j} + \sum_{s=o}^{K} R_{i-r,s} A_{(\tau-i)m-s} \}$$

$$+ B_{im} \sum_{s=o}^{K} R_{o,s} A_{(\tau-i)m-s} \quad .$$

For the sake of simplicity we have thereby put as in section 2:

$$A_j := 0 \qquad \text{for all} \quad j \in Z \setminus \{o,\dots,H\}.$$

This work has been supported by the Deutsche Forschungsgemeinschaft at the University of Bonn.

References.

1 Ben-Israel, A./ Greville, T.N.E. (1974). Generalized Inverses: Theory and Applications. Wiley-Interscience Publication, New York.

2 Geweke, J. (1978). Temporal Aggregation in the Multiple Regression Model, Econometrica 46, pp. 643-661.

3 Mundlak, Y. (1961). Aggregation over Time in Distributed Lag Models. International Economic Review 2, pp. 154-163.

4 Schönfeld, P. (198o). Fehlspezifikationen dynamischer Modelle durch temporale Aggregation. In: Empirische Wirtschaftsforschung, Festschrift für R.Krengel, ed.: J.Frohn/ R.Stäglin, pp. 253-266. Duncker & Humblot, Berlin.

5 Sims, Chr. A. (1971). Discrete Approximation to Continuous Time Distributed Lags in Econometrics. Econometrica 39, pp. 545-563.

6 Telser, L. (1967). Discrete Samples and Moving Sums in Stationary Stochastic Processes. Journal of the American Statistical Association 62, pp. 484-499.

7 Theil, H. (1955). Linear Aggregation of Economic Relations. North-Holland Publication, Amsterdam.

8 Werner, H.J. (1979). On the Matrix Monotonicity of Generalized Inversion. Linear Algebra and its Applications 27, pp. 141-145.

9 Werner, H.J. (198o). Zur temporalen Aggregation diskreter dynamischer Systeme. Discussion paper, University of Bonn.

GRAPHICAL TECHNIQUES USED FOR A DYNAMIC CHEMICAL PROCESS SIMULATION

Keiji Yajima, Junkichi Tsunekawa
Computation Center
Institute of JUSE
(Japanese Union of
Scientists and Engineers)
Tokyo 151/Japan

1. Introduction

A model of dynamic system which consists of algebraic and ordinary differential equations can be handled effectively by graphical techniques. By eliminating unnecessary variables, a size of simultaneous equations can be reduced and the diagnosis to detect the inadequacy of variable dependencies can be realized. These techniques that were used for static simulation program JUSE - GIFS and reported in the reference [1] have been expanded into dynamic simulation program DPS.

2. Equation handling in DPS

In DPS, a process is represented in terms of connecting flows and equipments. A flow includes streams representing material flow through pipes, and heat flow and information flow, representing flow of heat and information respectively. The equipment on the other hand consists of elements or the smallest mathematical models.

FORTRAN-like statements provide the function description of the element. The mathematical descriptions consist of the following equations:

(a) Algebraic equation

$$y = f(x_1, x_2, \ldots, x_n), \tag{1}$$

(b) Ordinary differential equation

$$y = \text{integral } (y'). \tag{2}$$

This signifies determination of the variable y by solving the orinary differential equation $y'=dy/dt$ in respect of time, namely y is an integral variable and y' is a differential variable.

(c) Delay relation

$$y = x(t - \tau), \tag{3}$$

where τ stands for the dalay time, and we may call y as delayed variable and x as delay variable.

DPS system constructs a simultaneous algebraic and ordinary differential equations in accordance with the following input data sections.

(a) Topology section

The whole structure of the process is described in this section by designating the interconnections of flows and elements.

(b) Variable section

The values and the characterizations of variables of following categories are designated in this section:

* known values of variables to be specified
* initial values of integral variables
* initial guess of variables to solve nonlinear equations
* candidates of tearing variables which relate to loop in a process.

The terminating condition of simulation run can be specified by values of variables.

(c) Output section

The variable names which are required to be output are described in this section. These variables are called required variables.

(d) Operation section

In the chemical processes there are operations such as open and close of valves, changes of valves etc, and these invoke events. This section serves to provide all event information. The conditions for starting and terminating the execution of the event are also designated.

In DPS system the necessary equations in the model are selected and the order of computation sequence is constructed internally. Thus for DPS users it is not necessary to consider the computation sequence, by contrast in the so-called modular type simulators users are not to be free from the sorting of equations in advance.

3. Graphical representation of equations

Dependence relation of variables in algebraic equation can be represented by graphical form. This graph is called static in contrast with dynamic graph mentioned later. In equation (1), it is possible to assign the vertices corresponding to all variables (y, x_1, x_2, ..., x_n) and to these vertices we may build the directed arcs from right-hand side variables (x_1, x_2, ..., x_n) to left-hand side variable y

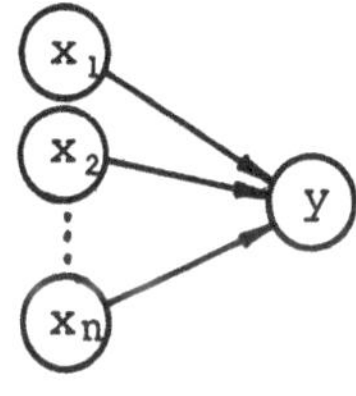

Fig. 1 Graph of algebraic equation

(fig. 1). It is possible to construct a graph with no loss of generality in such way that the variables in left-hand side of equations appear only once. Under this assumption, the graph may have the vertices corresponding one-to-one to the variables.

The variables are subject to the following internal classification according to their individual graphical characteristics resulting from the specification of the input data.

(a) S (set)-type variables

These represent variables which are assigned to specific values. These variables cause no computation. All equations in which an S-type variable appears on the left-hand side are disregarded.

(b) G (given)-type variables

A variable assigned a specific value of left-hand side will form a G-type variable, unlike in the case of the S-type variable, the values of the right hand side variables will be suitably adjusted so as to preserve the equality of the two sides. If a variable of this nature appears on the right-hand of a relation, it will be treated as an S-type variables.

(c) A (assumed)-type variables

An initial vertex on the graph (i.e. one not appearing on the left-hand side of an equation) without any value assigned to it, will be referred to as an A-type variable. Since the value of this variable is unknown, it is regarded as dependent variable which is adjusted according to the value of connecting G-type variable.

(d) DD (divided)-type variables

In a graph representing a directed cycle, the value of a variable in it may be assumed, and all the variables including itself may be evaluated on the basis of this value till the computed and assumed values of the variable first taken are reached to equal. A variable of this type is referred to as a DD-type variable. Unless the input data specifies which variable is to be treated as the DD-type variable, the system itself will internally determine it.

(e) X-type variables

All variables other than those of internal types (a) - (d) constitute X-type variables. S-type, G-type variables are indicated externally by input data, but other type variables can be determined automatically. Candidates of DD-type variable may be assigned externally.

As X-type variables are successively evaluated by substitution, provided the following reduced system of equations consists of A-type, G-type and DD-type variables, is constructed:

$$c_1 = F_1(x_1, x_2, \ldots, z_1, z_2, \ldots)$$
$$c_2 = F_2(x_1, x_2, \ldots, z_1, z_2, \ldots)$$
$$\ldots \tag{4}$$
$$z_1 = G_1(x_1, x_2, \ldots, z_1, z_2, \ldots)$$
$$z_2 = G_2(x_1, x_2, \ldots, z_1, z_2, \ldots)$$
$$\ldots$$

where $c_1, c_2, \ldots, x_1, x_2, \ldots$ and $z_1, z_2, \ldots$ represent G-, A- and DD-type variables respectively.

A system of equations including integral and delay functions is represented by dynamic graph. A dynamic graph is defined at each time step of integration and to these graphs direct arcs are attached from the previous static graph to succeeding static graph.

In an explicit method to solve the ordinary differential equations, an integral value is evaluated by the corresponding differential value and integral value at previous time. As the values of variable at previous time were already evaluated, integral variables on a static graph are treated as S-type variables.

In an implicit method for solving ordinary differential equations, an integral variable is evaluated in consideration of the corresponding differential values at present time, and integral and differential values at previous time. In this method, the static graph had a cyclic directed path including the integral variable and differential variable, namely this cosists the algebraic equations. In static graph, delayed variables are treated as S-type.

To obtain a steady state of process, two kinds of approaches are available. One is a dynamic simulation approach by which integration continues till all values of differential variables reach to almost steady state. The other is a static simulation approach by solving algebraic equation.

A differential variable can be assigned to G-type (or S-type according to an appearence of that variable in the right-hand side) and to value of 0, and an integral variable is assigned to A-type variable (there is a case in which user should input S-type specification).

For examle, if the mathematical model consists of following equations

$$x_1 = x_1(t) \qquad (x_1\text{: delay variable})$$
$$x_2 = x_1(t - \tau) \qquad (x_2\text{: delayed variable})$$
$$x_3 = x_2 - x_4 \qquad (x_3\text{: differential valiable})$$
$$x_4 = f(x_5) \tag{5}$$
$$x_5 = \text{integral } (x_3), \qquad (x_5\text{: integral variable})$$

then fig. 2 shows dynamic graphs applying Euler and backward Euler method and fig. 3 shows static graphs for Euler, backward Euler method and static simulation for solving steady state.

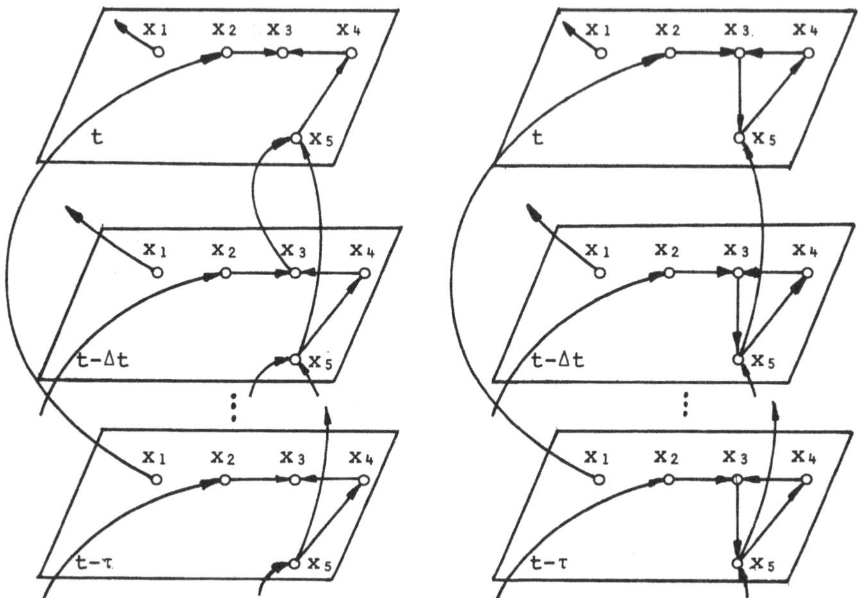

(a) Euler method (b) Backward Euler method

Fig. 2 Dynamic graph of integration

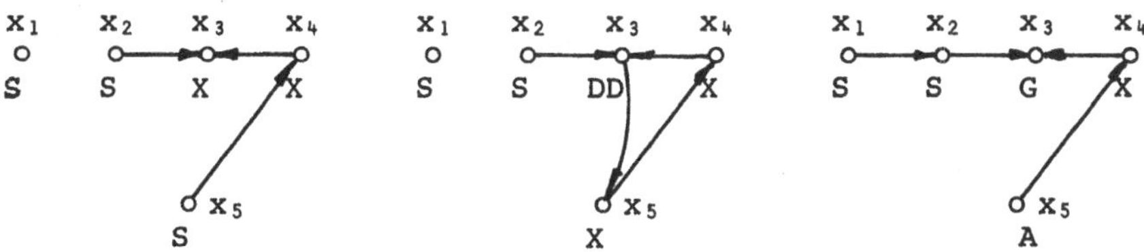

(a) Euler method (b) Backward Euler method (c) Solving steady state by algebraic equation

Fig. 3 Static graph of integration

4. Graph theoretic algorithms

(1) Elimination of equipments

Valve operation has an effect to the equipment functions now operating. The equations related to the equipments not operating are necessarily eliminated according to the operation data. This elimination is determined by using topological graph which represents the connection of equipments and flows.

(2) Elimination of variables

The variables required to be evaluated are called as output variables and those related to conditions of procedure termination, termed as, in either case, the R (required)-type variables. In other words, all internally classified variables are also assigned to R or non R-type variables. The program preserves both the R-type variables and the variables necessary to compute them.

(3) Consistency of variables and equations in number

In reduced equations, the number of equations and the number of independent variables should be coincident. Moreover, a secondary condition that must be satisfied here for the system of equations to be generally solvable is given by the following theorem [2], where if the number of equations and the number of unknown variables in a system are both n, and the number of equations involving any number k of variables ($n \geq k \geq 1$) is greater than or equal to k, the system is said to be generally solvable.

Theorem: Draw as many vertex-disjoint directed paths as possible from the initial A-vertices to the terminal G-vertices in the original graph. Then, "each A-vertex being connected to a G-vertex and each G-vertex being connected from an A-vertex by a path" constitutes a necessary and sufficient condition for the system of equation to be generally solvable.

The construction of a set of paths with the properties required in the above theorem may be efficiently carried out by means of the network-flow algorithm [3]. Thus, we may consider the problem as one of shipping along the paths in the graph as much commodity as possible from the A-vertices to the G-vertices, where each vertex as well as each arc is assigned a unit capacity. DPS system checks this theorem condition.

(4) Selection of DD-type variables

Some vertices in the cycles must be split up. Wherever necessary, variables, if any, specified as divisible in the input data are assigned to a high priority. Furthermore, if cycles occur in the graph, DPS system assigns variables of type DD. It is desirable that the number of DD-type variables is at its smallest, but no efficient algorithm is known to achieve this.

(5) Block-triangularization

If a reduced graph corresponding to reduced system of equations can be divided into several subgraphs by the block-triangularization algorithm, the reduced equations can be succesively solved with less calculations in an appropriate order. The solution of these equations of smaller size allows computations to achieve economy in both time and memory utilization. In the DPS system following decomposition algorithm is carried out for every connected component of the graph. (a) A reduced graph is constructed so that the vertices in it are constituted of the left- and right-hand side variables (a DD-type variable is interpreted as constituting two vertices of type A- and G-respectively), and a directed arc is set up from the A-vertex to the G-vertex if a path exists in the original graph from the A- to the G-vertices. (b) A one-to-one correspondence is established between the A-vertices and the G-vertices through directed arcs on the reduced graph. The algorithm here is familiarly referred to in graph theory as matching algorithm. (c) The reduced graph is reconstucted by drawing directed arcs from the G- to the corresponding A-vertices. (d) If a reduced graph has cycles in it, all G-vertices in such cycles will be taken as belonging to the same block. (e) Computation sequence is established by taking the order of the directed paths (higher to lower).

(6) Rearrangement of computation order

Differential equations are solved at the time intervals decided, but solution of algebraic equations is not needed on all occasions. To arrange these situations properly, it is natural to classify the variables as follows: (a) values of which are necessary at the initial point of time in an event, (b) necessary at every point in time, (c) required to be printed out in respect of prescribed time points, (d) necessary at the final point of time in an event.

We wish to express our deep appreciation to the members of Reserach Committee on DPS, in particular we thank to Prof. Masao Iri for his dedication to our project.

Reference

[1] M. Iri, J. Tsunekawa and K. Yajima, The Graphical Techniques Used for A Chemical Process Simulator "JUSE-GIFS", IFIP Proceedings (1971), Vol. 2, 1150-1155, North-Holland.

[2] M. Iri, J. Tsunekawa and K. Murota, Graph-Theoretic Approach to Large-Scale Systems of Equations, Research Memorandum RMI 81-05, Department of Mathematical Engineering and Instrumentation Physics, Faculty of Engineering, University of Tokyo, pp. 27, 1981.

[3] M. Iri, Network Flow, Transportation and Scheduling-Theory and Algorithms, Academic Press, New York, 1969.

SCHEDULING MAINTENANCE OPERATIONS WHICH CAUSE AGE-DEPENDENT FAILURE RATE CHANGES*

L. Shaw, B. Ebrahimian, J. Chan
Polytechnic Institute of New York
Brooklyn, NY 11201

ABSTRACT

We consider optimization of schedules for maintenance or repairs, in order to minimize long term average operating cost or to maximize availability. The novelty here is that the failure rate after a maintenance operation is a function of the system's previously expended lifetime. This generalizes earlier work by others on the simpler case where the future rate depends only on the number of previous repairs, but not on the times when they took place. The underlying lifetime distributions are assumed to have the Weibull form and two classes of maintenance strategies are considered.

The first case optimizes a set of successive maintenance intervals $T_1, T_2, \ldots, T_N$, and the number N, where a replacement by a new system is made at $t_N = \sum_{i=1}^{N} T_i$. We show among other results, that the optimal times T_i exist and are ordered as $T_1 \geq T_2 > \ldots \geq T_N$.

In the second case, the period of periodic maintenance is optimized numerically. The main contribution in that case is the formulation of the new failure rate model, and the efficient organization of the optimization calculations.

1. INTRODUCTION

This work extends the maintenance models and policy optimizations described in the references. In particular, the ideas in Nguyen and Murthy (1981) and Goldstein (1980) were strong influences.

We consider optimization of the schedule of times at which preventive maintenance will be carried out on a system subject to stochastic failures. Several different criteria are used to judge the quality of a schedule, including long-term average operating costs per unit time, and various definitions of availability. The effect of maintenance is represented as a reduction in the subsequent failure rate. One of our contributions is an approach for incorporating the influence of past operating time on the new failure rate function which applies after a maintenance operation.

A component or system with a failure time $\underset{\sim}{t}$ (random variable) can be described by its failure rate or hazard rate r(t): $r(t)dt = P[t < \underset{\sim}{t} \leq t + dt \mid \underset{\sim}{t} > t]$. Figure 1 shows two forms used later for the reduction in failure rate at maintenance times t_i. In case a), maintenance reduces $r(t_i)$ to zero, but subsequent growth rate is great-

* This work was partially supported by Grant No. N00014-75-C-0858 from the Office of Naval Research.

er than before the maintenance. That growth rate will be a function of the previous operating time t_i. In case b), the growth rate of r(t) is constant, but maintenance reduces r at t_i by an amount dependent on t_i.

We define $r_i(t)$ between maintenance times as: $r(t) = r_i(t - t_{i-1})$; and $t_{i-1} < t \le t_i$. We consider $r_i(t)$ which are increasing in t, with numerical results based on the Weibull forms $r_i(t) = \lambda_i t^\alpha$ or $r_i(t) = \lambda_1(t + t_{i-1})^\alpha - \Delta_i$

$$(\text{equivalent to } r(t) = \lambda_1 t^\alpha - \Delta_i) \tag{1}$$

These $r_i(t)$ are also increasing functions of the maintenance intervals $T_j = (t_j - t_{j-1})$ for $j < i$.

In some models we permit minor repairs when the system fails. After such a repair, the failure rate r(t) evolves as if no failure had occurred. This model is often used for complex equipment in which the next failure is unlikely to be related to the component which was just repaired. From this point of view, the number of failures and minor repairs between maintenance times, $N(t_{i-1}, t_i)$, has

$$E[N(t_{i-1}, t_i)] = \int_{t_{i-1}}^{t_i} r_i(t)\, dt \tag{2}$$

For both cases in Fig. 1, the system wears out ($r(t) \to \infty$ as $t \to \infty$) so complete replacement (renewal) is eventually required.

Section 2 considers a problem where minor repairs are permitted between maintenance times, and costs are assigned to renewal, minor repair and maintenance operations. With replacement at t_N, we seek the best N and $t_1, t_2, \dots, t_{N-1}$ to minimize the average cost per unit time when renewal cycles are repeated indefinitely. This is mathematically equivalent to maximizing the steady-state availability, by suitable reinterpretation of the costs.

Section 3 assumes that each failure requires a complete renewal, and that the maintenance times are equally spaced (often the case due to work schedules). Cycle availability A_c is defined as the fraction of time that the system is up (not in maintenance or replacement operations) during one renewal cycle. The maintenance period is sought to maximize $P[A_c > \delta]$, $0 < \delta < 1$.

2. SCHEDULES WITH MINOR REPAIRS

We assume that renewal occurs at t_N with a cost C_R; maintenance occurs at $T_1, T_2, \dots, T_{N-1}$ at a cost of C_M each time; and minor repairs occur at each failure with a cost C_r. The expected cost per renewal cycle is then

$$L(N, \bar{T}) = C_R + (N-1)\, C_M + C_r \sum_{i=1}^{N} \int_0^{T_j} r_j(t)\, dt \tag{3}$$

where $\bar{T} = (T_1, T_2, \ldots, T_N)$.

For these purposes, all three corrective operations are assumed to be instantaneous, corresponding to practical durations which are small compared to system uptime. The renewal cycle duration is then

$$D(N, \bar{T}) = t_N = \sum_1^N T_j \tag{4}$$

and the long-term average cost per renewal cycle is

$$C(N, \bar{T}) = L(N, \bar{T})/D(N, \bar{T}) \tag{5}$$

Theorem 1 If i) $r_i(t)$ is strictly increasing in $t, T_1, \ldots, T_{i-1}$

ii) $\dfrac{\partial^2 r_i(t)}{\partial T_m \, \partial T_n} \geq 0$ all i, t, m, n

Then the necessary conditions

$$L_i(N, \bar{T}) \triangleq \frac{\partial L(N,\bar{T})}{\partial T_i} = C^*(N, \bar{T}) \ , \quad i = 1,2,\ldots, N \tag{6}$$

have a solution corresponding to a global minimum of $C(N, \bar{T})$ with respect to the T_i. Saddle points, but not a relative maximum, are also possible solutions to (6).

Proof: Evaluation of the Hessian matrix at the stationary point shows that a relative maximum cannot exist for any $\bar{T}$. $C(N, \bar{T})$ is a continuous function of the T_i and it approaches infinity as $\|\bar{T}\| \to 0$ or $\to \infty$.

Theorem 2 If the $r_i(t)$ have the Weibull form $r_i(t) = \lambda_0[1 + \varepsilon \sum_{j=1}^{i-1} T_j] \, t^{\alpha}$ then the optimal T_i are ordered as $T_1^* \geq T_2^* \geq \ldots \geq T_N^*$.

Proof: This result is derived by showing that if $T_i = a < T_{i+1} = b$, the average cost will be reduced by changing to $T_i = b$, $T_{i+1} = a$, while keeping all other T_j unchanged. The comparison is facilitated by the fact that this interchange does not affect t_N or $r_k(t)$ for $k < i$ or $k > i+1$.

Another optimal-solution property which aids in computational problems is summarized in

Theorem 3: If a) $r_i(t)$ are strictly increasing in t, and b) $\partial r_i/\partial T_m = \partial r_j/\partial T_n$ all $n < i$, $m < j$. Then the optimal T_i satisfy $r_i(T_i^*) < r_j(T_j^*)$; $i < j$. The Weibull $r_i(t)$ in Theorem 2 satisfies the conditions of Theorem 3. The result is exemplified in Fig. 1(a) where the peaks get higher going to the right.

Proof: This follows from examination of the relation $L_i(N, \bar{T}) - L_{i+1}(N, \bar{T}) = 0$ based on (6).

The preceding theorems all deal with the case of fixed N. We also want to find the optimum number of maintenance operations (N-1). Clearly, interesting cases will require $C_R > C_M$ and $C_R > C_r$. The following properties have been observed in all numerical examples we have studied, but sufficient conditions and proofs have not yet been found.

Conjecture 1. $t_N^* = \sum_1^N T_i^*$ is an increasing function of N.

Conjecture 2. The minimal cost as a function of N is convex, as in Fig. 2, so that an optimal N* can be found by getting $\bar{T}^*_{(1)}, \bar{T}^*_{(2)}, \ldots,$ until $C(N+1, \bar{T}^*_{(N+1)}) > C(N, \bar{T}^*_{(N)})$.

The model considered here is closely related to one in Nguyen and Murthy (1981). They have $r_i(t)$ as functions of i only (not on t_i). In that simpler case, equation the inequality in Theorem 3 is replaced by $r_i(T_i^*) = r_j(T_j^*)$ and Conjecture 2 has been proved.

We now turn to a slightly different problem, using (5) for the criterion, but with case (b) of Fig. 1 to represent the maintenance effect. In particular

$$r_i(t) = \lambda(t + t_{i-1}) - \gamma\, t_{i-1} ; \qquad 0 < \gamma < \lambda \tag{7}$$

Fig. 1(b) has $\gamma = \lambda/2$. Expressing (7) in terms of the intervals T_j we have

$$r_i(t) = \lambda\, t + (\lambda-\gamma) \sum_1^{i-1} T_j.$$

so that $r_i(t)$ is an increasing function of t and the T_j $(j < i)$.

Here, a necessary condition for the best T_i's, based on partial derivatives of (5), is $\gamma(T_i - T_j) = 0$; $i, j = 1,2,\ldots, N$. Thus, for this model $T_1^* = T_2^* = \ldots T_N^* = T^*$ and

$$T^* = [-(\lambda-\gamma)(N-1) + \sqrt{(\lambda-\gamma)^2(N-1)^2 + 8\lambda(C_R - (N-1)C_M)/C_r}]/2\lambda \tag{8}$$

In this example we have also found an optimal N in every numerical example. Although the joint optimization of N and $\bar{T}$ reduces here to a two parameter optimization (N and T*), existence and uniqueness of the overall minimum have not yet been proved.

All of the preceding results can be interpreted with respect to an availability criterion. If C_R, C_M and C_r are viewed as the mean times required, respectively, for replacement, maintenance and minor repair, then the long-term availability is

$$A[N, \bar{T}] = \frac{D[N, \bar{T}]}{L[N, \bar{T}] + D[N, \bar{T}]} = [1 + C(N, \bar{T})]^{-1}$$

Thus, A is maximized here by minimizing C using the techniques described earlier.

3. PERIODIC MAINTENANCE -- NO MINOR REPAIRS

In this model every failure is followed by a replacement (renewal). The replacement time has an exponential distribution with mean value $1/\mu$, and maintenance operations have fixed duration d. The operating time between maintenances is u, and failure may occur during either system operation or maintenance. Failure rate reduction, of the form of case (b) in Fig. 1, occurs at the end of each maintenance. The use of equal operating intervals between maintenances is motivated by work-rule convenience and the results in the previous section for a related problem where that structure is optimal.

The maintenance effect is introduced differently here, but for the example of linearly increasing hazard rate, there is a direct connection with the models of Section 2. Here we say that maintenance reduces r(t) by a factor: $r(t_i^+) = g\ r(t_i^-)$; $0 < g < 1$ Thus, for $t_{i-1} < t < t_i$, $r(t) = r_1(t) - \Delta_i$ and $\Delta_{i+1} = (1-g)\ [r_1(t_i) - \Delta_i]$; $\Delta_1 = 0$. For the case of $r_1(t) = \lambda t$, the maintenance effect becomes $\Delta_{i+1} = (1-g)[i\ \lambda s - \Delta_i]$ where $s = (u + d)$. Using the notation of (1), this is equivalent to

$$r_i(t) = \lambda t + \theta_{i-1}\ ;\quad i = 2,3,\ldots \text{ where } \theta_i = \lambda\ s \sum_{j=1}^{i} g^j$$

and the θ_i are increasing functions of u (period between maintenances).

We want to find u to maximize the probability that cycle availability is above an acceptable level. With T_u and T_d representing the total up and down times in a renewal cycle, we have $A_c = T_u/(T_u + T_d)$ and $P[A_c > \delta]$ is to be maximized. T_d includes all of the time spent in maintenance before the failure, plus the replacement time T_R after the failure.

It is convenient to define E_n^u and E_n^d as the events that failure occurs, respectively, during the nth operating interval or the nth maintenance interval. Then

$$P[A_c \geq \delta] = \sum_{n=1}^{\infty} \left[\frac{U_n\ P_n(\delta) + D_n\ Q_n(\delta)}{U_n + D_n}\right] \tag{9}$$

where $U_n = P[E_n^u]$, $D_n = P[E_n^d]$

$$P_n(\delta) = P[A_c \geq \delta | E_n^u]\ ,\quad Q_n(\delta) = P[A_c \geq \delta | E_n^d] \tag{10}$$

Goldstein (1980) found analytical expressions for (9) in the simpler case where the durations of each operating period, maintenance period and replacement were independent random variables with exponential distributions. Only numerical evaluation seems possible for the present model.

The piecewise definition of r(t), given above can be converted to a corresponding definition for the lifetime probability density which can be used for numerical integration to evaluate the quantities in (9) for each n. (The summation is truncated when terms become small.) For $t_{n-1} < t < t_n$ we have $f(t) = r(t)\ R(t)$;

$$R(t) = R_n(t) \prod_{j=1}^{n-1} R_j(t_j)$$

$$R_j(t) = \exp[\int_{t_{j-1}}^{t} r(\tau)d\tau] \ : \ t_{j-1} < t \leq t_j$$

Also, we note that using T_R for the replacement time after failure

$$P_n(\delta) = \int_{t_{n-1}}^{t_{n-1}+u} f(t)\ P\{T_R < [t(1-\delta)/\delta - (n-1)d/\delta]\}dt$$

and a similar expression applies for $Q_n(\delta)$.

Numerical examples have shown that this approach is practical to get $P[A_c > \delta]$ for fixed u, and that a maximum can be found by varying the value of u. Very small u will introduce excessive time spent on maintenance; large u will delay maintenance until it is too late to be effective. Figure 3 shows examples for $P[A_c > 0.9]$ where the optimum maintenance period is 0.07 of the mean lifetime, or 0.18 of the mean lifetime, for different parameter choices.

4. CONCLUSIONS

Some useful properties of optimal solutions have been demonstrated for problems where maintenance reduces the subsequent failure rate by an amount depending on the previous operating time of the system. The analysis is more difficult than in similar problems where the effect of maintenance depends only on the number of previous maintenances.

REFERENCES

1. R.E. Barlow, L.C. Hunter, "Optimum Preventive Maintenance Policies," Operations Research, Vol, 8, 1960, Jan.-Feb., pp. 90-100.
2. R.E. Barlow, F. Proschan, Mathematical Theory of Reliability, John Wiley & Son, Inc., New York, 1965.
3. W.P. Pierskalla, J.E. Voelker, "A Survey of Maintenance Models: The Control and Surveillance of Deteriorating Systems," Naval Research Logistics Quarterly, Vol. 23, 1976, Sep., pp. 353-388.
4. S.M. Ross, Applied Probability Models with Optimization Applications, Holden-Day, San Francisco, 1970.
5. A. Tahara, T. Nishida, "Optimal Replacement Policies for a Repairable System with Markovian Transition of States," J. Operations Research Soc. of Japan, Vol. 16, 1973, June, pp. 78-103.
6. R.S. Gottfield, T. Weisman, "Introduction to Optimization Theory," Prentice Hall, Englewood Cliffs, NJ.
7. D.G. Nguyen and D.N.P. Murthy, "Optimal Preventive Maintenance Policies for Repairable Systems," to appear in Operations Research, Vol 29, No. 6, Nov.-Dec., 1981.
8. J. Goldstein, "A Theory of Cycle Availability for Repairable Redundant Systems," Ph.D. Dissertation, Polytechnic Institute of New York, June 1980.

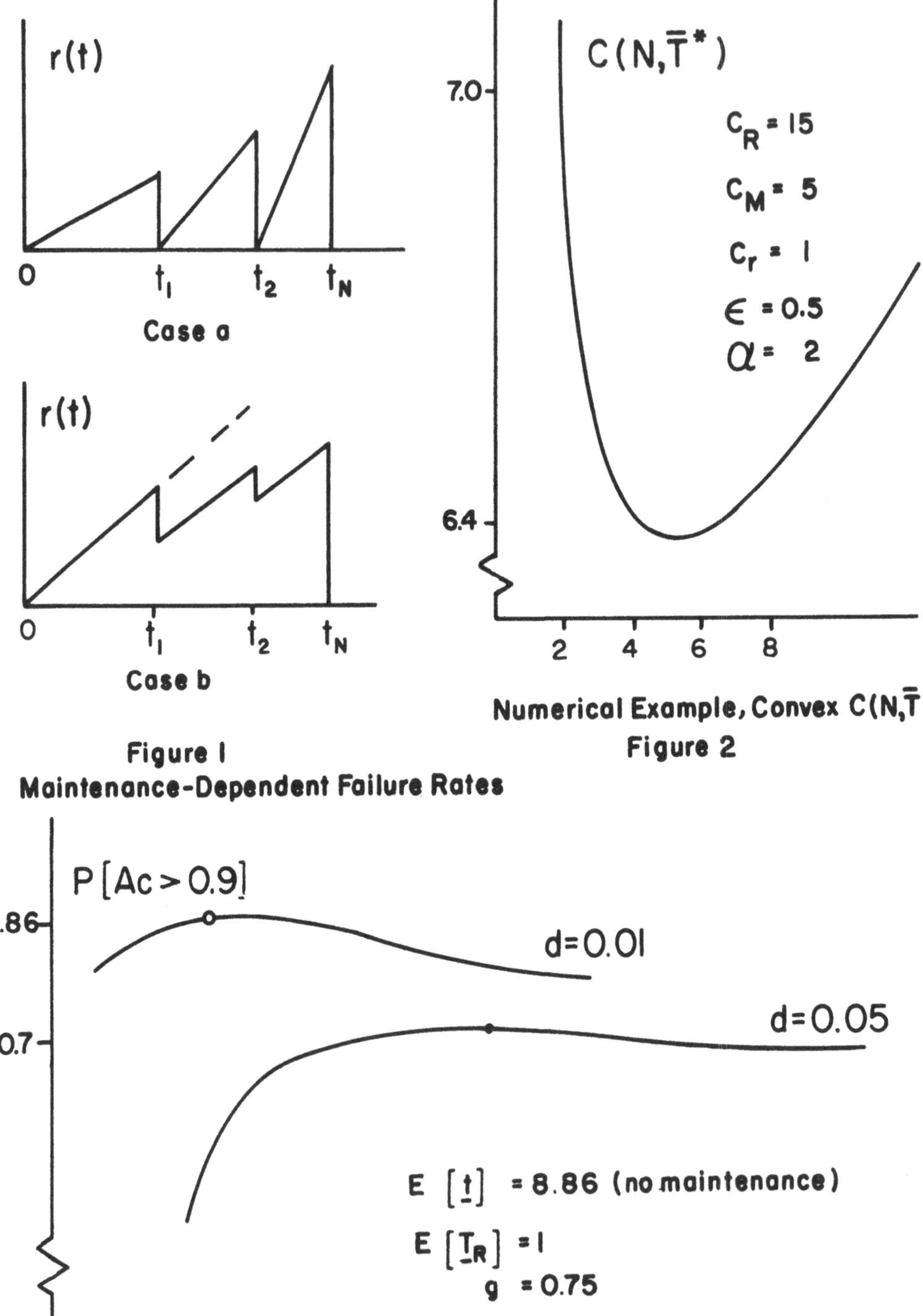

Figure 1
Maintenance-Dependent Failure Rates

Numerical Example, Convex $C(N,\bar{T})$
Figure 2

Figure 3 - Optimization of Periodic Maintenance Interval u

OPTIMAL MAINTENANCE POLICY AND SALE DATE FOR A MACHINE WITH RANDOM DETERIORATION AND SUBJECT TO RANDOM CATASTROPHIC FAILURE

Ilkka Virtanen
University of Vaasa
School of Business Studies
Vaasa, Finland

Abstract. The problem of providing optimal maintenance for a machine during its service life and simultaneously selecting its optimal sale date is considered from a control-theoretic viewpoint. Both the deterioration and the life time of the machine are considered as random processes. The stochastic maximum principle is applied to derive the conditions for the optimal maintenance policy and for the optimal planned sale date which maximize the expected net present value of the machine. An explicit solution is found analytically for the problem in the special case when some of the random processes of the model are independent of time and thus simply random variables. The case of one particular life-time distribution, the exponential case, is analyzed in full detail.

1 Introduction

When a machine is used for production purposes and it ages, it suffers one of the two fates - either there is a gradual deterioration or a sudden failure. The first situation means more frequent repairs, a decrease in performance of the machine etc., the machine produces decreasing net receipts over time. This deterioration can be partially offset via preventive maintenance, and there also exists, of course, the possibility of selling the machine at any time, although its salvage value declines over time. The second situation makes the machine unusable for production and it has to be junked and replaced by a new machine.

Since Näslund [7] had initiated the control theory approach to the problem of simultaneous maintenance and sale date optimization, Thompson [10] first formulated for the problem an explicit model and solved it in detail. Thompson's model is deterministic: the machine cannot fail and its deterioration obeys a given mathematical law. Other formulations for the deterministic problem have been later presented e.g. by Arora and Lele [3], Bensoussan et al. [5] and Scott and Jefferson [9]. Kamien and Schwartz [6] developed a stochastic model where the failure part of the problem was included but the degradation of the machine with age was not considered. Due to Alam and Sarma [2] is a model where both these features have been incorporated in a single model. The deterioration is taken as deterministic, whereas the machine is subject to random failure. Also the author of this paper has recently presented a generalized model [11] for the problem considered in [2].

In this paper we consider a model where the random nature of both the deterioration and the life time of the machine are taken into account. Alam et al. [1] have earlier presented a model where both these aspects have been included, but that model contains the unsatisfactory aspect that the sale date of the machine is not as an object of optimization: the machine is kept as long as it is operable, even though its use would not be profitable any more. This may lead to an unprofitable use of the machine and to an inproper optimum for the problem. Therefore, we generalize model [1] and make it more realistic by taking also the sale date of the machine (called the planned sale date due to the possibility of machine failing before that time) as a tool of optimization.

2 Review of the models of Thompson and Alam et al.

Thompson considers the following problem: find the optimal maintenance policy $u(t)$ and the optimal sale date T for a machine to maximize the present value $V(T)$ of the machine given by

$$(1)\qquad V(T) = S(T)\exp(-rT) + \int_0^T [pS(t) - u(t)]\exp(-rt)dt$$

where the salvage value $S(t)$ is affected by the deterioration factor and the amount and the effectiveness of preventive maintenance according to the differential equation

$$(2)\qquad \frac{dS(t)}{dt} = -\delta(t) + f(t)u(t)\ , \qquad S(0) = S_0\ .$$

In (1) and (2) r is the discount rate, $\delta(t)$ the deterioration rate, $f(t)$ the maintenance effectiveness function, and p is the (constant) production rate. The maintenance function $u(t)$ is the control variable satisfying the requirement

$$(3)\qquad 0 \leq u(t) \leq U\ , \qquad 0 \leq t \leq T\ ,$$

and $V(t)$ and $S(t)$ are the state variables. The solution of the problem can be found by a direct application of the maximum principle (see [10], pp. 545-547).

Alam et al. [1] take Thompson's model as the starting point and they model deterioration as a random process whereas machine failing is not considered in the first phase. We call this model Alam I. The deterioration rate is described by the following stochastic differential equation

$$(4)\qquad d\underline{\delta}(t)/dt = \underline{\alpha}(t) - \beta(t)u(t)$$

with the stochastic boundary condition

(5) $\underline{\delta}(0) = \underline{\delta}_0$.

The stochastic processes $\underline{\alpha}(t)$ and $\underline{\delta}(t)$ as well as the random variable $\underline{\delta}_0$ are assumed to be defined on a certain sample space Ω, the probability measure joining with Ω being P (generally speaking, we use the notation $\underline{z}$ or $\underline{z}(t)$ to indicate that the quantity z or z(t) is a random variable or a stochastic process, respectively). Because of (1) and (2), also the salvage value S(t) and the present value V(T) will now be stochastic processes on Ω, denoted by $\underline{S}(t)$ and $\underline{V}(T)$, respectively.

The problem (Alam I) is now to choose $u^*(t)$ and T so as to maximize

(6) $\bar{V}(T) = E\{\underline{V}(T)\} = \int_\Omega \underline{V}(T)dP$, where

(7) $\underline{V}(T) = \underline{S}(T)\exp(-rT) + \int_0^T [\underline{p}(t)\underline{S}(t) - u(t)]\exp(-rt)dt$

subject to the state equations

(8) $d\underline{S}(t)/dt = -\underline{\delta}(t) + f(t)u(t)$, $0 \leq t \leq T$; $\underline{S}(0) = S_0$

and (4) with boundary condition (5), and to the control constraint (3).

Applying the stochastic maximum principle, the solution of the problem can be derived. An analytic solution is possible in the special case when $\underline{\alpha}(t)$ and $\underline{p}(t)$ don't depend on time, they are simply random variables: $\underline{\alpha}(t) \equiv \underline{\alpha}$ and $\underline{p}(t) \equiv \underline{p}$ (for the solution of problem Alam I see [1], pp. 1073-1074).

In the second phase Alam et al. [1] take also the probability of machine failure into account and derive now the optimal maintenance policy for the machine, whereas the sale date of the machine is not considered, but the machine is assumed to be kept as long as it is operable. We call this model Alam II. Let τ denote the random life time of the machine and let $p_\tau(t;u(s), 0 \leq s \leq t)$, $P_\tau(t;u(s), 0 \leq s \leq t)$ and $Q_\tau(t;u(s), 0 \leq s \leq t)$ denote its density function, cumulative distribution function and reliability function, respectively. Further, let $p_\tau(t;u)$, $P_\tau(t;u)$ and $Q_\tau(t;u)$ compactly represent these quantities. Assuming the deterioration and failure processes mutually independent, the following model (Alam II) can be stated: choose an optimal policy $u^*(t)$ so as to satisfy the state equations (4), (5) and (8) and the control constraint (3) and to maximize the expectation

(9) $E_P\{E_\tau[\underline{V}(\tau)]\} = E_P\{\underline{V}_F\} = \int_\Omega \underline{V}_F dP$, where

(10) $\underline{V}_F = E_\tau[\underline{V}(\tau)] = \int_0^\infty \{[Q_\tau(t;u)\underline{p}(t) + p_\tau(t;u)]\underline{S}(t) - Q_\tau(t;u)u(t)\}\exp(-rt)dt$

is the expectation of $\underline{V}(\tau)$, the quantity in (7) with T considered as the random variable τ, and the expectation being taken with respect to τ. The second expectation E_p in (9) is with respect to the probability measure P.

Again, the solution of the problem can be found via application of the stochastic maximum principle. An analytic solution becomes possible when $\underline{\alpha}(t)$ and $\underline{p}(t)$ are simply random variables: $\underline{\alpha}(t) \equiv \underline{\alpha}$ and $\underline{p}(t) = \underline{p}$, and when failure probability is independent of maintenance: $p_\tau(t;u) = p_\tau(t)$ and, hence, $Q_\tau(t;u) = Q_\tau(t)$ (see [1], pp. 1076-1077).

3 The generalized model

Both the models Alam I and Alam II contain deficiencies in their formulation. The former represents an unrealistic situation in practice by assuming the machine as unbreakable, the latter may lead to an unprofitable use of the machine by forcing the owner to use the machine until it fails, regardless of its ever declining quality and productivity. In this paper, we provide for the problem a generalized formulation in which the above disadvantages are not included. We seek a planned sale date T and a planned maintenance policy $u^*(t)$, $0 \leq t \leq T$, for machine until it is sold or it fails and must be junked, whichever comes first, so as to maximize the expected present value of the machine. The machine is assumed to suffer random deterioration as well as to be subject to random catastrophic failure.

The state equations considered are again (8) and (4) with (5). The control constraint is (3). In the expressions above T now denotes the planned sale date of the machine, i.e. T is the time at which the machine will be sold provided it has not failed and been junked before that time.

The present value of the machine at time t is, provided the machine is still operable, according to (7),

$$(11) \quad \underline{V}(t) = \underline{S}(t)\exp(-rt) + \int_0^t [\underline{p}(t)\underline{S}(t) - u(t)]\exp(-rt)dt \,.$$

Let $\underline{V}_0(T)$ denote the present value which will be really obtained when the planned sale date of the machine is T. By assuming the junk value of the machine equal to its salvage value at the failure time, we get

$$(12) \quad \underline{V}_0(T) = \begin{cases} \underline{V}(T)\,, & \text{if } \tau > T \\ \underline{V}(\tau)\,, & \text{if } \tau < T\,. \end{cases}$$

Now, taking the expectation of $\underline{V}_0(T)$ with respect to the random variable τ and assuming mutual independence between the deterioration and failure processes, we get

$$(13) \quad \underline{V}_F(T) = E_\tau\{\underline{V}_0(T)\} = \int_0^T \underline{V}(t)p_\tau(t;u)dt + \int_T^\infty \underline{V}(T)p_\tau(t;u)dt = \int_0^T \underline{V}(t)p_\tau(t;u)dt + Q_\tau(T;u)\underline{V}(T) .$$

Substituting (11) in (13) we get after some labour (for details, see [12])

$$(14) \quad \underline{V}_F(T) = Q_\tau(T;u)\underline{S}(T)\exp(-rT) + \int_0^T \{[\underline{p}(t)Q_\tau(t;u) + p_\tau(t;u)]\underline{S}(t) - Q_\tau(t;u)u(t)\}\exp(-rt)dt .$$

Our problem is now to choose a (planned) optimal maintenance policy $u^*(t)$ and a (planned) optimal sale date T so as to maximize the expectation

$$(15) \quad \overline{V}_F(T) = E_P\{\underline{V}_F(T)\} = \int_\Omega \underline{V}_F(T)dP$$

where $\underline{V}_F(T)$ is given by (14) and the expectation is taken with respect to the probability measure P over the sample space Ω.

We can readily see that our generalized model is of the same form as the model Alam I, only with the coefficients of $\underline{S}(t)$ and $u(t)$ modified. The generalized model coincides with the model Alam I, when we only set $p_\tau(t;u) \equiv 0$ (the failure part of the model is omitted). We can also see that our generalized model coincides with the model Alam II, if we in (14) set $T = \infty$ to give (10) (the sale date optimization is omitted). Our model thus contains both the models Alam I and Alam II as its special cases.

4 Solution by stochastic maximum principle

The solution of the problem needs an application of the stochastic maximum principle. For this we must assume certain smoothness and regularity conditions: f, β, and u are piecewise continuous, $\underline{\alpha}(t)$ and $\underline{p}(t)$ and, hence, $\underline{\delta}(t)$, $\underline{S}(t)$ and $\underline{V}_F(t)$ are stochastic processes and $\underline{\delta}_0$ a random variable on a sample space Ω which is assumed to be a compact subset of an Euclidean space. The stochastic processes and the random variable are random quantities with respect to the probability measure P on Ω (for a detailed and strict description of the assumptions for the stochastic maximum principle see [4], pp. 876-878).

To solve the problem, we first form the Hamiltonian, which is now a random variable

$$(16)\quad \underline{H} = H(\underline{S}, \underline{\delta}, u, \underline{\lambda}_1, \underline{\lambda}_2, t)$$
$$= - \{[\underline{p}(t)Q_\tau(t;u) + p_\tau(t;u)]\underline{S}(t) - Q_\tau(t;u)u(t)\}\exp(-rt)$$
$$+ \underline{\lambda}_1(t)[-\underline{\delta}(t) + f(t)u(t)] + \underline{\lambda}_2(t)[\underline{\alpha}(t) - \beta(t)u(t)] ,$$

where the adjoint variables $\underline{\lambda}_1(t)$ and $\underline{\lambda}_2(t)$ also are stochastic processes on Ω and satisfy the stochastic differential equations

$$(17)\quad \begin{cases} \dfrac{d\underline{\lambda}_1(t)}{dt} = - \dfrac{\partial \underline{H}}{\partial \underline{S}} = [\underline{p}(t)Q_\tau(t;u) + p_\tau(t;u)]\exp(-rt) \\[2ex] \dfrac{d\underline{\lambda}_2(t)}{dt} = - \dfrac{\partial \underline{H}}{\partial \underline{\delta}} = \underline{\lambda}_1(t) \end{cases}$$

with the boundary conditions

$$(18)\quad \begin{cases} \underline{\lambda}_1(T) = - \dfrac{\partial}{\partial \underline{S}}[Q_\tau(T;u)\underline{S}(T)\exp(-rT)] = - Q_\tau(T;u)\exp(-rT) \\[2ex] \underline{\lambda}_2(T) = - \dfrac{\partial}{\partial \underline{\delta}}[Q_\tau(T;u)\underline{S}(T)\exp(-rT)] = 0 . \end{cases}$$

To find the solution for our problem we should proceed as follows. First we consider T as fixed and apply the stochastic maximum principle (i.e. minimize $E_p\{\underline{H}\}$ with respect to u, see [4], pp. 879-880) to obtain the optimal maintenance policy $u^*(t)$ for $0 \le t \le T$. Then we choose T so as to maximize $\overline{V}_F(T)$.

There exist, however, two reasons, why an analytic solution for this general case is not possible, and, in order to find out the solution, we had to use one of the iterative computational techniques. First, equations (17) are general stochastic differential equations, and secondly, the failure probability $p_\tau(t;u)$ depends on the maintenance performed. Here we present the solution for the problem in the special case where an analytic solution is possible. We make the following additional assumptions. First we assume that $\underline{\alpha}(t)$ and $\underline{p}(t)$ are independent of t, $\underline{\alpha}(t) \equiv \underline{\alpha}$ and $p(t) \equiv \underline{p}$ are simply random variables. The assumption makes it possible to obtain an explicit solution for the co-state equations (17) with (18) and, hence, for the state equations. The solution is achieved by replacing the required stochastic quantities with their expected values. The second assumption is that the failure probability is independent of maintenance: $p_\tau(t;u) = p_\tau(t)$ and, hence $Q_\tau(t;u) = Q_\tau(t)$. With this assumption, an analytic application of the stochastic maximum principle is possible.

Applying the stochastic maximum principle, i.e. minimizing $E_p\{\underline{H}\}$ with respect to u for the above special problem, the following condition for the optimal maintenance policy $u^*(t)$ is obtained

$$(19) \qquad u^*(t) = \begin{cases} U, \text{ if } E_p\{G(Q_\tau, \underline{\lambda}_1, \underline{\lambda}_2, t)\} < 0 \\ \text{arbitrary} \in [0,U], \text{ if } E_p\{G(Q_\tau, \underline{\lambda}_1, \underline{\lambda}_2, t)\} = 0 \\ 0, \text{ if } E_p\{G(Q_\tau, \underline{\lambda}_1, \underline{\lambda}_2, t)\} > 0 . \end{cases}$$

In (19) we have denoted

$$(20) \qquad G(Q_\tau, \underline{\lambda}_1, \underline{\lambda}_2, t) = Q_\tau(t)\exp(-rt) + \underline{\lambda}_1(t)f(t) - \underline{\lambda}_2(t)\beta(t) .$$

Equation (19) shows that the optimal maintenance policy is bang-bang. The possible switching point(s) T', where the level of maintenance is changed from U to 0 or vice versa, satisfy the switching equation $E_p\{G(Q_\tau, \underline{\lambda}_1, \underline{\lambda}_2, T')\} = 0$ or

$$(21) \qquad f(T') = [\beta(T')\overline{\lambda}_2(T') - Q_\tau(T')\exp(-rT')]/\overline{\lambda}_1(T')$$

In (21), $\overline{\lambda}_1(t)$ and $\overline{\lambda}_2(t)$ denote the expectations of $\underline{\lambda}_1(t)$ and $\underline{\lambda}_2(t)$, respectively, the expectations being with respect to the probability measure P.

Thus far we have considered the planned sale date T as fixed. We still have to choose T so as to maximize the expected present value $\overline{V}_F(T)$. Using similar reasoning as Thompson ([10], p. 546), and assuming mutual independence between the random variables $\underline{\alpha}$ and $\underline{p}$ we get the following condition for the optimal planned sale date T:

$$(22) \qquad \overline{S}(T) = \{\overline{\delta}(T) - [f(T)-1]u^*(T)\}/(\overline{p}-r) .$$

5 A particular case: exponentially distributed life time

We shall now demonstrate an explicit calculation of the optimal maintenance policy (19) for a machine with exponentially distributed life time. Therefore, let $p_\tau(t) = \sigma\exp(-\sigma t)$ and, hence, $Q_\tau(t) = \exp(-\sigma t)$ (for $t \geq 0$). As it is well known, the parameter of the distribution (=σ) corresponds to the constant failure rate of the machine.

By using the above expressions for $p_\tau(t)$ and $Q_\tau(t)$ the optimal maintenance policy (19) becomes

$$(23)\qquad u^*(t) = \begin{cases} U\,, & \text{if } \bar{G}(t) < 0 \\ \text{arbitrary} \in [0,U], & \text{if } \bar{G}(t) = 0 \\ 0\,, & \text{if } \bar{G}(t) > 0\,, \end{cases}$$

where

$$\begin{aligned}(24)\qquad \bar{G}(t) &= E_P\{G(Q_\tau,\underline{\lambda}_1,\underline{\lambda}_2,t)\} = G(Q_\tau,\bar{\lambda}_1,\bar{\lambda}_2,t) \\ &= \exp\{-(r+\sigma)t\}\Big\{(r+\sigma) - f(t)\Big[(\bar{p}+\sigma) - (\bar{p}-r)\exp\{-(r+\sigma)(T-t)\}\Big] \\ &\quad - \beta(t)\Big[(\bar{p}+\sigma)[1 - \exp -(r+\sigma)(T-t)\}]/(r+\sigma) \\ &\quad - (\bar{p}-r)\exp\{-(r+\sigma)(T-t)\}(T-t)\Big]\Big\}/(r+\sigma)\ .\end{aligned}$$

The bang-bang optimal policy (23) may have none, one or more switching points. For a switching point T' we have $\bar{G}(T') = 0$.

In section 3 we showed that our generalized model contains the prior models Alam I and Alam II as its special cases. As we now in the exponential case have obtained an explicit solution for the problem, we can also compare the results.

Comparing the optimal maintenance policy (23) with the optimal policy of the model Alam I (see eg. (10) in [1]), we see that they are of the same form. If we instead of the discount rate r in the model Alam I use the 'risk-adjusted' discount rate $r+\sigma$ and instead of the mean production rate $\bar{p}$ use the 'risk-adjusted' mean production rate $\bar{p}+\sigma$, we get (23). Or on the contrary, if we in our model ignore the possibility of random failure and set $\sigma = 0$, the two models coincide. The failure rate σ may thus be interpreted as a risk premium which is to be used to adjust both the discount rate and the mean production rate to the level of those in a certainty-equivalent problem.

In the model Alam II, instead of optimizing also the sale date of the machine, the machine was assumed to be kept until it fails and becomes junked, or in our terms, the planned sale date was fixed to infinity. Setting $T = \infty$ in (24) we obtain, that (23) coincides with the optimal policy for the model Alam II (see eg. (29) in [1]).

In this paper we have pointed out the importance of the sale date optimization also in the case of random machine life. The effect of this optimization can very clearly be demonstrated by a numerical example (see [12], pp. 17-19).

References

[1] Alam, M., Lynn, J.W. and Sarma, V.V.S., "Optimal maintenance policy for equipment subject to random deterioration and random failure. A modern control theory approach", Int. J. Systems Sci., 1976, Vol. 7, No. 9, 1971-1080

[2] Alam, M. and Sarma, V.V.S., "Optimum maintenance policy for an equipment subject to deterioration and random failure", IEEE Trans. Syst. Man Cybernet., 1974, Vol. SMC-4, No. 2, 172-175

[3] Arora, S.R. and Lele, P.T., "A note on optimal maintenance policy and sale date of a machine", Mgmt Sci., 1970, Vol. 17, No. 3, 170-173

[4] Baum, R.F., "Optimal control systems with stochastic boundary conditions and state equations", Ops Res., 1972, Vol. 20, 875-887

[5] Bensoussan, A., Hurst, E.G. Jr. and Näslund, B., Management applications of modern control theory, Amsterdam 1974

[6] Kamien, M.I. and Schwartz, N.L., "Optimal maintenance and sale age for a machine subject to failure", Mgmt Sci., 1971, Vol. 17B, No. 8, 495-504

[7] Näslund, B., "Simultaneous determination of optimal repair policy and service life", Swedish J. Econ., 1966, Vol. 68, No. 2, 63-73

[8] Saaty, T.L., Modern nonlinear equations, New York 1967

[9] Scott, C.H. and Jefferson, T.R., "A bilinear control model for optimal maintenance", Int. J. Control, 1979, Vol. 30, No. 2, 323-330

[10] Thompson, G.L., "Optimal maintenance policy and sale date of a machine", Mgmt Sci., 1968, Vol. 14, No. 9, 543-550

[11] Virtanen, I., "Optimal maintenance policy and planned sale date for a machine subject to deterioration and random failure", European J. Operational Res. (to appear)

[12] Virtanen, I., "Optimal maintenance policy and sale date for a machine with random deterioration and subject to random catastrophic failure", Lappeenranta University of Technology, Dept. of Physics and Mathematics, Report 2/1980

ANALYSIS OF A DISTRIBUTION TRANSFORMER INVENTORY SYSTEM

Robert L. Williams, Ph.D., P.E.
Department of Industrial & Systems Engineering
Ohio University
Athens, Ohio 45701

John K. Helbling, P.E.
Columbus & Southern Ohio Electric Company
Columbus, Ohio 43215

I. INTRODUCTION

I.1 Problem Setting

Transformers are used to convert electrical energy from high distribution voltages to lower voltages required by customers. A public utility must have on hand an adequate number of transformers to meet the varied requirements of customers. Annual capital requirements may approach $3,000,000. It is important that capital devoted to inventory be controlled as closely as possible to strike an appropriate balance between inventory costs and the risk of stock outs.

This paper describes a mathematical simulation model of an electrical transformer inventory/distribution system. The model addresses the organizational structure, material movement, delays in information about decisions and actions, and capital commitment resulting from various inventory policies. A system of ordinary differential equations, using Schiesser's Differential Systems Simulator, Version 2 (DSS/2), is used to describe and simulate the transfer of material, orders and information about the status of the two level distribution system.

The effect of information delays, inventory stock policies, demand forecasting techniques and repair policies are assessed under varying usage scenarios. Four demand scenarios were tested that represent possible system perturbations with two significant ones reported here: increasing average demand, stepped instantaneous demand.

I.2 Dynamic Models

In 1961 Forrester introduced a text book on the behavior of industrial systems entitled "Industrial Dynamics" (1). This text brought information-feedback theory, decision making, the experimental design approach to the design of social systems and the digital computer together to cope with the dynamics of industrial and economic systems. The creation of the DYNAMO compiler at MIT in conjunction with Forrester's dynamic modeling efforts accelerated the use of this method to study organizational behavior. In 1969 Forrester presented his analysis of a computer-simulated model of an urban area that initiated discussion and much controversy.

The use of the dynamic model was extended to planning of patient care city-suburban interactions regional air quality entire economic sectors and, by Forrester, to the world.

II MODEL DESIGN

II.1 Description of Model

The model was designed and implemented using the DSS/2 (3) system which implements numerical integration of ordinary and partial differential equations in FORTRAN. DSS/2 requires that the differential equations be ordinary first order differential equations of the initial value type. Higher order derivatives must be defined as a set of first-order equations. In this model, a system of 81 first-order differential equations was developed to model material, information, and order flows.

II.2 Material Network

The model was designed around the material flow through the system. The installation and removal of transformers in the field is the driving force of the material system and model. There are many different divisions handling distribution transformers. But because of their similarity it is possible to reduce the number needed for analysis to four areas: Operations, Local Stores, Stores, and Substation Maintenance. Figure II.1 shows the material flow among the four areas and the locations where inventories are held.

II.2.1 Operations Area:

The operations area generates the demand for installation in the field that necessitates the movement of transformers throughout the whole system. Inventories are maintained so that units are available for installation in the field when needed. The demand for transformers is dependent on several factors including: industrial, commercial, or residential development; load saturation; and weather damage. All inventory locations can be aggregated but two areas were used in the model in order to allow for differences in information flow delays.

The number of units in the field, FIP, is dependent on the number of transformers installed less any transformers that may be removed due to damage or planned replacements. The time rate of change in the number of units installed in the field is therefore:

$$\frac{DFIP(I)}{DT} = IRATE(I)-RUR(I)-FRR(I), \ I = AREA\ NUMBER$$

where:

$$IRATE(I) = \begin{cases} LSI(I) & \text{if } LSI(I) < WOTR(I) \\ WOTR(I) & \text{if } LSI(I) >= WOTR(I) \end{cases}$$

(See Figure II.1 for identification of variables.)

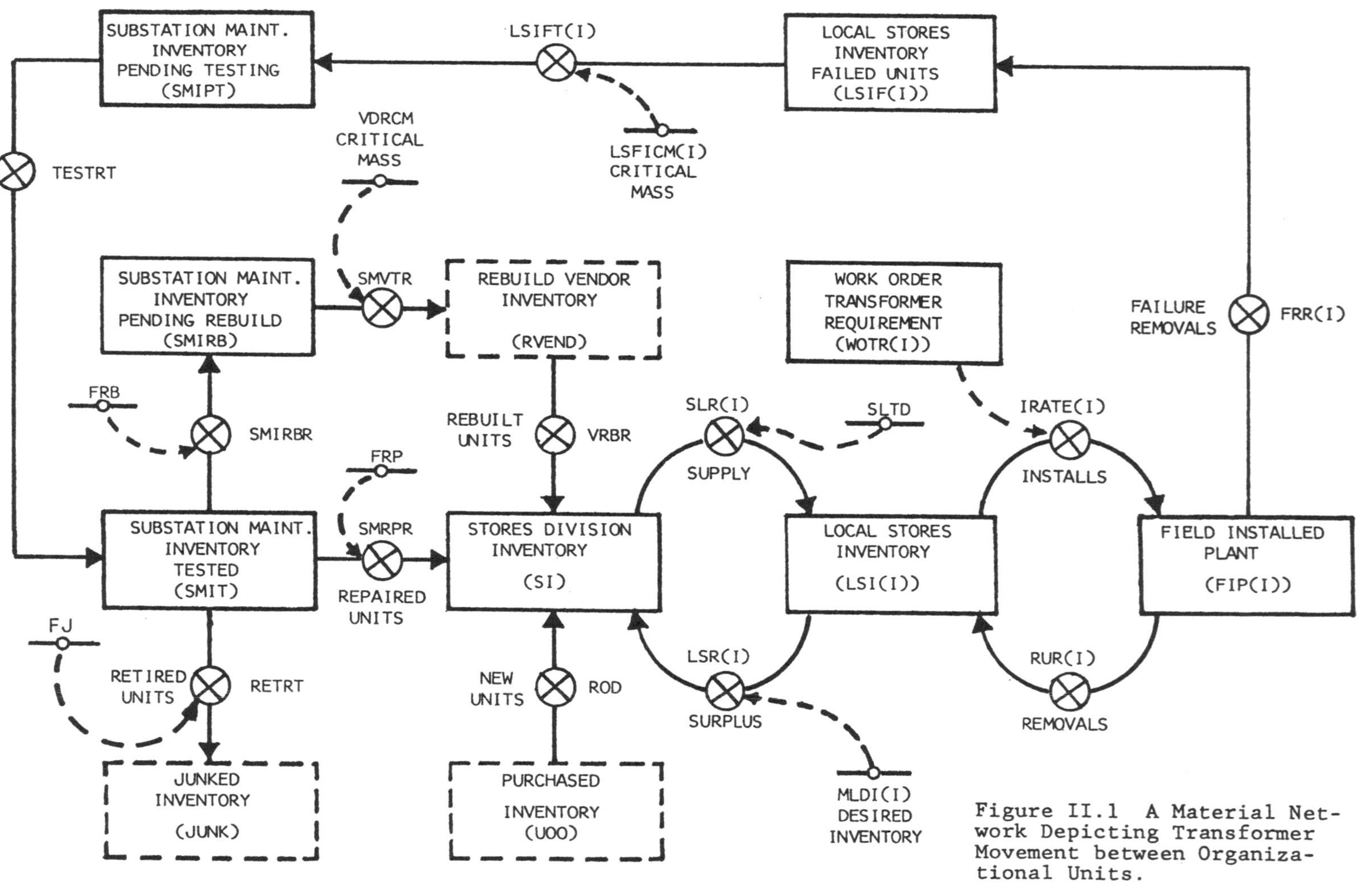

Figure II.1 A Material Network Depicting Transformer Movement between Organizational Units.

The total installation rate, IRATE(I), is equal to the work order transformer requirements rate, WOTR(I), if the supply is great enough.

II.2.2 Local Stores Area

To simplify the model, it was assumed that all operations areas utilize the same stock and requisitioning policies. The aggregate stock equalled approximately six months of average usage plus a contingency supply of low usage units. Each local area has an inventory of transformers which is actually two different stocks of transformers; those that have failed and those that are usable. The change in the level of local stores inventory is defined as:

$$\frac{DLSI(I)}{DT} = RUR(I) + SLR(I) - IRATE(I) - LSR(I)$$

where: $LSR(I) = LSI(I) - MLDI(I)$ if $MLDI(I) < LSI(I)$
$= 0.0$ if $MLDI(I) >= LSI(I)$

The change in the level of failed units in the local stores area is:

$$DLSIF(I) = FRR(I) - LSIFT(I)$$

where: $LSIFT(I) = LSIF(I)$ if $LSIF(I) >= LSFICM(I)$
$= 0$ if $LSIF(I) < LSFICM(I)$

In the above equations a critical mass level, LSFICM(I), is introduced for the transfer of failed units and surplus units.

II.2.3 Stores Division

The stores division receives new, repaired, and rebuilt transformers, and distributes them on request to the local stores areas. It also receives surplus units from the local stores areas. It is the stores division that monitors the usage so that the purchase of new units is timed properly to avoid stockout and overstock conditions.

The change in the stores inventory level is:

$$\frac{DSI}{DT} = ROD + SMRPR + VRBR + NETLSR - NETSLR$$

where: ROD = sixth-order delay of order placement rate with average delivery delay.

$$NETLSR = \sum_{I=1}^{N} LSR(I)$$

$$NETSLR = \sum_{I=1}^{N} SLR(I), \quad N = \text{number of local stores.}$$

$SLR(I) = URS(I)/SLTD$ $SI>=TOTURS$
$= (URS(I)/TOTURS)*SI/SLTD$ if $SI< TOTURS$

URS(I) = number of units on requisition for transfer from stores to local stores.

$$TOTURS = \sum_{I=1}^{N} URS(I)$$

The sixth-order delay was used because it proved to be closer than others to the empirical delay distribution.

II.2.4 Substation Maintenance Area

The substation maintenance area tests damaged units to determine the extent of the damage. There are three possible outcomes of the testing: (1) minor damage, repaired immediately; (2) major damage requiring unit to be rewound; (3) unit is junk. The distribution of units within the three categories was assumed to be a fixed percentage of the total units.

The three possible outcomes of the testing were modeled through the use of four different inventory levels. First, the change in the inventory pending testing is:

$$\frac{DSMIPT}{DT} = TLSIFT - TESTRT$$

where: $TLSIFT = \sum_{I=1}^{N} LSIFT(I)$

TESTRT = AVLAB/LPUT if SMIPT> AVLAB/LPUT
= SMIPT if SMIPT<= AVLAB/LPUT
AVLAB = available labor in manhours.
LPUT = average labor required for testing and repair.

Secondly, the derivative of the number of units tested is:

$$\frac{DSMIT}{DT} = TESTRT - SMIRBR - SMRPR - RETRT$$

where: SMIRBR = FRB * SMIT
FRB = fraction of total tested units rebuildable.
SMRPR = FRP * SMIT
FRP = fraction of total tested units repairable.
RETRT = FJ * SMIT
FJ = fraction of total tested units junk.

Thirdly, the change in the number of units junked is simply the number of units tested out as junk during the previous time period.

$$\frac{DJUNK}{DT} = RETRT$$

Lastly, the units that are found to be in need of rebuild are held until a critical mass is attained and then sent to a vendor. The change in the number of units held for rebuild is:

$$\frac{DSMIRB}{DT} = SMIRBR - SMVTR$$

where: SMVTR = SMIRB if SMIRB > = VDRCM
= 0.0 if SMIRB < 0

Rebuilt units are delivered to the stores division. The time that the units are within the control of the vender is assumed to be a function of the transfer rate and a third-order delay represents this process.

The change in the vendor stock level of units pending rebuild is:

$$\frac{DRVEND}{DT} = SMVTR - VRBR$$

where: VRBR = a third-order exponential delay.

II.3 Information Network-Transformer Master File

The life cycle of a distribution transformer is recorded and monitored in an on-line record system. Each transfer, installation, and removal is recorded. Due to clerical delays, errors in filling out forms, and the amount of time required to forward the forms from remote locations, the masterfile represents only an approximation of the actual current inventory and lags the actual transfers by 5 to 20 days. The differential equations are similar to those in the previous section but are not given here due to space limitations.

II.4 Order Network

This is a two level inventory distribution system requiring two inventory stock level policies and two ordering policies. They are discussed separately.

II.4.1 Stores Inventory Policies

The existing stores policy is to have on hand or on order a five to six month supply of transformers. The six month usage is based on a twelve-month moving average of the number of transformers issued from the stores area based on the masterfile (and not actual usage.)

The desired inventory is:

SID = TMA(3) * IUR + CSL
SID = desired stores inventory.
TMA(3) = twelve month moving average of stores usage.
IUR = number of months usage desired in inventory.
CSL = contingency stock level.

The recommended order quantity is based on the masterfile quantity indicated to be in stock except when there is actually no inventory on hand.

ROQ = SID + OJR - (SILIST + UOO + URQ) if SI> 0;
= SID + OJR - URQ - UOO if SI<=0;

where: ROQ = recommended order quantity.
OJR = outstanding job requirements.
SILIST = masterfiles stores inventory at first of month.
URQ = units on requisition.
UOO = units on order.
SI = stores inventory.

The number of units that can be ordered is limited by the level of approved funding. If the recommended number of units represent an expenditure in excess of the available uncommitted funds then the requisition rate is normally limited to those units which can be purchased with the available funds.

II.4.2 Local Stores Area Inventory Policies

The local stores policies are based only on the usage requirements. The model uses a three month average usage with some contingency stock as a buffer as a typical policy. The desired inventory of the local stores is:

LDI(I) = 3 * TMA(I) + LSCSL(I)

where: LDI(I) = local stores desired inventory.
TMA(I) = twelve month moving average for stores(I).
LSCSL(I) = local stores contingency stock levels.

The local stores requisition rate is:

LSRR(I) = LDI(I) - MFLSI(I) - URS(I); if LSI(I)> 0.0
= LDI(I) - URS(I); if LSI(I)<=0.0

where: MFLSI(I) = masterfile record of local stores inventory.
URS(I) = units on requisition for local stores(I).
LSI(I) = local stores inventory of usable transformers.

II.4.3 Division Operations

Each planned installation is initiated by a work order. The model determines the install rate by the number of pending work orders. Any backlog is accumulated in the work order transformer requirements. The change in work order requirements is:

$$\frac{DWOTR(I)}{DT} = NIR(I) + FRR(I) + ORR(I) - IRATE(I)$$

NIR(I) = rate of new installations.
FRR(I) = rate of failure removal.
ORR(I) = rate of planned replacement.
IRATE(I) = rate of total installations.

III MODEL IMPLEMENTATION

III.1 Demand Submodels

An initial value for the monthly average was determined from three years of transformer history upon which monthly adjustment factors were applied. The monthly adjustment factors were obtained by taking the three year average usage for each month of the year and dividing it by the total three year monthly average usage (AVEI) in order to reduce it to a dimensionless proportionality factor (AMONTI(N)). The twelve factors are then applied to the average monthly usage to predict the total installations for the month (TOTINS). The installation activity is affected by the time of year. Winter activity is reduced while late summer and fall activity is higher than the average.

The removal activity was predicted much the same as the installation activity. Weather also has a substantial bearing on the removal rate.

The model was run using two sets of monthly proportionality values. The first set assumes that there was no change in usage due to weather or other seasonal factors. The second set of proportionality values account for weather and the construction season.

The model was run under three demand conditions and three stock level policies. The demand conditions reported here are:

1) 10% annual compounded growth in the average monthly installation rate.
2) 37% step increase in the average monthly installation rate.

The stock level policy is:

> Existing stores policy of 5 months average usage plus a contingency stock in the "pipeline" and local stores policy of 3 months average usage plus a contingency stock in local "pipeline".

III.2 Case of Constant Average Demand, 10% Annual Growth for One Year

Figure III.1 depicts the response of the system to a compounded 10% annual increase in the installation rate. The seasonal proportionality factors are all unity giving a flat response curve at equilibrium. The increase in usage is seen to affect the stock levels in varying degrees and with different rates of response. The net result of the perturbation (10% growth) is an increase in the target levels for all three inventories shown. The local stores areas ability to perceive an increase and to establish new stock levels lagged the stimulus by 4 to 5 months.

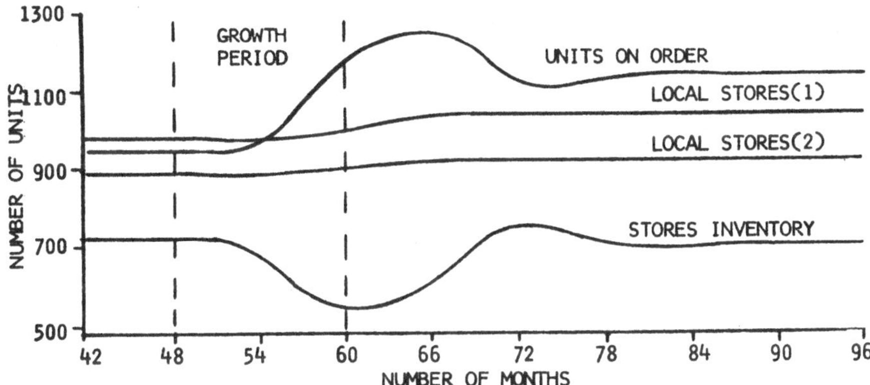

Figure III.1 Constant Demand with 10% Annual Growth for One Year.

The existing stores division policy may be only sufficient for short periods of growth and that sustained growth will require additional quantities ordered to establish the desired quantity on hand. This problem could be compounded by an increasing average delivery time which would shift more and more inventory into quantity held at the factory and not available for use.

During the growth period there was not an increase in the backlog of work orders that would result from insufficient quantities of transformers. Therefore the order policy and the stock level policy are adequate for small periods of growth provided that sufficient inventories exist prior to the growth period.

III.3 Case With Constant Average Demand, 200 Unit Step Increase

The worst case situation for the inventory system is a substantial step increase in demand. This type of increase puts severe demands on the stock policy and order policies. Unlike the annual growth case where the moving twelve month average usage has more time to respond, the step function has a sharp increase in demand that is not truely reflected in the desired target levels for several months. A 37% (200 unit) increase in demand was imposed at the 48th month.

Figure III.2 depicts the results of the stepped increase in demand. The perturbation resulted in a much sharper and drastic system response, as would be expected. The local stores areas took nearly two years to recover and establish a new equilibrium. The stores division was affected more drastically and could not recover until the local stores areas had re-established their states of equilibrium. The stores division inventory was reduced and the number of units on order stabilized at a significantly larger quantity compensating for the growth and the decrease in stock in-hand.

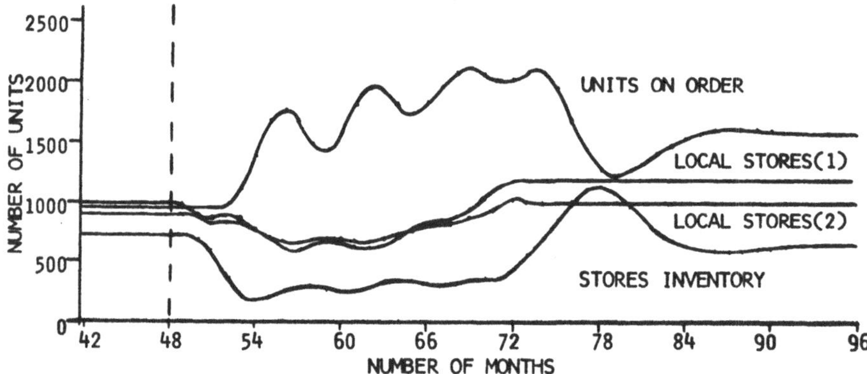

Figure III.2 Constant Demand with a 200 Unit Step Increase.

With sufficient inventory on-hand at the time of the increase, the set of stock level and order policies was sufficient to prevent a backlog of work orders. However, as in the previous case, the stores division inventory was the buffer between the long leadtime of the factory and the short leadtime required by the local areas. Thus an insufficient stores inventory would prevent absorbing the shock and increase the backlog.

III.3.3 Variations in Stock Level Policies

Each of the cases presented previously utilized the same stock level and ordering policy. The results indicated that the stock policy, though not optimal, was acceptable and did not result in undue delays in fulfilling customer demand. In order to determine how sensitive the system is to stock level policies two additional policies were tested which varied the amount of desired inventory by one to two months. The results were unacceptable which implies that current order policies of the company are quite good.

CONCLUSIONS

This dynamic simulation model can be used to assist management in establishing economical stock level and reorder policies. The results of the model were consistent with the observed behavior of the real system. The amount of hard data about actual stock levels is very limited and the real system has not maintained the same policies long enough to collect such data. Abnormal situations have caused management to adjust stock levels in contradiction with normal ordering and stocking policies. The authors' experience with specifying stock levels and reorder quantities indicates that the system is never allowed to reach equilibrium due to the lack of stability in the economy. However the model's initial response to stimulus is representative of the real system and in this respect it is of value to management.

The model indicated that the existing stock level and ordering policies were, if not optimal, at least satisfactory in their ability to avoid stock-out conditions. However the stock level at the local stores area could be reduced significantly as long as no major increase in usage was experienced. This was evident from the cases with the step increase in demand which demonstrated some delays in installation.

A more complete description of the model and results is available from the authors (2).

REFERENCES

(1) Forrester, J.W., INDUSTRIAL DYNAMICS, MIT Press, Cambridge, MASS., 1961.

(2) Helbling, J.K., & Williams, R.L. "Analysis of a Distribution Transformer Inventory System", Paper presented at the 10th IFIP Conference on System Modeling and Optimization, New York City, September 4, 1981.

(3) Schiesser, W.E., DSS/2 (DIFFERENTIAL SYSTEMS SIMULATOR, VERSION 2) INTRODUCTORY PROGRAMMING MANUAL, Lehigh University, 1976.

EXPERIMENTAL, ANALYTICAL AND COMPUTATIONAL STUDY OF A SIMPLIFIED MAXIMAL HEIGHT JUMP

William S. Levine and Felix E. Zajac
Dept. of E.E. RER and D Center (153)
University of Maryland VA Medical Center
College Park, MD 20742 Palo Alto, CA 94304

Abstract

Humans were asked to jump as high as possible subject to the constraint that they keep their knees and hips locked. In essence, the experimental subjects were asked to imitate a two segment inverted pendulum.

The experiment is modelled by a two segment inverted pendulum controlled by an "activation". That is, torque is exerted by a model of muscle activation in which the maximum torque depends on the joint angle and torque lags activation (activation is the control). This muscle model is based on the physiology. The resulting mathematical optimal control problem is solved and it is shown that the results are largely independent of the muscle model. The agreement between experiment and analysis is very good, although not perfect. Some suggestions are made regarding further sources of error.

I. INTRODUCTION

Several people have realized that optimal control techniques have great potential for the elucidation of the control of the musculoskeletal system [1-5]. However, the application of such techniques to musculoskeletal control problems is difficult. Perhaps the greatest difficulty is that the performance criterion is usually ambiguous. Another problem is that the system controlled usually has the dynamics of multi-segment pendula. Such systems are non-linear, have unusual state constraints, and often involve controls that are neither bang-bang nor linear.

At the moment, we have chosen to theoretically and experimentally study human subjects who were instructed to jump as high as possible while keeping their legs and body straight (i.e., knee and hip joints fully extended). This problem has the advantages of an unambiguous performance criterion and somewhat simplified dynamics. In this study, our theoretical objective is to formulate and solve the mathematical optimization problem with forethought to the extrapolation of these results to the study of more normal jumps. Our biological objective is to develop neuromuscular and musculoskeletal models for the lower leg and foot of complexity compatible with experimental resolution.

The paper is organized as follows: Section 2 describes the formulation of a mathematical model that approximates the physiological jumper. This is followed by a description of the technique whereby the resulting mathematical optimization problem is solved. In Section 4 we give the experimental and theoretical results. We conclude with a number of suggestions for further research.

This paper is a substantially abridged version of [6] where many details which we have omitted here can be found.

2. PROBLEM FORMULATION

Our approach to the study of movement is to add complexity to our model step by step. In this way the importance of each additional component can be determined, and the sensitivity of performance to each component can be used to guide future experiments. In such an approach, as long as the model does not fit the experiments to within experimental accuracy, the limit of resolution of the approach has not been reached and there is value in continuing to refine the model.

It should also be noted that our model is basically an optimal control problem. We test our model by solving the optimal control problem and then comparing the mathematical solution with the data obtained from human experimental subjects.

The model is based on the two segment inverted pendulum shown in Fig. 1. The incongruency between ankle and heel is accounted for by the triangular "foot". Note the approximation that the center of mass lies on the straight line connecting the toe joint and the ankle. Assume the pendulum is not fastened to the ground so that a "jump" is possible. The masses, lengths and inertias will be chosen to approximate those of the experimental subject. Since our subjects were instructed to keep their arms over their heads in a fixed position and to keep their legs and torso straight this model is reasonable.

The torque, u(t) in Fig. 1, is created by a simplified model for the torque generation properties of human muscle. We will describe this muscle model in detail shortly. For the moment, it is more convenient to treat u(t) as the control. Thus, given this physical structure, the problem is to find the u(t) ($0<t<t_f$) which will cause this mechanism to "jump" to its maximum possible height. This must be accomplished without violating two constraints: (1) neither segment can go through the ground and (2) $0<\Theta_1+\Theta_2<\pi$.

We have made a large number of simplifying assumptions and approximations in arriving at this physical model. Human joints are not pin-joints, our experimental subjects are not able to keep their knees and hips perfectly locked, and body segments do not have fixed inertia and center of mass.

Next we translate the above physical problem into a meaningful mathematical optimization problem. The mathematical problem requires four unambiguous components: (a) dynamics, (b) initial and terminal conditions, (c) constraints and (d) performance criterion.

(2a) Dynamics

There are two distinct cases that are important. The first is illustrated in Fig. 1 and corresponds to only the toe on the ground. The second corresponds to the whole foot on the ground. One might think that it is also necessary to consider the situation where the entire device is airborne. We have shown, for a one segment jumper [7], there is very little difference between maximizing the peak height reached by the upper tip and maximizing the peak height attained by the C.M. Similar results have been found for competitive high jumpers [8]. Thus, we assume that the performance objective is to maximize the peak height reached by the subject's center of mass. This height is completely determined at the instant of lift-off. Hence there is no need to consider the dynamics once the subject is airborne.

Case 1: Only the "toe joint" on the ground. The dynamics are given by (see Fig. 1 for notation)

$$\dot{\underline{\theta}}(t) = \underline{\omega}(t) \tag{1}$$

$$\dot{\underline{\omega}}(t)=A^{-1}(\underline{\theta})\{b(\theta)\begin{bmatrix}\omega_2^2\\ \omega_1^2\end{bmatrix} -\underline{C}(\underline{\theta}) +\begin{bmatrix}1\\ 1\end{bmatrix} u(t)\} \tag{2}$$

where $\underline{A}(\theta) = \begin{bmatrix}a_1 & a_2\cos(\theta_1+\theta_2)\\ a_2\cos(\theta_1+\theta_2) & a_2\end{bmatrix}$

$$b(\underline{\theta}) = a_2 \sin(\theta_1+\theta_2)$$

$$\underline{C}(\underline{\theta}) = \begin{bmatrix}c_1\, g\cos\theta_1\\ c_2\, g\cos\theta_2\end{bmatrix}$$

and the coefficients a_i, c_i are determined by the subject's masses, lengths, etc. (See [6] for details).

It is convenient to include a third equation here. This is the equation for F_v, the force exerted by the toe joint on the ground.

$$F_v = (m_1+m_2)g + m_1\ddot{y}_1 + m_2\ddot{y}_2 \tag{3}$$

where

$y_1(t)$ = vertical position of c.m. of foot

$y_2(t)$ = vertical position of c.m. of trunk

thus

$$F_v=(m_1+m_2)g + c_1(-\sin\theta_1\omega_1^2+\cos\theta_1\dot{\omega}_1)+c_2(-\sin\theta_2\omega_2^2+\cos\theta_2\dot{\omega}_2) \tag{4}$$

Case 2: Foot on the ground. The dynamics simplify to

$$\dot{\omega}_2(t) = \frac{u(t)-c_2 g\cos\theta_2(t)}{a_3} \tag{5}$$

$$\theta_1(t) = 34.31^0 = \theta_{10},\ \omega_1(t) = \dot{\omega}_1(t) = 0 \tag{6}$$

$$F_v = (m_1+m_2)g + c_2(\cos\theta_2\dot{\omega}_2 - \sin\theta_2\omega_2^2) \tag{7}$$

It is important to know the boundary between Case 1 and Case 2. This can be expressed in several ways. We use the center of pressure, denoted ℓcp, (see [6].

$$\ell cp = (\ell_1+\ell_2)\cos\theta_{10} + \frac{[F_h(\ell_1+\ell_2)\sin\theta_{10}-(m_1 g\ell_2\cos\theta_{10}+u(t))]}{F_v} \tag{8}$$

$$\text{where } F_h = -c_2(\cos\theta_2\omega_2^2 + \sin\theta_2\dot{\omega}_2) \tag{9}$$

The "foot" stays flat on the ground while $(\ell_1+\ell_2)>\ell cp>0$. The "heel" leaves the ground when $\ell cp<0$.

Notice that the controller, at certain states, can choose between the dynamics represented by (1) and (2) and the dynamics represented by (5) and (6).

(2b) Initial & Terminal Conditions

Experimentally, the subjects begin by standing with their feet flat on the ground and their trunk approximately vertical. Thus, the initial condition is an equilibrium point ($\theta_1 = 34.31^0, \theta_2 = 90^o$). All stationary initial conditions (states for which $\underline{\omega} = \underline{0}$) can be reached from this one, provided we perturb θ_2 by $\pm\varepsilon$.

The terminal time is the instant that the mechanism leaves the ground. Since the toe, or foot, rests on the ground, this corresponds to the first instant at which $F_v < 0$. Obviously, one does not want an open set for the target set. For reasonable jumps there are no optimal trajectories along which $F_v = 0$. Thus, we can take the terminal condition to be

$$F_v(\underline{\theta},\underline{\omega},u) \leq 0 \tag{10}$$

(2c) Constraints

The constraint is imposed by the need to keep the toe on the ground until Eq. (10) is satisfied. Thus, there is a constraint

$$F_v(\underline{\theta}(t),\underline{\omega}(t),u(t)) \geq 0 \tag{11}$$

Note that the physical constraint imposed by the ground does not appear among the constraints. Instead, it appears as a complicated nonlinearity in the dynamics.

(2d) Performance Criterion

Finally, the performance criterion is to maximize

$$J(t_f,u) = y_c(t_f) + \frac{1}{2g}\dot{y}_c^2(t_f) \tag{12}$$

where t_f = the instant of lift-off

$y_c(t)$ = the vertical position of the subject's center of mass

To complete the problem formulation we have to specify the muscle model which creates u(t). The modelling of muscle has been extensively studied with the result that there are a multitude of competing muscle models (see 6 for references). Out of this collection we elected to choose a model with the following characteristics:

(1) A small number of parameters characterizes the model.

(2) These parameters are "measureable" in the sense that they could be determined by reasonable tests on intact human subjects.

(3) The model is mathematically tractable.

Thus, the model was developed by replacing u(t) with

$$u(t) = d(\theta_1(t) + \theta_2(t))\xi(t) \tag{13}$$

where

$$\dot{\xi}(t) = \alpha\xi(t) + \alpha u_1(t) \tag{14}$$

$u_1(t)$ is the new control variable corresponding to muscular "activation" of the calf muscles with the constraint

$$0 \leq u_1(t) \leq 1. \tag{15}$$

$d(\theta_1(t)+\theta_2(t))$ is the maximum ankle extensor torque as a function of ankle angle and corresponds to the force-length property of muscle. It is known that maximum ankle torque also depends on the rate of change of ankle angle ($\omega_1(t)+\omega_2(t)$). Based on the work of Fugl-Meyer et al [9], however, this dependence is almost non-existent over the range of angles ($116^o < \Theta_1 < 142^o$) relevant to our experiment; thus this dependence is neglected. The exact form of $d(\Theta_1+\Theta_2)$ is shown in Fig. 2. Thus through the multiplication of $d(\Theta_1+\Theta_2)$ with $\xi(t)$, the basic muscle characteristic modelled is its force-length property scaled by the level of muscular activation after a lag.

Implicit in (13) and (14) are the assumptions that ankle torque is linear with respect to ξ, and linear with a lag in u_1 and that flexor muscles can be neglected. Virtually every muscle model in the literature is linear in the control variable. However, our model is too simple to account for all of the known experimental results on muscle mechanics. It will be shown that our muscle model is still rich enough to account for the significant characteristics of the jump.

3. Solution

Because of space limitations we present only an outline of the solution technique. The details can be found in [6].

First, we show that $u_1(t)=1$ is optimal near lift-off. This is done by posing and solving the problem of jumping to a maximum height starting from initial conditions near lift-off. Second, we calculate the complete set of states that are reachable from the actual initial condition of the human jumper while keeping his feet flat on the ground. Third, we show that $u_1(t)=1$ (the optimal solution near lift-off) is in fact optimal in a region of the state space that extends to the boundary of the set of states reachable with feet flat on the ground. This allows us to determine the optimal state for the heel to leave the ground.

The complete optimal control is thus composed of two basic epochs:

(1) Go from the initial condition to the optimal state for heel lift-off.

(2) Once the optimal heel lift-off state is reached set $u_1(t)=1$ and keep it there until lift-off of the whole body.

We give slightly more detail regarding each of the three components of the problem below.

(3.1) $u_1=1$ AT STATES NEAR LIFT-OFF.

To demonstrate this it is convenient to formulate a new optimal control problem that is valid only in a subset of the state space. In particular, assume that once the jumper's heel leaves the ground it is not advantageous to return the heel to the ground before lift-off.

We thus temporarily restrict our attention to states in which the heel is off the ground ($\Theta_1>0$). The performance criterion is still given by (12), the terminal condition by (10) and the control constraint by (15). The mixed constraint given by (11) still applies, but, we are fairly sure that (11) is not an active constraint just prior to lift-off. Thus, we apply the minimum principle to the "relaxed" problem obtained by ignoring the constraint (11). As long as the optimal control for this "relaxed" problem does not violate the constraint (11) we know it is the optimal control for the real problem as well.

(3.2) SET OF STATES REACHABLE FROM INITIAL STATE WITH FOOT FLAT ON THE GROUND

This set of states is shown in Fig. 3. The derivation of Fig. 3 is based on (5)-(8), the equations describing the dynamics when the foot is flat on the ground. The crucial requirement is that ℓcp (see (8)) remain greater than or equal to zero so the foot stays flat on the ground. As long as the foot is flat on the ground the dynamics are given by (5) and, the phase plane trajectories can be calculated analytically. The details can be found in [6].

(3.3) SHOW THAT $u_1=1$ FOR THE REGION FROM THE SET OF FLAT FOOT STATES TO LIFT-OFF

This is done by the following algorithm, which verifies the conjecture that $u_1=1$ from heel lift-off to lift-off and determines the optimal heel lift-off state.

Step 0: Choose any state among the set of states reachable with heel on the ground. Call this $\underline{x}_r^0$. Set k=0.

Step 1: Apply the minimum principle to verify that $u_1^*(t)=1$ from $\underline{x}_r^k$ until body lift-off. Remember that this also gives the costate at $t_0, \underline{p}^*(t_0)$ which is also the gradient of performance with respect to $\underline{x}_r^*$.

Step 2: Compute the projection of $\underline{p}^*(t_0)$ onto the set of reachable states with the heel on the ground. Call this projection $\underline{d}$. If $\underline{d}=\underline{0}$ stop. $\underline{x}_r^k$ is the optimal state for heel lift-off. If $\underline{d}\neq\underline{0}$, set $\hat{\underline{x}}_r=\underline{x}_r^k+\beta\underline{d}$ where $\beta>0$ and "small".

Step 3: Set $\underline{x}_r^k=\hat{\underline{x}}_r$, k=k+1 and return to Step 1.

It is clear that the algorithm will either converge to a locally optimal $\underline{x}_r^*$ or demonstrate that our conjectured optimal control is incorrect. We have used the algorithm to compute $\underline{x}_r^*$ and to verify that $u_1^*(t)=1$ until lift-off for several choices of the parameters. In every case the algorithm has converged and verified the conjecture. Additional calculations as well as further details of the algorithm performance are given in [6].

4. Results

Fig. 3 shows the set of reachable states while the foot is flat on the ground. The optimal trajectory is

(i) move from the initial equilibrium state Q to the minimum equilibrium state $A=(34.3^o\ \ 82.40^o\ \ 0\ \ 0)'$

(ii) then move from A to the optimal state for heel lift-off, $\underline{x}_r^*=(34.31^o\ \ 81.76^o\ \ 0\ \ -1.71^o/\text{sec})'$ (filled circle in Fig. 3). Notice that $\underline{x}_r^*$ lies on the boundary of the "forward moving" state feedback trajectory A-B. The corresponding optimal controls are

(i) not unique for moving from Q to A

(ii) unique for moving from A to $\underline{x}_r^*$ but very nearly equal to the torque required to maintain equilibrium at point A.

Our experimental results are slightly different. Our best estimate derived from experimental data of the minimum equilibrium state (the theoretical state A) is $A_e=(34.7^o\ \ 81^o\ \ 0\ \ 0)$ and is indicated by a Δ in Fig. 3. In contrast, there is considerable ambiguity with regard to the experimental state corresponding to heel lift-off. We are quite certain that heel lift-off has occurred by 790 msec. This is the time indicated by downward pointing arrows in Fig. 4. The experimentally deter-

FIG. 1 **FIG. 3**

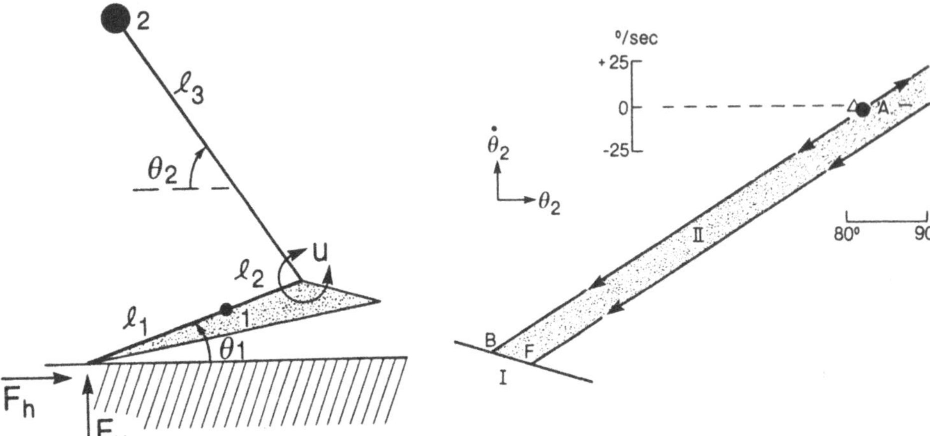

Fig. 1 Two-segment, triangular foot model. In this model $\theta_1(0)=34.31^o$ when the foot is flat on the ground.

Fig. 3 Plot of reachable states when the foot is flat on the ground. Point A, the minimum equilibrium angle, is $\theta_2=82.40^o$, $\dot{\theta}_2=0$. The optimal state at heel lift-off, lying along the boundary curve AB, is $\theta_2=81.76^o$, $\dot{\theta}_2=-1.71^o/$sec. Open triangle (Δ) corresponds to both the most forward equilibrium state and the heel lift-off state derived from experimental data.

FIG. 2

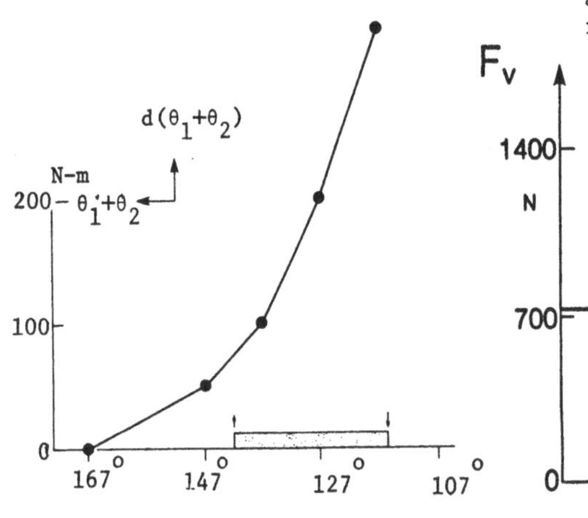

Fig. 2 Plot of maximum ankle extensor torque $d(\theta_1+\theta_2)$ vs. $\theta_1+\theta_2$. The "shaded bar" on the abscissa indicates the range of ankle angles observed in our human jumper from heel lift-off (↓) to body lift-off (↑).

FIG. 4

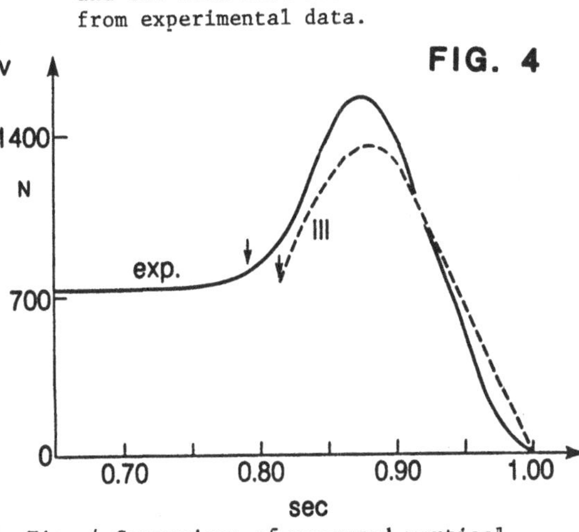

Fig. 4 Comparison of measured vertical force F_v from our human jumper ("exp", solid curve) with that computed from our model ("broken" curve). Downward-pointing arrows indicate the time at which heel lift-off occurs.

mined state at this time is $(36^\circ\ 81^\circ\ 36^\circ/\text{sec}\ 0^\circ/\text{sec})'$. However, it is possible that heel lift-off occurs as early as 725 msec. since Θ_1 and ω_1 are monotone increasing after this time. The state at this time is identical to the minimum equilibrium state $(34.7^\circ\ 81^\circ\ 0\ 0)'$. The true time and state at the instant of heel lift-off is probably somewhere between these two extremes. These extreme values are not very far apart relative to the experimental accuracy. The differences between theory and experiment apparent in Fig. 3 are believed insignificant and within experimental resolution.

Heel Lift-Off to Body Lift-Off

For this second part of the jump, experimental data is compared with theoretical results in Fig. 4. It should be noted that both curves are synchronized at the instant at which $F_v=0$ (body lift-off), since this is the least ambiguous point in the jump. We believe F_v, the vertical force as measured by the force plate, is the most important experimental datum. Since we actually measure F_v and F_v is an accurate measurement.

It is obvious from Fig. 4 that there is good agreement with both the experimentally derived F_v and the time for the subject to go from heel lift-off to body lift-off. There are several discrepancies between experimental and theoretical results.

One reason for these discrepancies is that the human jumper may use a non-optimal control $u_1(t)$. Remember that optimality is achieved with $u_1(t)=1$ for all times after heel lift-off. Thus ankle extensor muscles should be fully activated and ankle flexor muscles inactivated. We recorded EMG activity from an ankle extensor muscle, medial gastrocnemius (MG), and from an ankle flexor muscle, tibialis anterior (TA). Maximum EMG activity from (MG) muscle and no EMG activity from (TA) muscle is expected from heel lift-off to body lift-off. Indeed, we do find large (MG) and no (TA) EMG activity in this epoch, except that these EMG states terminate before body lift-off and switch to the opposite (presumably bang-bang) state. Specifically, MC muscle is inactivated for the final 68 msec of propulsion and TA muscle is activated for the final 50 msec. While such inactivation of ankle extensor muscles and activation of flexor muscles is clearly sub-optimal, performance may not be greatly affected since it probably takes MG muscle a long time (relative to 68 msec) to mechanically relax and it takes time after TA activation begins for its force to develop. Reasons for this unexpected EMG activity may be either to protect the ankle from exceeding full, yet comfortable, extension or to prepare for landing, an event occurring only 175 msec after body lift-off.

Another reason for the observed deceleration near the end of propulsion is that our model for torque does not account for the force-velocity property of muscle, as we discussed earlier.

Other possible reasons for the observed discrepancies are that we have accounted neither for passive ligamentous torques that arise as the ankle approaches full extension, nor for the toes, nor for the 50 Hz low-pass filtering of F_V inherent in our experimental recordings.

5. Suggestions for Future Research

The main thrust of our own future research is twofold. First, we have obtained experimental data from several other types of maximum-height jumps. These jumps include normal jumps by cats as well as several types of human jumps. We are currently trying to extend our analytical techniques to these other jumps. The key problem is to be able to calculate the set of reachable states when the foot is flat on the ground. This calculation is much harder for more normal jumps because modelling these jumps requires three or more segments and two or more controllers. We could then add toes to our current model for the jump studied in this report to test the speculations mentioned in previous sections.

Second, we believe that there is considerable potential gain from better estimates of experimental quantities and in particular of joint angular velocities, forces and torques. We are therefore developing techniques to obtain such estimates using force-plate and kinematic experimental data and various types of biomechanical models.

6. References

1. Chow, D. and D. Jacobson (1971). Studies of human locomotion via optimal programming. Math. Biosci. 10:239-306.
2. Hatze, H. (1976). The complete optimization of a human motion. Math. Biosci. 28:99-135.
3. Hardt, D.E. (1980). Optimal solutions for muscle forces. Proc. Joint Auto. Control Conf., paper TA10-E, San Francisco.
4. Zajac, F.E. and Levine, W.S. (1979). Novel experimental and theoretical approaches to study the neural control of locomotion and jumping, in Posture and Movement (edited by R.E. Talbott and D.R. Humphrey), pp. 259-279, Raven Press: N.Y..
5. Levine, W.S., F.E. Zajac, M.R. Zomlefer and M.R. Belzer (1980). Experimental, analytical and computational study of maximum height jumps. Proc. Joint Auto. Control Conf., paper TA10-D, San Francisco.
6. Levine, W.S., F.E. Zajac, M.R. Belzer and M.R. Zomlefer (1981). Ankle controls that produce a maximal vertical jump when other joints are locked. (submitted to) IEEE Trans. on AC.
7. Levine, W.S., Christodoulou, M. and F.E. Zajac (1981). On propelling a rod to a maximum vertical or horizontal distance. (Submitted to) Automatica.
8. Hay, J.G., (1978), The Biomechanics of Sports Techniques, 2nd Ed., Prentice-Hall Inc., Englewood Cliffs, N.J.
9. Fugl-Meyer, A.R., M. Sjostrom and L. Wahlby (1979). Human plantar flexion strength and structure. Acta Physiol. Scan. 107:47-56.

Acknowledgement:

This research has been supported in part by N.I.H. grant #NS11971 and in part by the Computer Science Center of the University of Maryland.

AN INNOVATIONS APPROACH TO CARDIAC HEMODYNAMICS MODELING

Abdel-H. Rashwan , Abdalla Sayed Ahmed.

Systems & Biomedical Engineering Department,
Faculty of Engineering,Cairo University,Giza,
EGYPT.

ABSTRACT

A new approach to the problem of modeling the blood vessels attached to the heart especially pulmonary artery is presented. The analyses are based on a dynamic nonlinear compartmental model that describes the transfer of pressure energy between two sites in that vessel.The system kernels are represented by two paths,each path contains a non-linear network in cascade with a linear dynamics followed by a nonlinear transformation.The first path reflects the dynamic compliance of the vessel under pulstile action.The second path represents the variations of input impedance along the cardiac cycle. An experiment with(20)patients was performed to validate the model output especially in cases with aneurysm.

I-INTRODUCTION

The anatomical description of the cardiovascular system gives the fact that the pulmonary artery is attached to the right ventricle (RV)as well as the attachment of the aorta to the left ventricle (LV).It receives the output of RV and transfers it to the respiratory system similar to aorta which receives the output of LV and transfers it to the systemic arterial system.But there are major anatomical and physiological differences between both pulmonary artery and the aortic arc as great vessels attached to the output of the heart.Compared with the aortic arc,the pulmonary artery has great variations in geometric shape and the developed pressure in the systolic period [1].In many respects,the analysis of basic function of the arterial part of the pulmonary system has followed the same paths as that for systemic arterial system.Frank[2]et.al.pioneered in applying Hale's idea which became known as the "Windkessel"theory.It conceives the arteries as system of interconnected tubes with fluid storage capacity(reservoir).Fluid is pumped in at one end in an intermettent fashion(ventricular ejection)while out flow at the other end through the peripheral resistance is approximately constant and poiseuillean.Apeter [3]avoides the windkessel drawbacks especially reflected waves concept by showing theoretically the dependence of peripheral resistance and pressure controlled by barorecptors.The windkessel theory was applied to the pulmonary circulation by Engelberg and DuBois[4]in spite of its dissipative aspects.When transmission lines theory had acquired enough status to replace the whole conglomerate of windkessel, Caro and McDonald[5]formulated a model based on wave transmission in a single uniform elastic tube. Both the windkessel and tube technology imply that the pulmonary arterial system would be a strong resonating system.This problem was eliminated in the

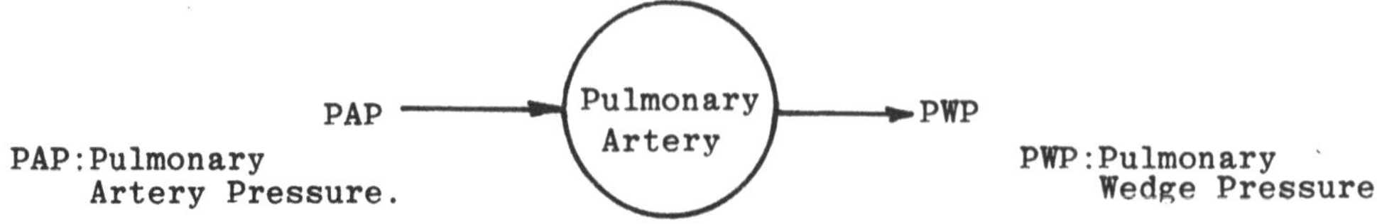

Fig.(1) Pulmonary Artery Compartment for Normal Cases.

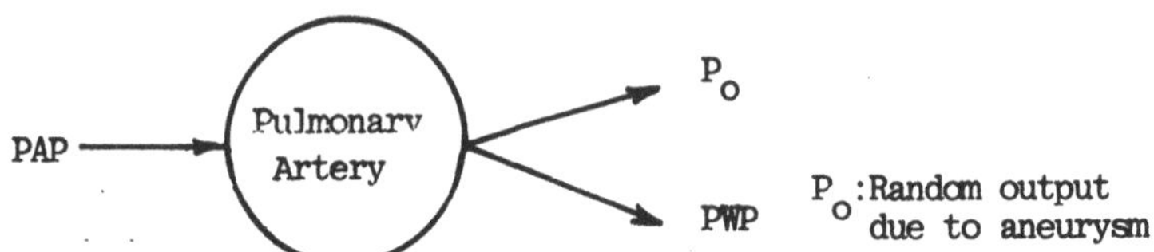

Fig.(2) Pulmonary Artery Compartment for the casees with aneurysm.

electrical analog model developed by Pollack et.al.[6]under several assumptions to keep the model network simple enough to be solved especially in steady state conditions.

The approach here is based on considering the pulmonary artery(PA)as a nonlinear compartment inside which the intracardiac pressure energy is transferred in unidirectional way for normal cases as shown in Fig.(1).For the abnormal,the compartment depicted in Fig.(1) would take another organization if aneurysm exists as illustrated in Fig.(2).The structure of the model is given in section-II.In section-III the procedure of data measurement and preprocessing is explained.The algorthim for parameter estimation is dipected in section-IV.

II-THE MODEL

Consider the segment of pulmonary artery behined the pulmonary valve to wedge before branching as a compartment which contains a definite form of matter or energy.The objective of the proposed model is to find the mathematical operator which transforms the compartment input domain to the range of its output:

$$Z(t) = G\left[T;\ U(k), k \leqslant T\right] \qquad (1)$$

Assuming the system under study(PA)is causual,continuous,bounded,and nonlinear over the period of data measurements.It is required to find out a set of functions that defines the transformation(G)together with the space U,and Z the domain and range of (G).Volterra[7]suggested that the input/output relationship of nonlinear system could be represented as the power series with functional terms:

$$Z(t_k) = \sum_{i=o}^{\infty} z_i(t_k) \qquad (2)$$

$$z_i(t_k) = \sum_{r_1=o}^{\infty} \sum_{r_2=o}^{\infty} \cdots \sum_{r_i=o}^{\infty} f_i(r_1, r_2, \ldots, r_i) \cdot \prod_{j=1}^{i} u(t_{k-r_j}) \qquad (3)$$

where:

$f_i(r_1, r_2, \ldots, r_i) \quad i=1,\ldots n$

are called the system kernels.

$t_k \leqslant T$

T is the period of measurements.

Truncating the series to L terms will simplify the result of equation(3)for practical application and will restrict the input past informations to a finite number of periods before t_k[8].The functional relationships of equation(3) could be characterized by a nonlinear network with no memory of past informations cascaded with linear network[9]. The proposed modelhas different structure from those where a nonlinear transfom ation following the linear network is present.The model consists of a set of parallel paths, each path contains a nonlinear network represented by a

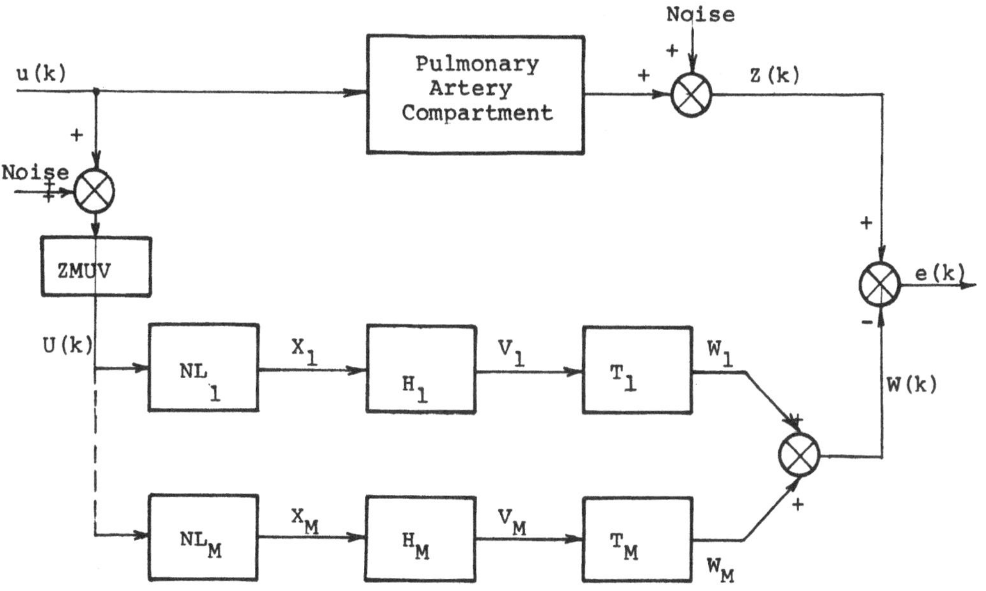

Fig.(3) The Model Structure For Pulmonary Artery Compartment.

polynomial of order(n)in cascade with a linear autoregressive moving average ARMA digital filter of order(m)following by a transformation to achieve zero mean unit variance output.The model output is the sum of the path's outputs.The number of paths is determined from a criterion of minimum error between system and model outputs as shown in Fig.(3). The model output is given by:

$$W(k)=\sum_{i=1}^{M}\left[\sum_{k=1}^{N}\left(X_i(u_k)\,\frac{N_i(z^{-1})}{1+D_i(z^{-1})}\right)v_i + m_i\right] \tag{4}$$

where:

- $X_i(u_K)$ is the output of nonlinear network in response to actual input u_k.
- N_i, D_i are polynomial coefficients of the ARMA filter.
- m_i the mean value of the i-th path output.
- v_i the variance of the i-th path output.
- z^{-1} delay operator.
- M No. of paths.
- N No. of points of measured data.

III-DATA ACQUISITION

The experiment was performed on(2o)patients with right ventricular hypertrophyRVH under cardiac catheterization at Kaser El-Eni hospital,Cairo university. The blood pressure measurements were done at two anatomical sites, behind the pulmonary valve and pulmonary wedge pressure PWP.The pressure associated with electrocardiogram lead-II were recorded by Mingograaf 42B.With the catheter of length loocm,size 7F,luman 1D.o58" introduced to pulmonary artery,a radiopaque dye(hypaque) was injected and a film of the pulmonary artery was obtained. This film contains a fluorscope image for at least three cardiac cycles using an enhanced X-ray

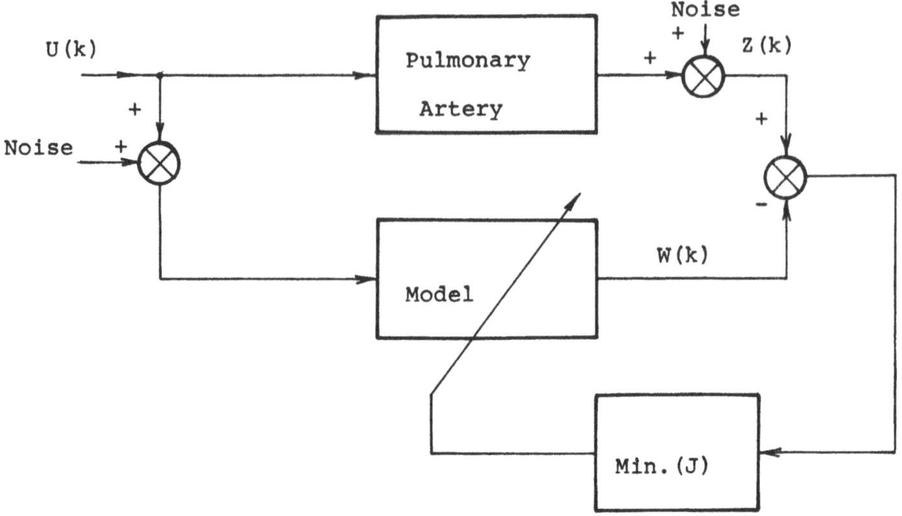

Fig.(4) Parameter Estimation Algorithm.

source, image intensifier and a 35mm camera run at a speed 64 frames/sec. The resulting single plane angiogram was used to judge -by physcians- the presence of aneurysm. The (2o) patients of the experiment were classified into two groups. The first group (1o) has RVH due to pulmonary hypertension. The second group has a pulmonary aneurysm as a late stage of pulmonary hypertension.

The baseline shift of the measured data due to muscular artifacts and respiration was removed by first order bandlimited Markov preprocessor [1o]. Using the R-peak of ECG trace as a time reference for each beat, five or more beats over several respiratory cycles were then averaged. Prior to use, the measured data were smoothed by a more complex but more satisfactory type which determines succesively the parabola that best fit a series of pressure samples. Such a parabola is defined as that which minimizes the mean square difference between the parabola and the value of data samples at corresponding points. The seven-points least squares parabola was used [11]. This way of fitting the data gives directly the first and second derivatives at each point.

IV-PARAMETER ESTIMATION

It is shown that the nonlinear compartment of pulmonary artery is represented by the interconnection of (NL-L-T) networks as depicted in Fig.(3). The model output W(k) is compared with the system output Z(k) to produce an error which is reduced in a root mean square sence RMS to determine the model parameters. This is achieved by minimizing the performance index J as shown in Fig.(4).

$$J=\sqrt{\frac{1}{N}\sum_{k=1}^{N}\left[Z(k)-W(k)\right]^{2}} \quad (5)$$

As the polynomial of NL is defined by :

$$X(u)=C_1U(k)+C_2U^2(k)+\ldots+C_nU^n(k) \quad (6)$$

and the ARMA digital filter,

$$H(z)=\frac{a_o+a_1z^{-1}+\ldots+a_{m-1}z^{-m+1}}{1+b_1z^{-1}+\ldots+b_mz^{-m}} \quad (7)$$

the following vectors are defined

by the following forms:

$$A^T=\left[a_o..a_{m-1},C_2a_o..C_2a_{m-1},\ldots, C_na_o..C_na_{m-1},-b_1,\ldots,-b_m\right]$$

$$B^T(k)=\left[U(k)..U(k-m+1),U^2(k)..U^2(k-m+1),\ldots U^n(k)..U^n(k-m+1),W_1(k-m+1)\ldots\ldots,W_1(k-m)\right]$$

where:

$$U(k)=\frac{u(k)-m_u}{v_u} \quad (8)$$

$$W_1(k)=\frac{Z(k)-m_z}{v_z}$$

m_u, v_u are the mean value and of the input u(k).

m_z,v_z are the mean value and variance of the output Z(k).

When only a finite number of data measurements is provided and no knowledge of the data statistics is available(other than stationary assumption),an alternative approach is to use least squares analysis.

For the first path,the error $e^{(1)}$ is given by:

$$e^{(1)}(k)=B^T(k).A - W_1(k) \quad (9)$$

Hence,the vector A which minimizes J is given by:

$$A=\left[\sum_{k=1}^{N} B(k).B^T(K)\right]^{-1}\sum_{k=1}^{N} B(k).W_1(k) \quad (1o)$$

The coefficients of the vector A will have a dimension(m.n+n)with certain amount of redundancy of the polynomial coefficients C_i, i=1,...n.Choose the set of C_i coefficients which gives minimum performance index J.For the i-th path,the error $e_n^{(i)}(k)$ replaces the system output $W_1(k)$ in vector B(k).

where:

$$e_n^{(i)}(k)=\frac{e^{(i)}(k)-m_e^{(i)}}{v_e^{(i)}} \quad (11)$$

and;

$$e^{(i)}(k)=W_i(k)-e^{(i-1)}(k) \quad (12)$$

i=2,...M

The procedure is repeated until the difference between two successive values of J is less than a prescribed accuracy and the value of that index also is lower than the permisible error;i.e.

$$\left|\frac{J^{i+1}-J^i}{J^i}\right| \leqslant \varepsilon \qquad J^i \leqslant \varepsilon \quad (13)$$

V-RESULTS

A comarison between the measured data PWP and the mean value of model output for both groups is illustrated in Fig.(5),(6).The corresponding model parameters for each group associated with the RMS are listed in table-I. All computations were done on PDP-11/34 digital computer of the department.Also,the programs were written in FORTRANlIV.A comparison between the response of the first path and an index reflecting the vessel elasticity(compliance)is depicted in Fig.(7),(8)for both groups respectively.Also,the response of the second path is compared with the pressure gradient along the vessel for both groups is illustrated in Fig.(9),(1o). The model wave velocity loops for both groups are shown in Fig.(11)(12).

VI-DISCUSSION

It has been demonstrted that the physical characteristics of the pulmonary artery can be determined from the blood pressure measurements at one site only; behind the valve.This is because the model predicts easily the wedge pressure which is at a distance from the first site.

An index I is introduced analogous to that used in evaluating the myocardium contractility[12] to detect the presence of aneurysm and consequently the dynamic variations of the vessel's compliance.

Evaluation of that index I is based on the assumptions:

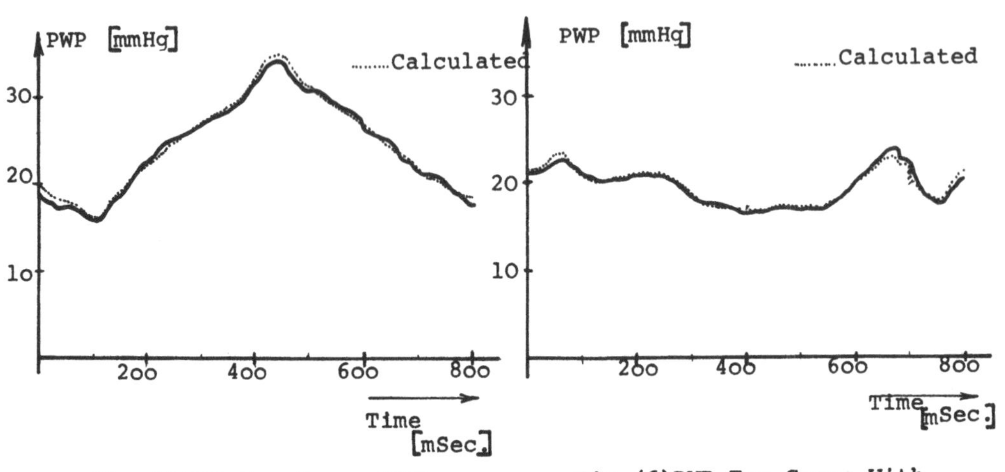

Fig.(5) PWP For Cases With Pulmonary Hypertension.

Fig.(6) PWP For Cases With Pulmon. Hyp.and Aneurysm

Fig.(7) Comparison Between First Path Output And Index Of Elasticity For Pul.Hyp.

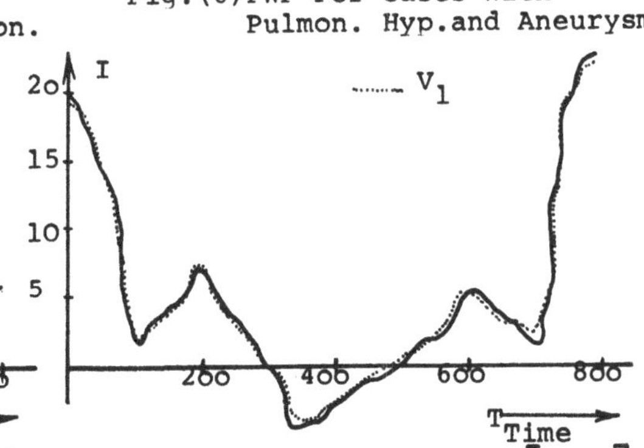

Fig.(8) Comparison Between First Path Output And Index Of Elasticity For Aneurysm.

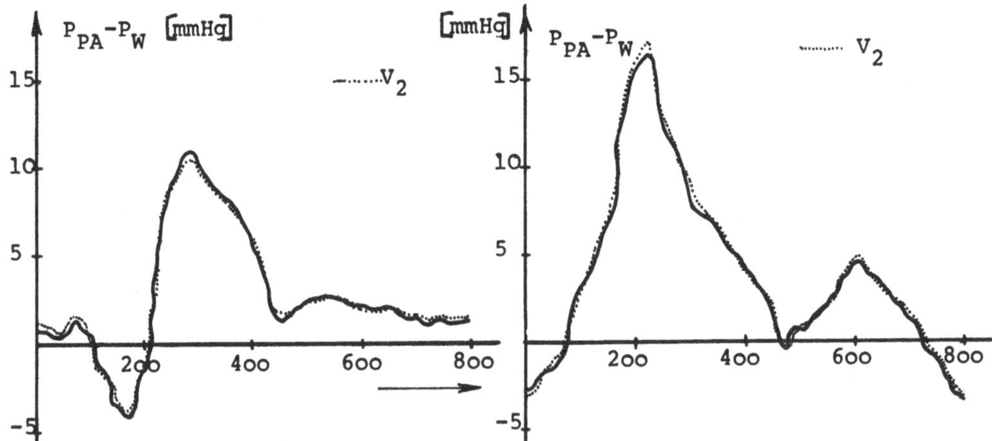

Fig.(9) Comparison Between Second Path Output and Pressure Gradient For Cases With Pulmon.Hyp.

Fig.(10) Comparison Between Second Path Output And Pressure Gradient For Cases With Aneurysm.

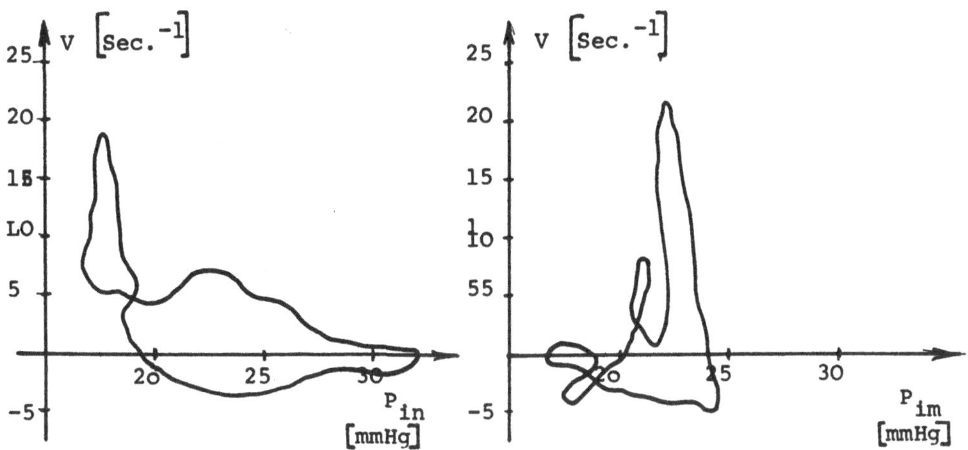

Fig.(11) Wave Velocity Loop For Pulmonary Hyp.

Fig.(12) Wave Velocity For Pulmonary Hyp. With Aneurysm.

TABLE-I

Case Type	Pulmonary Hyp. without An.			Pulmon. Hyp. with AN.		
Parameters	No. of Paths			No. of paths		
	1	2	3	1	2	3
C_1	1.00000	1.00000		1.00000	1.00000	
C_2	0.86645	1.265945		-0.9346	-0.77488	
C_3	0.00000	-2.19304		0.000000	0.66612	
a_o	1.00000	1.00000		1.00000	1.00000	
a_1	-1.00283	-1.00199		-3.19997	-1.01045	
a_2	0.251502	0.240999		2.56677	0.255256	
b_1	1.000000	1.000000		1.000000	1.000000	
b_2	-0.65645	-0.63751		-1.60167	-0.65290	
b_3	0.107800	0.108082		0.64140	0.106570	
RMS	0.375E-1	0.222E-5	0.375E-6	0.477E-1	0.423E-5	0.4117E-6

The presence of aneurysm will lead to additional output p_o-see Fig.(2)-for the instants of ventricular ejection.
Due to the vortexes of the pumped blood,the absorbed amount of blood will be fed again to the vessel.
The presence of negative I reflects the presence of aneurysm. Therefore,no need to do angiogram for checking the aneurysm and avoiding the side effects of angiogram protocol;see Fig.(8).
The direction of the returned blood from aneurysm is random and depends on the degree of aneurysm but it will affect the pressure gradient before and after the site of aneurysm.This effect is reflected on an occurrence of peak followed by a rapid decay on pressure gradient patteren as depicted in Fig.(10).
Plotting the index I against the input pressure to the vessel will produce a loop reflecting wave velociety propagation along the vessel.The presence of multi-loops reflects the presence of aneurysm as illustrated in Fig.(12).It is clear from model structure that the input is the actual physical input without the assumption of special form sinusoidal,or white Gaussian.But it is required to know the order of both polynomial and linear ARMA filter;n,m respectively.The calculation of RMS gives;during iteration;limitation for $n \leq 3$,and $m \leq 3$ [13].Also,the number of paths is limited to two paths where equation(13)yields a value of less than 0.0005.
Comparing with other models,it seems to be similar to Hammerstein model [14] , [15],but the presence of nonlinear transformation T yields an output with zero mean unit variance.The Hammerstein model is restricted to the use of fixed poles of the linear network which decreases the generality of selecting the order m.The multi-path choice seems to be similar to Uryson model [16],while the Uryson model consists of set of Hammerstein models in parallel and is based on the choice of Hammerstein polynomial order increasing with the number of paths.The solution of Uryson model is based on solving set of linear equations equals to number of paths.The vector of independent variables is the correlation function of the polynomial in each path,while the vector of dependant variables is the correlation of system output and the polynomial in each path.
The drawbacks of this approach are the need of large data to be processed to compute the requist correlations.If the linear dynamic namics response is alslowly varying,the whole procedure must be periodically repeated.But in the proposed model,in each path there are at most four iterations to select the polynomial and linear network orders to yields minimum J.Moreover,convergence of Uryson model iteration by that way cannot be guarnteed.

ACKNOWLEDGMENTS

We appreciate the program and effort done by Prof.Galal M.El-Said and his associates from the department of cardiology,Cairo university;for providing the clinical data.

REFERENCES

[1] E.O.Attinger,(1963), Pressure Transmission In Pulmonary Arteries Related To Frequency And Geometry, Circ.Res.Vol.12pp623.

[2] O.Frank,(1899), Die Grundform Des Arterien Pulses, Z.Biol. Vol.37,pp483.

[3] J.T.Apter,(1965), An Analysis Of Aortic Pressure Curves Taking Into Account Viscoelastic Properties Of The Aorta And Variations In Peripheral Resistance, Bull.Math.Biophys.Vol.27, pp 27 .

[4] J.Engelberg,and A.B.DuBios, (1959), Mechanics Of Pulmonary

Circulation In Isolated Rabbit Lungs, Am.J.Physiol. Vol.196, pp4ol.

[5] C.G.Caro,and D.A.McDonald, (1961), The Relation Of Pulstile Pressure And Flow In The Pulmonary Vascular Bed, J.Physiol.(LONDON)Vol.157, pp 426.

[6] G.H.Pollack,R.V.Reddy,and A.Noordergraaf,(1968), Input Impedence,Wave Travel,and Reflection In The Human Pulmonary Artery Tree:Studies Using Electrical Analog, IEEE Trans. On Biomd.Eng. Vol.BME-15,pp 151.

[7] John H.Seinfeld,and L.Lapidus,(1974), Mathematical Methods In Chemical Engineering, Vol.3,Prentice-Hall Inc.

[8] A.G.Bose,(1959), A Theory Of Nonlinear System, Res.Lab.Of Electronics, M.I.T.Cambridge Tech. Rep.309.

[9] Francis H.I.Chang,and Rein Luus,(1971 , A Noniterative Method For Identification Using Hammerstein Model, IEEE Trans. On Auto. Control Vol.AC-16,pp464.

[10] Abdel Hamid Rashwan,Galal M. El-Said,and Abdalla S Sayed Ahmed,(1980), Cardiac Signal Filtering By Markov Precessor, The First National Conf. Biophysics&Bioeng.Sci., Dec. Cairo,EGYPT.

[11] A.Steinberg,S.Abraham,and C.Caceres,(1962), Clinical Electrocardiogram Pattern Recognition, IRE Trans. On Biomed. Electronics,Vol.BME-9, pp 23-30.

[12] W.Grossman,H.Brook,S.Meister, H.Sherman,andLL.Dexter(1971), New Technique For Determining Instantaneous Myocardial Force-Velocity Relations In The Intact Heart, AmCirc.Res.Vol.XXVIII,Feb. pp 290-297.

[13] Abdel-H.Rashwan,and Abdalla Sayed Ahmed,(1982), A Unifying Approach To The Modelling Of Cardiac Hemodynamics, IFAC Sympsium On Theory and Application Of Digital Control,NewDelhi Jan.

[14] K.S.Narendra,and P.G.Gallman, (1966), An Iterative Method For The Identification Of Nonlinear System Using Hammerstein Model, IEEE Trans. On Auto.Control Vol.AC-11 pp546.

[15] T.C.Hsia,andA.L.Bailey, (1968), Learning Model Approach For Nonlinear System Identification, IEEE Trans.On System Sci. and Cybernetics Conf.Rec. pp228-232.

[16] P.G.Gallman,(1975), An Iterative Method For The Identification Of Nonlinear Systems, IEEE Trans. On Auto.Cont. Vol.AC-20 pp771-775.

MODELIZATION OF A MULTIPROCESSOR ARCHITECTURE

Jean René MENAND, Monique BECKER *

Université PARIS VI, 4 place Jussieu
LA 248 - Tour 55-65
75230 PARIS CEDEX 05 - France

ABSTRACT

Instead of using a very expensive and very powerful central unit, the Hypercube F8 consists of independent microprocessors F8 which work simultaneously. They are slow and inexpensive.

This architecture is modelised in order to choose the number of the microprocessors and to validate the system.

The model is hierarchical. Different levels of models are considered. Each model corresponds to a subsystem of hypercube F8. Most of the usual methods are used and direct calculations are made. For each level several methods are compared and the approximations are validated.

The global model gives the throughput of the system, the rate utilisations of the processors and the service times. It is possible to discuss the values of the parameters of the system.

* - Professor Roger DUPUY created this architecture. Let him receive our thanks for his great help.

I - DESCRIPTION OF THE MACHINE

Most of the usual architectures consist of a central unit which is very expensive and powerful and which is connected with many peripheral units. This structure implies a high complexity of the system ; i.e. multiprogrammation, shared memory, high cost and low efficiency. A lot of works study the optimisation of shared memory systems and of paging policies.

But nowadays there exist processors which are not very powerful but very inexpensive. So it seems interesting to consider a multiprocessor system which avoids multiprogramming. The processors being cheap we shall not mind if their utilisation rate is lower.

In this paper such a multiprocessor is studied [2]. The basic idea is the following : we replace the usual virtual units by real ones. Parallel microprocessors are used. They are slow and have a low capacity but a large number of them can be used. Here the expensive parts of the system are the peripheral units : disks and memory. Again, for economical reasons, it is interesting to share them.

n processors are sharing the disk and the I/O resources.

The processors : are F8 microprocessors. The cycle time est 2 μs. The local memory is 64 K octets. Each processor is executing one task at the time (instead of having one processor and multiprogrammation we have several processors and for each of them there is monoprogrammation). The I/O operations are performed through a buffer which is connected to the transfer processor. There are I/O processors and computing processors.

The transfer processor is taking care of the connections between the processors.

So the architecture is the following :

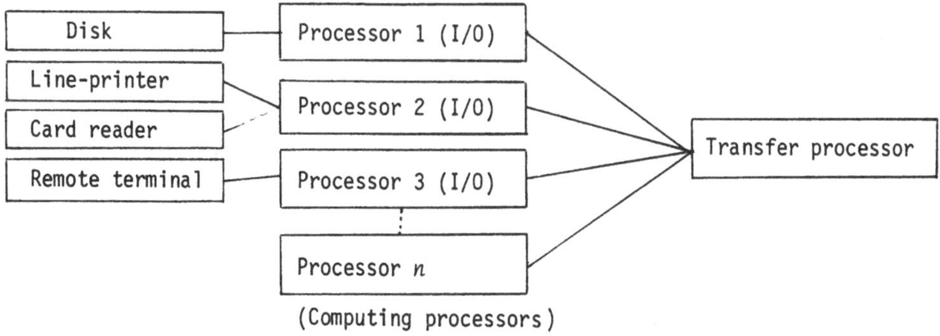

FIGURE 1

This processor is a Sigmetics 8 X 300 microprocessor, which transfers 1 K octet blocks.

There is a polling which means that the priority is the following : the transfer processor searches whether processor i is asking for a transfer. If yes it transfers the block. If no it searches whether processor $i + 1$ is asking for a transfer. After processor n it goes to processor 1.

The transfer time of a block is 4 ms.

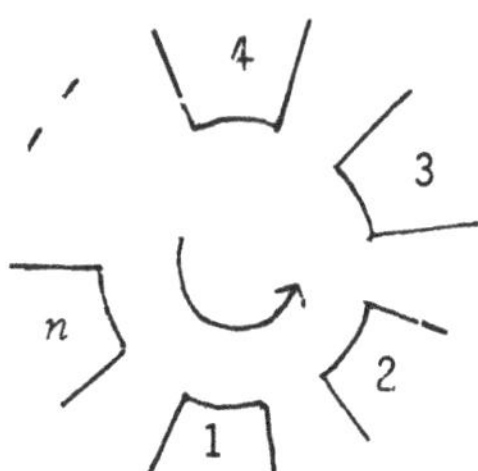

FIGURE 2 : the polling

The disk : there is one Control Data 97 M octets disk. The average service time is 30 ms. It is connected to the transfer processor through an F8 microprocessor, and a controller, which consists of two alternated buffers. The average service time of the system including disk, buffers and controller is : 40 ms.

The model is a queue network. Let us describe it.

When the jobs enter the system they first ask for several writings on the disk. When it is finished the job waits till a processor is free. Let us tell that the job waits in the allocation queue. Then the job is processed by a processor. We shall say that the job is active. This processor will not take care of any other job until the processing is finished. It is monoprogrammation. We shall model this by using a lock : a server which is before each processor.

While being processed by the processor, the job will ask for reading or writing on the disk. The transfer processor will choose the request with a polling priority. During a reading request, a given processor is blocked until the result of the request is transfered back through the polling which works twice.

When the processing is finished, the processor is free, and the job goes to the out queue, waiting for the reading of several results before having them printed on a line printer.

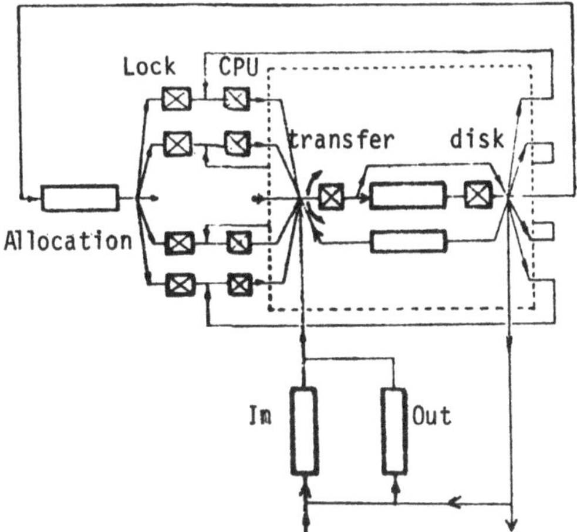

FIGURE 3 : the whole network

II - THE MODELLING METHOD [4]

The general method is hierarchical. We shall consider different level models. The first one is the transfer processor. It is an open network.

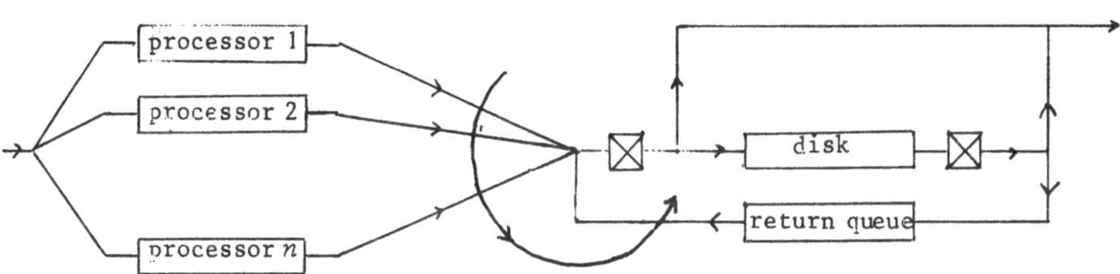

FIGURE 4 : the level 1 network

It is valid to isolate this subsystem since :

- the transfer service time is much smaller than the calculation service times
- the frequence of requests to disk is much higher than the frequence of job arrivals

The results of the study of this model will show that it is possible to replace this subsystem by a set of three queues : the transfer to the disk, the disk, and again the transfer. We shall consider a closed network with those three queues and one queue which models the processors.

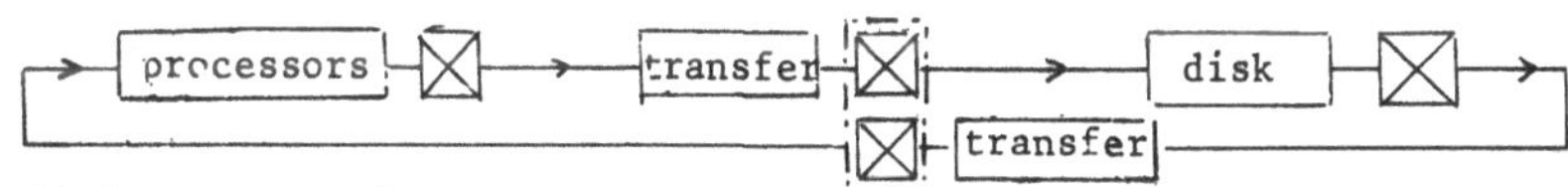

FIGURE 5 : the level 2 network

We shall consider a closed network with n_0 active jobs and for each value of n_0 we shall get the processing time. Of course : $n_0 \leq n$
n is the number of processors.

Then the level 3 model will be the wole system whose scheme is the following :

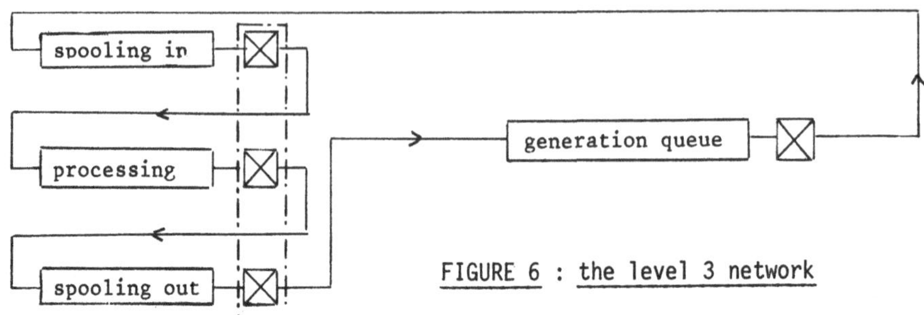

FIGURE 6 : the level 3 network

The processing queue is a queue which is assumed to be equivalent to the subsystem of level 2, for each value of n_0.
The dotted line on figure 6 shows which is the part of the network that is equivalent to the level 1 network. The dotted line on figure 7 also shows the queue which is equivalent to the level 2 network.

The whole set of approximations is validated by a simulation of the whole system.

III - THE RESULTS

III-1.- The level 1 model

Because of the polling priority this network is difficult.

1.- For low traffic if we neglect the polling priority, it is easy to solve this network by B.C.M.P. [3]

2.- We derived a direct method, using the conservation laws. We are actually improving this method which is a general solution of the polling problem. This work shall be published soon.

3.- For high traffic we used the diffusion approximation, with Reiser and Kobayashi way of calculating the constants and also with Gelenbe way of calculating the constants [8]

4.- A simulation was also performed

The results for the polling itself :
These methods were used for different values of the arrival rate of the requests : λ (λ is the number of requests by second).

We did the calculation for each value of λ from 16 to 224. We also calculated the limit value for saturation :

	B.C.M.P.	Diffusion	Direct Method	Simulation
limit λ	248	296	213	240

Figure 7 shows that the B.C.M.P. results are very bad for very high traffic but are good for low and medium traffic. The direct method results look like those of the simulation. The diffusion approximation is good for high traffic.

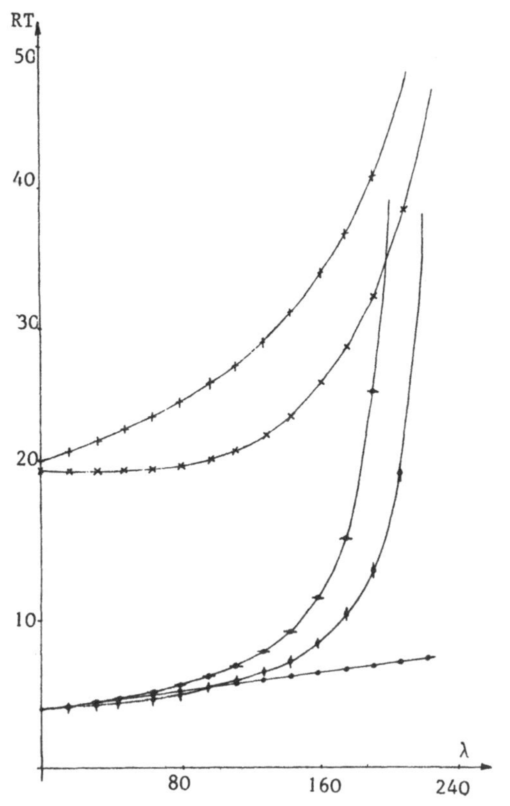

We then made the calculations for the whole network of figure 4. It appears that the transfer processor is very much faster than the disk. The bottleneck is the disk. So only low values for λ can be used, then diffusion approximation is not possible. The calculation of the transfer time of returns of reading requests shows that it is right to approximate the return queue by an M/M/I queue.

FIGURE 7

Throughput as a function of the arrival rate.

- Direct method
- Simulation
- B.C.M.P.
- + Diffusion Gelenbe
- x Diffusion Reiser and Kobayashi

III-2.- The level 2 model

We used approximation by B.C.M.P. and direct solution of transition state matrix. We study a closed network with N customers, for any value of N (N = number of active processors, N ≤ number of processors). Reading requests block the processors, writing request dont. When considering the active period of the processors, we modelize the reading requests. Writing requests are not modelised but the service time is iteratively calculated so as to take care of reading and writing request :

- first the arrival rate of reading requests λ_R is evaluated without taking care of writing. Then λ_W, the arrival rate of writing request is derived and the service times are modified
- the same calculation is iterated till convergence

III-3.- The level 3 model

The level 3 model is solved by an iterative method.

The throughput : Figure 8 gives the throughput as a function of μ the service time of the generation queue. The throughput first increases nearly linearly, to about 5.10^{-3} jobs/s ; we shall see that this corresponds to the saturation of the disk. Then it decreases and tends to a value which depends of the number of customers in the system.

Let us notice, that there is a saturation of the system, but no thrashing. This is due to the monoprogrammation of the processors. There is no swapping, and very little overhead. The processors have no system processing to perform, their efficiency nearly is independent of the number of customers in the system.

We conclude that the optimal number of customers in the system is more than 18.

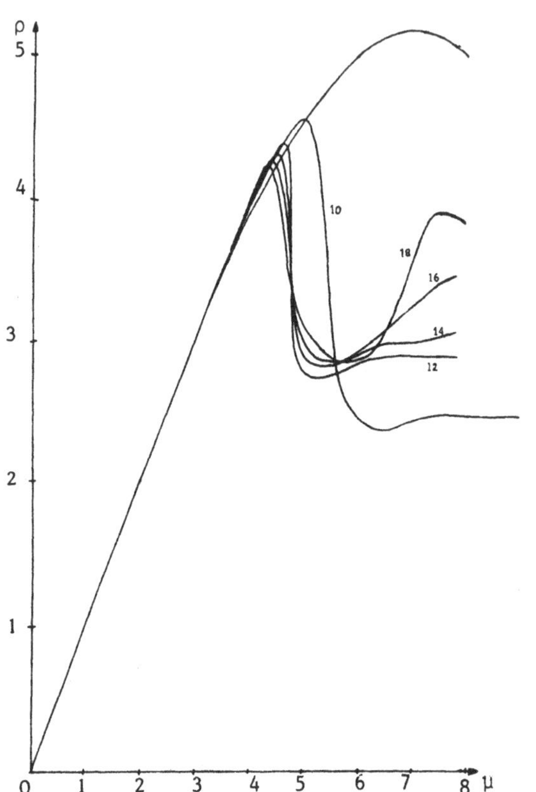

FIGURE 8

Throughput

The busy rate of the disk

It increases very quickly.Figure 9 gives this rate as a function of μ. The disk is the bottleneck. The service rate of the disk is the important parameter of the system. If this service rate increases, the throughput increases.

The mean number of active and non blocked processors

It increases, with the arrival rate of jobs, until the saturation of the disk. The upper value is 3.4. This proves that it is not useful to have much more than 4 processors for this kind of disk. But let us repeat that this is not very important since the processors are cheap and the disk is expensive.

The efficiency of an active processor, and the ratio of active processors

The upper efficiency is 85 % for 1 active processor, and 36 % for 8 active processors.

It can be seen that to get a 60 % efficiency for the active processors, it is enough to have 5 processors. Then the disk is saturated and the throughput cannot be improved very much.

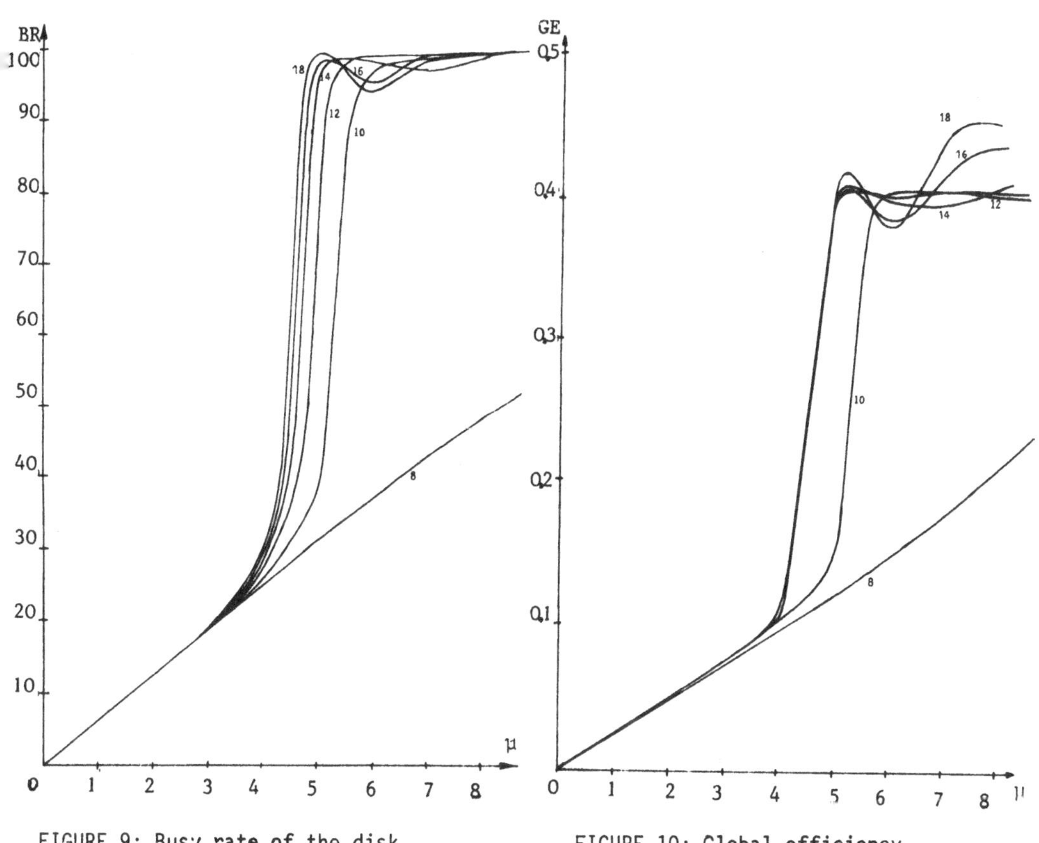

FIGURE 9: Busy rate of the disk.

FIGURE 10: Global efficiency.

The efficiency of the processors

Figure 10 gives the global efficiency as a function of the arrival rate of admission queue : μ, for different values of the number of active processors. The global efficiency is the product of the efficiency of an active processor and of the ratio of active processors; it gets to 40% when the disk is saturated.

The throughput as a function of the number of processors

It is given on figure 11, for a mean access time to the disk of 30 ms, and $\mu = 0.09$ 10^{-3} customers/s. The maximum throughput corresponds to the case of 4 or 5 processors. This result had already being obtained when studying the efficiency of processors. For more than 5 processors, the congestion appears. (For this multiprocessor system the number of processors corresponds to the number of customers for a classical multiprogrammed system).

This curve was confirmed by the simulation of the whole system, since it appeared that the results on figure 11 are within the confidence interval of the simulation.

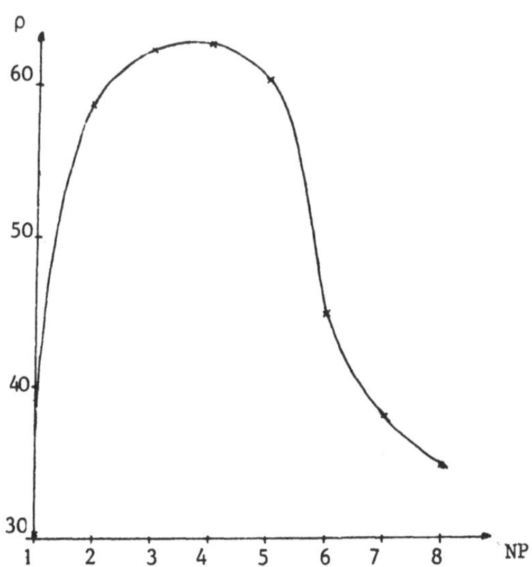

FIGURE 11

Throughput as a function of the number of processors.

CONCLUSION

The architecture has being validated. It appears that the bottleneck is the disk, it is not the transfer processor. It appears that the number of processors has to be less that 5, otherwise there is congestion. Of course the performance can be improved, if the speed of the disk is improved or if several disks are connected to the system.

The model : the results depend on the values of the parameters. The curves which are given in this paper are only valid for the values of the parameters that were chosen. Yet, the model is very general, and those values could be easily modified. It would only be necessary to run again the programs, with, for example different number of customers, of processors, a different speed for the disk, or for the processors, or a different arrival rate for the jobs...

Practical aspects : this system is presently under commercialization. It can replace management computers. It uses modular computers. It is a good example of distributed architecture programming. Of course it can be connected with a distant computing center (though Transpac network).

REFERENCES

[1] J.R. MENAND
"Etude et Modélisation d'une architecture multiprocesseur". Thèse de docteur-ingénieur. Université Paris VI, 23.06.1980

[2] R. DUPUY
"Architecture de machine à système dispersé". Rapport final de la convention SESORI n° 78053

[3] BASKETT, CHANDY, MUNTZ, PALACIOS
"Open, closed and mixed networks of Queues with different classes of customers" JACM 22, 1975

[4] P.J. COURTOIS
"Decomposability : queueing and computer system applications" ACM MONOGRAPH SERIES, 1977

[5] Y. BARD
"An analytic model of the VM/370 SYSTEM". IBM J. RES. DEVELOP. vol. 22, n° 5, september 1978

[6] M. BECKER, R. FORTET
"Projector method and iterative method to solve a packet switching network node. Validation by simulation". Measuring, Modelling and Evaluating Computer Systems, H. Beilner and E. Gelenbe (eds), North-Holland Publishing Company, 1977

[7] L. KLEINROCK
"Queueing Systems". vol. 2 : Computer Application. WILEY INTERSCIENCE, 1976

[8] E. GELENBE
"On Approximate Computer System Models"; J.A.C.M. vol. 22, n.2, april 1975

A NEW METHOD TO HANDLE THE LEFT-RECURSIONS FOR TOP-DOWN PARSING IN COMPILER DESIGN

Che Chang Chen
Department of Electrical Engineering and Computer Science
Stevens Institute of Technology
Hoboken, New Jersey 07030, U.S.A.

Abstract. A new method for handling the left-recursions for top-down parsing is presented. This method preserves the structure of the parse tree specified by the given grammar. Besides, the parsers written based on this method are usually smaller than the parsers written based on the traditional method.

I. INTRODUCTION

Top-down parsing strategy can reduce the effort of compiler design when it is applicable. In many practical cases, the need for backtracking, which is considered as the main shortcoming of top-down parsing, can be avoided by rearranging the grammar. When it does not require backtracking, it is called recursive-decent parsing. Recursive-decent parsing is efficient when the high-level system programming language is available. Even in the situation that the high-level programming language is not available, the technique of bootstrapping can be applied , and we still can use recursive-decent parsing for efficient compiler design.

Besides backtracking, the other problem of recursive-decent parsing is the left-recursions. When there are left-recursions in the grammar, direct application of recursive-decent parsing will result in infinite loops. The traditional way of handling this problem requires introducing new nonterminal symbols and productions. The key point of that method is to change the left-recursions into right-recursions. By introducing new nonterminal symbols and productions, it increases the size of the parser. By changing the left-recursions into right-recursions, it alters the structure of the parse tree. If what we want to accomplish is to decide whether a given character string belongs to the language specified by the given grammar, we do not have to worry about preserving the structure of the parse tree. However, in compiler writing, we are not merely interested in the recognition problem. The parsing is only the first step of the entire translation process. Therefore, it is desired that the structure of the parse tree can be preserved.

In this paper, a new method is proposed. This method does not introduce new nonterminal symbols or productions. It also preserves the structure of the parse tree as specified by the given grammar.

II. REVIEW OF THE TRADITIONAL METHOD

Given a context-free grammar $G=(N,\Sigma,P,S)$. Let $A\in N$, and $A\to A\alpha_1$, $A\to A\alpha_2,\ldots\ldots$, $A\to A\alpha_n$, $A\to\beta_1$, $A\to\beta_2,\ldots\ldots\ldots$, $A\to\beta_m$ be all the productions with the nonterminal symbol A on the left-hand side, where $\alpha_1, \alpha_2,\ldots, \alpha_n \in \Sigma^+$, $\beta_1, \beta_2,\ldots, \beta_m \in \Sigma^*$, and β_i does not start with A for any $1\leq i\leq m$.

Let A' be a new nonterminal symbol. Use the following new productions to replace those productions mentioned above, $A\to\beta_1A'$, $A\to\beta_2A'$, $\ldots$, $A\to\beta_mA'$, $A'\to\alpha_1A'$, $A'\to\alpha_2A'$, $\ldots$, $A'\to\alpha_nA'$, $A'\to\lambda$, where λ denotes the empty string. It can be proved that the new grammar generates the same language as the old grammar. Now, the left-recursions have been changed into the right-recursions. A recursive-decent recognizer can be written based on this new grammar.

In the transformation of the grammar, the left-recursions are changed into the right-recursions. Therefore, for the same sentence, the structure of the parse tree according to the transformed grammar may not be tha same as the structure of the parse tree according to the given grammar.

III. THE NEW METHOD

Treating the productions as the regular expressions over $N\cup\Sigma$, we can express those productions with A on the left-hand side mentioned in the beginning of Section II as follows according to Arden's theorem on regular expressions.

$$A := (\beta_1|\beta_2|\ldots\ldots|\beta_m)(\alpha_1|\alpha_2|\ldots\ldots|\alpha_n)^*$$

Let F_1 and F_2 be the state-transition-diagrams for $(\beta_1|\beta_2|....|\beta_m)$ and $(\alpha_1|\alpha_2|....|\alpha_n)$ respectively. Then the state-transition-diagram for A, denoted as F can be constructed as follows.

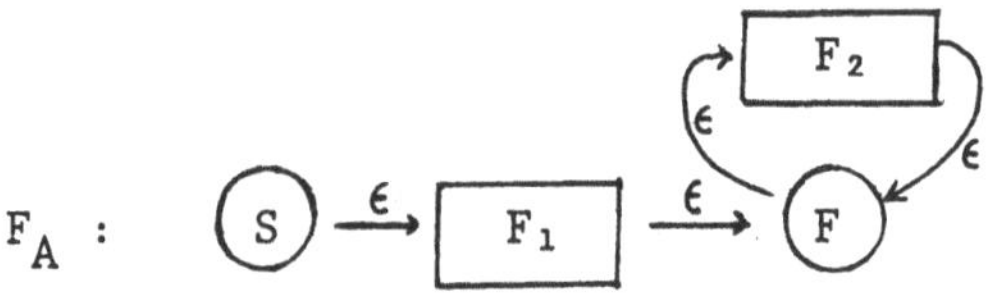

Fig. 1. State-transition-diagram for A

Where S is the starting state, F is the final state, and ε denotes the trivial transition.

Now, we want to construct a processor P_A which will simulate F_A and produce a parse tree. Let P_1 be a processor which simulates F_1 and produces a parse forest (in which the trees are ordered) according to $(\beta_1|\beta_2 ....|\beta_m)$. Let P_2 be a processor which simulates F_2 and produces a parse forest (in which the trees are ordered) according to $(\alpha_1|\alpha_2| ....|\alpha_n)$. Then, the operation of P_A can be described as follows.

Step 1 : Apply P_1. Let G be the parse forest produced by P_1.

Step 2 : Build a new node with label A. Let the component trees of G be the subtrees of this new node according to their order in G. Let this new tree be denoted as T_0.

Step 3 : If the simulation of F_2 fails, then return T_0; otherwise, apply P_2, let the ordered set of trees $(T_1, T_2, ..., T_d)$ be the parse forest produced by P_2, let $G \leftarrow (T_0, T_1, T_2, ..., T_d)$, go to Step 2.

It is easy to see that the processor P_A produces a parse tree of which the structure is according to the specification of the given grammar.

IV. CONCLUSION

A new method for handling the left-recursions for top-down parsing has heen presented in this paper. This method preserves the structure of the parse tree specified by the given grammar. Besides, the parsers written based on this method are usually of smaller size than the parsers written based on the traditional method.

REFERENCES

1. A. V. Aho and J. D. Ullman, _Principles of Compiler Design_, Addison-Wesley, Reading, Mass., 1978.

2. A. V. Aho and J. D. Ullman, _The Theory of Parsing, Translation, and Compiling_, Vol. II: _Compiling_, Prentice-Hall, Englewood Cliffs, N.J., 1973.

3. H. R. Lewis and C. H. Papadimitriou, _Elements of the Theory of Computation_, Prentice-Hall, Englewood Cliffs, N.J., 1981.

INDEX OF AUTHORS

INDEX OF AUTHORS

Lecture Notes in Control and Information Sciences

Edited by A. V. Balakrishnan and M. Thoma

Printed in the United States
By Bookmasters